Topic	Symbol	Meaning	Page
	i_A	identity function on A	136
	f^{-1}	inverse of f	153
	$f \circ g$	composition of g and f	151
	$a \bmod b$	remainder when a is divided by b	132
	$a \text{ div } b$	quotient when a is divided by b	132
	$\lfloor x \rfloor$	floor of x	126
	$\lceil x \rceil$	ceiling of x	127
	s_n	nth term the sequence $\{s_n\}$	157
	$\sum_{i=m}^{n} a_i$	$a_m + a_{m+1} + \cdots + a_n$	158
	$\sum_{i \in S} a_i$	sum the terms a_i, where $i \in S$	160
	$\prod_{i=m}^{n} a_i$	$a_m a_{m+1} \cdots a_n$	163
	$n!$	n factorial	241
	$f(x) = O(g(x))$	$f(x)$ is big-oh of $g(x)$	237
	$f(x) = \Omega(g(x))$	$f(x)$ is big-omega of $g(x)$	243
	$f(x) = \Theta(g(x))$	$f(x)$ is big-theta of $g(x)$	245
	$\min\{x, y\}$	minimum of x and y	252
	$\max\{x, y\}$	maximum of x and y	251
MATRICES	$(a_{ij})_{m\times n}$		165
	$\mathbf{A} + \mathbf{B}$		166
	$\mathbf{A} -$		167
	$k\mathbf{A}$		167
	$\mathbf{AB}$		170
	$\mathbf{I}_n$		166
	$\mathbf{A}^{\mathrm{T}}$		174
INTEGERS	$a \mid b$	a divides b	189
	$a \nmid b$	a is not a factor of b	189
	$\gcd\{a, b\}$	greatest common divisor a and b	191
	$\operatorname{lcm}\{a, b\}$	least common divisor of a and b	
	$(a_1 a_2 \cdots a_n)_b$	base-b representation	197
COMBINATORICS	$P(n, r)$	number of r-permutations of a set with n elements	352
	D_n	number of derangements of n elements	361
	$C(n, r)$	number of r-combinations of a set with n elements	366
	$\binom{n}{r}$	binomial coefficient: coefficient of x^n in expansion of $(1+x)^n$	366
	$N(P_{i_1} P_{i_2} \ldots P_{i_n})$	number of elements having each of the properties P_{i_j}, $j = 1, 2, \ldots, n$	400
	$N(P'_{i_1} P'_{i_2} \cdots P'_{i_n})$	number of elements not having any of the properties P_{i_j}, $j = 1, 2, \ldots, n$	400
PROBABILITY	$P(E)$	probability of event E	410
	$P(E \mid F)$	conditional probability of E given F	417
BOOLEAN MATRICES	$\mathbf{A} \vee \mathbf{B}$	join of $\mathbf{A}$ and $\mathbf{B}$	439
	$\mathbf{A} \wedge \mathbf{B}$	meet of $\mathbf{A}$ and $\mathbf{B}$	439
	$\mathbf{A} \odot \mathbf{B}$	boolean product of $\mathbf{A}$ and $\mathbf{B}$	439
	$\mathbf{A}^{[n]}$	nth boolean power of $\mathbf{A}$	441

continued

Topic	*Symbol*	*Meaning*	*Page*
RELATIONS	$S \circ R$	composite of relations S and R	463
	R^n	nth power of relation R	466
	R^{-1}	inverse relation	469
	$a \equiv b (\text{mod } m)$	a is congruent to b modulo m	484
	$a \not\equiv b (\text{mod } m)$	a is not congruent to b modulo m	484
	$[a]_R$	equivalence class of a with respect to R	486
	$a \prec b$	a is less than b	495
	$a \preceq b$	a is less than or equal to b	495
	$a \succ b$	a is greater than b	
	$a \succeq b$	a is greater than or equal to b	
GRAPHS	$\{u, v\}$ or $\{u - v\}$	undirected edge	517
	$G = (V, E)$	graph with vertex set V and edge set E	517
	$\deg(v)$	degree of vertex v	520
	K_n	complete graph with n vertices	524
	C_n	cycle with n vertices	526
	W_n	wheel with $n + 1$ vertices	526
	$K_{m,n}$	complete bipartite graph	527
	Q_n	n-cube	566
	$a\text{-}v_1\text{-} \cdots \text{-} v_{n-1}\text{-}b$	path from a to b	546
	$G_1 \cup G_2$	union of G_1 and G_2	532
	$\deg(R)$	degree of region R	581
BOOLEAN ALGEBRA	B	$\{0, 1\}$	807
	x, y	boolean variables	813
	x'	complement of x	
	$x + y$	boolean sum of x and y	814
	$x \cdot y$ (or xy)	boolean product of x and y	814
	$x \uparrow y$	x NAND y	820
	$x \downarrow y$	x NOR y	820
	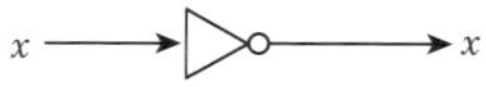	inverter	827
	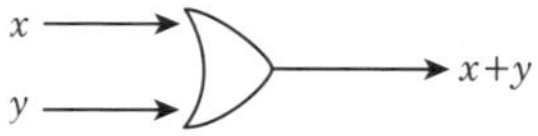	OR gate	826
	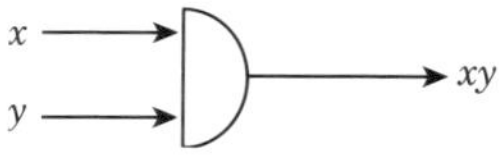	AND gate	825
LANGUAGES	xy	concatenation of words x and y	76
	λ	empty string	75
	$\lVert x \rVert$	length of the string x	75
	w^R	reversal of w	113
	AB	concatenation of languages A and B	736
	A^*	Kleene closure	739
	Σ	alphabet	75
FINITE-STATE MACHINES	σ	start symbol	744
	$w \rightarrow w'$	production	744
	$w \Rightarrow w'$	w' is directly derivable from w	746
	$\langle A \rangle$	nonterminal symbol	748
	$\langle A \rangle ::= \langle B \rangle \mid x$	Backus–Naur form	748
	(S, I, O, f, g, s_0)	finite-state machine with output	772
	s_0	start state	760
	(S, I, f, s_0, F)	finite-state machine with no output	761

Discrete Mathematics with Applications

Discrete Mathematics with Applications

Thomas Koshy
Framingham State College

ELSEVIER
ACADEMIC
PRESS

Amsterdam Boston Heidelberg London New York Oxford Paris
San Diego San Francisco Singapore Sydney Tokyo

Senior Editor	Barbara A. Holland
Project Manager	Marcy Barnes-Henrie
Associate Editor	Tom Singer
Marketing Manager	Linda Beattie
Cover Design	Hannus & Associates
Composition	Cepha
Printer	Quebecor World

Elsevier Academic Press
200 Wheeler Road, Burlington, MA 01803, USA
525 B Street, Suite 1900, San Diego, California 92101-4495, USA
84 Theobald's Road, London WC1X 8RR, UK

This book is printed on acid-free paper. ∞

Library of Congress Cataloging-in-Publication Data
Application Submitted

British Library Cataloguing in Publication Data
A catalogue record for this book is available from the British Library

ISBN: 0-12-421180-1

For all information on all Academic Press publications visit our website at www.academicpressbooks.com

Printed in the United States of America
03 04 05 06 07 08 9 8 7 6 5 4 3 2 1

Dedicated to

the memory of a close friend and an outstanding colleague Kenneth J. Preskenis (1940–2002)

Table of Contents

Preface xiii

A Word to the Student xxi

1 ***The Language of Logic*** 1

1.1 Propositions 2

1.2 Logical Equivalences 20

1.3 Quantifiers 32

***1.4** Arguments (optional) 38

1.5 Proof Methods 49

Chapter Summary 56

Review Exercises 58

Supplementary Exercises 62

Computer Exercises 63

Exploratory Writing Projects 63

Enrichment Readings 64

2 ***The Language of Sets*** 67

2.1 The Concept of a Set 67

2.2 Operations with Sets 78

***2.3** Computer Operations with Sets (optional) 94

2.4 The Cardinality of a Set 98

2.5 Recursively Defined Sets 104

Chapter Summary 109

Review Exercises 111

Supplementary Exercises 113

Computer Exercises 114

Exploratory Writing Projects 114

Enrichment Readings 115

3 ***Functions and Matrices*** 117

3.1 The Concept of a Function 117

3.2 Special Functions 125

3.3 Properties of Functions 136

3.4 The Pigeonhole Principle 144

3.5 Composition of Functions 150

3.6 Sequences and the Summation Notation 157

3.7 Matrices ... 164
Chapter Summary ... 175
Review Exercises ... 177
Supplementary Exercises ... 179
Computer Exercises ... 181
Exploratory Writing Projects ... 182
Enrichment Readings ... 183

4 ***Induction and Algorithms*** ... 185
4.1 The Division Algorithm ... 185
4.2 Divisibility Properties ... 189
4.3 Nondecimal Bases ... 197
4.4 Mathematical Induction ... 207
4.5 Algorithm Correctness ... 224
4.6 The Growth of Functions ... 237
***4.7** Complexities of Algorithms (optional) ... 247
Chapter Summary ... 252
Review Exercises ... 254
Supplementary Exercises ... 256
Computer Exercises ... 257
Exploratory Writing Projects ... 259
Enrichment Readings ... 259

5 ***Recursion*** ... 261
5.1 Recursively Defined Functions ... 262
5.2 Solving Recurrence Relations ... 278
5.3 Solving Recurrence Relations Revisited ... 286
5.4 Generating Functions ... 298
5.5 Recursive Algorithms ... 307
5.6 Correctness of Recursive Algorithms ... 316
***5.7** Complexities of Recursive Algorithms (optional) ... 319
Chapter Summary ... 333
Review Exercises ... 334
Supplementary Exercises ... 339
Computer Exercises ... 340
Exploratory Writing Projects ... 342
Enrichment Readings ... 342

6 ***Combinatorics and Discrete Probability*** ... 343
6.1 The Fundamental Counting Principles ... 344
6.2 Permutations ... 351
6.3 Derangements ... 360
6.4 Combinations ... 365
6.5 Permutations and Combinations with Repetitions ... 375
6.6 The Binomial Theorem ... 386

***6.7** The Generalized Inclusion–Exclusion Principle (GIEP) (optional) 399
***6.8** Discrete Probability (optional) ... 409
***6.9** Additional Topics in Probability (optional) ... 417
Chapter Summary ... 427
Review Exercises ... 429
Supplementary Exercises ... 432
Computer Exercises ... 434
Exploratory Writing Projects ... 434
Enrichment Readings ... 435

7 ***Relations*** ... 437
7.1 Boolean Matrices ... 438
7.2 Relations and Digraphs ... 443
***7.3** Computer Representations of Relations (optional) ... 449
7.4 Properties of Relations ... 454
7.5 Operations on Relations ... 461
***7.6** The Connectivity Relation (optional) ... 471
***7.7** Transitive Closure (optional) ... 475
7.8 Equivalence Relations ... 482
7.9 Partial and Total Orderings ... 493
Chapter Summary ... 506
Review Exercises ... 508
Supplementary Exercises ... 511
Computer Exercises ... 512
Exploratory Writing Projects ... 513
Enrichment Readings ... 514

8 ***Graphs*** ... 515
8.1 Graphs ... 516
***8.2** Computer Representations of Graphs (optional) ... 538
8.3 Isomorphic Graphs ... 541
8.4 Paths, Cycles, and Circuits ... 546
8.5 Eulerian and Hamiltonian Graphs ... 556
8.6 Planar Graphs ... 576
8.7 Graph Coloring ... 586
Chapter Summary ... 598
Review Exercises ... 601
Supplementary Exercises ... 604
Computer Exercises ... 606
Exploratory Writing Projects ... 607
Enrichment Readings ... 608

9 ***Trees*** ... 609
9.1 Trees ... 610
9.2 Spanning trees ... 614

9.3 Minimal Spanning Trees 626
9.4 Rooted Trees 635
9.5 Binary Trees 646
9.6 Binary Search Trees 664
***9.7** Huffman Trees (optional) 670
***9.8** Decision Trees (optional) 676
Chapter Summary 680
Review Exercises 681
Supplementary Exercises 686
Computer Exercises 687
Exploratory Writing Projects 688
Enrichment Readings 688

10 ***Digraphs*** 691
10.1 Digraphs 691
10.2 Dags 707
10.3 Weighted Digraphs 715
Chapter Summary 726
Review Exercises 727
Supplementary Exercises 730
Computer Exercises 731
Exploratory Writing Projects 732
Enrichment Readings 732

11 ***Formal Languages and Finite-State Machines*** 733
11.1 Formal Languages 734
11.2 Grammars 743
11.3 Finite-State Automata 759
11.4 Finite-State Machines 771
11.5 Deterministic Finite-State Automata and Regular Languages 779
11.6 Nondeterministic Finite-State Automata 782
11.7 Automata and Regular Languages 787
Chapter Summary 792
Review Exercises 794
Supplementary Exercises 798
Computer Exercises 800
Exploratory Writing Projects 801
Enrichment Readings 802

12 ***Boolean Algebra and Combinatorial Circuits*** 803
12.1 Boolean Algebra 804
12.2 Boolean functions 813
12.3 Logic Gates 824
12.4 Combinatorial Circuits 830
12.5 Minimization of Combinatorial Circuits 840

12.6 Don't Care Conditions ... 851
Chapter Summary ... 857
Review Exercises ... 859
Supplementary Exercises ... 862
Computer Exercises ... 863
Exploratory Writing Projects ... 864
Enrichment Readings ... 864

Appendix A ... 867
A.1 ASCII Character Set ... 867
A.2 Determinants ... 867
A.3 Exponential and Logarithmic Functions ... 874
A.4 Generating Permutations and Combinations ... 883
A.5 The Multinomial Theorem ... 888
A.6 The Greek Alphabet ... 894
A.7 Web Sites ... 895

References ... 899

Solutions to Odd-Numbered Exercises ... 907

Credits ... 1029

Index ... 1031

Preface

A journey of a thousand miles
begins with a single step.

—A Chinese proverb

People often ask: What is discrete mathematics? It's the mathematics of discrete (distinct and disconnected) objects. In other words, it is the study of discrete objects and relationships that bind them. The geometric representations of discrete objects have gaps in them. For example, integers are discrete objects, therefore (elementary) number theory, for instance, is part of discrete mathematics; so are linear algebra and abstract algebra.

On the other hand, calculus deals with sets of connected (without any gaps) objects. The set of real numbers and the set of points on a plane are two such sets; they have continuous pictorial representations. Therefore, calculus does not belong to discrete mathematics, but to continuous mathematics. However, calculus is relevant in the study of discrete mathematics. The sets in discrete mathematics are often finite or countable, whereas those in continuous mathematics are often uncountable.

Interestingly, an analogous situation exists in the field of computers. Just as mathematics can be divided into discrete and continuous mathematics, computers can be divided into digital and analog. Digital computers process the discrete objects 0 and 1, whereas analog computers process continuous data—that is, data obtained through measurement. Thus the terms *discrete* and *continuous* are analogous to the terms *digital* and *analog*, respectively.

The advent of modern digital computers has increased the need for understanding discrete mathematics. The tools and techniques of discrete mathematics enable us to appreciate the power and beauty of mathematics in designing problem-solving strategies in everyday life, especially in computer science, and to communicate with ease in the language of discrete mathematics.

The Realization of a Dream

This book is the fruit of many years of many dreams; it is the end-product of my fascination for the myriad applications of discrete mathematics to a variety of courses, such as *Data Structures*, *Analysis of Algorithms*, *Programming Languages*, *Theory of Compilers*, and *Databases*. *Data structures* and *Discrete Mathematics* compliment each other. The information in this book is applicable to quite a few areas in mathematics; discrete mathematics is also an excellent preparation for *number theory* and *abstract algebra*.

A logically conceived, self-contained, well-organized, and a user-friendly book, it is suitable for students and amateurs as well; so the language employed is, hopefully, fairly simple and accessible. Although the book features a well-balanced mix of conversational and formal writing style, mathematical rigor has not been sacrificed. Also great care has been taken to be attentive to even minute details.

Audience

The book has been designed for students in computer science, electrical engineering, and mathematics as a one- or two-semester course in discrete mathematics at the sophomore/junior level. Several earlier versions of the text were class-tested at two different institutions, with positive responses from students.

Prerequisites

No formal prerequisites are needed to enjoy the material or to employ its power, except a very strong background in college algebra. A good background in pre-calculus mathematics is desirable, but not essential. Perhaps the most important requirement is a bit of sophisticated mathematical maturity: a combination of patience, logical and analytical thinking, motivation, systematism, decision-making, and the willingness to persevere through failure until success is achieved.

Although no programming background is required to enjoy the discrete mathematics, knowledge of a structured programming language, such as Java or C++, can make the study of discrete mathematics more rewarding.

Coverage

The text contains in-depth coverage of all major topics proposed by professional associations for a discrete mathematics course. It emphasizes problem-solving techniques, pattern recognition, conjecturing, induction, applications of varying nature, proof techniques, algorithm development, algorithm correctness, and numeric computations.

Recursion, a powerful problem-solving strategy, is used heavily in both mathematics and computer science. Initially, for some students, it can be a bitter-sweet and demanding experience, so the strategy is presented with great care to help amateurs feel at home with this fascinating and frequently used technique for program development.

This book also includes discussions on Fibonacci and Lucas numbers, Fermat numbers, and figurate numbers and their geometric representations, all excellent tools for exploring and understanding recursion.

A sufficient amount of theory is included for those who enjoy the beauty in the development of the subject, and a wealth of applications as well for those who enjoy the power of problem-solving techniques. Hopefully, the student will benefit from the nice balance between theory and applications.

Optional sections in the book are identified with an asterisk (*) in the left margin. Most of these sections deal with interesting applications or discussions. They can be omitted without negatively affecting the logical development of the topic. However, students are strongly encouraged to pursue the optional sections to maximize their learning.

Historical Anecdotes and Biographies

Biographical sketches of about 60 mathematicians and computer scientists who have played a significant role in the development of the field are threaded into the text. Hopefully, they provide a human dimension and attach a human face to major discoveries. A biographical index, keyed to page, appears on the inside of the back cover for easy access.

Examples and Exercises

Each section in the book contains a generous selection of carefully tailored examples to clarify and illuminate various concepts and facts. The backbone of the book is the 560 examples worked out in detail for easy understanding.

Every section ends with a large collection of carefully prepared and well-graded exercises (more than 3700 in total), including thought-provoking true-false questions. Some exercises enhance routine computational skills; some reinforce facts, formulas, and techniques; and some require mastery of various proof techniques coupled with algebraic manipulation. Often exercises of the latter category require a mathematically sophisticated mind and hence are meant to challenge the mathematically curious.

Most of the exercise sets contain optional exercises, identified by the letter o in the left margin. These are intended for more mathematically sophisticated students.

Exercises marked with one asterisk (*) are slightly more advanced than the ones that precede them. Double-starred (**) exercises are more challenging than the single-starred; they require a higher level of mathematical maturity.

Exercises identified with the letter **c** in the left margin require a calculus background; they can be omitted by those with no or minimal calculus.

Answers or partial solutions to all odd-numbered exercises are given at the end of the book.

Foundation

Theorems are the backbones of mathematics. Consequently, this book contains the various proof techniques, explained and illustrated in detail.

They provide a strong foundation in problem-solving techniques, algorithmic approach, verification and analysis of algorithms, as well as in every discrete mathematics topic needed to pursue computer science courses such as *Data Structures*, *Analysis of Algorithms*, *Programming Languages*, *Theory of Compilers*, *Databases*, and *Theory of Computation*.

Proofs

Most of the concepts, definitions, and theorems in the book are illustrated with appropriate examples. Proofs shed additional light on the topic and enable students to sharpen their problem-solving skills. The various proof techniques appear throughout the text.

Applications

Numerous current and relevant applications are woven into the text, taken from computer science, chemistry, genetics, sports, coding theory, banking, casino games, electronics, decision-making, and gambling. They enhance understanding and show the relevance of discrete mathematics to everyday life. A detailed index of applications, keyed to pages, is given at the end of the book.

Algorithms

Clearly written algorithms are presented throughout the text as problem-solving tools. Some standard algorithms used in computer science are developed in a straightforward fashion; they are analyzed and proved to enhance problem-solving techniques. The computational complexities of a number of standard algorithms are investigated for comparison.

Algorithms are written in a simple-to-understand pseudocode that can easily be translated into any programming language. In this pseudocode:

- Explanatory comments are enclosed within the delimeters (* and *).
- The body of the algorithm begins with a **Begin** and ends in an **End**; they serve as the outermost parentheses.
- Every compound statement begins with a **begin** and ends in an **end**; again, they serve as parentheses. In particular, for easy readability, a while (for) loop with a compound statement ends in **endwhile** (**endfor**).

Chapter Summaries

Each chapter ends with a summary of important vocabulary, formulas, and properties developed in the chapter. All the terms are keyed to the text pages for easy reference and a quick review.

Review and Supplementary Exercises

Each chapter summary is followed by an extensive set of well-constructed review exercises. Used along with the summary, these provide a comprehensive review of the chapter. Chapter-end supplementary exercises provide additional challenging opportunities for the mathematically sophisticated and curious-minded for further experimentation and exploration. The book contains about 950 review and supplementary exercises.

Computer Assignments

Over 150 relevant computer assignments are given at the end of chapters. They provide hands-on experience with concepts and an opportunity to enhance programming skills. A computer algebra system, such as Maple, Derive, or Mathematica, or a programming language of choice can be used.

Exploratory Writing Projects

Each chapter contains a set of well-conceived writing projects, for a total of about 600. These expository projects allow students to explore areas not pursued in the book, as well as to enhance research techniques and to practice writing skills. They can lead to original research, and can be assigned as group projects in a real world environment.

For convenience, a comprehensive list of references for the writing projects, compiled from various sources, is provided in the *Student's Solutions Manual*.

Enrichment Readings

Each chapter ends with a list of selected references for further exploration and enrichment. Most expand the themes studied in this book.

Numbering System

A concise numbering system is used to label each item, where an *item* can be an algorithm, figure, example, exercises, section, table, or theorem. Item $m.n$ refers to item n in Chapter "m". For example, Section 3.4 is Section 4 in Chapter 3.

Special Symbols

Colored boxes are used to highlight items that may need special attention. The letter o in the left margin of an exercise indicates that it is optional, whereas a **c** indicates that it requires the knowledge of calculus. Besides, every theorem is easily identifiable, and the end of every proof and example is marked with a solid square (■). An asterisk (∗) next to an exercise indicates that it is challenging, whereas a double-star (∗∗) indicates that it is even more challenging. While "=" stands for equality, the closely related symbol "≈" means *is approximately equal to*:

o	optional exercises
c	requires a knowledge of calculus
■	end of a proof or a solution
∗	a challenging exercise
∗∗	a more challenging exercise
=	is equal to
≈	is approximately equal to

Abbreviations

For the sake of brevity, four useful abbreviations are used throughout the text: LHS, RHS, PMI, and IH:

LHS	Left-Hand Side
RHS	Right-Hand Side
PMI	Principle of Mathematical Induction
IH	Inductive Hypothesis

Symbols Index

An index of symbols used in the text and the page numbers where they occur can be found inside the front and back covers.

Web Links

The World Wide Web can be a useful resource for collecting information about the various topics and algorithms. Web links also provide biographies and discuss the discoveries of major mathematical contributors. Some Web sites for specific topics are listed in the Appendix.

Student's Solutions Manual

The *Student's Solutions Manual* contains detailed solutions of all odd-numbered exercises. It also includes suggestions for studying mathematics, and for preparing to take an math exam. The Manual also contains a comprehensive list of references for the various writing projects and assignments.

Instructor's Manual

The *Instructor's Manual* contains detailed solutions to all even-numbered exercises, two sample tests and their keys for each chapter, and two sample final examinations and their keys.

Acknowledgments

A number of people, including many students, have played a major role in substantially improving the quality of the manuscript through its development. I am truly grateful to every one of them for their unfailing encouragement, cooperation, and support.

To begin with, I am sincerely indebted to the following reviewers for their unblemished enthusiasm and constructive suggestions:

Gerald Alexanderson	Santa Clara University
Stephen Brick	University of South Alabama
Neil Calkin	Clemson University
Andre Chapuis	Indiana University
Luis E. Cuellar	McNeese State University
H. K. Dai	Oklahoma State University
Michael Daven	Mt. St. Mary College
Henry Etlinger	Rochester Institute of Technology
Jerrold R. Griggs	University of South Carolina
John Harding	New Mexico State University
Nan Jiang	University of South Dakota
Warren McGovern	Bowling Green State University
Tim O'Neil	University of Notre Dame
Michael O'Sullivan	San Diego State University
Stanley Selkow	Worcester Polytechnic Institute

Thanks also go to Henry Etlinger of Rochester Institute of Technology and Jerrold R. Griggs of the University of South Carolina for reading the entire manuscript for accuracy; to Michael Dillencourt of the University of California at Irvine, and Thomas E. Moore of Bridgewater State College for preparing the solutions to the exercises; and to Margarite Roumas for her excellent editorial assistance.

My sincere thanks also go to Senior Editor, Barbara Holland, Production Editor, Marcy Barnes-Henrie, Copy Editor, Kristin Landon, and Associate Editor, Thomas Singer for their devotion, cooperation, promptness, and patience, and for their unwavering support for the project.

Finally, I must accept responsibility for any errors that may still remain. I would certainly appreciate receiving comments about any unwelcome surprises, alternate or better solutions, and exercises, puzzles, and applications you have enjoyed.

Framingham, Massachusetts
September 19, 2003

Thomas Koshy
tkoshy@frc.mass.edu

A Word to the Student

Tell me and I will forget.
Show me and I will remember.
Involve me and I will understand.

— Confucius

The SALT of Life

Mathematics is a **s**cience; it is an **a**rt; it is a precise and concise **l**anguage; and it is a great problem-solving **t**ool. Thus mathematics is the **SALT** of life.

To learn a language, such as Greek or Russian, first you have to learn its alphabet, grammar, and syntax; you also have to build up a decent vocabulary to speak, read, or write. Each takes a lot of time and practice.

The Language of Mathematics

Because mathematics is a concise language with its own symbolism, vocabulary, and properties (or rules), to be successful in mathematics, you must know them well and be able to apply them.

For example, it is important to know that there is a difference between perimeter and area, area and volume, factor and multiple, divisor and dividend, hypothesis and hypotenuse, algorithm and logarithm, reminder and remainder, computing and solving, disjunction and destruction, conjunction and construction, and negation and negative. So you must be fluent in the language of mathematics, just like you need to be fluent in any foreign language. So keep speaking the language of mathematics.

Although mathematics is itself an unambiguous language, algebra is the language of mathematics. Studying algebra develops confidence, improves logical and critical thinking, and enhances what is called mathematical maturity, all needed for developing and establishing mathematical facts, and for solving problems.

This book is written in a clear and concise language that is easy to understand and easy to build on. It presents the essential (discrete) mathematical tools needed to succeed in all undergraduate computer science courses.

Theory and Applications

This book features a perfect blend of both theory and applications. Mathematics does not exist without its logically developed theory; in fact, theorems are like the steel beams of mathematics. So study the various proof techniques, follow the various proofs presented, and try to reproduce them in your own words. Whenever possible, create your own proofs. Try to feel at home with the various methods and proofs. Besides developing a working vocabulary, pay close attention to facts, properties, and formulas, and enjoy the beautiful development of each topic.

This book also draws on a vast array of interesting and practical applications to several disciplines, especially to computer science. These applications are spread throughout the book. Enjoy them, and appreciate the power of mathematics that can be applied to a variety of situations, many of which are found in business, industry, and scientific discovery in today's workplace.

Problem-Solving Strategies

To master mathematics, you must practice it; that is, you must apply and do mathematics. You must be able to apply previously developed facts to solve problems. For this reason, this book emphasizes problem-solving techniques. You will encounter two types of exercises in the exercise sets: The first type is computational, and the second type is algebraic and theoretical. Being able to do computational exercises does not automatically imply that you are able to do algebraic and theoretical exercises. So do not get discouraged, but keep trying until you succeed.

Of course, before you attempt the exercises in any section, you will need to first master the section; know the definitions, symbols, and facts, and redo the examples using your own steps.

Since the exercises are graded in ascending order of difficulty, always do them in order; save the solutions and refine them as you become mathematically more sophisticated.

The chapter-end review exercises give you a chance to re-visit the chapter. They can be used as a quick review of important concepts.

Recursion

Recursion is an extremely powerful problem-solving strategy, used often in mathematics and computer science. Although some students may need a lot of practice to get used to it, once you know how to approach problems recursively, you will certainly appreciate its great power.

Stay Actively Involved

Professional basketball players Magic Johnson, Larry Bird, and Michael Jordan didn't become superstars overnight by reading about basketball or by watching others play on television. Besides knowing the rules and the skills needed to play, they underwent countless hours of practice, hard work, a lot of patience and perseverance, willingness to meet failures, and determination to achieve their goal.

Likewise, you cannot master mathematics by reading about it or by simply watching your professor do it in class; you have to get involved and stay involved by doing it every day, just as skill is acquired in a sport. You can learn mathematics only in small, progressive steps, building on skills you have already mastered. Remember the saying: *Rome wasn't built in a day*.

Keep using the vocabulary and facts you have already studied. They must be fresh in your mind; review them every week.

A Few Suggestions for Learning Mathematics

- Read a few sections before each class. You might not fully understand the material, but you'll follow it far better when your professor discusses it in class. In addition, you will be able to ask more questions in class and answer more questions.
- Whenever you study the book, make sure you have a pencil and enough paper to write down definitions, theorems, and proofs, and to do the exercises.
- Return to review the material taught in class later in the same day. Read actively; do not just read as if it was a novel or a newspaper. Write down the definitions, theorems, and properties in your own words, without looking in your notes or the book. Good note-taking and writing aid retention. Re-write the examples, proofs, and exercises done in class, all in your own words. If you find them too challenging, study them again and try again; continue until you succeed.
- Always study the relevant section in the text and do the examples there; then do the exercises at the end of the section. Since the exercises are graded in order of difficulty, do them in order. Don't skip steps or write over previous steps; this way you'll progress logically, and you can locate and correct your errors. If you can't solve a problem because it involves a new term, formula, or some property, then re-study the relevant portion of the section and try again. Don't assume that you'll be able to do every problem the first time you try it. Remember, practice is the only way to success.

Solutions Manual

The *Student's Solutions Manual* contains additional helpful tips for studying mathematics, and preparing for and taking an examination in mathematics. It also gives detailed solutions to all odd-numbered exercises and a comprehensive list of references for the various exploratory writing projects.

A Final Word

Mathematics is no more difficult than any other subject. If you have the motivation, and patience to learn and do the work, then you will enjoy the beauty and power of discrete mathematics; you will see that discrete mathematics is really fun.

Keep in mind that learning mathematics is a step-by-step process. Practice regularly and systematically; review earlier chapters every week, since things must be fresh in your mind to apply and build on them. In this way, you will enjoy the subject, feel confident, and to explore more. The name of the game is practice, so practice, practice, practice.

I look forward to hearing from you with your comments and suggestions. In the meantime, enjoy the beauty and power of mathematics.

Thomas Koshy

Chapter 1

The Language of Logic

Symbolic logic has been disowned by many logicians on the plea that its interest is mathematical and by many mathematicians on the plea that its interest is logical.

—A. N. WHITEHEAD

Logic is the study of the principles and techniques of reasoning. It originated with the ancient Greeks, led by the philosopher Aristotle, who is often called the father of logic. However, it was not until the 17th century that symbols were used in the development of logic. German philosopher and mathematician Gottfried Leibniz introduced symbolism into logic. Nevertheless, no significant contributions in symbolic logic were made until those of George Boole, an English mathematician. At the age of 39, Boole published his outstanding work in symbolic logic, *An Investigation of the Laws of Thought*.

Logic plays a central role in the development of every area of learning, especially in mathematics and computer science. Computer scientists, for example, employ logic to develop programming languages and to establish the correctness of programs. Electronics engineers apply logic in the design of computer chips.

This chapter presents the fundamentals of logic, its symbols, and rules to help you to think systematically, to express yourself in precise and concise terms, and to make valid arguments.

Here are a few interesting problems we shall pursue in this chapter:

- Consider the following two sentences, both true:
  There are more residents in New York City than there are hairs on the head of any resident. No resident is totally bald. What is your conclusion: Is it true that at least two residents have the same number of hairs? (R. M. Smullyan, 1978)
- There are two kinds of inhabitants, "knights" and "knaves," on an island. Knights always tell the truth, whereas knaves always lie. Every inhabitant is either a knight or a knave. Tom and Dick are two residents. Tom says, "At least one of us is a knave." What are Tom and Dick?

Aristotle *(384–322* B.C.*), one of the greatest philosophers in Western culture, was born in Stagira, a small town in northern Greece. His father was the personal physician of the king of Macedonia. Orphaned young, Aristotle was raised by a guardian.*

At the age of 18, Aristotle entered Plato's Academy in Athens. He was the "brightest and most learned student" at the Academy which he left when Plato died in 347 B.C.

About 342 B.C.*, the king of Macedonia invited him to supervise the education of his young son, Alexander, who later became Alexander the Great. Aristotle taught him until 336* B.C.*, when the youth became ruler following the assassination of his father.*

Around 334 B.C.*, Aristotle returned to Athens and founded a school called the Lyceum. His philosophy and followers were called* peripatetic, *a Greek word meaning* "walking around," *since Aristotle taught his students while walking with them.*

The Athenians, perhaps resenting his relationship with Alexander the Great, who had conquered them, accused him of impiety soon after the Emperor's death in 323 B.C. *Aristotle, knowing the fate of Socrates, who had been condemned to death on a similar charge, fled to Chalcis, so the Athenians would not "sin twice against philosophy." He died there the following year.*

What are they if Tom says, "Either I'm a knave or Dick is a knight"? (R. M. Smullyan, 1978)

- Are there positive integers that can be expressed as the sum of two different cubes in two different ways?
- Does the formula $E(n) = n^2 - n + 41$ yield a prime number for every positive integer n?

1.1 Propositions

A declarative sentence that is either true or false, but *not* both, is a **proposition** (or a **statement**), which we will denote by the lowercase letter p, q, r, s, or t. The variables p, q, r, s, or t are **boolean variables** (or **logic variables**).

EXAMPLE 1.1 The following sentences are propositions:

(1) Socrates was a Greek philosopher.
(2) $3 + 4 = 5$.
(3) $1 + 1 = 0$ and the moon is made of green cheese.
(4) If $1 = 2$, then roses are red.

The following sentences are *not* propositions:

- Let me go! (exclamation)
- $x + 3 = 5$ (x is an unknown.)

Baron Gottfried Wilhelm Leibniz *(1646–1716), an outstanding German mathematician, philosopher, physicist, diplomat, and linguist, was born into a Lutheran family. The son of a professor of philosophy, he "grew up to be a genius with encylopedic knowledge."*

He had taught himself Latin, Greek, and philosophy before entering the University of Leipzig at age 15 as a law student. There he read the works of great scientists and philosophers such as Galileo, Francis Bacon, and René Descartes. Because of his youth, Leipzig refused to award him the degree of the doctor of laws, so he left his native city forever.

During 1663–1666, he attended the universities of Jena and Altdorf, and receiving his doctorate from the latter in 1666, he began legal services for the Elector of Mainz.

After the Elector's death, Leibniz pursued scientific studies. In 1672, he built a calculating machine that could multiply and divide and presented it to the Royal Society in London the following year.

In late 1675, Leibniz laid the foundations of calculus, an honor he shares with Sir Isaac Newton. He discovered the fundamental theorem of calculus, and invented the popular notations—d/dx *for differentiation and* $\int$ *for integration. He also introduced such modern notations as dot for multiplication, the decimal point, the equal sign, and the colon for ratio.*

From 1676, until his death, Leibniz worked for the Duke of Brunswick at Hanover and his estate after the duke's death in 1680. He played a key role in the founding of the Berlin Academy of Sciences in 1700.

Twelve years later, Leibniz was appointed councilor of the Russian Empire and was given the title of baron by Peter the Great.

Suffering greatly from gout, Leibniz died in Hanover. He was never married.

His works influenced such diverse disciplines as theology, philosophy, mathematics, the natural sciences, history, and technology.

- Close the door! (command)
- Kennedy was a great president of the United States. (opinion)
- What is my line? (interrogation) ■

Truth Value

The truthfulness or falsity of a proposition is called its **truth value**, denoted by T(true) and F(false), respectively. (These values are often denoted by 1 and 0 by computer scientists.) For example, the truth value of statement (1) in Example 1.1 is T and that of statement (2) is F.

Consider the sentence, *This sentence is false*. It is certainly a valid declarative sentence, but is it a proposition? To answer this, assume the sentence is true. But the sentence says it is false. This contradicts our assumption. On the other hand, suppose the sentence is false. This implies the sentence

***George Boole** (1815–1864), the son of a cobbler whose main interests were mathematics and the making of optical instruments, was born in Lincoln, England. Beyond attending a local elementary school and briefly a commercial school, Boole was self-taught in mathematics and the classics. When his father's business failed, he started working to support the family. At 16, he began his teaching career, opening a school of his own four years later in Lincoln.*

In his leisure time, Boole read mathematical journals at the Mechanics Institute. There he grappled with the works of English physicist and mathematician Sir Isaac Newton and French mathematicians Pierre-Simon Laplace and Joseph-Louis Lagrange.

In 1839, Boole began contributing original papers on differential equations to The Cambridge Mathematics Journal *and on analysis to the Royal Society. In 1844, he was awarded a Royal Medal by the Society for his contributions to analysis; he was elected a fellow of the Society in 1857.*

Developing novel ideas in logic and symbolic reasoning, he published his first contribution to symbolic logic, The Mathematical Analysis of Logic, *in 1847. His publications played a key role in his appointment as professor of mathematics at Queen's College, Cork, Ireland, in 1849, although he lacked a university education.*

In 1854, he published his most important work, An Investigation to the Laws of Thought, *in which he presented the algebra of logic now known as boolean algebra (see Chapter 12). The next year he married Mary Everest, the niece of Sir George Everest, for whom the mountain is named.*

In addition to writing about 50 papers, Boole published two textbooks, Treatise on Differential Equations *(1859) and* Treatise on the Calculus of Finite Differences; *both were used as texts in the United Kingdom for many years.*

A conscientious and devoted teacher, Boole died of pneumonia in Cork.

is true, which again contradicts our assumption. Thus, if we assume that the sentence is true, it is false; and if we assume that it is false, it is true. It is a meaningless and self-contradictory sentence, so it is *not* a proposition, but a **paradox**.

The truth value of a proposition may not be known for some reason, but that does not prevent it from being a proposition. For example, around 1637, the French mathematical genius Pierre-Simon de Fermat conjectured that the equation $x^n + y^n = z^n$ has *no* positive integer solutions, where $n \geq 3$. His conjecture, known as **Fermat's Last "Theorem,"** was one of the celebrated unsolved problems in number theory, until it was proved in 1993 by the English mathematician Andrew J. Wiles (1953–) of Princeton University. Although the truth value of the conjecture eluded mathematicians for over three centuries, it was still a proposition!

Here is another example of such a proposition. In 1742 the Prussian mathematician Christian Goldbach conjectured that every even integer greater than 2 is the sum of two primes, not necessarily distinct. For example, $4 = 2 + 2, 6 = 3 + 3$, and $18 = 7 + 11$. It has been shown true for every

Fermat *(1601–1665) was born near Toulouse as the son of a leather merchant. A lawyer by profession, he devoted his leisure time to mathematics. Although he published almost none of his discoveries, he did correspond with contemporary mathematicians.*

Fermat contributed to several branches of mathematics, but he is best known for his work in number theory. Many of his results appear in margins of his copy of the works of the Greek mathematician Diophantus (250 A.D.?). He wrote the following about his famous conjecture: "I have discovered a truly wonderful proof, but the margin is too small to contain it."

Christian Goldbach *(1690–1764) was born in Königsberg, Prussia. He studied medicine and mathematics at the University of Königsberg and became professor of mathematics at the Imperial Academy of Sciences in St. Petersburg in 1725. In 1728, he moved to Moscow to tutor Tsarevich Peter II and his cousin Anna of Courland. From 1729 to 1763, he corresponded with Euler on number theory. He returned to the Imperial Academy in 1732, when Peter's successor Anna moved the imperial court to St. Petersburg.*

In 1742, Goldbach joined the Russian Ministry of Foreign Affairs, and later became privy councilor and established guidelines for the education of royal children.

Noted for his conjectures in number theory and work in analysis, Goldbach died in Moscow.

even integer less than 4×10^{14}, but no one has been able to prove or disprove his conjecture. Nonetheless, the Goldbach conjecture is a proposition.

Propositions (1) and (2) in Example 1.1 are **simple propositions**. A **compound proposition** is formed by combining two or more simple propositions called **components**. For instance, propositions (3) and (4) in Example 1.1 are compound. The components of proposition (4) are *1* $= 2$ and *Roses are red*. The truth value of a compound proposition depends on the truth values of its components.

Compound propositions can be formed in several ways, and they are presented in the rest of this section.

Conjunction

The **conjunction** of two arbitrary propositions p and q, denoted by $p \wedge q$, is the proposition p *and* q. It is formed by combining the propositions using the word *and*, called a **connective**.

EXAMPLE 1.2 Consider the statements

p: *Socrates was a Greek philosopher*

and q: *Euclid was a Chinese musician.*

Their conjunction is given by

$p \wedge q$: *Socrates was a Greek philosopher and Euclid was a Chinese musician.* ■

To define the truth value of $p \wedge q$, where p and q are arbitrary propositions, we need to consider four possible cases:

- p is true, q is true.
- p is true, q is false.
- p is false, q is true.
- p is false, q is false.

(See the **tree diagram** in Figure 1.1 and Table 1.1.) If both p and q are true, then $p \wedge q$ is true; if p is true and q is false, then $p \wedge q$ is false; if p is false and q is true, then $p \wedge q$ is false; and if both p and q are false, then $p \wedge q$ is also false.

Figure 1.1

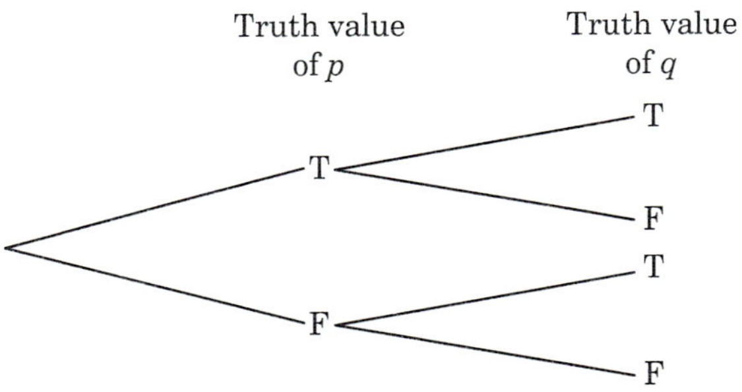

Table 1.1

p	q	$p \wedge q$
T	T	
T	F	
F	T	
F	F	

This information can be summarized in a table. In the third column of Table 1.1, enter the truth value of $p \wedge q$ corresponding to each pair of truth values of p and q. The resulting table, Table 1.2, is the **truth table** for $p \wedge q$.

Table 1.2

Truth table for $p \wedge q$

p	q	$p \wedge q$
T	T	T
T	F	F
F	T	F
F	F	F

Expressions that yield the value true or false are **boolean expressions**, and they often occur in both mathematics and computer science. For instance, $3 < 5$ and $5 < 5$ are boolean expressions. *If-statements* and *while-loops* in computer programs often use such expressions, and their values determine whether or not *if-statements* and *while-loops* will be executed, as the next example illustrates.

EXAMPLE 1.3 Determine whether the assignment statement, sum $\leftarrow$ sum $+ i + j$,* will be executed in the following sequence of statements:

```
i ← 3
j ← 5
sum ← 0
if (i < 4) and (j ≤ 5) then
  sum ← sum + i + j
```

SOLUTION:

The assignment statement will be executed if the truth value of the boolean expression $(i < 4)$ and $(j \leq 5)$ is T. So, let us evaluate it. Since $i \leftarrow 3$, $i < 4$ is true; since $j \leftarrow 5$, $j \leq 5$ is also true. Therefore, $(i < 4)$ and $(j \leq 5)$ is true (see row 1 of Table 1.2). Consequently, the given assignment statement will be executed. ■

Disjunction

A second way of combining two propositions p and q is by using the connective *or*. The resulting proposition p *or* q is the **disjunction** of p and q and is denoted by $p \vee q$.

EXAMPLE 1.4 Consider the statements

p: *Harry likes pepperoni pizza for lunch*

and

q: *Harry likes mushroom pizza for lunch.*

---

*The statement $x \leftarrow y$ means the value of the expression y is assigned to x, where $\leftarrow$ is the **assignment operator**. The general form of an **assignment statement** is *variable* $\leftarrow$ *expression*.

Their disjunction is given by

$p \vee q$: *Harry likes pepperoni pizza for lunch or Harry likes mushroom pizza for lunch.*

This sentence, however, is often written as

$p \vee q$: *Harry likes pepperoni or mushroom pizza for lunch.* ■

An interesting observation: In this example, Harry could like pepperoni pizza or mushroom pizza, or both, for lunch. In other words, the connective *or* is used in the inclusive sense *and/or* to mean *at least one, maybe both*. Such a disjunction is an **inclusive disjunction**.

Table 1.3 gives the truth table for an inclusive disjunction.

Table 1.3

Truth table for $p \vee q$

p	q	$p \vee q$
T	T	T
T	F	T
F	T	T
F	F	F

The disjunction of two propositions is true if at least one component is true; it is false only if both components are false.

EXAMPLE 1.5 Consider the statements

r: *Bernie will play basketball at 3 P.M. today*

and

s: *Bernie will go to a matinee at 3 P.M. today.*

Then $r \vee s$: *Bernie will play basketball or go to a matinee at 3 P.M. today.* ■

In this example, Bernie cannot play basketball and go to a matinee at the *same* time, so the word *or* is used in the exclusive sense to mean at least one, but *not* both. Such a disjunction is an **exclusive disjunction**. (See Exercise 31.) Throughout our discussion, we will be concerned with only inclusive disjunction, so the word "disjunction" will mean "inclusive disjunction."

Negation

The **negation** of a proposition p is *It is not the case that p*, denoted by $\sim p$. You may read $\sim p$ as *the negation of p* or simply *not p*.

EXAMPLE 1.6 Let p: Paris is the capital of France
and q: Apollo is a Hindu god.
The negation of p is

$\sim p$: It is not the case that Paris is the capital of France.

This sentence, however, is often written as

$\sim p$: Paris is not the capital of France.

Likewise, the negation of q is

$\sim q$: Apollo is not a Hindu god. ■

If a proposition p is true, then $\sim p$ is false; if p is false, then $\sim p$ is true. This definition is summarized in Table 1.4.

Table 1.4

Truth table for $\sim p$

p	$\sim p$
T	F
F	T

EXAMPLE 1.7 Evaluate each boolean expression, where $a = 3, b = 5$, and $c = 6$.

(1) $[\sim(a > b)] \wedge (b < c)$ (2) $\sim[(a \leq b) \vee (b > c)]$

SOLUTION:

(1) Since $a > b$ is false, $\sim(a > b)$ is true. Also, $b < c$ is true. Therefore,

$$[\sim (a > b)] \wedge (b < c) \text{ is true. (See row 1 of Table 1.2.)}$$

(2) $a \leq b$ is true; but $b > c$ is false. So $(a \leq b) \vee (b > c)$ is true. (See row 2 of Table 1.3.) Therefore, $\sim[(a \leq b) \vee (b > c)]$ is false. ■

Next we present another way of constructing new propositions.

Implication

Two propositions p and q can be combined to form statements of the form: *If p, then q*. Such a statement is an **implication**, denoted by $p \to q$. Since it involves a condition, it is also called a **conditional statement**. The component p is the **hypothesis** (or **premise**) of the implication and q the **conclusion**.

EXAMPLE 1.8 Let p: $\triangle$ABC is equilateral

and

q: $\triangle$ABC is isosceles.

Then

$p \to q$: *If $\triangle ABC$ is equilateral, then it is isosceles.*

Likewise,

$$q \to p: \textit{If } \triangle ABC \textit{ is isosceles, then it is equilateral.}$$

(*Note*: In the implication $q \to p$, q is the hypothesis and p is the conclusion.) ■

Implications occur in a variety of ways. The following are some commonly known occurrences:

- If p, then q.
- If p, q.
- p implies q.
- p only if q.
- q if p.
- p is sufficient for q.
- q is necessary for p.

Accordingly, the implication $p \to q$ can be read in one of these ways. For instance, consider the proposition

$$p \to q: \text{If } \triangle\text{ABC is equilateral, then it is isosceles.}$$

It means exactly the same as any of the following propositions:

- If △ABC is equilateral, it is isosceles.
- △ABC is equilateral implies it is isosceles.
- △ABC is equilateral only if it is isosceles.
- △ABC is isosceles if it is equilateral.
- That △ABC is equilateral, is a sufficient condition for it to be isosceles.
- That △ABC is isosceles, is a necessary condition for it to be equilateral.

Warning: The statement *p only if q* is often misunderstood as having the same meaning as the statement *p if q*. Remember, *p if q* means *If q, then p*. So be careful. Think of *only if* as one phrase; do not split it.

To construct the truth table for an implication *If p, then q*, we shall think of it as a conditional promise. If you do p, then I promise to do q. If the promise is kept, we consider the implication true; if the promise is not kept, we consider it false. We can use this analogy to construct the truth table, as shown below.

Consider the following implication:

$$p \to q: \text{If you wax my car, then I will pay you \$25.}$$

If you wax my car (p true) and if I pay you \$25 (q true), then the implication is true. If you wax my car (p true) and if I do not pay you \$25 (q false), then the promise is violated; hence the implication is false. What if you do not wax my car (p false)? Then I may give you \$25 (being generous!) or not. (So q may be true or false). In either case, my promise has not been tested and hence has not been violated. Consequently, the implication has not been proved false. If it is not false, it must be true. In other words, if p is false, the implication $p \to q$ is true by default. (If p is false, the implication is said to be **vacuously true**.)

This discussion is summarized in Table 1.5.

Table 1.5

Truth table for $p \to q$

p	q	$p \to q$
T	T	T
T	F	F
F	T	T
F	F	T

In the ordinary use of implications in the English language, there is a relationship between hypothesis and conclusion, as in the car waxing example. This relation, however, does not necessarily hold for formal implications. For instance, in the implication, *If the power is on, then* $3 + 5 = 8$, the conclusion $3 + 5 = 8$ is not even related to the hypothesis; however, from a mathematical point of view, the implication is true. This is so because the conclusion is true regardless of whether or not the power is on.

From an implication we can form three new implications — its converse, inverse, and contrapositive — as defined below.

Converse, Inverse, and Contrapositive

The **converse** of the implication $p \to q$ is $q \to p$ (switch the premise and the conclusion in $p \to q$). The **inverse** of $p \to q$ is $\sim p \to \sim q$ (negate the premise and the conclusion). The **contrapositive** of $p \to q$ is $\sim q \to \sim p$ (negate the premise and the conclusion, and then switch them).

EXAMPLE 1.9 Let $p \to q$: If △ABC is equilateral, then it is isosceles.

Its converse, inverse, and contrapositive are given by:

Converse $q \to p$: If △ABC is isosceles, then it is equilateral.

Inverse $\sim p \to \sim q$: If △ABC is not equilateral, then it is not isosceles.

Contrapositive $\sim q \to \sim p$: If △ABC is not isosceles, then it is not equilateral. ■

A word of caution: Many people mistakenly think that an implication and its converse mean the same thing; they usually say one to mean the other. In fact, they need not have the same truth value. You will, however, learn in Example 1.18 that an implication and its contrapositive have the same truth value, and so do the converse and the inverse.

Thus far, we have presented four **boolean operators**: $\wedge$, $\vee$, $\rightarrow$, and $\sim$. The first three enable us to combine two propositions; accordingly, they are **binary operators**. On the other hand, we need only one proposition to perform negation, so $\sim$ is a **unary operator**. These operators can be employed to construct more complex statements, as the next example demonstrates.

EXAMPLE 1.10 Construct a truth table for $(p \rightarrow q) \wedge (q \rightarrow p)$.

SOLUTION:

We construct the proposition $(p \rightarrow q) \wedge (q \rightarrow p)$ step-by-step. From the propositions p and q, we can form $p \rightarrow q$ and $q \rightarrow p$; then take their conjunction to yield the given statement. Thus, the truth table for $(p \rightarrow q) \wedge (q \rightarrow p)$ requires five columns: $p, q, p \rightarrow q, q \rightarrow p$, and $(p \rightarrow q) \wedge (q \rightarrow p)$ in that order (see Table 1.6). As before, first enter the possible pairs of truth values for p and q in columns 1 and 2. Then use the truth tables for implication (Table 1.5) and conjunction (Table 1.2) to complete the remaining columns. The resulting table is displayed in Table 1.6. It follows from the table that $(p \rightarrow q) \wedge (q \rightarrow p)$ is true if both p and q have the *same* truth values. ■

Table 1.6

p	q	$p \rightarrow q$	$q \rightarrow p$	$(p \rightarrow q) \wedge (q \rightarrow p)$
T	T	T	T	T
T	F	F	T	F
F	T	T	F	F
F	F	T	T	T

The Island of Knights and Knaves

The next two examples* illustrate the power of truth tables in decision-making and in arriving at logical conclusions in the midst of seemingly confusing and contradictory statements.

EXAMPLE 1.11 Faced with engine problems, Ellen Wright made an emergency landing on the beach of the Island of Knights and Knaves. The island is inhabited by two distinct groups of people, knights and knaves. Knights always tell the truth and knaves always lie. Ellen decided that her best move was to reach the capital and call for service.

*Based on C. Baltus, "A Truth Table on the Island of Truthtellers and Liars," *Mathematics Teacher*, Vol. 94 (Dec. 2001), pp. 730–732.

Walking from the beach, she came to an intersection, where she saw two men, A and B, working nearby. After hearing her story, A told Ellen, "The capital is in the mountains, or the road on the right goes to the capital." B then said, "The capital is in the mountains, and the road on the right goes to the capital." Then A looked up and said, "That man is a liar." Shrugging his shoulders, B then said, "If the capital is in the mountains, then the road to the right goes to the capital." Ellen then made a table on the back of her guidebook, thanked the two men, and walked down the road on the left. Did Ellen make the correct decision?

SOLUTION:

Let c: The capital is in the mountains

and

r: The road on the right goes to the capital.

Now we build a truth table, as Table 1.7 shows. Since B could not give both false and true statements (see rows 3 and 4 in columns 4 and 5), the last two rows of the table do not fit the given scenario; so they can be ignored.

Table 1.7

c	r	A: $c \vee r$	B: $c \wedge r$	B: $c \to r$
T	T	T	T	T
T	F	T	F	F
F	T	T	F	T
F	F	F	F	T

It now follows from the rest of column 3 that A is a knight. So his statement that "B is a liar" is true; thus B is a knave. Consequently, we can ignore row 1 also. This leaves us with row 2. Therefore, the statement r is false; that is, the road on the left goes to the capital. Thus Ellen made the correct decision. ■

The following example is a continuation of Ellen's saga.

EXAMPLE 1.12 Walking up the road to the left, Ellen encountered a group of people gathered at what she thought to be a bus stop. She approached three women, C, D, and E, and asked them whether the road went to the capital and whether the location was indeed a bus stop. She received three different responses:

C: "The road goes to the capital, and the bus stop is not here."

D: "The road does not go to the capital, and the bus stop is here."

E: "The road does not go to the capital, and the bus stop is not here."

Confused and somewhat perplexed, Ellen asked them whether they are knights or knaves. To this they all answered, "Two of us are knights, and one is a liar." How many of the three women are knights? Does the road go to the capital? Is the location where Ellen met them a bus stop?

SOLUTION:
Once again, we build a truth table. To this end, we let

g: The road goes to the capital

and

b: The bus stop is here.

Table 1.8 shows the resulting table, where only some columns are shown for convenience.

Table 1.8

g	b	C: $g \wedge \sim b$	D: $\sim g \wedge b$	E: $\sim g \wedge \sim b$
T	T	F	F	F
T	F	T	F	F
F	T	F	T	F
F	F	F	F	T

Since all three women made the same statement, "Two of us are knights, and one is a liar," they must all be knaves. Consequently, we can discard rows 2–4 in Table 1.8. (It now follows from row 1 that the three women are all knaves.) So the road does indeed go to the capital and the location is a bus stop. ■

Ellen's story is continued further in the exercises. See Exercises 76–78. Next we present yet another method of combining propositions.

Biconditional Statement

Two propositions p and q can be combined using the connective *if and only if*. The resulting proposition, p *if and only if* q, is the conjunction of two implications: (1) p only if q, and (2) p if q, that is, $p \to q$ and $q \to p$. Accordingly, it is called a **biconditional statement**, symbolized by $p \leftrightarrow q$.

EXAMPLE 1.13

Let p: $\triangle$ABC is equilateral

and q: $\triangle$ABC is equiangular.

Then the biconditional statement is given by

$p \leftrightarrow q$: $\triangle$ABC is equilateral if and only if it is equiangular. ■

Since the biconditional $p \leftrightarrow q$ means exactly the same as the statement $(p \to q) \wedge (q \to p)$, they have the same truth value in every case. We can use this fact, and columns 1, 2, and 5 of Table 1.6 to construct the truth table for $p \leftrightarrow q$, as in Table 1.9.

Notice that the statement $p \leftrightarrow q$ is true if both p and q have the same truth value; conversely, if $p \leftrightarrow q$ is true, then p and q have the same truth value.

Table 1.9

Truth table for $p \leftrightarrow q$

p	q	$p \leftrightarrow q$
T	T	T
T	F	F
F	T	F
F	F	T

Here is a simple application of this fact with which you are already familiar (see Section 7.8).

EXAMPLE 1.14 Let S denote the sum of the digits in 2034. If 3 is a factor of S, then 3 is a factor of 2034 also. Conversely, if 3 is a factor of 2034, then 3 is a factor of S also. Thus the biconditional, 2034 is divisible by 3 if and only if S is divisible by 3, is a true proposition. Consequently, if one component — say, S is divisible by 3 — is true, then the other component is also true. ■

Order of Precedence

To evaluate complex logical expressions, you must know the order of precedence of the logical operators. The order of precedence from the highest to the lowest is: (1) $\sim$ (2) $\wedge$ (3) $\vee$ (4) $\rightarrow$ (5) $\leftrightarrow$. Note that parenthesized subexpressions are always evaluated first; if two operators have equal precedence, the corresponding expression is evaluated from left to right. For example, the expression $(p \rightarrow q) \wedge \sim q \rightarrow \sim p$ is evaluated as $[(p \rightarrow q) \wedge (\sim q)] \rightarrow (\sim p)$, and $p \rightarrow q \leftrightarrow \sim q \rightarrow \sim p$ is evaluated as $(p \rightarrow q) \leftrightarrow [(\sim q) \rightarrow (\sim p)]$.

The next example involves constructing a truth table for a conditional statement and we shall use it shortly to make a few definitions.

EXAMPLE 1.15 Construct a truth table for $(p \rightarrow q) \leftrightarrow (\sim p \vee q)$.

SOLUTION:

We need columns for p, q, $p \rightarrow q$, $\sim p$, $\sim p \vee q$, and $(p \rightarrow q) \leftrightarrow (\sim p \vee q)$. First, fill in the first two columns with the four pairs of truth values for p and q. Then use the truth tables for implication, negation, disjunction, and biconditional to complete the remaining columns. Table 1.10 shows the resulting table.

Table 1.10

p	q	$p \rightarrow q$	$\sim p$	$\sim p \vee q$	$(p \rightarrow q) \leftrightarrow (\sim p \vee q)$
T	T	T	F	T	T
T	F	F	F	F	T
F	T	T	T	T	T
F	F	T	T	T	T

↑ always true!

Tautology, Contradiction, and Contingency

An interesting observation: It is clear from Table 1.10 that the compound statement $(p \to q) \leftrightarrow (\sim p \vee q)$ is always true, regardless of the truth values of its components. Such a compound proposition is a **tautology**; it is an eternal truth. For example, $p \vee \sim p$ is a tautology. (Verify this.)

On the other hand, a compound statement that is always false is a **contradiction**. For instance, $p \wedge \sim p$ is a contradiction (Verify this). A compound proposition that is neither a tautology nor a contradiction is a **contingency**. For example, $p \vee q$ is a contingency.

Next we show that there is a close relationship between symbolic logic and switching networks.

Switching Network (optional)

A **switching network** is an arrangement of wires and switches connecting two terminals. A switch that permits the flow of current is said to be **closed**; otherwise, it is **open**. Likewise, a switching network is **closed** if current can flow from one end of the network to the other; otherwise, it is **open**.

Two switches A and B can be connected either in **series** (see Figure 1.2) or in **parallel** (see Figure 1.3). The switching network in Figure 1.2 is closed if and only if both A and B are closed; accordingly, it is symbolically denoted by A $\wedge$ B. The network in Figure 1.3 is closed if and only if at least one of the two switches is closed; consequently, it is denoted by A $\vee$ B.

Figure 1.2

Switches connected in series, A $\wedge$ B.

Figure 1.3

Switches connected in parallel, A $\vee$ B.

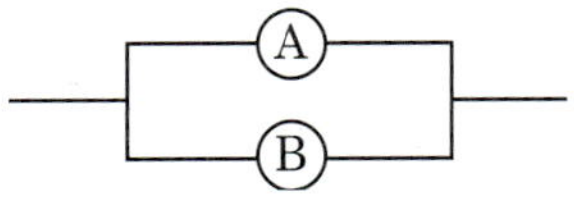

An electrical network may contain two switches A and A′ (A prime) such that if one is closed, then the other is open, and vice versa. (The operator ′ corresponds to the logical operator $\sim$.)

A switching network, in general, consists of series and parallel connections and hence can be described symbolically using the operators $\wedge$, $\vee$, and ′, as the following example illustrates.

EXAMPLE 1.16 Find a symbolic representation of the switching network in Figure 1.4.

SOLUTION:

Switches A and B′ are connected in series; the corresponding portion of the circuit is symbolized by A $\wedge$ B′. Switch B is in parallel with A $\wedge$ B′; so we have (A $\wedge$ B′) $\vee$ B. Since A′ and C are connected in series, the corresponding portion of the network is described by A′ $\wedge$ C. The circuits (A $\wedge$ B′) $\vee$ B and (A′ $\wedge$ C) are connected in parallel. Therefore, the given network is

Figure 1.4

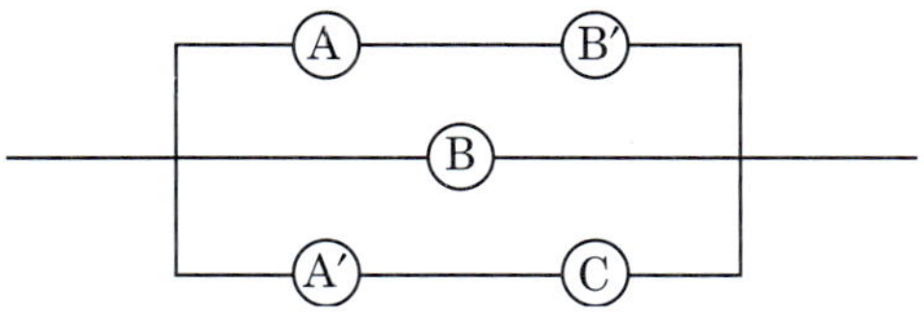

symbolized by $[(A \wedge B') \vee B] \vee (A' \wedge C)$. Since the operation $\vee$ is associative (see Table 1.13), this expression can be rewritten as $(A \wedge B') \vee B \vee (A' \wedge C)$. ■

Exercises 1.1

Which of the following are propositions?

1. The earth is flat.

2. Toronto is the capital of Canada.

3. What a beautiful day!

4. Come in.

Find the truth value of each compound statement.

5. $(5 < 8)$ and $(2 + 3 = 4)$

6. Paris is in France or $2 + 3 = 4$.

7. If $1 = 2$, then $3 = 3$.

8. $\triangle$ABC is equilateral if and only if it is equiangular.

Negate each proposition.

9. $1 + 1 = 0$.

10. The chalkboard is black.

Let x, y, and z be any real numbers. Represent each sentence symbolically, where p: $x < y$, q: $y < z$, and r: $x < z$.

11. $(x \geq y)$ or $(y < z)$

12. $(y \geq z)$ or $(x \geq z)$

13. $(x \geq y)$ and $[(y < z)$ or $(z > x)]$

14. $(x < y)$ or $[(y \geq z)$ and $(z > x)]$

Evaluate each boolean expression, where $a = 2, b = 3, c = 5$, and $d = 7$.

15. $[\sim(a > b)] \vee [\sim(c < d)]$

16. $[\sim(b < c)] \wedge [\sim(c < d)]$

17. $\sim[(a > b) \vee (b \leq d)]$

18. $\sim\{(a \leq b) \wedge [\sim(c > d)]\}$

Let t be a tautology and p an arbitrary proposition. Give the truth value of each proposition.

19. $\sim p \vee t$

20. $\sim p \wedge \sim t$

21. $\sim t \wedge p$

22. $\sim (\sim p \wedge \sim t)$

Construct a truth table for each proposition.

23. $\sim p \vee \sim q$

24. $\sim(\sim p \vee q)$

25. $(p \vee q) \vee (\sim q)$

26. $p \wedge (q \wedge r)$

Give the truth value of each proposition, using the given information.

27. $p \wedge q$, where q is not true.

28. $p \wedge q$, where $\sim q$ is not false.

29. $p \vee q$, where $\sim p$ is false.

30. $p \vee q$, where $\sim p$ is not true.

31. The **exclusive disjunction** of two propositions p and q is denoted by p **XOR** q. Construct a truth table for p XOR q.

Write each sentence in *if–then* form.

32. An equiangular triangle is isosceles.

33. Lines perpendicular to the same line are parallel.

34. $x^2 = 16$ is necessary for $x = 4$.

35. $x = 1$ is sufficient for $x^2 = 1$.

Write the converse, inverse, and contrapositive of each implication.

36. If the calculator is working, then the battery is good.

37. If London is in France, then Paris is in England.

Let x, y, and z be any real numbers. Represent each sentence symbolically, where p: $x < y$, q: $y < z$, and r: $x < z$.

38. If $x \geq y$ and $x < z$, then $y < z$.

39. If $z \geq y$ and $x < y$, then $z > x$.

40. $x < z$ if and only if $x < y$ and $y < z$.

41. $x \geq y$ and $y \geq z$ if and only if $x \geq z$.

Determine whether or not the assignment statement $x \leftarrow x + 1$ will be executed in each sequence of statements, where $i \leftarrow 2, j \leftarrow 3$, $k \leftarrow 6$, and $x \leftarrow 0$.

42. If $(i < 3) \wedge (j < 4)$ then
    $x \leftarrow x + 1$
  else
    $y \leftarrow y + 1$

43. If $(i < j) \vee (k > 4)$ then
    $x \leftarrow x - 1$
  else
    $x \leftarrow x + 1$

44. While $\sim(i \leq j)$ do
  begin
    $x \leftarrow x + 1$
    $i \leftarrow i + 1$
  endwhile

45. While $\sim(i + j \geq k)$ do
    $x \leftarrow x + 1$

Let t be a tautology and p an arbitrary proposition. Find the truth value of each.

46. $(\sim t) \rightarrow p$

47. $p \rightarrow t$

48. $(p \vee t) \rightarrow t$

49. $(p \vee t) \rightarrow (\sim t)$

50. $(p \wedge t) \rightarrow p$

51. $p \rightarrow (p \wedge t)$

52. $t \leftrightarrow (p \vee t)$

53. $p \leftrightarrow (p \wedge t)$

Construct a truth table for each proposition.

54. $p \to (p \vee q)$ **55.** $(p \wedge q) \to \sim p$

56. $(p \wedge q) \to (p \vee q)$ **57.** $(p \vee q) \leftrightarrow (p \wedge q)$

Determine whether or not each is a tautology.

58. $p \vee (\sim p)$ **59.** $[p \wedge (p \to q)] \to q$

60. $[(p \to q) \wedge (\sim q)] \to \sim p$ **61.** $[(p \vee q) \wedge (\sim q)] \to p$

Determine whether or not each is a contradiction.

62. $p \wedge (\sim p)$ **63.** $p \leftrightarrow \sim p$

64. $\sim(p \vee \sim p)$ **65.** $\sim p \leftrightarrow (p \vee \sim p)$

Indicate the order in which each logical expression is evaluated by properly grouping the operands using parentheses.

66. $p \vee q \wedge r$ **67.** $p \wedge q \to \sim p \vee \sim q$

68. $p \vee q \leftrightarrow \sim p \wedge \sim q$ **69.** $p \to q \leftrightarrow \sim p \vee q$

○ Represent each switching network symbolically.

70.

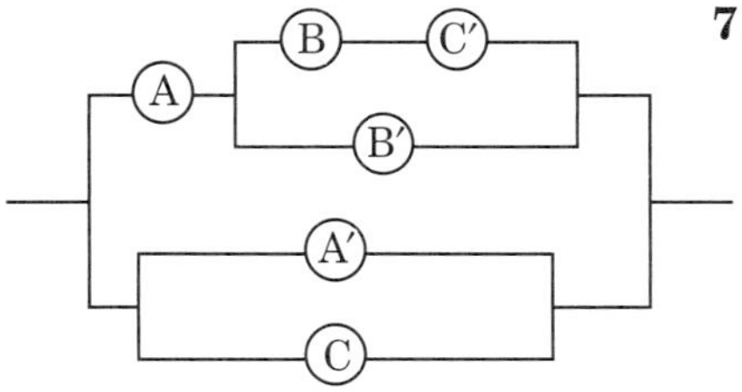

71.

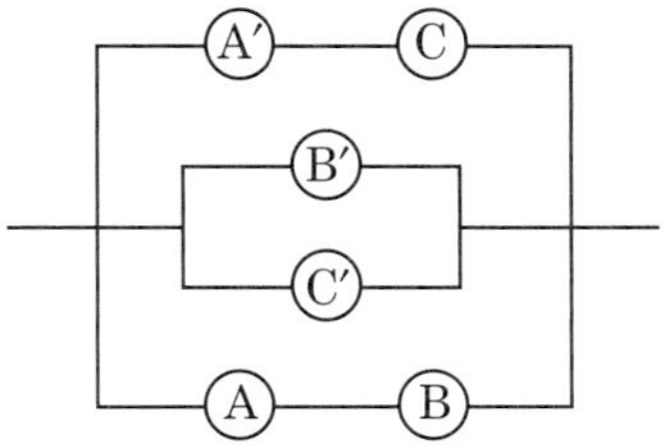

○ Draw a switching network with each representation.

72. $(A \vee B) \wedge (A \vee C)$ **73.** $(A \vee B') \vee (A \vee B)$

74. $(A \wedge B') \vee (A' \wedge B)$ **75.** $(A \wedge B) \vee (A' \wedge B) \vee (B' \wedge C)$

76. (Examples 1.11 and 1.12 continued) At the bus stop, Ellen noticed signs for three buses, B_1, B_2, and B_3, and approached another trio of women, F, G, and H. A conversation ensued:

Ellen: Where do the buses go?

F: At least one of B_1 and B_2 goes to the capital.

G: B_1 goes to the capital.

H: B_2 and B_3 go to the capital.

F: B_3 goes to the beach.

G: B_2 and B_3 go to the beach.

H: B_1 goes to the beach.

Which bus did Ellen take?

77. After reaching the bus terminal at the capital, Ellen saw three personal computers. She asked a young woman, I, whether the computers had Internet connections. She replied, "Computer 1 is not connected to the Internet. Ask that man, J; he is a knight." When Ellen approached the man, he told her, "Computer 2 has an Internet connection, but computer 3 does not." A second man, K, who overheard the conversation, then said, "If computer 2 has an Internet connection, then so does computer 1. Computer 3 is not connected to the Internet." Which computer had an Internet connection?

78. At the bus terminal, Ellen overheard the following conversation between two baseball fans, L and M:

L: I like the Yankees.

M: You do not like the Yankees. You like the Dodgers.

L: We both like the Dodgers.

Does fan L like the Yankees? Who likes the Dodgers?

1.2 Logical Equivalences

Two compound propositions p and q, although they may look different, can have identical truth values for all possible pairs of truth values of their components. Such statements are **logically equivalent**, symbolized by $p \equiv q$; otherwise, we write $p \not\equiv q$. If $p \equiv q$, the columns headed by them in a truth table are identical.

The next two examples illustrate this definition.

EXAMPLE 1.17 Verify that $p \to q \equiv \sim p \vee q$.

SOLUTION:
Construct a truth table containing columns headed by $p \to q$ and $\sim p \vee q$, as in Table 1.11. Use the truth tables for implication, negation, and disjunction to fill in the last three columns. Since the columns headed by $p \to q$ and $\sim p \vee q$ are identical, the two propositions have identical truth values. In other words, $p \to q \equiv \sim p \vee q$.

Table 1.11

p	q	$p \to q$	$\sim p$	$\sim p \vee q$
T	T	T	F	T
T	F	F	F	F
F	T	T	T	T
F	F	T	T	T

↑ identical columns ↑

(*Note*: This example shows that an implication can be expressed in terms of negation and disjunction.)

EXAMPLE 1.18 Show that $p \to q \equiv {\sim}q \to {\sim}p$; that is, an implication is logically equivalent to its contrapositive.

SOLUTION:
Once again, construct a truth table, with columns headed by $p, q, p \to q$, ${\sim}q$, ${\sim}p$, and ${\sim}q \to {\sim}p$. Use the truth tables for implication and negation to complete the last four columns. The resulting table (see Table 1.12) shows that the columns headed by $p \to q$ and ${\sim}q \to {\sim}p$ are identical; therefore, $p \to q \equiv {\sim}q \to {\sim}p$.

Table 1.12

p	q	$p \to q$	${\sim}q$	${\sim}p$	${\sim}q \to {\sim}p$
T	T	T	F	F	T
T	F	F	T	F	F
F	T	T	F	T	T
F	F	T	T	T	T

↑ identical columns ↑ ■

This is an extremely useful, powerful result that plays an important role in proving theorems, as we will see in Section 1.5.

It follows by this example that $q \to p \equiv {\sim}p \to {\sim}q$ (Why?); that is, the converse of an implication and its inverse have identical truth values.

Using truth tables, the laws of logic in Table 1.13 can be proved. We shall prove one of the De Morgan laws and leave the others as routine exercises.

Table 1.13

Laws of Logic

Let p, q, and r be any three propositions. Let t denote a tautology and f a contradiction. Then:

Idempotent laws

1. $p \wedge p \equiv p$	2. $p \vee p \equiv p$

Identity laws

3. $p \wedge t \equiv p$	4. $p \vee f \equiv p$

Inverse laws

5. $p \wedge ({\sim}p) \equiv f$	6. $p \vee ({\sim}p) \equiv t$

Domination laws

7. $p \vee t \equiv t$	8. $p \wedge f \equiv f$

Continued

Table 1.13

(continued)

Commutative laws	
9. $p \wedge q \equiv q \wedge p$	10. $p \vee q \equiv q \vee p$
Double negation	
11. $\sim(\sim p) \equiv p$	
Associative laws	
12. $p \wedge (q \wedge r) \equiv (p \wedge q) \wedge r$	13. $p \vee (q \vee r) \equiv (p \vee q) \vee r$
Distributive laws	
14. $p \wedge (q \vee r) \equiv (p \wedge q) \vee (p \wedge r)$	15. $p \vee (q \wedge r) \equiv (p \vee q) \wedge (p \vee r)$
De Morgan's laws	
16. $\sim(p \wedge q) \equiv \sim p \vee \sim q$	17. $\sim(p \vee q) \equiv \sim p \wedge \sim q$
Implication conversion law	
18. $p \to q \equiv \sim p \vee q$	
Contrapositive law	
19. $p \to q \equiv \sim q \to \sim p$	
Reductio ad absurdum law	
20. $p \to q \equiv (p \wedge \sim q) \to f$	

We can make a few observations about some of the laws. The commutative laws imply that the order in which we take the conjunction (or disjunction) of two propositions does not affect their truth values. The associative laws say that the way we group the components in a conjunction (or disjunction) of three or more propositions does not alter the truth value of the resulting proposition; accordingly, parentheses are *not* needed to indicate the grouping. In other words, the expressions $p \wedge q \wedge r$ and $p \vee q \vee r$ are no longer ambiguous, but do make sense.

Nonetheless, parentheses are essential to indicate the groupings in the distributive laws. For instance, if we delete the parentheses in law 14, then its left-hand side (LHS) becomes $p \wedge q \vee r \equiv (p \wedge q) \vee r$, since $\wedge$ has higher priority than $\vee$. But $(p \wedge q) \vee r \not\equiv p \wedge (q \vee r)$. (You may verify this.)

We now verify De Morgan's law 16 in the following example.

EXAMPLE 1.19 Verify that $\sim(p \wedge q) \equiv \sim p \vee \sim q$.

SOLUTION:
Construct a truth table with columns headed by $p, q, p \wedge q$, $\sim(p \wedge q)$, $\sim p$, $\sim q$, and $\sim p \vee \sim q$. Since columns 4 and 7 in Table 1.14 are identical, it follows that $\sim(p \wedge q) \equiv \sim p \vee \sim q$. ■

Augustus De Morgan *(1806–1871) was born in Madurai, Tamil Nadu, India, where his father was a colonel in the Indian army. When the young De Morgan was 7 months old, the family moved to England. He attended private schools, where he mastered Latin, Greek, and Hebrew and developed a strong interest in mathematics. After graduating in 1827 from Trinity College, Cambridge, he pondered a career either in medicine or law, but pursued mathematics. His professional career began in 1828 at University College, London. Three years later, however, when the college dismissed a colleague in anatomy without explanation, De Morgan resigned on principle. He returned to Trinity in 1836 when his successor died and remained there until a second resignation in 1866.*

A fellow of the Astronomical Society and a founder of the London Mathematical Society, De Morgan greatly influenced the development of mathematics in the 19th century. He exuded his passion for the subject in his teaching, stressing principles over techniques.

An incredibly prolific writer, De Morgan authored more than 1000 articles in more than 15 journals, as well as a number of textbooks, all characterized by clarity, logical presentation, and meticulous detail.

De Morgan's original contributions to mathematics were mainly in analysis and logic. In 1838, he coined the term mathematical induction *and gave a clear justification to this proof method, although it had been in use. His* The Differential and Integral Calculus *(1842) gives the first precise definition of a limit and some tests for convergence of infinite series.*

De Morgan was also interested in the history of mathematics. He wrote biographies of Sir Isaac Newton and Edmund Halley. His wife wrote De Morgan's biography in 1882.

His researches into all branches of knowledge and his prolific writing left him little time for social or family life, but he was well-known for his sense of humor.

Table 1.14

p	q	$p \wedge q$	$\sim(p \wedge q)$	$\sim p$	$\sim q$	$\sim p \vee \sim q$
T	T	T	F	F	F	T
T	F	F	T	F	T	F
F	T	F	T	T	F	T
F	F	F	T	T	T	T

↑ identical columns ↑

De Morgan's laws, although important, can be confusing, so be careful when you negate a conjunction or a disjunction. The negation of a conjunction (or disjunction) of two statements is the disjunction (or conjunction) of their negations.

The next two examples illustrate De Morgan's laws.

EXAMPLE 1.20 Let

p: Peter likes plain yogurt

and

q: Peter likes flavored yogurt.

Then

$$p \wedge q\text{: Peter likes plain yogurt and flavored yogurt.}$$

$$p \vee q\text{: Peter likes plain yogurt or flavored yogurt.}$$

By De Morgan's laws,

$\sim(p \wedge q) \equiv \sim p \vee \sim q$: Peter does not like plain yogurt or does not like flavored yogurt

and

$\sim(p \vee q) \equiv \sim p \wedge \sim q$: Peter likes neither plain yogurt nor flavored yogurt. ■

De Morgan's laws can be used in reverse order also; that is, $\sim p \vee \sim q \equiv \sim(p \wedge q)$ and $\sim p \wedge \sim q \equiv \sim(p \vee q)$. For instance, the sentence, *Claire does not like a sandwich or does not like pizza for lunch* can be rewritten as *It is false that Claire likes a sandwich and pizza for lunch*. Likewise, the sentence, *The earth is not flat and not round* can be restated as *It is false that the earth is flat or round.*

EXAMPLE 1.21 Determine whether or not the statement $x \leftarrow x + 1$ will be executed in the following sequence of statements:

```
a ← 7; b ← 4
if ~[(a < b) ∨ (b ≥ 5)] then
  x ← x + 1
```

SOLUTION:

The statement $x \leftarrow x + 1$ will be executed if the value of the boolean expression $\sim[(a < b) \vee (b \geq 5)]$ is true. By De Morgan's law,

$$\begin{aligned}\sim[(a < b) \vee (b \geq 5)] &\equiv \sim(a < b) \wedge \sim(b \geq 5)\\ &\equiv (a \geq b) \wedge (b < 5)\end{aligned}$$

Since $a = 7$ and $b = 4$, both $a \geq b$ and $b < 5$ are true; so, $(a \geq b) \wedge (b < 5)$ is true. Therefore, the assignment statement will be executed. ■

One of the elegant applications of the laws of logic is employing them to simplify complex boolean expressions, as the next example illustrates.

EXAMPLE 1.22 Using the laws of logic simplify the boolean expression $(p \wedge \sim q) \vee q \vee (\sim p \wedge q)$.

SOLUTION:

[The justification for every step is given on its right-hand-side (RHS).]

$$\begin{aligned}(p \wedge \sim q) \vee q \vee (\sim p \wedge q) &\equiv [(p \wedge \sim q) \vee q] \vee (\sim p \wedge q) && \text{assoc. law}\\ &\equiv [q \vee (p \wedge \sim q)] \vee (\sim p \wedge q) && \text{comm. law}\end{aligned}$$

$\equiv [(q \vee p) \wedge (q \vee \sim q)] \vee (\sim p \wedge q)$	dist. law
$\equiv [(q \vee p) \wedge t] \vee (\sim p \wedge q)$	$q \vee \sim q \equiv t$
$\equiv (q \vee p) \vee (\sim p \wedge q)$	$r \wedge t \equiv r$
$\equiv (\sim p \wedge q) \vee (p \vee q)$	comm. law
$\equiv [\sim p \vee (p \vee q)] \wedge [q \vee (p \vee q)]$	dist. law
$\equiv [(\sim p \vee p) \vee q] \wedge [q \vee (p \vee q)]$	assoc. law
$\equiv (t \vee q) \wedge [q \vee (p \vee q)]$	$\sim p \vee p \equiv t$
$\equiv t \wedge [q \vee (p \vee q)]$	$t \vee q \equiv t$
$\equiv q \vee (p \vee q)$	$t \wedge r \equiv r$
$\equiv q \vee (q \vee p)$	comm. law
$\equiv (q \vee q) \vee p$	assoc. law
$\equiv q \vee p$	idem. law
$\equiv p \vee q$	comm. law ■

For any propositions p, q, and r, it can be shown that $p \to (q \vee r) \equiv (p \wedge \sim q) \to r$ (see Exercise 12). We shall employ this result in Section 1.5.

Here are two elementary but elegant applications of this equivalence.

Suppose a and b are any two real numbers, and we would like to prove the following theorem: *If* $a \cdot b = 0$, *then either* $a = 0$ *or* $b = 0$. By virtue of the above logical equivalence, we need only prove the following proposition: *If* $a \cdot b = 0$ *and* $a \neq 0$, *then* $b = 0$ (see Exercise 43 in Section 1.5).

Second, suppose a and b are two arbitrary positive integers, and p a prime number. Suppose we would like to prove the following fact: *If* $p|ab$,* *then either* $p|a$ *or* $p|b$. Using the above equivalence, it suffices to prove the following equivalent statement: *If* $p|ab$ *and* $p \nmid a$, *then* $p|b$ (see Exercise 37 in Section 4.2).

We shall now show how useful symbolic logic is in the design of switching networks.

Equivalent Switching Networks (optional)

Two switching networks A and B are **equivalent** if they have the same electrical behavior, either both open or both closed, symbolically described by A ≡ B. One of the important applications of symbolic logic is to replace an electrical network, whenever possible, by an equivalent simpler network to minimize cost, as illustrated in the following example. To this end,

*$x|y$ means "x is a factor of y."

let A be any circuit, T a closed circuit, and F an open circuit. Then $A \wedge T \equiv A$, $A \wedge A' \equiv F$, $A \vee T \equiv T$, and $A \vee A' \equiv T$ (see laws 3 through 8). Likewise, laws 1 through 11 can also be extended to circuits in an obvious way.

EXAMPLE 1.23 Replace the switching network in Figure 1.5 by an equivalent simpler network.

Figure 1.5

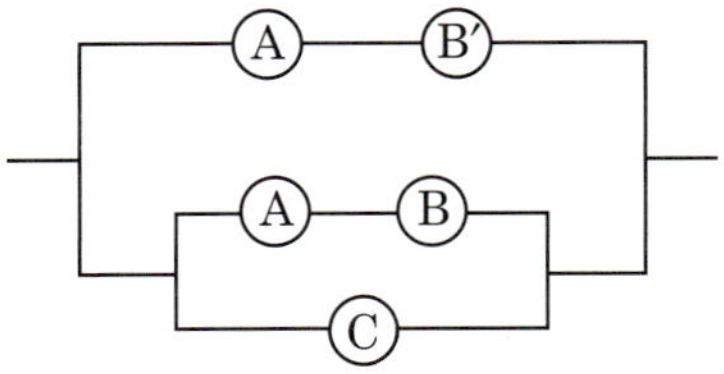

SOLUTION:

The given network is represented by $(A \wedge B') \vee [(A \wedge B) \vee C]$. Let us simplify this expression using the laws of logic. (The reason for each step is given on its RHS.)

$$\begin{aligned}(A \wedge B') \vee [(A \wedge B) \vee C] &\equiv [(A \wedge B') \vee (A \wedge B)] \vee C && \text{assoc. law}\\ &\equiv [A \wedge (B' \vee B)] \vee C && \text{dist. law}\\ &\equiv (A \wedge T) \vee C && B' \vee B \equiv T\\ &\equiv A \vee C && A \wedge T \equiv A\end{aligned}$$

Consequently, the given circuit can be replaced by the simpler circuit in Figure 1.6.

Figure 1.6

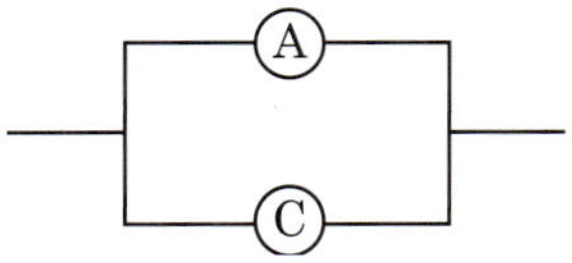

■

We close this section with a brief introduction to fuzzy logic.

Fuzzy Logic (optional)

"The binary logic of modern computers," wrote Bart Kosko and Satoru Isaka, two pioneers in the development of fuzzy logic systems, "often falls short when describing the vagueness of the real world. Fuzzy logic offers more graceful alternatives." Fuzzy logic, a branch of artificial intelligence, incorporates the vagueness or value judgements that exist in everyday life, such as "young," "smart," "hot," and "cold."

The first company to use a fuzzy system was F. L. Smidth and Co., a contracting company in Copenhagen, Denmark, which in 1980 used it to run a

__Bart Kosko__ holds degrees in philosophy and economics from the Universityof Southern California, an M.S. in applied mathematics, and a Ph.D. in electrical engineering from the University of California, Irvine. Currently, he is on the faculty in electrical engineering at the University of Southern California.

__Satoru Isaka__ received his M.S. and Ph.D. in systems science from the University of California, San Diego. He specializes in fuzzy information processing at Omron Advanced Systems at Santa Clara, and in the application of machine learning and adaptive control systems to biomedical systems and factory automation.

cement kiln. Eight years later, Hitachi used a fuzzy system to run the subway system in Sendai, Japan. Since then Japanese and American companies have employed fuzzy logic to control hundreds of household appliances, such as microwave ovens and washing machines, and electronic devices, such as cameras and camcorders. (See Figure 1.7.) It is generally believed that fuzzy, common-sense models are far more useful and accurate than standard mathematical ones.

Figure 1.7

In fuzzy logic, the truth value $t(p)$ of a proposition p varies from 0 to 1, depending on the degree of its truth; so $0 \leq t(p) \leq 1$. For example, the statement "The room is cool" may be assigned a truth value of 0.4; and the statement "Sarah is smart" may be assigned a truth value of 0.7.

Let $0 \leq x, y \leq 1$. Then the operations $\wedge$, $\vee$, and $'$ are defined as follows:

$$x \wedge y = \min\{x, y\}$$
$$x \vee y = \max\{x, y\}$$
$$x' = 1 - x$$

where $\min\{x, y\}$ denotes the minimum of x and y, and $\max\{x, y\}$ denotes the maximum of x and y.

Not all properties in propositional logic are valid in fuzzy logic. For instance, the **law of excluded middle**, $p \vee \sim p$ is true, does not hold in fuzzy logic. To see this, let p be a simple proposition with $t(p) = 0.3$. Then $t(p') = 1 - 0.3 = 0.7$; so $t(p \vee p') = t(p) \vee t(p') = 0.3 \vee 0.7 = \max\{0.3, 0.7\} = 0.7 \neq 1$. Thus $p \vee p'$ is not a tautology in fuzzy logic. [In propositional logic, $t(p \vee p') = 1$; so $p \vee p'$ is a tautology. Think of 1 representing a T and 0 representing an F.]

Likewise, $t(p \wedge p') = t(p) \wedge t(p') = 0.3 \wedge 0.7 = \min\{0.3, 0.7\} = 0.3 \neq 0$; so $p \wedge p'$ is not a contradiction, unlike in propositional logic.

Next we present briefly an interesting application* of fuzzy logic to decision making. It is based on the Yager method, developed in 1981 by Ronald R. Yager of Iona College, and employs fuzzy intersection and implication $\rightarrow$, defined by $p \rightarrow q \equiv \sim p \vee q$.

Fuzzy Decisions

Suppose that from among five U.S. cities—Boston, Cleveland, Miami, New York, and San Diego—we would like to select the best city to live in. We will use seven categories C1 through C7 to make the decision; they are *climate*, *cost of housing*, *cost of living*, *outdoor activities*, *employment*, *crime*, and *culture*, respectively, and are judged on a scale 0–6: 0 = *terrible*, 1 = *bad*, 2 = *poor*, 3 = *average*, 4 = *fairly good*, 5 = *very good*, and 6 = *excellent*. Table 1.15 shows the relative importance of each criterion on a scale 0–6 and the rating for each city in each category.

Table 1.15

Category	Importance	Boston	Cleveland	Miami	New York	San Diego
C1	6	3	2	5	1	6
C2	3	1	5	4	0	1
C3	2	3	4	3	1	5
C4	4	5	3	6	2	6
C5	4	4	3	3	4	3
C6	5	2	4	0	1	3
C7	4	6	3	3	6	5

*Based on M. Caudill, "Using Neural Nets: Fuzzy Decisions," *AI Expert*, Vol. 5 (April 1990), pp. 59–64.

The ideal city to live in will score high in the categories considered most important. In order to choose the finest city, we need to evaluate each city by each criterion, weighing the relative importance of each category. Thus, given a particular category's importance, we must check the city's score in that category; in other words, we must compute the truth value of $i \to s \equiv {\sim}i \vee s$ for each city, where i denotes the importance ranking for a particular category and s the city's score for that category. Table 1.16 shows the resulting data.

Now we take the conjunction of all scores for each city, using the *min* function (see Table 1.16). The lowest combined score determines the city's overall ranking. It follows from the table that San Diego is clearly the winner.

Table 1.16

Category	$\sim i$	Boston		Cleveland		Miami		New York		San Diego	
		s	$\sim i \vee s$	s	$\sim i \vee s$	s	$\sim i \vee s$	s	$\sim i \vee s$	s	$\sim i \vee s$
C1	0	3	3	2	2	5	5	1	1	6	6
C2	3	1	3	5	5	4	4	0	3	1	3
C3	4	3	4	4	4	3	4	1	4	5	5
C4	2	5	5	3	3	6	6	2	2	6	6
C5	2	4	4	3	3	3	3	4	4	3	3
C6	1	2	2	4	4	0	1	1	1	3	3
C7	2	6	6	3	3	3	3	6	6	5	5
Intersection			2		2		1		1		3

next best choices (Boston, Cleveland) — winner (San Diego) ↑

Finally, suppose we add a sixth city, say, Atlanta, for consideration. Then the Yager method ensures that the revised choice will be the existing choice (San Diego) or Atlanta; it can't be any of the others. Thus the procedure allows incremental decision making, so manageable subdecisions can be combined into an overall final choice.

Exercises 1.2

Give the truth value of p in each case.

1. $p \equiv q$, and q is not true.

2. $p \equiv q$, $q \equiv r$, and r is true.

Verify each, where f denotes a contradiction. (See Table 1.14.)

3. ${\sim}({\sim}p) \equiv p$

4. $p \wedge p \equiv p$

5. $p \vee p \equiv p$

6. $p \wedge q \equiv q \wedge p$

7. $p \vee q \equiv q \vee p$

8. ${\sim}(p \vee q) \equiv {\sim}p \wedge {\sim}q$

9. ${\sim}(p \to q) \equiv p \wedge {\sim}q$

10. $p \to q \equiv (p \wedge \sim q) \to f$

11. $p \wedge (q \wedge r) \equiv (p \wedge q) \wedge r$

12. $p \to (q \vee r) \equiv (p \wedge \sim q) \to r$

13. $(p \vee q) \to r \equiv (p \to r) \wedge (q \to r)$

Use De Morgan's laws to evaluate each boolean expression, where $x = 2$, $y = 5$, and $z = 3$.

14. $\sim[(x < z) \wedge (y < z)]$

15. $\sim[(y < x) \wedge (z < x)]$

16. $\sim[(x \geq y) \vee (y < z)]$

17. $\sim[(x < z) \vee (z < y)]$

Determine whether the assignment statement $c \leftarrow c + 1$ will be executed by the *if-statement* or *while-loop*, where $x \leftarrow 5$, $y \leftarrow 3$, and $z \leftarrow 7$.

18. If $\sim[(x < y) \wedge (y \leq z)]$ then $c \leftarrow c + 1$

19. If $\sim[(x = z) \wedge (x \leq y)]$ then $c \leftarrow c + 1$

20. If $\sim[(x \geq y) \vee (x < z)]$ then $c \leftarrow c + 1$

21. If $\sim[(x \leq z) \vee (y = z)]$ then $c \leftarrow c + 1$

22. While $\sim[(x \leq z) \vee (x < 3)]$ do $c \leftarrow c + 1$

23. While $\sim[(x > 6) \wedge (y = 4)]$ do $c \leftarrow c + 1$

24. While $\sim[(x = y) \vee (y = z)]$ do $c \leftarrow c + 1$

25. While $\sim[(x = 6) \vee (y = z)]$ do $c \leftarrow c + 1$

The logical operators **NAND** (not and) and **NOR** (not or) are defined as follows:

$$p \text{ NAND } q \equiv \sim(p \wedge q)$$
$$p \text{ NOR } q \equiv \sim(p \vee q)$$

Construct a truth table for each proposition.

26. p NAND q

27. p NOR q

Mark each sentence as true or false, where p, q, and r are arbitrary statements, t a tautology, and f a contradiction.

28. $p \wedge q \equiv q \wedge p$

29. $p \vee q \equiv q \vee p$

30. $p \wedge t \equiv p$

31. $p \vee f \equiv p$

32. $p \vee \sim p \equiv t$

33. $p \wedge \sim p \equiv f$

34. $\sim(p \wedge q) \equiv \sim p \wedge \sim q$

35. $\sim(p \vee q) \equiv \sim p \vee \sim q$

36. $p \to q \equiv q \to p$

37. $p \equiv p$

38. If $p \equiv q$, then $q \equiv p$.

39. If $p \equiv q$ and $q \equiv r$, then $p \equiv r$.

40. If $p \wedge q \equiv p \wedge r$, then $q \equiv r$.

41. If $p \vee q \equiv p \vee r$, then $q \equiv r$.

Use De Morgan's laws to verify each. (*Hint*: $p \to q \equiv \sim p \vee q$).

42. $\sim(\sim p \wedge \sim q) \equiv p \vee q$

43. $\sim(\sim p \vee q) \equiv p \wedge \sim q$

44. $\sim(\sim p \vee \sim q) \equiv p \wedge q$

45. $\sim(p \wedge \sim q) \equiv \sim p \vee q$

46. $\sim(p \to q) \equiv p \wedge \sim q$

47. $p \to \sim q \equiv \sim(p \wedge q)$

48. Show that the connectives $\wedge$, $\rightarrow$, and $\leftrightarrow$ can be expressed in terms of $\vee$ and $\sim$. (*Hint*: Use Exercise 44, law 18, and Tables 1.6 and 1.7.)

Simplify each boolean expression.

49. $p \wedge (p \wedge q)$

50. $p \vee (p \vee q)$

51. $p \vee (\sim p \wedge q)$

52. $(p \wedge \sim q) \vee (p \wedge q) \vee r$

***53.** $p \wedge (p \vee \sim q) \wedge (\sim p \vee \sim q)$

***54.** $(p \wedge \sim q) \vee (\sim p \wedge q) \vee (\sim p \wedge \sim q)$

Construct an equivalent, simpler network for each switching network.

55.

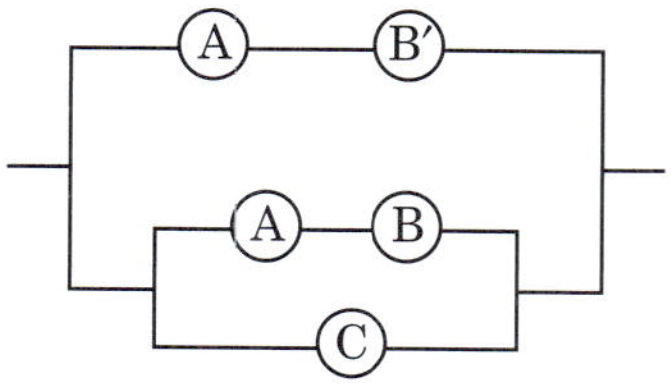

56.

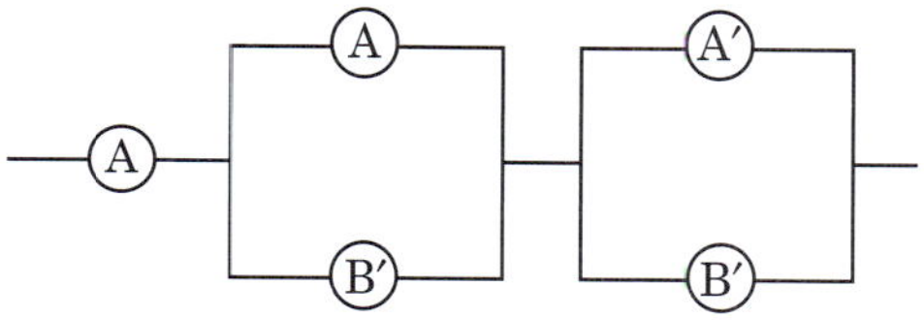

The **Sheffer stroke** | is a binary operator** defined by the following truth table.

p	q	$p \mid q$
T	T	F
T	F	T
F	T	T
F	F	T

(*Note*: On page 25 we used the vertical bar | to mean *is a factor of*. The actual meaning should be clear from the context. So be careful.) Verify each. (*Note*: Exercise 58 shows that the logical operators | and NAND are the same.)

57. $\sim p \equiv p \mid p$

58. $p \mid q \equiv \sim(p \wedge q)$

59. $p \wedge q \equiv (p \mid q) \mid (p \mid q)$

60. $p \vee q \equiv (p \mid p) \mid (q \mid q)$

61. $p \rightarrow q \equiv ((p \mid p) \mid (p \mid p)) \mid (q \mid q)$

62. $\sim(p \vee q) \equiv ((p \mid p) \mid (q \mid q)) \mid ((p \mid p) \mid (q \mid q))$

***63.** Express p XOR q in terms of the Sheffer stroke. (*Hint*: p XOR $q \equiv [(p \vee q) \wedge \sim(p \wedge q)]$.)

**The Sheffer stroke, named after the American logician Henry M. Sheffer (1883–1964), was devised by the American logician Charles S. Peirce (1839–1914).

***64.** Express $p \leftrightarrow q$ in terms of the Sheffer stroke. (*Hint*: $p \leftrightarrow q \equiv (p \to q) \wedge (q \to p)$.) [*Note*: Exercises 57–64 indicate that all boolean operators can be expressed in terms of the Sheffer stroke!]

Exercises 65–78 deal with propositions in fuzzy logic.

○ Let p, q, and r be simple propositions with $t(p) = 1, t(q) = 0.3$, and $t(r) = 0.5$. Compute the truth value of each, where s' denotes the negation of the statement s.

65. $(p')'$	**66.** $p \wedge q$	**67.** $p \vee r$	**68.** $q \vee q'$
69. $q \wedge q'$	**70.** $p' \vee q$	**71.** $(p \wedge q)'$	**72.** $p' \vee q'$
73. $(p \vee q)'$	**74.** $p \wedge q'$	**75.** $q \vee r'$	**76.** $(p \vee q) \wedge (p' \vee q)$

Let p be a simple proposition with $t(p) = x$ and p' its negation. Find each.

77. $t(p \vee p')$ **78.** $t(p \wedge p')$

1.3 Quantifiers

We now investigate a class of propositions different from those presented in the preceding sections.

Take a good look at the following propositions:

- All people are mortal.
- Every computer is a 16-bit machine.
- No birds are black.
- Some people have blue eyes.
- There exists an even prime number.

Each contains a word indicating quantity such as *all, every, none, some,* and *one*. Such words, called **quantifiers**, give us an idea about how many objects have a certain property.

There are two different quantifiers. The first is *all*, the **universal quantifier**, denoted by $\forall$, an inverted A. You may read $\forall$ as *for all, for each,* or *for every*. The second quantifier is *some*, the **existential quantifier**, denoted by $\exists$, a backward E. You may read $\exists$ *for some, there exists a,* or *for at least one*. Note that the word *some* means *at least one*.

The next two examples demonstrate how to write quantified propositions symbolically.

EXAMPLE 1.24 Let x be any apple. Then the sentence *All apples are green* can be written as *For every x, x is green*. Using the universal quantifier $\forall$, this sentence can be represented symbolically as $(\forall x)(x \text{ is green})$ or $(\forall x)P(x)$ where $P(x)$: x is green. (*Note*: x is just a dummy variable.) ■

Predicate

Here $P(x)$, called a **predicate**, states the property the object x has. Since $P(x)$ involves just one variable, it is a **unary** predicate. The set of all values x can have is called the **universe of discourse** (UD). In the above example, the UD is the set of all apples.

Note that $P(x)$ is *not* a proposition, but just an expression. However, it can be transformed into a proposition by assigning values to x. The truth value of $P(x)$ is predicated on the values assigned to x from the UD.

The variable x in the predicate $P(x)$ is a **free** variable. As x varies over the UD, the truth value of $P(x)$ can vary. On the other hand, the variable x in $(\forall x)P(x)$ is a **bound** variable, bound by the quantifier $\forall$. The proposition $(\forall x)P(x)$ has a fixed truth value.

EXAMPLE 1.25 Rewrite the sentence *Some chalkboards are black*, symbolically.

SOLUTION:
Choose the set of all chalkboards as the UD. Let x be an arbitrary chalkboard. Then the given sentence can be written as:

There exists an x such that x is black.

Using the existential quantifier, this can be symbolized as $(\exists x)\mathrm{b}(x)$, where $\mathrm{b}(x)$: x is black. ■

The next example illustrates how to find the truth values of quantified propositions.

EXAMPLE 1.26 The **absolute value** of a real number x, denoted by $|x|$, is defined by

$$|x| = \begin{cases} x & \text{if } x \geq 0 \\ -x & \text{if } x < 0 \end{cases}$$

Determine the truth value of each proposition, where the UD = set of all real numbers:
(1) $(\forall x)\,(x^2 \geq 0)$ (2) $(\forall x)\,(|x| > 0)$

SOLUTION:

(1) Since the square of every real number is nonnegative, the truth value of $(\forall x)\,(x^2 \geq 0)$ is T.
(2) It is *not* true that the absolute value of every number is positive, since $|0| = 0$, *not* greater than zero. So the truth value of $(\forall x)\,(|x| > 0)$ is F. ■

A predicate may contain two or more variables. A predicate that contains two variables is a **binary** predicate. For instance, $P(x, y)$ is a binary predicate. If a predicate contains n variables, it is an ***n*-ary** predicate.

The next two examples involve binary predicates.

EXAMPLE 1.27 Rewrite each proposition symbolically, where UD = set of real numbers.

(1) For each integer x, there exists an integer y such that $x + y = 0$.
(2) There exists an integer x such that $x + y = y$ for every integer y.
(3) For all integers x and y, $x \cdot y = y \cdot x$.
(4) There are integers x and y such that $x + y = 5$.

SOLUTION:

(1) $(\forall x)((\exists y)(x + y = 0))$, which is usually written as $(\forall x)(\exists y)(x + y = 0)$.
(2) $(\exists x)(\forall y)(x + y = y)$
(3) $(\forall x)(\forall y)(x \cdot y = y \cdot x)$
(4) $(\exists x)(\exists y)(x + y = 5)$ ■

The order of the variables x and y in $(\forall x)(\forall y)$ and $(\exists x)(\exists y)$ can be changed without affecting the truth values of the propositions. For instance, $(\forall x)(\forall y)(xy = yx) \equiv (\forall y)(\forall x)(xy = yx)$. Nonetheless, the order is important in $(\forall x)(\exists y)$ and $(\exists y)(\forall x)$. For example, let $P(x,y)$: $x < y$ where x and y are integers. Then $(\forall x)(\exists y)P(x,y)$ means *For every integer x, there is a suitable integer y such that x < y*; $y = x + 1$ is such an integer. Therefore, $(\forall x)(\exists y)P(x,y)$ is true. But $(\exists y)(\forall x)P(x,y)$ means *There exists an integer y, say, b, such that* $(\forall x)P(x,b)$; that is, *every integer x is less than b*. Clearly, it is false. Moral? The proposition $(Q_1x)(Q_2y)P(x,y)$ is evaluated as $(Q_1x)[(Q_2y)P(x,y)]$, where Q_1 and Q_2 are quantifiers.

A graphical approach can be helpful in finding the truth values of propositions in the form $(Q_1x)(Q_2y)P(x,y)$, where x and y are real numbers, as the next example illustrates.

EXAMPLE 1.28 (optional*) Determine the truth value of each proposition, where $P(x,y)$: $y < x^2$, and x and y are real numbers.

(1) $(\forall x)(\forall y)P(x,y)$ (2) $(\exists x)(\exists y)P(x,y)$ (3) $(\forall x)(\exists y)P(x,y)$
(4) $(\forall y)(\exists x)P(x,y)$ (5) $(\exists x)(\forall y)P(x,y)$ (6) $(\exists y)(\forall x)P(x,y)$

SOLUTION:

The graph of the equation $y = x^2$ is a parabola, as shown in Figure 1.8. The parabola is shown as a broken graph since no points on it satisfy the inequality $y < x^2$. The shaded region represents the solutions of the inequality.

(1) Is $y < x^2$ for all x and y? In other words, is the entire plane shaded? Since this is not the case, proposition (1) is false.
(2) $(\exists x)(\exists y)P(x,y)$ is true if $y < x^2$ for some real numbers x and y; that is, if and only if some portion of the cartesian plane is shaded. Since this is true, proposition (2) is true.

*Based on E. A. Kuehls, "The Truth-Value of $\{\forall, \exists, P(x,y)\}$: A Graphical Approach," *Mathematics Magazine*, Vol. 43 (Nov. 1970), pp. 260–261.

Figure 1.8

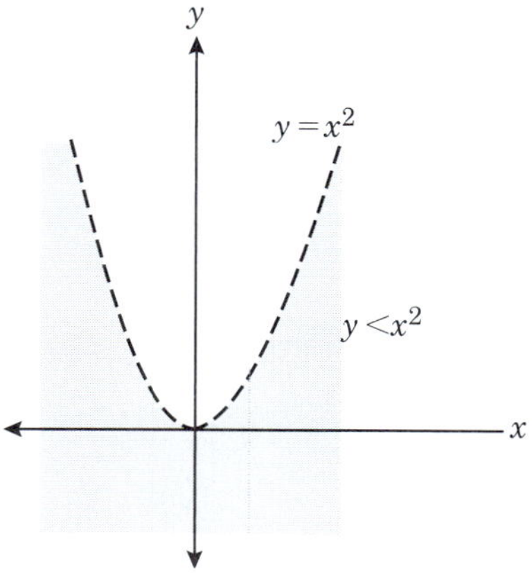

(3) $(\forall x)(\exists y)P(x,y)$ is true if there is a point (x,y) in the shaded area corresponding to every x; that is, it is true if every vertical line intersects the shaded area. Since this is the case, the proposition is true.

(4) $(\forall y)(\exists x)P(x,y)$ is true since every horizontal line intersects the shaded area.

(5) $(\exists x)(\forall y)P(x,y)$ means there is an x such that $y < x^2$ for all y. Therefore, the proposition is true if there is a vertical line which lies wholly within the shaded region. Since no such line exists, the proposition is false.

(6) $(\exists y)(\forall x)P(x,y)$ is true if there is a horizontal line which lies wholly within the shaded area. Since there are such lines, the proposition is true.

> *Note*: This graphical approach elucidates the difference between $(\forall x)(\exists y)$ and $(\exists x)(\forall y)$, and also between $(\forall y)(\exists x)$ and $(\exists x)(\forall y)$. $(\exists x)(\forall y)$ demands a *fixed* x, whereas $(\forall x)(\exists y)$ does *not* demand such a fixed x.

Next we discuss how to negate quantified propositions.

Recall from Example 1.24 that the proposition *All apples are green* can be symbolized as $(\forall x)P(x)$, where $P(x)$: x is green. Its negation is: It is false that all apples are green. That is, there exists an apple that is not green. In symbols, this can be written as $(\exists x)(\sim P(x))$. Thus, $\sim[(\forall x)P(x)] \equiv (\exists x)[\sim P(x)]$. Similarly, $\sim[(\exists x)P(x)] \equiv (\forall x)[\sim P(x)]$. These two properties are De Morgan's laws for negating quantifiers.

De Morgan's laws

- $\sim[(\forall x)P(x)] \equiv (\exists x)[\sim P(x)]$
- $\sim[(\forall x)P(x)] \equiv (\forall x)[\sim P(x)]$

By virtue of these laws, be careful when negating quantified propositions. When you negate the universal quantifier $\forall$, it becomes the existential quantifier $\exists$; when you negate the existential quantifier, it becomes the universal quantifier. In Section 1.5, we discuss a nice application of the first law to disproving propositions.

EXAMPLE 1.29 Negate each proposition, where the UD = set of integers.

(1) $(\forall x)\ (x^2 = x)$ (2) $(\exists x)\ (|x| = x)$

SOLUTION:

- $\sim[(\forall x)(x^2 = x)] \equiv (\exists x)[\sim(x^2 = x)]$
  $\equiv (\exists x)(x^2 \neq x).$
- $\sim[(\exists x)(|x| = x)] \equiv (\forall x)[\sim(|x| = x)]$
  $\equiv (\forall x)(|x| \neq x).$

EXAMPLE 1.30 Negate each quantified proposition.

(1) Every computer is a 16-bit machine.
(2) Some girls are blondes.
(3) All chalkboards are black.
(4) No person has green eyes.

SOLUTION:
Their negations are:

(1) Some computers are not 16-bit machines.
(2) No girls are blondes.
(3) Some chalkboards are not black.
(4) Some people have green eyes. ■

In closing, we should point out that what we discussed in Sections 1.1 and 1.2 is **propositional logic**; it deals with unquantified propositions. However, as we saw throughout this section, not all propositions can be symbolized in propositional logic, so quantifiers are needed. The area of logic that deals with quantified propositions is **predicate logic.**

Exercises 1.3

Determine the truth value of each proposition, where the UD consists of the numbers ±1, ±2, and 0.

1. $(\forall x)(x^2 = 4)$ **2.** $(\exists x)(x^3 + 2x^2 = x + 2)$ **3.** $(\forall x)(x^5 + 4x = 5x^3)$

4. $(\forall y)(y^4 + 3y^2 = 2)$ **5.** $\sim(\forall x)(x^3 = x)$ **6.** $(\forall x)[\sim(x^5 = 4x)]$

Let P(x): $x^2 > x$, Q(x): $x^2 = x$, and the UD = set of integers. Determine the truth value of each proposition.

7. $(\forall x)[\sim P(x)]$ **8.** $(\exists x)[\sim P(x)]$ **9.** $(\exists x)[P(x) \wedge Q(x)]$

10. $(\forall x)[P(x) \wedge Q(x)]$ **11.** $(\exists x)[P(x) \vee Q(x)]$ **12.** $(\forall x)[P(x) \vee Q(x)]$

Rewrite each sentence symbolically, where P(x): x is a 16-bit machine, Q(x): x uses the ASCII** character set, and the UD = set of all computers.

13. There is a computer that is a 16-bit machine and uses the ASCII character set as well.

14. We can find a 16-bit computer that does not use the ASCII character set.

15. We can find a computer that is either a 16-bit machine or does not use the ASCII character set.

16. There exists a computer that is neither a 16-bit machine nor uses the ASCII character set.

Negate each proposition, where x is an arbitrary integer.

17. $(\forall x)(x^2 > 0)$

18. $(\exists x)(x^2 \neq 5x - 6)$

19. Every supercomputer is manufactured in Japan.

20. There are no white elephants.

Rewrite each sentence symbolically, where the UD consists of real numbers.

21. The product of any two real numbers x and y is positive.

22. There are real numbers x and y such that $x = 2y$.

23. For each real number x, there is some real number y such that $x \cdot y = x$.

24. There is a real number x such that $x + y = y$ for every real number y.

25–28. Find the truth value of each proposition in Exercises 21–24.

Rewrite each in words, where UD = set of integers.

29. $(\forall x)(x^2 \geq 0)$ **30.** $\sim(\exists x)(x^2 = 2)$ **31.** $(\exists x)(\exists y)(x + y = 7)$

32. $(\forall x)(\exists y)(xy = 3)$ **33.** $(\exists x)(\forall y)(y - x = y)$ **34.** $(\forall x)(\forall y)(x + y = y + x)$

35–40. Find the truth value of each proposition in Exercises 29–34.

Let UD = set of integers, P(x,y): x is a multiple of y, and Q(x,y) : $x \geq y$. Determine the truth value of each proposition.

**ASCII is the acronym for American Standard Code for Information Interchange.

41. $(\exists x)P(15, x)$ **42.** $(\forall x)P(x, 2)$ **43.** $\sim(\exists x)P(x, 5)$

44. $(\exists x)[P(x, 3) \wedge Q(x, 3)]$ **45.** $(\exists x)[P(x, 2) \vee Q(x, 6)]$

46. $(\forall x)(\exists y)P(x, y)$ **47.** $(\forall x)(\exists y)Q(x, y)$

48. $(\forall x)[P(x, 3) \rightarrow Q(x, 3)]$ **49.** $(\exists x)[Q(x, 3) \rightarrow P(x, 3)]$

Let UD = set of real numbers and $P(x, y)$: $y^2 < x$. Determine the truth value of each proposition.

50. $(\forall x)(\forall y)P(x, y)$ **51.** $(\exists x)(\exists y)P(x, y)$ **52.** $(\forall x)(\exists y)P(x, y)$

53. $(\forall y)(\exists x)P(x, y)$ **54.** $(\exists x)(\forall y)P(x, y)$ **55.** $(\exists y)(\forall x)P(x, y)$

A third useful quantifier is the **uniqueness quantifier** $\exists!$. The proposition $(\exists!x)P(x)$ means *There exists a unique (meaning exactly one) x such that P(x)*. Determine the truth value of each proposition, where UD = set of integers.

56. $(\exists!x)(x + 3 = 3)$ **57.** $(\exists!x)(x^2 = 1)$ **58.** $(\exists!x)(\exists!y)(xy = 1)$

59. $(\exists!x)(\forall y)(x + y = y)$ **60.** $(\exists!x)(\exists!y)(2x = 3y)$ **61.** $(\forall x)(\exists!y)(x + y = 4)$

Determine the truth value of each, where P(s) denotes an arbitrary predicate.

62. $(\exists x)P(x) \rightarrow (\exists!x)(P(x)$ **63.** $(\exists!x)P(x) \rightarrow (\exists x)P(x)$

64. $(\forall x)P(x) \rightarrow (\exists!x)P(x)$ **65.** $(\exists x)P(x) \rightarrow (\forall x)P(x)$

66. $(\forall x)P(x) \rightarrow (\exists x)P(x)$ **67.** $(\exists!x)P(x) \rightarrow (\exists!y)P(y)$

***68.** Define the quantifier $\exists!$ in terms of the quantifiers $\exists$ and $\forall$.

*1.4 Arguments (optional)

Suppose we are given a finite set of propositions (called **hypotheses**) $H_1, H_2, \ldots, H_n$, all assumed true. Also assume that from these **premises**, we can arrive at a conclusion C through reasoning (or argument). Such a discussion can be written in **inferential form** as follows, where the symbol $\therefore$ means *therefore*:

$$\left.\begin{array}{l} H_1 \\ H_2 \\ \vdots \\ H_n \end{array}\right\} \text{hypotheses}$$
$$\overline{\therefore C} \quad \leftarrow \text{conclusion}$$

What does it mean to say that our reasoning in such a discussion is logical—that is, that the argument is valid?

Valid and Invalid Arguments

An argument is **valid** if the conjunction of the hypotheses $H_1, H_2, \ldots, H_n$ logically implies the conclusion C: that is, the implication $H_1 \wedge H_2 \wedge \cdots \wedge H_n \rightarrow C$ is a tautology. Otherwise, the argument is **invalid**, a **fallacy**.

Thus, an argument is valid if and only if the conclusion is a logical consequence of the hypotheses. In other words, if the hypotheses are assumed true, then the conclusion must follow logically from them. True hypotheses always lead to a true conclusion by a valid argument.

We begin checking the validity of arguments by using a well-known logic puzzle, due to R. M. Smullyan.

EXAMPLE 1.31 Test the validity of the following argument.

H_1 : There are more residents in New York City than there are hairs in the head of any resident.
H_2 : No resident is totally bald.

$\therefore$ At least two residents must have the same number of hairs on their heads.

SOLUTION:

(The argument contains two hypotheses. We always assume they are true and need to check whether the given conclusion follows logically from them.)

Suppose there are n residents in New York City. By H_1, the number of hairs on the head of every resident is less n; by H_2 every resident has at least one hair on his head. If each person has a different number of hairs, there must be n positive integers less than n, which is impossible. Therefore, at least two residents must have the same number of hairs on their heads.

Since the logical conclusion agrees with the given conclusion, the argument is valid. (This example is an application of the pigeonhole principle presented in Section 3.4.) ■

The next example presents another well-known logic puzzle, again due to Smullyan.

EXAMPLE 1.32 There are two kinds of inhabitants, knights and knaves, on an island. Knights always tell the truth, whereas knaves always lie. Every inhabitant is either a knight or a knave.

One day three inhabitants—A, B, and C—were standing together in a garden. A nomad came by and asked A, "Are you a knight or a knave?" Since A answered rather indistinctly, the stranger could make nothing out of his reply. So he asked B, "What did A say?" B replied, "A said, he is

Henry Ernest Dudeney *(1857–1930), England's greatest puzzlist and perhaps the greatest puzzlist who ever lived, was born in Mayfield, Sussex, England. Dudeney and Sam Loyd, the American puzzle genius, used to exchange puzzles and collaborate on puzzle articles in magazines and newspapers. Dudeney authored six books on puzzles, beginning with* The Canterbury Puzzles *(1907). Three of his collections were published posthumously.*

a knave." At this point C jumped into the conversation and said, "Don't believe B; he is lying." What are B and C?

SOLUTION:

A knight would never say, "I'm a knave," since he never lies. A knave would not say that either since he never tells the truth. Therefore, A did not say he was a knave. So B lied to the nomad and hence is a knave. Consequently, C was telling the truth, so C is a knight.

Thus B is a knave and C is a knight. (This example is pursued further in the exercises.) ■

The next puzzle* is a variation of a brainteaser developed by the English puzzlist, Henry Dudeney. Its solution does not employ any logic variables, but illustrates a clever problem-solving technique.

EXAMPLE 1.33 Smith, Jones, and Robinson are the brakeman, engineer, and fireman on a train, not necessarily in that order. Riding on the train are three passengers with the same last names who are identified by a "Mr." before their names. Assuming the following premises are true, determine who the engineer is.

H_1 : No two passengers live in the same city.
H_2 : Mr. Robinson lives in New York.
H_3 : The brakeman lives in Dallas.
H_4 : Mr. Jones has forgotten all the algebra he learned in high school.
H_5 : The passenger whose last name is the same as the brakeman's lives in Los Angeles.
H_6 : The brakeman and one of the passengers, a mathematical genius, attend the same local church.
H_7 : Smith beats the fireman in golf.

*Based on M. Gardner, *Mathematical Puzzles and Diversions*, The University of Chicago Press, Chicago, 1987.

SOLUTION:

We begin with two three-by-three arrays of empty cells with labels as in Figure 1.9. Use the premises to fill in the cells with 0's and 1's; enter a 1 in a cell with headings x and y if x has property y and enter a 0 otherwise.

Premise H_7 implies Smith is not a fireman, so enter a 0 in the upper right cell in Figure 1.9b. Since Mr. Robinson lives in New York (H_2), place a 1 in the lower left cell in Figure 1.9a and 0's in the remaining cells of the same row and column (why?).

It now follows that either Mr. Smith or Mr. Jones lives in Dallas. Does Mr. Jones live there?

Figure 1.9

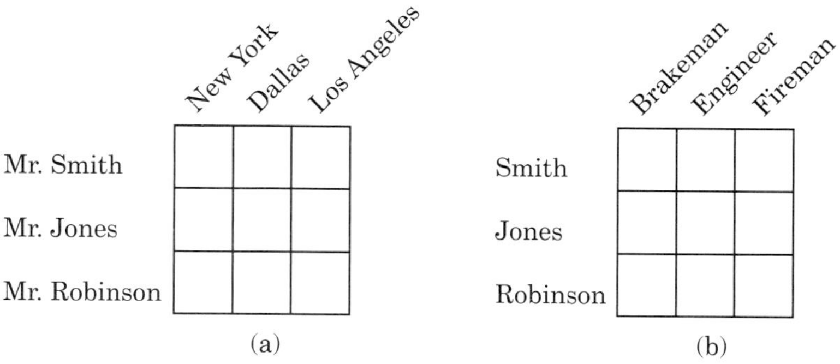

Since Mr. Jones cannot be a mathematical genius by H_4, the passenger genius must be Mr. Smith. By H_6, Mr. Smith and the brakeman live in the same city. Since it must be Dallas by H_3, enter a 1 in the upper middle cell in Figure 1.9a and 0's in the remaining cells of the same row and column.

It now follows that Mr. Jones lives in Los Angeles, so place a 1 in the middle cell of the third column in Figure 1.9a and 0's everywhere else. Figure 1.10a displays the resulting array.

By premise H_5, the brakeman and the passenger who lives in Los Angeles have the same last name, so the brakeman must be Jones; therefore, put a 1 in the first cell of the middle row in Figure 1.9b and 0's in the remaining cells of the same row and column.

By now, the top row in Figure 1.9b contains two 0's, so the middle cell must occupy a 1 (why?); so the middle cell in the bottom row a 0; hence the lower right cell must occupy a 1. Figure 1.10b shows the resulting array.

Figure 1.10

Mr. Smith	0	1	0
Mr. Jones	0	0	1
Mr. Robinson	1	0	0

(a)

Smith	0	1	0
Jones	1	0	0
Robinson	0	0	1

(b)

It follows from Figure 1.10b that Smith is the engineer. ■

***Bertrand Arthur William Russell** (1872–1970), a British philosopher and mathematician, was born into a prominent, aristocratic, and progressive-minded family near Trelleck, Wales. His mother died in 1874 and his father two years later; so the young Russell was brought up by his father's parents.*

Russell was home-educated by tutors. In 1890 he entered Trinity College, Cambridge, where he excelled in both mathematics and the moral sciences. In 1895, he was awarded a fellowship for his original dissertation on the foundations of geometry, published in 1897. After graduation, he worked briefly in the British embassy in Paris and then he went to Germany, where he wrote his first book, German Social Democracy *(1896). In 1910, Trinity appointed him a lecturer in logic and the philosophy of mathematics.*

Russell's outspokenness and liberal views often landed him in controversies. Around 1907, Russell fought hard for women's suffrage in the United Kingdom. During World War I, he was dismissed by Trinity for his protests and pacifist views. In 1918, he was imprisoned for 6 months for an article that was branded seditious. While in prison, he wrote Introduction to Mathematical Philosophy. *When he was about 90 years old, he was imprisoned again for campaigning for nuclear disarmament.*

In 1925, Trinity, realizing that the 1916 dismissal was excessively harsh, invited Russell back. He served there as a fellow from 1944 until his death.

Russell wrote more than 40 books on diverse subjects, including philosophy and physics; his greatest work is the three-volume Principia Mathematica *(1910–1913), which he coauthored with the Cambridge philosopher Alfred North Whitehead (1861–1947). It describes the logical construction of the foundations of mathematics from a set of primitive axioms.*

Russell won the 1950 Nobel prize for literature "as a defender of humanity and freedom of thought."

The following example is yet another well-known puzzle, the **barber paradox**, presented by the British mathematician and philosopher Bertrand Russell in 1918.

EXAMPLE 1.34 There is a male barber in a certain town. He shaves all those men and only those men who do not shave themselves. Does the barber shave himself?

SOLUTION:

Suppose the barber shaves himself. Then he belongs to the class of men who shave themselves. But no one in this class is shaved by the barber, so the barber does not shave himself, which is a contradiction.

On the other hand, suppose the barber does not shave himself. Since the barber shaves all those men who do not shave themselves, he shaves himself, again a contradiction.

Thus either case leads to a paradox: If the barber shaves himself, he does not shave himself; and conversely if he does not shave himself, then he shaves himself. So, logically, no such barber exists. ■

The symbols and the laws of logic can often be applied to check the validity of an argument, as the next two examples illustrate. To this end, follow the steps below:

- Rewrite the hypotheses symbolically.

- Assume the hypotheses are true.
- If the **inference rules** in Table 1.17 and/or the laws of logic can be used to reach the given conclusion, then the given argument is valid; otherwise, it is invalid; that is, the argument contains a flaw.

Table 1.17

Inference Rules

1. $p \wedge q \to (p \wedge q)$	conjunction
2. $p \wedge q \to p$	simplification
3. $p \to p \vee q$	addition
4. $[p \wedge (p \to q)] \to q$	law of detachment
5. $[(p \to q) \wedge (\sim q)] \to \sim p$	law of the contrapositive
6. $[(p \vee q) \wedge (\sim p)] \to q$	disjunctive syllogism
7. $[(p \to q) \wedge (q \to r)] \to (p \to r)$	hypothetical syllogism

A few words of explanation about each rule: The **conjunction rule** says that if both p and q are true, then $p \wedge q$ is true — a fact you already knew. According to the **simplification rule**, if $p \wedge q$ is true, then p is true. The **addition rule** says that if p is true, then $p \vee q$ is true regardless of the truth value of q. By the **law of detachment**, if an implication $p \to q$ is true and the premise p is true, then you can always conclude that q is also true; in other words, a true premise leads to a true conclusion logically. The **law of the contrapositive** says that if an implication $p \to q$ is true, but the conclusion q is false, then the premise p must be false. The two syllogisms can be interpreted similarly.

It is obvious that the inference rules play a central role in determining the validity of an argument. These rules, which are tautologies, can be established using truth tables. Try a few.

Each of the inference rules can be written in inferential form. For instance, the law of detachment can be rewritten as follows:

$$\begin{array}{l} p \\ p \to q \\ \hline \therefore \quad q \end{array}$$

EXAMPLE 1.35 Check the validity of the following argument.

If the computer was down Saturday afternoon, then Mary went to a matinee.
Either Mary went to a matinee or took a nap Saturday afternoon.
Mary did not take a nap that afternoon.

∴ The computer was down Saturday afternoon.

SOLUTION:

To avoid our emotions' playing any role in the way we reason, first translate the discussion into symbols. Let

p: The computer was down Saturday afternoon.
q: Mary went to a matinee Saturday afternoon.
r: Mary took a nap Saturday afternoon.

Then the given argument can be symbolized as follows:

$$\left.\begin{array}{l} H_1 : p \to q \\ H_2 : q \vee r \\ H_3 : \sim r \end{array}\right\} \quad \text{hypotheses}$$

$$\therefore p \qquad \leftarrow \text{conclusion}$$

Every step in our logical reasoning and the corresponding justification are given below:

1. $\sim r$ is true. — hypothesis H_3
2. $q \vee r$ is true. — hypothesis H_2
3. q is true. — step 1, step 2, and disjunctive syllogism
4. $p \to q$ is true. — hypothesis H_1
5. Then p may be true or false. — step 4, step 5, and definition of implication.

Since our logical conclusion does *not* agree with the given conclusion, the given argument is invalid. (Using a truth table you may verify that $[(p \to q) \wedge (q \vee r) \wedge (\sim r)] \to p$ is *not* a tautology. This provides an alternate demonstration that this argument is invalid.)

Note: Trivial steps may be omitted from such a reasoning without jeopardizing the logical progression. ■

We conclude this section with an example from Lewis Carroll's famous book *Symbolic Logic*. Two additional examples appear in the exercises.

EXAMPLE 1.36 Check the validity of the following argument.

Babies are illogical.
Nobody is despised who can manage a crocodile.
Illogical persons are despised.

∴ Babies cannot manage crocodiles.

SOLUTION:

First translate the sentences into *if-then* form using symbols, so we let

p: Harry is a baby.
q: Harry is illogical.
r: Harry can manage a crocodile.
s: Harry is despised.

Lewis Carroll *(1832–1898) (a pseudonym of Charles Lutwidge Dodgson) was the son of a clergyman and was born in Daresbury, England. He graduated from Christ Church College, Oxford University, in 1854 and began teaching mathematics at his alma mater in 1855, where he spent most of his life. He became a deacon in the Church of England in 1861.*

Carroll's famous Alice in Wonderland *and its sequel,* Through the Looking-Glass and What Alice Found There, *have provided a lot of pleasure to both children and adults all over the world.* Alice in Wonderland *is available in more than 30 languages, including Arabic and Chinese, and also in braille. The character is named for Alice Liddell, a daughter of the dean of Christ Church College.*

Then the argument can be written as:

$$\begin{array}{l} H_1: p \to q \\ H_2: r \to \sim s \\ H_3: q \to s \\ \hline \therefore\ p \to \sim r \end{array}$$

Every step of our logical reasoning is given below:

1. $(p \to q) \wedge (q \to s)$ is true. — conjunction rule
2. $p \to s$ is true. — hypothetical syllogism
3. $s \to \sim r$ is true. — law of the contrapositive
4. $(p \to s) \wedge (s \to \sim r)$ is true. — conjunction rule
5. $p \to \sim r$ is true. — hypothetical syllogism

Since the given conclusion agrees with the logical conclusion, the argument is valid. ■

Exercises 1.4

Rewrite each implication in inferential form.

1. $[(p \to q) \wedge (\sim q)] \to \sim p$

2. $[(p \to q) \wedge (q \to r)] \to (p \to r)$

Verify that each inference rule is a tautology.

3. $p \to (p \vee q)$

4. $[(p \to q) \wedge (q \to r)] \to (p \to r)$

Test the validity of each argument.

5.
$$\begin{array}{l} p \vee q \\ q \vee r \\ \sim r \\ \hline \therefore\ p \end{array}$$

6.
$$\begin{array}{l} p \leftrightarrow q \\ \sim p \vee r \\ \sim r \\ \hline \therefore\ \sim q \end{array}$$

7. If Bill likes cats, he dislikes dogs.
Bill likes dogs.

$\therefore$ Bill dislikes cats.

8. If Pat passes this course, she will graduate this year.
Pat does not pass this course.

$\therefore$ Pat will not graduate this year.

9. Frank bought a personal computer or a video cassette recorder (VCR).
If he bought a VCR, then he likes to watch movies at home.
He does not like to watch movies at home.

$\therefore$ Frank bought a personal computer.

10. If Peter is married, he is happy.
If he is happy, then he does not read the computer magazine.
He does read the computer magazine.

$\therefore$ Peter is unmarried.

(Exercises 11 and 12 come from Lewis Carroll's Symbolic Logic.)

***11.** All philosophers are logical.
An illogical person is always obstinate.

$\therefore$ Some obstinate persons are not philosophers.

***12.** No ducks waltz.
No officers ever decline to waltz.
All my poultry are ducks.

$\therefore$ My poultry are not officers.

Give the simplest possible conclusion in each argument. Assume each premise is true.

13. $p \leftrightarrow q$
$\sim p \vee r$
$\sim r$

14. $p \rightarrow q$
$p \vee \sim r$
r

15. $p \rightarrow \sim q$
$\sim r \rightarrow q$
p

16. $p \rightarrow q$
$\sim r \rightarrow \sim q$
$\sim r$

17. The program is running if and only if the computer is working.
The computer is working or the power is off.
The power is on.

18. Linda has a video cassette recorder (VCR).
If she has a personal computer, then she does not have a VCR.
If she does not have a personal computer, then she has a calculator.

19. Carol is a baby if and only if she is illogical.
Either she is illogical or unhappy.
But she is happy.

20. Three persons took a room for \$30 at a hotel. Soon after they checked out, the room clerk realized she had overcharged them since the room rents for \$25. She sent a bellhop to them with a \$5 reimbursement, but he returned to them only \$3, keeping \$2 for himself. Thus the room cost \$30 − \$3 = \$27 and \$27 + \$2 = \$29, so what happened to the extra dollar?

21. Aaron, Benjamin, Cindy, and Daphne are all friends. They are 34, 29, 28, and 27 years old, not necessarily in that order. Cindy is married to the oldest person. Aaron is older than Cindy, but younger than Daphne. Who is married to whom and how old are they? (*Mathematics Teacher*, 1990)

22. A family party consisted of one grandfather, one grandmother, two fathers, two mothers, four children, three grandchildren, one brother, two sisters, two sons, two daughters, one father-in-law, one mother-in-law, and one daughter-in-law. A total of 23 people, apparently. But no; there were only seven people at the party. How could this be possible? (B. Hamilton, 1992)

23. Three gentlemen—Mr. Blue, Mr. Gray, and Mr. White—have shirts and ties that are blue, gray, and white, but not necessarily in that order. No person's clothing has the same color as his last name. Mr. Blue's tie has the same color as Mr. Gray's shirt. What color is Mr. White's shirt? (*Mathematics Teacher*, 1986)

24. Three men and their wives were given \$5400. The wives together received \$2400. Sue had \$200 more than Jan, and Lynn had \$200 more than Sue. Lou got half as much as his wife, Bob the same as his wife, and Matt twice as much as his wife. Who is married to whom? (*Mathematics Teacher*, 1986)

There are seven lots, 1 through 7, to be developed in a certain city. A builder would like to build one bank, two hotels, and two restaurants on these lots, subject to the following restrictions by the city planning board (*The Official LSAT PrepBook*, 1991):

> If lot 2 is used, lot 4 cannot be used. If lot 5 is used, lot 6 cannot be used. The bank can be built only on lot 5, 6, or 7. A hotel cannot be built on lot 5. A restaurant can be built only on lot 1, 2, 3, or 5.

25. Which of the following is a possible list of locations for building them?

A. The bank on lot 7, hotels on lots 1 and 4, and restaurants on lots 2 and 5.

B. The bank on lot 7, hotels on lots 3 and 4, and restaurants on lots 1 and 5.

C. The bank on lot 7, hotels on lots 4 and 5, and restaurants on lots 1 and 3.

26. If a restaurant is built on lot 5, which of the following is *not* a possible list of locations?

A. A hotel on lot 2 and lot 4 is left undeveloped.

B. A restaurant on lot 2 and lot 4 is left undeveloped.

C. A hotel on lot 2 and lot 3 is left undeveloped.

Exercises 27–31 refer to Example 1.32 and are based on Smullyan's *What is the name of this book?*
A and B are inhabitants of the island. What are they if A says each of the following?

27. "At least one of us is a knave."

28. "Either I'm a knave or B is a knight."

29. A, B, and C are inhabitants of the island. Two residents are of the *same type* if they are both knights or both knaves. A says, "B and C are of the same type." Someone then asks C, "Are A and B of the same type?" What does C answer?

30. A says, "All of us are knaves," and B says, "Exactly one of us is a knight." What are A, B, and C?

31. A says, "All of us are knaves," and B says, "Exactly one of us is a knave." What is C?

Every inhabitant on a mysterious planet is either red or green. In addition, each inhabitant is either male or female. Every red man always tells the truth, whereas every green man always lies. The women, on the other hand, are opposite: every green woman tells the truth and every red woman lies. Since the natives always disguise their voices, and wear masks and gloves, it is impossible to identify their sex or color. But a clever anthropologist from Mathland met a native who made a statement from which he was able to deduce that the native was a green woman. (R. Smullyan, *Discover*, 1993)

32. What could the native have said? Justify your answer.

33. The second native the anthropologist interviewed also made a statement from which he was able to conclude that the native was a man (but not his color). Give a statement that would work. Again, justify your answer.

Four women, one of whom was known to have committed a serious crime, made the following statements when questioned by the police: (B. Bissinger, *Parade Magazine*, 1993)

Fawn:	"Kitty did it."
Kitty:	"Robin did it."
Bunny:	"I didn't do it."
Robin:	"Kitty lied."

34. If exactly one of these statements is true, identify the guilty woman.

35. If exactly one of these statements is false, identify the guilty woman.

***36.** "How is it, Professor Whipple," asked a curious student, "that someone as notoriously absentminded as you are manages to remember his telephone number?" "Quite simple, young man" replied the professor. "I simply keep in mind that it is the only seven-digit number such that the number obtained by reversing its digits is a factor of the number." What is Professor Whipple's telephone number? (A. J. Friedland, 1970)

***37.** Five angry cowgirls, standing in a field, accuse each other of rustling. No two distances between every two women are the same. Each has one bullet in her gun. At the count of ten, each shoots the nearest person in the toe. Will any cowgirl be shot or will at least one escape injury? (M. Gardner, *Parade Magazine*, 1993)

1.5 Proof Methods

Proofs, no matter how simple or complicated they are, are the heart and soul of mathematics. They play a central role in the development of mathematics and guarantee the correctness of mathematical results and algorithms (see Chapters 4 and 5). No mathematical results or computer algorithms are accepted as correct unless they are proved using logical reasoning.

A **theorem** in mathematics is a true proposition. Many theorems are implications $H_1 \wedge H_2 \wedge \cdots \wedge H_n \to C$. **Proving** such a theorem means verifying that the proposition $H_1 \wedge H_2 \wedge \cdots \wedge H_n \to C$ is a tautology. This section presents six standard methods for proving theorems: **vacuous proof, trivial proof, direct proof, indirect proof, proof by cases**, and **existence proof**. Vacuous and trivial proofs are, in general, parts of larger and complicated proofs, as will be seen in Chapters 4 and 5.

Vacuous Proof

Suppose the hypothesis H of the implication $H \to C$ is false. Then the implication is true regardless of whether C is true or false. Thus if the hypothesis H can be shown to be false, the theorem $H \to C$ is true by default; such a proof is a **vacuous proof**. Vacuous proofs, although rare, are necessary to handle special cases, as will be seen in Chapter 5.

EXAMPLE 1.37 Since the hypothesis of the statement *If* $1 = 2$, *then* $3 = 4$ is false, the proposition is vacuously true. ■

Trivial Proof

Suppose the conclusion c of the implication $H \to C$ is true. Again, the implication is true irrespective of the truth value of H. Consequently, if C can be shown to be true, such a proof is a **trivial proof**.

EXAMPLE 1.38 Let $P(n)$: If x is a positive real number and n any nonnegative integer, then $(1+x)^n \geq 1+nx$. Since $(1+x)^0 \geq 1+0\cdot x$ always, the proposition $P(0)$ is true. Thus the theorem is trivially true when $n = 0$. In this trivial proof we did not use the premise that $x > 0$. ■

Next we pursue another proof method.

Direct Proof

In the **direct proof** of the theorem $H_1 \wedge H_2 \wedge \cdots \wedge H_n \to C$, assume the given hypotheses H_i are true. Using the laws of logic or previously known facts, establish the desired conclusion C as the final step of a chain of implications: $H \to C_1$, $C_1 \to C_2, \ldots, C_n \to C$. Then, by the repeated application of the hypothetical syllogism, it follows that $H \to C$. The next example illustrates this method.

Often, theorems are stated in terms of sentences, so we need to first rewrite them symbolically and then work with the symbols, as the next example demonstrates.

EXAMPLE 1.39 Prove directly that the product of any two odd integers is an odd integer.

PROOF:
Let x and y be any two odd integers. Then there exist integers m and n such that $x = 2m+1$ and $y = 2n+1$. Thus,

$$\begin{aligned} x\cdot y &= (2m+1)\cdot(2n+1) \\ &= 4mn+2m+2n+1 \\ &= 2(2mn+m+n)+1 \\ &= 2k+1 \end{aligned}$$

where $k = 2mn+m+n$ is an integer. Therefore, xy is an odd integer. This concludes the proof. (Can you rewrite this proof as a chain of implications?) ■

Indirect Proof

There are two kinds of **indirect proofs** for the theorem $H_1 \wedge H_2 \wedge \cdots \wedge H_n \to C$: **proof of the contrapositive** and **proof by contradiction**. The first method is based on the law of the contrapositive, $H_1 \wedge H_2 \wedge \cdots \wedge H_n \to C \equiv {\sim}C \to {\sim}(H_1 \wedge H_2 \wedge \cdots \wedge H_n)$. [You may recall, by De Morgan's law, that ${\sim}(H_1 \wedge H_2 \wedge \cdots \wedge H_n) \equiv {\sim}H_1 \vee {\sim}H_2 \vee \cdots \vee {\sim}H_n$.] In this method, assume the desired conclusion C is false; then using the laws of logic, establish that some hypothesis H_i is also false. Once you have done this, the theorem is proved. The next example enlightens this method.

EXAMPLE 1.40 Prove indirectly: If the square of an integer is odd, then the integer is odd.

PROOF OF THE CONTRAPOSITIVE

Let x be any integer such that x^2 is odd. We would like to prove that x must be an odd integer. In the indirect method, we assume the conclusion is false; that is, x is *not* odd; in other words, assume x is an even integer. Let $x = 2k$ for some integer k. Then $x^2 = (2k)^2 = 4k^2 = 2(2k^2)$, which is an even integer. This makes our hypothesis that x^2 is an odd integer false. Therefore, by the law of the contrapositive, our assumption must be wrong; in other words, x must be an odd integer. Thus, if x^2 is an odd integer, then x is also an odd integer. ∎

Proof by contradiction, the other variation of indirect proof, is based on the law of *reductio ad absurdum*: $H_1 \wedge H_2 \wedge \cdots \wedge H_n \to C \equiv [H_1 \wedge H_2 \wedge \cdots \wedge H_n \wedge ({\sim}C)] \to F$. In this method, assume the given hypotheses H_i are true, but the conclusion C is false. Then argue logically and reach a contradiction F. The next example illustrates this method, where a **prime number** p is a positive integer with exactly two positive factors, 1 and p.

EXAMPLE 1.41 Prove by contradiction: There is no largest prime number; that is, there are infinitely many prime numbers.

PROOF BY CONTRADICTION

(*Notice that the theorem has no explicit hypothesis.*) Suppose the given conclusion is false; that is, there is a largest prime number p. So the prime numbers we have are $2, 3, 5, \ldots, p$; assume there are k such primes, $p_1, p_2, \ldots,$ and p_k.

Let x denote the product of all of these prime numbers plus one: $x = (2 \cdot 3 \cdot 5 \ldots p) + 1$. Clearly, $x > p$. When x is divided by each of the primes $2, 3, 5, \ldots, p$, we get 1 as the remainder. So x is *not* divisible by any of the primes. Hence either x must be a prime, or if x is composite then x is divisible by a prime $q \neq p_i$. In either case, there are more than k primes.

But this contradicts the assumption that there are k primes, so our assumption is false. In other words, there is no largest prime number. ∎

Now we turn to yet another proof technique.

Proof by Cases

Suppose we would like to prove a theorem of the form $H_1 \vee H_2 \vee \cdots \vee H_n \to C$. Since $H_1 \vee H_2 \vee \cdots \vee H_n \to C \equiv (H_1 \to C) \wedge (H_2 \to C) \wedge \cdots \wedge (H_n \to C)$, the statement $H_1 \vee H_2 \vee \cdots \vee H_n \to C$ is true if and only if each implication $H_i \to C$ is true. Consequently, we need only prove that each implication is true. Such a proof is a **proof by cases**, as illustrated in the following example, due to R. M. Smullyan.

EXAMPLE 1.42 Let A, B, and C be three inhabitants of the island described in Example 1.32. Two inhabitants are of the *same* type if they are both knights or both knaves. Suppose A says, "B is a knave," and B says, "A and C are of the same type." Prove that C is a knave.

PROOF BY CASES

Although this theorem is not explicitly of the form $H_1 \vee H_2 \vee \cdots \vee H_n \to C$, we artificially create two cases, namely, A is a knight and A is a knave.

Case 1 Suppose A is a knight. Since knights always tell the truth, his statement that B is a knave is true. So B is a knave and hence B's statement is false. Therefore, A and C are of different types; thus C is a knave.

Case 2 Suppose A is a knave. Then his statement is false, so B is a knight. Since knights always tell the truth, B's statement is true. So A and C are of the same type; thus C is a knave.

Thus in both cases, C is a knave. ■

Existence Proof

Finally, theorems of the form $(\exists x)P(x)$ also occur in mathematics. To prove such a theorem, we must establish the existence of an object a for which $P(a)$ is true. Accordingly, such a proof is an **existence proof**.

There are two kinds of existence proofs: the **constructive existence proof** and the **nonconstructive existence proof**. If we are able to find a mathematical object b such that $P(b)$ is true, such an existence proof is a **constructive proof**. The following example elucidates this method.

EXAMPLE 1.43 Prove that there is a positive integer that can be expressed in two different ways as the sum of two cubes.

CONSTRUCTIVE PROOF

By the discussion above, all we need is to produce a positive integer b that has the required properties. Choose $b = 1729$. Since $1729 = 1^3 + 12^3 = 9^3 + 10^3$, 1729 is such an integer.* ■

---

*A fascinating anecdote is told about the number 1729. In 1919, when the Indian mathematical genius Srinivasa Ramanujan (1887–1920) was sick in a nursing home in England, the eminent

A **nonconstructive** existence proof of the theorem $(\exists x)P(x)$ does not provide us with an element a for which $P(a)$ is true, but rather establishes its existence by an indirect method, usually contradiction, as illustrated by the next example.

EXAMPLE 1.44 Prove that there is a prime number > 3.

NONCONSTRUCTIVE PROOF

Suppose there are no primes > 3. Then 2 and 3 are the only primes. Since every integer ≥ 2 can be expressed as a product of powers of primes, 25 must be expressible as a product of powers of 2 and 3, that is, $25 = 2^i 3^j$ for some integers i and j. But neither 2 nor 3 is a factor of 25, so 25 cannot be written in the form $2^i 3^j$, a contradiction. Consequently, there must be a prime >3. ■

We invite you to give a constructive proof of the statement in the example. We conclude this section with a brief discussion of counterexamples.

Counterexample

Is the statement *Every girl is a brunette* true or false? Since we can find at least one girl who is not a brunette, it is false!

More generally, suppose you would like to show that the statement $(\forall x)P(x)$ is false. Since $\sim[(\forall x)P(x)] \equiv (\exists x)[\sim P(x)]$ by De Morgan's law, the statement $(\forall x)P(x)$ is false if there exists an item x in the UD for which the predicate $P(x)$ is false. Such an object x is a **counterexample**. Thus, to disprove the proposition $(\forall x)P(x)$, all we need is to produce a counterexample c for which $P(c)$ is false, as the next two examples demonstrate.

EXAMPLE 1.45 Number theorists dream of finding formulas that generate prime numbers. One such formula was found by the Swiss mathematician Leonhard Euler (see Chapter 8), namely, $E(n) = n^2 - n + 41$. It yields a prime for $n = 1, 2, \ldots, 40$. Suppose we claim that the formula generates a prime for every positive integer n. Since $E(41) = 41^2 - 41 + 41 = 41^2$ is not a prime, 41 is a counterexample, thus disproving the claim. ■

EXAMPLE 1.46 Around 1640, Fermat conjectured that numbers of the form $f(n) = 2^{2^n} + 1$ are prime numbers for all nonnegative integers n. For instance, $f(0) = 3$, $f(1) = 5$, $f(2) = 17$, $f(3) = 257$, and $f(4) = 65{,}537$ are all primes. In 1732, however, Euler established the falsity of Fermat's conjecture by producing a counterexample. He showed that $f(5) = 2^{2^5} + 1 = 641 \times 6700417$, a composite number. (Prime numbers of the form $2^{2^n} + 1$ are called **Fermat primes**.) ■

English mathematician Godfrey Harold Hardy (1877–1947) visited him. He told Ramanujan that the number of the cab he came in, 1729, was "a rather dull number" and hoped that it wasn't a bad omen. "No, Hardy," Ramanujan responded, "It is a very interesting number. It is the smallest number expressible as the sum of two cubes in two different ways."

Exercises 1.5

Determine if each implication is vacuously true for the indicated value of n.

1. If $n \geq 1$, then $2^n \geq n; n = 0$

2. If $n \geq 4$, then $2^n \geq n^2; n = 0, 1, 2, 3$

Determine if each implication is trivially true.

3. If n is a prime number, then $n^2 + n$ is an even integer.

4. If $n \geq 41$, then $n^3 - n$ is divisible by 3.

Prove each directly.

5. The sum of any two even integers is even.

6. The sum of any two odd integers is even.

7. The square of an even integer is even.

8. The product of any two even integers is even.

9. The square of an odd integer is odd.

10. The product of any two odd integers is odd.

11. The product of any even integer and any odd integer is even.

12. The square of every integer of the form $3k + 1$ is also of the same form, where k is an arbitrary integer.

13. The square of every integer of the form $4k + 1$ is also of the same form, where k is an arbitrary integer.

14. The **arithmetic mean** $\frac{a+b}{2}$ of any two nonnegative real numbers a and b is greater than or equal to their **geometric mean** $\sqrt{ab}$. [*Hint*: consider $(\sqrt{a} - \sqrt{b})^2 \geq 0$.]

Prove each using the law of the contrapositive.

15. If the square of an integer is even, then the integer is even.

16. If the square of an integer is odd, then the integer is odd.

17. If the product of two integers is even, then at least one of them must be an even integer.

18. If the product of two integers is odd, then both must be odd integers.

Prove by contradiction, where p is a prime number.

19. $\sqrt{2}$ is an irrational number.

20. $\sqrt{5}$ is an irrational number.

21. $\sqrt{p}$ is an irrational number.

***22.** $\log_{10} 2$ is an irrational number.

Prove by cases, where n is an arbitrary integer and $|x|$ denotes the absolute value of x.

23. $n^2 + n$ is an even integer. **24.** $2n^3 + 3n^2 + n$ is an even integer.

25. $n^3 - n$ is divisible by 3. (*Hint*: Assume that every integer is of the form $3k$, $3k + 1$, or $3k + 2$.)

26. $|-x| = |x|$ **27.** $|x \cdot y| = |x| \cdot |y|$ **28.** $|x + y| \leq |x| + |y|$

Prove by the existence method.

29. There are integers x such that $x^2 = x$.

30. There are integers x such that $|x| = x$.

31. There are infinitely many integers that can be expressed as the sum of two cubes in two different ways.

32. The equation $x^2 + y^2 = z^2$ has infinitely many integer solutions.

Give a counterexample to disprove each statement, where $\text{P}(x)$ denotes an arbitrary predicate.

33. The absolute value of every real number is positive.

34. The square of every real number is positive.

35. Every prime number is odd.

36. Every month has exactly 30 days.

37. $(\exists x)\text{P}(x) \to (\exists!x)\text{P}(x)$

38. $(\exists x)\text{P}(x) \to (\forall x)\text{P}(x)$

39. Find the flaw in the following "proof":

> Let a and b be real numbers such that $a = b$. Then $ab = b^2$.
>
> Therefore, $a^2 - ab = a^2 - b^2$
>
> Factoring, $a(a - b) = (a + b)(a - b)$
>
> Cancel $a - b$ from both sides:
>
> $$a = a + b$$
>
> Since $a = b$, this yields $a = 2a$.
>
> Cancel a from both sides.
>
> Then we get $1 = 2$.

Let a, b, and c be any real numbers. Then $a < b$ if and only if there is a positive real number x such that $a + x = b$. Use this fact to prove each.

40. If $a < b$ and $b < c$, then $a < c$. (**transitive property**)

41. If $a < b$, then $a + c < b + c$.

42. If $a + c < b + c$, then $a < b$.

43. Let a and b be any two real numbers such that $a \cdot b = 0$. Then either $a = 0$ or $b = 0$. [*Hint*: $p \to (q \vee r) \equiv (p \wedge \sim q) \to r$.]

***44.** The formula $f(n) = n^2 - 79n + 1601$ yields a prime for $0 \le n \le 10$. Give a counterexample to disprove the claim that the formula yields a prime for every nonnegative integer n.

***45.** Prime numbers of the form $f(n) = 2^n - 1$, where n is a positive integer, are called **Mersenne primes**, after the Franciscan monk Marin Mersenne (1588–1648). For example, $f(2) = 3, f(3) = 7$, and $f(5) = 31$ are Mersenne primes. Give a counterexample to disprove the claim that if n is a prime, then $2^n - 1$ is a prime.

Chapter Summary

This chapter presented the fundamentals of symbolic logic and the standard techniques of proving theorems.

Proposition

- A **proposition** is a declarative sentence that is either true or false, but not both (page 2).
- A **compound proposition** can be formed by combining two or more simple propositions, using logical operators: $\wedge$, $\vee$, $\sim$, $\to$, and $\leftrightarrow$ (page 5).
- The **conjunction** of two propositions is true if and only if both components are true; their **disjunction** is true if at least one component is true. An **implication** is false only if the premise is true and the conclusion is false. A **biconditional** is true if and only if both components have the same truth value (pages 5–14).
- The truth tables for the various logical operations can be combined into a single table, as in Table 1.18.

Table 1.18

p	q	$p \wedge q$	$(p \vee q)$	$\sim p$	$p \to q$	$p \leftrightarrow q$
T	T	T	T	F	T	T
T	F	F	T	F	F	F
F	T	F	T	T	T	F
F	F	F	F	T	T	T

- Three new implications can be constructed from a given implication: **converse, inverse**, and **contrapositive** (page 11).

- Various types of sentences and propositions can be summarized in a tree diagram, as in Figure 1.11.

Figure 1.11

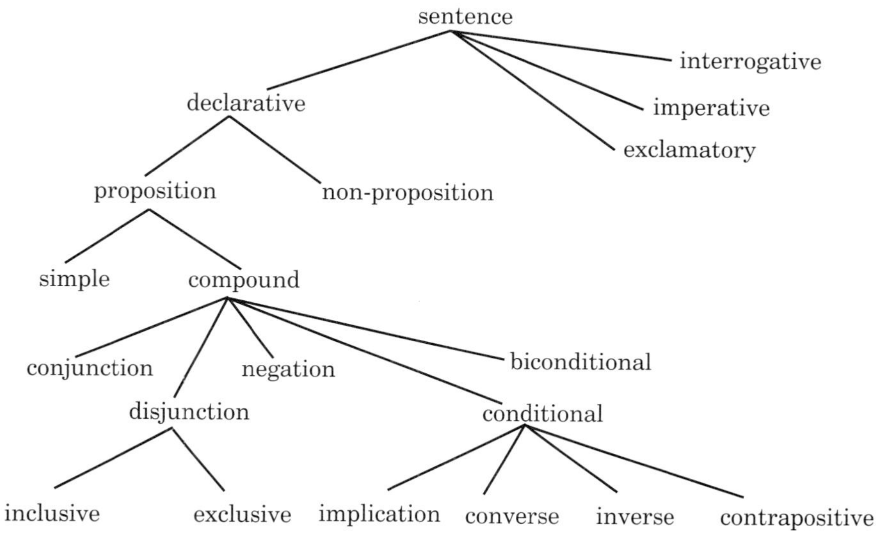

- A **tautology** is a compound statement that is always true. A **contradiction** is a compound statement that is always false. A **contingency** is a proposition that is neither a tautology nor a contradiction (page 16).
- Two compound propositions, p and q, are **logically equivalent** if they have identical truth values, symbolized by $p \equiv q$ (page 20).
- The important laws of logic are listed in Table 1.13 on page 21.

Argument

- An argument $\mathrm{H}_1 \wedge \mathrm{H}_2 \wedge \cdots \wedge \mathrm{H}_n \to \mathrm{C}$ is **valid** if the implication is a tautology; otherwise, it is **invalid** (page 39).
- The important inference rules are listed in Table 1.15 on page 43.

Quantifiers

- There are two quantifiers: **universal quantifier** ($\forall$) and **existential quantifier** ($\exists$) (page 32).
- A **predicate** $\mathrm{P}(x)$ is a sentence about the properties of the object x. The set of all values of x is the **universe of discourse** (UD) (page 33).
- **De Morgan's laws:** $\sim[(\forall x)\mathrm{P}(x)] \equiv (\exists x)[\sim \mathrm{P}(x)]$ (page 35)

  $\sim[(\exists x)\mathrm{P}(x)] \equiv (\forall x)[\sim \mathrm{P}(x)]$

Proof Methods

- There are six commonly used proof techniques: **vacuous proof, trivial proof, direct method, indirect method**, **proof by cases**, and **existence proof** (pages 49–53).

Figure 1.12

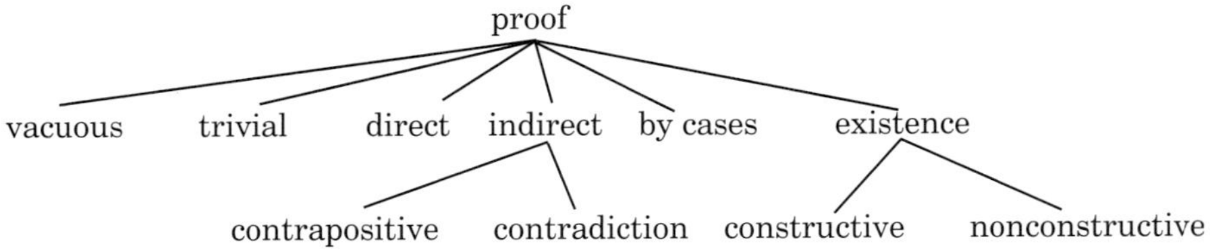

- In a direct proof, assume the given hypotheses are true. Then try to reach the given conclusion logically (page 50).
- For indirect proof by contrapositive, assume the given conclusion is false. Then establish directly that the given hypothesis is also false (page 51).
- For indirect proof by contradiction, assume the given hypothesis is true, but the given conclusion is false. Then try to reach a contradiction (page 51).
- For a constructive existence proof of a theorem $(\exists x)P(x)$, produce an element b such that $P(b)$ is true (page 52). In a nonconstructive existence proof, establish the existence of such an element b by an indirect method (page 53).

Counterexample

- To disprove the proposition $(\forall x)P(x)$, it suffices to produce an object c for which $P(c)$ is false (page 53).

Review Exercises

Construct a truth table for each proposition.

1. $(p \vee q) \wedge (\sim q)$ **2.** $(p \to q) \to r$ **3.** $p \to (q \to r)$ **4.** $(p \leftrightarrow q) \vee r$

Evaluate each boolean expression, where $a = 3, b = 7, c = 2$, and $d = 11$.

5. $(a \leq d) \wedge [(a > c) \vee (b > d)]$ **6.** $[(a > b) \wedge (b \leq c)] \vee (c < d)$

7. $(c > d) \vee [(b \leq c) \wedge (d < b)]$ **8.** $(b < c) \vee \sim[(a < c) \wedge (c < d)]$

Represent each sentence symbolically, where w, x, y, and z are real numbers.

9. If $w < x$ and $y < z$, then $w + y < x + z$.

10. If $w = x$ and $y = z$, then $w \cdot y = x \cdot z$.

Determine if the assignment statement $x \leftarrow y + z$ will be executed in each sequence of statements, where $i \leftarrow 5$, $j \leftarrow 3$, and $k \leftarrow 7$.

11. If $(i \leq j) \vee (j \leq k)$ then
$x \leftarrow y + z$

12. If $(i > j) \wedge (j \geq k)$ then
$x \leftarrow y + z$

13. If $\sim[(i > j) \vee (j < k)]$ then
$x \leftarrow y - z$
else
$x \leftarrow y + z$

14. Odd $\leftarrow 0$
while (odd ≤ 2) $\wedge$ $(i \leq 4)$ do
$x \leftarrow y + z$

Represent each network symbolically.

15.

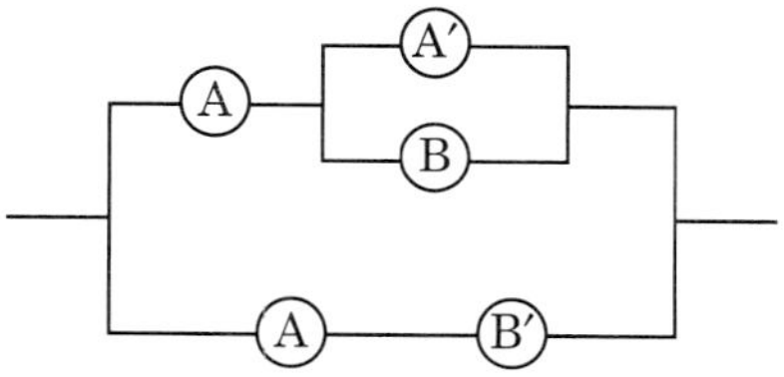

16.

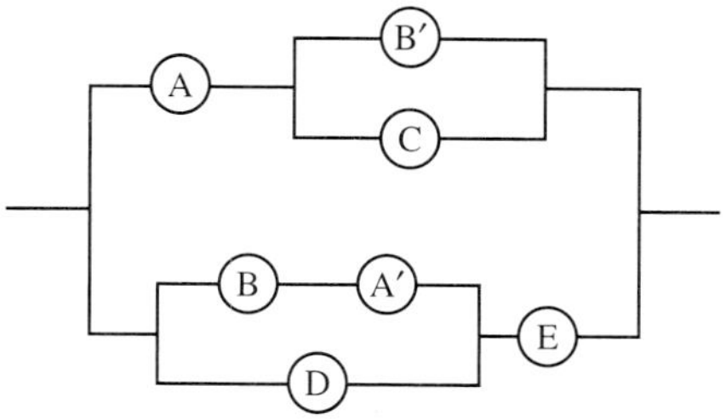

Let t be a true statement and p an arbitrary statement. Find the truth value of each.

17. $p \vee \sim t \to p$ **18.** $p \vee t \leftrightarrow \sim t$ **19.** $t \wedge (p \vee t)$ **20.** $p \wedge t \to p \vee t$

Use the given information to determine the truth value of each statement.

21. $p \to q$, if $p \vee q$ is false.

22. $p \to q$, if $\sim p \vee q$ is false.

23. $p \to \sim q$, if $q \to \sim p$ is false.

24. $p \wedge q$, if $p \to q$ is false.

25. $p \vee q$, if $p \to q$ is false.

26. $p \leftrightarrow q$, if $p \wedge q$ is true.

27. $p \leftrightarrow q$, if $p \wedge \sim q$ is true.

28. $p \vee \sim q$, if $q \wedge \sim p$ is true.

29. $p \wedge (q \wedge r)$, if $r \equiv s$ and s is not true.

Give the converse, inverse, and contrapositive of each implication.

30. If Pat is a girl, then she has green eyes.

31. If $x < y$, then $x + z < y + z$.

Write the contrapositive of each implication.

32. If $|x| < 3$, then $x < 3$ and $x > -3$.

33. If $|x| > 3$, then either $x > 3$ or $x < -3$.

Determine if each is a logical equivalence.

34. $p \wedge q \equiv \sim(p \to \sim q)$

35. $\sim(p \wedge \sim q) \equiv \sim p \vee q$.

36. $p \wedge (p \vee q) \equiv p \vee (p \wedge q)$

37. $p \to (q \to r) \equiv (p \to q) \to r$

Determine if each is a tautology.

38. $p \vee (p \wedge q) \leftrightarrow p$

39. $p \wedge (p \vee q) \leftrightarrow p$

40. $(p \wedge q) \leftrightarrow \sim(p \leftrightarrow \sim q)$

41. $(p \to \sim q) \leftrightarrow (q \to \sim p)$

Mark true or false, where p, q, r, and s are arbitrary statements.

42. If $q \equiv r$, then $p \wedge q \equiv p \wedge r$.

43. If $q \equiv r$, then $p \vee q \equiv p \vee r$.

44. If $p \equiv q$, then $p \to r \equiv q \to r$.

45. If $q \equiv r$, then $p \to q \equiv p \to r$.

46. If $p \leftrightarrow q$ is a tautology, then $p \equiv q$.

47. If $p \to q \equiv p \to r$, then $q \equiv r$.

Construct an equivalent simpler switching network for each circuit.

48.

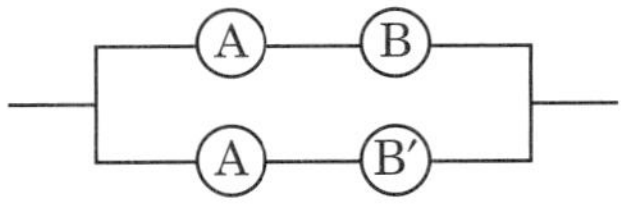

49.

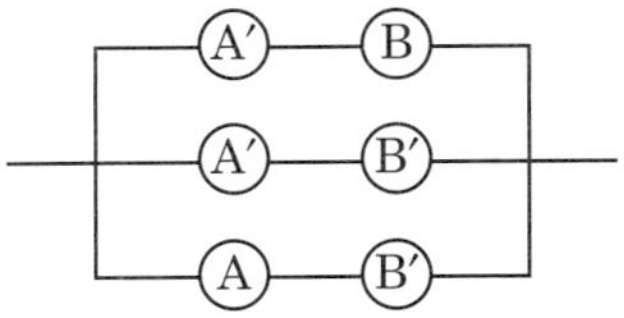

Test the validity of each argument.

50.
$$\begin{array}{l} p \vee q \\ q \to r \\ \sim r \\ \hline \therefore\ \sim p \end{array}$$

51.
$$\begin{array}{l} p \to \sim q \\ (q \wedge r) \to \sim s \\ r \wedge s \\ \hline \therefore\ \sim p \end{array}$$

52. Either Jane is not John's sister or Mary is not Harry's wife.
Mary is Harry's wife or Jane is not married.
John goes to school if and only if Jane is not married.
John does not go to school.

∴ Jane is not John's sister.

Determine the truth value of each, where the UD consists of the integers 0 and 1.

53. $(\exists x)(x^3 \neq x)$

54. $(\forall x)[(x+1)^2 = x^2 + 1]$

55. $(\exists y)[(y-1)^2 \neq y^2 - 1]$

56. $(\forall x)(\forall y)[(x+y)^2 = x^2 + y^2]$

Let UD = set of integers, $P(x)$: $x < 3$, and $Q(x)$: $x \geq 3$. Determine the truth value of each.

57. $(\forall x)[P(x) \wedge Q(x)]$ **58.** $(\forall x)[P(x) \vee Q(x)]$ **59.** $(\exists y)[P(y) \wedge Q(y)]$

60. $(\exists z)[P(z) \vee Q(z)]$ **61.** $(\forall x)[\sim P(x)]$ **62.** $(\exists z)[\sim Q(z)]$

Prove each, where a, b, c, d, and n are any integers.

63. The product of two consecutive integers is even.

64. $n^3 + n$ is divisible by 2.

65. $n^4 - n^2$ is divisible by 3.

66. If $a < b$ and $c < d$, then $a + c < b + d$.

67. If $a + b > 12$, then either $a > 6$ or $b > 6$.

68. If $ab = ac$, then either $a = 0$ or $b = c$. [*Hint*: $p \to (q \vee r) \equiv (p \wedge \sim q) \to r$.]

69. If $a^2 = b^2$, then either $a = b$ or $a = -b$.
[*Hint*: $p \to (q \vee r) \equiv (p \wedge \sim q) \to r$.]

70. Give a counterexample to disprove the following statement: If n is a positive integer, then $n^2 + n + 41$ is a prime number.
[Note: In 1798 the eminent French mathematician Adrien-Marie Legendre (1752–1833) discovered that the formula $L(n) = n^2 + n + 41$ yields distinct primes for 40 consecutive values of n. Notice that $L(n) = E(-n)$; see Example 1.45.]

○ *The propositions in Exercises 71–81 are fuzzy logic.*

Let p, q, and r be simple propositions with $t(p) = 1$, $t(q) = 0.3$, and $t(r) = 0.5$.
Compute the truth value of each, where s' denotes the negation of the statement s.

71. $p \wedge (q \vee r)$ **72.** $p \vee (q \wedge r)$

73. $(p \wedge q) \vee (p \wedge r)$ **74.** $(p \vee q) \wedge (p \vee r)$

75. $p' \wedge q'$ **76.** $(p \vee q') \vee (p \wedge q)$

77. $(p \vee q')' \vee q$ **78.** $(p \wedge q)' \wedge (p \vee q)$

79. Let p be a simple proposition with $t(p) = x$ and p' its negation. Show that $t(p \vee p') = 1$ if and only if $t(p) = 0$ or 1.

Let p and q be simple propositions with $t(p) = x$ and $t(q) = y$, where $0 \leq x, y \leq 1$. Verify each.

80. $(p \wedge q)' \equiv p' \vee q'$ [*Hint*: Show that $t((p \wedge q)') = t(p') \vee t(q')$.]

81. $(p \vee q)' \equiv p' \wedge q'$ [*Hint*: Show that $t((p \vee q)') = t(p') \wedge t(q')$.]

Supplementary Exercises

Write the converse, inverse, and contrapositive of each implication.

1. If $|x| < a$, then $-a < x < a$.

2. If $|x| > a$, then $x < -a$ or $x > a$.

Simplify each boolean expression.

3. $(p \vee \sim q) \wedge \sim(p \wedge q)$

***4.** $[p \vee q \vee (\sim p \wedge \sim q)] \vee (p \wedge \sim q)$

***5.** $(p \wedge \sim q) \vee (\sim p \wedge q) \vee (\sim p \wedge \sim q)$

***6.** $(p \vee q) \wedge \sim(p \wedge q) \wedge (\sim p \vee q)$

7. Let $p \equiv q$ and $r \equiv s$. Determine if $p \to (p \wedge r) \equiv q \to (q \wedge s)$.

Negate each proposition, where UD = set of real numbers.

8. $(\forall x)(\exists y)(xy \geq 1)$

9. $(\forall x)(\forall y)(xy = yx)$

10. $(\forall x)(\forall y)(\exists z)(x + y = z)$

11. $(\forall x)(\exists y)(\exists z)(x + y = z)$

Prove each.

12. The equation $x^3 + y^3 = z^3$ has infinitely many integer solutions.

***13.** Let n be a positive integer. Then $n(3n^4 + 7n^2 + 2)$ is divisible by 12.

***14.** Let n be a positive integer. Then $n(3n^4 + 13n^2 + 8)$ is divisible by 24.

***15.** In 1981 O. Higgins discovered that the formula $h(x) = 9x^2 - 471x + 6203$ generates a prime for 40 consecutive values of x. Give a counterexample to show that not every value of $h(x)$ is a prime.

***16.** The formula $g(x) = x^2 - 2999x + 2248541$ yields a prime for 80 consecutive values of x. Give a counterexample to disprove that every value of $g(x)$ is a prime.

In a **three-valued logic**, developed by the Polish logician Jan Lukasiewicz (1878–1956), the possible truth values of a proposition are 0, u, and 1, where 0 represents F, u represents *undecided*, and 1 represents T. The logical connectives $\wedge$, $\vee$, $'$, $\to$, and $\leftrightarrow$ are defined as follows:

$\wedge$	0	u	1
0	0	0	0
u	0	u	u
1	0	u	1

$\vee$	0	u	1
0	0	u	1
u	u	u	1
1	1	1	1

	$'$
0	1
u	u
1	0

$\to$	0	u	1
0	1	1	1
u	u	1	1
1	0	u	1

$\leftrightarrow$	0	u	1
0	1	u	0
u	u	1	u
1	0	u	1

Let p and q be arbitrary propositions in a three-valued logic, where r' denotes the negation of statement r and $t(r)$ denotes the truth value of r.

17. If $t(p \vee p') = 1$, show that $t(p) = 0$ or 1.

18. Show that $p \wedge q \to p \vee q$ is a three-valued tautology.

19. Show that $(p \to q) \leftrightarrow (p' \vee q)$ is not a three-valued tautology.

20. Show that $(p \to q) \leftrightarrow ({\sim}q \to {\sim}p)$ is a three-valued tautology.

21. Determine if $[p \wedge (p \to q)] \to q$ is a three-valued tautology.

Verify each.

22. $(p \wedge q)' \equiv p' \vee q'$ [*Hint*: Show that $t((p \wedge q)') = t(p') \vee t(q')$.]

23. $(p \vee q)' \equiv p' \wedge q'$ [*Hint*: Show that $t((p \vee q)') = t(p') \wedge t(q')$.]

Computer Exercises

Write a program to perform each task.
Construct a truth table for each proposition.

1. $(p \vee q) \wedge {\sim}q$ **2.** p NAND q **3.** p NOR q

4. $(p \to q) \leftrightarrow ({\sim}p \vee q)$ **5.** $(p \to q) \to r$ **6.** $(p \to q) \leftrightarrow ({\sim}q \to {\sim}p)$

Determine if each proposition is a tautology, by constructing a truth table.

7. $p \wedge (p \to q) \to q$ **8.** $(p \vee q) \wedge ({\sim}q) \to p$

9. $p \wedge (p \vee q) \leftrightarrow p$ **10.** $(p \to q) \wedge ({\sim}q) \to {\sim}p$

11. $p \wedge q \to p \vee q$ **12.** $(p \to q) \wedge (q \to r) \to (p \to r)$

Determine if the given propositions are logically equivalent, by constructing truth tables.

13. ${\sim}(p \wedge q)$, ${\sim}p \wedge {\sim}q$ **14.** $p \to q$, ${\sim}q \to {\sim}p$

15. $p \wedge (q \wedge r)$, $(p \wedge q) \wedge r$ **16.** $p \wedge (q \vee r)$, $(p \wedge q) \vee (p \wedge r)$

17. $(p \to q) \to r$, $p \to (q \to r)$ **18.** $p \to (q \vee r)$, $p \wedge ({\sim}q) \to r$

Exploratory Writing Projects

Using library and Internet resources, write a team report on each of the following in your own words. Provide a well-documented bibliography.

1. Write an essay on the contributions of G. Boole and W. Leibniz to mathematical logic.
2. Explain how (symbolic) logic helps you in everyday life. Give concrete examples.
3. Explain why proofs are important in mathematics and computer science. Do they help you in everyday life? In problem-solving? In a work environment? Give examples of proofs in computer science.

4. Give a detailed history of Fermat's last theorem. Include biographies of mathematicians who have worked on the problem.

5. Collect a number of well-known conjectures from number theory and explain recent advances toward establishing them or disproving them.

Study a number of puzzles from R. M. Smullyan's *Alice in Puzzle-land*, *What is the name of this book?*, and *The Lady or the Tiger*.

6. Write each as an argument and test the validity of it.

7. Write each as a theorem and establish it.

8. List a number of applications of fuzzy logic to everyday life. How do they enrich our lives?

9. Write a biography of H. M. Sheffer, C. S. Peirce, and A. M. Legendre.

10. Collect several examples on arguments from Lewis Carroll's *Symbolic Logic* and test the validity of each. Explain with examples the use of Euler diagrams in the analysis of arguments.

11. Collect several logic puzzles from recent issues of *Discover* magazine and *Parade* magazine. Solve each.

12. Study logic problems in the recent edition of *The official LSAT PrepBook* and solve them.

13. List several attempts to develop formulas for generating prime numbers.

14. Write an account of Fermat primes, Mersenne primes, the infinitude of each family, and their applications.

15. Investigate the *pentomino puzzle*, developed in 1954 by S. W. Golomb of the University of Southern California.

Enrichment Readings

1. L. Carroll, *Symbolic Logic and the Game of Logic*, Dover, New York, 1958.

2. L. Carroll, *Alice's Adventures in Wonderland and Through the Looking-Glass and What Alice Found There*, Oxford, New York, 1982.

3. N. Falletta, *The Paradoxicon*, Wiley, New York, 1990.

4. M. Gardner, *Mathematical Puzzles and Diversions*, The University of Chicago Press, Chicago, 1987.

5. J. T. Johnson, "Fuzzy Logic," *Popular Science*, Vol. 237 (July 1990), pp. 87–89.

6. B. Kosko and S. Isaka, "Fuzzy Logic," *Scientific American*, Vol. 269 (July 1993), pp. 76–81.

7. H. T. Nguyen and E. A. Walker, *A First Course in Fuzzy Logic*, 2nd ed., Chapman and Hall/CRC, Boca Raton, FL, 2000.

8. R. M. Smullyan, *Alice in Puzzle-land*, Penguin, New York, 1984.

9. R. M. Smullyan, *What is the name of this book?*, Prentice Hall, Englewood Cliffs, NJ, 1978.

10. R. M. Smullyan, *The Lady or the Tiger?*, Random House, New York, 1992.

11. D. Solow, *How to Read and Do Proofs*, Wiley, New York, 1982.

Chapter 2

The Language of Sets

The essence of mathematics lies in its freedom.

—GEORG CANTOR

The concept of a set is so fundamental that it unifies mathematics and its cognates. It has revolutionized mathematical thinking, enabling us to express ourselves in clear and concise terms.

The foundation of set theory was laid by the eminent German mathematician Georg Cantor during the latter part of the 19th century. "Today, Cantor's set theory has penetrated into almost every branch of mathematics," as the mathematical historian Howard Eves writes in *An Introduction to the History of Mathematics*.

In this chapter we present the language of sets. We introduce the concept of a set, the various ways of describing a set and of constructing new sets from known sets, a variety of applications, and a brief introduction to fuzzy sets.

The following are some of the problems we shall pursue in this chapter:

- Find the number of positive integers $\leq N$ and divisible by a, b, or c.
- How many subsets does a finite set with n elements have?
- How would you define the set of legally paired parentheses?
- How many sequences of legally paired parentheses can be formed using n pairs of left and right parentheses?

2.1 The Concept of a Set

This section introduces the concept of a set, various methods of defining sets, and relationships between sets.

Georg Cantor *(1845–1918) was born in St. Petersburg, Russia, where his father was a successful merchant and broker. Cantor showed great interest in mathematics from early childhood. In 1856, the family moved to Germany. Six years later, he entered the University of Zurich, but in the following year he moved to the University of Halle to study mathematics, physics, and philosophy. There he was greatly influenced by the eminent mathematician Karl Weierstrass (1815–1897). Although his father wanted him to become an engineer, Cantor relentlessly pursued his interest in mathematics and received his doctorate of philosophy at 22 from the University of Berlin for his work in number theory.*

In 1869, Cantor began his professional career as an unsalaried lecturer at the University of Halle. Five years later, he published his revolutionary work on set theory. Cantor developed an arithmetic of transfinite numbers analogous to that of finite numbers, thus creating another area of mathematical study. He proved that the set of real numbers is uncountable and he also established the existence of infinitely many different transfinite cardinal numbers by ingenious methods. He also made significant contributions to indeterminate equations and trigonometric series. Deeply religious, Cantor was also interested in art, music, and philosophy.

Being unhappy with his low salary at the University, Cantor tried to secure a better-paid position at the University of Berlin, but was sabotaged by Leopold Kronecker (1823–1891), an eminent mathematician at the University, who severely criticized Cantor's views on sets.

Relentless attacks by contemporary mathematicians intensified the manic depression he suffered from. Cantor died in a mental hospital in Halle in 1918.

Cantor was "one of the greatest intellects of the nineteenth century," according to Bertrand Russell. He "was an imaginative genius whose work has inspired [every aspect of] mathematical thought," Hazel Perfect of the University of Sheffield wrote in 1994.

Set

A **set** is a collection of well-defined objects,* called **elements** (or **members**) of the set.

There should be no ambiguity in determining whether or not a given object belongs to the set. For example, the vowels of the English alphabet form a (well-defined) set, whereas beautiful cities in the United States do not form a set since its membership would be debatable.

Sets are denoted by capital letters and their elements by lowercase letters. If an object x is an element of a set A, we write $x \in A$; otherwise $x \notin A$. For example, let A be the set of New England states. Then *Connecticut* $\in A$, whereas *Michigan* $\notin A$.

There are two methods of defining sets.

Listing Method

A set can sometimes be described by listing its members within braces. For instance, the set B of New England states can be described as $B = \{\text{ME}, \text{VT}, \text{NH}, \text{MA}, \text{CT}, \text{RI}\}$.

---

*To be precise, this is a circular definition; *set* is an undefined term, like *point* and *line* in geometry.

The order in which the elements are enumerated is immaterial. Thus B can also be written as {VT, RI, MA, CT, NH, ME}. If an element is repeated, it is not counted more than once. For example, $\{x, x, y, x, y, z\} = \{x, y, z\}$.

A set with a large number of elements that follow a definite pattern is often described using ellipses (...) by listing a few elements at the beginning. For example, the set of letters of the alphabet can be written as {a, b, c, ..., z} and the set of odd positive integers as $\{1, 3, 5, \ldots\}$.

Set-Builder Notation

Another way of describing a set is by using the **set-builder notation**. Its general form is $\{x \mid \mathrm{P}(x)\}$, where $\mathrm{P}(x)$ is a predicate indicating the property (or properties) the object x has. You may read $\{x \mid \mathrm{P}(x)\}$ as *the set consisting of all objects x such that x has the property P(x)*. Here the vertical bar "|" means *such that*. (Again, the meaning of the vertical bar should be clear from the context.)

EXAMPLE 2.1 Let B be the set of all months of the year with exactly 30 days. Then

$$B = \{x \mid x \text{ is a month of the year with exactly 30 days }\}$$
$$= \{\text{September, April, June, November}\}$$ ■

Next we present another of Russell's paradoxes introduced in 1901, which is quite similar to the barber paradox.

Russell's Paradox

Let $S = \{X \mid X \notin X\}$; that is, S consists of all sets that do not belong to themselves as elements. Does $S \in S$? If $S \in S$, then, by definition, $S \notin S$; on the other hand, if $S \notin S$, then, again by definition, $S \in S$. Thus, in either case, we have a contradiction. This paradox shows, not every predicate defines a set; that is, there is no set of all sets.

Next we present several relationships between sets.

Subset

If every element of A is also an element of B, A is a **subset** of B, denoted by $A \subseteq B$. In symbols, $(A \subseteq B) \leftrightarrow (\forall x)(x \in A \rightarrow x \in B)$. If $A \subseteq B$, we also say that B **contains** A and write $B \supseteq A$. If A is not a subset of B, we write $A \nsubseteq B$; thus $(A \nsubseteq B) \leftrightarrow (\exists x)(x \in A \wedge x \notin B)$.

EXAMPLE 2.2 Let A = set of states in the United States, B = set of New England states, and C = set of Canadian provinces. Then $B \subseteq A$, but $B \nsubseteq C$ and $A \nsubseteq C$. ■

To show that a set X is a subset of Y, select an arbitrary element x in X; then using the laws of logic and known facts, show that x is in Y also. We shall apply this technique in later sections. To show that $X \nsubseteq Y$, all you need is to find an element $x \in X$ which does not belong to Y.

Equal Sets

Two sets A and B are **equal**, denoted by $A = B$, if they contain the same elements. In other words, $A = B$ if $(A \subseteq B) \wedge (B \subseteq A)$. (We shall use this property to prove the equality of sets.) If $A \subseteq B$ and $A \neq B$, then A is a **proper subset** of B, denoted by $A \subset B$.

EXAMPLE 2.3 Consider the sets $A = \{x \mid x$ is a vowel of the alphabet$\}$, $B = \{\mathrm{a, e, i, o, u}\}$, $C = \{2, 3, 4\}$, and $D = \{x \mid x$ is a digit in the numeral 23432.$\}$ Then $A = B$, and $C = D$. ■

Does a set have to contain any element? Can there be a set with no elements? Suppose Fred went hunting in a nearby jungle and returned home with great tales, but no animals. The set of animals he caught is null. This leads us to the following definition.

Empty Set

The set containing no elements is the **empty** (or **null) set**; it is denoted by $\emptyset$ or $\{\}$.

EXAMPLE 2.4 The set of pink elephants is empty. So are the set of mountains on the earth that are 50,000 feet tall and the set of prime numbers between 23 and 28. ■

Many people mistakenly believe that $\{\emptyset\} = \emptyset$; this is *not* true, since $\{\emptyset\}$ contains an element $\emptyset$, whereas $\emptyset = \{\}$ contains no elements. Thus $\{\emptyset\} \neq \emptyset$.

Logically, it can be proved that $\emptyset$ is a subset of every set; that is, $\emptyset \subseteq A$ for every set A. Besides, although many people think that there are many empty sets, it can be proved that it is unique, meaning there is exactly one empty set. (See Exercises 53 and 54.)

Universal Set

It is always possible to choose a special set $U (\neq \emptyset)$ such that every set under discussion is a subset of U. Such a set is called a **universal set**, denoted by U. Thus $A \subseteq U$ for every set A.

EXAMPLE 2.5 Suppose we wish to discuss something about the sets $\{\mathrm{a}\}$, $\{\mathrm{b, c, d}\}$, and $\{\mathrm{b, d, e, f}\}$. Then $U = \{\mathrm{a, b, c, d, e, f}\}$ may be chosen as a valid universal set. (There are other valid choices also.) ■

EXAMPLE 2.6 (optional) Programming languages such as *Pascal* support the data type SET, although the implementations have a limit on the number of elements on the base-type of the set, that is, on the size of the universal set. For example, consider the Pascal declarations:

```
TYPE
   MONTHS = (JAN, FEB, MAR, APR, MAY, JUN, JUL, AUG, SEP, OCT, NOV, DEC);
   SETOFMONTHS = SET OF MONTHS;
VAR
   SPRING,SUMMER,FALL,WINTER:  SETOFMONTHS;
```

Here the universal set is

```
SETOFMONTHS = {JAN,FEB,MAR,APR,MAY,JUN,JUL,AUG,SEP,OCT,NOV,DEC}.
```

The above variable declarations define four set variables, namely, SPRING, SUMMER, FALL, and WINTER. The set values assigned to them must be subsets of SETOFMONTHS. For instance,

```
SPRING := [JAN,FEB,MAR];
```

is a legal Pascal assignment, although it is preposterous.

The set membership operator in Pascal is **IN** and can be used to determine if an element belongs to a set. For example, `FEB IN SPRING` is a legal boolean expression. Likewise, the set inclusion and containment operators are <= and >=, respectively. ■

Disjoint Sets

Sets need not have common elements. Two such sets are **disjoint** sets.

For example, the sets {Ada, BASIC, FORTRAN} and {C++, Java} are disjoint; so are the sets $\{+, -, *, /\}$ and $\{\wedge, \vee, \rightarrow, \leftrightarrow\}$.

Venn Diagrams

Relationships between sets can be displayed using Venn diagrams, named after the English logician John Venn. In a Venn diagram, the universal set U is represented by the points inside a rectangle and sets by the points enclosed by simple closed curves inside the rectangle, as in Figure 2.1. Figure 2.2 shows $A \subseteq B$, whereas Figure 2.3 shows they are not disjoint.

Figure 2.1

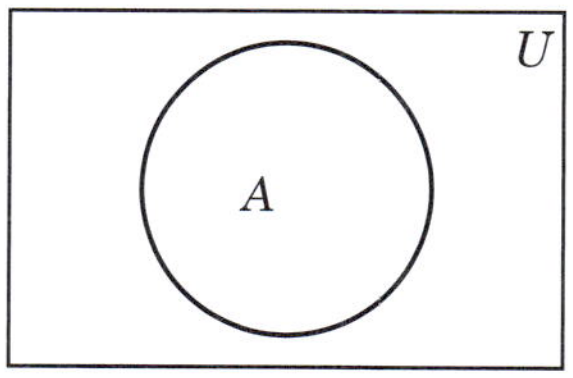

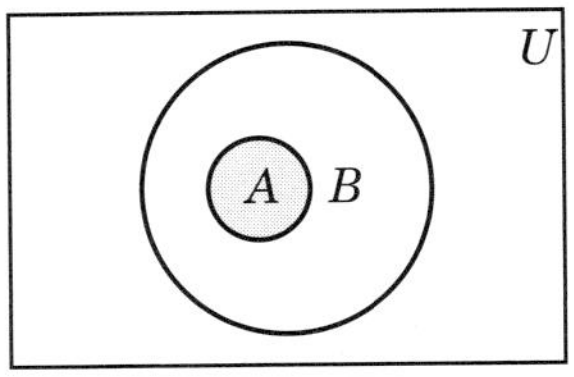

Figure 2.2

$A \subseteq B$.

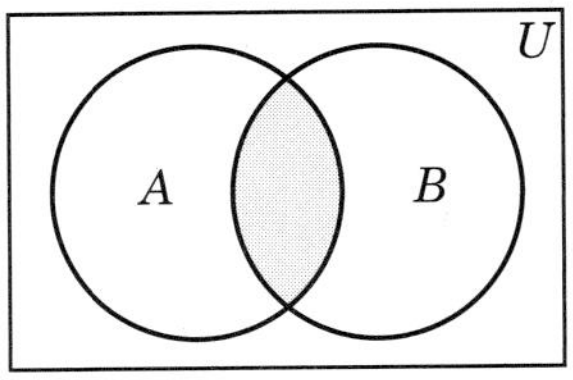

Figure 2.3

A and B may have common elements.

John Venn *(1834–1923) was born into a philanthropic family in Hull, England. After attending the high schools at Highgate and Islington, in 1853 he entered Gonville and Caius College, Cambridge, and graduated in mathematics three years later. He was elected a fellow of the College, a position he held until his death.*

In 1859 Venn was ordained in the Church of England, but after a brief period of church work, he returned to Cambridge as a lecturer on moral sciences. In 1883 he gave up his priesthood. The same year, he received a D.Sc. from Cambridge and was elected a fellow of the Royal Society of London.

Venn was greatly influenced by Boole's work in symbolic logic. Venn's masterpiece, Symbolic Logic *(1881), clarifies the inconsistencies and ambiguities in Boole's ideas and notations. He employed geometric diagrams to represent logical arguments, a technique originated by Leibniz and developed further by Euler. Venn added a rectangle to represent the universe of discourse.*

Venn published two additional books, The Logic of Chance *(1866) and* The Principles of Empirical Logic *(1889).*

Can the elements of a set be sets? Certainly. $\{\{a\}, \{b, c\}\}$, and $\{\emptyset, \{\emptyset\}, \{a, b\}\}$ are two such sets. In fact, the subsets of a set can be used to build a new set.

Power Set

The family of subsets of a set A is the **power set** of A, denoted by $P(A)$.

EXAMPLE 2.7 Find the power set $P(A)$ of the set $A = \{a, b\}$.

SOLUTION:
Since $\emptyset$ is a subset of every set, $\emptyset \in P(A)$. Also, {a} and {b} are subsets of A. Further, every set is a subset of itself, so $A \in P(A)$. Thus, the various elements of $P(A)$ are $\emptyset$, {a}, {b}, and A; that is, $P(A) = \{\emptyset, \{a\}, \{b\}, A\}$. ■

Sets can be classified as finite and infinite sets, as defined below.

Finite and Infinite Sets

A set with a definite number of elements is a **finite set**. A set that is not finite is **infinite**.

EXAMPLE 2.8 The sets {a,b,c} and the set of computers in the world are finite, but the set of integers and the set of points on a line are infinite. ■

> It may sometimes be difficult to know the exact number of elements in a finite set. But that does not affect its finiteness. For example, the set of residents in California at a given time is finite, although it is difficult to determine the actual count.

It is impossible to list all the elements of an infinite set. Consequently, the enumeration method with ellipsis or the set-builder notation is used to define infinite sets. In the former case, the ellipsis would come at the end of the list, for example, $\mathbb{N} = \{1, 2, 3, \ldots\}$.

The following are some special infinite sets we will be using frequently:

$$\mathbf{Z} = \text{set of integers} = \{\ldots, -2, -1, 0, 1, 2, \ldots\}$$

$$\mathbb{N} = \mathbf{Z}^+ = \text{set of positive integers} = \{1, 2, 3, \ldots\}$$

$$\mathbf{Z}^- = \text{set of negative integers} = \{\ldots, -3, -2, -1\}$$

$$\mathbf{W} = \text{set of whole numbers} = \{0, 1, 2, 3, \ldots\}$$

$$\mathbf{Q} = \text{set of rational numbers} = \{p/q | p, q \in \mathbf{Z} \wedge q \neq 0\}$$

$$\mathbb{R} = \text{set of real numbers}$$

$$\mathbb{R}^+ = \text{set of positive real numbers} = \{x \in \mathbb{R} | x > 0\}$$

$$\mathbb{R}^- = \text{set of negative real numbers} = \{x \in \mathbb{R} | x < 0\}$$

A few additional subsets of $\mathbb{R}$, called **intervals**, will prove useful in our discussions. They are given below, where $a < b$:

closed interval	$[a, b] = \{x \in \mathbb{R} \| a \leq x \leq b\}$
closed–open interval	$[a, b) = \{x \in \mathbb{R} \| a \leq x < b\}$
open–closed interval	$(a, b] = \{x \in \mathbb{R} \| a < x \leq b\}$
open interval	$(a, b) = \{x \in \mathbb{R} \| a < x < b\}$

David Hilbert *(1862–1943) was born and educated in Königsberg, Germany (now in Russia). He made significant contributions to algebra, analysis, geometry, and mathematical physics. He described the importance of set theory in the development of mathematics: "No one shall expel us from the paradise which Cantor has created for us."*

A bracket at an endpoint indicates it is included in the set, whereas a parenthesis indicates it is not included.

The set $\{x \in \mathbb{R} | x \geq a\}$ is denoted by $[a, \infty)$ using the **infinity symbol** ∞. Likewise, the set $\{x \in \mathbb{R} | x \leq a\}$ is denoted by $(-\infty, a]$.

Next we present two interesting paradoxes related to infinite sets and proposed in the 1920s by the German mathematician David Hilbert.

The Hilbert Hotel Paradoxes

Imagine a grand hotel in a major city with an infinite number of rooms, all occupied. One morning a visitor arrives at the registration desk looking for a room. "I'm sorry, we are full," replies the manager, "but we can certainly accommodate you." How is this possible? Is she contradicting herself?

To give a room to the new guest, Hilbert suggested moving the guest in Room 1 to Room 2, the guest in Room 2 to Room 3, the one in Room 3 to Room 4, and so on; Room 1 is now vacant and can be given to the new guest. The clerk is happy that she can accommodate him by moving each guest one room down the hall.

The second paradox involves an infinite number of conventioneers arriving at the hotel, each looking for a room. The clerk realizes that the hotel can make a fortune if she can somehow accommodate them. She knows she can give each a room one at a time as above, but that will involve moving each guest constantly from one room to the next, resulting in total chaos and frustration.

So Hilbert proposed the following solution: move the guest in Room 1 to Room 2, the guest in Room 2 to Room 4, the one in Room 3 to Room 6, and so on. This puts the old guests in even-numbered rooms, so the new guests can be checked into the odd-numbered rooms.

Notice that in both cases the hotel could accommodate the guests only because it has infinitely many rooms.

A third paradox: Infinitely many hotels with infinitely many rooms are leveled by an earthquake. All guests survive and come to Hilbert Hotel. How can they be accommodated? See Example 3.23 for a solution.

We close this section by introducing a special set used in the study of formal languages.

Every word in the English language is an arrangement of the letters of the alphabet $\{A, B, \ldots, Z, a, b, \ldots, z\}$. The alphabet is finite and not every arrangement of the letters need make any sense. These ideas can be generalized as follows.

Alphabet

A finite set Σ of symbols is an **alphabet**. (Σ is the uppercase Greek letter *sigma*.) A **word** (or **string**) **over** Σ is a finite arrangement of symbols from Σ.

For instance, the only alphabet understood by a computer is the **binary alphabet** $\{0,1\}$; every word is a finite and unique arrangement of 0's and 1's. Every zip code is a word over the alphabet $\{0, \ldots, 9\}$.

Sets such as $\{a, b, c, ab, bc\}$ are *not* considered alphabets since the string ab, for instance, can be obtained by juxtaposing, that is, placing next to each other, the symbols a and b.

Length of a Word

The **length** of a word w, denoted by $\|w\|$, is the number of symbols in it. A word of length zero is the **empty word** (or the **null word**), denoted by the lowercase Greek letter λ (**lambda**); It contains no symbols.

For example, $\|ab\| = 2$, $\|aabba\| = 5$, and $\|\lambda\| = 0$.

The set of words over an alphabet Σ is denoted by Σ^*. The empty word λ belongs to Σ^* for every alphabet Σ. In particular, if Σ denotes the English alphabet, then Σ^* consists of all words, both meaningful and meaningless. Consequently, the English language is a subset of Σ^*. More generally, we make the following definition.

Language

A **language** over an alphabet Σ is a subset of Σ^*.

The following two examples illustrate this definition.

EXAMPLE 2.9 The set of zip codes is a finite language over the alphabet $\Sigma = \{0, \ldots, 9\}$. ■

EXAMPLE 2.10 Let $\Sigma = \{a, b\}$. Then $\Sigma^* = \{\lambda, a, b, aa, ab, ba, bb, aaa, aab, aba, abb, baa, \ldots\}$, an infinite set. Notice that $\{aa, ab, ba, bb\}$ is a finite language over Σ, whereas $\{a, aa, aba, bab, aaaa, abba, \ldots\}$ is an infinite language. ■

Words can be combined to create new words, as defined below.

Concatenation

The **concatenation** of two words x and y over an alphabet, denoted by xy, is obtained by appending the word y at the end of x. Thus if $x = x_1 \dots x_m$ and $y = y_1 \dots y_n$, $xy = x_1 \dots x_m y_1 \dots y_n$.

For example, let Σ be the English alphabet, $x =$ CAN, and $y =$ ADA; then $xy =$ CANADA. Notice that concatenation is *not* a commutative operation; that is, $xy \neq yx$. It is, however, associative; that is, $x(yz) = (xy)z = xyz$.

Two interesting properties are satisfied by the concatenation operation:

- The concatenation of any word x with λ is itself; that is, $\lambda x = x = x\lambda$ for every $x \in \Sigma^*$.
- Let $x, y \in \Sigma^*$. Then $\|xy\| = \|x\| + \|y\|$. (See Section 5.1 for a proof.)

For example, let $\Sigma = \{\text{a,b}\}$, $x =$ aba, and $y =$ bbaab. Then $xy =$ ababbaab and $\|xy\| = 8 = 3 + 5 = \|x\| + \|y\|$.

A useful notation: As in algebra, the exponential notation can be employed to eliminate the repeating of symbols in a word. Let x be a symbol and n an integer ≥ 2; then x^n denotes the concatenation $xx \dots x$ to $n - 1$ times. Using this compact notation, the words aaabb and ababab can be abbreviated as a^3b^2 and $(\text{ab})^3$, respectively. Notice, however, that $(\text{ab})^3 =$ ababab $\neq \text{a}^3\text{b}^3 =$ aaabbb.

Exercises 2.1

Rewrite each set using the listing method.

1. The set of months that begin with the letter A.
2. The set of letters of the word GOOGOL.
3. The set of months with exactly 31 days.
4. The set of solutions of the equation $x^2 - 5x + 6 = 0$.

Rewrite each set using the set-builder notation.

5. The set of integers between 0 and 5.
6. The set of January, February, May, and July.
7. The set of all members of the United Nations.
8. {Asia, Australia, Antarctica}

Determine if the given sets are equal.

9. $\{x, y, z\}, \{x, z, y\}$ **10.** $\{x|x^2 = 1\}, \{x|x^2 = x\}$

11. $\{x|x^2 = x\}, \{0, 1\}$ **12.** $\{x, \{y\}\}, \{\{x\}, y\}$

Mark each as true or false.

13. $\text{a} \in \{\text{alfa}\}$ **14.** $\text{b} \subseteq \{\text{a,b,c}\}$ **15.** $\{\text{x}\} \subseteq \{\text{x, y, z}\}$

16. $\{0\} = \emptyset$ **17.** $0 \in \emptyset$ **18.** $\{\emptyset\} = 0$

19. $\{\emptyset\} = \emptyset$ **20.** $\emptyset \subseteq \emptyset$ **21.** $\emptyset \in \{\emptyset\}$

22. $\{x|x \neq x\} = \emptyset$ **23.** $\{x, y\} = \{y, x\}$ **24.** $\{x\} \in \{\{x\}, y\}$

25. $\emptyset$ is a subset of every set. **26.** Every set is a subset of itself.

27. Every nonempty set has at least two subsets.

28. The set of people in the world is infinite.

29. The set of words in a dictionary is infinite.

Find the power set of each set.

30. $\emptyset$ **31.** $\{\mathrm{a}\}$ **32.** $\{\mathrm{a}, \mathrm{b}, \mathrm{c}\}$

33. Using Exercises 30–32, predict the number of subsets of a set with n elements.

In Exercises 34–37, n denotes a positive integer less than 10. Rewrite each set using the listing method.

34. $\{n|n$ is divisible by $2\}$ **35.** $\{n|n$ is divisible by $3\}$

36. $\{n|n$ is divisible by 2 and $3\}$ **37.** $\{n|n$ is divisible by 2 or $3\}$

Find the family of subsets of each set that do *not* contain consecutive integers.

38. $\{1, 2\}$ **39.** $\{1, 2, 3\}$

40. Let a_n denote the number of subsets of the set $S = \{1, 2, \ldots, n\}$ that do not contain consecutive integers, where $n \geq 1$. Find a_3 and a_4.

In Exercises 41–46, a language L over $\Sigma = \{a, b\}$ is given. Find five words in each language.

41. $L = \{x \in \Sigma^* | x$ begins with and ends in $b.\}$

42. $L = \{x \in \Sigma^* | x$ contains exactly one $b.\}$

43. $L = \{x \in \Sigma^* | x$ contains an even number of a's.$\}$

44. $L = \{x \in \Sigma^* | x$ contains an even number of a's followed by an odd number of b's.$\}$

Compute the length of each word over $\{\mathrm{a}, \mathrm{b}\}$.

45. aab **46.** aabbb

47. ab^4 **48.** $\mathrm{a}^3\mathrm{b}^2$

Arrange the binary words of the given length in increasing order of magnitude.

49. Length two. **50.** Length three.

A **ternary word** is a word over the alphabet $\{0, 1, 2\}$. Arrange the ternary words of the given length in increasing order of magnitude.

51. Length one. **52.** Length two.

Prove each.

***53.** The empty set is a subset of every set.
(*Hint*: Consider the implication $x \in \emptyset \rightarrow x \in A$.)

***54.** The empty set is unique.
(*Hint*: Assume there are two empty sets, $\emptyset_1$ and $\emptyset_2$. Then use Exercise 53.)

***55.** Let A, B, and C be arbitrary sets such that $A \subseteq B$ and $B \subseteq C$. Then $A \subseteq C$.
(**transitive property**)

***56.** If Σ is a nonempty alphabet, then Σ^* is infinite.
(*Hint*: Assume Σ^* is finite. Since $\Sigma \neq \emptyset$, it contains an element a. Let $x \in \Sigma^*$ with largest length. Now consider xa.)

2.2 Operations with Sets

Just as propositions can be combined in several ways to construct new propositions, sets can be combined in different ways to build new sets. You will find a close relationship between logic operations and set operations.

Union

The **union** of two sets A and B, denoted by $A \cup B$, is obtained by merging them; that is, $A \cup B = \{x | (x \in A) \vee (x \in B)\}$.

Notice the similarity between union and disjunction.

EXAMPLE 2.11 Let $A = \{\text{a, b, c}\}$, $B = \{\text{b, c, d, e}\}$, and $C = \{x, y\}$. Then $A \cup B = \{\text{a, b, c, d, e}\} = B \cup A$ and $B \cup C = \{\text{b, c, d, e, x, y}\} = C \cup B$. ■

The shaded areas in Figure 2.4 represent the set $A \cup B$ in three different cases.

Intersection

The **intersection** of two sets A and B, denoted by $A \cap B$, is the set of elements common to both A and B; that is, $A \cap B = \{x | (x \in A) \vee (x \in B)\}$.

Figure 2.4

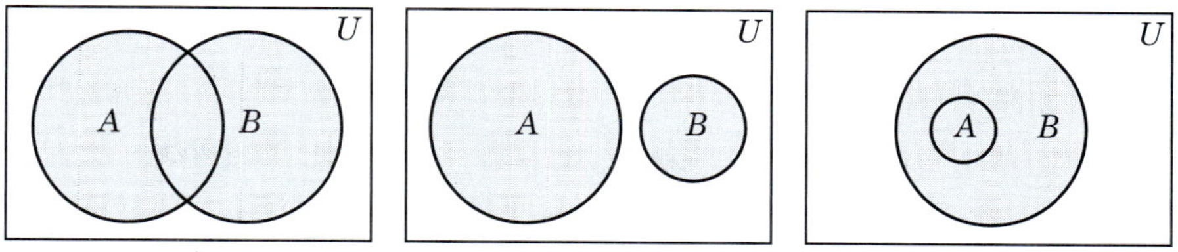

$A \cup B$ | $A \cup B$, where A and B are disjoint | $A \cup B = B$

Notice the relationship between intersection and conjunction.

EXAMPLE 2.12 Let $A = \{\text{Nov}, \text{Dec}, \text{Jan}, \text{Feb}\}$, $B = \{\text{Feb}, \text{Mar}, \text{Apr}, \text{May}\}$, and $C = \{\text{Sept}, \text{Oct}, \text{Nov}, \text{Dec}\}$. Then $A \cap B = \{\text{Feb}\} = B \cap A$ and $B \cap C = \emptyset = C \cap B$. (Notice that B and C are disjoint sets. More generally, two sets are disjoint if and only if their intersection is null.) ■

Figure 2.5

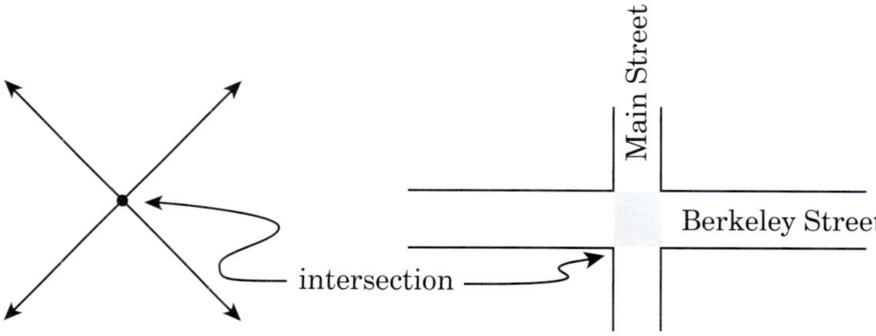

Figure 2.5 shows the intersection of two lines and that of two streets, and Figure 2.6 displays the set $A \cap B$ in three different cases.

Figure 2.6

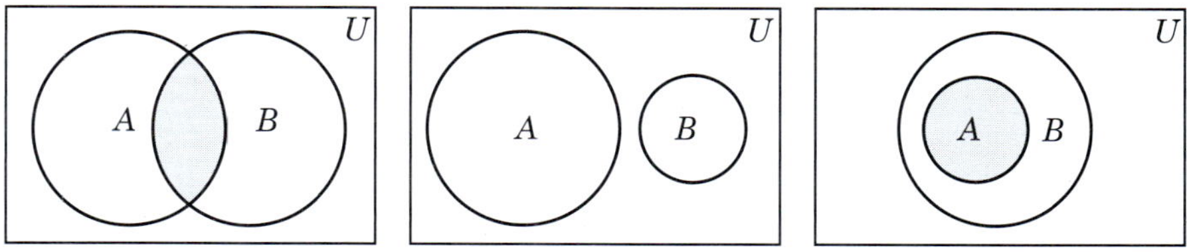

$A \cap B$ | $A \cap B = \emptyset$ | $A \cap B = A$

EXAMPLE 2.13 Let $A = \{a, b, c, d, g\}$, $B = \{b, c, d, e, f\}$, and $C = \{b, c, e, g, h\}$. Find $A \cup (B \cap C)$ and $(A \cup B) \cap (A \cup C)$.

SOLUTION:

(1)
$$B \cap C = \{\text{b}, \text{c}, \text{e}\}$$
$$A \cup (B \cap C) = \{\text{a}, \text{b}, \text{c}, \text{d}, \text{e}, \text{g}\}$$

(2)
$$A \cup B = \{\text{a}, \text{b}, \text{c}, \text{d}, \text{e}, \text{f}, \text{g}\}$$
$$A \cup C = \{\text{a}, \text{b}, \text{c}, \text{d}, \text{e}, \text{g}, \text{h}\}$$
$$(A \cup B) \cap (A \cup C) = \{\text{a}, \text{b}, \text{c}, \text{d}, \text{e}, \text{g}\}$$
$$= A \cup (B \cap C)$$

See the Venn diagram in Figure 2.7.

Figure 2.7

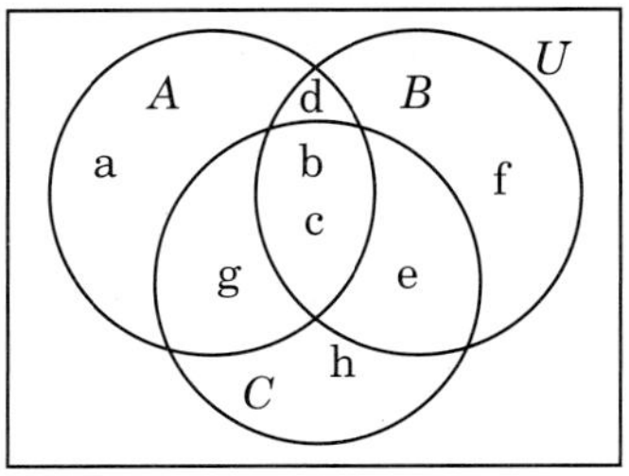

■

A third way of combining two sets is by finding their difference, as defined below.

Difference

The **difference** of two sets A and B (or the **relative complement** of B in A), denoted by $A - B$ (notice the order), is the set of elements in A that are *not* in B. Thus $A - B = \{x \in A | x \notin B\}$.

EXAMPLE 2.14 Let $A = \{\text{a}, \ldots, \text{z}, 0, \ldots, 9\}$, and $B = \{0, \ldots, 9\}$. Then $A - B = \{\text{a}, \ldots, \text{z}\}$ and $B - A = \emptyset$.

The shaded areas in Figure 2.8 represent the set $A - B$ in three different cases.

Figure 2.8

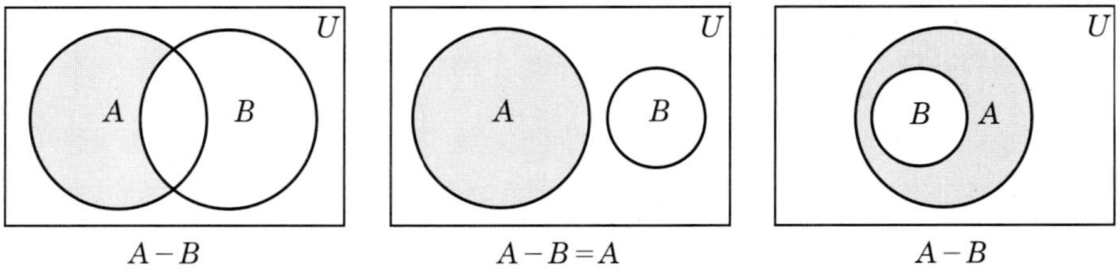

■

For any set $A \neq U$, although $A - U = \emptyset$, the difference $U - A \neq \emptyset$. This shows yet another way of obtaining a new set.

Complement

The difference $U - A$ is the (absolute) **complement** of A, denoted by A' (A prime). Thus $A' = U - A = \{x \in U | x \notin A\}$.

Figure 2.9 represents the complement of a set A. (Complementation corresponds to negation.)

Figure 2.9

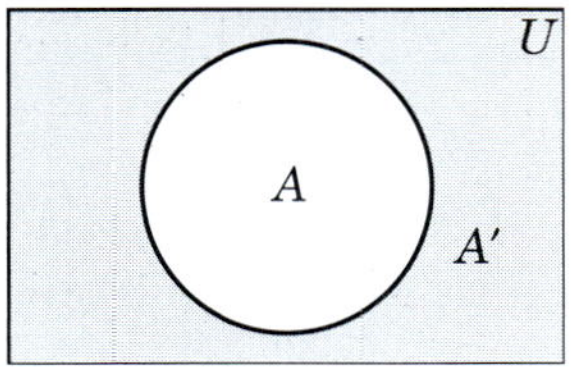

EXAMPLE 2.15 Let $U = \{a, \ldots, z\}$. Find the complements of the sets $A = \{a, e, i, o, u\}$ and $B = \{a, c, d, e, \ldots, w\}$. Then $A' = U - A =$ set of all consonants in the alphabet, and $B' = U - B = \{b, x, y, z\}$. ■

EXAMPLE 2.16 Let $A = \{a, b, x, y, z\}$, $B = \{c, d, e, x, y, z\}$, and $U = \{a, b, c, d, e, w, x, y, z\}$. Find $(A \cup B)'$ and $A' \cap B'$.

SOLUTION:

(1)
$$A \cup B = \{a, b, c, d, e, x, y, z\}$$
$$(A \cup B)' = \{w\}$$

(2)
$$A' = \{c, d, e, w\}$$
$$B' = \{a, b, w\}$$
$$A' \cap B' = \{w\} = (A \cup B)'$$

See Figure 2.10.

Figure 2.10

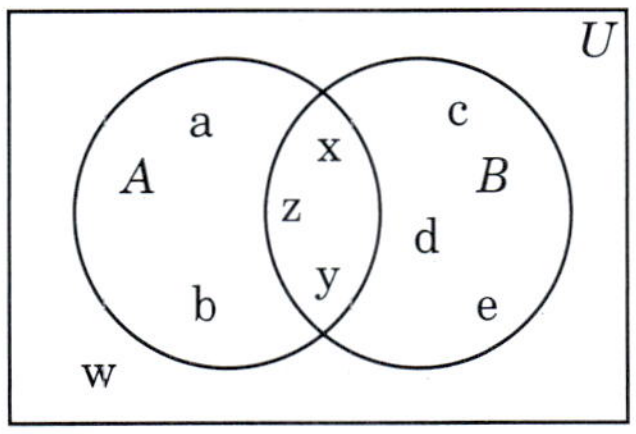

■

Since as a rule, $A - B \neq B - A$, by taking their union we can form a new set.

Symmetric Difference

The **symmetric difference** of A and B, denoted by $A \oplus B$, is defined by $A \oplus B = (A - B) \cup (B - A)$.

EXAMPLE 2.17 Let $A = \{\text{a}, \dots, \text{z}, 0, \dots, 9\}$ and $B = \{0, \dots, 9, +, -, *, /\}$. Then $A - B = \{\text{a}, \dots, \text{z}\}$ and $B - A = \{+, -, *, /\}$. So $A \oplus B = (A - B) \cup (B - A) = \{\text{a}, \dots, \text{z}, +, -, *, /\}$. ■

The symmetric difference of A and B is pictorially displayed in Figure 2.11 in three different cases.

Figure 2.11

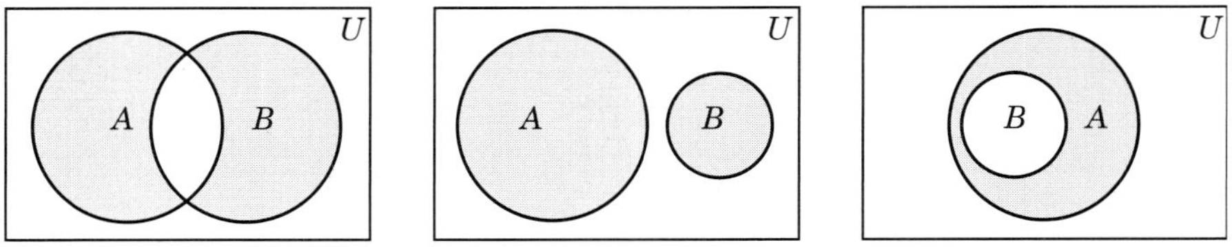

$A \oplus B = A \cup B - A \cap B$ $A \oplus B = A \cup B$ $A \oplus B = A - B$

Set and Logic Operations

> Set operations and logic operations are closely related, as Table 2.1 shows.

Table 2.1

Set operation	Logic operation
$A \cup B$	$p \vee q$
$A \cap B$	$p \vee q$
A'	$\sim p$
$A \oplus B$	p XOR q

The important properties satisfied by the set operations are listed in Table 2.2. (Notice the similarity between these properties and the laws of logic in Section 1.2.) We shall prove one of them. Use its proof as a model to prove the others as routine exercises.

We shall prove law 16. It uses De Morgan's law in symbolic logic, and the fact that $X = Y$ if and only if $X \subseteq Y$ and $Y \subseteq X$.

PROOF:

In order to prove that $(A \cup B)' = A' \cap B'$, we must prove two parts: $(A \cup B)' \subseteq A' \cap B'$ and $A' \cap B' \subseteq (A \cup B)'$.

- To prove that $(A \cup B)' \subseteq (A' \cap B')$:

  Let x be an arbitrary element of $(A \cup B)'$. Then $x \notin (A \cup B)$. Therefore, by De Morgan's law, $x \notin A$ and $x \notin B$; that is, $x \in A'$ and $x \in B'$. So $x \in A' \cap B'$. Thus every element of $(A \cup B)'$ is also an element of $A' \cap B'$; that is, $(A \cup B)' \subseteq A' \cap B'$.

Table 2.2

Laws of Sets

Let A, B, and C be any three sets and U the universal set. Then:

Idempotent laws

1. $A \cup A = A$	2. $A \cap A = A$

Identity laws

3. $A \cup \emptyset = A$	4. $A \cap U = A$

Inverse laws

5. $A \cup A' = U$	6. $A \cap A' = \emptyset$

Domination laws

7. $A \cup U = U$	8. $A \cap \emptyset = \emptyset$

Commutative laws

9. $A \cup B = B \cup A$	10. $A \cap B = B \cap A$

Double complementation law

11. $(A')' = A$

Associative laws

12. $A \cup (B \cup C) = (A \cup B) \cup C$	13. $A \cap (B \cap C) = (A \cap B) \cap C$

Distributive laws

14. $A \cup (B \cap C) = (A \cup B) \cap (A \cup C)$	15. $A \cap (B \cup C) = (A \cap B) \cup (A \cap C)$

De Morgan's laws

16. $(A \cup B)' = A' \cap B'$	17. $(A \cap B)' = A' \cup B'$

Absorption laws

18. $A \cup (A \cap B) = A$	19. $A \cap (A \cup B) = A$

(Note: The following laws have no names.)

20. If $A \subseteq B$, then $A \cap B = A$.	21. If $A \subseteq B$, then $A \cup B = B$.
22. If $A \subseteq B$, then $B' \subseteq A'$.	23. $A - B = A \cap B'$
24. $A \oplus B = A \cup B - A \cap B$	

- To prove that $A' \cap B' \subseteq (A \cup B)'$:

  Let x be any element of $A' \cap B'$. Then $x \in A'$ and $x \in B'$. Therefore, $x \notin A$ and $x \notin B$. So, by De Morgan's law, $x \notin (A \cup B)$. Consequently, $x \in (A \cup B)'$. Thus, since x is arbitrary, $A' \cap B' \subseteq (A \cup B)'$.

Thus, $(A \cup B)' = A' \cap B'$. See the Venn diagrams in Figure 2.12 also.

Note: Law 23 is a very useful result and will be used in the next section.

A few words of explanation: The commutative laws imply that the order in which the union (or intersection) of two sets is taken is irrelevant. The associative laws imply that when the union (or intersection) of three or more

Figure 2.12

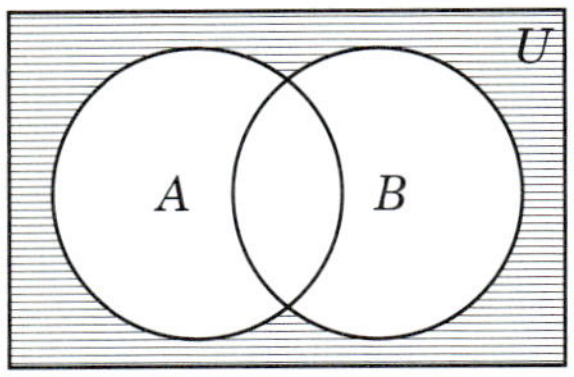

$(A \cup B)' =$ shaded area

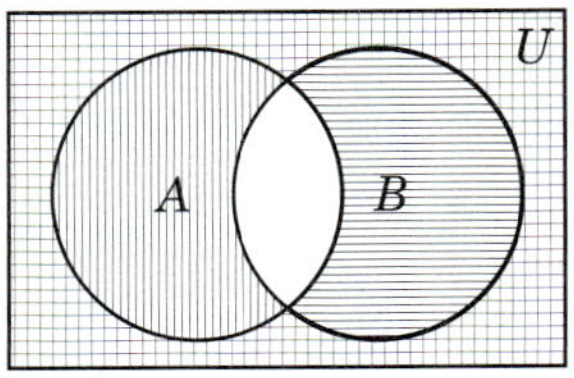

$A' \cap B' =$ cross-shaded area

sets is taken, the way the sets are grouped is immaterial; in other words, such expressions without parentheses are perfectly legal. For instance, $A \cup B \cup C = A \cup (B \cup C) = (A \cup B) \cup C$ is certainly valid. The two De Morgan's laws in propositional logic play a central role in deriving the corresponding laws in sets.

> Again, as in propositional logic, parentheses are essential to indicate the groupings in the distributive laws. For example, if you do not parenthesize the expression $A \cap (B \cup C)$ in law 15, then the LHS becomes $A \cap B \cup C = (A \cap B) \cup C = (A \cup C) \cap (B \cup C) \neq (A \cap B) \cup (A \cap C)$.

> Notice the similarity between the set laws and the laws of logic. For example, properties 1 through 19 and 22 have their counterparts in logic. Every corresponding law of logic can be obtained by replacing sets A, B, and C with propositions p, q, and r, respectively, the set operators $\cap$, $\cup$, and $'$ with the logic operators $\wedge$, $\vee$, and $\sim$ respectively, and equality $(=)$ with logical equivalence $(\equiv)$.

Using this procedure, the absorption law $A \cup (A \cap B) = A$, for instance, can be translated as $p \vee (p \wedge q) \equiv p$, which is the corresponding absorption law in logic.

Just as truth tables were used in Chapter 1 to establish the logical equivalence of compound statements, they can be applied to verify set laws as well. The next example illustrates this method.

EXAMPLE 2.18 Using a truth table, prove that $(A \cup B)' = A' \cap B'$.

SOLUTION:

Let x be an arbitrary element. Then x may or may not be in A. Likewise, x may or may not belong to B. Enter this information, as in logic, in the first two columns of the table, which are headed by $x \in A$ and $x \in B$.

The table needs five more columns, headed by $x \in (A \cup B)$, $x \in (A \cup B)'$, $x \in A'$, $x \in B'$, and $x \in (A' \cap B')$ (see Table 2.3). Again, as in logic, use the entries in the first two columns to fill in the remaining columns, as in the table.

Table 2.3

$x \in A$	$x \in B$	$x \in (A \cup B)$	$x \in (A \cup B)'$	$x \in A'$	$x \in B'$	$x \in (A' \cap B')$
T	T	T	F	F	F	F
T	F	T	F	F	T	F
F	T	T	F	T	F	F
F	F	F	T	T	T	T

Note: The shaded columns are identical

Since the columns headed by $x \in (A \cup B)'$ and $x \in (A' \cap B')$ are identical, it follows that $(A \cup B)' = A' \cap B'$. ■

Using truth tables to prove set laws is purely mechanical and elementary. It does not provide any insight into the development of a mathematical proof. Such a proof does not build on previously known set laws, so we shall not resort to such proofs in subsequent discussions.

Just as the laws of logic can be used to simplify logic expressions and derive new laws, set laws can be applied to simplify set expressions and derive new laws. In order to be successful in this art, you must know the laws well and be able to apply them as needed. So, practice, practice, practice.

EXAMPLE 2.19 Using set laws, verify that $(X - Y) - Z = X - (Y \cup Z)$.

PROOF:

$$\begin{aligned}
(X - Y) - Z &= (X - Y) \cap Z' && A - B = A \cap B' \\
&= (X \cap Y') \cap Z' && A - B = A \cap B' \\
&= X \cap (Y' \cap Z') && \text{associative law 13} \\
&= X \cap (Y \cup Z)' && \text{De Morgan's law 16} \\
&= X - (Y \cup Z) && A - B = A \cap B'
\end{aligned}$$
■

EXAMPLE 2.20 Simplify the set expression $(A \cap B') \cup (A' \cap B) \cup (A' \cap B')$.

SOLUTION:
(You may supply the justification for each step.)

$$\begin{aligned}
(A \cap B') \cup (A' \cap B) \cup (A' \cap B') &= (A \cap B') \cup [(A' \cap B) \cup (A' \cap B')] \\
&= (A \cap B') \cup [A' \cap (B \cup B')] \\
&= (A \cap B') \cup (A' \cap U) \\
&= (A \cap B') \cup A'
\end{aligned}$$

$$
\begin{aligned}
&= A' \cup (A \cap B') \\
&= (A' \cup A) \cap (A' \cup B') \\
&= U \cap (A' \cup B') \\
&= A' \cup B'
\end{aligned}
$$ ■

Often subscripts are used to name sets, so we now turn our attention to such sets.

Indexed Sets

Let I, called the **index set**, be the set of subscripts i used to name the sets A_i. Then the union of the sets A_i as i varies over I is denoted by $\bigcup_{i \in I} A_i$. Similarly, $\bigcap_{i \in I} A_i$ denotes the intersection of the sets A_i as i runs over I. In particular, let $I = \{1, 2, \dots, n\}$. Then $\bigcup_{i \in I} A_i = A_1 \cup A_2 \cup \cdots \cup A_n$, which is often written as $\bigcup_{i=1}^{n} A_i$ or simply $\bigcup_{1}^{n} A_i$. Likewise, $\bigcap_{i \in I} A_i = \bigcap_{i=I}^{n} A_i = \bigcap_{1}^{n} A_i = A_1 \cap A_2 \cap \cdots \cap A_n$. If $I = \mathbb{N}$, the expression $\bigcup_{i \in \mathbb{N}} A_i$ is written as $\bigcup_{i \in 1}^{\infty} A_i = \bigcup_{1}^{\infty} A_i$, using the infinity symbol ∞; similarly, $\bigcap_{i \in \mathbb{N}} A_i = \bigcap_{i=1}^{\infty} A_i = \bigcap_{1}^{\infty} A_i$.

Before we proceed to define a new binary operation on sets $\bigcap_{i=1}^{\infty}$, we define an ordered set.

Ordered Set

Recall that the set $\{a_1, a_2, \dots, a_n\}$ is an unordered collection of elements. Suppose we assign a position to each element. The resulting set is an **ordered set** with n elements or an ***n*-tuple**, denoted by $(a_1, a_2, \dots, a_n)$. (Notice the use of parentheses versus braces.) The set (a_1, a_2) is an **ordered pair.**

Two n-tuples are **equal** if and only if their corresponding elements are equal. That is, $(a_1, a_2, \dots, a_n) = (b_1, b_2, \dots, b_n)$ if and only if $a_i = b_i$ for every i.

EXAMPLE 2.21 Every numeral and word can be considered an n-tuple. For instance,

$345 = (3,\ 4,\ 5)$

- 5 ← ones
- 4 ← tens
- 3 ← hundreds

computer = (c, o, m, p, u, t, e, r)

$1001011 = (1, 0, 0, 1, 0, 1, 1)$ ← ASCII* code for letter K

$11010010 = (1, 1, 0, 1, 0, 0, 1, 0)$ ← EBCDIC** code for letter K ■

*American Standard Code for Information Interchange.

**Extended Binary Coded Decimal Interchange Code.

René Descartes *(1596–1650) was born near Tours, France. At eight, he entered the Jesuit school at La Fleche, where because of poor health he developed the habit of lying in bed thinking until late in the morning; he considered those times the most productive. He left the school in 1612 and moved to Paris, where he studied mathematics for a brief period.*

After a short military career and travel through Europe for about 5 years, he returned to Paris and studied mathematics and philosophy. He then moved to Holland, where he lived for 20 years writing several books. In 1637 he wrote Discours, *which contains his contributions to analytic geometry.*

In 1649 Descartes moved to Sweden at the invitation of Queen Christina. There he contracted pneumonia and died.

We are now ready to define the next and final operation on sets.

Cartesian Product

The **cartesian product** of two sets A and B, denoted by $A \times B$, is the set of all ordered pairs (a, b) with $a \in A$ and $b \in B$. Thus $A \times B = \{(a,b) | a \in A \wedge b \in B\}$. $A \times A$ is denoted by A^2.

It is named after the French philosopher and mathematician René Descartes.

EXAMPLE 2.22 Let $A = \{\mathrm{a, b}\}$ and $B = \{\mathrm{x, y, z}\}$. Then

$$A \times B = \{(\mathrm{a, x}), (\mathrm{a, y}), (\mathrm{a, z}), (\mathrm{b, x}), (\mathrm{b, y}), (\mathrm{b, z})\}$$

$$B \times A = \{(\mathrm{x, a}), (\mathrm{x, b}), (\mathrm{y, a}), (\mathrm{y, b}), (\mathrm{z, a}), (\mathrm{z, b})\}$$

$$A^2 = A \times A = \{(\mathrm{a, a}), (\mathrm{a, b}), (\mathrm{b, a}), (\mathrm{b, b})\}$$

(Notice that $A \times B \neq B \times A$.) ■

The various elements of $A \times B$ in Example 2.22 can be displayed in a rectangular fashion, as in Figure 2.13, and pictorially, using dots as in Figure 2.14. The circled dot in row a and column y, for instance, represents the element (a, y). The pictorial representation in Figure 2.14 is the **graph** of $A \times B$.

Figure 2.13

Elements of A				
	a	(a, x)	(a, y)	(a, z)
	b	(b, x)	(b, y)	(b, z)
		x	y	z
		Elements of B		

Figure 2.14

Pictorial representation of $A \times B$.

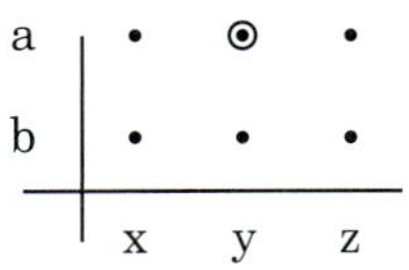

Figure 2.15 shows the graph of the infinite set $\mathbb{N}^2 = \mathbb{N} \times \mathbb{N}$. The circled dot in column 4 and row 3, for instance, represents the element (4,3). The horizontal and vertical dots indicate that the pattern is to be continued indefinitely in both directions.

Figure 2.15

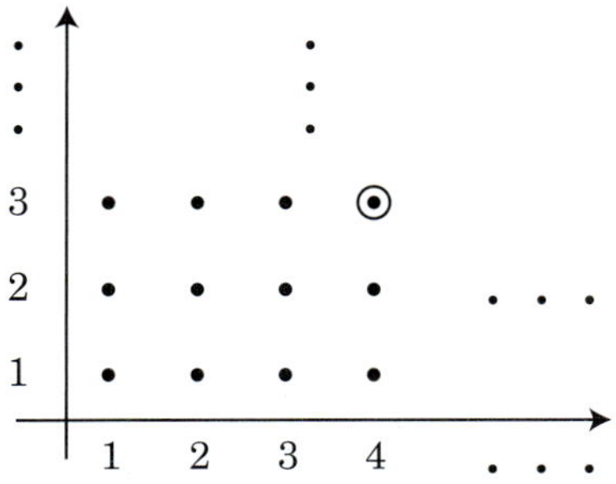

More generally, $\mathbb{R}^2 = \mathbb{R} \times \mathbb{R}$ consists of all possible ordered pairs (x, y) of real numbers. It is represented by the familiar ***xy*-plane** or the **cartesian plane** used for graphing (see Figure 2.16).

Figure 2.16

The cartesian plane $\mathbb{R}^2$.

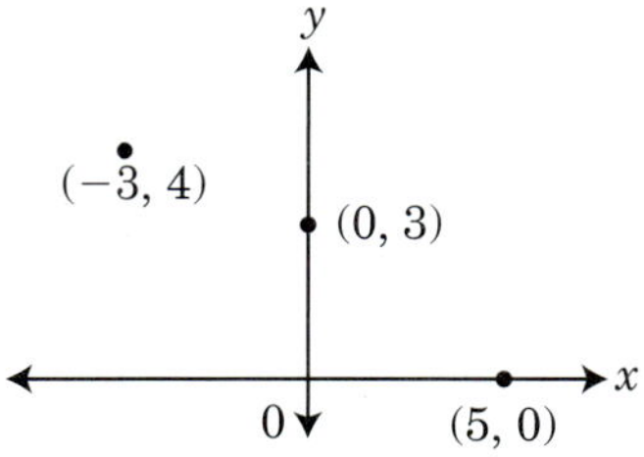

The following example presents an application of cartesian product.

EXAMPLE 2.23 Linda would like to make a trip from Boston to New York and then to London. She can travel by car, plane, or ship from Boston to New York, and by plane or ship from New York to London. Find the set of various modes of transportation for the entire trip.

SOLUTION:

Let A be the set of means of transportation from Boston to New York and B the set from New York to London. Clearly $A = \{\text{car, plane, ship}\}$ and $B = \{\text{plane, ship}\}$. So the set of possible modes of transportation is given by

Figure 2.17

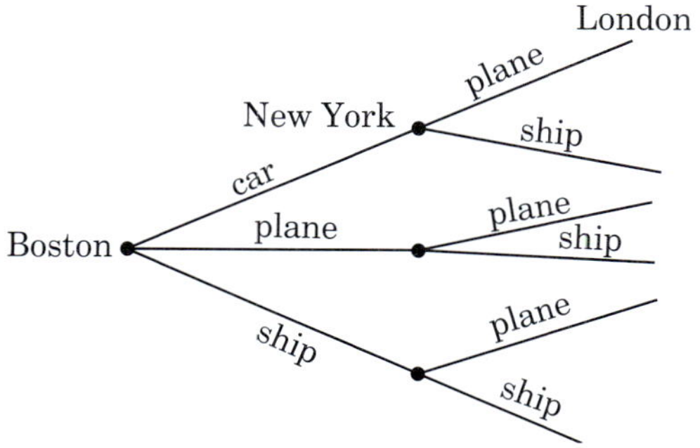

$A \times B =$ {(car, plane), (car, ship), (plane, plane), (plane, ship), (ship, plane), (ship, ship)}. See Figure 2.17. ■

The definition of the product of two sets can be extended to n sets. The **cartesian product of *n* sets** $A_1, A_2, \ldots, A_n$ consists of all possible n-tuples $(a_1, a_2, \ldots, a_n)$, where $a_i \in A_i$ for every i; it is denoted by $A_1 \times A_2 \times \cdots \times A_n$. If all A_i's are equal to A, the product set is denoted by A^n.

EXAMPLE 2.24 Let $A = \{\text{x}\}$, $B = \{\text{y}, \text{z}\}$, and $C = \{1, 2, 3\}$. Then

$$\begin{aligned} A \times B \times C &= \{(a, b, c) | a \in A, b \in B, \text{ and } c \in C\} \\ &= \{(\text{x}, \text{y}, 1), (\text{x}, \text{y}, 2), (\text{x}, \text{y}, 3), (\text{x}, \text{z}, 1), (\text{x}, \text{z}, 2), (\text{x}, \text{z}, 3)\} \end{aligned}$$

Finally, take a look at the map of the continental United States in Figure 2.18. It provides a geographical illustration of partitioning, a concept that can be extended to sets in an obvious way.

Figure 2.18

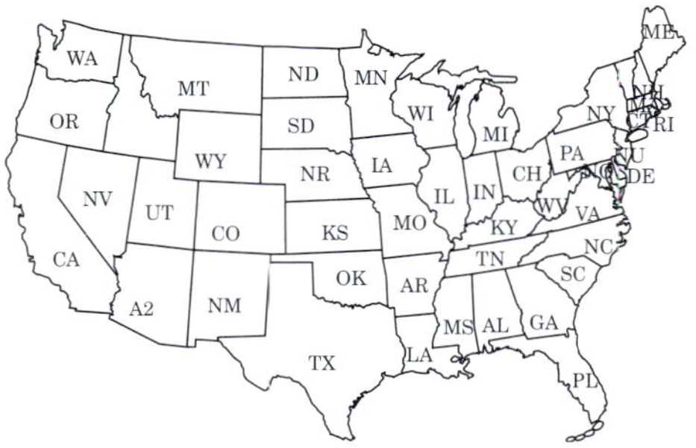

■

Partition

Consider the set $S = \{\text{a, b, c, d, e, f, g, h, i}\}$ and the subsets $S_1 = \{\text{a, b}\}$, $S_2 = \{\text{c}\}$, $S_3 = \{\text{d, e, f}\}$, $S_4 = \{\text{g, h}\}$, and $S_5 = \{\text{i}\}$. Notice that these subsets have three interesting properties: (1) They are nonempty; (2) they are pairwise disjoint; that is, no two subsets have any common elements; (3) their union is S. (See Figure 2.19.) The set $P = \{S_1, S_2, S_3, S_4, S_5\}$ is called a **partition** of S.

Figure 2.19

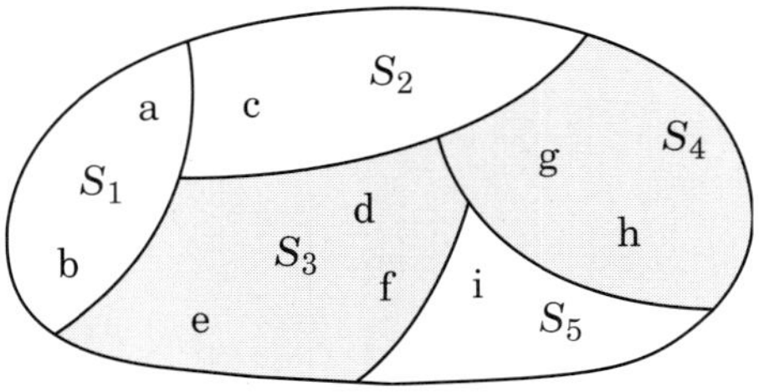

More generally, let I be an index set and P a family of subsets S_i of a nonempty set S, where $i \in I$. Then P is a **partition** of S if:

- Each set S_i is nonempty.
- The subsets are pairwise disjoint; that is, $S_i \cap S_j = \emptyset$ if $i \neq j$.
- The union of the subsets S_i is S; that is, $\bigcup_{i \in I} S_i = S$.

(Each subset S_i is a **block** of the partition.) Thus a partition of S is a collection of nonempty, pairwise disjoint subsets of S whose union is S.

EXAMPLE 2.25 Let Z_r denote the set of integers which, when divided by 5, leave r as the remainder. Then $0 \leq r < 5$ (see Section 4.1):

$$\mathbf{Z}_0 = \{\ldots, -5, 0, 5, \ldots\}$$
$$\mathbf{Z}_1 = \{\ldots, -4, 1, 6, \ldots\}$$
$$\mathbf{Z}_2 = \{\ldots, -3, 2, 7, \ldots\}$$
$$\mathbf{Z}_3 = \{\ldots, -2, 3, 8, \ldots\}$$
$$\mathbf{Z}_4 = \{\ldots, -1, 4, 9, \ldots\}$$

$P = \{\mathbf{Z}_0, \mathbf{Z}_1, \mathbf{Z}_2, \mathbf{Z}_3, \mathbf{Z}_4\}$ is a partition of the set of integers. See Figure 2.20. (This example is discussed in more detail in Section 7.4.) ■

Figure 2.20

Set of integers **Z**.

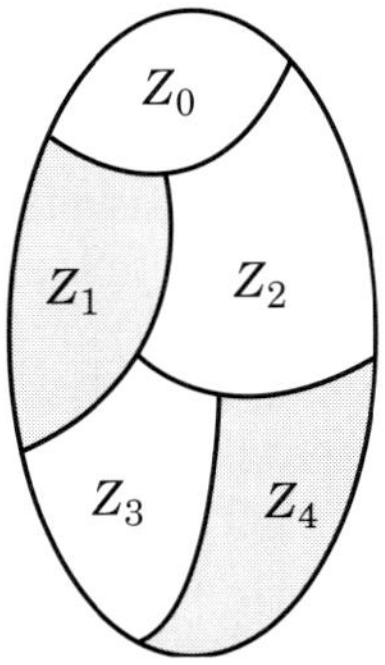

The sports pages of newspapers provide fine examples of partitions, as the next example illustrates.

EXAMPLE 2.26 In 2003, the set of teams S in the National Football League was divided into two conferences, American and National, and each conference into four divisions—East, South, North, and West. Let E_1, S_1, N_1, and W_1 denote the set of teams in East, South, North, and West Divisions in the American Conference, respectively, and E_2, S_2, N_2, and W_2 the corresponding sets in the National Conference. Then:

$$E_1 = \{\text{Buffalo, Miami, New England, NY Jets}\}$$
$$S_1 = \{\text{Indianapolis, Tennessee, Houston, Jacksonville}\}$$
$$N_1 = \{\text{Baltimore, Cincinnati, Cleveland, Pittsburgh}\}$$
$$W_1 = \{\text{Denver, Kansas City, Oakland, San Diego}\}$$
$$E_2 = \{\text{Washington, Philadelphia, Dallas, NY Giants}\}$$
$$S_2 = \{\text{Atlanta, Tampa Bay, Carolina, New Orleans}\}$$
$$N_2 = \{\text{Chicago, Detroit, Minnesota, Green Bay}\}$$
$$W_2 = \{\text{Arizona, Seattle, St. Louis, San Francisco}\}$$

Clearly, $P = \{E_1, S_1, N_1, W_1, E_2, S_2, N_2, W_2\}$ is a partition of S.

We close this section with a brief introduction to fuzzy sets.

Fuzzy Sets (optional)

Fuzzy sets, a generalization of ordinary sets, were introduced in 1965 by Lotfi A. Zadeh of the University of California at Berkeley. They have applications to human cognition, communications, decision analysis, psychology, medicine, law, information retrieval, and, of course, artificial intelligence. Like fuzzy logic, they model the fuzziness in the natural language—for example, in terms like *young, healthy, wealthy*, and *beautiful*.

In fuzzy set theory, every element x in the universal set U has a certain degree of membership $d_U(x)$, where $0 \leq d_U(x) \leq 1$; $d_U(x)$ indicates the degree of fuzziness. Accordingly, a fuzzy set S is denoted by listing its elements along with their degrees of membership; an element with zero degree of membership is not listed.

For example, let U be the fuzzy set of wealthy people and $S = \{\text{Tom } 0.4, \text{Dick } 0.7, \text{Harry } 0.6\}$. Then Harry belongs to S with degree of membership 0.6; $d_S(\text{Harry}) = 0.6$ measures Harry's degree of wealthiness.

The concept of an ordinary subset can be extended to fuzzy sets also.

Fuzzy Subset

Let A and B be fuzzy sets. Then A is a **fuzzy subset** of B if $A \subseteq B$ and $d_A(x) \leq d_B(x)$ for every element x in A.

Lotfi A. Zadeh *(1921–) was born in Baku, Azerbaijan. An alumnus of the University of Tehran (1942) and the Massachusetts Institute of Technology (1946), he received his Ph.D. from Columbia University in 1949 for his dissertation on frequency analysis of time-varying networks. He began his professional career in the Department of Electrical Engineering at Columbia. In 1959, he joined the Department of Electrical Engineering and Computer Science at the University of California, Berkeley, serving as its chair during the years 1963–1968. Currently, he is a professor at Berkeley and Director of Berkeley Initiative in Soft Computing.*

Zadeh's earlier "work was centered on systems analysis, decision analysis, and information systems. Since then his current research has shifted to the theory of fuzzy sets and its applications to artificial intelligence (AI). His research interest now is focused on fuzzy logic, soft computing, computing with words, and the newly developed computational theory of perceptions and precisiated natural language," according to the University of California Web site.

A truly gifted mind and an expert on AI, Zadeh has authored about 200 journal articles on a wide variety of subjects relating to the conception, design, and analysis of information/intelligent systems. He serves on the editorial boards of more than 50 journals and on the advisory boards of a number of institutions related to AI.

Zadeh is a recipient of numerous awards and medals, including the IEEE Education Medal, IEEE Richard W. Hamming Medal, IEEE Medal of Honor, the ASME Rufus Oldenburger Medal, B. Bolzano Medal of the Czech Academy of Sciences, Kampe de Feriet Medal, AACC Richard E. Bellman Central Heritage Award, the Grigore Moisil Prize, Honda Prize, Okawa Prize, AIM Information Science Award, IEEE-SMC J. P. Wohl Career Achievement Award, SOFT Scientific Contribution Memorial Award of the Japan Society for Fuzzy Theory, IEEE Millennium Medal, and the ACM 2000 Allen Newell Award. He has received honorary doctorates from many universities from around the world.

For example, let $S = \{\text{Betsey } 0.6, \text{Mat } 0.5\}$ and $T = \{\text{Betsey } 0.8, \text{Jonathan } 0.3, \text{Mary } 0.5, \text{Mat } 0.7\}$ by fuzzy sets of smart people. Then S is a fuzzy subset of T.

Operations on ordinary sets can be extended to fuzzy sets as well.

Operations on Fuzzy Sets

Let A and B be any fuzzy set. The **union** of A and B is $A \cup B$, where $d_{A\cup B}(x) = \max\{d_A(x), d_B(x)\}$; their **intersection** is $A \cap B$, where $d_{A\cap B}(x) = \min\{d_A(x), d_B(x)\}$; and the **complement** of A is A', where $d_{A'}(x) = 1 - d_A(x)$; in A' only the degrees of membership change.

Using the sets S and T above,

$$S \cup T = \{\text{Betsey } 0.8, \text{Jonathan } 0.3, \text{Mary } 0.5, \text{Mat } 0.7\}$$

$$S \cap T = \{\text{Betsey } 0.6, \text{Mat } 0.5\}$$

$$S' = \{\text{Betsey } 0.4, \text{Mat } 0.5\}$$

Additional opportunities to practice the various operations are given in the exercises.

Exercises 2.2

Let $A = \{a, e, f, g, i\}$, $B = \{b, d, e, g, h\}$, $C = \{d, e, f, h, i\}$, and $U = \{a, b, \ldots, k\}$. Find each set.

1. C' **2.** $B \cap C'$ **3.** $C \cap A'$ **4.** $(A \cup B)'$

5. $(B \cap C)'$ **6.** $(A \cup C')'$ **7.** $(B \cap C')'$ **8.** $A \oplus B$

9. $(A - B) - C$ **10.** $A - (B - C)$ **11.** $(A \cup B) - C$ **12.** $(A \cap B) - C$

Using the Venn diagram in Figure 2.21 find each set.

13. $(A \cup B) \cap C$ **14.** $A \cap (B \cup C)$ **15.** $A - (B - C)$

16. $(A \oplus B) \cup C$ **17.** $A \cap (B \oplus C)$ **18.** $A - (B \oplus C)$

Figure 2.21

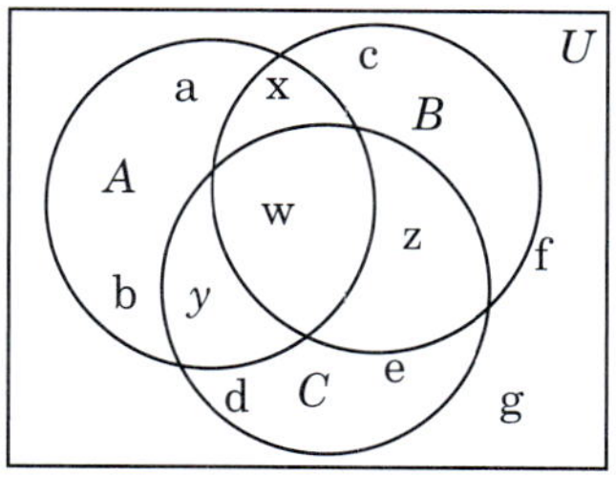

Let $A = \{b, c\}$, $B = \{x\}$, and $C = \{x, z\}$. Find each set.

19. $A \times B$ **20.** $B \times A$ **21.** $A \times \emptyset$ **22.** $A \times B \times \emptyset$

23. $A \times (B \cup C)$ **24.** $A \times (B \cap C)$ **25.** $A \times B \times C$ **26.** $A \times C \times B$

Mark each as true or false, where A, B, and C are arbitrary sets and U the universal set.

27. $A - \emptyset = A$ **28.** $\emptyset - A = -A$ **29.** $\emptyset - \emptyset = 0$

30. $A - A = 0$ **31.** $A - B = B - A$ **32.** $A - A' = \emptyset$

33. $(A')' = A$ **34.** $(A \cap B)' = A' \cap B'$ **35.** $(A \cup B)' = A' \cup B'$

36. $A \subseteq A \cup B$ **37.** $A \subseteq A \cap B$ **38.** $B \cap (A - B) = \emptyset$

Give a counterexample to disprove each proposition.

39. $(A - B) - C = A - (B - C)$ **40.** $A \cup (B - C) = (A \cup B) - (A \cup C)$

41. $A \cup (B \oplus C) = (A \cup B) \oplus (A \cup C)$ **42.** $A \oplus (B \cap C) = (A \oplus B) \cap (A \oplus C)$

Determine if each is a partition of the set $\{a, \ldots, z, 0, \ldots, 9\}$.

43. $\{\{a, \ldots, z\}, \{0, \ldots, 9\}, \emptyset\}$

44. $\{\{a, \ldots, j\}, \{i, \ldots, t\}, \{u, \ldots, z\}, \{0, \ldots, 9\}\}$

45. $\{\{a, \ldots, l\}, \{n, \ldots, t\}, \{u, \ldots, z\}, \{0, \ldots, 9\}\}$

46. $\{\{a, \ldots, u\}, \{v, \ldots, z\}, \{0, 3\}, \{1, 2, 4, \ldots, 9\}$

Prove each, where A, B, and C are any sets.

47. $(A')' = A$

48. $A \cup (A \cap B) = A$

49. $A \cap (A \cup B) = A$

50. $(A \cap B)' = A' \cup B'$

51. $A \oplus A = \emptyset$

52. $A \oplus U = A'$

53. $A \oplus B = B \oplus A$

54. $A - B = A \cap B'$

55. $(A \cup B \cup C)' = A' \cap B' \cap C'$

56. $(A \cap B \cap C)' = A' \cup B' \cup C'$

Simplify each set expression.

57. $A \cap (A - B)$

58. $(A - A') \cup (B - A)$

59. $(A - B') - (B - A')$

60. $(A \cup B) \cup (A \cap B')'$

61. $(A \cup B) - (A \cap B)'$

62. $(A \cup B)' \cap (A \cap B')$

63. $(A \cap B)' \cup (A \cup B')$

64. $(A \cup B')' \cap (A' \cap B)$

65. $(A' \cup B')' \cup (A' \cap B)$

***66.** State De Morgan's laws for sets A_i, $i \in I$. (I is an index set.)

***67.** State the distributive laws using the sets A and B_i, $i \in I$.

○ The **sum** of two fuzzy sets A and B is the fuzzy set $A \oplus B$, where $d_{A \oplus B}(x) = 1 \wedge [d_A(x) + d_B(x)]$; their **difference** is the fuzzy set $A - B$, where $d_{A-B}(x) = 0 \vee [d_A(x) - d_B(x)]$; and their **cartesian product** is the fuzzy set $A \times B$, where $d_{A \times B}(x, y) = d_A(x) \wedge d_B(x)$. Use the fuzzy sets $A = \{$Angelo 0.4, Bart 0.7, Cathy 0.6$\}$ and $B = \{$Dan 0.3, Elsie 0.8, Frank 0.4$\}$ to find each fuzzy set.

68. $A \cup B$

69. $A \cap B$

70. A'

71. $A \cup B'$

72. $A \cap B'$

73. $A \cap A'$

74. $A \oplus B$

75. $A - B$

76. $B - A$

77. $A \times B$

78. $B \times A$

79. $A \times A$

○ Let A and B be any fuzzy sets. Prove each.

***80.** $(A \cup B)' = A' \cap B'$

***81.** $(A \cap B)' = A' \cup B'$

*2.3 Computer Operations with Sets (optional)

Sets and the various set operations can be implemented in a computer in an elegant manner.

Computer Representation

Although the elements of a set have no inherent order, when the set is represented in a computer, an order is imposed upon them to permit implementation. The universal set U with n elements is represented as an array with n cells, each containing a 1:

	$n-1$				2	1	0
U	1	1	...	1	1	1	1

The elements are represented by the **binary digits** (or **bits**) 0 and 1 in the right-to-left fashion.

Subsets of U are represented by assigning appropriate bits to the various cells. A bit 1 in a cell indicates the corresponding element belongs to the set, whereas a 0 would indicate the element does *not* belong to the set.

EXAMPLE 2.27 Using $U = \{a, b, \dots, h\}$, represent the sets $A = \{a, b, g\}$ and $B = \{c, e, h\}$ as 8-bit strings.

SOLUTION:
Remember, the elements are represented in the right-to-left order. Thus:

	h	g	f	e	d	c	b	a
U	1	1	1	1	1	1	1	1
A	0	1	0	0	0	0	1	1
B	1	0	0	1	0	1	0	0

■

Next we discuss how the various subsets of a finite set can be found methodically.

Table 2.4

Subset	Bit String
Ø	000
{x}	001
{y}	010
{x, y}	011
{z}	100
{x, z}	101
{y, z}	110
{x, y, z}	111

Interestingly enough, there is a close relationship between sets and bit strings. Table 2.4, for instance, lists the various subsets of the set {x, y, z}. Notice that the table contains all possible three-bit strings and their decimal values increase from 0 to 7. (See Section 4.3 for a discussion of nondecimal bases.)

Next we present a systematic procedure to find the bit string of the subset that "follows" a given subset with bit string $b_2b_1b_0$. Such a recipe for solving a problem in a finite number of steps is called an **algorithm**.*

Next-Subset Algorithm

Take a good look at each string in Table 2.4. Can you find a rule to obtain each, except 000, from the preceding string? It is fairly simple: From right to left, locate the first 0. Change it to 1 and the 1's to its right to 0's.

For example, suppose you would like to find the subset following {x, y} with bit string $b_2b_1b_0 = 011$. From right to left, the first 0 is b_2. Change it to 1, and b_1 and b_0 to 0's. The resulting string is 100 and the corresponding subset is {z}.

This rule can be generalized and translated into an algorithm. See Algorithm 2.1. Use it to find the subsets following {z} and {y, z}.

```
Algorithm next-subset (b_{n-1}b_{n-2} ... b_0)
(* This algorithm finds the bit string of the subset that
   follows a given subset of an n-element set S. *)
Begin (* next-subset *)
  find the first 0 from the right
  change it to 1
  replace the bits to its right with 0's
End (* next-subset *)
```

Algorithm 2.1

The next-subset algorithm can be employed to find all subsets of a finite set S. Algorithm 2.2 shows the steps involved. Use it to find the subsets of {x, y, z}.

```
Algorithm subsets (S)
(* Using the next-subset algorithm, this algorithm finds the bit
   representations of all subsets of an n-element set S. *)
Begin (* subsets *)
  b_{n-1}b_{n-2} ... b_0 ← 00 ... 0  (* initialize string *)
  done ← false (* boolean flag *)
  while not done do
  begin (* while *)
    find the subset following b_{n-1}b_{n-2} ... b_0.
```

*The word *algorithm* is derived from the last name of the ninth-century Arabian astronomer and mathematician Abu-Abdullah Muhammed ibn-Musa al-Khowarizmi (Muhammed, the father of Abdullah and the son of Moses of Khwarizm). He was a teacher in the mathematical school in Baghdad, Iraq. His last name indicates he or his family originally came from Khwarizm (now called Khiva) in Uzbekistan.

His books on algebra and Indian numerals had a significant influence in Europe in the 12th century through their Latin translations. The term *algebra* is derived from the title of his algebra book *Kitab al-jabr w'al-muqabalah*.

```
      if every bit b_i = 1 then (* terminate the loop *)
        done ← true
    endwhile
  End (* subsets *)
```

Algorithm 2.2

Next we show how the set operations can be implemented in a computer.

Computer Operations

The representation of sets as n-bit strings allows us to use logic operations to perform set operations. They are implemented through the bit operations—AND, OR, XOR, COMP—defined by Table 2.5, where COMP indicates *one's complement*: comp(1) = 0 and comp(0) = 1.

Table 2.5 logic operators

bit (y \ x)	AND 0	AND 1	OR 0	OR 1	XOR 0	XOR 1	COMP
0	0	0	0	1	0	1	1
1	0	1	1	1	1	0	0

The various set operations are accomplished by performing the corresponding logic operations, as shown in Table 2.6. Notice that the logic operation corresponding to $A - B$ makes sense since $A - B = A \cap B'$, by law 23 in Table 2.2.

Table 2.6

Set operations	Logic operations
$A \cap B$	A AND B
$A \cup B$	A OR B
A'	COMP(A)
$A \oplus B$	A XOR B
$A - B$	A AND (COMP(B))

EXAMPLE 2.28 Let $U = \{a, b, \ldots, h\}$, $A = \{a, b, c, e, g\}$, and $B = \{b, e, g, h\}$. Using bit representations, find the sets $A \cap B$, $A \cup B$, $A \oplus B$, B', and $A - B$ as 8-bit words.

SOLUTION:

$$A = 0\ 1\ 0\ 1\ 0\ 1\ 1\ 1$$

$$B = 1\ 1\ 0\ 1\ 0\ 0\ 1\ 0$$

Using Tables 2.5 and 2.6, we have:

(1) $A \cap B = 0\ 1\ 0\ 1\ 0\ 0\ 1\ 0$ (2) $A \cup B = 1\ 1\ 0\ 1\ 0\ 1\ 1\ 1$

(3) $A \oplus B = 1\ 0\ 0\ 0\ 0\ 1\ 0\ 1$ (4) $B' = 0\ 0\ 1\ 0\ 1\ 1\ 0\ 1$

(5) $A = 0\ 1\ 0\ 1\ 0\ 1\ 1\ 1$

$B' = 0\ 0\ 1\ 0\ 1\ 1\ 0\ 1$

So $A - B = A$ AND (COMP(B))

$= 0\ 0\ 0\ 0\ 0\ 1\ 0\ 1$

Using the bit representations, you may verify that $A \cap B = \{\text{b, e, g}\}$, $A \cup B = \{\text{a, b, c, e, g, h}\}$, $A \oplus B = \{\text{a, c, h}\}$, $B' = \{\text{a, c, d, f}\}$, and $A - B = \{\text{a, c}\}$. ■

Exercises 2.3

Using the universal set $U = \{\text{a}, \ldots, \text{h}\}$, represent each set as an 8-bit word.

1. $\{\text{a, c, e, g}\}$ **2.** $\{\text{b, d, f}\}$ **3.** $\{\text{a, e, f, g, h}\}$ **4.** $\emptyset$

Use Algorithm 2.1 to find the subset of the set $\{s_0, s_1, s_2, s_3\}$ that follows the given subset.

5. $\{s_3\}$ **6.** $\{s_0, s_3\}$ **7.** $\{s_2, s_3\}$ **8.** $\{s_0, s_2, s_3\}$

Using Algorithm 2.2, find the subsets of each set.

9. $\{s_0, s_1\}$ **10.** $\{s_0, s_1, s_2, s_3\}$

Using the sets $A = \{\text{a, b, e, h}\}$, $B = \{\text{b, c, e, f, h}\}$, $C = \{\text{c, d, f, g}\}$, and $U = \{\text{a}, \ldots, \text{h}\}$, find the binary representation of each set.

11. $A \cap B$ **12.** $A \cup B$ **13.** B' **14.** $A - B$

15. $C - B$ **16.** $A \oplus B$ **17.** $B \oplus C$ **18.** $C \oplus A$

19. $A \cap C'$ **20.** $A \cup B'$ **21.** $A \cap (B \cap C)$ **22.** $A \cup (B \cap C)$

23. $A - (B \oplus C)$ **24.** $(A \oplus B) - C$ **25.** $A \oplus (B \oplus C)$ **26.** $(A \oplus B) \oplus C$

2.4 The Cardinality of a Set

This section presents four formulas involving finite sets, which we shall use frequently. Recall that every finite set has a fixed number of elements, so we make the following definition.

Cardinality

The **cardinality** of A, denoted by $|A|$, is the number of elements in it.*

*It should be clear from the context whether the symbol "| |" refers to absolute value or cardinality.

For example, $|\emptyset| = 0$, $|\{\emptyset\}| = 1$, and $|\{a, b, c\}| = 3$.

Let A and B be any two finite sets. How is $|A \cup B|$ related to $|A|$ and $|B|$? First, let's study an example.

EXAMPLE 2.29 Let $A = \{a, b, c\}$ and $B = \{b, c, d, e, f\}$. Clearly, $|A| = 3$, $|B| = 5$, $|A \cup B| = 6$, and $|A \cap B| = 2$, so $|A \cup B| = |A| + |B| - |A \cap B|$. ■

More generally, we have the following result:

THEOREM 2.1 (**Inclusion–Exclusion Principle**) Let A and B be two finite sets. Then $|A \cup B| = |A| + |B| - |A \cap B|$.

PROOF:

Suppose $|A \cap B| = k$. Since $A \cap B \subseteq A$ and $A \cap B \subseteq B$, we can assume that $|A| = k + m$ and $|B| = k + n$ for some nonnegative integers m and n (see Figure 2.22). Then:

$$\begin{aligned} |A \cup B| &= m + k + n \\ &= (m + k) + (n + k) - k \\ &= |A| + |B| - |A \cap B| \end{aligned}$$

Figure 2.22

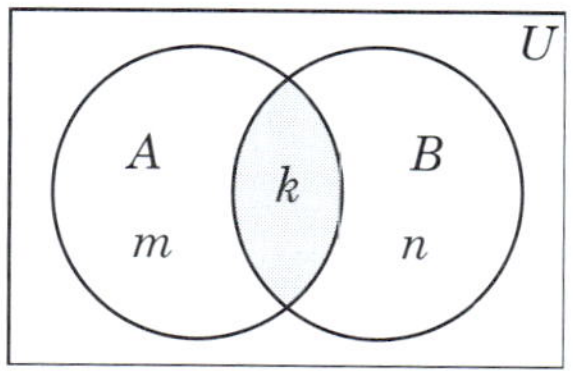

This completes the proof. ■

In addition, if A and B are disjoint sets, then $|A \cap B| = |\emptyset| = 0$, so $|A \cup B| = |A| + |B|$. Thus we have the following result.

COROLLARY 2.1
(**Addition Principle**) Let A and B be finite disjoint sets. Then $|A \cup B| = |A| + |B|$.

The next example demonstrates the inclusion–exclusion principle.

EXAMPLE 2.30 Find the number of positive integers ≤ 300 and divisible by 2 or 3.

SOLUTION:

Let $A = \{x \in \mathbb{N} | x \leq 300$ and is divisible by $2\}$ and $B = \{x \in \mathbb{N} | x \leq 300$ and is divisible by $3\}$. Then $A \cap B$ consists of positive integers ≤ 300 that are divisible by 2 and 3, that is, divisible by 6. Thus $A = \{2, 4, \ldots, 300\}$,

$B = \{3, 6, \ldots, 300\}$, and $A \cap B = \{6, 12, \ldots, 300\}$. Clearly, $|A| = 150$, $|B| = 100$, and $|A \cap B| = 50$, so by Theorem 2.1,

$$\begin{aligned} |A \cup B| &= |A| + |B| - |A \cap B| \\ &= 150 + 100 - 50 = 200 \end{aligned}$$

Thus there are 200 positive integers ≤ 300 and divisible by 2 or 3. (See Examples 3.11, and 3.12 in Section 3.2.)

Theorem 2.1 can be extended to any finite number of finite sets. For instance, the next example derives the formula for three finite sets.

EXAMPLE 2.31 Let A, B, and C be three finite sets. Prove that

$$|A \cup B \cup C| = |A| + |B| + |C| - |A \cap B| - |B \cap C| - |C \cap A| + |A \cap B \cap C|.$$

PROOF:

$$\begin{aligned} |A \cup B \cup C| &= |A \cup (B \cup C)| \\ &= |A| + |B \cup C| - |A \cap (B \cup C)| && \text{by Theorem 2.1} \\ &= |A| + |B \cup C| - |(A \cap B) \cup (A \cap C)| && \text{by the distributive law} \\ &= |A| + (|B| + |C| - |B \cap C|) - [|A \cap B| + |A \cap C| \\ &\quad - |(A \cap B) \cap (A \cap C)|] \\ &= |A| + |B| + |C| - |A \cap B| - |B \cap C| - |C \cap A| + |A \cap B \cap C|, \end{aligned}$$

since $A \cap C = C \cap A$ and $(A \cap B) \cap (A \cap C) = A \cap B \cap C$. ■

The next example shows how useful sets are in data analysis.

EXAMPLE 2.32 A survey among 100 students shows that of the three ice cream flavors vanilla, chocolate, and strawberry, 50 students like vanilla, 43 like chocolate, 28 like strawberry, 13 like vanilla and chocolate, 11 like chocolate and strawberry, 12 like strawberry and vanilla, and 5 like all of them. Find the number of students surveyed who like each of the following flavors.

(1) Chocolate but not strawberry.
(2) Chocolate and strawberry, but not vanilla.
(3) Vanilla or chocolate, but not strawberry.

SOLUTION:

Let V, C, and S symbolize the set of students who like vanilla, chocolate, and strawberry flavors, respectively. Draw three intersecting circles to represent them in the most general case, as in Figure 2.23.

Our first goal is to distribute the 100 students surveyed into the various regions. Since five students like all flavors, $|V \cap C \cap S| = 5$. Twelve students like both strawberry and vanilla, so $|S \cap V| = 12$. But five of them like chocolate also. Therefore, $|(S \cap V) - C| = 7$. Similarly, $|(V \cap C) - S| = 8$ and $|(C \cap S) - V| = 6$.

Figure 2.23

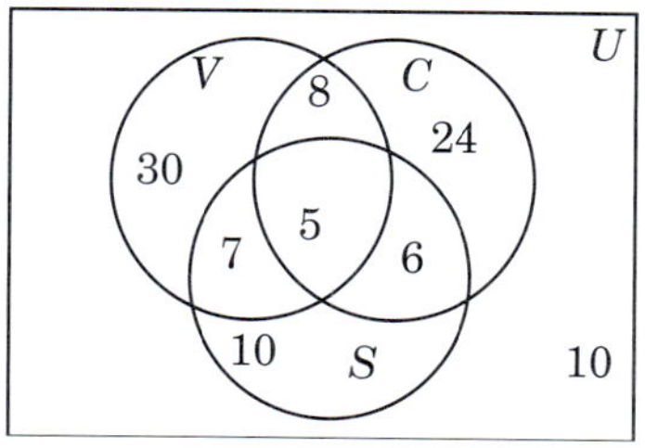

Of the 28 students who like strawberry, we have already accounted for $7 + 5 + 6 = 18$. So the remaining 10 students belong to the set $S - (V \cup C)$. Similarly, $|V - (C \cup S)| = 30$ and $|C - (S \cup V)| = 24$.

Thus far, we have accounted for 90 of the 100 students. The remaining 10 students lie outside the region $V \cup S \cup C$, as in Figure 2.23. The required answers can now be directly read from this Venn diagram:

(1) $|C - S| = 24 + 8 = 32$. So 32 students like chocolate but not strawberry.

(2) $|(C \cap S) - V| = 6$. Therefore, 6 students like both chocolate and strawberry, but not vanilla.

(3) $30 + 8 + 24 = 62$ students like vanilla or chocolate, but not strawberry. They are represented by the region $(V \cup C) - S$. ■

Finally, suppose a set contains n elements. How many subsets does it have? Before we answer this partially, let us study the next example, which uses the addition principle.

EXAMPLE 2.33 Let s_3 denote the number of subsets of the set $S = \{a, b, c\}$. Let $S^* = S - \{b\}$. We shall use the subsets of S^* in a clever way to find s_3 and all subsets of S. Let A denote the subsets of S^*. Then $A = \{\emptyset, \{a\}, \{c\}, \{a, c\}\}$. Clearly every element of A is also a subset of S.

Now add b to every element in A. Let B denote the resulting set: $B = \{\{b\}, \{b, a\}, \{b, c\}, \{b, a, c\}\}$. Every subset of S either contains b or does not contain b; so, by the addition principle, $s_3 = |A| + |B| = 4 + 4 = 8$. ■

More generally, we have the following result.

THEOREM 2.2 Let s_n denote the number of subsets of a set S with n elements. Then $s_n = 2s_{n-1}$, where $n \geq 1$.

PROOF:

Let $x \in S$. Let $S^* = S - \{x\}$. Then S^* contains $n - 1$ elements and hence has s_{n-1} subsets by definition. Each of them is also a subset of S. Now insert x in each of them. The resulting s_{n-1} sets are also subsets of S. Since every subset of S either contains x or does not contain x, the addition principle indicates a total of $s_{n-1} + s_{n-1} = 2s_{n-1}$ subsets of S. (Notice that $s_0 = 1$. Why?) ■

Consequently, if you know the number of subsets of a set with $n-1$ elements, this theorem can be employed to compute the number of subsets of a set with n elements. For instance, by Example 2.33, a set with three elements has eight subsets; therefore, a set with four elements has $2 \cdot 8 = 16$ subsets.

The technique used in the proof of Theorem 2.2 can be applied to write an algorithm for finding the power set of a set S. See Algorithm 2.3. It uses the fact that if A is a subset of S and $s \in S$, then $A \cup \{s\}$ is also a subset of S.

```
Algorithm subsets(S)
(* This algorithm finds the power set of a set S with n elements
   s1, s2, ..., sn. Sj denotes the jth element in the power set. *)
   Begin (* subsets *)
     power set ← {Ø}     (* initialize power set *)
     numsubsets ← 1 (* initialize the number of subsets *)
     for i = 1 to n do (* si denotes the ith element in S *)
     begin (* for *)
       j ← 1              (* j-th element in P(S) *)
       temp ← numsubsets (* temp is a temporary variable*)
       while j ≤ temp do  (* construct a new subset *)
       begin (* while *)
         add Sj ∪ {si} to the power set
         j ← j + 1
         numsubsets ← numsubsets + 1
       endwhile
     endfor
   End (* subsets *)
```

Algorithm 2.3 ■

Although Theorem 2.2 does not give us an explicit formula for the number of subsets, it can be used to find the formula. The next theorem gives us the explicit formula, which we shall prove in Section 4.4 (see Example 4.18).

THEOREM 2.3 A set with n elements has 2^n subsets, where $n \geq 0$. ■

For example, a set with four elements has $2^4 = 16$ subsets!

Exercises 2.4

Find the cardinality of each set.

1. The set of letters of the English alphabet.
2. The set of letters of the word TWEEDLEDEE.
3. The set of months of the year with 31 days.
4. The set of identifiers in Java that begin with 3.

Let A and B be two sets such that $|A| = 2a - b$, $|B| = 2a$, $|A \cap B| = a - b$, and $|U| = 3a + 2b$. Find the cardinality of each set.

5. $A \cup B$ **6.** $A - B$ **7.** B' **8.** $A - A'$

9. Find $|A|$ if $|A| = |B|$, $|A \cup B| = 2a + 3b$, and $|A \cap B| = b$.

10. Find $|A \cap B|$ if $|A| = a + b = |B|$ and $|A \cup B| = 2a + 2b$.

11. Find $|A \cap B|$ if $|A| = 2a$, $|B| = a$, and $|A \cup B| = 2a + b$.

Let A and B be finite sets such that $A \subseteq B$, $|A| = b$, $|B| = a + b$. Find the cardinality of each set.

12. $A \cup B$ **13.** $A - B$ **14.** $B - A$ **15.** $A \cap B$

Let A and B be finite disjoint sets, where $|A| = a$, and $|B| = b$. Find the cardinality of each set.

16. $A \cup B$ **17.** $A - B$ **18.** $B - A$

19–21. Find the cardinality of each set in Exercises 16–18, where $A \subseteq B$, B is finite, $|A| = a$, and $|B| = b$.

22. A survey conducted recently among 300 adults in Omega City shows 160 like to have their houses painted green, and 140 like them blue. Seventy-five adults like both colors. How many do not like either color?

23. A survey was taken to determine the preference between two laundry detergents, Lex and Rex. It was found that 15 people liked Lex only, 10 liked both, 20 liked Rex only, and 5 liked neither of them. How many people were surveyed?

Find the number of positive integers ≤ 500 and divisible by:

24. Two or three. **25.** Two, three, or five.

26. Two or three, but not six. **27.** Neither two, three, nor five.

Find the number of positive integers ≤ 1776 and divisible by:

28. Two, three, or five. **29.** Two, three, or five, but not six.

30. Two, three, or five, but not 15. **31.** Two, three, or five, but not 30.

According to a survey among 160 college students, 95 students take a course in English, 72 take a course in French, 67 take a course in German, 35 take a course in English and in French, 37 take a course in French and in German, 40 take a course in German and in English, and 25 take a course in all three languages. Find the number of students in the survey who take a course in:

32. English, but not German. **33.** English, French, or German.

34. English or French, but not German.

35. English and French, but not German.

36. English, but neither French nor German.

37. Neither English, French, nor German.

A recent survey by the MAD corporation indicates that of the 700 families interviewed, 220 own a television set but no stereo, 200 own a stereo but no camera, 170 own a camera but no television set, 80 own a television set and a stereo but no camera, 80 own a stereo and a camera but no television set, 70 own a camera and a television set but no stereo, and 50 do not have any of these. Find the number of families with:

38. Exactly one of the items.

39. Exactly two of the items.

40. At least one of the items.

41. All of the items.

Using Algorithm 2.3, find the power set of each set. List the elements in the order obtained.

42. $\{a, b\}$

43. $\{a, b, c\}$

A finite set with a elements has b subsets. Find the number of subsets of a finite set with the given cardinality.

44. $a+1$

45. $a+2$

46. $a+5$

47. $2a$

Let A, B, and C be subsets of a finite set U. Derive a formula for each.

48. $|A' \cap B'|$

49. $|A' \cap B' \cap C'|$

***50.** State the inclusion–exclusion principle for four finite sets A_i, $1 \le i \le 4$. (The formula contains 15 terms.)

***51.** Prove the formula in Exercise 50.

****52.** State the inclusion–exclusion principle for n finite sets A_i, $1 \le i \le n$.

2.5 Recursively Defined Sets

A new way of defining sets is using recursion. (It is a powerful problem-solving technique discussed in detail in Chapter 5.)

Notice that the set of numbers $S = \{2, 2^2, 2^{2^2}, 2^{2^{2^2}}, \ldots\}$ has three interesting characteristics:

(1) $2 \in S$.

(2) If $x \in S$, then $2^x \in S$.

(3) Every element of S is obtained by a finite number of applications of properties 1 and 2 only.

Property 1 identifies explicitly the primitive element in S and hence ensures that it is nonempty. Property 2 establishes a systematic procedure to construct new elements from known elements. How do we know, for instance, that $2^{2^2} \in S$? By property 1, $2 \in S$; then, by property (2), $2^2 \in S$; now choose $x = 2^2$ and apply property 2 again; so $2^{2^2} \in S$. Property 3 guarantees that in no other way can the elements of S be constructed. Thus the various elements of S can be obtained systematically by applying the above properties.

These three characteristics can be generalized and may be employed to define a set S implicitly. Such a definition is a recursive definition.

Recursively Defined Set

A **recursive definition** of a set S consists of three clauses:

- The **basis clause** explicitly lists at least one primitive element in S, ensuring that S is nonempty.
- The **recursive clause** establishes a systematic recipe to generate new elements from known elements.
- The **terminal clause** guarantees that the first two clauses are the only ways the elements of S can be obtained.

The terminal clause is generally omitted for convenience.

EXAMPLE 2.34 Let S be the set defined recursively as follows.

(1) $2 \in S$. (2) If $x \in S$, then $x^2 \in S$.

Describe the set by the listing method.

SOLUTION:

- $2 \in S$, by the basis clause.
- Choose $x = 2$. Then by the recursive clause, $4 \in S$.
- Now choose $x = 4$ and apply the recursive clause again, so $16 \in$ S. Continuing like this, we get $S = \{2, 4, 16, 256, 65536, \ldots\}$. ■

The next three examples further elucidate the recursive definition.

EXAMPLE 2.35 Notice that the language $L = \{\text{a, aa, ba, aaa, aba, baa, bba}, \ldots\}$ consists of words over the alphabet $\Sigma = \{\text{a, b}\}$ that end in the letter a. It can be defined recursively as follows.

- $\text{a} \in L$.
- If $x \in L$, then ax, b$x \in L$.

For instance, the word aba can be constructed as follows:

- $\text{a} \in L$. Choosing $x = \text{a}$, bx = ba $\in L$.

- Now choose $x =$ ba. Then a$x =$ aba $\in L$.

The tree diagram in Figure 2.24 illustrates systematically how to derive the words in L.

Figure 2.24

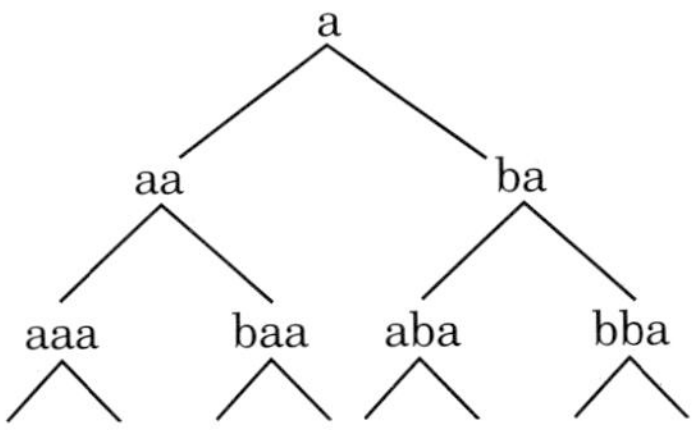

■

EXAMPLE 2.36 **(Legally Paired Parentheses)** An important problem in computer science is to determine whether or not a given expression is legally parenthesized. For example, (( )), ( ) ( ), and (( ) ( )) are validly paired sequences of parentheses, but) ( ), ( ) (, and) ( ) (are not. The set S of sequences of legally paired parentheses can be defined recursively as follows:

- ( ) $\in S$.
- If $x, y \in S$, then xy and (x) belong to S.

The tree diagram in Figure 2.25 shows the various ways of constructing the elements in S.

Figure 2.25

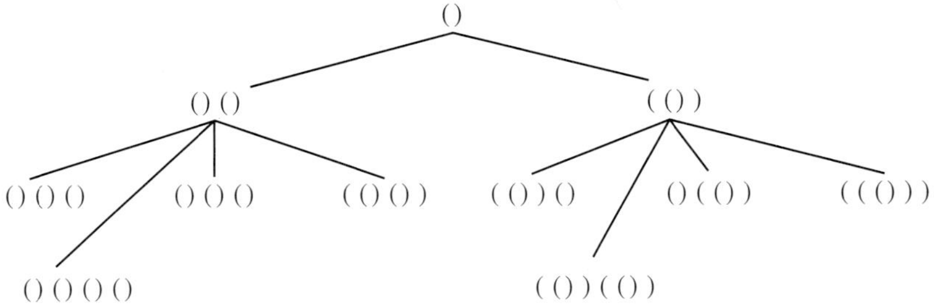

■

A simplified recipe to determine if a sequence of parentheses is legally paired is given in Algorithm 2.4.

```
Algorithm Legally Paired Sequence
(* This algorithm determines if a nonempty sequence of parentheses is
   legally paired. Count keeps track of the number of parentheses. It is
   incremented by 1 if the current parenthesis is a left parenthesis.
   and decremented by 1 if it is a right parenthesis. *)
   Begin (* algorithm *)
     count ← 0                     (* initialize *)
     read a symbol
     if symbol = left paren then
```

```
    while not the end of the sequence do
    begin (* while *)
      if symbol = left paren then
        count ← count + 1
      else  (* symbol = right parenthesis *)
        count ← count - 1
      read the next symbol
    endwhile
    if count = 0 then
      legal sequence
    else
      invalid sequence
else
  invalid sequence (* begins with a right paren *)
End (* algorithm *)
```

Algorithm 2.4

This example is studied further in Chapters 6 and 9. ■

EXAMPLE 2.37 A legal expression in propositional logic is called a **well-formed formula** (wff). For convenience, we restrict our discussion to the logical variables p, q, and r, and the operators $\wedge$, $\vee$, and $\sim$. Then the set of well-formed formulas can be defined recursively:

- The logic variables are wffs.
- If x and y are wffs, then so are (x), $\sim(x)$, $(x \wedge y)$, and $(x \vee y)$.

For instance, the expression $((p) \wedge ((\sim(q)) \vee (r)))$ is a wff, but $(q \wedge (\sim r))$ is not (why?). (Parentheses are often omitted when ambiguity is impossible.) ■

Exercises 2.5

In Exercises 1–6, a set S is defined recursively. Find four elements in each case.

1. i) $1 \in S$
ii) $x \in S \rightarrow 2x \in S$

2. i) $1 \in S$
ii) $x \in S \rightarrow 2^x \in S$

3. i) $e \in S$
ii) $x \in S \rightarrow e^x \in S$

4. i) $3 \in S$
ii) $x \in S \rightarrow \lg x \in S$†

5. i) $\lambda \in L$
ii) $x \in L \rightarrow x\text{bb} \in L$

6. i) $\lambda \in L$
ii) $x \in L \rightarrow \text{a}x\text{b} \in L$

†lg x means $\log_2 x$.

In Exercises 7–10, identify the set S that is defined recursively.

7. i) $1 \in S$
ii) $x, y \in S \to x + y \in S$

8. i) $1 \in S$
ii) $x, y \in S \to x \pm y \in S$

9. i) $2 \in S$
ii) $x, y \in S \to x \pm y \in S$

10. i) $\emptyset \in S$
ii) $x \in X, A \in S \to \{\mathrm{x}\} \cup A \in S$

Define each language L over the given alphabet recursively.

11. $\{0, 00, 10, 100, 110, 0000, 1010, \ldots\}$, $\Sigma = \{0, 1\}$.

12. $L = \{1, 11, 111, 1111, 11111, \ldots\}$, $\Sigma = \{0, 1\}$.

13. $L = \{x \in \Sigma^* | x = \mathrm{b}^n\mathrm{ab}^n, n \geq 0\}$, $\Sigma = \{\mathrm{a}, \mathrm{b}\}$.

14. The language L of all palindromes over $\Sigma = \{\mathrm{a}, \mathrm{b}\}$. (A **palindrome** is a word that reads the same both forwards and backwards. For instance, abba is a palindrome.)

***15.** $\{\mathrm{b}, \mathrm{bb}, \mathrm{bbb}, \mathrm{bbbb}, \ldots\}$, $\Sigma = \{\mathrm{a}, \mathrm{b}\}$.

***16.** $\{\mathrm{b}, \mathrm{aba}, \mathrm{aabaa}, \mathrm{aaabaaa}, \ldots\}$, $\Sigma = \{\mathrm{a}, \mathrm{b}\}$.

***17.** $\{\mathrm{a}, \mathrm{aaa}, \mathrm{aaaaa}, \mathrm{aaaaaaa}, \ldots\}$, $\Sigma = \{\mathrm{a}, \mathrm{b}\}$.

***18.** $\{1, 10, 11, 100, 101, \ldots\}$, $\Sigma = \{0, 1\}$.

Determine if each sequence of parentheses is legal.

19. (()()) **20.** (())(**21.** (()() **22.** (()())()

The nth **Catalan number** C_n, named after the Belgian mathematician, Eugene Charles Catalan (1814–1894), is defined by

$$\mathrm{C}_n = \frac{(2n)!}{n!(n+1)!}, \quad n \geq 0$$

where $n!$ (n **factorial**) is defined by $n! = n(n-1)\ldots 3 \cdot 2 \cdot 1$ and $0! = 1$. Catalan numbers have many interesting applications in computer science. For example, the number of well-formed sequences of n pairs of left and right parentheses is given by the nth Catalan number. Compute the number of legally paired sequences with the given pairs of left and right parentheses.

23. Three **24.** Four **25.** Five **26.** Six

27. List the well-formed sequences of parentheses with three pairs of left and right parentheses.

28. Redo Exercise 27 with four pairs of left and right parentheses.

Using Example 2.37, determine if each is a wff in propositional logic.

29. $(p \wedge ((\sim(q)) \vee r))$

30. $((\sim(p)) \vee ((q) \wedge (\sim r))$

31. $(((\sim p) \vee q) \wedge (\sim q) \vee (\sim p)))$

32. $((p \vee q) \wedge ((\sim(q)) \vee (\sim(r))))$

33. Determine if the following recursive definition yields the set S of legally paired parentheses. If not, find a validly paired sequence that cannot be generated by this definition.

i) $() \in S$.

ii) If $x \in S$, then $()x$, (x), $x() \in S$.

34. Define the set of words S over an alphabet Σ recursively. Assume $\lambda \in S$.
(*Hint*: use concatenation.)

35. Let Σ be an alphabet. Define Σ^* recursively.
(*Hint*: use concatenation.)

***36.** Define the language L of all binary representations of nonnegative integers recursively.

Chapter Summary

This chapter presented the concept of a set, different ways of describing a set, relations between sets, operations with sets and their properties, and formal languages. How sets and set operations work in a typical computer were also discussed.

Set

- A **set** is a well-defined collection of objects (page 68).
- A set can be described using words, listing the elements, or by the set-builder notation (page 69).
- $A \subseteq B$ if and only if every element of A is also an element of B (page 69).
- $(A = B) \leftrightarrow (A \subseteq B) \wedge (B \subseteq A)$ (page 70).
- The **null set** $\emptyset$ contains no elements (page 70).
- The **universal set** U contains all elements under discussion (page 70).
- A and B are **disjoint sets** if $A \cap B = \emptyset$ (page 71).
- The **power set** $P(A)$ of a set A is the family of all subsets of A (page 72).
- A set with a definite number of elements is **finite**; if a set is not finite, it is **infinite** (page 73).

Formal Language

- An **alphabet** Σ is a finite set of symbols; $\{0,1\}$ is the binary alphabet (page 75).
- A **word** over Σ is a finite arrangement of symbols from Σ. A word of length zero is the empty word λ (page 75).
- Σ^* consists of all possible words over Σ (page 75).
- A **formal language** over Σ is a subset of Σ^* (page 75).
- The **concatenation** of two words x and y is the word xy. (page 76).

Set Operations

- **Union** $A \cup B = x|(x \in A) \vee (x \in B)$ (page 78).
- **Intersection** $A \cap B = \{x|(x \in A) \wedge (x \in B)\}$ (page 78).
- **Difference** $A - B = \{x \in A|x \notin B\}$ (page 80).
- **Complement** $A' = U - A = \{x \in U|x \notin A\}$ (page 81).
- **Symmetric difference** $A \oplus B = (A - B) \cup (B - A)$ (page 82).
- **Cartesian product** $A \times B = \{(a, b)|(a \in A) \wedge (b \in B)\}$ (page 87).
- The fundamental properties of set operations are listed in Table 2.2 (page 83).

Partition

- A **partition** of a set S is a finite collection of nonempty, pairwise disjoint subsets of S whose union is S (page 90).

Computer Implementation

- Set operations are implemented in a computer using the bit operations in Table 2.5 and the logic operations in Table 2.6. (page 97).

Cardinality

- **Inclusion–exclusion principle** $|A \cup B| = |A| + |B| - |A \cap B|$ (page 99).
- **Addition principle** $|A \cup B| = |A| + |B|$, where $A \cap B = \emptyset$ (page 99).
- A set with n elements has 2^n subsets (page 102).

Recursion

- The **recursive** definition of a set consists of a basis clause, recursive clause, and a terminal clause (page 105).

Review Exercises

Using the Venn diagram in Figure 2.26, find each.

1. $A - (B \cap C)$ **2.** $(A \cup B) - C$ **3.** $A - (B - C)$

4. $(A - B) - C$ **5.** $A \oplus B$ **6.** $(A - B) \times (B - C)$

7. $A - (B \oplus C)$ **8.** $A \cup (B \oplus C')$

Figure 2.26

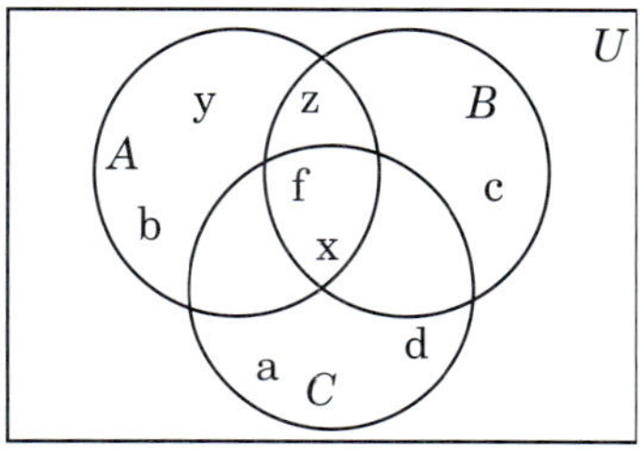

9. Find the sets A and B if $A \cap B' = \{\text{a, c}\}$, $B \cap A' = \{\text{b, e, g}\}$, $A \cap B = \{\text{d, f}\}$, and $A' \cap B' = \{\text{i}\}$.

Let A, B, and C be sets such that $A - (B \cup C) = \{\text{b,e}\}$, $B - (C \cup A) = \{\text{k}\}$, $C - (A \cup B) = \{\text{h}\}$, $A \cap B = \{\text{f, g}\}$, $B \cap C = \{\text{j}\}$, $C \cap A = \{\text{i}\}$, and $A \cap B \cap C = \emptyset$. Find each set.

10. $A - (B \cap C)$ **11.** $(A \oplus B) - C$ **12.** $A \oplus (B \oplus C)$

Find the power set of each set.

13. $\{\emptyset, \{\emptyset\}\}$ ***14.** $\{2, \{3\}, \{2, 3\}\}$

15. Let $A = \{n \in \mathbb{N} | n < 20$ and n is divisible by $2\}$, $B = \{n \in \mathbb{N} | n < 20$ and n is divisible by $3\}$, and $C = \{n \in \mathbb{N} | n < 20$ and n is divisible by $5\}$. Determine if they form a partition of the set $\{n \in \mathbb{N} | n < 20\}$.

○ Let $U = \{1, \ldots, 8\}$, $A = \{1, 3, 5, 7, 8\}$, and $B = \{2, 3, 6, 7\}$. Find the binary representation of each set.

16. $A - (A \cap B)$ **17.** $A - B'$ **18.** $A - (A \oplus B)$ **19.** $A \oplus (A \oplus B)$

A survey found that 45% of women like plain yogurt, 55% like flavored yogurt, and 23% like both. Compute the percentage of women who like each.

20. Plain yogurt, but not flavored.

21. Plain or flavored yogurt, but not both.

A survey was taken among the students on campus to find out whether they prefer vanilla or strawberry ice cream and whether they prefer chocolate or

Table 2.7

		Pudding			
		Chocolate	**Tapioca**	**Neither**	**Total**
	Vanilla	68	53	12	133
Ice Cream	Strawberry	59	48	9	116
	Neither	23	21	7	51
	Total	150	122	28	300

tapioca pudding. The results are summarized in Table 2.7. Find the number of students who:

22. Like strawberry ice cream and tapioca pudding.

23. Do not like pudding.

24. Like at least one of the ice cream flavors.

25. Like neither ice cream nor pudding.

Find the number of positive integers ≤ 4567 and divisible by:

26. Two, three, or five.

27. Two, five, or seven, but not 35.

Find four elements in each set S defined recursively.

28. i) $1 \in S$
ii) $x \in S \rightarrow 1 + x \in S$

29. i) $3 \in S$
ii) $x \in S \rightarrow \lg x \in S$

30. i) $\sqrt{2} \in S$
ii) $x \in S \rightarrow \sqrt{2 + x} \in S$

31. i) $1 \in S$
ii) $x \in S \rightarrow \sqrt{1 + 2x} \in S$

Define each set S recursively.

32. $\{2, 4, 16, 256, \ldots\}$

33. $\{1, 3, 7, 15, 31, \ldots\}$

34. $\{\mathrm{b}, \mathrm{ba}^2, \mathrm{ba}^4, \mathrm{ba}^6, \ldots\}$

35. $\{\lambda, ba, b^2a^2, b^3a^3, \ldots\}$

Find five words in each language L over the alphabet $\Sigma = \{\mathrm{a}, \mathrm{b}\}$.

36. $\{x \in \Sigma^* | x \text{ contains exactly one } a\}$

37. $\{x \in \Sigma^* | x \text{ contains an odd number of } a\text{'s}\}$

Define each language L over the given alphabet recursively.

38. $\{x \in \Sigma^* | x \text{ contains exactly one } a\}$, $\Sigma = \{\mathrm{a}, \mathrm{b}\}$.

39. $\{x \in \Sigma^* | x \text{ ends in ab}\}$, $\Sigma = \{a, b\}$.

40. $\{2, 3, 4, 5, 6, \ldots\}$, $\Sigma = \{2, 3\}$.

41. $\{1, 010, 00100, 0001000, 000010000, \ldots\}$, $\Sigma = \{0, 1\}$.

Determine if each is a well-formed formula.

42. $(p \wedge ((\sim(q)) \vee (r)))$

43. $(((p) \wedge (q)) \vee (\sim(q) \wedge (r)))$

Let A, B, and C be any sets. Prove each.

***44.** $A \cap (B \cup C) = (A \cap B) \cup (A \cap C)$

***45.** $A \cup (B \cap C) = (A \cup B) \cap (A \cup C)$

***46.** $A \cup (B - C) = (A \cup B) - (C - A)$

***47.** $A \cap (B - C) = (A \cap B) - (A \cap C)$

Simplify each set expression.

48. $(A' \cup B')' \cup (A' \cap B)$

***49.** $[A - (B \cup C)] \cap [(B \cap C) - A]$

○ Consider the fuzzy sets, where $A = \{\text{Mike } 0.6, \text{ Andy } 0.3, \text{ Jeff } 0.7\}$ and $B = \{\text{Jean } 0.8, \text{ June } 0.5\}$. Find each fuzzy set.

50. $A \cup B'$

51. $A' \cap B$

52. $A \oplus B'$

53. $A \times B$

○ Let A and B be any fuzzy sets. Prove each.

***54.** $(A \cup B)' = A' - B$

***55.** $(A - B)' = A' \cup B$

Supplementary Exercises

Prove each, where A, B, and C are arbitrary sets.

1. $A - (B \cup C) = (A - B) \cap (A - C)$

2. $[A \cap (A - B)] \cup (A' \cup B)' = A - B$

***3.** $A \cap (B \oplus C) = (A \cap B) \oplus (A \cap C)$

***4.** $A \oplus (B \oplus C) = (A \oplus B) \oplus C$

Simplify each set expression.

5. $(A \cap B) \cap (B \cap C) \cap (C \cap A)$

***6.** $[(A \cup B) \cap C] \cup [A \cap (B \cup C)]$

***7.** $(A \cup B') \cap (A' \cup B) \cap (A' \cup B')$

***8.** $[(A \cup B') \cup (A' \cup B)]' \cap (A' \cap B')$

Find the number of positive integers ≤ 1000 and *not* divisible by:

9. 2, 3, or 5.

***10.** 2, 3, 5, or 7.

11. Define recursively the language $\{0^n 1^n | n \geq 0\}$ over $\Sigma = \{0, 1\}$.

12. Define recursively a word w over a finite alphabet Σ.

Let $x = x_1 x_2 \ldots x_n \in \sum^*$. Then the string $x_n \ldots x_2 x_1$ is called the **reverse** of x, denoted by x^{R}. For example, the reverse of the binary word 01101 is 10110. Let $x, y \in \sum^n$. Prove each.

13. $(xy)^{\mathrm{R}} = y^{\mathrm{R}} x^{\mathrm{R}}$

14. The string x is palindromic if and only if $x^{\mathrm{R}} = x$.

15. The word xx^{R} is palindromic.

Computer Exercises

Write a program to do each task, where n denotes a positive integer ≤ 20.

1. Read in k subsets of the set $S = \{1, 2, \ldots, n\}$ and determine if the subsets form a partition of S.

2. Read in two sets A and B, where $U = \{1, 2, 3, \ldots, n\}$. Print the bit-representations of A and B. Use them to find the elements in $A \cup B$, $A \cap B$, A', $A - B$, $A \oplus B$, and $A \times B$, and their cardinalities.

3. Find all subsets of the set $\{1, 2, \ldots, n\}$.

4. Read in sequences of left and right parentheses, each containing at most 25 symbols. Determine if each word consists of legally paired parentheses.

5. Print the Catalan numbers C_0 through C_n.

Exploratory Writing Projects

Using library and Internet resources, write a team report on each of the following in your own words. Provide a well-documented bibliography.

1. Write an essay on the life and contributions of G. Cantor.

2. Explain the various occurrences of the ordered pair notation in everyday life.

3. Explain how the addition principle is used to define the addition of positive integers. Give concrete examples.

4. Explain how the concept of partitioning is used in everyday life. In sports. In computer science. Give concrete examples.

5. Study a number of mathematical paradoxes and explain them.

6. Discuss the various string operations and list the programming languages that support them.

7. Describe fuzzy sets and their applications, and L. A. Zadeh's contributions to them.

8. Write a biography of Abu-Abdullah Muhammed ibn-Musa al-Khowarizmi and the origin of the word *algorithm*.

9. Extend the concept of the cardinality of a finite set to infinite sets. Describe the arithmetic of transfinite cardinal numbers.

10. Discuss the halting problem.

Enrichment Readings

1. R. R. Christian, *Introduction to Logic and Sets*, Blaisdell, Waltham, MA, 1965.
2. M. Guillen, *Bridges to Infinity: The Human Side of Mathematics*, J. P. Tarcher, Inc., Los Angeles, 1983, pp. 41–60.
3. P. R. Halmos, *Naive Set Theory*, Van Nostrand, New York, 1960.
4. S. Sahni, *Concepts in Discrete Mathematics*, 2nd ed., Camelot, Fridley, MN, 1985, pp. 93–111.
5. P. C. Suppes, *Axiomatic Set Theory*, Van Nostrand, New York, 1960.
6. R. L. Wilder, *Evolution of Mathematical Concepts: An Elementary Study*, Wiley, New York, 1968.

Chapter 3

Functions and Matrices

To know him [Sylvester] was to know one of the historic figures of all time, one of the immortals; and when he was really moved to speak, his eloquence equaled his genius.

—G. B. HALSTED

This chapter presents two mathematical entities in some detail: functions and matrices. The concept of a function is central to every branch of mathematics and to many other areas of learning as well. We will look at the notion of a function and study a few exotic functions. In addition, we will discuss a few important properties of special functions and a few techniques for constructing new functions from known ones.

Matrices find their applications in diverse fields such as computer science, engineering, the natural sciences, and the social sciences.

A few of the interesting problems we shall study in this chapter are:

- Find the number of leap years beyond 1600 and not exceeding a given year N.
- Find the first day and the number of Friday-the-thirteenths in a given year, and the date for Easter Sunday of the year.
- If we select 367 students from a campus, will at least two of them have the same birthday?
- Suppose every pair of nonadjacent vertices of a hexagon is joined by a line segment, and each line segment is colored red or blue. Will the line segments form at least one monochromatic triangle?

3.1 The Concept of a Function

The concept of a function is so fundamental that it plays the role of a unifying thread that intertwines every branch of mathematics.

It is used in your everyday life as well. For example, when you compute your electric or water bill, you are using the concept of a function, perhaps unknowingly.

Here is an example of a function from the academic world. Consider five mathematics majors—Benson, Goldberg, Hall, Rawlings, and Wilcox. Their quality-point averages (QPA) on a 0–4 scale are 3.56, 3.80, 2.65, 3.56, and 2.23, respectively. Each element in the set $A = \{\text{Benson, Goldberg, Hall, Rawlings, Wilcox}\}$ is assigned a unique element from the set $B = \{2.23, 2.65, 3.56, 3.80\}$, as shown in Figure 3.1.

Figure 3.1

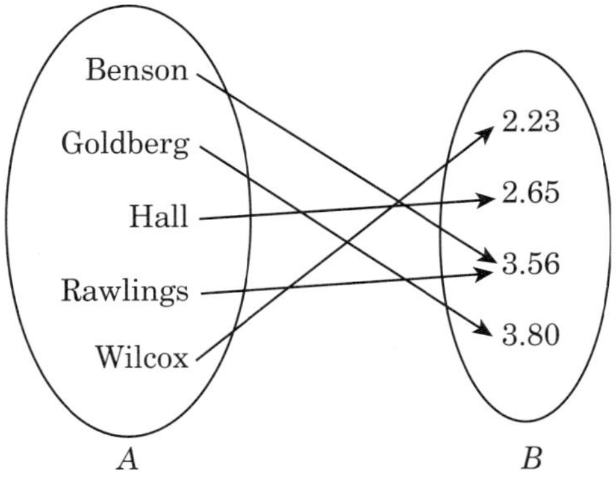

This assignment has two interesting properties:

- Every major is assigned a QPA.
- Every major has a *unique* (meaning exactly one) QPA. Such an assignment is a function.

More generally, we make the following definition.

Function

Let X and Y be any two nonempty sets. A **function** from X to Y is a rule that assigns to each element $x \in X$ a unique element $y \in Y$. Functions are usually denoted by the letters f, g, h, i, etc. If f is a function from X to Y, we write $f : X \to Y$. The set X is the **domain** of the function f and Y the **codomain** of f, denoted by dom(f) and codom(f), respectively. If $X = Y$, then f is said to be a function **on** X.

The next example elucidates these definitions.

EXAMPLE 3.1 Determine whether or not the assignments in Figures 3.2–3.4 are functions.

Figure 3.2

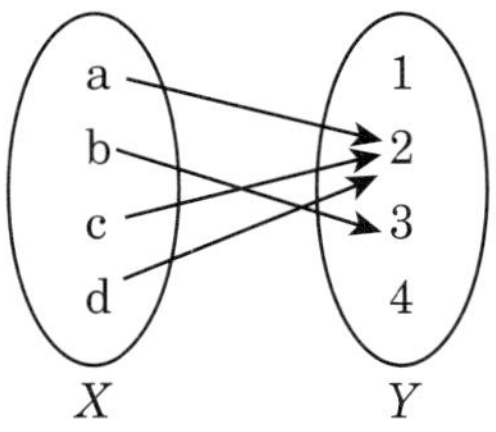

Figure 3.3

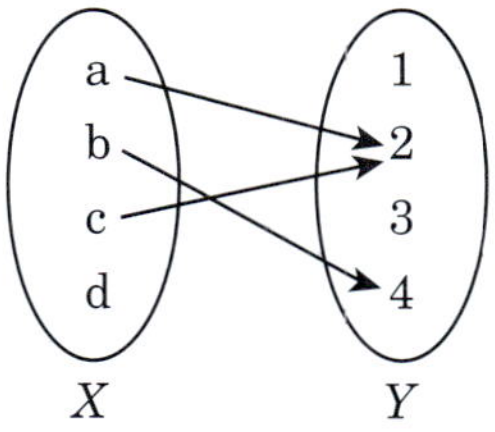

Figure 3.4

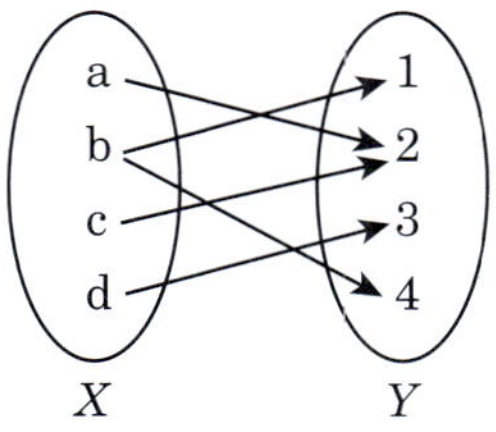

SOLUTION:

- The assignment in Figure 3.2 describes a function f from X to Y, since every element in X is assigned to exactly one element in Y. $\text{Dom}(f) = X$ and $\text{codom}(f) = Y$. (Notice that the definition does not prohibit two or more distinct elements in X being paired with the same element in Y. Also, it does not require that every element of Y be used.)
- On the other hand, the assignment in Figure 3.3 is *not* a function since not every element in X is assigned an element in Y.
- The "pairing" in Figure 3.4 is also not a function since $b \in X$ is not assigned a unique element in Y. ■

Let $f : X \to Y$, so every element $x \in X$ is paired with a unique element $y \in Y$, as in Figure 3.5. Then y is the **value** (or **image**) of the function f at x, denoted by $f(x)$, and x is a **pre-image** of y under f (see Figure 3.5); y is also known as the **output** corresponding to the **input** (or **argument**) x. Thus $y = f(x)$.* Read $f(x)$ as *f of x*.

Figure 3.5

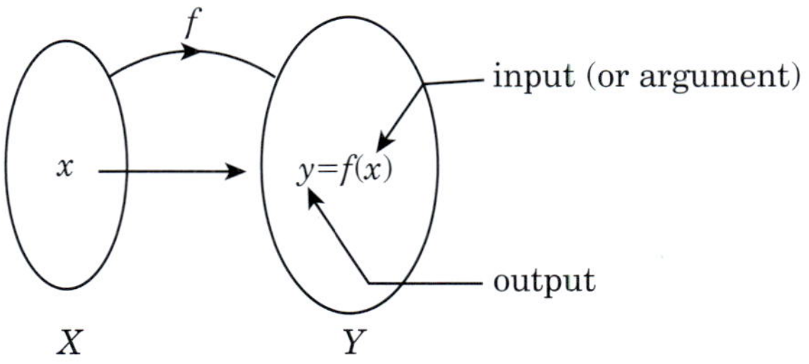

*This functional notation is due to Euler. See Chapter 8 for a biography of Euler.

> *Warning*: (1) $f(x)$ does *not* mean f times x. It simply denotes the item $y \in Y$ that $x \in X$ is paired with. (2) Let $f : X \to Y$ and x any element in X. Then, for convenience, we may call $f(x)$ the function, although it is *incorrect*. Remember, f is the function and $f(x)$ is just a value!

> There is an alternate way of defining a function $f : X \to Y$. Since every $x \in X$ determines a unique element $y = f(x)$ in Y, we can form the ordered pair (x,y) which belongs to $X \times Y$. The set of all such pairs (x, y) can be used to define f.

Next we define a useful subset of the codomain of a function.

Range of a Function

Let $f : X \to Y$ and $A \subseteq X$. Then $f(A)$ denotes the set $\{f(a) \mid a \in A\}$. In particular, $f(X)$ is the **range** of f, denoted by range(f). Thus range(f) $= f(X) = \{f(x) \mid x \in X\}$. Notice that range($f$) $\subseteq Y$.

EXAMPLE 3.2 Consider the function f in Figure 3.2. Then $f(\mathrm{a}) = f(\mathrm{c}) = f(\mathrm{d}) = 2$ and $f(\mathrm{b}) = 3$. Let $A = \{\mathrm{b}, \mathrm{c}\}$. Then $f(A) = \{f(\mathrm{b}), f(\mathrm{c})\} = \{3, 2\}$. Also, range($f$) $= \{2, 3\} \neq$ codom(f). ■

Programming languages provide built-in functions. For example, **ROUND** and **TRUNC** are two such functions. Both are functions from $\mathbb{R}$ to **Z**; ROUND rounds off a real number to the nearest integer, whereas TRUNC chops off the fractional part. The FORTRAN, C++, and Java functions **MAX** and **MIN** select the largest and the smallest of n integers (or real numbers); they are functions from $\mathbf{Z^n}$ to **Z** (or $\mathbb{R}^n$ to $\mathbb{R}$).

Functions are often defined using formulas; that is, by stating their general behavior. Many of the formulas you are familiar with are, in fact, examples of functions. For instance, the formula $C(r) = 2\pi r$ (circumference of a circle of radius r) defines a function.

We now present a few examples of abstract functions defined by formulas.

EXAMPLE 3.3 Let $\Sigma = \{\mathrm{a}, \mathrm{b}, \mathrm{c}\}$. Let $f : \Sigma^* \to \mathbf{W}$, defined by $f(x) = \|x\|$. Then $f(\lambda) = 0$, $f(\mathrm{abc}) = 3$, and $f(\mathrm{a}^i\mathrm{b}^j\mathrm{c}^k) = i + j + k$. ■

EXAMPLE 3.4 Let S be the set of binary words defined recursively as follows:

i) $1 \in S$.

ii) If $x \in S$ then $x0, x1 \in S$.

S consists of binary representations of positive integers with no leading zeros. (See Section 4.3 for a discussion of binary numbers.) Let $g : S \to \mathbb{N}$ defined by $g(x)$ = decimal value of x. Then $g(100) = 4$, $g(110) = 6$, and $g(101001) = 41$. ■

EXAMPLE 3.5 The character sets ASCII, multinational 1, box drawing, typographical symbols, math/scientific symbols, and Greek symbols, used by *WordPerfect** are denoted by the character set numbers 0, 1, 2, 3, 4, 6, and 8, respectively (see Appendix A.1). Let $A = \{0, 1, 2, 3, 4, 6, 8\}$. Each character in a character set is associated with a unique decimal number, called its **ordinal number** (or its relative position). For example, the ordinal number of the character '&' in ASCII is 38.** Let $B = \{32, 33, 34, \ldots, 60\}$, the set of ordinal numbers. Then we can define a function $f : A \times B \to C$ defined by $f(i,j) = c$, where c is the character with ordinal number j in the character set i. For example, $f(0,36)$ = '\$' and $f(8,38) = $ 'Σ'. ■

Piecewise Definition

The above definitions of functions consist of just one formula. In fact, the definitions of many of the real-world functions consist of more than one formula. Such a definition is a **piecewise definition**.

EXAMPLE 3.6 A town in Massachusetts charges each household a minimum of \$75 for up to 4000 cubic feet (ft^3) of water every 6 months. In addition, each household has to pay 60¢ for every 100 ft^3 of water in excess of 4000 ft^3. Express the water bill $f(x)$ as a function of the number of cubic feet of water x used for 6 months.

SOLUTION:

The minimum charge is \$75 for up to 4000 ft^3 of water, so $f(x) = 75$ if $0 \le x \le 4000$.

Suppose you used more than 4000 ft^3 of water. Then

$$\text{Cost for the excess} = \frac{x - 4000}{100}(0.60) = 0.006(x - 4000)$$

Then

$$\text{Total cost} = \text{minimum charge} + \text{cost for the excess} = 75 + 0.006(x - 4000)$$

Thus, the water bill $f(x)$ in dollars can be computed using the piecewise definition:

$$f(x) = \begin{cases} 75 & \text{if } 0 \le x \le 4000 \\ 75 + 0.006(x - 4000) & \text{if } x > 4000 \end{cases}$$ ■

***WordPerfect** is a wordprocessing program marketed by Corel Corporation.

**A character within single quotes indicates a literal character.

EXAMPLE 3.7 Let $A = \{0, 1, \ldots, 127\}$, the set of ordinal numbers in ASCII. Let $f : A \to$ ASCII be defined by

$$f(n) = \begin{cases} \text{nonprintable control character} & \text{if } 0 \le n \le 31 \text{ or } n = 127 \\ \text{uppercase letter} & \text{if } 65 \le n \le 90 \\ \text{lowercase letter} & \text{if } 97 \le n \le 122 \\ \text{other printable character} & \text{otherwise} \end{cases}$$

Clearly, f is defined piecewise. ■

Functions defined piecewise are written as **if-then-else** statements in most programming languages. For example, the function in Example 3.6 can be written as follows:

```
if (x ≥ 0) and (x ≤ 4000) then
  f(x) ← 50
else
  f(x) ← 50 + 0.006(x − 4000)
```

The geometrical representation of a function, called a **graph**, is often used to study functions. Remember, a picture is worth a thousand words. Since every function $f : X \to Y$ is a set of ordered pairs (x, y), the graph of f consists of points corresponding to the ordered pairs in f, as the next example illustrates.

EXAMPLE 3.8 Graph each function.

(1) Let $f : \mathbf{Z} \to \mathbf{Z}$ defined by $f(x) = x^2$.
(2) Let $g : \mathbb{R} \to \mathbb{R}$ defined by

$$g(x) = \begin{cases} x^2 & \text{if } x \ge 0 \\ -1 & \text{if } -2 \le x < 0 \\ 3x + 4 & \text{otherwise} \end{cases}$$

SOLUTION:
The graphs of the functions are displayed in Figures 3.6 and 3.7, respectively. Notice that the graph of f is a discrete collection of points.

Figure 3.6

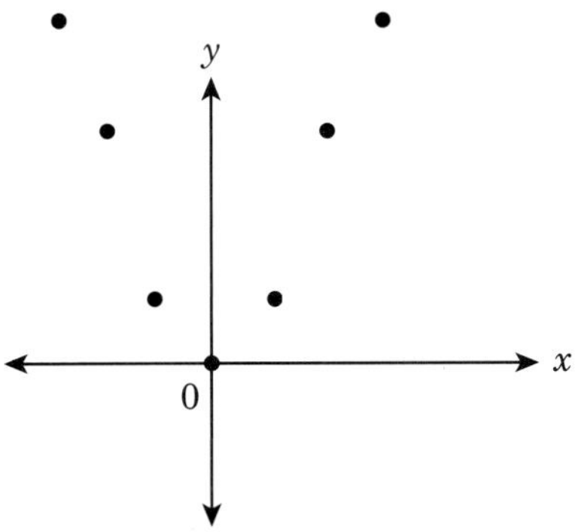

Figure 3.7

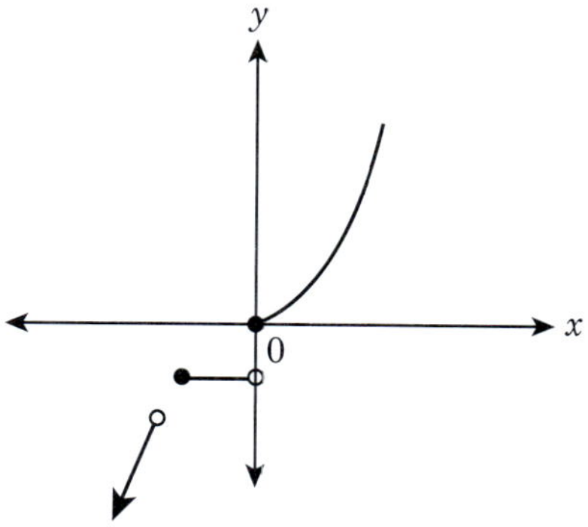

■

Next we define and illustrate two ways to construct new functions from known ones.

Sum and Product

Let $f : X \to \mathbb{R}$ and $g : Y \to \mathbb{R}$. They can be combined to construct new functions. The **sum** and **product** of f and g, denoted by $f + g$ and fg, respectively, are defined as follows:

$$(f+g)(x) = f(x) + g(x)$$
$$(fg)(x) = f(x) \cdot g(x)$$

The functions $f + g$ and fg are defined wherever *both* f and g are defined. Thus $\text{dom}(f+g) = \text{dom}(fg) = \text{dom}(f) \cap \text{dom}(g)$.

EXAMPLE 3.9 Let $f(x) = x^2$ and $g(x) = \sqrt{x-1}$, where $\text{dom}(f) = (-\infty,\infty)$ and $\text{dom}(g) = [1,\infty)$. Then

$$(f+g)(x) = f(x) + g(x) = x^2 + \sqrt{x-1}$$

and

$$(fg)(x) = f(x) \cdot g(x) = x^2\sqrt{x-1}$$

Since $\text{dom}(f) \cap \text{dom}(g) = [1,\infty)$, both $f(x)$ and $g(x)$ are defined only when $x \geq 1$, so $\text{dom}(f+g) = \text{dom}(fg) = [1,\infty)$. ■

Finally, two functions $f : A \to B$ and $g : C \to D$ are **equal** if $A = C$, $B = D$, and $f(x) = g(x)$ for every $x \in A$. We shall use this definition in the next section.

Exercises 3.1

The Celsius and Fahrenheit scales are related by the formula $F = \frac{9}{5}C + 32$.

1. Express $-40°C$ on the Fahrenheit scale.

2. Express 131°F on the Celsius scale.

$$\text{Let } g(x) = \begin{cases} 2|x| + 3 & \text{if } x \le 0 \\ 5 & \text{if } 0 < x \le 3. \\ -x^2 & \text{otherwise} \end{cases} \quad \text{Compute each.}$$

3. $g(-3.4)$ **4.** $g(0)$ **5.** $g(0.27)$ **6.** $g(4.5)$

Using Example 3.6 compute the water bill for each amount of water.

7. 1000 ft^3 **8.** 4000 ft^3 **9.** 5600 ft^3 **10.** 7280 ft^3

Let $\Sigma = \{0, 1\}$. Let $f : \Sigma^* \to \mathbf{W}$ defined by $f(x) = \|x\|$. Evaluate $f(x)$ for each value of x.

11. 000101 **12.** 1010100 **13.** 0001000 **14.** 00110011

Let Σ denote the English alphabet. Let $f : \Sigma^* \times \Sigma^* \to \Sigma^*$ defined by $f(x, y) = xy$, the concatenation of x and y. Find $f(x,y)$ for each pair of words x and y.

15. combi, natorics **16.** net, work

Let $A = \{32, 33, \ldots, 126\}$. Let $f : A \to$ ASCII defined by $f(n) =$ character with ordinal number n. Find $f(n)$ for each value of n.

17. 38 **18.** 64 **19.** 90 **20.** 123

Let g : ASCII $\to A$ defined by $g(c) = n$, where $A = \{32, 33, \ldots, 126\}$ and n denotes the ordinal number of the character c. Find $g(c)$ for each character c.

21. ‘+’ **22.** ‘<’ **23.** ‘z’ **24.** ‘{’

Let $f : \mathbf{Z} \times \mathbf{Z} \to \mathbf{Z}$ defined by $f(x,y) = 2x + 3y - 6xy$. Compute the following.

25. $f(2,3)$ **26.** $f(-3,0)$ **27.** $f(-2,3)$ **28.** $f(-3,-5)$

Let Σ denote the English alphabet. Let $g : \Sigma^* \to \Sigma^*$ defined by $f(w) = awa$. The function prefixes and suffixes each word with a. Find $f(w)$ for each word w.

29. zale **30.** mbrosi **31.** rom **32.** nesthesi

Using the function in Example 3.4 evaluate each, if defined.

33. $f(101)$ **34.** $f(1010)$ **35.** $f(001)$ **36.** $f(11011)$

Let $n \in \mathbb{N}$. A positive integer d is a **proper factor** of n if d is a factor of n and $d < n$. For example, the proper factors of 12 are 1, 2, 3, 4, and 6. Let $\sigma : \mathbb{N} \to \mathbb{N}$ defined by $\sigma(n) =$ sum of the proper factors of n. (σ is the lowercase Greek letter, sigma.) Compute $\sigma(n)$ for each value of n, where p and q are distinct primes. [A positive integer n such that $\sigma(n) = n$ is a **perfect number**.]

37. 6 **38.** 12 **39.** pq **40.** p^2

Using the functions $f(x) = 2x + 3$ and $g(x) = x^2 - 1$, find the following.

41. $(f+g)(-3)$ **42.** $(fg)(2)$ **43.** $(f+g)(x)$ **44.** $(fg)(x)$

Let $f : X \to Y$ and $A, B \subseteq X$†. Prove each.

***45.** $f(A \cup B) = f(A) \cup f(B)$ ***46.** $f(A \cap B) \subseteq f(A) \cap f(B)$

***47.** If $B \subseteq A \subseteq X$, then $f(A) - f(B) \subseteq f(A - B)$.

3.2 Special Functions

Here we turn our attention to some important functions used in discrete mathematics.

Polynomial Function

A function $f : \mathbb{R} \to \mathbb{R}$ defined by $f(x) = a_n x^n + a_{n-1}x^{n-1} + \cdots + a_1 + a_0$, where $a_0, a_1, \ldots, a_n \in \mathbb{R}$, $n \in \mathbf{W}$, and $a_n \neq 0$, is a **polynomial function**. The expression $a_n x^n + a_{n-1}x^{n-1} + \cdots + a_1 x + a_0$ is a **polynomial of degree** n in x. When $n = 1$, f is a **linear function**; when $n = 2$, f is a **quadratic function**.

Exponential and Logarithmic Functions

Let $a \in \mathbb{R}^+$, $a \neq 1$, and x any real number. The function $f: \mathbb{R} \to \mathbb{R}^+$ defined by $f(x) = a^x$ is an **exponential function** with base a. The most frequently used base in computer science is two. Figure 3.8 shows the graph of the exponential function $f(x) = 2^x$.

Figure 3.8

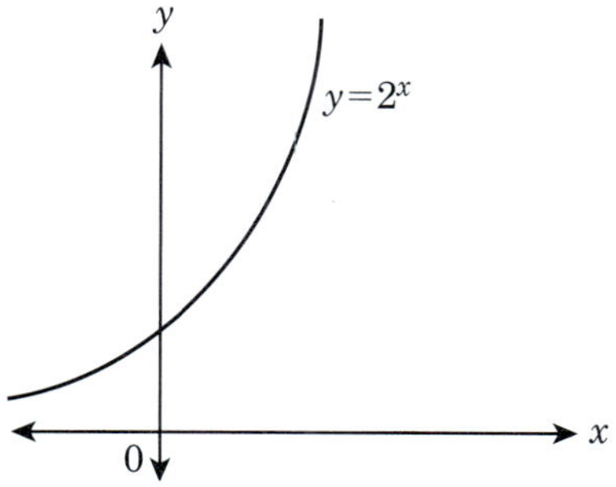

Let $a \in \mathbb{R}^+$, $a \neq 1$, and x and y any real numbers such that $y = a^x$. Then x is called the **logarithm** of y to the base a, denoted by $\log_a y$. Thus $(\log_a y = x) \leftrightarrow (y = a^x)$. Accordingly, the function $f: \mathbb{R}^+ \to \mathbb{R}$ defined

† $S, T \subseteq X$ means $S \subseteq X$ and $T \subseteq X$.

by $f(x) = \log_a x$ is the **logarithmic function** with base a. (See Appendix A.3 for a brief discussion of exponential and logarithmic functions.)

Remember that the most commonly used base in computer science is two. The corresponding logarithm is denoted by lg. Thus $\lg x = \log_2 x$, and Figure 3.9 shows the graph of the logarithmic function $f(x) = \lg x$.

Figure 3.9

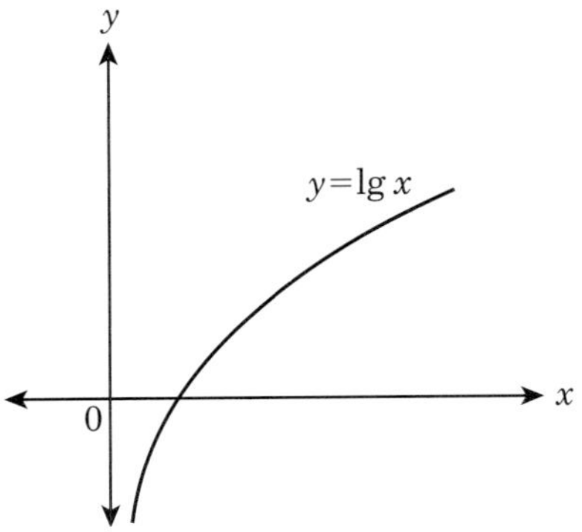

Absolute Value Function

The **absolute value function** is a function $f: \mathbb{R} \to \mathbb{R}$ defined by $f(x) = |x|$. Its graph is displayed in Figure 3.10. (Languages such as FORTRAN and Java provide a built-in function, **ABS**, for finding absolute values.)

Figure 3.10

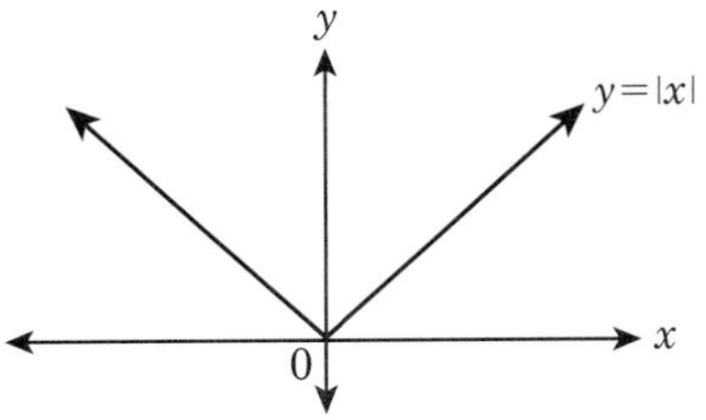

Floor and Ceiling Functions

The floor and ceiling functions are often used in the study of algorithms, as will be seen in the next two chapters. The **floor** of a real number x, denoted by $\lfloor x \rfloor$, is the greatest integer $\leq x$. The **ceiling** of x, denoted by $\lceil x \rceil$, is the least integer $\geq x$. The floor of x *rounds down* x while the ceiling of x *rounds up*. Accordingly, if $x \notin \mathbf{Z}$, the floor of x is the nearest integer to the *left of* x on the number line and the ceiling of x is the nearest integer to the *right of* x, as shown in Figure 3.11. The **floor function** $f(x) = \lfloor x \rfloor$ and the

Figure 3.11

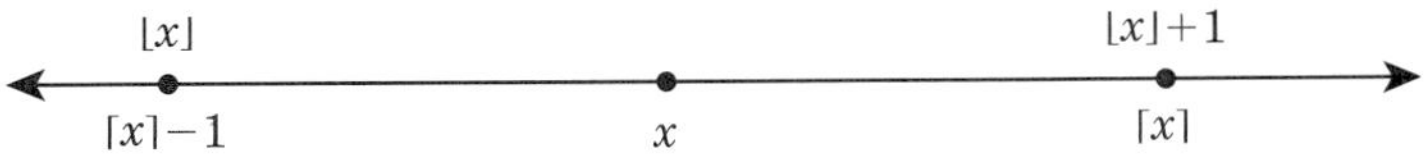

Kenneth E. Iverson *(1920–) was born at Camrose, Alberta, Canada. After graduating from Queen's University, Kingston, Ontario, in 1950, he received his M.A. from Harvard University in the following year. Three years later, he received his Ph.D. in applied mathematics from Harvard, where he taught until 1960.*

The programming language APL (A Programming Language) was his brainchild. He conceived the idea while a student at Harvard. After joining IBM in 1960, Iverson and Adin D. Falkoff developed APL into a full-fledged programming language.

Iverson's many honors include the Harry Goode Award from American Federation of Information Processing Society (1975), the Turing Award from the Association of Computing Machinery (1979), and an honorary degree from York University, Toronto (1997).

ceiling function $g(x) = \lceil x \rceil$ are also known as the **greatest integer function** and the **least integer function**, respectively.

For example, $\lfloor \pi \rfloor = 3$, $\lfloor \lg 3 \rfloor = 1$, $\lfloor -3.5 \rfloor = -4$, $\lfloor -2.7 \rfloor = -3$, $\lceil \pi \rceil = 4$, $\lceil \lg 3 \rceil = 2$, $\lceil -3.5 \rceil = -3$, $\lceil -2.7 \rceil = -2$, and $\lfloor 3 \rfloor = 3 = \lceil 3 \rceil$.

These two notations and the names *floor* and *ceiling* were introduced by Kenneth E. Iverson in the early 1960s. Both notations are variations of the notation $[x]$, which was used in number theory.

Figures 3.12 and 3.13 show the graphs of the floor and ceiling functions.

The programming languages PL/1, C, and Java provide the floor and ceiling functions as built-in functions, namely, **FLOOR** and **CEIL**. BASIC supports an intrinsic function called **INT**, which is in fact the floor function.

The floor function comes in handy when real numbers are to be truncated or rounded off to a desired number of decimal places. For example, the real number $\pi = 3.1415926535\ldots$ truncated to three decimal places is given by $\lfloor 1000\pi \rfloor/1000 = 3141/1000 = 3.141$; on the other hand, π rounded to three decimal places is $\lfloor 1000\pi + 0.5 \rfloor/1000 = 3.142$.

Figure 3.12

Graph of the floor function.

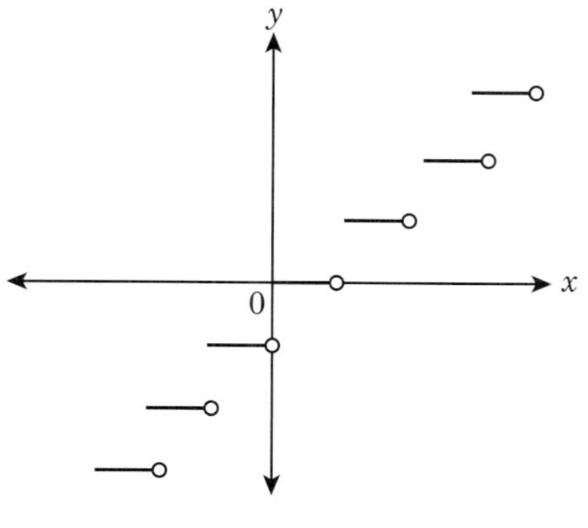

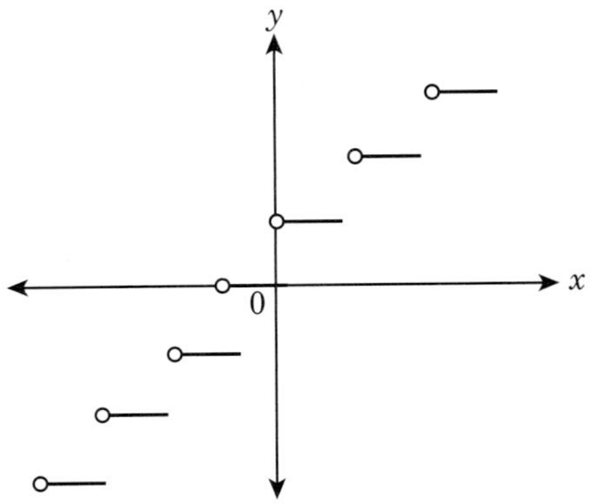

Figure 3.13

Graph of the ceiling function.

The next example presents an application of the ceiling function to everyday life.

EXAMPLE 3.10 **(The post-office function)** In 2003, the postage rate for a first-class letter of weight x not more than 1 ounce was 37¢; the rate for each additional ounce or a fraction thereof up to 11 ounces was an additional 23¢. Thus, the postage $p(x)$ for a first-class letter is given by $p(x) = 0.37 + 0.23\lceil x - 1\rceil$, where $0 < x \leq 11$.

For instance, the postage for a letter weighing 7.8 ounces is $p(7.8) = 0.37 + 0.23\lceil 7.8 - 1\rceil = \$1.98$. ■

Some properties of the floor and ceiling functions are listed below. We shall prove part 3, leaving the other parts as routine exercises.

THEOREM 3.1° Let x be any real number and n any integer. Then:

(1) $\lfloor n\rfloor = n = \lceil n\rceil$

(2) $\lceil x\rceil = \lfloor x\rfloor + 1$ $(x \notin \mathbf{Z})$

(3) $\lfloor x + n\rfloor = \lfloor x\rfloor + n$

(4) $\lceil x + n\rceil = \lceil x\rceil + n$

(5) $\left\lfloor \frac{n}{2}\right\rfloor = \frac{n-1}{2}$ if n is odd

(6) $\left\lceil \frac{n}{2}\right\rceil = \frac{n+1}{2}$ if n is odd

PROOF:
3) Every real number x can be written as $x = k + x'$ where $k = \lfloor x\rfloor$ and $0 \leq x' < 1$. Then

$$x + n = k + n + x' = (n + k) + x'$$

So
$$\lfloor x + n\rfloor = n + k, \text{ since } 0 \leq x' < 1$$
$$= \lfloor x\rfloor + n$$

as desired. ■

The floor function can be used to determine the number of positive integers less than or equal to a positive integer a and divisible by a positive integer b, as the next theorem shows.

THEOREM 3.2 Let a and b be any positive integers. Then the number of positive integers $\leq a$ and divisible by b is $\lfloor a/b \rfloor$.

PROOF:
Suppose there are k positive integers $\leq a$ and divisible by b. We need to show that $k = \lfloor a/b \rfloor$. The positive multiples of b less than or equal to a are $b, 2b, \ldots, kb$. Clearly, $kb \leq a$ or $k \leq a/b$. Further, $(k+1)b > a$. So $k + 1 > a/b$ or $a/b - 1 < k$. So

$$\frac{a}{b} - 1 < k \leq \frac{a}{b}$$

Thus k is the largest integer contained in a/b, so $k = \lfloor a/b \rfloor$. ■

For example, the number of positive integers ≤ 1776 and divisible by 13 is $\lfloor 1776/13 \rfloor = \lfloor 136.615\ldots \rfloor = 136$.

The next two examples employ Theorem 3.2 and the inclusion–exclusion principle.

EXAMPLE 3.11 Find the number of positive integers ≤ 3000 and *not* divisible by 7 or 8.

SOLUTION:
Let $A = \{x \in \mathbb{N} \mid x \leq 3000$ and divisible by $7\}$ and $B = \{x \in \mathbb{N} \mid x \leq 3000$ and divisible by $8\}$. We need to find $|A' \cap B'|$:

$$\begin{aligned}
|A' \cap B'| &= |(A \cup B)'| \\
&= |U| - |A \cup B| \\
&= |U| - |A| - |B| + |A \cap B| \\
&= 3000 - \lfloor 3000/7 \rfloor - \lfloor 3000/8 \rfloor + \lfloor 3000/56 \rfloor \\
&= 3000 - 428 - 375 + 53 = 2250
\end{aligned}$$

■

EXAMPLE 3.12 Find the number of positive integers ≤ 2076 and divisible by 3, 5, or 7.

SOLUTION:
Let A, B, and C denote the sets of positive integers ≤ 2076 and divisible by 3, 5, and 7, respectively. By the inclusion–exclusion principle,

$$\begin{aligned}
|A \cup B \cup C| &= |A| + |B| + |C| - |A \cap B| - |B \cap C| - |C \cap A| + |A \cap B \cap C| \\
&= \left\lfloor \frac{2076}{3} \right\rfloor + \left\lfloor \frac{2076}{5} \right\rfloor + \left\lfloor \frac{2076}{7} \right\rfloor - \left\lfloor \frac{2076}{15} \right\rfloor - \left\lfloor \frac{2076}{35} \right\rfloor \\
&\quad - \left\lfloor \frac{2076}{21} \right\rfloor + \left\lfloor \frac{2076}{105} \right\rfloor \\
&= 692 + 415 + 296 - 138 - 59 - 98 + 19 = 1127
\end{aligned}$$

■

In October 1582, Fr. Christopher Clavius and Aloysius Giglio introduced the Gregorian calendar at the request of Pope Gregory XIII to rectify the errors of the Julian calendar. In the Gregorian calendar, which is universally accepted, a nonleap year contains 365 days and a leap year contains 366 days. (A year is a **leap year** if it is a century divisible by 400, or if it is a noncentury and divisible by 4. For example, 1600 and 1976 were leap years, whereas 1778 and 1900 were not.)

The next example shows how to derive a formula to compute the number of leap years beyond 1600 and not exceeding a given year y.

EXAMPLE 3.13 Prove that the number of leap years ℓ after 1600 and not exceeding a given year y is given by

$$\ell = \left\lfloor \frac{y}{4} \right\rfloor - \left\lfloor \frac{y}{100} \right\rfloor + \left\lfloor \frac{y}{400} \right\rfloor - 388$$

PROOF:

Let n be a year such that $1600 < n \leq y$. To derive the formula for ℓ we proceed step-by-step:

- *To find the number of years n in the range divisible by 4*:
  Let $4n_1$ be such a year. Then $1600 < 4n_1 \leq y$; that is, $400 < n_1 \leq \frac{y}{4}$.
  Therefore, there are $n_1 = \left\lfloor \frac{y}{4} \right\rfloor - 400$ such years.
- *To find the number of centuries in the range* $1600 < n \leq y$:
  Let $100n_2$ be a century such that $1600 < 100n_2 \leq y$.
  Then $16 < n_2 \leq \frac{y}{100}$.
  Therefore, there are $n_2 = \left\lfloor \frac{y}{100} \right\rfloor - 16$ centuries beyond 1600 and $\leq y$.
- *To find the number of centuries in the range divisible by 400:*
  Since they are of the form $400n_3$, we have $1600 < 400n_3 \leq \mathrm{y}$. Then $4 < n_3 \leq \frac{y}{400}$, so $n_3 = \left\lfloor \frac{y}{400} \right\rfloor - 4$.
- Therefore,

$$\begin{aligned}
\ell &= n_1 - n_2 + n_3 \\
&= \left\lfloor \frac{y}{4} \right\rfloor - 400 - \left\lfloor \frac{y}{100} \right\rfloor + 16 \left\lfloor \frac{y}{400} \right\rfloor - 4 \\
&= \left\lfloor \frac{y}{4} \right\rfloor - \left\lfloor \frac{y}{100} \right\rfloor - \left\lfloor \frac{y}{400} \right\rfloor - 388
\end{aligned}$$

■

The technique employed for computer representation of sets (Section 2.3) is a consequence of the next function.

Characteristic Function

Let U be a universal set and S an arbitrary subset of U. Then we can define a function $f_S : U \to \{0, 1\}$ as follows:

$$f_S(x) = \begin{cases} 1 & \text{if } x \in S \\ 0 & \text{otherwise} \end{cases}$$

The function f_S is called the **characteristic function** of S.

The following example illustrates this definition.

EXAMPLE 3.14 Let $U = \{\text{a, b, c, d, e, f}\}$, $A = \{\text{a, c, d, e}\}$, and $B = \{\text{a, b, d}\}$. Then

$$f_A(x) = \begin{cases} 1 & \text{when } x = \text{a, c, d, e} \\ 0 & \text{otherwise} \end{cases}$$

In other words, $f_A(\text{a}) = f_A(\text{c}) = f_A(\text{d}) = f_A(\text{e}) = 1$ and $f_A(\text{b}) = f_A(\text{f}) = 0$ (see Figure 3.14). Similarly, $f_B(\text{a}) = f_B(\text{b}) = f_B(\text{d}) = 1$ and $f_B(\text{c}) = f_B(\text{e}) = f_B(\text{f}) = 0$. ■

Figure 3.14

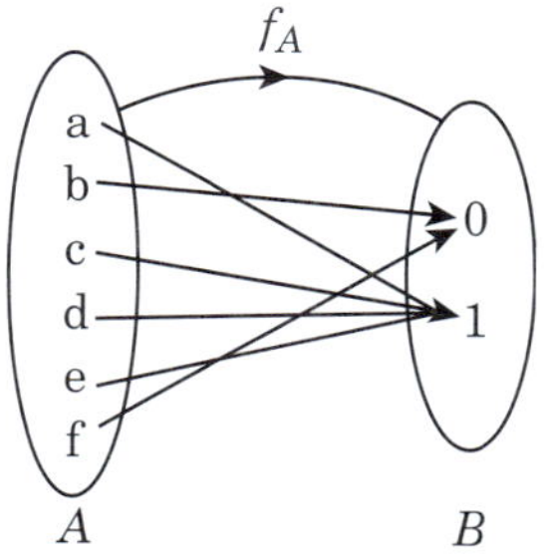

The characteristic function f_S assigns the value 1 or 0 to each element of the universe. So f_S and hence the set S can be uniquely identified by an n-bit word and vice versa, where $|U| = n$. This fact enabled us to represent sets as n-bit words in Section 2.3.

For example, the characteristic function f_A and hence the set A in the above example uniquely determine the 6-bit word 011101, where we have listed the bits from right to left for consistency. Similarly, f_B determines the word 001011.

The characteristic function satisfies the following properties.

THEOREM 3.3 Let A and B be any two sets, and U the universe. Let f_S denote the characteristic function of a subset S of U and x an arbitrary element in U. Then:

(1) $f_{A\cap B}(x) = f_A(x) \cdot f_B(x)$
(2) $f_{A\cup B}(x) = f_A(x) + f_B(x) - f_{A\cap B}(x)$
(3) $f_{A'}(x) = 1 - f_A(x)$
(4) $f_{A\oplus B}(x) = f_A(x) + f_B(x) - 2f_{A\cap B}(x)$

PROOF:

We shall prove part 1 and leave the other parts as exercises.

Case 1 Let $x \in A \cap B$.
Then $f_{A \cap B}(x) = 1$, by the definition of the characteristic function. Since $x \in A \cap B$, $x \in A$ and $x \in B$. Therefore, $f_A(x) = 1 = f_B(x)$; so, $f_A(x) \cdot f_B(x) = 1 \cdot 1$. Thus $f_{A \cap B}(x) = 1 = f_A(x) \cdot f_B(x)$.

Case 2 Let $x \notin A \cap B$.
Then $f_{A \cap B}(x) = 0$. Since $x \notin A \cap B, x \notin A$ or $x \notin B$. Therefore, either $f_A(x) = 0$ or $f_B(x) = 0$. So, in any case, $f_A(x) \cdot f_B(x) = 0$. Thus $f_{A \cap B}(x) = 0 = f_A(x) \cdot f_B(x)$.

Thus $f_{A \cap B}(x) = f_A(x) \cdot f_B(x)$ for every $x \in U$. ∎

Mod and Div Functions

The **mod function** $f(x, y) = x \bmod y$ denotes the remainder when an integer x is divided by a positive integer y. The **div function** $g(x,y) = x \operatorname{div} y$ denotes the quotient when x is divided by y. Programming languages often provide two such built-in operators, **mod** and **div**; in C++, the mod operator is denoted by the percent symbol %, and the div operator by the forward slash /.

For example, 23 mod 5 = 3, 18 mod 6 = 0, 23 div 5 = 4, and 5 div 6 = 0.

A scientific calculator, such as the TI-86, can be used to compute $x \bmod y$ using the keys [MATH], [Num], [MORE], and [F_4]. Consult the manual for your calculator to check if it supports the operator.

The mod function can determine the day of the week in n days from a given day. In 7 days, 14 days, and so on from a given day, it will again be the same day. Consequently, all we need do is remove the maximum number of 7's from n. Let r be the remainder when n is divided by 7. Then the rth day from the given day is the day we are looking for, as the next example demonstrates.

EXAMPLE 3.15 Today is Thursday. What day of the week will it be in 100 days from today?

SOLUTION:

100 mod 7 = 2. Two days from Thursday is Saturday, so it will be Saturday in 100 days from Thursday. ∎

The following example is a simple application of both div and mod operators.

Card Dealing

Consider a standard deck of 52 playing cards. They are originally assigned the numbers 0 through 51 in order. Use the suit labels 0 = clubs, 1 = diamonds, 2 = hearts, and 3 = spades to identify each suit, and the card labels 0 = ace, 1 = deuce, 2 = three, . . . , and 12 = king to identify the cards in each suit. Suppose card x is drawn at random from a well-shuffled deck, where $0 \leq x \leq 51$. How do we identify the card?

First, we need to determine the suit to which the card belongs. It is given by x div 13. Next, we need to determine the card within the suit; this is given by x mod 13. Thus card x is card (x mod 13) in suit (x div 13).

For example, let $x = 50$. Since 50 div 13 = 3, the card is a spade. Now 50 mod 13 = 11, so it is a queen. Thus card 50 is the queen of spades. ■

Next we pursue an intriguing application of the floor function and the mod operator to the game of chess.

The Two Queens Puzzle

There are two queens on an 8×8 chessboard. One can capture the other if they are on the same row, column, or diagonal. The 64 squares on the board are numbered 0 through 63. Suppose one queen is in square x and the other in square y, where $0 \leq x, y \leq 63$. Can one queen capture the other?

Since the squares are labeled 0 through 63, we can label each row with the numbers 0 through 7, and each column with the same numbers 0 through 7. In fact, each row label = $\lfloor r/8 \rfloor$ and each column label = c mod 8, where $0 \leq r, c \leq 63$. See Figure 3.15. Thus, the queen in square x lies in row $\lfloor x/8 \rfloor$ and column x mod 8, and that in square y lies in row $\lfloor y/8 \rfloor$ and column y mod 8. Consequently, the two queens will be in the same row if and only if $\lfloor x/8 \rfloor = \lfloor y/8 \rfloor$ and in the same column if and only if x mod 8 = y mod 8. For example, if $x = 41$ and $y = 47$, the two queens lie on the same row.

Figure 3.15

	0	1	2	3	4	5	6	7
0	0	1	2	3	4	5	6	7
1	8	9	10	11	12	13	14	15
2	16	17	18	19	20	21	22	23
3	24	25	26	27	28	29	30	31
4	32	33	34	35	36	37	38	39
5	40	41	42	43	44	45	46	47
6	48	49	50	51	52	53	54	55
7	56	57	58	59	60	61	62	63

← column label

↑ row label

How do we determine if they lie on the same diagonal? There are 15 northeast diagonals and 15 southeast diagonals. With a bit of patience, we can show that the queens lie on the same diagonal if and only if the absolute value of the difference of their row labels equals that of the difference of their column labels; that is, if and only if $|\lfloor x/8 \rfloor - \lfloor y/8 \rfloor| = |x \text{ mod } 8 - y \text{ mod } 8|$.

For example, let $x = 51$ and $y = 23$; see Figure 3.15. Then $|\lfloor 51/8 \rfloor - \lfloor 23/8 \rfloor| = |6-2| = 4 = |3-7| = |51 \bmod 8 - 23 \bmod 8|$, so one queen captures the other. On the other hand, if $x = 49$ and $y = 13$, then $|\lfloor 49/8 \rfloor - \lfloor 13/8 \rfloor| \neq |49 \bmod 8 - 13 \bmod 8|$; so one queen cannot capture the other. ■

Exercises 3.2

Evaluate each, where n is an integer.

1. $\lfloor n + 1/2 \rfloor$ **2.** $\lfloor n/2 \rfloor$ **3.** $\lceil n + 1/2 \rceil$ **4.** $\lceil n/2 \rceil$

Let $x = 3.456$ and $y = 2.789$. Compute each.

5. $\lfloor x + y \rfloor$ **6.** $\lfloor x \rfloor + \lfloor y \rfloor$ **7.** $\lfloor xy \rfloor$ **8.** $\lfloor x \rfloor \lfloor y \rfloor$

9. $\lfloor -x \rfloor$ **10.** $-\lfloor x \rfloor$ **11.** $\lceil x + y \rceil$ **12.** $\lceil x \rceil + \lceil y \rceil$

Find the range of each function on $\mathbb{R}$.

13. $f(x) = \lfloor x \rfloor + \lfloor -x \rfloor$ **14.** $f(x) = \lceil x \rceil + \lceil -x \rceil$

Find the number of positive integers ≤ 3076 and divisible by:

15. 3 or 4 **16.** 3, 5, or 7 **17.** 3, 5, or 6 **18.** Neither 3 nor 5

Compute the number of leap years after 1600 and not beyond each year.

19. 2000 **20.** 2020 **21.** 3076 **22.** 4050

Let $U = \{\text{a}, \ldots, \text{g}\}$. Define the characteristic function h of each set.

23. $\{\text{a, c, d, f}\}$ **24.** $\{\text{a, e, g}\}$ **25.** $\{\text{b, c, g}\}$ **26.** $\{\text{a, c, d, f, g}\}$

Let $U = \{\text{a}, \ldots, \text{h}\}$. In Exercises 27–30, a characteristic function f_S is given as an 8-bit word. Find the corresponding set S.

27. 11010100 **28.** 00101101 **29.** 10101010 **30.** 01010101

Find the day of the week in each case.

31. 234 days from Monday **32.** 365 days from Friday

33. 1776 days from Wednesday **34.** 2076 days from Saturday

Let $S = \{\text{true, false}\}$. Define a **boolean function** $f : \mathbb{N} \to S$ by $f(n) = \text{true}$ if year n is a leap year and false otherwise. Find $f(n)$ for each year n.

35. 1996 **36.** 2020 **37.** 2076 **38.** 3000

39. January 1, 2000, falls on a Saturday. What day of the week will January 1, 2020, be?
(*Hint*: Look for leap years.)

40. January 1, 1990, was a Monday. What day of the week was January 1, 1976?
(*Hint*: Again, look for leap years.)

Each day of the week, beginning with Sunday, can be identified by a code x, where $0 \le x \le 6$. January 1 of any year y can be determined using the following formula**.

$$x \equiv \left(y + \left\lfloor \frac{y-1}{4} \right\rfloor - \left\lfloor \frac{y-1}{100} \right\rfloor + \left\lfloor \frac{y-1}{400} \right\rfloor\right) \bmod 7 \qquad (3.1)$$

Using this formula determine the first day in each year.

41. 2000 **42.** 2020 **43.** 2076 **44.** 3000

The number of Friday-the-thirteenths in a given year y can be computed using formula (1) above and Table 3.1. For example, suppose that January 1 of a year y falls on a Sunday(0). If it is not a leap year, there will be two Friday-the-thirteenths: January 13 and October 13; if it is a leap year, there will be three: January 13, April 13, and July 13. Compute the number of Friday-the-thirteenths in each year.

45. 2000 **46.** 2020 **47.** 2076 **48.** 3076

Table 3.1

Code x	**January 1**	**Nonleap year** y	**Leap year** y
0	Sunday	January, October	January, April, July
1	Monday	April, July	September, December
2	Tuesday	September, December	June
3	Wednesday	June	March, November
4	Thursday	February, March, November	February, August
5	Friday	August	May
6	Saturday	May	October

(**Easter Sunday**) The date for Easter Sunday in any year y can be computed as follows. Let $a = y \bmod 19$, $b = y \bmod 4$, $c = y \bmod 7$, $d = (19a + 24) \bmod 30$, $e = (2b + 4c + 6d + 5) \bmod 7$, and $r = (22 + d + e)$. If $r \le 31$, then Easter Sunday is March r; otherwise, it is April $[r \pmod{31}]$. Compute the date for Easter Sunday in each year.

49. 1996 **50.** 2000 **51.** 2076 **52.** 3000

Prove each, where $x \in \mathbb{R}$ and $n \in \mathbf{Z}$.

53. $\left\lfloor \frac{n}{2} \right\rfloor = \frac{n-1}{2}$ if n is odd. **54.** $\left\lceil \frac{n}{2} \right\rceil = \frac{n+1}{2}$ if n is odd.

**Based on G. L. Ritter *et al.*, "An Aid to the Superstitious," *Mathematics Teacher*, Vol. 70, May 1977, pp. 456–457.

55. $\left\lfloor \frac{n^2}{4} \right\rfloor = \frac{n^2 - 1}{4}$ if n is odd.

56. $\left\lceil \frac{n^2}{4} \right\rceil = \frac{n^2 + 3}{4}$ if n is odd.

57. $\left\lfloor \frac{n}{2} \right\rfloor + \left\lceil \frac{n}{2} \right\rceil = n$

58. $\lceil x \rceil = \lfloor x \rfloor + 1 \quad (x \notin \mathbf{Z})$

59. $\lceil x \rceil = -\lfloor -x \rfloor$

60. $\lceil x + n \rceil = \lceil x \rceil + n$

Let A and B be any two sets, and U the universe. Let f_S denote the characteristic function of a subset S of U and x an arbitrary element in U. Prove each.

***61.** $f_{A \cup B}(x) = f_A(x) + f_B(x) - f_{A \cap B}(x)$

***62.** $f_{A'}(x) = 1 - f_A(x)$

***63.** $f_{A \oplus B}(x) = f_A(x) + f_B(x) - 2f_{A \cap B}(x)$

Let $x, y \in \mathbb{R}$. Let $\max\{x,y\}$ denote the maximum of x and y, and $\min\{x,y\}$ denote the minimum of x and y. Prove each.

***64.** $\max\{x,y\} + \min\{x,y\} = x + y$

***65.** $\max\{x,y\} - \min\{x,y\} = |x - y|$

3.3 Properties of Functions

Functions satisfy a number of properties and we begin with the identity function.

Identity Function

A function f on X is the **identity function** if $f(x) = x$ for every x in X. It is denoted by 1_X and leaves every input unchanged. The graph of the identity function on $\mathbb{R}$ is the 45°-line $y = x$.

EXAMPLE 3.16 Let S be an ordered set. **ORD**(x) denotes the ordinal number of each element x in S, the first ordinal number being 0. For example, using ASCII, ORD('<')[†] = 60 and ORD('$') = 36. (Pascal, for instance, provides such a built-in function.) If the argument x, however, is an integer n, ORD(n) = n. Thus ORD is the identity function on **W**. ■

Injection

A function $f : X \to Y$ is an **injection** (or **one-to-one function**) if different input values yield different output values. Thus f is injective, if $x_1 \neq x_2$ implies $f(x_1) \neq f(x_2)$; equivalently, f is injective if $f(x_1) = f(x_2)$ implies $x_1 = x_2$ (why?).

The next two examples illustrate this definition.

[†] x within single quotes indicates the character x.

EXAMPLE 3.17 Let $A = \{0, 1, 2, \ldots, 127\}$. Let **CHR**: $A \to$ ASCII defined by CHR(n) = ASCII character with ordinal number n. For example, CHR(59) = ';' and CHR(43) = '+'. Since distinct ordinal numbers correspond to different characters, CHR is injective. (C, C++, and Java, for instance, provide such a built-in function.) ■

EXAMPLE 3.18 Let $f : \Sigma^* \to W$ defined by $f(x) = \|x\|$, where $\Sigma = \{a, b, c\}$. The words *aaa* and *bab* are distinct words with the same length, so f is not injective. ■

How do you characterize the graph of an injective function $f : \mathbb{R} \to \mathbb{R}$? The function is not injective if there are two distinct input elements $a, b \in \mathbb{R}$ such that $f(a) = f(b)$, that is, if a horizontal line intersects its graph in two distinct points $(a, f(a))$ and $(b, f(b))$. Thus f is injective if and only if no horizontal line intersects the graph in more than one point. For example, the function $f(x) = x^2$ is not injective (see Figure 3.16).

Figure 3.16

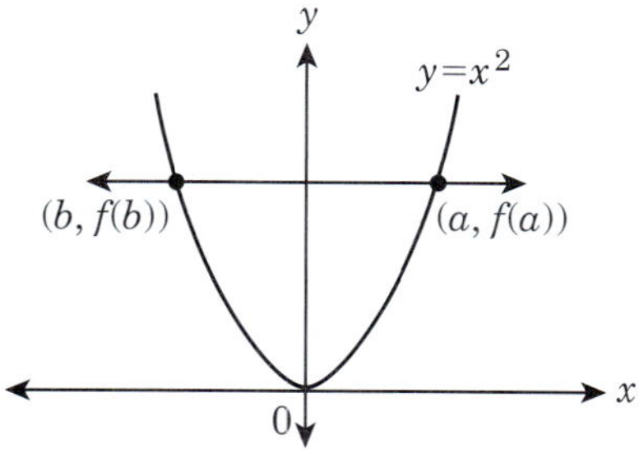

Surjection

A function $f : X \to Y$ is a **surjection** (or an **onto function**) if for every y in Y there exists an x in X such that $f(x) = y$, that is, if every element in Y has at least one pre-image in X. In other words, f is surjective if range(f) = Y.

The following two examples clarify this definition.

EXAMPLE 3.19 Let Σ be a nonempty alphabet. Let $f : \Sigma^* \to \mathbf{W}$ defined by $f(x) = \|x\|$. Let $n \in \mathbf{W}$. Then $x^n \in \Sigma^*$ and $\|x^n\| = n$. Thus, given any $n \in \mathbf{W}$, there exists an element $u = x^n \in \Sigma^*$ such that $f(u) = n$. Consequently, f is surjective. ■

EXAMPLE 3.20 Determine if the function $f(x) = x^2$ on $\mathbb{R}$ is surjective.

SOLUTION:
For every y in $\mathbb{R}$, does there exist a real number x such that $x^2 = y$? No, for instance, there is no real number x such that $x^2 = -1$, so f is *not* surjective. ■

Bijection

A function $f : X \to Y$ is a **bijection** (or **one-to-one correspondence**) if it is both injective and surjective.

EXAMPLE 3.21 Let A be the set of printable ASCII characters and $B = \{32, 33, \ldots, 126\}$. Let $f : A \to B$ defined by $f(c)$ = ordinal number of character c. Since f is both injective and surjective, f is bijective. (Notice the deliberate choice of B to make the function surjective.) ■

Notice that $23 \bmod 5 = 3 = 48 \bmod 5$, but $23 \neq 48$; therefore, the mod function is not injective. However, when an integer a is divided by m, there are m possible remainders, namely, $0, 1, 2, \ldots, m-1$ (see Section 4.1); so, given any nonnegative integer r less than m, we can always find an integer a such that $r = a \bmod m$; thus the mod function is surjective.

Hashing

We are now ready to examine an interesting application of the mod function in everyday life.

EXAMPLE 3.22 Banks use nine-digit account numbers to create and maintain customer accounts. Customer records are stored in an array in a computer and can be accessed fairly easily and quickly using their unique **keys**, which in this case are the account numbers. Access is often accomplished using the **hashing function** $h(x) = x \bmod m$, where x denotes the key (account number) and m the number of cells in the array; $h(x)$ denotes the **hash address** of the customer record with key x. See Figure 3.17.

Figure 3.17

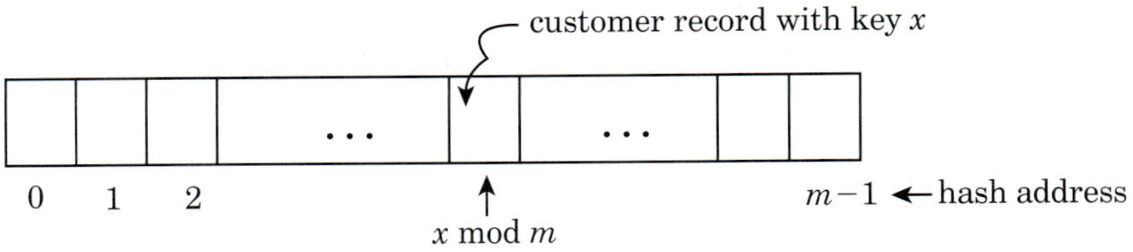

In particular, let $m = 1009$ and $x = 207630764$. The corresponding record is stored in location

$$h(207630764) = 207630764 \bmod 1009 = 762$$

Likewise,

$$h(307620765) = 307620765 \bmod 1009 = 881$$

Since the hashing function is not injective (why?), theoretically different customer records can be assigned to the same location. For example,

$$h(207630764) = 762 = h(208801204)$$

This results in a **collision**.

One simple way to resolve a collision is to do a sequential search for the next available cell, beginning with the cell where the collision has occurred. Then we store the item in the available cell. If we come to the end of the array without any success, then we would continue the search back at the beginning of the array, as if the array were circular. This way of resolving a collision is called **linear probing**. So we would store the data with the account number 208801204 in location 763 (assuming that is available). ■

Obviously, the technique illustrated in this example can be adapted to a variety of situations. For example, the various identifiers in a computer program can be stored in a *symbol table* using their first letters as keys; student records can be stored in a hash table using their social security numbers; and patients' medical records can be maintained in a table using their social security numbers as keys.

Next we present a few simple and useful properties of functions associated with finite sets.

THEOREM 3.4 Let X and Y be any two finite sets with $|X| = |Y| = n$. A function $f : X \to Y$ is injective if and only if f is surjective.

PROOF:

Let $X = \{x_1, x_2, \ldots, x_n\}$ and $Y = \{y_1, y_2, \ldots, y_n\}$. Suppose f is injective. Then $f(x_1), \ldots, f(x_n)$ are n distinct elements. So they must be the same elements $y_1, \ldots, y_n$ in some order. Therefore, f is surjective.

Conversely, suppose f is surjective. Then $f(x_1), \ldots, f(x_n) = Y$. Since $|Y| = n$, the elements $f(x_1), \ldots, f(x_n)$ must be different, so f must be injective. ■

Let A and B be two finite sets with the same cardinality. Suppose we would like to show that a function $f : A \to B$ is bijective. Then, by Theorem 3.4, it suffices to show that f is either injective or surjective.

THEOREM 3.5 Two finite sets have the same cardinality if and only if there exists a bijection between them.

PROOF:

Let X and Y be two finite sets with $|X| = m$ and $|Y| = n$. Let $X = \{x_1, \ldots, x_m\}$. Let $f : X \to Y$ be bijective. Since f is injective, $f(x_1), \ldots, f(x_m)$ are m distinct elements in Y. Consequently, $m \leq n$. Since f is surjective, every element y in Y has at least one input in X, so $n \leq m$. Thus $|X| = |Y|$.

Conversely, suppose $m = n$ and $Y = \{y_1, \ldots, y_m\}$. Define a function $f : X \to Y$ by $f(x_i) = y_i$ for every i. We will now show that f is injective. Let x_j and x_k be two elements in X such that $f(x_j) = f(x_k)$. Then, by definition, $y_j = y_k$; so, $j = k$ and hence $x_j = x_k$. Therefore, f is injective and hence, by Theorem 3.4, f is bijective. ■

Let $f : X \to Y$, where X and Y are finite sets and $|X| > |Y|$. Then what can we say about the function f? (Obviously, f can't be bijective.) This is answered in the next section.

Cardinality of an Infinite Set (optional)

Before closing this section, we extend the concept of cardinality of finite sets to infinite sets, a topic of great importance to theoretical computer science and certainly to mathematics. Recall that two finite sets have the same cardinality if there is a bijection between them. This leads to the following definition.

Two sets X and Y have the same **cardinality** if there exists a bijection from X to Y, denoted by $|X| = |Y|$.

This definition can be used to partition the family of infinite sets into two disjoint classes. To this end, we make the following definition.

Countable and Uncountable Sets

A set S is **countably infinite** if there exists a bijection between S and $\mathbb{N}$. A set that is finite or countably infinite is **countable**. A set that is not countable is **uncountable**.

The cardinality of $\mathbb{N}$ is denoted by $\aleph_0$ (read "aleph-naught," "aleph" being the first letter of the Hebrew alphabet. This symbol was introduced by Cantor). Thus a set S is countably infinite if $|S| = \aleph_0 = |\mathbb{N}|$.

If A and B are finite sets such that $A \subset B$, then $|A| < |B|$. This, however, need not be true in the case of infinite sets. For example, $E \subset \mathbb{N}$, where E denotes the set of even positive integers; nonetheless, $|E| = |\mathbb{N}| = \aleph_0$, as shown by the pairings in Figure 3.18.

Figure 3.18

EXAMPLE 3.23 Show that $\mathbb{N} \times \mathbb{N}$ is countably infinite.

PROOF:
Although we shall not give a formal proof, the arrows in Figure 3.19 show how the various elements of $\mathbb{N} \times \mathbb{N}$ can be listed as the first, second, third, and so on in a systematic way, showing that $\mathbb{N} \times \mathbb{N}$ is countably infinite.

Figure 3.19

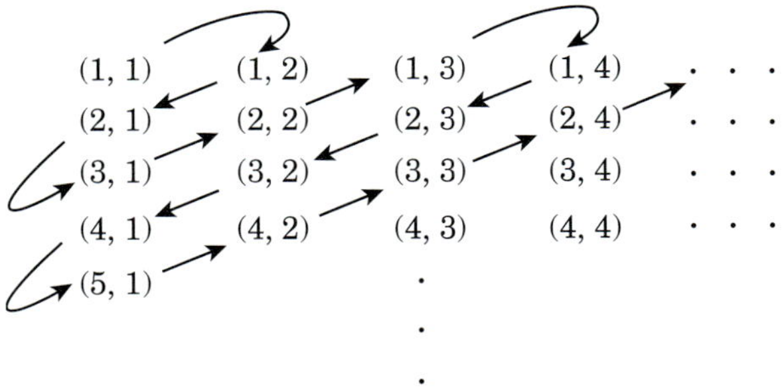

■

It follows by Example 3.22 that the set of positive rational numbers is countable. Consequently, the set of negative rational numbers is also countable. Since the union of two countable sets is countable (see Exercise 49), it follows that **Q** is countable.

It may seem improbable that there exist infinite sets that are uncountable. For instance, the open interval (0,1) is such a set, as the next example shows.

EXAMPLE 3.24 Show that the open interval (0,1) is uncountable.

PROOF (by contradiction)**:** Assume that the interval (0,1) is countable. Then every real number between 0 and 1 can be listed as $a_1, a_2, a_3, \ldots$ Each a_i has a unique decimal expansion (for numbers with two different decimal expansions, choose the expansion with trailing 9's. For example, although $0.5 = 0.5000\ldots = 0.4999\ldots$, select $0.4999\ldots$ for our discussion.):

$$\begin{aligned} a_1 &= 0.\,a_{11}a_{12}a_{13}a_{14}\ldots \\ a_2 &= 0.\,a_{21}a_{22}a_{23}a_{24}\ldots \\ a_3 &= 0.\,a_{31}a_{32}a_{33}a_{34}\ldots \\ a_4 &= 0.\,a_{41}a_{42}a_{43}a_{44}\ldots \\ &\vdots \end{aligned}$$

where each a_{ij} is a digit.

Now construct a real number $b = 0.b_1b_2b_3b_4\ldots$ as follows:

$$b_i = \begin{cases} 1 & \text{if } a_{ii} \neq 1 \\ 2 & \text{if } a_{ii} = 1 \end{cases}$$

Clearly, $0 < b < 1$; therefore b must be one of the numbers in the above list $a_1, a_2, a_3, a_4, \ldots$. However, since $b_i \neq a_{ii}$ for every i, b cannot be in the list. This leads to a contradiction. Therefore, the real numbers between 0 and 1 cannot be listed and hence the interval (0,1) is uncountable. (The technique employed is called **Cantor's diagonalization procedure**.) ■

Since the interval (0,1) is uncountable, it follows that $\mathbb{R}$ is also uncountable. So, although both $\mathbb{N}$ and $\mathbb{R}$ are infinite, $|\mathbb{R}| > \aleph_0$.

Exercises 3.3

Determine if each function is the identity function.

1.

x	a	b	c	d
$f(x)$	a	b	c	d

2.

x	a	b	c	d
$f(x)$	b	c	d	a

3.

x	a	b	c	d
$f(x)$	a	b	c	c

Determine if each function is injective, where **trunc(x)** denotes the integral part of the real number of x.

4. $f(x) = |x|, x \in \mathbb{R}$

5. $g(x) = 2^x, x \in \mathbb{R}$

6. $h(x) = \lg x, x \in \mathbb{R}^+$

7. $f(x) = \lfloor x \rfloor, x \in \mathbb{R}$

8. $g(x) = \lceil x \rceil, x \in \mathbb{R}$

9. $h(x) = \text{trunc}(x), x \in \mathbb{R}$

10. $f : S \to \mathbf{W}$ defined by $f(A) = |A|$, where S is the family of all finite sets.

Determine if each function from $\mathbb{R}$ to $\mathbf{Z}$ is surjective.

11. $f(x) = |x|$

12. $g(x) = \lfloor x \rfloor$

13. $h(x) = \lceil x \rceil$

14. $h(x) = \lg |x|, x \neq 0$

15. ORD: ASCII $\to$ **W** defined by ORD(c) $=$ ordinal number of the character c.

16. Let $f : \mathbb{R} \to \mathbb{R}$ defined by $f(x) = ax + b$, where $a,b \in \mathbb{R}$ and $a \neq 0$. Show that f is surjective; that is, find a real number x such that $f(x) = c$.

Determine if each function $f : A \to B$ is bijective.

17. $f(x) = x^2, A = B = \mathbb{R}$

18. $f(x) = \sqrt{x}, A = \mathbb{R}^+, B = \mathbb{R}$

19. $f(x) = |x|, A = B = \mathbb{R}$

20. $f(x) = \lfloor x \rfloor, A = B = \mathbb{R}$

21. $f(x) = \lceil x \rceil, A = B = \mathbb{R}$

22. $f(x) = 2^{|x|}, A = B = \mathbb{R}$

Determine if the functions in Exercises 23–30 are bijective. If they are not bijective, explain why.

23. $f : \Sigma^* \to \mathbf{W}$ defined by $f(x) =$ decimal value of x, where $\Sigma = \{0,1\}$.

24. $f : \Sigma^* \times \Sigma^* \to \Sigma^*$ defined by $f(x,y) = xy$, where Σ denotes the English alphabet.

25. $g : \Sigma^* \to \Sigma^*$ defined by $g(w) = awa$, where $\Sigma = \{\text{a,b,c}\}$.

26. $f : \mathbb{R} \times \mathbb{R} \to \mathbb{R} \times \mathbb{R}$ defined by $f(x,y) = (x,-y)$.

27. The ORD function on ASCII.

28. The **predecessor function** (**PRED**) and **successor function** (**SUCC**) are two important functions used in computer science. They are defined on ordered sets. If c is a printable ASCII character, PRED(c) denotes the predecessor of c and SUCC(c) denotes the successor of c; for example, PRED('?') = '@' and SUCC(':') = ';'. Determine if PRED and SUCC are bijective.

Using the hash function in Example 3.2, compute the location corresponding to the given key.

29. 012398745

30. 430358856

Student records are maintained in a table using the hashing function $h(x) = x \bmod 9767$, where x denotes the student's social security number.

Compute the location in the table corresponding to the given key, where the record is stored.

31. 012-34-5678 **32.** 876-54-3210

33–34. Redo Exercises 31 and 32 if $h(x)$ = first part in x mod 13.

35. Store the following two-letter abbreviations of states in the United States in a hash table with 26 cells, using the hashing function $h(x)$ = first letter in x:

NY, OH, FL, AL, MA, CA, MI, AZ

36. Redo Exercise 35 with the following state abbreviations: MD, CT, ID, MA, NB, NJ, MI, WI, CA, IA, WA, MN, NH, IN, NC, WY, NM, MS, MO, CO, NY, IL, NV, WV, ND, MT

Two sets A and B are **equivalent**, denoted by $A \sim B$, if there exists a bijection between them. Prove each.

37. $A \sim A$ (**reflexive property**)

38. $A \sim A \times \{1\}$

39. If $A \sim B$, then $A \times \{1\} \sim B \times \{2\}$

40. $\mathbf{Z} \sim \mathrm{O}$, the set of odd integers

Prove each.

41. A bijection exists between any two closed intervals $[a, b]$ and $[c, d]$, where $a < b$ and $c < d$.
(*Hint*: Find a suitable function that works.)

42. The set of odd positive integers is countably infinite.

43. The set of integers is countably infinite.

44. Any subset of a countable set is countable.

45. A set A is infinite if and only if there exists a bijection between A and a proper subset of itself.

46. The open interval (a,b) is uncountable.
[*Hint*: Find a suitable bijection from $(0,1)$ to (a,b).]

47. The set Q^+ of positive rational numbers is countable.

48. The set of irrational numbers is uncountable.
(*Hint*: Prove by contradiction.)

***49.** A countable union of countable sets is countable.

***50.** The cartesian product of two countable sets is countable.

***51.** If Σ is a finite alphabet, then Σ^* is countable.

3.4 The Pigeonhole Principle

Suppose m pigeons fly into n pigeonholes to roost, where $m > n$. Then obviously at least two pigeons must roost in the same pigeonhole (see Figures 3.20 and 3.21). This property, called the pigeonhole principle, can be stated in terms of functions, as the next theorem shows.

Figure 3.20

Figure 3.21

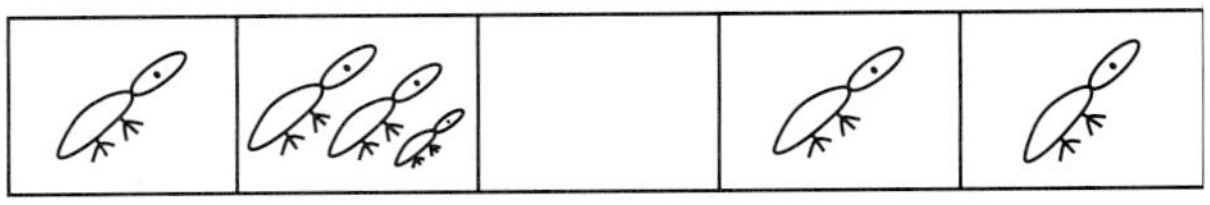

THEOREM 3.6 **(The Pigeonhole Principle)** Let $f : X \to Y$, where X and Y are finite sets, $|X| = m$, $|Y| = n$, and $m > n$. Then there exist at least two distinct elements x_1 and x_2 in X such that $f(x_1) = f(x_2)$.

PROOF:
Let $X = \{x_1, \ldots, x_m\}$. Suppose f is injective. Then $f(x_1), \ldots, f(x_m)$ are distinct elements in Y. So $m \leq n$. But this contradicts the assumption that $m > n$. Therefore, f is not injective and there must be at least two distinct elements x_1 and x_2 such that $f(x_1) = f(x_2)$. Hence the theorem. ■

The pigeonhole principle is a simple but important counting principle that we shall use in Chapters 4, 7, and 8.

The pigeonhole principle, which can be applied in a variety of situations, can be restated as follows: If m objects are placed into n boxes, then at least one box must contain two or more objects, where $m > n$. Accordingly, the pigeonhole principle is also called the **Dirichlet Box Principle** after the German mathematician Peter Gustav Lejeune Dirichlet, who used it extensively in his work on number theory.

Although the principle looks simple and straightforward, to apply it successfully you must choose the pigeons and pigeonholes appropriately, as the next few examples illustrate.

EXAMPLE 3.25 Suppose we select 367 students from campus. Show that at least two of them must have the same birthday.

SOLUTION:
The maximum number of days in a year is 366, and this occurs in a leap year. Think of students as pigeons and days of the year as pigeonholes. Let A be the set of students and B the set of days, where $|A| = m = 367$ and $|B| = n = 366$. Let $f : A \to B$ defined by $f(a) =$ birthday of student a.

***Gustav Peter Lejeune Dirichlet** (1805–1859) was born in Duren, Germany. The son of a postmaster, he first attended a public school and then a private school that emphasized Latin. After attending the Gymnasium in Bonn for 2 years, Dirichlet entered a Jesuit college in Cologne where he received a strong background in theoretical physics under the physicist Georg Simon Ohm. In May 1822, he moved to the University of Paris.*

In 1826, Dirichlet returned to Germany and taught at the University of Breslau. Three years later, he moved to the University of Berlin where he spent the next 27 years.

Dirichlet's primary interest in mathematics was number theory, inspired by Gauss' masterpiece, Disquisitiones Arithmeticae *(1801). He established* Fermat's Last Theorem *for* n = *14. Among the many results he discovered include the proof of a theorem presented to the Paris Academy of Sciences on algebraic number theory in 1837:* The sequence $\{an + b\}$ contains infinitely many primes, where a and b are relatively prime.

In 1855, when Gauss died, Dirichlet moved to the University of Göttingen. Three years later, he went to Montreaux, Switzerland, to deliver a speech in honor of Gauss. While there, he suffered a heart attack and was barely able to return home. During his illness his wife succumbed to a stroke, and Dirichlet died.

Since $m > n$, by the pigeonhole principle, there should be at least two students a_1 and a_2 such that $f(a_1) = f(a_2)$; that is, at least two students have the same birthday. ■

The next example* is geometric, demonstrating that the pigeonhole principle can pop up in seemingly unusual situations.

EXAMPLE 3.26 Suppose five **lattice points**, that is, points with integer coordinates, are selected on the cartesian plane and each pair of points is joined by a line segment. Show that at least one of the line segments must contain a lattice point between its endpoints.

SOLUTION:

The set of lattice points can be partitioned into four nonempty disjoint classes according to the **parity** (evenness or oddness) of their coordinates: (odd,odd), (odd,even), (even,odd), and (even,even). Since there are five points (pigeons) and four classes (pigeonholes), by the pigeonhole principle, at least two of them — say, A(a,b) and B(c,d)—must belong to the same class.

By the **midpoint formula** in analytic geometry, the midpoint M of the line segment $\overline{AB}$ is $\left(\frac{a+c}{2}, \frac{b+d}{2}\right)$. Since the sum of any two odd or even integers is an even integer, it follows that M is also a lattice point. Thus $\overline{AB}$ contains a lattice point M different from its endpoints. ■

*Based on C. T. Long, "On Pigeons and Problems," *Mathematics Teacher*, Vol. 81 (January 1988), pp. 28–30, 64.

It is well known that the decimal expansions of rational numbers are periodic. Using the pigeonhole principle, we shall establish this, but first a few words of explanation may be helpful.

Using the familiar long division method, you may verify that

$$\frac{4111}{33300} = 0.12345345345345\ldots$$

Although the decimal expansion is nonterminating, it is **periodic**; that is, a certain block of digits, namely, 345, gets repeated. Accordingly, the expansion is usually written as $0.12\overline{345}$, using a bar over the first repeating block. The number of digits in the smallest repeating block is the **period** of the expansion; here it is 3. We are now ready to prove the above proposition.

EXAMPLE 3.27 Prove that the decimal expansion of a rational number is periodic.

PROOF:

Consider, for convenience, a positive rational number $\frac{a}{b}$, where $0 < a < b$. Let $\frac{a}{b} = 0.d_1d_2d_3\ldots$ where, by the division algorithm (see Section 4.1), we have:

$$\begin{aligned} 10a &= bd_1 + r_1 \\ 10r_1 &= bd_2 + r_2 \\ 10r_2 &= bd_3 + r_3 \\ &\vdots \\ 10r_j &= bd_{j+1} + r_{j+1} \\ &\vdots \end{aligned} \tag{3.2}$$

and $0 \le r_i < b$ for every i. (*Note*: The digits $d_1, d_2, \ldots$ in the decimal expansion are the quotients when $10a$, $10r_1, \ldots$ are divided by b. Since a remainder has only b choices, by the pigeonhole principle, two of the remainders $r_1, r_2, \ldots, r_{b+1}$ must be equal; that is, $r_j = r_k$ for some j and k, where $1 \le j < k \le b+1$. Consequently, $d_{k+1} = d_{j+1}$, $d_{k+2} = d_{j+2}, \ldots$, $d_{2k-j} = d_k, d_{2k-j+1} = d_{j+1}$, and so on. Thus $d_{j+1}\ldots d_k$ is the smallest block getting repeated and the period of the decimal expansion is $k - j$. ■

The next example, a rather sophisticated application of the pigeonhole principle, is due to the Hungarian mathematician Paul Erdös.

EXAMPLE 3.28 **(Erdös Theorem)** If $n+1$ integers are selected from the set $\{1, 2, \ldots, 2n\}$, one of them divides another integer that has been selected.

PROOF:

Let $a_1, a_2, \ldots, a_{n+1}$ denote the integers selected. Write each of them as a product of a power of 2 and an odd integer; that is, $a_i = 2^{e_i}b_i$, where $1 \le i \le n+1$ and $e_i \ge 0$. The integers $b_1, b_2, \ldots, b_{n+1}$ are odd positive integers $\le 2n$.

Paul Erdös *(1913–1996) was born in Budapest, Hungary. Except for about three years in schools, Erdös (pronounced air-dosh) was taught at home, mostly by his father, who had returned from a Siberian prison after 6 years.*

A child prodigy, Erdös, at age 3, discovered negative numbers for himself. In 1930 Erdös entered the Peter Pazmany University in Budapest. Three years later, he discovered a beautiful proof of the celebrated Chebyshev theorem that there is a prime between any positive integer n *and* 2n. *In 1934 he received his Ph.D. from the university.*

An author of about 1500 articles and coauthor of about 500, Erdös was one of the most prolific writers in mathematics. A tribute in 1983 described him as "the prince of problem-solvers and the absolute monarch of problem-posers." As "the Euler of our time," he contributed extensively to number theory, combinatorics, function theory, complex analysis, set theory, group theory, and probability, the first two areas being closest to his heart.

"Always searching for mathematical truths," he deemed worldly possessions a nuisance, so he never had a home, a car, checks, or even an address. Always traveling from meeting to meeting, carrying a half-empty suitcase, he would stay with mathematicians wherever he went and donate the honoraria he earned as prizes to students.

A recipient of many honors, Erdös died of a heart attack while attending a mathematics meeting in Warsaw.

Since there are exactly n odd positive integers $\leq 2n$, by the pigeonhole principle, two of the elements $b_1, b_2, \ldots, b_{n+1}$ must be equal, say, $b_i = b_j$. That is, $a_j = 2^{e_j} b_j = 2^{e_j} b_i$. Thus, if $e_i < e_j$ then $a_i | a_j$, and if $e_j < e_i$ then $a_j | a_i$.* ■

The pigeonhole principle tells us that if m pigeons are distributed into n pigeonholes, where $m > n$, at least two pigeons must share the same pigeonhole. In fact, if more than $2m$ pigeons are assigned to m pigeonholes, then at least three pigeons must share the same pigeonhole. Thus the pigeonhole principle can be generalized as follows.

THEOREM 3.7 **(The Generalized Pigeonhole Principle)** If m pigeons are assigned to n pigeonholes, there must be a pigeonhole containing at least $\lfloor (m-1)/n \rfloor + 1$ pigeons.

PROOF (by contradiction):

Suppose *no* pigeonhole contains more than $\lfloor (m-1)/n \rfloor$ pigeons. Then:

$$\begin{aligned} \text{maximum number of pigeons} &= n \cdot \lfloor (m-1)/n \rfloor \\ &\leq n \cdot \frac{m-1}{n} \\ &= m - 1 \end{aligned}$$

*a/b means a is a factor of b. See Section 4.2.

This contradicts our assumption that there are m pigeons. Thus, one pigeonhole must contain at least $\lfloor (m-1)/n \rfloor + 1$ pigeons. ■

This generalized version of the pigeonhole principle is illustrated in the following examples.

EXAMPLE 3.29 If we select any group of 1000 students on campus, show that at least three of them must have the same birthday.

SOLUTION:

The maximum number of days in a year is 366. Think of students as pigeons and days of the year as pigeonholes. Then, by the generalized pigeonhole principle, the minimum number of students having the same birthday is $\lfloor (1000-1)/366 \rfloor + 1 = 2 + 1 = 3$. ■

The next example provides an interesting application of the generalized version to geometry. We shall revisit it in Chapter 8.

EXAMPLE 3.30 Suppose every pair of vertices of a hexagon is joined by a line segment, which is colored red or blue. Prove that the line segments form at least one monochromatic triangle, that is, a triangle with all its sides having the same color.

PROOF:

Let the letters A through F denote the vertices of a hexagon. Five line segments (pigeons) emanate from each vertex (see Figure 3.22). Without loss of generality, consider the line segments at A. Since there are exactly two colors (pigeonholes), by the generalized pigeonhole principle, at least three of the line segments at A must be monochromatic, say, red. Suppose they are $\overline{AB}$, $\overline{AD}$, and $\overline{AF}$, indicated by the solid line segments in Figure 3.23.

Figure 3.22

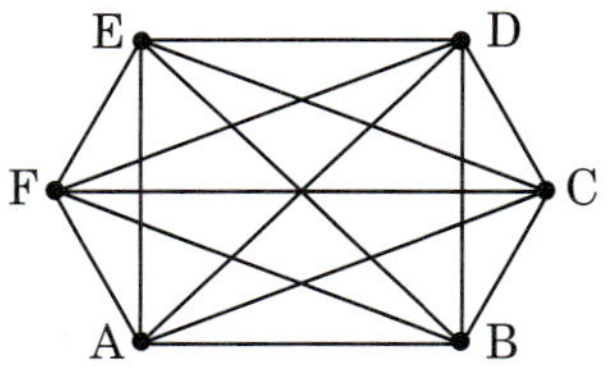

Figure 3.23

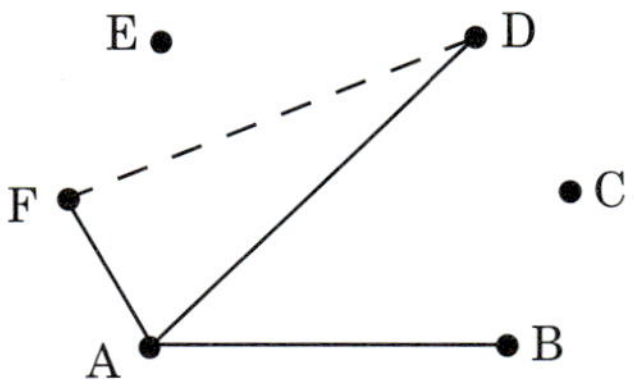

Case 1 Suppose $\overline{DF}$ is colored red. Then $\triangle$ADF is monochromatic.

Case 2 Suppose $\overline{DF}$ is not red. Then it is blue, indicated by the broken line segment in Figure 3.23. If $\overline{BD}$ is red, the ΔABD is monochromatic. If $\overline{BD}$ is not red, consider $\overline{BF}$. If $\overline{BF}$ is red, the ΔABF is a red triangle. If $\overline{BF}$ is blue, then ΔBDF is a blue triangle.

Thus the line segments form at least one monochromatic triangle. ■

Additional examples of the pigeonhole principle are presented in Section 3.7, as well as Chapters 4, 6, 7, and 8. Look for them.

Exercises 3.4

1. Show that in any 11-digit integer, at least two digits are the same.
2. Show that in any 27-letter word, at least two letters are the same.
3. Six positive integers are selected. Show that at least two of them will have the same remainder when divided by five.
4. ○ A C++ identifier contains 37 alphanumeric characters. Show that at least two characters are the same.
5. Show that in any group of eight people, at least two must have been born on the same day of the week.
6. Show that in any group of 13 people, at least two must have been born in the same month.
7. There are six matching pairs of gloves. Show that any set of seven gloves will contain a matching pair.
8. The sum of nine integers in the range 1–25 is 83. Show that one of them must be at least 10.
9. The total cost of 13 refrigerators at a department store is $12,305. Show that one refrigerator must cost at least $947.
10. Mrs. Zee has 19 skirts and would like to arrange them in a chest that has four drawers. Show that one drawer must contain at least five skirts.
11. Show that the repeating decimal $0.a_1a_2\ldots a_i\overline{b_1b_2\ldots b_j}$ is a rational number.
12. Let $n \in \mathbb{N}$. Suppose n elements are selected from the set $\{1, 2, \ldots, 2n\}$. Find a pair of integers in which one is not a factor of another integer.

Use the pigeonhole principle to prove the following.

13. If five points are chosen inside a unit square, then the distance between at least two of them is no more than $\sqrt{2}/2$.
14. Five points are chosen inside an equilateral triangle of unit side. The distance between at least two of them is no more than 1/2.

15. If 10 points are selected inside an equilateral triangle of unit side, then at least two of them are no more than 1/3 of a unit apart.

16. Let $f : X \rightarrow Y$ and $y \in Y$. Define $f^{-1}(y) = \{x \in X | f(x) = y\}$. In other words, $f^{-1}(y)$ consists of all pre-images of y. Use Example 3.1 to find $f^{-1}(y)$ for every $y \in Y$.

***17.** Prove the following alternate version of the generalized pigeonhole principle: Let $f : X \rightarrow Y$, where X and Y are finite sets, $|X| > k \cdot |Y|$, and $k \in \mathbb{N}$. Then there is an element $t \in Y$ such that $f^{-1}(t)$ contains more than k elements.

18. Prove that any set S of three integers contains at least two integers whose sum is even.
(*Hint*: Define a suitable function $f : S \rightarrow \{0, 1\}$ and use Exercise 17.)

***19.** Using the pigeonhole principle, prove that the cardinality of a finite set is unique.

3.5 Composition of Functions

Besides adding and multiplying functions, there is a very fundamental way of constructing new functions.

Consider the functions $f, g : \mathbb{R} \rightarrow \mathbb{R}$ defined by $f(x) = 2x+3$ and $g(x) = x^2$. Let x be an input into f. Then $f(x) = 2x+3$ is a real number and hence can be considered an input into g. The resulting output is $g(f(x))$ (see Figure 3.24). Thus the functions f and g can be employed to define a new function, called the **composite** of f and g, as shown in Figure 3.25.

Figure 3.24

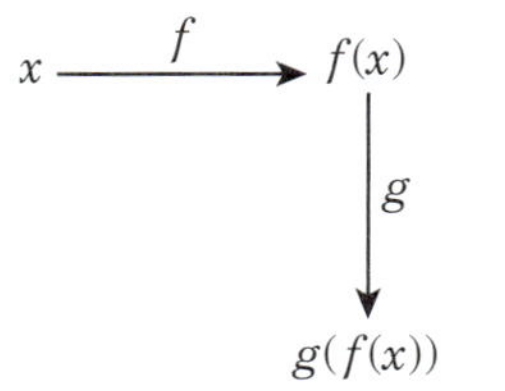

Figure 3.25

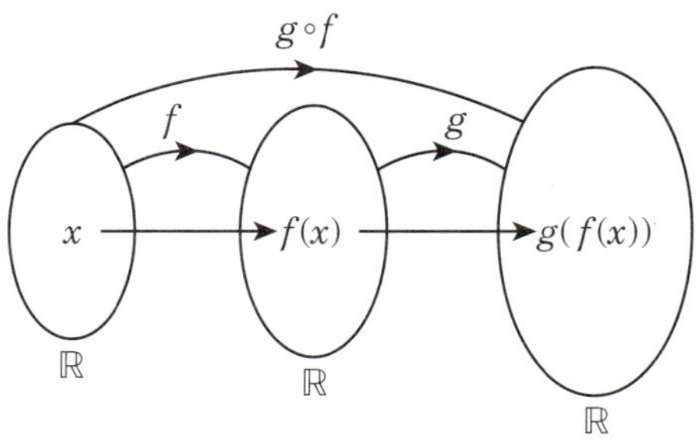

This leads us to the following definition.

Composition

Let $f : X \to Y$ and $g : Y \to Z$. The **composition** of f and g, denoted by $g \circ f$ (notice the order of the functions), is a function from X to Z, defined by $(g \circ f)(x) = g(f(x))$. Read $g \circ f$ as *g circle f* or *the composition of f and g*. [In general, dom(g) need not be the same as codom(f); all that is needed is that range $(f) \subseteq$ dom(g).]

EXAMPLE 3.31 Let $f, g : \mathbb{R} \to \mathbb{R}$[†] defined by $f(x) = 2x + 3$ and $g(x) = x^2$. Find $(g \circ f)(x)$ and $(f \circ g)(x)$.

SOLUTION:

- $$f(x) = 2x + 3$$

  Then
  $$\begin{aligned}(g \circ f)(x) &= g(f(x)) \\ &= g(2x + 3) = (2x + 3)^2\end{aligned}$$

- $$g(x) = x^2$$

  So
  $$\begin{aligned}(f \circ g)(x) &= f(g(x)) \\ &= f(x^2) = 2(x^2) + 3 \\ &= 2x^2 + 3\end{aligned}$$ ■

> It follows from Example 3.31 that, in general, $f \circ g \neq g \circ f$; in other words, composition is *not* a commutative operation. For instance, putting clothes in a washing machine and then in a dryer does not yield the same result as putting them in a dryer and then in a washing machine!

EXAMPLE 3.32 (optional) Composition is easily accomplished in computer science. To illustrate this, study the following algorithm fragment, where $x \in \mathbb{R}$:

```
if x ≤ 4 then
  x ← x + 2
else
  x ← x - 3
if x ≤ 5 then
  x ← x²
else
  x ← 2x - 1
```

Find the value of x resulting from the execution of this fragment with the initial values of $x = 2$ and $x = 5$.

[†]$f, g : X \to Y$ is an abbreviation for $f : X \to Y$ and $g : X \to Y$.

SOLUTION:

- Suppose $x = 2$. Since $2 \leq 4$, line 2 is executed and hence $x \leftarrow 4$. Then the condition in line 5 is tested. Since $4 \leq 5$, line 6 is executed. So x gets the value 16 from line 6.
- Suppose $x = 5$ initially; then line 2 is skipped. So $x \leftarrow 2$ by line 4. Since $2 \leq 5$, line 6 is executed. Therefore, $x \leftarrow 4$. ■

To see that this example employs composition, let $f(x)$ and $g(x)$ denote the functions defined by the **if-then-else** statements in the above algorithm fragment. Then:

$$f(x) = \begin{cases} x+2 & \text{if } x \leq 4 \\ x-3 & \text{if } x > 4 \end{cases} \quad \text{and} \quad g(x) = \begin{cases} x^2 & \text{if } x \leq 5 \\ 2x-1 & \text{if } x > 5 \end{cases}$$

The output resulting from the fragment is given by the composition of f and g. You may verify that $(g \circ f)\,(2) = 16$ and $(g \circ f)(5) = 4$; in fact $g \circ f$ is defined by

$$(g \circ f)(x) = \begin{cases} (x+2)^2 & \text{if } x \leq 3 \\ 2x+3 & \text{if } 3 < x \leq 4 \\ (x-3)^2 & \text{if } 4 < x \leq 8 \\ 2x-7 & \text{otherwise} \end{cases}$$

(See Exercise 54.)

A few simple properties satisfied by the composition operation follow. Their proofs are fairly straightforward; we shall prove part 3 and leave the others as exercises.

THEOREM 3.8 Let $f : X \to Y$ and $g : Y \to Z$. Then:

(1) $f \circ 1_X = f$ (2) $1_Y \circ f = f$
(3) If f and g are injective, then $g \circ f$ is injective.
(4) If f and g are surjective, then $g \circ f$ is surjective.
(5) If f and g are bijective, then $g \circ f$ is bijective.

PROOF:

(3) Let $x_1, x_2 \in X$ such that $(g \circ f)(x_1) = (g \circ f)(x_2)$. Then

$$g(f(x_1)) = g(f(x_2)), \text{ by definition.}$$

Then $$f(x_1) = f(x_2), \text{ since } g \text{ is injective.}$$

Consequently, $$x_1 = x_2, \text{ since } f \text{ is injective.}$$

Thus, if $(g \circ f)(x_1) = (g \circ f)(x_2)$, then $x_1 = x_2$, so $g \circ f$ is injective. (Exercises 44–46 provide partial converses to the results 3 through 5.) ■

Before we define the inverse of a function and discuss its properties, let us study the next example.

EXAMPLE 3.33 Let $f(x) = ax + b$ and $g(x) = \frac{x-b}{a}$ on $\mathbb{R}$, where $a \neq 0$. Find $(g \circ f)(x)$ and $(f \circ g)(x)$.

SOLUTION:
Let $x \in \mathbb{R}$. Then:

$$(1)\ (g \circ f)(x) = g(f(x)) = g(ax+b) = \frac{(ax+b)-b}{a} = x$$

$$(2)\ (f \circ g)(x) = f(g(x)) = f\left(\frac{x-b}{a}\right) = a\left(\frac{x-b}{a}\right) + b = x$$

In this example, $(g \circ f)(x) = x = (f \circ g)(x)$ for all x. That is, $g \circ f = f \circ g = 1_{\mathbb{R}}$, the identity function. In other words, one function *undoes* what the other has done. This leads to the following definition. ■

Inverse Function

Let $f: X \to Y$. Suppose there is a function $g: Y \to X$ such that $(g \circ f)(x) = x$ for every $x \in X$ and $(f \circ g)(y) = y$ for every $y \in Y$; it is called the **inverse** of f, denoted by f^{-1}; that is, $g = f^{-1}$. [*Note*: dom(f) = codom(f^{-1}) and codom(f) = dom(f^{-1}); also $f^{-1}(x) \neq 1/f(x)$.] It can be shown that the inverse of f is unique (see Exercise 52). The function f^{-1} does just the opposite of what f has done, as illustrated in Figure 3.26. A function that has an inverse is said to be **invertible**.

Figure 3.26

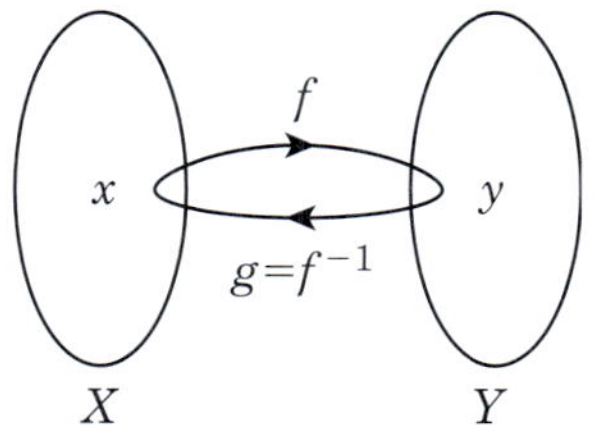

The next two examples illustrate this definition.

EXAMPLE 3.34 With the functions f and g in Figure 3.27, notice that: $(g \circ f)(a) = g(f(a)) = g(3) = a$, $(g \circ f)(b) = g(f(b)) = g(2) = b$, $(g \circ f)(c) = g(f(c)) = g(1) = c$, and $(g \circ f)(d) = g(f(d)) = g(0) = d$. Thus, $(g \circ f)(x) = x$ for every x in X and, similarly, $(f \circ g)(y) = y$ for every y in Y. So $g = f^{-1}$. ■

EXAMPLE 3.35 Consider the functions ORD : ASCII $\to \{0, 1, \ldots, 127\}$ and CHR: $\{0, 1, \ldots, 127\} \to$ ASCII. ORD(c) gives the ordinal number of the character c in ASCII, whereas CHR(n) returns the character with ordinal number n.

Figure 3.27

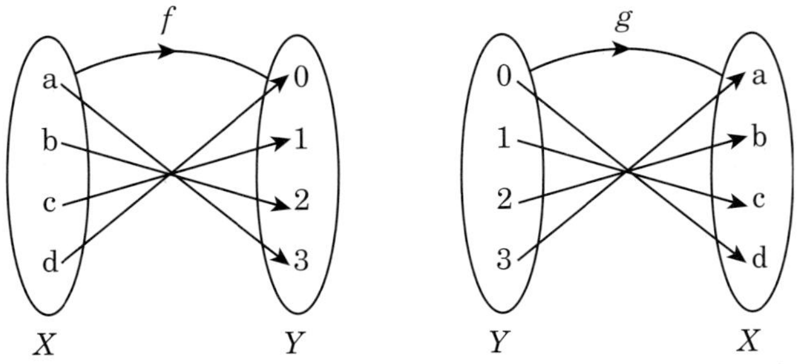

Let CH denote a character variable and n a valid ordinal number. Then CHR(ORD(CH)) = CH and ORD(CHR(n)) = n. Thus CHR and ORD are inverse functions. ■

Unfortunately, *not* every function is invertible. The next theorem gives a necessary and sufficient condition for invertibility.

THEOREM 3.9 A function $f : X \to Y$ is invertible if and only if it is bijective.

PROOF:
Suppose f is invertible. We would like to show f is bijective.

- *To prove that f is injective*:
  Let x_1 and x_2 be any two elements in X such that $f(x_1) = f(x_2)$. Since f is invertible, f^{-1} exists. Then

$$f^{-1}(f(x_1)) = f^{-1}(f(x_2))$$
$$(f^{-1} \circ f)(x_1) = (f^{-1} \circ f)(x_2)$$
$$x_1 = x_2$$

  Therefore, f is injective.

- *To prove that f is surjective:*
  Let y be any element in Y. We have to produce a suitable element x in X such that $f(x) = y$. Choose $x = f^{-1}(y)$ (see Figure 3.26). Then $f(x) = f(f^{-1}(y)) = (f \circ f^{-1})(y) = y$.
  Thus f is both injective and surjective; therefore, it is bijective.
  Conversely, suppose f is bijective. Then every element x in X is paired with a unique element y in Y and vice versa. Define a function $g : Y \to X$ as follows: $g(y)$ = the unique element x in X such that $f(x) = y$. Then $(g \circ f)(x) = g(f(x)) = g(y) = x$ and $(f \circ g)(y) = f(g(y)) = f(x) = y$. Therefore, $g = f^{-1}$ and hence f is invertible. ■

The next two examples use Theorem 3.9 to determine the invertibility of a function.

EXAMPLE 3.36 The exponential function $f : \mathbb{R} \to \mathbb{R}^+$ defined by $f(x) = 2^x$ is bijective, so it is invertible. Its inverse is the logarithmic function $g : \mathbb{R}^+ \to \mathbb{R}$ defined by $g(x) = \lg x$. ■

EXAMPLE 3.37 Let $f : \Sigma^* \to \mathbf{W}$ defined by $f(x) = \|x\|$, where Σ denotes the English alphabet. Since f is not bijective (why?), f is not invertible. ■

We close this section with a list of additional properties satisfied by the inverse of a function and leave their proofs as exercises for you to pursue.

THEOREM 3.10 Let $f : X \to Y$ and $g : Y \to Z$ be invertible functions. Then:

- $f^{-1} \circ f = 1_X$
- $f \circ f^{-1} = 1_Y$
- f^{-1} is bijective.
- $(f^{-1})^{-1} = f$
- $(g \circ f)^{-1} = f^{-1} \circ g^{-1}$ ■

Exercises 3.5

Let $f, g : \mathbb{R} \to \mathbb{R}$ be defined by $f(x) = 2x - 1$ and $g(x) = x^2 + 1$. Find:

1. $(g \circ f)(2)$ **2.** $(f \circ g)(-1)$ **3.** $(g \circ f)(x)$ **4.** $(f \circ g)(x)$

Let $f(x) = \lfloor x \rfloor$ and $g(x) = \lceil x \rceil$, where $x \in \mathbb{R}$. Compute each.

5. $(g \circ f)(-2.3)$ **6.** $(f \circ g)(-2.3)$ **7.** $(g \circ f)(-4.1)$ **8.** $(f \circ g)(-3.9)$

Let $f, g : \mathbf{W} \to \mathbf{W}$ defined by $f(x) = x \bmod 5$ and $g(x) = x \operatorname{div} 7$. Evaluate each.

9. $(g \circ f)(17)$ **10.** $(f \circ g)(23)$ **11.** $(g \circ f)(97)$ **12.** $(f \circ g)(78)$

Determine if the function g is the inverse of the corresponding function f.

13. $f(x) = x^2, x \geq 0;\ g(x) = \sqrt{x}, x \geq 0$

14. $f(x) = x^2, x \leq 0;\ g(x) = -\sqrt{x}, x \geq 0$

Define the inverse g of each function f.

15.

x	a	b	c	d
$f(x)$	4	1	3	2

16.

x	a	b	c	d
$f(x)$	b	c	d	a

Determine if the given function is invertible. If it is not invertible, explain why.

17. ORD on **Z**.

18. $f :$ ASCII $\to \mathbf{W}$ defined by $f(c) =$ ordinal number of the character c.

19. $f : \mathbf{W} \to \mathbf{W}$ defined by $f(n) = n \pmod 5$.

20. $f : \Sigma^* \to \Sigma^*$ defined by $f(w) = awa$, where $\Sigma = \{a, b, c\}$.

21. $f : S \to \mathbb{N}$ defined by $f(x) =$ decimal value of x, where S is the set of binary representations of positive integers with no leading zeros.

22. $f : \Sigma^* \to \mathbf{W}$ defined by $f(x) =$ decimal value of x, where $\Sigma = \{0,1\}$.

23. Let $f : \Sigma^n \to \mathbf{W}$ defined by $f(x) = \sum_{i=1}^{n} x_i$, where Σ^n denotes the set of words of length n over $\Sigma = \{0, 1, 2\}$ and $x = x_1x_2 \cdots x_n$. [$f(x)$ is the **weight** of x; for example, $f(10211) = 5$.]

Mark each sentence as true or false. Assume the composites and inverses are defined:

24. The composition of two injections is injective.

25. The composition of two surjections is surjective.

26. The composition of two bijections is a bijection.

27. Every function is invertible.

28. Every injective function is invertible.

29. Every invertible function is injective.

30. Every invertible function is surjective.

31. Every invertible function is bijective.

32. Every bijection is invertible.

33. The composition of two invertible functions is invertible.

Using the algorithm fragment in Example 3.32, compute the output resulting from each initial value of x.

34. -5 **35.** 0 **36.** 3 **37.** 7

Let $f : X \to Y$ and $g : Y \to Z$. Prove each.

38. $f \circ 1_X = f$ **39.** $1_Y \circ f = f$

40. If f and g are injective, then $g \circ f$ is injective.

41. If f and g are surjective, then $g \circ f$ is surjective.

42. If f and g are bijective, then $g \circ f$ is bijective.

43. The identity function 1_X is bijective.

44. If $g \circ f$ is injective, then f is injective.

45. If $g \circ f$ is surjective, then g is surjective.

46. If $g \circ f$ is bijective, then f is injective and g is surjective.

Let $f : X \to Y$ and $g : Y \to Z$ be invertible functions. Prove each.

47. $f^{-1} \circ f = 1_X$ **48.** $f \circ f^{-1} = 1_Y$

49. f^{-1} is bijective. **50.** $(f^{-1})^{-1} = f$

51. $(g \circ f)^{-1} = f^{-1} \circ g^{-1}$ ***52.** The inverse of f is unique.

53. Let $f : A \to B$, $g : B \to C$, and $h : C \to D$. Prove that $h \circ (g \circ f) = (h \circ g) \circ f$ (**associative property**).
[*Hint*: Verify that $(h \circ (g \circ f))(x) = ((h \circ g) \circ f)(x)$ for every x in A.]

○ ***54.** Let f and g denote the functions defined by the **if-then-else** statements in Example 3.31. Show that $g \circ f$ is defined as given in the example.
(*Hint*: Consider the cases $x \leq 4$ and $x > 4$, and then two subcases in each case.)

Prove each, where $X \sim Y$ implies set X is equivalent to set Y.

***55.** If $A \sim B$, then $B \sim A$ (**symmetric property**).

***56.** If $A \sim B$ and $B \sim C$, then $A \sim C$ (**transitive property**).

Let $f : X \to Y$ be bijective. Let S and T be subsets of Y. Prove each.

***57.** $f^{-1}(S \cup T) = f^{-1}(S) \cup f^{-1}(T)$ ***58.** $f^{-1}(S \cap T) = f^{-1}(S) \cap f^{-1}(T)$

3.6 Sequences and the Summation Notation

Sequences and the summation notation play a key role in the next three chapters, so we present them here.

Let a be a whole number and $X = \{a, a+1, a+2, \ldots\}$. A function s with domain X or a subset of X is called a **sequence**. Let $n \in X$. Then $s(n)$ is called a **term** of the sequence, denoted by s_n. The various terms of the sequence can be listed as s_a, s_{a+1}, $s_{a+2}, \ldots$ in increasing order of subscripts.

In particular, let $X = \mathbb{N}$. Then the terms of the sequence are:

$$s_1, s_2, s_3, \ldots, s_n, \ldots$$

$\uparrow$ general term

The nth term s_n is the **general term** of the sequence; the sequence is often denoted by $\{s_n\}_1^\infty$ or simply $\{s_n\}$. (It should be clear from the context whether the braces indicate a set or a sequence.) The general term is often used to define a sequence.

EXAMPLE 3.38 Consider the sequence $\{s_n\}$, where $s_n = 2n - 1$. The various terms of the sequence are $1, 3, 5, 7, \ldots$. Formally, the sequence is the function $s : \mathbb{N} \to \mathbb{N}$ defined by $s(n) = 2n - 1$. ■

EXAMPLE 3.39 Let a_n be the binary representation of the positive integer n with no leading zeros. The various terms of the sequence $\{a_n\}$ are 1, 10, 11, 100, 101, 110, 111, ... ■

Sequences can be classified as finite or infinite, as the next definition shows.

Finite and Infinite Sequences

A sequence is **finite** if its domain is finite; otherwise, it is **infinite**.

Thus, a finite sequence is made up of a finite number of terms, and an infinite sequence contains infinitely many terms. Both types are useful in mathematics and computer science as well.

Every word over an alphabet can be considered a finite sequence. For instance, the binary word 010110111 is a finite sequence containing nine terms. The elements of a finite language form a finite sequence; for example, the words of length ≤ 2 over the alphabet $\{a, b, c\}$ form a finite sequence, namely, λ, a, b, c, aa, ab, ac, ba, bb, bc, ca, cb, cc; on the other hand, λ, a, a^2, a^3, ... is an infinite sequence.

We now turn to the summation notation you will find very useful throughout the remainder of the book.

The Summation Notation

Often we need to work with sums of terms of number sequences $\{a_n\}$. Sums such as $a_k + a_{k+1} + \cdots + a_m$ can be written in a compact form using the **summation symbol** $\sum$, which denotes the word *sum*. The summation notation was introduced in 1772 by the brilliant French mathematician Joseph Louis Lagrange. (Recall that Σ denoted an alphabet in Chapter 2; its actual meaning should be clear from the context.)

A typical term in the above sum can be denoted by a_i, so it is the sum of the terms a_i as i runs from k to m. It is denoted by $\sum_{i=k}^{i=m} a_i$. Thus

$$\sum_{i=k}^{i=m} a_i = a_k + a_{k+1} + \cdots + a_m$$

The variable i is the **summation index**. The values k and m are the **lower** and **upper limits** of the index i. The "$i =$" above the $\sum$ is usually omitted; in fact, the indices above and below the $\sum$ are also omitted when there is no confusion. Thus

$$\sum_{i=k}^{i=m} a_i = \sum_{i=k}^{m} a_i = \sum_{k}^{m} a_i$$

For example,

$$\sum_{i=1}^{6} i = 1 + 2 + 3 + 4 + 5 + 6 = 21$$

Joseph Louis Lagrange *(1736–1813) ranks with Leonhard Euler (see Chapter 8) as one of the greatest mathematicians of the 18th century. The eldest of 11 children in a wealthy family in Turin, Italy, Lagrange was forced to pursue a profession after his father, an influential cabinet official, lost all his wealth by engaging in unsuccessful financial speculations.*

While studying the classics at the College of Turin, the 17-year-old Lagrange found his interest in mathematics kindled by an essay by the astronomer Edmund Halley on the superiority of the analytical methods of calculus over geometry in the solution of optical problems. In 1754, he began corresponding with several outstanding mathematicians in Europe. The following year, he was appointed professor of mathematics at the Royal Artillery School in Turin. Three years later, he helped to found a society that later became the Turin Academy of Sciences. While at Turin, Lagrange developed revolutionary results in the calculus of variations, mechanics, sound, and probability, winning the prestigious Grand Prix of the Paris Academy of Sciences in 1764 and 1766.

In 1766, when Euler left the Berlin Academy of Sciences, Frederick the Great wrote to Lagrange that "the greatest king in Europe" would like to have "the greatest mathematician of Europe" at his court. Accepting the invitation, Lagrange moved to Berlin to head the Academy and remained there for 20 years. When Frederick died in 1786, Lagrange moved to Paris at the invitation of Louis XVI. He was appointed professor at the École Normale and then at the École Polytechnique, where he taught until 1799. He died in Paris.

Lagrange made significant contributions to analysis, analytic mechanics, calculus, probability, and number theory, as well as helping to establish the French metric system.

$$\sum_{i=-1}^{2} i(i-1) = (-1)(-1-1) + 0(0-1) + 1(1-1) + 2(2-1) = 4$$

The index i is a dummy variable; you can use any variable as the index without affecting the value of the sum, so

$$\sum_{i=\ell}^{m} a_i = \sum_{j=\ell}^{m} a_j = \sum_{k=\ell}^{m} a_k$$

EXAMPLE 3.40 Evaluate $\sum_{i=-2}^{3} i^2$.

SOLUTION:

$$\sum_{i=-2}^{3} i^2 = (-2)^2 + (-1)^2 + 0^2 + 1^2 + 2^2 + 3^2 = 19$$ ■

The following results are extremely useful in evaluating finite sums. They can be proved using mathematical induction (Section 4.4).

THEOREM 3.11 Let $n \in \mathbf{N}$ and $c \in \mathbb{R}$. Let $a_1, a_2, \ldots,$ and $b_1, b_2, \ldots,$ be any two number sequences. Then:

- $$\sum_{i=1}^{n} c = nc \tag{3.3}$$
- $$\sum_{i=1}^{n}(ca_i) = c\left(\sum_{i=1}^{n} a_i\right) \tag{3.4}$$
- $$\sum_{i=1}^{n}(a_i + b_i) = \left(\sum_{i=1}^{n} a_i\right) + \left(\sum_{i=1}^{n} b_i\right) \tag{3.5}$$

(These results can be extended for any lower limit $k \in \mathbf{Z}$.) ■

The next example illustrates this theorem.

EXAMPLE 3.41 Evaluate $\sum_{j=1}^{2}[(5j)^3 - 2j]$.

SOLUTION:

$$\begin{aligned}
\sum_{j=-1}^{2}[(5j)^3 - 2j] &= \left(\sum_{j=-1}^{2}(5j)^3 - 2\sum_{j=-1}^{2} j\right)\\
&= 125\left(\sum_{j=-1}^{2} j^3\right) - 2\sum_{j=-1}^{2} j\\
&= 125[(-1)^3 + 0^3 + 1^3 + 2^3] - 2(-1 + 0 + 1 + 2)\\
&= 996
\end{aligned}$$
■

Indexed Summation

The summation notation can be extended to sequences with index sets I as their domains. For instance, $\sum_{i \in I} a_i$ denotes the sum of the values a_i as i runs over the various values in I.

As an example, let $I = \{0, 1, 3, 5\}$. Then $\sum_{i \in I}(2i + 1)$ represents the sum of the values of $2i + 1$, so

$$\sum_{i \in I}(2i + 1) = (2 \cdot 0 + 1) + (2 \cdot 1 + 1) + (2 \cdot 3 + 1) + (2 \cdot 5 + 1) = 22$$

Often we need to evaluate sums of the form $\sum_{P} a_{ij}$, where the subscripts i and j satisfy certain properties P. (Such summations are used in Chapters 4 and 6.)

For example, let $I = \{1, 2, 3, 4\}$. Then $\sum_{1 \le i < j \le 4} (2i + 3j)$ denotes the sum of the values of $2i + 3j$, where $1 \le i < j \le 4$. This can be abbreviated as $\sum_{i<j} (2i + 3j)$ provided the index set is obvious from the context. To find this sum, we must consider every possible pair (i,j), where $i,j \in I$ and $i < j$. Thus:

$$\begin{aligned}\sum_{i<j}(2i + 3j) &= (2 \cdot 1 + 3 \cdot 2) + (2 \cdot 1 + 3 \cdot 3) + (2 \cdot 1 + 3 \cdot 4) \\ &\quad + (2 \cdot 2 + 3 \cdot 3) + (2 \cdot 2 + 3 \cdot 4) + (2 \cdot 3 + 3 \cdot 4) \\ &= 80\end{aligned}$$

EXAMPLE 3.42 Evaluate $\sum_{\substack{d \ge 1 \\ d|6}} d$, where $d|6$ indicates that d is a factor of 6.

SOLUTION:

$$\begin{aligned}\sum_{\substack{d \ge 1 \\ d|6}} d &= \text{sum of positive integers } d, \text{ where } d \text{ is a factor of 6.} \\ &= \text{sum of positive factors of 6} \\ &= 1 + 2 + 3 + 6 = 12\end{aligned}$$

■

Multiple summations arise often in mathematics. They are evaluated in the right-to-left fashion. For example, the double summation $\sum_i \sum_j a_{ij}$ is evaluated as $\sum_i \left(\sum_j a_{ij}\right)$ and the triple summation $\sum_i \sum_j \sum_k a_{ijk}$ as $\sum_i \left[\sum_j \left(\sum_k a_{ijk}\right)\right]$.

We close this section with an example of a double summation.

EXAMPLE 3.43 Evaluate $\sum_{i=-1}^{1} \sum_{j=0}^{2} (2i + 3j)$.

SOLUTION:

$$\begin{aligned}\sum_{i=-1}^{1} \sum_{j=0}^{2}(2i + 3j) &= \sum_{i=-1}^{1} \left[\sum_{j=0}^{2}(2i + 3j)\right] \\ &= \sum_{i=-1}^{1} [(2i + 3 \cdot 0) + (2i + 3 \cdot 1) + (2i + 3 \cdot 2)] \\ &= \sum_{i=-1}^{1} (6i + 9) \\ &= [6 \cdot (-1) + 9] + (6 \cdot 0 + 9) + (6 \cdot 1 + 9) \\ &= 27\end{aligned}$$

■

Exercises 3.6

Evaluate each sum.

1. $\sum_{i=1}^{6} i$ **2.** $\sum_{k=0}^{4} (3+k)$ **3.** $\sum_{j=0}^{4} (j-1)$

4. $\sum_{i=-1}^{4} 3$ **5.** $\sum_{n=0}^{4} (3n-2)$ **6.** $\sum_{j=-2}^{2} j(j-2)$

7. $\sum_{k=-2}^{4} 3k$ **8.** $\sum_{k=-2}^{3} 3(k^2)$ **9.** $\sum_{k=-1}^{3} (3k)^2$

10. $\sum_{k=1}^{5} (3-2k)k$ **11.** $\sum_{j=-1}^{4} (j-2)^2$ **12.** $\sum_{i=0}^{5} (0.1)^i(0.9)^{5-i}$

Rewrite each sum using the summation notation.

13. $1+3+5+\cdots+23$ **14.** $3^1+3^2+\cdots+3^{10}$

15. $1\cdot 2+2\cdot 3+\cdots+11\cdot 12$ **16.** $1(1+2)+2(2+2)+\cdots+5(5+2)$

Determine if each is true or false.

17. $\sum_{i=m}^{n} i = \sum_{i=m}^{n} (n+m-i)$ **18.** $\sum_{i=m}^{n} x^i = \sum_{i=m}^{n} x^{n+m-i}$

19. Sums of the form $S = \sum_{i=m+1}^{n} (a_i - a_{i-1})$ are **telescoping** sums. Show that $S = a_n - a_m$.

20. Using Exercise 19 and the identity $\frac{1}{i(i+1)} = \frac{1}{i} - \frac{1}{i+1}$, derive a formula for $\sum_{i=1}^{n} \frac{1}{i(i+1)}$.

21. Using Exercise 19 and the identity $(i+1)^2 - i^2 = 2i+1$, find a formula for $\sum_{i=1}^{n} i$.

Evaluate each sum, where δ_{ij} is defined as follows.

$$\delta_{ij} = \begin{cases} 1 & \text{if } i = j \\ 0 & \text{otherwise} \end{cases}$$

[δ_{ij} is called **Kronecker's delta**, after the German mathematician Leopold Kronecker (1823–1891).]

22. $\sum_{i=1}^{5}\sum_{j=1}^{6}(2i+3j)$

23. $\sum_{i=1}^{3}\sum_{j=1}^{i}(j+3)$

24. $\sum_{j=1}^{6}\sum_{i=1}^{5}(2i+3j)$

25. $\sum_{i=1}^{6}\sum_{j=1}^{5}(i^2-i)$

26. $\sum_{i=1}^{5}\sum_{j=1}^{6}(i^2-j+1)$

27. $\sum_{j=1}^{6}\sum_{i=1}^{5}(i^2-j+1)$

28. $\sum_{i=1}^{5}\sum_{j=1}^{5}\delta_{ij}$

29. $\sum_{i=1}^{3}\sum_{j=1}^{5}(2+3\delta_{ij})$

30. $\sum_{i=1}^{6}\sum_{j=1}^{7}(i^2-3i+\delta_{ij})$

Just as $\sum$ is used to denote sums, the product $a_k a_{k+1} \dots a_m$ is denoted by $\prod_{i=k}^{m} a_i$. The **product symbol** Π is the Greek capital letter *pi*. For example, $n! = \prod_{i=1}^{n} i$. Evaluate each product.

31. $\prod_{i=1}^{3}(i+1)$ **32.** $\prod_{j=3}^{5}(j^2+1)$ **33.** $\prod_{j=-5}^{50} 1$ **34.** $\prod_{k=0}^{50}(-1)^k$

Evaluate each sum and product, where p is a prime and $I = \{1, 2, 3, 5\}$.

35. $\sum_{k=0}^{3} k!$

36. $\sum_{p\le 10} p$

37. $\prod_{p\le 10} p$

38. $\prod_{i\in I}(3i-1)$

39. $\sum_{\substack{d\ge 1\\ d\mid 12}} d$

40. $\sum_{\substack{d\ge 1\\ d\mid 12}} \left(\frac{12}{d}\right)$

41. $\sum_{\substack{d\ge 1\\ d\mid 18}} 1$

42. $\sum_{p\le 25} 1$

43. $\prod_{\substack{i,j\in I\\ i<j}} (i+2j)$

44. $\prod_{\substack{i,j\in I\\ i\le j}} i^j$

45. $\sum_{\substack{i,j\in I\\ i\mid j}} (2^i+3^j)$

46. $\sum_{j=1}^{4}(3^j-3^{j-1})$

Expand each.

47. $\sum_{i=1}^{3} a_{ij}$

48. $\sum_{j=1}^{2} a_{ij}$

49. $\sum_{i=1}^{3}\sum_{j=1}^{2} a_{ij}$

50. $\sum_{j=1}^{2}\sum_{i=1}^{3} a_{ij}$

51. $\sum_{1\le i<j\le 3}(a_i+a_j)$

52. $\sum_{1\le i\le j<3}(a_i+a_j)$

Arthur Cayley *(1821–1895) was born in Richmond, England. At 14 he entered King's College, London. His teachers, recognizing his superb mathematical talents, encouraged him to be a mathematician.*

At 17, Cayley entered Trinity College, Cambridge, where he was rated to be in a class by himself, "above the first." By age 25, he had published 25 papers, the first one at age 20.

In 1846, he left his position at Cambridge to study law and became a successful lawyer. Feeling unfulfilled, he left the law after 14 years, although during this period he had published more than 200 papers.

In 1863 Cayley rejoined the faculty at Cambridge University. He pursued his mathematical interests, until his death.

James Joseph Sylvester *(1814–1897) attended Cambridge University, which for several years denied him the degrees he earned, because he was Jewish.*

At 24, he became professor of natural philosophy at the University of London. Three years later, he taught at the University of Virginia for a year and then returned to England to become an actuary while continuing his mathematical investigations.

Sylvester was professor of mathematics at Johns Hopkins University from 1876 to 1883. In 1878 he founded The American Journal of Mathematics.

53. $\sum_{1 \le i < j \le 3} |A_i \cap A_j|$

54. $\sum_{1 \le i < j < k \le 3} |A_i \cap A_j \cap A_k|$

3.7 Matrices

Matrices were discovered jointly by two English mathematicians, Arthur Cayley and James Joseph Sylvester. Matrix notation allows data to be summarized in a very compact form and manipulated in a convenient way.

The sports pages of every newspaper provide fine examples of matrices. For example, during the National Hockey League 2001–2002 regular season, the Boston Bruins won 43 games, lost 24 games, tied 6 games, and had 9 overtime losses; the New York Rangers won 36 games, lost 38 games, tied 4 games, and had 4 overtime losses; the Detroit Red Wings won 51 games, lost 17 games, tied 10 games, and had 4 overtime losses; and the Los Angeles Kings won 40 games, lost 27 games, tied 11 games, and had 4 overtime losses. These data can be arranged in a compact form:

	won	lost	tied	overtime loss
Boston	43	24	6	9
New York	36	38	4	4
Detroit	51	17	10	4
Los Angeles	40	27	11	4

Suppose you know that the first row refers to Boston, the second row to New York, and so on, and the first column refers to the number of wins, the second column to the number of ties, and so on. Then the row and column headings can be deleted. Call the resulting arrangement A:

$$A = \begin{bmatrix} 43 & 24 & 6 & 9 \\ 37 & 11 & 36 & 4 \\ 39 & 10 & 35 & 4 \\ 51 & 7 & 26 & 4 \end{bmatrix}$$

Such a rectangular arrangement of numbers is called a matrix. More generally, we have the following definition.

Matrix

A **matrix** is a rectangular arrangement of numbers enclosed by brackets. A matrix with m rows and n columns is an $m \times n$ (read m by n) matrix, its **size** being $m \times n$. If $m = 1$, it is a **row vector**; and if $n = 1$, it is a **column vector**. If $m = n$, it is a **square matrix of order** n. Each number in the arrangement is an **element** of the matrix. Matrices are denoted by uppercase letters.

For example, let

$$A = \begin{bmatrix} 3 & -5 & 6 \\ 1 & 0 & 4 \end{bmatrix} \quad \text{and} \quad B = \begin{bmatrix} 1 & 0 & -3 \\ -6 & 4 & -2 \\ 2 & 7 & -1 \end{bmatrix}$$

A is a 2×3 matrix, whereas B is a square matrix of order 3. The elements of the row vector $[0\ 3\ {-7}]$ are 0, 3, and -7.

The double subscript notation is extremely useful in naming the elements of an $m \times n$ matrix A. Let a_{ij} denote the element in row i and column j of A. Then the matrix has the form

$$A = \begin{bmatrix} a_{11} & a_{12} & \cdots & a_{1j} & \cdots & a_{1n} \\ a_{21} & a_{22} & \cdots & a_{2j} & \cdots & a_{2n} \\ \cdot & \cdot & \cdots & \cdot & \cdots & \cdot \\ a_{i1} & a_{i2} & \cdots & a_{ij} & \cdots & a_{in} \\ \cdot & \cdot & \cdots & \cdot & \cdots & \cdot \\ a_{m1} & a_{m2} & \cdots & a_{mj} & \cdots & a_{mn} \end{bmatrix} \longleftarrow \text{row } i$$

$$\uparrow\!\_\_\_\_ \quad \text{column } j$$

For convenience, it is abbreviated as $A = (a_{ij})_{m \times n}$, or simply (a_{ij}) if the size is clear from the context.

How do we determine if two matrices are equal? This is answered by the next definition.

Equality of Matrices

Two matrices $A = (a_{ij})$ and $B = (b_{ij})$ are **equal** if they have the same size and $a_{ij} = b_{ij}$ for every i and j. For example, if

$$\begin{bmatrix} 1 & x & -3 \\ 2 & 0 & y \end{bmatrix} = \begin{bmatrix} 1 & 0 & -3 \\ z & 0 & -1 \end{bmatrix}$$

then $x = 0, y = 1$, and $z = 2$.

The following definition presents two special matrices.

Zero and Identity Matrices

If every element of a matrix is zero, then it is a **zero matrix**, denoted by O.

Let $A = (a_{ij})_{n\times n}$. Then the elements $a_{11}, a_{22}, \ldots, a_{nn}$ form the **main diagonal** of the matrix A. Suppose

$$a_{ij} = \begin{cases} 1 & \text{if } i = j \\ 0 & \text{otherwise} \end{cases}$$

Then A is the **identity matrix** of order n; it is denoted by I_n, or I when there is no ambiguity.

For example, $[0\ 0\ 0]$ and $\begin{bmatrix} 0 & 0 \\ 0 & 0 \end{bmatrix}$ are zero matrices.

$A = (a_{ij})_{n\times n}$ is the identity matrix I_n if every element on its main diagonal is 1, and every element above and below it is 0. For example, $\begin{bmatrix} 1 & 0 \\ 0 & 1 \end{bmatrix}$ is the identity matrix of order 2, namely, I_2.

Just as propositions and sets can be combined to construct new propositions and new sets, matrices also can be combined to produce new matrices. The various matrix operations are presented and illustrated below.

Matrix Addition

The **sum** of the matrices $A = (a_{ij})_{m\times n}$ and $B = (b_{ij})_{m\times n}$ is defined by $A+B = (a_{ij} + b_{ij})_{m\times n}$. (We can add only matrices of the same size.)

EXAMPLE 3.44 Let

$$A = \begin{bmatrix} 2 & -3 & 7 \\ 0 & 1 & 1 \end{bmatrix} \quad \text{and} \quad B = \begin{bmatrix} 1 & 5 & 0 \\ 2 & 0 & -1 \end{bmatrix}$$

Then

$$A + B = \begin{bmatrix} 2+(-1) & (-3)+5 & 7+0 \\ 0+2 & 1+0 & 1+(-1) \end{bmatrix} = \begin{bmatrix} 1 & 2 & 7 \\ 2 & 1 & 0 \end{bmatrix}$$ ■

Negative of a Matrix

The **negative** or (**additive inverse**) of a matrix $A = (a_{ij})$, denoted by $-A$, is defined by $-A = (-a_{ij})$.

For instance, the negative of

$$A = \begin{bmatrix} 2 & 3 & -4 \\ 0 & -5 & 6 \end{bmatrix} \quad \text{is} \quad -A = \begin{bmatrix} -2 & -3 & 4 \\ 0 & 5 & -6 \end{bmatrix}$$

You may verify that $A + (-A) = O$.

Matrix Subtraction

The **difference** $A - B$ of the matrices $A = (a_{ij})_{m\times n}$ and $B = (b_{ij})_{m\times n}$ is defined by $A - B = (a_{ij} - b_{ij})_{m\times n}$. (We can subtract only matrices of the same size.)

For example, using the matrices A and B in Example 3.44,

$$A - B = \begin{bmatrix} 2-(-1) & (-3)-5 & 7-0 \\ 0-2 & 1-0 & 1-(-1) \end{bmatrix} = \begin{bmatrix} 3 & -8 & 7 \\ -2 & 1 & 2 \end{bmatrix}$$

The next example introduces us to the fourth matrix operation.

EXAMPLE 3.45 Suppose you bought 12 coconut donuts, 15 butternut donuts, and 6 cinnamon donuts from shop I, and you bought 9 coconut donuts, 12 butternut donuts, and 16 cinnamon donuts from shop II. Then the number of donuts of each kind you bought from each shop is given by the matrix

$$A = \begin{array}{l} \\ \text{shop I} \\ \text{shop II} \end{array} \begin{array}{c} \begin{array}{ccc} \text{coconut} & \text{butternut} & \text{cinnamon} \end{array} \\ \begin{bmatrix} 12 & 15 & 6 \\ 9 & 12 & 16 \end{bmatrix} \end{array}$$

Suppose each donut costs 75¢. Then the cost of each type of donut at each shop is obtained by multiplying each entry of A by 75. The resulting matrix is denoted by $75A$. Thus

$$75A = \begin{bmatrix} 75\cdot 12 & 75\cdot 15 & 75\cdot 6 \\ 75\cdot 9 & 75\cdot 12 & 75\cdot 16 \end{bmatrix} = \begin{bmatrix} 900 & 1125 & 450 \\ 675 & 900 & 1200 \end{bmatrix} \quad \blacksquare$$

This example leads to the next definition.

Scalar Multiplication

Let $A = (a_{ij})$ be any matrix and k any real number (called a **scalar**). Then $kA = (ka_{ij})$.

The fundamental properties of the various matrix operations are stated in the following theorem. We shall prove two of them, and leave the others as routine exercises.

THEOREM 3.12 Let A, B, and C be any $m \times n$ matrices, O the $m \times n$ zero matrix, and c and d any real numbers. Then:

- $A + B = B + A$
- $A + (B + C) = (A + B) + C$
- $A + O = A = O + A$
- $A + (-A) = O = (-A) + A$
- $(-1)A = -A$
- $c(A + B) = cA + cB$
- $(c + d)A = cA + dA$
- $(cd)A = c(dA)$

PROOF:
Let A $= (a_{ij})_{m\times n}$.

- *To prove that* $A + (-A) = O = (-A) + A$:

$$\begin{aligned} A + (-A) &= (a_{ij})_{m\times n} + (-a_{ij})_{m\times n} && \text{negative of } A \\ &= (a_{ij} + (-a_{ij}))_{m\times n} && \text{matrix addition} \\ &= (0)_{m\times n} && a_{ij} + (-a_{ij}) = 0 \\ &= O && \text{zero matrix} \end{aligned}$$

Similarly, $(-A) + A = O$. Thus $A + (-A) = O = (-A) + A$.

- *To prove that* $(c + d)A = cA + dA$:

$$\begin{aligned} (c + d)A &= (c + d)(a_{ij})_{m\times n} && \text{definition of } A \\ &= ((c + d)a_{ij})_{m\times n} && \text{scalar multiplication} \\ &= (ca_{ij} + da_{ij})_{m\times n} && \text{dist. prop. of numbers} \\ &= (ca_{ij})_{m\times n} + (da_{ij})_{m\times n} && \text{matrix addition} \\ &= c(a_{ij})_{m\times n} + d(a_{ij})_{m\times n} && \text{scalar multiplication} \\ &= cA + dA && \text{definition of } A \end{aligned}$$

This concludes the proofs. ■

Before we define matrix multiplication, let us study the next example.

EXAMPLE 3.46 *Discount Gas* sells regular, unleaded, and premium gasoline at two gasoline stations X and Y. Matrix A shows the price (in dollars) of a gallon of each kind of gasoline; matrix B, the average number of gallons sold at each location:

$$\begin{array}{cccc} & \text{regular} & \text{unleaded} & \text{premium} \\ A = & [2.50 & 2.75 & 3.00] \end{array}$$

$$B = \begin{array}{c} \text{regular} \\ \text{unleaded} \\ \text{premium} \end{array} \overset{\begin{array}{cc} X \quad & \quad Y \end{array}}{\begin{bmatrix} 3000 & 3500 \\ 4000 & 3750 \\ 1500 & 2000 \end{bmatrix}}$$

SOLUTION:
Notice that:

$$\begin{aligned} \text{Revenue from location X} &= 2.50(3000) + 2.75(4000) + 3.00(1500) \\ &= \$23{,}000.00 \\ \text{Revenue from location Y} &= 2.50(3500) + 2.75(3750) + 3.00(2000) \\ &= \$25{,}062.50 \end{aligned}$$

These two values can be used to form the matrix

$$\begin{array}{cc} X & Y \\ [23{,}000.00 & 25{,}062.50] \end{array}$$

Each of its elements can be obtained by multiplying each element of A by the corresponding element in each column of B and adding them up, as shown below:

$$\begin{aligned} &[2.50 \quad 2.75 \quad 3.00] \begin{bmatrix} 3000 & 3500 \\ 4000 & 3750 \\ 1500 & 2000 \end{bmatrix} \\ &= 2.50(3000) + 2.75(4000) + 3.00(1500) \\ &= 23{,}000.00 \end{aligned}$$

$$\begin{aligned} &[2.50 \quad 2.75 \quad 3.00] \begin{bmatrix} 3000 & 3500 \\ 4000 & 3750 \\ 1500 & 2000 \end{bmatrix} \\ &= 2.50(3500) + 2.75(3750) + 3.00(2000) \\ &= 25{,}062.50 \end{aligned}$$

The matrix [23,000.00 25,062.50] is the **product** of the matrices A and B, denoted by AB. Thus

$$AB = \begin{bmatrix} 2.50 & 2.75 & 3.00 \end{bmatrix} \begin{bmatrix} 3000 & 3500 \\ 4000 & 3750 \\ 1500 & 2000 \end{bmatrix} = \begin{bmatrix} 23{,}000.000 & 25{,}062.50 \end{bmatrix}$$

More generally, we define the product of two matrices as follows. ■

Matrix Multiplication

The **product** AB of the matrices $A = (a_{ij})_{m\times n}$ and $B = (b_{ij})_{n\times p}$ is the matrix $C = (c_{ij})_{m\times p}$, where c_{ij} is the sum of the products of the corresponding elements in row i of A and column j of B, as shown below:

$$\begin{bmatrix} a_{11} & a_{12} & \cdots & a_{1n} \\ a_{21} & a_{22} & \cdots & a_{2n} \\ \cdot & \cdot & \cdots & \cdot \\ a_{i1} & a_{i2} & \cdots & a_{in} \\ \cdot & \cdot & \cdots & \cdot \\ a_{m1} & a_{m2} & \cdots & a_{mn} \end{bmatrix} \begin{bmatrix} b_{11} & b_{12} & \cdot & b_{1j} & \cdot & b_{1p} \\ b_{21} & b_{22} & \cdot & b_{2j} & \cdot & b_{2p} \\ \cdot & \cdot & \cdots & \cdot & \cdot & \cdot \\ b_{i1} & b_{i2} & \cdot & b_{ij} & \cdot & b_{ip} \\ \cdot & \cdot & \cdots & \cdot & \cdot & \cdot \\ b_{n1} & b_{n2} & \cdot & b_{nj} & \cdot & b_{np} \end{bmatrix} = \begin{bmatrix} & \cdot & \\ & \cdot & \\ \cdots & c_{ij} & \cdots \\ & \cdot & \\ & \cdot & \end{bmatrix}$$

where $c_{ij} = a_{i1}b_{1j} + a_{i2}b_{2j} + \cdots + a_{in}b_{nj} = \sum_{k=1}^{n} a_{ik}b_{kj}$.

> The product $C = AB$ is defined only if the number of columns in A equals the number of rows in B. The size of the product is $m \times p$.

The next example illustrates this definition.

EXAMPLE 3.47 Let

$$A = \begin{bmatrix} 1 & -2 & 3 \\ 0 & 4 & -1 \end{bmatrix} \quad \text{and} \quad B = \begin{bmatrix} 3 & -2 \\ 0 & 1 \\ -1 & 0 \end{bmatrix}$$

Find AB and BA, if defined.

SOLUTION:

Since the number of columns of A equals the number of rows of B, the product AB is defined. Furthermore, the size of AB is 2×2:

$$AB = \begin{bmatrix} 1 & -2 & 3 \\ 0 & 4 & -1 \end{bmatrix} \begin{bmatrix} 3 & -2 \\ 0 & 1 \\ -1 & 0 \end{bmatrix}$$

$$= \begin{bmatrix} 1 \cdot 3 + (-2) \cdot 0 + 3 \cdot (-1) & 1 \cdot (-2) + (-2) \cdot 1 + 3 \cdot 0 \\ 0 \cdot 3 + 4 \cdot 0 + (-1) \cdot (-1) & 0 \cdot (-2) + 4 \cdot 1 + (-1) \cdot 0 \end{bmatrix}$$

$$= \begin{bmatrix} 0 & -4 \\ 1 & 4 \end{bmatrix}$$

The product BA is also defined (why?) and its size is 3×3:

$$BA = \begin{bmatrix} 3 & -2 \\ 0 & 1 \\ -1 & 0 \end{bmatrix} \begin{bmatrix} 1 & -2 & 3 \\ 0 & 4 & -1 \end{bmatrix}$$

$$= \begin{bmatrix} 3 \cdot 1 + (-2) \cdot 0 & 3 \cdot (-2) + (-2) \cdot 4 & 3 \cdot 3 + (-2) \cdot (-1) \\ 0 \cdot 1 + 1 \cdot 0 & 0 \cdot (-2) + 1 \cdot 4 & 0 \cdot 3 + 1 \cdot (-1) \\ (-1) \cdot 1 + 0 \cdot 0 & (-1) \cdot (-2) + 0 \cdot 4 & (-1) \cdot 3 + 0 \cdot (-1) \end{bmatrix}$$

$$= \begin{bmatrix} 3 & -14 & 11 \\ 0 & 4 & -1 \\ -1 & 2 & -3 \end{bmatrix}$$

You may notice that $AB \neq BA$. ■

We can use the definition of matrix multiplication to develop an algorithm to find the product of two matrices in an obvious way, as Algorithm 3.1 shows.

```
Algorithm product (A,B,C)
(* Let A = (aij)m×n and B = (bij)n×p. This algorithm shows
   how to find their product C = (cij)m×p. *)
   Begin (* product *)
     for i = 1 to m do
       for j = 1 to p do
       begin (* for j *)
         cij ← 0 (* initialize *)
         for k = 1 to n do
           cij ← cij + aikbkj (* update cij *)
       endfor
   End (* product *)
```

Algorithm 3.1

The fundamental properties of matrix multiplication are stated in the next theorem. They can be proved without much difficulty using the summation notation.

THEOREM 3.13 Let A, B, and C be three matrices. Then:

(1) $A(BC) = (AB)C$ (2) $AI = A = IA$
(3) $A(B+C) = AB + AC$ (4) $(A+B)C = AC + BC$

provided the indicated sums and products are defined. ■

We close this section with an example to illustrate part 3 of Theorem 3.13.

EXAMPLE 3.48 Let

$$A = \begin{bmatrix} 2 & -3 \\ 5 & 0 \end{bmatrix}, \quad B = \begin{bmatrix} 1 & 0 & -1 \\ 2 & -3 & 5 \end{bmatrix}, \quad \text{and} \quad C = \begin{bmatrix} 0 & -2 & 1 \\ -3 & 0 & 4 \end{bmatrix}$$

Show that $A(B+C) = AB + AC$.

SOLUTION:
First notice that both B and C are the same size, so $B+C$ is defined and is of size 2×3. Furthermore, since A is 2×2 and $B+C$ is 2×3, $A(B+C)$ is defined. Similarly, $AB + AC$ is also defined.

$$B + C = \begin{bmatrix} 1 & 0 & -1 \\ 2 & -3 & 5 \end{bmatrix} + \begin{bmatrix} 0 & -2 & 1 \\ -3 & 0 & 4 \end{bmatrix} = \begin{bmatrix} 1 & -2 & 0 \\ -1 & -3 & 9 \end{bmatrix}$$

$$A(B+C) = \begin{bmatrix} 2 & -3 \\ 5 & 0 \end{bmatrix}\begin{bmatrix} 1 & -2 & 0 \\ -1 & -3 & 9 \end{bmatrix} = \begin{bmatrix} 5 & 5 & -27 \\ 5 & -10 & 0 \end{bmatrix}$$

$$AB = \begin{bmatrix} 2 & -3 \\ 5 & 0 \end{bmatrix}\begin{bmatrix} 1 & 0 & -1 \\ 2 & -3 & 5 \end{bmatrix} = \begin{bmatrix} -4 & 9 & -17 \\ 5 & 0 & -5 \end{bmatrix}$$

$$AC = \begin{bmatrix} 2 & -3 \\ 5 & 0 \end{bmatrix}\begin{bmatrix} 0 & -2 & 1 \\ -3 & 0 & 4 \end{bmatrix} = \begin{bmatrix} 9 & -4 & -10 \\ 0 & -10 & 5 \end{bmatrix}$$

$$AB + AC = \begin{bmatrix} -4 & 9 & -17 \\ 5 & 0 & -5 \end{bmatrix}\begin{bmatrix} 9 & -4 & -10 \\ 0 & -10 & 5 \end{bmatrix} = \begin{bmatrix} 5 & 5 & -27 \\ 5 & -10 & 0 \end{bmatrix}$$

$$= A(B+C)$$ ■

Exercises 3.7

Solve the following equations.

1. $\begin{bmatrix} x-1 & 2 & 0 \\ 0 & y+3 & 4 \\ -3 & 1 & z+2 \end{bmatrix} = \begin{bmatrix} -2 & 2 & 0 \\ 0 & -1 & 4 \\ -3 & 1 & -2 \end{bmatrix}$

2. $\begin{bmatrix} x-y & -1 & 0 \\ -3 & y-z & 2 \\ 4 & -5 & z-x \end{bmatrix} = \begin{bmatrix} 3 & -1 & 0 \\ -3 & -4 & 2 \\ 4 & -5 & 1 \end{bmatrix}$

Find the additive inverse of each matrix.

3. $\begin{bmatrix} 2 & -3 \\ 0 & 4 \end{bmatrix}$ **4.** $\begin{bmatrix} 1 & -2 & 3 \\ 3 & 3 & -1 \end{bmatrix}$ **5.** $\begin{bmatrix} 0 & -3 & -2 \\ 1 & -2 & 4 \\ 2 & -5 & 6 \end{bmatrix}$

Let $A = \begin{bmatrix} 1 & 0 & -1 \\ 0 & 2 & 3 \end{bmatrix}$, $B = \begin{bmatrix} 0 & -2 & 5 \\ 0 & 0 & 1 \end{bmatrix}$, and $C = \begin{bmatrix} -3 & 0 & 0 \\ 0 & 1 & 2 \end{bmatrix}$. Find each.

6. $A - B$ **7.** $B + C$ **8.** $A + 2C$ **9.** $-2B$

10. $2B - C$ **11.** $2A + 3B$ **12.** $3B - 2C$ **13.** $3A + (-2)B$

Let A be an $m \times n$ matrix, B a $p \times q$ matrix, and C an $r \times s$ matrix. Under what conditions is each defined? Find the size of each when defined. (*Note*: A^2 means AA.)

14. $A + B$ **15.** $B - C$ **16.** BC **17.** A^2

18. $A(B + C)$ **19.** $AB - AC$ **20.** $AB + C$ **21.** $A(BC)$

22. A team in the NHL earns 2 points, 1 point, or 0 points for a win, tie or overtime loss, or a loss, respectively. Using matrices, find the number of points earned by each team listed at the beginning of this section.

A summer vacation lodge in sunny California expects four guests: A, B, C, and D. They plan to stay at the lodge for 7, 14, 21, and 28 days, respectively. Each of them has diabetes. Since the nearest drugstore is several miles away, the manager of the lodge decides to store three different types of insulin — semi-lente, lente, and ultra — needed by these guests. Their daily insulin requirements are summarized in Table 3.2.

Table 3.2

	Guests			
Insulin	**A**	**B**	**C**	**D**
Semi-lente	25	40	35	0
Lente	20	0	15	15
Ultra	20	0	30	40

Each gram of insulin of the three types costs 10, 11, and 12 cents, respectively[†]. Using matrices, compute each:

[†]Based on R. F. Baum, "Insulin Requirements as a Linear Process in Time," in *Some Mathematical Models in Biology*, R. M. Thrall, ed., The University of Michigan Press, Ann Arbor, MI, 1967, pp. 0L2.1–0L2.4.

23. The number of grams of each type of insulin needed.

24. The total cost of the insulin.

25. The insulin requirements if the guests decide to stay an additional 3, 5, 8, and 13 days, respectively.

26. The insulin requirements if the guests decide to stay three times their original time.

Let A, B, and C be any $m \times n$ matrices, O the $m \times n$ zero matrix, and c and d any real numbers. Prove each (see Theorem 3.12).

27. $A + B = B + A$

28. $A + (B + C) = (A + B) + C$

29. $A + O = A = O + A$

30. $(-1)A = -A$

31. $c(A + B) = cA + cB$

32. $(cd)A = c(dA)$

Let A, B, and C be any square matrices of order 2. Prove each.

33. $A(BC) = (AB)C$

34. $A(B + C) = AB + AC$

35. $(A + B)C = AC + BC$

The **transpose** of a matrix $A = (a_{ij})_{m\times n}$, denoted by A^{T}, is defined as $A^{\mathrm{T}} = (a_{ji})_{n\times m}$. Find the transpose of each.

36. $\begin{bmatrix} 1 & 2 & 3 \\ 2 & 0 & -1 \\ -2 & 1 & 0 \end{bmatrix}$

37. $\begin{bmatrix} a & b & c \\ d & e & f \\ f & g & h \end{bmatrix}$

38. A square matrix A is **symmetric** if $A^{\mathrm{T}} = A$. What can you say about the elements of a symmetric matrix A?

39. Let A be a square matrix. Prove that $(A^{\mathrm{T}})^{\mathrm{T}} = A$.

Let A, B, and C be square matrices of order 2. Prove each.

40. $(A + B)^{\mathrm{T}} = A^{\mathrm{T}} + B^{\mathrm{T}}$

41. $(AB)^{\mathrm{T}} = B^{\mathrm{T}}A^{\mathrm{T}}$

42. $(AA^{\mathrm{T}})^{\mathrm{T}} = AA^{\mathrm{T}}$

43. $(ABC)^{\mathrm{T}} = C^{\mathrm{T}}B^{\mathrm{T}}A^{\mathrm{T}}$

A square matrix A of order n is **invertible** if there is a matrix B such that $AB = I_n = BA$. Then B is the **inverse** of A, denoted by A^{-1}. In Exercises 44 and 45, verify that $B = A^{-1}$. Assume that $k = ad - bc \neq 0$.

44. $A = \begin{bmatrix} a & b \\ c & d \end{bmatrix}$, $B = \dfrac{1}{k}\begin{bmatrix} d & -b \\ -c & a \end{bmatrix}$

45. $A = \begin{bmatrix} 1 & -2 & 0 \\ 3 & 1 & -1 \\ 1 & 2 & -3 \end{bmatrix}, B = \frac{1}{17}\begin{bmatrix} 1 & 6 & -2 \\ -8 & 3 & -1 \\ -5 & 4 & -7 \end{bmatrix}$

Find each product.

46. $\begin{bmatrix} 1 & 2 \\ 2 & 3 \end{bmatrix}\begin{bmatrix} x \\ y \end{bmatrix}$ **47.** $\begin{bmatrix} 1 & 0 \\ 0 & 1 \end{bmatrix}\begin{bmatrix} x \\ y \end{bmatrix}$ **48.** $\begin{bmatrix} 1 & 1 & 1 \\ 1 & -2 & 3 \\ 2 & -3 & 4 \end{bmatrix}\begin{bmatrix} x \\ y \\ z \end{bmatrix}$

Rewrite each linear system as a matrix equation $AX = B$.

49. $2x + 3y = 4$
$4x + 5y = 6$

50. $x - 2y = 4$
$3x + y - z = -5$
$x + 2y - 3z = 6$

51–52. Using Exercises 44 and 45, solve the linear systems in Exercises 49 and 50, respectively.

Let M denote the set of 2×2 matrices over **W**. Let $f : \mathbb{N} \to M$ defined by $f(n) = \begin{bmatrix} 1 & 1 \\ 1 & 0 \end{bmatrix}^n$. Compute $f(n)$ for each value of n.

53. 2 **54.** 3 **55.** 4 **56.** 5

Prove each.

***57.** The inverse of a square matrix A is unique.
(*Hint*: Assume A has two inverses B and C. Show that $B = C$.)

***58.** If A is an invertible matrix, then $(A^{-1})^{-1} = A$.

***59.** If A and B are two invertible matrices of order n, then $(AB)^{-1} = B^{-1}A^{-1}$.

60. Write an algorithm to compute the sum of the matrices $A = (a_{ij})_{m\times n}$ and $B = (b_{ij})_{m\times n}$.

Chapter Summary

This chapter presented the concept of a function, the summation notation, and matrices. Functions can be defined by the ordered pair notation, tables, or graphs. Several properties of functions and some exotic functions were examined, including how to construct new functions from known ones.

Function

- A **function** $f: X \to Y$ is a pairing of every element x in X with a unique element y in Y. Dom(f) $= X$, codom(f) $= Y$, and range(f) $= \{f(x) \in Y | x \in X\}$ (page 118).

Special Functions

- Polynomial function $f(x) = \sum_{l=0}^{n} a_i x^i \quad (a_n \neq 0)$ (page 125).
- Exponential function $f(x) = a^x$ $(a > 0, a \neq 1)$ (page 125).
- Logarithmic function $f(x) = \log_a x$ $(a > 0,\ a \neq 1)$ (page 126).
- Absolute value function $f(x) = |x|$ (page 126).
- Floor function $f(x) = \lfloor x \rfloor$ (page 126).
- Ceiling function $f(x) = \lceil x \rceil$ (page 126).
- Characteristic function $f_A(x) = \begin{cases} 1 & \text{if } x \in A \\ 0 & \text{otherwise} \end{cases}$ (page 131).
- Mod function $f(x, y) = x \bmod y$ (page 132).
- Div function $g(x, y) = x \operatorname{div} y$ (page 132).

Properties of Functions

- A function $f : X \to X$ is the **identity function** on X if $f(x) = x$ for every $x \in X$ (page 136).
- $f : X \to Y$ is **injective**, if $x_1 \neq x_2 \to f(x_1) \neq f(x_2)$ (page 136).
- $f : X \to Y$ is **surjective**, if $\text{range}(f) = Y$ (page 137).
- $f : X \to Y$ is **bijective**, if it is both injective and surjective (page 137).
- If X and Y are finite sets with $|X| = |Y|$, $f : X \to Y$ is injective if and only if it is surjective (page 139).
- Two sets have the same cardinality if and only if a bijection exists between them (page 140).

The Pigeonhole Principle

- **Simple version** If m pigeons fly into n pigeonholes to roost, where $m > n$, then at least two pigeons must roost in the same pigeonhole (page 147).
- **Generalized version** If m pigeons fly into n pigeonholes to roost, where $m > n$, one pigeonhole must contain at least $\lfloor (m - 1)/n \rfloor + 1$ pigeons (page 147).

Composition

- The **composition** of the functions $f : X \to Y$ and $g : Y \to Z$ is given by $(g \circ f)(x) = g(f(x))$ for every x in X (page 151).

- The composition of two bijections is bijective (page 152).
- The function $g : Y \to X$ is the inverse of the function $f : X \to Y$ if $g \circ f = 1_X$ and $f \circ g = 1_Y$ (page 153).
- A function is invertible if and only if it is bijective (page 154).

Sequences and the Summation Notation

- A sequence $\{s_n\}$ is a function with domain $X = \{a, a+1, a+2, \ldots\}$ or a finite subset of X, where $a \in \mathbf{W}$ (page 157).
- Using the summation symbol $\sum$, the sum $a_k + a_{k+1} + \cdots + a_m$ is written as $\sum_{i=k}^{i=m} a_i$ (page 158).

Matrix

- An $\boldsymbol{m} \times \boldsymbol{n}$ **matrix** $(a_{ij})_{m \times n}$ is a rectangular arrangement of elements, where a_{ij} denotes the element in row i and column j (page 165).

Review Exercises

Find the number of positive integers ≤ 1776 and divisible by each:

1. 5 or 7
2. 5 but not 7
3. 5, 6, or 7
4. 3 or 5, but not 7

Find the day of the week in each case.

5. 1024 days from Sunday
6. 1948 days from Thursday

Find the month of the year in each case.

7. 256 months from March
8. 1976 months from August

Using formula (3.1) in Exercises 3.2, determine the first day in each year.

9. 2048
10. 4076
11. 7776
12. 7997

Using the formula in Exercises 3.2, compute the date for Easter Sunday in each year.

13. 2550
14. 3443
15. 4076
16. 6666

Determine if each function is injective.

17. $f(x) = -|x|, x \in \mathbb{R}$
18. $g(x) = \sqrt{x}, x \in \mathbb{R}^+$

Determine if each function $f : A \to B$ is surjective.

19. $f(x) = -\sqrt{x}$, $A = \mathbb{R}^+$, $B = \mathbb{R}^-$
20. $f(x) = 2^x$, $A = \mathbb{R}$, $B = \mathbb{R}^+$

Determine if each function $f : A \to B$ is bijective.

21. $f(x) = 2^{|x|}, A = B = \mathbb{R}$

22. $f = \mathrm{ORD}, A = B = \mathbf{Z}$

Let A and B be finite sets with $|A| = 3$ and $|B| = 2$. Find the number of:

23. Functions that can be defined from A to B.

24. Constant functions that can be defined from A to B.

25. Injections that can be defined from A to B.

26. Surjections that can be defined from A to B.

27. Bijections that can be defined from A to B.

Student records are maintained in a table using the hashing function $h(x) = x \bmod 9767$, where x denotes the student's social security number. Compute the location in the table corresponding to the given key, where the record is stored.

28. 011-53-1212

29. 212-44-7557

30–31. Redo Exercises 28 and 29 if $h(x) =$ first part in $x \bmod 23$.

32. The confirmation number for flight reservations made with an airline over the Internet consists of three letters followed by a digit and then two letters. Store the following confirmation numbers in a hash table of 26 cells using the hashing function $h(x) =$ first letter in x:

VPS3SL, NBC4GK, CBS1AA, AQX5CD, CBA3BA, NCR4SK,

CNN1TK, ABC5ZZ

33. Redo Exercise 32 if $h(x) =$ last letter in x.

34. Redo Exercise 32 using a hash table of 10 cells and $h(x) =$ digit in x.

35. Redo Exercise 32 using a hash table of 10 cells and $h(x) =$ digit in $x \bmod 5$.

36. Show that in any group of seven positive integers, at least two of them leave the same remainder when divided by six.

37. The total cost of mailing six letters is \$19. Show that the mailing charge for at least one letter is \$3 or more.

Let $f, g : \mathbb{R} \to \mathbb{R}$ defined by $f(x) = 2\lceil x \rceil + 1$ and $g(x) = 3\lfloor x \rfloor - 2$. Compute each.

38. $(g \circ f)(2.56)$

39. $(f \circ g)(-3.45)$

40. $(g \circ f)(-4.67)$

41. $(f \circ g)(-5.73)$

Let $f, g : \mathbf{W} \to \mathbf{W}$ defined by $f(x) = x \bmod 6$ and $g(x) = x \text{ div } 6$. Evaluate each.

42. $(g \circ f)(31)$

43. $(f \circ f)(49)$

44. $(f \circ g)(176)$

45. $(g \circ g)(1331)$

Mark each sentence as true or false, where x and y are arbitrary real numbers.

46. $\lfloor x + y \rfloor = \lfloor x \rfloor + \lfloor y \rfloor$

47. $\lceil x + y \rceil = \lceil x \rceil + \lceil y \rceil$

48. $\lfloor xy \rfloor = \lfloor x \rfloor \lfloor y \rfloor$

49. $\lceil xy \rceil = \lceil x \rceil \lceil y \rceil$

Give a counterexample to disprove each proposition, where $x, y \in \mathbb{R}$ and $n \in \mathbf{Z}$.

50. $\lfloor x + y \rfloor = \lfloor x \rfloor + \lfloor y \rfloor$

51. $\lceil x + y \rceil = \lceil x \rceil + \lceil y \rceil$

52. $\lfloor xy \rfloor = \lfloor x \rfloor \lfloor y \rfloor$

53. $\lceil xy \rceil = \lceil x \rceil \lceil y \rceil$

54. $\lfloor nx \rfloor = n \lfloor x \rfloor$

55. $\lceil nx \rceil = n \lceil x \rceil$

Find the first four terms of the sequence with the given general term, where $\alpha = (1 + \sqrt{5})/2$ and $\beta = (1 - \sqrt{5})/2$. (The number α is the **golden ratio**.)

56. $a_n = \left\lfloor \frac{\alpha^n}{\sqrt{5}} + \frac{1}{2} \right\rfloor, n \geq 1$

57. $b_n = \left\lceil \frac{\alpha^n}{\sqrt{5}} - \frac{1}{2} \right\rceil, n \geq 1$

58. $L_n = \alpha^n + \beta^n\ , n \geq 1$

59. $f_n = \frac{1}{\sqrt{5}}(\alpha^n - \beta^n)$

60. Arrange the terms of the sequence of ternary words of length ≤ 2 over the alphabet $\{0,1,2\}$ in increasing order of their numeric values.

Evaluate each.

61. $\sum_{i=1}^{10} \left\lfloor \frac{i}{2} \right\rfloor$

62. $\sum_{i=1}^{10} \left\lceil \frac{i}{2} \right\rceil$

63. $\sum_{i=1}^{17} \left\lfloor \frac{i}{2} \right\rfloor$

64. $\sum_{i=1}^{17} \left\lceil \frac{i}{2} \right\rceil$

65. Let $f : A \to B$ and $g : B \to A$ such that $f \circ g = 1_B$. Prove that g is injective.

66. Let $f : A \to B$ and $g : B \to A$ such that $g \circ f = 1_A$. Prove that g is surjective.

Supplementary Exercises

Let $A = (a_{ij})_{n \times n}$ and $B = (b_{ij})_{n \times n}$. Prove each.

1. $(A+B)^{\mathrm{T}} = A^{\mathrm{T}} + B^{\mathrm{T}}$

2. $(AB)^{\mathrm{T}} = B^{\mathrm{T}} A^{\mathrm{T}}$

3. AA^{T} is symmetric.

4. Let $G = \{0, 1\}$ and d: $G^n \times G^n \to \mathbf{W}$ defined by $d(x,y)$ = number of components in which the words x and y differ. $d(x,y)$ is called the **Hamming distance** between the n-bit words x and y. Is d bijective?

Richard Wesley Hamming *(1915–1998) was born in Chicago, graduated from the University of Chicago in 1937, and received an M.S. from the University of Nebraska 2 years later. After receiving his Ph.D. in mathematics in 1942 from the University of Illinois, he began his teaching career at the university and moved to the university of Louisville until 1945. After a year working on the Manhattan project at Los Alamos Science Laboratory, he joined the technical staff at Bell Telephone Labs in 1946; he headed the numerical methods research department from 1964 to 1967, and then the computer science research department until 1977. He left Bell in 1977 and became an adjunct professor in computer science at the Naval Postgraduate School, Monterey, California.*

Recipient of numerous awards, Hamming made significant contributions to algebraic coding theory, numerical methods, statistics, and digital filters.

5. Let $a_1, a_2, \ldots, a_n \in \mathbb{R}^+$. Prove that at least one of them is greater than or equal to their average $\frac{1}{n}(\sum_{1}^{n} a_i)$.

Evaluate each sum and product.

6. $\sum_{i=0}^{2} \sum_{j=1}^{3} \sum_{k=0}^{1} (i+j+k)$

7. $\sum_{i=0}^{1} \sum_{j=-1}^{1} \sum_{k=0}^{2} (i+2j+3k)$

8. $\sum_{i=1}^{3} \sum_{j=2}^{4} \sum_{k=1}^{5} 2^{i+j-k}$

9. $\prod_{i=1}^{3} \prod_{j=0}^{2} 2^{i+j}$

10. $\prod_{i=0}^{2} \prod_{j=1}^{3} (i-j)$

11. $\prod_{i=2}^{5} \prod_{j=0}^{3} 2$

An **arithmetic sequence** is a number sequence in which every term except the first is obtained by adding a fixed number, called the **common difference**, to the preceding term. For example, 1, 3, 5, 7, ... is an arithmetic sequence with common difference 2. Let a_n denote the nth term of the arithmetic sequence with first term a and common difference d.

12. Find a formula for a_n.

13. Let S_n denote the sum of the first n terms of the sequence. Prove that $S_n = \frac{n}{2}[2a + (n-1)d]$.

A **geometric sequence** is a number sequence in which every term except the first is obtained by multiplying the previous term by a constant, called the **common ratio**. For example, 2, 6, 18, 54, ... is a geometric sequence with common ratio 3. Let a_n denote the nth term of the geometric sequence with first term a and common ratio r.

14. Find a formula for a_n.

15. Let S_n denote the sum of the first n terms of the sequence. Prove that $S_n = \frac{a(r^n-1)}{r-1} \quad (r \neq 1)$.

***16.** Let $f \colon X \to Y$ be injective and $A \subseteq B \subseteq X$. Prove that $f(A \cap B) = f(A) \cap f(B)$.

***17.** Let $f \colon X \to Y$ and $A \subseteq B \subseteq X$ such that $f(A \cap B) = f(A) \cap f(B)$. Give a counterexample to show that f need not be injective.

***18.** Suppose $a_1 + a_2 + \cdots + a_n - n + 1$ pigeons occupy n pigeonholes $H_i, 1 \leq i \leq n$. Prove that either H_1 contains $\geq a_1$ pigeons, or H_2 contains $\geq a_2$ pigeons, ..., or H_n contains $\geq a_n$ pigeons.

Use Exercise 18, prove each.

19. The pigeonhole principle.

20. The generalized pigeonhole principle.

21. The sequence $\{a_n\}_1^\infty$ satisfies the property that $a_n = \sum\limits_{i=n+1}^{\infty} a_i$ for every $n \geq 1$. Show that $a_{n+1} = \frac{1}{2}a_n, n \geq 1$. (T. Fletcher, 1978).

Computer Exercises

Write a program to perform each task.

1. Read in the amount of water used by a household for 6 months and compute the water bill, using the rate in Example 3.6.

2. Read in a positive integer $n \leq 1000$ and print all perfect numbers $\leq n$. (There are three perfect numbers ≤ 1000. See Exercises 3.1.)

3. Read in a year $y > 1600$ and determine each:

- Whether or not it is a leap year.
- The number of leap years > 1600 and $\leq y$.

4. Read in a year y and find the following (see Exercises 3.2):

- The day of January 1 in year y and year $y + 1$.
- The number of Friday-the-thirteenths in year y.

5. January 1, 2000, fell on a Saturday. Determine the day of the week of January 1, 1776, and January 1, 3000. Print the calendar for January in each year.

6. Read in a series of years greater than 2000 and determine the Easter date in each year. (See Exercises 3.2.)

7. The discrete probability $p(r)$ that two people in a group of r people selected at random have the same birthday is given by

$$p(r) = 1 - \frac{365.364 \ldots (365 - r + 1)}{365^r}$$

assuming 365 days in a year. Compute the probability for each value of r: 10, 20, 30, ..., 100, including 23. (You will see that if 23 people are selected at random, there is a better than 50% chance that two have the same birthday. This is known as the **birthday paradox**.)

8. Assign the numbers 0–51 in order to the 52 playing cards in a standard deck. Read in a number x, where $0 \leq x \leq 51$. Identify the card numbered x. Use the suit labels 0 = clubs, 1 = diamonds, 2 = hearts, and 3 = spades, and the card labels 0 = ace, 1 = deuce, 2 = three, ..., in each suit.

9. Assign the numbers 0–63, row by row, to the various squares on an 8×8 chessboard. Read in two numbers x and y, where $0 \leq x, y \leq 63$. Determine if the queen at square x can capture the queen at square y.

10. Read in n customers' nine-digit account numbers at a bank, where n is a positive integer ≤ 100. Store them in a hash table using the hashing function $h(x) = x \bmod 113$, where x denotes an account number. Print the hash table.

11. Read in n students' social security numbers and store them in a hash table using the hashing function $h(x) = x \bmod 109$, where x denotes a social security number and n is a positive integer ≤ 100. Print the hash table.

12. Read in the two-letter abbreviations of all states in the United States. Store them in a hash table of 26 cells, using the hash function $h(x) =$ first letter in x.

13. Read in a positive integer $n \leq 15$ and a square matrix A of order n. Determine if it is symmetric.

14. Read in an $m \times n$ matrix A and a $p \times q$ matrix B. Find $A + B$, $A - B$, and AB if they are defined.

15. The **trace** of a matrix $(a_{ij})_{n \times n}$ is $\sum_{i=1}^{n} a_{ii}$. Read in a positive integer $n \leq 20$ and an $n \times n$ matrix, and print the trace of the matrix.

Exploratory Writing Projects

Using library and Internet resources, write a team report on each of the following in your own words. Provide a well-documented bibliography.

1. Discuss the development of the concept of a function.

2. Collect the various rates for water and electricity consumption from neighboring towns and cities. Write each as a word problem and then define each as a function.
3. Describe the history of perfect numbers and their relationship to Mersenne primes. Comment on the existence of odd perfect numbers.
4. Give a number of applications of the floor and ceiling functions to everyday life.
5. Investigate the two-queens puzzle on an $n \times n$ chessboard, where $2 \leq n \leq 10$.
6. Examine the various classes of infinite sets, their properties, and their cardinalities.
7. Describe the origin of the pigeonhole principle and give several applications to everyday life. Comment on the power of the principle.
8. List the various built-in functions in your favorite programming language. Determine whether each is injective, surjective, or bijective.
9. Investigate the origin of the summation notation.
10. Describe the various mathematical structures and operations hidden on the sports pages of a national newspaper.
11. Give a brief introduction to coding theory.
12. Discuss the *Leontief input-output model*, developed by Wassily Leontief, who won the Nobel Prize in Economic Science in 1973.

Enrichment Readings

1. W. T. Bailey, "Friday-the-Thirteenth," *Mathematics Teacher*, Vol. 62 (May 1969), pp. 363–364.
2. D. R. Camp, "Secret Codes with Matrices," *Mathematics Teacher*, Vol. 78 (Dec. 1985), pp. 676–680.
3. D. I. A. Cohen, *Basic Techniques of Combinatorial Theory*, Wiley, New York, 1978, pp. 144–178.
4. R. L. Graham *et al.*, *Concrete Mathematics*, Addison-Wesley, Reading, MA, 1989, pp. 397–424.
5. E. Horowitz and S. Sahni, *Fundamentals of Computer Algorithms*, Computer Science Press, Rockville, MD, 1978.
6. F. S. Koltz, "When Is a Program Like a Function?" *Mathematics Teacher*, Vol. 79 (Nov. 1986), pp. 648–651.
7. Z. Usiskin, "The Greatest Integer Symbol," *Mathematics Teacher*, Vol. 70 (Dec. 1977), pp. 739–743.

Chapter 4

Induction and Algorithms

God created the natural numbers; all else is the work of man.

—L. KRONECKER

This chapter presents the well-ordering principle, the division algorithm with which you are already familiar, and some fundamental divisibility properties. In addition, through the well-ordering principle we will establish an additional proof technique, the principle of mathematical induction. Interesting applications of this principle, as well as the pigeonhole principle from Chapter 3, will be investigated.

Some of the intriguing problems pursued in this chapter lie below:

- Are there integers between 0 and 1?
- If n is a positive integer ≥ 2 and $a_1, a_2, \ldots, a_n \in \mathbf{Z}$, are there consecutive elements $a_{k+1}, a_{k+2}, \ldots, a_\ell$ such that $a_{k+1} + a_{k+2} + \cdots + a_\ell$ is divisible by n, where $k < \ell$?
- If $a_1, a_2, \ldots, a_n$ are the first n positive integers in some order, arranged around a circle, is it true that there must be a set of k consecutive elements in the cyclic arrangement whose sum is greater than $\lfloor [kn(n+1) - 2]/2n \rfloor$?
- Can any postage of $n \geq 2$ cents be paid using two- and three-cent stamps?

4.1 The Division Algorithm

The division algorithm, with which you are already familiar, is often employed to verify the correctness of a division problem. Its proof is based on the following cardinal fact, which is accepted as an axiom. (An **axiom** is a proposition that is accepted as true. It is usually a self-evident proposition and is consistent with known facts.)

The Well-Ordering Principle
Every nonempty set of positive integers has a least element. ∎

For example, the set $\{13, 5, 8, 23\}$ has a least element, 5. The well-ordering principle applies to any nonempty subset S of $T = \{n \in \mathbf{Z} \mid n \geq n_0\}$, where n_0 is any integer. To see this, let $S^* = \{n - n_0 + 1 \mid n \in S\}$ and $T^* = \{n - n_0 + 1 \mid n \in T\}$. Since $S^* \subseteq T^*$ and $T^* \subseteq N$, by the well-ordering principle, S^* contains a least element ℓ^*. Then $n_0 + \ell^* - 1$ is a least element of S (why?).

For example, let $S = \{-3, -1, 0, 1, 3, 5\}$ and $T = \{n \in \mathbf{Z} \mid n \geq -5\}$. Then $S^* = \{3, 5, 6, 7, 9, 11\}$ has a least element $\ell^* = 3$, so $n_0 + \ell^* - 1 = -5 + 3 - 1 = -3$ is the least element of S.

Next we present the division algorithm. Its proof is a bit complicated, so we omit it here; but a proof, using the well-ordering principle, can be established (see, for instance, the author's number theory book).

The Division Algorithm

When an integer a is divided by a positive integer b, we get a unique (integer) **quotient** q and a unique (integer) **remainder** r, where $0 \leq r < b$. The integer a is the **dividend** and b the **divisor**. This is formally stated as follows.

THEOREM 4.1 **(The Division Algorithm)** Let a be any integer and b any positive integer. Then there exist unique integers q and r such that

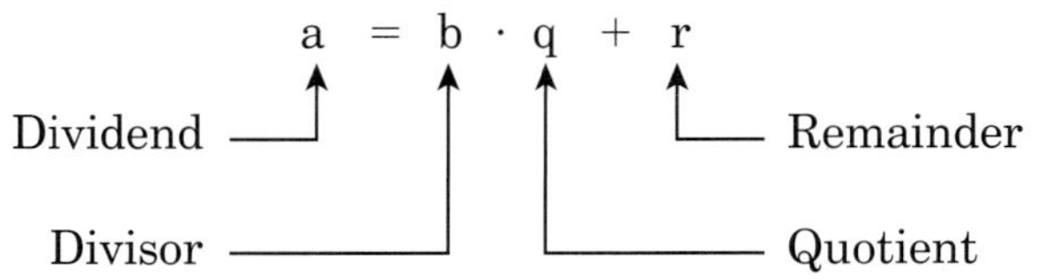

where $0 \leq r < b$. ∎

Although this theorem does not present an algorithm for finding q and r, it has been traditionally called the division algorithm. The values of q and r can be found using the familiar long division method.

Notice that the equation $a = bq + r$ can be written as

$$\frac{a}{b} = q + \frac{r}{b}$$

so $q = a \text{ div } b = \lfloor a/b \rfloor$ and $r = a - bq = a \bmod b$.

The next example shows that we should be careful in finding the quotient and the remainder when the dividend is negative.

EXAMPLE 4.1 Find the quotient q and the remainder r when -23 is divided by 5.

SOLUTION:
Since $-23 = 5 \cdot (-4) + (-3)$, you might be tempted to say that $q = -4$ and $r = -3$. Recall that the remainder can *never* be negative, so we rewrite -23 as $-23 = 5 \cdot (-5) + 2$, where $0 \leq r(= 2) < 5$ (see the number line in Figure 4.1). Thus $q = -5$ and $r = 2$; in other words, -23 div $5 = -5$ and -23 mod $5 = 2$.

Figure 4.1

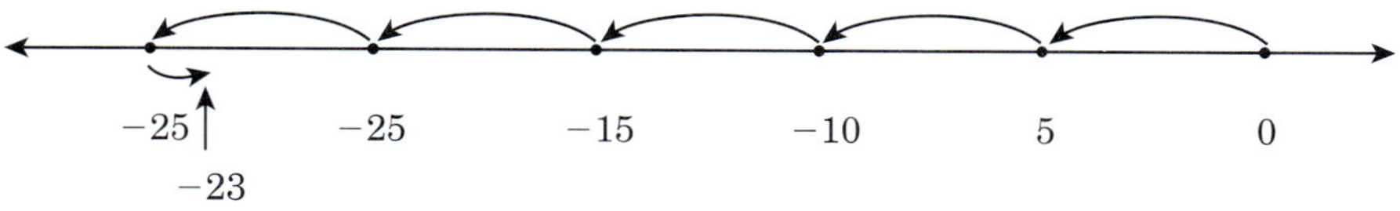

We close this section with two applications of the division algorithm and the pigeonhole principle. ■

EXAMPLE 4.2 Let b be an integer ≥ 2. If $b + 1$ distinct integers are randomly selected, prove that the difference of some two of them must be divisible by b.

PROOF
Let q be the quotient and r the remainder when an integer a is divisible by b. Then, by the division algorithm, $a = bq + r$ where $0 \leq r < b$. The $b+1$ distinct integers yield $b+1$ remainders (pigeons); but there are only b possible remainders (pigeonholes). Therefore, by the pigeonhole principle, two of the remainders must be equal.

Let x and y be the corresponding integers. Then $x = bq_1 + r$ and $y = bq_2 + r$ for some quotients q_1 and q_2. Then

$$\begin{aligned} x - y &= (bq_1 + r) - (bq_2 + r) \\ &= b(q_1 - q_2) \end{aligned}$$

Thus, $x - y$ is divisible by b. ■

EXAMPLE 4.3 Let n be an integer ≥ 2 and let $a_1, a_2, \ldots, a_n \in \mathbf{Z}$. Prove that there exist integers k and ℓ such that $a_{k+1} + a_{k+2} + \cdots + a_\ell$ is divisible by n, where $1 \leq k < \ell \leq n$; that is, there exist consecutive elements $a_{k+1}, a_{k+2}, \ldots, a_\ell$ whose sum is divisible by n.

PROOF (by cases):
Consider the n sums $S_i = a_1 + a_2 + \cdots + a_i$, where $1 \leq i \leq n$.

Case 1 If any of the sums S_i is divisible by n, then the statement is true.

Case 2 Suppose none of the sums S_i is divisible by n. When S_i is divided by n, the remainder must be nonzero. So, by the division algorithm, the possible remainders are $1, 2, \ldots, (n-1)$. Since there are n sums and $n-1$ possible remainders, by the pigeonhole principle, two of the sums S_k and S_ℓ must yield the same remainder r when divided by n, where $k < \ell$.

Therefore, there must exist integers q_1 and q_2 such that $a_1 + a_2 + \cdots + a_k = nq_1 + r$ and $a_1 + a_2 + \cdots + a_\ell = nq_2 + r$, where $k < \ell$. Subtracting, we get $a_{k+1} + a_{k+2} + \cdots + a_\ell = n(q_1 - q_2)$. Thus $a_{k+1} + a_{k+2} + \cdots + a_\ell$ is divisible by n. ■

To cite a specific example, consider the seven integers 2, 3, 8, 15, 23, 29, and 57. Then $S_1 = a_1 = 2 = 0 \cdot 7 + 2$ and $S_5 = a_1 + a_2 + a_3 + a_4 + a_5 = 2 + 3 + 8 + 15 + 23 = 51 = 7 \cdot 7 + 2$. Then $S_5 - S_1 = a_2 + a_3 + a_4 + a_5 = 3 + 8 + 15 + 23 = 49$ is divisible by 7. Here $k = 1$ and $\ell = 5$. (You may notice that $S_4 = a_1 + a_2 + a_3 + a_4 = 2 + 3 + 8 + 15$ is also divisible by 7.)

Exercises 4.1

1. Is the set of positive odd integers well-ordered?

2. Is the set of positive even integers well-ordered?

In Exercises 3–6, find the quotient and the remainder when the first integer is divided by the second.

3. 137, 11 **4.** 15, 23 **5.** −43, 16 **6.** −37, 73

Find the set of possible remainders when an integer is divided by the given integer.

7. Two **8.** Five **9.** Seven **10.** Twelve

11. Prove that there exists no integer between 0 and 1.

12. Let $a \in \mathbf{Z}$. Prove that no integer exists between a and $a + 1$.

13. Let $n_0 \in \mathbf{Z}, S$ be a nonempty subset of the set $T = \{n \in \mathbf{Z} \mid n \geq n_0\}$, and ℓ^* be a least element of the set $T^* = \{n - n_0 + 1 \mid n \in T\}$. Prove that $n_0 + \ell^* - 1$ is a least element of S.

14. Using the well-ordering principle, prove that 1 is the smallest positive integer.
(*Hint*: Prove by contradiction.)

***15.** Let $a \in \mathbf{Z}, S = \{a, a+1, \ldots\}, T \subseteq S$, and $a \in T$. Let k be any element of S such that whenever $k \in T$, $k + 1 \in T$. Prove that $S = T$.

***16.** Let $a \in \mathbf{Z}$ and $S = \{a, a+1, \ldots\}$. Let $\mathrm{P}(n)$ be a predicate on S such that the following conditions are satisfied: (1) $\mathrm{P}(a)$ is true; (2) If $\mathrm{P}(a)$,

P($a+1$), ..., P(k) are true for any $k \geq a$, then P($k+1$) is also true. Prove that P(n) is true for every $n \geq a$.

4.2 Divisibility Properties

The celebrated euclidean algorithm can be used to find the greatest common divisor of two positive integers, but first a very few properties of prime and composite numbers, and some divisibility properties.

Let a and b ($\neq 0$) be any two integers. If there is an integer q such that $a = bq$, we say b **divides** a, b is a **factor** of a, a is **divisible** by b, or a is a **multiple** of b. We then write $b \mid a$; otherwise, $b \nmid a$. (Again, the meaning of the vertical bar should be clear from the context.) For instance, $3 \mid 6$, $8 \mid 24$, but $6 \nmid 14$.

A positive factor b of a positive integer a is a **proper factor** of a if $b \neq a$. For example, the proper factors of 6 are 1, 2, and 3.

There are positive integers with exactly two positive factors. Accordingly, we make the following definition.

Prime Numbers and Composite Numbers

A positive integer > 1 is a **prime number** (or simply a **prime**) if its only positive factors are 1 and itself. A positive integer > 1 is a **composite number** if it is not a prime.

For example, 2 and 19 are primes, whereas 6 and 21 are composite numbers (why?).

There is a systematic procedure for determining whether or not a positive integer $n \geq 2$ is a prime. It is based on the next theorem.

THEOREM 4.2 Any composite number n has a prime factor $\leq \lfloor \sqrt{n} \rfloor$.

PROOF (by contradiction):
Since n is composite, there are positive integers a and b such that $n = ab$ where $1 < a < n$ and $1 < b < n$. Suppose $a > \sqrt{n}$ and $b > \sqrt{n}$. Then $n = ab > \sqrt{n} \cdot \sqrt{n} = n$, which is impossible. Therefore, either $a \leq \sqrt{n}$ or $b \leq \sqrt{n}$. Since both a and b are integers, it follows that either $a \leq \lfloor \sqrt{n} \rfloor$ or $b \leq \lfloor \sqrt{n} \rfloor$.

By the fundamental theorem of arithmetic (see Theorem 4.13), every positive integer has a prime factor. Any such factor of a or b is also a factor of $a \cdot b = n$, so n must have a prime factor $\lfloor \sqrt{n} \rfloor$. ■

> It follows from Theorem 4.2 that if n has *no* prime factors $\leq \lfloor \sqrt{n} \rfloor$, then n is a prime; otherwise, it is a composite number.

This fact can be used to determine whether or not an integer $n \geq 2$ is a prime, as the next example illustrates.

EXAMPLE 4.4 Determine if 1601 is a prime number.

SOLUTION:
First list all primes $\leq \lfloor\sqrt{1601}\rfloor$. They are 2, 3, 5, 7, 11, 13, 17, 19, 23, 29, 31, and 37. None of them is a factor of 1601 (verify); so 1601 is a prime. ■

An algorithm for determining the primality of a positive integer n ≥ 2 is given in Algorithm 4.1.

```
Algorithm prime number(n)
(* This algorithm determines if a positive integer n≥2 is
   prime or not using Theorem 4.2. *)
 Begin (* algorithm *)
   list all primes ≤ ⌊√n⌋
   if any of them is a factor of n then
     n is not a prime
   else
     n is a prime
 End (* algorithm *)
```

Algorithm 4.1

In the remainder of this section we discuss some useful divisibility properties. We begin with a simple and straightforward property.

THEOREM 4.3 If a and b are positive integers such that $a \mid b$ and $b \mid a$, then $a = b$. ■

> Notice that this theorem does *not* hold if a and b are any integers. For example, $3 \mid (-3)$ and $(-3) \mid 3$, but $3 \neq -3$.

THEOREM 4.4 Let a, b, and c be any integers. Then:

(1) If $a \mid b$ and $b \mid c$, then $a \mid c$ (**transitive property**).
(2) If $a \mid b$ and $a \mid c$, then $a \mid (b + c)$.
(3) If $a \mid b$ and $a \mid c$, then $a \mid (b - c)$.
(4) If $a \mid b$, then $a \mid bc$.

PROOF:
We shall prove properties 1 and 2, and leave the others as exercises.

(1) Since $a \mid b$, there exists an integer q_1 such that $b = aq_1$. Similarly, there exists an integer q_2 such that $c = bq_2$. Then $c = bq_2 = (aq_1)q_2 = a(q_1q_2)$. Thus, there exists an integer $q = q_1q_2$ such that $c = aq$. Therefore, $a \mid c$.

(2) As above, we have $b = aq_1$ and $c = aq_3$. Then $b + c = aq_1 + aq_3 = a(q_1 + q_3)$. Since $q_1 + q_3$ is an integer, it follows that $a \mid (b + c)$. ■

The Greatest Common Divisor

A positive integer can be a factor of two positive integers a and b. Such a positive integer is a **common factor** of a and b. The largest such common factor is the **greatest common divisor** (gcd) of a and b, denoted by $\gcd\{a, b\}$.

For instance, $\gcd\{6, 9\} = 3$, $\gcd\{12, 24\} = 12$, and $\gcd\{6, 35\} = 1$.

This definition of gcd, although simple and clear, is not practical, so we give an alternate, equivalent definition below.

An Alternate Definition of GCD

A positive integer d is the **gcd** of two positive integers a and b if:

- $d \mid a$ and $d \mid b$; and
- if $d' \mid a$ and $d' \mid b$, then $d' \mid d$, where d' is a positive integer.

Thus, d is $\gcd\{a, b\}$ if (1) d is a common divisor of both a and b; and (2) any common divisor of a and b is also a divisor of d.

The next theorem, an extremely useful and powerful result, can be applied to develop an algorithm to compute $\gcd\{a, b\}$.

THEOREM 4.5 Let a and b be any positive integers, and r the remainder when a is divided by b. Then $\gcd\{a, b\} = \gcd\{b, r\}$.

PROOF

Let $\gcd\{a, b\} = d$ and $\gcd\{b, r\} = d'$. To prove that $d = d'$, it suffices to show that $d \mid d'$ and $d' \mid d$. By the division algorithm, a unique quotient q exists such that

$$a = bq + r \tag{4.1}$$

To show that $d \mid d'$:

Since $d = \gcd\{a, b\}$, $d \mid a$ and $d \mid b$. Therefore, $d \mid bq$, by Theorem 4.4. Then $d \mid (a - bq)$, again by Theorem 4.4. In other words, $d \mid r$, by Equation (4.1). Thus, $d \mid b$ and $d \mid r$. Therefore, $d \mid \gcd\{b, r\}$; that is, $d \mid d'$.

Similarly, it can be shown that $d' \mid d$. (See Exercise 33.) Thus, by Theorem 4.3, $d = d'$; that is, $\gcd\{a, b\} = \gcd\{b, r\}$. ■

EXAMPLE 4.5 Illustrate Theorem 4.5, using $a = 108$ and $b = 20$.

SOLUTION:

$\gcd\{108, 20\} = 4$ (verify). When 108 is divided by 20, the remainder is 8. $\gcd\{20, 8\} = 4$ (verify). Thus, $\gcd\{108, 20\} = \gcd\{20, 8\}$. ■

Euclidean Algorithm

Among several procedures for finding the gcd of two positive integers, one efficient algorithm is the **euclidean algorithm**, named after the

Little is known about Euclid's life. He taught at the University of Alexandria and founded the Alexandrian School of Mathematics. When the Egyptian ruler King Ptolemy I asked Euclid if there were an easier way to learn geometry than by studying The Elements, *he replied, "There is no royal road to geometry."* *Euclid is called the* father of geometry.

No work, except for the Bible, has been more widely read, studied, or edited," according to J. E. Lightner of Western Maryland College, Westminister, Maryland. "More than 2000 editions of the work have appeared since the first printed one in 1482; however, no extant copy of The Elements *dates from Euclid's own time."*

Greek mathematician Euclid (330?–275 B.C.), who included it in his extraordinary work *The Elements*. The algorithm repeatedly applies the division algorithm and Theorem 4.5. Before formally discussing the algorithm, we illustrate it in the next example.

EXAMPLE 4.6 Find $\gcd\{1976, 1776\}$.

SOLUTION:
Apply the division algorithm with 1976 (the larger of the two numbers) as the dividend and 1776 as the divisor:

$$1976 = 1 \cdot 1776 + 200$$

Apply the division algorithm again with 1776 and 200, using 1776 as the dividend and 200 as the divisor:

$$1776 = 8 \cdot 200 + 176$$

Continue this procedure until a zero remainder is obtained:

$$\begin{aligned} 1976 &= 1 \cdot 1776 + 200 \\ 1776 &= 8 \cdot \ \ 200 + 176 \\ 200 &= 1 \cdot \ \ 176 + 24 \\ 176 &= 7 \cdot \ \ \ \ 24 + 8 && \longleftarrow \text{last nonzero remainder} \\ 24 &= 3 \cdot \ \ \ \ \ \ 8 + 0 \end{aligned}$$

The last nonzero remainder in this procedure is the gcd. Thus $\gcd\{1976, 1776\} = 8$. ■

Will this method work for any two positive integers a and b? If $a = b$, then $\gcd\{a, b\} = a$. So assume, for convenience, $a > b$. (If this is not true, simply switch them.) Let $r_0 = b$. Then by successive application of the division algorithm, we get a sequence of equations:

$$\begin{aligned} a &= q_0 r_0 + r_1 & 0 \le r_1 < r_0 \\ r_0 &= q_1 r_1 + r_2 & 0 \le r_2 < r_1 \\ r_1 &= q_2 r_2 + r_3 & 0 \le r_3 < r_2 \\ &\vdots \end{aligned}$$

Continuing like this, we get the following sequence of remainders:

$$b = r_0 > r_1 > r_2 > r_3 > \cdots \ge 0$$

Since the remainders are nonnegative and getting smaller and smaller, this sequence must eventually terminate with remainder $r_n = 0$. Thus, the last two equations in the above procedure are:

$$r_{n-2} = q_{n-1} r_{n-1} + r_n \qquad 0 \le r_n < r_{n-1}$$

and

$$r_{n-1} = q_n r_n$$

It then follows that $\gcd\{a, b\} = \gcd\{a, r_0\} = \gcd\{r_0, r_1\} = \gcd\{r_1, r_2\} = \cdots = \gcd\{r_{n-1}, r_n\} = r_n$, the last nonzero remainder. (This can be established by using mathematical induction; see Exercise 56 in Section 4.4.)

EXAMPLE 4.7 Apply the euclidean algorithm to find $\gcd\{2076, 1024\}$.

SOLUTION:
By the successive application of the division algorithm, we get:

$$\begin{aligned} 2076 &= 2 \cdot 1024 + 28 \\ 1024 &= 36 \cdot \ \ 28 + 16 \\ 28 &= 1 \cdot \ \ 16 + 12 \\ 16 &= 1 \cdot \ \ 12 + 4 & \longleftarrow \text{last nonzero remainder} \\ 12 &= 3 \cdot \ \ \ 4 + 0 \end{aligned}$$

Since the last nonzero remainder is 4, $\gcd\{2076, 1024\} = 4$. ■

The euclidean algorithm is formally presented in Algorithm 4.2.

```
Algorithm Euclid(x,y,divisor)
(* This algorithm returns gcd{x,y} in divisor, where
   x ≥ y > 0. *)
 0. Begin (* algorithm *)
 1.   dividend ← x
 2.   divisor ← y
```

```
    remainder ← dividend mod divisor
    while remainder > 0 do (* update dividend,
                              divisor, and remainder *)
    begin (* while *)
      dividend ← divisor
      divisor ← remainder
      remainder ← dividend mod divisor
    endwhile
End (* algorithm *)
```

Algorithm 4.2

The euclidean algorithm provides a procedure for expressing the gcd of two positive integers in terms of themselves, as the next example shows.

EXAMPLE 4.8 Example 4.7 showed that $\gcd\{2076, 1024\} = 4$. Express the gcd in terms of 2076 and 1024.

SOLUTION:
We use the equations in Example 4.7 in the reverse order:

$$\begin{aligned} 4 &= 16 - 1 \cdot 12 && = 16 - 1 \cdot (28 - 1 \cdot 16) \\ &= 2 \cdot 16 - 1 \cdot 28 && = (1024 - 36 \cdot 28) - 1 \cdot 28 \\ &= 2 \cdot 1024 - 72 \cdot 28 - 1 \cdot 28 && = 2 \cdot 1024 - 73 \cdot 28 \\ &= 2 \cdot 1024 - 73(2076 - 2 \cdot 1024) && = 2 \cdot 1024 - 73 \cdot 2076 + 146 \cdot 1024 \\ &= (-73) \cdot 2076 + 148 \cdot 1024 \end{aligned}$$

(You may verify this by direct computation.) ■

Example 4.8 can be generalized as in the following theorem. We omit its proof.

THEOREM 4.6 Let a and b be any positive integers, and $d = \gcd\{a, b\}$. Then there exist integers s and t such that $d = sa + tb$. ■

Note: (1) The expression $sa + tb$ is called a **linear combination** of a and b. (2) The integers s and t are *not* unique. For example, $\gcd\{28, 12\} = 4$ and $4 = 1 \cdot 28 + (-2) \cdot 12 = (-2) \cdot 28 + 5 \cdot 12$. (3) The integers s and t can be found by using the various equations in the euclidean algorithm, or by trial and error especially when a and b are fairly small.

Theorem 4.6 can be used to derive other divisibility properties. To this end, we define two positive integers to be **relatively prime** if their gcd is 1. For example, 6 and 35 are relatively prime, whereas 12 and 18 are not relatively prime.

THEOREM 4.7 Let a and b be relatively prime numbers. If $a \mid bc$, then $a \mid c$.

PROOF:
Since a and b are relatively prime, Theorem 4.6 indicates integers s and t exist such that $sa + tb = 1$. Then $sac + tbc = c$. By Theorem 4.4, $a \mid (sac)$ and $a \mid (tbc)$. Therefore, by Theorem 4.4, $a \mid (sac + tbc)$; that is, $a \mid c$. ■

The following exercises offer additional divisibility properties to verify; again, consult a number theory book.

Exercises 4.2

Determine if each positive integer is a prime.

1. 727 **2.** 1001 **3.** 1681 **4.** 1723

5. Prove or disprove: Every prime is a perfect number.

Using the euclidean algorithm, find the gcd of the given integers.

6. 2024, 1024 **7.** 2076, 1076 **8.** 2076, 1776 **9.** 3076, 1976

In Exercises 10–13, express the gcd of the given integers as a linear combination of them.

10. 12, 9 **11.** 18, 28 **12.** 12, 29 **13.** 28, 15

14. Two prime numbers that differ by 2 are called **twin primes**. For example, 5 and 7 are twin primes. Prove that one more than the product of two twin primes is a perfect square. (Twin primes played a key role in 1994 in establishing a flaw in the Pentium chip, manufactured by Intel Corporation.)

Evaluate each sum, where d is a positive integer.

15. $\sum_{d \mid 6} d$ **16.** $\sum_{d \mid 12} 1$ **17.** $\sum_{d \mid 18} \left(\frac{1}{d}\right)$ **18.** $\sum_{d \mid 18} \left(\frac{18}{d}\right)$

Disprove each statement, where a, b, and c are arbitrary integers.

19. If $a \mid (b + c)$, then $a \mid b$ and $a \mid c$. **20.** If $a \mid bc$, then $a \mid b$ and $a \mid c$.

(Easter Sunday) Here is a second method* for determining Easter Sunday in a given year N. Let $a = N \bmod 19$, $b = N \text{ div } 100$, $c = N \bmod 100$, $d = b \text{ div } 4$, $e = b \bmod 4$, $f = (b+8) \text{ div } 25$, $g = (b-f+1) \text{ div } 3$, $h = (19a + b - d - g + 15) \bmod 30$, $i = c \text{ div } 4$, $j = c \bmod 4$, $k = (32 + 2e + 2i - h - j) \bmod 7$, $\ell = (a + 11h + 22k) \text{ div } 451$, $m = (h + k - 7\ell + 114) \text{ div } 31$, and $n = (h + k - 7\ell + 114) \bmod 31$. Then Easter Sunday falls on the $(n + 1)$st

*Based on "To Find Easter," *Nature* (April 20, 1876). For bringing this method to his attention, the author would like to thank Thomas Moore of Bridgewater State College.

day of the mth month of the year. Compute the date for Easter Sunday in each year.

21. 2000 **22.** 2076 **23.** 3000 **24.** 3663

Euler's phi-function φ is another important number-theoretic function on $\mathbb{N}$, defined by $\varphi(n) =$ number of positive integers $\leq n$ and relatively prime to n. For example, $\varphi(1) = 1 = \varphi(2), \varphi(3) = 2 = \varphi(4)$, and $\varphi(5) = 4$. Evaluate $\varphi(n)$ for each value of n.

25. 10 **26.** 15 **27.** 17 **28.** 24

29. Compute $\sum_{d|n} \varphi(d)$ for $n = 5$, 6, 10, and 12.

30. Using Exercise 29, predict a formula for $\sum_{d\,|\,n} \varphi(d)$.

Let a, b, c, and n be any positive integers and p be any prime. Prove each.

31. If $a \mid b$ and $a \mid c$, then $a \mid (b - c)$.

32. If $a \mid b$, then $a \mid bc$.

33. Let r be the remainder when a is divided by b. Let $d = \gcd\{a, b\}$ and $d' = \gcd\{b, r\}$. Then $d' \mid d$.

34. Let $a > b$. Then $\gcd\{a, b\} = \gcd\{a, a - b\}$.

35. Let $a > b$. Then $\gcd\{a, b\} = \gcd\{b, a + b\}$.

36. The gcd of a and b is unique.
(*Hint*: Assume two gcd's d and d'; show that $d = d'$.)

37. If $p \mid ab$, then $p \mid a$ or $p \mid b$.
[*Hint*: Assume $p \mid ab$ and $p \nmid a$. Since $p \nmid a$, $\gcd\{p, a\} = 1$.]

38. Any two consecutive integers are relatively prime.

39. Let $d = \gcd\{a, b\}$. Then a/d and b/d are relatively prime.

40. $\gcd\{na, nb\} = n \cdot \gcd\{a, b\}$ **41.** $\gcd\{\gcd\{a, b\}, c\} = \gcd\{a, \gcd\{b, c\}\}$

42. Let $a \mid c$ and $b \mid c$, where a and b are relatively prime numbers. Then $ab \mid c$.

43. 2 and 3 are the only two consecutive integers that are primes.

44. 3, 5, and 7 are the only three consecutive odd integers that are primes.

45. If p and $p^2 + 8$ are primes, then $p^3 + 4$ is also a prime. (D. L. Silverman, 1968)

46. If p and $p + 2$ are twin primes, then p must be odd.

47. Suppose p and q are primes such that $p - q = 3$. Then $p = 5$.

48. Every odd prime is of the form $4n + 1$ or $4n + 3$.

Disprove each statement.

49. If $\gcd\{a,b\}=1$ and $\gcd\{b,c\}=1$, then $\gcd\{a,c\}=1$, where a, b, and c are positive integers.

50. $n!+1$ is a prime for every $n \geq 0$.

51. $E_n = p_1p_2\cdots p_n + 1$ is a prime, where p_i denotes the ith prime and $i \geq 1$.

52. Let n be a positive integer. Prove that $(n+1)!+2$, $(n+1)!+3, \dots,$ $(n+1)!+(n+1)$ are n consecutive composite numbers.

4.3 Nondecimal Bases

In everyday life we use the decimal notation, base ten, to represent any real number. For example, $234 = 2(10^2)+3\ (10^1)+4(10^0)$, which is the **decimal expansion** of 234. Likewise, $23\cdot 45 = 2(10^1)+3(10^0)+4(10^{-1})+5(10^{-2})$. Computers use base two (*binary*), and very long binary numbers are often handled by humans (as opposed to computers) using bases eight (*octal*) and sixteen (*hexadecimal*).

Actually, any positive integer $b \geq 2$ is a valid choice for a base. This is a consequence of the following fundamental result.

THEOREM 4.8 Let b be a positive integer ≥ 2. Then every positive integer a can be expressed uniquely in the form $a = a_kb^k + a_{k-1}\,b^{k-1} + \cdots + a_1b + a_0$, where $a_0, a_1, \dots, a_k$ are nonnegative integers less than b, $a_k \neq 0$, and $k \geq 0$. ■

This leads us to the following definition.

Base-*b* Representation

The expression $a_kb^k + a_{k-1}b^{k-1} + \cdots + a_1b + a_0$ is the **base-*b* expansion** of the integer a. Accordingly, we write $a = (a_ka_{k-1}\cdots a_1a_0)_b$ in base b. The base is omitted when it is 10.

For example, $234 = 234_{\text{ten}}$ and $22 = 10110_{\text{two}}$ (see Example 4.9).

When the base is greater than 10, to avoid confusion we use the letters A, B, C, ... to represent the *digits* 10, 11, 12, ..., respectively. It is easy to find the decimal value of an integer from its base-b representation, as the next example illustrates.

EXAMPLE 4.9 Express 10110_{two} in base 10.

SOLUTION:

$$\begin{aligned} 10110_{\text{two}} &= 1(2^4)+0(2^3)+1(2^2)+1(2^1)+0(2^0) \quad \longleftarrow \text{binary expansion} \\ &= 16+0+4+2+0 \\ &= 22 \end{aligned}$$

■

Conversely, suppose we are given a decimal integer. How do we express it in another base b? By Theorem 4.8, all we have to do is express it as a sum of powers of b, then simply collect the coefficients in the right order. Always remember to account for missing coefficients.

A Brainteaser

Take a look at the tablets A, B, C, D, and E in Figure 4.2. Assuming you are under 32 years old, identify the tablets on which your age appears; we can then easily tell your age. For example, if your age appears on tablets A, B, C, and E, then you must be 23. Can you explain how this puzzle works?

Figure 4.2

A		B		C		D		E	
1	17	2	18	4	20	8	24	16	24
3	19	3	19	5	21	9	25	17	25
5	21	6	22	6	22	10	26	18	26
7	23	7	23	7	23	11	27	19	27
9	25	10	26	12	28	12	28	20	28
11	27	11	27	13	29	13	29	21	29
13	29	14	30	14	30	14	30	22	30
15	31	15	31	15	31	15	31	23	31

Returning to nondecimal representations, a simple algorithm expresses an integer a in any nondecimal base b: divide a, and its successive quotients by b until a zero quotient is reached, then pick the remainders in the reverse order. These steps can be translated into the elegant algorithm given in Algorithm 4.3.

```
Algorithm nondecimal base(n,b)
(* This algorithm finds the base-b representation (a_m a_{m-1} ... a_1 a_0)_b
   of a positive integer n. The variables q and r denote the quotient
   of the remainder of the division algorithm, and i is a subscript. *)
  Begin (* algorithm *)
    (* initialize the variables q, r, and i *)
    q ← n; i ← 0
    while q > 0 do
    begin (* while *)
      r ← q mod b
      a_i ← r
      q ← q div b
```

```
    i ← i+1
  endwhile
End (* algorithm *)
```

Algorithm 4.3

The next example illustrates this algorithm.

EXAMPLE 4.10 Represent 15,036 in the hexadecimal system, that is, in base 16.

SOLUTION:

Applying Algorithm 4.3 we have:

$$\begin{aligned} 15036 &= 939 \cdot 16 + 12 \\ 939 &= 58 \cdot 16 + 11 \quad \uparrow \\ 58 &= 3 \cdot 16 + 10 \quad \text{read up} \\ 3 &= 0 \cdot 16 + 3 \end{aligned}$$

Thus $15{,}036 = 3ABC_{\text{sixteen}}$. ■

Addition in Base b

Before we discuss how to add nondecimal numbers, let us examine the familiar addition algorithm in base 10.

To find the sum of any two decimal digits a and b, we find the remainder $r = (a+b) \bmod 10$ and the quotient $q = (a+b) \operatorname{div} 10$. Then $a+b = (qr)_{\text{ten}}$; q is the **carry** resulting from the addition of a and b. Using this idea we can add any two decimal integers.

Fortunately, the addition algorithm can be extended to any nondecimal base b in an obvious way. For example, let $x = (x_m \ldots x_0)_b$ and $y = (y_n \ldots y_0)_b$ where $m \geq n$. If $m > n$, we could assume that $y_{n+1} = \cdots = y_m = 0$. We add the corresponding digits in x and y in a right-to-left fashion. Let $s_i = (x_i + y_i + c_i) \bmod b$ and $c_{i+1} = (x_i + y_i + c_i) \operatorname{div} b$, where $c_0 = 0$. Then $x + y = (s_{m+1} s_m \ldots s_0)_b$ where s_{m+1} may be 0 or 1. (Leading zeros are deleted from the answer.)

These steps translate into a straightforward algorithm, as in Algorithm 4.4.

```
Algorithm addition (x,y,s,b)
(* This algorithm computes the sum s = (s_{m+1} s_m ... s_0) of two
   integers x = x_m ... x_0 and y = y_n ... y_0 in base b, where m ≥ n. *)
  Begin (* algorithm *)
    carry ← 0 (* initialize carry *)
    for i = 0 to n do
    begin (* for *)
      s_i ← (x_i + y_i + carry) mod b
      carry ← (x_i + y_i + carry) div b
    endfor
    for i = n+1 to m do
```

```
    begin (* for *)
      si ← (xi + carry) mod b
      carry ← (xi + carry) div b
    endfor
    if carry > 0 then
      sm+1 ← carry
  End (* algorithm *)
```

Algorithm 4.4

This algorithm is illustrated in the next two examples.

EXAMPLE 4.11 Add the binary integers 10110_{two} and 1011_{two}.

SOLUTION:
First write the integers one below the other in such a way that the corresponding bits are vertically aligned. See Figure 4.3. (For convenience, the base two is not shown.)

Figure 4.3

	1	0	1	1	0
+		1	0	1	1

Figure 4.4

				⓪	
	1	0	1	1	0
+		1	0	1	1
					1

Figure 4.5

			①	⓪	
	1	0	1	1	0
+		1	0	1	1
				0	1

Figure 4.6

	①	①	①	①	⓪	
		1	0	1	1	0
+			1	0	1	1
	1	0	0	0	0	1

Add the corresponding bits from right to left, beginning with the one's column: $0+1=1$. Since $1 \bmod 2 = 1$, enter 1 as the one's bit in the sum. Since $1 \text{ div } 2 = 0$, the resulting carry is 0, shown circled in Figure 4.4. (In practice when the carry is 0, it is simply ignored.) Now add the bits 0, 1, and 1 in the twos column: $0+1+1=2$. Since $2 \bmod 2 = 0$ and $2 \text{ div } 2 = 1$, enter 0 in the twos column and the new carry is 1 (see Figure 4.5). Continuing like this, we get the sum 100001_{two}. See Figure 4.6. ■

> The addition of binary numbers can be made easy by observing that $0+0=0$, $0+1=1=1+0$, and $1+1=10$, all in base two.

Next we illustrate the multiplication algorithm in base b.

Multiplication in Base b

The traditional algorithm for multiplying two integers x and y works for any base in an obvious way: multiply every digit in x by every digit in y as in base b and add up the partial products, as in Example 4.12.

EXAMPLE 4.12 Multiply 1011_{two} and 101_{two}.

SOLUTION:
The various steps unfold in Figures 4.7–4.9. The product is 110111_{two}.

Figure 4.7

$$\begin{array}{rcccc} & 1 & 0 & 1 & 1 \\ \times & & 1 & 0 & 1 \\ \hline & 1 & 0 & 1 & 1 \end{array} \quad \leftarrow \text{multiply 1011 by 1}$$

Figure 4.8

$$\begin{array}{cccccc} & & 1 & 0 & 1 & 1 \\ & \times & & 1 & 0 & 1 \\ \hline & & 1 & 0 & 1 & 1 \\ & 0 & 0 & 0 & 0 & \\ 1 & 0 & 1 & 1 & & \end{array} \quad \begin{array}{l} \\ \\ \\ \leftarrow \text{multiply 1011 by 0} \\ \leftarrow \text{multiply 1011 by 1} \end{array}$$

Figure 4.9

$$\begin{array}{cccccc} & & 1 & 0 & 1 & 1 \\ & \times & & 1 & 0 & 1 \\ \hline & & 1 & 0 & 1 & 1 \\ & 0 & 0 & 0 & 0 & \\ 1 & 0 & 1 & 1 & & \\ \hline 1 & 1 & 0 & 1 & 1 & 1 \end{array} \quad \text{add the partial products}$$

■

Shifting and Binary Multiplication

If you found these two examples confusing, don't be discouraged. Fortunately, most computers do binary multiplications using a technique called **shifting**, as discussed below.

Consider the binary number $x = (x_m x_{m-1} \ldots x_1 x_0)_{\text{two}} = \sum_{i=0}^{m} x_i 2^i$. What is the effect of multiplying x by 2^j? Since

$$x2^j = \sum_{i=0}^{m} x_i 2^{i+j} = x_m \ldots x_1 x_0 \underbrace{00\ldots 0}_{j \text{ zeros}}{}_{\text{two}},$$

every bit in x is shifted to the left by j columns.

More generally, let a be any bit. Then

$$x(a2^j) = \sum_{i=0}^{m} (ax_i) 2^{i+j} = (ax_m)\ldots(ax_0)\underbrace{00\ldots 0}_{j \text{ zeros}}{}_{\text{two}}$$

The bit ax_i equals x_i if $a = 1$ and equals 0 if $a = 0$. Thus, the effect of multiplying the number $x = (x_m \ldots x_0)_{\text{two}}$ by the bit y_j in the multiplicand $y = (y_n \ldots y_j \ldots y_0)_{\text{two}}$ is the same as multiplying each bit x_i by y_j and shifting the result to the left by j columns. Then add the partial products to get the desired product, as illustrated below.

EXAMPLE 4.13 Evaluate $1011_{\text{two}} \times 101_{\text{two}}$.

SOLUTION:

The various steps are displayed in Figures 4.10–4.13. It follows from Figure 4.13 that the resulting product is 110111_{two}.

Figure 4.10

```
    1 0 1 1
×     1 0 1
  ---------
    1 0 1 1   ← multiply 1011 by 1; no shifting.
```

Figure 4.11

```
    1 0 1 1
×     1 0 1
  ---------
    1 0 1 1
  0 0 0 0     ← multiply 1011 by 0; shift by one column.
```

Figure 4.12

```
      1 0 1 1
  ×     1 0 1
    ---------
      1 0 1 1
    0 0 0 0
  1 0 1 1       ← multiply 1011 by 1; shift by 2 columns.
```

Figure 4.13

$$
\begin{array}{rrrrrr}
 & & 1 & 0 & 1 & 1 \\
\times & & & 1 & 0 & 1 \\
\hline
 & & 1 & 0 & 1 & 1 \\
 & 0 & 0 & 0 & 0 & \\
1 & 0 & 1 & 1 & & \\
\hline
1 & 1 & 0 & 1 & 1 & 1
\end{array}
\quad \text{add the partial products.}
$$

■

The shifting method of multiplication leads to Algorithm 4.5 for multiplying two binary numbers.

```
Algorithm binary multiplication (x, y, p)
(* This algorithm computes the product p = (p_{m+n}p_{m+n-1}...p_0)_two
   of the binary numbers x = (x_m x_{m-1}...x_1 x_0)_two and
   y = (y_n y_{n-1}...y_1 y_0)_two, using shifting. *)
  Begin (* algorithm *)
    for j = 0 to n do
      begin (* for *)
        multiply each bit x_i by y_j
        shift the resulting binary word to the left
        by j columns
        w_j ← resulting binary word
      endfor
    add the partial products w_j
    p ← resulting sum
  End (* algorithm *)
```

Algorithm 4.5

Binary Subtraction

We can subtract binary numbers without the bother of "borrows," using one's complement and addition. The **one's complement** x' of a binary number x is obtained by replacing each 0 in x with a 1 and vice versa. For example, the one's complement of 1011_{two} is 0100_{two} and that of 1001_{two} is 0110_{two}. The **two's complement** of x is $x' + 1$. For instance, the two's complement of 1011_{two} is $0100_{\text{two}} + 1 = 0101_{\text{two}}$.

The next example illustrates this new technique step-by-step before it is justified in a formal discussion.

EXAMPLE 4.14 Subtract 1011_{two} from 100001_{two}.

SOLUTION:

For convenience, we shall drop the base two.

Step 1 Find the one's complement of the subtrahend 1011. Since the minuend 100001 contains six bits, keep the same number of bits in the subtrahend by padding it with two 0's at the beginning. The one's complement of $1011 = 001011$ is 110100.

Step 2 Find the two's complement by adding 1 to the one's complement: $110100 + 1 = 110101$.

Step 3 Add the two's complement in step 2 to the minuend 100001:

$$\begin{array}{r} 1\ 0\ 0\ 0\ 0\ 1 \\ +\ 1\ 1\ 0\ 1\ 0\ 1 \\ \hline \textcircled{1}\ 0\ 1\ 0\ 1\ 1\ 0 \end{array}$$

delete ⟶ (the circled leading 1)

Step 4 Delete the leading carry 1. The resulting number $010110 = 10110$ is the desired answer.

Thus $100001_{\text{two}} - 1011_{\text{two}} = 10110_{\text{two}}$. (To check this, you may verify that $1011_{\text{two}} + 10110_{\text{two}} = 100001_{\text{two}}$.) ■

How can this technique work? To justify the algorithm illustrated, first notice that $x - y = x + (-y)$; that is, subtracting y from x is equivalent to adding the additive inverse $-y$ of y to x. This is the basic idea behind the binary subtraction algorithm.

Now how to find $-y$? First, assume that $\|x\| = \|y\| = n$. (If $\|y\| < \|x\|$, pad y with enough 0's at the beginning so the length of the resulting word is n.) Let y' denote the one's complement of y. Then $y + y'$ is an n-bit word w containing all 1's:

	$n-1$	$n-2$					
$w =$	1	1	1	...	1	1	1

For example, let $y = 10110$. Then $y' = 01001$, so $y + y' = 11111$.

The value of the n-bit word w is $2^n - 1$ (see Section 4.4). Thus $y + y' = w = 2^n - 1$, so $-y = y' + 1 - 2^n = y'' - 2^n$, where $y'' = y' + 1$ denotes the two's complement of y. Therefore, $x + (-y) = x + y'' - 2^n = (x + y'') - 2^n$. Thus, to subtract y from x, it suffices to add y'' to x and drop the leading carry 1. This explains why the above subtraction algorithm works.

The algorithm for the case $\|x\| < \|y\|$ is complicated, so we omit its discussion here.*

We close this section with an intriguing numeric puzzle that will test your mastery of both nondecimal addition and subtraction.

---

*For a discussion of negative binary numbers, see A. S. Tanerbaum, *Structured Computer Organization*, Prentice Hall, Englewood, NJ, 1976, pp. 420–423.

A Nondecimal Puzzle

Write down a three-digit number in base eight. Reverse its digits. Subtract the smaller number from the other (in base eight); save all leading zeros. Reverse its digits. Add the last two numbers. Is your answer 1067_{eight}? Now redo the puzzle in base 12; your answer should be $10AB_{\text{twelve}}$.

Exercises 4.3

Express each number in base 10.

1. 1101_{two} **2.** 11011_{two} **3.** 1776_{eight} **4.** 1976_{sixteen}

Express each decimal number as required.

5. $1076 = (\quad)_{\text{two}}$ **6.** $676 = (\quad)_{\text{eight}}$

7. $1776 = (\quad)_{\text{eight}}$ **8.** $2076 = (\quad)_{\text{sixteen}}$

The binary representation of an integer can conveniently be used to find its octal representation. Group the bits in threes from right to left and replace each group with the corresponding octal digit. For example,

$$243 = 11110011_{\text{two}} = 011\ \ 110\ \ 011_{\text{two}} = 363_{\text{eight}}$$

Using this short cut, rewrite each binary number as an octal integer.

9. 1101_{two} **10.** 11011_{two} **11.** 111010_{two} **12.** 10110101_{two}

The binary representation of an integer can also be used to find its hexadecimal representation. Group the bits in fours from right to left and then replace each group with the equivalent hexadecimal digit. For instance,

$$243 = 11110011_{\text{two}} = 1111\ \ 0011_{\text{two}} = \text{F}3_{\text{sixteen}}$$

Using this method express each binary number in base 16.

13. 11101_{two} **14.** 110111_{two} **15.** 1110101_{two} **16.** 10110101_{two}

The techniques explained in Exercises 9–12 are reversible; that is, the octal and hexadecimal representations of integers can be used to find their binary representations. For example,

$$345_{\text{eight}} = 011\ \ 100\ \ 101_{\text{two}} = 11100101_{\text{two}}$$

Using this technique, rewrite each number in base two.

17. 36_{sixteen} **18.** 237_{eight} **19.** 237_{sixteen} **20.** $3\text{AD}_{\text{sixteen}}$

In Exercises 21–28, perform the indicated operations.

21. $\begin{array}{r} 1111_{\text{two}} \\ +\ 1011_{\text{two}} \end{array}$ **22.** $\begin{array}{r} 1076_{\text{eight}} \\ +\ 2076_{\text{eight}} \end{array}$ **23.** $\begin{array}{r} 3076_{\text{sixteen}} \\ +\ 5776_{\text{sixteen}} \end{array}$ **24.** $\begin{array}{r} 101101_{\text{two}} \\ -\ 10011_{\text{two}} \end{array}$

25. $\begin{array}{r} 11000_{\text{two}} \\ -\ 100_{\text{two}} \end{array}$ **26.** $\begin{array}{r} 10111_{\text{two}} \\ \times\ 1101_{\text{two}} \end{array}$ **27.** $\begin{array}{r} 1024_{\text{eight}} \\ \times\ 2776_{\text{eight}} \end{array}$ **28.** $\begin{array}{r} 3ABC_{\text{sixteen}} \\ \times\ 4CBA_{\text{sixteen}} \end{array}$

29. Arrange the binary numbers 1011, 110, 11011, 10110, and 101010 in order of increasing magnitude.

30. Arrange the hexadecimal numbers 1076, 3056, 3CAB, 5ABC, and CACB in order of increasing magnitude.

31. What can you say about the ones bit in the binary representation of an even integer? An odd integer?

Find the value of the base b in each case.

32. $54_b = 64$ **33.** $1001_b = 9$ **34.** $1001_b = 126$ **35.** $144_b = 49$

36. Suppose a space investigative team to Venus sends back the picture of an addition problem scratched on a wall, as shown in Figure 4.14. The Venusian numeration system is a place value system, just like ours. The base of the system is the same as the number of fingers on a Venusian hand. Determine the base of the Venusian numeration system. (This puzzle is due to H. L. Nelson.**)

Figure 4.14

The sum in "Venusian" notation.

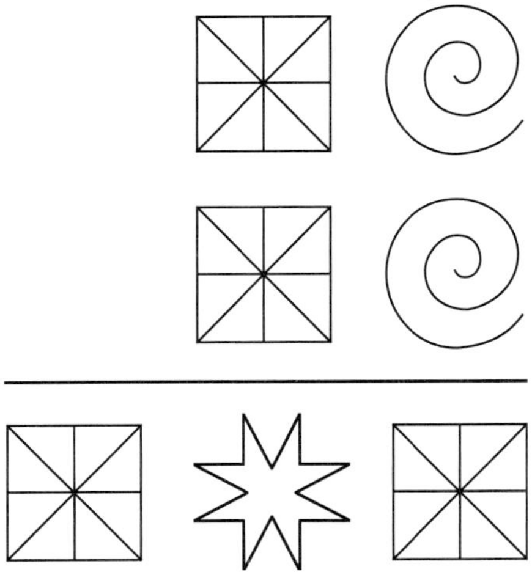

Define recursively each set S of binary words.

37. Set of binary words that represent even positive integers.

38. Set of binary words that represent odd positive integers.

**M. Gardner, "Mathematical Games," *Scientific American*, Vol. 219, Sept. 1968, pp. 218–230

39. Set of binary words that represent positive integers with no leading zeros.

40. Set of palindromic binary words.

Polynomials can be evaluated efficiently using the technique of **nested multiplication**, called **Horner's method**. [This method is named after the English schoolmaster, William G. Horner (1786–1837), who published it in 1819.] For instance, the polynomial $f(x) = 4x^3 + 5x^2 + 6x + 7$ can be evaluated as $f(x) = ((4x + 5)x + 6)x + 7$. Using this method, express each integer as a decimal integer.

41. 245_{eight} **42.** 101101_{two} **43.** 1100101_{two} **44.** $43\text{BC}_{\text{sixteen}}$

***45.** Let x be a three-digit hexadecimal number with distinct digits. Reverse the digits. Subtract the smaller number from the other number (save all the digits in your answer). Reverse the digits in the difference. Add this number to x. Find the sum.

4.4 Mathematical Induction

The principle of mathematical induction† (PMI) is a frequently used proof technique in both mathematics and computer science, as will be seen shortly.

Many interesting results in mathematics hold true for all positive integers. For example, the following statements are true for every positive integer n, where x, y, and x_i are any positive real numbers:

- $(x \cdot y)^n = x^n \cdot y^n$
- $\log(x_1 \dots x_n) = \sum_{i=1}^{n} \log x_i$
- $\sum_{i=1}^{n} i = \frac{n(n+1)}{2}$
- $\sum_{i=0}^{n-1} r^i = \frac{r^n - 1}{r - 1} (r \neq 1)$

How do we prove that these results hold for every positive integer n? Obviously, it is impossible to substitute each positive integer for n and verify that the formula holds. The principle of induction can establish the validity of such formulas.

To begin with, suppose the orange cans in a collection can be arranged as in Figure 4.15. Row 1 contains one can, row 2 contains two cans, ..., row n contains n cans. Can you predict a formula for the total number of cans in the collection? See Example 4.15 for a formula.

†Although the Venetian scientist Francesco Maurocylus (1491–1575) applied it in proofs in a book he wrote in 1575, the term *mathematical induction* was coined by De Morgan.

Figure 4.15

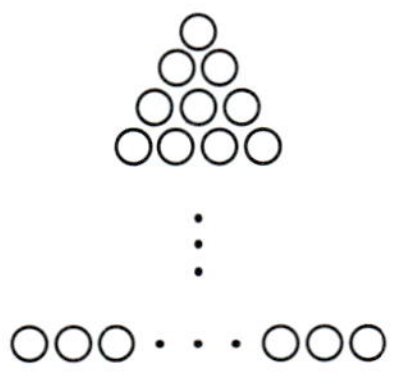

The next result is the cornerstone of the principle of induction. Its proof, as we shall see shortly, follows by the well-ordering principle in Section 4.1.

THEOREM 4.9 Let S be a subset of $\mathbb{N}$ satisfying the following properties:

(1) $1 \in S$.
(2) If k is an arbitrary positive integer in S, then $k + 1 \in S$. Then $S = \mathbb{N}$.

PROOF (by contradiction):
Suppose $S \neq \mathbb{N}$. Let $S' = n \in N \mid n \notin S$. Since $S' \neq \emptyset$, by the well-ordering principle, S' contains a least element ℓ'. Then $\ell' > 1$ by condition 1. Since ℓ' is the least element in S', $\ell' - 1 \notin S'$; so $\ell' - 1 \in S$. Consequently, by condition 2, $(\ell' - 1) + 1 = \ell' \in S$. This contradiction establishes the theorem. ■

This theorem can be generalized as in Theorem 4.10. We leave its proof as an exercise.

THEOREM 4.10 Let n_0 be a fixed integer. Let S be a subset of $\mathbf{Z}$ satisfying the following conditions:

- $n_0 \in S$.
- If k is an arbitrary integer $\geq n_0$ such that $k \in S$, then $k + 1 \in S$.

Then $S \supseteq \{n \in \mathbf{Z} \mid n \geq n_0\}$. ■

Weak Version of Induction

Before we formalize the principle of induction, let's look at a trivial example. Consider an infinite number of dominoes arranged in a row (see Figure 4.16a). Suppose we knock down the first domino.

What happens to the rest of the dominoes? Do they all fall? Not necessarily; see Figures 4.16b and c.

So let's further assume the following: If the kth domino is knocked down, then the $(k + 1)$st domino also falls down. If we topple the first domino, what would happen to the rest? They all would fall; see Figure 4.16d.

This illustration can be expressed in symbols. Let $\mathrm{P}(n)$ denote the predicate that the nth domino falls. (*Note*: UD $= \mathbb{N}$.) Assume the following propositions are true:

- $\mathrm{P}(1)$.
- $\mathrm{P}(k) \rightarrow \mathrm{P}(k + 1)$ for every positive integer k.

Figure 4.16

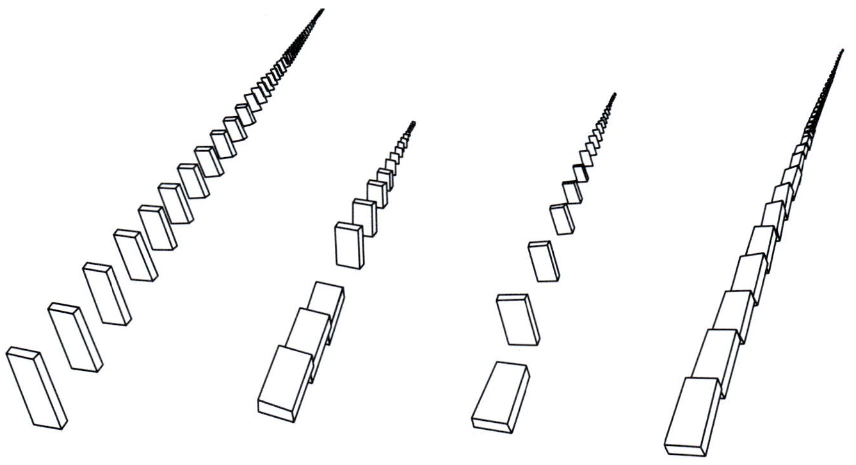

Then P(n) is true for every positive integer n; that is, every domino would fall. This leads us to the **weak version** of the principle.

THEOREM 4.11 **(The Principle of Mathematical Induction)** Let P(n) be a predicate satisfying the following conditions, where n is an integer:

(1) P(n_0) is true for some integer n_0.
(2) If P(k) is true for an arbitrary integer $k \geq n_0$, then P($k + 1$) is also true.

Then P(n) is true for every integer $n \geq n_0$.

PROOF:
Let S denote the set of integers $\geq n_0$ for which P(n) is true. Since P(n_0) is true, $n_0 \in S$. By condition 2, whenever $k \in S$, $k+1 \in S$. Therefore, by Theorem 4.10, S consists of all integers $\geq n_0$. Consequently, P(n) is true for every integer $n \geq n_0$. This establishes the validity of the principle. ∎

Condition 1 assumes the proposition P(n) is true when $n = n_0$. Look at condition 2: If P(n) is true for an arbitrary integer $k \geq n_0$, it is also true for $n = k + 1$. Then, by the repeated applications of condition 2 and the law of detachment, it follows that P($n_0 + 1$), P($n_0 + 2$), ... all hold true. In other words, P(n) holds for every $n \geq n_0$.

Proving a result by PMI involves two key steps:

1. **Basis step** Verify that P(n_0) is true.
2. **Induction step** Assume P(k) is true for an arbitrary integer $k \geq n_0$ (**inductive hypothesis**).
   Then verify that P($k + 1$) is also true.

> *A word of caution*: A question frequently asked is, "Isn't this cyclic reasoning? Are you not assuming what you are asked to prove?" The confusion stems from misinterpreting step 2 for the conclusion. The induction step involves showing that the implication $P(k) \rightarrow P(k+1)$ is a tautology; that is, if $P(k)$ is true, then so is $P(k+1)$. The conclusion is "$P(n)$ is true for every $n \geq n_0$." So be careful.

A variety of interesting examples will show how useful this important proof technique is.

The next example gives a nice formula for computing the total number of cans in the collection in Figure 4.15.

EXAMPLE 4.15 Using PMI, prove that, for every positive integer n,

$$1 + 2 + 3 + \cdots + n = \frac{n(n+1)}{2}$$

PROOF (by induction):

Let $P(n)$: $\sum_{i=1}^{n} i = \dfrac{n(n+1)}{2}$.

Basis step To verify that $P(1)$ is true (*Note*: Here $n_0 = 1$):

When $n = 1$, RHS $= \frac{1(1+1)}{2} = 1 = \sum_{i=1}^{1} i =$ LHS; so $P(1)$ is true.

Induction step Let k be an arbitrary positive integer. We would like to show that $P(k) \rightarrow P(k+1)$: Assume $P(k)$ is true; that is,

$$\sum_{i=1}^{k} i = \frac{k(k+1)}{2} \qquad \longleftarrow \text{inductive hypothesis}$$

To establish that $P(k) \rightarrow P(k+1)$ is true, that is,

$$\sum i = \frac{(k+1)(k+2)}{2}$$

we start with the LHS of this equation:

$$\text{LHS} = \sum_{i=1}^{k+1} i = \sum_{i=1}^{k} i + (k+1) \qquad \left(\text{Note}: \sum_{i=1}^{k+1} x_i = \sum_{i=1}^{k} x_i + x_{k+1}\right)$$

$$= \frac{k(k+1)}{2} + (k+1), \qquad \text{by the inductive hypothesis}$$

$$= \frac{(k+1)(k+2)}{2}$$

$$= \text{RHS}$$

Thus, if P(k) is true, then P($k+1$) is also true.

Therefore, by PMI, P(n) is true for every $n \geq 1$; that is, the formula holds for every positive integer n. ■

Figure 4.17 provides a geometric proof of this formula without words.

Figure 4.17

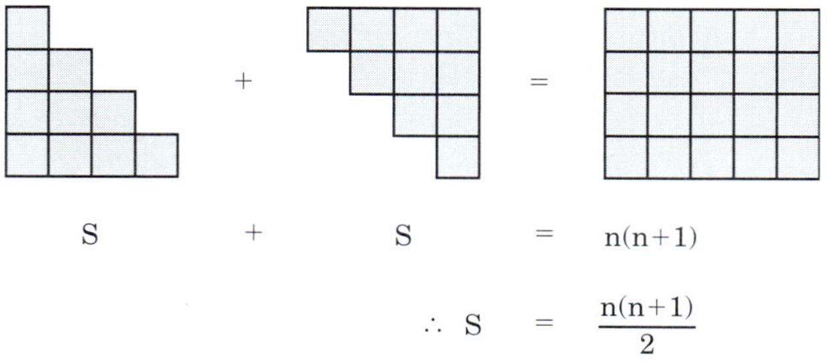

The next example, again an application of induction, employs a divisibility property, so we follow it in some detail.

EXAMPLE 4.16 Prove that $2n^3 + 3n^2 + n$ is divisible by 6 for every integer $n \geq 1$.

PROOF (by PMI):
Let P(n): $2n^3 + 3n^2 + n$ is divisible by 6.

Basis step When $n=1$, $2n^3 + 3n^2 + n = 2(1) + 3(1) + 1 = 6$ is clearly divisible by 6. Therefore, P(1) is true.

Induction step Assume P(k) is true, that is, $2k^3 + 3k^2 + k$ is divisible by 6 for any $k \geq 1$. Then $2k^3 + 3k^2 + k = 6m$ for some integer m (inductive hypothesis). We must show that P($k+1$) is true; that is, $2(k+1)^3 + 3(k+1)^2 + (k+1)$ is divisible by 6. Notice that

$$\begin{aligned} &2(k+1)^3 + 3(k+1)^2 + (k+1) \\ &\quad = 2(k^3 + 3k^2 + 3k + 1) + 3(k^2 + 2k + 1) + (k+1) \\ &\quad = (2k^3 + 3k^2 + k) + 6(k^2 + 2k + 1) \\ &\quad = 6m + 6(k^2 + 2k + 1) \qquad \text{by the inductive hypothesis} \\ &\quad = 6(m + k^2 + 2k + 1), \end{aligned}$$

which is clearly divisible by 6. Thus P($k+1$) is true.

Thus, by induction, the given statement is true for every $n \geq 1$. ■

Notice that in the above examples, $n_0 = 1$, but it need not always be 1, as the next example shows.

***Jacob I. Bernoulli** (1654–1705), a member of the most distinguished family of mathematicans (see the family tree in Section 9.1), was born in Basel, Switzerland. His grandfather, a pharmacist in Amsterdam, had become a Swiss through marriage, and his father was a town councilor and a magistrate.*

Bernoulli received his M.A. in philosophy in 1671 and a theological degree 5 years later. During this time, he studied mathematics and astronomy against his father's will. He spent the next 2 years tutoring in Geneva. In 1687 he became professor of mathematics at the University of Basel, remaining there until his death. His brother Johann succeeded him at Basel.

In May 1690 he used the term integral *in the calculus sense known today. Bernoulli's most famous work,* Ars Conjectandi, *was published posthumously in 1713. It contains significant contributions to probability theory, the theory of series, and gravitational theory.*

EXAMPLE 4.17 **(Bernoulli's Inequality)** Let x be any real number greater than -1. Prove that $(1+x)^n \geq 1+nx$ for every $n \geq 0$.

PROOF (by PMI):
Let x be any real number > -1. Let P(n): $(1+x)^n \geq 1+nx$. (*Note*: The induction is on the discrete variable n and not on the "continuous" variable x.)

Basis step To verify that P(0) is true: Notice that

$$\begin{aligned}(1+x)^0 &= 1\\ &\geq 1+0x\end{aligned}$$

So P(0) is true. (*Note*: Here $n_0 = 0$.)

Induction step Assume P(k) is true; that is, $(1+x)^k \geq 1+kx$ for an arbitrary integer $k \geq 0$. We need to show that P($k+1$) is true; that is, $(1+x)^{k+1} \geq 1+(k+1)x$.

By the inductive hypothesis, we have $(1+x)^k \geq 1+kx$. Then

$$\begin{aligned}(1+x)^{k+1} &= (1+x)(1+x)^k,\\ &\geq (1+x)(1+kx), && \text{by IH and since } 1+x>0\\ &= 1+(k+1)x+kx^2\\ &\geq 1+(k+1)x, && \text{since } kx^2 \geq 0\end{aligned}$$

Therefore, P($k+1$) is also true.
Thus, by PMI, $(1+x)^n \geq 1+nx$ for every $n \geq 0$. ■

The next example inductively establishes Theorem 2.3 from Chapter 2.

EXAMPLE 4.18 A finite set A with n elements has exactly 2^n subsets.

PROOF (by PMI):

Basis step When $n = 0$, $A = \emptyset$, so A has exactly $1 = 2^0$ subset. Thus the result is true when $n = 0$.

Induction step Assume any finite set with k elements has 2^k subsets, where $k \geq 0$. Let A be a set with $k + 1$ elements. We would like to show that A has 2^{k+1} subsets.

To this end, let $x \in A$. Let $B = A - \{x\}$. Since $|B| = k$, B has 2^k subsets by the inductive hypothesis. Each of the subsets of B is a subset of A. Now add x to each of them. The resulting 2^k sets are also subsets of A. Since every subset of A either contains x or does not contain x, by the addition principle, A has $2^k + 2^k = 2^{k+1}$ subsets.

Thus, by the principle of induction, the result holds for every finite set. ■

> Both the basis and the induction steps are essential in the principle of induction, as the next two examples illustrate.

EXAMPLE 4.19 Let $g(n)$ denote the maximum number of nonoverlapping regions formed inside a circle by joining n distinct points on it. Figures 4.18–4.22 show the cases $n = 1, 2, 3, 4$, and 5, where the various regions are numbered 1, 2, 3, etc. The results are summarized in Table 4.1.

Figure 4.18

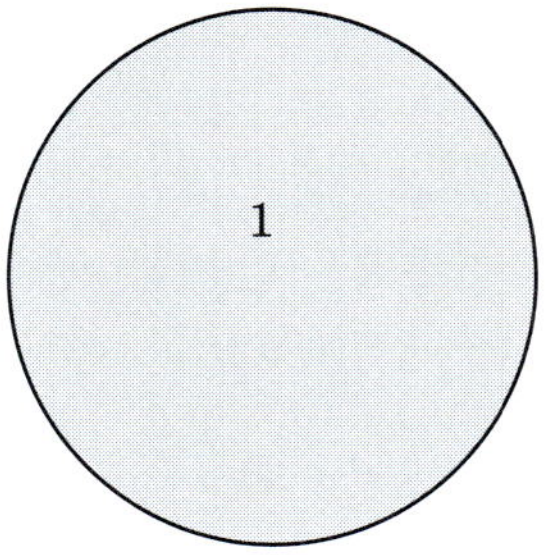

Figure 4.19

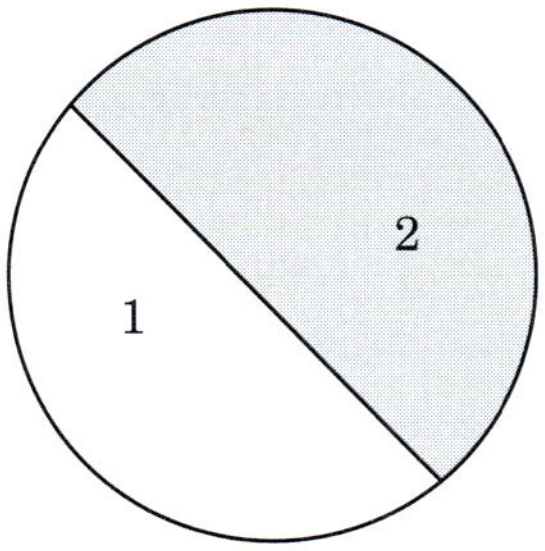

Figure 4.20

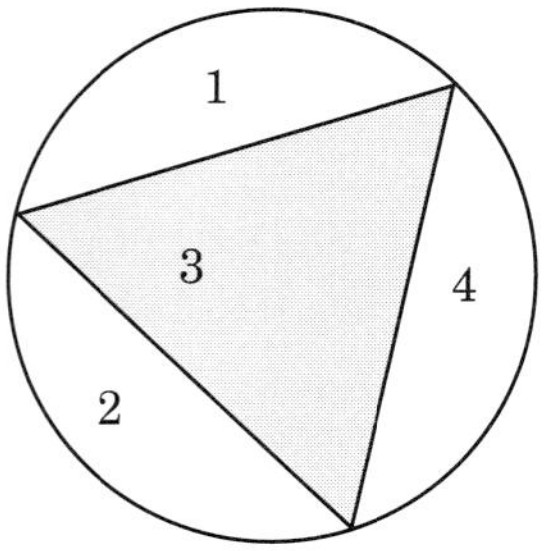

Figure 4.21

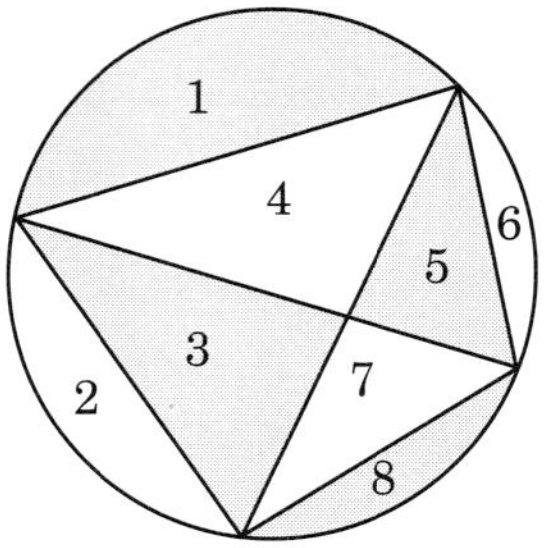

Figure 4.22

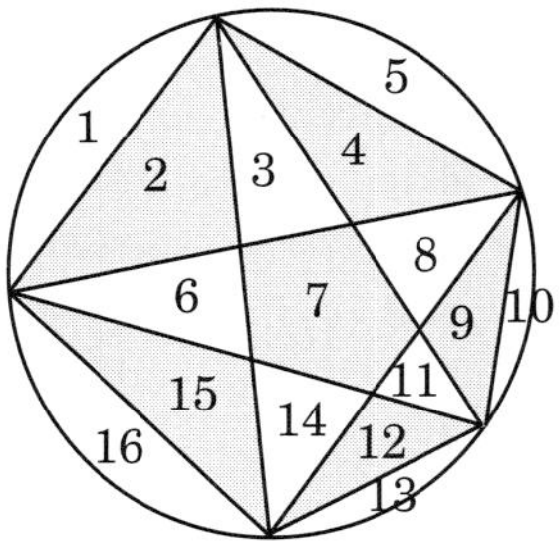

Figure 4.23

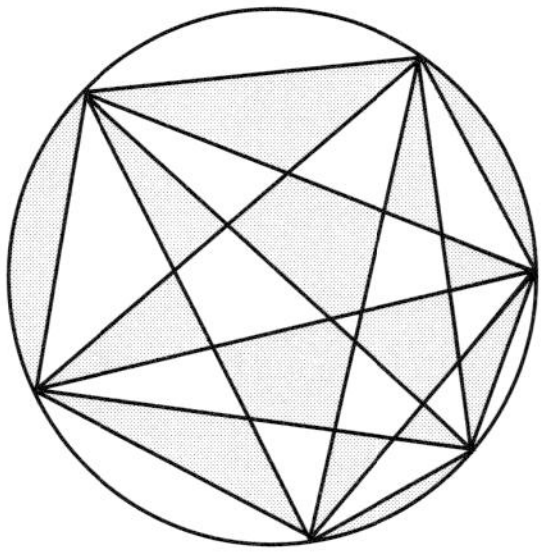

It appears from the table that $g(n) = 2^{n-1}$. Then $g(1) = 2^0 = 1$, which is true (basis step). Nonetheless, this does not guarantee that $g(n) = 2^{n-1}$ for every $n \geq 1$. If the formula were true, there would be $g(6) = 2^5 = 32$ nonoverlapping regions with six points. Unfortunately, there are only 31 such regions (see Figure 4.23) We shall derive the correct formula in Chapter 6.

Table 4.1

Number of points n	**1**	**2**	**3**	**4**	**5**	**6**
Maximum number of nonoverlapping regions $g(n)$	1	2	4	8	16	?

We can conclude that the truthfulness of the basis step and an apparent pattern do *not* ensure that P(n) is true for every n. ■

The following example shows that the validity of the induction step is necessary, but not sufficient, to guarantee that P(n) is true for all integers in the UD.

EXAMPLE 4.20 Consider the "formula" $\text{P}(n): 1+3+5+\cdots+(2n-1)=n^2+1$. Suppose P($k$) is true: $\sum_{i=1}^{k}(2i-1)=k^2+1$. Then:

$$\begin{aligned}\sum_{i=1}^{k+1}(2i-1) &= \sum_{i=1}^{k}(2i-1)+(2k+1)\\ &= (k^2+1)+(2k+1)\\ &= (k+1)^2+1\end{aligned}$$

So if P(k) is true, P($k+1$) is true. Nevertheless, the formula does not hold for any positive integer n. Try P(1) (see Exercise 5). ■

Using induction, the next example "proves" that every person is of the same sex.

EXAMPLE 4.21 "Prove" that every person in a set of n people is of the same sex.

"PROOF":
Let P(n): Everyone in a set of n people is of the same sex. Clearly, P(1) is true. Let k be a positive integer such that P(k) is true; that is, everyone in a set of k people is of the same sex. To show that P($k+1$) is true, consider a set $A=\{a_1,a_2,\ldots,a_{k+1}\}$ of $k+1$ people. Partition A into two overlapping sets, $B=\{a_1,a_2,\ldots,a_k\}$ and $C=\{a_2,\ldots,a_{k+1}\}$, as in Figure 4.24. Since $|B|=k=|C|$, by the inductive hypothesis, everyone in B is of the same sex and everyone in C is of the same sex. Since B and C overlap, everyone in $B\cup C$ must be of the same sex; that is, everyone in A is of the same sex.

Thus, by PMI, P(n) is true for every $n\geq 1$. This concludes the "proof." ■

> *Note*: The assertion that everyone is of the same sex is clearly false. Can you find the flaw in the "proof"? See Exercise 46.

Figure 4.24

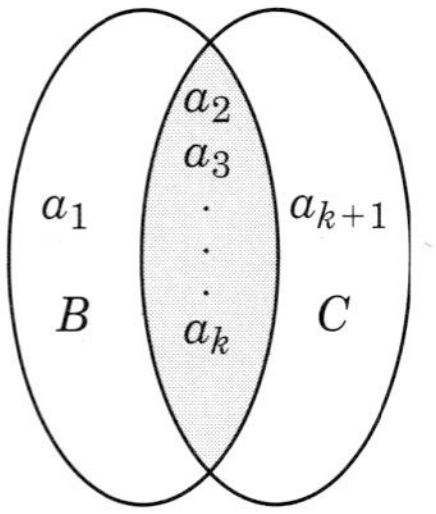

Before discussing the second version of the principle of induction, we will look at a few applications of the formula in Example 4.15. First a definition.

Polygonal Number

A **polygonal number** is a positive integer n that can be represented by n dots in a polygonal array in a systematic fashion. For example, the integers 1, 3, 6, 10, ... are **triangular numbers** since they can be represented by triangular arrays, as shown in Figure 4.25; the number of pins in a bowling alley and that of balls in the game of pool are triangular numbers. Let t_n denote the nth triangular number. Then

$$t_n = 1 + 2 + 3 + \cdots + n = \frac{n(n+1)}{2}$$

Figure 4.25

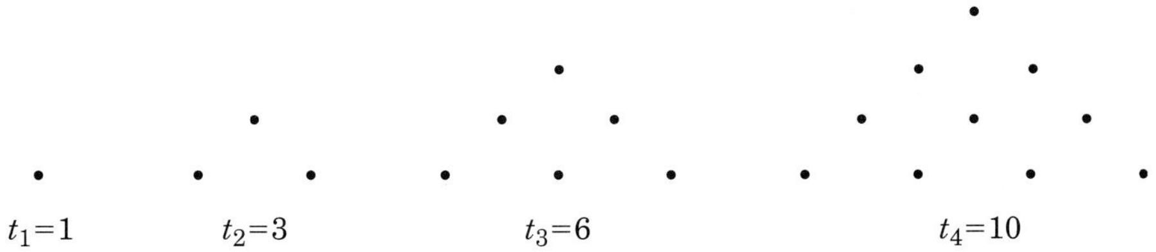

Triangular numbers manifest delightful properties. For example, $t_n + t_{n-1} = n^2$; Figures 4.26 and 4.27 provide a nonverbal, geometric proof of this result. See Exercises 47–50.

Figure 4.26

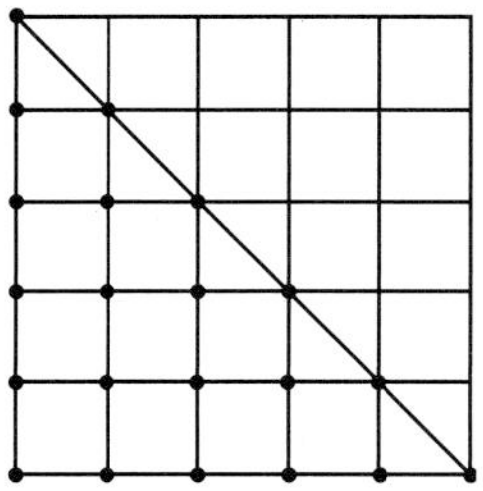

Figure 4.27

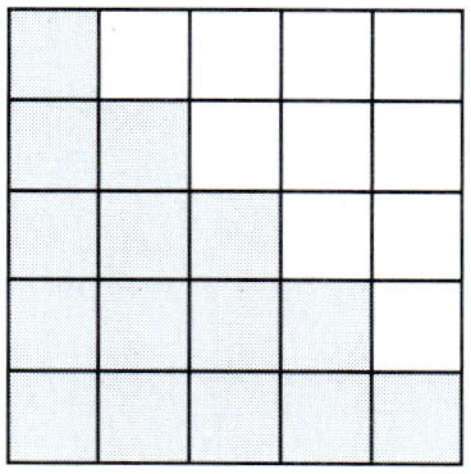

The next example is another application of the formula in Example 4.15 and the generalized pigeonhole principle.

EXAMPLE 4.22 Let $a_1, a_2, \ldots, a_n$ be the first n positive integers in some order. Suppose they are arranged around a circle (see Figure 4.28). Let k be any positive integer $\leq n$. Prove that there exists a set of k consecutive elements in the arrangement with a sum $\lfloor [kn(n+1) - 2]/2n \rfloor$, where $\lfloor x \rfloor$ denotes the floor of x.

Figure 4.28

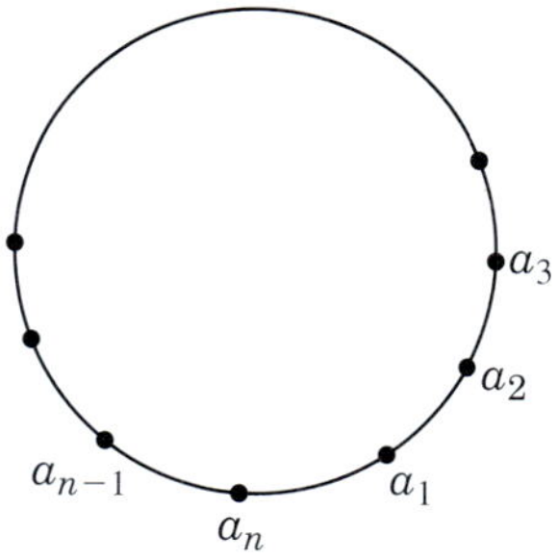

PROOF:

Consider the following sums:

$$
\begin{aligned}
S_1 &= a_1 + a_2 + \cdots + a_k \\
S_2 &= a_2 + a_3 + \cdots + a_{k+1} \\
&\vdots \\
S_n &= a_n + a_1 + \cdots + a_{k-1}
\end{aligned}
$$

Each of the first n positive integers appears k times in this set of sums. Then

$$\sum_{i=1}^{n} S_i = k\left(\sum_{i=1}^{n} a_i\right) = k\left(\sum_{i=1}^{n} i\right) = \frac{kn(n+1)}{2}, \text{ by Example 4.15}$$

Consider $kn(n+1)/2$ pigeons. We would like to distribute them among n pigeonholes, called $S_1, S_2, \ldots, S_n$. By the generalized pigeonhole principle, at least one of the pigeonholes S_i must contain more than $\lfloor kn(n+1)/2n - 1/n \rfloor = \lfloor [kn(n+1-2)]/2n \rfloor$ pigeons. In other words, $s_i > \lfloor kn(n+1) - 2/2n \rfloor$, as desired. ■

In particular, if numbers 1 through 10 are randomly placed around a circle, at least three consecutive integers in the arrangement must have a sum exceeding $\lfloor [3 \cdot 10 \cdot 11 - 2]/(2 \cdot 10) \rfloor = 16$.

We now discuss the strong version of the principle of induction.

Strong Version of Induction

Sometimes the truth of P(k) might not be enough to establish that of P($k+1$). In other words, the truthfulness of P($k+1$) may require more than that of P(k). In such cases, we have to assume a stronger inductive hypothesis that P(n_0), P(n_0+1), ..., P(k) are all true; then verify that P($k+1$) is also true. This **strong version,** which can be proved using the weak version (see Exercise 57), is stated as follows.

THEOREM 4.12 **(The Second Principle of Mathematical Induction)** Let P(n) be a predicate satisfying the following conditions, where n is any integer:

- P(n_0) is true for some integer n_0.
- If k is an arbitrary integer $\geq n_0$ such that $P(n_0) \wedge P(n_0+1) \wedge \cdots \wedge P(k)$ is true, then P($k+1$) is also true. Then P(n) is true for every $n \geq n_0$.

The next theorem illustrates this proof technique. ■

THEOREM 4.13 **(The Fundamental Theorem of Arithmetic)** Every positive integer $n \geq 2$ either is a prime or can be written as a product of primes.

PROOF (by strong induction):
Let P(n) denote the given predicate.

Basis step Choose $n_0 = 2$. Since 2 is itself a prime, P(2) is true.

Inductive step Let k be a positive integer ≥ 2 such that P(2), P(3), ..., P(k) are true; that is, assume that integers 2 through k are primes or can be written as products of primes. We would like to show that P($k+1$) is also true; that is, integer $k+1$ is a prime or can be expressed as a product of primes.

If $k+1$ is itself a prime, then we are done. If $k+1$ is not a prime, it must be the product of two positive integers x and y, where $1 < x, y < k+1$. By the inductive hypothesis, both x and y are primes or products of primes. Therefore, $k+1 = x \times y$ is also a product of two or more prime numbers. In other words, P($k+1$) also holds:

Thus, by the strong version of induction, P(n) is true for every $n \geq 2$. ■

We now present an interesting application of the fundamental theorem of arithmetic, which is the cornerstone of number theory, and the floor function.

EXAMPLE 4.23 Find the number of trailing zeros in 123!

SOLUTION:
By the fundamental theorem of arithmetic, 123! can be factored as $2^a 5^b c$, where c denotes the product of primes other than 2 and 5. Clearly $a > b$. Each trailing zero in 123! corresponds to a factor of 10 and vice versa.

$$\begin{aligned}\therefore \text{ Number of trailing zeros} &= \begin{pmatrix}\text{Number of products of the form}\\ 2\cdot 5 \text{ in the prime factorization}\end{pmatrix}\\ &= \text{minimum of } a \text{ and } b\\ &= b, \text{ since } a > b\end{aligned}$$

We proceed to find b:

Number of positive integers ≤ 123 and divisible by $5 = \lfloor 123/5 \rfloor = 24$

Each of them contributes a 5 to the prime factorization of 123!

Number of positive integers ≤ 123 and divisible by $25 = \lfloor 123/25 \rfloor = 4$

(See Figure 4.29.) Each of them contributes an additional 5 to the prime factorization. Since no higher power of 5 contributes a 5 in the prime factorization of 123!, the total number of 5's in the prime factorization equals $24 + 4 = 28$. Thus the total number of trailing zeros in 123! is 28.

Figure 4.29

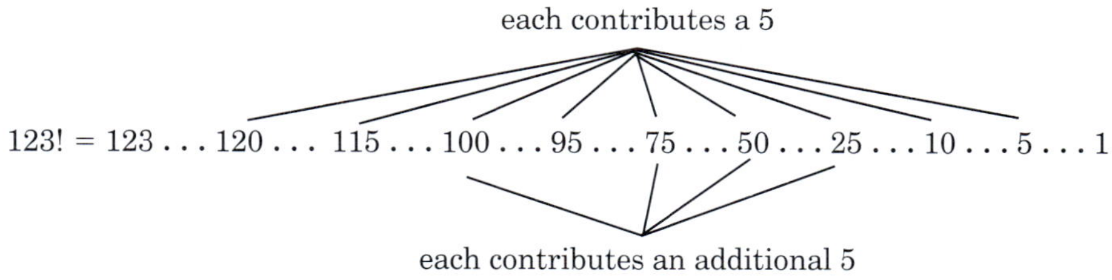

■

The next example is another interesting application of the floor function. It employs the following facts from number theory:

- *Every positive integer that is not a square has an even number of positive factors.* For example, 18 has six positive factors: $1, 2, 3, 6, 9, 18$; 21 has four: $1, 3, 7, 21$; 19 has two: $1, 19$.
- *Every perfect square has an odd number of positive factors.* For example, 25 has three positive factors, namely, $1, 5$, and 25; 64 has seven: $1, 2, 4, 8, 16, 32$, and 64.

- *There are* $\lfloor\sqrt{n}\rfloor$ *perfect squares* $\leq n$.
- For example, there are $\lfloor\sqrt{27}\rfloor = 5$ perfect squares not exceeding $27 : 1, 4, 9, 16, 25$; there are $\lfloor\sqrt{68}\rfloor = 8$ perfect squares $< 68 : 1, 4, 9, 16, 25, 36, 49, 64$.

EXAMPLE 4.24 There are 1000 rooms in a hotel and every room is occupied by a guest. The first guest opens the door to every room. The nth guest closes every nth door if it is open and opens it otherwise, where $2 \leq n \leq 1000$. How many doors will be open at the end?*

SOLUTION:
Before applying these results to solve the puzzle, let us study a mini-version with 10 tenants and 10 apartments. The first tenant opens all 10 doors; the second tenant closes the 2nd, 4th, 6th, 8th, and 10th doors; the third closes the 3rd door, opens the 6th door, and closes the 9th door; the fourth tenant opens the 4th and 8th doors. Continuing like this, the 10th tenant closes the 10th door. These data are summarized in Table 4.2, where O indicates the door is open and C indicates the door is closed.

Table 4.2

	Door									
Tenant	**1**	**2**	**3**	**4**	**5**	**6**	**7**	**8**	**9**	**10**
1	O	O	O	O	O	O	O	O	O	O
2	.	C	.	C	.	C	.	C	.	C
3	.	.	C	.	.	O	.	.	C	.
4	.	.	.	O	.	.	.	O	.	.
5	.	.	.	.	C	.	.	.	.	O
6	.	.	.	.	.	C	.	.	.	.
7	.	.	.	.	.	.	C	.	.	.
8	.	.	.	.	.	.	.	C	.	.
9	.	.	.	.	.	.	.	.	O	.
10	.	.	.	.	.	.	.	.	.	C

It follows from the table that doors 1, 4, and 9 remain open at the end, so the number of such doors is three. (Notice that $3 = \lfloor\sqrt{10}\rfloor$; so can you predict the answer to the given problem? Construct tables like Table 4.2 for 13 tenants and 13 apartments, 18 tenants and 18 apartments, and 25 tenants and 25 apartments, and look for a pattern.)

Let us now return to the original problem. The first tenant opens all doors. Consider the kth tenant, where $2 \leq k \leq 1000$.

Case 1 Let n be a perfect square, where $n^2 \leq 1000$. Since n has an odd number of positive factors, the last person to touch the door will open it. Thus every nth door will remain open if n is a perfect square. The number

*Based on M. vos Savant, *Ask Marilyn*, St. Martin Press, New York, 1992, p. 228.

of such doors equals the number of perfect squares ≤ 1000, namely, $\lfloor\sqrt{1000}\rfloor = 31$.

Case 2 Suppose n is not a perfect square, where $n^2 \le 1000$. Since n has an even number of positive factors, the last person to touch the door will close it. In other words, every nth door will remain closed if n is not a perfect square.

Thus, by the addition principle, $31 + 0 = 31$ doors will remain open. They are doors numbered $1, 4, 9, 16, 25, \ldots, 900$, and 961. ■

More generally, suppose there are m tenants and m apartments, and the first tenant opens all doors. The jth tenant closes every jth door if it is open, and opens it otherwise, where $2 \le j \le m$. How many doors will remain open at the end?

Exercises 4.4

1. Compute the 36th triangular number. (It is the so-called *beastly number*.)

2. Prove that the sum of two consecutive triangular numbers is a perfect square.

(**Twelve Days of Christmas**) Suppose you sent your love 1 gift on the first day of Christmas, $1 + 2$ gifts on the second day, $1 + 2 + 3$ gifts on the third day and so on.

3. How many gifts did you send on the 12th day of Christmas?

4. How many gifts did your love receive in the 12 days of Christmas?

Using PMI, prove each for every integer $n \ge 1$.

5. $\sum_{i=1}^{n} (2i - 1) = n^2$

6. $\sum_{i=1}^{n} i^2 = \frac{(n+1)(2n+1)}{6}$

7. $\sum_{i=1}^{n} i^3 = \left[\frac{n(n+1)}{2}\right]^2$

8. $\sum_{i=1}^{n} ar^{i-1} = \frac{a(r^n - 1)}{r - 1} \quad (r \ne 1)$

9. $n^2 + n$ is divisible by 2.

10. $n^4 + 2n^3 + n^2$ is divisible by 4.

11. The number of lines formed by joining n (≥ 2) distinct points in a plane, no three of which being collinear, is $n(n-1)/2$.

12. The number of diagonals of a convex n-gon* is $n(n-1)/2 \ge 3$.

13. Let a be a positive integer and p a prime number such that $p \mid a^n$. Then $p \mid a$, where $n \ge 1$.
(*Hint*: Use Exercise 37 in Section 4.2.)

*An n-gon is a polygon with n sides. An n-gon such that the line segment joining any two points inside it lies within it is a **convex polygon.**

14. Prove that $1 + 2 + \cdots + n = n(n+1)/2$ by considering the sum in the reverse order.* (Do not use induction.)

Evaluate each sum.

15. $\sum_{k=1}^{30} (3k^2 - 1)$ **16.** $\sum_{k=1}^{50} (k^3 + 2)$ **17.** $\sum_{i=1}^{n} \lfloor i/2 \rfloor$ **18.** $\sum_{i=1}^{n} \lceil i/2 \rceil$

Find the value of x resulting from executing each algorithm fragment.

19.

```
x ← 0
for i = 1 to n do
  x ← x + (2i − 1)
```

20.

```
x ← 0
for i = 1 to n do
  x ← x + i(i + 1)
```

21.

```
x ← 0
for i = 1 to n do
  for j = 1 to i do
  x ← x + 1
```

Evaluate each sum and product.

22. $\sum_{i=1}^{n} \sum_{j=1}^{i} i$ **23.** $\sum_{i=1}^{n} \sum_{j=1}^{i} j$ **24.** $\sum_{i=1}^{n} \sum_{j=1}^{i} j^2$ **25.** $\sum_{i=1}^{n} \sum_{j=1}^{i} (2j-1)$

26. $\prod_{i=1}^{n} 2^{2i}$ **27.** $\prod_{i=1}^{n} i^2$ **28.** $\prod_{i=1}^{n} \prod_{j=1}^{n} i^j$ **29.** $\prod_{i=1}^{n} \prod_{j=1}^{n} 2^{i+j}$

30. A **magic square** of order n is a square arrangement of the positive integers 1 through n^2 such that the sum of the integers along each row, column, and diagonal is a constant k, called the **magic constant**. Figure 4.30 shows two magic squares, one of order 3 and the other of order 4. Prove that the magic constant of a magic square of order n is $n(n^2 + 1)/2$.

Figure 4.30

8	1	6
3	5	7
4	9	2

$k = 15$

1	14	15	4
12	7	6	9
8	11	10	5
13	2	3	16

$k = 34$

*An interesting anecdote is told about Karl Frederich Gauss (1777–1855), one of the great mathematicians. When he was a child, his teacher asked his pupils to compute the sum of the first 100 positive integers. According to the story, the teacher did so to get some time to grade his papers. To the teacher's dismay, Gauss found the answer in a few moments by pairing the numbers from both ends:

$$1 + 2 + 3 + \cdots + 50 + 51 + \cdots + 98 + 99 + 100$$

The sum of each pair is 101 and there are 50 pairs. So the total sum is $50 \cdot 101 = 5050$.

Let p, q, and r be prime numbers, and i, j, and k whole numbers. Find the sum of the positive divisors of each.

31. p^i **32.** $p^i q^j$ **33.** $p^i q^j r^k$

34. Let p be a prime and $n \in \mathbb{N}$. Prove that p^n is not a perfect number. (*Hint*: Prove by contradiction.)

Find the number of times the statement x ← x + 1 is executed by each loop.

35.
```
for i=1 to n do
  for j=1 to i do
    x ← x + 1
```

36.
```
for i = 1 to n do
  for j = 1 to i do
    for k = 1 to i do
      x ← x + 1
```

37.
```
for i = 1 to n do
  for j = 1 to i do
    for k = 1 to j do
      x ← x + 1
```

38.
```
for i = 1 to n do
  for j = 1 to i do
    for k = 1 to i do
      for l = 1 to i do
        x ← x + 1
```

According to legend, King Shirham of India was so pleased with the invention of chess that he offered to reward its inventor Sissa Ben Dahir with anything he wished. His request was a seemingly modest one: one grain of wheat on the first square of a chessboard, two on the second, four on the third, and so on. The king was delighted with this simple request, but soon realized he could not fulfill it. The last square alone would take $2^{63} = 9{,}223{,}372{,}036{,}854{,}775{,}808$ grains of wheat. Find each for an $n \times n$ chessboard.

39. The number of grains on the last square.

40. The total number of grains on the chessboard.

41. Let a_n denote the number of times the statement $x \leftarrow x + 1$ is executed in the following loop:

```
for i = 1 to n do
  for j = 1 to ⌊i/2⌋ do
    x ← x + 1
```

Show that

$$a_n = \begin{cases} \dfrac{n^2}{4} & \text{if } n \text{ is even} \\[2ex] \dfrac{n^2 - 1}{4} & \text{if } n \text{ is odd} \end{cases}$$

Find the number of trailing zeros in the decimal value of each.

42. 100! **43.** 378! **44.** 500! **45.** 1000!

46. Find the flaw in the "proof" in Example 4.21.

Prove each, where t_n denotes the nth triangular number and $n \geq 2$.

47. $8t_n + 1 = (2n+1)^2$

48. $8t_{n-1} + 4n = (2n)^2$

49. $t_{n-1}^2 + t_n^2 = t_{n^2}$

50. $\sum_{i=1}^{n} t_i = \dfrac{n(n+1)(n+2)}{6}$

Let $A, A_1, A_2, \ldots, A_n, B_1, B_2, \ldots, B_n$ be any sets, and $p_1, p_2, \ldots, p_n, q, q_1, q_2, \ldots, q_n$ be any propositions. Using induction prove each.

***51.** $A \cup \left(\bigcap_{i=1}^{n} B_i\right) = \bigcap_{i=1}^{n} (A \cup B_i)$

***52.** $A \bigcap_{i=1}^{n} (\cup B_i) = \bigcup_{i=1}^{n} (A \cap B_i)$

***53.** $\sim (p_1 \wedge p_2 \wedge \cdots \wedge p_n) \equiv (\sim p_1) \vee (\sim p_2) \vee \cdots \vee (\sim p_n)$

***54.** $\sim (p_1 \vee p_2 \vee \cdots \vee p_n) \equiv (\sim p_1) \wedge (\sim p_2) \wedge \cdots \wedge (\sim p_n)$

***55.** Prove that any postage of n (≥ 2) cents can be made using two- and three-cent stamps. (*Hint*: Use the division algorithm and induction.)

***56.** Let a and b be any two positive integers with $a \geq b$. Using the sequence of equations in the euclidean algorithm prove that $\gcd\{a, b\} = \gcd\{r_{n-1}, r_n\}$, $n \geq 1$.

***57.** Prove the strong version of mathematical induction, using the weak version.

***58.** Prove the weak version of induction, using the well-ordering principle.

****59.** Let S_n denote the sum of the elements in the nth set of the sequence of sets of squares $\{1\}$, $\{4, 9\}$, $\{16, 25, 36\}, \ldots$. Find a formula for S_n. (J. M. Howell, 1989)

****60.** Redo Exercise 59 using the sequence of triangular numbers $\{1\}$, $\{3, 6\}$, $\{10, 15, 21\}, \ldots$. (J. M. Howell, 1988)

4.5 Algorithm Correctness

Suppose we wrote an algorithm to solve a problem and translated the algorithm into a computer program. Since it is impossible to test the program for all sets of input values, we rely on a mathematical proof to ensure that the program will always yield the correct output. The principle of induction can certify the correctness of algorithms.

Correct Program

A **correct** program yields the correct result for all legal input values, assuming the program contains no compilation and execution errors.

Proving the correctness of a program, especially a complex one, is not at all an easy task. It consists of two steps:

(1) Proving that the program will always terminate; and
(2) proving that it will always produce the correct result. The second step constitutes the **partial correctness** of the program.

Loop Invariant

First, we will establish the partial correctness of simple **while** loops. Let n denote the number of iterations of a **while** loop. Assume a predicate P(n). A relationship among the variables holds true before the loop is executed and after each iteration of the loop, no matter how large n is. As the algorithm execution progresses, the values of the variables in the loop may vary, but the relationship remains unaffected. Such a predicate is a **loop invariant.**

To prove that P(n) is a loop invariant, we apply PMI, as the next two examples demonstrate.

EXAMPLE 4.25 Algorithm 4.6 computes the product of two positive integers x and y. Notice that the values of the variables x and y are not affected by the loop in lines 3–7. But the values of i and *answer* do get changed during each iteration of the loop.

```
Algorithm multiplication(x,y)
(* This algorithm computes the product of the positive integers x and y,
   and prints the answer. *)
0.  Begin (* algorithm *)
1.    answer ← 0 (* initialize answer *)
2.    i ← 0     (* counter *)
3.    while i < x do
4.    begin (* while *)
5.      answer ← answer + y
6.      i ← i + 1
7.    endwhile
8.  End (* algorithm *)
```

Algorithm 4.6

Let a_n and i_n denote the values of *answer* and i at the end of n iterations. Let P(n): $a_n = i_n \cdot y$. We shall prove that the predicate P(n) is a loop invariant.

PROOF (by PMI):
Let P(n): $a_n = i_n \cdot y$, $n \geq 0$.

Basis step The value $n = 0$ means zero iterations; it corresponds to the situation before the loop is entered. When $n = 0, a_0 = 0$ and $i_0 = 0$. Therefore, $a_0 = i_0 \cdot y$; so, P(0) is true.

Induction step Assume P(k) is true; that is, $a_k = i_k \cdot y$ after k iterations. Then $a_{k+1} = a_k + y$ and $i_{k+1} = i_k + 1$, by lines 5 and 6. Thus:

$$\begin{aligned} a_{k+1} &= i_k \cdot y + y, \text{ by the inductive hypothesis} \\ &= (i_k + 1)y \\ &= i_{k+1} \cdot y \end{aligned}$$

So P($k + 1$) is true.

Thus, by PMI, P(n) is true for every $n \geq 0$; that is, P(n) is a loop invariant. ■

How is the property that P(n) is a loop invariant useful? Since $a_n = i_n \cdot y$ after n iterations, it must be true even when we exit the loop. The loop is terminated when $i_n = x$. Then answer $= a_n = x \cdot y$, as expected. Since P(n) is a loop invariant, the algorithm does indeed work correctly.

What exactly is the iteration method? Suppose we would like to compute the value $f(n)$ of a function f at an integer $n \geq n_0$. In the **iteration method**, we use $f(n_0)$ to compute $f(n_0 + 1)$, then use the successive values $f(n_0 + 2)$, $f(n_0 + 3), \ldots$ to evaluate $f(n)$. For instance, to evaluate $n!$ by iteration, we successively evaluate $0!, 1!, 2!, \ldots, (n - 1)!$ and then evaluate $n!$.

EXAMPLE 4.26 Algorithm 4.7 is an iterative algorithm for computing $n!$, where $n \geq 0$. Let fact(n) be the value of *factorial* at the end of n iterations of the loop. Prove that *P(n): fact(n) = n!* is a loop invariant.

```
Algorithm factorial (n)
(* This algorithm computes and prints the value of
   n! for every n ≥ 0. *)
0.   Begin (* algorithm *)
1.     factorial ← 1            (* initialize *)
2.     i ← 1                    (* counter *)
3.     while i < n do
4.     begin (* while *)
5.       i ← i + 1
6.       factorial ← factorial * i
7.     endwhile
9.   End (* algorithm *)
```

Algorithm 4.7

PROOF (by PMI):
Let P(n): fact(n) $= n!$, $n \geq 0$.

Basis step When $n = 0$, fact(0) $= 1 = 1!$ by line 1; so P(0) is true.

Induction step Assume P(k) is true: fact(k) $= k!$. Then:

$$\begin{aligned} \text{fact}(k+1) &= \text{fact}(k) \cdot (k+1), \text{ by line 6} \\ &= k! \cdot (k+1), \text{ by the inductive hypothesis} \\ &= (k+1)! \end{aligned}$$

Therefore, P($k+1$) is true.

Thus, by induction, P(n) holds true for every $n \geq 0$; that is, P(n) is a loop invariant and hence the algorithm correctly computes the value of $n!$, for every $n \geq 0$. ■

Searching and Sorting Algorithms

The remainder of this section establishes the partial correctness of a few standard searching and sorting algorithms. We begin with two searching algorithms, linear and binary.

Linear Search Algorithm

Let $X = [x_1, x_2, \ldots, x_n]$ be an unordered list (also known as a one-dimensional array or simply an array) of n distinct items. We would like to search the list for a specific item, called *key*. If *key* exists in the list, the algorithm should return the location of *key*.

We search the list from right to left for convenience. Compare x_n and *key*. If $x_n = key$, *key* occurs and location $= n$. Otherwise, compare x_{n-1} and *key*. If they are equal, we are done. Otherwise, continue the search until it is successful or the list is empty. This algorithm is the **linear search algorithm**.

For example, let $X = $ [Dallas, Boston, Nashville, Albany, Portland] and *key* = Albany. Then *key* occurs in the list at location 4.

In general, we cannot assume *key* occurs in the list. To make the search process always successful, we store *key* in location 0: $x_0 \leftarrow key$. So if the search routine returns the value zero for location, it implies *key* does not occur in the list.

An iterative version of the linear search algorithm is given in Algorithm 4.8.

```
Algorithm linear search (X,n,key,location)
(* This algorithm searches a list by the linear search method for
   a key and returns its location in the list. To make the search
   always successful, we store key in x0. If the algorithm returns
   the value 0 for location, key does not occur in the list. *)
Begin (* algorithm *)
  x0 ← key
  i ← n
  while xi ≠ key do
    i ← i − 1
```

```
5.     location ← i
6. End (* algorithm *)
```

Algorithm 4.8

EXAMPLE 4.27 Prove that the linear search algorithm in Algorithm 4.8 works correctly for every $n \geq 0$.

PROOF (by PMI):
Let P(n): The algorithm returns the correct location for every list of size $n \geq 0$.

Basis step When $n = 0$, the **while** loop is skipped. The algorithm returns the value 0 in location by line 5, which is correct. So P(0) is true.

Induction step Assume P(k) is true for an arbitrary integer $k \geq 0$; that is, the algorithm works when the list contains k items.

To show that P($k + 1$) is true, consider a list X with $k + 1$ elements.

Case 1 If $x_{k+1} = key$ in line 3, the **while** loop will not be entered and the algorithm returns the correct value $k + 1$ for location in line 5.

Case 2 If $x_{k+1} \neq key$, $i = k$ at the end of the first iteration. This restricts us to a sublist with k elements. By the inductive hypothesis, the algorithm works correctly for such a list.

In both cases, P($k + 1$) holds. Thus, by induction, P(n) is true for $n \geq 0$. In other words, the algorithm returns the correct location for every list with $n \geq 0$ elements.

Binary Search Algorithm

The **binary search algorithm** searches for a given *key* if the list X is ordered. The technique employed is **divide and conquer**. First compute the *middle* (*mid*) of the list, where $mid = \lfloor (1 + n)/2 \rfloor$. The middle item is x_{mid}.

Now partition the list into three disjoint sublists: $[x_1, \ldots, x_{\text{mid}-1}]$, $[x_{\text{mid}}]$, and $[x_{\text{mid}+1}, \ldots, x_n]$. If $x_{\text{mid}} = key$, the search is successful and $location = mid$. If they are not equal, we search only the lower half or the upper half of the list. If $key < x_{\text{mid}}$, search the sublist $[x_1, \ldots, x_{\text{mid}-1}]$; otherwise, search the sublist $[x_{\text{mid}+1}, \ldots, x_n]$. Continue like this until the search is successful or the sublist is empty.

EXAMPLE 4.28 Use the binary search algorithm to search the list $X = [3, 5, 8, 13, 21, 34, 55, 89]$ for $key = 5$.

SOLUTION:
Let x_i denote the ith element of the list X, where $1 \leq i \leq n$ and $n = 8$.

Step 1 Compute *mid* for the list X:

$$mid = \lfloor (1+n)/2 \rfloor = \lfloor (1+8)/2 \rfloor = 4.$$

Therefore, the middle term is $x_{mid} = 13$.

Step 2 Compare x_{mid} and *key*:
Since $x_4 \neq 5$, *key*, if it occurs, must exist in the lower sublist $[x_1, x_2, x_3] = [3, 5, 8]$ or in the upper sublist $[x_5, x_6, x_7, x_8] = [21, 34, 55, 89]$. Since $key < x_4$, search the first sublist and continue steps 1 and 2 until either *key* is located or the sublist becomes empty.

Step 3 Compute *mid* for the list $[x_1, x_2, x_3]$:

$$mid = \lfloor (1+3)/2 \rfloor = 2$$

So $x_{mid} = x_2 = 5$.

Step 4 Compare x_{mid} and *key*:
Since $x_{mid} = key$, the search is successful. *Key* occurs at location 2 and we are done. (As an exercise, use the algorithm to search the list *X* with $key = 23$.) ■

The steps in this example can be translated into an algorithm. See Algorithm 4.9.

```
Algorithm binary search(X,1,n,key,mid)
(* This algorithm searches an ordered list X of n elements for a special
   item (key). It returns the location of key if the search is
   successful and zero otherwise. The variable mid returns such a value.
   The variables low and high denote the lower and upper indices of the
   list being searched. *)
Begin (* algorithm *)
low ← 1
high ← n
while low ≤ high do        (* list is nonempty *)
begin (* while *)
  mid ← ⌊(low + high)/2⌋
  if key = x_mid then       (* key exists in the list*)
    exit the loop
  else if key < x_mid then        (* search lower half*)
       high ← mid − 1
     else   (* search the upper half *)
       low ← mid + 1
endwhile
  if low > high then        (* search is unsuccessful *)
    mid ← 0
End (* algorithm *)
```

Algorithm 4.9

The next example establishes the partial correctness of this algorithm using strong induction.

EXAMPLE 4.29 Prove that the binary search algorithm (Algorithm 4.9) works correctly for every ordered list of size $n \geq 0$.

PROOF (by strong induction):
Let P(n): The algorithm works for every ordered list of size n.

Basis step When $n = 0$, $low = 1$ and $high = 0$. Since $low \leq high$ is false in line 3, the **while** loop is not executed. So the algorithm returns the correct value 0 from line 14, as expected, and P(0) is true.

Induction step Assume P(i) holds for every $i \leq k$, where $k \geq 0$; that is, the algorithm returns the correct value for any list of size $i \leq k$.

To show that P($k + 1$) is true, consider an ordered list X of size $k + 1$. Since $high = k + 1 \geq 1 = low$, the loop is entered and the middle index is computed in line 5.

Case 1 If $key = x_{\text{mid}}$, we exit the loop (line 7) and the value of mid is returned, so the algorithm works.

Case 2 If $key < x_{\text{mid}}$, search the sublist $x_1, \ldots, x_{\text{mid}-1}$; otherwise, search the sublist $x_{\text{mid}+1}, \ldots, x_n$. In both cases, the sublists contain fewer than $k + 1$ elements, so the algorithm works in either case by the inductive hypothesis.

Thus P($k + 1$) is true. So, by PMI, P(n) is true for $n \geq 0$; that is, the algorithm works correctly for every ordered list of zero or more items. ■

Next we present two standard sorting algorithms and prove their correctness.

Sorting Algorithms

Suppose we are given a list of n items and would like to sort them in "ascending order." Several methods are available. Two algorithms that can do the job are bubble sort and selection sort.

Bubble Sort

Bubble sort is a simple, elegant algorithm for sorting a list of n items. It "bubbles up" smaller items to the top and pushes larger items to the bottom: Compare consecutive elements, beginning with the first pair. Swap them if they are out of order. Compare the next pair and swap them if necessary. Continue like this to the end of the list. This ends the first pass. Now place the largest element at the end of the list. Repeat these steps with all but the largest element until the resulting sublist consists of one element. The list is now ordered.

The following example demonstrates this method.

EXAMPLE 4.30 Using bubble sort, sort the list $X = [34, 13, 21, 3, 89]$.

SOLUTION:
Let x_i denote the ith element in the list, where $1 \le i \le 5$. The given list is

	1	2	3	4	5
X	34	13	21	3	89

Step 1 Compare x_1 and x_2. Since $x_1 > x_2$, swap them. This yields the list

	1	2	3	4	5
X	13	34	21	3	89

Now compare x_2 and x_3. Since $x_2 > x_3$, interchange x_2 and x_3. This produces the list

	1	2	3	4	5
X	13	21	34	3	89

Since $x_3 > x_4$, switch them, yielding the list

	1	2	3	4	5
X	13	21	3	34	89

Compare x_4 and x_5. Since $x_4 < x_5$, they are in the correct order and no interchanging is needed. This completes the first pass. At the end of the first pass, the largest element in the list is placed in proper position:

	1	2	3	4	5
X	13	21	3	34	89
	to be sorted				in correct position

Step 2 In the second pass, compare the elements x_1 through x_4 and swap them if necessary. This results in the two largest elements being placed correctly:

	1	2	3	4	5
X	13	3	21	34	89
	to be sorted			correctly sorted	

Step 3 The third pass involves the elements x_1 through x_3. At the end of this pass, the three largest elements are correctly placed:

	1	2	3	4	5
X	3	13	21	34	89

to be sorted (elements 1–2) in correct order (elements 3–5)

Step 4 At the end of the fourth pass the list is completely sorted:

	1	2	3	4	5
X	3	13	21	34	89

all in correct order

> *Two important observations*:
>
> - At the end of the ith pass, the i largest elements are correctly placed at the end of the list, where $1 \le i \le n$. So the $(i + 1)$st pass involves the elements x_1 through x_{n-i}.
> - Bubble sort takes $n - 1$ passes to sort a list of n items, even if the list becomes ordered at the end of the ith pass, where $i < n - 1$. Once the list is sorted, it makes no sense to go through the remaining passes, so the additional passes can be avoided with a boolean variable.

The various steps in Example 4.30 can be developed into an algorithm for bubble sort, as presented in Algorithm 4.10.

```
Algorithm bubble sort(X,n)
(* This algorithm sorts a list X of n elements
   using the bubble algorithm. *)
Begin (* algorithm *)
  for i = 1 to n - 1 do
    for j = 1 to n - i do
      if x_j > x_{j+1} then
        swap x_j and x_{j+1}
End (* algorithm *)
```

Algorithm 4.10

EXAMPLE 4.31 Establish the correctness of the bubble sort algorithm.

PROOF (by PMI):
Let P(n): The algorithm sorts every list of size $n \geq 1$.

Basis step When $n = 1$, the list contains just one element and hence is clearly sorted, so P(1) is true.

Induction step Assume P(k) is true; that is, the algorithm sorts correctly every list of k (≥ 1) items.

To show that P($k + 1$) is true, consider a list $X = [x_1, x_2, \ldots, x_{k+1}]$. Since $k + 1 \geq 2$, the **for** loop in line 1 is entered. When $i = 1$, j runs from 1 through $n - 1$. Lines 3 and 4 are executed: the consecutive elements x_j and x_{j+1} are compared and swapped if out of order. The inner **for** loop places the largest of the elements $x_1, x_2, \ldots, x_{k+1}$ in position $k + 1$. This leaves a sublist of k elements, $[x_1, x_2, \ldots, x_k]$. By the inductive hypothesis, the algorithm correctly sorts it. It follows that the algorithm correctly sorts the entire list X; that is, P($k + 1$) is true.

Thus, by the principle of induction, P(n) is true for $n \geq 1$; that is, the bubble sort algorithm always works. ■

Selection Sort

Unlike bubble sort, **selection sort** finds the largest element and swaps it with x_n if x_n is not the largest element. Find the largest of the remaining elements $x_1, x_2, \ldots, x_{n-1}$, and switch it with x_{n-1} if it isn't x_{n-1}. Continue like this until the list is completely sorted.

> In each pass, unlike in bubble sort, if two elements are out of order, we do not swap them right away but wait to find the largest element of the sublist. At the end of the ith pass, the largest of the elements $x_1, x_2, \ldots, x_{n-i+1}$ is swapped with x_{n-i+1}, where $1 \leq i < n$.

This outline of the selection sort algorithm can be a bit refined. In the ith pass, initially assume x_{n-i+1} is the largest element. Find the largest of the elements $x_1, x_2, \ldots, x_{n-i}$. Swap it with x_{n-i+1} if necessary. Algorithm 4.11 results.

```
Algorithm selection sort(X,n)
(* This algorithm sorts a list X of n items using the iterative version
   of selection sort. Maxindex denotes the index of the largest element
   in a given pass. *)
Begin (* algorithm *)
  if n > 1 then(* list contains at least two elements *)
   for i = 1 to n − 1 do
```

```
3.      begin (* for *)
4.        maxindex ← n − i + 1 (* assume x_{n−i+1} is the
              largest element; save its index. *)
5.        for j=1 to n − i do
6.          if x_j > x_maxindex, then        (* update maxindex *)
7.            maxindex ← j
8.          if maxindex ≠ n − i + 1, then (* found a larger
              element; swap the corresponding elements *)
9.            swap x_maxindex and x_{n−i+1}
10.       endfor
11.   End (* algorithm *)
```

Algorithm 4.11

EXAMPLE 4.32 Establish the correctness of Algorithm 4.11.

PROOF (by PMI):
Let P(n): The algorithm works correctly for every list of size $n \geq 1$.

Basis step When $n = 1$, the list contains one element and is clearly sorted, so P(1) is true.

Induction step Assume P(k) is true; that is, the algorithm sorts correctly every list of size $k \geq 1$.

To show that P(k + 1) is true, consider a list $X = [x_1, x_2, \ldots, x_{k+1}]$ with $k + 1$ elements, where $k + 1 \geq 2$. Since $k + 1 \geq 2$, the condition in line 1 is satisfied, and we enter the loop in line 2. When $i = 1$, $maxindex = (k + 1) - 1 + 1 = k + 1$. The **for** loop in lines 5–7 compares each of the elements $x_1, x_2, \ldots, x_k$ with x_{maxindex} and updates it as needed. Line 8 updates *maxindex* if we have found an element larger than x_{k+1}. If maxindex $\neq k + 1$, then the elements x_{k+1} and $\mathrm{x}_{\text{maxindex}}$ are swapped. This stores the largest of the $k + 1$ elements in position $k + 1$, leaving a sublist of k elements, namely, $x_1, x_2, \ldots, x_k$ to be sorted.

Therefore, by the inductive hypothesis, the algorithm sorts correctly the list X containing $k + 1$ elements.

Thus, by induction, P(n) is true for every $n \geq 1$; that is, the algorithm correctly sorts every list of size n. ■

These searching and sorting algorithms are pursued again in Section 4.7. Additional sorting algorithms appear in the exercises.

Exercises 4.5

Prove that the given predicate P(n) in each algorithm is a loop invariant.

1.
```
Algorithm exponential(x,n)
(* This algorithm computes
x^n, where x ∈ ℝ+ and
n ∈ W. *)
```

2.
```
Algorithm division(x,y)
(* This algorithm computes
the quotient and the
remainder when a positive
```

```
0. Begin (* algorithm *)
1.   answer ← 1
2.   while n > 0 do
3.   begin (* while *)
4.     answer ← answer · x
5.     n ← n − 1
6.   endwhile
7. End  (* algorithm *)
```

P(n): $a_n = x^n$, where a_n denotes the value of answer after n iterations of the **while** loop.

3. **Algorithm Euclid(x,y,divisor)**
(* See Algorithm 4.2 *)
P(n): $\gcd\{x_n, y_n\} = \gcd\{x, y\}$ where x_n and y_n denote the values of x = dividend and y = divisor after n iterations.

integer x is divided by a positive integer y using addition and subtraction. *)

```
 0. Begin (* algorithm *)
 1.   dividend ← x
 2.   divisor ← y
 3.   quotient ← 0
 4.   remainder ← dividend
 5.   while dividend ≥
         divisor do
 6.   begin (* while *)
 7.     dividend ← dividend
           − divisor
 8.     quotient ← quotient + 1
 9.     remainder ← dividend
10.   endwhile
11. End  (* algorithm *)
```

4. **Algorithm gcd(x,y)**
(* This algorithm computes the gcd of two positive integers x and y. *)

```
0. Begin (* algorithm *)
1.   while x ≠ y do
2.     if x > y then
3.       x ← x − y
4.     else
5.       y ← y − x
6.     gcd ← x
7. End (* algorithm *)
```

P(n): $\gcd\{x_n, y_n\} = \gcd\{x, y\}$, where x_n and y_n denote the values of x and y at the end of n iterations of the loop.

5. **Algorithm sum** (x,y) (* This algorithm prints the sum of two nonnegative integers x and y. *)

```
0. Begin (* algorithm *)
1.   sum ← x
2.   count ← 0 (* counter *)
3.   while count < y do
4.   begin (* while *)
5.     sum ← sum + 1
6.     count ← count + 1
7.   endwhile
8. End (* algorithm *)
```

P(n): $x = q_n y + r_n$, where q_n and r_n denote the quotient and the remainder after n iterations.

6. **Algorithm square (x)** (* This algorithm prints the square of $x \in \mathbf{W}$. *)

```
0.   Begin (* algorithm *)
1.   answer ← 0
2.   i ← 0 (* counter *)
3.   While i < x do
4.   begin (* while *)
5.     answer ← answer + (2i + 1):
6.   i ← i + 1
7.   endwhile
8. End (* algorithm *)
```

P(n): $a_n = n^2$, where a_n denotes the value of answer at the end of n iterations.

Using the algorithm in Exercise 4, compute the gcd of each pair of integers.

7. $18, 3$ **8.** $28, 12$ **9.** $28, 48$ **10.** $24, 112$

Sort the following lists using the bubble sort algorithm.

11. 23, 7, 18, 19, 53 **12.** 19, 17, 13, 8, 5

13–14. Sort each list in Exercises 11 and 12 using the selection sort algorithm.

Write an iterative algorithm to do the tasks in Exercises 15–17.

15. Compute $n!$, $n \geq 0$.

16. Determine if two $n \times n$ matrices A and B are equal.

17. Compute the product of two $n \times n$ matrices A and B.

18. Let $A = (a_{ij})_{n \times n}$ and $B = (b_{ij})_{n \times n}$. A is **less than or equal to** B, denoted by $A \leq B$, if $a_{ij} \leq b_{ij}$ for every i and j. Write an algorithm to determine if $A \leq B$.

Consider a list X of n numbers $x_1, x_2, \ldots, x_n$. Write iterative algorithms to do the tasks in Exercises 19–25.

19. Find the sum of the numbers.

20. Find the product of the numbers.

21. Find the maximum of the numbers.

22. Find the minimum of the numbers.

23. Print the numbers in the given order $x_1, x_2, \ldots, x_n$.

24. Print the numbers in the reverse order $x_1, x_2, \ldots, x_n$.

25. Write an algorithm to determine if a string S of n characters is a palindrome.

26–36. Establish the correctness of each algorithm in Exercises 15–25.

Use the **insertion sort** algorithm in Algorithm 4.12 to answer Exercises 37–39.

```
Algorithm insertion sort(X,n)
(* This algorithm sorts a list X of n elements into
   ascending order by inserting a new element in the
   proper place at the end of each pass. *)
Begin (* algorithm *)
  for i = 2 to n do
  begin (* for *)
    temp ← x_i    (* temp is a temporary variable *)
    j ← i − 1
    while j ≥ 1 do
```

```
    begin (* while *)
     if x_j > temp then
       x_{j+1} ← x_j
     j ← j - 1
    endwhile
    x_{j+1} ← temp
  endfor
End (* algorithm *)
```

Algorithm 4.12

Sort each list.

37. 3, 13, 8, 6, 5, 2

38. 11, 7, 4, 15, 6, 2, 9

39. Establish the correctness of the algorithm.

4.6 The Growth of Functions

The growth of functions can be investigated using three important notations: the big-oh (O), the big-omega (Ω), and the big-theta (Θ) notations.* We will employ it in Sections 4.7 and 5.7 to analyze some standard algorithms.

Suppose we have developed two algorithms to solve a problem. To determine if one is better than the other, we need some type of yardstick to measure their efficiency. Since the complexity of an algorithm is a function of the input size n, we measure efficiency in terms of n. To this end, we begin with the big-oh notation, introduced in 1892 by the German mathematician Paul Gustav Heinrich Bachmann. The big-oh symbol is also known as the **Landau symbol** after the German mathematician Edmund Landau who popularized it.

The Big-Oh Notation

Let $f, g: \mathbb{N} \to \mathbb{R}$. Then $f(n)$ is of **order at most** $g(n)$, if a positive constant C and a positive integer n_0 exist such that $|f(n)| \le C|g(n)|$ for every $n \ge n_0$. In symbols, we write $f(n) = O(g(n))$. (Read this as *f(n) is big-oh of g(n).*)

In this definition, if we can find one value for C, any value greater than that also will work, so the value of C is not unique.

When we say the time needed to execute an algorithm is $O(g(n))$, it simply means the time needed is not more than some constant times $|g(n)|$ when n is sufficiently large. For instance, let c_n denote the maximum number of element comparisons required in line 3 of the linear search algorithm (Algorithm 4.8), where n denotes the input size. Using c_n as an

*Ω and Θ are the uppercase Greek letters *omega* and *theta*, respectively.

Paul Gustav Heinrich Bachmann *(1837–1920), the son of a Lutheran minister, was born in Berlin. He inherited a pious attitude and a great love for music. During his early years, he had difficulties in mathematical studies, but his talent was discovered by one of his teachers.*

After recovering from tuberculosis in Switzerland, Bachmann studied mathematics, first at the University of Berlin and then at the University of Göttingen, where he attended Dirichlet's lectures. In 1862 he received his doctorate from Berlin under the guidance of the famous German mathematician Ernst Kummer, for a thesis on group theory. He became a professor at Breslau and later at Munster.

Around 1890, he resigned his position and moved to Weimar, Germany, where he continued his mathematical writing, composed music, played the piano, and wrote music criticism for newspapers. His writings include several volumes on number theory and a book on Fermat's Last Theorem. Bachmann died in Weimar.

Edmund Landau *(1877–1938), the son of a gynecologist, was born in Berlin. After attending high school, he studied mathematics at the University of Berlin, receiving his doctorate under the German mathematician Georg Frobenius in 1899. He taught at Berlin until 1909 and then moved to the University of Göttingen, where both David Hilbert and Felix Klien were colleagues. After the Nazis forced him to quit teaching, he never gave another lecture in Germany.*

Landau's principal contributions were to analytic number theory, especially to the distribution of primes. He wrote several books and more than 250 papers, and exercised tremendous influence on the development of number theory. Landau died suddenly in Berlin.

estimate of the execution of the algorithm, it can be shown that $c_n = O(n)$ (see Example 4.44). This means c_n grows no faster than n, when n is sufficiently large.

Before we analyze the execution times of algorithms, we will study a few simple examples to show how to use the big-oh notation.

EXAMPLE 4.33 Let $f(n) = 50n^3 - 6n + 23$. Show that $f(n) = O(n^3)$.

SOLUTION:

$$f(n) = 50n^3 - 6n + 23$$

Therefore,

$$|f(n)| = |50n^3 - 6n + 23|$$

$$\leq |50n^3| + |-6n| + |23|, \text{ by the triangle inequality}$$

$$
\begin{aligned}
&= 50n^3 + 6n + 23 \\
&\le 50n^3 + 6n^3 + 23n^3, \text{ when } n \ge 1 \quad (\textit{Note}: n_0 = 1) \\
&= 79n^3
\end{aligned}
$$

Thus, by taking $C = 79$, it follows that $f(n) = O(n^3)$. ■

More generally, we have the following result.

THEOREM 4.14 Let $f(n) = \sum_{i=0}^{m} a_i n^i$ be a polynomial in n of degree m. Then $f(n) = O(n^m)$.

PROOF:
$f(n) = a_m n^m + a_{m-1} n^{m-1} + \cdots + a_1 n + a_0$. By the triangle inequality, we have:

$$
\begin{aligned}
|f(n)| &\le |a_m| n^m + |a_{m-1}| n^{m-1} + \cdots + |a_1| n + |a_0| \\
&\le |a_m| n^m + |a_{m-1}| n^m + \cdots + |a_1| n^m + |a_0| n^m, \quad n \ge 1 \\
&= \left(\sum_{i=1}^{m} |a_i| \right) n^m = C n^m, \text{ where } C = \sum_{i=1}^{m} |a_i| \\
&= O(n^m)
\end{aligned}
$$

Thus, when n is sufficiently large, the leading term dominates the value of the polynomial. ■

In Example 4.33, although $f(n) = O(n^3)$, it is also true that $f(n) \le 79n^5$ and $f(n) \le 79n^6$. So we could say correctly, but meaninglessly, that $f(n) = O(n^5)$ and also $f(n) = O(n^6)$. To make comparisons meaningful, however, we shall always choose the smallest possible order of magnitude.

Commonly Used Order Functions

The most common order functions and their names are listed below, arranged in increasing order of magnitude:

- Constant $O(1)$
- Logarithmic $O(\lg n)$
- Linear $O(n)$
- (no name exists) $O(n \lg n)$
- Quadratic $O(n^2)$
- Cubic $O(n^3)$
- Polynomial $O(n^m)$

- Exponential $O(2^n)$
- Factorial $O(n!)$

When we say that the order of magnitude of an algorithm is a constant, we mean that the execution time is bounded by a constant; that is, it is independent of the input size n. If the order is linear, the execution time grows linearly; it is directly proportional to the input size.

Approximate values of some of the order functions are given in Table 4.3 for comparison; the graphs of a few of them are given in Figure 4.31.

Table 4.3

lg n	n	n lg n	n^2
3	10	30	100
6	100	600	10,000
9	1,000	9,000	100,000
13	10,000	130,000	100,000,000
16	100,000	1,600,000	10,000,000,000
19	1,000,000	19,000,000	one trillion

Figure 4.31

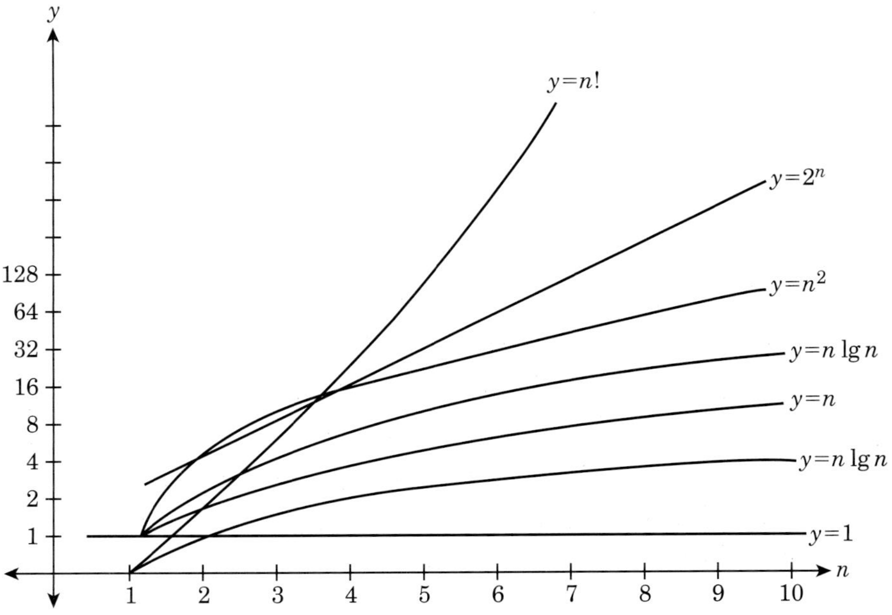

The order functions satisfy the following relationships among the frequently used execution times, when n is sufficiently large: $O(1) < O(\lg n) < O(n) < O(n \lg n) < O(n^2) < O(n^3) < O(2^n) < O(n!)$. They give us an idea of how long algorithms of varying orders will take to execute jobs.

For instance, if two algorithms solve a problem, one with $O(n)$ and the other with $O(\lg n)$, then (other things being equal) the second algorithm will work faster.

The next two examples also illustrate how to estimate the growth of functions.

EXAMPLE 4.34 Show that $n! = O(n^n)$ and $\lg n! = O(n \lg n)$.

SOLUTION:

- $$\begin{aligned} n! &= n(n-1)\cdots 3\cdot 2\cdot 1 \\ &\le n\cdot n\cdots n\cdot n\cdot n, \quad \text{where } n \ge 1 \\ &= n^n \\ &= O(n^n) \quad (\textit{Note}\text{: Use } C = 1.) \end{aligned}$$
- Since $n! \le n^n$ from above,
$$\begin{aligned} \lg n! &\le n \lg n \quad (\textit{Note}\text{: If } 0 < x \le y, \text{ then } \lg x \le \lg y.) \\ &= O(n \lg n) \end{aligned}$$

■

The following example shows how to estimate in a nested **for** loop the growth of the number of times an assignment statement is executed.

EXAMPLE 4.35 Estimate $f(n)$, the number of times the statement $x \leftarrow x + 1$, is executed in the following **for** loop.

```
for i = 1 to n do
  for j = 1 to i do
    x ← x + 1
```

SOLUTION:

Since the statement $x \leftarrow x + 1$ is executed i times for each value of i, where $1 \le i \le n$,

$$f(n) = \sum_{i=1}^{n} i = \frac{n(n+1)}{n} = O(n^2)$$

As n increases, $f(n)$ grows as n^2. ■

The Growth of a Sum of Two Functions

Imagine an algorithm consisting of two subalgorithms. Suppose the orders of execution times of the subalgorithms are given by $f_1(n) = O(g_1(n))$ and $f_2(n) = O(g_2(n))$. The next theorem shows how to compute the order of the algorithm.

THEOREM 4.15 Let $f_1(n) = O(g_1(n))$ and $f_2(n) = O(g_2(n))$. Then $(f_1 + f_2)(n) = O(max\{|g_1(n)|, |g_2(n)|\})$.

PROOF

By definition, there exist positive constants C_1, C_2, n_1, and n_2 such that $|f_1(n)| \leq C_1 \ |g_1(n)|$ for $n \geq n_1$, and $|f_2(n)| \leq C_2|g_2(n)|$ for $n \geq n_2$. Let $C = \max\{C_1, C_2\}$, $n_0 = \max\{n_1, n_2\}$, and $g(n) = \max\{|g_1(n)|, |g_2(n)|\}$. Then:

$$\begin{aligned} |f_1(n) + f_2(n)| &\leq C_1|g_1(n)| + C_2|g_2(n)| \\ &\leq C|g(n)| + C|g(n)|, \text{ where } n \geq n_0 \\ &= 2C|g(n)| \end{aligned}$$

Thus $f_1(n) + f_2(n) = O(g(n))$; that is, $(f_1 + f_2)(n) = O(\max\{|g_1(n)|, |g_2(n)|\})$. ■

It follows by this theorem that if $f_1(n) = O(g(n))$ and $f_2(n) = O(g(n))$, then $(f_1 + f_2)(n) = O(g(n))$. Why?

The Growth of a Product of Two Functions

The next theorem helps us to estimate the growth of $(f_1 \cdot f_2)(n)$, the product of the functions f_1 and f_2.

THEOREM 4.16 Let $f_1(n) = O(g_1(n))$ and $f_2(n) = O(g_2(n))$. Then $(f_1 \cdot f_2)(n) = O(g_1(n) \cdot g_2(n))$.

PROOF

Again, by definition, there are constants C_1, C_2, n_1, and n_2 such that $|f_1(n)| \leq C_1|g_1(n)|$ for $n \geq n_1$, and $|f_2(n)| \leq C_2|g_2(n)|$ for $n \geq n_2$. Let $C = C_1C_2$ and $n_0 = \max\{n_1, n_2\}$. Then:

$$\begin{aligned} |(f_1 \cdot f_2)(n)| &= |f_1(n) \cdot f_2(n)| \\ &= |f_1(n)| \cdot |f_2(n)| \\ &\leq C_1|g_1(n)| \cdot C_2|g_2(n)| \\ &= C|g_1(n)g_2(n)|, \text{ where } n \geq n_0 \end{aligned}$$

Thus $(f_1 \cdot f_2)(n) = O(g_1(n)g_2(n))$. ■

The next two examples employ this handy theorem along with the earlier theorems.

EXAMPLE 4.36 Let $f(n) = 6n^2 + 5n + 7 \lg n!$. Estimate the growth of $f(n)$.

SOLUTION:

Since $6n^2 = O(n^2)$ and $5n = O(n)$, $6n^2 + 5n = O(n^2)$ by Theorem 4.15. Furthermore, $7 = O(1)$, and $\lg n! = O(n \lg n)$ by Example 4.34. So

$$\begin{aligned} 7 \lg n! &= O(1) \cdot O(n \lg n) \\ &= O(1 \cdot n \lg n), \text{ by Theorem 4.16} \\ &= O(n \lg n) \end{aligned}$$

Since lg $n \leq n$, n lg $n \leq n^2$ for $n \geq 1$ (see Figure 4.31), it follows by Theorem 4.15 that $f(n) = O(n^2) + O(n \lg n) = O(n^2)$. ■

EXAMPLE 4.37 Let $f(n) = (3n^2 + 4n - 5) \lg n$. Estimate the growth of $f(n)$.

SOLUTION:
$3n^2 + 4n - 5 = O(n^2)$, by Theorem 4.14
Clearly,

$$\lg n = O(\lg n)$$

So

$$\begin{aligned} f(n) &= (3n^2 + 4n - 5) \lg n \\ &= O(n^2) \cdot O(\lg n) \\ &= O(n^2 \lg n), \text{ by Theorem 4. 16} \end{aligned}$$ ■

We now turn to the big-omega and the big-theta notations for investigating the growth of functions.

The Big-Omega and Big-Theta Notations

The big-oh notation has been widely used in the study of the growth of functions; however, it does not give us an exact order of growth. For instance, $f(n) = O(g(n))$ just implies that the function f does not grow any faster than g. In other words, it simply provides an upper bound for the size of $f(n)$ for large values of n, but no lower bound.

When we need the lower bound, we employ the big-omega notation. When we need both bounds to estimate the growth of f, we use the big-theta notation. Both notations were introduced in the 1970s by Donald Knuth of Stanford University.

We now pursue the big-omega notation. As you could imagine by now, its definition closely resembles that of the big-oh notation; it can be obtained by simply changing $\leq$ to $\geq$.

The Big-Omega Notation

Let $f, g: \mathbb{N} \to \mathbb{R}$. Suppose there is a positive constant C and a positive integer n_0 such that $|f(n)| \geq C|g(n)|$ for every $n \geq n_0$. Then $f(n)$ is $\Omega(g(n))$; that is, $f(n) = \Omega(g(n))$. [As above, read this as $f(n)$ is big-omega of $g(n)$.]

The following example illustrates this definition.

EXAMPLE 4.38 Let $f(n) = 50n^3 - 6n + 23$. When $n \geq 0$, $50n^3 - 6n + 23 \geq 50n^3$. So, with $C = 50$ and $g(n) = n^3$, it follows that $f(n) \geq C \cdot g(n)$ for every $n \geq 0$. Thus $f(n) = \Omega(n^3)$. (Notice that here $n_0 = 0$.) ■

Donald Ervin Knuth *(1938–), a pioneer in the development of the theory of compilers, programming languages, and the analysis of algorithms, is also a prolific writer in computer science. He was born in Milwaukee, Wisconsin, where his father, the first college graduate in the Knuth family, taught bookkeeping at a Lutheran high school; his talent for mathematics and music played a significant role in the intellectual development and pursuit of the young Knuth.*

As a youngster, Knuth had a marvelous gift for solving complex problems. As an eighth grader, he entered the Ziegler's Candies Contest to find the number of words that can be formed from the letters in Ziegler's Giant Bar. *Knuth listed 4500 such words, 2000 more than in Ziegler's master list. This won a television set for the school and enough Ziegler candy for the entire student body.*

In high school, Knuth entered the prestigious Westinghouse Science Talent Search (now Intel Science Talent Search) with his project, The Prtrzebie System of Weights and Measures, *that would replace the cumbersome British system. His project won an honorable mention, and $25 from* MAD Magazine *for publishing it. When he graduated from high school, he was already an accomplished mathematician, musician, and writer.*

He majored in physics at the Case Institute of Technology (now Case Western Reserve University) and was introduced to an IBM 650 computer, one of the earliest mainframes. After studying the manual from cover to cover, he decided that he could do better and wrote assembler and compiler code for the school's IBM 650.

In 1958, Knuth developed a system for analyzing the value of a basketball player, which the coach then used to help the team win a league championship. Newsweek *wrote an article about Knuth's system and Walter Cronkite carried it on the* CBS Evening News.

In his sophomore year, Knuth switched his major to mathematics. His work at Case was so distinguished that when he was awarded his B.S. in 1960, the faculty made an unprecedented decision to grant him an M.S. concurrently.

Knuth then entered the California Institute of Technology for graduate work and received his Ph.D. in mathematics 3 years later. He joined the faculty there, also consulting for the Burroughs Corporation writing compilers for various programming languages, including ALGOL 58 and FORTRAN II.

From 1968–1969, he worked at the Institute for Defense Analyses, Princeton, New Jersey. In 1969, Knuth joined the faculty at Stanford University.

Knuth's landmark project, The Art of Computer Programming, *was initiated by Addison-Wesley Publishing Co. in early 1962, while he was still in graduate school. Dedicated to the study of algorithms, it would be a seven-volume series when completed. A revered work, it was the pioneer textbook in the 1970s and continues to be an invaluable resource. Knuth developed two computer languages to deal with mathematics typography,* TEX, *a typesetting program, and* Metafont, *a program to develop the shapes of letters.*

He has received numerous honorary degrees from universities around the world: the Grace Murray Hopper Award (1971), the Alan M. Turing Award (1974), the Lester R. Ford Award (1975), the National Medal of Science (1979), the McDowell Award (1980), the Computer Pioneer Award (1982), and the Steele Prize (1987).

An accomplished church organist and composer of music for the organ, Knuth retired from Stanford in 1992.

We now make an interesting observation. To this end, let $f(n) = \Omega(g(n))$; so $|f(n)| \geq C|g(n)|$ for $n \geq n_0$. Then $|g(n)| \leq C'|f(n)|$ for some positive constant $C' = 1/C$; so $g(n) = O(f(n))$. Conversely, let $g(n) = O(f(n))$. By retracing these steps, it follows that $f(n) = \Omega(g(n))$. Thus $f(n) = \Omega(g(n))$ if and only if $g(n) = O(f(n))$.

We now define the big-theta notation, using the big-oh and big-omega notations.

The Big-Theta Notation

Let $f, g : \mathbb{N} \to \mathbb{R}$ such that $f(n) = O(g(n))$ and $f(n) = \Omega(g(n))$. Then $f(n)$ is said to be of **order** $g(n)$. We then write $f(n) = \Theta(g(n))$; read this as *$f(n)$ is big-theta of $g(n)$*.

The next two examples illustrate this definition.

EXAMPLE 4.39 Let $f(n) = (3n^2 + 4n - 5)\lg n$. By Example 4.37, $f(n) = O(n^2 \lg n)$. When $n \geq 1$, we also have:

$$(3n^2 + 4n - 5)\lg n \geq 3n^2 \lg n$$

That is,

$$f(n) \geq 3(n^2 \lg n)$$

So

$$f(n) = \Omega(n^2 \lg n)$$

Thus $f(n) = O(n^2 \lg n) = \Omega(n^2 \lg n)$, so $f(n) = \Theta(n^2 \lg n)$. ■

EXAMPLE 4.40 Let $f(n)$ show the number of times the assignment statement $x \leftarrow x + 1$ is executed by the nested for loops in Example 4.35. Recall that $f(n) = n(n+1)/2 = O(n^2)$.

Since $n + 1 \geq n$ for every $n \geq 1$, it follows that $n(n+1)/2 \geq n^2/2$; so $f(n) = \Omega(n^2)$. Thus $f(n) = \Theta(n^2)$. ■

We now make two interesting observations from Examples 4.39 and 4.40:

- If $f(n)$ is a polynomial in n of degree m, then $f(n) = \theta(n^m)$.
- $f(n) = \Theta(g(n))$ if and only if $A|g(n)| \leq |f(n)| \leq B \mid g(n) \mid$ for some constants A and B.

See Exercises 50 and 51.

Before closing this section, we add that the definitions of the big-oh, big-omega, and big-theta notations remain valid even if the domain of f consists of real numbers.

Exercises 4.6

Using the big-oh notation, estimate the growth of each function.

1. $f(n) = 2n + 3$ **2.** $f(n) = 4n^2 + 2n - 3$ **3.** $f(n) = 2n^3 - 3n^2 + 4n$

4. $f(n) = 3 + \lg n$ **5.** $f(n) = 3 \lg n + 2$ **6.** $f(n) = (3n)!$

7. $f(n) = \lg (5n)!$ **8.** $f(n) = 23$ **9.** $f(n) = \sum_{k=1}^{n} k^2$

10. $f(n) = \sum_{k=1}^{n} k^3$ **11.** $f(n) = \sum_{i=1}^{n} \lfloor i/2 \rfloor$ **12.** $f(n) = \sum_{i=1}^{n} \lceil i/2 \rceil$

Verify each.

13. $2^n = O(n!)$ **14.** $\sum_{i=0}^{n-1} 2^i = O(2^n)$ **15.** $\sum_{i=0}^{n} i^k = O(n^{k+1})$

16. $\sum_{i=1}^{n} \frac{1}{i(i+1)} = O(1)$ **17.** $\sum_{i=1}^{n} i(i+1) = O(n^3)$ **18.** $\sum_{i=1}^{n} (2i-1)^2 = O(n^3)$

19–22. Let a_n denote the number of times the statement $x \leftarrow x + 1$ is executed by each loop in Exercises 35–38 in Section 4.4. Using the big-oh notation, estimate the growth of a_n in each case.

23–32. Using the big-omega notation, estimate the growth of each function in Exercises 1–5 and 8–12.

Verify each.

33. $(3n)! = \Omega(6^n)$ **34.** $\sum_{i=1}^{n} i(i+1) = \Omega(n^3)$

35. $\sum_{i=1}^{n} (2i-1) = \Omega(n^2)$ **36.** $\sum_{i=1}^{n} (2i-1)^2 = \Omega(n^3)$

37. $2n + 3 = \Omega(n)$ **38.** $4n^2 + 2n - 3 = \Omega(n^2)$

39. $2n^3 - 3n^2 + 4n = \Omega(n^3)$ **40.** $3 + \lg n = \Omega(\lg n)$

41. $3 \lg n + 2 = \Omega(\lg n)$ **42.** $23 = \Omega(1)$

43. $\sum_{i=1}^{n} \lfloor i/2 \rfloor = \Omega(n^2)$ **44.** $\sum_{i=1}^{n} \lceil i/2 \rceil = \Omega(n^2)$

45. Let $f_1(n) = O(g(n))$ and $f_2(n) = kf_1(n)$, where k is a positive constant. Show that $f_2(n) = O(g(n))$.

46. Consider the constant function $f(n) = k$. Show that $f(n) = O(1)$.

Let $f(n) = O(h(n))$ and $g(n) = O(h(n))$. Verify each.

47. $(f + g)(n) = O(h(n))$

48. $(f \cdot g)(n) = O((h(n))^2)$

49. Let f, g, and h be three functions such that $f(n) = O(g(n))$ and $g(n) = O(h(n))$. Show that $f(n) = O(h(n))$.

50. Let $f(n) = \sum_{i=0}^{m} a_i n^i$, where each a_i is a real number and $a_m \neq 0$. Prove that $f(n) = \Theta(n^m)$.

51. Let $f, g : \mathbb{N} \to \mathbb{R}$. Prove that $f(n) = \Theta(g(n))$ if and only if $A|g(n)| \leq |f(n)| \leq B|g(n)|$ for some constants A and B.

4.7 Complexities of Algorithms (optional)

The time complexities of standard algorithms can be used to estimate theoretically using the big-oh and big-theta notations. Before beginning to code an algorithm we should make sure it will do its job. Why is analyzing the algorithm important? Several routines can perform the same task, but not necessarily with the same efficiency, so we should employ the one that is most efficient.

Two norms are used to measure the efficiency of an algorithm: space complexity and time complexity.

Space Complexity

Space complexity refers to how much storage space the algorithm needs. Since this depends on factors such as the computer used and methods of data storage, we restrict our discussion to time complexity.

Time Complexity

The **time complexity** of an algorithm refers to the time it takes to run the algorithm. It is often measured by the number of fundamental operations performed by the algorithm. In the case of a sorting or searching algorithm, we shall use element-comparison as the basic operation. Since the time required by an algorithm depends on the input size n, we measure time complexity in terms of n.

Often we are interested in three cases:

- The **best-case time** is the minimum time needed to execute an algorithm for an input of size n.

- The **worst-case-time** is the maximum time needed to execute the algorithm for an input of size n.
- The **average-case-time** is the average time needed to execute the algorithm for an input of size n. Estimating the average time is often a difficult task, involving probability.

We begin our analysis with the algorithm for matrix multiplication.

EXAMPLE 4.41 Estimate the number a_n of operations (additions and multiplications) needed to compute the product C of two matrices A and B of order n.

SOLUTION:
Let $A = (a_{ij})_{n\times n}$, $B = (b_{ij})_{n\times n}$, and $C = (c_{ij})_{n\times n}$. Since $c_{ij} = \sum_{k=1}^{n} a_{ik}b_{kj}$, it takes n multiplications and $n - 1$ additions to compute each c_{ij}. There are n^2 elements in C and each takes a total of $n + (n - 1) = 2n - 1$ operations. Therefore, $a_n = n^2(2n - 1) = O(n^3) = \Theta(n^3)$. Thus the product takes $O(n^3) = \Theta(n^3)$ operations. ■

Next we estimate the number of operations required to compute the product of two binary integers.

EXAMPLE 4.42 Use Algorithm 4.5 to estimate the maximum number a_n of operations (shifting and additions) required to compute the product of two binary integers $x = (x_n \dots x_0)_{\text{two}}$ and $y = (y_n \dots y_0)_{\text{two}}$.

SOLUTION:
The worst case occurs when $y_j = 1$ for every j. Each y_j contributes a shift of j places to the left. Therefore, the total number of shifts $= \sum_{j=0}^{n} j = n(n + 1)/2$, by Example 4.15.

There are $n + 1$ partial products. Adding them involves an $(n + 1)$-bit integer, an $(n + 2)$-bit integer, ..., a $(2n + 1)$-bit integer. Therefore, the total number of bit additions required is $2n + 1$. Thus:

$$\begin{aligned} a_n &= (\text{maximum no. of shifts}) + (\text{maximum no. of additions}) \\ &= \frac{n(n+1)}{2} + 2n + 1 \\ &= O(n^3) = \Theta(n^3) \end{aligned}$$

■

Next, we estimate the number of comparisons required by the bubble sort algorithm, so review it before proceeding any further.

EXAMPLE 4.43 Let c_n denote the number of comparisons required in line 3 of the bubble sort algorithm (see Algorithm 4.10). Estimate the order of magnitude of c_n.

SOLUTION:
In line 3 of the algorithm, the consecutive elements x_j and x_{j+1} are compared for every value of j. Since j varies from 1 to $n - i$, the

number of comparisons is $n - i$, by virtue of the inner loop, where $1 \leq i \leq n - 1$. So

$$\begin{aligned} c_n &= \sum_{i=1}^{n-1}(n-i) = \sum_{i=1}^{n-1} n - \sum_{i=1}^{n-1} i \\ &= n(n-1) - \frac{(n-1)n}{2}, \text{ by Example 4.15} \\ &= \frac{n(n-1)}{2} \\ &= O(n^2) = \Theta(n^2) \end{aligned}$$

Thus the bubble sort algorithm takes $O(n^2) = \Theta(n^2)$ comparisons. ■

We turn our analysis to the search algorithms presented in Section 5. Review them before proceeding any further.

EXAMPLE 4.44 Use the linear search algorithm (Algorithm 4.8) to estimate the best time, the worst time, and the average time required to search for a *key* in a list X of n elements.

SOLUTION:
Let a_n, b_n, and c_n denote the number of element comparisons needed in line 3 in the average case, the best case, and the worst case, respectively.

- The best case is realized if $x_n = key$. Since this takes only one comparison for all inputs of size n, $b_n = 1$. So $b_n = O(1)$ and the execution time is a constant.

- To compute c_n, notice that the worst case occurs when *key* does not exist in the list, in which case the **while** loop is executed $n + 1$ times. Therefore,

$$\begin{aligned} c_n &= n + 1 \\ &\leq n + n, \text{ when } n \geq 1 \\ &= 2n = O(n) \end{aligned}$$

Thus, in the worst case, the linear search algorithm takes $O(n)$ comparisons. The run time varies linearly with input size.

- To compute the average time a_n, we need to consider two cases: *key* occurs or does not occur in the list. If *key* occurs in position i, $n-i+1$ element comparisons will be required, where $1 \leq i \leq n$. If *key* does not occur in the list, $n + 1$ comparisons will be needed. So the average time taken is given by

$$\begin{aligned} a_n &= \frac{(1+2+\cdots+n)+(n+1)}{n+1} \\ &\le \frac{(n+1)(n+2)}{2(n+1)} \\ &= \frac{n}{2}+1 = O(n) \end{aligned}$$

Again, it takes $O(n)$ element comparisons. Thus, the average case, from the complexity point, is no better than the worst case in linear search. ■

Note: In the average case analysis, we assumed *key* could occur in any of the n positions with an equal chance. We also assumed that it had the same chance of not occurring in the list. If that were not the case, we would need to apply the concept of expected value in probability theory to compute a_n.

Next we examine the complexity of the binary search algorithm.

EXAMPLE 4.45 Let c_n denote the maximum number of comparisons in lines 6 through 8 of the binary search algorithm (Algorithm 4.9). Show that $c_n = O(\lg n)$.

SOLUTION:

Case 1 Let n be a power of 2, say, $n = 2^k$ where $k \ge 0$. Initially, $mid = \lfloor (low + high)/2 \rfloor = \lfloor (1+2^k)/2 \rfloor = 2^{k-1}$, so the lower sublist contains $2^{k-1} - 1$ elements and the upper sublist 2^{k-1} elements. By now two comparisons have taken place, one in line 6 and the other in line 8. Since the upper sublist contains more elements, partition it into three sublists. This time the maximum number of elements in a sublist is 2^{k-2} and two more comparisons are needed. At the next stage, two more comparisons are needed. Continue like this until the list contains one element, when $k = 0$. Again, two more comparisons ensue.

Thus, in the worst case, two comparisons are needed for each power i of 2, where $0 \le i \le k$. Therefore,

$$\begin{aligned} c_n &= 2(k+1) = 2k+2 \\ &= 2\lg n + 2, \text{ since } n = 2^k \\ &= O(\lg n) \end{aligned}$$

Case 2 Suppose n is not a power of 2. Let n be an integer such that $2^j < n < 2^{j+1}$. Then $j < \lg n$. Let $N = 2^{j+1}$. Clearly, $c_n < c_N$. By the above analysis, $c_N = 2(j+2)$. Thus:

$$\begin{aligned} c_n &< c_N \\ &= 2(j+2) \end{aligned}$$

$$
\begin{aligned}
&< 2(\lg n + 2)\\
&\leq 2(\lg n + \lg n), \text{ when } n \geq 4\\
&= 4 \lg n\\
&= \mathrm{O}(\lg n)
\end{aligned}
$$

Thus, whether or not n is a power of 2, $c_n = \mathrm{O}(\lg n)$, so the algorithm takes $\mathrm{O}(\lg n)$ comparisons in the worst case. ■

Additional examples of analyzing the complexities of algorithms appear in the exercises and the next chapter.

Exercises 4.7

1. Show that it takes $\mathrm{O}(n^2)$ additions to compute the sum of two square matrices of order n.
2. Let A and B be two square matrices of order n. Let c_n denote the number of comparisons needed to determine whether or not $A \leq B$. Show that $c_n = \mathrm{O}(n^2)$.

Let A be a square matrix of order n. Let s_n denote the number of swappings of elements needed to find the transpose A^{T} of A.

3. Find a formula for s_n.
4. Show that $s_n = \mathrm{O}(n^2)$.
5. Show that the number of additions of two n-bit integers is $\mathrm{O}(n)$.

Let a_n denote the number of additions (lines 5 and 6) required to compute the square of an integer using the algorithm in Exercise 6 of Section 5.

6. Find a formula for a_n.
7. Show that $a_n = \mathrm{O}(n)$.

Algorithm 4.13 finds the maximum value in a list X of n items. Use it to answer Exercises 8 and 9.

```
Algorithm findmax(X,n,max)
(* This algorithm returns the largest item in a list X of n
   items in a variable called max. *)
 Begin (* algorithm *)
   max ← x1      (* initialize max *)
   i ← 2
   while i ≤ n do
   begin (* while *)
     if xi > max then     (* update max *)
       max ← xi
     i ← i + 1
   endwhile
 End (* algorithm *)
```

Algorithm 4.13

8. Establish the correctness of the algorithm.

9. Let c_n denote the number of comparisons needed in line 5. Show that $c_n = O(n)$.

10. Let c_n denote the number of element-comparisons in line 6 of the insertion sort algorithm in Algorithm 4.12. Show that $c_n = O(n^2)$.

Use the minmax algorithm in Algorithm 4.14 to answer Exercises 11–13.

```
Algorithm iterative minmax(X,n,min,max)
(* This algorithm returns the minimum and the maximum
   of a list X of n elements. *)
 0. Begin (* algorithm *)
 1.    if n ≥ 1 then
 2.    begin  (* if *)
 3.       min ← x1
 4.       max ← x1
 5.       for i = 2 to n do
 6.       begin  (* for *)
 7.          if xi < min then
 8.             min ← xi
 9.          if xi > max then
10.             max ← xi
11.       endfor
12.    endif
13. End (* algorithm *)
```

Algorithm 4.14

11. Find the maximum and the minimum of the list 12, 23, 6, 2, 19, 15, 37.

12. Establish the correctness of the algorithm.

13. Using the big-oh notation, estimate the number c_n of comparisons in lines 7 and 9 of the algorithm.

14. Let c_n denote the maximum number of comparisons in lines 6 through 8 of the binary search algorithm (Algorithm 4.9). Show that $c_n = \Theta(\lg n)$.

Chapter Summary

This chapter provided a quick introduction to number theory, one of the oldest branches of mathematics. By accepting the well-ordering principle as an axiom, we established the principle of induction. We saw many examples of how pivotal induction is in proving loop invariants.

We also illustrated how to add and multiply any two nondecimal numbers, and how to subtract binary integers using complements.

Finally, we established the partial correctness of algorithms and discussed the time complexities of some standard algorithms using the big-oh and big-theta notations.

The Well-Ordering Principle

- Every nonempty subset of N has a least element (page 186).

The Division Algorithm

- The **division algorithm** When an integer a is divided by a positive integer b, there exist a unique quotient q and a unique remainder r such that $a = bq + r$, where $0 \leq x < b$ (page 186).
- An integer $p \geq 2$ is a **prime** if its only positive factors are 1 and p (page 189).

The Greatest Common Divisor (gcd)

- A positive integer d is the gcd of two positive integers a and b if:
  - $d \mid a$ and $d \mid b$; and
  - if $d' \mid a$ and $d' \mid b$, then $d' \mid d$. (page 191).
- The **euclidean algorithm**, which uses successive applications of the division algorithm, provides a procedure to compute $\gcd\{a,b\}$ (page 193).
- Two positive integers a and b are **relatively prime** if $\gcd\{a,b\} = 1$ (page 194).
- Every decimal integer has a unique nondecimal representation in a given base and every nondecimal integer has a unique decimal value (page 197).
- Binary subtraction can be performed using two's complement (page 203).

Mathematical Induction

- **Weak version** Let P(n) be a predicate such that
  - P(n_0) is true; and
  - for every $k \geq n_0$, if P(k) is true, $P(k+1)$ is also true.

  Then P(n) is true for every $n \geq n_0$ (page 209).
- **Strong version** Let P(n) be a predicate such that
  - P(n_0) is true; and

- for every $k \geq n_0$, if P(n_0), P($n_0 + 1$), ..., P(k) are true, P($k + 1$) is also true. Then P(n) is true for $n \geq n_0$ (page 218).

- **The Fundamental Theorem of Arithmetic** Every positive integer ≥ 2 is either a prime or can be expressed as a product of primes (page 218).

Algorithm Correctness

- Using induction, we verified the partial correctness of several standard algorithms: linear search (page 228), binary search (page 230), bubble sort (page 233), and selection sort (page 234).

The Big-Oh Notation

- $f(n) = O(g(n))$, if there are positive constants C and n_0 such that $|f(n)| \geq C|g(n)|$ for every $n \geq n_0$ (page 237).
- $f(n) = \Omega(g(n))$, if $|f(n)| \geq C|g(n)|$ for every $n \geq n_0$ (page 243).
- $f(n) = \Theta g(n)$, if $f(n) = O(g(n))$ and $f(n) = \Omega(g(n))$ (page 245).
- The time complexity of an algorithm is the execution time of the algorithm (page 245).

Review Exercises

Using the euclidean algorithm, find the gcd of each pair of integers.

1. 18, 28 **2.** 36, 12 **3.** 15, 24 **4.** 1024, 3076

Express each number in base 10.

5. 2000_{eight} **6.** 2345_{sixteen} **7.** $\text{BAD}_{\text{sixteen}}$ ***8.** $\text{BAD.CA}_{\text{sixteen}}$

Rewrite each number in the indicated base b.

9. 245, $b = 2$ **10.** 348, $b = 8$ **11.** 1221, $b = 8$ **12.** 1976, $b = 16$

In Exercises 13–16, perform the indicated operation.

13. 11010_{two} $+111_{\text{two}}$

14. 5768_{sixteen} $+78\text{CB}_{\text{sixteen}}$

15. $5\text{AB}8_{\text{sixteen}}$ $\times\text{BAD}_{\text{sixteen}}$

16. 110110_{two} -11011_{two}

Rewrite each binary integer in base eight.

17. 10110101 **18.** 1101101101 **19.** 100110011 **20.** 10011011001

21–24. Rewrite the binary integers in Exercises 17–20 in base 16.

Find the value of x resulting from the execution of each algorithm fragment.

25.
```
x ← 0
for i = 1 to n do
  for j = 1 to n do
    x ← x + 1
```

26.
```
x ← 0
for i = 1 to n do
  for j = 1 to i do
    for k = 1 to j do
      x ← x + 1
```

27. Find a formula for the number a_n of times the statement $x \leftarrow x + 1$ is executed by the following loop:

```
for i = 1 to n do
  for j = 1 to ⌈i/2⌉ do
    x ←  x + 1
```

28. Let $a,b,c,d \in \mathbb{N}$. Let $d \mid ab$, $d \mid ac$, and b and c be relatively prime numbers. Prove that $d \mid a$.

29. Let $a,b \in \mathbb{N}$ and $\gcd\{a,b\} = 1$. Prove that $\gcd\{a - b, a + b\} = 1$ or 2.

Using induction prove each, where n is a positive integer.

30. $n^2 - n$ is divisible by 2.

31. $n^3 - n$ is divisible by 3.

32. $\sum_{i=1}^{n}(2i-1)^2 = \frac{n(4n^2-1)}{3}$

33. $\sum_{i=1}^{n}\frac{1}{(2i-1)(2i+1)} = \frac{n}{2n+1}$

34. The product of any two consecutive positive integers is even.

35. Suppose you have an unlimited supply of identical black and white socks. Using induction and the pigeonhole principle, show that you must select at least $2n + 1$ socks in order to ensure n matching pairs. (C. T. Long)

Evaluate each sum and product.

36. $\sum_{i=1}^{n} i(i+1)$

37. $\sum_{i=1}^{n}\sum_{j=1}^{n}(2i+3j)$

38. $\sum_{i=1}^{n}\sum_{j=1}^{n}2^i3^j$

39. $\sum_{i=1}^{n}\sum_{j=1}^{n}2^j$

40. $\prod_{i=1}^{n}\prod_{j=1}^{n}2^i3^j$

41. $\prod_{i=1}^{n}\prod_{j=1}^{n}3^{2j}$

42. $\prod_{i=1}^{n}\prod_{j=1}^{n}2^i$

***43.** $\sum_{i=1}^{n} i \prod_{j=1}^{i} j$

44. Let S_n denote the value of *sum* after n iterations of the while loop in Algorithm 4.15. Prove that P(n): $S_n = n(n+1)$ is a loop invariant.

```
Algorithm evensum (n)
(* This algorithm computes the sum of the first x
   positive even integers. *)
0. Begin (* algorithm *)
1.    sum ← 0
```

```
2.     i ← 0 (* counter *)
3.     while i < n do
4.     begin (* while *)
5.       i ← i + 1
6.       sum ← sum + 2 * i
7.     endwhile
8.   End (* algorithm *)
```

Algorithm 4.15

45. Using Example 4.23 predict a formula for the number of trailing zeros in $n!$, where $n \geq 1$.

46. Let a_n denote the number of operations (additions and multiplications) in line 6 of the algorithm in Exercise 44. Find the order of magnitude of a_n.

47. Add two lines to the following number pattern, where t_n denotes the nth triangular number.

$$t_1 + t_2 + t_3 = t_4$$
$$t_5 + t_6 + t_7 + t_8 = t_9 + t_{10}$$
$$t_{11} + t_{12} + t_{13} + t_{14} + t_{15} = t_{16} + t_{17} + t_{18}$$

Prove each, where t_n denotes the nth triangular number.

48. $t_n^2 - t_{n-1}^2 = n^3$ **49.** $t_n^2 = t_n + t_{n-1}t_{n+1}$ **50.** $2t_n t_{n-1} = t_{n^2-1}$

Supplementary Exercises

1. Prove that $(m^2 - n^2, 2mn, m^2 + n^2)$ is a solution of the equation $x^2 + y^2 = z^2$.

2. Prove that the product of the sums of two squares of two integers can be written as a sum of two squares.

3. Let t_k denote the kth triangular number and n any triangular number. Prove that $(2k + 1)^2 n + t_k$ is also a triangular number. (R. F. Jordan, 1991)

4. In 1950, P. A. Piza discovered the following formula about sums of powers of triangular numbers t_i: $[3 \sum_{i=1}^{n} t_i]^3 = \sum_{i=1}^{n} t_i^3 + 2 \sum_{i=1}^{n} t_i^4$. Verify it for $n = 3$ and $n = 4$.

5. Show that 111 cannot be a square in any base.

***6.** Prove that one more than the product of four consecutive integers is a perfect square, and the square root of the resulting number is the average of the product of the smaller and larger numbers and the product of the two middle integers. (W. M. Waters, 1990)

A composite number n is **Duffinian** if none of its positive factors, except 1, is a factor of the sum s of its proper factors. For example, let $n = 21$. The sum of its proper factors $= 1 + 3 + 7 = 11$. Since both 3 and 7 are not factors of 11, 21 is Duffinian. (You may verify that 10 is not Duffinian.)

7. Determine if 18, 25, 36, and 43 are Duffinian.

8. Let p be a prime and k a positive integer ≥ 2. Prove that p^k is Duffinian.

9. Prove that n is Duffinian if and only if none of the factors of n, except 1, is a factor of n.

10. Prove or disprove: The product of two Duffinian numbers is Duffinian.

Prove each, where n is a positive integer.

***11.** $n(3n^4 + 7n^2 + 2)$ is divisible by 12.

***12.** $n(3n^4 + 13n^2 + 8)$ is divisible by 24.

****13.** Let S_n denote the sum of the elements in the nth set in the sequence of sets of positive integers $\{1\}, \{3, 5\}, \{7, 9, 11\}, \{13, 15, 17, 19\}, \ldots$. Find a formula for S_n. (R. Euler, 1988)

****14.** Let S_n denote the sum of the elements in the nth set in the sequence of positive integers $\{1\}, \{2, 3, \ldots, 8\}, \{9, 10, \ldots, 21\}, \{22, 23, \ldots, 40\}, \ldots$. Find a formula for S_n. (C. W. Trigg, 1980)

****15.** Three schools in each state, Alabama, Georgia, and Florida, enter one person in each of the events in a track meet. The number of events and the scoring system are unknown, but the number of points for the third place is less than that for the second place, which in turn is less than the number of points for the first place. Georgia scored 22 points, and Alabama and Florida tied with 9 each. Florida won the high jump. Who won the mile run? (M. vos Savant, 1993)

Computer Exercises

Write a program to perform each task.

1. Read in an integer $b \geq 2$ and select $b + 1$ integers at random. Find two integers in the list such that their difference is divisible by b.

2. Read in an integer $n \geq 2$ and select n positive integers at random. Find a sequence of integers from the list whose sum is divisible by n.

3. Read in a positive integer ≥ 2 and determine if it is a prime.

4. Determine if each value of $f(n) = n^2 - n + 41$ is a prime, where $0 \leq n \leq 41$.

5. Redo Program 4 with $f(n) = n^2 - 79n + 1601$, where $0 \leq n \leq 80$.

6. Determine if the nth **Fermat number** $f(n) = 2^{2^n} + 1$ is a prime, where $0 \leq n \leq 4$.

7. Find all perfect numbers ≤ 1000. (There are three such numbers.)

8. Find the gcd$\{x, y\}$ using the euclidean algorithm.

9. Read in a sequence of pairs of integers n and b. For each integer n, determine its base-b representation and use this representation to compute the corresponding decimal value. Print each integer n, base b, base-b representation, and its decimal value in a tabular form.

10. Read in a positive integer n and find the number of trailing zeros in $n!$.

11. A **palindrome** is a positive integer that reads the same backward and forward. Find the eight palindromic triangular numbers < 1000.

12. Compute the total number of grains of wheat needed for each of the squares on an 8×8 chessboard, as in Exercises 39 and 40 in Section 4.4. (*Hint*: The answer is 18,446,744,073,709,551,615 grains, which may be too large for an integer variable to hold, so think of a suitable data structure.)

13. Read in a positive integer $N \leq 1000$. Using Example 4.24, determine how many doors will remain open at the end. Do *not* use the fact that there are $\lfloor \sqrt{n} \rfloor$ perfect squares $\leq n$.

14. Print the ages 1–31 on five tablets A, B, C, D, and E, as in Figure 4.2. Read in some tablets at random and compute the corresponding age. Extend the puzzle to six tablets to include ages through 63.

15. Read in a positive integer n and determine if it is a prime.

16. Construct a table of values of the function $E(n) = n^2 - n + 41$, where $0 \leq n \leq 41$, and identify each value as prime or composite.

17. Redo program 16 with $L(n) = n^2 + n + 41$, where $0 \leq n \leq 41$, and identify each value as prime or composite.

18. Redo program 16 with $H(n) = 9n^2 - 471n + 6203$, where $0 \leq n \leq 39$, and identify each value as prime or composite.

19. Redo program 16 with $G(n) = n^2 - 2999n + 2248541$, where $1460 \leq n \leq 1539$, and identify each value as prime or composite.

20. Read in a positive integer n, and list all primes $\leq n$ and are of the form $k^2 + 1$.

21. Read in a positive integer n and find a prime between:

(a) n and $2n$. (b) n^2 and $n^2 + 1$.

Exploratory Writing Projects

Using library and Internet resources, write a team report on each of the following in your own words. Provide a well-documented bibliography.

1. Describe how twin primes were used in 1994 by Thomas Nicely of Lynchburg College, Virginia, to detect defects in the Pentium chip.
2. Explain how to construct Tables A–E in Figure 4.2 and how the puzzle works. Extend the puzzle to cover ages through 63.
3. Describe the origin of mathematical induction. Include biographies of those who developed this proof technique. Comment on its importance in computer science.
4. Describe the origin of *figurate numbers*. Explain the various types and their properties. Include the relationships between the 12 days of Christmas puzzle, and polygonal numbers and tetrahedral numbers.
5. Explore the history of magic squares. Do they have any practical applications?
6. Describe the origin of the big-oh, big-omega, and big-theta notations. Include biographies of mathematicians who developed them.
7. Investigate the various classes of prime numbers.
8. Describe the history of finding larger and larger primes, and their practical applications. Comment on the Greatest Internet Mersenne Prime Search (GIMPS), founded in 1996 by George Woltman.
9. Discuss the game of *Nim* and its relationship to binary numbers.
10. Discuss *Eleusis*, a card game devised by R. Abbott of New York.

Enrichment Readings

1. R. G. Archibald, *An Introduction to the Theory of Numbers*, Merrill, Columbus, OH, 1970, pp. 1–95.
2. G. Brassard and P. Bratley, *Algorithmics: Theory & Practice*, Prentice Hall, Englewood Cliffs, NJ, 1988.
3. J. Dugle, "The Twelve Days of Christmas and Pascal's Triangle," *Mathematics Teacher*, Vol. 75 (Dec. 1982), pp. 755–757.
4. G. H. Hardy, *A Mathematician's Apology*, Cambridge University Press, Cambridge, 1941.

5. T. Koshy, *Elementary Number Theory with Applications*, Harcourt/Academic Press, Boston, 2002, pp. 1–189.

6. C. Oliver, "The Twelve Days of Christmas," *Mathematics Teacher*, Vol. 70 (Dec. 1977), pp. 752–754.

7. H. S. Wilf, *Algorithms and Complexity*, Prentice-Hall, Englewood Cliffs, NJ, 1986, pp. 8–22, 137–175.

Chapter 5

Recursion

It is common sense to take a method and try it. If it fails, admit it frankly and try another. But above all, try something.

— FRANKLIN ROOSEVELT

Recursion is an elegant and powerful problem-solving technique, used extensively in both discrete mathematics and computer science. Many programming languages, such as ALGOL, FORTRAN 90, C++, and Java, support recursion. This chapter investigates this powerful method in detail.

In addition, we will study three simple methods for solving recurrence relations: iteration, characteristic equations, and generating functions.

We also will establish the validity of recursive algorithms using induction and analyze their complexities using the big-oh and big-theta notations.

Some of the interesting problems we pursue in this chapter are:

- There are three pegs X, Y, and Z on a platform and 64 disks of increasing sizes at X. We would like to move them from X to Z using Y as an auxiliary peg subject to the following conditions:

  Only one disk can be moved at a time.
  No disk can be placed on the top of a smaller disk.

  If it takes one second to transfer a disk from one peg to another, how long will it take to solve the puzzle?

- Is there a formula for the number of n-bit words containing no two consecutive 1's?
- Suppose we introduce a mixed pair (male and female) of 1-month-old rabbits into a large enclosure on January 1. By the end of each month, the rabbits become mature, and each pair produces $k - 1$ mixed pairs of offspring at the beginning of the following month. Find the average age of the rabbit pairs at the beginning of the nth month.
- Can we estimate the number of divisions required to compute $\gcd\{a, b\}$ by the euclidean algorithm?
- What is a divide-and-conquer algorithm? If $f(n)$ denotes the number of operations required by such an algorithm, what can you say about its order of complexity?

5.1 Recursively Defined Functions

Recall that in Section 2.5 we employed recursion to define sets; we invoked the recursive clause to construct new elements from known elements. The same idea can be applied to define functions, and hence sequences as well.

This section illustrates how powerful a problem-solving technique recursion is. We begin with a simple problem:

> There are n guests at a sesquicentennial ball. Each person shakes hands with everybody else exactly once. How many handshakes are made?

Suppose you would like to solve a problem such as this. (See Example 5.3.) The solution may not be obvious. However, it may turn out that the problem could be defined in terms of a simpler version of itself. Such a definition is a **recursive definition**. Consequently, the given problem can be solved provided the simpler version can be solved. This idea is pictorially represented in Figure 5.1.

Figure 5.1

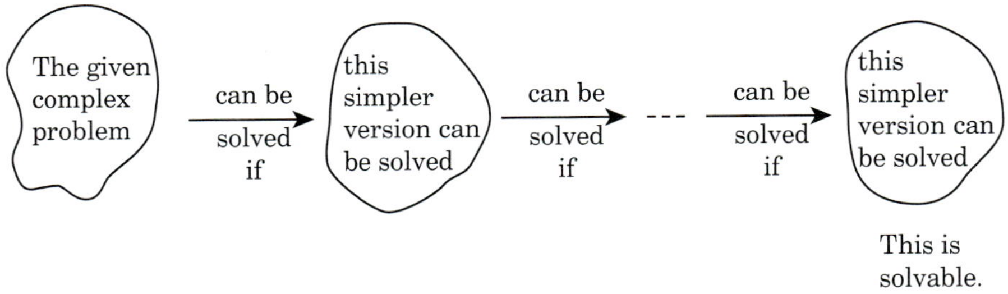

Recursive Definition of a Function

Let $a \in \mathbf{W}$ and $X = \{a, a+1, a+2, \ldots\}$. The **recursive definition** of a function f with domain X consists of three parts, where $k \geq 1$:

- **Basis clause** A few initial values of the function $f(a), f(a+1), \ldots, f(a+k-1)$ are specified. An equation that specifies such initial values is an **initial condition**.
- **Recursive clause** A formula to compute $f(n)$ from the k preceding functional values $f(n-1), f(n-2), \ldots, f(n-k)$ is made. Such a formula is a **recurrence relation** (or **recursion formula**).
- **Terminal clause** Only values thus obtained are valid functional values. (For convenience, we drop this clause from our recursive definition.)

Thus the recursive definition of f consists of one or more (a finite number of) initial conditions, and a recurrence relation.

Is the recursive definition of f a valid definition? In other words, if the k initial values $f(a), f(a+1), \ldots, f(a+k-1)$ are known and $f(n)$ is defined in terms of k of its predecessors $f(n-1), f(n-2), \ldots, f(n-k)$, where $n \geq a+k$, is $f(n)$ defined for $n \geq a$? Fortunately, the next theorem comes to our rescue. Its proof uses strong induction and is complicated, so we omit it.

THEOREM 5.1 Let $a \in \mathbf{W}$, $X = \{a, a+1, a+2, \ldots\}$, and $k \in \mathbb{N}$. Let $f : X \to \mathbb{R}$ such that $f(a), f(a+1), \ldots, f(a+k-1)$ are known. Let n be any positive integer $\geq a+k$ such that $f(n)$ is defined in terms of $f(n-1), f(n-2), \ldots$ and $f(n-k)$. Then $f(n)$ is defined for every $n \geq a$. ■

> By virtue of this theorem, recursive definitions are also known as **inductive definitions**.

The following examples illustrate the recursive definition of a function.

EXAMPLE 5.1 Define recursively the factorial function f.

SOLUTION:

Recall that the factorial function f is defined by $f(n) = n!$, where $f(0) = 1$. Since $n! = n(n-1)!$, f can be defined recursively as follows:

$$f(0) = 1 \qquad \leftarrow \text{initial condition}$$

$$f(n) = n \cdot f(n-1), \quad n \geq 1 \qquad \leftarrow \text{recurrence relation}$$

■

Suppose we would like to compute $f(3)$ using this recursive definition. We then continue to apply the recurrence relation until the initial condition is reached, as shown below:

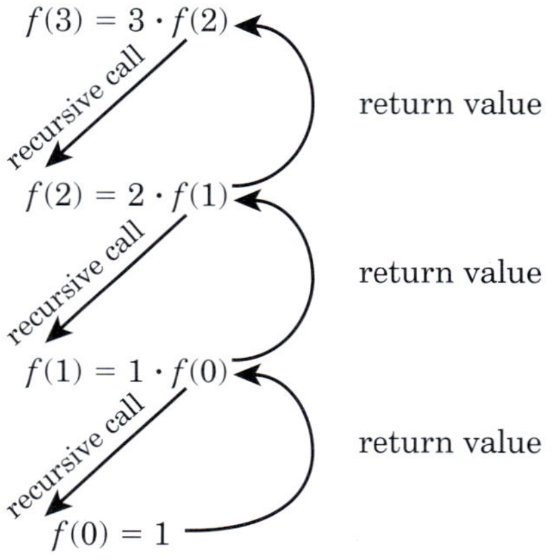

(5.1)

(5.2)

(5.3)

(5.4)

Since $f(0) = 1$, 1 is substituted for $f(0)$ in Equation (5.3) and $f(1)$ is computed: $f(1) = 1 \cdot f(0) = 1 \cdot 1 = 1$. This value is substituted for $f(1)$ in Equation (5.2) and $f(2)$ is computed: $f(2) = 2 \cdot f(1) = 2 \cdot 1 = 2$. This value is now returned to Equation (5.1) to compute $f(3)$: $f(3) = 3 \cdot f(2) = 3 \cdot 2 = 6$, as expected.

EXAMPLE 5.2 Judy deposits \$1000 in a local savings bank at an annual interest rate of 8% compounded annually. Define recursively the compound amount $A(n)$ she will have in her account at the end of n years.

SOLUTION:
Clearly, $A(0) =$ initial deposit $=$ \$1000. Let $n \geq 1$. Then:

$$\begin{aligned} A(n) &= \begin{pmatrix} \text{compound amount} \\ \text{at the end of the} \\ (n-1)\text{st year} \end{pmatrix} + \begin{pmatrix} \text{interest earned} \\ \text{during the} \\ n\text{th year} \end{pmatrix} \\ &= A(n-1) + (0.08)A(n-1) \\ &= 1.08A(n-1) \end{aligned}$$

Thus $A(n)$ can be defined recursively as follows:

$$\begin{aligned} &A(0) = 1000 && \leftarrow \text{initial condition} \\ &A(n) = 1.08A(n-1), \quad n \geq 1 && \leftarrow \text{recurrence relation} \end{aligned}$$ ■

For instance, the compound amount Judy will have at the end of three years is

$$\begin{aligned} A(3) &= 1.08A(2) \\ &= 1.08\,[1.08A(1)] = 1.08^2A(1) \\ &= 1.08^2[1.08A(0)] = 1.08^3(1000) \\ &\approx \$1259.71^* \end{aligned}$$

The next two examples illustrate an extremely useful problem-solving technique, used often in discrete mathematics and computer science.

EXAMPLE 5.3 **(The handshake problem)** There are n guests at a sesquicentennial ball. Each person shakes hands with everybody else exactly once. Define recursively the number of handshakes $h(n)$ that occur.

SOLUTION:
Clearly, $h(1) = 0$, so let $n \geq 2$. Let x be one of the guests. By definition, the number of handshakes made by the remaining $n - 1$ guests among themselves is $h(n-1)$. Now person x shakes hands with each of these

---

*The symbol $\approx$ means is *approximately equal to*.

$n - 1$ guests, yielding $n - 1$ additional handshakes. So the total number of handshakes made equals $h(n-1) + (n-1)$, where $n \geq 2$.

Thus $h(n)$ can be defined recursively as follows:

$$h(1) = 0 \qquad \leftarrow \text{initial condition}$$

$$h(n) = h(n-1) + (n-1), \quad n \geq 2 \qquad \leftarrow \text{recurrence relation} \quad ■$$

EXAMPLE 5.4 **(Tower of Brahma*)** According to a legend of India, at the beginning of creation, God stacked 64 golden disks on one of three diamond pegs on a brass platform in the temple of Brahma at Benares† (see Figure 5.2). The priests on duty were asked to move the disks from peg X to peg Z using Y as an auxiliary peg under the following conditions:

- Only one disk can be moved at a time.
- No disk can be placed on the top of a smaller disk.

The priests were told that the world would end when the job was completed.

Figure 5.2

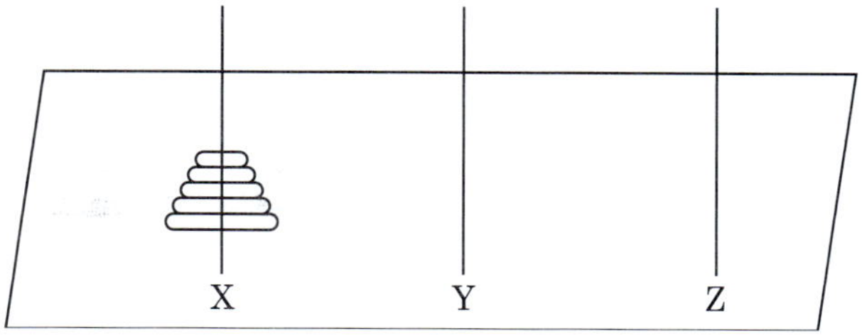

Suppose there are n disks on peg X. Let b_n denote the number of moves needed to move them from peg X to peg Z, using peg Y as an intermediary. Define b_n recursively.

SOLUTION:

Clearly $b_1 = 1$. Assume $n \geq 2$. Consider the top $n - 1$ disks on peg X. By definition, it takes b_{n-1} moves to transfer them from X to Y using Z as an auxiliary. That leaves the largest disk at peg X; it takes one move to transfer it from X to Z. See Figure 5.3.

Now the $n - 1$ disks at Y can be moved from Y to Z using X as an intermediary in b_{n-1} moves, so the total number of moves needed is $b_{n-1} + 1 + b_{n-1} = 2b_{n-1} + 1$. Thus b_n can be defined recursively as follows:

$$b_n = \begin{cases} 1 & \text{if } n = 1 \quad \leftarrow \text{initial condition} \\ 2b_{n-1} + 1 & \text{otherwise} \quad \leftarrow \text{recurrence relation} \end{cases} \quad ■$$

*A puzzle based on the Tower of Brahma was marketed in 1883 under the name **Tower of Hanoi**.

†Benares is now known as Varanasi.

Figure 5.3

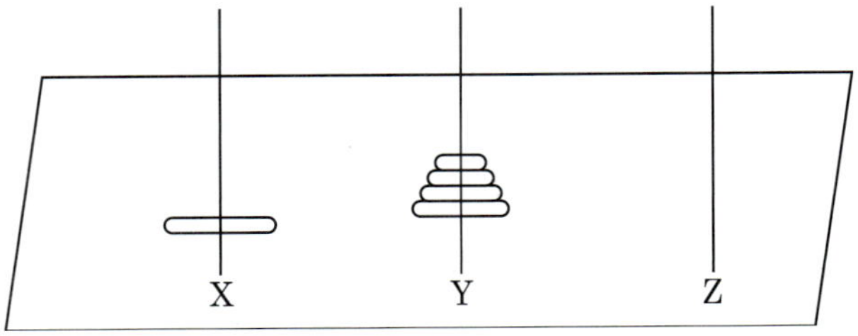

For example,

$$
\begin{aligned}
b_4 &= 2b_3 + 1 & &= 2[2b_2 + 1] + 1 \\
&= 4b_2 + 2 + 1 & &= 4[2b_1 + 1] + 2 + 1 \\
&= 8b_1 + 4 + 2 + 1 & &= 8(1) + 4 + 2 + 1 \\
&= 15
\end{aligned}
$$

so it takes 15 moves to transfer 4 disks from X to Z, by this strategy.

The next example also illustrates the same technique. We will take it a step further in Chapter 6.

EXAMPLE 5.5 Imagine n lines in a plane such that no two lines are parallel, and no three are concurrent.* Let f_n denote the number of distinct regions into which the plane is divided by them. Define f_n recursively.

SOLUTION:
If there is just one line ℓ_1 in the plane, then $f_1 = 2$ (see Figure 5.4). Now consider a second line ℓ_2; it is intersected at exactly one point by ℓ_1. Each half of ℓ_2 divides an original region into two, adding two more regions (see Figure 5.5). Thus $f_2 = f_1 + 2 = 4$. Suppose we add a third line ℓ_3. It is

Figure 5.4

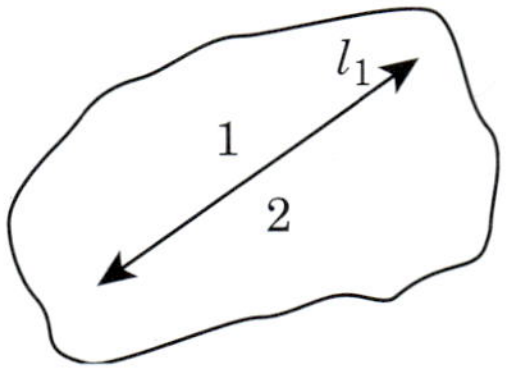

Figure 5.5

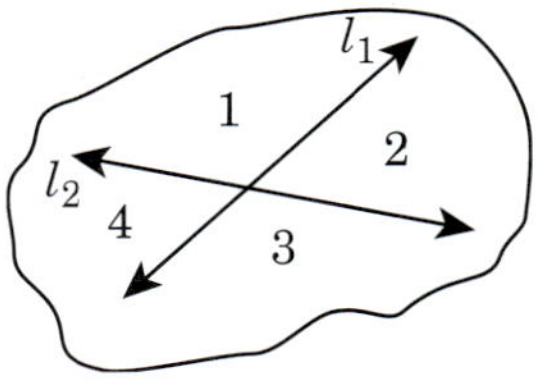

*Three or more lines in a plane are **concurrent** if they intersect at a point.

intersected by ℓ_1 and ℓ_2 in two points; in other words, line ℓ_3 is divided by ℓ_1 and ℓ_2 into three parts. Each portion divides an existing region into two, yielding three new regions (see Figure 5.6). So $f_3 = f_2 + 3 = 7$.

Figure 5.6

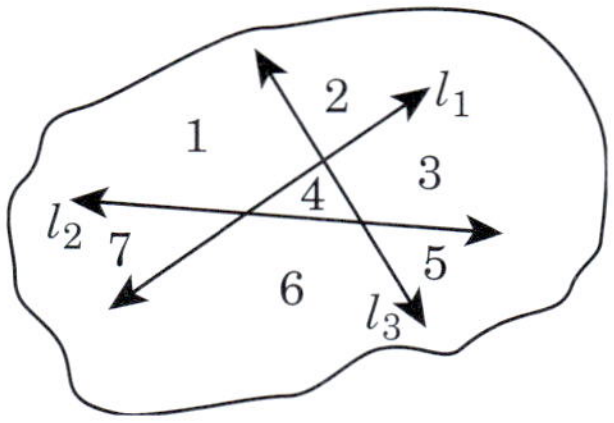

More generally, suppose there are $n-1$ lines $\ell_1, \ell_2, \ldots, \ell_{n-1}$ in the plane. They divide the plane into f_{n-1} disjoint regions, by definition. Now add one more line ℓ_n (see Figure 5.7). Since no three lines are concurrent, line ℓ_n must intersect lines $\ell_1, \ell_2, \ldots, \ell_{n-1}$ at new points and hence is divided by

Figure 5.7

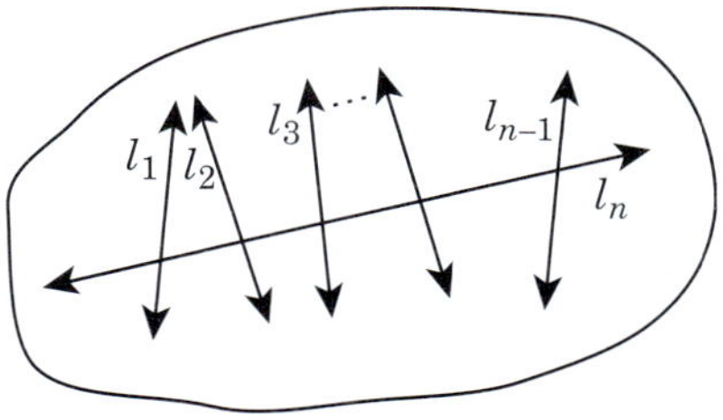

them into n segments. Each segment divides an existing region into two subregions, contributing n more regions, so $f_n = f_{n-1} + n$. Thus f_n can be defined recursively as follows:

$$f_n = \begin{cases} 1 & \text{if } n = 0 \\ f_{n-1} + n & \text{otherwise} \end{cases}$$ ■

The next example illustrates how to define recursively the number of times an assignment is executed by nested **for** loops.

EXAMPLE 5.6 Let a_n denote the number of times the assignment statement $x \leftarrow x + 1$ is executed by the following nested **for** loops. Define a_n recursively.

```
for i = 1 to n do
  for j = 1 to i do
    for k = 1 to j do
      x ← x + 1
```

SOLUTION:

- First, we must find the initial condition satisfied by a_n. When $n = 1$, $i = j = k = 1$, so the assignment statement is executed exactly once. Thus $a_1 = 1$.

- *To find the recurrence relation satisfied by a_n:*
  Let $n \geq 2$. As i runs from 1 through $n-1$, by definition, the statement is executed a_{n-1} times.
  When $i = n$, the inner loops become:

```
for j = 1 to n do
  for k = 1 to j do
    x ← x + 1
```

For each value of j, where $1 \leq j \leq n$, the innermost loop executes the statement j times. So these nested loops execute it $\sum_{j=1}^{n} j = \frac{n(n+1)}{2}$ times. Therefore,

$$a_n = \begin{pmatrix}\text{no. of times the statement}\\ \text{is executed as } i \text{ runs from}\\ 1 \text{ through } n-1\end{pmatrix} + \begin{pmatrix}\text{no. of times the}\\ \text{statement is executed}\\ \text{when } i = n\end{pmatrix}$$

$$= a_{n-1} + \frac{n(n+1)}{2}$$

Thus a_n can be defined as follows:

$$a_1 = 1$$

$$a_n = a_{n-1} + \frac{n(n+1)}{2}, \quad n \geq 2$$

(We shall pursue this definition in Example 5. 11.) ■

The next example provides a recursive definition with two initial conditions. We shall use it often in the following sections and in the next chapter.

EXAMPLE 5.7 **(Fibonacci)** Leonardo Fibonacci, the most outstanding Italian mathematician of the Middle Ages, proposed the following problem around 1202:

Suppose there are two newborn rabbits, one male and the other female. Find the number of rabbits produced in a year if:

- Each pair takes one month to become mature.
- Each pair produces a mixed pair every month, from the second month.
- No rabbits die.

Suppose, for convenience, that the original pair of rabbits was born on January 1. They take a month to become mature. So there is still only one pair on February 1. On March 1, they are 2 months old and produce a new mixed pair, a total of two pairs. Continuing like this, there will be three pairs on April 1, five pairs on May 1, and so on. See the last row of Table 5.1.

Table 5.1

No. of pairs	Jan	Feb	March	April	May	June	July	Aug
Adults	0	1	1	2	3	5	8	13
Babies	1	0	1	1	2	3	5	8
Total	1	1	2	3	5	8	13	21

←

Leonardo Fibonacci *(1170?–1250?), also known as Leonardo of Pisa, was born in the commercial center of Pisa, Italy, into the Bonacci family. His father, a customs manager, expected the son to become a merchant and took him to Bougie, Algeria, to receive good training in arithmetic with Indian numerals. Leonardo's subsequent business trips to Egypt, Syria, Greece, and Sicily brought him closer to Indian mathematics.*

In 1202, shortly after his return, convinced of the elegance of the Indian methods of computation, Fibonacci published his famous work, Liber Abaci. *(The word* abaci *in the title does not refer to the old abacus, but to computation in general.) This book, devoted to arithmetic and elementary algebra, introduced the Indian notation and arithmetic algorithms to Europe.*

Fibonacci wrote three additional books: Practica Geometriae, *a collection of results in geometry and trigonometry;* Liber Quadratorum, *a major work on number theory; and* Flos, *also on number theory.*

Fibonacci's importance and usefulness to Pisa and its citizenry through his teaching and services were honored by Emperor Frederick II of Pisa.

The numbers 1, 1, 2, 3, 5, 8, ... are **Fibonacci numbers**.* They have a fascinating property: Any Fibonacci number, except the first two, is the sum of the two immediately preceding Fibonacci numbers. (At the given rate, there will be 144 pairs of rabbits on December 1.)

This yields the following recursive definition of the nth Fibonacci number F_n:

$$F_1 = F_2 = 1 \qquad \leftarrow \text{initial conditions}$$

$$F_n = F_{n-1} + F_{n-2}, \quad n \geq 3 \qquad \leftarrow \text{recurrence relation}$$

■

The next example illustrates recursion and also shows that Fibonacci numbers occur in quite unexpected places.

EXAMPLE 5.8 Let a_n denote the number of n-bit words containing no two consecutive 1's. Define a_n recursively.

*See author's *Fibonacci and Lucas Numbers with Applications* for a thorough discussion of Fibonacci numbers.

SOLUTION:

First, let us find the n-bit words containing no two consecutive 1's corresponding to $n = 1, 2, 3$, and 4 (see Table 5.2). It follows from the table that $a_1 = 2$, $a_2 = 3$, $a_3 = 5$, and $a_4 = 8$.

Table 5.2

$n = 1$	$n = 2$	$n = 3$	$n = 4$
0	00	000	0000
1	01	010	0100
	10	100	1000
		001	0010
		101	1010
			0001
			0101
			1001

Now, consider an arbitrary n-bit word. It may end in 0 or 1.

Case 1 Suppose the n-bit word ends in 0. Then the $(n-1)$st bit can be a 0 or a 1, so there are no restrictions on the $(n-1)$st bit:

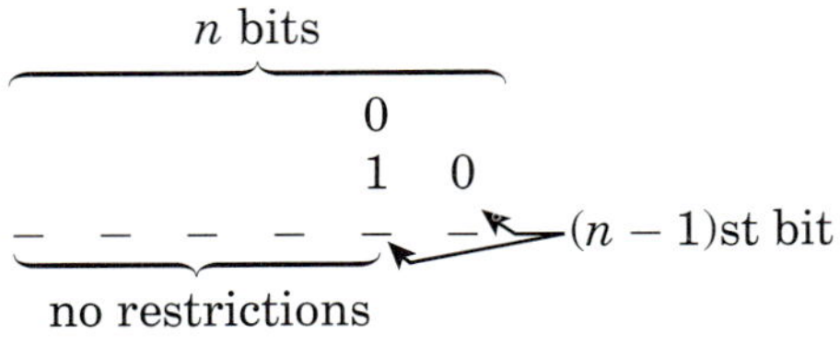

Therefore, a_{n-1} n-bit words end in 0 and contain no two consecutive 1's.

Case 2 Suppose the n-bit word ends in 1. Then the $(n-1)$st bit must be a zero. Further, there are no restrictions on the $(n-2)$nd bit:

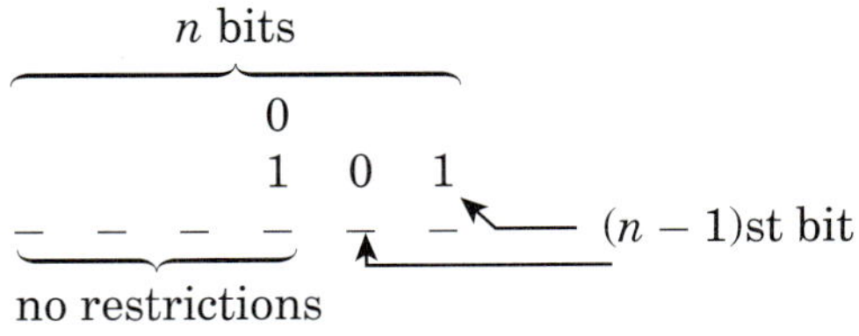

Thus a_{n-2} n-bit words end in 1 and contain no two consecutive 1's.

Since the two cases are mutually exclusive, by the addition principle, we have:

$$a_1 = 2, \quad a_2 = 3 \qquad \leftarrow \text{initial conditions}$$

$$a_n = a_{n-1} + a_{n-2}, \quad n \geq 3 \qquad \leftarrow \text{recurrence relation}$$

Notice that the above recurrence relation is exactly the same as the Fibonacci recurrence relation, but with different initial conditions! The resulting numbers are the Fibonacci numbers 2, 3, 5, 8, 13, ... ■

Notice that this example does *not* provide a constructive method for systematically listing all n-bit words with the required property. It is given in Exercise 19.

Interestingly enough, the delightful Fibonacci numbers occur in numerous totally unexpected places. For instance, the numbers of spiral arrays of seeds in mature sunflowers in the clockwise and counterclockwise directions are often consecutive Fibonacci numbers, usually 34 and 55, or 55 and 89. See Figures 5.8 and 5.9.

Figure 5.8

Figure 5.9

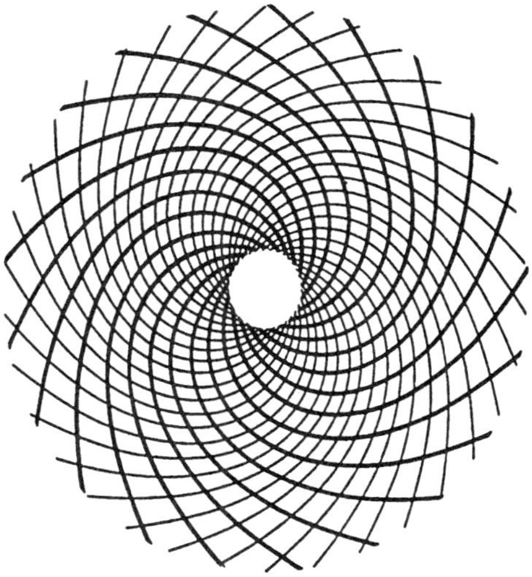

Before closing this section, we establish an important result from the theory of formal languages. First, recall that Σ^* denotes the set of words over an alphabet Σ. Also Σ^* can be defined recursively as follows (see Exercise 35 in Section 2.6):

- $\lambda \in \Sigma^*$.
- If $w \in \Sigma^*$ and $s \in \Sigma$, then $ws \in \Sigma^*$.

Furthermore, the length $\|w\|$ of a word w over Σ can be defined recursively as follows:

- $\|\lambda\| = 0$.
- If $w \in \Sigma^*$ and $s \in \Sigma$, then $\|ws\| = \|w\| + 1$.

Using these definitions and induction, we prove below that $\|xy\| = \|x\| + \|y\|$ for any two words x and y in Σ^*.

EXAMPLE 5.9 Let x and y be any two words over an alphabet Σ. Prove that $\|xy\| = \|x\| + \|y\|$.

PROOF (by induction):
Let x be any element in Σ^*. Let P(y) denote the predicate that $\|xy\| = \|x\| + \|y\|$, where $y \in \Sigma^*$. Since $y \in \Sigma^*$, y can be the null word λ or a nonempty word.

Basis step *To show that* P(λ) *is true;* that is, $\|x\lambda\| = \|x\| + \|\lambda\|$:

$$\text{Since } x\lambda = x,\ \|x\lambda\| = \|x\| = \|x\| + 0 = \|x\| + \|\lambda\|. \text{ So } P(\lambda) \text{ is true.}$$

Induction step Assume P(y) is true, that is, $\|xy\| = \|x\| + \|y\|$ (inductive hypothesis). We must show that P(ys) is true, that is, $\|xys\| = \|x\| + \|ys\|$. Notice that:

$$xys = (xy)s \qquad \text{assoc. prop. of concatenation}$$

Then

$$\begin{aligned}
\|xys\| &= \|(xy)s\| && \text{length is a function}\\
&= \|xy\| + 1 && \text{recursive def. of length}\\
&= (\|x\| + \|y\|) + 1 && \text{inductive hypothesis}\\
&= \|x\| + (\|y\| + 1) && \text{assoc. prop. of addition}\\
&= \|x\| + \|ys\| && \text{recursive def. of length}
\end{aligned}$$

Therefore, P(ys) is true. Thus P(y) implies P(ys).

Therefore, by induction, P(y) is true for every $y \in \Sigma^*$; that is, $\|xy\| = \|x\| + \|y\|$ for every $x, y \in \Sigma^*$. ■

Finally, we emphasize that the immediate predecessor f_{n-1} need not appear in the recursive definition of a function f at n. For example, consider the function f: **W** → **W** defined by

$$f_0 = 1, \quad f_1 = 0, \quad f_2 = 1$$
$$f_n = f_{n-2} + 2f_{n-3}, \quad n \geq 3$$

Clearly, f_{n-1} is not needed to compute f_n, when $n \geq 3$. Try f_6 as an exercise.

Exercises 5.1

In Exercises 1–6, a_n denotes the nth term of a number sequence satisfying the given initial condition(s) and the recurrence relation. Compute the first four terms of the sequence.

1. $a_1 = 1$

$a_n = a_{n-1} + 3, n \geq 2$

2. $a_0 = 1$

$a_n = a_{n-1} + n, n \geq 1$

3. $a_1 = 1$

$a_n = \dfrac{n}{n-1} a_{n-1}, n \geq 2$

4. $a_1 = 1, a_2 = 2$

$a_n = a_{n-1} + a_{n-2}, n \geq 3$

5. $a_1 = 1, a_2 = 1, a_3 = 2$

$a_n = a_{n-1} + a_{n-2} + a_{n-3}, n \geq 4$

6. $a_1 = 1, a_2 = 2, a_3 = 3$

$a_n = a_{n-1} + a_{n-2} + a_{n-3}, n \geq 4$

7. The nth **Lucas number** L_n, named after the French mathematician François-Edouard-Anatole Lucas, is defined recursively as follows:

$$L_1 = 1, \quad L_2 = 3$$
$$L_n = L_{n-1} + L_{n-2}, n \geq 3$$

(The Lucas sequence and the Fibonacci sequence satisfy the same recurrence relation, but have different initial conditions.) Compute the first six Lucas numbers.

The gcd of two integers x (> 0) and y ($\geq$ 0) can be defined recursively as follows:

$$\gcd\{x, y\} = \begin{cases} \gcd\{y, x\} & \text{if } y > x \\ x & \text{if } y \leq x \text{ and } y = 0 \\ \gcd\{y, x \bmod y\} & \text{if } y \leq x \text{ and } y > 0 \end{cases}$$

Using this definition, compute the gcd of each pair of integers.

8. 28, 18

9. 24, 75

François-Edouard-Anatole Lucas *(1842–1891) was born in Amiens, France. After completing his studies at the École Normale in Amiens, he worked as an assistant at the Paris Observatory. He served as an artillery officer in the Franco-Prussian war and then became professor of mathematics at the Lycée Saint-Louis and Lycée Charlemagne, both in Paris. A gifted and entertaining teacher, Lucas died of a freak accident at a banquet: His cheek was gashed by a piece of a plate that was accidentally dropped, and he died from infection within a few days.*

Lucas loved computing and developed plans for a computer that never materialized. Besides his contributions to number theory, he is known for his four-volume classic on recreational mathematics. Best known among the problems he developed is the Tower of Brahma.

A person deposits \$1000 in a bank at an annual interest rate of 6%. Let $A(n)$ denote the compound amount she will receive at the end of n interest periods. Define $A(n)$ recursively if interest is compounded:

10. Semiannually **11.** Quarterly **12.** Monthly

Ned deposits a certain amount A_0 in a bank at an annual interest rate of 12% compounded annually. The compound amount he would receive at the end of n years is given by $A_n = 1.12A_{n-1}$, where $n \geq 1$. Determine the initial deposit A_0 if he would receive:

13. \$1804.64 at the end of 5 years. **14.** \$3507.00 at the end of 6 years.

Define recursively each sequence of numbers. (*Hint*: Look for a pattern and define the nth term a_n recursively.)

15. 1, 4, 7, 10, 13 … **16.** 3, 8, 13, 18, 23 …

17. 0, 3, 9, 21, 45 … **18.** 1, 2, 5, 26, 677 …

19. An n-bit word containing no two consecutive ones can be constructed recursively as follows: Append a 0 to such $(n-1)$-bit words or append a 01 to such $(n-2)$-bit words. Using this procedure construct all 5-bit words containing no two consecutive ones. There are 13 such words.

Define each recursively, where $n \geq 0$.

20. The nth power of a positive real number x.

21. The union of n sets.

22. The intersection of n sets.

23. The number S_n of subsets of a set with n elements.

24. The nth term a_n of an arithmetic sequence with first term a and common difference d.

***John McCarthy** (1927–), one of the fathers of artificial intelligence (AI), was born in Boston. He graduated in mathematics from the California Institute of Technology, receiving his Ph.D. from Princeton in 1951. After teaching at Princeton, Stanford, Dartmouth, and MIT, he returned to Stanford as a full professor. While at Princeton, he was named a Proctor Fellow and later the Higgins Research Instructor in mathematics. At Stanford, he headed the Artificial Intelligence Laboratory.*

During his tenure at Dartmouth, McCarthy coined the term artificial intelligence *(AI). He developed LISP (*LIS*t* Programming*), one of the most widely used programming languages in AI. He also helped develop ALGOL 58 and ALGOL 60. In 1971 he received the prestigious Alan M. Turing award for his outstanding contributions to data processing.*

25. The nth term a_n of a geometric sequence with first term a and common ratio r.

26. Let $f : X \to X$ be bijective. Define f^n recursively, where $f^2 = f \circ f$.

The **91-function** f, invented by John McCarthy, is defined recursively on **W** as follows.

$$f(x) = \begin{cases} x - 10 & \text{if } x > 100 \\ f(f(x + 11)) & \text{if } 0 \le x \le 100 \end{cases}$$

Compute each

27. $f(99)$ **28.** $f(98)$ **29.** $f(f(99))$ **30.** $f(f(91))$

31. Show that $f(99) = 91$.

32. Prove that $f(x) = 91$ for $90 \le x \le 100$.

33. Prove that $f(x) = 91$ for $0 \le x < 90$.

(Triangulation of convex polygons) The nth Catalan number C_n denotes the number of ways to divide a convex $(n + 2)$-gon into triangles by drawing nonintersecting diagonals. For instance, there are five ways of triangulating a convex pentagon, as shown in Figure 5.10; therefore, $C_3 = 5$. C_n is given recursively by $C_{n+1} = \sum_{i=0}^{n} C_i C_{n-i}$, where $C_0 = 1$. Compute each.

Figure 5.10

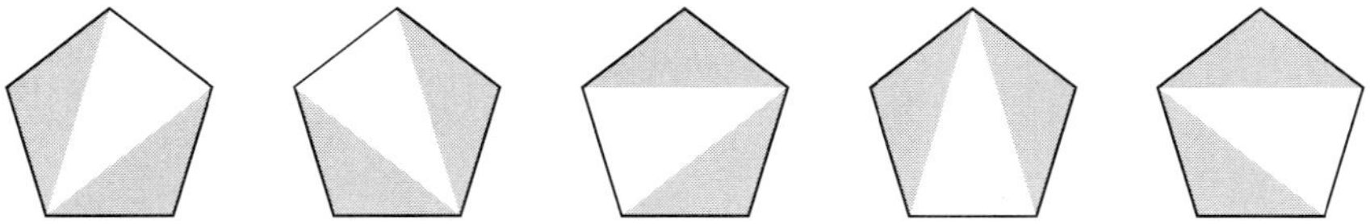

34. C_6 **35.** C_7

36. The sequence defined by $a_{n+1} = \frac{1}{2}(a_n + \frac{N}{a_n})$ can be used to approximate $\sqrt{N}$ to any desired degree of accuracy, where a_1 is an estimate of $\sqrt{N}$. Use this fact to compute $\sqrt{19}$ correct to six decimal places. Use $a_1 = 4$.

37. Let F_n denote the nth Fibonacci number. Compute $\frac{F_{n+1}}{F_n}$ correct to eight decimal places for $1 \leq n \leq 10$. Compare each value to $(1+\sqrt{5})/2$ correct to eight decimal places.

38. (For those familiar with the concept of limits) Use Exercise 37 to predict $\lim_{n\to\infty} \frac{F_{n+1}}{F_n}$.

Prove each, where F_n is the nth Fibonacci number, L_n the nth Lucas number, and $\alpha = (1+\sqrt{5})/2$, the **golden ratio**.

39. $F_n = 2F_{n-2} + F_{n-3}, n \geq 4$

40. $F_n^2 - F_{n-1}F_{n+1} = (-1)^{n-1}, n \geq 2$

41. F_{5n} is divisible by 5, $n \geq 1$.

42. $F_n < \alpha^{n-1}, n \geq 3$

43. $F_n \leq 2^n, n \geq 1$

44. Let $A = \begin{bmatrix} 1 & 1 \\ 1 & 0 \end{bmatrix}$. Then $A^n = \begin{bmatrix} F_{n+1} & F_n \\ F_n & F_{n-1} \end{bmatrix}$, $n \geq 1$. Assume $F_0 = 0$.

45. Using Exercise 44, deduce that $F_{n+1}F_{n-1} - F_n^2 = (-1)^n$.

(*Hint*: Let A be a square matrix. Then $|A^n| = |A|^n$, where $|A|$ denotes the determinant of A.)

46. $L_n = F_{n+1} + F_{n-1}, n \geq 2$

47. $L_{2n} = 3 + \sum_{k=1}^{2n-2} L_k$

The nth term b_n of a number sequence is defined by $b_n = \frac{\alpha^n - \beta^n}{\alpha - \beta}$, where $\alpha = (1+\sqrt{5})/2$ and $\beta = (1-\sqrt{5})/2$ are solutions of the equation $x^2 = x+1$. Verify each.

48. $b_1 = 1$

49. $b_2 = 1$

50. $b_n = b_{n-1} + b_{n-2}, n \geq 3$

(It follows from Exercises 48–50 that $b_n = F_n$. It is called the **Binet form** of the nth Fibonacci number, after the French mathematician Jacques-Phillipe-Marie Binet.)

With α and β as above, let $u_n = \alpha^n + \beta^n, n \geq 1$. Verify each.

51. $u_1 = 1$

52. $u_2 = 3$

53. $u_n = u_{n-1} + u_{n-2}, n \geq 3$

[These exercises indicate that $u_n = L_n$, the nth Lucas number. Accordingly, $u_n = \alpha^n + \beta^n$ is the Binet form of L_n.]

Jacques Phillippe Marie Binet *(1786–1865), a French mathematician and astronomer, was born at Rennes, Brittany. In 1804, he entered the Ecole Polytechnique in Paris, graduated 2 years later, and took a job in the Department of Bridges and Roads of the French government. In 1807, Binet became a teacher at the École Polytechnique, and the following year became assistant to the professor of applied analysis and descriptive geometry. In 1814, he was appointed examiner of descriptive geometry, then professor of mechanics (1815), and then inspector general of studies (1816). In 1821, he was awarded the Chevalier de la Légion d'Honneur. Two years later, Binet was appointed chair of astronomy at the Collège de France.*

But the July 1830 revolution was not kind to him. A strong supporter of Charles X, Binet became a victim of Charles' abdication; he was dismissed from École Polytechnique by King Louis-Phillipe in November, 1830.

Binet made many contributions to mathematics, physics, and astronomy. In 1812, he discovered the rule for matrix multiplication and, in 1840, discovered the explicit formula for the n*th Fibonacci number. In 1843, he was elected to the Academy of Sciences and later became its president. A devout Catholic, Binet died in Paris.*

54. Let $a_1, a_2, \ldots, a_n \in \mathbb{N}$, where $n \geq 2$. Prove that $\gcd\{a_1, a_2, \ldots, a_n\} = \gcd\{\gcd\{a_1, a_2, \ldots, a_{n-1}\}, a_n\}$.

Using Exercise 54 compute the gcd of each set of numbers.

55. 6, 12, 20, 38

56. 12, 28, 48, 104, 252

Let a_n denote the number of times the assignment statement $x \leftarrow x + 1$ is executed by each nested **for** loop. Define a_n recursively.

57.
```
for i = 1 to n do
  for j = 1 to i do
    x ← x + 1
```

58.
```
for i = 1 to n do
  for j = 1 to i do
    for k = 1 to i do
      x ← x + 1
```

59. Let a_n denote the number of rectangles that can be formed on a $1 \times n$ rectangular board. Find the recurrence relation satisfied by a_n. (*Hint*: Look for a pattern. Every square is also a rectangle.)

A subset of the set $S = \{1, 2, \ldots, n\}$ is **alternating** if its elements, when arranged in increasing order, follow the pattern odd, even, odd, even, etc. For example, $\{3\}$, $\{1,2,5\}$, and $\{3,4\}$ are alternating subsets of $\{1,2,3,4,5\}$, whereas $\{1,3,4\}$ and $\{2,3,4,5\}$ are not; $\emptyset$ is considered alternating.* Let a_n denote the number of alternating subsets of S.

60. Define a_n recursively.

61. Prove that $a_n = F_{n+2}$, where F_n denotes the nth Fibonacci number.

*Proposed by Olry Terquem (1782–1862).

Stirling numbers of the second kind, denoted by $S(n, r)$ and used in combinatorics, are defined recursively as follows, where $n, r \in \mathbb{N}$:

$$S(n,r) = \begin{cases} 1 & \text{if } r = 1 \text{ or } r = n \\ S(n-1, r-1) + rS(n-1, r) & \text{if } 1 < r < n \\ 0 & \text{if } r > n \end{cases}$$

They are named after the English mathematician James Stirling (1692–1770). Compute each Stirling number.

62. $S(2, 2)$

63. $S(5, 2)$

A function of theoretical importance in the study of algorithms is the **Ackermann's function**, named after the German mathematician and logician Wilhelm Ackermann (1896–1962). It is defined recursively as follows, where $m, n \in \mathbf{W}$:

$$A(m,n) = \begin{cases} n+1 & \text{if } m = 0 \\ A(m-1, 1) & \text{if } n = 0 \\ A(m-1,\ A(m, n-1)) & \text{otherwise} \end{cases}$$

Compute each.

64. $A(0, 7)$

65. $A(1, 1)$

66. $A(4, 0)$

67. $A(2, 2)$

Prove each for $n \geq 0$.

68. $A(1, n) = n + 2$

69. $A(2, n) = 2n + 3$

***70.** Predict a formula for $A(3, n)$.

***71.** Prove the formula in Exercise 70, where $n \geq 0$.

5.2 Solving Recurrence Relations

The recursive definition of a function f does not provide us with an explicit formula for $f(n)$, but establishes a systematic procedure for finding it. This section illustrates the iterative method of finding a formula for $f(n)$ for a simple class of recurrence relations.

Solving the recurrence relation for a function f means finding an explicit formula for $f(n)$. The **iterative method** of solving it involves two steps:

- Apply the recurrence formula iteratively and look for a pattern to predict an explicit formula.
- Use induction to prove that the formula does indeed hold for every possible value of the integer n.

The next example illustrates this method.

EXAMPLE 5.10 (The handshake problem continued) By Example 5.3, the number of handshakes made by n guests at a dinner party is given by

$$\begin{aligned} h(1) &= 0 \\ h(n) &= h(n-1) + (n-1), n \geq 2 \end{aligned}$$

Solve this recurrence relation.

SOLUTION:

Step 1 To predict a formula for $h(n)$:

Using iteration,
$$\begin{aligned} h(n) &= h(n-1) + (n-1) \\ &= h(n-2) + (n-2) + (n-1) \\ &= h(n-3) + (n-3) + (n-2) + (n-1) \\ &\vdots \\ &= h(1) + 1 + 2 + 3 + \cdots + (n-2) + (n-1) \\ &= 0 + 1 + 2 + 3 + \cdots + (n-1) \\ &= \frac{n(n-1)}{2} \end{aligned}$$

Step 2 To prove, by induction, that $h(n) = \dfrac{n(n-1)}{2}$, where $n \geq 1$:

Basis step When $n = 1, h(1) = \dfrac{1 \cdot 0}{2} = 0$, which agrees with the initial condition. So the formula holds when $n = 1$.

Induction step Assume $h(k) = \dfrac{k(k-1)}{2}$ for any $k \geq 1$. Then:

$$h(k+1) = h(k) + k, \qquad \text{by the recurrence relation}$$

$$= \frac{k(k-1)}{2} + k, \qquad \text{by the induction hypothesis}$$
$$= \frac{k(k+1)}{2}$$

Therefore, if the formula holds for $n = k$, it also holds for $n = k + 1$. Thus, by PMI, the result holds for $n \geq 1$. ■

More generally, using iteration we can solve the recurrence relation

$$a_n = a_{n-1} + f(n) \tag{5.5}$$

as follows:

$$\begin{aligned}
a_n &= a_{n-1} + f(n) \\
&= [a_{n-2} + f(n-1)] + f(n) = a_{n-2} + f(n-1) + f(n) \\
&= [a_{n-3} + f(n-2)] + f(n-1) + f(n) \\
&= a_{n-3} + f(n-2) + f(n-1) + f(n) \\
&\vdots \\
&= a_0 + \sum_{i=1}^{n} f(i)
\end{aligned} \tag{5.6}$$

You can verify that this is the actual solution of the recurrence relation (5.5).

For example, in the handshake problem $f(n) = n - 1$ and $h(0) = 0$, so the solution of the recurrence relation is

$$\begin{aligned}
h(n) &= h(0) + \sum_{i=1}^{n} f(i) = 0 + \sum_{i=1}^{n} (i-1) \\
&= \sum_{i=1}^{n-1} i = \frac{n(n-1)}{2}, \quad n \geq 1
\end{aligned}$$

which is exactly the solution obtained in the example.

EXAMPLE 5.11 Solve the recurrence relation in Example 5.6.

SOLUTION:

Notice that a_n can be redefined as

$$a_n = a_{n-1} + \frac{n(n+1)}{2}, \quad n \geq 1$$

where $a_0 = 0$. Comparing this with recurrence relation (5.5), we have $f(n) = \dfrac{n(n+1)}{2}$. Therefore, by Equation (5.6),

$$\begin{aligned}
a_n &= a_0 + \sum_{i=1}^{n} f(i) \\
&= a_0 + \sum_{i=1}^{n} \frac{i(i+1)}{2} = 0 + \frac{1}{2}\sum_{i=1}^{n}(i^2 + i) \\
&= \frac{1}{2}\left(\sum_{i=1}^{n} i^2 + \sum_{i=1}^{n} i\right) \\
&= \frac{1}{2}\left[\frac{n(n+1)(2n+1)}{6} + \frac{n(n+1)}{2}\right] \\
&= \frac{n(n+1)}{2}\left(\frac{2n+1}{6} + \frac{1}{2}\right) = \frac{n(n+1)}{2} \cdot \frac{2n+4}{6} \\
&= \frac{n(n+1)(n+2)}{6}, \quad n \geq 0
\end{aligned}$$

■

The following illustration of the iterative method brings us again to the Tower of Brahma puzzle.

EXAMPLE 5.12 Recall from Example 5.4 that the number of moves needed to transfer n disks from peg X to peg Z is given by

$$\begin{aligned}
b_1 &= 1 \\
b_n &= 2b_{n-1} + 1, n \geq 2
\end{aligned}$$

Solve this recurrence relation.

SOLUTION:

Step 1 To predict a formula for b_n:

Using iteration,

$$\begin{aligned}
b_n &= 2b_{n-1} + 1 \\
&= 2[2b_{n-2} + 1] + 1 = 2^2 b_{n-2} + 2 + 1 \\
&= 2^2[2b_{n-3} + 1] + 2 + 1 = 2^3 b_{n-3} + 2^2 + 2 + 1 \\
&\vdots \\
&= 2^{n-1} b_1 + 2^{n-2} + \cdots + 2^2 + 2 + 1 \\
&= 2^{n-1} + 2^{n-2} + \cdots + 2 + 1 \\
&= 2^n - 1, \quad \text{by Exercise 8 in Section 4.4.}
\end{aligned}$$

Step 2 You may prove by induction that $b_n = 2^n - 1$, where $n \geq 1$. ■

More generally, you may verify that the solution of the recurrence relation $a_n = ca_{n-1} + 1$, where c is a constant ($\neq 1$), is

$$a_n = c^n a_0 + \frac{c^n - 1}{c - 1}$$

For instance, in Example 5.12, $b_0 = 0$ and $c = 2$, so

$$b_n = 2^n \cdot 0 + \frac{2^n - 1}{2 - 1} = 2^n - 1$$

as expected.

Let us pursue Example 5.12 a bit further. Suppose there are 64 disks at peg X, as in the original puzzle, and it takes 1 second to move a disk from one peg to another. Then it takes a total of $2^{64} - 1$ seconds to solve the puzzle.

To get an idea how incredibly large this total is, notice that there are about $365 \cdot 24 \cdot 60 \cdot 60 = 31{,}536{,}000$ seconds in a year. Therefore,

$$\begin{aligned}\text{Total time taken} &= 2^{64} - 1 \text{ seconds}\\ &\approx 1.844674407 \times 10^{19} \text{ seconds}\\ &\approx 5.84942417 \times 10^{11} \text{ years}\\ &\approx 600 \text{ billion years!}\end{aligned}$$

Intriguingly, according to some estimates, the universe is only about 18 billion years old.

Exercises 5.2

Using the iterative method, predict a solution to each recurrence relation satisfying the given initial condition.

1. $s_0 = 1$

$s_n = 2s_{n-1}, n \geq 1$

2. $a_1 = 1$

$a_n = a_{n-1} + n, n \geq 2$

3. $a_0 = 1$

$a_n = a_{n-1} + n, n \geq 1$

4. $a_1 = 1$

$a_n = a_{n-1} + (2n - 1), n \geq 2$

5. $a_0 = 0$

$a_n = a_{n-1} + 4n, n \geq 1$

6. $s_1 = 1$

$s_n = s_{n-1} + n^3, n \geq 2$

7. $s_1 = 1$

$s_n = s_{n-1} + n^2,\ n \geq 2$

8. $a_1 = 1$

$a_n = 2a_{n-1} + (2^n - 1),\ n \geq 2$

9–16. Using induction, verify the solutions to Exercises 1–8.

17. Using the data in Example 5.2, show that the compound amount Judy will receive at the end of n years is given by $A(n) = 1000(1.08)^n$, where $n \geq 0$.

Use the recursive definition of f_n in Example 5.5 to answer Exercises 18 and 19.

18. Predict a formula for f_n.

19. Prove that the formula holds for $n \geq 1$.

20. Using induction, establish the explicit formula for b_n in Example 5.12.

Using induction, prove that each is a solution to the corresponding recurrence relation, where c is a constant and $f(n)$ a function of n.

21. $a_n = a_0 + \sum_{i=1}^{n} f(i),\ a_n = a_{n-1} + f(n)$

22. $a_n = c^n a_0 + \dfrac{c^n - 1}{c - 1},\ a_n = ca_{n-1} + 1$ (assume $c \neq 1$)

23. $a_n = c^n a_0 + \sum_{i=1}^{n} c^{n-i} f(i),\ a_n = ca_{n-1} + f(n)$

Let a_n denote the number of times the statement $x \leftarrow x + 1$ is executed by the following loops.

```
for i = 1 to n do
  for j = 1 to ⌊i/2⌋ do
    x ← x + 1
```

24. Define a_n recursively.

25. Show that $a_n = \begin{cases} 0 & \text{if } n = 1 \\ a_{n-1} + n/2 & \text{if } n > 1 \text{ and even} \\ a_{n-1} + (n-1)/2 & \text{if } n > 1 \text{ and odd} \end{cases}$

26. Solve the recurrence relation satisfied by a_n.

Let a_n denote the number of times the statement $x \leftarrow x + 1$ is executed by the following **for** loops:

```
for i = 1 to n do
  for j = 1 to ⌈i/2⌉ do
    x ← x + 1
```

27. Define a_n recursively.

28. Show that $a_n = \begin{cases} 1 & \text{if } n = 1 \\ a_{n-1} + n/2 & \text{if } n > 1 \text{ and even} \\ a_{n-1} + (n+1)/2 & \text{if } n > 1 \text{ and odd} \end{cases}$

29. Solve the recurrence relation satisfied by a_n.

Let a_n denote the number of times the statement $x \leftarrow x + 1$ is executed by the nested **for** loops in Exercise 35 in Section 4.4.

30. Define a_n recursively.

31. Solve the recurrence relation satisfied by a_n.

32–33. Redo Exercises 30 and 31 using the loops in Exercise 36 in Section 4.4.

34–35. Redo Exercises 30 and 31 using the loops in Exercise 37 in Section 4.4.

36–37. Redo Exercises 30 and 31 using the loops in Exercise 38 in Section 4.4.

Let t_n denote the nth triangular number.

38. Define t_n recursively.

39. Find an explicit formula for t_n.

40. Prove that $8t_n + 1$ is a perfect square.

The nth **pentagonal number** p_n is obtained from its predecessor by adding three rows of dots plus one. The first four pentagonal numbers are represented pictorially in Figure 5.11.

Figure 5.11

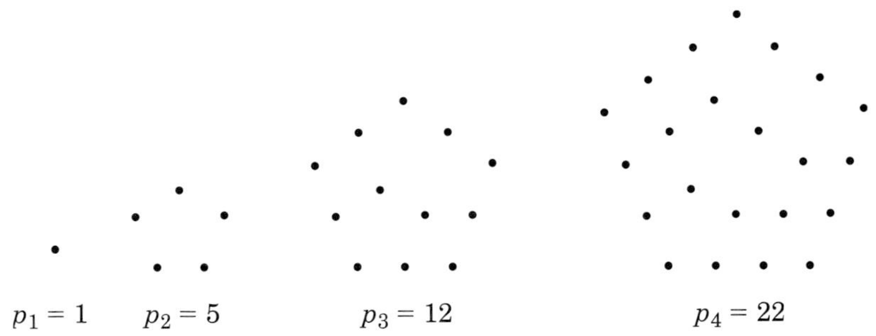

41. Represent p_5 pictorially.

42–43. Redo Exercises 38 and 39 using p_n.

The nth **hexagonal number** h_n is obtained from its predecessor by adding four rows of dots plus one dot. The first four hexagonal numbers are shown pictorially in Figure 5.12.

44–46. Redo Exercises 41–43 using h_n.

47. Prove that $h_n = p_n + t_n - n$, using the explicit formulas for p_n and t_n.

48. Prove that $h_n = p_n + t_n - n$, using the recurrence relations for p_n and t_n.

Figure 5.12

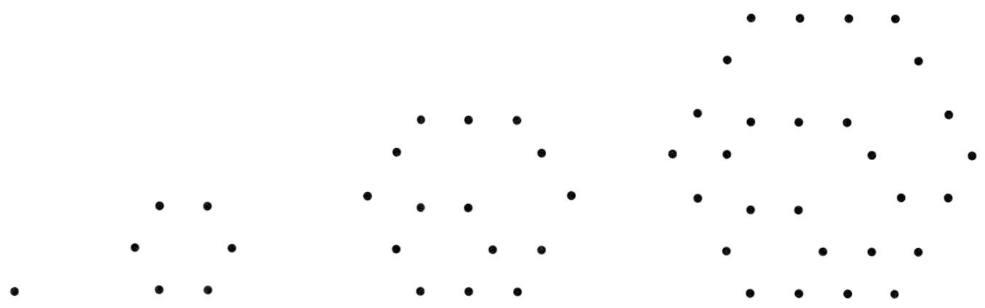

Triangular pyramidal numbers T_n (or **tetrahedral numbers**) are positive integers that can be represented by triangular pyramidal shapes. The first four tetrahedral numbers are 1, 4, 10, and 20; see Figure 5.13.

Figure 5.13

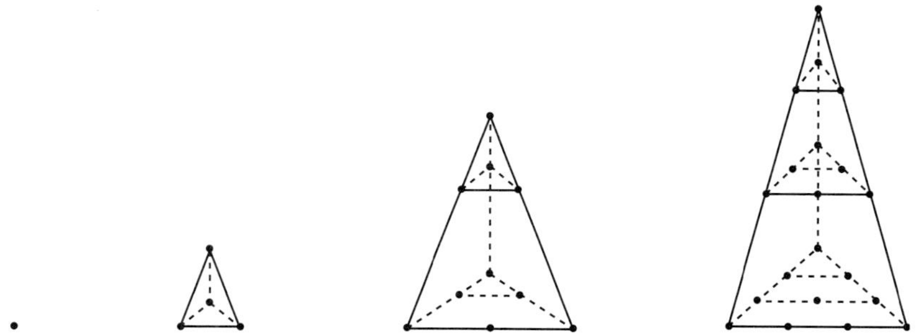

49. Define T_n recursively.

50. Conjecture an explicit formula for T_n.

51. Establish the formula in Exercise 50.

Square pyramidal numbers S_n are positive integers that can be represented by pyramidal shapes, where the base is a square. The first four square pyramidal numbers are 1, 5, 14, and 30; see Figure 5.14.

52–54. Redo Exercises 49–51 with S_n.

Let a_n denote the number of subsets of the set $S = \{1, 2, \ldots, n\}$ that contain no consecutive integers, where $n \geq 0$. When $n = 0$, $S = \emptyset$.† Compute each.

55. a_0 **56.** a_1 **57.** a_2 **58.** a_3

†Proposed by Irving Kaplansky of The University of Chicago.

Figure 5.14

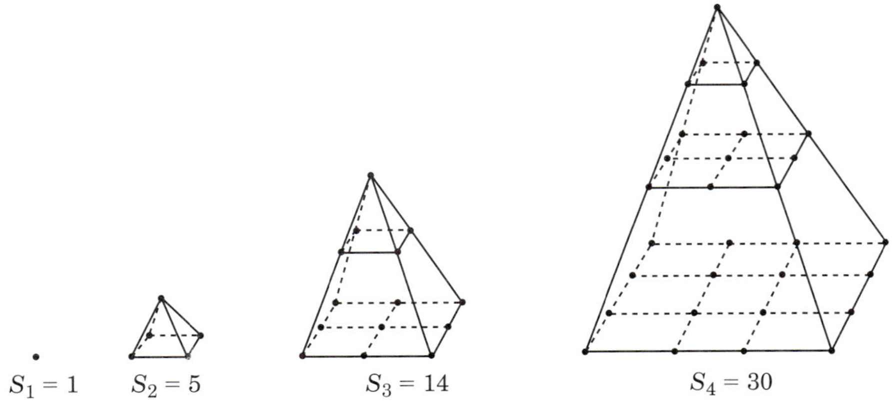

59. Define a_n recursively.

60. Solve the recurrence relation satisfied by a_n.

Suppose we introduce a mixed pair of 1-month-old rabbits into a large enclosure on the first day of a certain month. By the end of each month, the rabbits become mature and each pair produces $k-1$ mixed pairs of offspring at the beginning of the following month. (*Note*: $k \geq 2$.) For instance, at the beginning of the second month, there is one pair of 2-month-old rabbits and $k-1$ pairs of 0-month-olds; at the beginning of the third month, there is one pair of 3-month-olds, $k-1$ pairs of 1-month-olds, and $k(k-1)$ pairs of 0-month-olds. Assume the rabbits are immortal. Let a_n denote the average age of the rabbit pairs at the beginning of the nth month. (P. Filipponi, 1990)

****61.** Define a_n recursively.

****62.** Predict an explicit formula for a_n.

****63.** Prove the formula in Exercise 64.

64. (For those familiar with the concept of limits) Find $\lim_{n\to\infty} a_n$.

5.3 Solving Recurrence Relations Revisited

Unfortunately, the iterative method illustrated in the preceding section can be applied to only a small and simple class of recurrence relations. The present section develops a method for solving two large, important classes of recurrence relations.

Linear Homogeneous Recurrence Relations with Constant Coefficients (LHRRWCCs)

A ***k*****th-order linear homogeneous recurrence relation with constant coefficients** is a recurrence relation of the form

$$a_n = c_1 a_{n-1} + c_2 a_{n-2} + \cdots + c_k a_{n-k} \tag{5.7}$$

where $c_1, c_2, \ldots, c_k \in \mathbb{R}$ and $c_k \neq 0$.

First, a few words of explanation: The term *linear* means that every term on the RHS of Equation (5.7) contains at most the first power of any predecessor a_i. A recurrence relation is *homogeneous* if every term on the RHS is a multiple of some a_i; in other words, the relation is satisfied by the sequence $\{0\}$; that is, $a_n = 0$ for every n. All coefficients c_i are constants. Since a_n depends on its k immediate predecessors, the *order* of the recurrence relation is k. Accordingly, to solve a kth-order LHRRWCC, we will need k initial conditions, say, $a_0 = C_0$, $a_1 = C_1$, ..., $a_{k-1} = C_{k-1}$.

The next example illustrates in detail the various terms in this definition.

EXAMPLE 5.13

- The recurrence relation $s_n = 2s_{n-1}$ is a LHRRWCC. Its order is one.
- The recurrence relation $a_n = na_{n-1}$ is linear and homogeneous. But the coefficient on the RHS is not a constant. Therefore, it is not a LHRRWCC.
- $h_n = h_{n-1} + (n-1)$ is a linear recurrence relation. But it is not homogeneous because of the term $n - 1$.
- The recurrence relation $a_n = a_{n-1}^2 + 3a_{n-2}$ is homogeneous. But it is not linear since the power of a_{n-1} is 2.
- $a_n = a_{n-1} + 2a_{n-2} + 3a_{n-6}$ is a LHRRWCC of order six. ■

Before we discuss solving second-order LHRRWCCs, notice that the solution of the recurrence relation $s_n = 2s_{n-1}$, where $s_0 = 1$, is $s_n = 2^n$, $n \geq 0$ (see Exercise 1 in Section 5.2). More generally, you may verify that the solution of the recurrence relation $a_n = \alpha a_{n-1}$, where $a_0 = c$, is $a_n = c\alpha^n$, $n \geq 0$.

We now turn our attention to the second-order LHRRWCC

$$a_n = aa_{n-1} + ba_{n-2} \tag{5.8}$$

where a and b are nonzero constants. If it has a nonzero solution of the form $c\alpha^n$, then $c\alpha^n = ac\alpha^{n-1} + bc\alpha^{n-2}$. Since $c\alpha \neq 0$, this yields $\alpha^2 = a\alpha + b$; that is, $\alpha^2 - a\alpha - b = 0$, so α must be a solution of the **characteristic equation**

$$x^2 - ax - b = 0 \tag{5.9}$$

of the recurrence relation (5.8). The roots of Equation (5.9) are the **characteristic roots** of recurrence relation (5.8).

Theorems 5.2 through 5.4 show how characteristic roots help solve LHRRWCCs.

THEOREM 5.2 Let α and β be the distinct (real or complex) solutions of the equation $x^2 - ax - b = 0$, where $a, b \in \mathbb{R}$ and $b \neq 0$. Then every solution of the LHRRWCC $a_n = aa_{n-1} + ba_{n-2}$, where $a_0 = C_0$ and $a_1 = C_1$, is of the form $a_n = A\alpha^n + B\beta^n$ for some constants A and B.

PROOF:
The proof consists of two parts:

- First, we will show that $a_n = A\alpha^n + B\beta^n$ is a solution of the recurrence relation for any constants A and B.
- We will then find the values of A and B satisfying the given initial conditions.

First, notice that since α and β are solutions of equation (5.9), $\alpha^2 = a\alpha + b$ and $\beta^2 = a\beta + b$.

- *To show that* $a_n = A\alpha^n + B\beta^n$ *is a solution of the recurrence relation*:

$$\begin{aligned} aa_{n-1} + ba_{n-2} &= a(A\alpha^{n-1} + B\beta^{n-1}) + b(A\alpha^{n-2} + B\beta^{n-2}) \\ &= A\alpha^{n-2}(a\alpha + b) + B\beta^{n-2}(a\beta + b) \\ &= A\alpha^{n-2} \cdot \alpha^2 + B\beta^{n-2} \cdot \beta^2 \\ &= A\alpha^n + B\beta^n \\ &= a_n \end{aligned}$$

  Thus $a_n = A\alpha^n + B\beta^n$ is a solution of the recurrence relation (5.8).

- Secondly, let $a_n = A\alpha^n + B\beta^n$ be a solution of (5.8). To find the values of A and B, notice that the conditions $a_0 = C_0$ and $a_1 = C_1$ yield the following linear system:

$$C_0 = A + B \tag{5.10}$$

$$C_1 = A\alpha + B\beta \tag{5.11}$$

  Solving this system, we get (Verify.)

$$A = \frac{C_1 - C_0\beta}{\alpha - \beta} \text{ and } \frac{C_0\alpha - C_1}{\alpha - \beta} \text{ (Remember, } \alpha \neq \beta.)$$

With these values for A and B, a_n satisfies the initial conditions and the recurrence relation. Since the recurrence relation and the initial conditions determine a unique sequence, $\{a_n\}$, $a_n = A\alpha^n + B\beta^n$ is indeed the unique solution of the recurrence relation. ■

A few interesting observations:

- The solutions α and β are nonzero, since $\alpha = 0$, for instance, would imply that $b = 0$.
- Theorem 5.2 *cannot* be applied if $\alpha = \beta$. However, it works even if α and β are complex numbers.
- The solutions α^n and β^n are the **basic solutions** of the recurrence relation. In general, the number of basic solutions equals the order of the recurrence relation. The **general solution** $a_n = A\alpha^n + B\beta^n$ is a **linear combination** of the basic solutions. The particular solution is obtained by selecting A and B in such a way that the initial conditions are satisfied, as in Theorem 5.2.

The next three examples illustrate how to solve second-order LHRRWCCs using their characteristic equations.

EXAMPLE 5.14 Solve the recurrence relation $a_n = 5a_{n-1} - 6a_{n-2}$, where $a_0 = 4$ and $a_1 = 7$.

SOLUTION:

- *To find the general solution of the recurrence relation*:
The characteristic equation of the recurrence relation is $x^2 - 5x + 6 = 0$; the characteristic roots are 2 and 3. Therefore, by Theorem 5.2, the general solution of the recurrence relation is $a_n = A \cdot 2^n + B \cdot 3^n$. (This solution is used in Examples 5.19 and 5.20.)
- *To find the values of A and B*:
Using the initial conditions we find:

$$a_0 = A + B = 4$$
$$a_1 = 2A + 3B = 7$$

Solving this linear system yields $A = 5$ and $B = -1$ (Verify this.).

Thus the solution of the recurrence relation satisfying the given conditions is $a_n = 5 \cdot 2^n - 3^n$, $n \geq 0$. ■

The next example finds an explicit formula for the nth Fibonacci number F_n, which we have been waiting for.

EXAMPLE 5.15 Solve the Fibonacci recurrence relation $F_n = F_{n-1} + F_{n-2}$, where $F_1 = 1 = F_2$.

SOLUTION:

The characteristic equation of the recurrence relation is $x^2 - x - 1 = 0$, and its solutions are $\alpha = \dfrac{1+\sqrt{5}}{2}$ and $\beta = \dfrac{1-\sqrt{5}}{2}$. You may verify $\alpha + \beta = 1$ and $\alpha\beta = -1$.

The general solution is $F_n = A\alpha^n + B\beta^n$. To find A and B, we have:

$$F_1 = A\alpha + B\beta = 1$$
$$F_2 = A\alpha^2 + B\beta^2 = 1$$

Solving these two equations, we get (Verify):

$$A = \frac{\alpha}{1+\alpha^2} = \frac{(1+\sqrt{5})/2}{(5+\sqrt{5})/2} = \frac{1+\sqrt{5}}{5+\sqrt{5}}$$
$$= \frac{(1+\sqrt{5})(5-\sqrt{5})}{(5+\sqrt{5})(5-\sqrt{5})} = \frac{5+5\sqrt{5}-\sqrt{5}-5}{25-5} = \frac{1}{\sqrt{5}}$$

and similarly $B = \dfrac{\beta}{1+\beta^2} = -\dfrac{1}{\sqrt{5}}$ (Verify this.).

Thus the solution of the recurrence relation satisfying the given conditions is

$$a_n = \frac{\alpha^n - \beta^n}{\sqrt{5}} = \frac{\alpha^n - \beta^n}{\alpha - \beta}$$

which is the *Binet form* for the nth Fibonacci number F_n. (See Example 5.26 for a different method.) ■

The next example, proposed by Irving Kaplansky of The University of Chicago, also illustrates solving second order LHRRWCCs and is closely related to Example 5.15.

EXAMPLE 5.16 Let a_n denote the number of subsets of the set $S = \{1, 2, \ldots, n\}$ that do not contain consecutive integers, where $n \geq 0$. When $n = 0$, $S = \emptyset$. Find an explicit formula for a_n.

SOLUTION:
To get an idea about a_n, let us find its value for $n = 0$, 1, 2, 3, and 4 by constructing a table, as in Table 5.3. It appears from the table that a_n is a Fibonacci number and $a_n = F_{n+2}$.

Table 5.3

n	**Subsets of S that do not contain consecutive integers**	a_n
0	$\emptyset$,	1
1	$\emptyset$, $\{1\}$	2
2	$\emptyset$, $\{1\}$, $\{2\}$	3
3	$\emptyset$, $\{1\}$, $\{2\}$, $\{3\}$, $\{1,3\}$	5
4	$\emptyset$, $\{1\}$, $\{2\}$, $\{3\}$, $\{4\}$, $\{1,3\}$, $\{1,4\}$, $\{2,4\}$	8

↑ F_{n+2}

We shall, in fact, prove that $a_n = F_{n+2}$ in two steps: First we shall define a_n recursively and then solve the recurrence relation to obtain this explicit formula.

- *To define a_n recursively*:
  From Table 5.3, $a_0 = 1$ and $a_1 = 2$. So let $n \geq 2$. Let A be a subset of S that does not contain two consecutive integers. Then either $n \in A$ or $n \notin A$.

Case 1 Suppose $n \in A$. Then $n - 1 \notin A$. By definition, $S^* = \{1, 2, \ldots, n-2\}$ has a_{n-2} subsets not containing two consecutive integers. Add n to each of the subsets. The resulting sets are subsets of S satisfying the desired property, so S has a_{n-2} such subsets.

Case 2 Suppose $n \notin A$. By definition, there are a_{n-1} such subsets of S having the required property.
Since these two cases are mutually exclusive, by the addition principle, $a_n = a_{n-1} + a_{n-2}$.

Thus a_n can be defined recursively as

$$a_0 = 1, a_1 = 2$$
$$a_n = a_{n-1} + a_{n-2}, \quad n \geq 2.$$

- *To solve the recurrence relation*:

  This recurrence relation is exactly the same as the Fibonacci one with the initial conditions $a_0 = 1$, $a_1 = 2$. So instead of going through a complete solution, as in Example 5.15, notice that this definition yields the Fibonacci numbers 1, 2, 3, 5, 8, ... It follows that $a_n = F_{n+2}$, $n \geq 0$.

Using the values of α and β from Example 5.15,

$$a_n = F_{n+2} = \frac{\alpha^{n+2} - \beta^{n+2}}{\alpha - \beta}, \quad n \geq 0$$

(Verify this. See Exercise 13.) ■

Theorem 5.2 does not work if the characteristic roots α and β are equal, that is, if α is a root with degree of multiplicity two. The following theorem, however, comes to our rescue. It shows that, in addition to α^n, $n\alpha^n$ is a basic solution.

THEOREM 5.3 Let $a, b \in \mathbb{R}$ and $b \neq 0$. Let α be a real or complex solution of the equation $x^2 - ax - b = 0$ with degree of multiplicity two. Then $a_n = A\alpha^n + Bn\alpha^n$ is the general solution of the LHRRWCC $a_n = aa_{n-1} + ba_{n-2}$.

PROOF:
Since α is a root of the equation $x^2 - ax - b = 0$ with degree of multiplicity two,

$$x^2 - ax - b = (x - \alpha)^2$$
$$= x^2 - 2\alpha x + \alpha^2$$

Therefore,

$$a = 2\alpha \quad \text{and} \quad b = -\alpha^2 \tag{5.12}$$

- *To show that $a_n = n\alpha^n$ satisfies the recurrence relation*:
  Notice that

$$\begin{aligned} aa_{n-1} + ba_{n-2} &= a[(n-1)\alpha^{n-1}] + b[(n-2)\alpha^{n-2}] \\ &= 2\alpha[(n-1)\alpha^{n-1}] + (-\alpha^2)[(n-2)\alpha^{n-2}] \qquad \text{by (5.12)} \\ &= \alpha^n[2(n-1) - (n-2)] \\ &= n\alpha^n = a_n \end{aligned}$$

  Therefore, $n\alpha^n$ is a solution of the recurrence relation.

Then $a_n = A\alpha^n + Bn\beta^n$ is the general solution of the given recurrence relation, where A and B are selected in such a way that the initial conditions are satisfied. (The values of A and B can be found using initial conditions, as in Theorem 5.2.) ■

The next example illustrates Theorem 5.3.

EXAMPLE 5.17 Solve the recurrence relation $a_n = 6a_{n-1} - 9a_{n-2}$, where $a_0 = 2$ and $a_1 = 3$.

SOLUTION:
The characteristic equation of the recurrence relation is $x^2 - 6x + 9 = 0$; its solution is 3 with degree of multiplicity two. Therefore, by Theorem 5.3, the general solution of the recurrence relation is $a_n = A \cdot 3^n + B \cdot n3^n$. (We use this in Example 5.21.)

The initial conditions $a_0 = 2$ and $a_1 = 3$ yield the equations

$$A \cdot 3^0 + B \cdot 0 \cdot 3^0 = 2$$

and

$$A \cdot 3 + B \cdot 1 \cdot 3 = 3.$$

Solving these equations, we get $A = 2$ and $B = -1$. (Verify).

Thus the solution of the recurrence relation satisfying the given conditions is $a_n = 2 \cdot 3^n - n \cdot 3^n$, $n \geq 0$. ■

Theorems 5.2 and 5.3 can be combined to yield the following general result.

THEOREM 5.4 Let α be a characteristic root of the LHRRWCC (5.7).

- If the degree of multiplicity of α is 1, then α^n is a basic solution of the LHRRWCC.
- If the degree of multiplicity of α is m, then $\alpha^n, n\alpha^n, \ldots, n^{m-1}\alpha^n$ are basic solutions of the LHRRWCC. (*Note*: A kth-order LHRRWCC has k basic solutions.)
- The general solution of the LHRRWCC is a linear combination of all basic solutions. ■

The following example illustrates this general theorem.

EXAMPLE 5.18 Solve the recurrence relation $a_n = 7a_{n-1} - 13a_{n-2} - 3a_{n-3} + 18a_{n-4}$, where $a_0 = 5, a_1 = 3, a_2 = 6$, and $a_3 = -21$.

SOLUTION:
The characteristic equation of the LHRRWCC is $x^4 - 7x^3 + 13x^2 + 3x - 18 = 0$. Since $x^4 - 7x^3 + 13x^2 + 3x - 18 = (x+1)(x-2)(x-3)^2$, the characteristic roots are:

-1 and 2 with degree of multiplicity one each

and 3 with degree of multiplicity two

Since 3 is a root with degree of multiplicity two, it yields two basic solutions, 3^n and $n3^n$. Thus the general solution of the LHRRWCC is a linear combination of the basic solutions $(-1)^n, 2^n, 3^n$, and $n3^n$; that is, $a_n = A(-1)^n + B2^n + C3^n + Dn3^n$.

To find the values of A, B, C, and D:
Since $a_0 = 5, a_1 = 3, a_2 = 6$, and $a_3 = -21$, we have

$$A + B + C = 5$$

$$-A + 2B + 3C + 3D = 3$$

$$A + 4B + 9C + 18D = 6$$

and

$$-A + 8B + 27C + 81D = -21$$

Solving this linear system, we get $A = 2 = C, B = 1$, and $D = -1$ (Verify this.). Thus the solution of the LHRRWCC satisfying the initial conditions is $a_n = 2(-1)^n + 2^n + 2 \cdot 3^n - n3^n, n \geq 0$. ■

The technique of solving LHRRWCCs cannot be applied to the seemingly simple recurrence relations $f_n = f_{n-1} + n$ (Example 5.5) and $b_n = 2b_{n-1} + 1$ (Example 5.4), which are linear, but nonhomogeneous. So we now turn to solving **linear nonhomogeneous recurrence relations with constant coefficients** (LNHRRWCCs).

LNHRRWCCs

The *general form* of a LNHRRWCC is

$$a_n = c_1 a_{n-1} + c_2 a_{n-2} + \cdots + c_k a_{n-k} + f(n) \tag{5.13}$$

where $c_1, c_2, \ldots, c_k \in \mathbb{R}, c_k \neq 0$, and $f(n)$ is *not* identically zero. Its solution depends on that of the **associated linear homogeneous recurrence relation with constant coefficients** (ALHRRWCCs)

$$a_n = c_1 a_{n-1} + c_2 a_{n-2} + \cdots + c_k a_{n-k} \tag{5.14}$$

we studied earlier.

Solving LNHRRWCCs

To solve the LNHRRWCCs (5.13), let $a_n^{(h)}$ denote the general solution of the ALHRRWCCs (5.14). Suppose we know some solution $a_n^{(p)}$ of the recurrence relation (5.13); $a_n^{(p)}$ is a **particular solution** of the LNHRRWCCs (5.13). Then the **general solution** of (5.13) is given by

$$a_n = a_n^{(h)} + a_n^{(p)}$$

This fact is confirmed by the following theorem; we leave its proof as an exercise (see Exercise 44).

THEOREM 5.5 Let $a_n^{(h)}$ denote the general solution of the ALHRRWCCs (5.14) and $a_n^{(p)}$ a particular solution of the LNHRRWCC (5.13). Then $a_n = a_n^{(h)} + a_n^{(p)}$ is the general solution of the LNHRRWCCs (5.13). ■

It follows from this theorem that solving the LNHRRWCCs (5.13) depends on finding a particular solution $a_n^{(p)}$. Although no general algorithm exists for solving an arbitrary LNHRRWCCs, two special cases can be handled fairly easily. When $f(n)$ is a polynomial in n or is of the form $C\alpha^n$, a particular solution can be extracted with ease, as the next two examples demonstrate, where C and α are constants. The techniques we employ are similar to those used to solve linear nonhomogeneous differential equations.

EXAMPLE 5.19 Solve the LNHRRWCCs $a_n = 5a_{n-1} - 6a_{n-2} + 8n^2$, where $a_0 = 4$ and $a_1 = 7$.

SOLUTION:
It follows from Example 5.14 that the general solution of the ALHRRWCCs $a_n = 5a_{n-1} - 6a_{n-2}$ is given by $a_n^{(h)} = A \cdot 2^n + B \cdot 3^n$. Since $f(n) = 8n^2$ is a quadratic polynomial in n, it seems reasonable to look for a particular solution of the same form, say, $a_n = an^2 + bn + c$. Then the given recurrence relation yields

$$\begin{aligned} an^2 + bn + c &= 5[a(n-1)^2 + b(n-1) + c] - 6[a(n-2)^2 + b(n-2) + c] + 8n^2 \\ &= (8-a)n^2 + (14a - b)n - 19a + 7b - c \end{aligned}$$

Equating the coefficients of like terms, we get the linear system:

$$\begin{aligned} a &= 8 - a \\ b &= 14a - b \\ c &= -19a + 7b - c \end{aligned}$$

Solving the system, we get $a = 4, b = 28$, and $c = 60$ (Verify). We now claim that $a_n^{(p)} = 4n^2 + 28n + 60$ is a particular solution (Verify).

Thus, by Theorem 5.5, the general solution of the given recurrence relation is

$$\begin{aligned} a_n &= a_n^{(h)} + a_n^{(p)} \\ &= A \cdot 2^n + B \cdot 3^n + 4n^2 + 28n + 60 \end{aligned}$$

Using the two given initial conditions, this yields the linear system:

$$\begin{aligned} A + B &= -56 \\ 2A + 3B &= -85 \end{aligned}$$

This yields $A = -83$ and $B = 27$ (Verify this also.).

Thus the desired solution is

$$a_n = (-83) \cdot 2^n + 27 \cdot 3^n + 4n^2 + 28n + 60, \quad n \geq 0$$ ■

The next example illustrates how to solve the LNHRRWCCs (5.13) when $f(n)$ is of the form $C\alpha^n$, where C and α are constants.

EXAMPLE 5.20 Solve the LNHRRWCCs $a_n = 5a_{n-1} - 6a_{n-2} + 3 \cdot 5^n$, where $a_0 = 4$ and $a_1 = 7$.

SOLUTION:

As in Example 5.19, the general solution of the ALHRRWCCs $a_n = 5a_{n-1} - 6a_{n-2}$ is given by $a_n^{(h)} = A \cdot 2^n + B \cdot 3^n$. Since $f(n) = 3 \cdot 5^n$, we search for a particular solution of the form $a_n = c \cdot 5^n$. Then we must have

$$c \cdot 5^n = 5(c \cdot 5^{n-1}) - 6(c \cdot 5^{n-2}) + 3 \cdot 5^n$$

Canceling 5^{n-2} from both sides, the resulting equation yields $c = 25/2$. We now claim that $a_n = (25/2)5^n$ is a particular solution of the recurrence relation (Verify this.).

Thus the general solution of the LNHRRWCCs is

$$a_n = A \cdot 2^n + B \cdot 3^n + (25/2)5^n$$

Using the initial conditions, we get the linear system:

$$\begin{aligned} A + B &= -17/2 \\ 2A + 3B &= -111/2 \end{aligned}$$

Solving this system, we get $A = 30$ and $B = -77/2$ (Verify this.).

Thus the solutions of the given recurrence relation are given by

$$a_n = (30) \cdot 2^n - (77/2) \cdot 3^n + (25/2) \cdot 5^n, \quad n \geq 0$$

(Verify this also.) ■

An important observation: In this example, notice that the 5 in $f(n)$ is *not* a characteristic root of the ALHRRWCCs. If it were, we would have needed to make adjustments in our search for a particular solution, as in Theorem 5.3. We shall pursue this case shortly.

The following theorem justifies the techniques demonstrated in these two examples; we omit its proof in the interest of brevity.

THEOREM 5.6 In the LNHRRWCCs (5.13), suppose $f(n) = (b_k n^k + b_{k-1}n^{k-1} + \cdots + b_1 n + b_0)\alpha^n$. If α is *not* a characteristic root of the ALHRRWCCs (5.14), then a particular solution is of the form $(d_k n^k + d_{k-1}n^{k-1} + \cdots + d_1 n + d_0)\alpha^n$. If α is a characteristic root with multiplicity m, then a particular solution is of the form $n^m(e_k n^k + e_{k-1}n^{k-1} + \cdots + e_1 n + e_0)\alpha^n$. ■

We conclude this section with the following example, which illustrates this theorem when α is a characteristic root of the ALHRRWCCs.

EXAMPLE 5.21 Solve the LNHRRWCCs $a_n = 6a_{n-1} - 9a_{n-2} + 4(n+1)3^n$, where $a_0 = 2$ and $a_1 = 3$.

SOLUTION:
From Example 5.17, the general solution of the ALHRRWCCs is $a_n^{(h)} = A\cdot 3^n + B\cdot n3^n$, where $n \geq 0$. Since 3 is a characteristic root with multiplicity 2, we search for a particular solution of the form $n^2(cn+d)3^n$, where the constants c and d are to be determined. Then we must have

$$\begin{aligned} n^2(cn+d)3^n &= 6\{(n-1)^2[c(n-1)+d]3^{n-1}\} \\ &\quad - 9\{(n-2)^2[c(n-2)+d]3^{n-2}\} + 4(n+1)3^n \end{aligned}$$

Equating the coefficients of like terms, this yields $c = 2/3$ and $d = 4$ (Verify); so $a_n^{(p)} = 2n^2(n+6)3^{n-1}$.

Thus the general solution of the recurrence relation is

$$a_n = A\cdot 3^n + B\cdot n3^n + 2n^2(n+6)3^{n-1}, \quad n \geq 0$$

Using the initial conditions, this yields

$$a_n = (6-19n)\cdot 3^{n-1} + 2n^2(n+6)3^{n-1}, \quad n \geq 0$$ ■

(You can confirm this.)

Exercises 5.3

Determine if each recurrence relation is a LHRRWCC.

1. $L_n = L_{n-1} + L_{n-2}$

2. $D_n = nD_{n-1} + (-1)^n$

3. $a_n = 1.08a_{n-1}$

4. $b_n = 2b_{n-1} + 1$

5. $a_n = a_{n-1} + n$

6. $a_n = 2a_{n-1} + (2^n - 1)$

7. $a_n = a_{n-1} + 2a_{n-2} + 3a_{n-5}$

8. $a_n = a_{n-1} + 2a_{n-3} + n^2$

Solve each LHRRWCC.

9. $a_n = a_{n-1} + 2a_{n-2}, a_0 = 3, a_1 = 0$

10. $a_n = 5a_{n-1} - 6a_{n-2}, a_0 = 4, a_1 = 7$

11. $a_n = a_{n-1} + 6a_{n-2}, a_0 = 5, a_1 = 0$

12. $a_n = 4a_{n-2}, a_0 = 2, a_1 = -8$

13. $a_n = a_{n-1} + a_{n-2}, a_0 = 1, a_1 = 2$

14. $a_n = a_{n-1} + a_{n-2}, a_0 = 2, a_1 = 3$

15. $L_n = L_{n-1} + L_{n-2}, L_1 = 1, L_2 = 3$

16. $a_n = 4a_{n-1} - 4a_{n-2}, a_0 = 3, a_1 = 10$

17. $a_n = 6a_{n-1} - 9a_{n-2}, a_0 = 2, a_1 = 3$

18. $a_n = 3a_{n-1} + 4a_{n-2} - 12a_{n-3}, a_0 = 3, a_1 = -7, a_2 = 7$

19. $a_n = 8a_{n-1} - 21a_{n-2} + 18a_{n-3}, a_0 = 0, a_1 = 2, a_2 = 13$

20. $a_n = 7a_{n-1} - 16a_{n-2} + 12a_{n-3}, a_0 = 0, a_1 = 5, a_2 = 19$

21. $a_n = -a_{n-1} + 16a_{n-2} + 4a_{n-3} - 48a_{n-4}, a_0 = 0, a_1 = 16, a_2 = -2, a_3 = 142$

22. $a_n = 13a_{n-2} - 36a_{n-4}, a_0 = 7, a_1 = -6, a_2 = 38, a_3 = -84$

23. $a_n = 9a_{n-1} - 30a_{n-2} + 44a_{n-3} - 24a_{n-4}, a_0 = 5, a_1 = 12, a_2 = 38, a_3 = 126$

24. $a_n = 8a_{n-1} - 24a_{n-2} + 32a_{n-3} - 16a_{n-4}, a_0 = 1, a_1 = 4, a_2 = 44, a_3 = 272$

Find the general form of a particular solution of the LNHRRWCCs (5.13) corresponding to each function $f(n)$.

25. $f(n) = n$

26. $f(n) = 1$

27. $f(n) = 3n^2$

28. $f(n) = 3^n$

29. $f(n) = n2^n$

30. $f(n) = 43n^2 5^n$

Find the general form of a particular solution of the LNHRRWCCs $a_n = 4a_{n-1} - 4a_{n-2} + f(n)$ corresponding to each function $f(n)$.

31. $f(n) = 3 \cdot 2^n$

32. $f(n) = n2^n$

33. $f(n) = 23n^2 2^n$

34. $(17n^3 - 1)2^n$

Solve each LNHRRWCCs.

35. $a_n = 2a_{n-1} + 1, a_0 = 1$

36. $a_n = 7a_{n-1} - 10a_{n-2} + n^2, a_0 = 0, a_1 = 1$

37. $a_n = 7a_{n-1} - 12a_{n-2} + 3^n, a_0 = 0, a_1 = 2$

38. $a_n = 7a_{n-1} - 12a_{n-2} + 3n4^n, a_0 = 0, a_1 = 2$

***39.** $a_n = a_{n-1} + n, a_0 = 1$

***40.** $a_n = a_{n-1} + n - 1, a_1 = 0$

41. Let r_n and s_n be two solutions of the recurrence relation (5.8). Prove that $a_n = r_n + s_n$ is also a solution.

42. Let α be a solution of the equation $x^k - c_1x^{k-1} - \cdots - c_k = 0$. Show that α^n is a solution of LHRRWCC (5.7).

43. Let α be a characteristic root of the LHRRWCC $a_n = aa_{n-1} + ba_{n-2} + ca_{n-3}$ with degree of multiplicity three. Show that $\alpha^n, n\alpha^n, n^2\alpha^n$ are solutions of LHRRWCC.

44. Let $a_n^{(h)}$ denote the general solution of the ALHRRWCCs (5.14) and $a_n^{(p)}$ a particular solution of the LNHRRWCCs (5.13). Prove that $a_n = a_n^{(h)} + a_n^{(p)}$ is the general solution of the LNHRRWCCs (5.13).

5.4 Generating Functions

Generating functions provide a powerful tool for solving LHRRWCCs, as will be seen shortly. They were invented in 1718 by the French mathematician Abraham De Moivre, when he used them to solve the Fibonacci recurrence relation (see Example 5.26). Generating functions can also solve combinatorial problems, as the next chapter shows.

To begin with, notice that the polynomial $1 + x + x^2 + x^3 + x^4 + x^5$ can be written as $\dfrac{x^6 - 1}{x - 1}$. You may verify this by either cross-multiplication, the familiar long division method, or Exercise 8 in Section 4.4. Accordingly, $f(x) = \dfrac{x^6 - 1}{x - 1}$ is called the **generating function** of the sequence of coefficients 1, 1, 1, 1, 1, 1 in the polynomial.

More generally, we make the following definition.

Generating Function

Let $a_0, a_1, a_2, \ldots$ be a sequence of real numbers. Then the function

$$g(x) = a_0 + a_1x + a_2x^2 + \cdots + a_nx^n + \cdots \tag{5.15}$$

is the **generating function** for the sequence $\{a_n\}$. Generating functions for the finite sequence $a_0, a_1, \ldots, a_n$ can also be defined by letting $a_i = 0$ for $i > n$; thus $g(x) = a_0 + a_1x + a_2x^2 + \cdots + a_nx^n$ is the generating function for the finite sequence $a_0, a_1, \ldots, a_n$.

Abraham De Moivre *(1667–1754), son of a surgeon, was born in Vitry-le-Francois, France. His formal education began at the Catholic village school, and then continued at the Protestant Academy at Sedan and later at Saumur. He did not receive good training in mathematics until he moved to Paris in 1684, where he studied Euclid's later books and other texts.*

Around 1686, De Moivre emigrated to England, where he began his life-long profession, tutoring in mathematics, and mastered Newton's Principia Mathematica. *In 1695 he presented a paper, his first, on Newton's theory of fluxions to the Royal Society of London and 2 years later he was elected a member of the Society. Unfortunately, despite his influential friends, he could not find an academic position. He had to earn a living as a tutor, author, and expert on applications of probability to gambling and annuities.*

He dedicated his first book, a masterpiece, The Doctrine of Chances, *to Newton. His most notable discovery concerns probability theory:* The binomial probability distribution can be approximated by the normal distribution.

De Moivre died in London.

For example,

$$g(x) = 1 + 2x + 3x^2 + \cdots + (n+1)x^n + \cdots$$

is the generating function for the sequence of positive integers and

$$f(x) = 1 + 3x + 6x^2 + \cdots + \frac{n(n+1)}{2}x^2 + \cdots$$

is the generating function for the sequence of triangular numbers. Since

$$\frac{x^n - 1}{x - 1} = 1 + x + x^2 + \cdots + x^{n-1}$$

$g(x) = \dfrac{x^n - 1}{x - 1}$ is the generating function for the sequence of n ones.

> *A word of caution*: The RHS of Equation (5.15) is a **formal power series** in x. The letter x does not represent anything. The various powers x^n of x are simply used to keep track of the corresponding terms a_n of the sequence. In other words, think of the powers x^n as placeholders. Consequently, unlike in calculus, the convergence of the series is of no interest to us.

Equality of Generating Functions

Two generating functions $f(x) = \sum_{n=0}^{\infty} a_n x^n$ and $g(x) = \sum_{n=0}^{\infty} b_n x^n$ are **equal** if $a_n = b_n$ for every $n \geq 0$.

For example, let $f(x) = 1 + 3x + 6x^2 + 10x^3 + \cdots$ and $g(x) = 1 + \frac{2 \cdot 3}{2}x + \frac{3 \cdot 4}{2}x^2 + \frac{4 \cdot 5}{2}x^3 + \cdots$. Then $f(x) = g(x)$.

A generating function we will use frequently is

$$\frac{1}{1-ax} = 1 + ax + a^2x^2 + \cdots + a^nx^n + \cdots \tag{5.16}$$

Then

$$\frac{1}{1-x} = 1 + x + x^2 + \cdots + x^n + \cdots \tag{5.17}$$

Can we add and multiply generating functions? Yes! Such operations are performed exactly the same way as polynomials are combined.

Addition and Multiplication of Generating Functions

Let $f(x) = \sum_{n=0}^{\infty} a_n x^n$ and $g(x) = \sum_{n=0}^{\infty} b_n x^n$ be two generating functions. Then

$$f(x) + g(x) = \sum_{n=0}^{\infty}(a_n + b_n)x^n \text{ and } f(x)g(x) = \sum_{n=0}^{\infty}\left(\sum_{i=0}^{n} a_i b_{n-i}\right)x^n$$

For example,

$$\begin{aligned}
\frac{1}{(1-x)^2} &= \frac{1}{1-x} \cdot \frac{1}{1-x} \\
&= \left(\sum_{i=0}^{\infty} x^i\right)\left(\sum_{i=0}^{\infty} x^i\right) = \sum_{n=0}^{\infty}\left(\sum_{i=0}^{n} 1 \cdot 1\right)x^n \\
&= \sum_{n=0}^{\infty}(n+1)x^n \\
&= 1 + 2x + 3x^2 + \cdots + (n+1)x^n + \cdots
\end{aligned} \tag{5.18}$$

and

$$\begin{aligned}
\frac{1}{(1-x)^3} &= \frac{1}{1-x} \cdot \frac{1}{(1-x)^2} \\
&= \left(\sum_{n=0}^{\infty} x^n\right)\left[\sum_{n=0}^{\infty}(n+1)x^n\right] \\
&= \sum_{n=0}^{\infty}\left[\sum_{i=0}^{n} 1 \cdot (n+1-i)\right]x^n \\
&= \sum_{n=0}^{\infty}[(n+1) + n + \cdots + 1]x^n
\end{aligned}$$

$$= \sum_{n=0}^{\infty} \frac{(n+1)(n+2)}{2} x^n$$

$$= 1 + 3x + 6x^2 + 10x^3 + \cdots \tag{5.19}$$

Before exploring how valuable generating functions are in solving LHRRWCCs, we illustrate how the technique of **partial fraction decomposition**, used in integral calculus, enables us to express the quotient $\frac{p(x)}{q(x)}$ of two polynomials $p(x)$ and $q(x)$ as a sum of proper fractions, where $\deg p(x) < \deg q(x)$.†

For example,

$$\frac{6x+1}{(2x-1)(2x+3)} = \frac{1}{2x-1} + \frac{2}{2x+3}$$

Partial Fraction Decomposition Rule for $\frac{p(x)}{q(x)}$, where $\deg p(x) < \deg q(x)$

If $q(x)$ has a factor of the form $(ax+b)^m$, then the decomposition contains a sum of the form

$$\frac{A_1}{ax+b} + \frac{A_2}{(ax+b)^2} + \cdots + \frac{A_m}{(ax+b)^m}$$

where A_i is a rational number.

Examples 5.22–5.24 illustrate the partial fraction decomposition technique. We use their results to solve the recurrence relations in Examples 5.25–5.27.

EXAMPLE 5.22 Express $\frac{x}{(1-x)(1-2x)}$ as a sum of partial fractions.

SOLUTION:

Since the denominator contains two linear factors, we let

$$\frac{x}{(1-x)(1-2x)} = \frac{A}{1-x} + \frac{B}{1-2x}$$

To find the constants A and B, multiply both sides by $(1-x)(1-2x)$:

$$x = A(1-2x) + B(1-x)$$

Now give convenient values to x. Setting $x = 1$ yields $A = -1$ and setting $x = 1/2$ yields $B = 1$. (The values of A and B can also be found by equating

†$\deg f(x)$ denotes the degree of the polynomial $f(x)$.

coefficients of like terms from either side of the equation and solving the resulting linear system.)

$$\frac{x}{(1-x)(1-2x)} = \frac{-1}{1-x} + \frac{1}{1-2x}$$

(You may verify this by combining the sum on the RHS into a single fraction.) We use this result in Example 5.25. ■

EXAMPLE 5.23 Express $\dfrac{x}{1-x-x^2}$ as a sum of partial fractions.

SOLUTION:
First, factor $1-x-x^2$:

$$1-x-x^2 = (1-\alpha x)(1-\beta x)$$

where $\alpha = \dfrac{1+\sqrt{5}}{2}$ and $\beta = \dfrac{1-\sqrt{5}}{2}$. (Notice that $\alpha+\beta = 1, \alpha\beta = -1$, and $\alpha - \beta = \sqrt{5}$.)

Let

$$\frac{x}{1-x-x^2} = \frac{A}{1-\alpha x} + \frac{B}{1-\beta x}$$

Then

$$x = A(1-\beta x) + B(1-\alpha x)$$

Equating coefficients of like terms, we get:

$$A + B = 0$$
$$-\beta A - \alpha B = 1$$

Solving this linear system yields $A = \dfrac{1}{\sqrt{5}} = -B$ (Verify this.).

Thus

$$\frac{x}{(1-x-x^2)} = \frac{1}{\sqrt{5}}\left[\frac{1}{1-\alpha x} - \frac{1}{1-\beta x}\right]$$

We use this result in Example 5.26. ■

EXAMPLE 5.24 Express $\dfrac{2-9x}{1-6x+9x^2}$ as a sum of partial fractions.

SOLUTION:
Again, factor the denominator:

$$1-6x+9x^2 = (1-3x)^2$$

By the decomposition rule, let

$$\frac{2-9x}{1-6x+9x^2} = \frac{A}{1-3x} + \frac{B}{(1-3x)^2}$$

Then

$$2-9x = A(1-3x)+B$$

This yields $A = 3$ and $B = -1$ (Verify this.).

Thus

$$\frac{2-9x}{1-6x+9x^2} = \frac{3}{1-3x} - \frac{1}{(1-3x)^2}$$

We use this result in Example 5.27. ■

Now we are ready to use partial fraction decompositions and generating functions to solve recurrence relations in the next three examples.

EXAMPLE 5.25 Use generating functions to solve the recurrence relation $b_n = 2b_{n-1} + 1$, where $b_1 = 1$.

SOLUTION:

First, notice that the condition $b_1 = 1$ yields $b_0 = 0$. To find the sequence $\{b_n\}$ that satisfies the recurrence relation, consider the corresponding generating function

$$g(x) = b_0 + b_1x + \quad b_2x^2 + \quad b_3x^3 + \cdots + \quad b_nx^n + \cdots$$

Then

$$2xg(x) = \qquad 2b_1x^2 + 2b_2x^3 + \cdots + 2b_{n-1}x^n + \cdots$$

Also,

$$\frac{1}{1-x} = 1 + x + \quad x^2 + \quad x^3 + \cdots + \quad x^n + \cdots$$

Then

$$\begin{aligned} g(x) - 2xg(x) - \frac{1}{1-x} &= -1 + (b_1 - 1)x + (b_2 - 2b_1 - 1)x^2 + \cdots \\ &\quad + (b_n - 2b_{n-1} - 1)x^n + \cdots \\ &= -1 \end{aligned}$$

since $b_1 = 1$ and $b_n = 2b_{n-1} + 1$ for $n \geq 2$. That is,

$$(1-2x)g(x) = \frac{1}{1-x} - 1 = \frac{x}{1-x}$$

Then

$$g(x) = \frac{x}{(1-x)(1-2x)}$$

$$= -\frac{1}{1-x} + \frac{1}{1-2x}, \text{ by Example 5.22}$$

$$= -\left(\sum_{n=0}^{\infty} x^n\right) + \left(\sum_{n=0}^{\infty} 2^n x^n\right), \text{ by (5.16)}$$

$$= \sum_{n=0}^{\infty} (2^n - 1)x^n$$

But $g(x) = \sum_{n=0}^{\infty} b_n x^n$, so $b_n = 2^n - 1$, $n \geq 1$. (Notice that this is the same solution obtained in Example 5.12.) ■

EXAMPLE 5.26 Using generating functions, solve the Fibonacci recurrence relation $F_n = F_{n-1} + F_{n-2}$, where $F_1 = 1 = F_2$.

SOLUTION:

Notice that the two initial conditions yield $F_0 = 0$. Let

$$g(x) = F_0 + F_1 x + F_2 x^2 + \cdots + F_n x^n + \cdots$$

be the generating function of the Fibonacci sequence. Since the orders of F_{n-1} and F_{n-2} are 1 and 2 less than the order of F_n, respectively, we find $xg(x)$ and $x^2 g(x)$:

$$xg(x) = F_1 x^2 + F_2 x^3 + F_3 x^4 + \cdots + F_{n-1} x^n + \cdots$$

$$x^2 g(x) = F_1 x^3 + F_2 x^4 + F_3 x^5 + \cdots + F_{n-2} x^n + \cdots$$

Then

$$g(x) - xg(x) - x^2 g(x) = F_1 x + (F_2 - F_1)x^2 + (F_3 - F_2 - F_1)x^3 + \cdots + (F_n - F_{n-1} - F_{n-2})x^n + \cdots$$

$$= x$$

since $F_2 = F_1$ and $F_n = F_{n-1} + F_{n-2}$.

That is,

$$(1 - x - x^2)g(x) = x$$

$$g(x) = \frac{x}{1 - x - x^2}$$

$$= \frac{1}{\sqrt{5}}\left[\frac{1}{1-\alpha x} - \frac{1}{1-\beta x}\right], \text{ by Example 5.23}$$

where $\alpha = \dfrac{1+\sqrt{5}}{2}$ and $\beta = \dfrac{1-\sqrt{5}}{2}$

Then

$$\sqrt{5}g(x) = \frac{1}{1-\alpha x} - \frac{1}{1-\beta x}$$
$$= \sum_{n=0}^{\infty} \alpha^n x^n - \sum_{n=0}^{\infty} \beta^n x^n = \sum_{n=0}^{\infty} (\alpha^n - \beta^n) x^n$$

So

$$g(x) = \sum_{n=0}^{\infty} \frac{(\alpha^n - \beta^n)}{\sqrt{5}} x^n$$

Therefore, by the equality of generating functions,

$$F_n = \frac{\alpha^n - \beta^n}{\sqrt{5}} = \frac{\alpha^n - \beta^n}{\alpha - \beta}$$

(Recall that this is the **Binet form** of F_n.) ■

We close this section with the following example.

EXAMPLE 5.27 Using generating functions, solve the recurrence relation $a_n = 6a_{n-1} - 9a_{n-2}$, where $a_0 = 2$ and $a_1 = 3$.

SOLUTION:
Let

$$g(x) = a_0 + a_1 x + a_2 x^2 + \cdots + a_n x^n + \cdots$$

Then

$$6xg(x) = 6a_0 x + 6a_1 x^2 + 6a_2 x^3 + \cdots + 6a_{n-1} x^n + \cdots$$
$$9x^2 g(x) = 9a_0 x^2 + 9a_1 x^3 + 9a_2 x^4 + \cdots + 9a_{n-2} x^n + \cdots$$

Then

$$g(x) - 6xg(x) + 9x^2 g(x) = a_0 + (a_1 - 6a_0)x + (a_2 - 6a_1 + 9a_0)x^2 + \cdots + (a_n - 6a_{n-1} + 9a_{n-2})x^n + \cdots$$
$$= 2 - 9x$$

using the given conditions. Thus

$$(1 - 6x + 9x^2)g(x) = 2 - 9x$$

Therefore,

$$g(x) = \frac{2 - 9x}{1 - 6x + 9x^2}$$

$$= \frac{3}{1-3x} - \frac{1}{(1-3x)^2}, \text{ by Example 5.24}$$

$$= 3\left(\sum_{n=0}^{\infty} 3^n x^n\right) - \sum_{n=0}^{\infty}(n+1)3^n x^n$$

$$= \sum_{n=0}^{\infty}[3^{n+1} - (n+1)3^n]x^n$$

$$= \sum_{n=0}^{\infty} 3^n(2-n)x^n$$

Thus

$$a_n = (2-n)3^n, \quad n \geq 0$$ ■

The following exercises provide ample practice in this problem-solving technique.

Exercises 5.4

Express each quotient as a sum of partial fractions.

1. $\dfrac{x+7}{(x-1)(x+3)}$

2. $\dfrac{4x^2-3x-25}{(x+1)(x-2)(x+3)}$

3. $\dfrac{5}{1-x-6x^2}$

4. $\dfrac{2+4x}{1+8x+15x^2}$

5. $\dfrac{x(x+2)}{(2+3x)(x^2+1)}$

6. $\dfrac{-2x^2-2x+2}{(x-1)(x^2+2x)}$

7. $\dfrac{x^3+x^2+x+3}{x^4+5x^2+6}$

8. $\dfrac{-x^3+2x^2+x}{x^4+x^3+x+1}$

9. $\dfrac{3x^3-x^2+4x}{x^4-x^3+2x^2-x+1}$

***10.** $\dfrac{x^3+x^2+5x-2}{x^4-x^2+x-1}$

Using generating functions, solve each LHRRWCC.

11. $a_n = 2a_{n-1}, a_0 = 1$

12. $a_n = a_{n-1} + 1, a_1 = 1$

13. $a_n = a_{n-1} + 2, a_1 = 1$

14. $a_n = a_{n-1} + 2a_{n-2}, a_0 = 3, a_1 = 0$

15. $a_n = 4a_{n-2}, a_0 = 2, a_1 = -8$

16. $a_n = a_{n-1} + 6a_{n-2}, a_0 = 5, a_1 = 0$

17. $a_n = 5a_{n-1} - 6a_{n-2}, a_0 = 4, a_1 = 7$

18. $a_n = a_{n-1} + a_{n-2}, a_0 = 1, a_1 = 2$

19. $a_n = a_{n-1} + a_{n-2}, a_0 = 2, a_1 = 3$

20. $L_n = L_{n-1} + L_{n-2}, L_1 = 1, L_2 = 3$

21. $a_n = 4a_{n-1} - 4a_{n-2}, a_0 = 3, a_1 = 10$

22. $a_n = 6a_{n-1} - 9a_{n-2}, a_0 = 2, a_1 = 3$

23. $a_n = 3a_{n-1} + 4a_{n-2} - 12a_{n-3}, a_0 = 3, a_1 = -7, a_2 = 7$

24. $a_n = 8a_{n-1} - 21a_{n-2} + 18a_{n-3}, a_0 = 0, a_1 = 2, a_2 = 13$

25. $a_n = 7a_{n-1} - 16a_{n-2} + 12a_{n-3}, a_0 = 0, a_1 = 5, a_2 = 19$

26. $a_n = 3a_{n-1} + 4a_{n-2} - 12a_{n-3}, a_0 = 3, a_1 = -7, a_2 = 7$

27. $a_n = 6a_{n-1} - 12a_{n-2} + 8a_{n-3}, a_0 = 0, a_1 = 2, a_2 = -2$

28. $a_n = 13a_{n-2} - 36a_{n-4}, a_0 = 7, a_1 = -6, a_2 = 38, a_3 = -84$

29. $a_n = -a_{n-1} + 3a_{n-2} + 5a_{n-3} + 2a_{n-4}, a_0 = 0, a_1 = -8, a_2 = 4, a_3 = -42$

5.5 Recursive Algorithms

Recall that the recursive definition of the factorial function f expresses $f(n)$ in terms of itself with a smaller argument $n-1$. Accordingly, it can be employed to write a simple algorithm to compute $n!$ This algorithm has the interesting property that it invokes itself with a smaller argument. Such an algorithm is a recursive algorithm.

Recursive Algorithm

An algorithm is **recursive** if it invokes itself with a smaller argument; that is, if it invokes a reduced version of itself. (See Figure 5.1.)

Recursive definitions invariably lead to recursive algorithms. This section translates some of the examples discussed in Section 5.1 into recursive algorithms and presents a few new ones—gcd, binary search, and merge sort.

EXAMPLE 5.28 Write a recursive algorithm to compute $n!$, where $n \geq 0$.

SOLUTION:

When $n = 0$, the algorithm must terminate and yield the value 1. When $n > 0$, the recurrence relation $f(n) = n \cdot f(n-1)$ must be applied: the algorithm must invoke itself with $n-1$ as the new argument. The recursive algorithm is given in Algorithm 5.1.

```
Algorithm factorial(n)
(* This algorithm computes n! using recursion *)
Begin (* algorithm *)
   if n = 0 then      (* base case *)
      factorial ← 1
   else               (* invoke the algorithm *)
      factorial ← n · factorial(n - 1)
End (* algorithm *)
```

Algorithm 5.1 ■

Figure 5.15 shows the result of invoking the factorial algorithm with $n = 3$, where f means *factorial*.

Figure 5.15

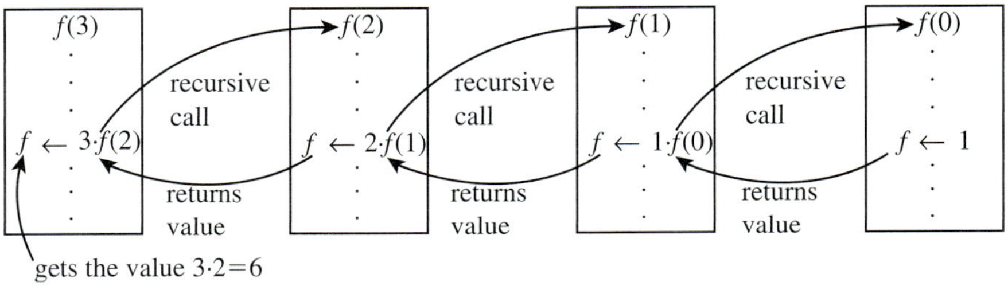

Every recursive algorithm has two important characteristics, or cases:

- The **base case** ensures the sequence of recursive calls will terminate after a finite number of steps. This case corresponds to the initial condition(s) of a recursive definition.
- The **general case** continues to call itself so long as the base case is not satisfied.

The next example presents an algorithm for computing the number of handshakes made by n guests, discussed in Example 5.3.

EXAMPLE 5.29 Using Example 5.3 write a recursive algorithm to compute the number of handshakes made by n guests.

SOLUTION:

Base case The algorithm terminates when $n = 1$, in which case the number of handshakes made is zero.

General case When $n \geq 2$, the algorithm invokes itself using the recurrence relation $h(n) = h(n-1) + (n-1)$.

These two cases lead to Algorithm 5.2.

```
Algorithm handshake(n)
(* This algorithm computes the number of handshakes made
   by n guests at a party by recursion. *)
Begin (* algorithm *)
```

```
  if n = 1 then      (* basis case *)
    handshake ← 0
  else               (* general case *)
    handshake ← handshake(n - 1) + (n - 1)
End (* algorithm *)
```

Algorithm 5.2 ■

EXAMPLE 5.30 Write a recursive algorithm to print the moves and the total number of moves needed to transfer the n disks from peg X to peg Z in the Tower of Brahma puzzle in Example 5.4.

SOLUTION:

Recall that solving the puzzle involves three steps:

- Move the top $n-1$ disks from X to Y using Z as an auxiliary peg;
- Move disk n from X to Z; and
- Move the $n-1$ disks from Y to Z using X as an auxiliary.

We also must count the moves made. The resulting Algorithm 5.3 follows.

```
Algorithm tower (X,Z,Y,n,count)
(* This algorithm, using recursion, prints the various moves
   needed to solve the Tower of Brahma puzzle and returns
   the total number of moves needed in the global variable count.
   Count must be initialized to 0 in the calling module. *)
Begin (* algorithm *)
  if n = 1 then (* base case *)
    begin (* if *)
      move disk 1 from X to Z
      count ← count + 1
    endif
  else (* general case *)
    begin (* else *)
      tower(X,Y,Z,n - 1,count) (* move the top n - 1 disks *)
      move disk n from X to Z
      count ← count + 1
      tower(Y,Z,X,n - 1,count)
    endelse
End (* algorithm *)
```

Algorithm 5.3 ■

Suppose we invoke this algorithm by tower (X,Y,Z,3,count). The tree diagram in Figure 5.16 illustrates the various recursive calls, where t stands for *tower* and c for *count*. Seven moves are needed:

> move 1 from X to Z; move 2 from X to Y; move 1 from Z to Y; move 3 from X to Z; move 1 from Y to X; move 2 from Y to Z; move 1 from X to Z.

You may verify this.

Figure 5.16

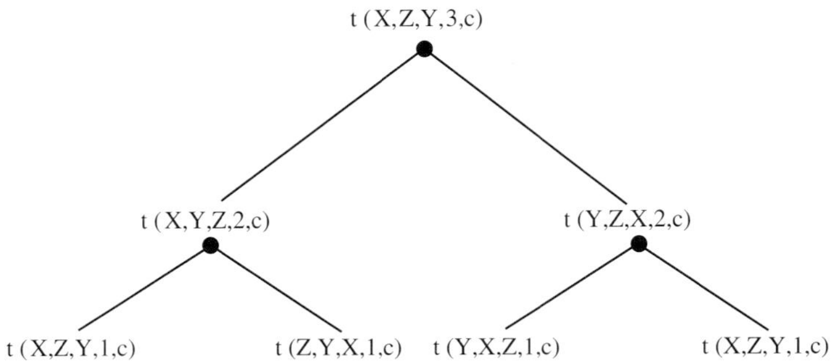

The next example displays a Fibonacci algorithm.

EXAMPLE 5.31 Write a recursive algorithm to compute the nth Fibonacci number F_n.

SOLUTION:

Recall from Example 5.7 that the recursive definition of F_n involves two initial conditions $F_1 = 1 = F_2$, and the recurrence relation $F_n = F_{n-1} + F_{n-2}$, where $n \geq 3$. These two cases can be combined into straightforward Algorithm 5.4.

```
Algorithm Fibonacci(n)
(* This algorithm computes the nth Fibonacci number
   using recursion. *)
Begin (* algorithm *)
   if n = 1 or n = 2 then      (* base cases *)
     Fibonacci ← 1
   else                        (* general case *)
     Fibonacci ← Fibonacci(n - 1) + Fibonacci(n - 2)
End (* algorithm *)
```

Algorithm 5.4 ■

The tree diagram in Figure 5.17 illustrates the recursive computing of F_5, where each dot represents an addition.

Figure 5.17

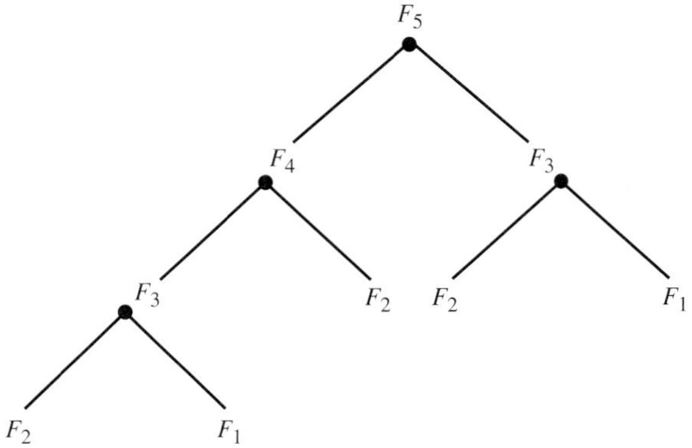

The next example shows how we can use recursion to compute the gcd of two positive integers x and y.

EXAMPLE 5.32 Write a recursive algorithm to compute the gcd of two positive integers x and y.

SOLUTION:

If $x > y$, $\gcd\{x,y\} = \gcd\{x - y, y\}$. (See Exercise 34 in Section 4.2.) We use this fact to write Algorithm 5.5.

```
Algorithm gcd(x,y)
(* This algorithm computes the gcd of two positive
   integers x and y using recursion. *)
 Begin (* algorithm *)
   if x > y then
      gcd ← gcd{x - y, y}
   else if x < y then
           gcd ← gcd{y, x}
        else
           gcd ← x
 End (* algorithm *)
```

Algorithm 5.5

(As an exercise, use this algorithm to compute $\gcd\{x,y\}$ with $x = 28$ and $y = 12$, $x = 13$ and $y = 20$, and $x = 17$ and $y = y$.) ■

We now turn our attention to the recursive version of the binary search algorithm, presented in Example 4.28 in Section 4.5. Recall that binary search, a divide-and-conquer technique, is an efficient method for searching an ordered list for a *key* (say, for example, a certain name in your local telephone directory).

EXAMPLE 5.33 **(Binary Search Algorithm)** Write a recursive algorithm to search an ordered list X of n items and determine if a certain item (*key*) occurs in the list. Return the location of *key* if the search is successful.

SOLUTION:

Because the algorithm is extremely useful, we first outline it:

```
compute the middle index.
if key = middle value then
  we are done and exit
else if key < middle value then
        search the lower half
     else
        search the upper half.
```

The algorithm is given in Algorithm 5.6.

```
Algorithm binary search(X,low,high,key,found,mid)
(* The algorithm returns the location of key in the
   variable mid in the list X if the search is successful.
```

```
   Low, mid, and high denote the lowest, middle, and highest
   indices of the list. Found is a boolean variable;
   it is true if key is found and false otherwise. *)
Begin (* algorithm *)
  if low ≤ high then (* list is nonempty *)
    begin (* if *)
      found ← false (* boolean flag *)
      mid ← ⌊(low + high)/2⌋
      if key = x_mid then
        found ← true (* we are done. *)
      else
        if key < x_mid then (* search the lower half *)
          binary search(X,low,mid - 1,key,found,mid)
        else (* search the upper half *)
          binary search(X,mid + 1,high,key,found,mid)
  endif
End (* algorithm *)
```

Algorithm 5.6

(As an exercise, use this algorithm to search the list $[3, 5, 8, 13, 21, 34, 55, 89]$ with $key = 5$ and $key = 23$.) ■

The Merge Algorithm

Before presenting the merge sort algorithm that sorts a list into ascending order, we show how the **merge algorithm** works. It combines two ordered lists A and B into an ordered list C, eliminating all duplicate elements.

Consider the two lists A and B:

	1	2	3
A	2	3	5

	1	2	3	4	5
B	1	3	5	8	13

Clearly, the combined sorted list contains at most 8 elements.

Let a_i denote the ith element of A, b_j the jth element of B, and c_k the kth element of C, where $1 \leq i \leq 3, 1 \leq j \leq 5$, and $1 \leq k \leq 8$.

Step 1 Initially, compare a_1 and b_1. Since $b_1 < a_1$, store b_1 in c_1. This yields the following

	1	2	3	4	5	6	7	8
C	1							

Step 2 Compare a_1 and b_2. $a_1 < b_2$. So store a_1 in c_2:

	1	2	3	4	5	6	7	8
C	1	2						

Step 3 Compare a_2 and b_2. Since they are equal, store a_2 in c_3:

	1	2	3	4	5	6	7	8
C	1	2	3					

Step 4 Since $a_3 = b_3$, store a_3 in c_4:

	1	2	3	4	5	6	7	8
C	1	2	3	5				

Step 5 There are no more elements left in A, so copy the remaining elements of B into C. This yields the following sorted list:

	1	2	3	4	5	6	7	8
C	1	2	3	5	8	13		

We now explore the merge sort algorithm, which uses both recursion and the merge algorithm.

The Merge Sort Algorithm

The **merge sort algorithm** sorts a list X of n elements into increasing order. First, partition the list into one-element sublists by successively dividing lists in two. Then invoke the merge algorithm successively to merge the sublists, a pair at a time, into increasing order until the entire list is sorted.

For instance, suppose the one-element sublists after successive division are $x_1, x_2, \ldots$, and x_n; then merge the sublists x_1 and x_2, x_3 and x_4, etc., to form new sublists x_{12}, x_{34}, etc.; now merge the sublists $x_{12}, x_{34}, \ldots$ pair by pair; continue like this until there is a single ordered list.

The following example illustrates this method.

EXAMPLE 5.34 Using the merge sort algorithm, sort the list 13, 8, 3, 5, 2 into ascending order.

SOLUTION:

Divide the given list into two sublists of equal or about the same size: $[13, 8, 3]$ and $[5, 2]$. Split each sublist into two sublists, resulting in four sublists: $[13, 8]$, $[3]$, $[5]$, $[2]$. Now divide the first sublist into two sublists, resulting in five one-element sublists: $[13]$, $[8]$, $[3]$, $[5]$, $[2]$.

The tree diagram in Figure 5.18 illustrates this splitting process.

Now the merge algorithm combines them successively in pairs into sorted sublists until the original list is sorted, as shown by the upside-down tree in Figure 5.19.

The recursive merge sort algorithm is given in Algorithm 5.7. Use it to sort the list $[13, 55, 3, 8, 34, 5, 2, 31, 29, 6]$.

Figure 5.18

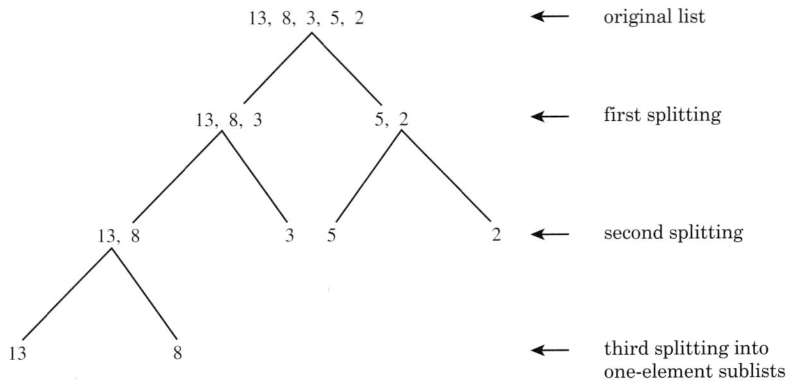

Figure 5.19

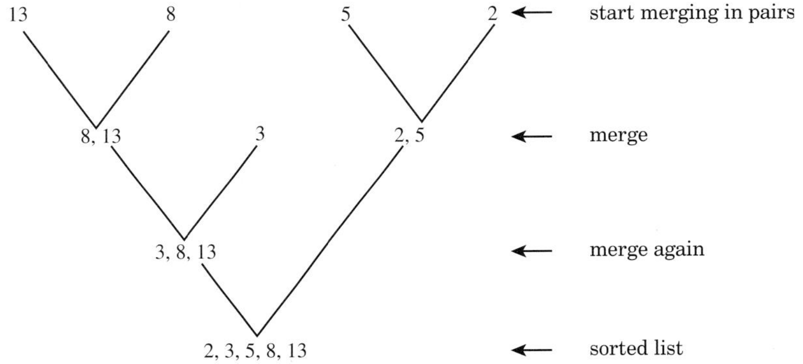


```
Algorithm merge sort(X,low,high)
(* This recursive algorithm   successively divides a list X of
   high - low + 1 elements into sublists of one element. Then it
   continues to merge sublists in pairs into ordered sublists by
   invoking the merge algorithm until the whole list is ordered. *)
 Begin (* algorithm *)
   if low < high then(*list contains more than one element* )
   begin (* if *)
     middle ← ⌊(low + high)/2⌋
     merge sort(X,low,middle) (* sort the lower sublist *)
     merge sort(X,middle + 1,high) (* sort the upper list*)
     merge the two sublists
   endif
 End (* algorithm *)
```

Algorithm 5.7

Exercises 5.5

Using Algorithm 5.4, compute the nth Fibonacci number for each value of n.

1. 3 **2.** 6 **3.** 7 **4.** 10

Using Algorithm 5.4, find the number of computations needed to compute the nth Fibonacci number F_n for each value of n. (*Hint*: Draw a tree diagram.)

5. 4 **6.** 5 **7.** 6 **8.** 7

9. Let a_n denote the number of additions needed to compute F_n using recursion. Use Exercises 5–8 to predict a formula for a_n.

10. Using induction, prove the formula in Exercise 9 for every $n \geq 1$.

11. Write an iterative algorithm to compute the nth Fibonacci number.

12. Mrs. Zee deposits A dollars at a bank at an annual interest rate of $r\%$ compounded semiannually. Write a recursive algorithm to compute the compound amount she will receive at the end of n years.

Using the recursive binary search algorithm in Example 5.33, determine if the given key occurs in the corresponding list. Show the successive values of *low*, *high*, and *mid*.

13. 2, 3, 5, 8, 13, 21; *key* = 13 **14.** 3, 5, 7, 8, 10; *key* = 9

Using the merge sort algorithm, arrange each list into ascending order.

15. 9, 5, 2, 7, 19, 17, 3, 11 **16.** 9, 11, 6, 2, 12, 3, 8, 5, 31, 13

17. Write an algorithm to compute the nth Lucas number L_n using recursion.

18. Let x be a positive real number and n a nonnegative integer. Write a recursive algorithm to compute x^n.

Let $X = [x_1, x_2, \ldots, x_n]$ and $Y = [y_1, y_2, \ldots, y_n]$ be two lists of numbers. Write a recursive algorithm to accomplish the tasks in Exercises 19–31.

19. Find the sum of the numbers from left to right.

20. Find the sum of the numbers from right to left.

21. Compute the product of the numbers from left to right.

22. Compute the product of the numbers from right to left.

23. Find the maximum of the numbers in the list.

24. Find the minimum of the numbers in the list.

25. Print the numbers in the given order $x_1, x_2, \ldots, x_n$.

26. Print the numbers in the reverse order $x_n, x_{n-1}, \ldots, x_2, x_1$.

27. (**Linear search**) Search the list for a specific item (*key*). Return the location of *key* if the search is successful.

28. Determine if two lists X and Y of n items of the same type are identical.

29. Determine if a word of n alphanumeric characters is a palindrome.

30. Evaluate Ackermann's function $A(x,y)$, where x and y are nonnegative integers. See Exercises 5.1 for a definition of $A(x,y)$.

31. Sort the list X using bubble sort.

32. Use the recursive bubble sort algorithm to sort the list 13, 5, 2, 8, 3.

Quicksort, invented in 1962 by C. Anthony R. Hoare of Oxford University, is an extremely efficient technique for sorting a large list X of n items $x_1, x_2, \ldots, x_n$. It is based on the fact that it is easier to sort two small lists than one large list. Choose the first element x_1 as the *pivot*. To place the pivot in its final *resting place*, compare it to each element in the list. Move the elements less than x_1 to the front of the list and those greater than x_1 to the rear. Now place pivot in its final position. Partition the list X into two sublists such that the elements in the first sublist are less than x_1 and the elements in the second sublist are greater than x_1. Continue this procedure recursively with the two sublists.

***33.** Use quicksort to sort the list 7, 8, 13, 11, 5, 6, 4.

***34.** Use quicksort to write a recursive algorithm to sort a list X of n elements.

5.6 Correctness of Recursive Algorithms

We now use induction to establish the correctness of two well-known recursive algorithms, linear search and bubble sort. We begin with the linear search algorithm.

Recall that the linear search algorithm searches a list X of n elements for a given *key*. If the search is successful, the algorithm returns the location of *key*; otherwise, it returns zero. A recursive version is given in Algorithm 5.8. Again, as an exercise, use it to search the list $X = [13, 5, 47, 7, 11, 8, 3]$ for $key = 11$.

```
Algorithm linear search (X,n,key,location)
(* This algorithm returns the position of key in the
   variable location. If location = 0, then key does
   not exist in the list. *)
 Begin (* algorithm *)
    if n = 0 then (* unsuccessful search *)
      location ← 0
    else if x_n = key then
           location ← n
         else
           linear search(X,n - 1,key,location)
  End (* algorithm *)
```

Algorithm 5.8

EXAMPLE 5.35 Establish the correctness of Algorithm 5.8.

PROOF (by PMI):
To prove the correctness of the algorithm, we must show that it works correctly for $n \geq 0$. Let P(n): The algorithm returns the correct value of location for every list of size n.

Basis step When $n = 0$, lines 3 through 6 in the algorithm are skipped and the algorithm returns the value 0 from line 2. So the algorithm works correctly when $n = 0$.

Induction step Let k be an arbitrary integer $k \geq 0$ such that P(k) is true; that is, assume the algorithm works correctly for a list of arbitrary size $k \geq 0$. To prove that P($k + 1$) is true, invoke the algorithm for a list X of size $k + 1$. Note that $k + 1 \geq 1$.

Case 1 If $x_{k+1} = key$, the algorithm returns the value $k + 1$ from line 4.

Case 2 If $x_{k+1} \neq key$, line 6 is executed; so the algorithm is invoked for a list with k elements. By our inductive hypothesis, the algorithm works for such a list.

Thus in both cases, the algorithm returns the correct value of location. Therefore, P($k + 1$) is true.

Consequently, P(n) holds for $n \geq 0$ by induction; that is, the algorithm works correctly for every list. ■

Next we verify the correctness of the recursive version of the bubble sort algorithm, given in Algorithm 5.9. To get used to it, you may use it to sort the list $X = [13, 5, 47, 7, 11, 8, 3]$.

```
Algorithm Bubble Sort(X,n)
(* This algorithm sorts a list X of n items using recursion. *)
Begin (* algorithm *)
  if n > 1 then (* list contains at least two elements *)
  begin (* if *)
    for i = 1 to n - 1 do
     if x_i > x_{i+1} then (* they are out of order *)
      swap x_i and x_{i+1}
    bubble sort(X,n - 1)
  endif
End (* algorithm *)
```

Algorithm 5.9

EXAMPLE 5.36 Establish the correctness of Algorithm 5.9.

PROOF (by PMI):
Let P(n): The algorithm works for every list of size n.

Basis step When $n = 0$, the list contains no elements. So the algorithm works by default. Thus, P(0) is true.

Induction step Assume P(k) is true for an arbitrary integer $k \geq 0$; that is, the algorithm correctly sorts every list of k (≥ 0) elements. To prove that P($k+1$) is true, invoke the algorithm for a list X with $k+1$ elements, where $k+1 \geq 1$.

If $k+1 = 1$, the **for** loop is not entered. So P($k+1$) is true, by default.

If $k+1 > 1$, the **for** loop is entered. Consecutive elements x_i and x_{i+1} are compared in line 4 and switched in line 5 if necessary. When we exit the loop, the largest of the $k+1$ elements is placed in the correct position, in location $k+1$.

This leaves a sublist of k elements, $x_1, \ldots, x_k$. By the inductive hypothesis, the algorithm correctly sorts such a list.

Thus if P(k) is true, then P($k+1$) is also true.

Therefore, by induction, P(n) is true for every $n \geq 0$: the algorithm sorts every list of every size $n \geq 0$. ■

The following exercises provide additional opportunities to establish the correctness of recursive algorithms.

Exercises 5.6

Establish the correctness of each algorithm.

1. The factorial algorithm in Example 5.28.

2. The handshake algorithm in Example 5.29.

3. The Tower of Brahma algorithm in Example 5.30.

4. The Fibonacci algorithm in Example 5.31.

5. The binary search algorithm in Example 5.33.

6. The merge sort algorithm in Algorithm 5.7.

7–17. The algorithms in Exercises 19–29 of Section 5.5.

Algorithm 5.10 computes the nth power of a positive real number x, where $n \geq 0$. Use it to answer Exercises 18–24.

```
Algorithm exponentiation(x,n)
(* This algorithm computes the nth power of x using recursion
   and returns the value in the variable answer. *)
 0. Begin (* algorithm *)
 1.   if n = 0 then
 2.     answer ← 1
 3.   else if n = 1 then
 4.           answer ← x
 5.   else
```

```
6.       begin (* else *)
7.        value ← exponentiation(x,⌊n/2⌋)
8.        answer ← value · value
9.        if n is odd then
10.        answer ← answer · x
11.      endelse
12. End (* algorithm *)
```

Algorithm 5.10

Let a_n denote the number of multiplications (lines 7–10) required by the algorithm to compute x^n. Compute each.

18. a_0 **19.** a_1 **20.** a_4 **21.** a_5

22. Find the recurrence relation satisfied by a_n.

23. Solve the recurrence relation in Exercise 22, where $n = 2^k$.

24. Establish the correctness of Algorithm 5.10.

25. Prove the correctness of the iterative Fibonacci algorithm in Exercise 11 of Section 5.5.

*5.7 Complexities of Recursive Algorithms (optional)

Using the big-oh and big-theta notations, we now investigate the complexities of a few standard recursive algorithms: linear search, Fibonacci, selection sort, binary search, and merge sort. In addition, using Fibonacci numbers, we estimate the number of divisions needed to compute $\gcd\{a, b\}$ using the euclidean algorithm.

We begin our analysis with the recursive linear search algorithm.

EXAMPLE 5.37 Use the recursive linear search in Algorithm 5.8 to estimate the worst time required to search for a *key* in a list X of n items.

SOLUTION:

Let c_n denote the maximum number of element comparisons needed in line 3 of the algorithm. To find a big-oh estimate of c_n, first define it recursively.

Clearly, $c_0 = 0$. When $n \geq 1$

$$\begin{aligned} c_n &= \begin{pmatrix}\text{maximum number of calls}\\ \text{from the recursive call in}\\ \text{line 6}\end{pmatrix} + \begin{pmatrix}\text{number of}\\ \text{comparisons}\\ \text{in line 3}\end{pmatrix} \\ &= c_{n-1} + 1 \end{aligned}$$

Thus

$$c_0 = 0$$
$$c_n = c_{n-1} + 1, \quad n \geq 1$$

Solving this recurrence relation (try) yields $c_n = n, n \geq 0$; so $c_n = O(n) = \Theta(n)$. Thus, in the worst case, the algorithm takes $O(n) = \Theta(n)$ comparisons to locate the key, the same as the iterative version. ■

Next we analyze the recursive and iterative Fibonacci algorithms.

EXAMPLE 5.38 Using the recursive algorithm in Example 5.31, estimate the number of additions a_n needed to compute the nth Fibonacci number.

SOLUTION:
By Exercises 9 and 10 in Section 5.5, $a_n = F_n - 1, n \geq 1$. But, by Exercise 43 in Section 5.1, $F_n \leq 2^n$, where $n \geq 1$. Therefore,

$$\begin{aligned} a_n &\leq 2^n - 1 \\ &< 2^n \\ &= O(2^n) \end{aligned}$$

Thus, the recursive Fibonacci algorithm takes $O(2^n)$ additions. ■

For comparison, we now study the complexity of the iterative version of the Fibonacci algorithm.

EXAMPLE 5.39 Estimate the number of additions a_n required in line 5 to compute the nth Fibonacci number F_n by Algorithm 5.11.

```
Algorithm iterative Fibonacci(n)
(* This iterative algorithm uses the values of the
   variables of the last and the current Fibonacci
   numbers to compute the next Fibonacci number. *)
0.  Begin (* algorithm *)
1.    last ← 1
2.    current ← 1
3.    for i = 2 to n do
4.    begin (* for *)
5.          next ← last + current
6.          last ← current
7.          current ← next
8.    endfor
9.  End (* algorithm *)
```

Algorithm 5.11

SOLUTION:
The first two Fibonacci numbers need no computations; therefore, $a_1 = 0 = a_2$. Suppose $n > 2$. It takes one addition to compute the next item

F_n from the current term F_{n-1}. So $a_n = a_{n-1} + 1$. Solving this recurrence relation (try), we get

$$\begin{aligned} a_n &= n - 2, \quad n \geq 2 \\ &= \Theta(n) \end{aligned}$$

Thus the iterative version takes $\Theta(n)$ additions to compute F_n. ■

The time it takes to compute F_n by the recursive algorithm grows exponentially with n, whereas by the iterative algorithm it grows only linearly. As n gets larger and larger, it takes more time to compute F_n by recursion than by iteration. Thus, by dividing and conquering the problem, we have made it complicated.

Should we prefer the iterative method to the recursive method? Since every recursive algorithm has a nonrecursive version, if the algorithm makes just one recursive call to itself, as in the factorial algorithm, the iterative approach will, in general, save time. On the other hand, if the problem has a recursive definition, it will be easy to write a recursive algorithm for the problem. Writing the nonrecursive version of a recursive algorithm is often a painful task and the resulting algorithm is often much longer, complicated, and difficult to understand. For instance, the nonrecursive version of the Tower of Brahma algorithm is longer and that of quicksort is rather complicated.

Next we estimate the number of element-comparisons required by the recursive selection sort algorithm presented in Algorithm 5.12. (See Algorithm 4.11 in Chapter 4 for an iterative version.)

```
Algorithm selection sort(X,n)
(* This algorithm invokes a subalgorithm called swap
   which switches two elements. Maxindex denotes the
   index of the largest of the n elements. *)
0. Begin (* algorithm *)
1.   maxindex ← n (*initialize maxindex at each pass *)
2.   for i = 1 to n - 1 do
3.     if x_i > x_maxindex then
4.       maxindex ← i
5.   if maxindex ≠ n then (* swap the corresponding items *)
6.     swap x_maxindex and x_n
7.   selection sort(X,n - 1)
8. End (* algorithm *)
```

Algorithm 5.12

EXAMPLE 5.40 Estimate the number c_n of comparisons (lines 3 and 5) required by Algorithm 5.12.

SOLUTION:

To estimate c_n, first define it recursively.

If the list contains just one element, lines 3 and 5 are not executed; therefore, $c_1 = 0$.

Suppose $n \geq 2$. Since the **for** loop is executed $n - 1$ times, line 3 is executed $n - 1$ times. Furthermore, line 5 is executed once. Therefore,

$$\begin{aligned} c_n &= c_{n-1} + (n-1) + 1 \\ &= c_{n-1} + n, \quad n \geq 2 \end{aligned}$$

Solving the recurrence relation by the iterative method, we get

$$\begin{aligned} c_n &= \frac{n(n+1)}{2} - 1, \quad n \geq 1 \\ &= \Theta(n^2) \end{aligned}$$

Thus the algorithm takes $\Theta(n^2)$ comparisons to sort a list of n items, as in the iterative version. ■

Example 5.41 investigates one of the many properties of Fibonacci numbers. Example 5.42 uses the property to estimate the number of divisions in the euclidean algorithm.

EXAMPLE 5.41 Let F_n denote the nth Fibonacci number and $\alpha = \dfrac{1+\sqrt{5}}{2}$. Prove that $\alpha^{n-2} < F_n < \alpha^{n-1}$, $n \geq 3$.

PROOF (by strong induction):
(We shall prove that $\alpha^{n-2} < F_n$ and leave the other half as an exercise.) You may verify that α is a solution of the equation $x^2 = x + 1$, so $\alpha^2 = \alpha + 1$. Let P(n): $\alpha^{n-2} < F_n$, where $n \geq 3$.

Basis step Since the induction step below uses the recurrence relation $F_{k+1} = F_k + F_{k-1}$, the basis step involves verifying that both P(3) and P(4) are true.

- *To show that P(3) is true*: When $n = 3$,

$$\alpha^{n-2} = \alpha = \frac{1+\sqrt{5}}{2} < \frac{1+3}{2} = 2 = F_3$$

So P(3) is true.

- *To show that P(4) is true*:

$$\begin{aligned} \alpha^2 &= \left(\frac{1+\sqrt{5}}{2}\right)^2 = \frac{3+\sqrt{5}}{2} \\ &< \frac{3+3}{2} = 3 = F_4 \end{aligned}$$

Thus P(4) is also true.

Induction step Assume P(3), P(4), ..., P(k) are true; that is, assume $\alpha^{i-2} < F_i$ for $3 \leq i \leq k$. We must show that $P(k+1)$ is true; that is, $\alpha^{k-1} < F_{k+1}$.

We have

$$\alpha^2 = \alpha + 1.$$

Multiplying both sides by α^{k-3},

$$\begin{aligned}
\alpha^{k-1} &= \alpha^{k-2} + \alpha^{k-3} && \text{(Note: } k-3 \geq 2.\text{)} \\
&< F_k + F_{k-1}, && \text{by the inductive hypothesis} \\
&= F_{k+1}, && \text{by the recurrence relation}
\end{aligned}$$

Thus P($k+1$) is true.

Therefore, by the strong version of induction, P(n) is true for $n \geq 3$; that is, $\alpha^{n-2} < F_n$ for every $n \geq 3$. ■

Now we can estimate the number of divisions required by the euclidean algorithm to compute $\gcd\{a, b\}$.

EXAMPLE 5.42 **(Lamé's Theorem)** The number of divisions needed to compute $g\{a, b\}$ by the euclidean algorithm is no more than five times the number of decimal digits in b, where $a \geq b \geq 2$.

PROOF:

Let F_n denote the nth Fibonacci number, $a = r_0$, and $b = r_1$. By the repeated application of the division algorithm we have:

$$\begin{aligned}
r_0 &= r_1 q_1 + r_2 && 0 \leq r_2 < r_1 \\
r_1 &= r_2 q_2 + r_3 && 0 \leq r_3 < r_2 \\
&\vdots \\
r_{n-2} &= r_{n-1} q_{n-1} + r_n && 0 \leq r_n < r_{n-1} \\
r_{n-1} &= r_n q_n
\end{aligned}$$

Clearly, it takes n divisions to evaluate $\gcd\{a, b\} = r_n$. Since $r_i < r_{i-1}$, $q_i \geq 1$ for $1 \leq i \leq n$. In particular, since $r_n < r_{n-1}$, $q_n \geq 2$; so $r_n \geq 1$ and $r_{n-1} \geq 2 = F_3$. Consequently, we have:

$$\begin{aligned}
r_{n-2} &= r_{n-1} q_{n-1} + r_n \\
&\geq r_{n-1} + r_n \\
&\geq F_3 + 1 \\
&= F_3 + F_2 = F_4
\end{aligned}$$

$$\begin{aligned} r_{n-3} &= r_{n-2}q_{n-2} + r_{n-1} \\ &\geq r_{n-2} + r_{n-1} \\ &\geq F_4 + F_3 = F_5 \end{aligned}$$

Continuing like this,

$$\begin{aligned} r_1 &= r_2 q_2 + r_3 \\ &\geq r_2 + r_3 \\ &\geq F_n + F_{n-1} = F_{n+1} \end{aligned}$$

That is,

$$b \geq F_{n+1}$$

By Example 5.41, $F_{n+1} > \alpha^{n-1}$, where $\alpha = \dfrac{1+\sqrt{5}}{2}$. Therefore,

$$b > \alpha^{n-1}$$

Then

$$\log b > (n-1)\log \alpha$$

Since $\alpha = \dfrac{1+\sqrt{5}}{2} \approx 1.618033989$, $\log \alpha \approx 0.2089876403 > \dfrac{1}{5}$. So

$$\log b > \frac{n-1}{5}$$

Suppose b contains k decimal digits. Then $b < 10^k$. Therefore, $\log b < k$ and hence $k > \dfrac{n-1}{5}$. Thus $n < 5k+1$ or $n \leq 5k$. That is, the number of divisions needed by the algorithm is no more than five times the number of decimal digits in n. ■

Let us pursue this example a bit further. Since $\log b > \dfrac{n-1}{5}$, $n < 1 + 5\log b$. Also, since $b \geq 2$,

$$\begin{aligned} 5\log b &\geq 5\log 2 \\ &> 1 \end{aligned}$$

Thus

$$\begin{aligned} n &< 1 + 5\log b \\ &< 5\log b + 5\log b \\ &= 10\log b \\ &= O(\log b) \end{aligned}$$

Thus it takes $O(\log b)$ divisions to compute $\gcd\{a, b\}$ by the euclidean algorithm.

Gabriel Lamé *(1795–1870) was born in Tours, France. After graduating from the École Polytechnique in 1817, he continued his studies at the École des Mines, from which he graduated in 1820.*

The same year Lamé was appointed director of the School of Highways and Transportation in St. Petersburg, Russia. There he taught mathematics, physics, and chemistry and planned roads and bridges in and around the city. In 1832, he returned to Paris to form an engineering firm. Within a few months, however, he left it to become the chair of physics at the École Polytechnique, where he remained until 1844. While teaching, he served as a consulting engineer, becoming the chief engineer of mines in 1836. He helped build the railroads from Paris to Versailles and to St. Germain.

In 1844, Lamé became graduate examiner for the University of Paris in mathematical physics and probability, and professor 7 years later. In 1862, he became deaf and resigned his positions. He died in Paris.

Although Lamé did original work in number theory and mathematical physics, his greatest contribution was the development of the curvilinear coordinates and their applications. His work on the curvilinear system led him to number theory. In 1840, he proved Fermat's Last Theorem for n = 7.

Gauss considered Lamé the foremost French mathematician of his time. French mathematicians, however, considered him too practical, and French scientists, too theoretical.

The next example, due to S. H. Friedberg, explores the number of multiplications needed to compute the determinant of an $n \times n$ matrix by cofactor expansion. (It may be omitted by those not familiar with determinants and calculus.)

○ **EXAMPLE 5.43** (optional) Let f_n denote the number of multiplications needed to compute $\det A$, the determinant of an arbitrary $n \times n$ matrix $A = (a_{ij})$ by cofactor expansion. Estimate f_n.

SOLUTION:

We estimate f_n in three steps:

- Define f_n recursively.
- Solve the recurrence relation.
- Use the solution to estimate f_n.
- *To define f_n recursively*:
  Let C_{ij} denote the $(n-1) \times (n-1)$ determinant obtained from $\det A$ by deleting its ith row and jth column. By expanding $\det A$ with respect to the first row, we have

$$\det A = \sum_{j=1}^{n} (-1)^{j+1} a_{1j} C_{1j} \quad \leftarrow \text{cofactor expansion by row 1}$$

In particular, let $A = \begin{bmatrix} a & b \\ c & d \end{bmatrix}$. Then $\det A = aC_{11} - bC_{12} = ad - bc$. Clearly, two multiplications are needed to evaluate $\det A$ and hence $f_2 = 2$. Also $f_1 = 0$.
Suppose $n \geq 3$. Then, by definition, it takes f_{n-1} multiplications to compute C_{1j}. Therefore, it takes $f_{n-1} + 1$ multiplications to evaluate $a_{1j}C_{1j}$ and hence $n(f_{n-1} + 1)$ multiplications to compute $\det A$.

Thus f_n can be defined recursively as follows:

$$
\begin{aligned}
f_1 &= 0 \\
f_n &= n(f_{n-1} + 1), \quad n \geq 2
\end{aligned}
\tag{5.20}
$$

(This is a linear nonhomogeneous recurrence relation with nonconstant coefficients.)

- *To solve the recurrence relation (5.20)*:
Let $f_n = n!g_n$. Since $f_1 = 0$, $g_1 = 0$. Substituting for f_n in Equation (5.20), we get

$$
\begin{aligned}
n!g_n &= n[(n-1)!g_{n-1} + 1] \\
&= n!g_{n-1} + n
\end{aligned}
$$

So

$$
\begin{aligned}
(g_n - g_{n-1})n! &= n \\
g_n - g_{n-1} &= \frac{1}{(n-1)!} \quad \text{(Note: } g_1 = 0.\text{)}
\end{aligned}
$$

Solving this yields (see Exercise 64)

$$
g_n = \sum_{k=1}^{n-1} \frac{1}{k!}, \text{ since } g_1 = 0
$$

So,

$$
\begin{aligned}
f_n = n!g_n &= n!\left(\sum_{k=1}^{n-1} \frac{1}{k!}\right) \\
&= n!\left(\sum_{k=1}^{n} \frac{1}{k!}\right) - 1
\end{aligned}
$$

Therefore,

$$
\begin{aligned}
f_n &\leq n!\left(\sum_{k=1}^{\infty} \frac{1}{k!}\right) - 1 \\
&= n!(e-1) - 1, \text{ by calculus} \\
&\leq en! \\
&= O(n!)
\end{aligned}
$$

Thus the evaluation of $\det A$ by cofactor expansion takes $O(n!)$ multiplications. ■

Divide-and-Conquer Algorithms

We can now analyze the complexities of a special class of recursive algorithms called divide-and-conquer algorithms.

The binary search algorithm presented in Algorithm 5.6 is based on the divide-and-conquer approach. To search an ordered list of n items for a given key, we divide the list into two smaller and similar sublists of about the same size. If the middle value $\neq key$, then we search either the lower half or the upper half, and continue this procedure until we are done. This exemplifies a divide-and-conquer algorithm.

More generally, consider a problem of size n. Suppose the problem can be solved for a small initial value of n, and it can be broken up into a smaller and similar subproblems of approximately the same size, usually $\lfloor n/b \rfloor$ or $\lceil n/b \rceil$, where $a, b \in \mathbb{N}$, $1 \leq a < n$, and $1 < b < n$. Assume that we can solve each of the subproblems and employ their solutions to solve the original problem. Such an algorithm is a **divide-and-conquer algorithm**.

Let $f(n)$ denote the number of operations required to solve the original problem and $g(n)$ the number of operations resulting from the splitting. Then, assuming b is a factor of n,

$$f(n) = af(n/b) + g(n)$$

This is the **divide-and-conquer recurrence relation** resulting from the algorithm.

The binary search algorithm manifests the complexities of the divide-and-conquer technique.

EXAMPLE 5.44 **(binary search)** Using the recursive binary search in Algorithm 5.6, let c_n denote the maximum number of element comparisons needed to search for a given item (key) in an ordered list X of n items. If $n = 1$, then $low = high = mid = 1$ and the condition in line 5 is tested exactly once; so $c_1 = 1$.

Suppose $n > 1$. Then the middle term is $x_{\lfloor (n+1)/2 \rfloor}$. Compare *key* to $x_{\lfloor (n+1)/2 \rfloor}$. If they are *not* equal, search the lower sublist or the upper sublist, but *not* both. If n is even, $\left\lfloor \frac{n+1}{2} \right\rfloor = \left\lfloor \frac{n}{2} \right\rfloor$; so the upper half contains $\frac{n}{2} = \left\lfloor \frac{n}{2} \right\rfloor$ elements and the lower half contains $\frac{n}{2} - 1 \left(< \left\lfloor \frac{n}{2} \right\rfloor \right)$ elements. On the other hand, if n is odd, then $\left\lfloor \frac{n+1}{2} \right\rfloor = \frac{n+1}{2}$; so both sublists contain $\frac{n-1}{2} = \left\lfloor \frac{n}{2} \right\rfloor$ elements each. Thus, in any case, the maximum number of comparisons needed is $c_{\lfloor n/2 \rfloor} + 1$. So

$$\begin{aligned} c_1 &= 1 \\ c_n &= c_{\lfloor n/2 \rfloor} + 1, \quad n \geq 2 \end{aligned} \tag{5.21}$$

To solve this recurrence relation, assume, for convenience, that n is a power of 2, say $n = 2^k$, where $k \geq 0$. Let $c_n = a_k$. Then the recurrence relation (5.21) becomes $a_k = a_{k-1} + 1$, where $a_0 = 1$. Solving this recurrence relation yields $a_k = k + 1$, $k \geq 0$ (Verify.). Since $n = 2^k$, $k = \lg n$, so $c_n = 1 + \lg n$, $n \geq 1$. Thus, if n is a power of 2, then $c_n = \Theta(\lg n)$.

Suppose n is *not* a power of 2. Then, by induction, it can be shown that $c_n = 1 + \lfloor \lg n \rfloor$, where $n \geq 1$ (see Exercise 44), so $c_n = \Theta(\lg n)$.

Thus, in both cases, the algorithm takes $\Theta(\lg n)$ element comparisons in the worst case. ■

The preceding example is a special case of the following theorem. Since the proof is somewhat complicated, we skip it (see Exercises 65 and 66).

THEOREM 5.7 Let $a, b \in \mathbb{N}$ and $c, d \in \mathbb{R}^+$ with $b \geq 2$. Let f be a nondecreasing function* such that $f(n) = af(n/b) + c$ and $f(1) = d$. Then

$$f(n) = \begin{cases} O(\lg n) & \text{if } a = 1 \\ O(n^{\log_b a}) & \text{otherwise} \end{cases}$$ ■

For example, let f be a nondecreasing function such that $f(n) = 3f(n/2) + 5$ and $f(1) = 8$. Then, by Theorem 5.7, $f(n) = O(n^{\lg 3})$.

The next theorem is a generalization of Theorem 5.7. We state it without proof (see Exercises 67–69 for special cases of the theorem) and apply it in Example 5.45.

THEOREM 5.8 Let $a, b \in \mathbb{N}$ and $c, d \in \mathbb{R}^+$ with $b \geq 2$. Let f be a nondecreasing function such that $f(n) = af(n/b) + cn^d$. Then

$$f(n) = \begin{cases} O(n^d) & \text{if } a < b^d \\ O(n^d \lg n) & \text{if } a = b^d \\ O(n^{\log_b a}) & \text{if } a > b^d \end{cases}$$ ■

EXAMPLE 5.45 (optional) Let $A = (a_{ij})$ and $B = (b_{ij})$ be two $n \times n$ matrices. Let $C = (c_{ij})$ be their product where $c_{ij} = \sum_{k=1}^{n} a_{ik} b_{kj}$. Since C has n^2 entries and each takes n multiplications, the product C can be computed using $n^3 = O(n^3)$ multiplications; in fact, it can be computed using $O(n^3)$ computations (additions and multiplications), as Exercises 40 and 41 indicate. ■

We close this section with an analysis of the merge sort algorithm, a divide-and-conquer strategy.

EXAMPLE 5.46 **(merge sort)** The merge sort method in Algorithm 5.7 sorts a list of n elements. Assume, for convenience, that n is a power of 2, say, $n = 2^k$, $k \geq 0$.

---

*Let $S \subseteq \mathbb{R}$. A function $f : S \to \mathbb{R}^+$ is said to be **nondecreasing** if $x < y$ implies $f(x) \leq f(y)$.

Let c_n denote the maximum number of element comparisons needed in line 6. Show that $c_n = O(n \lg n)$.

SOLUTION:
When $n = 2$, one comparison is needed in line 6; therefore, $c_2 = 1$. So, let $n > 2$. The list is split into two, with each sublist containing $n/2$ elements. In the worst case, the number of comparisons resulting from line 4 is $c_{n/2}$, as it is from line 5. When the merge algorithm is invoked in line 6, each sublist contains $n/2$ elements; so the maximum number of comparisons from line 6 is $n - 1$. Thus

$$c_2 = 1$$
$$c_n = 2c_{n/2} + (n-1), \quad n \geq 3$$

Let $a_k = c_n$ where $n = 2^k$, $k \geq 0$. Then

$$a_1 = 1$$
$$a_k = 2a_{k-1} + (2^k - 1), \quad k \geq 2$$

This recurrence relation (see Exercise 8 in Section 5.2) yields

$$\begin{aligned} a_k &= (k-1)2^k + 1, \quad k \geq 1 \\ &= k \cdot 2^k - 2^k + 1 \end{aligned}$$

Thus

$$\begin{aligned} c_n &= (\lg n)n - n + 1 \\ &\leq n \lg n + 1 \\ &< 2n \lg n, \quad n \geq 2 \\ &= O(n \lg n) \end{aligned}$$

■

More generally, it can be shown that in the worst case the merge sort requires $O(n \lg n)$ element comparisons for a list of n elements. This time estimate is the best among all sorting algorithms.

Exercises 5.7

Find a big-oh estimate for each.

1. The number $h(n)$ of handshakes made by n guests at a party, using Example 5.3.
2. The number b_n of moves needed to transfer n disks in the Tower of Brahma puzzle in Example 5.4.
3. The number f_n of regions formed by n lines, using Example 5.5.

Estimate the solution f_n of each recurrence relation (see Exercises 5.2).

4. $f_1 = 1$

$f_n = f_{n-1} + (2n - 1),\ n \geq 2$

5. $f_0 = 0$

$f_n = f_{n-1} + 4n,\ n \geq 1$

6. $f_1 = 2$

$f_n = f_{n-1} + n,\ n \geq 2$

7. $f_1 = 1$

$f_n = 2f_{n-1} + (2^n - 1),\ n \geq 2$

Find the number of comparisons needed to search for *key* = 13 in each ordered list using the recursive binary search algorithm in Example 5.33.

8. 1, 2, 3, 5, 8, 13

9. 5, 8, 13, 21, 34

10. 3, 7, 8, 13, 21

11. 15, 16, 19, 21

Compute the maximum number of comparisons needed to search for a particular item in an ordered list containing the following number of items, using the recursive binary search algorithm.

12. 8

13. 20

14. 25

15. 31

Let b_n denote the number of multiplications needed to compute $n!$ using the recursive factorial algorithm in Example 5.1.

16. Define b_n recursively.

17. Solve the recurrence relation satisfied by b_n.

18. Show that $b_n = O(n)$.

19–22. Estimate the number of times a_n the assignment statement, $x \leftarrow x + 1$, is executed by the nested **for** loops in Exercises 35–38 of Section 4.4.

Estimate the number a_n of times the statement, $x \leftarrow x + 1$, is executed by each nested **for** loop.

23.

```
for i = 1 to n do
  for j = 1 to ⌊i/2⌋ do
    x ← x + 1
```

24.

```
for i = 1 to n do
  for j = 1 to ⌈i/2⌉ do
    x ← x + 1
```

***25.**

```
for i = 1 to n do
 for j = 1 to i do
   for k = 1 to j do
     for l = 1 to j do
       x ← x + 1
```

***26.**

```
for i = 1 to n do
 for j = 1 to i do
   for k = 1 to j do
     for l = 1 to k do
       x ← x + 1
```

Let b_n denote the number of element-comparisons needed by the bubble sort algorithm in Algorithm 5.9.

27. Define b_n recursively.

28. Solve the recurrence relation.

29. Find a big-oh estimate of b_n.

30. Let a_n denote the number of additions needed to compute the nth Fibonacci number F_n, using Algorithm 5.4. Prove that $a_n = \sum_{i=1}^{n-2} F_i$, $n \geq 3$.

Solve each recurrence relation.

31. $c_0 = 1$

$c_n = c_{n-1} + b,\ n \geq 1$

32. $a_2 = 0$

$a_n = a_{n-1} + b,\ n \geq 3$

33. $c_1 = 0$

$c_n = c_{n-1} + bn,\ n \geq 2$

34. $c_1 = a$

$c_n = c_{n-1} + bn^3,\ n \geq 2$

The number of operations $f(n)$ required by an algorithm is given by $f(n) = f(n-1) + (n-1) + (n-2)$, where $f(1) = 1$.

35. Find an explicit formula for $f(n)$.

36. Show that $f(n) = O(n^2)$.

Let $f(n)$ denote the number of bits in the binary representation of a positive integer n.

37. Find a formula for $f(n)$.

38. Show that $f(n) = O(\lg n)$.

39. Let $x \in \mathbb{R}^+$ and $n \in \mathbb{N}$. The technique of **successive squaring** can be applied to compute x^n faster than multiplying x by itself $n - 1$ times. For example, to find x^{43}, first evaluate x^2, x^4, x^8, x^{16}, and x^{32}; then multiply x^{32}, x^8, x^2, and x^1: $x^{43} = x^{32} \cdot x^8 \cdot x^2 \cdot x^1$. This process takes only $5 + 3 = 8$ multiplications instead of the conventional method's 42. The powers of x used in computing x^n are the place values of the bits in the binary representation of n; in fact, the number of powers of x used equals the number of nonzero bits in the binary representation of n. Let $f(n)$ denote the number of multiplications needed to compute x^n by successive squaring. Show that $f(n) = O(\lg n)$.

Let $A = (a_{ij})$ and $B = (b_{ij})$ be two $n \times n$ matrices. Let f_n denote the number of computations (additions and multiplications) to compute their product $C = (c_{ij})$, where $c_{ij} = \sum_{k=1}^{n} a_{ik}b_{kj}$.

40. Evaluate f_n.

41. Estimate f_n.

42. Solve the recurrence relation $C_n = 2C_{n/2} + 1$, where $C_1 = a$ and n is a power of 2.

43. Show that $C_n = O(n)$.

44. Let c_n denote the maximum number of comparisons needed to search for a *key* in an ordered list X of n elements, using the recursive binary search algorithm. Prove that $c_n = 1 + \lfloor \lg n \rfloor$, for every $n \geq 1$.

45. Let $a, b, k \in \mathbb{N}$, $b \geq 2$, and $n = b^k$. Consider the function f defined by $f(n) = af(n/b) + g(n)$. Show that $f(n) = a^k f(1) + \sum_{i=0}^{k-1} a^i g(n/b^i)$.

46. Solve the recurrence relation $a_n = 2a_{n/2} + n$, where $a_1 = 0$ and $n = 2^k$.

47. Use Exercise 46 to show that $a_n = O(n \lg n)$.

Let f be a function defined by $f(n) = af(n/b) + cn$, where $a, b \in \mathbb{N}$, $b \geq 2$, $c \in R^+$, and $f(1) = d$. Assume n is a power of b.

48. Solve the recurrence relation.

49. Let $a = b$ and $d = 0$. Show that $f(n) = O(n \lg n)$.

Consider the recurrence relation $c_n = c_{\lfloor n/2 \rfloor} + c_{\lfloor (n+1)/2 \rfloor} + 2$, where $c_1 = 0$.

50. Compute c_3 and c_4.

51. Solve the recurrence relation when n is a power of 2.

52. Find the order of magnitude of c_n when n is a power of 2.

Let t be a function defined by

$$t(n) = \begin{cases} a & \text{if } n = 1 \\ t(\lfloor n/2 \rfloor) + t(\lceil n/2 \rceil) + bn & \text{otherwise} \end{cases}$$

where $a, b \in \mathbb{R}^+$. (Such a function occurs in the analysis of merge sort.)

53. Evaluate $t(5)$ and $t(6)$.

54. Prove that $t(n)$ is a nondecreasing function; that is, $t(n) \leq t(n+1)$, where $n \geq 1$.

55. Show that $t(n) = O(n \lg n)$, where n is a power of 2.

Let $f(n) = 2f(n/2) + cn^2$, where $f(1) = d$ and n is a power of 2.

56. Solve the recurrence relation.

57. Show that $f(n) = O(n^2)$.

The number $h_n = \sum_{i=1}^{n} \left(\frac{1}{i}\right)$, called the **harmonic number**, occurs frequently in the analysis of algorithms.

58. Compute h_4 and h_5.

59. Define h_n recursively.

60. Prove that $\sum_{i=1}^{n} h_i = (n+1)h_n - n$, $n \geq 1$.

61. Prove that $h_{2^m} \geq 1 + \frac{m}{2}$, $m \geq 0$.

62. Prove that $h_n \leq \frac{n+1}{2}$.

***63.** (For those familiar with calculus) Let h_n denote the nth harmonic number $h_n = \sum_{i=1}^{n} \left(\frac{1}{i}\right)$. Show that $h_n = O(\lg n)$.
(*Hint*: Use integration.)

64. Solve the recurrence relation $g_n - g_{n-1} = 1/(n-1)!$, where $g_1 = 0$.

Let $a, b \in \mathbb{N}$ and $c, d \in \mathbb{R}^+$ with $b \geq 2$. Let f be a nondecreasing function such that $f(n) = af(n/b) + c$ and $f(1) = d$. Prove each.

****65.** If $a = 1$, then $f(n) = O(\lg n)$.

****66.** If $a \neq 1$, then $f(n) = O(n^{\log_b a})$.

Let $a, b, n \in \mathbb{N}$, $b \geq 2$, $c, d \in \mathbb{R}^+$, $f(1) = d$, and n is a power of b. Let f be a nondecreasing function such that $f(n) = af(n/b) + cn^2$. Prove each.

****67.** If $a = b^2$, then $f(n) = n^2 d + cn^2 \log_b n$.

****68.** If $a \neq b^2$, then $f(n) = An^2 + Bn^{\log_b a}$, where $A = \dfrac{b^2 c}{b^2 - a}$ and $B = d + \dfrac{b^2 c}{a - b^2}$.

****69.**

$$f(n) = \begin{cases} O(n^2) & \text{if } a < b^2 \\ O(n^2 \lg n) & \text{if } a = b^2 \\ O(n^{\log_b a}) & \text{if } a > b^2 \end{cases}$$

Chapter Summary

This chapter presented a new class of functions and hence sequences: recursively defined functions. The definitions of such functions can be translated into recursive algorithms. Just as the big-oh and big-theta notations worked well in analyzing the time complexities of algorithms, so does induction in proving the correctness of recursive algorithms.

Recursion

- The **recursive definition** of a function consists of one or more initial conditions and a recurrence relation (page 262).

Solving Recurrence Relations

- A simple class of recurrence relations can be solved using the iterative method (page 279).
- Every solution of the recurrence relation $a_n = a_{n-1} + f(n)$ is of the form $a_n = a_0 + \sum_{i=1}^{n} f(i)$ (page 280).

- Every solution of the recurrence relation $a_n = ca_{n-1} + 1$ is of the form $a_n = c^n a_0 + \frac{c^n - 1}{c - 1}$, where $c \neq 1$ (page 282).
- A ***k*th-order LHRRWCC** is of the form $a_n = \sum_{i=1}^{k} c_i a_{n-i}$, where $c_k \neq 0$ (page 287).
- The **characteristic equation** of this recurrence relation is $x^k - \sum_{i=1}^{k} c_i x^{k-i} = 0$ (page 287).
- The characteristic roots of a LHRRWCCs can be used to solve the LHRRWCCs (page 288).
- The general solution of a LNHRRWCCs is given by $a_n = a_n^{(h)} + a_n^{(p)}$ (page 294).

Generating Functions

- $g(x) = \sum_{n=0}^{\infty} a_n x^n$ is the **generating function** of the real number sequence $a_0, a_1, a_2, \ldots$ (page 298).
- Generating functions and the partial fraction decomposition rule can be used to solve LHRRWCCs (page 301).

Recursive Algorithms

- A **recursive algorithm** consists of two cases: base case(s) and a general case (page 307).
- **Lamé's Theorem** The euclidean algorithm for computing $\gcd\{a, b\}$ takes no more than five times the number of decimal digits in b, where $a \geq b \geq 2$ (page 323).

Divide-and-Conquer Algorithms

- The recurrence relation of a divide-and-conquer algorithm is of the form $f(n) = af(n/b) + g(n)$ (page 327).

Review Exercises

In Exercises 1 and 2, the nth term a_n of a number sequence is defined recursively. Compute a_5.

1. $a_1 = a_2 = 1, a_3 = 2$

$a_n = a_{n-1} + a_{n-2} + a_{n-3}, n \geq 4$

2. $a_1 = 0, a_2 = a_3 = 1$

   $a_n = a_{n-1} + 2a_{n-2} + 3a_{n-3}, n \geq 4$

3. The number of additions a_n needed to compute the nth Fibonacci number F_n by recursion is given by $a_n = F_n - 1$, $n \geq 1$. Find the recurrence relation satisfied by a_n.

(**A modified handshake problem**) Mr. and Mrs. Matrix hosted a party for n married couples. At the party, each person shook hands with everyone else, except the spouse. Let $h(n)$ denote the total number of handshakes made.

4. Define $h(n)$ recursively.
5. Predict an explicit formula for $h(n)$.
6. Prove the formula obtained in Exercise 5, where $n \geq 1$.

Using the iterative method, predict an explicit formula satisfied by each recurrence relation.

7. $a_1 = 1 \cdot 2$

   $a_n = a_{n-1} + n(n+1), n \geq 2$

8. $a_1 = 2 \cdot 3$

   $a_n = 3a_{n-1}, n \geq 2$

9. $a_1 = 1$

   $a_n = a_{n-1} + 2^{n-1}, n \geq 2$

10. $a_0 = 0$

    $a_n = a_{n-1} + (3n - 1), n \geq 1$

11–14. Using induction, prove the formulas obtained in Exercises 7–10.

Solve each recurrence relation.

15. $a_n = a_{n-1} + a_{n-2}, a_1 = 2, a_2 = 3$
16. $a_n = a_{n-1} + a_{n-2}, a_1 = a_2 = a$
17. $a_n = 2a_{n-1} + 7a_{n-2} - 8a_{n-3} - 12a_{n-4}, a_0 = 4, a_1 = 10, a_2 = 18, a_3 = 58$
18. $a_n = 4a_{n-1} + 2a_{n-2} - 12a_{n-3} - 9a_{n-4}, a_0 = 4, a_1 = 0, a_2 = 4, a_3 = -32$
19. $a_n = 10a_{n-1} - 21a_{n-2} + 5n, a_0 = 0, a_1 = 3$
20. $a_n = 8a_{n-1} - 15a_{n-2} + 4n5^n, a_0 = 1, a_1 = 3$
21. Let a_n denote the number of multiplications (lines 7–10) in Algorithm 5.10. Show that $a_n = O(n)$.

Let c_n denote the number of element comparisons made (line 4) by the recursive bubble sort algorithm in Algorithm 5.9.

22. Define c_n recursively.
23. Solve the recurrence relation.
24. Show that $c_n = O(n^2)$.

Algorithm 5.13 evaluates the polynomial $f(x) = \sum_{i=0}^{n} a_i x^i$ at $x = \alpha$. Use it for Exercises 25–29.

```
Algorithm evaluate poly(f,n,α,answer)
(* This algorithm returns the value of a polynomial f
   of degree n at α in the variable answer. *)
0. Begin (* algorithm *)
1.     answer ← a0
2.     power ← 1
3.     for i = 1 to n do
4.     begin (* for *)
5.         power ← power * α
6.         answer ← answer + ai * power
7.     endfor
8. End (* algorithm *)
```

Algorithm 5.13

Evaluate each polynomial at $x = -1$.

25. $f(x) = x^3 + 2x^2 - 3x + 4$

26. $f(x) = 2x^3 + 5x - 6$

Let c_n denote the number of operations (lines 5–6) required to evaluate a polynomial at $x = \alpha$.

27. Define c_n recursively.

28. Solve the recurrence relation.

29. Show that $c_n = O(n^2)$.

Use **Horner's algorithm** (Algorithm 5.14) to evaluate the polynomial $f(x) = \sum_{i=0}^{n} a_i x^i$ at $x = \alpha$ for Exercises 30–35.

```
Algorithm Horner(f,n,i,α)
(* This algorithm evaluates a polynomial f of degree n at
   x = α by recursion and is invoked by Horner(f,n,0,α). *)
0. Begin (* algorithm *)
1.   if i = n then
2.      Horner ← an
3.   else
4.      Horner ← Horner(f,n,i + 1,α) · α + ai
5. End (* algorithm *)
```

Algorithm 5.14

Evaluate each polynomial at $x = 2$.

30. $f(x) = 3x^2 + 4x - 5$

31. $f(x) = 2x^3 - 5x + 3$

Let b_n denote the number of operations (addition and multiplication) needed in line 4.

32. Define b_n recursively.

33. Solve the recurrence relation.

34. Show that $b_n = O(n)$.

35. Let a_n denote the number of n-bit words that do not contain the pattern 111. Define a_n recursively.

Let a_n denote the number of ways a $2 \times n$ rectangular board can be covered with 2×1 dominoes.

36. Define a_n recursively.
(*Hint*: Consider $2 \times (n-1)$ and $2 \times (n-2)$ boards.)

37. Find an explicit formula for a_n.

Write a recursive algorithm to compute each sum.

38. The sum of the first n even positive integers.

39. The sum of the first n odd positive integers.

40–41. Establish the correctness of the algorithms in Exercises 38 and 39.

42. Write an iterative algorithm to find the minimum and the maximum of a list X of n elements.

Let c_n denote the number of element comparisons made by the minmax algorithm in Exercise 42.

43. Define c_n recursively.

44. Solve the recurrence relation.

45. Show that $b_n = O(n)$.

Prove each, where α and β are the solutions of the equation $x^2 = x + 1$, F_n the nth Fibonacci number, and L_n the nth Lucas number. Identities in Exercises 46–53 were discovered in 1876 by Lucas.

46. $\sum_{i=1}^{n} F_i = F_{n+2} - 1$

47. $\sum_{i=1}^{n} F_{2i-1} = F_{2n}$

48. $\sum_{i=1}^{n} F_{2i} = F_{2n+1} - 1$

49. $\sum_{i=1}^{n} L_i = L_{n+2} - 3$

50. $\sum_{i=1}^{n} L_{2i-1} = L_{2n} - 2$

51. $\sum_{i=1}^{n} L_{2i} = L_{2n+1} - 1$

52. $F_{n+1}^2 + F_n^2 = F_{2n+1}$

53. $F_{n+1}^2 - F_{n-1}^2 = F_{2n}$

54. $\gcd\{F_n, F_{n+1}\} = 1, n \geq 1$

55. $x^n = F_n x + F_{n-1}, n \geq 2$

56. $F_n = \dfrac{\alpha^n - \beta^n}{\alpha - \beta}, n \geq 1$

Let $C(n)$ denote the number of comparisons needed by quicksort to sort a list of n items. In the worst case, $C(n) = C(n-1) + (n-1)$, where $C(0) = 0 = C(1)$.

57. Solve the recurrence relation.

58. Show that $C(n) = O(n^2)$.

(*Note*: This shows that, in the worst case, quicksort is as bad as selection sort.)

Let $A(n)$ denote the average number of comparisons needed by quicksort. Then $A(n) = (n+1) + \frac{1}{n}\sum_{i=1}^{n}[A(i-1) + A(n-i)]$, where $A(0) = 0 = A(1)$. Use this fact to answer Exercises 59 and 60.

***59.** Show that $\frac{A(n)}{n+1} = 2\sum_{i=3}^{n+1}\left(\frac{1}{i}\right)$.

c ***60.** Show that $A(n) = O(n \lg n)$.
(*Hint*: Use integration.)

Supplementary Exercises

A side of the equilateral triangle in Figure 5.20 is n units long. Let a_n denote the number of triangles pointing north.

Figure 5.20

1. Define a_n recursively.

2. Solve the recurrence relation.

The nth **Fermat number** f_n is defined by $f_n = 2^{2^n} + 1$, $n \geq 0$.

3. Prove that $f_{n+1} = f_n^2 - 2f_n + 2$. (J. M. Schram, 1983)

4. Using Exercise 3, compute f_1, f_2, f_3, and f_4.

5. Let a_n be an infinite sequence with $a_1 = 1$, $a_5 = 5$, $a_{12} = 144$, and $a_n + a_{n+3} = 2a_{n+2}$. Prove that $a_n = F_n$. (H. Larson, 1977)

6. Let $\alpha = \frac{1+\sqrt{5}}{2}$ and F_n the nth Fibonacci number. Prove that $\lim_{n\to\infty} \frac{F_{n+1}}{F_n} = \alpha$.

***7.** Let S_n denote the sum of the numbers in the nth term of the sequence of sets of pentagonal numbers $\{1\}$, $\{5, 12\}$, $\{22, 35, 51\}$, $\{70, 92, 117, 145\}, \ldots$. Find a formula for S_n.

***8.** Let S_n denote the sum of the numbers in the nth term of the sequence of sets of Fibonacci numbers $\{1\}$, $\{1, 2\}$, $\{3, 5, 8\}$, $\{13, 21, 34, 55\}, \ldots$. Find a formula for S_n.

Describe the behavior of each number sequence $\{a_n\}$, where $a_0 = a$, $a_1 = b$, and $a_2 = c$ are positive numbers. (R. L. Graham, 1991)

9. $a_{n+2} = (1 + a_{n+1})/a_n$

10. $a_{n+3} = (1 + a_{n+1} + a_{n+2})/a_n$

Let $n \in \mathbb{N}$ and φ Euler's phi-function. Define $\varphi^k = \varphi^{k-1} \circ \varphi$, where $\varphi^1 = \varphi$ and $\circ$ denotes composition. Let $f(n) = \varphi(n) + \varphi^2(n) + \varphi^3(n) + \cdots + \varphi(1)$. (D. L. Silverman, 1981)

11. Compute $f(5)$ and $f(8)$.

12. Prove that if $n = 2^k$, then $f(n) = n$.

13. Prove that $f(n)$ is even. [*Hint*: $\varphi(n)$ is even for $n > 2$.]

14. Consider the sequence of right triangles T_n, $n \geq 1$, with legs A_n and B_n, and hypotenuse C_n such that $A_{n+1} = B_n$ and $B_{n+1} = C_n$. Compute $\lim_{n\to\infty} \frac{B_n}{A_n}$ and $\lim_{n\to\infty} \frac{C_n}{B_n}$. (M. Flavio, 1980)

A set of integers A is **fat** if each of its elements is $\geq |A|$. For example, $\{5, 7, 91\}$ is a fat set, but $\{3, 7, 36, 41\}$ is not. Ø is considered a fat set. Let f_n denote the number of fat subsets of the set $\{1, 2, \ldots, n\}$. (G. F. Andrews)

***15.** Define f_n recursively.

***16.** Find an explicit formula for f_n.

Let $f(n, k)$ denote the number of k-element subsets of the set $S = \{1, 2, \ldots, n\}$ that do not contain consecutive integers. Let f_n denote the total number of subsets of S that do not contain consecutive integers. (I. Kaplansky)

***17.** Define $f(n, k)$ recursively.

***18.** Find an explicit formula for f_n.

Computer Exercises

Write a program to perform each task.

1. Read in a positive integer $n \leq 20$, and print the various moves and the number of moves needed to transfer n disks from peg X to peg Z, using the rules in Example 5.4.
2. Read in a positive integer n, and print the first n triangular and tetrahedral numbers.
3. Print the triangular numbers ≤ 1000 that are perfect squares.
4. Print the triangular numbers ≤ 1000 that are primes.
5. There are eight palindromic triangular numbers < 1000. Find them.
6. Search for two triangular numbers t_n such that t_n and n are palindromic, where $9 \leq n \leq 100$.
7. Read in a positive integer n and print the first n Fibonacci numbers, using recursion and iteration.

8. Read in a positive integer $n \leq 20$ and print the first n Lucas numbers.

9. Read in a positive integer $n \leq 25$ and print the values of $\frac{F_{n+1}}{F_n}$ and $\frac{L_{n+1}}{L_n}$ correct to 10 decimal places, where F_n denotes the nth Fibonacci number and L_n the nth Lucas number.

Read in a list of n positive integers. Use recursion to print each.

10. Their sum, product, maximum, and minimum.

11. The list in the given order.

12. The list in the reverse order.

13. Read in a *key* and search the list for *key*. Print the location if the search is successful; otherwise, print a suitable message.

14. Read in a *key* and a sorted list of n items; determine if *key* occurs in the list using recursion and iteration. Print the location of *key* if the search is successful.

15. Read in a list of n words and determine if each is a palindrome, using recursion.

16. Read in two lists of n integers. Determine if they are identical, using recursion.

17. Read in a nonnegative real number x and a nonnegative integer n; compute the nth power of x.

18. Read in a positive integer $n \leq 100$ and a positive real number $x \leq 2$. Use the binary representation of n and the technique of successive squaring to compute x^n. Print the number of multiplications needed to compute it.

19. Read in a number α, and a polynomial $\sum_{i=0}^{n} a_i x^i$ (that is, coefficients and the corresponding exponents); print the value of the polynomial at α, using Horner's method.

20. Read in n positive integers and print their minimum and maximum, using both iteration and recursion.

21. Read in a positive integer $n \leq 10$ and arrange the Stirling numbers of the second kind $S(n, r)$ in a triangular form, where $1 \leq r \leq n$.

22. Read in n positive integers and sort them using bubble sort, selection sort, and insertion sort. Print the number of element-comparisons needed by each algorithm.

23. Read in n four-letter words. Sort them, using merge sort and quicksort. Print the number of element comparisons needed by each sort.

Exploratory Writing Projects

Using library and Internet resources, write a team report on each of the following in your own words. Provide a well-documented bibliography.

1. Describe the properties of Fibonacci numbers, their occurrences in nature, applications to various disciplines, and relationships to Lucas numbers.
2. Explain how the golden ratio is related to Fibonacci and Lucas numbers. Describe its various occurrences in nature.
3. Describe the various forms of Ackermann's function. Investigate its importance in the study of recursive functions and the analysis of algorithms.
4. Investigate the *Josephus problem*, named for the first century Jewish historian Flavius Josephus (37?–100?).
5. Describe how, using Fibonacci numbers F_n ($n \geq 2$) as bases, nonnegative integers can be represented as binary numbers with no two adjacent 1's. Express the integers 1–25 as such binary numbers.
6. Define continued fractions and describe their relationship to Fibonacci numbers.
7. Describe the *Game of Life*, invented in 1970 by British mathematician John H. Conway, now at Princeton University.
8. Describe the *Game of Halma*, invented in 1883 by George H. Monks, a Harvard Medical School graduate.
9. Examine the history of Catalan numbers and their properties and applications. Include a biography of E. C. Catalan.
10. Write an essay on the Tower of Brahma (Hanoi).
11. Write an essay on Quicksort.
12. Discuss *the fifteen puzzle*, invented by American puzzlist Samuel Loyd (1841–1911).
13. Discuss *Markov chains*, named after Russian mathematician Andrei A. Markov (1856–1922), who developed the theory of stochastic processes, and their applications to business.

Enrichment Readings

1. G. Brassard and P. Bratley, *Algorithmics: Theory & Practice*, Prentice-Hall, Englewood Cliffs, NJ, 1986, pp. 26–34, 48–61.

2. R. P. Grimaldi, *Discrete and Combinatorial Mathematics: An Applied Introduction*, 4th edition, Addison-Wesley, Boston, MA, 1999, pp. 351–403.

3. B. W. Jackson and D. Thro, *Applied Combinatorics with Problem Solving*, Addison-Wesley, Reading, MA, 1990, pp. 226–252.

4. T. Koshy, *Fibonacci and Lucas Numbers with Applications*, Wiley, New York, 2001.

5. C. Oliver, "The Twelve days of Christmas," *Mathematics Teacher*, Vol. 70 (Dec. 1977), pp. 752–754.

6. S. Sahni, *Concepts in Discrete Mathematics*, 2nd ed., Camelot, Fridley, MN, 1985, pp. 205–335.

7. R. Sedgewick, *Algorithms*, 2nd ed., Addison-Wesley, Reading, MA, 1988, pp. 3–189.

8. K. B. Strangeman, "The Sum of *n* Polygonal Numbers," *Mathematics Teacher*, Vol. 67 (Nov. 1974), pp. 655–658.

9. C. W. Trigg, "Palindromic Triangular Numbers," *J. Recreational Mathematics*, Vol. 6 (Spring 1973), pp. 146–147.

10. A. Tucker, *Applied Combinatorics*, Wiley, New York, 1984, pp. 222–298.

11. H. S. Wilf, *Algorithms and Complexity*, Prentice-Hall, Englewood Cliffs, NJ, 1986, pp. 26–34, 48–61.

Chapter 6

Combinatorics and Discrete Probability

The theory of probability affords an excellent illustration of the application of the theory of permutations and combinations which is the fundamental part of the algebra of discrete quantity.

—G. CRYSTAL

Combinatorics is a fascinating branch of discrete mathematics, which deals with the art of counting. Very often we ask the question, *In how many ways can a certain task be done*? Usually combinatorics comes to our rescue. In most cases, listing the possibilities and counting them is the least desirable way of finding the answer to such a problem. Often we are *not* interested in enumerating the possibilities, but rather would like to know the total number of ways the task can be done.

For instance, consider the following combinatorial problem:

One type of automobile license plate number in Massachusetts consists of one letter and five digits.
Find the number of such license plate numbers possible.

• **5HO515** •
• **MASSACHUSETTS** •

Suppose you are willing to list all the possibilities and count them to find the answer. Assuming you know how to enumerate them systematically and that it would take a second to count each, it would take about 6 months to complete the counting alone. Obviously, this is an inefficient way to find the answer, especially when combinatorics can do the job in seconds. (See Example 6.6.)

A few other interesting problems we examine in this chapter are:

- A secretary types up 10 different letters and 10 envelopes. In how many different ways can she place each letter in an envelope so that no letter is placed in the correct envelope?

- Eleven guests would like to order soft drinks with their dinner. There are five choices for a soft drink: Coke Classic, Diet Coke, root beer, Pepsi, and Sprite. Find the number of different beverage selections possible.
- What is the maximum number of nonoverlapping regions formed inside a circle by joining n points on it?
- In how many ways can n married couples be seated at a round table so that men and women sit on alternate chairs and no wife sits next to her husband?

This chapter presents the fundamentals of combinatorics.* In several instances, you will find recursion and generating functions useful in solving combinatorial problems, so review them as needed.

6.1 The Fundamental Counting Principles

This section, a natural continuation of Section 2.4, presents three fundamental principles that form the foundation of combinatorics.

The addition and the inclusion–exclusion principles discussed in Section 2.4 have fine applications to combinatorics, as will be seen shortly.

EXAMPLE 6.1 Find the number of ways of drawing a red queen or a black king from a standard deck of playing cards.

SOLUTION:
Let A denote the set of red queens and B the set of black kings. Clearly, $|A| = 2 = |B|$. Since A and B are disjoint sets, by the addition principle, $|A \cup B| = |A| + |B| = 2 + 2 = 4$. Thus there are four different ways of drawing a red queen or a black king. ■

In this example, drawing a red queen can be considered a **task**, say, task A. Likewise, drawing a black king can be considered task B. Since the two tasks cannot occur simultaneously, they are **mutually exclusive**. So finding the number of ways of drawing a red queen or a black king is equivalent to finding the number of ways task A or task B can be performed.

Accordingly, the addition principle can be restated in terms of tasks as follows.

THEOREM 6.1 **(Addition Principle)** Let A and B be two mutually exclusive tasks. Suppose task A can be done in m ways and task B in n ways. Then task A or task B can take place in $m + n$ ways. ■

*The first book on combinatorics is *Ars Conjectandi*, written by the Swiss mathematician, Jakob Bernoulli (1654–1705) and published posthumously in 1713.

The addition principle can be extended to any finite number of pairwise mutually exclusive tasks, using induction. For instance, let $T_1, \dots, T_n$ be n pairwise mutually exclusive tasks. Suppose task T_i can be done in m_i ways, where $1 \le i \le n$. Then task $T_1, T_2, \dots$, or T_n can be done in $m_1 + m_2 + \cdots + m_n$ ways, as the next example illustrates.

EXAMPLE 6.2 A freshman has selected four courses and needs one more course for the next term. There are 15 courses in English, 10 in French, and 6 in German she is eligible to take. In how many ways can she choose the fifth course?

SOLUTION:
Let E be the task of selecting a course in English, F the task of selecting a course in French, and G that of selecting a course in German. These tasks can be done in 15, 10, and 6 ways, respectively, and are mutually exclusive, so, by the addition principle, the fifth course can be selected in $|E| + |F| + |G| = 15 + 10 + 6 = 31$ ways. ■

Like the addition principle, the inclusion–exclusion principle can be restated in terms of tasks in an obvious way. It can also be extended to a finite number of tasks (see Theorem 6.19).

THEOREM 6.2 **(Inclusion–Exclusion Principle)** Suppose a task A can be done in m ways, task B in n ways, and both can be accomplished in k different ways. Then task A or B can be done in $m + n - k$ ways. ■

The next problem exemplifies this.

EXAMPLE 6.3 In how many ways can you deal a king or a black card from a standard deck of cards?

SOLUTION:
A king can be selected in four different ways and a black card in 26 different ways. These two tasks can be done simultaneously in two ways, namely, by selecting a black king; so, by Theorem 6.2, a king or a black card can be selected in $4 + 26 - 2 = 28$ ways. ■

Before stating the next counting principle, let us return to Example 2.23 in Chapter 2. The task of selecting a mode of transportation for the trip from Boston to London via New York consists of two subtasks A and B: A is selecting a mode of transportation from Boston to New York — car, plane, or ship — and B is selecting a mode of transportation from New York to London — plane or ship. They can be done in $|A| = 3$ and $|B| = 2$ ways. Recall that the trip can be made in $6 = |A| \cdot |B|$ ways; that is, $|A \times B| = 6 = |A| \cdot |B|$.

More generally, we have the following result.

THEOREM 6.3 **(Multiplication Principle)** Suppose a task T is made up of two subtasks, subtask T_1 followed by subtask T_2. If subtask T_1 can be done in m_1 ways and subtask T_2 in m_2 different ways for each way subtask T_1 can be done, then task T can be done in $m_1 m_2$ ways. ■

The next four examples illustrate this principle.

EXAMPLE 6.4 Find the number of two-letter words that begin with a vowel—a, e, i, o, or u.

SOLUTION:
The task of forming a two-letter word consists of two subtasks T_1 and T_2: T_1 consists of selecting the first letter and T_2 selecting the second letter, as Figure 6.1 shows.

Figure 6.1

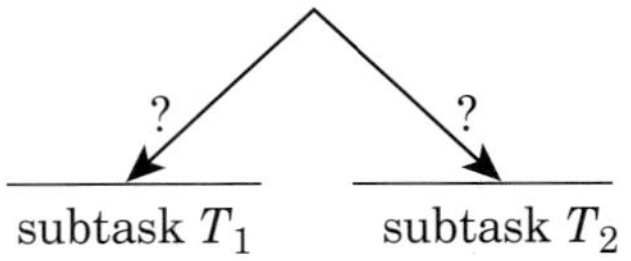

Since each word must begin with a vowel, T_1 can be accomplished in five ways. There are no restrictions on the choice of the second letter, so T_2 can be done in 26 ways (see Figure 6.2). Therefore, by the multiplication principle, the task can be performed in $5 \cdot 26 = 130$ different ways. In other words, 130 two-letter words begin with a vowel.

Figure 6.2

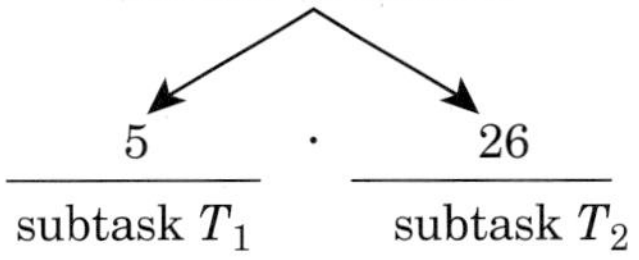

■

The various two-letter words in this example can be enumerated systematically by constructing a tree diagram, as in Figure 6.3. All desired words can be obtained by traversing the various branches of the tree, as indicated.

Figure 6.3

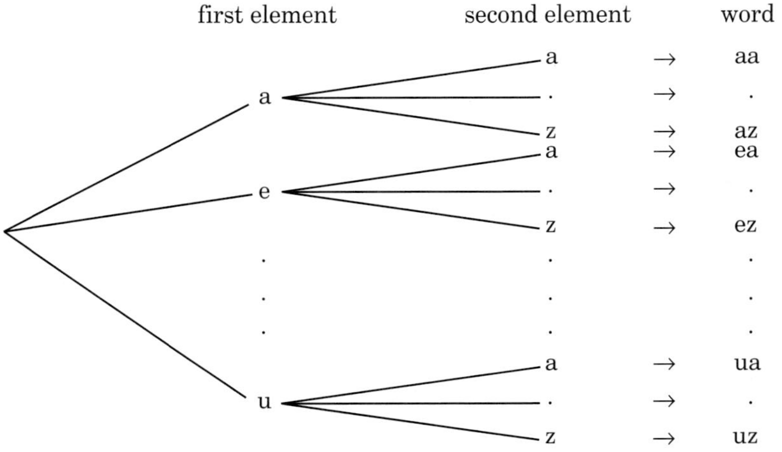

The multiplication principle can also be extended to any finite number of subtasks. Suppose a task T can be done by n successive subtasks, $T_1, T_2, \ldots, T_n$. If subtask T_i can be done in m_i different ways after T_{i-1} has been completed, where $1 \le i \le n$, then task T can be done in $m_1 m_2 \cdots m_n$ ways.

The multiplication principle can be applied to prove that a set with size n has 2^n subsets, as shown below.

EXAMPLE 6.5 Show that a set S with n elements has 2^n subsets.

SOLUTION:
Every subset of S can be uniquely identified by an n-bit word (see Figure 6.4). The task of forming an n-bit word can be broken down to n subtasks: selecting a bit for each of the n positions. Each position in the word

Figure 6.4

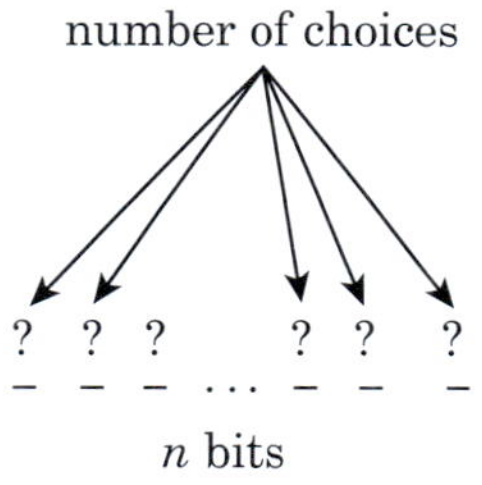

has two choices, 0 or 1; so, by the multiplication principle, the total number of n-bit words that can be formed is $\underbrace{2 \cdot 2 \cdot \cdots \cdot 2}_{n \text{ times}} = 2^n$ (see Figure 6.5). In other words, S has 2^n subsets.

Figure 6.5

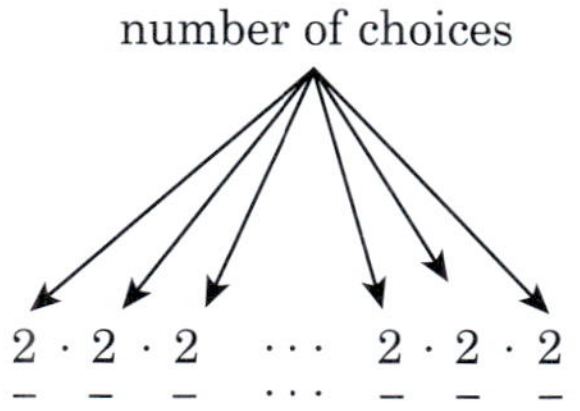

■

We now turn to solving the problem posed at the beginning of the chapter.

EXAMPLE 6.6 One type of automobile license plate number in Massachusetts consists of one letter and five digits. Compute the number of such license plate numbers possible.

SOLUTION:
For convenience, we decompose the task into three subtasks:

- Choosing a letter. It can be done in 26 ways.
- Choosing the position of the letter. It has six possible slots.

- Choosing the five digits. They can be selected in $10 \cdot 10 \cdot 10 \cdot 10 \cdot 10 = 100{,}000$ ways.

Now we are ready to find the final answer. By the multiplication principle, the total number of license plates is $26 \cdot 6 \cdot 100000 = 15{,}600{,}000$. ■

The next example depends on the multiplication and addition principles.

EXAMPLE 6.7 (optional) An identifier in a programming language consists of a letter followed by alphanumeric characters.* Find the number of legal identifiers of length at most 10.

SOLUTION:
Let S_i denote the set of identifiers of length i, where $1 \le i \le 10$. Then $|S_i| = 26 \cdot 36^{i-1}$ (see Figure 6.6). Since the subtasks $S_1, \ldots, S_{10}$ are mutually

Figure 6.6

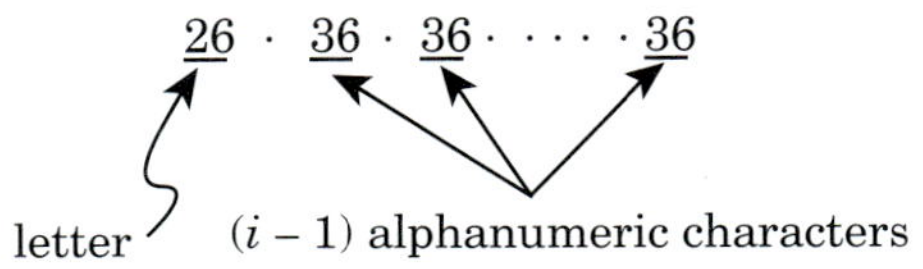

exclusive, by the addition principle, the total number of identifiers of length ≤ 10 is given by

$$\sum_{i=1}^{10} |S_i| = \sum_{i=1}^{10} 26 \cdot 36^{i-1} = 26\left(\sum_{i=0}^{9} 36^i\right)$$
$$= 26 \cdot \frac{(36^{10} - 1)}{36 - 1} = \frac{26(36^{10} - 1)}{35}$$
$$= 2{,}716{,}003{,}412{,}618{,}210$$
$$\approx 2.7 \text{ quadrillion!}$$
■

The final example in this opening section employs the multiplication and the inclusion–exclusion principles.

EXAMPLE 6.8 An eight-bit word is called a **byte**. Find the number of bytes with their second bit 0 or the third bit 1.

SOLUTION:

Number of bytes with second bit $0 = 2 \cdot 1 \cdot 2 \cdot 2 \cdot 2 \cdot 2 \cdot 2 \cdot 2 = 2^7$

Number of bytes with third bit $1 = 2 \cdot 2 \cdot 1 \cdot 2 \cdot 2 \cdot 2 \cdot 2 \cdot 2 = 2^7$

Since these two subtasks are *not* mutually exclusive, we cannot add these two partial answers and claim that the answer is $2^7 + 2^7 = 128 + 128 = 256$.

---

*An **alphanumeric** character is a letter or a digit.

So, we must find the number of bytes that have both properties. The number of bytes with second bit 0 and third bit 1 equals $2 \cdot 1 \cdot 1 \cdot 2 \cdot 2 \cdot 2 \cdot 2 \cdot 2 = 2^6 = 64$, so, by the inclusion–exclusion principle, the number of bytes with the given properties is $2^7 + 2^7 - 2^6 = 128 + 128 - 64 = 192$. ■

Exercises 6.1

Find the number of positive integers ≤ 1976 and divisible by:

1. 2 or 3. **2.** 3 or 5. **3.** 2, 3, or 5. **4.** 3, 5, or 7.

○ **5.** In one version of BASIC, a variable name consists of a letter, or a letter followed by a digit, or the dollar sign ($). Find the total number of possible variable names.

Find the number of terms in the expansion of each expression.

6. $(a+b)(c+d+e)(x+y)$ **7.** $(b+c)(d+e+f)(x+y+z)$

8. $\left(\sum_{i=0}^{2} a_i\right)\left(\sum_{i=1}^{4} b_i\right)\left(\sum_{i=2}^{5} c_i\right)$ **9.** $\left(\sum_{i=-2}^{5} a_i\right)\left(\sum_{i=-1}^{3} b_i\right)\left(\sum_{i=0}^{4} c_i\right)$

10. Find the number of palindromes of length n over the English alphabet.

○ **11.** Find the number of palindromic alphanumeric identifiers of length n. (See Example 6.7.)

12. Find the total number of bytes. (See Example 6.8.)

13. A word over the alphabet $\{0, 1, 2\}$ is called a **ternary** word. Find the number of ternary words of length n that can be formed.

A typical automobile license plate in New York contains three letters followed by three digits. Find the number of license plates of this kind that:

14. Can be formed. **15.** Begin with the word BAT.

16. Begin with a vowel. **17.** Begin with the digit 6.

18. Repeat no letters or digits.

19. Contain the same letters and the same digits.

20. Have the property that both words and numbers are palindromes.

An old zip code in the United States consists of five digits. Find the total number of possible zip codes that:

21. Have no repetitions. **22.** Begin with 0.

23. End in K. **24.** Are palindromes.

A zip code in Canada consists of three letters and three digits. Each zip code begins with a letter. The letters and digits alternate; for instance, A1B2C3. Find the number of zip codes that:

25. End in 6.

26. Begin with A and end in 3.

27. End in Z.

28. Are possible.

○ The password for a computer system consists of six alphanumeric characters and begins with a letter. Find the total number of passwords that:

29. Are possible.

30. End in RED.

31. Contain the word BAT.

32. End in 2076.

33. Every radio and television station in the United States has a unique call name. Each contains three or four letters, beginning with K or W. For example, KEY and WASP are legal call names. Find the number of possible call names.

Find the number of bytes that:

34. Begin with 101.

35. End with 110.

36. Begin with and end in the same bit.

37. Begin with and end in different bits.

38. Have the same third and fourth bits.

39. Have third bit or fifth bit 1.

40. Have second bit 0 or third bit 1.

41. Are palindromes.

Find the number of positive divisors of the following positive integers.

42. $2^i \cdot 3^j \cdot 5^k$

43. 600

44. A number-theoretic function used in the study of perfect numbers is the **tau function** τ on $\mathbb{N}$. (τ is the Greek letter, tau.) $\tau(n)$ denotes the number of positive divisors of $n \in \mathbb{N}$. Let $n = p_1^{e_1} p_2^{e_2} \cdots p_k^{e_k}$, where $p_1, p_2, \ldots, p_k$ are distinct primes and $e_1, e_2, \ldots, e_k \in \mathbf{W}$. Find $\tau(n)$.

45. Find the number of $n \times n$ matrices that can be constructed using bits.

Find the number of ternary words that have:

46. Length at most 3.

47. Length at most 5.

48. Length at most n.

49. Length 3 and are palindromes.

50. Length 4 and are palindromes.

51. $4 \leq$ length ≤ 6

In an alphabet of m characters, how many words have:

52. Length 3?

53. Length not more than 2?

54. Length at least 2, but not more than 4?

55. Length not more than n?

Let A and B be two finite sets with $|A| = m$ and $|B| = n$. How many:

56. Functions can be defined from A to B?

57. Bijections can be defined from A to B (assume $m = n$)?

58. Invertible functions can be defined from A to B (assume $m = n$)?

59. Injections can be defined from A to B (assume $m \leq n$)?

Let τ denote the tau function. Prove each.

60. $\tau(n)$ is odd if and only if n is a square.

***61.** If m and n are relatively prime numbers, then $\tau(mn) = \tau(m) \cdot \tau(n)$.

***62.** The **harmonic mean** m of the numbers $a_1, a_2, \ldots, a_n$ is the reciprocal of the arithmetic mean of their reciprocals; that is,

$$\frac{1}{m} = \frac{1}{n} \sum_{i=1}^{n} \left(\frac{1}{a_i}\right)$$

Prove that the harmonic mean of the positive factors of a perfect number N is an integer.
(Hint: If d is a factor of N, then so is N/d.) (R. Euler, 1987)

6.2 Permutations

The concept of ordered arrangements was familiar to Chinese mathematicians as early as 1100 B.C. This section introduces the concept of ordered arrangements and shows how to find the number of such arrangements.

Suppose a photographer would like to arrange 10 cats in a row for a television commercial. She can arrange them in any order she likes. In how many ways can she accomplish this? Although we could use the multiplication principle to arrive at an answer, we shall apply the concept of a permutation (see Example 6.11).

Recall that the elements of an ordered set are assigned unique positions. For convenience, let us denote the ordered set (a,b,c) as the arrangement *abc*. The words *acb* and *bac* are two different arrangements of the very same letters. (Remember, the order makes a difference.) Each of these arrangements is a **permutation** of the three letters taken all at a time, or a **3-permutation**.

Permutation

A **permutation** of a set of n (distinct) elements taken r $(0 \leq r \leq n)$ at a time is an arrangement of r elements of the set. For convenience, it is called an ***r*-permutation**. If $r = n$, then the r-permutation is called simply a **permutation**. The number of r-permutations of a set of size n is denoted by $P(n, r)$.

We begin our discussion with a simple example.

EXAMPLE 6.9 Find the number of permutations; that is, 3-permutations of the elements of the set $\{a, b, c\}$.

SOLUTION:
By the multiplication principle, the number of 3-permutations of three elements is $3 \cdot 2 \cdot 1 = 6$. Thus $P(3, 3) = 6$. ■

The various permutations in Example 6.9 can be obtained systematically using a tree diagram, as Figure 6.7 shows. They are abc, acb, bac, bca, cab, and cba.

Figure 6.7

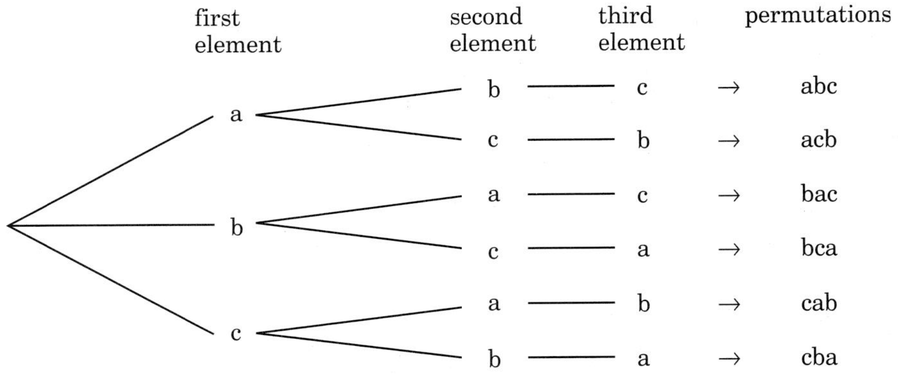

EXAMPLE 6.10 Find the number of 2-permutations of the elements of the set $\{a, b, c\}$.

SOLUTION:
Again, by the multiplication principle, the number of 2-permutations is $3 \cdot 2 = 6$; that is, $P(3, 2) = 6$. ■

The various 2-permutations in Example 6.10 are ab, ac, ba, bc, ca, and cb. They can be obtained using the tree diagram in Figure 6.8.

Examples 6.9 and 6.10 can be interpreted as follows: Suppose you have three books in your hands and would like to arrange them in a bookcase. If there is enough room for three books, they can be arranged in $3 \cdot 2 \cdot 1 = 6$ different ways. On the other hand, if there is room for only two books, they can be arranged in $3 \cdot 2 = 6$ different ways.

Figure 6.8

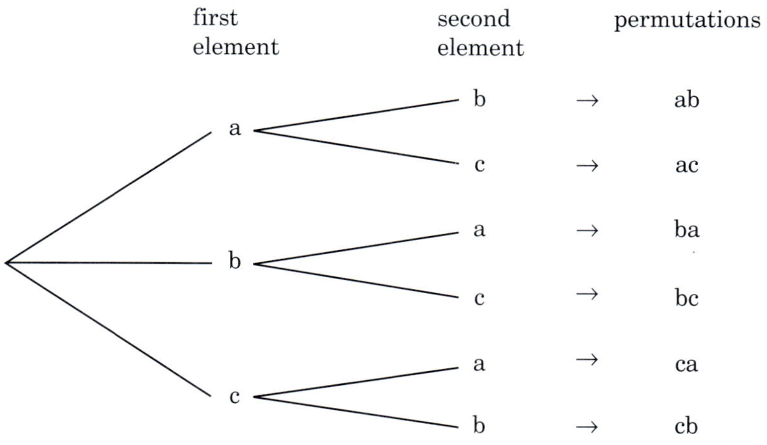

More generally, we have the following result.

THEOREM 6.4 The number of r-permutations of a set of n (distinct) elements is given by $P(n,r) = \dfrac{n!}{(n-r)!}$.

PROOF:
Since there are n elements, the first element can be chosen in n ways. Now $n-1$ elements are left; so the second element can be chosen in $n-1$ ways. Continue like this until the rth element is ready to be chosen. At this point there are $n-r+1$ elements left. Consequently, the rth element can be chosen in $n-r+1$ ways. Thus, by the multiplication principle,

$$\begin{aligned} P(n,r) &= n(n-1)(n-2)\cdots(n-r+1) \\ &= \frac{n(n-1)\cdots(n-r+1)(n-r)\cdots 2\cdot 1}{(n-r)\cdots 2\cdot 1} \\ &= \frac{n!}{(n-r)!} \end{aligned}$$

■

Although it is easy to remember the value of $P(n,r)$ using this formula, $P(n,r)$ is often computed using the formula $P(n,r) = n(n-1)(n-2)\cdots(n-r+1)$. The values $n!$ and $(n-r)!$ may be too large even for a calculator to compute. Then the value $n!/(n-r)!$ may not be exact. You will find the fact that $n! = n(n-1)!$ useful in computing $P(n,r)$. For example,

$$\begin{aligned} P(25,5) &= \frac{25!}{(25-5)!} = \frac{25!}{20!} = \frac{25\cdot 24\cdot 23\cdot 22\cdot 21\cdot 20!}{20!} \\ &= 25\cdot 24\cdot 23\cdot 22\cdot 21 = 6{,}375{,}600 \end{aligned}$$

Suppose we let $r = n$ in Theorem 6.4. Then

$$P(n,n) = \frac{n!}{(n-n)!} = \frac{n!}{0!} = \frac{n!}{1} = n!$$

Accordingly, we have the following result.

THEOREM 6.5 The number of permutations of a set of size n is given by $P(n,n) = n!$. That is, n elements can be arranged in $n!$ ways. ■

Interestingly enough, applications of this formula appear in the anonymous Hebrew book *Sefer Yetzirah (The Book of Creation)*, written between 300 and 600 A.D.

The next two examples illustrate this theorem.

EXAMPLE 6.11 A photographer would like to arrange 10 cats for a television commercial. How many ways can she arrange them in a row?

SOLUTION:
Since all the cats have to be in the commercial at the same time, $r = n = 10$. Therefore, the number of possible arrangements is $P(10,10) = 10! = 3{,}628{,}800$. ■

EXAMPLE 6.12 Find the number of words that can be formed by scrambling the letters of the word SCRAMBLE. (Remember, a word is just an arrangement of symbols; it need not make sense.)

SOLUTION:
The word SCRAMBLE contains eight distinct letters. Therefore, the number of words that can be formed equals the number of arrangements of the letters in the word, namely, $P(8,8) = 8! = 40{,}320$. ■

The next example uses the multiplication principle, as well as Theorem 6.5.

EXAMPLE 6.13 A salesperson at a computer store would like to display six models of personal computers, five models of computer monitors, and four models of keyboards. In how many different ways can he arrange them in a row if items of the same family are to be next to each other?

SOLUTION:
There are three types of items: personal computers, monitors, and keyboards. Think of the items in each family as *tied together* into one unit. These families can be arranged in $P(3,3) = 3!$ ways. Now the items within each family can be rearranged. The six models of personal computers can be arranged in $P(6,6) = 6!$ ways, the monitors in $P(5,5) = 5!$ ways, and the keyboards in $P(4,4) = 4!$ different ways. Thus, by the multiplication principle, the total number of possible arrangements is $3!\,6!\,5!\,4! = 12{,}441{,}600$. ■

Permutations of elements arranged in a circle work somewhat differently from the linear situations we have investigated up to this point.

Cyclic Permutation

In how many different ways can you place five beads on a necklace? The answer is *not* $5! = 120$, but far less, since it contains a lot of duplicate arrangements. For instance, the two circular arrangements shown in Figure 6.9 are identical. (Look at the relative positions of the beads p_1 through p_5.) Each circular arrangement is a **cyclic permutation**.

Figure 6.9

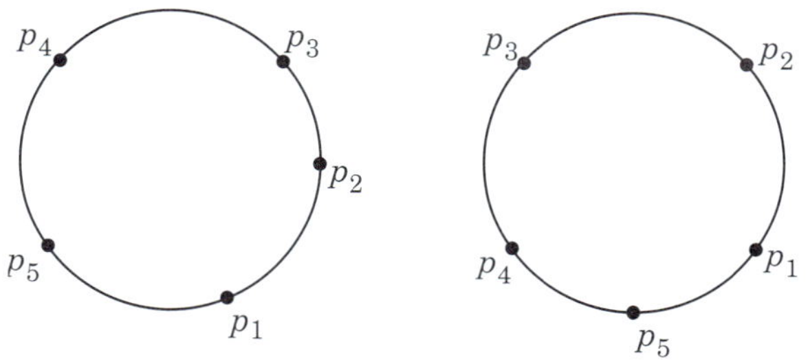

Before we find the number of cyclic permutations of the five beads in Example 6.14, the following general result will be useful to prove.

THEOREM 6.6 The number of cyclic permutations of n (distinct) items is $(n-1)!$.

PROOF:
To avoid duplicates, let us assign a fixed position to the first item a_1 around the circle (see Figure 6.10). Now $n-1$ positions are left. So the second item a_2 can be placed in any one of the $n-1$ positions. Now $n-2$ positions are left. Therefore, the third item a_3 can be placed in any of the $n-2$ positions. Continue like this until all items have been placed. Thus, by the multiplication principle, the number of cyclic permutations is $1 \cdot (n-1)(n-2) \cdots 2 \cdot 1 = (n-1)!$ ■

Figure 6.10

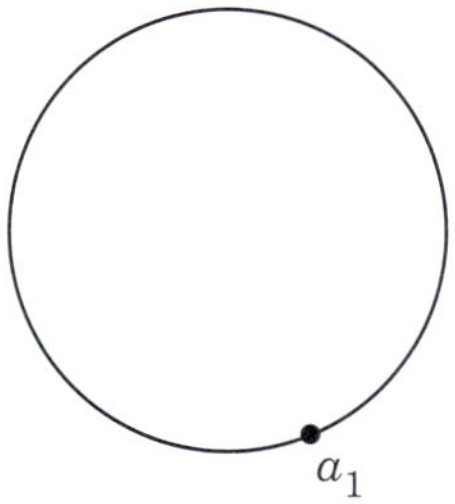

The next example illustrates this result.

EXAMPLE 6.14 Find the number of different ways five zinnias can be planted in a circle.

SOLUTION:

$$\begin{pmatrix}\text{Number of ways of planting}\\ \text{five zinnias in a circle}\end{pmatrix} = \begin{pmatrix}\text{Number of cyclic permutations}\\ \text{of five items}\end{pmatrix}$$
$$= (5-1)! = 24$$ ■

Let us return to "linear" permutations. We would like to find the recurrence relation satisfied by $P(n,r)$. We will use a combinatorial argument to find it, leaving the straightforward algebraic proof as a routine exercise (see Exercise 45).

THEOREM 6.7 The number of r-permutations of n distinct elements satisfies the recurrence relation $P(n,r) = P(n-1,r) + rP(n-1,r-1)$, where $0 < r < n$.

PROOF:
Let X be a set with n elements and x an arbitrary element in it. The set of r-permutations of X can be partitioned into two subsets: A, the set of permutations *not* containing x, and B, the set of permutations containing x.

- *To find the number of elements in* A: Since no permutations in A contain x, every element in A is an r-permutation of $n-1$ elements. The number of such permutations is $P(n-1,r)$.
- *To find the number of elements in* B: Since every permutation in B contains x, $n-1$ candidates are left in X for the remaining $r-1$ positions. They can be arranged in $P(n-1,r-1)$ ways. Now the position of x in a permutation has r choices. Therefore, by the multiplication principle, $rP(n-1,r-1)$ permutations contain x.

Since A and B are disjoint sets, by the addition principle,

$$P(n,r) = |A| + |B|$$
$$= P(n-1,r) + rP(n-1,r-1)$$ ■

> Note that it's much easier to compute $P(n,r)$ using the explicit formula in Theorem 6.4 rather than by using the recursive. Try $P(5,3)$ both ways to see the difference.

Fibonacci Numbers Revisited

The following example presents an interesting confluence of permutations and Fibonacci numbers.

EXAMPLE 6.15 Let p_n denote the number of permutations f of the set $S_n = \{1,2,\dots,n\}$ such that $|i - f(i)| \leq 1$ for all $1 \leq i \leq n$, where $n \geq 1$. So p_n counts the number of permutations that move each element no more than one position from its natural position.

Figure 6.11 shows the various such permutations for $n = 1, 2, 3$, and 4.

Figure 6.11

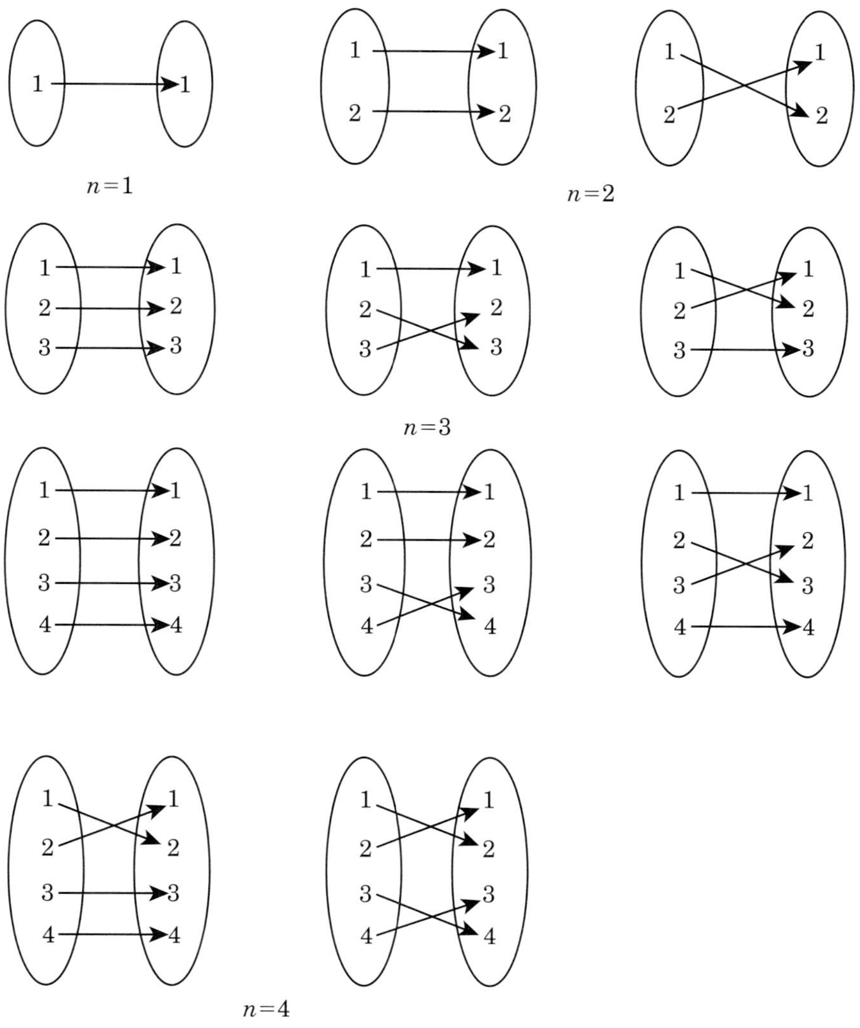

Table 6.1 summarizes these data. It appears from the table that $p_n = F_{n+1}$. We can in fact confirm this.

Table 6.1

n	1	2	3	4	...	n
p_n	1	2	3	5	...	?

Case 1 Let $f(n) = n$. Then the remaining $n - 1$ elements can be used to form p_{n-1} permutations such that $|i - f(i)| \leq 1$ for all i.

Case 2 Let $f(n) \neq n$. Then $f(n) = n - 1$ and $f(n - 1) = n$. The remaining $n - 2$ elements can be employed to form p_{n-2} permutations with the desired property.

Thus, by the addition principle, $p_n = p_{n-1} + p_{n-2}$, where $p_1 = 1$ and $p_2 = 2$. It now follows that $p_n = F_{n+1}$, where $n \geq 1$, as conjectured. ■

Since the total number of permutations of S_n is $n!$, it follows from this example that there are $n! - F_{n+1}$ permutations f of S_n such that $|i - f(i)| > 1$ for some integer i, where $1 \leq i \leq n$. Thus there are $n! - F_{n+1}$ permutations of S_n that move at least one element of S_n by two spaces from its natural position.

In particular, there are $3! - F_4 = 3$ such permutations of the set $\{1, 2, 3\}$, as Figure 6.12 depicts.

Figure 6.12

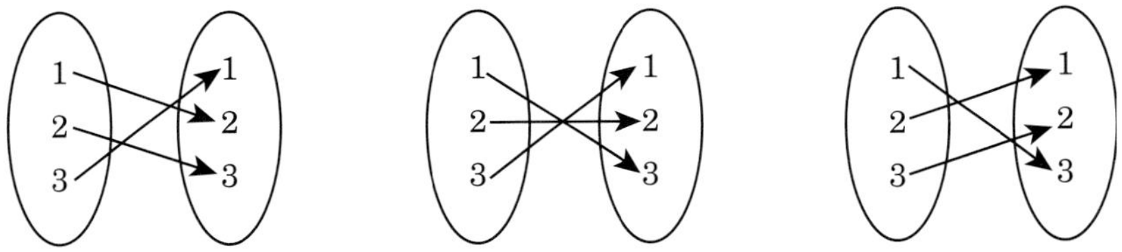

Exercises 6.2

Evaluate each.

1. $\dfrac{5!}{4!}$ **2.** $\dfrac{10!}{3!\,7!}$ **3.** $P(5,3)$ **4.** $P(6,6)$

Mark each sentence as true or false, where n is an arbitrary nonnegative integer and $0 \leq r \leq n$.

5. $0! = 0$

6. $1! = 1$

7. $5 \cdot 4! = 5!$

8. $(m + n)! = m! + n!$

9. $(2 + 3)! = 2! + 3!$

10. $(mn)! = m!\,n!$

11. $(2 \cdot 3)! = 2!\,3!$

12. $n(n - 1)! = n!$

13. $P(n, 0) = 0$

14. $P(n, 1) = P(n, n - 1)$

15. $P(n, r) = P(n, n - r)$

16. $n!$ is divisible by 10 if $n > 4$.

17. Find the number of two-digit numerals that can be formed using the digits 2, 3, 5, 6, and, 9 and that contain no repeated digits.

18. Find the number of three-digit numerals that can be formed using the digits 2, 3, 5, 6, and 9, if repetitions are not allowed.

Find the number of words that can be formed by scrambling the letters in each word.

19. algorithm **20.** word **21.** computer **22.** logic

○ The password for a computer system consists of eight distinct alphabetic characters. Find the number of passwords possible that:

23. End in the string MATH. **24.** Begin with the string CREAM.

25. Contain the word COMPUTER as a substring.

26. Contain the letters WORD together, but in any order.

27. Contain the string BLACK or the string WHITE.

28. Contain the strings BLACK and WHITE.

29. Do not contain the string SAMPLE.

A botanist would like to plant three coleus, four zinnias, and five dahlias in a row in her front garden. How many ways can she plant them if:

30. They can be planted in any order.

31. Plants of the same family must be next to each other.

32. The family of zinnias must be in between the other two families.

Find the number of ways seven boys and three girls can be seated in a row if:

33. A boy sits at each end of the row.

34. A girl sits at each end of the row.

35. The girls sit together at one end of the row.

36. Show that $P(n,0) = 1$.

Using the recursive definition of $P(n,r)$, evaluate each.

37. $P(5,4)$ **38.** $P(6,0)$ **39.** $P(3,2)$ **40.** $P(6,3)$

Solve each equation.

41. $P(n,1) = 6$ **42.** $P(n,2) = 42$

43. $P(n,n-1) = 5040$ **44.** $P(5,r) = 20$

45. Using Theorem 6.4, prove that $P(n,r) = P(n-1,r) + rP(n-1,r-1)$.

Verify each.

46. $(n+1)! + n! = (n+2)n!$ **47.** $(n+1)! - n! = n(n!)$

48. Prove by induction that $1 \cdot 1! + 2 \cdot 2! + \cdots + n \cdot n! = (n+1)! - 1$, $n \geq 1$.

49. Write an algorithm to compute $P(n,r)$, using Theorem 6.4.

50. Write a recursive algorithm to compute $P(n,r)$.

***51.** Show that $(n!)! > (2n)!$, if $n > 3$.

6.3 Derangements

At the beginning of the 18th century, the following problem was proposed:

> A secretary had written n different letters and addressed n different envelopes for them. Unfortunately, a wind storm mixed up the letters and the envelopes. After the storm was over, each letter was placed in an envelope. In how many ways can the letters be placed in the envelopes, so that every letter is in a wrong envelope?

This problem has several variations. One involves n couples attending a dance. In how many ways can the men dance with women other than their own wives?

A second variation involves n guests checking in their coats at the coat room of a fancy restaurant. In how many ways can the attendant return their coats, so no person gets the right coat?

Before answering these problems, we make the following definition.

Derangement

A permutation of n distinct items $a_1, a_2, \ldots, a_n$ in which no item a_i appears in its original position i for any i, $1 \leq i \leq n$, is called a **derangement.**

We would like to find the number of possible derangements of n items, so we begin with an example.

EXAMPLE 6.16 Find the number of derangements of the elements 1, 2, 3, and 4.

SOLUTION:
There are nine derangements of the four elements, namely:

2143	3142	4123
2341	3412	4312
2413	3421	4321

The permutation 2314 is not a derangement since 4 appears in its natural position. ■

Let D_n denote the number of derangements of n items. Then $D_0 = 1$, since there is one derangement with no elements. (This will be verified later.) There is no derangement with one element, so $D_1 = 0$. There is exactly one derangement of the elements 1 and 2, namely, 21; therefore, $D_2 = 1$. There are two derangements of the elements 1, 2, and 3: 231 and 312; therefore, $D_3 = 2$. It follows by Example 6.15 that $D_4 = 9$.

To find an explicit formula for D_n, first we derive a recurrence relation satisfied by D_n.

THEOREM 6.8 Let D_n denote the number of derangements of n distinct items. Then

$$D_n = (n-1)(D_{n-1} + D_{n-2}), \quad n \geq 2 \tag{6.1}$$

PROOF:

Let $a_1, a_2, \ldots, a_n$ denote the n items. Item a_n can be placed in any one of the positions $1, 2, \ldots, (n-1)$, so the location of a_n has $n-1$ choices.

Suppose it is placed in position 1 (see Figure 6.13). Let us now consider the following two cases:

Figure 6.13

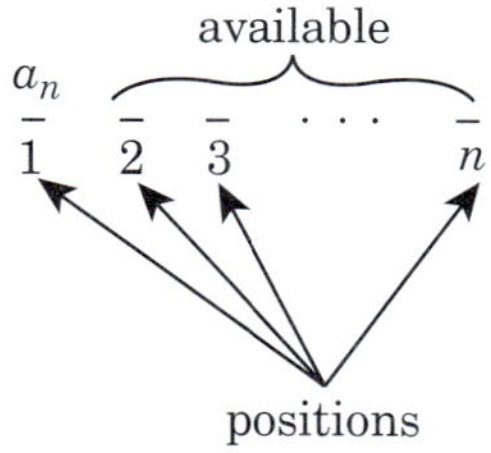

Case 1 Suppose a_1 is placed in position n (see Figure 6.14). Then $n-2$ items are left, namely, $a_2, a_3, \ldots, a_{n-1}$ and also $n-2$ positions, namely, positions 2 through $n-1$. The number of derangements of $n-2$ items, by definition, is D_{n-2}. Thus, if a_1 is placed in position n, there are D_{n-2} derangements.

Figure 6.14

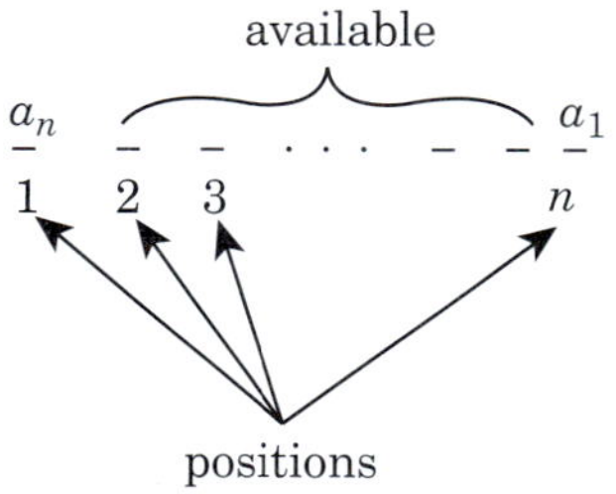

Case 2 Suppose a_1 is *not* placed in position n (see Figure 6.15). Then a_1 must occupy one of the positions 2 through $n - 1$ and one of the items a_2 through a_{n-1} must occupy position n. Thus, we have $n - 1$ items and $n - 1$ positions, yielding D_{n-1} derangements.

Figure 6.15

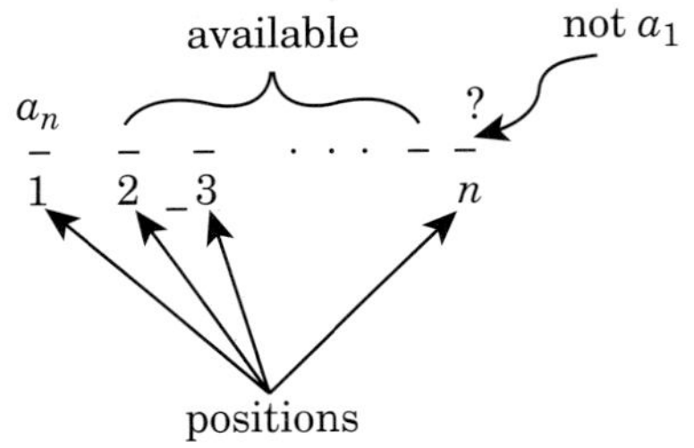

Thus, with a_n in position 1, a total of $D_{n-1} + D_{n-2}$ derangements are possible, by the addition principle. Since a_n may occupy any one of the $n - 1$ positions, the total number of derangements is $(n - 1)(D_{n-1} + D_{n-2})$. Thus $D_n = (n - 1)(D_{n-1} + D_{n-2})$, $n \geq 2$. ■

Using Theorem 6.8, D_n can be defined recursively as follows.

A Recursive Definition of D_n

$$D_0 = 1, \quad D_1 = 0$$

$$D_n = (n - 1)(D_{n-1} + D_{n-2}), \quad n \geq 2 \tag{6.1}$$

> The recurrence relation (6.1) can prove that D_0 must be 1. Using the recurrence relation, $D_2 = (2 - 1)(D_1 + D_0)$. Since $D_2 = 1$ and $D_1 = 0$, this yields $1 = 1(0 + D_0)$. For this to be true, D_0 must be 1, as in the above definition.

The next example illustrates this recursive definition.

EXAMPLE 6.17 Use the recursive definition of D_n to compute D_3 and D_4.

SOLUTION:

$$\begin{aligned} D_3 &= (3 - 1)(D_2 + D_1) \\ &= 2(1 + 0) \\ &= 2 \end{aligned} \qquad \begin{aligned} D_4 &= (4 - 1)(D_3 + D_2) \\ &= 3(2 + 1) \\ &= 9 \end{aligned}$$

Notice that these answers agree with those obtained earlier. ■

The above recursive definition can be used to derive an alternate definition of D_n.

Notice that the recurrence relation (6.1) can be rewritten as

$$D_n - nD_{n-1} = (-1)[D_{n-1} - (n-1)D_{n-2}]$$

To simplify this, we substitute $d_n = D_n - nD_{n-1}$. Then it becomes $d_n = -d_{n-1}$, where $d_1 = D_1 - 1D_0 = 0 - 1(1) = -1$.

Solving this recurrence relation (see Exercise 9),

$$d_n = (-1)^n, \quad n \geq 1$$

Thus

$$D_n - nD_{n-1} = (-1)^n$$

That is,

$$D_n = nD_{n-1} + (-1)^n, \quad n \geq 1 \tag{6.2}$$

Accordingly, D_n can be defined recursively as follows:

An Alternate Recursive Definition of D_n

$$D_0 = 1$$
$$D_n = nD_{n-1} + (-1)^n, \quad n \geq 1 \tag{6.3}$$

The next example uses this alternate definition.

EXAMPLE 6.18 With the alternate definition (6.3), compute D_3 and D_4.

SOLUTION:

$$\begin{aligned} D_3 &= 3D_2 + (-1)^3 \\ &= 3(1) + (-1) \\ &= 2 \end{aligned} \qquad \begin{aligned} D_4 &= 4D_3 + (-1)^4 \\ &= 4(2) + 1 \\ &= 9 \end{aligned}$$

■

The recurrence relation (6.2) can be solved using the iteration method (see Exercise 10). The solution is given in the following theorem.

THEOREM 6.9 The number of derangements of n distinct elements is

$$D_n = n!\left[1 - \frac{1}{1!} + \frac{1}{2!} - \frac{1}{3!} + \frac{1}{4!} - \cdots + \frac{(-1)^n}{n!}\right], \quad n \geq 0$$

■

An interesting observation: It is shown in calculus that $e^{-1} = \sum_{n=0}^{\infty} \frac{(-1)^n}{n!}$, so the expression inside the brackets in the formula is the sum of the first $(n+1)$ terms in the expansion of e^{-1}. See Exercise 40.

The next example illustrates this theorem.

EXAMPLE 6.19 Using Theorem 6.9, compute D_5.

SOLUTION:

$$\begin{aligned} D_5 &= 5!\left(1 - \frac{1}{1!} + \frac{1}{2!} - \frac{1}{3!} + \frac{1}{4!} - \frac{1}{5!}\right) \\ &= 120\left(1 - 1 + \frac{1}{2} - \frac{1}{6} + \frac{1}{24} - \frac{1}{120}\right) \\ &= 44 \end{aligned}$$

■

Returning to the 18th-century problem, n letters can be placed in wrong envelopes in D_n ways, where the value of D_n is given by Theorem 6.9.

Exercises 6.3

Using the recursive definition (6.1), compute the number of derangements D_n for each value of n.

1. 5 **2.** 6 **3.** 7 **4.** 10

Using Theorem 6.9, compute each.

5. D_2 **6.** D_4 **7.** D_6 **8.** D_7

9. Solve the recurrence relation $d_n = -d_{n-1}$, $n \geq 2$, where $d_1 = -1$.

10. Solve the recurrence relation (6.2).

Prove each.

11. D_n is even if n is an odd integer.

12. D_n is odd if n is an even integer.

Let b_n denote the number of computations (additions and multiplications) needed to find D_n, using the recursive definition (6.1). Compute the following.

13. b_2 **14.** b_3 **15.** b_4 **16.** b_5

17. Define b_n recursively.

Using the recursive definition of b_n, compute each.

18. b_4 **19.** b_5 **20.** b_6 **21.** b_8

22. Prove that b_n is an even integer for every $n \geq 0$.

Let c_n denote the number of computations (additions and multiplications) needed to find D_n, using the recursive definition (6.3). Compute each.

23. c_2 **24.** c_3 **25.** c_4 **26.** c_5

27. Define c_n recursively.

Using the recursive definition of c_n, compute each.

28. c_4 **29.** c_5 **30.** c_6 **31.** c_8

32. Solve the recurrence relation obtained in Exercise 27.

33. Show that $c_n = O(n)$.

34. Show that $D_n = 3 \cdot 4 \cdot \cdots \cdot n - 4 \cdot 5 \cdot \cdots \cdot n + \cdots + (-1)^{n-1}n + (-1)^n$.

35. Let a_n denote the number of multiplications needed to compute D_n, using the formula in Exercise 34. Show that $a_n = O(n^2)$.

Let $p_n = \dfrac{D_n}{n!}$. Compute the value of p_n correct to six decimal places for each value of n.

36. 5 **37.** 6 **38.** 7 **39.** 10

***40.** (For those familiar with the concept of a limit) Using Exercises 36–39, predict $\lim_{n\to\infty} p_n$.

41. Show that $p_n - p_{n-1} = \dfrac{(-1)^n}{n!}$.

42. Using Exercise 41, derive the explicit formula for D_n. (*Hint*: Solve the recurrence relation in Exercise 41.)

The formula in Exercise 41 can be derived in a slightly different way also. Verify each.

43. $p_n = \dfrac{n-1}{n}p_{n-1} + \dfrac{1}{n}p_{n-2}$ **44.** $p_n - p_{n-1} = \dfrac{1}{n} - (p_{n-1} - p_{n-2})$

45. $p_n - p_{n-1} = \dfrac{(-1)^n}{n!}$
(*Hint*: Let $g_n = p_n - p_{n-1}$ in Exercise 44.)

46. Write a nonrecursive algorithm to compute D_n.

47. Write a recursive algorithm to compute D_n.

6.4 Combinations

Recall that a permutation is an ordered arrangement of elements in a set. Sometimes, however, the order of elements is irrelevant; only their membership is important. We will investigate such unordered arrangements in this section.

For example, a committee such as A = {Costa, Shea, Weiss, Hall, Chen} is just a set, and the order in which the names are listed is immaterial. Suppose we would like to form a subcommittee of A

consisting of three members. Three such subcommittees are: {Costa, Shea, Weiss}, {Costa, Shea, Hall}, and {Costa, Shea, Chen} (see Example 6.20). Each is a **combination** of the five elements taken three at a time, or a **3-combination.**

More generally, we make the following definition.

Combination

An ***r*-combination** of a set of n elements, where $0 \leq r \leq n$, is a subset containing r elements.

The number of r-combinations of a set with n elements is denoted by $C(n, r)$ or $\binom{n}{r}$.* Both notations frequently appear in combinatorics. The number of combinations is also called the **binomial coefficient**** for reasons that will be clear from Section 6.6.

Before deriving a formula for $C(n, r)$, let us study the following example.

EXAMPLE 6.20 Find the number of r-combinations of the set $\{a, b, c\}$, when $r = 0$, 1, 2, or 3.

SOLUTION:

- Exactly one subset contains zero elements: the null set.
  Number of 0-combinations $= C(3, 0) = 1$.
- Three subsets contain one element each: $\{a\}$, $\{b\}$, and $\{c\}$.
  Number of 1-combinations $= C(3, 1) = 3$.
- Three subsets contain two elements each: $\{a, b\}$, $\{b, c\}$, and $\{c, a\}$.
  Number of 2-combinations $= C(3, 2) = 3$.
- Finally, exactly one subset contains three elements: the set itself.
  Number of 3-combinations $= C(3, 3) = 1$. ■

We now derive a formula for $C(n, r)$.

THEOREM 6.10 The number of r-combinations of a set of n elements is given by $C(n, r) = \dfrac{n!}{r!(n-r)!}$, $0 \leq r \leq n$.†

---

*This two-level parenthesis notation was introduced by German mathematician and physicist Baron Andreas von Ettinghausen (1796–1878). Von Ettinghausen was born in Heidelberg, attended the University of Vienna, and for two years worked as an assistant in mathematics and physics at the University. In 1821 he became professor of mathematics, and in 1835, professor of physics and director of the Physics Institute. Thirteen years later, he became the director of the Mathematical Studies and Engineering Academy in Vienna.

A pioneer in mathematical physics, von Ettinghausen worked in analysis, algebra, differential geometry, mechanics, optics, and electromagnetism.

**The term *binomial coefficient* was introduced by the German algebraist Michel Stifel (1486–1567). In *Arithmetica Integra* (1544), his best-known work, Stifel gives the binomial coefficients for $n \leq 17$.

†Rabbi Ibn Ezra was familiar with this formula in 1100, and shortly thereafter, it appeared in primitive form in Chinese, Indian, and Arabic works.

PROOF:

By definition, there are $C(n,r)$ r-combinations of a set of n elements. Each combination contains r elements and contributes $P(r,r) = r!$ r-permutations, so the total number of r-permutations is $r!C(n,r)$. But, by definition, there are $P(n,r) = \frac{n!}{(n-r)!}$ r-permutations. Therefore,

$$r!C(n,r) = \frac{n!}{(n-r)!}$$

That is,

$$C(n,r) = \frac{n!}{r!(n-r)!}$$

■

In particular, $C(n,0) = \frac{n!}{0!(n-0)!} = \frac{n!}{0!\,n!} = 1$. That is, the number of 0-combinations of a set with n elements is one (see Example 6.20). Also, $C(n,n) = \frac{n!}{n!(n-n)!} = \frac{n!}{n!\,0!} = 1$. That is, the number of n-combinations of a set with n elements is also one (see Example 6.20).

A word of caution: To compute $C(n,r)$ when n is fairly large, do *not* compute $n!$ and $r!(n-r)!$, and then divide. The value of $n!$ may be very large for your calculator to hold without approximating it, so you will find the following fact useful:

$$C(n,r) = \frac{n(n-1)\cdots(n-r+1)}{r!}$$

EXAMPLE 6.21 Compute the number of subcommittees of three members each that can be formed from a committee of 25 members.

SOLUTION:

$$\begin{aligned}\begin{pmatrix}\text{Number of subcommittees}\\ \text{of three people each}\end{pmatrix} &= \begin{pmatrix}\text{number of 3-combinations}\\ \text{of a set of 25 people}\end{pmatrix}\\ &= C(25,3)\\ &= \frac{25\cdot 24\cdot 23}{3!}\\ &= 2300\end{aligned}$$

■

The next example is an interesting close relative of Example 5.5 in Chapter 5.

EXAMPLE 6.22 **(The Pizza problem)** Let f_n denote the maximum number of places into which a pizza can be divided with n cuts. Find a formula for f_n.

SOLUTION:

Clearly, the maximum number of regions can be realized when every two chords, that is, cuts, intersect and no three chords are concurrent.

It follows by Example 5.5 in Chapter 5 that f_n can be defined recursively as

$$f_0 = 1$$
$$f_n = f_{n-1} + n, \quad n \geq 1$$

Solving this recurrence relation (see Exercise 3 in Section 5.2) yields

$$f_n = 1 + \frac{n(n+1)}{2} \quad \text{(Verify this.)}$$

This formula can be rewritten as

$$\begin{aligned} f_n &= 1 + n + \frac{n(n-1)}{2} \quad \text{(Verify this.)} \\ &= C(n,0) + C(n,1) + C(n,2), \quad n \geq 0 \end{aligned}$$

(You may prove this using induction and Pascal's identity in Theorem 6.12. We shall pursue this example in Section 6.6.) ■

Example 6.23 employs the multiplication principle, as well as Theorem 6.10.

EXAMPLE 6.23 How many committees of three blondes and four brunettes can be formed from a group of five blondes and six brunettes?

SOLUTION:

Three blondes can be selected from five blondes in $C(5,3) = 10$ different ways and four brunettes from six brunettes in $C(6,4) = 15$ different ways. Therefore, by the multiplication principle, the number of committees with three blondes and five brunettes is $10 \cdot 15 = 150$. ■

The following example is an interesting application of the pigeonhole principle.

EXAMPLE 6.24 Let A be a 10-element subset of the set $\{1, 2, \ldots, 15\}$. Let A_s be a subset of A containing three elements, where the subscript s denotes the sum of the elements in A_s. For example, the subset $\{2, 4, 5\}$ is denoted by A_{11}. Determine if each subset of A can be identified by a unique name A_s. In other words, does every sum $i + j + k$ have a unique value s, where $1 \leq i < j < k \leq 15$?

SOLUTION:

We let the pigeonhole principle do the job for us. The least value of s is $1 + 2 + 3 = 6$ and the largest value of s is $13 + 14 + 15 = 42$. Thus $6 \leq s \leq 42$; there are at most 37 possible values of s.

There are $C(10, 3) = 120$ three-element subsets (pigeons) of A and only 37 possible sums (pigeonholes), so, by the pigeonhole principle, at least two subsets must yield the same sum; that is, *not* every three-element subset of A can have a unique name.

For example, let $A = \{1, 2, \ldots, 10\}$. Since subsets, $\{1, 2, 5\}$ and $\{1, 3, 4\}$, yield the same sum, 8, they have the same name, A_8. ■

The next theorem will in many cases reduce your workload with combinations, as seen in Example 6.25.

THEOREM 6.11 $C(n, r) = C(n, n - r)$, where $0 \leq r \leq n$.

PROOF:

$$\begin{aligned} C(n, n-r) &= \frac{n!}{(n-r)![n-(n-r)]!} \\ &= \frac{n!}{(n-r)!r!} = C(n, r) \end{aligned}$$ ■

> According to Theorem 6.11, the number of r-combinations of a set with n elements equals that of the $(n - r)$-combinations of the set; for example, $C(5, 2) = C(5, 3) = 10$. This result can be used to cut down the amount of work needed to compute the number of combinations in several applications.

The next example takes advantage not only of this theorem, but also of the addition principle.

EXAMPLE 6.25 Find the number of groups that can be formed from a group of seven marbles if each group must contain at least three marbles.

SOLUTION:

Since each group must contain at least three marbles, it can contain three, four, five, six, or seven marbles.

Number of groups containing three marbles $= C(7, 3) = 35$

Number of groups containing four marbles $= C(7, 4) = C(7, 3) = 35$

Number of groups containing five marbles $= C(7, 5) = 21$

Number of groups containing six marbles $= C(7, 6) = C(7, 1) = 7$

Number of groups containing seven marbles $= C(7, 7) = 1$

Total number of groups $= 35 + 35 + 21 + 7 + 1 = 99$ ■

This problem can be done in a clever, shorter way as follows:

$$\begin{pmatrix}\text{number of groups containing}\\ \text{at most two marbles}\end{pmatrix} = \begin{pmatrix}\text{number of groups containing}\\ \text{0, 1, or 2 marbles}\end{pmatrix}$$
$$= C(7,0) + C(7,1) + C(7,2)$$
$$= 1 + 7 + 21 = 29$$

So

$$\begin{pmatrix}\text{number of groups}\\ \text{containing at least}\\ \text{three marbles}\end{pmatrix} = \begin{pmatrix}\text{total number}\\ \text{of possible}\\ \text{groups}\end{pmatrix} - \begin{pmatrix}\text{number of}\\ \text{groups}\\ \text{containing}\\ \text{at most two}\\ \text{marbles}\end{pmatrix}$$
$$= 2^7 - 29 = 99$$

We can now proceed to find a recurrence relation satisfied by $C(n,r)$. As before, we will give a combinatorial argument to establish the formula and leave the algebraic proof as a standard exercise (see Exercise 39).

THEOREM 6.12

$$C(n,r) = C(n-1, r-1) + C(n-1, r), \text{ where } 0 < r < n. \tag{6.4}$$

PROOF:

Let X be a set of n elements. There are $C(n,r)$ r-element subsets of X. Let x be any element of X. Let $Y = X - \{x\}$; it contains $n-1$ elements. The r-element subsets of X can be partitioned into two disjoint families: family A of r-element subsets that contain x, and family B of r-element subsets that do not contain x.

- *To find* $|A|$: Each subset in A contains x and hence contains $r-1$ elements excluding x. Therefore, they are the $(r-1)$-element subsets of Y containing x. There are $C(n-1, r-1)$ such subsets.
- *To find* $|B|$: The r-element subsets not containing x are the r-element subsets of Y. There are $C(n-1, r)$ such subsets.

Thus, by the addition principle, the total number of r-element subsets equals $|A| + |B| = C(n-1, r-1) + C(n-1, r)$. That is, $C(n,r) = C(n-1, r-1) + C(n-1, r)$. ■

The recurrence relation (6.4) is called **Pascal's identity**, after Blaise Pascal, an outstanding French mathematician and physicist.

The next example uses recursion to derive a formula for the maximum number of nonoverlapping regions formed by joining n distinct points on a circle, a problem presented in Example 4.19.

○ **EXAMPLE 6.26**

(**The Circle Problem**—optional) Let g_n denote the maximum number of nonoverlapping regions formed inside a circle by joining n distinct points on it. Derive a formula for g_n.

Blaise Pascal *(1623–1662) was born in Clermont-Ferrand, France. Although he showed phenomenal mathematical ability at an early age, he was encouraged by his father to pursue subjects such as ancient languages. His father even refused to teach him any sciences until he found that Pascal by himself at age 12 had discovered many theorems in elementary geometry. At 14 he attended the weekly meetings of a group of French mathematicians that later became the French Academy. At 16 he developed important results in conic sections and wrote a book on it.*

Observing that his father spent long hours auditing government accounts and feeling that intelligent people should not waste their time doing mundane things, Pascal at age 19 invented the first mechanical calculating machine.

In 1650, suffering from failing health, Pascal left his mathematical and scientific work to pursue religion. Three years later, he returned briefly to mathematics. During this period, working with Fermat, he laid the foundation for probability theory. Most of his life was spent in physical pain.

The programming language Pascal is named after him.

SOLUTION*

We derive an explicit formula for g_n, using recursion. Let $P_1, P_2, \ldots, P_n$ be n points on a circle such that no three chords $\overline{P_rP_s}$ are concurrent, where $1 \le r, s \le n$. Choose a new point Q on arc $\overset{\frown}{P_{n-1}P_n}$. Join Q to each of the points P_1 through P_n. The chord $\overline{QP_n}$ introduces an extra region (see Figure 6.16).

Figure 6.16

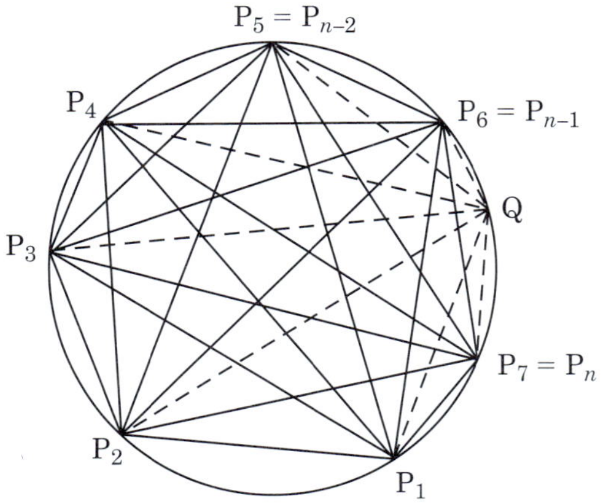

Now consider the chord $\overline{QP_i}$, $1 \le i \le n-1$. It intersects each of the chords obtained by joining one of the i points $P_n, P_1, \ldots, P_{i-1}$ to one of the $n-1-i$ points $P_{i+1}, P_{i+2}, \ldots, P_{n-1}$. Each intersection corresponds to a new region.

*Based on A. V. Boyd and M. J. Glencorss, "Dissecting a Circle by Chords through n Points," *Mathematics Teacher*, Vol. 84 (April 1991), pp. 318–319.

Consequently, the number of regions formed by the chord $\overline{QP_i}$ is one more than the total number of points of intersection, namely, $1 + i(n-1-i)$, $1 \le i \le n-1$.

Thus the total number of regions formed by the introduction of the $(n+1)$st point Q is

$$1 + \sum_{i=1}^{n-1}[1 + i(n-1-i)] = \left(\sum_{i=1}^{n} 1\right) + \sum_{i=1}^{n-1}[i(n-1-i)]$$

This yields the recurrence relation

$$g_{n+1} - g_n = \left(\sum_{i=1}^{n} 1\right) + \sum_{i=1}^{n-1}[i(n-1-i)]$$

The RHS of this equation can be simplified as follows:

$$\begin{aligned}
&= n + \sum_{i=1}^{n-1} i(n-1) - \sum_{i=1}^{n-1} i^2 \\
&= n + (n-1)\sum_{i=1}^{n-1} i - \sum_{i=1}^{n-1} i^2 \\
&= n + (n-1)\cdot\frac{(n-1)n}{2} - \frac{(n-1)n[2(n-1)+1]}{6} \\
&= n + \frac{n(n-1)^2}{2} - \frac{n(n-1)(2n-1)}{6} \\
&= n + \frac{n(n-1)(n-2)}{6} \\
&= C(n,1) + C(n,3) \qquad \text{(Verify this.)}
\end{aligned}$$

Solving this recurrence relation by iteration (see Exercise 40) yields $g_n = C(n,0) + C(n,2) + C(n,4)$, $n \ge 0$.

In particular, $g_5 = C(5,0) + C(5,2) + C(5,4) = 1 + 10 + 5 = 16$ and $g_6 = C(6,0) + C(6,2) + C(6,4) = 1 + 15 + 15 = 31$, as expected. We shall pursue this example a bit further in Section 6.6. ■

Exercises 6.4

A committee consists of nine members. Find the number of subcommittees that can be formed of each size.

1. Two **2.** Five **3.** Six **4.** Seven

5. Find the number of ways a committee of three students and five professors can be formed from a group of seven students and 11 professors.

6. Find the number of ways a committee of four students, four professors, and three administrators can be formed from a group of six students, eight professors, and five administrators.

7. Find the number of lines that can be drawn using 10 distinct points, no three being collinear.

8. Find the number of triangles that can be drawn using 10 points, no three being collinear.

9. Solve the recurrence relation in Example 6.22.

10. Prove the formula in Example 6.22, using induction.

Let A be a 10-element subset of the set $\{1, 2, \ldots, 20\}$.

11. Determine if A has two five-element subsets that yield the same sum of the elements.

12. Determine if A has two eight-element subsets that yield the same sum of the elements.

(**Twelve Days of Christmas**) Suppose that on the first day of Christmas you sent your love 1 gift, $1 + 2$ gifts on the second day, $1 + 2 + 3$ gifts on the third day, and so on.

13. Find the number of gifts sent on the 12th day.

14. Find the total number of gifts sent in 12 days.

15. Show that the number of gifts sent on the nth day is $C(n+1, 2)$, where $1 \le n \le 12$.

16. Show that the total number of gifts sent by the nth day is $C(n+2, 3)$, where $1 \le n \le 12$.

Solve each equation, where $n \ge 0$.

17. $C(n, 0) = 1$

18. $C(n, 1) = 10$

19. $C(n, 2) = 28$

20. $C(n, n-2) = 55$

21. Find the number of ways of dividing a set of size n into two disjoint subsets of sizes r and $n - r$.

A collection plate contains four nickels, five dimes, and seven quarters. In how many ways can you:

22. Choose three coins?

23. Form a sum of 40 cents?

Find the number of ways each sum can be formed from a collection of 10 nickels and 5 quarters.

24. 25 cents

25. 30 cents

Jane has two nickels, four dimes, three quarters, and two half-dollars in her handbag. Find the number of ways she can tip the waiter if she would like to give him:

26. Exactly three coins.

27. At least three coins, but not more than five coins.

28. Not more than three coins.

29. Exactly 50 cents.

30. Not more than 50 cents, using only one type of coin.

Find the number of ways a committee of five can be formed from a group of five boys and four girls, if each committee must contain:

31. Exactly two boys.

32. At least two boys.

33. At least two girls.

34. At least one boy and at least one girl.

35. At most one boy.

36. At most one girl.

37. Verify that $\binom{n}{r} = \frac{n}{r}\binom{n-1}{r-1}$, where $n \geq r \geq 1$.

38. Prove that $C(2n, n)$ is an even integer for every $n \geq 1$. (*Hint*: Use Exercise 37.)

39. Prove Pascal's identity algebraically.

40. Solve the recurrence relation in Example 6.26.

41. Prove the explicit formula for g_n in Example 6.26, using induction.

42. Using the explicit formula in Example 6.26, verify that $g_n = \sum_{k=0}^{4} C(n-1, k)$.

Let $A(n, r)$ denote the number of additions needed to compute $C(n, r)$ by its recursive definition. Compute each.

43. $A(3, 2)$

44. $A(5, 3)$

45. Define $A(n, r)$ recursively.

46. Prove that $A(n, r) = C(n, r) - 1$, using induction.

Recall that the nth Catalan number C_n is defined by $C_n = \frac{(2n)!}{n!(n+1)!}$, $n \geq 0$

47. Show that $C_n = C(2n, n) - C(2n, n-1)$.

Prove each.

48. $C_n = \frac{1}{n+1}C(2n, n), \quad n \geq 0$

49. $C_n = \frac{2(2n-1)}{n+1}C_{n-1}, \quad n \geq 1$

50. Define C_n recursively.

51. Show that $C_n = 3C_{n-1} + \left(C_{n-1} - \frac{6}{n+1}C_{n-1}\right), n \geq 1$

The nth Catalan number satisfies the recurrence relation $C_n = \sum_{i=0}^{n-1} C_i C_{n-1-i}$, $n \geq 2$. (*Note*: This relation can be used to compute C_n using n multiplications, $n-1$ additions, and no divisions.) Use it to compute each Catalan number.

52. C_4

53. C_5

54. Write an algorithm to compute $C(n, r)$ using Theorem 6.10.

Write a recursive algorithm to compute each.

55. $C(n, r)$

56. C_n

Stirling numbers of the second kind $S(n, r)$ are also given by the formula

$$S(n,r) = \frac{1}{r!}\sum_{k=0}^{r-1}(-1)^k \binom{r}{k}(r-k)^n$$

Compute each Stirling number.

57. $S(3, 2)$

58. $S(4, 2)$

The number of surjections that can be defined from a finite set A to a finite set B is given by $r!S(n, r)$, where $|A| = n$ and $|B| = r$. Compute the number of possible surjections from A to B if:

59. $|A| = 3$, $|B| = 2$.

60. $|A| = 4$, $|B| = 2$.

61. $|A| = n$, $|B| = 2$.

62. $|A| = n$, $|B| = 3$.

6.5 Permutations and Combinations with Repetitions

The permutations and combinations examined so far involved unrepeated items. If the items repeat, then computations become a bit more complicated. This section explores such permutations and combinations.

Permutations with Repetitions

Consider the word REFERENCE. If we swap the second E with the fourth E in the word, we do *not* get a new word. How can we compute the number of permutations in such cases?

EXAMPLE 6.27 Find the number of different arrangements of the letters of the word REFERENCE.

SOLUTION:
The word REFERENCE contains nine letters. If they were all distinct, the answer would be $9! = 362{,}880$. But, since duplicate letters exist, the answer is indeed much less.

Let N denote the number of different words. We shall find the value of N in an indirect way.

The word REFERENCE contains two R's and four E's; the remaining letters are distinct. Think of the two R's as two distinct letters, R_1 and R_2, and the four E's as four distinct letters, E_1 through E_4. The letters R_1 and R_2 can be arranged in 2! ways and the four E's in 4! ways. Therefore, if all the letters were distinct, there would be a total of $2!\,4!\,N$ different words. Thus $2!\,4!\,N = 9!$; so

$$N = \frac{9!}{2!\,4!} \tag{6.5}$$
$$= 7560$$

■

> *An interesting observation*: The number 9 in the numerator of Equation (6.5) indicates the number of letters in the word. Each number in the denominator indicates the frequency of each repeating letter.

This REFERENCE problem exemplifies the next theorem.

THEOREM 6.13 The number of permutations of n items of which n_1 items are of one type, n_2 are of a second type, ..., and n_k are of a kth type, is $n!/(n_1!\,n_2!\cdots n_k!)$.

PROOF:
Let N denote the total number of permutations. As in Example 6.27, we shall find the value of N indirectly.

Let $A_1, \ldots, A_{n_1}$ denote the items of the first type; $B_1, \ldots, B_{n_2}$ items of the second type; ...; and $Z_1, \ldots, Z_{n_k}$ items of the kth type. If all items were distinct, the total would be $n!$ permutations.

If items $A_1, \ldots, A_{n_1}$ are distinct, they can be arranged in $n_1!$ ways. Items $B_1, \ldots, B_{n_2}$, if distinct, can be arranged in $n_2!$ ways, and so on. Items $Z_1, \ldots, Z_{n_k}$, if distinct, can be arranged in $n_k!$ ways. Thus, by the multiplication principle, if all items are distinct, there would be $(n_1!\,n_2!\cdots n_k!)N$ permutations. So $n! = (n_1!\,n_2!\cdots n_k!)N$. Thus

$$N = \frac{n!}{n_1!\,n_2!\cdots n_k!}$$

■

This theorem works well in solving the next two problems.

EXAMPLE 6.28 Find the number of bytes containing exactly three 0's.

SOLUTION:

$$\begin{aligned}\begin{pmatrix}\text{number of bytes containing}\\ \text{exactly three 0's}\end{pmatrix} &= \begin{pmatrix}\text{number of bytes containing}\\ \text{three 0's and five 1's}\end{pmatrix}\\ &= \begin{pmatrix}\text{number of permutations of}\\ \text{eight symbols of which}\\ \text{three are alike (0's) and}\\ \text{five are alike (1's)}\end{pmatrix}\\ &= \frac{8!}{3!\,5!}, \quad \text{by Theorem 6.13}\\ &= 56\end{aligned}$$

■

The next example pertains to the layout of a planned city.

EXAMPLE 6.29 **(Lattice-Walking)** Figure 6.17 shows a portion of a city map.

Figure 6.17

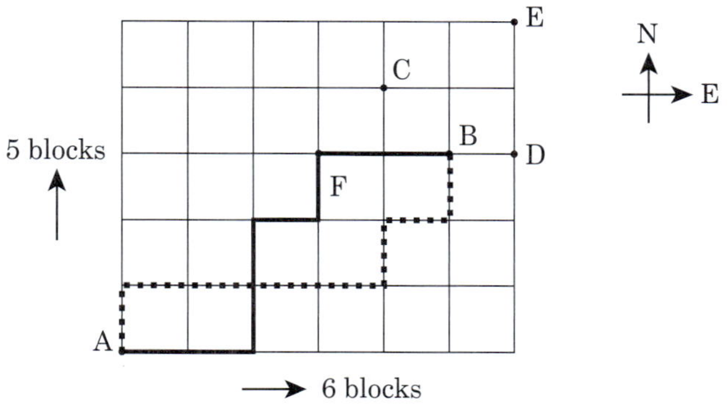

Suppose you would like to travel from point A to point B, covering exactly 8 blocks. You can travel in the easterly or northerly direction only. Two possible routes are shown in the figure. How many such routes are possible?

SOLUTION:

The heavy route in the figure can be represented by the string EENNENEE; it means, travel 2 blocks east, 2 blocks north, 1 block east, 1 block north, and 2 blocks east. The second route shown, NEEEENEN, can be interpreted similarly.

Every route from A to B can be represented by an eight-letter word, of which five letters are alike (E's) and three are alike (N's). Therefore,

$$\text{total number of paths from A to B} = \begin{pmatrix}\text{total number of 8-letter words}\\ \text{of which five letters are alike}\\ \text{and the other three are alike}\end{pmatrix}$$

$$= \frac{8!}{5!\,3!},$$
$$= 56 \qquad \blacksquare$$

Abracadabra

A related problem was proposed in 1966 by G. Polya of Stanford University: Find the number of ways the word ABRACADABRA can be read using the rhombic array in Figure 6.18, beginning at the apex A and ending at the bottommost A. It follows from Example 6.29 that this problem can be translated into a city-walking problem, walking around the blocks in the city in Figure 6.19. One-half of the streets run from northeast (NE) to southwest (SW) and the rest from northwest to southeast. Each path from S to E consists of 10 blocks: 5 blocks in the NW–SE direction and the rest

Figure 6.18

```
          A
        B   B
      R   R   R
    A   A   A   A
  C   C   C   C   C
A   A   A   A   A   A
  D   D   D   D   D
    A   A   A   A
      B   B   B
        R   R
          A
```

Figure 6.19

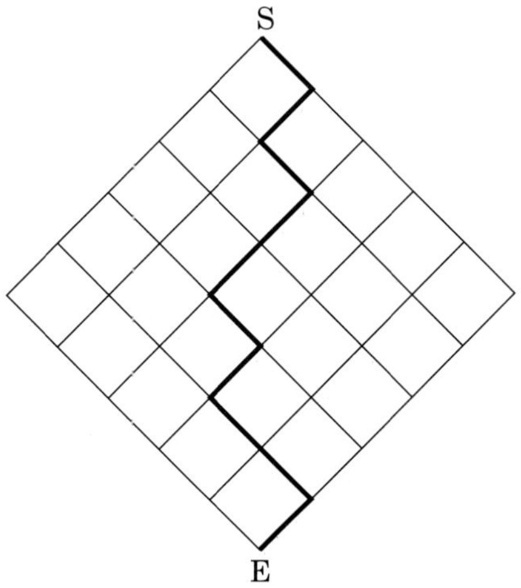

in the NE–SW direction. So the total number of paths from S to E equals the number of combinations of 10 items, of which five are alike and the other five are alike, namely, $\frac{10!}{5!\,5!} = 252$. Thus there are 252 different ways we can read the magic word in Figure 6.18. (We shall revisit this problem in Section 6.6.)

Combinations with Repetitions

Just as permutations can deal with repeated elements, so can combinations (called **selections**). For example, suppose five friends go to a local restaurant for beverages: iced tea, hot tea, or coffee. The waitress puts all five requests on the same order. How many different orders for the table are possible? The order in which the beverages are selected is immaterial and the same beverage can be selected by more than one person. Also, not every beverage need be selected.

Before returning to this problem in Example 6.32, let us study a couple of simple ones.

EXAMPLE 6.30 Find the number of 3-combinations of the set $S = \{a, b\}$.

SOLUTION:
S contains $n = 2$ elements. Since each combination must contain three elements, $r = 3$. Since $r > n$, the elements of each combination must be repeated. Consequently, a combination may contain three a's, two a's and one b, one a and two b's, or three b's. Using the set notation, the 3-combinations are $\{a, a, a\}$, $\{a, a, b\}$, $\{a, b, b\}$ and $\{b, b, b\}$. So there are four 3-combinations of a set of two elements. ■

EXAMPLE 6.31 Find the number of 3-combinations of the set $\{a, b, c\}$, where the elements may be repeated.

SOLUTION:
Again, using the set notation, the various 3-combinations are:

$$\{a,a,a\} \quad \{a,a,b\} \quad \{a,a,c\} \quad \{a,b,b\} \quad \{a,b,c\} \quad \{a,c,c\}$$
$$\{b,b,b\} \quad \{b,b,c\} \quad \{b,c,c\}$$
$$\{c,c,c\}$$

Thus the set $\{a, b, c\}$ has 10 3-combinations. ■

Before developing a formula for the number of combinations with repetitions, let us return to the beverage problem posed earlier.

EXAMPLE 6.32 Five friends would like to order beverages with their dinner at a local restaurant that serves iced tea, hot tea, or coffee. Find the number of beverage orders possible.

SOLUTION:
A convenient notation will prevent confusion.

Denote each type of beverage by a dash and separate them using two slashes, as shown below:

$$\underset{\text{iced tea}}{\underline{\qquad\qquad}} \;/\; \underset{\text{hot tea}}{\underline{\qquad\qquad}} \;/\; \underset{\text{coffee}}{\underline{\qquad\qquad}}$$

Mark each person's selection by an X in the appropriate area.

For instance, the distribution XX / X / XX indicates that two people selected iced tea, one selected hot tea, and two selected coffee; the distribution XXX / / XX means, three people selected iced tea, none ordered hot tea, and two selected coffee.

Thus the number of possible beverage orders equals the number of permutations of seven items (five X's and two /'s) of which five are alike (X's) and the other two are alike (/'s):

$$\frac{7!}{5!\,2!} = 21$$ ■

This solution strategy produces the following theorem.

THEOREM 6.14 The number of r-combinations with repetitions from a set of n elements is $C(n + r - 1, r)$.

PROOF:
Each r-combination with repeated elements from a set of n elements can be considered a string of r X's and $(n - 1)$ slashes, as in Example 6.32. Each string contains $r + n - 1 = n + r - 1$ symbols, of which r are alike (X's) and $n - 1$ are alike (slashes). Therefore, by Theorem 6.13, the number of such strings, that is, r-combinations, equals

$$\frac{(n + r - 1)!}{r!(n - 1)!} = C(n + r - 1, r)$$ ■

This theorem helps solve the next example.

EXAMPLE 6.33 There are five types of soft drinks at a fast food restaurant: Coke Classic, Diet Coke, root beer, Pepsi, and Sprite. Find the number of beverage orders 11 guests can make.

SOLUTION:
Since there are five types of soft drinks, $n = 5$. Each beverage order is a selection containing 11 items, that is, an 11-combination with repeating elements. Therefore, by Theorem 6.14, the number of possible beverage orders equals

$$\begin{aligned} C(n + r - 1, r) &= C(5 + 11 - 1, 11) \\ &= C(15, 11) \\ &= \frac{15!}{11!\,4!} = 1365 \end{aligned}$$ ■

This problem has a nice interpretation. Let x_i denote the number of guests ordering soft drink i, where $1 \leq i \leq 5$. Then $x_1+x_2+x_3+x_4+x_5 = 11$, where $x_i \geq 0$. The number of nonnegative integer solutions of this equation is the same as the number of possible beverage orders, so the number of integer solutions of this equation is $C(5+11-1, 11) = C(15, 11) = 1365$.

The next theorem generalizes this result, simply restating Theorem 6.14.

THEOREM 6.15 Let $x_1, x_2, \ldots, x_n$ be n nonnegative integer variables and r a nonnegative integer. The equation $x_1 + x_2 + \cdots + x_n = r$ has $C(n+r-1, r)$ integer solutions. ■

EXAMPLE 6.34 Find the number of solutions of the equation

$$x_1 + x_2 + x_3 = 5 \tag{6.6}$$

where x_1, x_2, and x_3 are nonnegative integer variables.

SOLUTION:

Here $r = 5$ and $n = 3$. By Theorem 6.15, the number of solutions is

$$\begin{aligned} C(n+r-1, r) &= C(3+5-1, 5) \\ &= C(7, 5) = 21 \end{aligned}$$

(Can you list all the solutions? See Example 6.35 also.) ■

Taking this example a step further, suppose you would like to find all solutions of Equation (6.6), where $x_1, x_2, x_3 \geq 1$. Make the substitution $y_i = x_i - 1, 1 \leq i \leq 3$. Clearly, $y_i \geq 0$. Equation (6.6) becomes

$$y_1 + y_2 + y_3 = 2$$

By Theorem 6.15, this equation has $C(n+r-1, r) = C(3+2-1, 2) = C(4, 2) = 6$ solutions: $(0,1,1), (1,0,1), (1,1,0), (2,0,0), (0,2,0)$, and $(0,0,2)$. Consequently, Equation (6.6) with $x_i \geq 1$ has six solutions: $(1,2,2)$, $(2,1,2), (2,2,1), (3,1,1), (1,3,1)$, and $(1,1,3)$.

For Loops Revisited

The following two examples provide applications of both these formulas to *for* loops.

EXAMPLE 6.35 Find the number of times the assignment statement $x \leftarrow x + 1$ is executed by the following nested *for* loops:

```
For i = 1 to n do
  For j = 1 to i do
    x ← x + 1
```

SOLUTION:
Notice that the assignment statement is executed for each pair (j, i), where $1 \leq j \leq i \leq n$. For example, the statement is executed when $j = 3, i = 5$; $j = 3, i = 3$; but not when $j = 3$ and $i = 2$. Thus the number of executions equals the number of pairs (j, i) with repetitions allowed; in other words, it equals the number of 2-selections of the set $\{1, 2, 3, \ldots, n\}$. The number of such 2-selections is $C(n + 2 - 1, 2)$, so the assignment is executed in $C(n + 2 - 1, 2) = C(n + 1, 2) = t_n$ different ways. ■

We now pursue this example with one more level added to the *for* loops.

EXAMPLE 6.36 Find the number of times the assignment statement $x \leftarrow x + 1$ is executed by the following nested *for* loops:

```
For i = 1 to n do
  For j = 1 to i do
    For k = 1 to j do
      x ← x + 1
```

SOLUTION:
The assignment statement is executed for each triplet (k, j, i), where $1 \leq k \leq j \leq i \leq n$ and where repetitions are allowed. The number of such 3-selections of the set $\{1, 2, 3, \ldots, n\}$ is $C(n + 3 - 1, 3)$, so the assignment is executed $C(n + 3 - 1, 3) = C(n + 2, 3) = T_n$ different ways. ■

A Generalization

More generally, consider the following nested sequence of *for* loops:

```
For i1 = 1 to n do
  For i2 = 1 to i1 do
    For i3 = 1 to i2 do
          ⋮
        For ir = 1 to ir - 1 do
          x ← x + 1
```

It follows from the combinatorial arguments in Examples 6.35 and 6.36 that the number of times the assignment statement $x \leftarrow x + 1$ is executed is given by the number of r-selections of the set $\{1, 2, 3, \ldots, n\}$ with repetitions allowed, that is, $C(n + r - 1, r)$.

Triangular and Tetrahedral Numbers Revisited

Returning to triangular numbers t_n and tetrahedral numbers T_n, we find that their formulas fit into the one in Theorem 6.14:

$$t_n = n(n + 1)/2 \qquad = C(n + 1, 2) = C(n + 2 - 1, 2)$$

$$T_n = n(n + 1)(n + 2)/6 = C(n + 2, 3) = C(n + 3 - 1, 3)$$

Generating Functions and Combinations

Generating functions, introduced in Chapter 5, are a valuable tool in solving combinatorial problems involving repeated elements. The next three examples illustrate this method.

EXAMPLE 6.37 Using generating functions, find the number of beverage orders possible in Example 6.32.

SOLUTION:

Suppose the three beverages are ordered by i, j, and k patrons. Then every beverage order can be considered a 3-tuple (i, j, k), where $i, j, k \geq 0$ and $i + j + k = 5$. Let x^i denote that iced tea was ordered by i customers. Since 0, 1, 2, 3, 4, or 5 people can order iced tea, we use the polynomial $1 + x + x^2 + x^3 + x^4 + x^5$ to represent the various possibilities. (Notice the exponents.) Both hot tea and coffee can also be ordered by 0, 1, 2, 3, 4, or 5 people; correspondingly, the polynomials $1 + y + y^2 + y^3 + y^4 + y^5$ and $1 + z + z^2 + z^3 + z^4 + z^5$ represent these possibilities. Consequently, we want products of the form $x^i y^j z^k$, where $i + j + k = 5$ in the product $\left(\sum_{i=0}^{5} x^i\right)\left(\sum_{j=0}^{5} y^j\right)\left(\sum_{k=0}^{5} z^k\right)$

For convenience, let $x = y = z$. Thus the total number of beverage orders possible is the coefficient of x^5 in the product $\left(\sum_{i=0}^{5} x^i\right)^3$.

You may verify that it is 21. Table 6.2 shows the various possible beverage orders.

Table 6.2

Iced tea	Hot tea	Coffee	Iced tea	Hot tea	Coffee
0	0	5	2	0	3
0	1	4	2	1	2
0	2	3	2	2	1
0	3	2	2	3	0
0	4	1	3	0	2
0	5	0	3	1	1
1	0	4	3	2	0
1	1	3	4	0	1
1	2	2	4	1	0
1	3	1	5	0	0
1	4	0			

■

The next problem resembles Example 6.34, with additional constraints on the variables.

EXAMPLE 6.38 Find the number of solutions of the equation $x_1 + x_2 + x_3 + x_4 = 11$, where $x_1 \geq 7, 1 \leq x_2, x_3 \leq 3$, and $0 \leq x_4 \leq 3$.

SOLUTION:
Again, generating functions can do the job for us. Since $x_1 \geq 7, x_1$ can be 7, 8, 9, 10, or 11, represented by the polynomial $x^7 + x^8 + \cdots + x^{11}$. The polynomials corresponding to the constraints $1 \leq x_2, x_3 \leq 3$ are $x + x^2 + x^3$ and $x + x^2 + x^3$. For $0 \leq x_4 \leq 3$, we have the polynomial $1 + x + x^2 + x^3$. Thus the number of solutions is the coefficient of x^{11} in the product

$$\left(\sum_{i=7}^{11} x^i\right)\left(\sum_{i=1}^{3} x^i\right)\left(\sum_{i=1}^{3} x^i\right)\left(\sum_{i=0}^{3} x^i\right) \tag{6.7}$$

that is, the coefficient of x^2 in the product

$$\left(\sum_{i=0}^{4} x^i\right)\left(\sum_{i=0}^{2} x^i\right)\left(\sum_{i=0}^{2} x^i\right)\left(\sum_{i=0}^{3} x^i\right)$$

You can verify that it is 10. Thus the equation has 10 solutions satisfying the given conditions. [The various solutions (i, j, k, l) can be obtained by picking the exponents in the products $x^i x^j x^k x^l$ that yield x^{11} in the product (6.7).] ■

We close this section with an example that is closely related to Example 6.38.

EXAMPLE 6.39 In how many ways can 11 cookies be distributed among four children—Amy, Betsy, Carol, and Daisy—so that Amy gets at least seven cookies, both Betsy and Carol get at least one cookie each but not more than three, and Daisy gets no more than three cookies?

SOLUTION:
Since Amy gets at least seven cookies, this case yields the polynomial $x^7 + x^8 + x^9 + x^{10} + x^{11}$. Similarly, the other three constraints yield the polynomials $x + x^2 + x^3, x + x^2 + x^3$, and $1 + x + x^2 + x^3$. Thus the number of ways of distributing 11 cookies under the given conditions is the coefficient of x^{11} in the product (6.7) above, namely, 10. ■

Exercises 6.5

Find the number of distinct words that can be formed by scrambling the letters in each word.

1. CALCULUS

2. TALLAHASSEE

Find the number of bytes that:

3. Contain exactly two 0's.

4. Contain exactly five 0's.

5. Contain at least five 0's.

6. Contain not more than two 0's.

7. Contain exactly eight 0's.

8. Contain exactly nine 0's.

Find the number of ternary words over the alphabet $\{0, 1, 2\}$ that are of length four and:

9. Contain exactly three 0's.

10. Contain at least six 0's.

11. Contain at most two 0's.

12. Contain two 0's, three 1's, and three 2's.

In Exercises 13–16, use Figure 6.17 to find the number of possible routes from A to the given point, traveling easterly or northerly for the given number of blocks.

13. Point F and 5 blocks.

14. Point C and 8 blocks.

15. Point D and 9 blocks.

16. Point E and 11 blocks.

List the 4-combinations of each set.

17. $\{a\}$

18. $\{a, b\}$

19. There are five types of desserts available at a restaurant. Find the number of ways eight people can select them, if order does not matter.

20. A restaurant offers six choices for the main dish. How many ways can a group of nine women select the main dish? Assume that order does not matter.

21. In how many ways can 10 quarters in a piggy bank be distributed among 7 people?

Find the number of solutions to each equation, where the variables are nonnegative integers.

22. $x_1 + x_2 + x_3 = 3$

23. $x_1 + x_2 + x_3 + x_4 = 7$

24. $x_1 + x_2 + x_3 + x_4 = 10$

25. $x_1 + x_2 + x_3 + x_4 + x_5 = 11$

Find the number of solutions to each equation, where $x_i \geq 1$.

26. $x_1 + x_2 + x_3 + x_4 = 11$

27. $x_1 + x_2 + x_3 + x_4 + x_5 = 13$

Use generating functions to solve the following counting problems.

28. Use generating function (6.7) to find the various solutions of the equation in Example 6.38.

Find the number of solutions to each equation.

29. $x_1 + x_2 + x_3 = 10,\ x_1 \geq 3,\ 1 \leq x_2 \leq 3,\ x_3 \geq 5$

30. $x_1 + x_2 + x_3 = 12,\ x_1,\ x_2 \geq 5,\ 1 \leq x_3 \leq 4$

31. $x_1 + x_2 + x_3 + x_4 = 10,\ x_1,\ x_2 \geq 2,\ x_3 \geq 0,\ x_4 \geq 5$

32. $x_1 + x_2 + x_3 + x_4 = 11,\ x_1,\ x_2 \geq 2,\ 2 \leq x_3 \leq 4,\ x_4 \geq 3$

33–34. Find the solutions to the equations in Exercises 29 and 30.

35. Find the number of ways 10 quarters can be distributed among three people—Aaron, Beena, and Cathy—so that both Aaron and Beena get at least one quarter, Beena gets no more than three, and Cathy gets at least two.

36. Find the number of ways 11 raisins can be distributed among four children—Daisy, Emily, Francis, Gail—so that Daisy, Emily, and Francis get at least two raisins, Francis gets no more than four, and Gail gets at least three.

6.6 The Binomial Theorem

The binomial coefficients satisfy a vast array of properties. We shall visit a few of them shortly.

Pascal's Triangle

The various binomial coefficients $\binom{n}{r}$, where $0 \le r \le n$, can be arranged in the form of a triangle, called **Pascal's triangle**,* as shown in Figures 6.20 and 6.21.

Figure 6.20

$$\begin{array}{ccccccccccl}
 & & & & \binom{0}{0} & & & & & & \leftarrow \text{row } 0 \\
 & & & \binom{1}{0} & & \binom{1}{1} & & & & & \leftarrow \text{row } 1 \\
 & & \binom{2}{0} & & \binom{2}{1} & & \binom{2}{2} & & & & \leftarrow \text{row } 2 \\
 & \binom{3}{0} & & \binom{3}{1} & & \binom{3}{2} & & \binom{3}{3} & & & \leftarrow \text{row } 3 \\
\binom{4}{0} & & \binom{4}{1} & & \binom{4}{2} & & \binom{4}{3} & & \binom{4}{4} & & \leftarrow \text{row } 4
\end{array}$$

Figure 6.21

$$\begin{array}{ccccccccccl}
 & & & & 1 & & & & & & \leftarrow \text{row } 0 \\
 & & & 1 & & 1 & & & & & \leftarrow \text{row } 1 \\
 & & 1 & & 2 & & 1 & & & & \leftarrow \text{row } 2 \\
 & 1 & & 3 & & 3 & & 1 & & & \leftarrow \text{row } 3 \\
1 & & 4 & & 6 & & 4 & & 1 & & \leftarrow \text{row } 4
\end{array}$$

Pascal's triangle has many intriguing properties:

- Every row begins with and ends in 1. This is no coincidence, since $C(n, 0) = 1 = C(n, n)$.

*Although Pascal's triangle is named after Pascal, it appeared in a 1303 work by the Chinese mathematician Chu Shi-Kie.

- Pascal's triangle is symmetric about a vertical line through the middle. This is so since $C(n, r) = C(n, n - r)$.
- Any interior number in each row is the sum of the numbers immediately to its left and to its right in the preceding row. This is so by virtue of Pascal's identity.
- Recall that, by Example 6.22, the maximum number of places f_n into which a pizza can be divided with n distinct cuts is given by $f_n = C(n, 0) + C(n, 1) + C(n, 2)$. It is the sum of the first three numbers in row n of Pascal's triangle (see Figure 6.22).

Figure 6.22

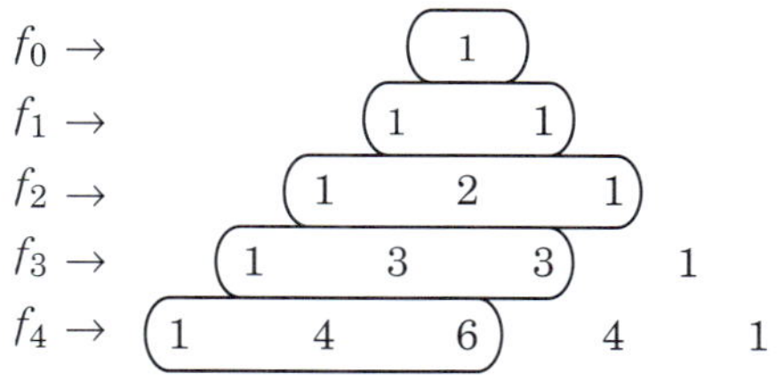

- By Example 6.26, the maximum number of nonoverlapping regions formed inside a circle by joining n distinct points on it is given by $g_n = C(n, 0) + C(n, 2) + C(n, 4)$. Using Pascal's identity, this formula can be rewritten as $g_n = C(n - 1, 0) + C(n - 1, 1) + C(n - 1, 2) + C(n - 1, 3) + C(n - 1, 4)$. (Verify this.) Consequently, the value of g_n can be obtained by adding the first five numbers in row $n-1$ of Pascal's identity (see Figure 6.23).

Figure 6.23

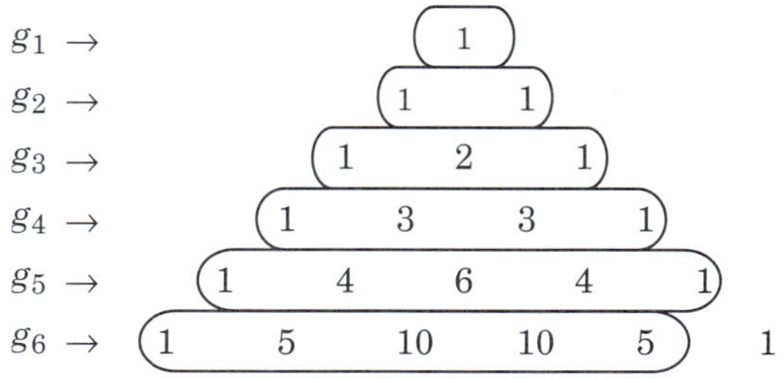

Pascal's Triangle and Abracadabra

Next we show how Polya's abracadabra problem is related to Pascal's triangle. Beginning with a 1 at the apex and using Pascal's identity, build the rhombic array in Figure 6.24. Each entry in Figure 6.24 gives the number of paths from the apex to the corresponding location. So there are $252 = C(10, 5)$ possible paths from the apex to the bottommost point in the array.

Figure 6.24

1

1 1

1 2 1

1 3 3 1

1 4 6 4 1

1 5 10 10 5 1

6 15 20 15 6

21 35 35 21

56 70 56

126 126

252

Next we show how Catalan numbers 1, 1, 2, 5, 14, 42, ... can be extracted from Pascal's triangle.

Pascal's Triangle and Catalan Numbers

In Chapter 2 we defined the nth *Catalan number* C_n as

$$C_n = \frac{(2n)!}{(n+1)!\,n!}, \qquad n \geq 0$$

Since $\dfrac{(2n)!}{n!\,n!} = \dbinom{2n}{n}$, this can be rewritten as

$$C_n = \frac{1}{n+1}\binom{2n}{n}, \qquad n \geq 0$$

Consequently, C_n can be obtained by dividing the *central binomial coefficient* $\dbinom{2n}{n}$ by $n+1$.

For example, $C_4 = \dfrac{1}{5}\dbinom{8}{4} = \dfrac{70}{5} = 14$; thus C_4 is obtained by dividing by 5 the central element 70 in row 8 in Pascal's triangle.

Figure 6.25 shows the first five central binomial coefficients; we can use them to compute the first five Catalan numbers.

Catalan numbers, like Fibonacci and Lucas numbers, have a propensity to appear in quite unexpected places. We shall pursue a few such delightful occurrences later in this Section and in Section 9.6.

The next theorem shows how the binomial coefficients and Theorem 6.13 are useful in finding the **binomial expansion** of $(x + y)^n$. We shall prove it using a combinatorial argument (see Exercise 43 for an algebraic method).

Figure 6.25

The central binomial coefficients.

$$
\begin{array}{ccccccccccccccccc}
 & & & & & & & & \textcircled{1} & & & & & & & & \\
 & & & & & & & 1 & & 1 & & & & & & & \\
 & & & & & & 1 & & \textcircled{2} & & 1 & & & & & & \\
 & & & & & 1 & & 3 & & 3 & & 1 & & & & & \\
 & & & & 1 & & 4 & & \textcircled{6} & & 4 & & 1 & & & & \\
 & & & 1 & & 5 & & 10 & & 10 & & 5 & & 1 & & & \\
 & & 1 & & 6 & & 15 & & \textcircled{20} & & 15 & & 6 & & 1 & & \\
 & 1 & & 7 & & 21 & & 35 & & 35 & & 21 & & 7 & & 1 & \\
1 & & 8 & & 28 & & 56 & & \textcircled{70} & & 56 & & 28 & & 8 & & 1
\end{array}
$$

THEOREM 6.16 **(The Binomial Theorem)*** Let x and y be any real numbers, and n any nonnegative integer. Then $(x+y)^n = \sum_{r=0}^{n} \binom{n}{r} x^{n-r}y^r$.

PROOF:

Since $(x+y)^n = (x+y)(x+y)\ldots(x+y)$ to n factors, $(x+y)^n$ is expanded by multiplying an x from some of the factors on the RHS and a y from the remaining factors. That is, every term is obtained by selecting an x from any of the $n-r$ factors and a y from the remaining r factors. Thus, every term in the expansion is of the form $Cx^{n-r}y^r$, where C denotes the coefficient and $0 \le r \le n$.

Notice that the coefficient of $x^{n-r}y^r$ is the number of ways of selecting an x from any $n-r$ of the n factors (and hence a y from the remaining r factors). Therefore,

$$
\begin{aligned}
\text{Coefficient of } x^{n-r}y^r &= \binom{n}{n-r} \\
&= \binom{n}{r}
\end{aligned}
$$

So, every term in the expansion is of the form $\binom{n}{r} x^{n-r}y^r$, where $0 \le r \le n$. Thus

$$(x+y)^n = \sum_{r=0}^{n} \binom{n}{r} x^{n-r}y^r \qquad \blacksquare$$

The next example illustrates the binomial theorem.

*The binomial theorem for $n = 2$ can be found in Euclid's work (ca. 300 B.C.).

EXAMPLE 6.40 Find the binomial expansion of $(2a - 3b)^4$.

SOLUTION:
Here $x = 2a$, $y = -3b$, and $n = 4$. Using the binomial theorem,

$$\begin{aligned}(2a-3b)^4 &= \binom{4}{0}(2a)^4(-3b)^0 + \binom{4}{1}(2a)^3(-3b)^1 + \binom{4}{2}(2a)^2(-3b)^2 \\ &\quad + \binom{4}{3}(2a)^1(-3b)^3 + \binom{4}{4}(2a)^0(-3b)^4 \\ &= (2a)^4 + 4(2a)^3(-3b) + 6(2a)^2(-3b)^2 + 4(2a)(-3b)^3 + (-3b)^4 \\ &= 16a^4 - 96a^3b + 216a^2b^2 - 216ab^3 + 81b^4\end{aligned}$$

■

The next example illustrates how to employ the binomial theorem to determine a particular coefficient in the expansion of $(ax + by)^n$.

EXAMPLE 6.41 Find the coefficient of x^3y^4 in the expansion of $(x + y)^7$.

SOLUTION:
By the binomial theorem, the coefficient of $x^{n-r}y^r$ is $\binom{n}{r}$, where $n = 7$ and $r = 4$. So

$$\text{Coefficient of } x^3y^4 = \binom{7}{4} = 35$$

■

Pascal's Triangle and the Nondecimal Bases Nine and Three

The binomial theorem and hence Pascal's triangle can be used to express any positive integer in bases nine and three. For example, by the binomial theorem,

$$\begin{aligned}10^n &= (9+1)^n \\ &= \sum_{r=0}^{n} \binom{n}{r} 9^{n-r}\end{aligned}$$

In particular,

$$\begin{aligned}10^3 &= \sum_{r=0}^{3} \binom{3}{r} 9^{3-r} \\ &= \binom{3}{0}9^3 + \binom{3}{1}9^2 + \binom{3}{2}9^1 \binom{3}{3}9^0 \\ &= 1\cdot 9^3 + 3\cdot 9^2 + 3\cdot 9 + 1\cdot 9^0 \\ &= 1331_{\text{nine}}\end{aligned}$$

Likewise, $10^4 = 14641_{\text{nine}}$ (Verify this.)

Since $9 = 3^2$, by the binomial theorem, we have

$$10^n = \sum_{r=0}^{n} \binom{n}{r} 3^{2n-2r}$$

For example,

$$\begin{aligned} 10^3 &= \binom{3}{0} 3^6 + \binom{3}{1} 3^4 + \binom{3}{2} 3^2 + \binom{3}{3} 3^0 \\ &= 1 \cdot 3^6 + 3 \cdot 3^4 + 3 \cdot 3^2 + 1 \cdot 3^0 \\ &= 1 \cdot 3^6 + 1 \cdot 3^5 + 0 \cdot 3^4 + 1 \cdot 3^3 + 0 \cdot 3^2 + 0 \cdot 3^1 + 1 \cdot 3^0 \\ &= 1101001_{\text{three}} \end{aligned}$$

This ternary expansion can be obtained from row 3 of Pascal's triangle, after converting each entry into base three and inserting zeros in between the entries.

The binomial theorem can conveniently establish several interesting combinatorial identities. Among them is one already seen in Chapter 2.

THEOREM 6.17

$$\sum_{r=0}^{n} \binom{n}{r} = 2^n$$

That is, the sum of the binomial coefficients is 2^n; in other words, a set with n elements has 2^n subsets.

PROOF:
By the binomial theorem,

$$(x + y)^n = \sum_{r=0}^{n} \binom{n}{r} x^{n-r} y^r$$

Let $x = y = 1$. Then

$$2^n = (1 + 1)^n = \sum_{r=0}^{n} \binom{n}{r} 1^{n-r} 1^r$$

That is,

$$\sum_{r=0}^{n} \binom{n}{r} = 2^n$$

■

Submatrices of a Matrix

The following example, an interesting application of Theorem 6.17, was proposed in 1943 by the well-known mathematics historian Howard Eves while at Syracuse University. The elegant solution, by A. Wayne of Flushing, New York, employs the addition and multiplication principles also.

EXAMPLE 6.42 Find the total number of submatrices of an $m \times n$ matrix.

SOLUTION:
Any r rows can be selected in $\binom{m}{r}$ ways. So, by Theorem 6.17, the total number of combinations of rows from m rows equals $\sum_{r=1}^{m}\binom{m}{r} = 2^m - 1$, Similarly, the total number of columns we can choose is $2^n - 1$. Thus there are $(2^m - 1)(2^n - 1)$ ways of choosing rows and columns; that is, there are $(2^m - 1)(2^n - 1)$ submatrices in an $m \times n$ matrix. ∎

Another identity emerging from the binomial theorem is given in the next theorem.

THEOREM 6.18

$$\binom{n}{0} + \binom{n}{2} + \binom{n}{4} + \cdots = \binom{n}{1} + \binom{n}{3} + \binom{n}{5} + \cdots \tag{6.8}$$

where $n \geq 1$. That is, the sum of the "even" binomial coefficients equals that of the "odd" binomial coefficients.

PROOF:
Again by the binomial theorem,

$$(x+y)^n = \sum_{r=0}^{n}\binom{n}{r}x^{n-r}y^r$$

Set $x = 1$ and $y = -1$. Then, for $n \geq 1$,

$$\begin{aligned} 0 = [1 + (-1)]^n &= \sum_{r=0}^{n}\binom{n}{r}1^{n-r}(-1)^r \\ &= \binom{n}{0} - \binom{n}{1} + \binom{n}{2} - \binom{n}{3} + \binom{n}{4} - \binom{n}{5} + \cdots \end{aligned}$$

That is,

$$\binom{n}{0} + \binom{n}{2} + \binom{n}{4} + \cdots = \binom{n}{1} + \binom{n}{3} + \binom{n}{5} + \cdots$$ ∎

We now make an interesting observation. Recall that the binomial coefficient $C(n, r)$ denotes the number of subsets of size r of a set with n elements, so the LHS of Equation (6.8) represents the total number of subsets with an even number of elements and the RHS represents that with an odd number of elements. Since the total number of subsets is 2^n, each equals 2^{n-1}, by Equation (6.8).

Next we turn to two interesting occurrences of Catalan numbers.

Catalan Numbers Revisited

The great Swiss mathematician Leonhard Euler (see Chapter 8) discovered Catalan numbers. He found them in his study of *triangulations* of convex polygons, that is, dividing the interior of a convex polygon into triangular areas by drawing nonintersecting diagonals. Let C_n denote the number of triangulations of convex $(n+2)$-gon, where $n \geq 1$. It follows from Figure 6.26 that $C_1 = 1$, $C_2 = 2$, $C_3 = 5$, and $C_4 = 14$. More generally, Euler established that

$$C_n = \frac{2 \cdot 6 \cdot 10 \cdots (4n-2)}{(n+1)!}$$

Figure 6.26

Triangulations of convex $(n+2)$-gons.

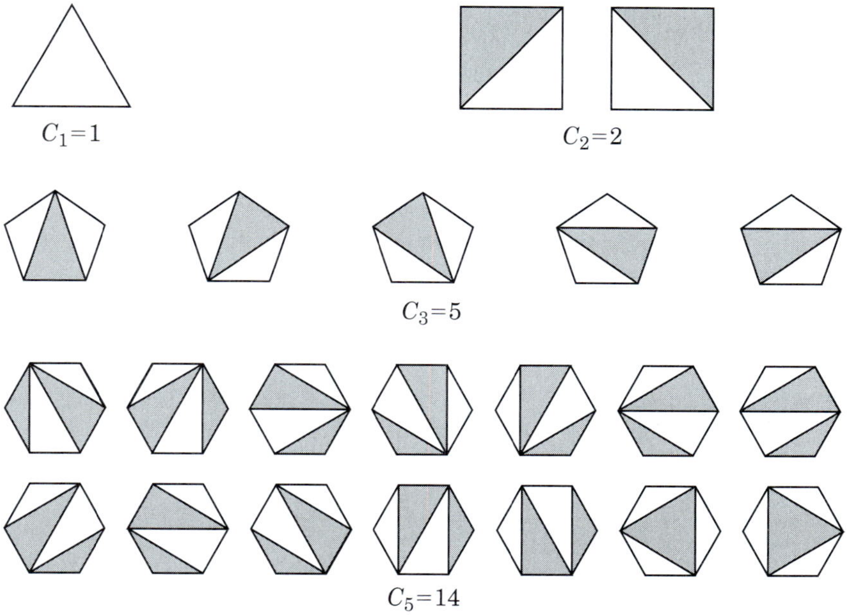

In 1759, the German mathematician Johann Andreas von Segner (1707–1777), a contemporary of Euler, established a recursive procedure to compute C_n:

$$C_n = C_0C_{n-1} + C_1C_{n-2} + \cdots + C_{n-1}C_0$$

where $C_0 = 1$. For example, $C_4 = 1 \cdot 5 + 1 \cdot 2 + 2 \cdot 1 + 5 \cdot 1 = 14$.

Catalan's Parenthesization Problem

Interestingly, Euler's triangulation problem is essentially the same as Catalan's parenthesization problem, which he solved in 1838: using n pairs of left and right parentheses, how many different ways we can parenthesize a sequence of $n + 1$ symbols for a binary nonassociative operation?

For example, with two symbols, there is one possibility: (ab); with three symbols, there are two possibilities: ((ab)c) and (a(ab)); and with four symbols, there are five ways: ((ab) (cd)), (((ab)c)d), (a(b(cd))), (a((bc)d), and ((a(bc))d); they yield the Catalan numbers 1, 2, and 5; and so on.

In 1961, H.G. Forder of the University of Auckland, New Zealand, showed that every triangulation of convex polygon yields a correctly parenthesized expression, and vice versa. This close relationship becomes clear if we consider the triangulation of the hexagon in Figure 6.27. We have labeled five of its sides a through e, leaving the base unlabeled. Label every diagonal spanning two adjacent sides with the concatentation of the corresponding labels in parentheses. Continue this algorithm until the base gets a label, as in Figure 6.28. Thus the triangulation in Figure 6.27 yields the correctly parenthesized expression (((ab)c)(de)). Retracing the steps, we can recover the triangulation from the expression.

Figure 6.27

Triangulation of a hexagon.

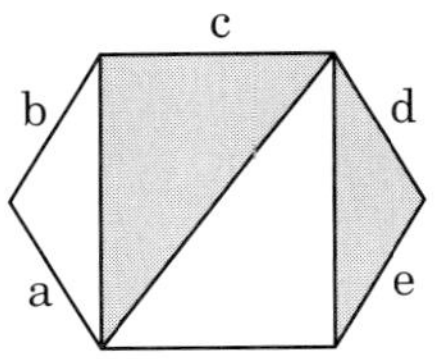

Figure 6.28

Parenthesized triangulation of a hexagon.

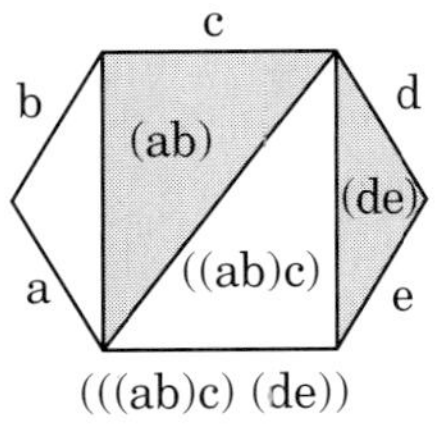

Triangulations, Parenthesized Expressions, and Binary Numbers

Each triangulation of a convex polygon and by extension the corresponding parenthesized expression, can be uniquely represented by a binary number. For instance, consider the expression (((ab)c)(de)) in Figure 6.28. If we replace each left parenthesis with a 1 and each letter with a 0 and ignore all right parentheses, the expression yields the binary number 111000100. We do not need to store the right parentheses, because if we know the left parentheses, we can insert the matching right parentheses correctly.

Thus such a binary number is a compact way of representing the expression and hence the triangulation.

We shall revisit this binary designation in Section 9.6 on binary trees.

Before closing this section, we derive the explicit formula for C_n. The proof employs the following generating function.

$$(1-4x)^{1/2} = 1 - 2\sum_{n=1}^{\infty} \frac{1}{n}\binom{2n-2}{n-1} x^n$$

An Explicit Formula for the nth Catalan Number (optional)

Let P_n denote the number of ways of parenthesizing $n+1$ symbols using n pairs of left and right parentheses. Then $P_1 = 1$. So, let $n \geq 2$. The first i symbols can be parenthesized in P_i ways and the next $n-i$ symbols in P_{n-i} ways, where $1 \leq i \leq n-1$. Using the multiplication and addition principles, we have the following recurrence relation for P_n:

$$P_n = P_1P_{n-1} + P_2P_{n-2} + \cdots + P_{n-1}P_1$$

(It follows from Segner's formula that $P_n = C_{n-1}$.)

Now consider the generating function

$$f(x) = P_1x + P_2x^2 + \cdots + P_nx^n + \cdots$$

Then

$$\begin{aligned}[f(x)]^2 &= P_1^2x^2 + (P_1P_2 + P_2P_1)x^3 + \cdots + (P_1P_{n-1} + P_2P_{n-2} \\ &\quad + \cdots + P_{n-1}P_1)x^n + \cdots \\ &= x^2 + P_3x^3 + \cdots + P_nx^n + \cdots \\ &= f(x) - x\end{aligned}$$

So

$$[f(x)]^2 - f(x) + x = 0$$

Solving,

$$f(x) = \frac{1 \pm \sqrt{1-4x}}{2}$$

Since $P_n > 0$ for every n, we take the minus sign, so

$$f(x) = \frac{1 - \sqrt{1-4x}}{2}$$

Using the above power series expansion, this yields

$$f(x) = \sum_{n=1}^{\infty} \frac{1}{n} \binom{2n-2}{n-1} x^n$$

Thus

$$P_n = \frac{1}{n} \binom{2n-2}{n-1} = C_{n-1}$$

So

$$C_n = \frac{1}{n+1} \binom{2n}{n}, \quad n \geq 0$$

as desired.

Exercises 6.6

Find the coefficient of each.

1. x^3y^5 in the expansion of $(x+y)^8$

2. x^4y^6 in the expansion of $(x-y)^{10}$

3. x^2y^6 in the expansion of $(2x+y)^8$

4. x^4y^5 in the expansion of $(2x-3y)^9$

Using the binomial theorem, expand each.

5. $(x+y)^4$ **6.** $(x-y)^5$ **7.** $(2x-1)^5$ **8.** $(x+2y)^6$

Find the middle term in the binomial expansion of each.

9. $\left(x+\frac{1}{x}\right)^4$ **10.** $\left(x-\frac{1}{x}\right)^6$

11. $\left(2x+\frac{2}{x}\right)^8$ **12.** $\left(x^2+\frac{1}{x^2}\right)^{10}$

Find the largest binomial coefficient in the expansion of each.

13. $(x+y)^5$ **14.** $(x+y)^6$ **15.** $(x+y)^7$ **16.** $(x+y)^8$

17. Using Exercises 13–16, predict the largest binomial coefficient in the expansion of $(x+y)^n$.

Use Pascal's triangle in Figure 6.21 to answer Exercises 18 and 19.

18. Find the sum of the numbers along the northeast diagonals.

19. What do you notice about them?

The nth Fibonacci number F_n is given by the sum of the numbers along the nth northeast diagonal of Pascal's triangle; that is,

$$F_n = \sum_{i=0}^{\lfloor (n-1)/2 \rfloor} \binom{n-i-1}{i}$$

Using this formula, compute each Fibonacci number.

20. F_1 **21.** F_2 **22.** F_5 **23.** F_6

The **Bell numbers** B_n, named after the English mathematician Eric T. Bell (1883–1960) and used in combinatorics, are defined recursively as follows:

$$B_0 = 1$$

$$B_n = \sum_{i=0}^{n-1} \binom{n-1}{i} B_i, \quad n \geq 1$$

Compute each Bell number.

24. B_2 **25.** B_3 **26.** B_4 **27.** B_5

Using the binomial theorem, prove each.

28. $2^{4n} + 3n - 1$ is divisible by 9. (*Hint*: $2 = 3 - 1$.)

29. $4^{2n} + 10n - 1$ is divisible by 25. (*Hint*: $4 = 5 - 1$.)

30. $\sum_{r=0}^{n} \binom{2n}{2r} = \sum_{r=1}^{n} \binom{2n}{2r-1}$ (*Hint*: Use Theorem 6.18.)

31. $\sum_{r=0}^{n} 2^r \binom{n}{r} = 3^n$

32. $\sum_{r=0}^{n} \binom{n}{r} \binom{n}{n-r} \binom{2n}{n}$

[*Hint*: Consider $(1+x)^{2n} = (1+x)^n(1+x)^n$. Equate the coefficients of x^n from either side.]

33. $\sum_{i=1}^{n} \binom{n}{i-1} \binom{n}{i} = \binom{2n}{n+1}$

[*Hint*: Consider $(1+x)^{2n} = (x+1)^n(1+x)^n$. Equate the coefficients of x^{n+1} from both sides.]

Evaluate each sum.

34. $1\binom{n}{1} + 2\binom{n}{2} + 3\binom{n}{3} + \cdots + n\binom{n}{n}$

(*Hint*: Let S denote the sum. Use S and the sum in the reverse order to compute $2S$.)

35. $a\binom{n}{0} + (a+d)\binom{n}{1} + (a+2d)\binom{n}{2} + \cdots + (a+nd)\binom{n}{n}$

(*Hint*: Use the same hint as in Exercise 34.)

36. Show that $C(n, r-1) < C(n, r)$ if and only if $r < \dfrac{n+1}{2}$, where $0 \le r < n$.

37. Using Exercise 36, prove that the largest binomial coefficient $C(n, r)$ occurs when $r = \lfloor n/2 \rfloor$.

Using induction, prove each.

38. $\binom{n}{0} + \binom{n+1}{1} + \binom{n+2}{2} + \cdots + \binom{n+r}{r} = \binom{n+r+1}{r}$

(*Hint*: Use Pascal's identity.)

39. $1\binom{n}{1} + 2\binom{n}{2} + \cdots + n\binom{n}{n} = n2^{n-1}$

40. $\binom{n}{0}^2 + \binom{n}{1}^2 + \binom{n}{2}^2 + \cdots + \binom{n}{n}^2 = \binom{2n}{n}$

From the binomial expansion $(1+x)^n = \sum_{r=0}^{n} \binom{n}{r} x^r$, it can be shown using calculus that $n(1+x)^{n-1} = \sum_{r=1}^{n} \binom{n}{r} r x^{r-1}$. Using this result, prove each.

41. $1\binom{n}{1} + 2\binom{n}{2} + 3\binom{n}{3} + \cdots + n\binom{n}{n} = n2^{n-1}$

42. $1\binom{n}{1} + 3\binom{n}{3} + 5\binom{n}{5} + \cdots = 2\binom{n}{2} + 4\binom{n}{4} + 6\binom{n}{6} + \cdots = n2^{n-2}$

43. Prove the binomial theorem, using mathematical induction.

***44.** Using a combinatorial argument prove that

$$\binom{n}{m}\binom{m}{r} = \binom{n}{r}\binom{n-r}{m-r} \qquad \textbf{(Newton's identity)}$$

(*Hint*: Select an r-element subset of an n-element set in two ways.)

45. Prove the result in Exercise 44 algebraically.

The following result is known as **Vandermonde's identity**, after the German mathematician Abnit-Theophile Vandermonde (1735–1796):

$$\binom{m+n}{r} = \binom{m}{0}\binom{n}{r} + \binom{m}{1}\binom{n}{r-1} + \binom{m}{2}\binom{n}{r-2} + \cdots + \binom{m}{r}\binom{n}{0}$$

***46.** Prove Vandermonde's identity, using a combinatorial argument. (*Hint*: Consider the ways of selecting r people from a group of m men and n women.)

47. Prove Vandermonde's identity algebraically. [*Hint*: Consider $(1+x)^m(x+1)^n = (1+x)^{m+n}$.]

48. Find a formula for $\sum_{i=2}^{n} \binom{i}{2}$.

49. Using induction, establish the formula guessed in Exercise 48.

50. Find a formula for $\sum_{i=3}^{n} \binom{i}{3}$.

51. Using induction, establish the formula guessed in Exercise 50.

52. Using Exercises 48–51, predict a formula for $\sum_{i=k}^{n} \binom{i}{k}$.

*6.7 The Generalized Inclusion–Exclusion Principle (GIEP) (optional)

The generalized version of the inclusion–exclusion principle has interesting applications to number theory, surjections, and derangements, as will be seen shortly.

THEOREM 6.19 **(GIEP)** Let $A_1, A_2, \ldots, A_n$ be n finite sets. Then

$$\left|\bigcup_{i=1}^{n} A_i\right| = \sum_{1\le i\le n} |A_i| - \sum_{1\le i<j\le n} |A_i \cap A_j| + \sum_{1\le i<j<k\le n} |A_i \cap A_j \cap A_k| - \cdots + (-1)^{n+1}\left|\bigcap_{i=1}^{n} A_i\right| \tag{6.9}$$

PROOF:

To prove this formula, we show that every element on the LHS is counted exactly once by the expression on the RHS of Equation (6.9).

Let x be an arbitrary element that occurs in exactly r of the sets $A_1, A_2, \ldots, A_n$. Then x is counted $C(r,1)$ times in $\sum |A_i|$, $C(r,2)$ times in $\sum |A_i \cap A_j|$, $C(r,3)$ times in $\sum |A_i \cap A_j \cap A_k|$, and so on. Therefore, the number of times x is counted on the RHS of Equation (6.9) is

$$C(r,1) - C(r,2) + C(r,3) - \cdots + (-1)^{r+1}C(r,r) = \sum_{k=1}^{r} (-1)^{k+1}C(r,k)$$

But

$$\sum_{k=0}^{r}(-1)^k C(r,k) = 0, \text{ by Equation (6.8)}$$

That is,

$$C(r,0) - \sum_{k=1}^{r}(-1)^{k+1} C(r,k) = 0$$

$$\sum_{k=1}^{r}(-1)^{k+1} C(r,k) = C(r,0)$$

That is,

$$\sum_{k=1}^{r}(-1)^{k+1} C(r,k) = 1$$

Consequently, every element x is counted exactly once on the RHS of equation (6.9). This completes the proof. ■

In many applications, an alternate form of the inclusion–exclusion principle works nicely. For instance, it can find the number of primes not exceeding a positive integer, as will be seen shortly.

An Alternate Inclusion–Exclusion Formula

Let S be a finite set. We would like to find the number of elements in S that have none of the properties $P_1, P_2, \ldots, P_n$. Let A_i be the set of elements in S that have property P_i. Let $N(P_{i_1}, P_{i_2}, \ldots, P_{i_k})$ denote the number of elements in S that have properties $P_{i_1}, P_{i_2}, \ldots, P_{i_k}$. Let $N(P'_1, P'_2, \ldots, P'_n)$ denote the number of elements in S that have none of the properties $P_1, P_2, \ldots, P_n$. Then

$$N(P_{i_1}, P_{i_2}, \ldots, P_{i_k}) = |A_{i_1} \cap A_{i_2} \cap \cdots \cap A_{i_k}|$$

So $N(P'_1 P'_2 \cdots P'_n) = |S| - |A_1 \cup A_2 \cup \cdots \cup A_n|$

$$= |S| - \left[\sum_{1 \le i \le n} N(P_i) - \sum_{1 \le i < j \le n} N(P_i P_j) + \sum_{1 \le i < j < k \le n} N(P_i P_j P_k) - \cdots \right.$$
$$\left. + (-1)^{n+1} N(P_1 P_2 \cdots P_n) \right]$$

$$= |S| - \sum_{1 \le i \le n} N(P_i) + \sum_{1 \le i < j \le n} N(P_i P_j) - \sum_{1 \le i < j < k \le n} N(P_i P_j P_k) + \cdots$$
$$+ (-1)^n N(P_1 P_2 \cdots P_n)$$

This is the **alternate inclusion–exclusion principle**.

This formula can find the number of primes not exceeding a positive integer n, as the next example demonstrates.

EXAMPLE 6.43 Find the number of primes ≤ 100.

SOLUTION:
Let $S = \{n \in \mathbb{N} \mid 1 < n \leq 100\}$. By Theorem 4.2, a positive integer n is a prime if and only if it has no prime factors $\leq \lfloor\sqrt{n}\rfloor$. Therefore, an element in S is prime if and only if it has no prime factors ≤ 10. There are four primes ≤ 10, namely, 2, 3, 5, and 7. Thus the primes ≤ 100 are these four primes, and those integers in S not divisible by 2, 3, 5, or 7.

Let P_2 be the property that an integer in S is divisible by 2, P_3 the property that an integer in S is divisible by 3, P_5 the property that an integer in S is divisible by 5, and P_7 the property that an integer in S is divisible by 7. Then $N(P'_2 P'_3 P'_5 P'_7)$ denotes the number of integers in S not divisible by 2, 3, 5, or 7. Thus there are $4 + N(P'_2 P'_3 P'_5 P'_7)$ primes in S.

To find $N(P'_2 P'_3 P'_5 P'_7)$: First notice that $|S| = 99$. Secondly, let $r, s, t \in \{2, 3, 5, 7\}$. Since r and s are primes, an integer has property $P_r P_s$ if it has both properties P_r and P_s. This process can be extended to $P_r P_s P_t$ and $P_2 P_3 P_5 P_7$. Therefore, the number of elements in S having property $P_r P_s$ is given by $\lfloor 100/rs \rfloor$, the number of elements in S having property $P_r P_s P_t$ is given by $\lfloor 100/rst \rfloor$, and those with $P_2 P_3 P_5 P_7$ by $\lfloor 100/2 \cdot 3 \cdot 5 \cdot 7 \rfloor$.

By the alternate inclusion–exclusion principle,

$$
\begin{aligned}
N(P'_1, P'_2, \ldots, P'_n) &= |S| - \sum N(P_i) + \sum N(P_i P_j) - \sum N(P_i P_j P_k) \\
&\quad + N(P_2 P_3 P_5 P_7) \\
&= 99 - [N(P_2) + N(P_3) + N(P_5) + N(P_7)] \\
&\quad + [N(P_2 P_3) + N(P_2 P_5) + N(P_2 P_7) + N(P_3 P_5) + N(P_3 P_7) \\
&\quad + N(P_5 P_7)] - [N(P_2 P_3 P_5) + N(P_2 P_3 P_7) + N(P_3 P_5 P_7)] \\
&\quad + N(P_2 P_3 P_5 P_7) \\
&= 99 - (\lfloor 100/2 \rfloor + \lfloor 100/3 \rfloor + \lfloor 100/5 \rfloor + \lfloor 100/7 \rfloor) + (\lfloor 100/2 \cdot 3 \rfloor \\
&\quad + \lfloor 100/2 \cdot 5 \rfloor + \lfloor 100/2 \cdot 7 \rfloor + \lfloor 100/3 \cdot 5 \rfloor + \lfloor 100/3 \cdot 7 \rfloor \\
&\quad + \lfloor 100/5 \cdot 7 \rfloor) - (\lfloor 100/2 \cdot 3 \cdot 5 \rfloor + \lfloor 100/2 \cdot 3 \cdot 7 \rfloor + \lfloor 100/3 \cdot 5 \cdot 7 \rfloor) \\
&\quad + \lfloor 100/2 \cdot 3 \cdot 5 \cdot 7 \rfloor \\
&= 99 - (50 + 33 + 20 + 14) + (16 + 10 + 7 + 6 + 4 + 2) \\
&\quad - (3 + 2 + 1) + 0 \\
&= 21
\end{aligned}
$$

Thus, there are $4 + 21 = 25$ primes ≤ 100. ■

We now present two delightful applications of the generalized inclusion–exclusion principle and the binomial theorem.

Counting Surjections

First, we develop an explicit formula for the number N of surjections f from a finite set A to a finite set B. To this end, let $|A| = m$ and $|B| = n$, where $|S|$ denotes the cardinality of the set S. If $m < n$, no surjections from A to B can be defined; so we let $m \geq n$.

Let a be any element in A. Since $f(a)$ has n choices, a total of n^m functions can be defined from A to B. It now follows that

$$N = n^m - (\text{number of functions that are } \textit{not} \text{ surjective})$$

So it suffices to count the number of functions from A to B that are not surjective; this is where we shall invoke the GIEP.

For convenience, let $A = \{1, 2, 3, \ldots, m\}$ and $B = \{b_1, b_2, b_3, \ldots, b_n\}$. Let S_i denote the set of all functions from A to B that do not produce b_i as an output, where $1 \leq i \leq n$. Then $S = S_1 \cup S_2 \cup \cdots \cup S_n$ denotes the set of all functions from A to B that do not output at least one element of B. In other words, $S = S_1 \cup S_2 \cup \cdots \cup S_n$ denotes the set of nonsurjections from A to B, so $|S| = |S_1 \cup S_2 \cup \cdots \cup S_n|$ denotes the number of nonsurjections from A to B. By Theorem 6.19,

$$|S| = \sum_{i=1}^{n} |S_i| - \sum_{i<j} |S_i \cap S_j| + \sum_{i<j<k} |S_i \cap S_j \cap S_k| - \cdots \pm |S_1 \cap S_2 \cap \cdots \cap S_n|$$

To begin with, consider S_1. It consists of all functions that do not output b_1; in other words, S_1 consists of all functions from A to $\{b_2, b_3, \ldots, b_n\}$. There are $(n-1)^m$ such functions; so $|S_1| = (n-1)^m$. Similarly, S_2 consists of all functions from A to $\{b_1, b_3, \ldots, b_n\}$. There are $(n-1)^m$ such functions; so $|S_2| = (n-1)^m$. More generally, $|S_i| = (n-1)^m$, where $1 \leq i \leq n$. Therefore,

$$\sum_{i=1}^{n} |S_i| = n(n-1)^m = \binom{n}{1}(n-1)^m$$

To compute $|S_i \cap S_j|$, where $i < j$, let us first investigate $S_1 \cap S_2$. It consists of all functions from A to B that do not output b_1 or b_2. So it consists of all functions from A to $\{b_3, b_4, \ldots, b_n\}$. There are $(n-2)^m$ such functions; so $|S_1 \cap S_2| = (n-2)^m$. Similarly, $S_2 \cap S_3$ consists of all functions from A to $\{b_1, b_4, \ldots, b_n\}$; so $|S_2 \cap S_3| = (n-2)^m$. Continuing like this, it follows that $|S_i \cap S_j| = (n-2)^m$, where $i < j$. Since there are $\binom{n}{2}$ pairs of sets S_1 through S_n, it follows that

$$\sum_{i<j} |S_i \cap S_j| = \binom{n}{2}(n-2)^m$$

Let us now compute $|S_i \cap S_j \cap S_k|$, where $i < j < k$. For convenience, we begin with $S_1 \cap S_2 \cap S_3$. It consists of all functions from A to B that do not output b_1, b_2, or b_3. Thus $S_1 \cap S_2 \cap S_3$ consists of all functions from A to $\{b_4, b_5, \ldots, b_n\}$; so $|S_1 \cap S_2 \cap S_3| = (n-3)^m$. Using the same argument, it follows that $|S_1 \cap S_2 \cap S_3| = (n-3)^m$, where $i < j < k$. Since there are $\binom{n}{3}$ triplets of the sets S_1 through S_n, it follows that

$$\sum_{i<j<k} |S_i \cap S_j \cap S_k| = \binom{n}{3}(n-3)^m$$

Slowly but surely, a pattern is emerging. In lieu of conjecturing the formula for the intersection $S_1 \cap S_2 \cap \cdots \cap S_j$ of j (distinct) sets S_1 through S_j, let us apply the same argument as above. As before, $|S_1 \cap S_2 \cap \cdots \cap S_j|$ denotes the number of functions from A to $\{b_{j+1}, b_{j+2}, \ldots, b_n\}$, so $|S_1 \cap S_2 \cap \cdots \cap S_j| = (n-j)^m$. Since there are $\binom{n}{j}$ such intersections of n sets, it follows that

$$\sum |S_1 \cap S_2 \cap \cdots \cap S_j| = \binom{n}{j}(n-j)^m$$

Finally, we would like to make sure that the same argument works for the last term $|S_1 \cap S_2 \cap \cdots \cap S_n|$. The set $S_1 \cap S_2 \cap \cdots \cap S_n$ consists of all functions that output none of the elements b_1 through b_n. But there are no such functions, so

$$|S_1 \cap S_2 \cap \cdots \cap S_n| = 0 = \binom{n}{n}(n-n)^m$$

Collecting all the pieces together, by Theorem 6.19, we have

$$\begin{aligned}|S| &= \binom{n}{1}(n-1)^m - \binom{n}{2}(n-2)^m + \binom{n}{3}(n-3)^m - \cdots \\ &\quad + (-1)^{n+1}\binom{n}{n}(n-n)^m \\ &= \sum_{r=1}^{n}(-1)^{r+1}\binom{n}{r}(n-r)^m\end{aligned}$$

Thus

$$\begin{aligned}N &= n^m - |S| \\ &= n^m - \sum_{r=1}^{n}(-1)^{r+1}\binom{n}{r}(n-r)^m \\ &= \sum_{r=0}^{n}(-1)^{r}\binom{n}{r}(n-r)^m \qquad (6.10)\end{aligned}$$

For example, let $A = \{1, 2, 3\}$ and $B = \{\mathrm{x}, \mathrm{y}\}$; so a total of $2^3 = 8$ functions can be defined from A to B. They are pictured in Figure 6.29.

Figure 6.29

Functions from $\{1, 2, 3\}$ to $\{x, y\}$.

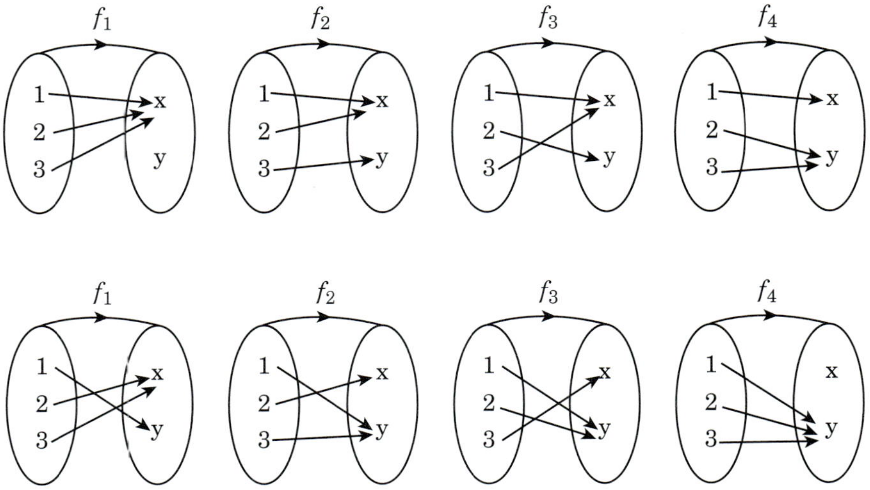

Of these functions, two are *not* surjective; they are displayed in Figure 6.30.

Figure 6.30

Nonsurjections from $\{1, 2, 3\}$ to $\{x, y\}$.

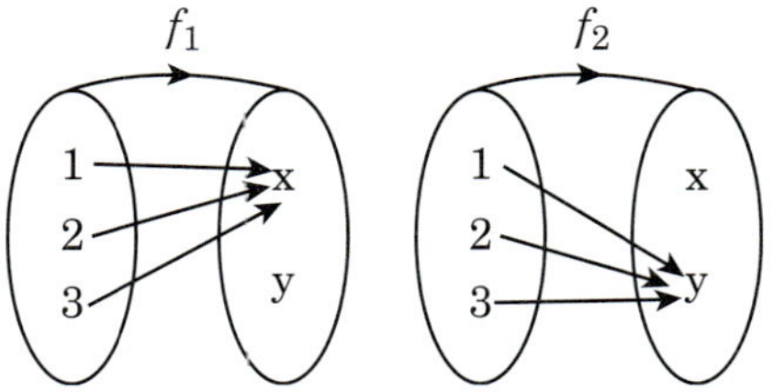

Thus there are

$$\sum_{r=0}^{2} (-1)^r \binom{2}{r} (2-r)^r = 8 - 2 + 0 = 6$$

surjections from A to B; they are displayed in Figure 6.31.

Counting Derangements

Next, we develop an explicit formula for the number of derangements D_n, again using Theorem 6.19. The technique employed is similar to the one we used for counting surjections.

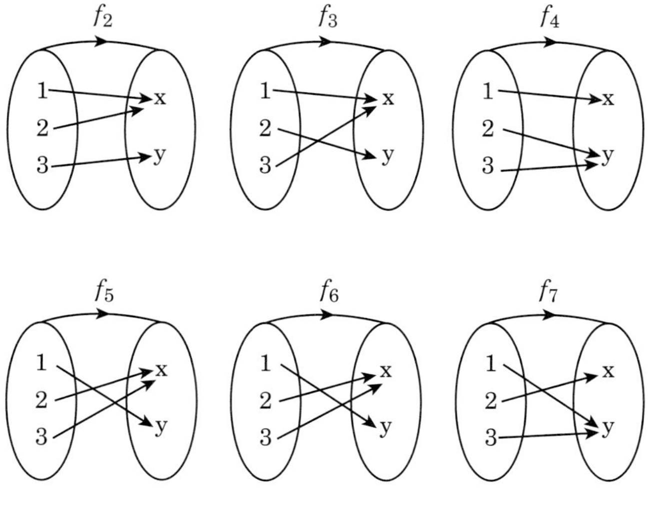

Figure 6.31

Surjections from $\{1, 2, 3\}$ to $\{x, y\}$.

Suppose the n elements are $1, 2, 3, \ldots, n$. They can be arranged in $n!$ ways, so

$$D_n = n! - \text{(number of permutations that leave at least one element fixed)}$$

To find the number of permutations that are *not* derangements, let S_i denote the set of permutations that leave the element i in position i, where $1 \le i \le n$. Then $S = \bigcup_{i=1}^{n} S_i$ denotes the set of all permutations of the elements 1 through n that are not derangements, so $D_n = n! - |S|$. To compute the value of $|S|$, as before, we apply Theorem 6.19.

To begin with, consider S_1. It consists of all permutations that leave 1 in position 1. The remaining $n - 1$ elements can be anywhere else and hence can be arranged in $(n - 1)!$ ways, so $|S_1| = (n - 1)!$ Similarly, S_2 consists of all permutations that leave 2; then also the remaining $n - 1$ elements can be arranged in (n − 1)! ways; so $|S_2| = (n - 1)!$ In general, $|S_i| = (n - 1)!$, where $1 \le i \le n$. Therefore,

$$\sum_{i=1}^{n} |S_i| = n(n-1)! = \binom{n}{1}(n-1)!$$

We now compute $|S_i \cap S_j|$, where $i < j$. $S_1 \cap S_2$, for instance, consists of all permutations that leave both 1 and 2 fixed. The remaining $n - 2$ elements can be rearranged in $(n - 2)!$ ways, so $|S_1 \cap S_2| = (n - 2)!$ More generally, $|S_i \cap S_j| = (n - 2)!$, where $i < j$. There are $\binom{n}{2}$ such pairs of intersections, so

$$\sum_{i<j} |S_i \cap S_j| = \binom{n}{2}(n-2)!$$

We now compute $|S_i \cap S_j \cap S_k|$, where $i < j < k$. The set $S_i \cap S_j \cap S_k$ consists of all permutations that leave the elements i, j, and k fixed. The remaining $n - 3$ elements can be permuted in $(n-3)!$ ways. There are $\binom{n}{3}$ such triplets of intersections, so

$$\sum_{i<j<k} |S_i \cap S_j \cap S_k| = \binom{n}{3} (n-3)!$$

More generally, $|S_1 \cap S_2 \cap \cdots \cap S_j| = (n-j)!$, since the elements $1, 2, \ldots, j$ remain fixed and the remaining $n - j$ elements can be rearranged in $(n-j)!$ ways. Since there are $\binom{n}{j}$ such intersections, it follows that

$$\sum |S_1 \cap S_2 \cap \cdots \cap S_j| = \binom{n}{j} (n-j)!$$

Finally, $S_1 \cap S_2 \cap \cdots \cap S_n$ consists of all permutations that leave every element fixed. There is only one such permutation, so

$$|S_1 \cap S_2 \cap \cdots \cap S_n| = 1 = \binom{n}{n} (n-n)!$$

Tying all the pieces together, we get

$$\begin{aligned}
|S| &= \sum_{i=1}^{n} |S_i| - \sum_{i<j} |S_i \cap S_j| + \sum_{i<j<k} |S_i \cap S_j \cap S_k| - \cdots \pm |S_1 \cap S_2 \cap \cdots \cap S_n| \\
&= \binom{n}{1}(n-1)! - \binom{n}{2}(n-2)! + \binom{n}{3}(n-3)! - \cdots + (-1)^{n+1}\binom{n}{n}(n-n)! \\
&= \sum_{r=1}^{n} (-1)^{r+1} \binom{n}{r} (n-r)!
\end{aligned}$$

Thus

$$\begin{aligned}
D_n &= n! - |S| \\
&= n! - \sum_{r=1}^{n} (-1)^{r+1} \binom{n}{r} (n-r)! \\
&= \binom{n}{0}(n-0)! - \sum_{r=1}^{n} (-1)^{r+1} \binom{n}{r} (n-r)! \\
&= \sum_{r=0}^{n} (-1)^{r} \binom{n}{r} (n-r)! \qquad (6.11)
\end{aligned}$$

For example,

$$\begin{aligned} D_4 &= \sum_{r=0}^{4} (-1)^r \binom{4}{r} (4-r)! \\ &= \binom{4}{0} 4! - \binom{4}{1} 3! + \binom{4}{2} 2! - \binom{4}{3} 1! + \binom{4}{3} 0! \\ &= 24 - 4 \cdot 6 + 6 \cdot 2 - 4 \cdot 1 + 1 \\ &= 9 \end{aligned}$$

as we found in Example 6.17.

Figures 6.32–6.34 represent the three major steps in the development of formula (6.11) for $n = 4$.

Figure 6.32

Permutations of four elements.

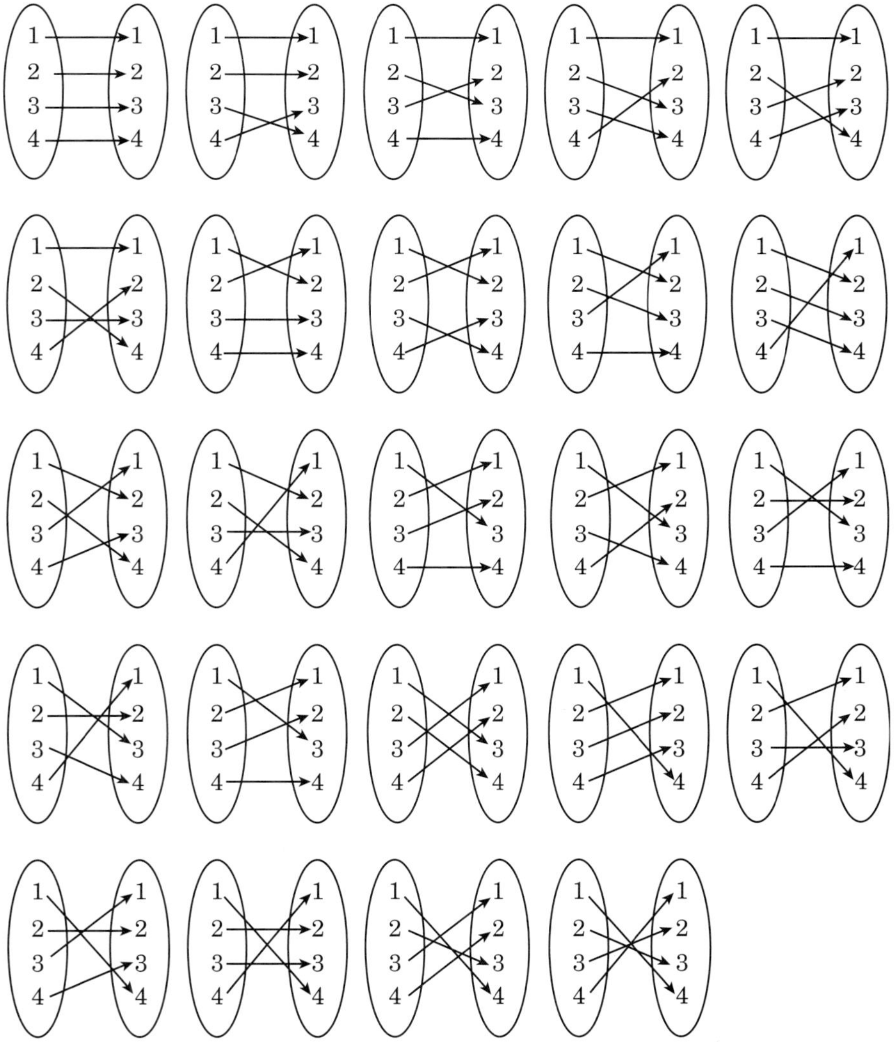

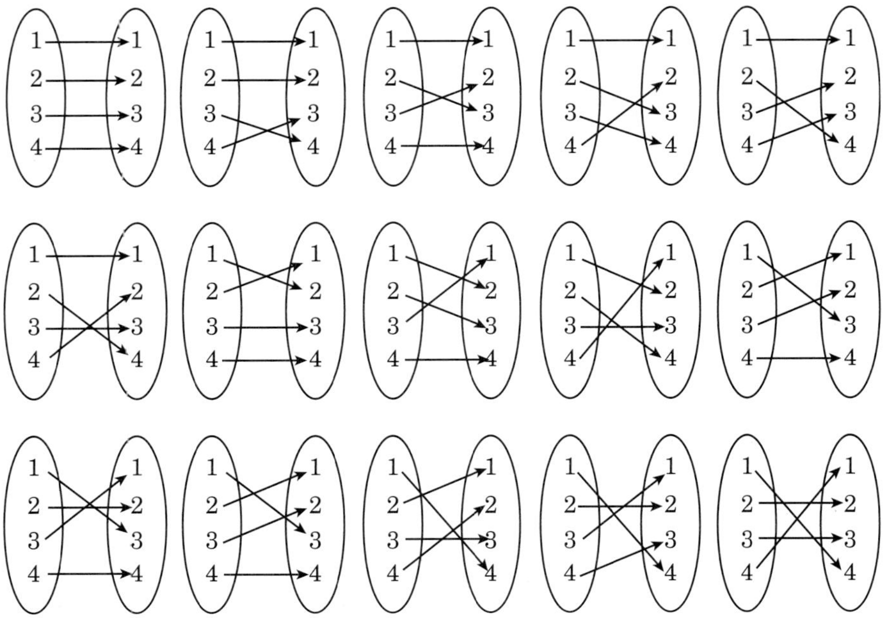

Figure 6.33

Permutations that are not derangements.

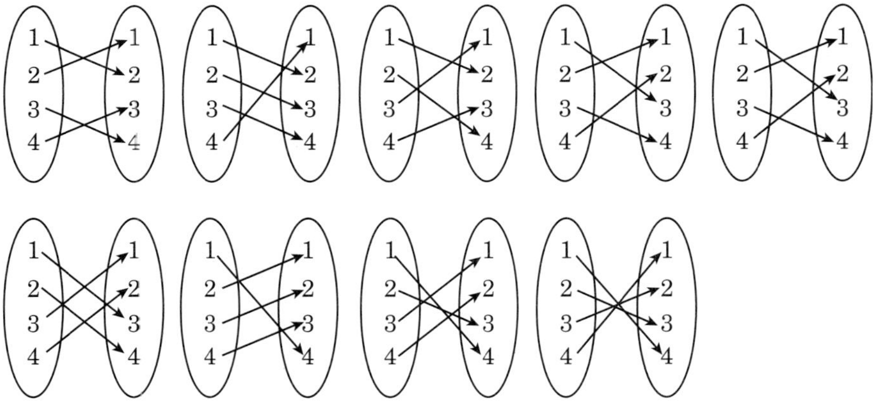

Figure 6.34

Derangements of four elements.

Notice that formula (6.11) can be rewritten as

$$D_n = n! \sum_{r=0}^{n} \frac{(-1)^r}{r!} \tag{6.12}$$

Exercises 6.7

1. A survey conducted among 300 adults shows that 160 like to have their houses painted white and 140 like blue. Seventy-four like both colors. How many do not like either color?

2. A survey among 100 consumers shows that of the two laundry detergents, Lex and Rex, 45 like Lex, 60 like Rex, and 20 like both. How many surveyed do not like either of them?

Find the number of positive integers ≤ 1000 and *not* divisible by:

3. 3 or 5
4. 5 or 6
5. 2, 3, or 5
6. 3, 5, or 7

Using the alternate inclusion–exclusion formula, find the number of primes *not* exceeding:

7. 75
8. 110
9. 125
10. 129

Find the number of solutions to each equation with nonnegative integer variables.

*11. $x + y + z = 11, x \leq 3, y \leq 4, z \leq 5$

*12. $w + x + y + z = 13, w \leq 3, x, y \leq 4, z \leq 7$

13. In 1984, E. T. H. Wang of Wilfrid Laurier University, Waterloo, Ontario, Canada, established that

$$\sum_{r=1}^{n} r^3 \binom{n}{r} D_{n-r} = 5n!$$

Verify the formula for $n = 5$ and $n = 6$.

6.8 Discrete Probability (optional)

The groundwork for probability theory was laid by chance in 1654 when an aristocratic gambler, Chevalier de Mere, asked Blaise Pascal the following question: *If two players of equal skill are forced to quit a game before it is over, how should the stakes be divided between them?* The problem sounds simple — the stakes should be divided so the person who had the greater chance to win the game when they stopped playing gets more than his opponent. Pascal communicated this problem to Pierre de Fermat and they solved it independently. This began probability theory.

Suppose you flip a coin. It can land heads up (H) or tails up (T) with equal likelihood. Each of them, H or T, is an **outcome** of the **experiment** of tossing the coin. The set {H, T} of possible outcomes of the experiment is the **sample space** of the experiment.

*Based on T. Koshy, *Finite Mathematics and Calculus with Applications*, Scott, Foresman Glenview, IL, 1979, pp. 87–134.

Sample Space and Event

The set of all possible outcomes of an experiment is the **sample space** of the experiment, denoted by S. (Throughout, we assume S is nonempty and finite.) An **event** E is a subset of the sample space. An outcome in E is a **favorable outcome** (or **success**); an outcome not in E is an **unfavorable outcome** (or **failure**). If $E = \emptyset$, E is an **impossible event**. If $|E| = 1$, then E is a **simple event**. The event $E' = S - E$ is the **complement** of event E.

For example, consider the experiment of tossing three coins. By the multiplication principle, the sample space S consists of $2 \cdot 2 \cdot 2 = 8$ possible outcomes: $S = \{\text{HHH}, \text{HHT}, \text{HTH}, \text{HTT}, \text{THH}, \text{THT}, \text{TTH}, \text{TTT}\}$. Let A be the event of obtaining exactly two heads, B that of obtaining at least two heads, and C that of obtaining four heads. Then $A = \{\text{HHT}, \text{HTH}, \text{THH}\}$ and $B = \{\text{HHT}, \text{HTH}, \text{THH}, \text{HHH}\}$. Clearly, $C = \emptyset$, an impossible event.

The definition of the probability of an event was given by the outstanding French mathematician Pierre-Simon Laplace.

Probability of an Event

Let E be an event of a finite sample space S consisting of equally likely outcomes. Then the **probability** of the event, denoted by $p(E)$, is defined by

$$p(E) = \frac{|E|}{|S|} = \frac{\text{number of ways } E \text{ can occur}}{\text{total number of possible outcomes}}$$

The next three examples illustrate this definition.

EXAMPLE 6.44 Suppose a card is drawn at random from a standard deck of playing cards. Find the probability that it will be a spade.

SOLUTION:

Since a standard deck contains 52 cards, the number of possible outcomes of drawing a card is 52. There are 13 spades in the deck; so a spade can be chosen in 13 ways. Hence, the probability that the card drawn is a spade is given by

$$\begin{aligned} p(E) &= \frac{|E|}{|S|} = \frac{\text{number of ways } E \text{ can occur}}{\text{total number of possible outcomes}} \\ &= \frac{13}{52} = \frac{1}{4} \end{aligned}$$

■

EXAMPLE 6.45 A Massachusetts state weekly lottery number is divided into three boxes, colored yellow, blue, and white. These boxes contain a three-digit number, a two-digit number, and a one-digit number, respectively. Assume that 000-00-0 is a valid lottery number and this week's winning number is

123 YELLOW	45 BLUE	6 WHITE

***Pierre-Simon Laplace** (1749–1827), one of the most influential mathematicians and scientists of all time, was born in Beaumont-en-Auge, Normandy, France, into a prosperous family of farmers and merchants. After graduating from school, he entered the University of Caen to study theology as his father had planned. During 2 years at Caen, Laplace discovered his mathematical talents and left for Paris to pursue a career in mathematics. With the help of mathematician and philosopher Jean le Rond d'Alembert, he became professor of mathematics at the École Militaire. At the age of 24, Laplace was elected a member of the Paris Academy of Sciences.*

Laplace is one of the founders of probability theory and is best known for his outstanding contributions to celestial mechanics, the study of motions of celestial bodies. In addition, he did significant work in applied mathematics and mathematical statistics.

Laplace published two monumental treatises, Traité de Méchanique Céleste *(five volumes, 1799–1825) and* Théorie Analytique des Probabilités *(1812). The former, one of the greatest scientific works of the 19th century, earned him the title* the Newton of France, *while the latter was the first comprehensive book on probability.*

Laplace was a strong advocate of the metric system and played an important role in reorganizing the École Polytechnique, a prestigious engineering school founded in 1795.

Indifferent to political conditions, Laplace remained loyal to every party in power, which enabled him to preserve his high scientific status throughout the turbulent era in which he lived.

He died in Paris.

Table 6.3 summarizes the payoffs. Each winning ticket is allowed only one prize, the largest for which it is eligible. Suppose you buy a ticket for 50¢. Find the probability that you will win \$50,000.

Table 6.3

Yellow	**Blue**	**White**	**Payoff**
—	—	6	Eligible for drawing
—	45	—	\$5
—	45	6	\$10
123	—	—	\$10
123	—	6	\$100
123	45	—	\$1000
123	45	6	\$50,000

SOLUTION:

Since each of the six digits could be any one of the 10 digits, the total number of possible outcomes is $10^6 = 1{,}000{,}000$. Of these, there is exactly one way of drawing the winning number; so, the probability of winning the Megabucks lottery is

$$\frac{1}{1{,}000{,}000} = 0.000001 \qquad \blacksquare$$

A few important observations arise from the definition of the probability of an event E. Since $0 \leq |E| \leq |S|$, $0 \leq p(E) \leq 1$ for any event E. Consequently, an event E is certain to occur if $p(E) = 1$, and will *not* occur if $p(E) = 0$.

Also, since $E' = S - E, p(E') = 1 - p(E)$ (Why?). The next two examples use this property.

EXAMPLE 6.46 Find the probability of obtaining at least one head when three coins are tossed.

SOLUTION:
Let E be the event of obtaining at least one head. Then E' denotes the event of obtaining no heads and $p(E') = 1/8$. Therefore,

$$p(E) = 1 - p(E') = 1 - \frac{1}{8} = \frac{7}{8} \qquad \blacksquare$$

EXAMPLE 6.47 **(The Birthday Paradox)** Suppose r people are selected at random. Find the probability that at least two of them have the same birthday. Do not distinguish between leap years and nonleap years.

SOLUTION:
Let E be the event that at least two of the r people have the same birthday. Then E' denotes the event that no two of them have the same birthday.

To compute $p(E)$, first find $p(E')$. Since there are 365 possibilities for a birthday, there are 365^r possibilities for the birthdays of the r people. Now the first person has 365 possibilities for his birthday, the second person has 364 possibilities, the third person has 363 possibilities, and so on; the rth person has $365 - (r - 1)$ possibilities for his birthday. Thus

$$p(E') = \frac{365 \cdot 364 \cdots (365 - r + 1)}{365^r}$$

So

$$\begin{aligned} p(E) &= 1 - p(E') \\ &= 1 - \frac{365 \cdot 364 \cdots (366 - r)}{365^r}, \quad 1 \leq r \leq 365 \end{aligned}$$

Table 6.4 gives the values of $p(E)$ for various values of r.

It follows from the table that if 23 people are selected at random, chances are better than 50% that at least two of them have the same birthday! This is known as the **birthday paradox**.

Table 6.4

Number of people selected r	$p(E)$
5	0.027135574
10	0.116948178
20	0.411438384
22	0.475695308
23	0.507297234
25	0.568699704
30	0.706316243
40	0.891231810
50	0.970373580
60	0.994122661
70	0.999159576
80	0.999914332
90	0.999993848
100	0.999999693

■

Combinatorics and the multiplication principle often help in computing probabilities, as the next example shows.

EXAMPLE 6.48 Five marbles are drawn at random from a bag of seven green marbles and four red marbles. Find the probability that three are green and two are red.

SOLUTION:
Since there are $7+4=11$ marbles, any five of them can be drawn in $C(11,5)$ ways.

Let A be the event that three marbles are green and two are red. Three green marbles can be selected in $C(7,3)$ ways and two red marbles in $C(4,2)$ ways; so, by the multiplication principle, the event A can occur in $C(7,3)\cdot C(4,2)$ ways. So,

$$p(A) = \frac{C(7,3)\cdot C(4,2)}{C(11,5)} = \frac{\frac{7!}{3!\,4!}\cdot\frac{4!}{2!\,2!}}{\frac{11!}{5!\,6!}}$$

$$= \frac{7!\,4!\,5!\,6!}{3!\,4!\,2!\,2!\,11!} = \frac{5}{11}. \qquad \text{(Verify this.)}$$ ■

If the outcomes of an experiment are not equally likely, Laplace's definition has to be modified.

A Modified Definition of the Probability of an Event

Let $E=\{a_1,a_2,\ldots,a_n\}$ be an event of a finite sample space consisting of *not* necessarily equally likely outcomes. Let $p(a_i)$ denote the probability that the outcome a_i will occur.

Then the probability of E is defined by $p(E) = \sum_{i=1}^{n} p(a_i)$. Thus, $p(E)$ is the sum of the probabilities of the outcomes in E.

EXAMPLE 6.49 Suppose the probability of obtaining a prime number is twice that of obtaining a non-prime number, when a certain loaded die is rolled. Find the probability of obtaining an odd number when it is rolled.

SOLUTION:

There are six possible outcomes when a die is rolled, of which three are primes: 2, 3, and 5. The probability of obtaining a prime is twice that of a nonprime; that is, $p(\text{prime}) = 2p(\text{nonprime})$. Since the sum of the probabilities of the various possible outcomes is $1, 3p(\text{prime}) + 3p(\text{nonprime}) = 1$. That is,

$$6p(\text{nonprime}) + 3p(\text{nonprime}) = 1$$

So

$$p(\text{nonprime}) = \frac{1}{9}$$

Thus

$$p(\text{prime}) = 2p(\text{nonprime}) = \frac{2}{9}$$

Then

$$\begin{aligned} p(\text{odd number}) &= p(1) + p(3) + p(5) \\ &= \frac{1}{9} + \frac{2}{9} + \frac{2}{9} = \frac{5}{9} \end{aligned}$$

■

We now proceed to the inclusion–exclusion and addition principles in probability. So we begin with a familiar definition.

Mutually Exclusive Events

Two events A and B are **mutually exclusive** if $A \cap B = \emptyset$, that is, if they cannot occur simultaneously.

For example, suppose a card is drawn from a standard deck of cards. Drawing a red queen and drawing a black king are mutually exclusive events.

The inclusion–exclusion principle and the addition principle in probability come into service for such circumstances. (Outcomes of any experiment are assumed equally likely, unless noted otherwise.)

THEOREM 6.20 **(Inclusion–Exclusion Principle)** If A and B are any two events of a finite sample space S, the probability that at least one of them will occur is given by $p(A \cup B) = p(A) + p(B) - p(A \cap B)$.

PROOF:

By the inclusion–exclusion principle on sets,

$$|A \cup B| = |A| + |B| - |A \cap B|$$

Then

$$\frac{|A \cup B|}{|S|} = \frac{|A|}{|S|} + \frac{|B|}{|S|} - \frac{|A \cap B|}{|S|}$$

That is,

$$p(A \cup B) = p(A) + p(B) - p(A \cap B)$$ ■

In particular, if A and B are mutually exclusive, $A \cap B = \emptyset$ and hence $p(A \cap B) = 0$. Therefore, $p(A \cup B) = p(A) + p(B)$, which can be stated formally as follows.

THEOREM 6.21 **(Addition Principle)** If A and B are two mutually exclusive events of a finite sample space, $p(A \cup B) = p(A) + p(B)$. ■

As in sets, these two results can be extended to any finite number of events of a finite sample space.

The following example uses the inclusion–exclusion principle.

EXAMPLE 6.50 A survey among 50 housewives about the two laundry detergents *Lex* (L) and *Rex* (R) shows that 25 like *Lex*, 30 like *Rex*, 10 like both, and 5 like neither. A housewife is selected at random from the group surveyed. Find the probability that she likes neither *Lex* nor *Rex*.

SOLUTION:

Using the Venn diagram in Figure 6.35, we have:

$$p(L) = \frac{25}{50} = \frac{1}{2},\ p(R) = \frac{30}{50} = \frac{3}{5},\ \text{and}\ p(L \cap R) = \frac{10}{50} = \frac{1}{5}$$

$$\begin{aligned} p(L \cup R) &= p(L) + p(R) - p(L \cap R) \\ &= \frac{1}{2} + \frac{3}{5} - \frac{1}{5} = \frac{9}{10} \end{aligned}$$

So

$$\begin{aligned} p(L' \cap R') &= 1 - p(L \cup R) \\ &= 1 - \frac{9}{10} = \frac{1}{10} \end{aligned}$$

Figure 6.35

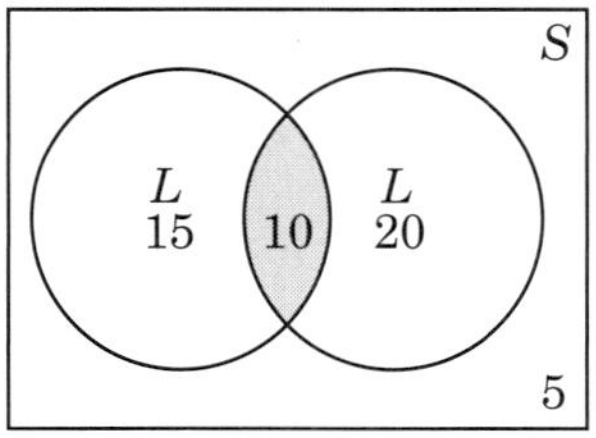

(We could read this answer directly from the Venn diagram.) ■

Exercises 6.8

A card is drawn at random from a standard deck of cards. Find the probability of obtaining:

1. A king.

2. A club.

3. A king or a queen.

4. A club or a diamond.

Two dice are rolled. Find the probability of obtaining:

5 Two fives.

6. A five and a six.

7. A sum of four.

8. A sum less than five.

Using Example 6.45, find the probability that you will:

9. Win \$1000.

10. Win \$100.

11. Win \$10.

12. Win \$5.

13. Be eligible for a drawing.

Two cards are drawn at random from a standard deck of cards. Find the probability that:

14. Both are kings.

15. Both are clubs.

16. One is a king and the other a queen.

17. One is a club and the other a diamond.

Five marbles are selected at random from a bag of seven white and six red marbles. Find the probability of each event.

18. All are white balls.

19. All are red balls.

20. Three are white and two are red.

21. Two are white and three are green.

Let $U = \{\mathrm{a, b, c, d, e}\}$ be the sample space of an experiment, where the outcomes are equally likely. Find the probability of each event.

22. $\{\mathrm{a}\}$

23. $\{\mathrm{a, b}\}$

24. $\{\mathrm{a, c, d}\}$

25. $\emptyset$

A survey of 475 customers at Chestnut Restaurant shows that of the three ice cream flavors — chocolate, strawberry, and vanilla — 65 like only chocolate, 75 like only strawberry, 85 like only vanilla, 100 like chocolate but not strawberry, 120 like strawberry but not vanilla, 140 like vanilla but not chocolate, and 65 like none of the flavors. A customer is selected at random from the survey. Find the probability that he likes:

26. All flavors.

27. Chocolate.

28. Exactly two flavors.

29. Exactly one flavor.

*6.9 Additional Topics in Probability (optional)

This section presents a few additional topics in probability, namely, conditional probability, expected value, Bernoulli trials, binomial probabilities, as well as the average-case complexity of the linear search algorithm.

The ensuing dice problem manifests conditional probability.

Let E be the event of rolling a sum of seven with two dice. Then $p(E) = 6/36 = 1/6$. Suppose a 3 comes up on one of the dice. This reduces the sample space to $\{(1,3),(2,3),(3,3),(4,3),(5,3),(6,3),(3,1),(3,2),(3,4),(3,5),(3,6)\}$. Consequently, a sum of 7 can be obtained in two ways: $(3,4)$ and $(4,3)$. Therefore, the probability of getting a sum of seven, knowing that a three has been rolled, is $2/11$. Thus the additional information has indeed affected the probability of E. Accordingly, we make the following definition.

Conditional Probability

The probability that an event A will occur, knowing that a certain other event B ($\neq \emptyset$) has already occurred, is the **conditional probability** of A, given B. It is denoted by $p(A \mid B)$.

EXAMPLE 6.51 In Example 6.50, find the probability that a housewife selected at random from the survey likes *Lex* knowing that she likes *Rex*.

SOLUTION:

Since the housewife likes *Rex*, the sample space has $10 + 20 = 30$ outcomes, of which 10 are favorable to the event L. Therefore,

$$p(L \mid R) = \frac{10}{30} = \frac{1}{3}$$ ■

In this example, you may note that

$$\frac{p(L \cap R)}{p(R)} = \frac{10/50}{30/50} = \frac{1}{3} = p(L|R)$$

A more general powerful result arises.

THEOREM 6.22 Let A and B be any two events of a finite sample space with $p(B) \neq 0$. Then $p(A|B) = \dfrac{p(A \cap B)}{p(B)}$.

PROOF:

Let $|A \cap B| = k$, $|A| = m + k$, and $|B| = n + k$ (see the Venn diagram in Figure 6.36). Then

$$P(A \mid B) = \frac{k}{n+k}$$

Figure 6.36

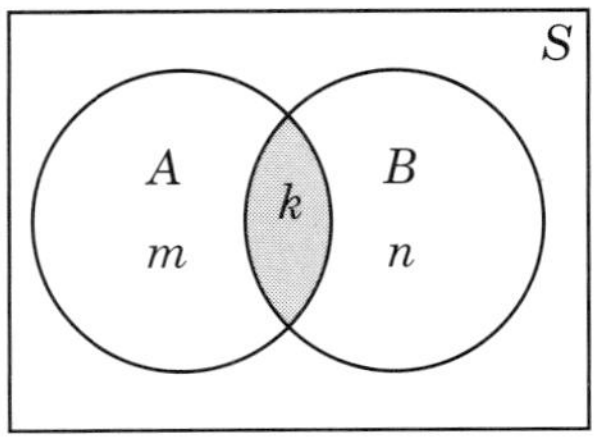

and

$$\frac{p(A \cap B)}{p(B)} = \frac{k/|S|}{(n+k)/|S|} = \frac{k}{n+k} = p(A \mid B) \qquad \blacksquare$$

EXAMPLE 6.52 A study conducted recently in a rural area shows that the probability of a randomly selected person being allergic to oak pollen is 7/24 and the probability of being allergic to both oak and birch pollen is 3/20. Find the probability that he is allergic to birch pollen, given that he is allergic to oak pollen.

SOLUTION:
Let B and K denote the events of being allergic to birch pollen and oak pollen, respectively. $p(K) = 7/24$ and $p(B \cap K) = 3/20$. Therefore,

$$p(B \mid K) = \frac{p(B \cap K)}{p(K)} = \frac{3/20}{7/24} = \frac{18}{35} \qquad \blacksquare$$

By Theorem 6.22, $p(B \mid A) = \dfrac{p(A \cap B)}{p(A)}$. That is, $p(A \cap B) = p(A) \cdot p(B \mid A)$. This result expresses the probability both A and B will occur, that is, the event $A \cap B$ in terms of $p(A)$ and $p(B \mid A)$. Accordingly, we have the following result.

THEOREM 6.23 **(Multiplication Theorem)** Let A and B be any two events of a finite sample space. Then the probability that both A and B will occur is given by $p(A \cap B) = p(A) \cdot p(B|A)$. ■

This result works for any finite number of events of a finite sample space. For example, if p_1 is the probability of an event A, p_2 the probability of an event B after A has occurred, and p_3 the probability of an event C after both A and B have occurred, the probability that the events A, B, and C will occur in that order is $p_1 p_2 p_3$.

EXAMPLE 6.53 Two marbles are drawn successively from a box of three black and four white marbles. Find the probability that both are black if the first marble is *not* replaced before the second drawing.

SOLUTION:
Let B_1 be the event of drawing the first black marble. Then $p(B_1) = 3/7$. Let B_2 be the event of drawing a second black marble. Since the first marble is not replaced before the second is drawn, there are only two black balls left in the box at the second drawing. Therefore, $p(B_2 \mid B_1) = 2/6$. Consequently, the probability of drawing two black balls successively without replacement is given by

$$p(B_1 \cap B_2) = p(B_1) \cdot p(B_2 \mid B_1) = \frac{3}{7} \cdot \frac{2}{6} = \frac{1}{7} \qquad \blacksquare$$

Dependent and Independent Events

Two events are **dependent** if the occurrence of one event affects the probability of the other event occurring; otherwise, they are **independent**.

In Example 6.53, the events B_1 and B_2 are dependent. On the other hand, if the first marble is replaced before the second is drawn, then two events would be independent.

If A and B are independent events, $p(A \mid B) = p(A)$ and $p(B \mid A) = p(B)$. Therefore, $p(A \cap B) = p(A) \cdot p(B \mid A) = p(A) \cdot p(B)$. Accordingly, we have the following result.

THEOREM 6.24 Let A and B be two independent events of a finite sample space. Then $p(A \cap B) = p(A) \cdot p(B)$. $\blacksquare$

EXAMPLE 6.54 Redo Example 6.53 with the first marble replaced before the second drawing.

SOLUTION:
Let B_1 be the event of drawing the first black marble. Then $p(B_1) = 3/7$. Let B_2 be the event of drawing the second black marble. Since the first marble is replaced before the second is drawn, $p(B_2) = 3/7$. Since B_1 and B_2 are independent events, the probability of drawing two black balls successively with replacement is given by

$$p(B_1 \cap B_2) = p(B_1) \cdot p(B_2) = \frac{3}{7} \cdot \frac{3}{7} = \frac{9}{49} \qquad \blacksquare$$

A concept very closely related to probability is **expected value**; it was introduced by the brilliant Dutch mathematician Christian Huygens (1629–1695). It can predict the number of occurrences of a possible outcome if an experiment is performed many times.

Suppose you toss a coin 100 times. How many times would you expect it come up heads? Intuitively, you would expect a head 50 times. In practice, however, you might get a head 59 times and a tail 41 times, or a head 43 times and a tail 57 times. Nevertheless, on an average, you would expect it to fall heads 50% of the time.

More generally, we make the following definition.

Expected Value

If $a_1, a_2, \ldots, a_n$ are the numerical values of the distinct outcomes of an experiment, and $p_1, p_2, \ldots, p_n$ are the corresponding probabilities of the corresponding outcomes, the **expected value** E of the experiment is given by $E = a_1 p_1 + a_2 p_2 + \cdots + a_n p_n$.

EXAMPLE 6.55 A coin is tossed four times. How many times would you expect it falls heads?

SOLUTION:
The possible outcomes when a coin is tossed four times are 0, 1, 2, 3, or 4 heads. The corresponding probabilities are $\frac{1}{16}, \frac{4}{16}, \frac{6}{16}, \frac{4}{16}$, and $\frac{1}{16}$, respectively. So the expected number of heads is $0 \cdot \frac{1}{16} + 1 \cdot \frac{4}{16} + 2 \cdot \frac{6}{16} + 3 \cdot \frac{4}{16} + 4 \cdot \frac{1}{16} = 2$, which seems intuitively right. ■

The concept of expected value plays a key role in gambling. For instance, to play the **numbers game**,* you bet a dollar on one of the whole numbers, 000 through 999. If yours is the winning number, you win \$700; otherwise, you lose your dollar. Clearly, $p(\text{winning}) = \frac{1}{1000}$ and $p(\text{losing}) = \frac{999}{1000}$. Therefore, the average expected profit of the game is

$$E = \$699\left(\frac{1}{1000}\right) + (-\$1)\left(\frac{999}{1000}\right) = -30¢$$

That $E = -30¢$ simply means if this game is played many times, you can expect to lose an average of 30¢ per game.

In **keno**, an unbiased machine selects 20 numbers without replacement from the positive integers 1 through 80. You predict a number in advance that will be selected by the machine. To play the game, you pay \$1. If your number is one of the 20 selected by the machine, you will win \$3.20.

$$p(\text{winning}) = \frac{20}{80} = \frac{1}{4} \text{ and } p(\text{losing}) = \frac{60}{80} = \frac{3}{4}$$

$$\therefore \quad \text{Average expected profit} = \$2.20\left(\frac{1}{4}\right) + (-1)\left(\frac{3}{4}\right) = -20¢$$

We now show how the concept of expected value is important in analyzing the average-case complexities of algorithms.

Average-Case Complexity of an Algorithm

To compute the average-case complexity of an algorithm, let $s_1, s_2, \ldots, s_n$ denote the input values to the algorithm. Then $\{s_1, s_2, \ldots, s_n\}$ denotes the

*Based on A. Sterrett, "Gambling Doesn't Pay!," *Mathematics Teacher*, Vol. 60 (March 1967), pp. 210–214.

sample space of an experiment. Let a_i denote the number of operations required by input s_i and p_i the probability assigned to a_i. The expected value $E = \sum_{i=1}^{n} a_i p_i$ measures the average-case time complexity of the algorithm.

To illustrate this definition, we turn to the linear search algorithm.

Average-Case Complexity of the Linear Search Algorithm

To compute the average-case complexity of the linear search algorithm, we return to Algorithm 4.8. Although the list contains n elements, there are $n+1$ cases for the algorithm: *key* occurs in the list (n cases) and *key* does not occur in the list (one case). Assume all the n items in the list are distinct and are equally likely with probability p; that is, if key occurs in the list, it can be any one of the n elements with probability p. Let q denote the probability that *key* does not occur in the list. Then $np + q = 1$.

If key occurs in position i, the algorithm takes i element comparisons to locate it with probability $p_i = p$ and $a_i = i$, where $1 \leq i \leq n$. Also, $p_{n+1} = q$ and $a_{n+1} = n$. Thus the average-case complexity of the algorithm is the average number of comparisons:

$$\begin{aligned}
E &= \sum_{i=1}^{n+1} a_i p_i = \sum_{i=1}^{n} a_i p_i + a_{n+1} p_{n+1} \\
&= \left(\sum_{i=1}^{n} ip\right) + nq = p\left(\sum_{i=1}^{n} i\right) + nq \\
&= p \cdot \frac{n(n+1)}{2} + nq
\end{aligned}$$

In particular, if $q = 0$, that is, if *key* occurs in the list, $np = 1$ and hence $E = \dfrac{n+1}{2} = \mathrm{O}(n)$. If $q = 1$, that is, if *key* does not occur in the list, $p = 0$ and $E = n = \mathrm{O}(n)$. If $0 < q < 1$, then $np < 1$ and

$$\begin{aligned}
E &< (1) \cdot \frac{n+1}{2} + n \\
&= \frac{3n+1}{2} \\
&= \mathrm{O}(n)
\end{aligned}$$

Thus, in all cases $0 \leq q \leq n$, $E = \mathrm{O}(n)$.

Note: In Example 4.44, we proved that the average-case complexity of the algorithm is

$$\begin{aligned}
a_n &= \frac{(1 + 2 + \cdots + n) + (n+1)}{n+1} \\
&= \mathrm{O}(n)
\end{aligned}$$

The analysis employed in Example 4.44 is a special case of the above analysis with $p = q = \dfrac{1}{n+1}$; in other words, all the $n+1$ cases were assumed equally likely with probability $\dfrac{1}{n+1}$.

We now turn to the discussion of probability of an event in a special class of experiments.

Bernoulli Trials

Notice that the experiment of tossing a coin three times consists of three repeated subexperiments: tossing the coin the first time, the second time, and the third time. Each is called a **trial**. The two outcomes in a trial are called **success** and **failure**, with probabilities p and $q = 1-p$, respectively. For instance, if getting a five is considered a success when a die is rolled, then not getting a five is a failure; here $p = 1/6$ and $q = 5/6$.

Repeated trials are called **Bernoulli trials**, after the outstanding Swiss mathematician Jakob Bernoulli (1654–1705), if:

- The trials are independent, and
- Each trial has exactly two outcomes, success or failure.

For example, the experiment of rolling a (fair) die three times consists of three independent trials. Let obtaining a five be a success in each trial. Each trial has exactly two outcomes: obtaining a five (success) and not obtaining a five (failure). In each trial, $p(\text{success}) = 1/5$ and $p(\text{failure}) = 5/6$. Thus the experiment consists of three Bernoulli trials.

EXAMPLE 6.56 A (fair) die is rolled three times. Let obtaining a five be a success (S) and a non-five a failure (F). Find the probability of obtaining exactly two successes.

SOLUTION:

Figure 6.37 indicates the sample space for the experiment: {SSS, SSF, SFS, SFF, FSS, FSF, FFS, FFF}. Each branch of the tree is labeled with its corresponding probability. These probabilities can be used to compute the probabilities of the various outcomes. For instance,

$$p(\text{SFS}) = \frac{1}{6}\cdot\frac{5}{6}\cdot\frac{1}{6} = \frac{5}{6}\cdot\left(\frac{1}{6}\right)^2$$

$$\begin{aligned} p(\text{exactly two successes}) &= p(\text{SSF}) + p(\text{SFS}) + p(\text{FSS}) \\ &= \frac{1}{6}\cdot\frac{1}{6}\cdot\frac{5}{6} + \frac{1}{6}\cdot\frac{5}{6}\cdot\frac{1}{6} + \frac{5}{6}\cdot\frac{1}{6}\cdot\frac{1}{6} \\ &= 3\cdot\left(\frac{1}{6}\right)^2\cdot\left(\frac{5}{6}\right) = \binom{3}{2}\left(\frac{1}{6}\right)^2\cdot\left(\frac{5}{6}\right) \end{aligned}$$

Figure 6.37

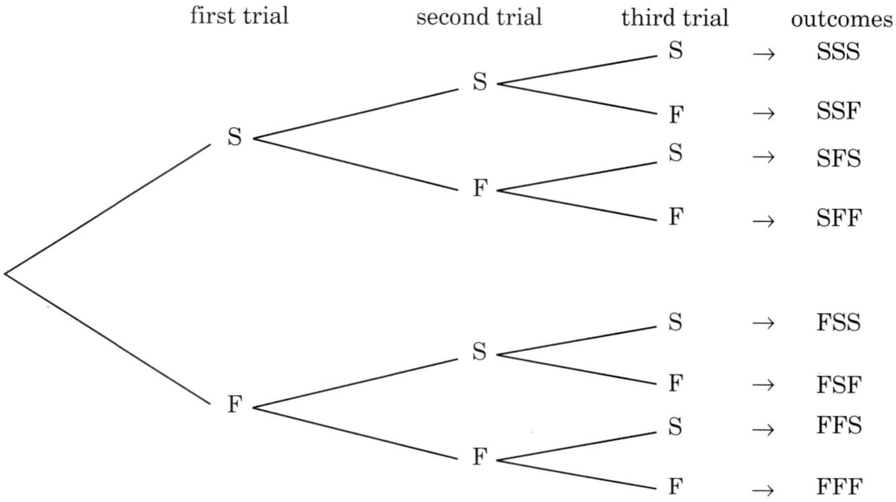

■

The next theorem generalizes these results.

THEOREM 6.25 Let p and q denote the probabilities of success and failure, respectively, in a Bernoulli trial. The probability of exactly k successes in a sequence of n Bernoulli trials is given by $C(n, k)p^k q^{n-k}$.

PROOF:
Since the trials are independent, the probability of obtaining k successes (and hence $n - k$ failures) is $p^k q^{n-k}$. But there are $C(n, k)$ ways of obtaining exactly k successes. Therefore, the probability of exactly k successes is $C(n, k)p^k q^{n-k}$. ■

The probability $C(n, k)p^k q^{n-k}$ is usually denoted by $b(n, k, p)$; it is called the **binomial probability**.

EXAMPLE 6.57 Ten percent of the population in a suburban town are allergic to ragweed. Five people of the town are selected at random. Find the probability that exactly two of them are allergic to ragweed.

SOLUTION:
The probability that a person selected at random is allergic to ragweed is $p = 0.1$. Therefore, by Theorem 6.25,

$$\begin{aligned} p(\text{exactly two are allergic}) &= C(5, 2)(0.1)^2(0.9)^3 \\ &= 0.0729 \end{aligned}$$

■

EXAMPLE 6.58 **Chuck-a-luck** is a dice game in which three dice in an enclosed cage are thoroughly mixed. You select one of the numbers 1, 2, 3, 4, 5, or 6 before the dice come to rest, and bet \$1 that this number will appear on at least one die. Suppose you choose five. If one or more fives appear, the dollar is returned, plus $1 for each five. Compute the expected profit of the game.*

SOLUTION:

First, compute the probabilities of obtaining no fives, exactly one five, exactly two fives, and exactly three fives. The game can be considered a sequence of three Bernoulli trials. Let getting a five be a success. Then $p(\text{success}) = \dfrac{1}{6}$ and $p(\text{failure}) = \dfrac{5}{6}$. Therefore, by Theorem 6.25, the binomial probability of exactly k successes is $C(3,k)\left(\dfrac{1}{6}\right)^k\left(\dfrac{5}{6}\right)^{3-k}$, where $k = 0, 1, 2,$ or 3. Thus

$$p(\text{no fives}) = C(3,0)\left(\frac{1}{6}\right)^0\left(\frac{5}{6}\right)^3 = \frac{125}{216}$$

$$p(\text{exactly one five}) = C(3,1)\left(\frac{1}{6}\right)\left(\frac{5}{6}\right)^2 = \frac{75}{216}$$

$$p(\text{exactly two fives}) = C(3,2)\left(\frac{1}{6}\right)^2\left(\frac{5}{6}\right) = \frac{15}{216}$$

$$p(\text{all fives}) = C(3,3)\left(\frac{1}{6}\right)^3\left(\frac{5}{6}\right)^0 = \frac{1}{216}$$

$$\begin{aligned}\therefore \quad \text{Expected profit} &= (-\$1)\left(\frac{125}{216}\right) + (\$1)\left(\frac{75}{216}\right) + (\$2)\left(\frac{15}{216}\right) \\ &\quad + (\$3)\left(\frac{1}{216}\right) \\ &\approx -7.9¢\end{aligned}$$

■

Finally, what can we say about the sum of the binomial probabilities in a sequence of Bernoulli trials? Since $p + q = 1$, we have

$$\begin{aligned}\sum_{k=0}^{n} b(n,k,p) &= \sum_{k=0}^{n} C(n,k)p^k q^{n-k} \\ &= (p+q)^n \quad \text{by the binomial theorem} \\ &= 1\end{aligned}$$

Thus the sum of the binomial probabilities is always 1.

*Based on A. Sterrett, "Gambling Doesn't Pay!," *Mathematics Teacher*, Vol. 60 (March 1967), pp. 210–214.

For example, the sum of the binomial probabilities in Example 6.58 is

$$\frac{125}{216} + \frac{75}{216} + \frac{15}{216} + \frac{1}{216} = 1$$

Exercises 6.9

Two dice are rolled. Find the probability of obtaining each event.

1. A sum of 11, knowing that a six has occurred on one die.

2. A sum of 11, knowing that one die shows an odd number.

It is found that 65% of the families in a town own a house, 25% own a house and a minivan, and 40% own a minivan. Find the probability that a family selected at random owns each of the following.

3. A house, given that it owns a minivan.

4. A minivan, given that it owns a house.

The Sealords have three children. Assuming that the outcomes are equally likely and independent, find the probability that they have three boys, knowing that:

5. The first child is a boy.

6. At least one child is a boy.

7. The second child is a boy.

8. The first two children are boys.

9. One child is a girl.

A survey of 475 customers at Chestnut Restaurant shows that of the three ice cream flavors—chocolate, strawberry, and vanilla—65 customers like only chocolate, 75 like only strawberry, 85 like only vanilla, 100 like chocolate but not strawberry, 120 like strawberry but not vanilla, 140 like vanilla but not chocolate, and 65 like none of the flavors. Find the probability that a customer selected at random from the survey:

10. Likes chocolate, given that she likes strawberry.

11. Likes strawberry, given that she likes vanilla.

12. Likes chocolate, given that she does not like strawberry.

13. Likes vanilla, given that she does not like chocolate.

14. Likes chocolate, given that she does not like strawberry or vanilla.

15. Does not like chocolate, given that she does not like strawberry or vanilla.

Two cards are drawn at random successively from a standard deck. The first card is replaced before the second is drawn. Find the probability that:

16. Both are queens.

17. Both are clubs.

18. The first is a club and the second a spade.

19. The first is a heart and the other a green ace.

20–23. Redo Exercises 16–19 assuming no replacement.

There are 15 rabbits in a cage. Five of them are injected with a certain drug. Three of the 15 rabbits are selected successively at random for an experiment. Find the probability that:

24. Only the first rabbit is injected with the drug.

25. Only the second rabbit is injected with the drug.

26. Only the first two rabbits are injected with the drug.

27. Only the last two rabbits are injected with the drug.

28. Ted and Ned are rolling a die. If a five appears Ned receives \$3 from Ted; otherwise, he loses 50¢ to Ted. Compute Ned's expected winnings.

29. Two dice are rolled. If a sum of six appears, Randy gets \$6 from Wanda; otherwise, he loses \$3 to her. Compute Randy's expected winnings.

30. A typical roulette wheel contains 38 slots. Two of them are numbered 0 and 00, and the others are numbered 1 through 36. The slots 0 and 00 are green; of the remaining 36 slots, 18 are red and the others black. A rolled ball has an equal chance of falling into any of the slots. If you bet \$1 on a number and the ball lands in that slot, then you receive \$35 in addition to the bet. Find your expected winnings.

31. In Exercise 30, if you bet \$1 on a red number and the ball lands there, you receive \$1 plus your bet. Find your expected profit.

A die is rolled four times. Find the probability of obtaining:

32. All sixes.

33. Exactly one six.

34. Exactly two sixes.

35. Exactly three sixes.

36. At least one six.

37. Not more than two sixes.

A survey shows that 20% of the adults in Simpleton have high blood pressure. A sample of four adults is selected at random. Find the probability that:

38. They all have high blood pressure.

39. Exactly one of them has high blood pressure.

40. Not more than two of them have high blood pressure.

41. Not all of them have high blood pressure.

42. For the casino game **football pools**, a list of 10 football games is printed on a ticket. If one team is considered weaker than its opponent by the people who run the pool, that team is given enough points to make the game a tossup. Thus the probability of picking a winning team is 0.5. You pay $1 to play the game and select all 10 winners. If all your selections win, you get $150; if nine win, you receive a consolation prize of $20; otherwise, you lose your dollar. Compute your expected profit. (A. Sterrett, 1967)

***43.** Let p denote the probability of success in a Bernoulli trial. Prove that the expected number of successes in a sequence of n Bernoulli trials is np. (*Hint*: Use the binomial theorem.)

Chapter Summary

This chapter presented the fundamentals of combinatorics and discrete probability. The counting principles in Section 6.1 are the cornerstone of combinatorics.

The Fundamental Counting Principles

- **Addition Principle** If A and B are two mutually exclusive tasks and can be done in m and n ways, respectively, task A or B can be done in $m + n$ ways (page 344).
- **Inclusion–Exclusion Principle** Suppose task A can be done in m ways and task B in n ways. If both can be done in k ways, task A or B can be done in $m + n - k$ ways (page 345).
- **Multiplication Principle** If task T_1 can be done in m_1 ways and task T_2 in m_2 ways corresponding to each way T_1 can occur, these two tasks can be done in that order in $m_1 m_2$ ways (page 345).

Permutations

- An r-permutation of a set of n distinct elements is an ordered arrangement of r elements of the set. The number of r-permutations of a set of size n is denoted by $P(n, r)$ (page 352).
- $P(n,r) = \dfrac{n!}{(n-r)!}$ (page 353); $P(n,n) = n!$ (page 353).
- A cyclic permutation is a circular arrangement. The number of cyclic permutations of n distinct items is $(n - 1)!$ (page 355).
- $P(n,r) = P(n-1,r) + rP(n-1, r-1)$ (page 356).

Derangements

- A **derangement** is a permutation of n distinct items $a_1, a_2, \ldots, a_n$ such that no item a_i occupies position i, where $1 \le i \le n$ (page 360).

- The number of derangements D_n of n items satisfies two recurrence relations:

  $D_n = (n-1)(D_{n-1} + D_{n-2}), \quad n \geq 2$
  where $D_0 = 1$ and $D_1 = 0$ (page 362)

  $D_n = nD_{n-1} + (-1)^n, \quad n \geq 1$
  where $D_0 = 1$ (page 363)

- $D_n = n!\left[1 - \frac{1}{1!} + \frac{1}{2!} \cdots + \frac{(-1)^n}{n!}\right], n \geq 0$ (page 363).

Combinations

- An r-combination of a set of n elements is a subset with size r, where $0 \leq r \leq n$. The number of r-combinations is denoted by $C(n,r)$ or $\binom{n}{r}$ (page 366).

- $C(n,r) = \dfrac{n!}{r!(n-r!)}$ (page 366) $C(n,r) = C(n, n-r)$ (page 369)

- **Pascal's identity** $C(n,r) = C(n-1, r-1) + C(n-1, r)$ (page 370).

- **Permutations with Repetitions** The number of permutations of n items where n_1 are alike, n_2 are alike, ..., and n_k are alike, is given by $\dfrac{n!}{n_1!n_2!\cdots n_k!}$ (page 376).

- **Combinations with Repetitions** The number of r-combinations with repetitions from a set with size n is $C(n+r-1, r)$ (page 380).

- Let $x_1, x_2, \ldots, x_n$ be n nonnegative integer variables. The equation $x_1 + x_2 + \cdots + x_n = r$ has $C(n+r-1, r)$ integer solutions (page 381).

- **The binomial theorem** Let x and y be real variables and n any nonnegative integer. Then

$$(x+y)^n = \sum_{r=0}^{n} \binom{n}{r} x^{n-r} y^r \qquad \text{(page 389)}$$

- $\sum_{r=0}^{n} C(n,r) = 2^n$ (page 391)

- The number of surjections N from a finite set A to a finite set B is given by

$$N = \sum_{r=0}^{n} (-1)^r \binom{n}{r} (n-r)^m$$

  where $|A| = m$, $|B| = n$, and $m \geq n$ (page 403).

- The number of derangements N of n items is given by

$$N = \sum_{r=0}^{n} (-1)^r \binom{n}{r} (n-r)! \qquad \text{(page 406)}$$

Discrete Probability

- Let E be an event of a finite sample space S consisting of equally likely outcomes. Then $p(E) = |E|/|S|$ (page 410).
- If the outcomes are not equally likely, $p(E) = \sum p(a_i)$, where $p(a_i)$ denotes the probability of outcome $a_i \in E$ (page 414).
- **Inclusion–Exclusion Principle** If A and B are any two events, then $p(A \cup B) = p(A) + p(B) - p(A \cap B)$ (page 414).
- **Addition Principle** If A and B are mutually exclusive events, then $p(A \cup B) = p(A) + p(B)$ (page 415).

Conditional Probability

- **Multiplication Theorem** $p(A \cap B) = p(A) \cdot p(B|A)$ (page 418).
- The **expected value** E of an experiment with numerical outcomes $a_1, \ldots, a_n$ is given by $E = \sum_{i=1}^{n} a_i p(a_i)$ (page 420).

Bernoulli Trials

- Repeated trials of an experiment are **Bernoulli trials** if:

  They are independent;

  Each trial has exactly two outcomes; and

  P(outcome) remains the same in every trial (page 422).
- The **binomial probability** of exactly k successes in a sequence of n Bernoulli trials is given by $b(n,k,p) = C(n,k)p^k q^{n-k}$, where $p = p(\text{success})$ and $q = p(\text{failure})$ (page 423).

Review Exercises

Find the number of positive integers ≤ 2076 and divisible by:

1. 3 or 4.

2. 3 or 4, but not 12.

3. 2, 3, or 5, but not 30.

4. 3, 4, or 5, but not 60.

○ **5.** In FORTRAN 77, a variable name consists of six alphanumeric characters, beginning with a letter. Find the total number of legal variable names possible.

○ In C++, an identifier consists of a letter followed by any number of alphanumeric characters. How many such identifiers of length six:

6. End in a letter?

7. End in a digit?

8. Are palindromes?

9. Are not palindromes?

A license plate number in Connecticut consists of two letters followed by four digits. Find each.

10. The total number of license plates possible.

11. The number of license plates that end in GREEN.

12. The number of license plates that end in JAVA.

Find the number of terms in the expansion of each.

13. $\left(\sum_{i=0}^{3} a_i\right)\left(\sum_{i=0}^{3} b_i\right)\left(\sum_{i=-1}^{3} c_i\right)$

14. $\left(\sum_{i=-2}^{3} a_i\right)\left(\sum_{i=-1}^{3} b_i\right)\left(\sum_{i=0}^{3} c_i\right)$

Master Mind is a fascinating logic game played by two players, the code maker and the code breaker. There are 72 pegs, available in six colors. The code maker creates a secret code pattern using four pegs in a row. The code breaker tries to figure out the code. Find the number of coding patterns possible if:

15. There are no restrictions on the use of colors.

16. The same color cannot be repeated.

17–18. Redo Exercises 15 and 16 if pegs are available in seven colors.

Find the number of ways three dahlias, four cannas, and five zinnias can be planted:

19. In a row such that all plants of the same family are next to each other.

20. In a row such that the zinnias are planted in between the other two families of plants.

21. In a circle.

22. In a circle such that all plants of the same family are next to each other.

Find the number of ternary words that can be formed of:

23. Length 10.

24. Length at most 10.

Find the number of subcommittees that can be formed from a 10-member committee such that each contains:

25. At least seven members.

26. At most five members.

A standard deck of playing cards contains 52 cards. Find the total number of:

27. Poker hands (five cards).

28. Flushes (all cards of the same suit).

29. Full houses (three of one kind and two of another).

Find the number of ways the letters of each word can be scrambled.

30. TENNESSEE

31. MISSISSIPPI

How many words of length six over the alphabet {a, b, c, d, e} contain:

32. Two a's, three b's, and one c.

33. Three a's and three b's.

(**A modified handshake problem**) Mr. and Mrs. Matrix hosted a party for n married couples. Each person shook hands with everyone else, excluding his or her spouse. No person shook hands with himself or herself. Let $h(n)$ denote the number of handshakes made.

34. Using induction, prove that $h(n) = C(2n - 1, 2) + (n - 1)$, $n \geq 0$.

35. Derive a recurrence relation satisfied by $h(n)$.

36. Show that $h(n) = 2n(n - 1)$, $n \geq 0$.

A local fast food restaurant serves four types of pizza—pepperoni, mushroom, cheese, and onion. Find the number of different pizza orders possible by:

37. Six students.

38. Seven students.

39–40. Redo Exercises 37 and 38, using generating functions.

Find the number of nonnegative integer solutions of each equation.

41. $x_1 + x_2 + x_3 + x_4 = 8$

42. $x_1 + x_2 + x_3 + x_4 + x_5 = 12$

Find the coefficient of each.

43. x^4y^7 in the expansion of $(x - y)^{11}$

44. x^5y^5 in the expansion of $(2x + 3y)^{10}$

Using the binomial theorem, expand each.

45. $(x + y)^6$

46. $(x - 2y)^4$

47. $(3x - y)^5$

48. $(2x + 3y)^5$

49. Prove that $C(n, r)$ is an integer for every integer $n \geq 0$, where $0 \leq r \leq n$.

50. Prove by induction that $\sum_{i=r}^{n} C(i, r) = C(n + 1, r + 1)$.

(**Twelve days of Christmas**) Suppose you sent your love 1 gift on the first day of Christmas, $1 + 2$ gifts on the second day, $1 + 2 + 3$ gifts on the third day, and so on. Show that the number of gifts:

51. Sent on the nth day is $C(n + 1, 2)$.

52. Received by your love in the first n days is $C(n + 2, 3)$.

Let D_n denote the number of derangements of n items. Prove each.

53. The number of permutations of n items with exactly k of them in their natural positions is $C(n,k)D_{n-k}$, where $0 \le k \le n$.

54. $C(n,0)D_0 + C(n,1)D_1 + \cdots + C(n,n)D_n = n!$, $n \ge 0$

55. Find the sum of the numbers in rows 0 through $n - 1$ of Pascal's triangle.

Let $A(n,r)$ denote the number of additions needed to compute $C(n,r)$ using recursion.

56. Find $A(n,0)$ and $A(n,n)$.

57. Find an explicit formula for $A(n,1)$.

***58.** Find the recurrence relation satisfied by $A(n,r)$.

***59.** Prove that $A(n,r) = C(n,r) - 1$.

○ ***60.** Find the probability of obtaining no two consecutive heads when a coin is tossed n times.

○ **61.** A contractor bids to construct a shopping plaza. It has been estimated that there is a 5/8 probability that she would make a profit of \$3 million, and a 3/8 probability that she would lose \$4 million. Compute her expected profit.

○ **62.** A construction company is bidding on two contracts to build apartment complexes in Janesville and Jeansville. The probabilities that the two bids will be accepted are 60% and 80%, respectively, and the bidding expenses are \$100,000 and \$150,000, respectively. It is estimated that the company will make profits of \$3 million from Janesville and \$2.3 million from Jeansville. In which location should the company be more interested?

○ It is found that 10% of the snow blowers manufactured by a certain company need repair before they can be sold. Five snow blowers are selected at random. Find the probability that:

63. At least one needs repair. **64.** Not all need repair.

Supplementary Exercises

Prove each, where C_n denotes the nth Catalan number.

1. The product of any r consecutive positive integers is divisible by $r!$.

2. $C(2n,n)$ is an integer for every $n \ge 0$.

3. $(n+1) \mid C(2n,n)$ for every $n \ge 0$.

4. $p \mid C(p,r)$, where p is a prime and $0 < r < p$.

5. $C_n = \dfrac{C(2n+1,n)}{2n+1}$

6. $C_n = \dfrac{2(2n-1)}{n+1}C_{n-1}, n \geq 1$

7. (**Probleme des Menages**) Let M_n denote the number of ways n married couples can be seated around a round table with men and women in alternate chairs and no wife next to her husband. It can be shown that

$$M_n = n! + \sum_{k=1}^{n}(-1)^k \frac{2n}{2n-k}\binom{2n-k}{k}(n-k)! \quad \text{(E. Lucas)}$$

Compute M_5 and M_6.

8. Prove that

$$\binom{n-i}{r-i}\binom{n}{r+i}\binom{n+i}{r} = \binom{n-i}{r}\binom{n+i}{r+i}\binom{n}{r-i}$$

Evaluate each.

9. $\sum_{k=0}^{n}\binom{n}{k}k^2$

***10.** $\sum_{k=0}^{n}\binom{n}{k}k^3$ (N. J. Kuenzi and B. Prielipp, 1985)

****11.** $\sum_{d=1}^{n-1}\sum_{j=0}^{d}\sum_{k=0}^{n-d-1}\binom{j+k}{j}$ (U.S.A Mathematical Olympiad, 1991)

***12.** Guess the number of odd binomial coefficients in row n of Pascal's triangle.
(*Hint*: Compare the number of odd binomial coefficients in row n and the binary expansion of n.)

13. Let $a, b \in \mathbf{W}$, $a = (a_n a_{n-1} \cdots a_0)_{\text{two}}$, and $b = (b_n b_{n-1} \cdots b_0)_{\text{two}}$. If $a_i \geq b_i$ for every i, we say a **implies** b and write $a \Rightarrow b$; otherwise $a \not\Rightarrow b$. Determine if $43 \Rightarrow 25$ and $47 \Rightarrow 29$.

14. The binomial coefficient $C(n,r)$ is odd if and only if $n \Rightarrow r$. Using this fact, determine the **parity** (oddness or evenness) of $C(25,18)$ and $C(29,19)$. (See Exercise 13.)

Let $f(n,k)$ denote the number of k-element subsets of the set $S = \{1,2,\ldots,n\}$ that do not contain consecutive integers. (I. Kaplansky)

15. Define $f(n,k)$ recursively.

16. Prove that $f(n,k) = C(n-k+1,k)$.

17. Prove that the total number of subsets of S that do not contain consecutive integers is F_{n+2}, where F_n denotes the nth Fibonacci number.

An ordered pair of subsets (A, B) of the set $S_n = \{1, 2, \dots, n\}$ is **admissible** if $a > |B|$ for every $a \in A$ and $b > |A|$ for every $b \in B$. For example, $(\{2, 3\}, \{4\})$ is an admissible pair of subsets of S_4.

18. Find the various admissible ordered pairs of subsets of the sets S_0, S_1, and S_2.

***19.** Predict the number of admissible ordered pairs of subsets of S_n.

Computer Exercises

Write a program to perform each task, where $n \in \mathbb{N}$ and $n \leq 20$.

1. Read in positive integers n and r, where $r \leq n \leq 10$, and compute $P(n, r)$, using Theorem 6.4 and recursion.

2. Print the number of derangements D_n of n distinct items using the recursive definition (6.1), the alternate recursive definition (6.3), and the explicit formula in Theorem 6.9.

3. Print the values of $p_n = \dfrac{D_n}{n!}$ and e^{-1} correct to 10 decimal places, for comparison.

4. Read in positive integers n and r, where $0 < r \leq n$, and compute $C(n, r)$ using Theorem 6.10 and recursion.

5. Print all solutions of the equation $x_1 + x_2 + x_3 = 5$, where $x_1, x_2, x_3 \in \mathbf{W}$. Also print all solutions if $x_1, x_2, x_3 \in \mathbb{N}$.

6. Compute the nth Catalan number, using recursion.

7. Print all bytes in increasing order.

8. Print in tabular form all binary words of length n and the corresponding subsets of the set $\{1, 2, 3, \dots, n\}$.

9. Print Pascal's triangle, as indicated below:

- As in Figure 6.21
- Right-justified
- Left-justified

Exploratory Writing Projects

Using library and Internet resources, write a team report on each of the following in your own words. Provide a well-documented bibliography.

1. Study the number of plates on automobiles in your state and neighboring states. Create a combinatorial problem that will generate such license plate numbers in each system.
2. Investigate the origin of binomial coefficients and the two-level parentheses notation.
3. Study the numerous properties of binomial coefficients, including their parity.
4. Examine the history of Pascal's triangle. Why is Pascal's triangle named after Pascal?
5. Investigate the numerous properties of Pascal's triangle. Explain how figurate numbers, Fibonacci numbers, and Catalan numbers can be extracted from it.
6. Extend Pascal's triangle upward, the definition of the binomial coefficient $C(n, r)$ to include negative and fractional values of n, and the binomial theorem to fractional exponents.
7. Discuss the number of ways n rooks of n different colors can be placed on an $n \times n$ chessboard such that no two rooks can attack each other.
8. Investigate the properties of Bell numbers and their applications to combinatorics. Include a biography of E. T. Bell.
9. Define Stirling numbers of the first and second kind. Include their applications to combinatorics and a biography of J. Stirling.
10. Discuss *Eulerian numbers* and their relationships to Stirling numbers.
11. Discuss *Bernoulli numbers*, named after Jakob Bernoulli (1654–1705), and their relationships with power series and the tangent function.
12. Write an essay on the inclusion–exclusion principle.
13. Study the origin of discrete probability.
14. Describe the various casino games and compute the expected value of each.
15. Examine the *Hardy–Weinberg probabilities* in population genetics, named for the English mathematician Godfrey. H. Hardy (1877–1947) and the German physician Wilhelm Weinberg (1862–1937) who published them independently in 1980.

Enrichment Readings

1. G. Berman and K. D. Fryer, *Introduction to Combinatorics*, Academic Press, New York, 1972, pp. 35–125.
2. K. P. Bogart, *Introductory Combinatorics*, Pitman, Boston, MA, 1983.

3. P. Z. Chinn, "Inductive Patterns, Finite Differences, and a Missing Region," *Mathematics Teacher*, Vol. 81 (Sept. 1988), pp. 446–449.

4. D. I. A. Cohen, *Basic Techniques of Combinatorial Theory*, Wiley, New York, 1978, pp. 13–178.

5. J. Dugle, "The Twelve Days of Christmas and Pascal's Triangle," *Mathematics Teacher*, Vol. 75 (Dec. 1982), pp. 755–757.

6. P. O. Eckhardt, "Discretely Deranged Squares," *Mathematics Teacher*, Vol. 83 (April 1990), pp. 318–320.

7. A. W. F. Edwards, *Pascal's Arithmetical Triangle*, The Johns Hopkins University Press, Baltimore, MD, 2002.

8. M. Eng and J. Casey, "Pascal's Triangle — A Serendipitous Source for Programming Activities," *Mathematics Teacher*, Vol. 76 (Dec. 1983), pp. 686–690.

9. B. W. Jackson and D. Thoro, *Applied Combinatorics with Problem Solving*, Addison-Wesley, Reading, MA, 1990, pp. 42–133.

10. B. H. Litwiller and D. R. Duncan, "Poker Probabilities," *Mathematics Teacher*, Vol. 70 (Dec. 1977), pp. 766–771.

11. C. Oliver, "The Twelve Days of Christmas," *Mathematics Teacher*, Vol. 70 (Dec. 1977), pp. 752–754.

12. J. Varnadore, "Pascal's Triangle and Fibonacci Numbers," *Mathematics Teacher*, Vol. 84 (April 1991), pp. 314–316, 319.

Chapter 7

Relations

The invention of the symbol ≡ by Gauss affords a striking example of the advantage which may be derived from an appropriate notation, and marks an epoch in the development of the science of arithmetic.

— G. B. MATHEWS

Functions are a special case of relations, which are also used in everyday life. Relations have applications to many disciplines, including biology, computer science, psychology, and sociology. The EQUIVALENCE statement in FORTRAN, for example, is based on the relation *has the same location as* (see Example 7.42). Graphs, digraphs, formal languages, finite state machines — all to be discussed in the next four chapters — are closely related to the theory of relations.

In this chapter we will examine the concept of a relation, its computer representations and properties, and different ways to construct new relations from known ones.

We will deal with the following problems, as well as others:

- Is it possible to arrange all n-bit words around a circle in such a way that any two adjacent words differ by exactly one bit?
- Can we determine the day corresponding to a given date $m/d/y$, where $y > 1582$, the year the Gregorian calendar was adopted?
- Five sailors and a monkey are marooned on a desert island. During the day they gather coconuts for food. They decide to divide them up in the morning and retire for the night. While the others sleep, one sailor gets up and divides them into equal piles, with one left over that he throws out for the monkey. He hides his share, puts the remaining coconuts together, and goes back to sleep. Later a second sailor gets up and divides the pile into five equal shares with one coconut left over, which he discards for the monkey. He also hides his share, puts the remaining coconuts together, and goes back to sleep. Later the remaining sailors repeat the process. Find the smallest possible number of coconuts in the original pile.

- The computer science courses required for a computer science major at a college are given in Table 7.1. In which order can a student take them?

Table 7.1

Number	Course	Prerequisite
CS 100	Computer Science I	None
CS 150	Computer Science II	CS 100
CS 200	Computer Organization	CS 150
CS 250	Data Structures	CS 150
CS 300	Computer Architecture	CS 200
CS 350	Programming Languages	CS 250
CS 400	Software Engineering	CS 250
CS 450	Operating Systems	CS 250, CS 300

A special class of matrices called boolean matrices is used to study relations, so we begin with a brief discussion of such matrices.

7.1 Boolean Matrices

(This section is closely related to Section 3.7 on matrices; you will probably find that section useful to review before reading further.)

A **boolean matrix** is a matrix with bits as its entries. Thus $A = (a_{ij})_{m \times n}$ is a boolean matrix if $a_{ij} = 0$ or 1 for every i and j. For instance, $\begin{bmatrix} 1 & 0 & 1 \\ 0 & 1 & 0 \end{bmatrix}$ is a boolean matrix, whereas $\begin{bmatrix} 1 & -3 \\ 0 & 2 \end{bmatrix}$ is *not*.

Boolean Operations and and or

The boolean operations *and* ($\wedge$) and *or* ($\vee$), defined by Table 2.1, signal the combining of boolean matrices to construct new ones. Listed below are several properties of these bit operations. They can be verified easily, so try a few.

THEOREM 7.1 Let a and b be arbitrary bits. Then:

- $a \wedge a = a$
- $a \vee a = a$
- $a \wedge (b \wedge c) = (a \wedge b) \wedge c$
- $a \vee (b \wedge c) = (a \vee b) \wedge (a \vee c)$
- $a \wedge b = b \wedge a$
- $a \vee b = b \vee a$
- $a \vee (b \vee c) = (a \vee b) \vee c$
- $a \wedge (b \vee c) = (a \wedge b) \vee (a \wedge c)$ ■

Using the two bit-operations, we now define two operations on boolean matrices.

Join and Meet

The **join** of the boolean matrices $A = (a_{ij})_{m \times n}$ and $B = (b_{ij})_{m \times n}$, denoted by $A \vee B$, is defined by $A \vee B = (a_{ij} \vee b_{ij})_{m \times n}$. Each element of $A \vee B$ is obtained by *oring* the corresponding elements of A and B. The **meet** of A and B, denoted by $A \wedge B$, is defined by $A \wedge B = (a_{ij} \wedge b_{ij})_{m \times n}$. Every element of $A \wedge B$ is obtained by *anding* the corresponding elements of A and B.

The following example illustrates these two definitions.

EXAMPLE 7.1 Let

$$A = \begin{bmatrix} 1 & 0 & 1 \\ 0 & 1 & 0 \end{bmatrix} \quad \text{and} \quad B = \begin{bmatrix} 0 & 0 & 1 \\ 1 & 0 & 1 \end{bmatrix}$$

Find $A \vee B$ and $A \wedge B$.

SOLUTION:

- $$A \vee B = \begin{bmatrix} 1 \vee 0 & 0 \vee 0 & 1 \vee 1 \\ 0 \vee 1 & 1 \vee 0 & 0 \vee 1 \end{bmatrix} = \begin{bmatrix} 1 & 0 & 1 \\ 1 & 1 & 1 \end{bmatrix}$$
- $$A \wedge B = \begin{bmatrix} 1 \wedge 0 & 0 \wedge 0 & 1 \wedge 1 \\ 0 \wedge 1 & 1 \wedge 0 & 0 \wedge 1 \end{bmatrix} = \begin{bmatrix} 0 & 0 & 1 \\ 0 & 0 & 0 \end{bmatrix}$$

■

Boolean Product

The **boolean product** of the boolean matrices $A = (a_{ij})_{m \times p}$ and $B = (b_{jk})_{p \times n}$, denoted by $A \odot B$, is the matrix $C = (c_{ij})_{m \times n}$, where $c_{ij} = (a_{i1} \wedge b_{1j}) \vee (a_{i2} \wedge b_{2j}) \vee \cdots \vee (a_{ip} \wedge b_{pj})$. (See Figure 7.1).

Figure 7.1

$$\begin{bmatrix} a_{11} & a_{12} & \dots & a_{1p} \\ \cdot & \cdot & \dots & \cdot \\ a_{i1} & a_{i2} & \dots & a_{ip} \\ \cdot & \cdot & \dots & \cdot \\ a_{m1} & a_{m2} & \dots & a_{mp} \end{bmatrix} \odot \begin{bmatrix} b_{11} & \dots & b_{1j} & \dots & b_{1n} \\ \cdot & \dots & \cdot & \dots & \cdot \\ b_{i1} & \dots & b_{ij} & \dots & b_{in} \\ \cdot & \dots & \cdot & \dots & \cdot \\ b_{p1} & \dots & b_{pj} & \dots & b_{pn} \end{bmatrix} = \begin{bmatrix} c_{11} & \dots & c_{1j} & \dots & c_{1n} \\ \cdot & \dots & \cdot & \dots & \cdot \\ c_{i1} & \dots & c_{ij} & \dots & c_{in} \\ \cdot & \dots & \cdot & \dots & \cdot \\ c_{m1} & \dots & c_{mj} & \dots & c_{mn} \end{bmatrix}$$

Notice the similarity between this definition and that of the usual product of matrices.

The next example clarifies this definition.

EXAMPLE 7.2 Let

$$A = \begin{bmatrix} 1 & 0 & 1 \\ 0 & 1 & 0 \end{bmatrix} \quad \text{and} \quad B = \begin{bmatrix} 1 & 1 \\ 1 & 0 \\ 0 & 0 \end{bmatrix}$$

Find $A \odot B$ and $B \odot A$, if defined.

SOLUTION:

(1) Since the number of columns in A equals the number of rows in B, $A \odot B$ is defined:

$$A \odot B = \begin{bmatrix} 1 & 0 & 1 \\ 0 & 1 & 0 \end{bmatrix} \odot \begin{bmatrix} 1 & 0 \\ 1 & 1 \\ 0 & 0 \end{bmatrix}$$

$$= \begin{bmatrix} (1 \wedge 1) \vee (0 \wedge 1) \vee (1 \wedge 0) & (1 \wedge 0) \vee (0 \wedge 1) \vee (1 \wedge 0) \\ (0 \wedge 1) \vee (1 \wedge 1) \vee (0 \wedge 0) & (0 \wedge 0) \vee (1 \wedge 1) \vee (0 \wedge 0) \end{bmatrix}$$

$$= \begin{bmatrix} 1 & 0 \\ 1 & 1 \end{bmatrix}$$

(2) Number of columns in $B = 2 =$ Number of rows in A. Therefore, $B \odot A$ is also defined:

$$B \odot A = \begin{bmatrix} 1 & 0 \\ 1 & 1 \\ 1 & 0 \end{bmatrix} \odot \begin{bmatrix} 1 & 0 & 1 \\ 0 & 1 & 0 \end{bmatrix}$$

$$= \begin{bmatrix} (1 \wedge 1) \vee (0 \wedge 0) & (1 \wedge 0) \vee (0 \wedge 1) & (1 \wedge 1) \vee (0 \wedge 0) \\ (1 \wedge 1) \vee (1 \wedge 0) & (1 \wedge 0) \vee (1 \wedge 1) & (1 \wedge 1) \vee (1 \wedge 0) \\ (0 \wedge 1) \vee (0 \wedge 0) & (0 \wedge 0) \vee (0 \wedge 1) & (0 \wedge 1) \vee (0 \wedge 0) \end{bmatrix}$$

$$= \begin{bmatrix} 1 & 0 & 1 \\ 1 & 1 & 1 \\ 0 & 0 & 0 \end{bmatrix}$$

■

The fundamental properties of the boolean matrix operations are listed in the following theorem. Their proofs being fairly straightforward, appear as routine exercises (see Exercises 36–43).

THEOREM 7.2 Let A, B, and C be three boolean matrices. Then:

- $A \vee A = A$
- $A \wedge A = A$
- $A \vee B = B \vee A$
- $A \wedge B = B \wedge A$
- $A \vee (B \vee C) = (A \vee B) \vee C$
- $A \wedge (B \wedge C) = (A \wedge B) \wedge C$
- $A \vee (B \wedge C) = (A \vee B) \wedge (A \vee C)$
- $A \wedge (B \vee C) = (A \wedge B) \vee (A \wedge C)$
- $A \odot (B \odot C) = (A \odot B) \odot C$

The sizes of the matrices are assumed compatible for the corresponding matrix operations. ■

Boolean Power of a Boolean Matrix

Let A be an $m \times m$ boolean matrix and n any positive integer. The nth **boolean power** of A, denoted by $A^{[n]}$, is defined recursively as follows:

$$A^{[0]} = I_m \qquad \text{(the identity matrix)}$$
$$A^{[n]} = A^{[n-1]} \odot A \quad \text{if } n \geq 1$$

The following example illustrates this definition.

EXAMPLE 7.3 Let

$$A = \begin{bmatrix} 1 & 1 & 0 \\ 1 & 1 & 0 \\ 0 & 0 & 0 \end{bmatrix}$$

Compute $A^{[2]}$ and $A^{[3]}$.

SOLUTION:

$$A^{[2]} = A^{[1]} \odot A = A \odot A = \begin{bmatrix} 1 & 1 & 0 \\ 1 & 1 & 0 \\ 0 & 0 & 0 \end{bmatrix} \odot \begin{bmatrix} 1 & 1 & 0 \\ 1 & 1 & 0 \\ 0 & 0 & 0 \end{bmatrix} = \begin{bmatrix} 1 & 1 & 0 \\ 1 & 1 & 0 \\ 0 & 0 & 0 \end{bmatrix}$$

$$A^{[3]} = A^{[2]} \odot A = \begin{bmatrix} 1 & 1 & 0 \\ 1 & 1 & 0 \\ 0 & 0 & 0 \end{bmatrix} \odot \begin{bmatrix} 1 & 1 & 0 \\ 1 & 1 & 0 \\ 0 & 0 & 0 \end{bmatrix} = \begin{bmatrix} 1 & 1 & 0 \\ 1 & 1 & 0 \\ 0 & 0 & 0 \end{bmatrix}$$

(You can verify that in this case, $A^{[n]} = A$ for every $n \geq 1$.) ■

You will find boolean matrices and their properties useful in the next few sections, so review them as needed.

Exercises 7.1

Using the boolean matrices

$$A = \begin{bmatrix} 1 & 1 \\ 0 & 0 \end{bmatrix}, \; B = \begin{bmatrix} 0 & 1 \\ 1 & 0 \end{bmatrix}, \text{ and } C = \begin{bmatrix} 0 & 0 \\ 1 & 0 \end{bmatrix}$$

find each.

1. $A \vee B$ **2.** $A \wedge B$ **3.** $A \odot C$ **4.** $C \odot A$

5. $A \vee (B \vee C)$ **6.** $A \wedge (B \wedge C)$ **7.** $A \odot (B \odot C)$ **8.** $(A \odot B) \odot C$

Using the boolean matrices

$$A = \begin{bmatrix} 1 & 0 & 1 \\ 0 & 0 & 0 \\ 1 & 1 & 0 \end{bmatrix}, \; B = \begin{bmatrix} 0 & 0 & 1 \\ 0 & 1 & 0 \\ 1 & 0 & 0 \end{bmatrix}, \text{ and } C = \begin{bmatrix} 0 & 0 & 0 \\ 0 & 0 & 1 \\ 0 & 1 & 0 \end{bmatrix}$$

find each.

9. $A \wedge (B \vee C)$ **10.** $A \vee (B \wedge C)$ **11.** $(A \wedge B) \vee (A \wedge C)$

12. $(A \vee B) \wedge (A \vee C)$ **13.** $(A \wedge B) \vee (A \vee C)$ **14.** $A \odot (B \odot C)$

15. $(A \odot B) \odot C$ **16.** $B \odot C \odot A$ **17.** $A \odot A \odot A$

18. Using the boolean matrix

$$A = \begin{bmatrix} 1 & 0 & 1 \\ 1 & 1 & 0 \\ 0 & 0 & 1 \end{bmatrix}$$

find $A^{[3]}$ and $A^{[5]}$.

Let A and B be any two $n \times n$ boolean matrices. Find the number of boolean operations needed to compute each.

19. $A \vee B$ **20.** $A \wedge B$ **21.** $A \odot B$

22. Find the number of $m \times n$ boolean matrices that can be defined.

23. Let A be an $m \times p$ boolean matrix and B a $p \times n$ boolean matrix. Find the number of boolean operations needed to compute $A \odot B$.

24. For the boolean matrix A in Example 7.3, prove that $A^{[n]} = A$ for every $n \geq 1$.

The **complement** of a boolean matrix A, denoted by A', is obtained by taking the one's complement of each element in A, that is, by replacing 0's with 1's and 1's with 0's. Use the boolean matrices A, B, and C in Exercises 1–8 to compute each.

25. A' **26.** B' **27.** $(A \vee B)'$ **28.** $A' \wedge B'$

29. $(A \wedge B)'$ **30.** $A' \vee B'$ **31.** $A \wedge (B' \vee C')$ **32.** $(A \odot B) \odot C'$

Let A and O be two $m \times n$ boolean matrices such that every entry of A is 1 and every entry of O is 0. Let B be any $m \times n$ boolean matrix. What can you say about each?

33. $A \vee B$ **34.** $A \wedge B$ **35.** A'

Let A, B, and C be any $n \times n$ boolean matrices. Prove each.

36. $A \vee A = A$ **37.** $A \wedge A = A$ **38.** $A \vee B = B \vee A$ **39.** $A \wedge B = B \wedge A$

40. $A \vee (B \vee C) = (A \vee B) \vee C$ **41.** $A \wedge (B \wedge C) = (A \wedge B) \wedge C$

42. $A \vee (B \wedge C) = (A \vee B) \wedge (A \vee C)$ **43.** $A \wedge (B \vee C) = (A \wedge B) \vee (A \wedge C)$

Write an algorithm to find each.

44. The join of two boolean matrices A and B.

45. The meet of two boolean matrices A and B.

46. The complement of a boolean matrix A.

47. The boolean product of two boolean matrices A and B.

48. The nth boolean power of an $m \times m$ boolean matrix A.

7.2 Relations and Digraphs

Clearly many relationships exist in the world around us. On the human level, they are parent–child, husband–wife, student–teacher, doctor–patient, and so on. Relationships exist between numbers also; the equality relation $(=)$ and the less-than relation $(<)$ are two such relationships. In fact, relationships can exist between any two sets; they are known as relations.

This section presents the concept of a relation and discusses how relations can be represented using matrices and graphs.

Before formally defining a binary relation, let us study an example.

EXAMPLE 7.4 Consider the sets $A = \{\text{Tom, Dick, Harry}\}$ and $B = \{\text{Amy, Betsy, Carol, Daisy}\}$. Suppose Tom is married to Daisy, Dick to Carol, and Harry to Amy. Let $R = \{(\text{Tom, Daisy}), (\text{Dick, Carol}), (\text{Harry, Amy})\}$. Using the set-builder notation, it can also be defined as

$$R = \{(a, b) \in A \times B \mid a \text{ is married to } b\}$$

Notice that $R \subseteq A \times B$. It is defined using the relation *is married to*. The set R is a **binary relation** from A to B. ■

More generally, we make the following definition.

Binary Relation

A **binary relation** R from a set A to a set B is a subset of $A \times B$. The **domain** of the relation consists of the first elements in R and the **range** consists of the second elements; they are denoted by **dom(R)** and **range(R)**, respectively. A binary relation from A to itself is a binary relation **on** A.

The following example illustrates these terms.

EXAMPLE 7.5 Let $A = \{2, 3, 5\}$ and $B = \{2, 3, 4, 6, 7\}$. Define a relation R from A to B as follows:

$$R = \{(a, b) \mid a \text{ is a factor of } b\}$$

Then $R = \{(2, 2), (2, 4), (2, 6), (3, 3), (3, 6)\}$, $\text{dom}(R) = \{2, 3\}$, and $\text{range}(R) = \{2, 3, 4, 6\}$. ■

Let R be a relation from A to B. If $(a,b) \in R$, we say a **is related to** b by the relation R; in symbols, we write aRb. If a is *not* related to b, we write $a \not R b$. For instance, $3 < 5$, but $7 \not< 6$. (Here the relation is $<$.) The next example illustrates this further.

EXAMPLE 7.6 Let A be the set of cities and B the countries in the world. Define a relation R from A to B, using the phrase *is the capital of.* So $R = \{(a,b) \in A \times B | a$ is the capital of $b\}$. Then Paris R France, but Toronto $\not R$ Canada. ■

Relations from a finite set to a finite set can be represented by boolean matrices, as defined below.

Adjacency Matrix of a Relation

A relation R from a set $\{a_1, a_2, \dots a_m\}$ to a set $\{b_1, b_2, \dots b_n\}$ can be represented by the $m \times n$ boolean matrix $M_R = (m_{ij})$, where

$$m_{ij} = \begin{cases} 1 & \text{if } a_i R b_j \\ 0 & \text{otherwise} \end{cases}$$

M_R is the **adjacency matrix** of the relation R.

EXAMPLE 7.7 Define a relation R from $A =$ {chicken, dog, cat} to $B =$ {fish, rice, cotton} by $R = \{(a,b) | a$ eats $b\}$. Then $R =$ {(chicken, fish), (chicken, rice), (dog, fish), (dog, rice), (cat, fish), (cat, rice)}. Its adjacency matrix is

$$M_R = \begin{array}{c} \\ \text{chicken} \\ \text{dog} \\ \text{cat} \end{array} \begin{array}{c} \begin{array}{ccc} \text{fish} & \text{rice} & \text{cotton} \end{array} \\ \begin{bmatrix} 1 & 1 & 0 \\ 1 & 1 & 0 \\ 1 & 1 & 0 \end{bmatrix} \end{array}$$

■

Figure 7.2

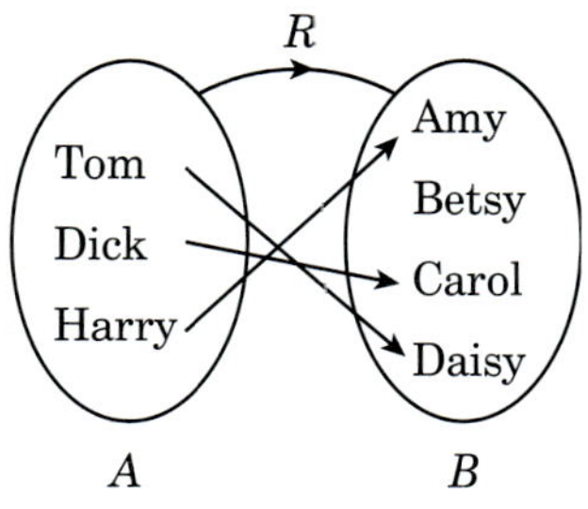

Relations can also be represented pictorially. For instance, the relation in Example 7.4 is displayed in Figure 7.2; an arrow from an element a in A to an element b in B indicates that a is related to b.

Figure 7.3

The circle $x^2 + y^2 = 9$.

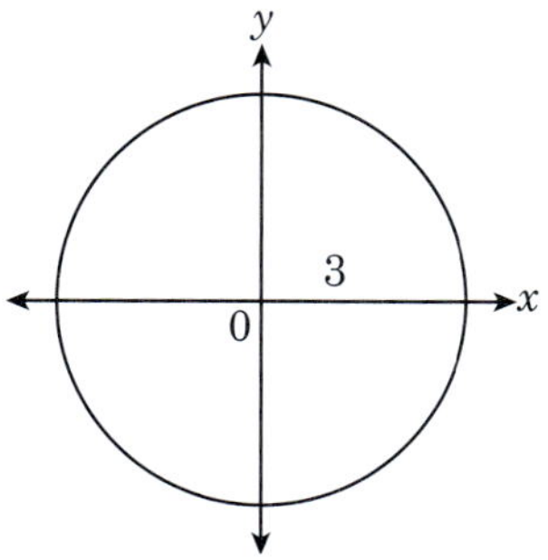

Relations can be displayed using familiar **graphs** as well. For example, the graph of the relation $\{(x, y) \in \mathbb{R} \times \mathbb{R} \mid x^2 + y^2 = 9\}$ is the circle $x^2 + y^2 = 9$ with center at the origin and radius 3 (see Figure 7.3).

Digraphs

Relations R on a finite set A can be represented pictorially in yet another way. We denote every element of A by a point, called a **vertex** (or **node**), and each ordered pair (a, b) in R by a directed arc or a directed line segment, called an **edge**, from a to b. The resulting diagram is a **directed graph** or simply a **digraph**. If an edge (a, b) exists, we say that vertex b is **adjacent to** vertex a. (Notice the order of the vertices.)

The next two examples illustrate these definitions.

EXAMPLE 7.8 Represent the relation R defined on $A = \{2, 3, 4, 6\}$ by the phrase *is a factor of* in a digraph.

SOLUTION:
Notice that

$$\begin{aligned} R &= \{(a, b) \in A \times A \mid a \text{ is a factor of } b\} \\ &= \{(2, 2), (2, 4), (2, 6), (3, 3), (3, 6), (4, 4), (6, 6)\} \end{aligned}$$

Figure 7.4 shows its digraph. It contains four vertices: 2, 3, 4, and 6. Since $3R6$, vertex 6 is adjacent to vertex 3.

Figure 7.4

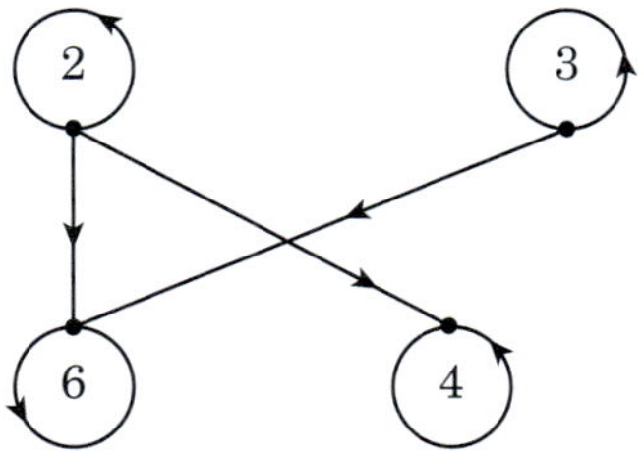

■

Notice that the digraph in Figure 7.4 contains an edge (x, x) leaving and terminating at the same vertex x. Such an edge is a **loop**. The digraph in Figure 7.4 contains four loops.

We now turn to the concept of a path in a relation, and hence in a digraph.

Paths in Digraphs and Relations

Let R be a relation on a set A, and let $a, b \in A$. A **path** in R, that is, in the digraph of R, from a to b is a finite sequence of edges $(a, x_1), (x_1, x_2), \ldots, (x_{n-1}, b)$; the vertices x_i's need *not* be distinct. The path from a to b is also denoted by a-x_1-x_2- $\cdots$ -x_{n-1}-b. The number of edges in the path is its **length**. A path that begins and terminates at the same vertex is a **cycle**. A cycle of length one is a **loop**.

The next example clarifies these terms.

EXAMPLE 7.9 Notice that the relation in Figure 7.5 contains a path of length three from a to b, namely, a-c-d-b. The path b-c-d-b is a cycle of length three. The cycle b-b is a loop.

Figure 7.5

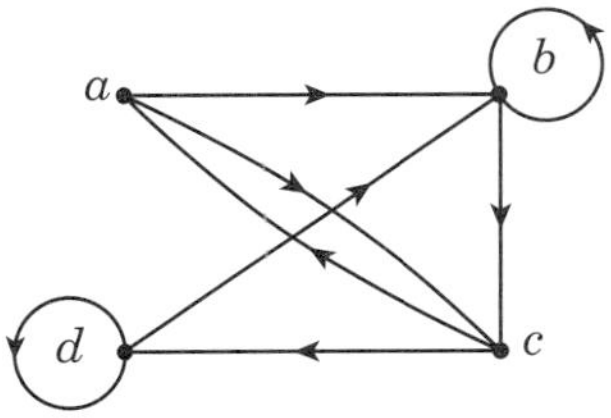

■

The next example presents an interesting relation in the language of binary words.

EXAMPLE 7.10 (**Gray Codes**) Suppose a switching network is composed of n switches a_i, where $1 \le i \le n$. Let $a_i = 1$ denote that switch a_i is closed and $a_i = 0$ denote that it is open. Every state of the network can be denoted by the n-bit word $a_1a_2 \ldots a_n$. Let Σ^n denote the set of n-bit words, that is, the set of all states of the network. For example, $\Sigma^3 = \{000, 001, 010, 100, 011, 101, 110, 111\}$. Naturally, we are tempted to ask: *Is it possible to test every state of the circuit by changing the state of exactly one switch? That is, is it possible to list every n-bit word by changing exactly one bit?*

Another definition can lead to rewording the question. Two n-bit words are **adjacent** if they differ in exactly one bit, that is, if the **Hamming distance** between them is one. For example, 010 and 011 are adjacent, whereas 001 and 110 are not.

Define a relation R on Σ^n as $\alpha R \beta$ if α and β are adjacent. We can rephrase this: Is it possible to arrange the elements α_i of Σ^n in such a way that $\alpha_i R \alpha_{i+1}$ where $1 \le i \le m - 1, \alpha_m R \alpha_1$, and $m = 2^n$? That is, is it possible to arrange the n-bit words around a circle in such a way that any two neighboring words are adjacent?

Figure 7.6

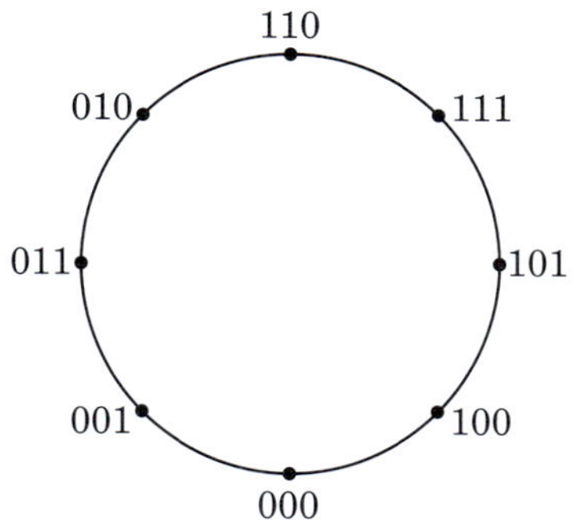

Interestingly enough, the elements of Σ^3 can be arranged in this fashion (see Figure 7.6): 000, 001, 011, 010, 110, 111, 101, 100. Such an ordering is called a **Gray code** for Σ^3. More generally, a **Gray code** for Σ^n is an arrangement of its elements $\alpha_1, \alpha_2, \ldots, \alpha_m$ such that $\alpha_i R \alpha_{i+1}$ and $\alpha_m R \alpha_1$, where $1 \leq i \leq m-1$. Gray codes are named for Frank Gray, who invented them in the 1940s at what was then AT&T Bell Labs.

We can restate our original question again: *Is there a Gray code for* Σ^n *for every* $n \geq 1$*?* Induction leads to an affirmative answer.

PROOF (by induction):
Let P(n): There exists a Gray code for every Σ^n.

Basis step When $n = 1$, $\{0, 1\}$ is clearly a Gray code; so P(1) is true.

Induction step Assume P(k) is true; that is, there is a Gray code for Σ^k. Suppose $\{\alpha_1, \alpha_2, \ldots, \alpha_r\}$ is a Gray code, where $r = 2^k$.

To show that P(k+1) is true:

Consider the $(k+1)$-bit words $0\alpha_1, 0\alpha_2, \ldots, 0\alpha_r, 1\alpha_r, 1\alpha_{r-1}, \ldots, 1\alpha_1$. Clearly they form the $2r = 2^{k+1}$ elements of Σ^{k+1}. Call them $\beta_1, \beta_2, \ldots, \beta_{2r}$, respectively, for convenience. Since $\alpha_i R \alpha_{i+1}$ and $\alpha_r R \alpha_1, \beta_j R \beta_{j+1}$ and $\beta_{2r} R \beta_1$, so $\{\beta_1, \beta_2, \ldots, \beta_{2r}\}$ is a Gray code; that is, P$(k+1)$ is true.

Thus, by induction, a Gray code exists for every Σ^n.

(Notice that the induction step provides a smooth method for constructing a Gray code for Σ^{k+1} from that of Σ^k. This example will be taken a bit further in Chapter 8.) ■

Finally, we will see how relations and functions are closely related, if we recall that a function $f : A \to B$ is a set of ordered pairs $(a, b) \in A \times B$ such that *every* element a in A is assigned a *unique* element b in B. Consequently, every function can be redefined as a relation, as follows.

An Alternate Definition of a Function

A **function** $f : A \to B$ is a relation from A to B such that:

- $\text{Dom}(f) = A$; and
- If $(a, b) \in f$ and $(a, c) \in f$, then $b = c$.

We close this section with an example that illustrates this definition.

EXAMPLE 7.11 Which of the relations R, S, and T in Figure 7.7 are functions?

Figure 7.7

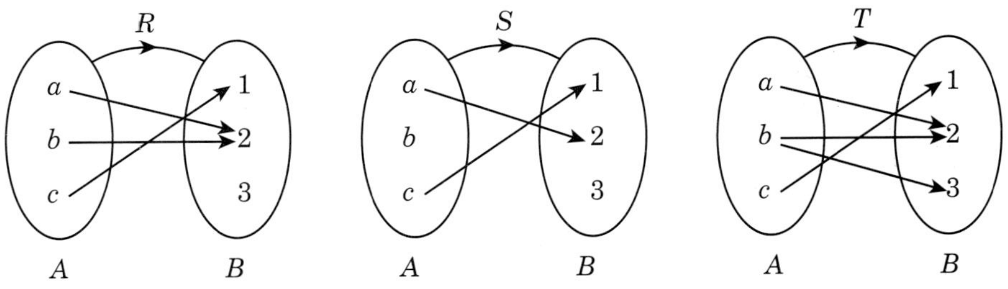

SOLUTION:
The relation R is a function, whereas S is *not* since $\text{dom}(S) \neq A$. T is also *not* a function since the same element b in A is paired with two distinct elements in B, namely, 2 and 3. ■

Exercises 7.2

List the elements in each relation from $A = \{1, 3, 5\}$ to $B = \{2, 4, 8\}$.

1. $\{(a, b) \mid a < b\}$ **2.** $\{(a, b) \mid b = a + 1\}$ **3.** $\{(a, b) \mid a + b = 5\}$

4. $\{(a, b) \mid a$ is a factor of $b\}$ **5.** $\{(a, b) \mid a + b \leq 3\}$ **6.** $\{(a, b) \mid a = b\}$

7–12. Find the domain and range of each relation in Exercises 1–6.

13–18. Find the adjacency matrix of each relation in Exercises 1–6.

Represent each relation R on the given set A in a digraph.

19. $\{(a, b) | a < b\}$, $\{2, 3, 5\}$ **20.** $\{(a, b) | a \leq b\}$, $\{2, 3, 5\}$

21. $\{(a, b) | a$ is a factor of $b\}$, $\{2, 4, 5, 8\}$

22. $\{(a, b) | b = a + 2\}$, $\{2, 4, 5, 6\}$

Using the relation $R = \{(x, y) | 2x + 3y = 12\}$ on $\mathbb{R}$, determine whether or not each is true.

23. $3R2$ **24.** $2R3$ **25.** $-3R5$ **26.** $-5R6$

Using the relation $R = \{(x, y) | x^2 + y^2 = 4\}$ on $\mathbb{R}$, determine if each is true.

27. $2R0$ **28.** $2R2$ **29.** $-2R0$ **30.** $4R0$

Define a relation R on $\mathbf{Z}$ by xRy if and only if $x - y$ is divisible by 5. Determine if:

31. $9R4$ **32.** $13R6$ **33.** $3R8$ **34.** $23R3$

List the elements in the relation R represented by each digraph.

35.

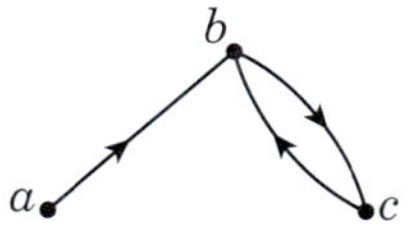

36.

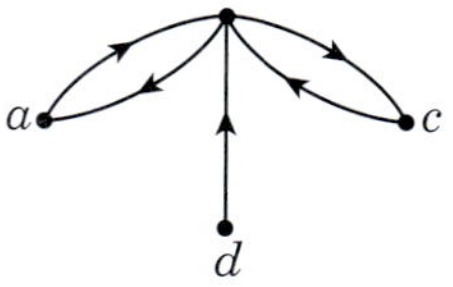

37–38. Find the adjacency matrix of each relation in Exercises 35 and 36.

39. Construct a Gray code for Σ^4, where $\Sigma = \{0, 1\}$.

Using the relation in Figure 7.5, find each.

40. Paths of length one starting at a.

41. Paths of length two starting at b.

42. Number of paths of length one. **43.** Number of paths of length two.

44. Number of cycles of length three. **45.** Number of loops.

Determine if each relation from $\{a, b, c, d\}$ to $\{0, 1, 2, 3, 4\}$ is a function.

46. $\{(a, 0), (b, 1), (c, 0), (d, 3)\}$ **47.** $\{(a, 3), (b, 3), (b, 4), (c, 1), (d, 0)\}$

48. $\{(a, 3), (b, 3), (c, 3), (d, 3)\}$ **49.** $\{(a, 1), (b, 2), (c, 3)\}$

Let A and B be finite sets with $|A| = m$ and $|B| = n$. Find the number of binary relations that can be defined:

50. From A to B. **51.** On A.

52. A relation R on the set $\{1, 2, \ldots, n\}$ is given in terms of its elements. Write an algorithm to find its adjacency matrix A.

53. Write an algorithm to print the elements of a relation R on $\{1, 2, \ldots, n\}$ using its adjacency matrix A.

*7.3 Computer Representations of Relations (optional)

Since relations from a finite set to a finite set can be represented by boolean matrices, the most straightforward way of implementing a relation and its digraph in a computer is by its adjacency matrix.

The second method involves **linked lists**. Since some programming languages such as FORTRAN do not support dynamic linked lists, the array representation of linked lists serves well. (*Note*: Arrays are nothing but matrices.) For example, the digraph in Figure 7.8 contains seven edges, arbitrarily numbered 1 through 7. Store the tails and the corresponding heads of each edge in two parallel one-dimensional arrays, $T = (t_i)$ and $H = (h_i)$, respectively (see Figure 7.9). Notice that $t_3 = 1$ and $h_3 = 2$, so an edge exists from vertex 1 to vertex 2, namely, edge 1. Since $t_7 = 3$ and $h_7 = 2$, there is also an edge from vertex 3 to vertex 2, namely, edge 5. The other edges can be read similarly.

Figure 7.8

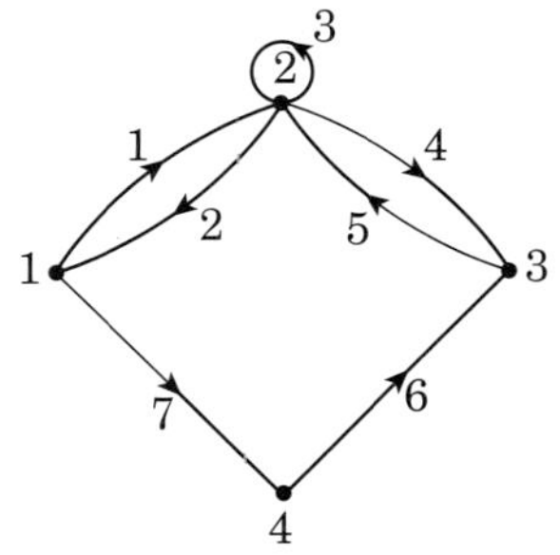

Figure 7.9

	TAIL	HEAD
1	2	2
2	4	3
3	1	2
4	2	3
5	2	1
6	1	4
7	3	2

The enumeration of the edges need not begin with edge 1. In this example, edge 1 is stored in t_3 and h_3. Accordingly, index 3 is stored in a variable called START (see Figure 7.10). Further, the edges can be stored in any order. To find the edge following each edge, an array N (for NEXT) is used. The element n_{i+1} locates the successor of edge n_i, $1 \leq i \leq 6$. We store 0 in n_6 to indicate the end of the linked list representation of the digraph, as in Figure 7.11.

Figure 7.10

START: 3 → 3

	T	H
1	2	2
2	4	3
3	1	2
4	2	3
5	2	1
6	1	4
7	3	2

Figure 7.11

Static linked list representation of the relation.

START: 3 → 3

	T	H	N
1	2	2	4
2	4	3	6
3	1	2	5
4	2	3	7
5	2	1	1
6	1	4	0
7	3	2	2

Most modern programming languages support dynamic data structures. In this type of language, a linked list consists of a set of **nodes** and each node contains (at least) two fields: a **data field** and a **link field** (or **pointer field**) (see Figure 7.12). The data field contains a data item, whereas the link field contains the address of the next node in the list. For instance, consider the linked list in Figure 7.13. HEADER contains the address of the first node in the list; it corresponds to START in the previous discussion. The link field of the last node contains a special pointer called the **nil pointer** that signals the end of the list. This pointer corresponds to 0 in the static representation; a slash (/) in the field signifies it.

Figure 7.12

A typical node.

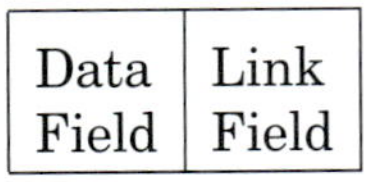

Figure 7.13

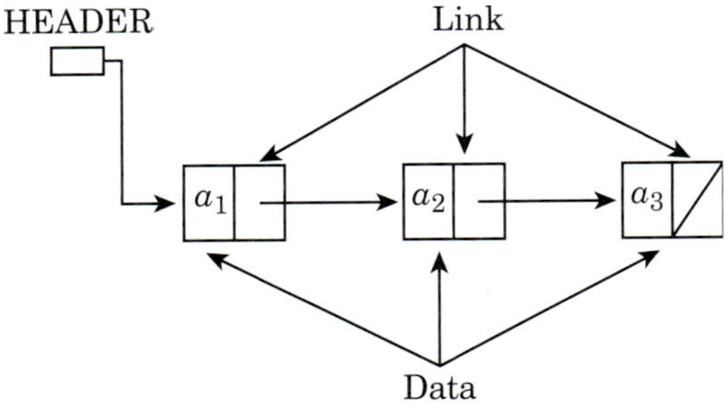

The relation in Figure 7.13 illustrates the dynamic linked list representation. First, for each vertex, create a linked list of vertices adjacent to it. Then store the header nodes in an array. The resulting linked representation appears in Figure 7.14.

Figure 7.14

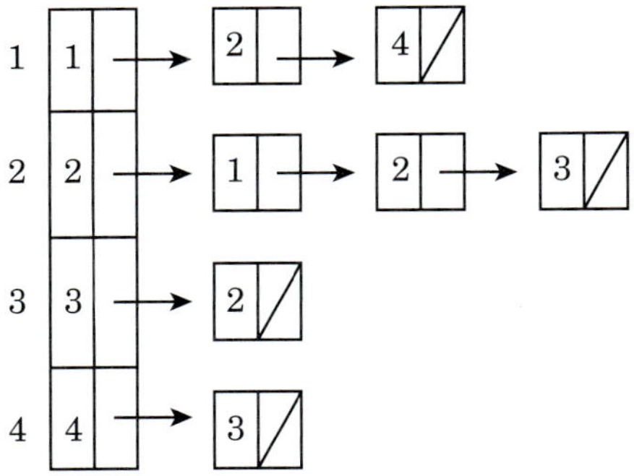

We can abbreviate this representation by storing the header nodes in an array of pointers, as in Figure 7.15. This simplified version is the **adjacency list representation** of the digraph and hence of the relation.

The next example shows how to find the adjacency matrix of a relation from its adjacency list representation.

EXAMPLE 7.12 Using the adjacency list representation of the relation in Figure 7.15, find its adjacency matrix.

SOLUTION:

The figure indicates vertex 1 is related to 2 and 4; vertex 2 is related to 1, 2, and 3; vertex 3 is related to 2; and vertex 4 is related to 3. Thus, the adjacency matrix of the relation is

$$M_R = \begin{array}{c} \\ 1 \\ 2 \\ 3 \\ 4 \end{array} \begin{array}{c} \begin{array}{cccc} 1 & 2 & 3 & 4 \end{array} \\ \begin{bmatrix} 0 & 1 & 0 & 1 \\ 1 & 1 & 1 & 0 \\ 0 & 1 & 0 & 0 \\ 0 & 0 & 1 & 0 \end{bmatrix} \end{array}$$

Figure 7.15

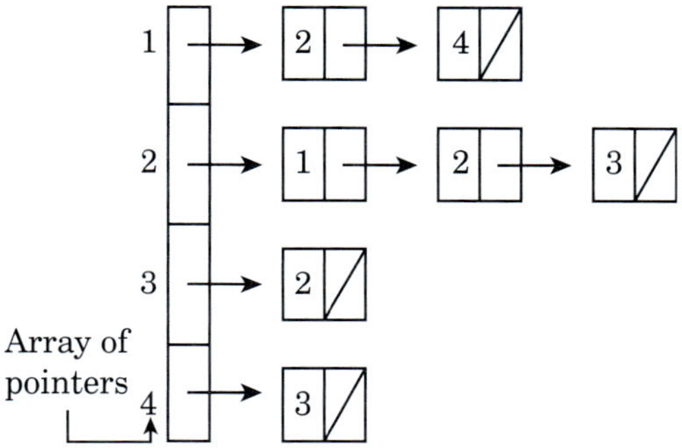

Exercise 7.3

Find the static linked list representation of each relation.

1.

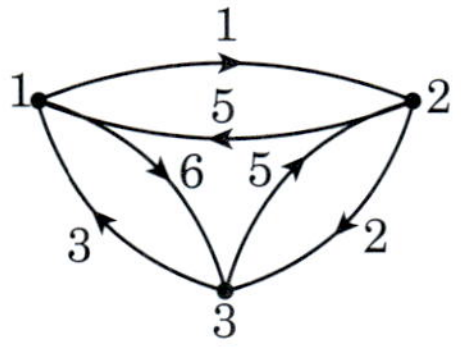

2.

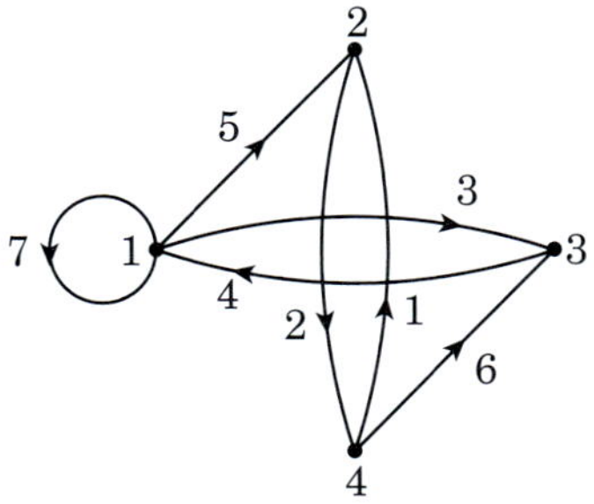

3–4. Find the adjacency list representation of the relations in Exercises 1 and 2.

Find the adjacency matrix of the relation with each adjacency list representation.

5.

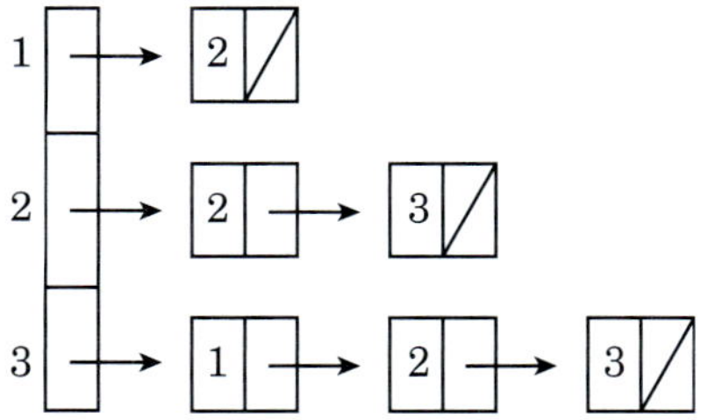

6.

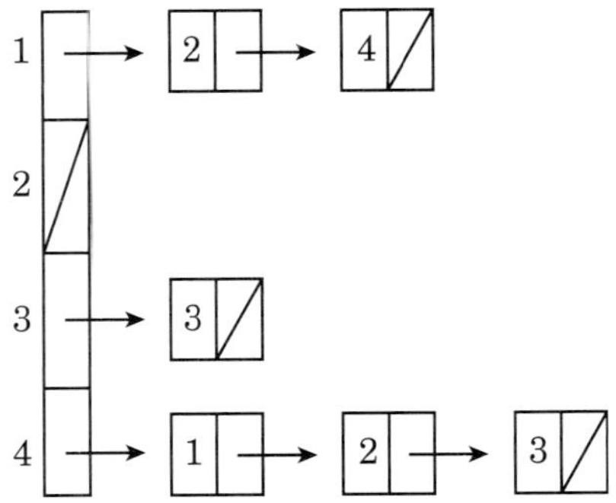

7–8. Draw the digraphs of the relations represented by the adjacency lists in Exercises 5 and 6.

Find the adjacency list representation of the relation with the given adjacency matrix.

9.
$$\begin{array}{c} \\ 1 \\ 2 \\ 3 \end{array}\begin{array}{c} \begin{array}{ccc} 1 & 2 & 3 \end{array} \\ \begin{bmatrix} 0 & 1 & 0 \\ 1 & 0 & 1 \\ 0 & 1 & 1 \end{bmatrix} \end{array}$$

10.
$$\begin{array}{c} \\ 1 \\ 2 \\ 3 \\ 4 \end{array}\begin{array}{c} \begin{array}{cccc} 1 & 2 & 3 & 4 \end{array} \\ \begin{bmatrix} 1 & 0 & 1 & 1 \\ 0 & 1 & 1 & 0 \\ 0 & 0 & 1 & 1 \\ 1 & 0 & 0 & 1 \end{bmatrix} \end{array}$$

Write an algorithm to find the adjacency list representation of a relation R on the set $\{1, 2, \dots, n\}$ using:

11. The relation, given in terms of ordered pairs.

12. Its adjacency matrix A.

13. Write an algorithm to find the adjacency matrix A of a relation on the set $\{1, 2, \dots, n\}$ from its adjacency list representation.

7.4 Properties of Relations

Since relations on finite sets can be represented by matrices, their properties can be identified from their adjacency matrices. In this section we will study the properties of reflexivity, symmetry, antisymmetry, and transitivity.

To begin with, consider the relation R, *is logically equivalent to*, on the set of propositions. Since every proposition is logically equivalent to itself, it has the property that xRx for every proposition x. Such a relation is reflexive.

Reflexive Relation

A relation R on a set A is **reflexive** if xRx for every element x in A, that is, if xRx for every $x \in A$.

Since every set A is a subset of itself, the relation *is a subset of* on its power set is reflexive. Similarly, the **equality relation** $(=)$ is also reflexive; it is denoted by Δ. Thus, a relation is reflexive if and only if $\Delta \subseteq R$.

The next two examples illustrate additional reflexive relations.

EXAMPLE 7.13 Since $x \leq x$ for every real number x, the relation $\leq$ on $\mathbb{R}$ is reflexive. No number is less than itself, so the less than relation is *not* reflexive. ■

EXAMPLE 7.14 Which of the following relations on $A = \{x, y, z\}$ are reflexive?

- $R_1 = \{(x,x), (x,y), (y,y), (z,z)\}$
- $R_2 = \{(x,x), (y,y), (y,z), (z,y)\}$
- $R_3 = \emptyset$ [the **empty relation**]
- $R_4 = \{(x,x), (y,y), (z,z)\}$

SOLUTION:

For a relation R on A to be reflexive, every element in A must be related to itself, that is, $(a,a) \in R$ for every $a \in A$. The element a has three choices, namely, x, y, and z; therefore, the ordered pairs (x,x), (y,y), and (z,z) must be in the relation for it to be reflexive. Consequently, the relations R_1 and R_4 are reflexive, whereas R_2 and R_3 are *not*. ■

How can we characterize the adjacency matrix $M = (m_{ij})$ of a reflexive relation on the set $A = \{a_1, a_2, \ldots a_n\}$? A relation R on A is reflexive if and only if $a_i R a_i$ for every a_i in A. Thus, R is reflexive if and only if $m_{ii} = 1$ for every ij that is, if and only if the main diagonal elements of M_R are all 1's, as Figure 7.16 shows.

Figure 7.16

$$M_R = \begin{bmatrix} 1 & & & & \\ & 1 & & & \\ & & \cdot & & \\ & & & \cdot & \\ & & & & \cdot \\ & & & & & 1 \end{bmatrix}$$

The digraph of a reflexive relation must contain a loop at each vertex, since every element of A is related to itself; see Figure 7.16.

Number of Reflexive Relations

We can use the adjacency matrix M_R of a relation R on a set A to compute the number of reflexive relations that can be defined on A, as the following example demonstrates.

EXAMPLE 7.15 Find the number of reflexive relations R that can be defined on a set with n elements.

SOLUTION:
Since R is reflexive, every element on the main diagonal of M_R is 1; there are n such elements. Since M_R contains n^2 elements, there are $n^2 - n = n(n-1)$ elements that do not lie on the main diagonal; each can be a 0 or 1; so each such element m_{ij} has two choices. Thus, by the multiplication principle, we can form $2^{n(n-1)}$ such adjacency matrices, that is, $2^{n(n-1)}$ reflexive relations on A. ■

For an exploration of symmetric and antisymmetric relations, again let R be the relation, *is logically equivalent to*, on the set of propositions. If x and y are any two propositions such that xRy, then yRx. Thus xRy implies yRx.

On the other hand, let x and y be any two real numbers such that $x \leq y$ and $y \leq x$. Then $x = y$. Thus the relation $R(\leq)$ has the property that if xRy and yRx, then $x = y$.

These two examples lead us to the next definitions.

Symmetric and Antisymmetric Relations

A relation R on a set A is **symmetric** if aRb implies bRa; that is, if $(a,b) \in R$, then $(b,a) \in R$. It is **antisymmetric** if aRb and bRa imply $a = b$.

By the law of the contrapositive, the definition of antisymmetry can be restated as follows: A relation R on A is **antisymmetric** if whenever $a \neq b$, either $a \not R b$ or $b \not R a$, that is, $\sim(aRb \wedge bRa)$. Thus R is antisymmetric if there are *no* pairs of distinct elements a and b such that aRb and bRa.

The next three examples demonstrate symmetric and antisymmetric relations.

EXAMPLE 7.16 Which of the following relations on $\{x,y,z\}$ are symmetric? Antisymmetric?

- $R_1 = \{(x,x), (y,y), (z,z)\}$
- $R_2 = \{(x,y)\}$
- $R_3 = \{(x,y), (y,x)\}$
- $R_4 = \{(x,x), (x,z), (z,x), (y,z)\}$

SOLUTION:
The relations R_1 and R_3 are symmetric. R_2 is *not* symmetric, since (y,x) is not in R_2. Similarly, R_4 is *not* symmetric. R_1 and R_2 are antisymmetric, but R_3 and R_4 are *not*. ■

EXAMPLE 7.17 The relation *is logically equivalent to* on the set of propositions is symmetric. Is it antisymmetric? Suppose $p \equiv q$ and $q \equiv p$; this does not imply that $p = q$, so the relation is *not* antisymmetric. ■

EXAMPLE 7.18 The relation $\leq$ on $\mathbb{R}$ is *not* symmetric, since $x \leq y$ does not imply that $y \leq x$. If, however, $x \leq y$ and $y \leq x$, then $x = y$, so the relation is antisymmetric. ■

> These two examples demonstrate that a symmetric relation need not be antisymmetric and vice versa.

> As for the adjacency matrix of a symmetric relation, a relation R on $\{a_1, a_2, \dots, a_n\}$ is symmetric only if $a_i R a_j$ implies $a_j R a_i$; that is, only if, $m_{ij} = m_{ji}$. Thus, R is symmetric if and only if M_R is symmetric; see Figure 7.17.

Figure 7.17

$$M_R = \begin{bmatrix} \ddots & 0 & & \\ 0 & \ddots & 1 & \\ & 1 & \ddots & 0 \\ & & 0 & \ddots \end{bmatrix}$$

Graphically, this means if a directed edge runs from a_i to a_j, then one should run from a_j to a_i. In other words, every edge must be bidirectional.

> For a relation R to be antisymmetric, if $a_i \neq a_j$ either $a_i \not R a_j$ or $a_j \not R a_i$. In other words, if $i \neq j$ and $m_{ij} = 1$, then $m_{ji} = 0$; that is, either $m_{ij} = 0$ or $m_{ji} = 0$; see Figure 7.18.

Figure 7.18

$$M_R = \begin{bmatrix} \ddots & 0 & & \\ 0 & \ddots & 0 & \\ & 1 & \ddots & 1 \\ & & 0 & \ddots \end{bmatrix}$$

Geometrically, if a directed edge runs from a_i to a_j, one should *not* run from a_j to a_i; that is, no edges are bidirectional.

EXAMPLE 7.19 Determine if the relation R on $\{a, b, c\}$ defined by

$$M_R = \begin{bmatrix} 1 & 1 & 0 \\ 0 & 0 & 0 \\ 0 & 1 & 1 \end{bmatrix}$$

is antisymmetric.

SOLUTION:
Consider the cases $i \neq j$ and $m_{ij} = 1$, where $1 \leq i,j \leq 3$. Clearly, $m_{12} = 1 \neq 0 = m_{21}$ and $m_{32} = 1 \neq 0 = m_{23}$. Thus, when $i \neq j$, either $m_{ij} = 0$ or $m_{ji} = 0$. Therefore, the relation is antisymmetric; see Figure 7.19. (Notice that $m_{11} = m_{33} = 1$ and $m_{22} = 0$, but this does not violate the condition for antisymmetry.)

Figure 7.19

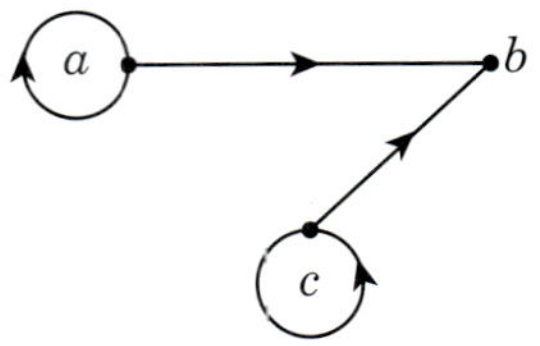

■

Number of Symmetric Relations

Again, the adjacency matrix of a relation on a set A can be effectively used to determine the number of symmetric relations that can be defined on A. The following example demonstrates this.

EXAMPLE 7.20 Find the number of symmetric relations that can be defined on a set with n elements.

SOLUTION:
Let R be a relation on the set and let $M_R = (m_{ij})_{n \times n}$. Then $m_{ij} = 1$ if and only if $m_{ji} = 1$ for every i and j. So each element m_{ij} below the main diagonal determines uniquely the corresponding element m_{ji} above the main diagonal; in other words, each m_{ji} has one choice (see Figure 7.20).

Figure 7.20

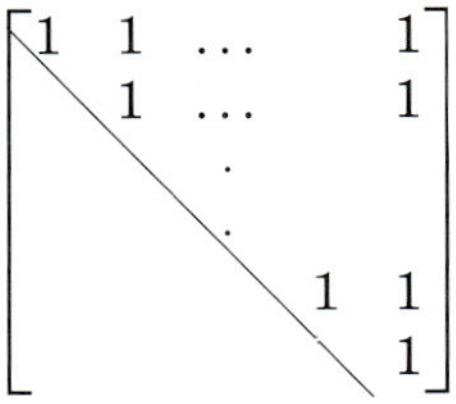

Figure 7.21

Each element has two choices.

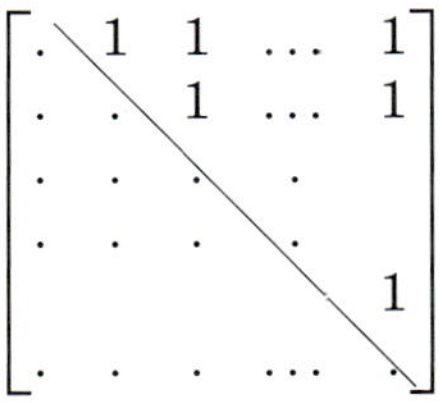

Now, each element on or below the main diagonal has two choices: 0 or 1 (see Figure 7.21). There are $1 + 2 + \cdots + n = n(n+1)/2$ such elements. So, by the multiplication principle, the number of such adjacency matrices equals $2^{n(n+1)/2}$; that is, we can define $2^{n(n+1)/2}$ symmetric relations on the set. ■

Notice that the less-than relation on $\mathbb{R}$ has the property that if $x < y$ and $y < z$, then $x < z$. Accordingly, the order relation $<$ is said to be transitive.

More generally, we make the following definition.

Transitive Relation

A relation R on A is **transitive** if aRb and bRc imply aRc; that is, whenever a is related to b and b is related to c, a is related to c.

The next three examples illuminate this definition.

EXAMPLE 7.21 Once again, consider the relation *is logically equivalent to* on the set of propositions. If $p \equiv q$ and $q \equiv r$, then $p \equiv r$, so the relation $\equiv$ is transitive. ■

EXAMPLE 7.22 Let A be the set of courses offered by a mathematics department. Define a relation R on A as follows: xRy if course x is a prerequisite for course y. The relation R is transitive (Why?). (R is the **precedence relation**.) ■

> Determining if a relation R is transitive can be time-consuming, especially if the relation contains many elements. We must look at all possible ordered pairs of the form (a,b) and (b,c), then ascertain if the element (a,c) is also in R, as the next example illustrates.

EXAMPLE 7.23 Which of the following relations on $\{a, b, c\}$ are transitive?

- $R_1 = \{(a,b), (b,c), (a,c)\}$
- $R_3 = \{(a,a), (b,b), (c,c)\}$
- $R_2 = \{(a,a), (a,b), (a,c), (b,a), (b,c)\}$
- $R_4 = \{(a,b)\}$

SOLUTION:

The relation R_1 is transitive; so are R_3 and R_4 by default. In relation R_2, $(b,a) \in R_2$ and $(a,b) \in R_2$, but $(b,b) \notin R_2$. So, R_2 is *not* transitive. ■

> As for the digraph of a transitive relation R, whenever there is a directed edge from a to b and one from b to c, one also runs from a to c.

Transitive relations are explored further in Section 7.7.

Exercises 7.4

Determine if the given relation on $\{a, b, c, d\}$ is reflexive, symmetric, antisymmetric, or transitive.

1. $\{(a,a), (b,b)\}$

2. $\{(a,a), (a,b), (b,b), (c,c), (d,d)\}$

3. $\emptyset$

4. $\{(a,b), (a,c), (b,c)\}$

Is the relation *has the same color hair as* on the set of people:

5. Reflexive? **6.** Symmetric?

7. Antisymmetric? **8.** Transitive?

9–12. Redo Exercises 5–8 using the relation *lives within 5 miles of* on the set of people.

13–16. Let Σ^n denote the set of n-bit words. Define a relation R on Σ^n as xRy if the Hamming distance between x and y is one. Redo Exercises 5–8 using the relation R.

In Exercises 17–19, the adjacency matrices of three relations on $\{a, b, c\}$ are given. Determine if each relation is reflexive, symmetric, or antisymmetric.

17. $\begin{bmatrix} 1 & 0 & 0 \\ 0 & 1 & 0 \\ 0 & 0 & 1 \end{bmatrix}$ **18.** $\begin{bmatrix} 0 & 1 & 1 \\ 1 & 0 & 1 \\ 1 & 0 & 0 \end{bmatrix}$ **19.** $\begin{bmatrix} 1 & 0 & 0 \\ 1 & 1 & 0 \\ 0 & 1 & 1 \end{bmatrix}$

When is a relation on a set A *not*:

20. Reflexive? **21.** Symmetric? **22.** Transitive?

Give an example of a relation on $\{a, b, c\}$ that is:

23. Reflexive, symmetric, and transitive.

24. Reflexive, symmetric, but not transitive.

25. Reflexive, transitive, but not symmetric.

26. Symmetric, transitive, but not reflexive.

27. Reflexive, but neither symmetric nor transitive.

28. Symmetric, but neither transitive nor reflexive.

29. Transitive, but neither reflexive nor symmetric.

30. Neither reflexive, symmetric, nor transitive.

31. Symmetric, but not antisymmetric.

32. Antisymmetric, but not symmetric.

33. Symmetric and antisymmetric.

34. Neither symmetric nor antisymmetric.

In Exercise 35–38, complete each adjacency matrix of a relation on $\{a, b, c\}$ in such a way that the relation has the given property.

35. $\begin{bmatrix} - & 1 & 0 \\ 0 & - & 1 \\ 1 & 0 & - \end{bmatrix}$, reflexive **36.** $\begin{bmatrix} 1 & - & 0 \\ 1 & 0 & 1 \\ - & - & 1 \end{bmatrix}$, symmetric

37. $\begin{bmatrix} 0 & - & 1 \\ 1 & 1 & - \\ - & 1 & 0 \end{bmatrix}$, antisymmetric

38. $\begin{bmatrix} - & 1 & - \\ - & 1 & 1 \\ 1 & - & - \end{bmatrix}$, transitive

39. When will a relation R on a set A be both symmetric and antisymmetric?

A relation R on a set A is **irreflexive** if no element of A is related to itself, that is, if $(a,a) \notin R$ for every $a \in A$. Determine if each relation is irreflexive.

40. The less-than relation on $\mathbb{R}$.

41. The relation *is a factor of* on $\mathbb{N}$.

42. The relation *is a parent of* on the set of people.

Determine if each relation on $\{a, b, c\}$ is irreflexive.

43. $\{(a, a)\}$

44. $\{(a, b), (b, b), (a, c)\}$

45. $\{(b, a), (c, a)\}$

46. $\emptyset$

Characterize each for an irreflexive relation on a finite set:

47. Its adjacency matrix.

48. Its digraph.

A relation R on a set A is **asymmetric** if whenever $aRb, b \not R a$. Determine if each relation is asymmetric.

49–51. The relations in Exercises 40–42.

52. $\{(a, a), (b, b), (c, c)\}$ on $\{a, b, c\}$

53. $\{(a, b), (a, c), (b, b)\}$ on $\{a, b, c\}$

54. $\{(a, b), (b, c), (c, a)\}$ on $\{a, b, c\}$

For an asymmetric relation on a finite set, characterize:

55. Its adjacency matrix.

56. Its digraph.

Find the number of binary relations that can be defined on a set of two elements that are:

***57.** Reflexive.

***58.** Symmetric.

***59.** Reflexive and symmetric.

***60.** Antisymmetric.

***61.** Irreflexive.

***62.** Asymmetric.

***63.** Prove: A relation R on a finite set is transitive if $M_R^{[2]} \le M_R$, where $(a_{ij}) \le (b_{ij})$ means $a_{ij} \le b_{ij}$ for every i and j.

7.5 Operations on Relations

Just as sets can be combined to construct new sets, relations can be combined to produce new relations. This section presents five such operations, three of which are analogous to the set operations of union, intersection, and complementation.

Union and Intersection

Let R and S be any two relations from A to B. Their **union** and **intersection**, denoted by $R \cup S$ and $R \cap S$, respectively, are defined as $R \cup S = \{(a,b) \mid aRb \vee aSb\}$ and $R \cap S = \{(a,b) \mid aRb \wedge aSb\}$. Thus $a(R \cup S)b$ if aRb or aSb. Likewise, $a(R \cap S)b$ if aRb and aSb.

The next two examples illustrate these definitions.

EXAMPLE 7.24 Consider the relations $R = \{(a,a),(a,b),(b,c)\}$ and $S = \{(a,a),(a,c),(b,b),(b,c),(c,c)\}$ on $\{a,b,c\}$ (see Figures 7.22 and 7.23). Then $R \cup S = \{(a,a),(a,b),(a,c),(b,b),(b,c),(c,c)\}$ and $R \cap S = \{(a,a),(b,c)\}$.

Figure 7.22

Digraph of R.

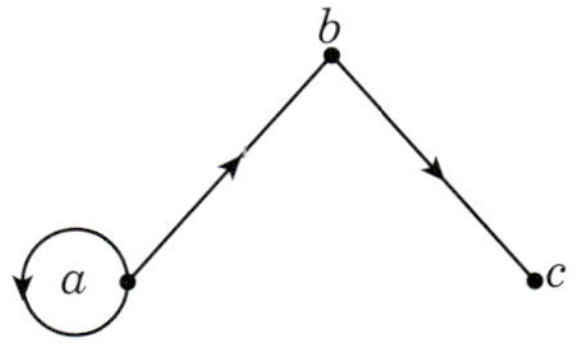

Figure 7.23

Digraph of S.

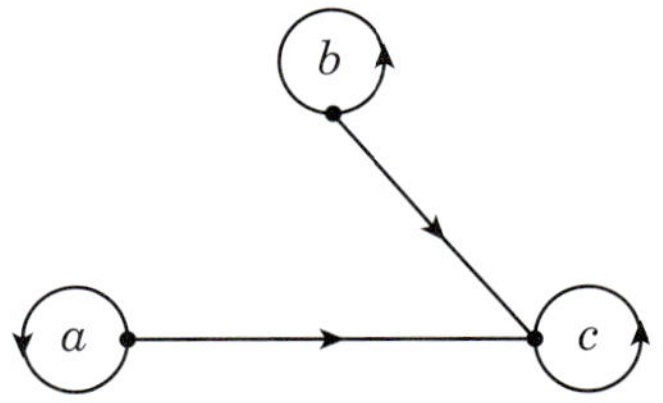

Figure 7.24

Digraph of $R \cup S$.

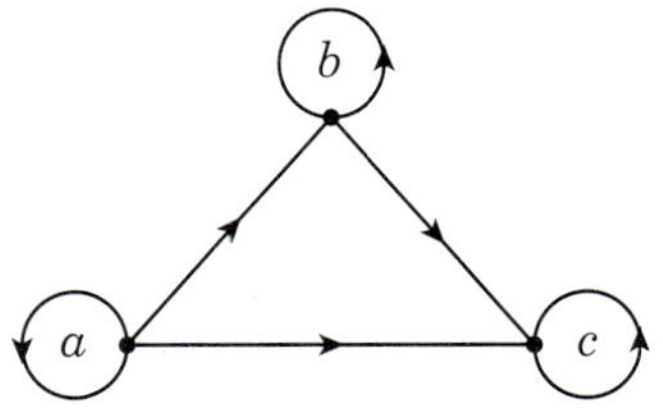

Figure 7.25

Digraph of $R \cap S$.

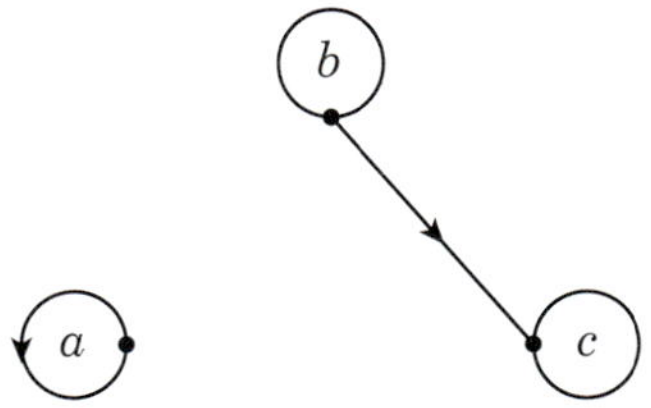

Graphically, $R \cup S$ consists of all edges in R together with those in S (see Figure 7.24), whereas $R \cap S$ consists of all common edges (see Figure 7.25). ■

EXAMPLE 7.25 Let R and S be the relations $\leq$ and $\geq$ on $\mathbb{R}$, respectively. Then $R \cup S$ consists of all possible ordered pairs $\mathbb{R} \times \mathbb{R}$ and $R \cap S$ is the equality relation. ■

We can use the adjacency matrices of relations R and S to find those of their union and intersection. By definition, an entry in $M_{R \cup S}$ is 1 if and only if the corresponding element of M_R or M_S is 1; that is, if and only if the corresponding element of their join, $M_R \vee M_S$, is 1. Since $M_{R \cup S}$ and $M_R \vee M_S$ are of the same size, $M_{R \cup S} = M_R \vee M_S$. Similarly, an element of $M_{R \cap S}$ is 1 if and only if the corresponding element of $M_R \wedge M_S$ is 1, so $M_{R \cap S} = M_R \wedge M_S$.

Theorem 7.3 summarizes these conclusions. We leave a formal proof as an exercise (see Exercise 62).

THEOREM 7.3 Let R and S be relations on a finite set. Then $M_{R \cup S} = M_R \vee M_S$ and $M_{R \cap S} = M_R \wedge M_S$. ■

The following example illustrates this theorem.

EXAMPLE 7.26 Using the adjacency matrices of the relations R and S in Example 7.24 find $M_{R \cup S} = M_R \vee M_S$ and $M_{R \cap S} = M_R \wedge M_S$.

SOLUTION:
We have

$$M_R = \begin{bmatrix} 1 & 1 & 0 \\ 0 & 0 & 1 \\ 0 & 0 & 0 \end{bmatrix} \quad \text{and} \quad M_S = \begin{bmatrix} 1 & 0 & 1 \\ 0 & 1 & 1 \\ 0 & 0 & 1 \end{bmatrix}$$

By Theorem 7.3,

$$M_{R \cup S} = M_R \vee M_S = \begin{bmatrix} 1 & 1 & 1 \\ 0 & 1 & 1 \\ 0 & 0 & 1 \end{bmatrix} \quad \text{and} \quad M_{R \cap S} = M_R \wedge M_S = \begin{bmatrix} 1 & 0 & 0 \\ 0 & 0 & 1 \\ 0 & 0 & 0 \end{bmatrix}$$

These matrices can recover the actual elements of $R \cup S$ and $R \cap S$ obtained in Example 7.24. ■

Another way to combine two relations is quite similar to the composition of functions we studied in Section 3.5.

Composition of Relations

Let R be a relation from A to B, and S a relation from B to C. The **composition** of R and S, denoted by $R \odot S$, is defined as follows. Let $a \in A$ and $c \in C$. Then $a(R \odot S)c$ if there exists an element b in B such that aRb and bSc, as in Figure 7.26.

Figure 7.26

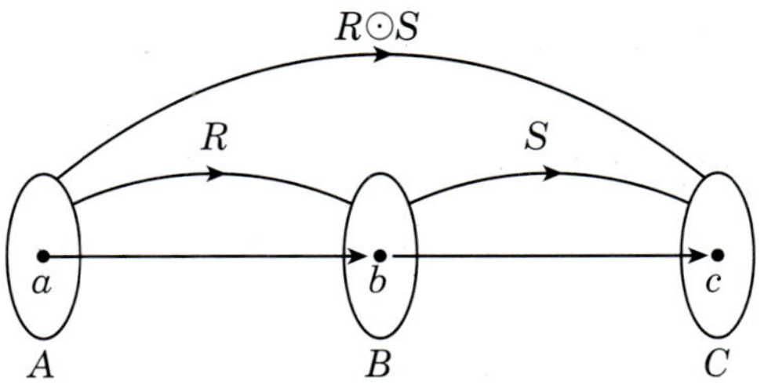

The next example illustrates this definition.

EXAMPLE 7.27 Let $A = \{a, b, c\}, B = \{1, 2, 3, 4\}$, and $C = \{w, x, y, z\}$. Using the relations $R = \{(a, 1), (a, 3), (b, 2)\}$ from A to B and $S = \{(1, x), (1, y), (2, w), (2, z), (4, y)\}$ from B to C (see Figure 7.27), find $R \odot S$.

SOLUTION:
Since $aR1$ and $1Sx$, $a(R \odot S)x$. Similarly, $a(R \odot S)y$, $b(R \odot S)w$, and $b(R \odot S)z$. Thus, $R \odot S = \{(a, x), (a, y), (b, w), (b, z)\}$.

Pictorially, all we need to do is simply follow the arrows from A to C in the figure. (Try this approach.)

Figure 7.27

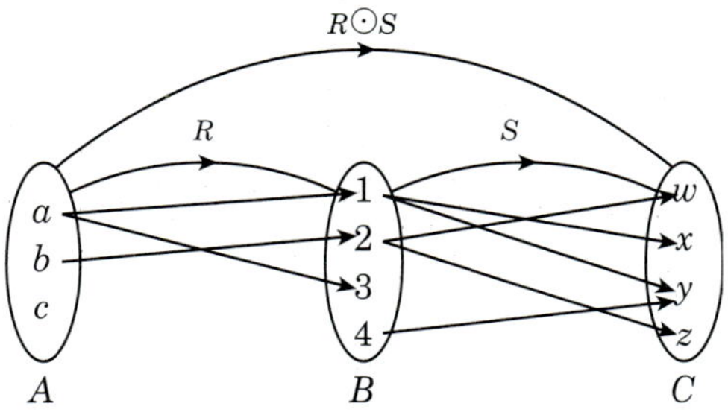

■

Databases

The next example gives an interesting application of the composition operation to the theory of databases.

EXAMPLE 7.28 Suppose a database consists of two files F_1 and F_2, given by Tables 7.2 and 7.3, respectively. File F_1 can be considered a relation from the set of names to the set of telephone numbers and file F_2 a relation from the set of telephone numbers to the set of telephone bills. Then $F_1 \odot F_2$ is a relation from the set of names to the set of telephone bills. In other words, $F_1 \odot F_2$ is a file of names and their corresponding telephone bills (see Table 7.4).

Table 7.2

Name	Telephone number
Hall	123-4567
Berkowitz	225-5061
Chand	124-3987
Benson	239-3883
Scott	534-3434
Abrams	345-5678

Table 7.3

Telephone number	Bill
123-4567	39.45
123-0011	25.00
243-1111	47.50
124-3987	23.35
124-8958	73.30
534-3434	95.65
345-5678	51.95
128-9876	64.85

Table 7.4

Name	Bill
Hall	39.45
Chand	23.15
Scott	95.65
Abrams	51.95

The adjacency matrices of the relations R, S, and $R \odot S$ display an intriguing connection. To see this, from Example 7.27, we have:

$$M_R = \begin{bmatrix} 1 & 0 & 1 & 0 \\ 0 & 1 & 0 & 0 \\ 0 & 0 & 0 & 0 \end{bmatrix} \quad M_S = \begin{bmatrix} 0 & 1 & 1 & 0 \\ 1 & 0 & 0 & 1 \\ 0 & 0 & 0 & 0 \\ 0 & 0 & 1 & 0 \end{bmatrix}, \quad \text{and}$$

$$M_{R \odot S} = \begin{bmatrix} 0 & 1 & 1 & 0 \\ 1 & 0 & 0 & 1 \\ 0 & 0 & 0 & 0 \end{bmatrix}$$

Then

$$M_R \odot M_S = \begin{bmatrix} 1 & 0 & 1 & 0 \\ 0 & 1 & 0 & 0 \\ 0 & 0 & 0 & 0 \end{bmatrix} \odot \begin{bmatrix} 0 & 1 & 1 & 0 \\ 1 & 0 & 0 & 1 \\ 0 & 0 & 0 & 0 \\ 0 & 0 & 1 & 0 \end{bmatrix}$$

$$= \begin{bmatrix} 0 & 1 & 1 & 0 \\ 1 & 0 & 0 & 1 \\ 0 & 0 & 0 & 0 \end{bmatrix} = M_{R \odot S}$$

■

More generally, we have the following result.

THEOREM 7.4 Let A, B, and C be finite sets. Let R be a relation from A to B, and S a relation from B to C. Then $M_{R \odot S} = M_R \odot M_S$.

PROOF:
Let $A = \{a_1, a_2, \ldots, a_m\}$, $B = \{b_1, b_2, \ldots, b_n\}$, and $C = \{c_1, c_2, \ldots, c_p\}$. Then the matrices $M_R, M_S, M_{R \odot S}$, and $M_R \odot M_S$ are of sizes $m \times n$, $n \times p$, $m \times p$, and $m \times p$, respectively.

Let $M_{R \odot S} = (x_{ij})$ and $M_R \odot M_S = (y_{ij})$. Then $x_{ij} = 1$ if and only if $a_i(R \odot S)c_j$. But $a_i(R \odot S)c_j$ if and only if a_iRb_k and b_kSc_j for some b_k in B. Thus, $x_{ij} = 1$ if and only if $y_{ij} = 1$, so $x_{ij} = y_{ij}$ for every i and j. Consequently, $M_{R \odot S} = M_R \odot M_S$. ■

The definition of composition can be extended to a finite number of relations. Accordingly, we now define the nth power of a relation using recursion.

Recursive Definition of R^n

Let R be a relation on a set A. The ***n*th power of *R***, denoted by R^n, is defined recursively as

$$R^n = \begin{cases} R & \text{if } n = 1 \\ R^{n-1} \odot R & \text{otherwise} \end{cases}$$

Geometrically, R^n consists of the endpoints of all possible paths of length n. Thus aR^nb if a path of length n exists from a to b.

The next two examples illuminate this definition

EXAMPLE 7.29 Using the relation $R = \{(a,b),(b,b),(c,a),(c,c)\}$ on $\{a,b,c\}$, find R^2 and R^3.

SOLUTION:

- $R^2 = R \odot R = \{(a,b),(b,b),(c,a),(c,b),(c,c)\}$
- $R^3 = R^2 \odot R = \{(a,b),(b,b),(c,a),(c,b),(c,c)\} = R^2$

The digraphs of the relations R and R^2 are displayed in Figures 7.28 and 7.29, respectively.

Figure 7.28

Digraph of R.

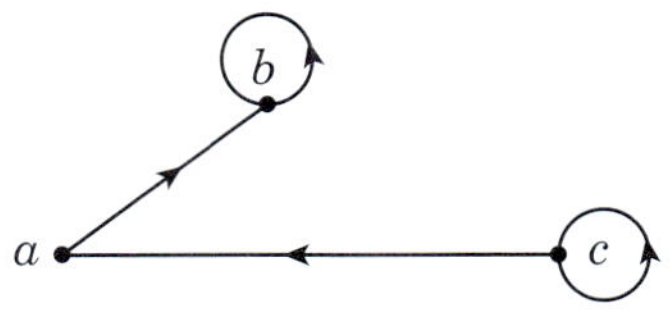

Figure 7.29

Digraph of $R^2 = R^3$.

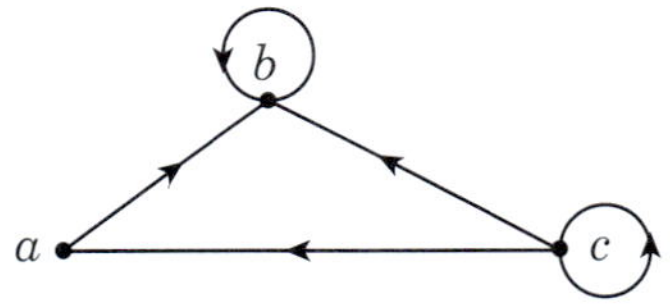

■

EXAMPLE 7.30 Define a relation R on the set of all U.S. cities as follows: xRy if there is a direct flight from city x to city y. Then xR^2y if there is a direct flight from city x to some city z and a direct flight from city z to city y. Thus R^2 consists of the endpoints of all airline routes in R passing through exactly one city. More generally, R^n consists of the endpoints of all airline routes in R passing through exactly $n-1$ cities. ■

Let R be a relation on a finite set. Then, by Theorem 7.4, $M_{R \odot R} = M_R \odot M_R$; that is, $M_{R^2} = (M_R)^{[2]}$.

More generally, we have the following result.

THEOREM 7.5 Let R be a relation on a finite set and n any positive integer. Then $M_{R^n} = (M_R)^{[n]}$. ■

EXAMPLE 7.31 For the relation R in Example 7.29, find M_{R^2} and M_{R^3}.

SOLUTION:

Notice that

$$M_R = \begin{bmatrix} 0 & 1 & 0 \\ 0 & 1 & 0 \\ 1 & 0 & 1 \end{bmatrix}$$

$$M_{R^2} = (M_R)^{[2]} = \begin{bmatrix} 0 & 1 & 0 \\ 0 & 1 & 0 \\ 1 & 0 & 1 \end{bmatrix} \odot \begin{bmatrix} 0 & 1 & 0 \\ 0 & 1 & 0 \\ 1 & 0 & 1 \end{bmatrix} = \begin{bmatrix} 0 & 1 & 0 \\ 0 & 1 & 0 \\ 1 & 1 & 1 \end{bmatrix}$$

$$M_{R^3} = (M_R)^{[3]} = M_R^{[2]} \odot M_R$$

$$= \begin{bmatrix} 0 & 1 & 0 \\ 0 & 1 & 0 \\ 1 & 1 & 1 \end{bmatrix} \odot \begin{bmatrix} 0 & 1 & 0 \\ 0 & 1 & 0 \\ 1 & 0 & 1 \end{bmatrix} = \begin{bmatrix} 0 & 1 & 0 \\ 0 & 1 & 0 \\ 1 & 1 & 1 \end{bmatrix}$$

Notice that M_{R^2} and M_{R^3} are the adjacency matrices of the relations R^2 and R^3, obtained in Example 7.29. ■

The next theorem tells us more about powers of transitive relations, and we will use it in Section 7.7.

THEOREM 7.6 Let R be a transitive relation on a set A. Then $R^n \subseteq R$ for every positive integer n.

PROOF (by PMI):
When $n = 1, R^1 \subseteq R$, which is true. Suppose $R^k \subseteq R$ for an arbitrary positive integer k.

To show that $R^{k+1} \subseteq R$:

Let $(x,y) \in R^{k+1}$. Since $R^{k+1} = R^k \odot R, (x,y) \in R^k \odot R$. Then, by definition, there is a z in A such that $(x,z) \in R^k$ and $(z,y) \in R$. But $R^k \subseteq R$, by the inductive hypothesis. Consequently, $(x,z) \in R$. Thus $(x,z) \in R$ and $(z,y) \in R$, so $(x,y) \in R$ by transitivity. Thus $R^{k+1} \subseteq R$.

Thus, by induction, $R^n \subseteq R$ for every $n \geq 1$. ■

We conclude this section with an example to illustrate this theorem.

EXAMPLE 7.32 Notice that the relation $R = \{(a,a),(a,b),(a,c),(b,c)\}$ on $\{a,b,c\}$ (see Figure 7.30) is transitive. You may verify that:

$$R^2 = R \odot R = \{(a,a),(a,b),(a,c)\} \subseteq R$$

$$R^3 = R^2 \odot R = \{(a,a),(a,b),(a,c)\} \subseteq R$$

$$R^4 = R^3 \odot R = \{(a,a),(a,b),(a,c)\} \subseteq R$$

(In fact, $R^n = R^2$ for every integer $n \geq 2$, so $R^n \subseteq R$ for every $n \geq 1$. See Exercise 7.38.)

Figure 7.30

Digraph of R.

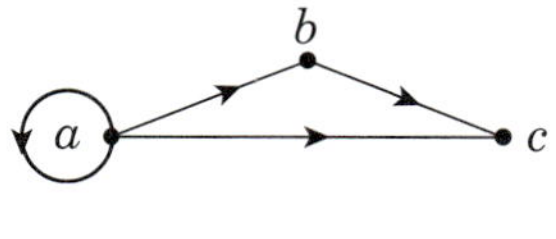

■

Exercises 7.5

1. Using the relations $R = \{(a,b),(a,c),(b,b),(b,c)\}$ and $S = \{(a,a),(a,b),(b,b),(c,a)\}$ on $\{a,b,c\}$, find $R \cup S$ and $R \cap S$.
2. Redo Exercise 1 using the relations $R = \{(a,a),(a,b),(b,c),(b,d)\}$ and $S = \{(a,b),(b,b),(b,c),(c,a),(d,a)\}$ on $\{a,b,c,d\}$.
3. Let R and S be the relations $<$ and $=$ on $\mathbf{R}$, respectively. Identify $R \cup S$ and $R \cap S$.
4. With the adjacency matrices of the relations R and S in Exercise 1, find those of the relations $R \cup S$ and $R \cap S$.
5. Redo Exercise 4 using the relations in Exercise 2.

Using the relations $R = \{(a,a),(a,b),(b,c),(c,c)\}$ and $S = \{(a,a),(b,b),(b,c),(c,a)\}$ on $\{a,b,c\}$, find each.

6. $R \odot S$
7. $S \odot R$
8. R^2
9. R^3

Let R be a relation from $\{a,b,c\}$ to $\{1,2,3,4\}$ and S a relation from $\{1,2,3,4\}$ to $\{x,y,z\}$. Find $R \odot S$ in each case.

10. $R = \{(a,2),(a,3),(b,1),(c,4)\}$ and $S = \{(1,x),(2,y),(4,y),(3,z)\}$
11. $R = \{(a,1),(b,2),(c,1)\}$ and $S = \{(3,x),(3,y),(4,z)\}$

Using the following adjacency matrices of relations R and S on $\{a,b,c\}$, find the adjacency matrices in Exercises 12–19.

$$M_R = \begin{bmatrix} 1 & 0 & 1 \\ 0 & 1 & 0 \\ 0 & 1 & 1 \end{bmatrix} \qquad M_S = \begin{bmatrix} 0 & 1 & 1 \\ 0 & 0 & 0 \\ 1 & 0 & 1 \end{bmatrix}$$

12. $M_{R \odot S}$
13. $M_{S \odot R}$
14. M_{R^4}
15. $(M_R)^{[4]}$
16. Define a relation R on the set of U.S. cities as follows: xRy if a direct communication link exists from city x to city y. How would you interpret R^2? R^n?
17. Redo Exercise 16 using the relation R on the set of all countries in the world, defined as follows: xRy if country x can communicate with country y directly.

The **complement** and **inverse** of a relation R from a set A to a set B, denoted by R' and R^{-1} respectively, are defined as follows: $R' = \{(a,b) \mid a \not{R} b\}$ and $R^{-1} = \{(a,b) \mid bRa\}$. So R' consists of all elements in $A \times B$ that are not in R, whereas R^{-1} consists of all elements (a,b), where $(b,a) \in R$. Using the relations $R = \{(a,a),(a,b),(b,c),(c,c)\}$ and $S = \{(a,a),(b,b),(b,c),(c,a)\}$ on $\{a,b,c\}$, find each.

18. R' **19.** R^{-1} **20.** M_{R^1} **21.** $(M_R)'$

22. $M_{R^{-1}}$ **23.** $(M_R)^T$ **24.** M_{S^1} **25.** $(M_S)^T$

Using the relations $R = \{(a,1),(b,2),(b,3)\}$ and $S = \{(a,2),(b,1),(b,2)\}$ from $\{a,b\}$ to $\{1,2,3\}$, find each.

26. R' **27.** R^{-1} **28.** $(R')'$ **29.** $(R^{-1})^{-1}$

30. $R' \cap S'$ **31.** $(R \cap S)'$ **32.** $R' \cup S'$ **33.** $(R \cup S)'$

34. $(R \cup S)^{-1}$ **35.** $R^{-1} \cup S^{-1}$ **36.** $R^{-1} \cap S^{-1}$ **37.** $(R \cap S)^{-1}$

38. For the relation R in Example 7.32, prove that $R^n = R^2$ for every $n \geq 2$.

Let R and S be relations on a finite set. Prove each.

39. $M_R = (M_{R^1})$ **40.** $M_{R^{-1}} = (M_R)^T$

Let R and S be relations from A to B. Prove each.

41. $(R^{-1})^{-1} = R$ **42.** If $R \subseteq S$, then $S' \subseteq R'$

43. If $R \subseteq S$, then $R^{-1} \subseteq S^{-1}$ **44.** $(R \cup S)' = R' \cap S'$

45. $(R \cap S)' = R' \cup S'$ **46.** $(R \cup S)^{-1} = R^{-1} \cup S^{-1}$

47. $(R \cap S)^{-1} = R^{-1} \cap S^{-1}$

Let R and S be relations on a set. Prove each.

48. R is reflexive if and only if R^{-1} is reflexive.

49. R is symmetric if and only if R' is symmetric.

50. R is symmetric if and only if R^{-1} is symmetric.

51. R is symmetric if and only if $R^{-1} = R$.

52. If R and S are symmetric, $R \cup S$ is symmetric.

53. If R and S are symmetric, $R \cap S$ is symmetric.

54. If R and S are transitive, $R \cap S$ is transitive.

55. Disprove: The union of two transitive relations on a set is transitive.

56. Let A, B, C, and D be any sets, R a relation from A to B, S a relation from B to C, and T a relation from C to D. Prove that $R\odot(S\odot T) = (R\odot S)\odot T$. (**associative property**)

Let R and S be two relations from A to B, where $|A| = m$ and $|B| = n$. Using their adjacency matrices, write an algorithm to find the adjacency matrix of each relation.

57. $R \cup S$ **58.** $R \cap S$ **59.** R' **60.** R^{-1}

61. Let $X = (x_{ij})$ be the adjacency matrix of a relation R from A to B and $Y = (y_{ij})$ that of a relation S from B to C, where $|A| = m$, $|B| = n$, and $|C| = p$. Write an algorithm to find the adjacency matrix $Z = (z_{ij})$ of the relation $R \odot S$.

***62.** Prove Theorem 7.3.

7.6 The Connectivity Relation (optional)

We can use the various powers R^n of a relation R to construct a new relation, called the connectivity relation. This section defines that new relation and then shows how to compute it.

Connectivity Relation

Let R be a relation on a set A. The **connectivity relation** of R, denoted by R^∞, is the union of all powers of R:

$$\begin{aligned} R^\infty &= R \cup R^2 \cup R^3 \cup R^4 \cup \ldots \cup R^n \cup \ldots \\ &= \bigcup_{n=1}^{\infty} R^n \end{aligned}$$

So $M_{R^\infty} = M_R \vee M_{R^2} \vee M_{R^3} \vee \ldots$

Geometrically, $aR^\infty b$ if there is a path of some length n from a to b. The connectivity relation consists of the endpoints of all possible paths in R.

The next two examples show how to find R^∞.

EXAMPLE 7.33 Find the connectivity relation R^∞ of the relation $R = \{(a,a),(a,b),(a,c),(b,c)\}$ on $\{a,b,c\}$.

SOLUTION:
From Example 7.32, $R^n = R^2$ for every integer $n \geq 2$. So

$$\begin{aligned} R^\infty &= R \cup R^2 \cup R^3 \cup R^4 \cup \ldots \\ &= R \cup R^2 \\ &= \{(a,a),(a,b),(a,c),(b,c)\} \end{aligned}$$ ■

EXAMPLE 7.34 Find the connectivity relation R^∞ of the relation $R = \{(a,b),(b,a),(b,b),(c,b)\}$ on $\{a,b,c\}$.

SOLUTION:
$R^2 = R \odot R = \{(a,a),(a,b),(b,a),(b,b),(c,a),(c,b)\}$ (see Figures 7.31 and 7.32.).

Figure 7.31

Digraph of R.

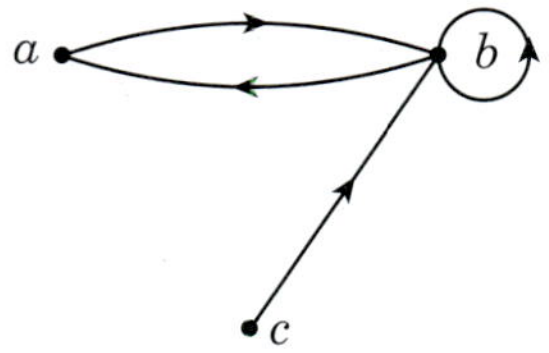

Figure 7.32

Digraph of R^2.

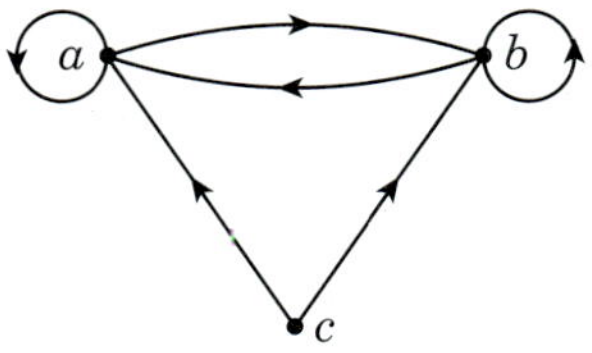

$$R^3 = R^2 \odot R = \{(a,a),(a,b),(b,a),(b,b),(c,a),(c,b)\} = R^2$$

$$R^4 = R^3 \odot R = \{(a,a),(a,b),(b,a),(b,b),(c,a),(c,b)\} = R^2$$

In fact $R^n = R^2$ for every $n \geq 2$. Thus,

$$R^\infty = R \cup R^2 = \{(a,a),(a,b),(b,a),(b,b),(c,a),(c,b)\}$$ ■

We can also determine connectivity using the adjacency matrix of a relation.

EXAMPLE 7.35 Using the adjacency matrix of the relation R in Example 7.34, find its connectivity relation.

SOLUTION:
Since $R = \{(a,b),(b,a),(b,b),(c,b)\}$,

$$M_R = \begin{bmatrix} 0 & 1 & 0 \\ 1 & 1 & 0 \\ 0 & 1 & 0 \end{bmatrix} \qquad M_{R^2} = M_R \odot M_R = \begin{bmatrix} 1 & 1 & 0 \\ 1 & 1 & 0 \\ 1 & 1 & 0 \end{bmatrix}$$

$$M_{R^3} = M_{R^2} \odot M_R = \begin{bmatrix} 1 & 1 & 0 \\ 1 & 1 & 0 \\ 1 & 1 & 0 \end{bmatrix} \qquad M_{R^4} = M_{R^3} \odot M_R = \begin{bmatrix} 1 & 1 & 0 \\ 1 & 1 & 0 \\ 1 & 1 & 0 \end{bmatrix}$$

Then

$$\begin{aligned} M_{R^\infty} &= M_R \vee M_{R^2} \vee M_{R^3} \vee \dots \\ &= M_R \vee M_{R^2} \\ &= \begin{bmatrix} 1 & 1 & 0 \\ 1 & 1 & 0 \\ 1 & 1 & 0 \end{bmatrix} \qquad \text{(Verify this.)} \end{aligned}$$

Thus $R^\infty = \{(a,a),(a,b),(b,a),(b,b),(c,a),(c,b)\}$, as in Example 7.34. ■

Theorem 7.7 comes in handy when computing R^∞. With the theorem, only the first n powers of R are needed to compute it, where $n = |A|$.

THEOREM 7.7 Let R be a relation on a set with size n. Then

$$R^\infty = R \cup R^2 \cup R^3 \cup \dots \cup R^n$$

$$\begin{aligned} M_{R^\infty} &= M_R \vee M_{R^2} \vee M_{R^3} \vee \dots \vee M_{R^n} \\ &= M_R \vee (M_R)^{[2]} \vee (M_R)^{[3]} \vee \dots \vee (M_R)^{[n]} \end{aligned}$$ ■

The next example illustrates this theorem.

EXAMPLE 7.36 Find R^∞ of the relation R on $\{a,b,c,d\}$ defined by

$$M_R = \begin{bmatrix} 0 & 1 & 0 & 0 \\ 1 & 0 & 1 & 0 \\ 0 & 0 & 0 & 1 \\ 1 & 0 & 0 & 0 \end{bmatrix}$$

SOLUTION:

$$M_{R^2} = M_R \odot M_R = \begin{bmatrix} 1 & 0 & 1 & 0 \\ 0 & 1 & 0 & 1 \\ 1 & 0 & 0 & 0 \\ 0 & 1 & 0 & 0 \end{bmatrix} \qquad M_{R^3} = M_{R^2} \odot M_R = \begin{bmatrix} 0 & 1 & 0 & 1 \\ 1 & 0 & 1 & 0 \\ 0 & 1 & 0 & 0 \\ 1 & 0 & 1 & 0 \end{bmatrix}$$

$$M_{R^4} = M_{R^3} \odot M_R = \begin{bmatrix} 1 & 0 & 1 & 0 \\ 0 & 1 & 0 & 1 \\ 1 & 0 & 1 & 0 \\ 0 & 1 & 0 & 1 \end{bmatrix}$$

By Theorem 7.7,

$$M_{R^\infty} = M_R \vee M_{R^2} \vee M_{R^3} \vee M_{R^4} = \begin{bmatrix} 1 & 1 & 1 & 1 \\ 1 & 1 & 1 & 1 \\ 1 & 1 & 1 & 1 \\ 1 & 1 & 1 & 1 \end{bmatrix}$$

Thus $R^\infty = \{(a,a),(a,b),(a,c),(a,d),(b,a),(b,b),(b,c),(b,d),(c,a),(c,b),(c,c),(c,d),(d,a),(d,b),(d,c),(d,d)\}$. (You may verify this using the digraph of R.) ■

We can use Theorem 7.7 to develop an algorithm for computing M_{R^∞}, which yields the connectivity relation of a relation R. It is given in Algorithm 7.1.

```
Algorithm Connectivity Relation (MR,MR∞)
(* This algorithm uses the adjacency matrix MR of a relation R on a set
   with size n and computes that of its connectivity relation, using
   Theorem 7.7. *)
0.  Begin (* algorithm *)
      (* Initialize MR∞ and B, where B denotes the ith boolean
         power of MR. *)
1.    MR∞ ← MR
      B ← MR
2.    for i = 2 to n do (* find the ith boolean power of MR*)
3.      begin (* for *)
4.        B ← B ⊙ MR
5.        MR∞ ← MR∞ ∨ B (* update MR *)
6.      endfor (* for *)
7.  End (* algorithm *)
```

Algorithm 7.1

We close this section with an analysis of the complexity of this algorithm. Let b_n denote the number of boolean operations needed to compute R^∞. Each element in line 4 takes n meets and $n-1$ joins, a total of $2n-1$ operations. Since the product contains n^2 elements, the total number of bit-operations in line 4 is $(2n-1)n^2$. The join of the two $n \times n$ matrices in line 5 takes n^2 boolean operations. Since the ***for*** loop is executed $n-1$ times, the total number of boolean operations is given by

$$\begin{aligned} b_n &= (n-1)[(2n-1)n^2 + n^2] \\ &= 2(n-1)n^3 \\ &= \Theta(n^4) \end{aligned}$$

Thus the connectivity algorithm takes $\Theta(n^4) =$ bit operations.

Exercises 7.6

Find the connectivity relation of each relation on $\{a, b, c\}$.

1. $\{(a,a)\}$ **2.** $\{(a,a),(b,b)\}$ **3.** $\{(a,a),(b,b),(c,c)\}$

4. $\{(a,a),(a,b),(c,a)\}$ **5.** $\emptyset$ **6.** $\{(a,b),(a,c),(b,a),(c,a)\}$

Find the connectivity relation of the relation on $\{a,b,c\}$ with each adjacency matrix.

7. $\begin{bmatrix} 1 & 0 & 0 \\ 0 & 1 & 0 \\ 0 & 0 & 1 \end{bmatrix}$ **8.** $\begin{bmatrix} 1 & 1 & 0 \\ 0 & 1 & 1 \\ 1 & 0 & 1 \end{bmatrix}$ **9.** $\begin{bmatrix} 1 & 0 & 1 \\ 1 & 1 & 1 \\ 1 & 1 & 0 \end{bmatrix}$

Find the connectivity relation of the relation on $\{a,b,c,d\}$ with each adjacency matrix.

10. $\begin{bmatrix} 0 & 1 & 1 & 0 \\ 0 & 0 & 0 & 1 \\ 1 & 0 & 0 & 0 \\ 0 & 0 & 1 & 0 \end{bmatrix}$ **11.** $\begin{bmatrix} 0 & 1 & 0 & 1 \\ 0 & 0 & 1 & 0 \\ 0 & 0 & 0 & 0 \\ 0 & 1 & 0 & 0 \end{bmatrix}$ **12.** $\begin{bmatrix} 1 & 0 & 0 & 0 \\ 0 & 1 & 1 & 0 \\ 0 & 0 & 1 & 1 \\ 1 & 0 & 0 & 0 \end{bmatrix}$

*7.7 Transitive Closure (optional)

The connectivity relation of a relation R is closely associated with its transitive closure. First, we define the closure of R.

A relation R may not have a desired property, such as reflexivity, symmetry, or transitivity. Suppose it is possible to find a relation containing R and having the desired property. The smallest such relation is the **closure** of R with respect to the property. Accordingly, we make the next definition.

Transitive Closure

Suppose a relation R on A is *not* transitive. The smallest transitive relation that contains R is the **transitive closure** of R, denoted by R^*.

> How do we find R^*? If R is not transitive, it should have ordered pairs (a,b) and (b,c) such that $(a,c) \notin R$; so add (a,c) to R. We can continue this with every such pair in the new relation. The resulting relation is transitive, the transitive closure of R.

The next example illustrates this method.

EXAMPLE 7.37 Find the transitive closures of the relations $R = \{(a,b),(b,a),(b,c)\}$, $S = \{(a,a),(b,b),(c,c)\}$, and $T = \emptyset$ on $\{a,b,c\}$.

SOLUTION:

- $R = \{(a,b),(b,c),(b,a)\}$. Since $(a,b) \in R$ and $(b,c) \in R$, it needs (a,c) to be transitive. So add (a,c) to R. The new relation is $R_1 = \{(a,b),(a,c),(b,c),(b,a)\}$. It contains both (a,b) and (b,a), but not (a,a) or (b,b). Add them to R_1: $R_2 = \{(a,a),(a,b),(a,c),(b,a),(b,b),(b,c)\}$. It is transitive and contains R, so it is the transitive closure of R.
- The relation S is transitive, by default, so $S^* = S$.
- The transitive closure of $\emptyset$ is itself. ■

The transitive closure R^* of the relation R in Example 7.37 has practical applications. Suppose the relation indicates the communication links in a network of computers a, b, and c, as in Figure 7.33. The transitive closure R^* shows the possible ways one computer can communicate with another, perhaps through intermediaries. For instance, computer a cannot communicate directly with c, but it can through b. Figure 7.34 displays the transitive closure R^*.

Figure 7.33

Digraph of R.

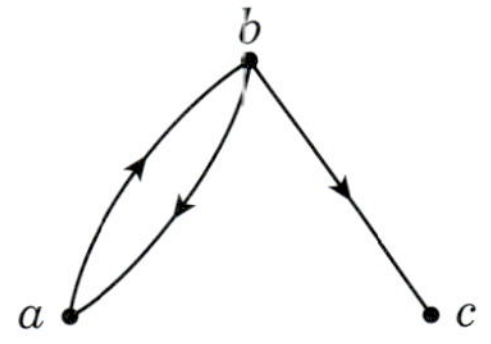

Figure 7.34

Digraph of R.*

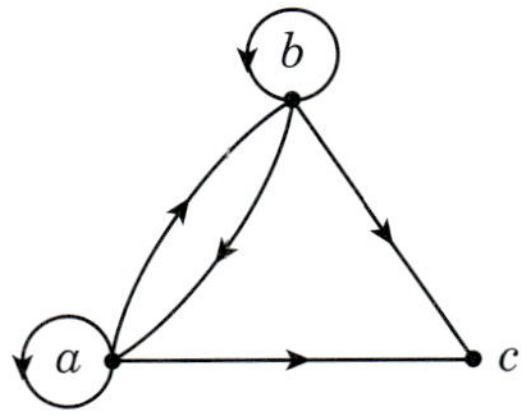

The close link between the transitive closure of a relation and its connectivity relation can be illustrated as follows.

THEOREM 7.8 The connectivity relation R^∞ of a relation R is its transitive closure R^*.

PROOF:

The proof unfolds in two parts. First, we must show that R^∞ is transitive and then show it is the smallest transitive relation containing R.

- *To prove that R^∞ is transitive*: Let $(a,b) \in R^\infty$ and $(b,c) \in R^\infty$. Since $(a,b) \in R^\infty$, a path runs from a to b. Similarly, one runs from b to c.

Combining these two paths produces a path from a to c. So $(a,c) \in R^\infty$ and R^∞ is transitive.

- *To prove that R^∞ is the smallest transitive relation containing R*: Suppose there is a transitive relation S such that $R \subseteq S \subseteq R^\infty$. We will show that $S = R^\infty$. Since S is transitive, by Theorem 7.6, $S^n \subseteq S$ for every $n \geq 1$. So

$$S^\infty = \bigcup_{n=1}^{\infty} S^n \subseteq S$$

Thus

$$S^\infty \subseteq S$$

By assumption, $R \subseteq S$; so $R^\infty \subseteq S^\infty$, since every path in R is also a path in S. Therefore, $R^\infty \subseteq S$.

Consequently, $S \subseteq R^\infty$ and $R^\infty \subseteq S$. Therefore, $S = R^\infty$. In other words, there are no transitive relations in between R and R^∞. So R^∞ is the smallest transitive relation containing R. ■

It follows by Theorems 7.7 and 7.8 that

$$R^* = R \cup R^2 \cup \cdots \cup R^n$$

and hence

$$M_{R^\infty} = M_R \vee M_{R^2} \vee \cdots \vee M_{R^n}$$

To illustrate this, using Example 7.36, the transitive closure of the relation $R = \{(a,b),(b,a),(b,c),(c,d),(d,a)\}$ on $\{a,b,c,d\}$ is $R^* = R^\infty = \{(a,a), (a,b),(a,c),(a,d),(b,a),(b,b),(b,c),(b,d),(c,a),(c,b),(c,c),(c,d),(d,a),(d,b), (d,c),(d,d)\}$.

Since $R^\infty = R^*$, the connectivity relation algorithm can be used to compute M_{R^*}, but it is not efficient, especially when M_{R^*} is fairly large. A better method to find R^* is **Warshall's algorithm**, named in honor of Stephen Warshall, who invented it in 1962.

Warshall's Algorithm

Let a-x_1-x_2-$\cdots$-x_m-b be a path in a relation R on a set $A = \{a_1, a_2, \ldots, a_n\}$. The vertices $x_1, x_2, \ldots, x_m$ are the **interior points** of the path. For instance, vertices c and d are the interior points on the path a-c-d-b of the digraph in Figure 7.5.

The essence of Warshall's algorithm lies in constructing a sequence of n boolean matrices $W_1, \ldots, W_n$, beginning with $W_0 = M_R$. Let $W_k = (w_{ij})$, where $1 \leq k \leq n$. Define $w_{ij} = 1$ if a path runs from a_i to a_j in R whose interior vertices, if any, belong to the set $\{a_1, a_2, \ldots, a_k\}$. Since the ijth element of W_n equals 1 if and only if a path exists from a_i to a_j whose interior points belong to the set $\{a_1, a_2, \ldots, a_n\}$, $W_n = W_{R^*}$.

In fact, the matrix $W_k = (w_{ij})$ can be constructed from its predecessor $W_{k-1} = (v_{ij})$ as follows. When can $w_{ij} = 1$? For $w_{ij} = 1$, there must be a path from a_i to a_j whose interior vertices belong to the set $\{a_1, a_2, \ldots, a_k\}$.

Case 1 If a_k is *not* an interior vertex, all interior vertices must belong to the set $\{a_1, a_2, \ldots, a_{k-1}\}$, so $v_{ij} = 1$.

Case 2 Suppose a_k is an interior vertex (see Figure 7.35). If a cycle exists at a_k, eliminate it to yield a shorter path. (This guarantees that the vertex a_k occurs exactly once in the path.) Therefore, all interior vertices of the paths a_i-$\cdots$-a_k and a_k-$\cdots$-a_j belong to the set $\{a_1, a_2, \ldots a_{k-1}\}$. In other words, $v_{ik} = 1$ and $v_{kj} = 1$.

Figure 7.35

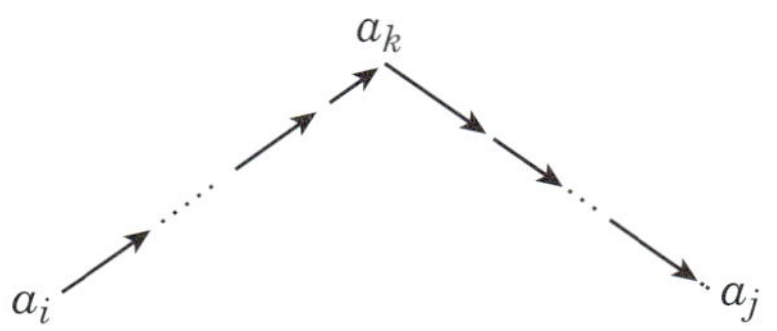

Consequently, $w_{ij} = 1$ only if $v_{ij} = 1$, or $v_{ik} = 1$ and $v_{kj} = 1$. This is the crux of Warshall's algorithm. Thus the ijth element of W_k is 1 if:

- The corresponding element of W_{k-1} is 1 or
- Both the ikth element and the kjth element of W_{k-1} are 1; that is, the ith element in column k of W_{k-1} and the jth element in row k of W_{k-1} are 1.

Use this property to construct W_1 from $W_0 = M_R$, W_2 from $W_1, \ldots$, and W_n from W_{n-1}. Since $W_n = M_{R^*}$, the actual elements of R^* can be read from W_n.

The next two examples clarify this algorithm.

EXAMPLE 7.38 Using Warshall's algorithm, find the transitive closure of the relation $R = \{(a, b), (b, a), (b, c)\}$ on $A = \{a, b, c\}$.

SOLUTION:

Step 1 Find W_0.

$$W_0 = M_R = \begin{bmatrix} 0 & 1 & 0 \\ 1 & 0 & 1 \\ 0 & 0 & 0 \end{bmatrix}$$

Step 2 Find W_1.

If the ijth element of W_0 is 1, the ijth element of W_1 is also 1. In other words, every 1 in W_0 stays in W_1. To find the remaining 1's in W_1, locate the 1's in column $1(=k)$; there is just one 1; it occurs in position $i = 2$. Now locate the 1's in row $1(=k)$. Again, there is just one 1, namely, in position $j = 2$. Therefore, the ijth entry in W_1 should be 1, where $i = 2$ and $j = 2$. Thus

$$W_1 = \begin{bmatrix} 0 & 1 & 0 \\ 1 & 1 & 1 \\ 0 & 0 & 0 \end{bmatrix}$$

Step 3 Find W_2.
Again, all the 1's in W_1 stay in W_2. To find the other 1's, if any, locate the 1's in column $2(=k)$ and row $2(=k)$. They occur in positions 1 and 2 of column 2 and in positions 1, 2, and 3 of row 2, so the ijth entry of W_2 must be 1, where $i = 1, 2$ and $j = 1, 2, 3$. So change the 0's in such locations of W_1 to 1's. Thus

$$W_2 = \begin{bmatrix} 1 & 1 & 1 \\ 1 & 1 & 1 \\ 0 & 0 & 0 \end{bmatrix}$$

Step 4 Find W_3.
All the 1's in W_2 remain in W_3. To find the remaining 1's, if any, locate the 1's in column 3—namely, positions 1 and 2—and the 1's in row 3. Because no 1's appear in row 3, we get no new 1's, so $W_3 = W_2$.

Since A contains three elements, $W_{R^*} = W_3$. Thus,

$$W_{R^*} = \begin{bmatrix} 1 & 1 & 1 \\ 1 & 1 & 1 \\ 0 & 0 & 0 \end{bmatrix}$$

which agrees with the transitive closure obtained in Example 7.37. ■

EXAMPLE 7.39 Using Warshall's algorithm, find the transitive closure of the relation $R = \{(a,a), (a,b), (a,d), (b,a), (c,b), (c,c), (d,b), (d,c), (d,d)\}$ on $\{a,b,c,d\}$.

SOLUTION:

Step 1 Find W_0.

$$W_0 = M_R = \begin{bmatrix} 1 & 1 & 0 & 1 \\ 1 & 0 & 0 & 0 \\ 0 & 1 & 1 & 0 \\ 0 & 1 & 1 & 1 \end{bmatrix}$$

Step 2 Find W_1.
Locate the 1's in column 1 and row 1; positions 1 and 2 in column 1; and positions 1, 2, and 4 in row 1. Therefore, W_1 should contain a 1 in locations (1,1), (1,2), (1,4), (2,1), (2,2), and (2,4):

$$W_1 = \begin{bmatrix} 1 & 1 & 0 & 1 \\ 1 & 1 & 0 & 1 \\ 0 & 1 & 1 & 0 \\ 0 & 1 & 1 & 1 \end{bmatrix}$$

(All the 1's in W_0 remain in W_1.)

Step 3 Find W_2.
Locate the 1's in column 2 and in row 2; positions 1, 2, 3, and 4 in column 2, and positions 1, 2, and 4 in row 2. So W_2 should contain a 1 in locations (1,1), (1,2), (1,4), (2,1), (2,2), (2,4), (3,1), (3,2), (3,4), (4,1), (4,2), and (4,4). Again, since all the 1's in W_1 stay in W_2,

$$W_2 = \begin{bmatrix} 1 & 1 & 0 & 1 \\ 1 & 1 & 0 & 1 \\ 1 & 1 & 1 & 1 \\ 1 & 1 & 1 & 1 \end{bmatrix}$$

Step 4 Find W_3.
The 1's of column 3 occur in positions 3 and 4; those of row 3 in positions 1, 2, 3, and 4. Consequently, W_3 should contain a 1 in locations (i,j) where $i = 3, 4$ and $j = 1, 2, 3, 4$:

$$W_3 = \begin{bmatrix} 1 & 1 & 0 & 1 \\ 1 & 1 & 0 & 1 \\ 1 & 1 & 1 & 1 \\ 1 & 1 & 1 & 1 \end{bmatrix}$$

Step 5 Find W_4.
The 1's of column 4 appear in positions 1, 2, 3, and 4; the 1's of row 4 in positions 1, 2, 3, and 4. So W_4 should contain a 1 in locations (i,j) where $i = 1, 2, 3, 4$ and $j = 1, 2, 3, 4$:

$$W_4 = \begin{bmatrix} 1 & 1 & 1 & 1 \\ 1 & 1 & 1 & 1 \\ 1 & 1 & 1 & 1 \\ 1 & 1 & 1 & 1 \end{bmatrix}$$

Since $M_{R^*} = W_4$, this is the adjacency matrix of the transitive closure. (Finding the connectivity relation of R will verify this.) ■

Warshall's algorithm is presented in Algorithm 7.2. It is based on the discussion preceding Example 7.38.

```
Algorithm Warshall(MR,W)
(* This algorithm employs the adjacency matrix of a relation R on
   finite set with n elements to find the adjacency matrix MR* of its
   transitive closure. *)
0. Begin (* algorithm *)
         (* Initialize W = (wij) *)
1.   W ← MR
2.   for k = 1 to n do (* compute Wk *)
3.     for i = 1 to n do
4.       for j = 1 to n do
5.         wij ← wij ∨ (wik ∧ wkj) (*compute the ij-th element *)
6.   MR ← W
7. End (* algorithm *)
```

Algorithm 7.2

A Comparison of Warshall's Algorithm with the Connectivity Algorithm

Why is this algorithm far more efficient than the connectivity relation algorithm? Notice that the number of boolean operations in line 5 is 2, so the total number of boolean operations in lines 2 through 5 (and hence in the algorithm) is $2 \cdot n \cdot n \cdot n = 2n^3 = \Theta(n^3)$, whereas the connectivity algorithm takes $\Theta(n^4)$ bit operations.

Exercises 7.7

Find the transitive closure of each relation on $A = \{a, b, c\}$.

1. $\{(a,b),(b,a)\}$

2. $\{(a,b),(b,c),(c,a)\}$

3. $\{(b,a),(b,c),(c,b)\}$

4. $\{(a,a),(a,c),(b,c),(c,a)\}$

Find the transitive closure of each relation on $A = \{a, b, c, d\}$.

5. $\{(a,a),(a,b)\}$

6. $\{(a,b),(b,c),(c,a)\}$

In Exercises 7–9, find the adjacency matrix of the transitive closure of each relation R on $\{a, b, c\}$ with the given adjacency matrix.

7. $\begin{bmatrix} 0 & 1 & 1 \\ 0 & 0 & 1 \\ 0 & 0 & 0 \end{bmatrix}$

8. $\begin{bmatrix} 1 & 0 & 1 \\ 0 & 1 & 0 \\ 0 & 1 & 1 \end{bmatrix}$

9. $\begin{bmatrix} 0 & 1 & 0 \\ 0 & 1 & 1 \\ 1 & 0 & 0 \end{bmatrix}$

10–12. Using the connectivity relation algorithm, find the transitive closure R^* of each relation in Exercises 7–9.

13–15. For the relation R on $\{a,b,c\}$ with each adjacency matrix in Exercises 7–9, compute the boolean matrix W_1 in Warshall's algorithm.

In Exercises 16–18, the adjacency matrix of a relation R on $\{a,b,c,d\}$ is given. In each case, compute the boolean matrices W_1 and W_2 in Warshall's algorithm.

16. $\begin{bmatrix} 0 & 1 & 0 & 0 \\ 0 & 0 & 1 & 0 \\ 1 & 0 & 1 & 0 \\ 0 & 1 & 0 & 1 \end{bmatrix}$ **17.** $\begin{bmatrix} 1 & 0 & 0 & 0 \\ 0 & 1 & 0 & 1 \\ 0 & 1 & 0 & 1 \\ 0 & 1 & 0 & 1 \end{bmatrix}$ **18.** $\begin{bmatrix} 0 & 1 & 0 & 1 \\ 1 & 0 & 1 & 0 \\ 0 & 0 & 0 & 1 \\ 1 & 0 & 0 & 1 \end{bmatrix}$

19–24. Using Warshall's algorithm, find the transitive closure of each relation in Exercises 7–9 and 16–18.

25–33. The **reflexive closure** of a relation on a set is the smallest reflexive relation that contains it. Find the reflexive closures of the relations in Exercises 1–9.

Find the reflexive closure of each relation on $\mathbb{R}$.

34. The less-than relation. **35.** The greater-than relation.

36–44. The **symmetric closure** of a relation on a set is the smallest symmetric relation that contains it. Find the symmetric closures of the relations in Exercises 1–9.

Let R be any relation on a set A. Prove each.

45. R is reflexive if and only if $\Delta \subseteq R$.

46. $R \cup \Delta$ is reflexive.

***47.** $R \cup \Delta$ is the smallest reflexive relation containing R .
(*Hint*: Assume there is a reflexive relation S such that $R \subseteq S \subseteq R \cup \Delta$. Prove that $S = R$ or $S = R \cup \Delta$.)

***48.** $R \cup R^{-1}$ is symmetric.
[*Hint*: Consider $(R \cup R^{-1})^{-1}$.]

***49.** $R \cup R^{-1}$ is the smallest symmetric relation that contains R.
(*Hint*: Suppose there is a symmetric relation S such that $R \subseteq S \subseteq R \cup R^{-1}$.)

7.8 Equivalence Relations

Section 7.4 introduced relations that are reflexive, symmetric, and transitive. Naturally we can now ask: Are there relations that simultaneously manifest all three properties? The answer is yes; for instance, the relation *is logically equivalent to* on the set of propositions has all these properties. Such a relation is an equivalence relation.

Equivalence Relation

A relation on a set is an **equivalence relation** if it is reflexive, symmetric, and transitive.

Examples 7.40–7.42 explore equivalence relations.

EXAMPLE 7.40 The relation *has the same color hair as* on the set of people is reflexive, symmetric, and transitive. So it is an equivalence relation. ■

EXAMPLE 7.41 Let Σ denote an alphabet. Define a relation R on Σ^* by xRy if $\|x\| = \|y\|$, where $\|w\|$ denotes the length of the word w. Is R an equivalence relation?

SOLUTION:

- Since every word has the same length as itself, R is reflexive.
- Suppose that xRy. Then $\|x\| = \|y\|$, so $\|y\| = \|x\|$. Consequently, yRx. Thus R is symmetric.
- If $xRy4$ and yRz, then $\|x\| = \|y\|$ and $\|y\| = \|z\|$. Therefore, $\|x\| = \|z\|$ and hence xRy. In other words, R is transitive.

Thus, R is an equivalence relation. ■

EXAMPLE 7.42 (optional) Is the relation *has the same memory location as* on the set of variables in a program an equivalence relation?

SOLUTION:

- Since every variable has the same location as itself, the relation is reflexive.
- If a variable x has the same location as a variable y, then y has the same location as x, so the relation is symmetric.
- Suppose x has the same location as y and y has the same location as z. Then x has the same location as z, so the relation is transitive.

Thus the relation is an equivalence relation. ■

FORTRAN provides an **equivalence statement**, so called since the relation *has the same location as* is an equivalence relation. We can see this in the following FORTRAN statement:

```
EQUIVALENCE (A,B),(C,D,E),(F,G,H)
```

It means the variables A and B share the same memory location; the variables C, D, and E share the same memory location; and so do the variables F, G, and H.

The congruence relation, an important relation in mathematics, is a classic example of an equivalence relation. It is closely related to the equality relation and partitions of a finite set, as will be seen shortly.

Karl Friedrich Gauss *(1777–1855), son of a laborer, was born in Brunswick, Germany. A child prodigy, he detected an error in his father's bookkeeping when he was 3. The Duke of Brunswick, recognizing his remarkable talents, sponsored his education. Gauss received his doctorate in 1799 from the University of Helmstedt. In his doctoral dissertation, he gave the first rigorous proof of the fundamental theorem of algebra, which states, "Every polynomial of degree n (≥ 1) with real coefficients has at least one zero." Newton and Euler, among other brilliant minds, had attempted to prove it, but failed.*

He made significant contributions to algebra, number theory, geometry, analysis, physics, and astronomy. His impressive work Disquisitiones Arithmeticae *of 1801 laid the foundation for modern number theory.*

From 1807 until his death, he was the director of the observatory and professor of mathematics at the University of Göttingen.

Called the "prince of mathematics" by his contemporary mathematicians, Gauss made the famous statement, "Mathematics is the queen of the sciences and the theory of numbers the queen of mathematics."

The congruence symbol ≡ was invented around 1800 by Karl Friedrich Gauss, the greatest mathematician of the 19th century.

Congruence Relation

Let $a, b, m \in \mathbf{Z}$, where $m \geq 2$. Then a is **congruent to *b* modulo *m***, denoted by $a \equiv b \pmod{m}$, if $a - b$ is divisible by m. The integer m is the **modulus** of the **congruence relation**. (This definition provides the basis of the **mod** operator we studied in Chapter 3.) If a is *not* congruent to b modulo m, we write $a \not\equiv b \pmod{m}$.

For example, since $5|(13-3)$, $13 \equiv 3 \pmod{5}$. Also, $-5 \equiv 3 \pmod{4}$ since $4|(-5-3)$. But $17 \not\equiv 4 \pmod{6}$, since $6 \nmid (17-4)$.

The congruence relation has several useful properties, some of which are given below.

THEOREM 7.9 Let $a, b, c, d, m \in \mathbf{Z}$ with $m \geq 2$. Then:

(1) $a \equiv a \pmod{m}$. (**reflexive property**)
(2) If $a \equiv b \pmod{m}$, then $b \equiv a \pmod{m}$. (**symmetric property**)
(3) If $a \equiv b \pmod{m}$ and $b \equiv c \pmod{m}$, then $a \equiv c \pmod{m}$. (**transitive property**)
(4) Let r be the remainder when a is divided by m. Then $a \equiv r \pmod{m}$.

PROOF:
We shall prove part 3 and leave the other parts as exercises.

(3) Suppose $a \equiv b \pmod{m}$ and $b \equiv c \pmod{m}$. Then $m|(a-b)$ and $m|(b-c)$. Consequently, $a - b = mq_1$ and $b - c = mq_2$ for some

integers q_1 and q_2. Then

$$\begin{aligned} a - c &= (a - b) + (b - c) \\ &= mq_1 + mq_2 \\ &= m(q_1 + q_2) \end{aligned}$$

Therefore, $m|(a - c)$ and $a \equiv c \pmod{m}$. ■

It follows by the theorem that the congruence relation is an equivalence relation.

The Congruence Relation and the Mod Operator

Suppose $a \equiv r \pmod{b}$, where $0 \le r \le b$. Then it can be shown that $r = a \bmod b$. Conversely, if $r = a \bmod b$, then $a \equiv r \pmod{b}$. Thus $a \equiv r \pmod{b}$ if and only if $r = a \bmod b$, where $0 \le r \le b$. See exercises 49 and 50.

For example, $43 \equiv 3 \pmod{5}$ and $0 \le 3 < 5$; clearly, $3 = 43 \bmod 5$. Let us digress briefly to look at an interesting application of congruences*.

Friday-the-13th

Congruences can be employed to find the number of Friday-the-13ths in a given year. Whether or not Friday-the-13th occurs in a given month depends on two factors: the day on which the 13th fell in the previous month and the number of days in the previous month.

Suppose that this is a non–leap year and that we would like to find the number of Friday-the-13ths in this year. Suppose also that we know the day the 13th occurred in December of last year. Let M_i denote each of the months December through November in that order and D_i the number of days in month M_i. The various values of D_i are $31, 31, 28, 31, 30, 31, 30, 31, 31, 30, 31$, and 30, respectively.

We label the days Sunday through Saturday by 0 through 6 respectively; so day 5 is a Friday.

Let $D_i \equiv d_i \pmod{7}$, where $0 \le d_i < 7$. The corresponding values of d_i are $3, 3, 0, 3, 2, 3, 2, 3, 3, 2, 3$, and 2, respectively. Each value of d_i indicates the number of days the day of the 13th in month M_i must be advanced to find the day the 13th falls in month M_{i+1}.

For example, December 13, 2000, was a Wednesday. So January 13, 2001, fell on day $(3 + 3) =$ day 6, which was a Saturday.

---

*T. Koshy, *Elementary Number Theory with Applications*, Harcourt/Academic Press, Boston, MA, 2002.

Let $t_i \equiv \sum_{j=1}^{i} d_j \pmod 7$, where $1 \le i \le 12$. Then t_i represents the total number of days the day of December 13 must be moved forward to determine the day of the thirteenth in month M_i.

For example, $t_3 \equiv d_1 + d_2 + d_3 = 3 + 3 + 0 \equiv 6 \pmod 7$. So, the day of December 13, 2000 (Wednesday) must be advanced by six days to determine the day of March 13, 2001; it is given by day $(3 + 6) =$ day $2 =$ Tuesday.

Notice that the various values of t_i modulo 7 are 3, 6, 6, 2, 4, 0, 2, 5, 1, 3, 6, and 1, respectively; they include all the least residues modulo 7. Given the day of December 13, they can be used to determine the day of the thirteenth of each month M_i in a non–leap year.

Table 7.5 summarizes the day of the 13th of each month in a non-leap year, corresponding to every choice of the day of December 13 of the previous year. You may verify this. Notice from the table that there can be at most three Friday-the-13ths in a non–leap year.

Table 7.5

Day of the 13th in Each Month in a Non–leap Year.

Dec. 13 \ t_i	**Jan. 3**	**Feb. 6**	**March 6**	**April 2**	**May 4**	**June 0**	**July 2**	**Aug. 5**	**Sept. 1**	**Oct. 3**	**Nov. 6**	**Dec. 1**
Sun	3	6	6	2	4	0	2	5	1	3	6	1
Mon	4	0	0	3	5	1	3	6	2	4	0	2
Tue	5	1	1	4	6	2	4	0	3	5	1	3
Wed	6	2	2	5	0	3	5	1	4	6	2	4
Thu	0	3	3	6	1	4	6	2	5	0	3	5
Fri	1	4	4	0	2	5	0	3	6	1	4	6
Sat	2	5	5	1	3	6	1	4	0	2	5	0

For a leap year, the various values of d_i are 3, 3, 1, 3, 2, 3, 2, 3, 3, 2, 3, and 2; and the corresponding values of t_i are 3, 6, 0, 3, 5, 1, 3, 6, 2, 4, 0, and 2. Using these, we can construct a similar table for a leap year.

Returning to the congruence relation, we now explore a close relationship between equivalence relations and partitions; but first we make the following definition.

Equivalence Class

Let R be an equivalence relation on a set A and let $a \in A$. The **equivalence class** of a, denoted by $[a]$, is defined as $[a] = \{x \in A \mid xRa\}$. It consists of all elements in A that are linked to a by the relation R. If $x \in [a]$, then x is a **representative** of the class $[a]$.

The next two examples explore equivalence relations.

EXAMPLE 7.43 The relation $R = \{(\mathrm{a,a}), (\mathrm{a,b}), (\mathrm{b,a}), (\mathrm{b,b}), (\mathrm{c,c})\}$ on $A = \{\mathrm{a,b,c}\}$ is an equivalence relation. Find the equivalence class of each element in A.

SOLUTION:

(1) $[a] = \{x \in A | xRa\}$
$= \{\text{a}, \text{b}\}$

(2) $[b] = \{x \in A | xRb\}$
$= \{\text{a}, \text{b}\}$
$= [a]$

(3) $[c] = \{x \in A | xRc\}$
$= \{c\}$

Two distinct equivalence classes exist, $[a]$ and $[c]$. Class $[a]$ has two representatives and class $[c]$ one representative. ■

EXAMPLE 7.44 The relation R on the set of words over the alphabet $\{\text{a}, \text{b}\}$, defined by $x\text{R}y$ if $\|x\| = \|y\|$, is an equivalence relation (see Example 7.41). Infinitely many equivalence classes exist for R, such as $\{\lambda\}$, $\{\text{a}, \text{b}\}$, and $\{\text{aa}, \text{ab}, \text{ba}, \text{bb}\}$. ■

EXAMPLE 7.45 Find all equivalence classes of the congruence relation mod 5 on the set of integers.

SOLUTION:
Let r be the remainder when an integer a is divided by 5. Then $a \equiv r (\text{mod } 5)$. Since the possible values of r, by the division algorithm, are $0, 1, 2, 3$, and 4, there are five distinct equivalence classes:

$$[0] = \{\ldots, -10, -5, 0, 5, 10, \ldots\}$$
$$[1] = \{\ldots, -9, -4, 1, 6, 11, \ldots\}$$
$$[2] = \{\ldots, -8, -3, 2, 7, 12, \ldots\}$$
$$[3] = \{\ldots, -7, -2, 3, 8, 13, \ldots\}$$
$$[4] = \{\ldots, -6, -1, 4, 9, 14, \ldots\}$$
■

These three examples lead us to the following observations:

- Every element belongs to an equivalence class.
- Any two distinct equivalence classes are disjoint.

These results can be stated more formally as follows.

THEOREM 7.10 Let R be an equivalence relation on a set A, with a and b any two elements in A. Then the following properties hold:

(1) $a \in [a]$.
(2) $[a] = [b]$ if and only if $a\text{R}b$.
(3) If $[a] \neq [b]$, then $[a] \cap [b] = \emptyset$.

PROOF:

(1) Since R is reflexive, aRa for every $a \in A$, so $a \in [a]$.
(2) Suppose $[a] = [b]$. Since $a \in [a]$ by part (1), $a \in [b]$. Therefore, by definition, aRb.

Conversely, let aRb. *To show that* $[a] \subseteq [b]$:
Let $x \in [a]$. Then xRa. Since xRa and aRb, xRb by transitivity. Therefore, $x \in [b]$ by definition. Thus $[a] \subseteq [b]$.
Similarly, $[b] \subseteq [a]$. Thus, $[a] = [b]$.

(3) We will prove the contrapositive of the given statement: If $[a] \cup [b] \neq \emptyset$, then $[a] = [b]$. Suppose $[a] \cup [b] \neq \emptyset$. Then an element x should be in $[a] \cap [b]$. Then $x \in [a]$ and $x \in [b]$. Since $x \in [a]$, xRa and hence aRx by symmetry. In addition, since $x \in [b]$, xRb. Thus aRx and xRb. Therefore, aRb by transitivity. Thus $[a] = [b]$, by part 2.

This concludes the proof. ■

It follows by Theorem 7.10 that any two equivalence classes are either identical or disjoint, but *not* both.

Notice that Example 7.43 has two disjoint equivalence classes, $[a]$ and $[c]$; their union is the whole set A. Therefore, $\{[a], [c]\}$ is a partition of A. In fact, every equivalence relation on a set induces a partition of the set, as given by the next theorem.

THEOREM 7.11 Let R be an equivalence relation on a set A. Then the set of distinct equivalence classes forms a partition of A. ■

The next four examples illuminate this theorem.

EXAMPLE 7.46 The relation *belongs to the same division as* is an equivalence relation on the set of teams in the American (National) League of major-league baseball. Let x denote a certain team in the American League. Then the class $[x]$ consists of all teams that belong to the same division as x. By Theorem 7.11, the set of teams in the league can be partitioned as {[Yankees], [White Sox], [Mariners]}. ■

EXAMPLE 7.47 By Example 7.41, the relation *has the same length as* on the set of words Σ^* over the alphabet $\Sigma = \{a, b\}$ is an equivalence relation. Then the set of equivalence classes formed is $\{[\lambda], [a], [aa], [aaa], \ldots\}$; it is a partition of Σ^*. ■

EXAMPLE 7.48 (optional) Suppose a FORTRAN program contains the variables A through J and the equivalence statement:

```
EQUIVALENCE (A,B),(C,D),(F,A,G),(C,J),(E,H)
```

By Example 7.42 the relation *shares the same memory location as* is an equivalence relation on the set of variables V. Let $V_1 = \{A,B,F,G\}$, $V_2 = \{C,D,J\}$, $V_3 = \{E,H\}$, and $V_4 = \{I\}$. The partition of V induced by this relation is $\{V_1, V_2, V_3, V_4\}$. See Figure 7.36.

Figure 7.36

Set of variables V.

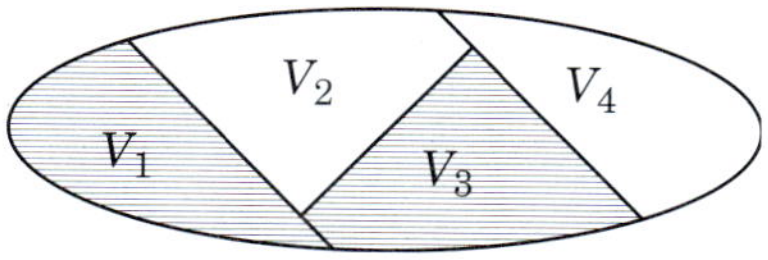

■

EXAMPLE 7.49 By Example 7.45, the distinct equivalence classes formed by the congruence relation modulo 5 on **Z** are [0], [1], [2], [3], and [4]. They form a partition of the set of integers, as shown in Figure 7.37.

Figure 7.37

A partition of the set of integers **Z**.

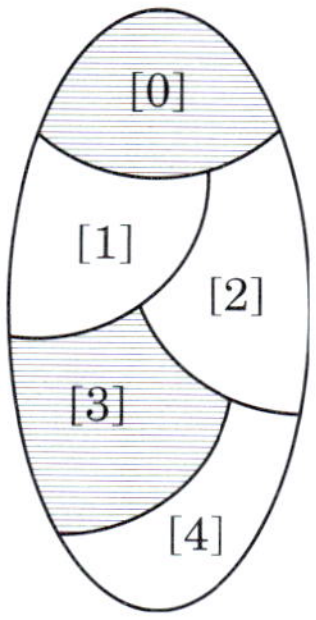

■

Conversely, does every partition yield an equivalence relation? The next theorem shows that every partition does.

THEOREM 7.12 Every partition of a set induces an equivalence relation on it.

PROOF:

Let $P = \{A_1, A_2, \ldots\}$ be a partition of a set A. Define a relation R on A as: aRb if a belongs to the same block as b. We shall show that R is indeed an equivalence relation.

- Since every element in A belongs to the same block as itself, R is reflexive.
- Let aRb. Then a belongs to the same block as b. So b belongs to the same block as a. Thus R is symmetric.
- Let aRb and bRc. Then a belongs to the same block as b and b to the same block as c. So a belongs to the same block as c. Therefore, R is transitive.

Thus R is an equivalence relation. ■

How can we find the equivalence relation corresponding to a partition of a set? The next example demonstrates how to accomplish this.

EXAMPLE 7.50 Find the equivalence relation on $A = \{a, b, c\}$ corresponding to the partition $\{\{a, b\}, \{c\}\}$.

SOLUTION:
Define a relation R on A as follows (see the above proof): xRy if x belongs to the same block as y. Since a and b belong to the same block, aRa, aRb, bRa, and bRb. Similarly, cRc. Thus $R = \{(a,a),(a,b,),(b,a),(b,b),(c,c)\}$. ■

Example 7.50 can serve to develop an algorithm for finding the equivalence relation corresponding to a partition P of a finite set A. It is given in Algorithm 7.3.

```
Algorithm Equivalence Relation (P,A,R)
(* This algorithm determines the equivalence relation R
   corresponding to a partition P of a finite set A. *)
Begin (* algorithm *)
    while P ≠ ø do
    begin (* while *)
    extract a block B
    pair each element in B with every element in B
    P ← P − B (* update P *)
    endwhile
  End (* algorithm *)
```

Algorithm 7.3

> Theorems 7.11 and 7.12 indicate a bijection between the family of partitions of a set and the family of equivalence relations on it.

Number of Partitions of a Finite Set

There is a delightful formula for computing the number of partitions (and hence the number of equivalence relations) of a set with size n. It is given by $\sum_{r=1}^{n} S(n,r)$, where $S(n,r)$ denotes a *Stirling number of the second kind*, defined by

$$S(n,1) = 1 = S(n,n)$$

$$S(n,r) = S(n-1,r-1) + rS(n-1,r), 1 < r < n$$

See Exercises 33–40.

Exercises 7.8

Determine if each is an equivalence relation.

1. The relation $\leq$ on $\mathbb{R}$.
2. The relation *is congruent to* on the set of triangles in a plane.
3. The relation *is similar to* on the set of triangles in a plane.

4. The relation *lives within 5 miles of* on the set of people.

5. The relation *takes a course with* on the set of students on campus.

Determine if each is an equivalence relation on $\{a,b,c\}$.

6. $\{(a,a),(b,b),(c,c)\}$

7. $\{(a,a),(a,c),(b,b),(c,a),(c,c)\}$

8. $\emptyset$

9. $\{(a,a),(b,b),(b,c),(c,b)\}$

Using the equivalence relation $\{(a,a),(a,b),(b,a),(b,b),(c,c),(d,d)\}$ on $\{a,b,c,d\}$, find each equivalence class.

10. $[a]$ **11.** $[b]$ **12.** $[c]$ **13.** $[d]$

A FORTRAN program contains 10 variables, A through J, and the following equivalence statement: EQUIVALENCE (A,B,C),(D,E),(F,B),(C,H). Find each class.

14. $[A]$ **15.** $[B]$ **16.** $[E]$ **17.** $[J]$

Using the equivalence relation in Example 7.47, find the equivalence class represented by:

18. a **19.** b **20.** aa **21.** aaa

Using the relation *has the same length as* on the set of words over the alphabet $\{a,b,c\}$, find the equivalence class with each representative.

22. λ **23.** a **24.** ab **25.** bc

26. Find the set of equivalence classes formed by the congruence relation modulo 4 on the set of integers.

Find the partition of the set $\{a,b,c\}$ induced by each equivalence relation.

27. $\{(a,a),(b,b),(c,c)\}$

28. $\{(a,a),(a,c),(b,b),(c,a),(c,c)\}$

A FORTRAN program contains the variables A through J. Find the partition of the set of variables induced by each equivalence statement.

29. EQUIVALENCE (A,B,C),(D,E),(F,B),(C,H)

30. EQUIVALENCE (A,B),(B,J),(C,J),(D,E,H)

Find the equivalence relation corresponding to each partition of the set $\{a,b,c,d\}$.

31. $\{\{a\},\{b,c\},\{d\}\}$

32. $\{\{a,b\},\{c,d\}\}$

The number of partitions of a set with size n is given by $\sum_{r=1}^{n} S(n,r)$, where $S(n,r)$ denotes a Stirling number of the second kind. Compute the number of partitions of a set with the given size.

33. Two **34.** Three **35.** Four **36.** Five

37–40. The number of partitions of a set with size n is also given by the **Bell number** B_n. Using Bell numbers, compute the number of partitions of a set with each of the sizes in Exercises 33–36.

Give a counterexample to disprove each.

41. The union of two equivalence relations is an equivalence relation.

42. The composition of two equivalence relations is an equivalence relation.

We can compute the day of the week corresponding to any date since 1582, the year the Gregorian calendar was adopted. The day d of the week for the rth day of month m in year y (> 1582) is given by

$$d = r + \lfloor 2.6m - 0.2 \rfloor - 2C + D + \lfloor C/4 \rfloor + \lfloor D/4 \rfloor \pmod 7$$

where $C = \lfloor y/100 \rfloor$ and $D = y \bmod 100$; $d = 0$ denotes Sunday; and $m = 1$ denotes March, $m = 11$ January, and $m = 12$ February. This formula is called **Zeller's formula**, after Christian Julius Johannes Zeller (1849–1899). Find the Christmas day of each year.

43. 2000 **44.** 2020 **45.** 2345 **46.** 3000

Let $a, b, c, d, m \in \mathbf{Z}$ with $m \geq 2$. Prove each.

47. If $a \equiv b \pmod m$ and $c \equiv d \pmod m$, then $a + c \equiv b + d \pmod m$.

48. If $a \equiv b \pmod m$ and $c \equiv d \pmod m$, then $ac \equiv bd \pmod m$.

49. Let r be the remainder when a is divided by m. Then $a \equiv r \pmod m$.

50. If $a \equiv r \pmod m$ and $0 \leq r < m$, r is the remainder when a is divided by m.

51. Let r_1 and r_2 be the remainders when a and b are divided by m, respectively. Then $a \equiv b \pmod m$ if and only if $r_1 \equiv r_2 \pmod m$.

52. A positive integer N is divisible by 3 if and only if the sum of its digits is divisible by 3. [*Hint*: $10 \equiv 1 \pmod 3$.]

53. A positive integer N is divisible by 9 if and only if the sum of its digits is divisible by 9. [*Hint*: $10 \equiv 1 \pmod 9$.]

Using the congruence relation, find the remainder when the first integer is divided by the second.

54. 256, 3 **55.** 657, 3 **56.** 1976, 9 **57.** 389, 276, 9

(*Hint*: Use Exercise 52 or 53.)

58. The United Parcel Service assigns to each parcel an identification number of nine digits and a check digit. The check digit is the remainder mod 9 of the 9-digit number. Compute the check digit for 359,876,015.

59. Every bank check has an 8-digit identification number $d_1d_2 \ldots d_8$ followed by a check digit d given by $d \equiv (d_1d_2, \ldots, d_8) \cdot (7,3,9,7,3,9,7,3) \bmod 10$, where $(x_1, x_2, \ldots, x_n) \cdot (y_1, y_2, \ldots, y_n) = \sum_{i=1}^{n} x_i y_i$. (It is the **dot product** of the two n-tuples.) Compute the check digit for 17,761,976.

60. Libraries use a sophisticated **code-a-bar system** to assign each book a 13-digit identification number $d_1, d_2 \ldots d_{13}$ and a check digit d. Let k denote the number of digits among d_1, d_3, d_5, d_7, d_9, d_{11}, and d_{13} greater than or equal to 5. Then d is computed as $d \equiv [-(d_1, d_2, \ldots, d_{13}) \cdot (2,1,2,1,2,1,2,1,2,1,2,1,2) - k] \bmod 10$, where the dot indicates the dot product. Compute the check digit for 2,035,798,008,938.

***61.** (**The coconuts and monkey problem**)* Five sailors and a monkey are marooned on a desert island. During the day they gather coconuts for food. They decide to divide them up in the morning and retire for the night. While the others are asleep, one sailor gets up and divides them into equal piles, with one left over that he throws out for the monkey. He hides his share, puts the remaining coconuts together, and goes back to sleep. Later a second sailor gets up and divides the pile into five equal shares with one coconut left over, which he discards for the monkey. Later the remaining sailors repeat the process. Find the smallest possible number of coconuts in the original pile.

7.9 Partial and Total Orderings

Just as we used the concepts of reflexivity, symmetry, and transitivity to define equivalence relations, we can use reflexivity, antisymmetry, and transitivity to introduce a new class of relations: partial orders. We begin this section with an example.

Building a house can be broken down into several tasks, as Table 7.6 shows. Define a relation R on the set of tasks as follows: Let x and y be any two tasks; then xRy if $x = y$ or must be done before y. This relation is reflexive, antisymmetric, and transitive (verify). Such a relation is a partial order.

---

*Writer Ben Ames Williams used this problem in a short story titled "Coconuts," which appeared in the October 9, 1926, issue of *The Saturday Evening Post*. The story concerned a contractor who wanted to bid on a large contract. Knowing of their competitor's strong passion for recreational mathematics, one of his employees gave him this problem. The competitor became so obsessed with solving the puzzle that he forgot to enter his bid before the deadline.

Table 7.6

Task	Requires the completion of
(1) Building the foundation (t_1)	None
(2) Framing (t_2)	t_1
(3) Subflooring (t_3)	t_1, t_2
(4) Partitioning into rooms (t_4)	t_1, t_2, t_3
(5) Roofing (t_5)	t_1, t_2
(6) Plumbing (t_6)	t_1, t_2
(7) Wiring (t_7)	t_1, t_2
(8) Siding (t_8)	t_1, t_2, t_5, t_9
(9) Flooring (t_9)	t_1, t_2, t_6
(10) Interior painting (t_{10})	t_1 through t_5, t_7
(11) Exterior painting (t_{11})	t_1, t_2, t_8, t_9
(12) Carpeting (t_{12})	t_1 through t_7, t_9, t_{10}
(13) Installing fixtures (t_{13})	t_1 through t_{11}

Partial Order

A relation R on a set A is a **partial order** if it is reflexive, antisymmetric, and transitive. The set A with its partial order R is a **partially ordered set** (or **poset**), denoted by (A, R). When the partial order is clear from the context, call the poset A.

The next three examples illustrate these definitions.

EXAMPLE 7.51 The relation $\leq$ on $\mathbb{R}$ is reflexive, antisymmetric, and transitive, so $\leq$ is a partial order on $\mathbb{R}$ and $(\mathbb{R}, \leq)$ a poset. Similarly, the divisibility relation $|$ on $\mathbb{N}$ is a partial order, so $(\mathbb{N}, |)$ is also a poset. ■

EXAMPLE 7.52 Let $\Sigma = \{a, b\}$. Define a relation R on Σ^* as: xRy if x is a prefix of y. Is R a partial order?

- Every word is a prefix of itself, so R is reflexive.
- Let xRy and yRx. Then $y = sx$ and $x = ty$ for some $s, t \in \Sigma^*$, so $x = t(sx) = (ts)x$.
  Consequently, $ts = \lambda$ and hence $t = s = \lambda$. So $x = y$ and the relation is antisymmetric.
- Suppose xRy and yRz. Then $y = sx$ and $z = ty$ for some $s, t \in \Sigma^*$. Therefore, $z = t(sx) = (ts)x$. Consequently xRz, and the relation is transitive.

Thus, R is a partial order on Σ^* and (Σ^*, R) is a poset. ■

EXAMPLE 7.53 The relation *has the same color hair as* on the set of people is reflexive, but not antisymmetric. Therefore it is not a partial order. ■

Just as an equivalence relation generalizes the equality relation, a partial order generalizes the relation $\leq$. Accordingly, a partial order is denoted by $\preceq$. $x \preceq y$ means x **precedes or equals** y. If $x \preceq y$ and $x \neq y$, we write $x \prec y$, meaning x **precedes** y.

Comparable Elements

Two elements x and y in a poset are **comparable** if either $x \preceq y$ or $y \preceq x$; otherwise, they are **noncomparable**.

EXAMPLE 7.54 Let x and y be any two real numbers. Then either $x \leq y$ or $y \leq x$. So any two real numbers can be compared using the relation $\leq$: they are comparable.

Using the divisibility relation | on $\mathbb{N}$, the positive integers 3 and 6 are comparable, since $3 \mid 6$. But 3 and 8 are not comparable, since $3 \nmid 8$ and $8 \nmid 3$. ■

> Example 7.54 indicates that a poset may contain noncomparable elements, which justifies the word *partial* in *partial order*. This leads us to the next definition.

Total Order

If any two elements in a poset are comparable, such a partial order is a **total order** or a **linear order**. The poset is then a **totally ordered set** or a **linearly ordered set**.

Notice that $\leq$ is a total order on $\mathbb{R}$, whereas the divisibility relation is *not* a total order on $\mathbb{N}$.

Just as sets can be used to construct new sets, posets can be combined to construct new posets. In order to do this, we first define a relation on the cartesian product of two posets.

Lexicographic Order

Let $(A, \preceq_1)$ and $(B, \preceq_2)$ be two posets. Define a relation $\preceq$ on $A \times B$ as $(a, b) \preceq (a', b')$ if $a \prec_1 a'$, or $a = a'$ and $b \preceq_2 b'$. The relation $\preceq$, an extension of the alphabetic order, is the **lexicographic order**.

The lexicographic order is a partial order on $A \times B$. If A and B are totally ordered sets, so is $A \times B$. The lexicographic order can be extended to the cartesian product $A_1 \times A_2 \times \cdots \times A_n$ of n posets and n totally ordered sets. The next two examples illustrate this.

EXAMPLE 7.55 Consider the cartesian product $\mathbb{N} \times \mathbb{N} \times \mathbb{N}$, where the partial order is the usual $\leq$. Then $(2, 5, 3) \leq (3, 2, 1)$ since the first element in the triplet $(2, 5, 3)$ is less than that in the second triplet $(3, 2, 1)$. Also, $(2, 4, 5) \leq (2, 4, 7)$. This ordering mirrors the familiar sequencing of three-digit numbers. ■

EXAMPLE 7.56 Let Σ be a partially ordered alphabet with the partial order $\preceq$ and Σ^n denote the set of words of length n over Σ. Since every word in Σ^n can be considered an n-tuple, the lexicographic order on the cartesian product on n posets can be applied to Σ^n also.

Let $x = a_1a_2\ldots a_n$ and $y = b_1b_2\ldots b_n$ be any two elements in Σ^n. Then $x \prec y$ if:

- Either $a_1 \preceq b_1$, or
- An integer i exists such that $a_1 = b_1, a_2 = b_2, \ldots, a_i = b_i$, and $a_{i+1} \prec b_{i+1}$.

In particular, let Σ denote the English alphabet, a totally ordered set: $a \prec b \prec c \prec \cdots \prec z$. Clearly, *computer*$\prec$ *demolish*, *compress* $\prec$ *computer*, *contend* $\prec$ *content*, and *content*$\prec$ *context*.

This lexicographic order can work for Σ^* in a familiar way. Let x and y be any two words over Σ. Then $x \prec y$ in lexicographic order if one of two conditions holds:

- $x = \lambda$, the empty word.
- If $x = su$ and $y = sv$, where s denotes the longest common prefix of x and y, the first symbol in u precedes that in v in alphabetic order.

For example, *marathon* $\prec$ *marble*, *margin* $\prec$ *market*, *limber* $\prec$ *timber*, and *creation* $\prec$ *discretion*. ■

Hasse Diagrams

We can simplify the digraph of a finite poset by omitting many of its edges. For instance, since a partial order is reflexive, each vertex has a loop, which we can delete. In addition, drop all edges implied by transitivity. For example, if the digraph contains the edges (a, b) and (b, c), it has the edge (a, c), which we can omit. Finally, draw the remaining edges upward and drop all arrows. The resulting is the **Hasse diagram**, named for the German mathematician Helmut Hasse.

Examples 7.57–7.60 generate Hasse diagrams.

EXAMPLE 7.57 Construct the Hasse diagram for the poset $(A, |)$, where $A = \{1, 2, 3, 6, 8, 12\}$ and $|$ denotes the divisibility relation.

SOLUTION:
The digraph of the poset is Figure 7.38.

Step 1 Delete the loop at each vertex. The result is Figure 7.39.

Step 2 Delete all edges implied by transitivity. Figure 7.40 shows the ensuing diagram.

Step 3 Omit all arrows and draw the edges "upward." The Hasse diagram appears in Figure 7.41.

Helmut Hasse *(1898–1979), a celebrated number theorist and dedicated teacher, was born in Kassel, Germany. His father was a judge. While studying at the gymnasiums in Kassel and later Berlin, he decided on a career in mathematics. After the gymnasiums, he entered the navy. While in the navy in the Baltic he studied number theory and then mathematics at the University of Kiel. Leaving the navy in December 1918, Hasse went to Göttingen to pursue his mathematical interest and then to Marburg, receiving his Ph.D. in 1921.*

His teaching career began in Kiel in 1922. Three years later, he became a professor at Halle, then moved to Marburg, Göttingen, Berlin, and finally Hamburg in 1950, where he remained until his retirement in 1966. Earlier he had been director of the Mathematics Institute at Göttingen. But he was dismissed by the British occupation authorities in September 1945.

Hasse was a member of several academies of science and author of numerous articles and books. Hasse received a number of awards including the German National prize for Science and Technology (1953) and the Cothenius Medal of the Academia Leopoldina (1968).

Figure 7.38

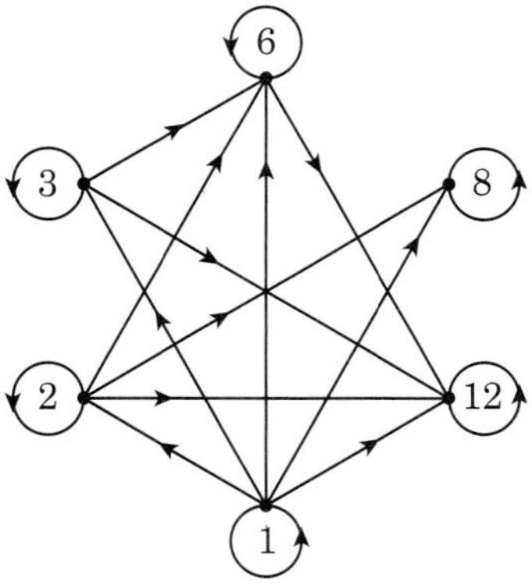

Figure 7.39

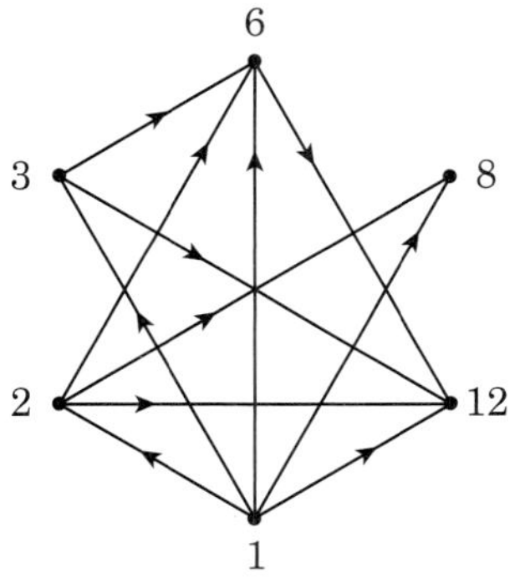

Figure 7.40

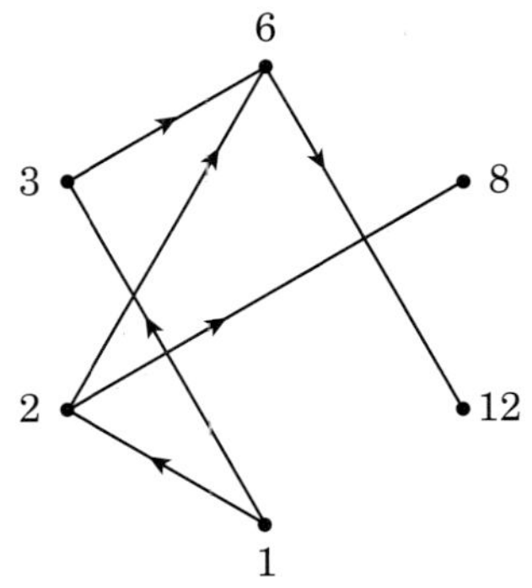

Figure 7.41

Hasse diagram for the poset.

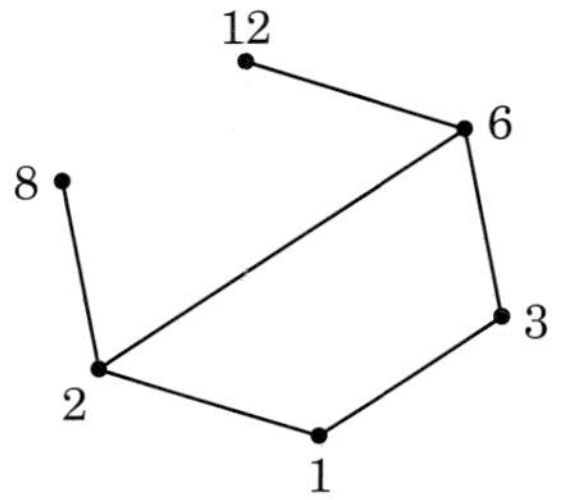

■

EXAMPLE 7.58 Draw the Hasse diagram for the poset $(A, \subseteq)$, where A denotes the power set of the set $\{a,b,c\}$.

SOLUTION:

The set $\{a,b,c\}$ has eight subsets: $\emptyset, \{a\}, \{b\}, \{c\}, \{a,b\}, \{b,c\}, \{c,a\}$, and $\{a,b,c\}$. Following steps 1–3, as in Example 7.57, produces the Hasse diagram in Figure 7.42.

Figure 7.42

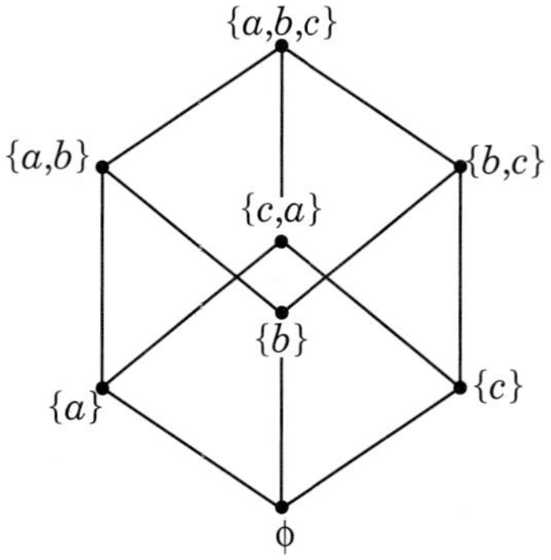

■

EXAMPLE 7.59 The relation $R = \{(a,a),(a,c),(a,e),(b,b),(b,c),(b,d),(b,e),(c,c),(c,e),(d,d),(d,e),(e,e)\}$ is a partial order on $\{a,b,c,d,e\}$. Figure 7.43 displays its Hasse diagram.

Figure 7.43

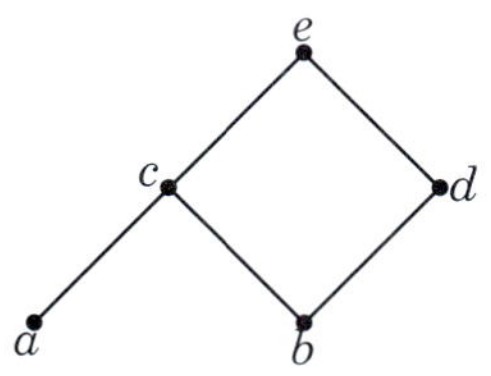

■

EXAMPLE 7.60 Consider the alphabet $\Sigma = \{a, b\}$. The relation $\preceq$ on Σ^*, defined by $x \preceq y$ if x is a prefix of y, is a partial order. The Hasse diagram for all words of maximum length two appears in Figure 7.44.

Figure 7.44

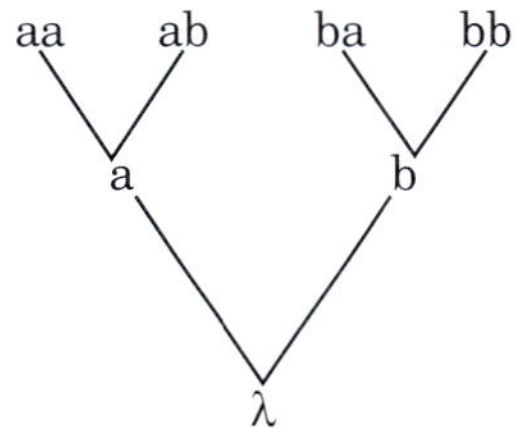

■

Extremal elements in a poset are important, especially in linear ordering.

Extremal Elements

An element a in a poset $(A, \preceq)$ is a **maximal element** if A has *no* element b such that $a \prec b$. Similarly, an element a in A is a **minimal element** if A has no element $b \prec a$.

The maximal and minimal elements in a finite poset can easily be read from its Hasse diagram, like the ones in Figures 7.41 and 7.42.

EXAMPLE 7.61 The poset in Figure 7.41 has two maximal elements, 8 and 12, and one minimal element, 1.

Figure 7.43 has one maximal element, e; it has two minimal elements, a and b. ■

A poset may exhibit the following properties:

- A poset may have more than one maximal element and more than one minimal element (see Example 7.61).
- A poset need not have any maximal or minimal elements. For instance, the poset $(\mathbf{Z}, \leq)$ has no maximal or minimal elements.
- A poset may have a maximal element but no minimal elements, or a minimal element but no maximal elements. For example, the poset $(\mathbf{Z}^-, \leq)$ has a maximal element but no minimal elements, whereas the poset $(\mathbf{Z}^+, \leq)$ has a minimal element but no maximal elements.

Two special extremal elements are the greatest and the least.

Greatest and Least Elements

If a poset A contains an element a such that $b \preceq a$ for every element b in A, a is the **greatest element** of the poset. If it contains an element a such that $a \preceq b$ for every b in A, a is the **least element**.

> The greatest element of a poset, if it exists, is unique; likewise, the least element. They are the topmost and the bottommost elements in the Hasse diagram.

For example, the poset in Figure 7.41 has no greatest element, but has a least element, 1. Figure 7.43, on the other hand, has a greatest element, e, but no least element.

Although an arbitrary poset need not have a minimal element, every nonempty finite poset has a minimal element, as Theorem 7.13 shows.

THEOREM 7.13 Every finite nonempty poset $(A, \preceq)$ has a minimal element.

PROOF:
Let a_1 be any element in A. If a_1 is not minimal, there must be an element a_2 in A such that $a_2 \prec a_1$. If a_2 is minimal, then we have finished. If a_2 is not minimal, A must have an element a_3 such that $a_3 \prec a_2$. If a_3 is not minimal, continue this procedure. Since A contains only a finite number of elements, it must terminate with some element a_n. Thus $a_n \prec a_{n-1} \prec \cdots \prec a_3 \prec a_2 \prec a_1$. Consequently, a_n is a minimal element. ■

This result forms the cornerstone of the topological sorting technique.

Topological Sorting

Study the tasks t_1 through t_{13} for building a house, given in Table 7.6. (Recall that the relation *precedes or is the same as* is a partial order on A). For these tasks to be entered in a computer, the elements of the poset must be arranged in a linear order consistent with the partial order. If $a \preceq b$, then enter task a before task b in linear order. This technique is called **topological sorting**.

To topologically sort a finite nonempty poset $(A, \preceq)$ with n elements, proceed as follows. By Theorem 7.13, the poset contains a minimal element, say, a_1. Exclude it from A. Then $A - \{a_1\}$ is also a finite poset. If it is nonempty, it contains a minimal element a_2. Delete a_2 from $A - \{a_1\}$. Then $A - \{a_1, a_2\}$ is a finite poset with minimal element a_3. Continue this procedure until the poset becomes null. This procedure yields the desired linear order, $a_1 \prec a_2 \prec a_3 \prec \cdots \prec a_n$.

A simple algorithm can handle this organizing (see Algorithm 7.4).

```
Algorithm Topological Sort (S)
(* This algorithm sorts a finite nonempty poset S into a linear order
   using topological sorting. *)
  Begin (* algorithm *)
    while S ≠ ø do
      begin (* while *)
        find a minimal element a in S
        S ← S − {a} (* delete a from S *)
    endwhile
  End (* algorithm *)
```

Algorithm 7.4

We can establish the validity of this algorithm using induction and Theorem 7.13. We leave its verification as an exercise.

EXAMPLE 7.62 Topologically sort the elements of the poset in Example 7.57.

SOLUTION:

The poset given by the Hasse diagram in Figure 7.41 has one minimal element, 1. Delete it from the poset and hence from the Hasse diagram. The diagram turns into Figure 7.45 with a poset of two minimal elements, 2 and 3. Delete one of them, say, 3. The resulting poset appears in Figure 7.46; it has two minimal elements, 2 and 6. Delete one of them, say, 2. The new poset in Figure 7.47 also has two minimal elements, 6 and 8. Extract, say, 8. The resulting poset is shown in Figure 7.48. Extract its minimal element, 6; this leaves just one element, 12 (see Figure 7.49). Deleting it yields the empty set, and the procedure terminates. Thus, we can sort the elements of the poset in a linear order compatible with the partial order: $1 \prec 3 \prec 2 \prec 8 \prec 6 \prec 12$.

Figure 7.45

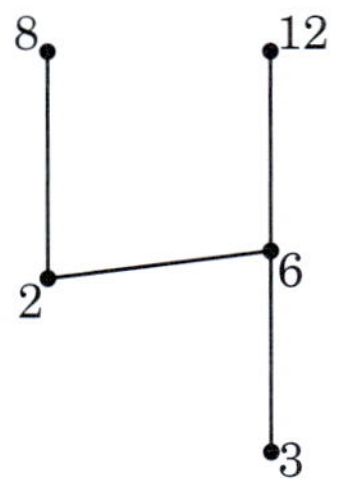

Figure 7.46

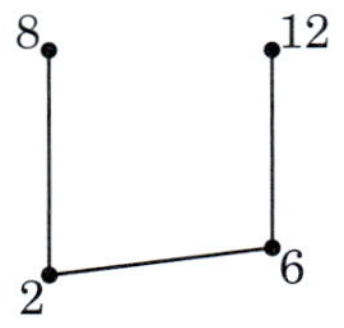

Figure 7.47

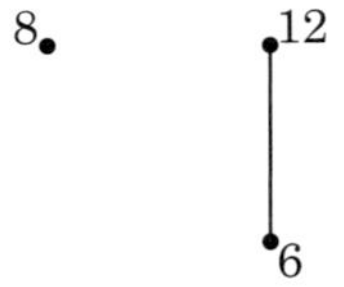

Figure 7.48

Figure 7.49

■

In this example, we could have chosen a minimal element in more than one way on three occasions; in other words, the output from the topological sorting need not be unique.

We close this section with another sorting example.

EXAMPLE 7.63 Topologically sort the elements of the poset in Example 7.59.

SOLUTION:
Figures 7.50–7.54 track the steps of the sorting algorithm. The resulting output is $a \prec b \prec d \prec c \prec e$.

Figure 7.50

Extract a.

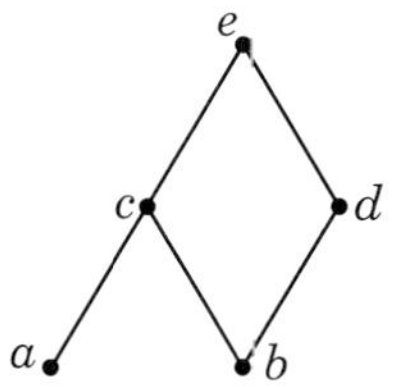

Figure 7.51

Extract b.

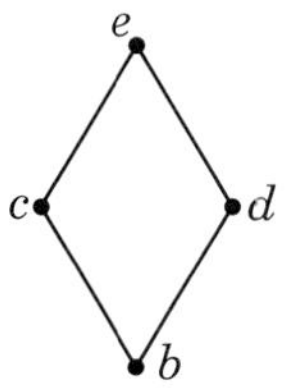

Figure 7.52

Extract c.

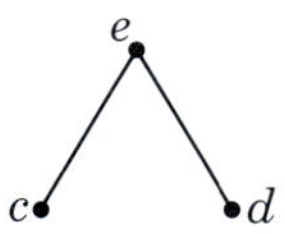

Figure 7.53

Extract d.

Figure 7.54

Extract e.

■

Exercises 7.9

Determine if each is a partial order.

1. The relation $<$ on $\mathbb{R}$

2. The relation $\geq$ on $\mathbb{R}$

3. The relation $\geq$ on $\mathbf{Z}$

4. The relation $|$ on $\mathbf{Z}$

Determine if each is a partial order on $\{a, b, c\}$.

5. $\{(a,a),(b,b),(c,c)\}$

6. $\{(a,a),(a,b),(b,a),(b,b),(c,c)\}$

7. $\{(a,a),(b,b),(b,c),(c,c)\}$

8. $\{(a,a),(a,b),(b,b),(b,c),(c,c)\}$

Determine if each is a partial order.

9.

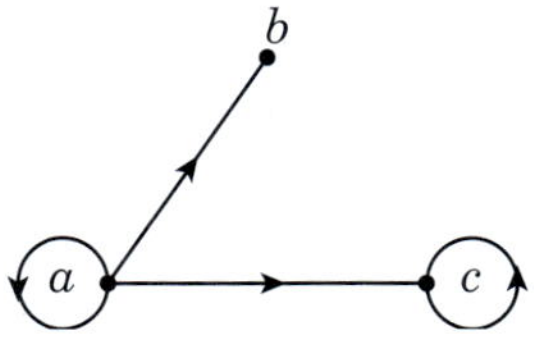

10.

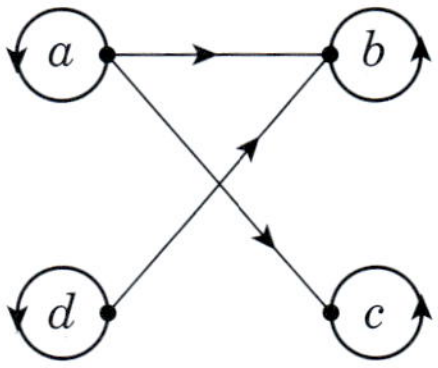

11.

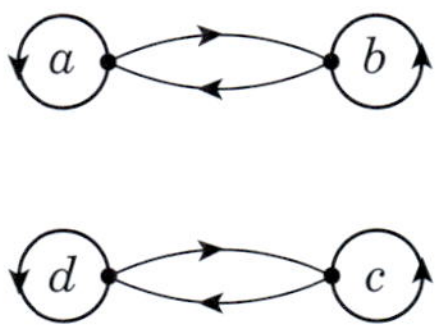

Determine if the given elements are comparable in the poset $(A, |)$, where $A = \{1, 2, 3, 6, 9, 18\}$ and $|$ denotes the divisibility relation.

12. $2, 3$

13. $2, 6$

14. $2, 9$

15. $3, 18$

Determine if the given elements are comparable in the poset $(A, \subseteq)$, where A denotes the power set of $\{a, b, c\}$ (see Example 7.58).

16. $\{a, b\}, \{b, c\}$

17. $\{a, b\}, \{b\}$

Arrange the following pairs from the poset $\mathbb{N} \times \mathbb{N}$ in lexicographic order.

18. $(3, 5), (2, 3)$

19. $(3, 5), (2, 6)$

20. Find three ordered pairs of positive integers that precede the pair $(2, 3)$ in lexicographic order.

21. Find three triplets of positive integers that precede the triplet $(2, 3, 5)$.

Arrange the following words over the English alphabet in lexicographic order.

22. *mat, rat, bat, cat, eat, fat*

23. *neighbor, neophyte, neglect, moment, luxury, maximum*

24. *custom, custody, custard, cushion, curtain, culvert*

25. *discreet, discrete, discount, discourse, diskette, discretion*

26. Arrange all words of length ≤ 2 over the alphabet $\{a, b\}$ in lexicographic order. Construct a Hasse diagram for each poset.

27. $(A, |)$, where $A = \{1, 2, 3, 6, 9, 18\}$ and $|$ denotes the divisibility relation.

28. $(A, |)$, where $A = \{1, 2, 3, 6, 8, 24\}$ and $|$ is the divisibility relation.

29. (A, R), where $A = \{a, b, c\}$ and $R = \{(a, a), (a, b), (b, b), (b, c), (c, c)\}$.

30. $(A, \subseteq)$, where A denotes the power set of the set $\{a, b\}$.

31. Let A denote the set of words of length ≤ 3 over the binary alphabet. The relation R, defined on A by xRy if x is a prefix of y, is a partial order. Draw a Hasse diagram for the poset (A, R).

Find the maximal and minimal elements in the poset with each Hasse diagram.

32.

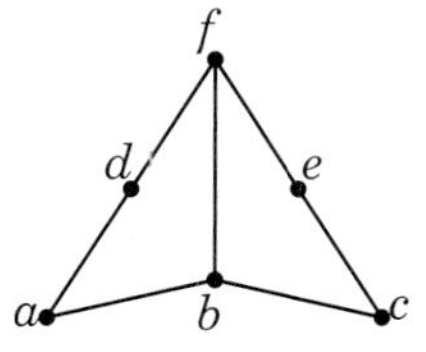

33.

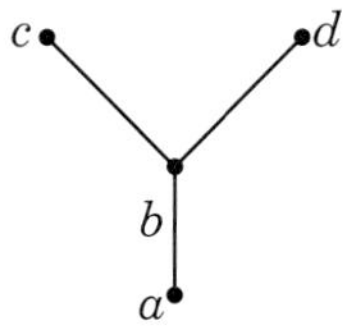

34.

Find the maximal and minimal elements, if they exist, in each poset.

35. $(A, \leq)$, where A denotes the set of positive even integers.

36. $(A, \leq)$, where A denotes the set of negative even integers.

37. $(A, |)$, where $A = \{1, 2, 3, 6, 9, 18\}$ **38.** $(A, |)$, where $A = \{1, 2, 3, 6, 8, 24\}$

39–42. Find the greatest and least elements, if they exist, in the posets of Exercises 35–38.

Mark each statement as true or false.

43. Every poset has a maximal element.

44. Every poset has a minimal element.

45. The maximal element in a poset, if it exists, is unique.

46. The minimal element in a poset, if it exists, is unique.

47. Every poset has a greatest element.

48. Every poset has a least element.

Give a counterexample to disprove each statement.

49. Every poset has a maximal element.

50. Every poset has a minimal element.

51. Every poset has a greatest element.

52. Every poset has a least element.

Topologically sort the elements of each poset.

53. The poset in Figure 7.43. **54.** The poset in Figure 7.44.

55. The poset in Exercise 32. **56.** The poset in Exercise 33.

57. $(A, |)$, where $A = \{1, 2, 3, 6, 9, 18\}$ **58.** $(A, |)$, where $A = \{1, 2, 3, 6, 8, 24\}$

59. Topologically sort the tasks t_1 through t_{13} in building a house, given by Table 7.6.

60. A project contains six subprojects, A through F. Results from some of the subprojects are needed by others, as Table 7.7 shows. Find the ways the subprojects can be sequentially arranged.

Table 7.7

Subproject	Requires results from
A	B, D
B	C
C	None
D	C, E
E	None
F	A

61. Seven tasks, A through G, comprise a project. Some of them can only be started after others are completed, as indicated by Table 7.8. How many ways can the tasks be arranged sequentially, so the prerequisites of each task will be completed before it is started? List one of them.

Table 7.8

Task	Requires the completion of
A	B, C
B	G
C	None
D	A, F
E	None
F	B, E
G	None

***62.** Let $(A, \preceq_1)$ and $(B, \preceq_2)$ be two posets. Define a relation $\preceq_3$ on $A \times B$ as follows: $(a, b) \preceq_3 (a', b')$ if $a \preceq_1 a'$ and $b \preceq_2 b'$. Prove that $\preceq_3$ is a partial order.

Prove each.

***63.** The greatest element of a poset $(A, \preceq)$, if it exists, is unique.

***64.** The least element of a poset $(A, \preceq)$, if it exists, is unique.

***65.** Every finite nonempty poset $(A, \preceq)$ contains a maximal element.

***66.** Establish the correctness of Algorithm 7.4.

Chapter Summary

We studied the fundamentals of the theory of relations and explored how relations on finite sets can be represented by graphs and boolean matrices.

Boolean Matrix

- A **boolean matrix** has bits for entries (page 438).
- The **join** $A \vee B$ and **meet** $A \vee B$ of two boolean matrices A and B are obtained by *oring* and *anding* the corresponding bits, respectively (page 439).
- The **boolean product** $A \odot B$ of two boolean matrices $A = (a_{ij})_{m \times p}$ and $B = (b_{jk})_{p \times n}$ is the matrix $C = (c_{ij})_{m \times n}$, where $c_{ij} = (a_{i1} \wedge b_{1j}) \vee (a_{i2} \wedge b_{2j}) \vee \cdots \vee (a_{ip} \wedge b_{pj})$ (page 439).

- The **complement** A' of a boolean matrix A results from swapping 0's and 1's (page 442).

Binary Relation

- A **binary relation** R from A to B is a subset of $A \times B$. If $(a,b) \in R$, we write aRb; otherwise, we write ab (page 443).
- A relation R from a finite set to a finite set can be represented by its **adjacency matrix**, M_R (page 444).
- A relation on a finite set can be represented by a **digraph** (page 445).
- Every function $f : A \to B$ is a binary relation from A to B such that (1) $\text{dom}(f) = A$; and (2) if $(a,b) \in f$ and $(a,c) \in f$, then $b = c$ (page 448).

Properties of Relations

- A relation R on A is **reflexive** if aRa for every $a \in A$ (page 455).
- A relation R on A is **symmetric** if aRb implies bRa (page 456).
- A relation R on A is **antisymmetric** if $aRb \wedge bRa$ implies $a = b$ (page 456).
- A relation R on A is **transitive** if $aRb \wedge bRc$ implies aRc (page 459).

Constructing New Relations

- The **union** and **intersection** of two relations R and S from A to B are $R \cup S = \{(a,b)|aRb \vee aSb\}; R \wedge S = \{(a,b)|aRb \vee aSb\}$ (page 462).
- If R and S are relations on a finite set, $M_{R \cup S} = M_R \vee M_S$ and $M_{R \cap S} = M_R \wedge M_S$ (page 463).
- Let R be a relation from A to B and S a relation from B to C. Their **composition** is $R \odot S = \{(a,c) \in A \times C \mid aRb \wedge bRc \text{ for some } b \text{ in } B\}$ (page 463).
- In particular, if A, B, and C are finite sets, then $M_{R \odot S} = M_R \odot M_S$ (page 466).
- For a relation R on a finite set, $M_{R^n} = (M_R)^{[n]}$ (page 467).
- For a transitive relation $R, R^n \subseteq R$ for every $n \geq 1$ (page 468).
- The **connectivity relation** R^∞ is the union of all powers of R:

$$R^\infty = \bigcup_{n=1}^{\infty} R^n; M_{R^\infty} = M_R \vee M_{R^2} \vee M_{R^3} \vee \ldots \quad \text{(page 471).}$$

- In particular, let R be a relation on a set with size n. Then

$$R^{\infty} = \bigcup_{i=1}^{n} R^i \text{ and } M_R = M_R \vee M_{R^2} \vee \cdots \vee M_{R^n} \quad \text{(page 473)}.$$

Transitive Closure

- The **transitive closure** R^* of a relation R is the smallest transitive relation containing it (page 475).
- $R^* = R^{\infty}$ (page 477).
- **Warshall's algorithm** systematically finds M_{R^*} (page 477).

Equivalence Relations and Partitions

- An **equivalence relation** is reflexive, symmetric, and transitive (page 483).
- An equivalence relation on a set induces a partition of the set and vice versa (page 488).

Partial and Total Orders

- A **partial order** $\preceq$ is reflexive, antisymmetric, and transitive. A set together with a partial order is a **poset** (page 494).
- Two elements, x and y, in a poset are **comparable** if either $x \preceq y$ or $y \preceq x$ (page 495).
- If any two elements in a poset are comparable, the partial order is a **total order** or **linear order** (page 495).
- The **lexicographic order** is an extension of the alphabetical order to posets (page 495).
- The **Hasse diagram** of a finite poset contains no loops, edges implied by transitivity, or arrows; its edges are drawn upward (page 496).
- The elements of a finite nonempty poset can be **sorted topologically** (page 500).

Review Exercises

Determine if each relation on $\{a, b, c\}$ is reflexive, symmetric, antisymmetric, or transitive.

1. $\{(a,a),(b,c),(c,b),(c,c)\}$

2. $\{(a,b),(b,a),(b,c),(c,b)\}$

Using the relations $R = \{(1,1),(1,2),(2,2),(3,2)\}$ and $S = \{(1,1),(2,2),(2,3),(3,2)\}$ on $\{1,2,3\}$, find each.

3. $(R \cup S)'$ **4.** $R' \cap S'$ **5.** $(R \cap S)'$ **6.** $R' \cup S'$

7. $(R \cup S)^{-1}$ **8.** $R^{-1} \cup S^{-1}$ **9.** $(R \cap S)^{-1}$ **10.** $R^{-1} \cap S^{-1}$

11. $R \odot S$ **12.** R^2 **13.** R^3 **14.** R^∞

With the adjacency matrices of the relations $R = \{(1,1),(1,2),(2,2),(3,2)\}$ and $S = \{(1,1),(2,2),(2,3),(3,2)\}$ on $\{1,2,3\}$, find each.

15. $(M_R)'$ **16.** $(M_R)^T$ **17.** $M_{R \odot S}$ **18.** M_{R^2}

19. $R \odot S$ **20.** R^2 **21.** M_{R^*} **22.** R^*

Find the transitive closure of the relation on $A = \{a,b,c\}$ with each adjacency matrix.

23. $\begin{bmatrix} 0 & 0 & 1 \\ 1 & 0 & 0 \\ 0 & 1 & 1 \end{bmatrix}$ **24.** $\begin{bmatrix} 1 & 0 & 0 \\ 0 & 1 & 1 \\ 1 & 0 & 1 \end{bmatrix}$

Since 1972 every book published commercially has a 10-digit identification number, its *International Standard Book Number* (**ISBN**). The ISBN consists of four parts: a group code (one digit), a publisher code (two digits), a book code (six digits), and a check digit. For instance, the ISBN of an earlier text by this author is 0-12-421171-2. The group code 0 indicates that the book was published in an English-speaking country. The publisher code (12) identifies the publisher, Academic Press, and the book code (421171) is assigned by the publisher to the book. The check digit d, where $0 \le d \le 10$ and 10 is denoted by X, is used to detect errors and is computed as follows: Let $x_1, x_2, \ldots, x_9$ denote the first nine digits in the ISBN. Let s denote the dot product of the 9-tuples $(x_1, x_2, x_3, x_4, x_5, x_6, x_7, x_8, x_9)$ and $(10, 9, 8, 7, 6, 5, 4, 3, 2)$. Then $d \equiv -s \pmod{11}$. Compute the check digit if the first 9 digits of the ISBN are:

25. 0-12-421171 **26.** 0-87-620321

Determine if each is an equivalence relation on $\{a,b,c\}$.

27. $\{(a,a),(a,b),(b,a),(c,c)\}$ **28.** $\{(a,a),(a,c),(b,b),(c,a),(c,c)\}$

Using the equivalence relation $\{(a,a),(a,c),(b,b),(b,d),(c,a),(c,c),(d,b),(d,d)\}$ on $A = \{a,b,c,d\}$, find each equivalence class.

29. $[a]$ **30.** $[b]$ **31.** $[c]$ **32.** $[d]$

33. Find the partition of A induced by the above relation.

Find the equivalence relation corresponding to each partition of the set $\{2,3,4,7\}$.

34. $\{\{2,4,7\},\{3\}\}$ **35.** $\{\{2,4\},\{3\},\{7\}\}$

Find the number of partitions of a set with the given size.

36. Two

37. Seven

Mark each statement as true or false, where A is an arbitrary set, R an arbitrary relation, and Δ the equality relation.

38. The null relation is reflexive.

39. The null relation is symmetric.

40. The null relation is transitive.

41. A relation R on A is reflexive if and only if $\Delta \subseteq R$.

42. The less-than relation on $\mathbb{R}$ is irreflexive.

43. The less-than relation on $\mathbb{R}$ is antisymmetric.

44. If R is transitive, $R^* = R$.

45. If $R^* = R$, R is transitive.

46. The less-than relation on $\mathbb{R}$ is a partial order.

47. The less-than relation on $\mathbb{R}$ is a total order.

48. Arrange all binary words of length 3 in lexicographic order, where $0 \prec 1$.

49. Arrange all binary words of length ≤ 3 in lexicographic order, where $0 \prec 1$.

The relation $\preceq$ on the set A of required courses given in Table 7.1 by $x \preceq y$ if x is a prerequisite of or the same as y is a partial order on A.

50. Draw the Hasse diagram for the poset.

51. Topologically sort the required computer science courses.

Use the poset in Figure 7.55 to find the following.

Figure 7.55

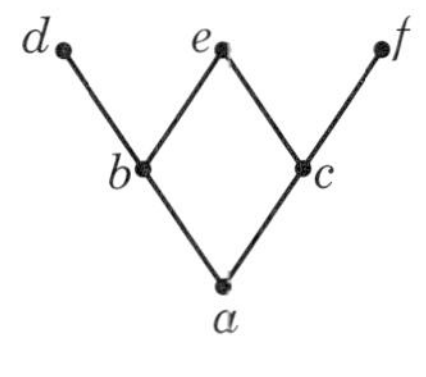

52. The maximal and minimal elements, if they exist.

53. The greatest and least elements, if they exist.

54. Topologically sort the elements in the poset.

Let R and S be any two relations on a set A. Prove each.

55. $(R \cap S)^2 \subseteq R^2 \cap S^2$

56. $(R \cap S)^n \subseteq R^n \cap S^n$, $n \geq 1$

57. R is antisymmetric if and only if $R \cap R^{-1} \subseteq \Delta$.

58. The intersection of two equivalence relations is an equivalence relation.

Supplementary Exercises

Let D denote any day of the week, where $0 \leq D \leq 6$ and $D = 0$ denotes Sunday. The day of the week corresponding to any day $(m/d/y)$ in the Gregorian calendar is given by

$$D \equiv \begin{cases} \left\lfloor \dfrac{23m}{9} \right\rfloor + d + 4 + y + \left\lfloor \dfrac{y-1}{4} \right\rfloor - \left\lfloor \dfrac{y-1}{100} \right\rfloor + \left\lfloor \dfrac{y-1}{400} \right\rfloor \pmod 7 & \text{if } m < 3 \\ \left\lfloor \dfrac{23m}{9} \right\rfloor + d + 4 + y + \left\lfloor \dfrac{y}{4} \right\rfloor - \left\lfloor \dfrac{y}{100} \right\rfloor + \left\lfloor \dfrac{y}{400} \right\rfloor - 2 \pmod 7 & \text{otherwise} \end{cases}$$

(M. Keith and T. Carver, 1990)

Compute the day of each date.

1. July 4, 1776

2. December 25, 2076

Prove each.

3. Let p be a prime. Then $p \mid \binom{p}{k}$ for $0 < k < p$.

4. (**Fermat's theorem**) Let $a \in \mathbb{N}$ and p a prime. Then $a^p \equiv a \pmod p$. (*Hint*: Use induction.)

Let $a, b \in \mathbb{R}$ and p a prime. Prove that $(a+b)^p \equiv a^p + b^p \pmod p$ using:

5. The binomial theorem and Exercise 3.

6. Fermat's theorem.

Evaluate each.

7. $5^{1000} \pmod 7$

8. $12^{4000} \pmod 5$

9. Prove that the product of any three consecutive integers is divisible by 3.

10. Let $n \in \mathbf{W}$. Prove that the number formed by concatenating the decimal values of 2^n and 2^{n+1} is divisible by 3. (For example, when $n = 5$, both 3264 and 6432 are divisible by 3.) (D. Burns, 1977)

11. Around 1760, John Wilson (1741–1793), an English mathematician, proved that $(p-1)! \equiv -1 \pmod{p}$; that is, the quotient

$$W(p) = \frac{(p-1)!+1}{p}$$

is an integer. This is known as **Wilson's theorem**.) p is a **Wilson prime** if $W(p) \equiv 0 \pmod{p}$; that is, if $(p-1)! \equiv -1 \pmod{p^2}$. Find the two Wilson's primes < 20. (The third and largest known Wilson prime is 563. It is not known whether or not there are infinitely many Wilson primes.)

12. (**Lucas' Theorem**) Let p be a prime, $n = (a_t a_{t-1} \dots a_0)_p$ and $k = (b_t b_{t-1} \dots b_0)_p$. Then $\binom{n}{k} \equiv \binom{a_t}{b_t}\binom{a_{t-1}}{b_{t-1}} \cdots \binom{a_0}{b_0} \pmod{p}$. Using Lucas' theorem, find the remainders when the binomial coefficients $C(234,19)$ and $C(3456,297)$ are divided by 5.

***13.** Let a and b be relatively prime integers. Prove that $a^{\varphi(b)} + b^{\varphi(a)} \equiv 1 \pmod{ab}$.
(*Hint*: Let $n \in \mathbb{N}$ and a an integer relatively prime to n. Then $a^{\varphi(n)} \equiv 1 \pmod{n}$. This is **Euler's theorem**.) (M. Charosh, 1983)

***14.** Show that a set with n elements must have at least 2^n relations with the same reflexive closure.
(*Hint*: Use the pigeonhole principle.)

***15.** Show that a set with size n must have at least $2^{n(n-1)/2}$ relations with the same symmetric closure.
(*Hint*: Use the pigeonhole principle.)

Computer Exercises

Write a program to perform each task, where n denotes a positive integer ≤ 20 and $A = \{1, 2, \dots, n\}$.

1. Read in two boolean matrices. Print their join, meet, complement, and boolean product, if defined.

2. Read in the elements of a relation R on A. Print its adjacency matrix M_R. Use M_R to enumerate the elements in the relation.

3. Read in the adjacency matrix of a relation on A. Determine if the relation is reflexive, symmetric, antisymmetric, or transitive.

4. Read in the adjacency matrices of two relations on A. Print the adjacency matrices of their union, intersection, complements, and inverses.

5. Read in the adjacency matrix of a relation R from A to B and that of a relation S from B to C. Print $M_{R \odot S}$.

6. Read in the adjacency matrix of a relation R on A. Print M_{R^*}, using the connectivity relation algorithm and Warshall's algorithm, and compare the number of bit operations required by them.

7. Read in the adjacency matrix of a relation on A. Determine if the relation is an equivalence relation.

8. Read in two positive integers r and n, where $r \leq n \leq 10$. Print the number of equivalence relations that can be defined on a set of size n, using Stirling numbers of the second kind and Bell numbers.

9. Read in the adjacency matrix of a partial order on a poset A.
   - Determine if it is a partial order.
   - Print the boolean matrix corresponding to its Hasse diagram.
   - Topologically sort the elements of the poset.

10. Determine the most likely day on which the 13th of a month will fall in the Gregorian calendar. Since the Gregorian calendar repeats every 400 years, you need only consider a period of 400 years.

11. Read in a positive integer $n \leq 1000$ and print all Wilson primes $\leq n$.

Exploratory Writing Projects

Using library and Internet resources, write a team report on each of the following in your own words. Provide a well-documented bibliography.

1. Explain and illustrate the various relational operations in the theory of databases.

2. Describe the algorithm employed by the United States Postal Service to encode the nine-digit zip code into barcodes, and decode the barcodes (62 bars) into zip codes.

3. Describe how modular arithmetic can be used to construct m-pointed stars.

4. Explain the coding scheme for creating *European Article Numbering* (EAN) barcodes to uniquely identify books. Extend it to include the five-digit add-on code to provide price information.

5. Describe the origins of the Julian and Gregorian calendars.

6. Develop a formula to determine the day d of the week for the rth day in a given month m of any given year y in the Gregorian calendar, where $y > 1600$.

7. Study the algorithms of assigning driver's license numbers in various states.

8. State and prove *the Chinese Remainder Theorem*. Illustrate it using ancient examples from China and India.

9. Write an essay on the various cryptosystems.

Enrichment Readings

1. A. V. Aho *et al.*, *Data Structures and Algorithms*, Addison-Wesley, Reading, MA, 1983.
2. W. T. Bailey, "Friday-the-Thirteenth,"*Mathematics Teacher,* Vol. 62 (May 1969), pp. 363–364.
3. J. A. Gallian, "Assigning Driver's License Numbers," *Mathematics Magazine*, Vol. 64 (Feb. 1991), pp. 13–22.
4. J. A. Gallian, "The Mathematics of Identification Numbers," *The College Mathematics Journal*, Vol. 22 (May 1991), pp. 194–202.
5. J. A. Gallian and S. Winters, "Modular Arithmetic in the Marketplace," *The American Mathematical Monthly*, Vol. 95 (June–July 1988), pp. 548–551.
6. D. W. Hardy and C. L. Walker, *Applied Algebra: Codes, Ciphers, and Discrete Algorithms*, Prentice-Hall, Upper Saddle River, NJ, 2003.
7. T. Koshy, *Elementary Number Theory with Applications*, Harcourt/ Academic Press, Boston, MA, 2002, pp. 210–436.
8. P. Lefton, "Number Theory and Public-Key Cryptology," *Mathematics Teacher*, Vol. 84 (Jan. 1991), pp. 54–62.
9. R. E. Lewand, *Cryptological Mathematics*, Math. Association of America, Washington, D.C., 2000.
10. J. E. Shockley, *Introduction to Number Theory*, Holt, Rinehart and Winston, New York, 1967, pp. 36–69.
11. J. R. Snow, "An Application of Number Theory to Cryptology," *Mathematics Teacher*, Vol. 82 (Jan. 1989), pp. 18–26.
12. P. M. Tuchinsky, "International Standard Book Number," *The UMAP Journal*, Vol. 5 (1985), pp. 41–54.

Chapter 8

Graphs

Euler calculated without effort, as men breathe,
or as eagles sustain themselves in the wind.

—FRANÇOIS ARAGO

Graph theory, a fascinating branch of mathematics, has numerous applications to such diverse areas as computer science, engineering, linguistics, and management science, as well as the natural and social sciences.

Like many important discoveries, graph theory grew out of an interesting physical problem, the celebrated Königsberg Bridge Puzzle (see Section 8.1). The outstanding Swiss mathematician Leonhard Euler solved the puzzle in 1736, thus laying the foundation for graph theory and earning his title as the father of graph theory.

This chapter presents the fundamentals of the field he created, with its assortment of new terms. Since graph terminology is not yet standard, definitions of basic terms can vary from book to book, an important fact to remember.

We will study the following interesting problems as well as others:

- The City of Königsberg (see Figure 8.1) comprises the river banks A and C and the islands B and D. These four land areas are connected by seven bridges. Could a Königsbergian take a walk through his beloved city, passing over each bridge exactly once? Could he take a walk through the city passing over each bridge exactly once and return home?
- At a sesquicentennial ball, there are n guests and each person shakes hands with everybody else exactly once. How would you represent the handshakes pictorially? How many handshakes are made?
- Three married couples want to cross a river in a rowboat which can carry only two people at a time. No husband will allow his wife to be in the boat or on the shore in the presence of another man unless he is also present. The women can, of course, row well. How can they all cross? (S. Gudder, 1976)

***Leonhard Euler** (1707–1783) was born in Basel, Switzerland. His father, a mathematician and a Calvinist pastor, wanted him also to become a pastor. Although Euler had different ideas, he followed his father's wishes, entering the University of Basel to study Hebrew and theology. His hard work at the University and remarkable ability brought him to the attention of the well-known mathematician Johann Bernoulli (1667–1748). Recognizing young Euler's talents, Bernoulli persuaded the boy's father to change his mind, and Euler was allowed to pursue his studies in mathematics.*

At age 19 Euler brought out his first paper. Although it failed to win the Paris Academy Prize in 1727, he eventually won the prize 12 times.

Euler was the most prolific mathematician, significantly contributing to every branch of mathematics. With his phenomenal memory, he had perfect recall for every formula. A genius, he could work anywhere and under any conditions. Euler belongs in a class by himself.

- A developer is building three new houses on one side of a street. If she would like to connect three utilities to each of them from the other side, can she lay the utility lines without any crossings?

8.1 Graphs

The Prussian city of Königsberg (called Kaliningrad during the era of the Soviet Union) lies on the Pregel river (see Figure 8.1). It consists of the two river banks A and C, and the two islands B and D. Seven bridges connect the four land areas of the city.

Figure 8.1

The City of Königsberg.

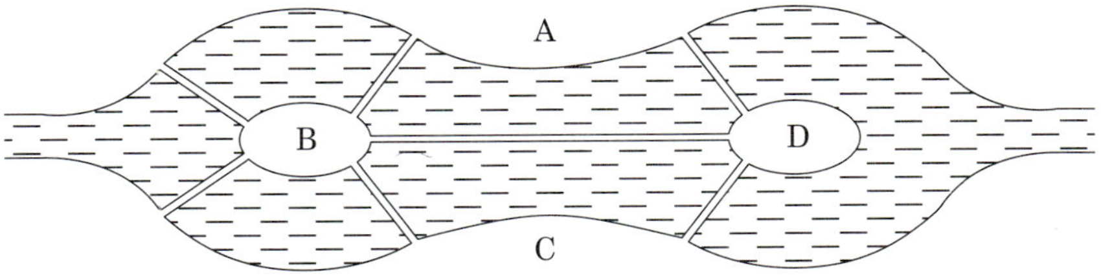

Residents of the city used to take evening walks from one part of the city to another. This, naturally, suggested the following question: *Is it possible to walk through the city, traversing each bridge exactly once*?

The problem sounds simple, and you might want to try a few possible paths before going any further. After all, by the multiplication principle, the maximum number of possible paths is $7! = 5040$.

In 1736, Euler, while at St. Petersburg Academy, published a solution to the problem: No such walk is possible. In fact, he proved a far more general result, of which the Königsberg bridge puzzle is a special case. Euler constructed a mathematical model like Figure 8.2 for the problem in which points A and C represent the two river banks; B and D the two islands. The arcs or line segments joining them represent the seven bridges.

Figure 8.2

A mathematical model for the Königsberg bridge puzzle.

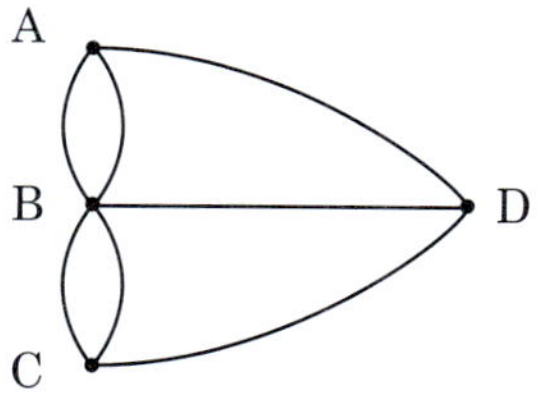

The Königsberg bridge problem can now be stated in layman's language as follows: *Beginning at one of the points A, B, C, or D, is it possible to trace the figure without lifting your pencil or traversing the same edge twice?* Section 8.5 will explore this further.

The Königsberg bridge model in Figure 8.2 consists of four points—A, B, C, and D—and the arcs or line segments joining them. Such a figure is called a graph.

Graph

A **graph** (or **undirected graph**) G consists of a nonempty finite set V of points (called **vertices** or **nodes**) and a set E of unordered pairs of elements in V (called **edges**). The graph G is the ordered pair $(V,E): G = (V,E)$. An edge connecting the vertices u and v is denoted by $\{u,v\}$, u-v, or some label. Geometrically, edges are denoted by arcs or line segments.

The next example uses these terms to apply graphs to the theory of communications.

EXAMPLE 8.1 A taxpayer files his federal tax return at an Internal Revenue Service (IRS) center located in the region in which he lives. Six of the IRS centers in the continental United States are (1) Atlanta, (2) Holtsville, (3) Philadelphia, (4) Cincinnati, (5) Austin, and (6) Fresno. A computer at each center can communicate with a computer at any other center over a telephone line. This network of computers can be modeled by a graph, as in Figure 8.3.

Figure 8.3

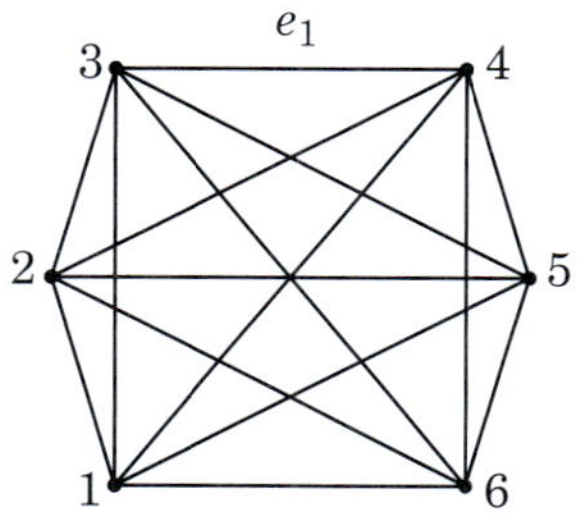

The graph has six vertices: 1, 2, 3, 4, 5, and 6. Each vertex represents a computer and each edge a telephone link. Since each computer can communicate with every other computer, an edge runs between any two vertices; e_1 denotes edge $\{3, 4\}$. ■

Figure 8.4 displays more graphs. The one in Figure 8.4a contains three vertices—a, b, and c: $V = \{a, b, c\}$. Its three edges are $e_1 = \{a, a\}$, $e_2 = \{a, b\}$, and $e_3 = \{b, c\} : E = \{e_1, e_2, e_3\} = \{\{a, a\}, \{a, b\}, \{b, c\}\}$.

Figure 8.4

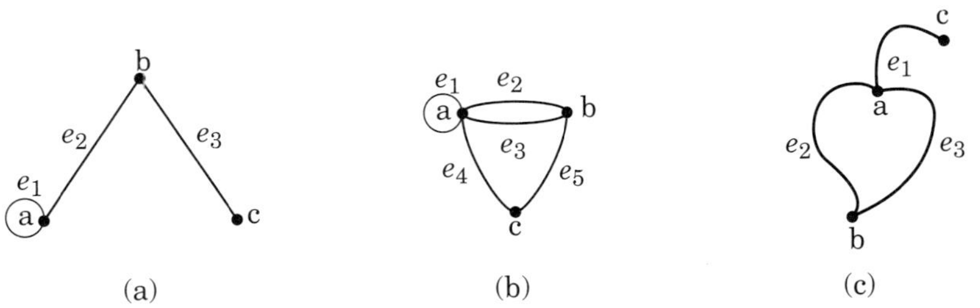

Airline route maps provide a fine paradigm of a graph, with each vertex representing a city and every edge a direct flight from one city to another.

We now introduce several special classes of graphs: simple, complete, bipartite, complete bipartite, and weighted graphs.

Simple Graph

An edge $\{a, a\}$ emanating from and terminating at the same vertex a is a **loop**.* **Parallel edges** have the same vertices. A **simple graph** contains no loops or parallel edges.

For example, the graphs in Figures 8.4a and b display a loop at a, while the ones in Figures 8.5 and 8.6 have two parallel edges, e_2 and e_3, connecting vertices a and b. Figure 8.3 is a simple graph, unlike the graphs in Figure 8.4 (why?).

By means of graphs, modular arithmetic can construct aesthetically pleasing designs, as the next example demonstrates.

EXAMPLE 8.2 Choose $V = \{0, 1, 2, 3, 4\}$, the set of integers modulo 5.

Figure 8.5

A pentagram.

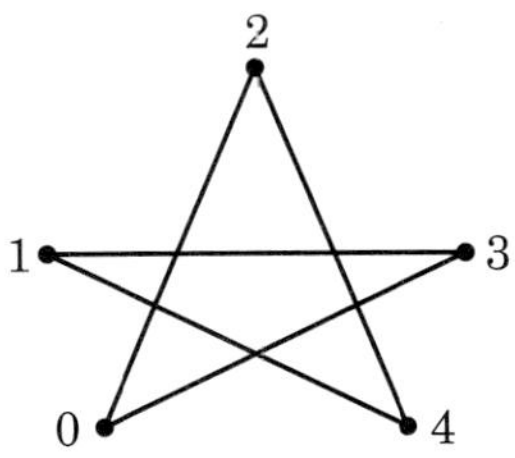

*Although $\{a, a\} = \{a\}$ as sets, the loop at a is denoted by $\{a, a\}$ or a-a.

Figure 8.6

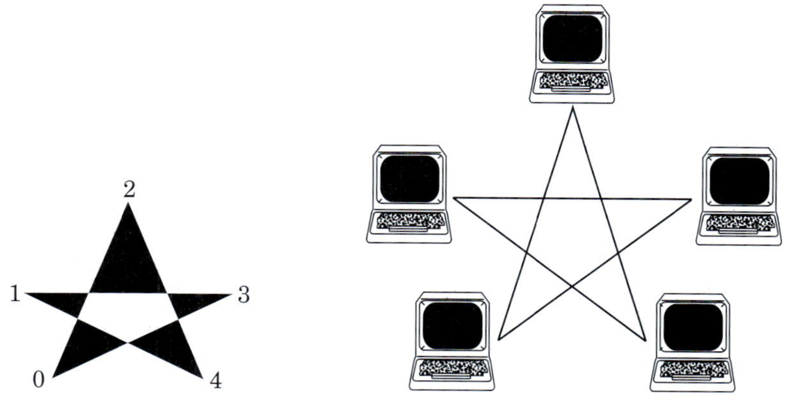

(a) (b)

Mark the vertices at equal intervals on a circle. An edge exists between the vertices x and y if $y \equiv x + 2 \pmod 5$. For instance, $1 \equiv 4 + 2 \pmod 5$; so an edge runs between the vertices 4 and 1. A simple graph, called a **pentagram,** materializes in Figure 8.5. Coloring the various wedge-shaped regions creates the pleasing design in Figure 8.6a. The pentagram reminds us of the point-to-point communication system in Figure 8.6b. ■

Graphs can also facilitate the study of hydrocarbons.

EXAMPLE 8.3

Arthur Cayley used graphs in studying isomers of hydrocarbons. A hydrocarbon molecule consists of carbon and hydrogen atoms. Each hydrogen atom (H) is bonded to a single carbon atom (C), whereas a carbon atom bonds with two, three, or four atoms which can be carbon or hydrogen.

Figure 8.7

Ethane, C_2H_6.

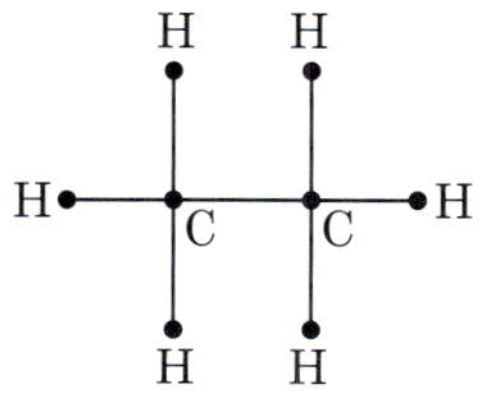

Figure 8.8

Cyclobutane, C_4H_8.

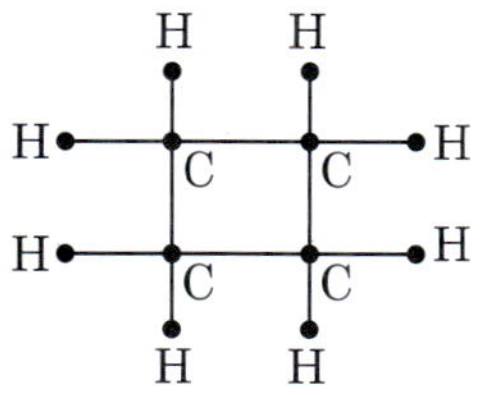

An ethane molecule, for instance, consists of two carbon atoms and six hydrogen atoms. Its **structural formula** appears as the graph in Figure 8.7, representing the **molecular formula** C_2H_6. Figures 8.8 and 8.9

Figure 8.9

Cyclopentane, C_5H_{10}.

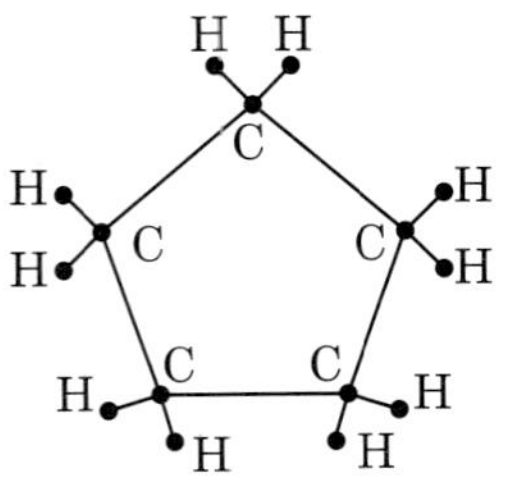

show the structural formulas of cyclobutane and cyclopentane molecules. ■

It is not necessary to have edges between any two vertices in a graph, so we make the following definition.

Adjacency and Incidence

Two vertices v and w in a graph are **adjacent** if an edge runs between them; if a loop occurs at v, v is adjacent to itself. An **isolated vertex** is not adjacent to any vertex. **Adjacent edges** have a common vertex. An edge is **incident** with a vertex v if v is an endpoint of the edge.

For example, in the graph in Figure 8.4a, vertices a and b are adjacent, but a and c are not. Edges $\{a, b\}$ and $\{b, c\}$ are adjacent. Edge e_2 is incident with vertices a and b. The graph contains no isolated vertices.

The concept of the degree of a vertex is important in the study of graphs, as will be seen later.

Degree of a Vertex

The **degree** of a vertex v in a graph is the number of edges meeting at v; it is denoted by $\deg(v)$.

> Clearly, a vertex v is isolated if $\deg(v) = 0$. In addition, a loop at v contributes two to its degree.

For example, in Figure 8.4b, $\deg(a) = 5$, $\deg(b) = 3$, and $\deg(c) = 2$. In Figure 8.4c, $\deg(a) = 3$, $\deg(b) = 2$, and $\deg(c) = 1$.

We have seen that digraphs can arise from matrices; graphs also can arise from them.

Adjacency Matrix

The **adjacency matrix** of a graph with n vertices $v_1, v_2, \ldots, v_n$ is an $n \times n$ matrix $A = (a_{ij})$, where a_{ij} = number of edges from v_i to v_j.

Because every edge in a graph is undirected, $a_{ij} = a_{ji}$ for every i and j, so the adjacency matrix of every graph is symmetric. If the graph is simple, A is a boolean matrix.

For example, the adjacency matrix of the Königsberg bridge model in Figure 8.2 is

$$A = \begin{array}{c} \\ A \\ B \\ C \\ D \end{array}\begin{array}{c} \begin{array}{cccc} A & B & C & D \end{array} \\ \begin{bmatrix} 0 & 2 & 0 & 1 \\ 2 & 0 & 2 & 1 \\ 0 & 2 & 0 & 1 \\ 1 & 1 & 1 & 0 \end{bmatrix} \end{array} \begin{array}{c} \text{row sum} \\ 3 \\ 5 \\ 3 \\ 3 \end{array} \begin{array}{c} \\ \\ \leftarrow \deg(B) \\ \\ \\ \end{array}$$

The sum of the numbers along each row gives the degree of the corresponding vertex. For instance, $\deg(A) = 3$ and $\deg(B) = 5$.

Theorem 8.1 shows a close relationship between the sum of the degrees of the vertices of a graph and the number of edges in it.

THEOREM 8.1 Let e denote the number of edges of a graph G with n vertices $v_1, v_2, \ldots, v_n$. Then $\sum_{i=1}^{n} \deg(v_i) = 2e$.

PROOF:

Every nonloop edge is incident with exactly two (distinct) vertices. On the other hand, every loop edge is incident with the vertex twice. Thus every edge, whether it is a loop or not, contributes a two to the sum of the degrees of the vertices; so $\sum_{i=1}^{n} \deg(v_i) = 2e$. ■

According to Theorem 8.1, the sum of the degrees in a graph is always an even integer. This fact can determine if a given number of edges and vertices with known degrees can generate a graph (and hence a hydrocarbon).

The next two examples employ this test.

EXAMPLE 8.4 Is a graph with four vertices a, b, c, and d with $\deg(a) = 4$, $\deg(b) = 5 = \deg(d)$, and $\deg(c) = 2$ possible?

SOLUTION:

Sum of the degrees $= 4 + 5 + 2 + 5 = 16$. Since the sum is even, there might be such a graph with $16/2 = 8$ edges. In fact, Figure 8.10 demonstrates one (see Exercise 65).

Figure 8.10

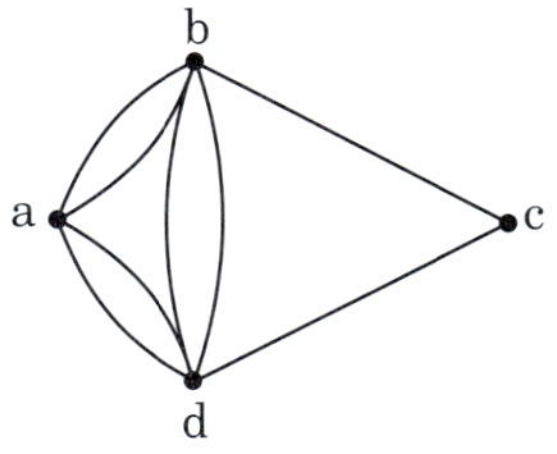

■

Example 8.5 applies Theorem 8.1 to hydrocarbon molecules.

○ **EXAMPLE 8.5** (optional) In a hydrocarbon molecule, a hydrogen atom is bonded to exactly one carbon atom. If a carbon atom bonds to four atoms, carbon or hydrogen, can a hydrocarbon molecule with three carbon atoms and five hydrogen atoms exist?

SOLUTION:
Suppose there is such a molecule. Its structural formula is a graph with vertices representing atoms and edges representing bonds. The sum of the degrees of the vertices of the graph is $3 \cdot 4 + 5 \cdot 1 = 17$. Since this sum is *not* an even integer, such a graph and hence such a hydrocarbon molecule cannot exist. ■

Another useful consequence of Theorem 8.1 lies below.

THEOREM 8.2 The number of odd degree vertices in a graph is an even integer.

PROOF:
Let G be a graph with e edges. Let x denote the sum of the degrees of even degree vertices and y the sum of the degrees of odd degree vertices. By Theorem 8.1, $x + y = 2e$. Since x is the sum of even integers, x is even, so $y = 2e - x$ is also an even integer. But y is the sum of odd integers, so for y to be even, the number of addends in the sum must be even. In other words, the number of odd degree vertices must be even. ■

For example, the Königsberg model in Figure 8.2 contains four vertices of odd degree.

Just as a set can have subsets, a graph can have subgraphs.

Subgraph of a Graph

A **subgraph** of a graph $G = (V, E)$ is a graph $G_1 = (V_1, E_1)$ where $V_1 \subseteq V$ and $E_1 \subseteq E$.

EXAMPLE 8.6 The graphs G_1 and G_2 in Figure 8.12 are subgraphs of the graph G in Figure 8.11 (Why?).

Figure 8.11

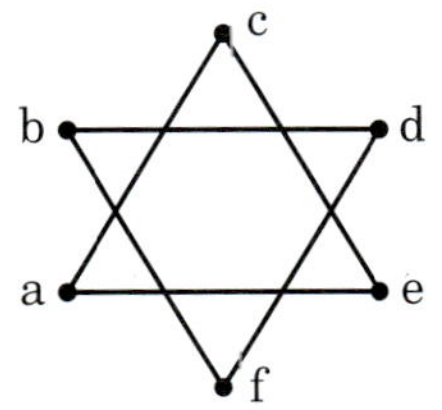

Graph G (**Star of David**)

Figure 8.12

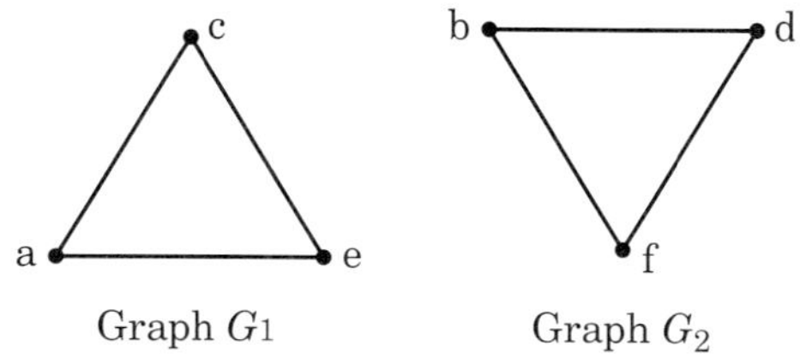

■

Next we present two fascinating occurrences of Fibonacci and Lucas numbers in the study of hydrocarbons.*

Fibonacci and Paraffins

Delete the hydrogen atoms from the structural formulas of saturated hydrocarbon molecules C_nH_{2n+2}. This yields graphs consisting of only carbon atoms and edges between adjacent vertices.

The **topological index** of such a graph G with n vertices is the total number of different ways the graph can be partitioned into disjoint subgraphs containing all vertices and exactly k edges, where $k \geq 0$. For example, Figure 8.13a shows the carbon atom skeleton for the paraffin, pentane C_5H_{12}, and Figure 8.13b shows its various possible partitionings. Consequently, the topological index of pentane is $1 + 4 + 3 = 8$.

Figure 8.13

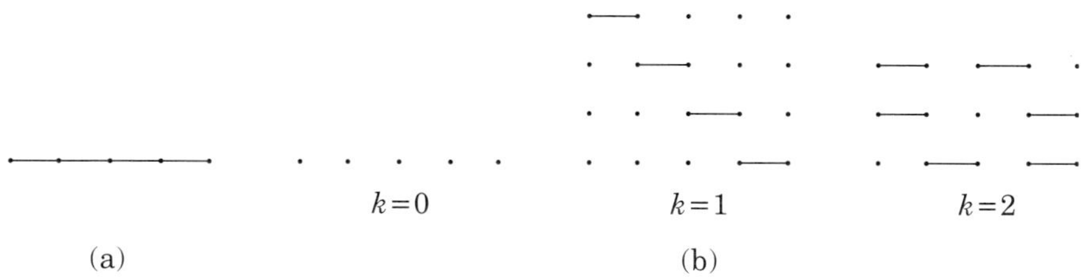

Table 8.1 shows the carbon atom graphs G_n and their topological indices of 10 paraffins $C_nH_{2n+2}, n \geq 1$. For a graph consisting of a single vertex, the index is defined as one. It appears from the table that the index of G_n is F_{n+1}.

To confirm this observation, let t_n denote the topological index of the carbon atom graph G_n of a paraffin with n vertices, as Figure 8.14 shows.

Case 1 Suppose the edge v_{n-1}-v_n is not included. Then the edge v_{n-2}-v_{n-1} may or may not be included. Consequently, the topological index of the remaining graph G_{n-1} is t_{n-1}.

*T. Koshy, *Fibonacci and Lucas Numbers with Applications,* Wiley, New York, 2001.

Table 8.1

Topological indices of paraffins C_nH_{2n+2}.

Paraffin	n	Graph	k = 0	1	2	3	4	5	Total
Methane	1		1						1
Ethane	2		1	1					2
Propane	3		1	2					3
Butane	4		1	3	1				5
Pentane	5		1	4	3				8
Hexane	6		1	5	6	1			13
Heptane	7		1	6	10	4			21
Octane	8		1	7	15	10	1		34
Nonane	9		1	8	21	20	5		55
Decane	10		1	9	28	35	15	1	89

↑ F_{n+1}

Figure 8.14

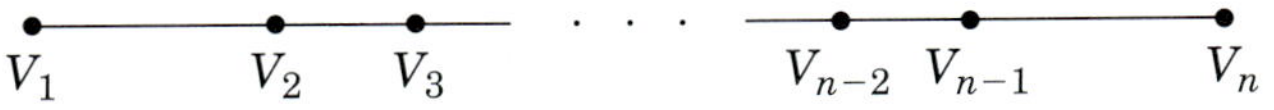

Case 2 Suppose the edge v_{n-1}-v_n is included. Then the edge v_{n-2}-v_{n-1} is not included. This yields the graph G_{n-2} and its index is t_{n-2}.

Thus, the addition principle, $t_n = t_{n-1} + t_{n-2}$. But $t_1 = 1$ and $t_2 = 2$, so, $t_n = F_{n+1}$.

Lucas and Cycloparaffins

Table 8.2 shows the carbon atom skeleton G_n of 10 cycloparaffins C_nH_{2n} and the corresponding indices. A similar argument shows that the index of G_n = index of G_n + index of $G_{n-2} = F_{n+1} + F_{n-1} = L_n$, where $n \geq 3$.

Five important classes of simple graphs receive attention below: complete, cycle, wheel, bipartite, and weighted.

Complete Graph

A simple graph with an edge between every two distinct vertices is a **complete graph**. A complete graph with n vertices is denoted by K_n.

The complete graphs K_1 through K_5 appear in Figure 8.15. Notice that Figure 8.3 displays K_6. The complete graph K_n has $C(n, 2)$ edges and each vertex has degree $n - 1$ (Why?).

The next example revisits the handshake problem from Chapter 5.

EXAMPLE 8.7 **(The handshake problem)** At a sesquicentennial ball, each of n guests shakes hands with everybody else exactly once. Find the number of handshakes h_n made.

Table 8.2

Topological indices of cycloparaffins C_nH_{2n}.

Cycloparaffin	n	Graph	k = 0	1	2	3	4	5	Total
·	1		1						1
·	2		1	2					3
Cyclopropane	3		1	3					4
Cyclobutane	4	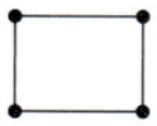	1	4	2				7
Cyclopentane	5		1	5	5				11
Cyclohexane	6		1	6	9	2			18
Cycloheptane	7		1	7	14	7			29
Cyclooctane	8		1	8	20	16	2		47
Cyclononane	9		1	9	27	30	9		76
Cyclodecane	10		1	10	35	50	25	2	123

↑ L_n

Figure 8.15

The complete graphs K_1 through K_5.

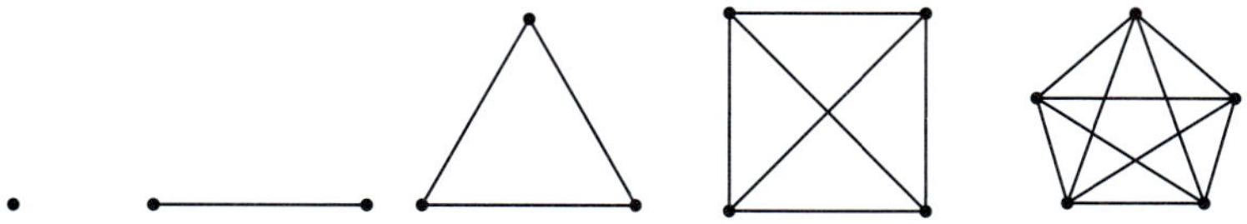

K_1 K_2 K_3 K_4 K_5

SOLUTION:
Represent each guest by a vertex of a graph and each handshake by an edge. Since each person shakes hands with everyone else exactly once, the complete graph K_n is generated. Therefore,

$$\begin{aligned} h_n &= \text{number of handshakes made} \\ &= \text{number of edges in } K_n \\ &= C(n, 2) \end{aligned}$$

■

Complete graphs are also useful in modeling **round-robin tournaments**, where every team plays every other team exactly once. With n teams entering the tournament, K_n provides the model.

Cycle and Wheel Graphs

The **cycle graph** C_n of length $n(\geq 3)$ consists of n vertices $v_1, \ldots, v_n$ and edges $\{v_i, v_{i+1}\}$, where $1 \leq i \leq n$ and $v_{n+1} = v_1$. Figure 8.16 displays cycle graphs C_3, C_4, and C_5.

Figure 8.16

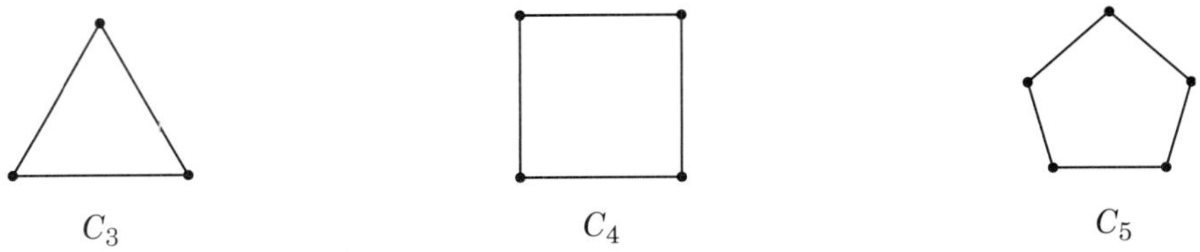

The **wheel graph** $W_n (n \geq 3)$ is obtained from C_n by adding a vertex v inside C_n and connecting it to every vertex in C_n. Figure 8.17 shows the wheel graphs W_3, W_4, and W_5.

Figure 8.17

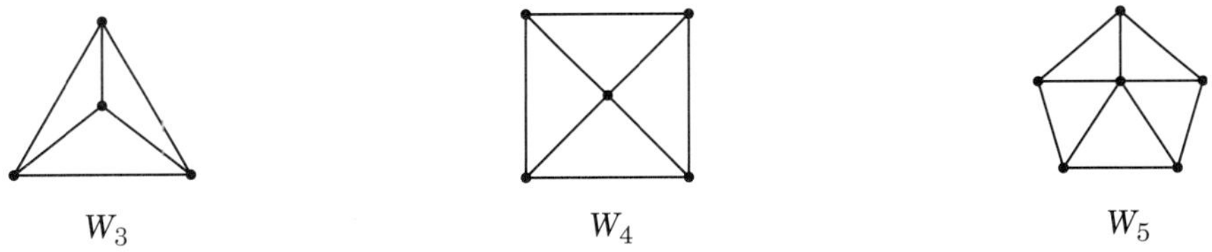

Next we present another type of simple graph.

Bipartite Graph

If the vertex set V of a simple graph $G = (V, E)$ can be partitioned into two disjoint (nonempty) sets V_1 and V_2, so every edge in G is incident with a vertex in V_1 and a vertex V_2, then G is **bipartite**.

The graph in Figure 8.18 is bipartite, since the vertex set $V = \{\mathrm{a, b, c, d, e, f}\}$ unites the disjoint sets $V_1 = \{\mathrm{a, b, c}\}$ and $V_2 = \{\mathrm{d, e, f}\}$, and every edge has one vertex in V_1 and the other in V_2. But the graph in Figure 8.19 is *not* bipartite (Why?).

Figure 8.18

A bipartite graph.

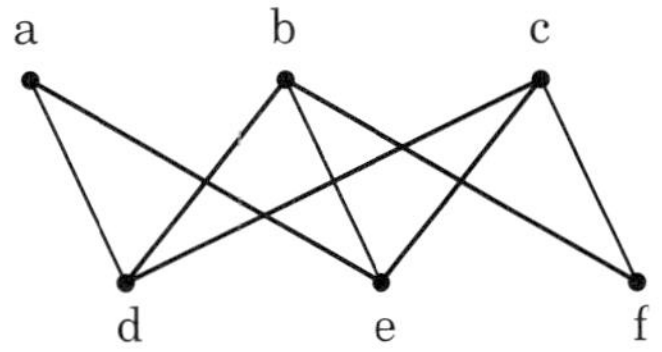

Figure 8.19

Not bipartite.

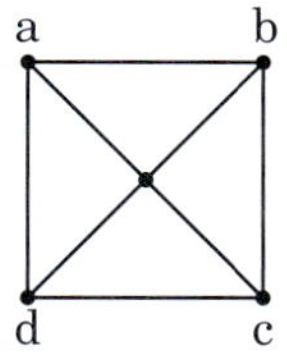

Complete Bipartite Graph

Let G be a bipartite graph with $|V_1| = m$ and $|V_2| = n$. If an edge runs between every vertex in V_1 and V_2, G is a **complete bipartite graph**, denoted by $K_{m,n}$.

The complete bipartite graph $K_{2,3}$ is shown in Figure 8.20. An interesting application of $K_{3,3}$ will be made in Example 8.36.

Figure 8.20

The complete bipartite graph $K_{2,3}$.

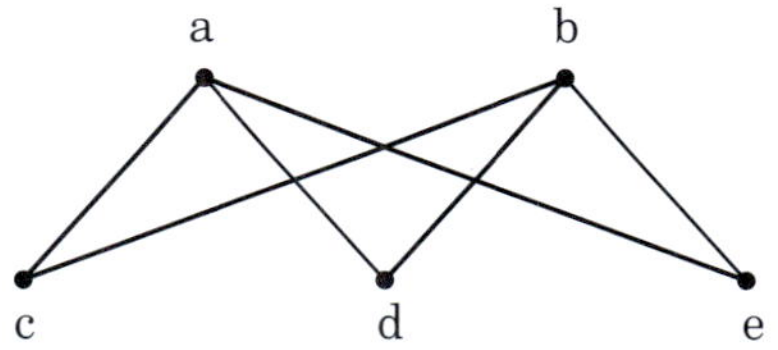

The last important class of simple graphs handled in this section follows.

Weighted Graph

A simple graph in which each edge e is assigned a positive real number w is a **weighted graph**. The number w is the **weight** of edge e. Thus a weighted graph may be considered an ordered triplet (V,E,f), where $f : E \to \mathbb{R}$ is a function.

Route maps often provide examples of weighted graphs, as the following example shows.

EXAMPLE 8.8 Figure 8.21 presents a weighted graph, where the weights represent distances in miles between the cities.

Figure 8.21

A weighted graph.

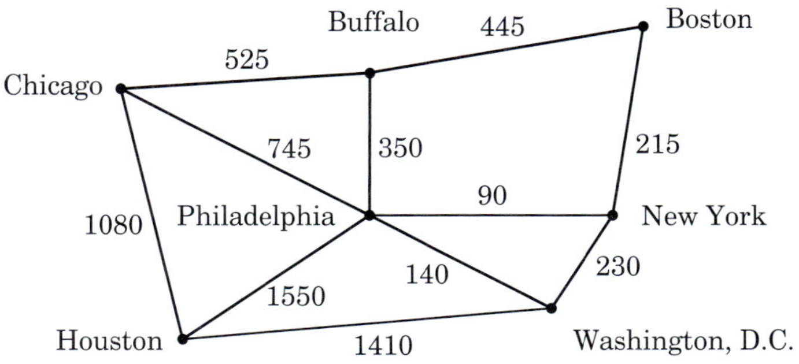

■

The weights in a weighted graph may represent distances, travel times, fares, transportation costs, etc. Every entry a_{ij} of the adjacency matrix A of a weighted graph denotes the weight of the edge $\{i, j\}$. Accordingly, A is the **weighted adjacency matrix** of the graph.

Graphs have interesting and useful applications in communications.

Graphs and Telecommunications

Suppose, for example, there are n telephones in a city and there is a dedicated line between every two telephones. Such an arrangement of communication lines, called a **network**, can be modeled by the complete graph K_n. Such a telephone network is inefficient and expensive. (How many cables are needed to link n customers? What if a new subscriber needs to be added to the network?)

So the local telephone company uses a network, called **star topology**, which links each subscriber to a central office switch (or exchange). It can be represented by the bipartite graph $K_{1,n}$, as in Figure 8.22, where the symbol □ indicates a central office switch. The star topology takes just n cables to connect n phones, much less than the $C(n, 2)$ cables needed for K_n.

To accommodate a large number of customers, two or more star topologies are linked by trunks in cities and large towns. Figure 8.23, for example, shows a network with two exchanges, 863 and 891.

Figure 8.22

A star topology.

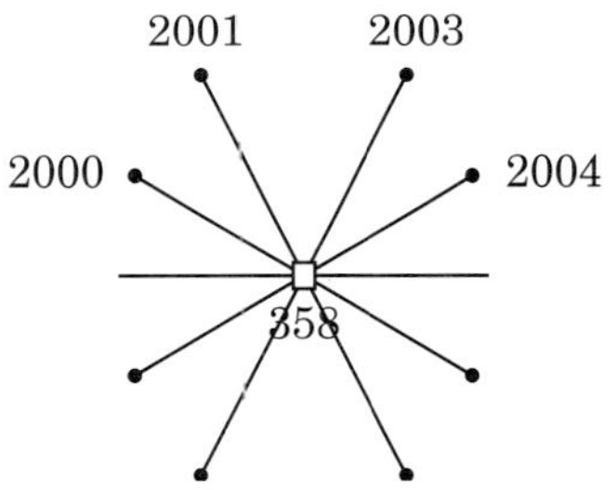

Figure 8.23

The star topologies linked by a trunk.

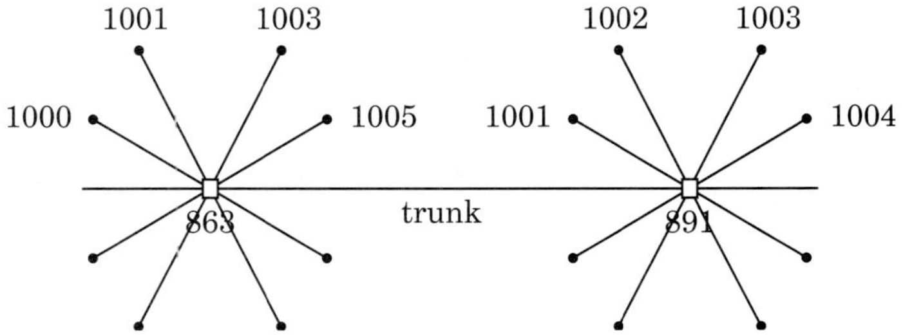

Graphs and Local Area Networks

Star topology is also used in the design of **local area networks** (LANs), which are arrangements of personal computers and peripheral devices such as printers and plotters in a building or on a campus. See Figure 8.24. All devices in the network are connected to a host computer at the center of the network, which directs the flow of communications between devices.

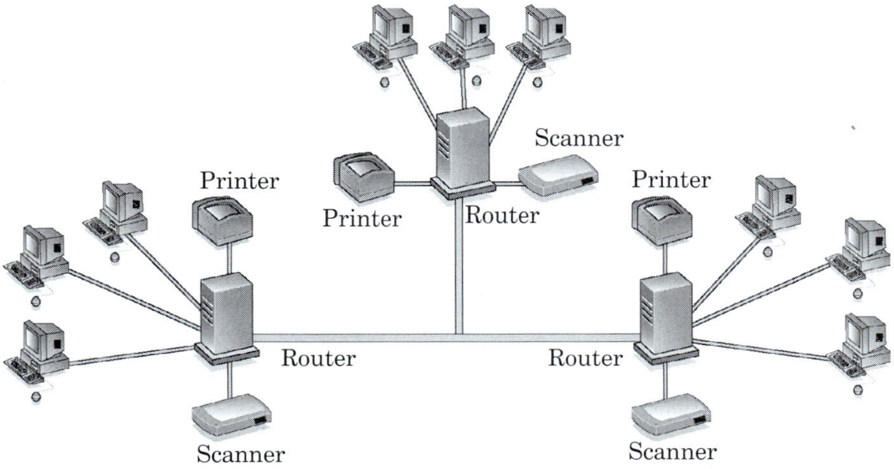

Figure 8.24

A local area network.

Bus topology, illustrated in Figure 8.25, links devices in the network via a single trunk. This topology is simple and takes the least amount of cable to install. Furthermore, it is easy to add a device to the network by simply tapping into the trunk.

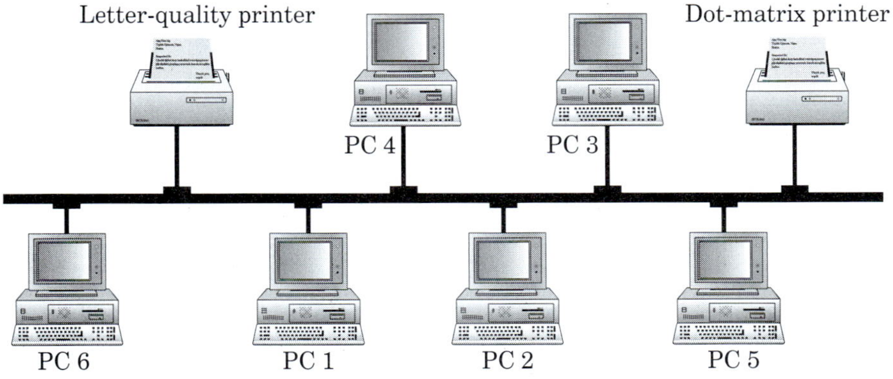

Figure 8.25

A bus topology.

Ring topology, no longer in use, connects computers in a circle, so each computer is linked to exactly two others. Such a LAN can be represented by the cycle graph C_n. See Figure 8.26.

Some LANs employ a **star-ring topology**, a hybrid of star and ring topologies, as in Figure 8.27. A star-ring topology can be modeled by the

Figure 8.26

A ring topology.

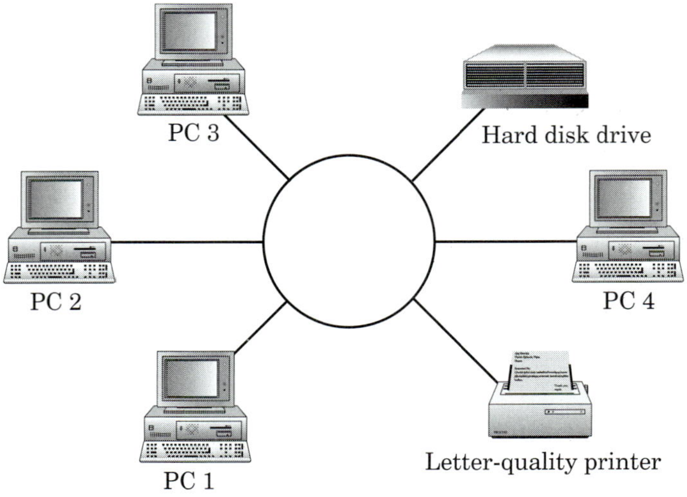

wheel graph W_n. In such a network, messages are sent around a circle or thorough a central host device.

Figure 8.27

A star-ring topology.

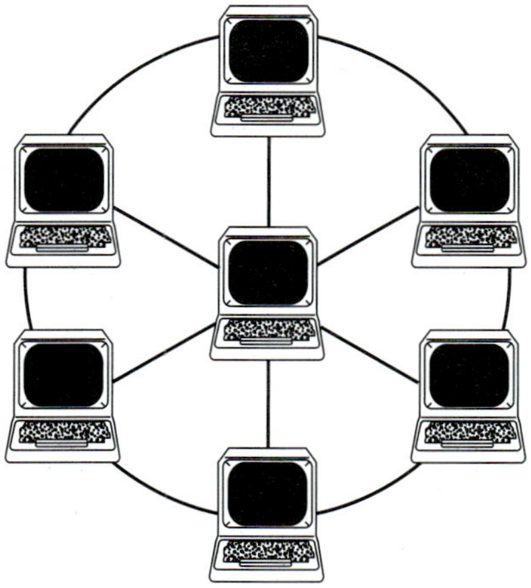

Among the sundry places graphs crop up are in games. Two such games follow.

○ **EXAMPLE 8.9** (optional) The game of **Yashima*** is played on the Yashima board (see Figure 8.28) by two players using two tokens, one red and the other white. (The graph is called the **Petersen graph**, after the Danish mathematician Julius Petersen, who developed it in 1891.) Initially, the tokens are placed

*Proposed by F. H. Kierstead, Jr., and solved by B. J. Schwartz, J. *Recreational Mathematics*. Vol. 14, 1981–1982, p. 309.

at two vertices of the graph. The players alternately move their tokens from one vertex to an unoccupied adjacent vertex, erasing the edge traversed. The game ends when a player lands at an isolated vertex. Find the maximum number of moves each player can make.

Figure 8.28

Petersen graph (the Yashima board).

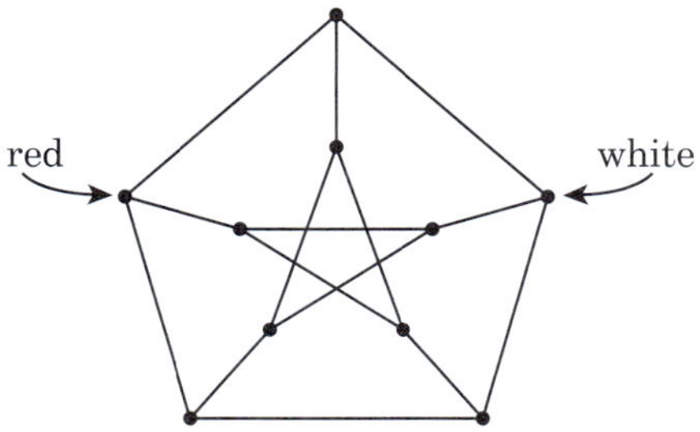

SOLUTION:

Notice that the degree of each vertex is three. Two edges are deleted at each vertex along the path taken by each player except at the initial and terminal vertices, where only one edge disappears. Thus a maximum of four vertices (two for each player) lose an odd number of edges, and when the game ends, at most four vertices can remain isolated; in other words, at most 12 of the 15 edges can be deleted. So each player can last six moves.

To achieve this maximum number, one player must travel along the pentagon and the other along the star. (Try this with a friend.) ■

The next example is related to Example 3.29, so review it before going any further.

EXAMPLE 8.10 The graph-theoretic game of **SIM**, invented by Gustavus J. Simmons of Sandia Corporation in 1969, is played by two players using the complete graph K_6. Player R has a red pencil and player B a blue one. They alternately color the edges of the graph. The first player to complete a monochromatic triangle loses the game.

Three nontrivial ways of playing the game are given in Figure 8.29, where solid edges indicate red edges and broken ones, blue.

How long can a game of SIM go on? Suppose player R uses his first three moves to color the edges of a triangle. By then player B will have made two moves, so player R can color a monochromatic triangle in five moves and the game ends. Such a sequence is clearly suicidal for R.

Can the game end in a draw? Since the edges form at least one monochromatic triangle, by Example 3.29, the game cannot end in a draw. In at most 15 moves, one player will lose by completing a monochromatic triangle. ■

This example has an intriguing application, as the next example shows.

Julius Petersen *(1839–1910) was born in Soro, Denmark. He graduated from Copenhagen University in 1866 and 5 years later received his Ph.D. from Copenhagen. In 1887, he became a professor at the university, a position he held until 1909.*

An author of several textbooks translated into several languages, Petersen strongly influenced mathematics education in Denmark. Although he is best known for his work in graph theory, he worked in algebra, analysis, geometry, mechanics, and number theory.

Figure 8.29

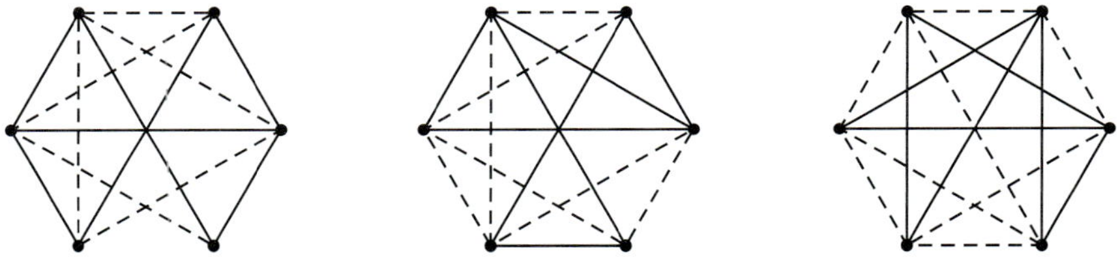

Player B losses in 13 moves

Player R losses in 14 moves

Player B losses in 15 moves

EXAMPLE 8.11 Show that any group of six people contains three mutual friends or three mutual strangers.

PROOF:

Represent each person in the group by a vertex of a graph. Draw a red edge between two vertices if the corresponding persons are friends; otherwise draw a blue edge. When all edges are drawn, we get the complete graph K_6. By Example 8.10, the edges form a monochromatic triangle, so the group contains three mutual friends or three mutual strangers. ■

We close this section with a technique of combining two graphs and an interesting application to the modified handshake problem.

Union of Simple Graphs

The **union** of two simple graphs, $G_1 = (V_1, E_1)$ and $G_2 = (V_2, E_2)$, is the simple graph $G = (V, E)$, where $V = V_1 \cup V_2$ and $E = E_1 \cup E_2$; it is denoted by $G_1 \cup G_2$.

For example, the *Star of David* in Figure 8.11 is the union of the graphs G_1 and G_2 in Figure 8.12.

We now present an application of this operation. To this end, we return to the handshake problem in Example 5.3.

Recall from Example 5.10 that the number of handshakes made by n people is given by $h(n) = n(n-1)/2 = C(n,2)$, where $n \geq 1$. Geometrically, $h(n)$ denotes the number of edges in the complete graph K_n.

A Generalized Handshake Problem

More generally, suppose a group of r (≥ 0) people hosts a party and there are n guests at the party. Everyone shakes hands with everybody else exactly once, except that no host shakes hands with any other host. Then

$$\begin{aligned} h(n) &= \begin{pmatrix}\text{number of handshakes}\\ \text{made by the } n \text{ guests}\end{pmatrix} + r\begin{pmatrix}\text{number of handshakes}\\ \text{made by a host } H\end{pmatrix} \\ &= c(n,2) + rn \end{aligned}$$

See Figure 8.30.

Figure 8.30

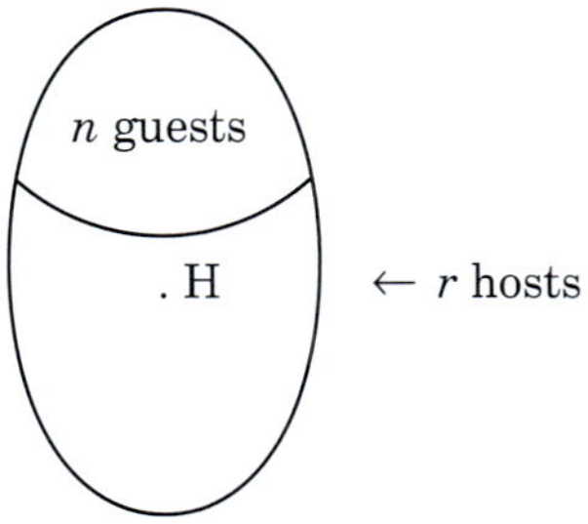

The formula $h(n) = C(n,2) + rn$ has an interesting graph-theoretic interpretation. As in the original simple version, the handshakes made among the guests can be represented by the edges in K_n. Likewise, rn handshakes between the guests and the hosts can be represented by the edges in the complete bipartite graph $K_{r,n}$.

Thus $h(n)$ denotes the number of edges in the union $K_n \cup K_{r,n}$ of the two graphs (see Figure 8.31), where $n = 4$ and $r = 3$. Consequently, $h(n) = e(K_n) + e(K_{r,n}) = e(K_n \cup K_{r,n})$, where $e(G)$ denotes the number of edges in a graph G.

Exercises 8.1

Determine if each graph is simple.

Figure 8.31

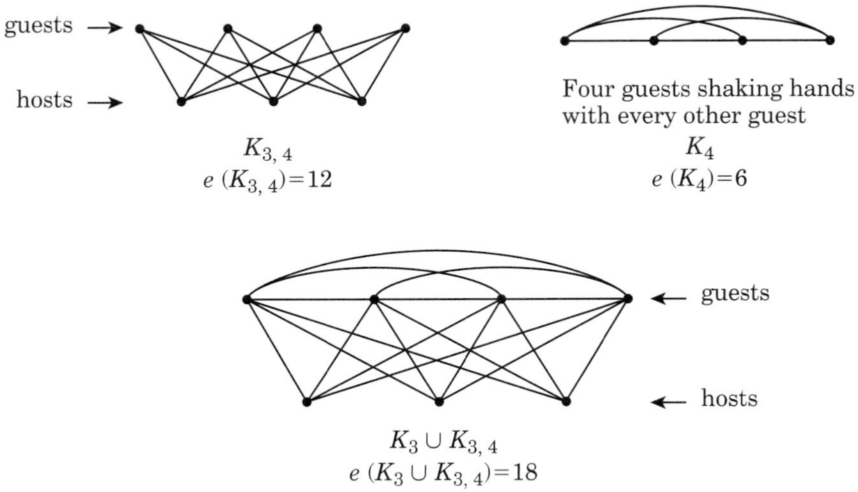

1.

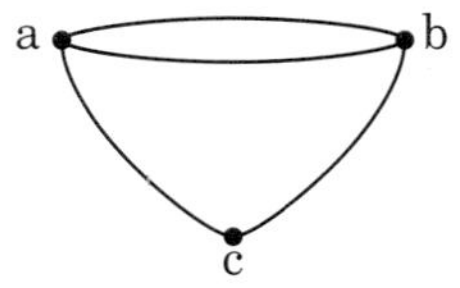

2.

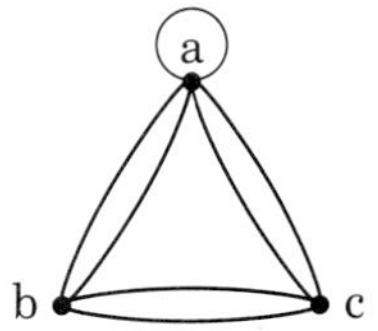

3.

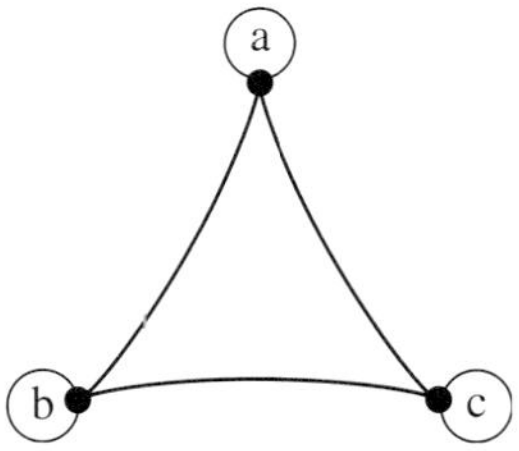

4.

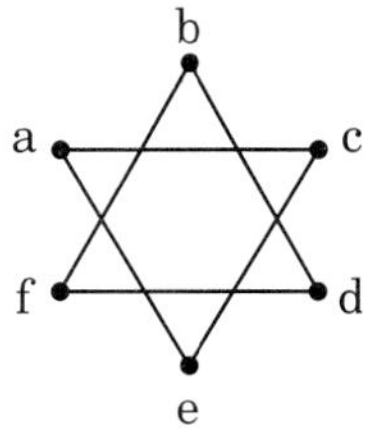

5–6. Find the number of vertices and edges of the graphs in Exercises 1–2.

7–8. Find the degrees of the vertices of the graphs in Exercises 3–4.

9–10. Find the adjacency matrix of the graphs in Exercises 1 and 2.

Draw the graph with the given adjacency matrix.

11.
$$\begin{array}{c} \\ a \\ b \\ c \\ d \end{array}\begin{array}{c} \begin{array}{cccc} a & b & c & d \end{array} \\ \begin{bmatrix} 0 & 0 & 1 & 1 \\ 0 & 0 & 1 & 1 \\ 1 & 1 & 0 & 0 \\ 1 & 1 & 0 & 0 \end{bmatrix} \end{array}$$

12.
$$\begin{array}{c} \\ a \\ b \\ c \\ d \end{array}\begin{array}{c} \begin{array}{cccc} a & b & c & d \end{array} \\ \begin{bmatrix} 1 & 1 & 1 & 0 \\ 1 & 1 & 0 & 1 \\ 1 & 0 & 1 & 1 \\ 0 & 1 & 1 & 1 \end{bmatrix} \end{array}$$

13. The adjacency matrix of a simple graph has the form

$$A = \left[\begin{array}{c|c} A_1 & 0 \\ \hline 0 & A_2 \end{array}\right]$$

What can you say about the graph?

14–15. Verify Theorem 8.1 for each graph in Exercises 1 and 2. Find the number of edges of a graph that has:

16. Exactly three vertices with degrees 1, 3, and 2.

17. Exactly five vertices with degrees 1, 1, 1, 1, and 4.

Could there could be a graph that has:

18. Three vertices with degrees 2, 3, and 4?

19. Four vertices with degrees 2, 2, 2, and 2?

Find the number of bonds in each hydrocarbon molecule. (Assume each carbon atom is bonded to four atoms.)

20. A propane molecule C_3H_8

21. A butane molecule C_4H_{10}

22. An ethylene molecule C_2H_4

23. A cyclobutane molecule C_4H_8

24. Find the number of subgraphs of the graph in Exercise 3. Compute the number of edges in each complete graph.

25. K_6

26. K_7

27. K_{10}

28. K_{11}

29. Characterize the adjacency matrix of the complete graph K_n.

Is each graph bipartite? If so, identify the vertex sets V_1 and V_2.

30.

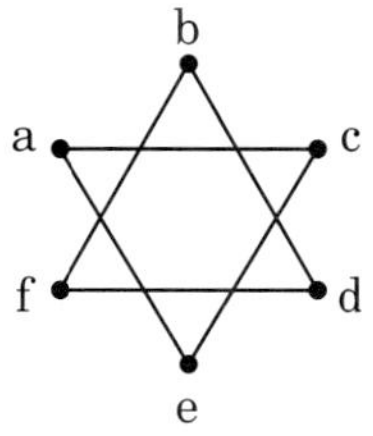

31.

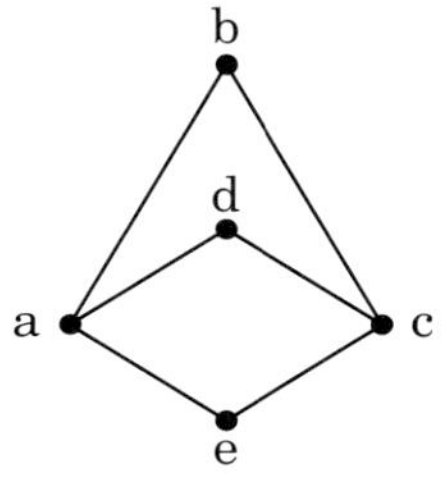

32. Find the number of vertices in the bipartite graph $K_{m,n}$.

33. Find the number of edges in the bipartite graph $K_{m,n}$.

34. Identify the general form of the adjacency matrix for $K_{m,n}$.

35. Let G be a graph with n vertices and e edges. Let M and m denote the maximum and minimum of the degrees of vertices in G, respectively. Prove that $m \leq 2e/n \leq M$.

A simple graph G is **regular** if every vertex has the same degree. If every vertex has degree r, G is r-**regular** with r the **degree** of the graph. Draw a regular graph with the given properties.

36. $r = 1$ and two vertices.

37. $r = 2$ and three vertices.

38. $r = 2$ and four vertices.

39. $r = 3$ and four vertices.

40. $r = 1$ and not complete.

41. $r = 2$ and not complete.

42. Is the Petersen graph in Figure 8.28 r-regular? If yes, find the value of r.

43. Is the complete graph K_n regular? If so, find its degree.

44. How many edges does an r-regular graph with n vertices have? (*Hint*: Use Exercise 35.)

45. Let G be an r-regular graph with n vertices. Prove that nr is even. (*Hint*: Use Exercise 44.)

46. Can there be a 1-regular graph with three vertices?

47. Can there be a 3-regular graph with five vertices?

The **complement** of simple graph G is a simple graph G' containing all vertices in G; two vertices are adjacent in G' if they are not adjacent in G. For example, the graphs in Figures 8.32 and 8.33 are complements of each other. Find the complements of the graphs in Exercises 48–50.

Figure 8.32

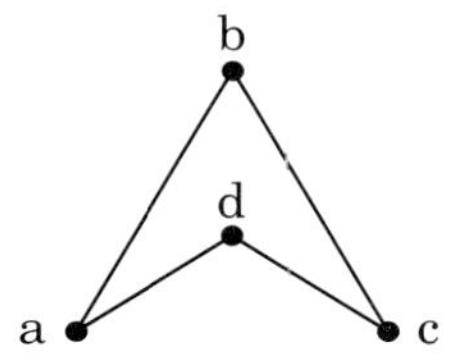

Figure 8.33

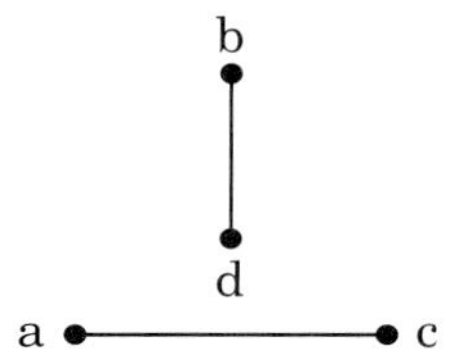

48.

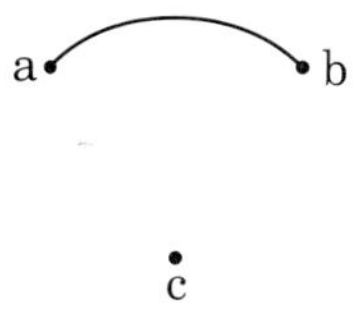

49.

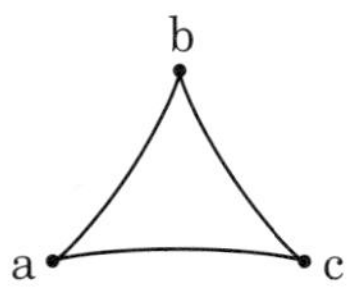

50.

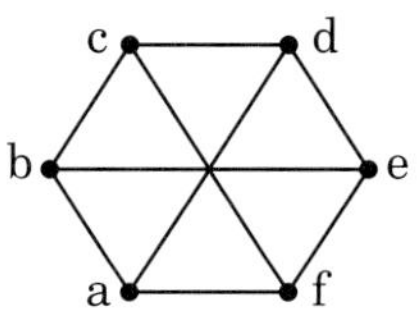

51. How are the adjacency matrices of G and G' related?

Characterize the complement of each graph.

52. K_n

53. $K_{m,n}$

54. Let n and e denote the numbers of vertices and edges in a simple graph G and e' the number of edges in G'. How are e and e' related?

Find the union of each pair of graphs.

55.

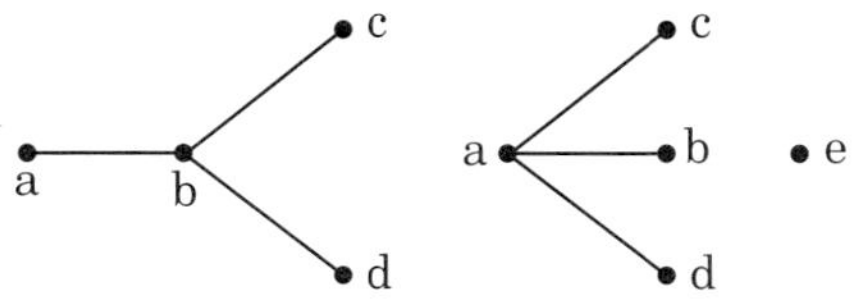

56.

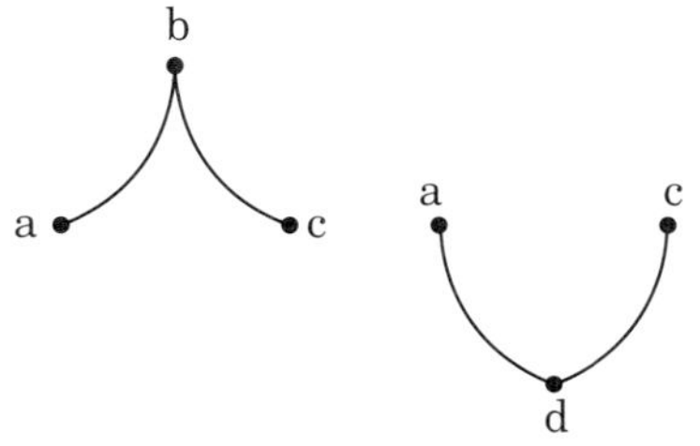

Characterize each graph.

57. $K_{2,3} \cup K'_{2,3}$

58. $K_{2,4} \cup K'_{2,4}$

59. $K_{3,3} \cup K'_{3,3}$

60. $K_{m,n} \cup K'_{m,n}$

61. Let G be a simple graph with n vertices. What can you say about $G \cup G'$?

62. A company wishes to schedule 1-hour meetings beginning at 7 A.M. between every two of its six regional managers—A, B, C, D, E, and F—so each can spend an hour with each of the other five for better acquaintance. Find the various possible schedule pairings. (This is a round-robin tournament in disguise.) (S. W. Golomb, 1993)

63. Five basketball teams, a through e, enter a round-robin tournament. Create a schedule so that every team plays every other team exactly once.
(*Hint*: Since the number of teams is odd, add a dummy team x. If a team is paired with x, the team draws a bye in that round.)

64. Show that any simple graph with two or more vertices has at least two vertices of the same degree.
(*Hint*: Use the pigeonhole principle.)

***65.** Let $v_1, \ldots, v_n$ be n vertices with degrees $\deg(v_1), \ldots, \deg(v_n)$, respectively, such that $\sum_{i=1}^{n} \deg(v_i)$ is even. Prove that there exists a graph satisfying these conditions.

[*Hint*: Let $\sum_{i=1}^{n} \deg(v_i) = 2e$. Use induction on e.]

*8.2 Computer Representations of Graphs (optional)

Like relations and digraphs, graphs can be computer-implemented by arrays and linked lists. A graph with n vertices can materialize by means of its adjacency matrix A.

Since many of the n^2 elements in A are frequently zero, a more efficient way to implement a graph is by linked lists. For each node, create a linked list of all nodes adjacent to it and store their header nodes in an array of pointers. The result is the **adjacency list representation** of the graph.

For example, the Königsberg bridge model in Figure 8.34, with the vertices labeled 1 through 4, has two edges from vertex 1 to vertex 2 and one to vertex 4. So the linked list headed by 1 contains three nodes and the lists headed by 2, 3, and 4 contain five, three, and three nodes, respectively. Figure 8.35 shows the adjacency list representation of the graph.

Figure 8.34

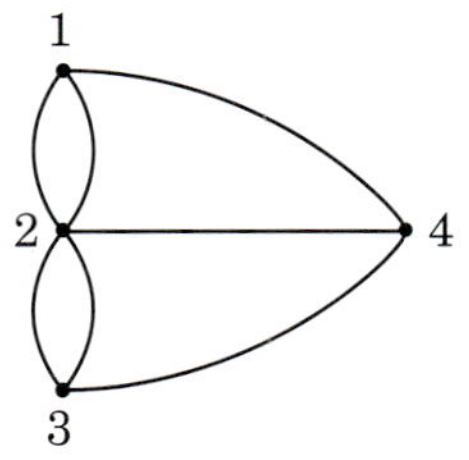

Figure 8.35

Adjacency list representation.

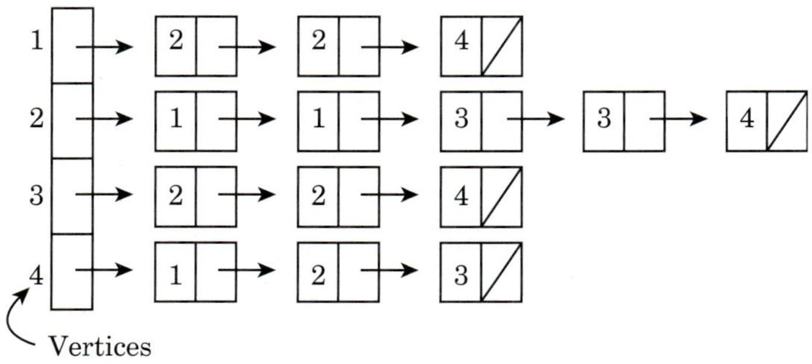

In the linked list representation of a weighted graph, each node contains three fields, as shown in Figure 8.36. For example, the weighted graph in Figure 8.37 yields the adjacency list representation in Figure 8.38.

Figure 8.36

A typical node in a weighted graph.

vertex	weight	pointer

Figure 8.37

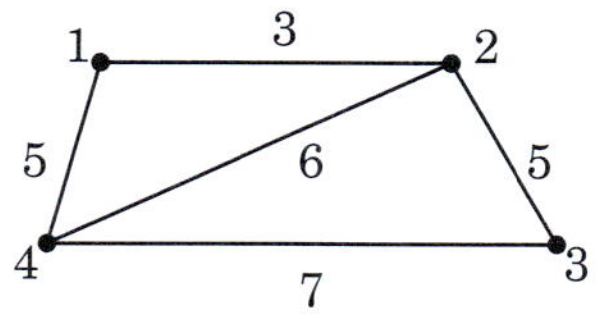

Figure 8.38

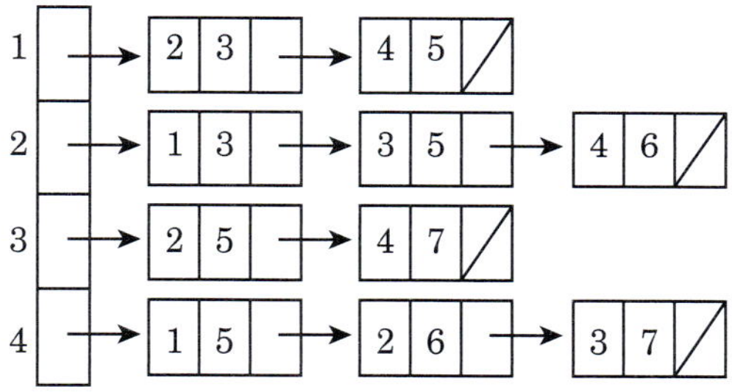

Exercises 8.2

Find the adjacency list representation of each graph.

1.

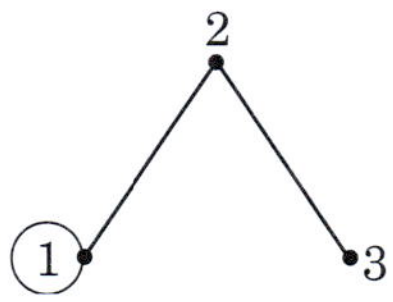

2.

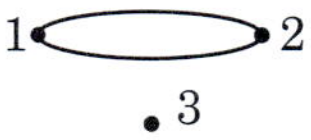

3.

4.

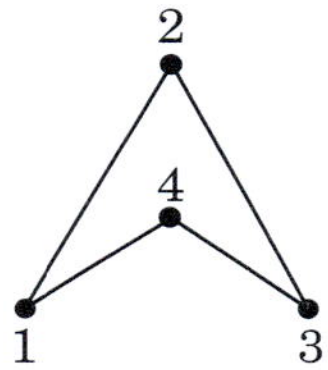

5.

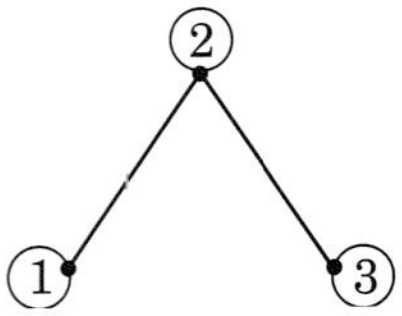

6.

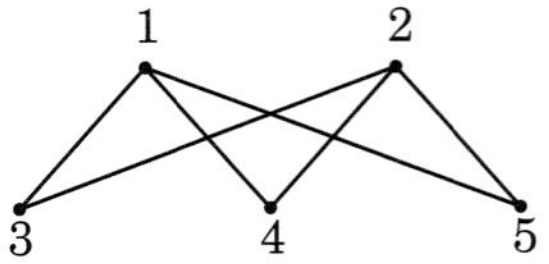

Find the adjacency list representation of each graph.

7. K_3 **8.** K_4 **9.** K_3' **10.** $K_{2,3'}$

11–16. Find the adjacency matrices of the graphs in Exercises 1-6.

Draw the graph with the given list representation.

17.

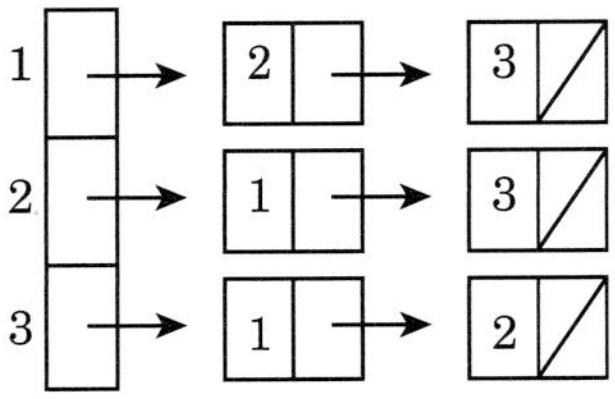

18.

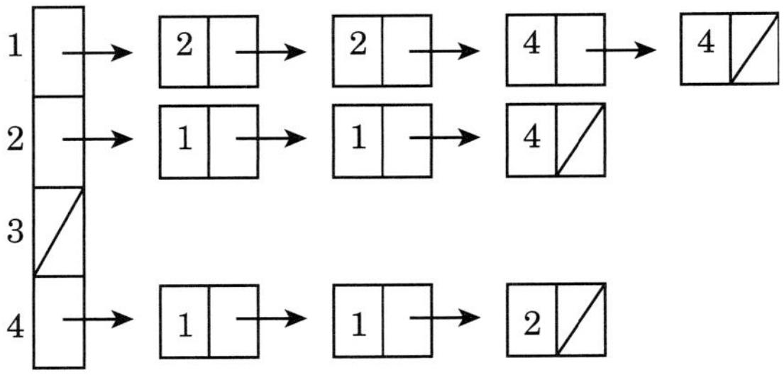

Find the adjacency list representation of the given weighted graph.

19.

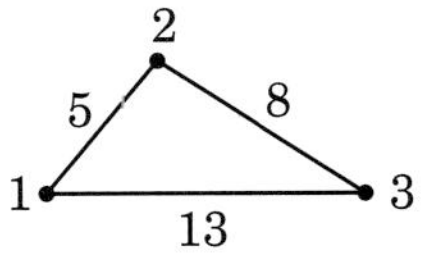

20.

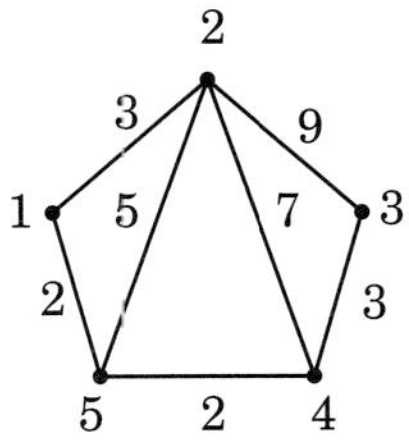

21.

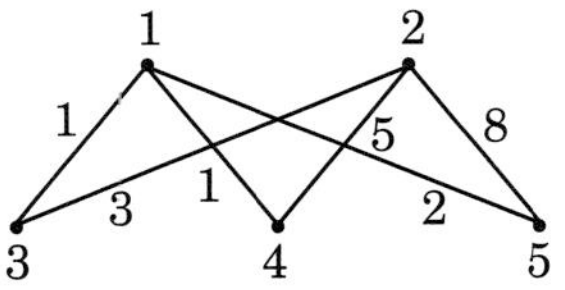

8.3 Isomorphic Graphs

Two simple graphs can have a closer relationship than just being complements. Isomorphic graphs, which we will study in this section, share identical properties.

Some graphs may seem to differ, but have essentially the same properties. Graphs G_1 and G_2 in Figure 8.39, for example, although they look different, have the same properties: both contain the same number of vertices and edges. Two vertices of degree two and two of degree three appear in both. Besides, we can redraw G_2 in such a way that it will look exactly like G_1. Where you place the vertices and how you draw the edges do *not* affect the structure of a graph and do not produce a different graph.

Figure 8.39

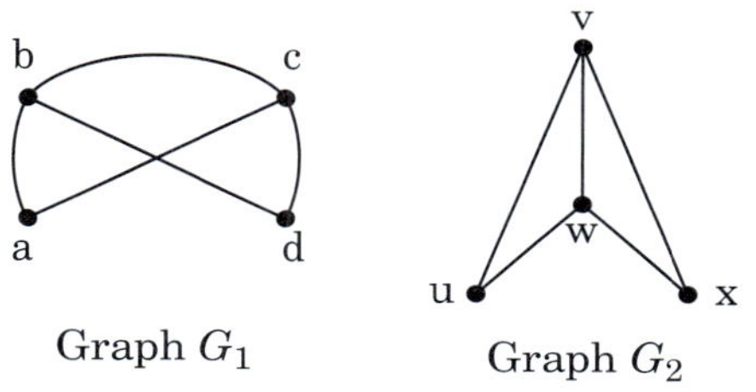

Isomorphic Graphs

Two simple graphs, $G_1 = (V_1, E_1)$ and $G_2 = (V_2, E_2)$, are **isomorphic** if a bijection $f : V_1 \to V_2$ exists such that $\{a, b\}$ is an edge in E_1 if and only if $\{f(a), f(b)\}$ is an edge in E_2, for any two elements a and b in V_1. The function f is an **isomorphism*** between G_1 and G_2.

For two graphs G_1 and G_2 to be isomorphic, the following conditions must be satisfied:

- $|V_1| = |V_2|$
- $|E_1| = |E_2|$
- A bijection $f : V_1 \to V_2$ should preserve the adjacency relationship: if $\{a, b\}$ is an edge in E_1 then $\{f(a), f(b)\}$ must be an edge in E_2, and vice versa. Consequently, the corresponding vertices in G_1 and G_2 will have the same degree.

EXAMPLE 8.12 The graphs $G_1 = (V_1, E_1)$ and $G_2 = (V_2, E_2)$ in Figure 8.39 clearly satisfy conditions 1 and 2. Define a function $f : V_1 \to V_2$ as follows: $f(a) = u$, $f(b) = v$, $f(c) = w$, and $f(d) = x$. Clearly, f is bijective. Does f preserve the

*The word ***isomorphism*** comes from the Greek words ***iso*** (the same as) and ***morphe*** (form).

adjacency relationship? {a, b} is an edge in E_1, so is $\{f(a), f(b)\} = \{u, v\}$ an edge in E_2? Yes. Similarly, Table 8.3 verifies the other correspondences of edges. Thus G_1 and G_2 are isomorphic.

Table 8.3

Edge $\{x, y\}$ in E_1	$\{f(x), f(y)\}$	**Is $\{f(x), f(y)\}$ an edge in E_2?**
{a, b}	$\{f(a), f(b)\} = \{u, v\}$	Yes
{a, c}	$\{f(a), f(c)\} = \{u, w\}$	Yes
{b, c}	$\{f(b), f(c)\} = \{v, w\}$	Yes
{b, d}	$\{f(b), f(d)\} = \{v, x\}$	Yes
{c, d}	$\{f(c), f(d)\} = \{w, x\}$	Yes

■

If two graphs are isomorphic, one can be obtained from the other by renaming its vertices (and edges) and redrawing it if necessary. In any case, they share exactly the same structure and hence the same properties. For instance, both have the same number of vertices and edges. Besides, $\deg(v) = \deg(f(v))$ for every $v \in V_1$. For the isomorphic graphs in Figure 8.39, $\deg(a) = 2 = \deg(f(a))$, $\deg(b) = 3 = \deg(f(b))$, $\deg(c) = 3 = \deg(f(c))$, and $\deg(d) = 2 = \deg(f(d))$. Such a common property is an **isomorphism invariant**.

Isomorphism Invariant

A property shared by isomorphic graphs is called an **isomorphism invariant**.

EXAMPLE 8.13 Are the pentagon $G_1 = (V_1, E_1)$ and the pentagram $G_2 = (V_2, E_2)$ in Figure 8.40 isomorphic?

Figure 8.40

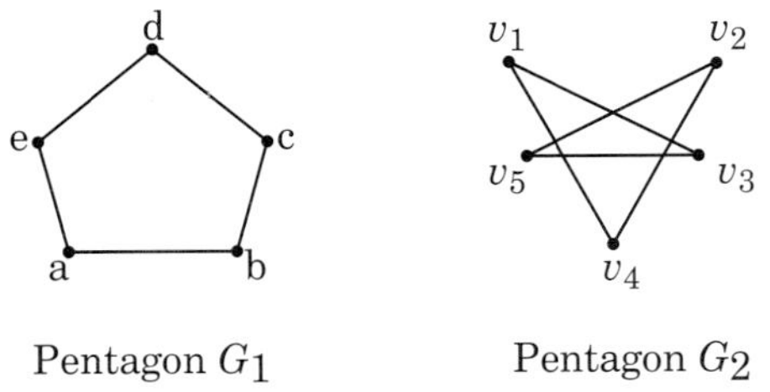

SOLUTION:
First $|V_1| = 5 = |V_2|$ and $|E_1| = 5 = |E_2|$. Define a function $f : V_1 \to V_2$ as $f(a) = v_1$, $f(b) = v_3$, $f(c) = v_5$, $f(d) = v_2$, and $f(e) = v_4$. It is certainly a bijection. To determine if f preserves the adjacency relationship, construct Table 8.4. The table shows that f preserves the adjacency relationship; so G_1 and G_2 are isomorphic.

Table 8.4

Edge $\{x, y\}$ in E_1	$\{f(x), f(y)\}$	Is $\{f(x), f(y)\}$ an edge in E_2?
{a, b}	$\{f(a), f(b)\} = \{v_1, v_3\}$	Yes
{b, c}	$\{f(b), f(c)\} = \{v_3, v_5\}$	Yes
{c, d}	$\{f(c), f(d)\} = \{v_5, v_2\}$	Yes
{d, e}	$\{f(d), f(e)\} = \{v_2, v_4\}$	Yes
{e, a}	$\{f(e), f(a)\} = \{v_4, v_1\}$	Yes

■

Isomorphism invariants can detect if two graphs $G_1 = (V_1, E_1)$ and $G_2 = (V_2, E_2)$ are not isomorphic. If $|V_1| \neq |V_2|$ or $|E_1| \neq |E_2|$, then G_1 and G_2 are not isomorphic. For example, the graphs in Figure 8.41 are *not* isomorphic (Why?).

Figure 8.41

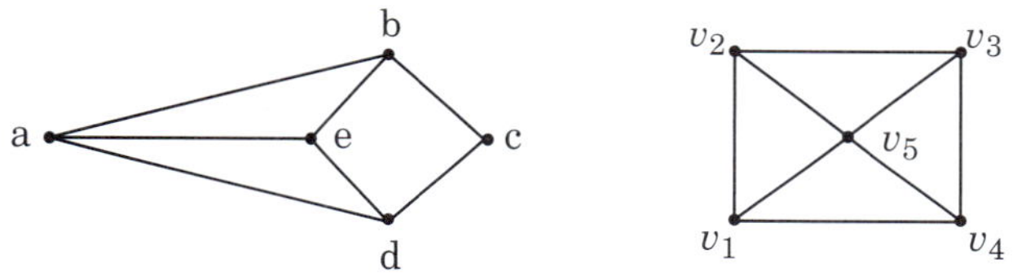

On the other hand, suppose two graphs share the same invariants. Unfortunately, this does not guarantee isomorphism, as the next example demonstrates.

EXAMPLE 8.14 Determine if graphs G_1 and G_2 in Figure 8.42 are isomorphic.

Figure 8.42

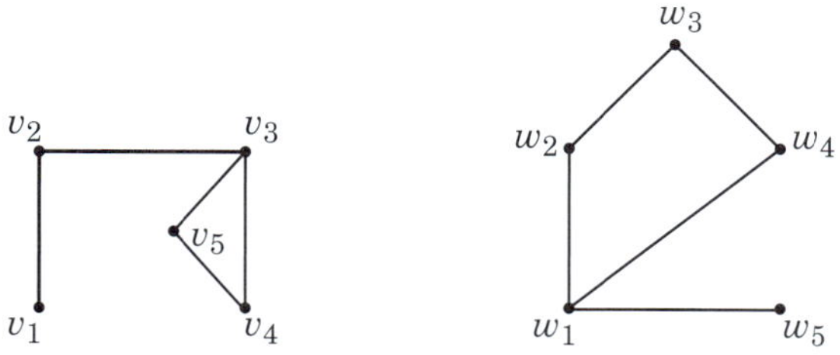

Graph G_1 Graph G_2

SOLUTION:

Both graphs have five vertices and five edges. Each contains one vertex of degree one, three vertices of degree two, and one vertex of degree three. Nonetheless, they are not isomorphic. Let us see why.

Suppose they are isomorphic. Since there is exactly one vertex of degree one in G_1 and G_2, namely, v_1 and w_5, v_1 must match w_5. Now v_1 is adjacent to v_2 and degree $(v_2) = 2$, so the vertex w_1 adjacent to w_5 must correspond to v_2. But deg $(w_1) = 3 \neq$ deg (v_2), which is a contradiction. Thus, G_1 and G_2 are *not* isomorphic. ■

> *Conclusion*? If the invariants of two graphs do not agree, they cannot be isomorphic. If they do agree, however, the graphs need not be isomorphic. Unfortunately, no simple tests exist for determining graph isomorphism.

Exercises 8.3

In Exercises 1–14, determine if the simple graphs are isomorphic. When they are, determine an isomorphism *f*.

1.

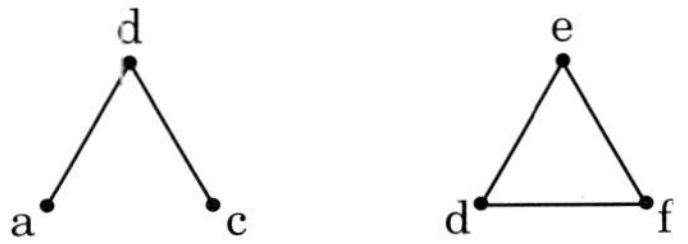

2.

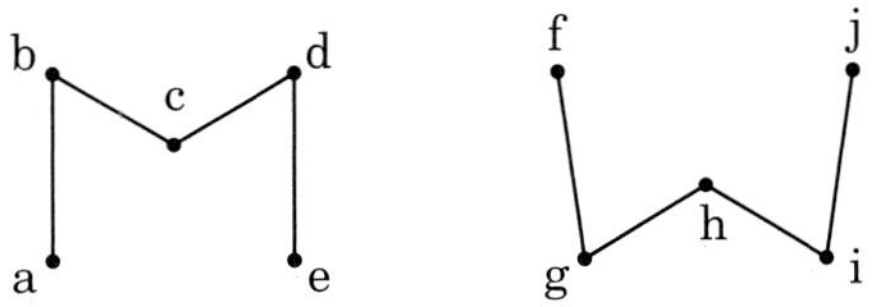

3.

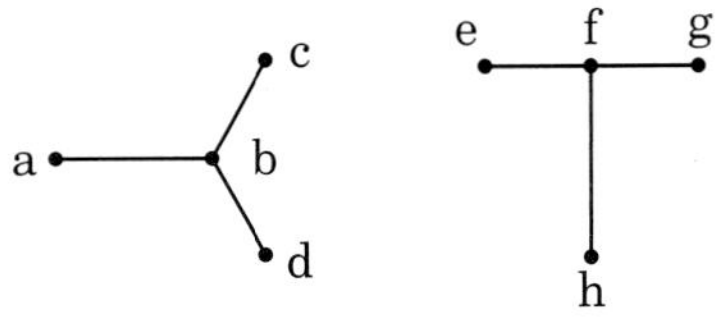

4.

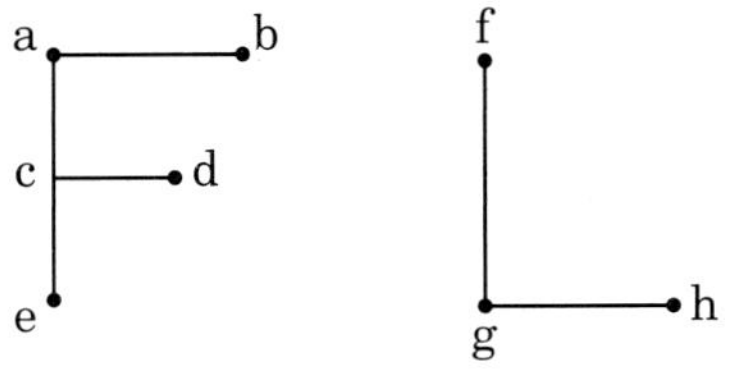

5.

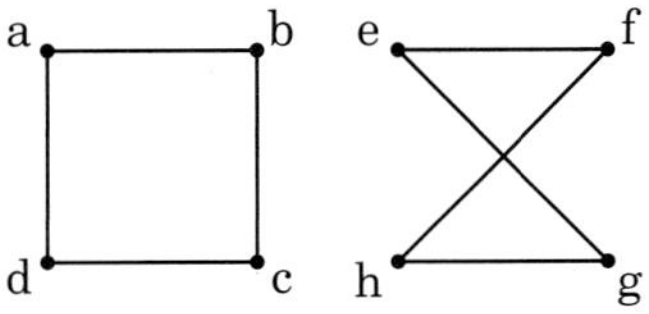

6.

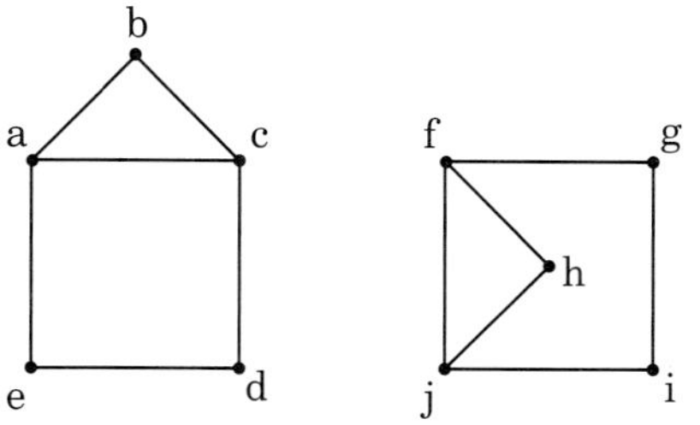
b
a
c
e
d
f
g
h
j
i

7.

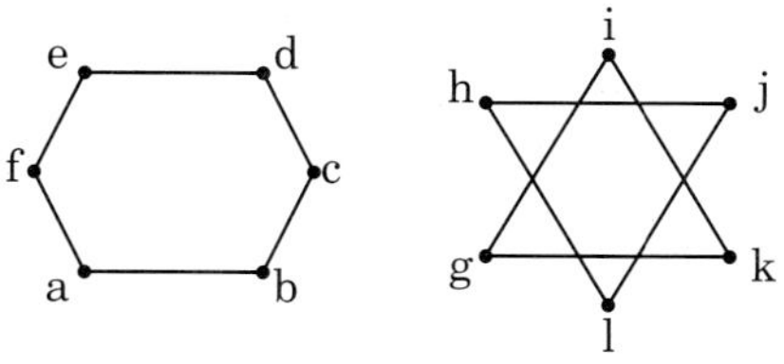
e
d
f
c
a
b
i
h
j
g
k
l

8.

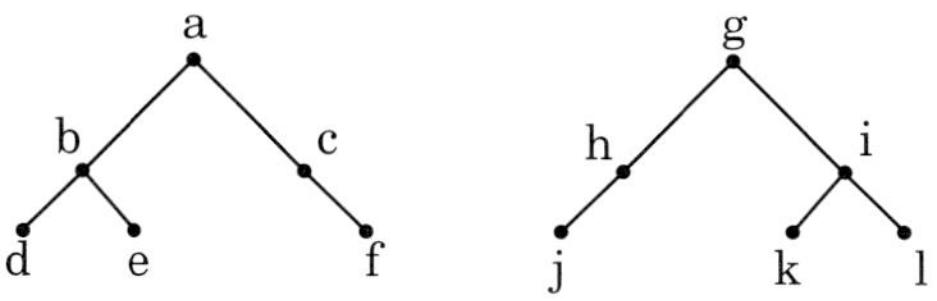
a
b
c
d
e
f
g
h
i
j
k
l

9.

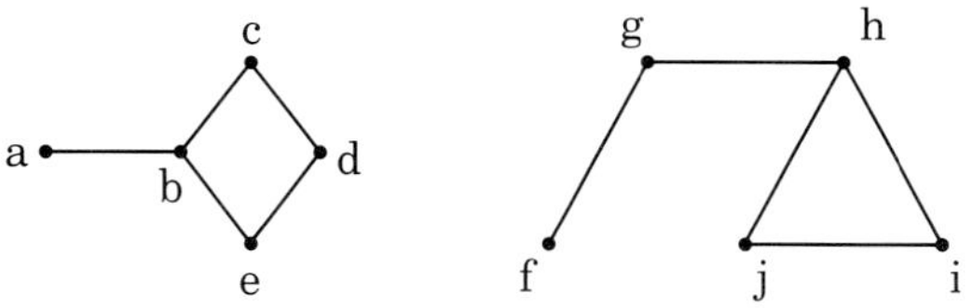
c
a
b
d
e
g
h
f
j
i

10.

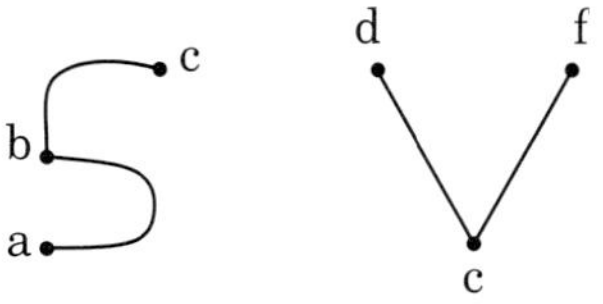
c
b
a
d
f
c

11.

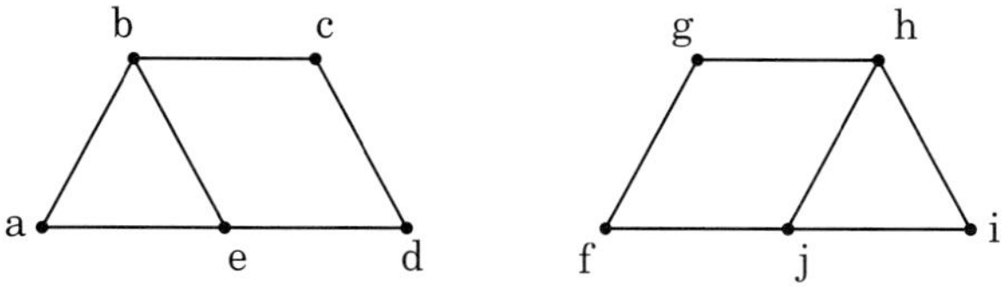
b
c
a
e
d
g
h
f
j
i

12.

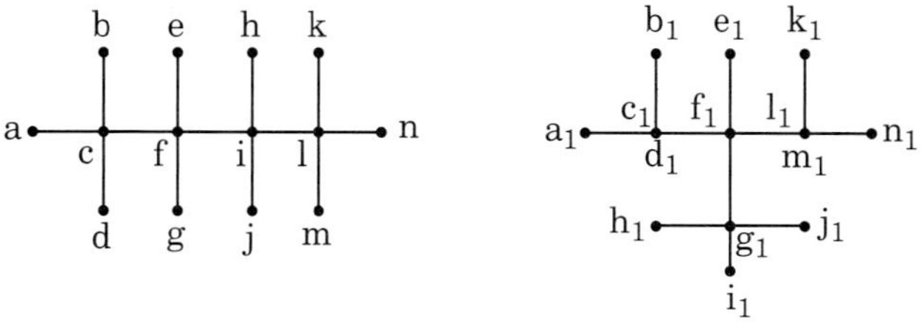
b
e
h
k
a
n
c
f
i
l
d
g
j
m
b1
e1
k1
c1
f1
l1
a1
n1
d1
m1
h1
j1
g1
i1

13.

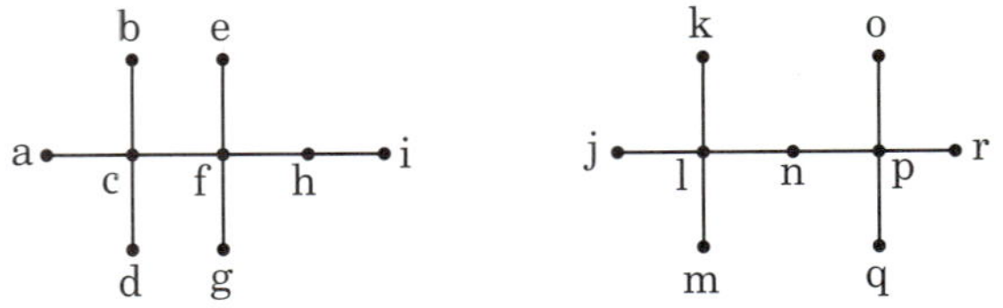

14.

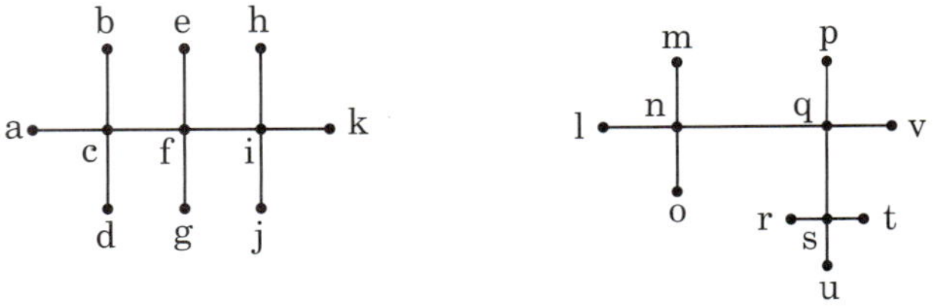

15. Show that isomorphism of simple graphs with n vertices is an equivalence relation.

8.4 Paths, Cycles, and Circuits

Reexamine the network of computers in Figure 8.3. The computer in Atlanta can communicate with that in Holtsville, and the computer in Holtsville can communicate with that in Philadelphia, so the computer in Atlanta can communicate with that in Philadelphia: in the graph model, a path runs from vertex 1 to vertex 3.

Paths, with their subclasses of cycles and circuits, carry great importance in graph theory since they can answer many questions about models of real-world situations, so we begin this section with the definition of a path.

Path

Let v_0 and v_n be two vertices in a graph. A **path of length** n from v_0 to v_n is a sequence of vertices v_i and edges e_i of the form v_0-e_1-v_1-e_2-$\cdots$-e_n-v_n where each edge e_i is incident with the vertices v_{i-1} and v_i, $1 \le i \le n$. The vertices v_0 and v_n are the **endpoints** of the path. If the graph is simple, the path is unique and is denoted by just listing the vertices along the path: v_0-v_1-$\cdots$-v_n. A **simple** path from v_0 to v_n contains no repeated vertices, with one possible exception: its endpoints could be the same.

The next example illustrates these definitions.

EXAMPLE 8.15 In the graph in Figure 8.43, the sequence a-e_1-b-e_4-c-e_5-b-e_3-d is a path of length 4 from a to d; so is the sequence a-e_1-b-e_5-c-e_4-b-e_3-d. Thus the path from a to d is not unique. The path a-e_1-b-e_3-d is even shorter, with a length of only two. (However, it is not the shortest path.) The path a-e_7-d-e_3-b is simple, because no one vertex reappears.

Figure 8.43

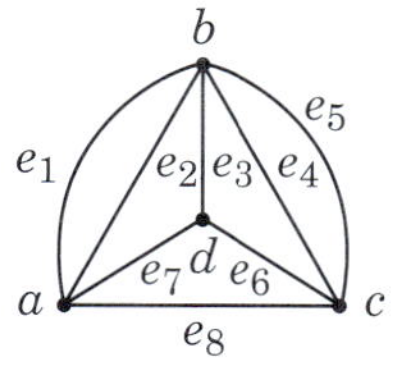

■

EXAMPLE 8.16 In Figure 8.44, the sequence a-e_1-b-e_3-c-e_4-d is a path of length three from a to d. Since the graph is simple, it can also be given by the

Figure 8.44

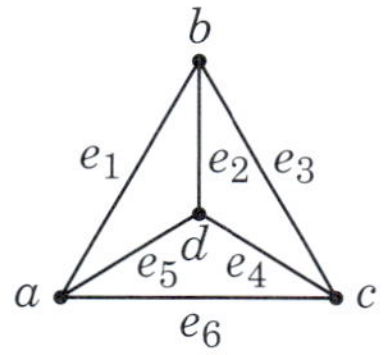

sequence a-b-c-d. (Notice that, in Figure 8.43, the sequence a-b-c-d does not define a unique path.) ■

Before we present yet another delightful occurrence of Fibonacci numbers, we make a simple definition.

Independent Subset of the Vertex Set

Let V denote the set of vertices of a graph. A subset S of V is **independent** if no two vertices in S are adjacent. Thus, S is independent if $x, y \in S$, but the edge x-y does not exist, where $x \neq y$.

For example, consider the pentagon G_1 in Figure 8.40. Then $S = \{\text{a}, \text{c}\}$ is an independent subset of the vertex set $V = \{\text{a}, \text{b}, \text{c}, \text{d}, \text{e}\}$; so is $\{\text{a}, \text{d}\}$, but not $\{\text{a}, \text{c}, \text{e}\}$.

We are now ready to present the application.*

EXAMPLE 8.17 Let P_n denote the path v_0-v_1-v_2-$\cdots$ -v_n of length n connecting the vertices v_0, v_1, $v_2, \ldots,$ and v_n in a simple graph, where $n \geq 0$. Let A_n denote the number of independent subsets of vertices in the path.

When $n = 0$, the path P_0 consists of a single point v_0, so there are two possible independent subsets: $\emptyset$, $\{v_0\}$.

When $n = 1$, the path is v_0-v_1. Then there are three independent subsets of $\{v_0, v_1\}$, namely, $\emptyset$, $\{v_0\}$, and $\{v_1\}$.

When $n = 2$, the path P_2 contains three vertices: v_0, v_1, and v_2. So, there are five independent subsets: $\emptyset$, $\{v_0\}, \{v_1\}, \{v_2\}$, and $\{v_0, v_2\}$.

*T. Koshy, *Fibonacci and Lucas Numbers with Applications*, Wiley, New York, 2001.

These data are summarized in Table 8.5. Clearly, a pattern emerges. It seems safe to conjecture that $A_n = F_{n+3}$, where $n \geq 0$. We invite you to confirm this.

Table 8.5

n	Path P_n	Independent subsets	A_n
0	v_0	$\emptyset$, $\{v_0\}$	2
1	v_0 — v_1	$\emptyset$, $\{v_0\}, \{v_1\}$	3
2	v_0 — v_1 — v_2	$\emptyset$, $\{v_0\}, \{v_1\}, \{v_2\}, \{v_0, v_2\}$	5
3	v_0 — v_1 — v_2 — v_3	$\emptyset$, $\{v_0\}, \{v_1\}, \{v_2\}, \{v_3\}$ $\{v_0, v_2\}, \{v_0, v_3\}, \{v_1, v_3\}$	8

↑ F_{n+3} ■

A simple path may repeat vertices if they be its endpoints. Such a phenomenon carries a special label.

Cycle and Circuit

A path with endpoints v_0 and v_n is **closed** if $v_0 = v_n$; otherwise, it is **open**. A simple closed path is a **cycle**; a closed path with no repeated edges is a **circuit**.

Table 8.6 summarizes and illustrates the basic terms introduced thus far in this section. Refer to it as often as needed.

Table 8.6

Term	Meaning	Example from Figure 8.43
Path	Sequence v_0-e_1-v_1-$\cdots$-e_n-v_n, where $e_i = \{v_{i-1}, v_i\}, 1 \leq i \leq n$	a-e_7-d-e_6-c-e_4-b-e_5-c
Simple path	All vertices are distinct; endpoints could be the same	a-e_7-d-e_6-c-e_4-b
Closed path	Endpoints are the same	a-e_2-b-e_4-c-e_5-b-d-e_7-a
Open path	Endpoints are not the same	a-e_8-c-e_4-b-e_5-c
Cycle	Simple closed path	a-e_1-b-e_4-c-e_8-a
Circuit	Closed path; no repeated edges	a-e_1-b-e_4-c-e_3-b-e_3-d-e_7-a

In a network of computers, every computer could communicate with every other computer either directly or indirectly. If a path runs from every vertex (computer) to every other vertex (computer), such a graph is said to be connected, as defined below.

Connected Graph

A graph is **connected** if there is a path between every two distinct vertices of the graph; otherwise, it is **disconnected**.

EXAMPLE 8.18 The Königsberg bridge model in Figure 8.2 is a connected graph, as are the graphs in Figures 8.43 and 8.44; but the *Star of David* in Figure 8.11 is not (Why?). ■

The next theorem shows a simple path exists between every two vertices in a connected graph.

THEOREM 8.3 There is a simple path between any two distinct vertices in a connected graph.

PROOF:
Let u and v be any two vertices in a connected graph G. Since G is connected, a path x_0-e_1-x_1-e_2-$\cdots$-x_{n-1}-e_n-x_n runs between $x_0 = \text{u}$ and $x_n = \text{v}$. If the path is not simple, some of the vertices $x_0, x_1, \ldots, x_n$ must be the same, say, for example, $x_i = x_j$, where $i < j$. Consequently, there must be a cycle x_i-e_{i+1}-x_{i+1}-$\cdots$-e_j-x_j, as in Figure 8.45. Eliminate this cycle to yield a shorter path.

Figure 8.45

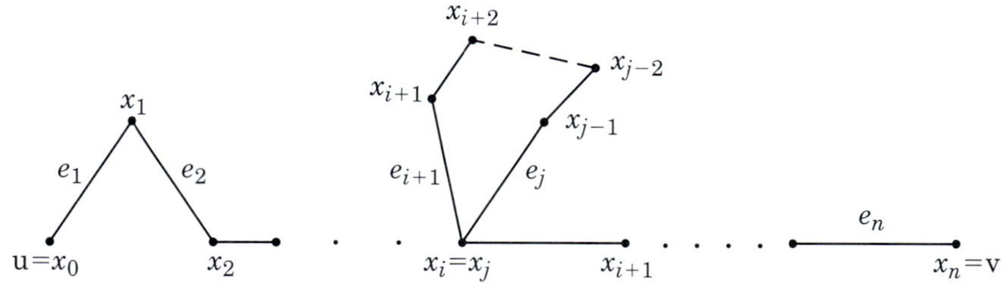

Similarly, eliminating all cycles in the path yields a shorter path from u to v. It contains no repeated vertices and therefore is simple. This completes the proof. ■

How long can a simple path between any two distinct vertices of a connected graph be? The next theorem provides a partial answer.

THEOREM 8.4 The length of a simple path between any two distinct vertices of a connected graph with n vertices is at most $n - 1$. ■

How can we compute the number of paths between any two vertices v_i and v_j in a connected graph? Theorem 8.5 answers this.

THEOREM 8.5 Let A be the adjacency matrix of a connected graph with n vertices $v_1, v_2, \ldots, v_n$ and k a positive integer $\leq n - 1$. The ijth entry of the matrix A^k gives the number of paths of length k from v_i to v_j.

PROOF (by induction):

Basis step By definition, the ijth entry in A is the number of edges from v_i to v_j. But an edge from v_i to v_j is a path of length one. Consequently, the ijth entry in A equals the number of paths of length 1 from v_i to v_j. Thus the result holds when $k = 1$.

Induction step Assume that the number of paths of length k from v_i to v_j equals the ijth entry in A^k (inductive hypothesis). We have $A^{k+1} = A^k \cdot A$. Let $A^{k+1} = (c_{ij}), A^k = (b_{ij})$, and $A = (a_{ij})$. Then $c_{ij} = \sum_{p=1}^{n} b_{ip}a_{pj}$. By the inductive hypothesis, b_{ip} paths of length k run from v_i to v_p; but there are a_{pj} paths of length 1 from v_p to v_j. Therefore, by the multiplication principle, $b_{ip}a_{pj}$ paths of length $k+1$ from v_i to v_j pass through v_p. Now v_p can be any one of the n vertices. Thus, by the addition principle, the total number of paths of length $k+1$ from v_i to v_j (passing through $v_1, v_2, \ldots,$ or v_n) is $\sum_{p=1}^{n} b_{ip}a_{pj}$, which equals c_{ij}. Thus, by induction, the result holds for every positive integer $\leq (n-1)$. ■

The following example demonstrates this theorem.

EXAMPLE 8.19 Figure 8.46 shows the direct telephone lines connecting cities a, b, c, and d. Each edge in the graph represents a direct telephone link.

Figure 8.46

A communication model.

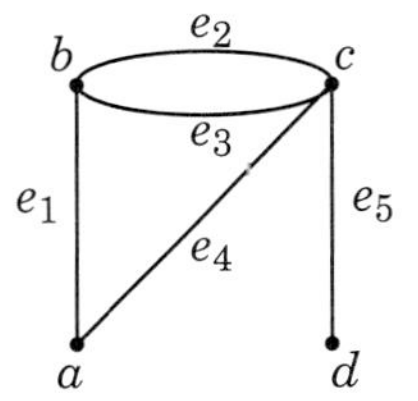

The adjacency matrix of the graph is

$$A = \begin{array}{c} \\ a \\ b \\ c \\ d \end{array} \begin{array}{c} \begin{matrix} a & b & c & d \end{matrix} \\ \begin{bmatrix} 0 & 1 & 1 & 0 \\ 1 & 0 & 2 & 0 \\ 1 & 2 & 0 & 1 \\ 0 & 0 & 1 & 0 \end{bmatrix} \end{array}$$

No direct lines run between a and d. However, a can communicate with d via c. This communication line, called a **2-stage communication link**, is the path a-e_4-c-e_5-d, a link between a and d passing through exactly one city, namely, c. Likewise, two 2-stage communication links run between b and d: b-e_2-c-e_5-d and b-e_3-c-e_5-d. By Theorem 8.5, the number of 2-stage links, lines passing through exactly one city, are the various entries in A^2:

$$A^2 = \begin{array}{c} \\ a \\ b \\ c \\ d \end{array} \begin{array}{c} \begin{array}{cccc} a & b & c & d \end{array} \\ \begin{bmatrix} 2 & 2 & 2 & 1 \\ 2 & 5 & 1 & 2 \\ 2 & 1 & 6 & 0 \\ 1 & 2 & 0 & 1 \end{bmatrix} \end{array}$$

There are five 2-stage links between city b and itself: b-e_1-a-e_1-b, b-e_2-c-e_2-b, b-e_2-c-e_3-b, b-e_3-c-e_2-b, and b-e_3-c-e_3-b. No 2-stage links connect c and d. ■

Theorems 8.4 and 8.5 lead us to the next result.

THEOREM 8.6 Let A be the adjacency matrix of a connected graph with n vertices and k a positive integer $\leq n-1$. The ijth entry of the matrix $A + A^2 + \cdots + A^k$ gives the number of paths of length $\leq k$ from vertex v_i to vertex v_j. ■

As an example, for the communication graph in Figure 8.46,

$$A + A^2 = \begin{array}{c} \\ a \\ b \\ c \\ d \end{array} \begin{array}{c} \begin{array}{cccc} a & b & c & d \end{array} \\ \begin{bmatrix} 2 & 3 & 3 & 1 \\ 3 & 5 & 3 & 2 \\ 3 & 3 & 6 & 1 \\ 1 & 2 & 1 & 1 \end{bmatrix} \end{array}$$

The entries of this matrix give the number of 1- or 2-stage communication links between two cities, links passing through *at most one* city. For instance, three links between cities b and c go through at most one city: b-e_2-c, b-e_3-c, and b-e_1-a-e_4-c. ■

The next example relates graph theory to combinatorics.

○ **EXAMPLE 8.20** (optional*) A **prime circle of order** n is a cyclic permutation of the integers 1 through n such that the sum of any two adjacent integers is a prime number. For instance, 143256 is a prime circle of order 6. Find all prime circles of order $n \leq 10$.

SOLUTION:

Since the sum of any two even or odd integers is even, the numbers in prime circles must alternate between odd and even. Consequently, there are no prime circles of odd order.

To find all prime circles of (even) order $n \leq 10$, draw a graph G with 10 vertices, labeled 1 through 10. Two vertices v and w in G are adjacent if

*Proposed by A. Filz and solved by B. Barwell, J. *Recreational Mathematics*, Vol. 15, 1982–1983, pp. 70–71.

$v + w$ is a prime. Such a graph is shown in Figure 8.47. To find the various prime circles of order n from the graph, delete vertices $n + 1, n + 2, \ldots, 10$ and edges incident with them. Now look for all distinct cycles that pass through the integers 1 through n in the resulting subgraph; the vertices along them form prime circles.

Figure 8.47

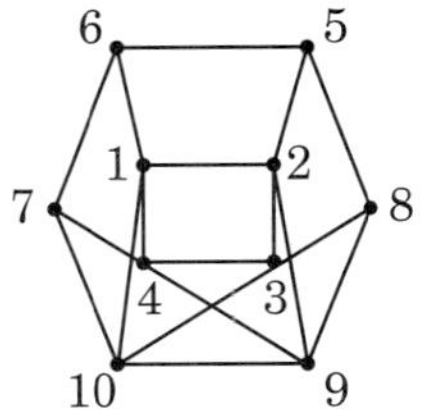

For example, to find the prime circles of order 8, delete vertices 9 and 10, and edges incident with them. The resulting subgraph in Figure 8.48 shows two distinct prime circles of order 8, namely, 12385674 and 12583476. (Without loss of generality, we list the permutations beginning with 1.)

Figure 8.48

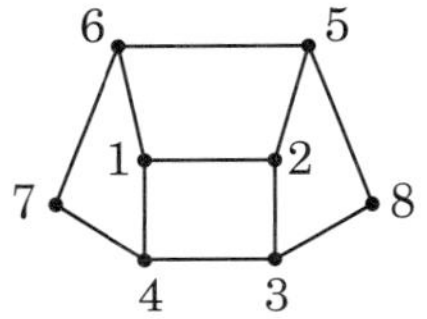

You may verify that there is exactly one prime circle of order two, one of order four, and one of order six. They are 12, 1234, and 143256, respectively.

There are 48 prime circles of order 10. Four of them are listed below, where X denotes 10:

123456789X 12347X9856 123498567X 123856749X

Can you find the others? ■

We conclude this section with two examples illustrating how useful graphs and their paths can be in solving familiar and interesting puzzles.

EXAMPLE 8.21 **(The cabbage–goat–wolf puzzle)** A farmer with a rowboat needs to transport a cabbage, a goat, and a wolf across a river. The rowboat has just enough room for him and either the cabbage, the goat, or the wolf. Since the wolf can eat the goat, they cannot be left alone in the absence of the farmer. Likewise, the goat and the cabbage also cannot be left alone. How can he transfer them across the river?

SOLUTION:

We shall represent the progress of a solution in a graph and update it at every step until a solution emerges. An edge indicates a transfer across the river.

Initially the cabbage (C) and the goat (G) or the goat and the wolf (W) cannot be left unattended by the farmer (F). The cabbage and the wolf, however, can be left alone. So the first step is to transfer the goat across the river. We represent this by an edge connecting the vertices "start" (S) and FG, meaning both the farmer and the goat must go across the river (see Figure 8.49). Now the man must return to the original side (see Figure 8.50) and has then two choices: transport the cabbage or the wolf to the other side (see Figure 8.51).

Figure 8.49

Figure 8.50

Figure 8.51

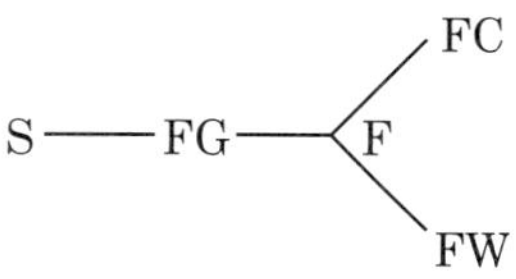

Case 1 Suppose the farmer moves the cabbage across the river. Then he cannot leave the cabbage and the goat on the same side, so he must return with the goat (see Figure 8.52). Continuing like this yields the graph in Figure 8.53.

Figure 8.52

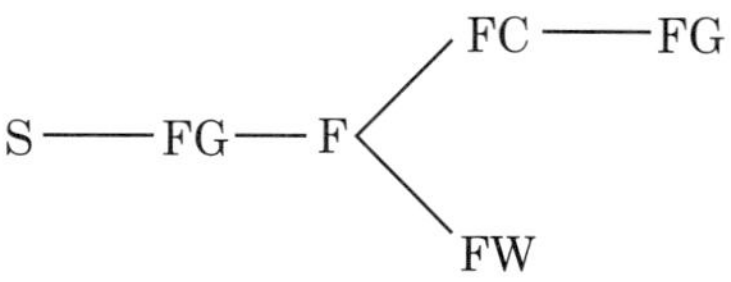

Figure 8.53

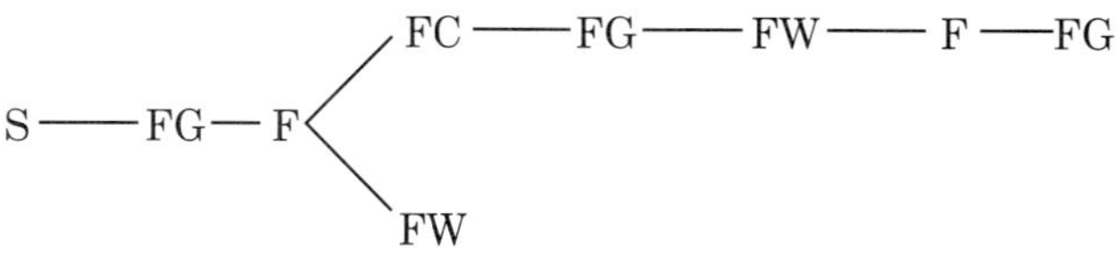

Case 2 Suppose the man transfers the wolf across the river. Continuing as in case 1 produces the graph in Figure 8.54.

Figure 8.54

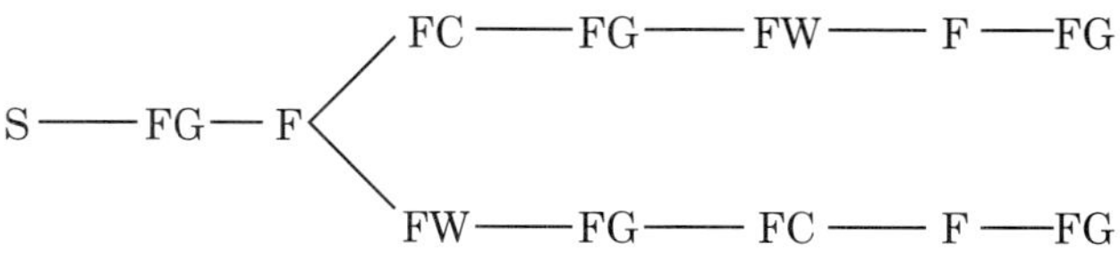

It follows from this graph that the puzzle has exactly two solutions, given by the two paths beginning at S. ■

EXAMPLE 8.22

Three married couples want to cross a river in a rowboat which can carry only two people at a time. No husband will allow his wife to be in the boat or stay ashore in the presence of another man unless he is also present. The women can, of course, row well. How can they cross the river?*

SOLUTION:

Let A, B, and C be the three husbands, and a, b, and c their wives, respectively. We shall denote the development of a solution by a graph in which each edge represents a strategy and each vertex indicates the person(s) crossing the river.

Since there are six people and only two can be in the boat at a time, there are $C(6, 2) = 15$ possibilities for the first move. Nine of them, however, are unacceptable: Ab, Ac, Ba, Bc, Ca, Cb, AB, AC, and BC. So the first acceptable strategy is Aa, Bb, Cc, ab, ac, or bc.

Figure 8.55 shows six of 12 possible solutions with ab first crossing the river. (*Note*: The path S-ab-a, for instance, indicates both a and b cross the river, b gets off the boat, and a returns to the original side.) With ac or bc making the first move, 24 additional solutions can be obtained. (Does a solution exist if initially a husband takes his wife across the river? See Exercise 36.) ■

Figure 8.55

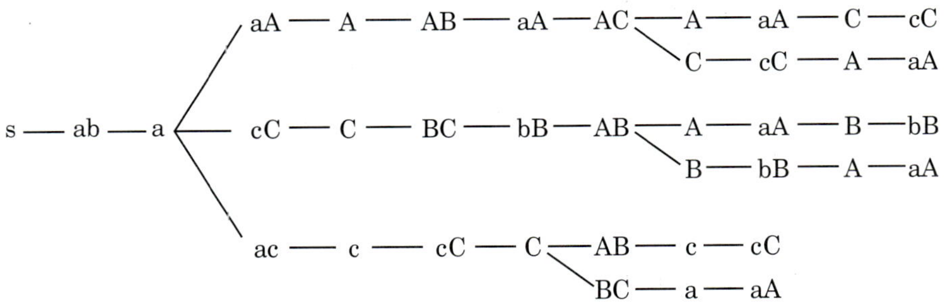

Exercises 8.4

Find the length of each path in the graph in Figure 8.43.

1. b-e_4-c-e_8-a

2. b-e_4-c-e_6-d-e_7-a

3. d-e_7-a-e_2-b-e_1-a-e_8-c

4. d-e_7-a-e_2-b-e_5-c-e_6-d

Use the graph in Figure 8.43 to find each.

5. The number of paths of length three from a to d.

*Based on S. Gudder, *A Mathematical Journey*, McGraw-Hill, New York, 1976, pp. 12,205.

6. All simple paths of length three from a to d.

7. The length of the shortest path from a to d.

8. All distinct cycles of length three beginning at a.

9. All distinct cycles of length three beginning at b.

10. All distinct circuits in the graph.

Determine if each is a connected graph.

11.

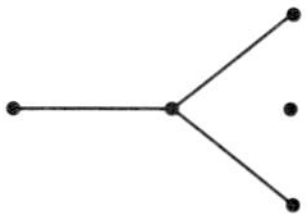

12.

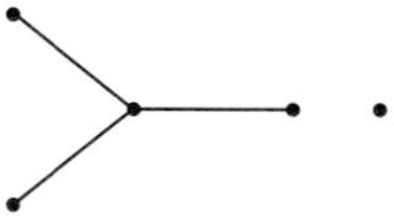

13.

14.

15.

16.

Use Figure 8.56 to answer Exercises 17 and 18.

Figure 8.56

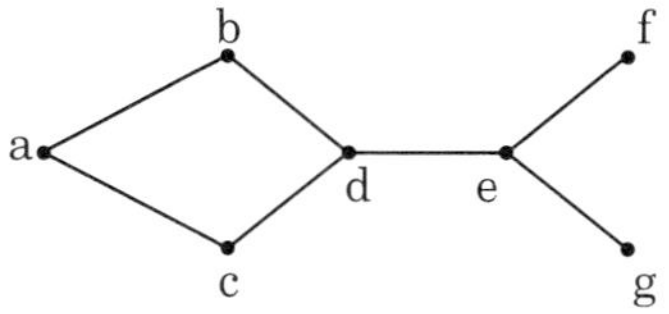

17. Find the length n of a longest open simple path.

18. Find all open simple paths of length n.

Find the number of distinct simple paths of length n in K_5, where n is:

19. 1 **20.** 2 **21.** 3 **22.** 4

23–26. Find the number of distinct cycles of length n in K_5 for each value in Exercises 19–22.

27–30. Find the number of distinct circuits of length n in K_5 for the values in Exercises 19–22.

The direct communication links between cities a, b, c, and d are represented in Figure 8.57. Find the number of communication links between a and d passing through:

Figure 8.57

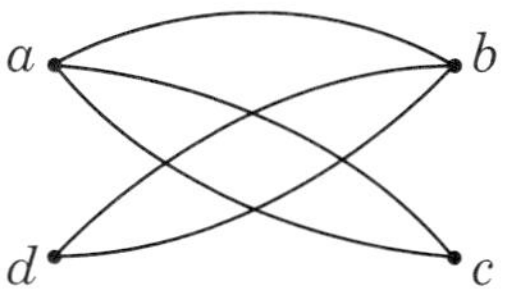

31. Exactly one city. **32.** At most one city.

33–34. Redo Exercises 31 and 32 for cities b and d.

35. Find two solutions to the puzzle in Example 8.22 if both a and b cross the river first.

36. Redo Exercise 35 if both A and a cross the river first.

37. A farmer has $2n + 1$ animals, $a_1, a_2, \ldots, a_{2n+1}$, and wishes to get them across a river in a boat that can hold only n animals at a time. Animals a_i and a_j cannot be left alone whenever $|i - j| = 1$. How can she accomplish this task?
(G. Gannon and M. Martelli, 1993)

***38.** Prove that a connected graph with n vertices has at least $n - 1$ edges. (*Hint*: Use induction.)

39. Using the adjacency matrix of a graph, write an algorithm to determine if it is connected.

8.5 Eulerian and Hamiltonian Graphs

The Königsberg bridge problem, considered earlier, raises two interesting questions:

- Can one walk through the city crossing each bridge exactly once?
- Can one walk through the city using each bridge exactly once and return home?

These two questions can be stated in terms of a connected graph:

- Does the Königsberg bridge model contain an open path that includes every edge exactly once?
- Does it contain a circuit that includes every edge?

We shall answer these questions a bit later, but first a few definitions.

Eulerian Graph

A path in a connected graph is an **Eulerian path** if it contains every edge exactly once. A circuit in a connected graph is an **Eulerian circuit** if it contains every edge of the graph. A connected graph with an Eulerian circuit is an **Eulerian graph**.

EXAMPLE 8.23 Study the graphs G_1, G_2, and G_3 in Figure 8.58. G_1 is Eulerian since it contains an Eulerian circuit, for example, *a-b-c-e-f-g-c-d-a*; in fact, it contains several Eulerian circuits. (Can you find another one?) G_2 has an Eulerian

Figure 8.58

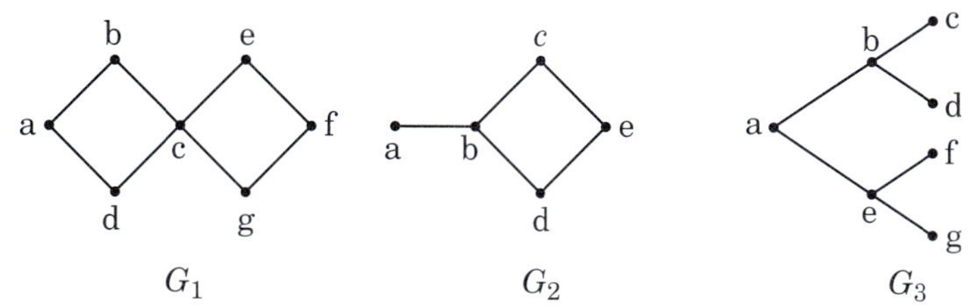

path, namely, a-b-c-e-d-b, but no Eulerian circuits. G_3 has no Eulerian paths or circuits. ■

Theorem 8.7 provides a necessary and sufficient condition for characterizing Eulerian graphs.

THEOREM 8.7 A connected graph G is Eulerian if and only if every vertex of G has even degree.

PROOF:

Suppose G is Eulerian. Then G contains an Eulerian circuit, say, from v_0 to v_0: v_0-e_1-v_1-e_2-$\cdots$-v_{n-1}-e_n-v_0. Both edges e_1 and e_n contribute a 1 to the degree of v_0; so deg (v_0) is at least two. Each time the circuit passes through a vertex (including v_0), the degree of the vertex is increased by 2. Consequently, the degree of every vertex, including v_0, is an even integer.

Conversely, suppose every vertex of G has even degree. Actually constructing an Eulerian circuit will prove G Eulerian. Let v_0 be an arbitrary vertex in G. Beginning with v_0, form a circuit $C_1 = v_0$-v_1-v_2-$\cdots$-v_{n-1}-v_0. (The longer the circuit, the smaller the number of steps involved.) This is possible since every vertex has even degree and a vertex ($\neq v_0$) can be left by an edge not used to enter it. If C_1 is Eulerian, G is too.

If C_1 is not Eulerian, consider the subgraph H obtained by deleting all the edges in C_1 and vertices *not* incident with the remaining edges. Note that all vertices of H have even degree. Since G is connected, H and C_1 must have a common vertex b. Beginning with b, construct a circuit C_2 for H.

Now combine C_1 and C_2 to form a larger circuit C. If it is Eulerian, then G is. If it is not, continue this procedure to form an Eulerian circuit. This procedure must terminate since the number of edges in G is finite. Thus G contains an Eulerian circuit and hence is Eulerian. ■

The next two examples illustrate this powerful theorem.

EXAMPLE 8.24 Recall that in Figure 8.58, G_1 is Eulerian. By Theorem 8.7, every vertex must have even degree: deg (a) = deg (b) = deg (d) = deg (e) = deg (g) = deg (f) = 2 and deg (c) = 4. Not every vertex in G_2 has even degree, so G_2, as expected, is not Eulerian. ■

EXAMPLE 8.25 In the Königsberg bridge puzzle, the degree of every vertex in Figure 8.2 is odd, so the graph is *not* Eulerian. Consequently, it is impossible to walk through the city crossing each bridge exactly once and return home. ■

Next we present a graph-theoretic solution to an ancient chess problem. The solution is ascribed to Dudeney, who knew little about graph theory and who called it the "buttons and string method." It exemplifies how valuable a tool a graph is in solving seemingly difficult problems. We find both beauty and elegance in the solution.*

EXAMPLE 8.26 Two white knights occupy the upper corner squares and two black knights occupy the lower corner squares of a 3×3 chessboard (see Figure 8.59). How can the white knights swap their places with the black knights in the smallest number of moves? (Try this yourself before studying the solution.)

Figure 8.59

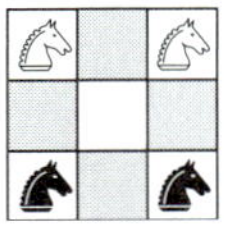

SOLUTION:

Knights move in an L-shaped pattern [a knight on a chessboard can move either two squares horizontally and one square vertically, or two squares vertically and one square horizontally, so a knight at position (i, j) has eight possible moves: $(i \pm 1, j \pm 2)$ and $(i \pm 2, j \pm 1)$. See Figure 8.60.

Figure 8.60

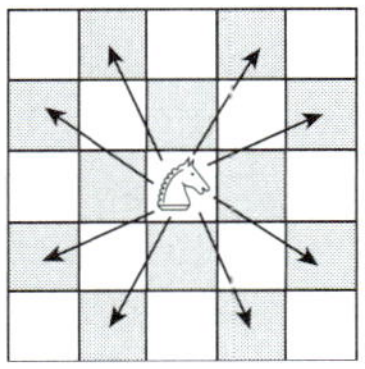

So *no* knight can move into the central square, and therefore we are interested only in the eight outside squares.

Number them 1 through 8, represent each by a vertex (button) and every possible move from square u to square v by an edge (string) from vertex u to vertex v. For example, the knight in square 1 can move to squares 5 and 7, so draw edges from vertex 1 to vertices 5 and 7; similarly, draw all other possible edges. Figure 8.61 shows the resulting graph.

Notice that every vertex has even degree, so the graph is Eulerian and 1-5-6-2-8-4-3-7-1 is an Eulerian circuit. As a result, the graph can be redrawn as (that is, it is isomorphic to) the one in Figure 8.62 that shows the puzzle has a solution (Why?).

*M. Gardner, *Mathematical Puzzles and Diversions*, University of Chicago Press, Chicago, 1987.

Figure 8.61

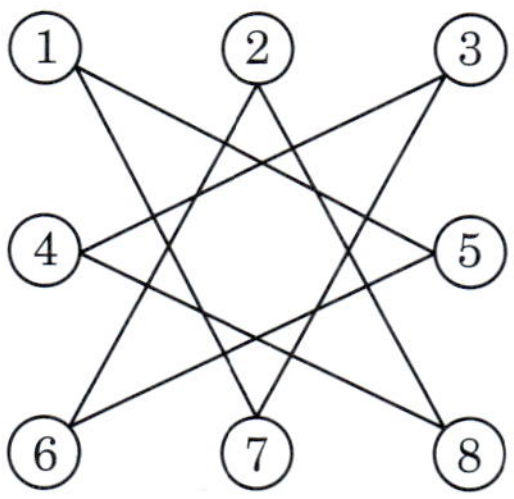

Figure 8.62

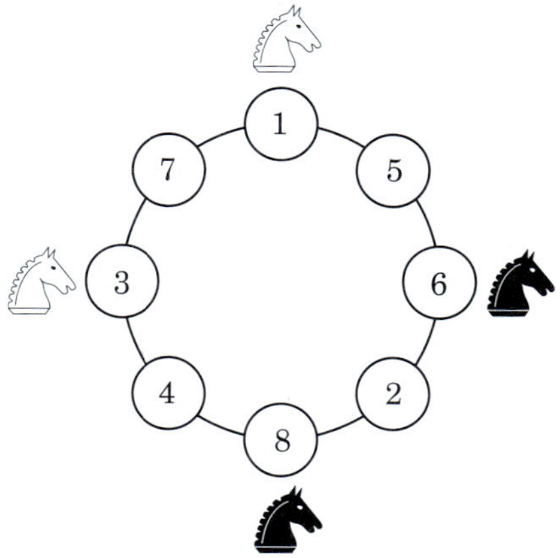

To find a solution, all you have to do now is move the knights to adjacent vertices cyclically (in either clockwise or counterclockwise direction) along the circuit until the black and white knights are swapped. (Try this, list all moves, and convince yourself.) ■

Using the adjacency matrix of a connected graph, Theorem 8.7 can produce an algorithm to determine whether or not the graph is Eulerian. It is given in Algorithm 8.1.

```
Algorithm Eulerian Graph(G,A)
(* Let G be a connected graph G with n vertices and adjacency matrix
   A = (aij)nxn. This algorithm, using Theorem 8.7, determines if G is
   Eulerian. Odd is a counter that keeps a count of the number of odd
   vertices in G.*)
Begin (* algorithm *)
  odd ← 0 (* initialize the counter *)
  for i = 1 to n do (* find each row sum *)
  begin (* for *)
    sum ← 0
    for j = 1 to n do (* compute deg(vi) *)
      sum ← sum + aij
    if sum is odd then (* update the counter *)
      odd ← odd + 1
  endfor
```

```
    if odd = 0 then
      graph is Eulerian
    else
      graph is not Eulerian
  End (* algorithm *)
```

Algorithm 8.1

The next example shows how an Eulerian circuit can be constructed in an Eulerian graph.

EXAMPLE 8.27 The vertices of the graph in Figure 8.63 represent the various cities a salesperson must visit, and the edges represent the various airline routes. Based at a, she would like to fly each route exactly once and return home. Find a sequence of routes she could take for the round trip. In other words, find an Eulerian circuit for the graph.

Figure 8.63

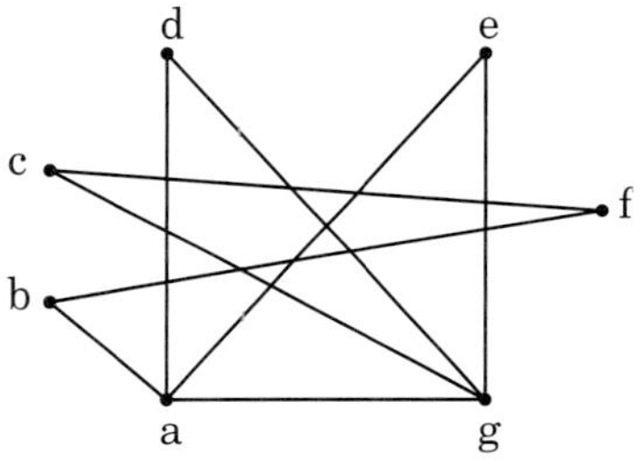

SOLUTION:

First, notice that the graph is Eulerian. To find an Eulerian circuit, use the construction method in the proof of Theorem 8.7.

Step 1

Find a circuit C_1 beginning with a, such as a-d-g-a, indicated by broken edges in Figure 8.64. C_1 is not Eulerian.

Figure 8.64

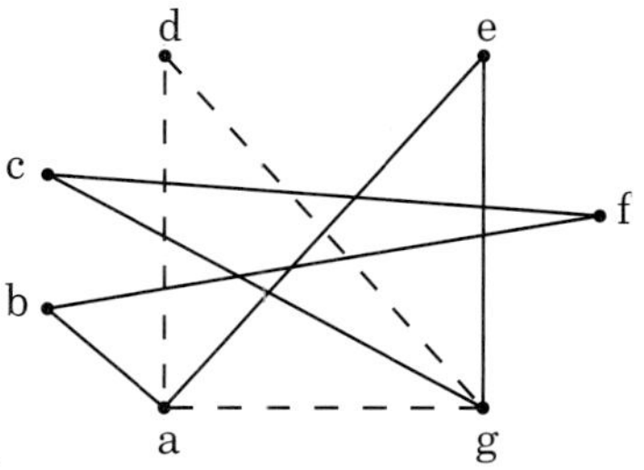

Step 2

Look for a vertex that lies on both C_1 and a solid edge. Vertices a and g are two such vertices. Beginning with one of them, say, g, build up a circuit C_2

using only solid edges: *g-e-a-b-f-c-g*. Indicate this also using broken edges, as in Figure 8.65.

Figure 8.65

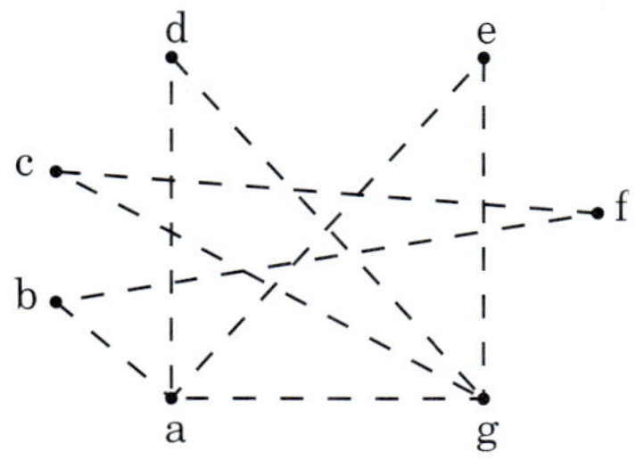

Figure 8.66

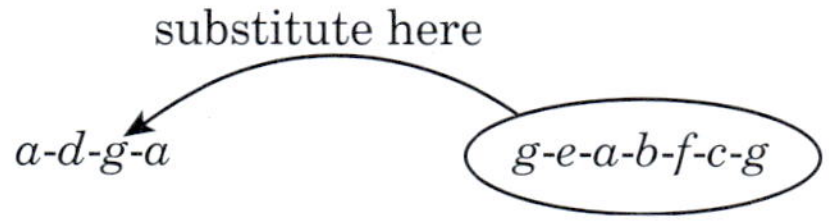

Since no more solid edges remain, the procedure stops here, with two circuits C_1 and C_2. Since C_2 begins with g, combine them as shown in Figure 8.66. The resulting is an Eulerian circuit, *a-d-g-e-a-b-f-c-g-a*, that the salesperson can fly and return home. ■

From Example 8.27 and the second half of the proof of Theorem 8.7, a simple algorithm evolves for finding an Eulerian circuit in an Eulerian graph. It is given in Algorithm 8.2.

```
Algorithm Eulerian Circuit (G,a)
(* Using Theorem 8.7, this algorithm finds an Eulerian circuit C in an
   Eulerian graph G, beginning with vertex a. *)
Begin (* algorithm *)
  find a circuit C beginning with vertex a.
  from G delete all edges in C and vertices incident only with edges
  in C. Call the resulting graph H.
  while there are edges in H do
  begin (* while *)
    select a vertex v that is common to C and H.
    find a circuit C_H in H beginning with vertex v.
    update the subgraph H
    update the circuit C using C_H.
  endwhile
End (* algorithm *)
```

Algorithm 8.2

For a non-Eulerian connected graph, the next theorem can be used to determine if it contains an Eulerian path.

THEOREM 8.8 A connected graph contains an Eulerian path, but *not* an Eulerian circuit, if and only if it has exactly two vertices of odd degree.

PROOF:

Let G be a connected graph. Suppose it contains an Eulerian path from v_0 to v_n, say, v_0-e_1-v_1-e_2-$\cdots$-v_{n-1}-e_n-v_n. Edges e_1 and e_n contribute a 1 to the degrees of v_0 and v_n, respectively. Every time the path passes through a vertex, it contributes a 2 to its degree. So every time the path passes through v_0, its degree is increased by 2; similarly for v_n. Consequently, the degrees of v_0 and v_n are odd. The degree of each internal vertex $v_1, v_2, \ldots, v_{n-1}$ remains even. Thus the graph contains exactly two vertices of odd degree.

Conversely, suppose G contains exactly two vertices of odd degree, say, v_0 and v_n. Adding a new edge $\{v_0, v_n\}$ to G results in a graph G_1 with all even degree vertices. Therefore, by Theorem 8.7, G_1 is Eulerian. Removing edge $\{v_0, v_n\}$ from G_1 shows that G contains an Eulerian path from v_0 to v_n. ■

> *A useful observation*: The proof of Theorem 8.8 implies that an Eulerian path in a connected graph must begin and end at a vertex of odd degree.

The following two examples illustrate Theorem 8.8.

EXAMPLE 8.28 The Königsberg bridge model in Figure 8.2 contains four vertices, all odd degree. Therefore, by Theorem 8.8, no Eulerian path exists. In other words, no walk through the city, traversing each bridge exactly once, is possible. ■

EXAMPLE 8.29 Determine if each graph in Figure 8.67 has an Eulerian path. If so, find it.

Figure 8.67

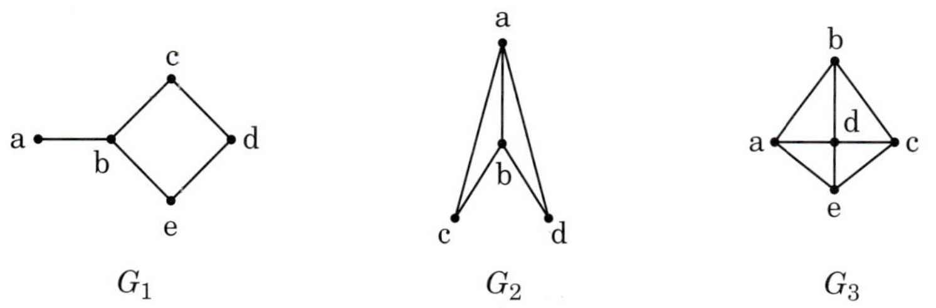

SOLUTION:

Both G_1 and G_2 contain exactly two vertices of odd degree, namely, a and b. Therefore, by Theorem 8.8, both contain Eulerian paths. For example, a-b-c-d-e-b is an Eulerian path in G_1 and a-d-b-c-a-b is an Eulerian path in G_2. G_3 contains four vertices of odd degree, So G_3 does not contain an Eulerian path. ■

The next example, although quite simple and straightforward, is a nice application of Eulerian paths.

EXAMPLE 8.30 Figure 8.68 shows the floor plan of a haunted house. Jeff enters it through the front door and the door locks behind him. In fact, every time he passes through a door, it locks behind him. Is it possible for him to visit every room exactly once and exit through the back door?*

Figure 8.68

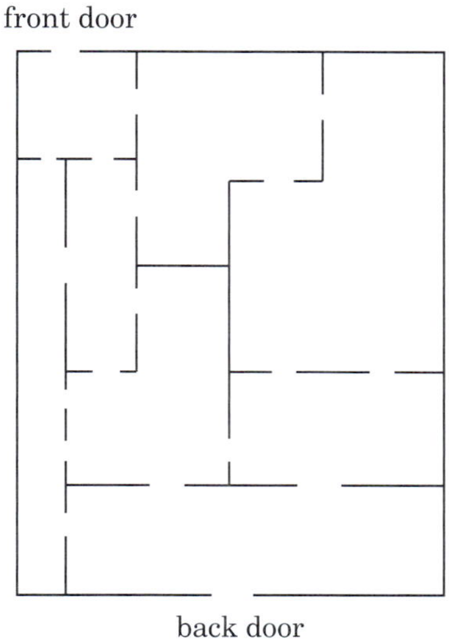

SOLUTION:

(The answer is obvious by inspection.) First, we construct a graph model, as in Figure 8.69, where F denotes the front yard, B the back yard, and every edge a door. Thus the problem can be restated as: *Does the graph in Figure 8.69 contain an Eulerian path?*

Figure 8.69

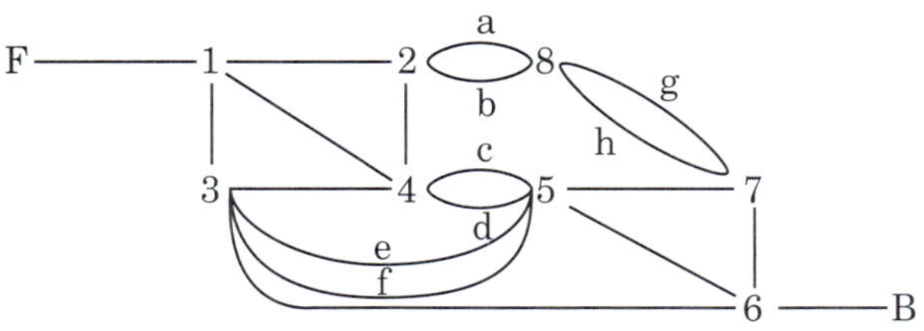

Since the graph has exactly two odd vertices, F and B, by Theorem 8.8, it contains an Eulerian path; one such path is F-1-2-4-1-3-e-5-f-3-6-5-7-g-8-a-2-b-8-h-7-6-B. (Can you find all Eulerian paths in the graph?) ■

*Based on "Calendar Problems," *Mathematics Teacher*, Vol. 85 (Dec. 1992), p. 736.

With the above set of necessary and sufficient conditions for a connected graph to have an Eulerian circuit and an Eulerian path, we examine two strikingly similar questions:

- Under what conditions will a connected graph have a cycle that includes every vertex of the graph?
- Under what conditions will it have a simple path that includes every vertex?

Before exploring answers, let us study the following puzzle.

EXAMPLE 8.31

(Around the World Puzzle) In 1857, the Irish mathematician Sir William Rowan Hamilton invented a puzzle. It consisted of a wooden regular dodecahedron (a polyhedron with 12 regular pentagonal faces), a peg at each of the 20 vertices of the dodecahedron, and a string (see Figure 8.70). Each peg represented a major city in the world and each edge of the decahedron a route between two cities. The problem was: *Beginning at an arbitrary city, find a route for a round trip that would take a person to every other city exactly once.* The route was identified using the pegs and the string.

Figure 8.70

Around the world puzzle.

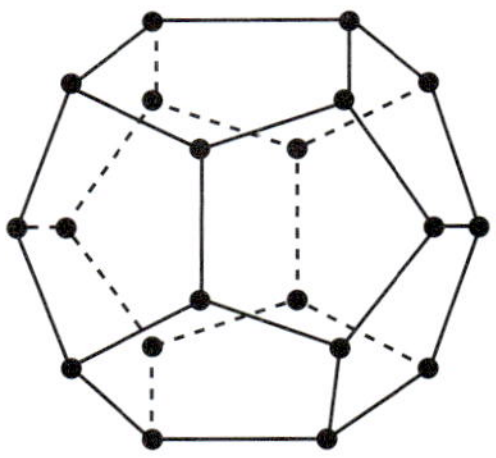

The same problem can be expressed as: *Does the graph model in Figure 8.71 have a cycle containing every vertex*? (Try to find such a cycle.) Hamilton's puzzle has a solution, indicated by the solid edges in Figure 8.72.

Figure 8.71

A graph model of Hamilton's puzzle.

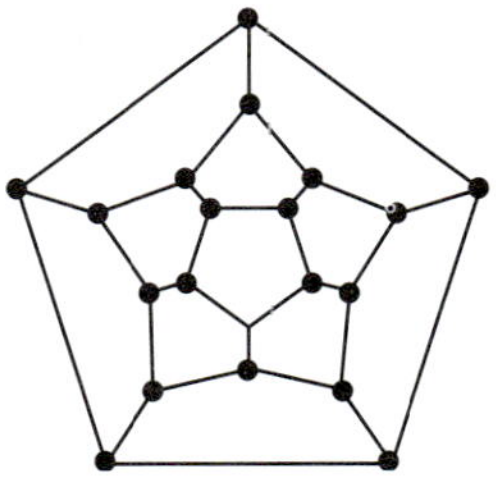

Figure 8.72

A solution to Hamilton's puzzle.

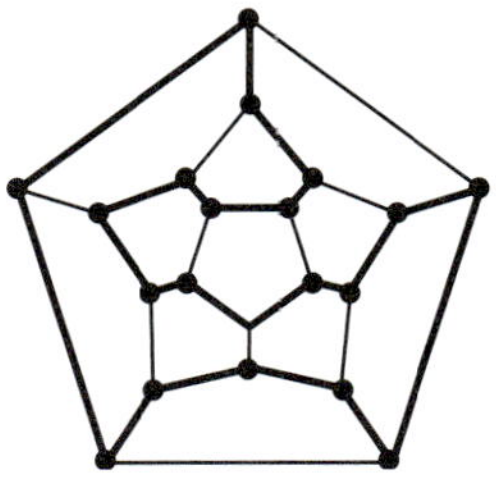

■

Sir William Rowan Hamilton *(1805–1865) was born in Dublin, Ireland. By the age of 3 he had mastered Greek, Hebrew, and Latin, and by 10 he had also learned Arabic, Persian, Bengali, Chaldee, Hindustani, Malay, Marathi, and Sanskrit. By 17 he had mastered calculus and astronomy without any formal training. At 18 he entered Trinity College, Dublin, where he excelled, winning extraordinary honors in classics and science. At 22 he was appointed the Astronomer Royal of Ireland, a position he held until his death. At age 30, he was knighted.*

Hamilton made significant contributions to abstract algebra, dynamics, and optics. He is well known for his development of the noncommutative **system of quarternions** *(or* **Icosian calculus** *as he called it).*

In 1875 Hamilton invented The Icosian Game, *based on his work on quarternions. He sold the idea to a dealer of games and puzzles for 25 pounds, but it was a bad investment for the dealer. The* **Around the World Puzzle** *is a variant of this game.*

This example and the two questions posed above naturally lead to the next definitions named in honor of Hamilton for obvious reasons.

Hamiltonian Graph

A simple path in a connected graph is **Hamiltonian** if it contains every vertex. A cycle in a connected graph that contains every vertex is **Hamiltonian**. A connected graph that contains a Hamiltonian cycle is a **Hamiltonian graph**.

The following example illuminates these definitions.

EXAMPLE 8.32 Are the graphs in Figure 8.73 Hamiltonian? If a graph is not Hamiltonian, does it contain a Hamiltonian path?

Figure 8.73

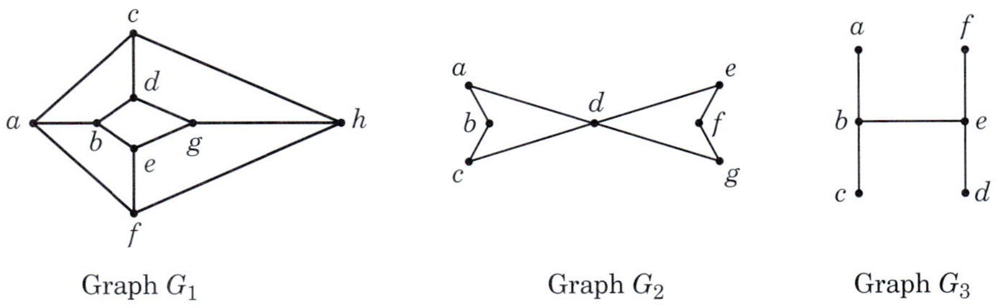

SOLUTION:

Graph G_1 contains a Hamiltonian cycle, for example, *a-c-d-g-h-f-e-b-a*, so it is Hamiltonian.

In G_2, suppose we start at a, b, or c. Then we have to pass through d to visit vertices e, f, and g. So, to return to the home vertex, we have to pass through d again. Thus G_2 contains no Hamiltonian cycles that begin at $a, b,$ or c. Suppose we start at d. Then again after visiting a, b, and c, we have to pass through d to visit $e, f,$ or g before returning to d. Consequently, G_2 contains no Hamiltonian cycles that begin at d either. Suppose we start at e, f, or g. This also yields no Hamiltonian cycles. Thus G_2 is not Hamiltonian. Nonetheless, it *does* contain a Hamiltonian path, *a-b-c-d-e-f-g*.

In G_3, to visit a, c, d, or f, we must enter and leave both b and c at least twice, so it is not Hamiltonian. By a similar argument, it has no Hamiltonian paths. ■

Example 8.33 uses Gray codes introduced in Chapter 7.

EXAMPLE 8.33

Let Σ^n denote the set of n-bit words. The graph with each vertex representing one of the n-bit words is called the ***n*-cube**, denoted by Q_n. Two edges in Q_n are adjacent if the Hamming distance between them is one. Figure 8.74 shows the n-cube for $n = 1$, 2, and 3.

Figure 8.74

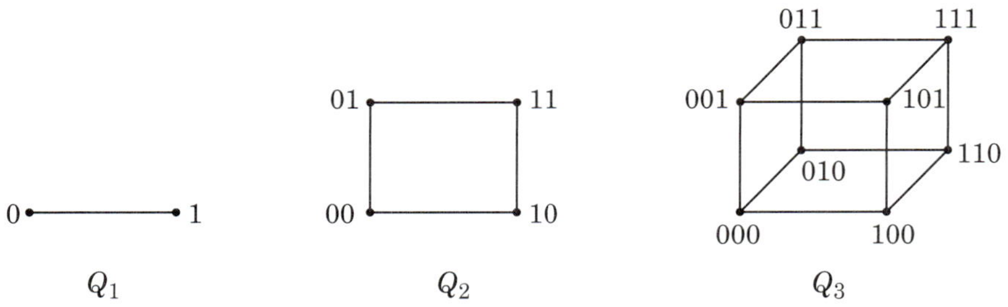

Is the n-cube Hamiltonian? By Example 7.10, a Gray code exists for every Σ^n; that is, a cycle with every vertex of the n-cube does exist, so Q_n is Hamiltonian. For instance, Figure 8.75 shows a Hamiltonian cycle for Q_3.

Figure 8.75

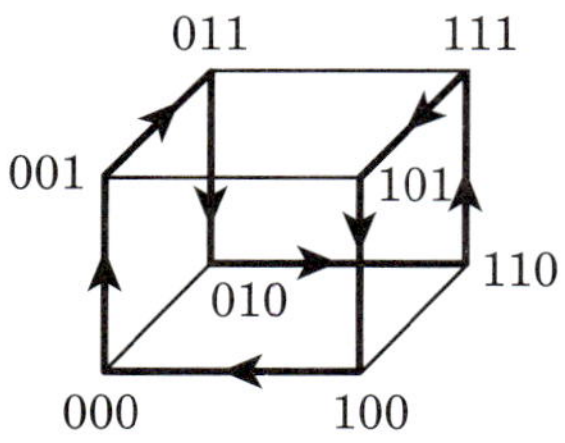

Using matrices, a Hamiltonian cycle H_n for Q_n can be defined recursively, where each element in H_n represents a vertex:

$$H_1 = \begin{bmatrix} 0 \\ 1 \end{bmatrix}$$

$$H_n = \begin{bmatrix} 0H_{n-1} \\ 1H^R_{n-1} \end{bmatrix}, \quad n \geq 2$$

where H^R_{n-1} denotes the list H_{n-1} in the reversed order.

For example,

$$H_2 = \begin{bmatrix} 00 \\ 01 \\ 11 \\ 10 \end{bmatrix}$$

Then

$$H_3 = \begin{bmatrix} 000 \\ 001 \\ 011 \\ 010 \\ 110 \\ 111 \\ 101 \\ 100 \end{bmatrix}$$

This clearly agrees with the Hamiltonian cycle in Figure 8.75. (Can you find such a cycle for Q_4 using recursion? See Exercise 60.) ■

Q_n serves as a fine model of parallel computation, where each vertex represents a processor. Since Q_n is Hamiltonian, it follows that every processor can communicate with every other processor, either directly or indirectly.

Hamiltonian cycles are a misnomer because Hamilton was certainly not the first person to look for them. One earlier reference to them can be found in the famous **Knight's tour problem** stated below.

Knight's Tour Problem

Is it possible for a knight to visit each square on a chessboard exactly once and return to the home square?

To see how this problem is related to finding Hamiltonian cycles in a graph, represent each square by a vertex and the two squares connecting a knight's move by an edge. Then a knight's tour is possible if and only if the corresponding graph is Hamiltonian.

Figure 8.76

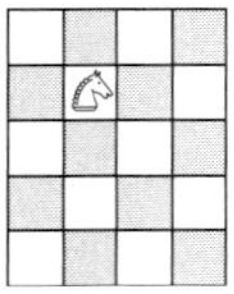

Figure 8.77

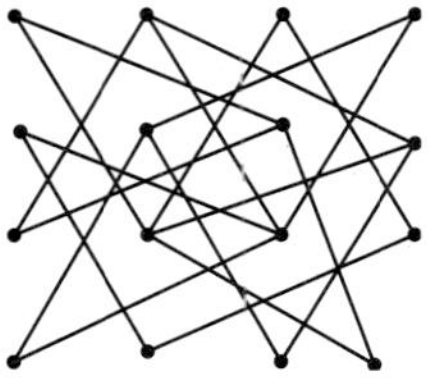

For example, Figure 8.76 shows a 4 × 4 chessboard and Figure 8.77 a graph representing a set of legal moves by a knight; unfortunately, it lands at some squares more than once; therefore, the graph is not Hamiltonian.

With a bit of patience, we will find that a knight's tour is impossible on a 4 × 4 board. Furthermore, no solution exists on an $n \times n$ board if n is odd.

In 1759, Euler published an algorithm for finding a tour on an 8 × 8 chessboard, and 12 years later Alexandre-Théophile Vandermonde presented another one. Figure 8.78 shows a solution. (Finding a knight's tour is a good exercise in "backtracking"; see Section 9.2.)

Figure 8.78

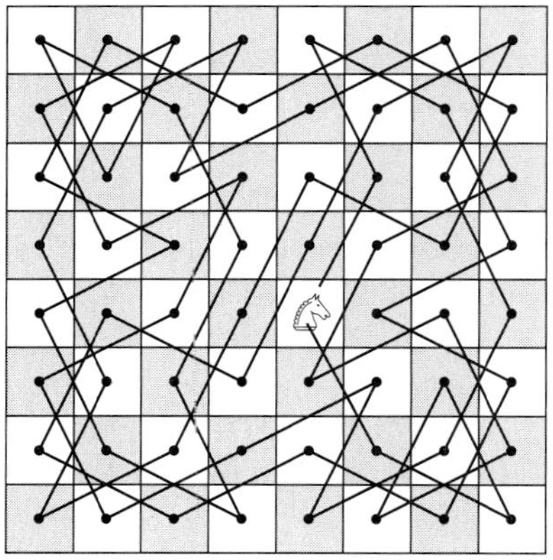

An interesting observation: Using the tour in Figure 8.78, suppose we number the squares 1 through 64 in the order they are visited. The resulting square (see Figure 8.79) is a pseudo-magic square: the row sums and column sums are equal, each being 260, but the diagonal sums are different.

Alexandre-Théophile Vandermonde *(1735–1796) was born in Paris. Music was his first love and he developed an interest in mathematics later in life. His entire mathematical contribution consists of four papers published in the Histoires of the Academy of Sciences during 1771–1772. They are concerned with the theory of equations, theory of determinants, factorials, and the knight's tour.*

Vandermonde died in Paris.

Figure 8.79

35	26	37	14	63	24	11	50
38	15	34	25	12	51	62	23
27	36	13	40	21	64	49	10
16	39	28	33	52	9	22	61
29	54	41	20	1	60	7	48
42	17	32	53	8	45	4	59
55	30	19	44	57	2	47	6
18	43	56	31	46	5	58	3

Hamiltonian paths have interesting applications to combinatorics, as the next example shows.

○ **EXAMPLE 8.34** (optional) A **power chain of order** n is a permutation of the first n (≥ 2) natural numbers such that the sum of every pair of adjacent elements is a power. For example, 81726354 is a power chain of order 8. Find all power chains of order 7.*

SOLUTION:

A systematic procedure for finding all power chains of order n follows. Draw a graph with n vertices, labeled 1 through n. Two vertices i and j in the graph are adjacent if $i + j$ is a power. Each Hamiltonian path yields a power chain.

In particular, consider the graph in Figure 8.80 with $n = 7$. Since the degree of vertex 4 is 1, it is easier to list all Hamiltonian paths beginning with 4. For convenience, redraw the graph as in Figure 8.81. It follows from this graph that there are exactly two power chains of order 7: 4531726 and 4536271.

*Proposed by H. L. Nelson and solved by B. Barwell, *J. Recreational Mathematics*, Vol. 12:1, 1979–1980, pp. 67–68.

Figure 8.80

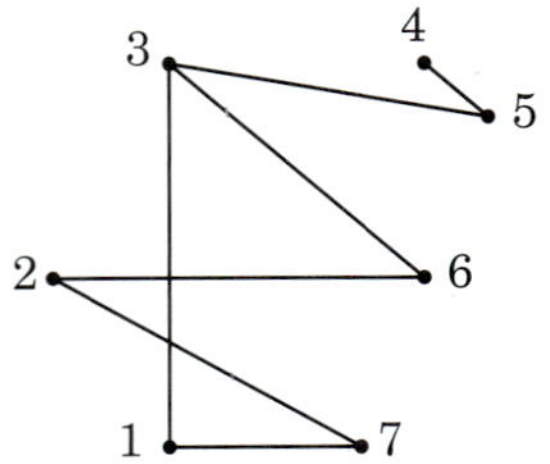

Figure 8.81

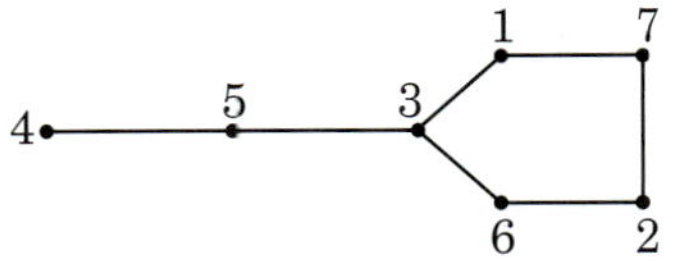

■

Although a simple test will determine whether or not a connected graph has an Eulerian circuit or an Eulerian path, nothing similar exists for determining if it contains a Hamiltonian cycle or path. Finding such a test, without an exhaustive search of all $n!$ possible paths, remains a major unsolved problem in graph theory.

A host of sufficient conditions exist, however, for a connected graph to be Hamiltonian. Two lie below, without proofs, but illustrated.

THEOREM 8.9 **(Dirac's Theorem)** A simple connected graph G with $n \geq 3$ vertices is Hamiltonian if $\deg(v) \geq \frac{n}{2}$ for every vertex v in G. ■

Notice that the condition in Theorem 8.9 does not apply to the graph G_1 in Figure 8.73 and the 3-cube in Figure 8.75. Nonetheless, both are Hamiltonian. So even if the condition is not satisfied, the graph may be Hamiltonian.

The next theorem provides another sufficient criterion for a graph to be Hamiltonian.

THEOREM 8.10 **(Ore's Theorem)** Let G be a simple connected graph with $n \geq 3$ vertices. If $\deg(u) + \deg(v) \geq n$ for every pair of nonadjacent vertices u and v, then G is Hamiltonian. ■

You may verify that $\deg(u) + \deg(v) \geq 5$ for every pair of nonadjacent vertices u and v of the graph G_3 in Figure 8.67. Consequently, by Theorem 8.10, G_3 is Hamiltonian, as expected. The condition in the theorem is not satisfied by the 3-cube in Figure 8.75; nonetheless, Q_3 is Hamiltonian.

Next we present the well-known traveling salesperson problem, which employs Hamiltonian cycles and has interesting applications to the theory of communications.

Gabriel Andrew Dirac *(1925–1984), a pioneer in graph theory, was born in Budapest. He moved to England in 1937 when his mother married Paul Adrien Maurice Dirac, a Nobel Laureate in physics. In 1942, he began his studies at Cambridge University. After a two-year interruption by the war, he continued his studies at Cambridge and London, receiving his Ph.D. in mathematics from the University of London (1951). That same year he was awarded the Rayleigh Prize by Cambridge University.*

Dirac taught at the universities of London, Toronto, Hamburg, Wales, and Aarhus. He was a member of the editorial board of the Journal of Graph Theory and the European Journal of Combinatorics.

Besides his work with graph theory, he made outstanding contributions to number theory and geometry. He was also a passionate art connoisseur.

Oystein Ore *(1899–1968), a Norwegian mathematician, was born in Oslo. He received his Ph.D. from Oslo University in 1924 and taught there for 2 years. He joined Yale University in 1927 and taught there until his retirement in 1967, holding the chair of the mathematics department from 1936 to 1945.*

Ore served on the board of American Relief for Norway from 1942 to 1947 and chaired the Relief Mission in 1945–1946.

Recognizing his contributions to Norway, King Haakon VII decorated him Knight of St. Olaf in 1947.

As an author of numerous articles and several books, Ore made outstanding contributions to graph theory, abstract algebra, number theory, and probability.

Traveling Salesperson Problem

Study the weighted graph in Figure 8.82. The vertices represent cities and the weights represent the distances between them. A salesperson assigned to city a would like to visit every other city exactly once and return to the home city so that the total distance traveled is a minimum. In other words, beginning at a, he would like to find a Hamiltonian cycle, so the sum of the weights along the cycle is a minimum. This is the celebrated **traveling salesperson problem**.

Figure 8.82

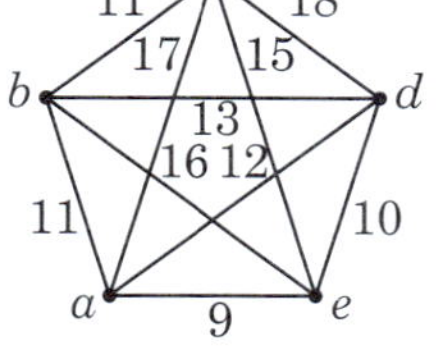

When the graph contains only a few vertices, you can list all possible cycles, compute the sum of the weights along each, and find the shortest Hamiltonian cycle.

For example, the graph in Figure 8.82, by the multiplication principle, has $4!/2 = 12$ distinct Hamiltonian cycles, each containing five edges. You may verify that the cycle with least weight is a-d-b-c-e-a; so the total distance traveled is 60 miles.

Finding a Hamiltonian cycle in a complete graph K_n without searching through all distinct $(n - 1)!/2$ Hamiltonian cycles remains an unsolved problem.

We close this section with another graph-theoretic game.

○ **EXAMPLE 8.35** (optional) The game of **Ham** (for **Ham**pton Court) is a graph-theoretic game quite similar to SIM, invented by B. Recaman in 1977. Like SIM, Ham is also a two-person game. Two players, R (for red) and B (for blue), take turns coloring an edge of the complete graph K_n. (A game on K_n is a game of **order** n.) Player R's objective is to obtain a Hamiltonian cycle made up entirely of red edges while B's goal is to prevent R from doing it. Whoever achieves his\her goal wins the game.

The game of order 3 has a trivial winning strategy for player B. She can always block R from completing a Hamiltonian cycle. (So B wins.) See Figure 8.83, where solid edges indicate red edges and broken edges, blue.

Figure 8.83

Player B wins.

Figure 8.84

Player B wins.

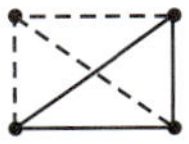

Figure 8.85

Player B wins.

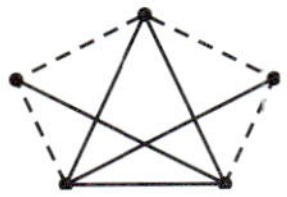

Player B can always win games of orders 4 and 5, as in Figures 8.84 and 8.85.

The game of order 6, although a bit more complicated, can also be won by player B. After R has colored the first edge $\{x, y\}$, B colors an edge $\{u, v\}$ not incident with x or y. Thereafter, B colors four more edges incident with x or y. At this stage, if R has at least two edges incident with both u and v, she can win (see Figure 8.86). But B can avoid this by coloring one of the two edges missing in the Hamiltonian cycle (see Figure 8.87).

Is player B favored to win if the game is of order 7 or more? It is conjectured so.

Figure 8.86

Player R wins.

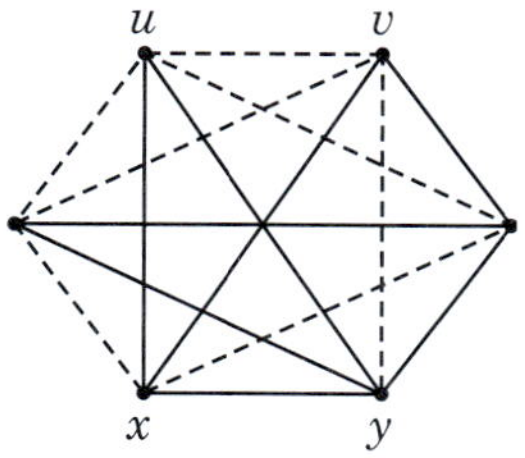

Figure 8.87

Player B wins.

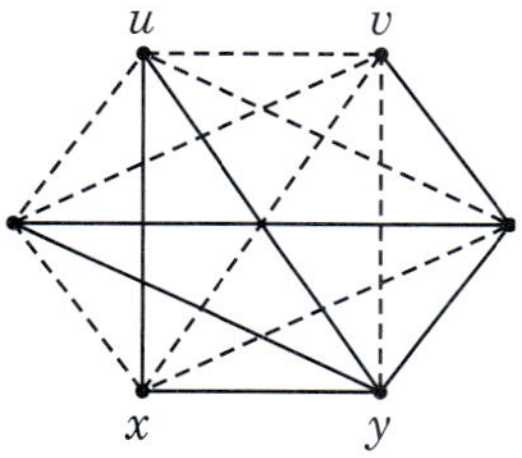

■

Exercises 8.5

Determine if each graph is Eulerian.

1.

2.

3.

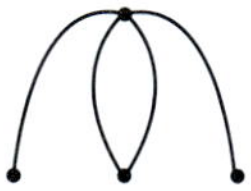

4.

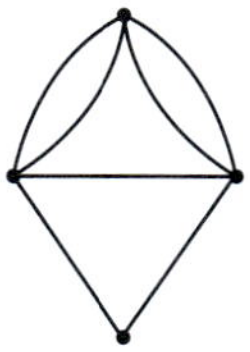

5.

6.

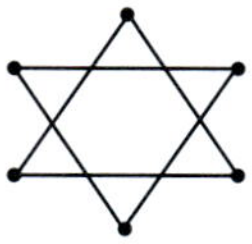

Construct an Eulerian circuit for each Eulerian graph.

7.

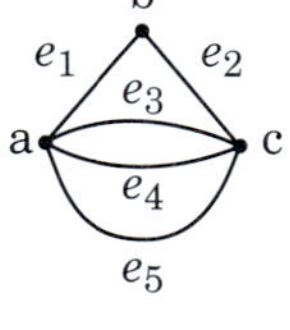

8.

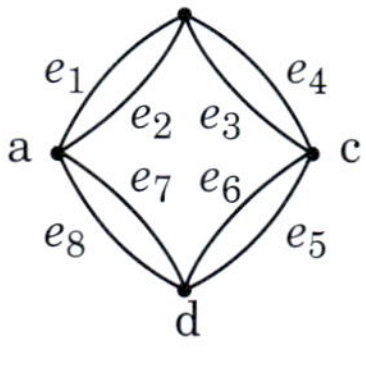

9.

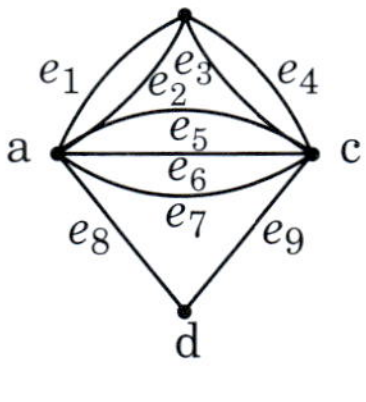

10.

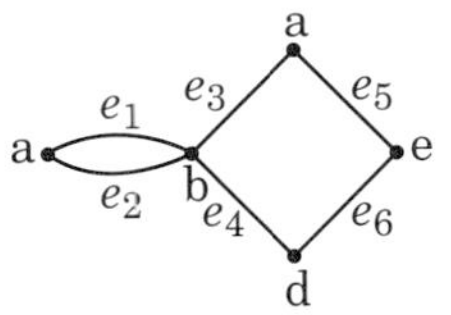

11.

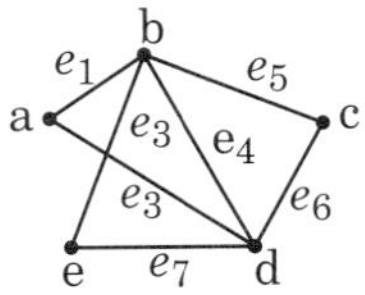

12.

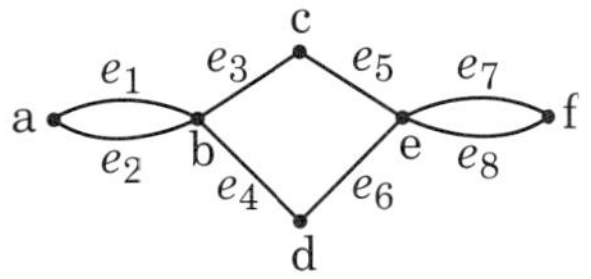

13. Is the Petersen graph in Figure 8.28 Eulerian?

14–19. Do the non-Eulerian graphs in Exercises 1–6 have an Eulerian path?

Find an Eulerian path in each graph, if possible.

20.

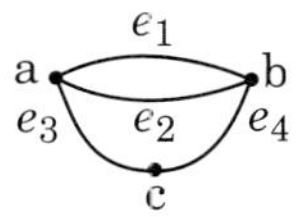

21.

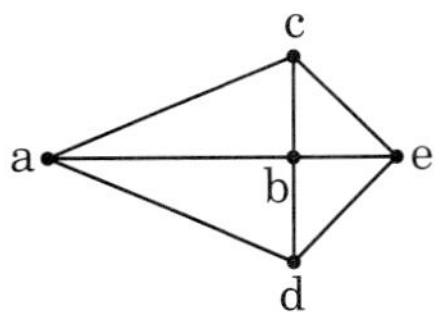

22.

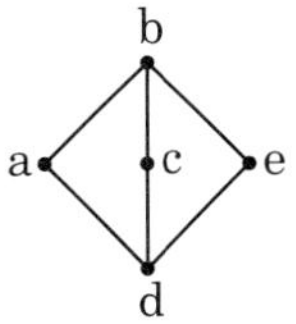

Under what conditions will each graph be Eulerian?

23. K_n **24.** $K_{m,n}$

25–26. Under what conditions will each graph in Exercises 23 and 24 contain an Eulerian path, but not an Eulerian circuit?

27. If a connected r-regular graph is Eulerian, what can you say about r?

28–42. Are the graphs in Exercises 1–12 and 20–22 Hamiltonian? If one is not Hamiltonian, determine if it has a Hamiltonian path; if so, find it.

43. Under what conditions will the complete graph K_n be Hamiltonian?

44. If G is a connected graph containing a vertex with degree 1, can it be Hamiltonian?

Determine if each complete bipartite graph $K_{m,n}$ is Hamiltonian. If a graph is not Hamiltonian, does it contain a Hamiltonian path?

45. $K_{2,3}$ **46.** $K_{3,3}$ **47.** $K_{2,4}$ **48.** $K_{3,4}$

49. For what values of m and n will $K_{m,n}$ be Hamiltonian?

50. Is the Petersen graph in Figure 8.28 Hamiltonian?

Consider the **triangle graph** T_n, where T_1, T_2, T_3 are shown in Figure 8.88.*

Figure 8.88

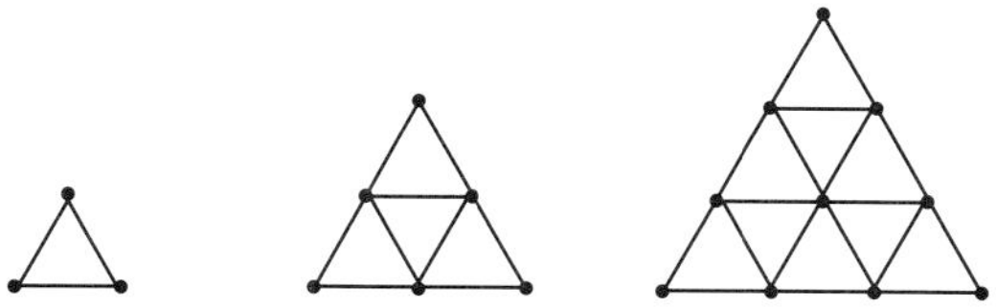

T_1 T_2 T_3

51. When will T_n be Eulerian?

52. When will T_n be Hamiltonian?

Consider the **square graph** S_n, where S_1, S_2, S_3 are pictured in Figure 8.89.

Figure 8.89

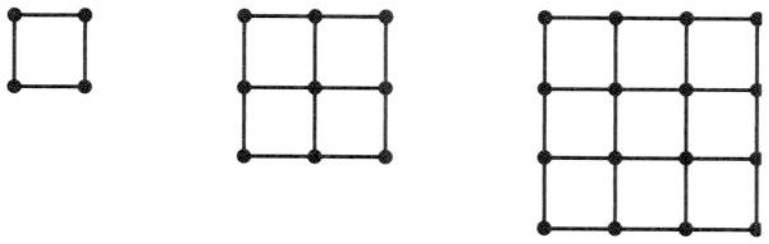

S_1 S_2 S_3

53. When will S_n be Eulerian?

54. When will S_n be Hamiltonian?

Give an example of a graph that is:

55. Both Eulerian and Hamiltonian.

56. Eulerian, but not Hamiltonian.

57. Hamiltonian, but not Eulerian.

58. Neither Eulerian nor Hamiltonian.

59. Figure 8.90 shows five cities, a through e, and the distances between them. A salesperson based at a would like to visit each city exactly once and return to a, covering the fewest miles. Find the route she should take and the minimum distance she would travel.

60. Display a Hamiltonian cycle for the 4-cube.

Find all power chains of order n, if they exist, for each value of n.

61. 6 **62.** 8 **63.** 9 **64.** 15

*Exercises 51–54 are based on A. Guckin *et al.*, *The Euler Circuit Project*, COMAP, Inc., Lexington, MA, 1989.

Figure 8.90

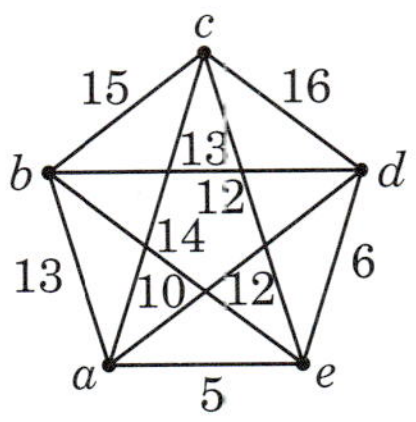

○ ***65.** A **power cycle of order** n is a cyclic permutation of the first n (≥ 2) positive integers such that the sum of every pair of adjacent elements is a power. Find a power cycle of order 17.

66. Write an algorithm to determine if a connected graph is Eulerian, using its adjacency list representation.

67. Write an algorithm to determine if a connected graph contains an Eulerian path, using its adjacency matrix.

8.6 Planar Graphs

Take a good look at every graph presented so far. In some cases, the edges meet only at the vertices, whereas in the other cases the edges meet at non-vertices. For example, the edges of the Königsberg bridge model (Figure 8.2) meet only at its vertices, whereas the edges of the pentagram in Figure 8.5 intersect at points that are not necessarily vertices. However, the confusion can be resolved. The pentagram can be redrawn in such a way that its edges intersect only at its vertices, as Figure 8.91 shows. The graphs in Figures 8.5 and 8.91 are clearly isomorphic. You are invited to verify this.

Figure 8.91

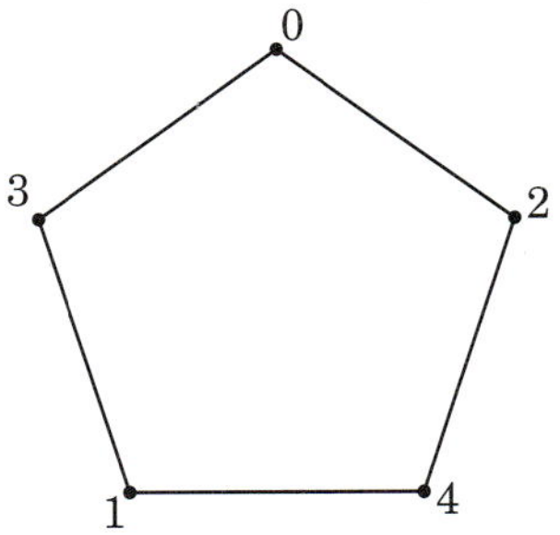

From these observations arises the following definition.

Planar Graph

A graph is **planar** if it can be drawn in the plane, so its edges meet only at the vertices. Such a drawing is a **planar representation** of the graph.

The previous discussion indicates the pentagram is planar. The graph in Figure 8.92 is planar since it can transform to Figure 8.93; they are isomorphic graphs. The graph in Figure 8.94 is also planar; Figure 8.95 shows its planar representation.

Figure 8.92

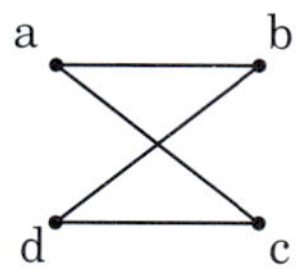

Figure 8.93

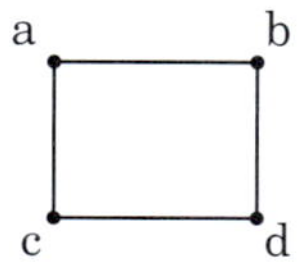

Figure 8.94

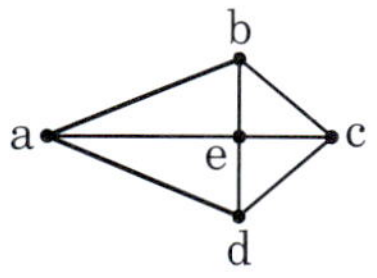

Figure 8.95

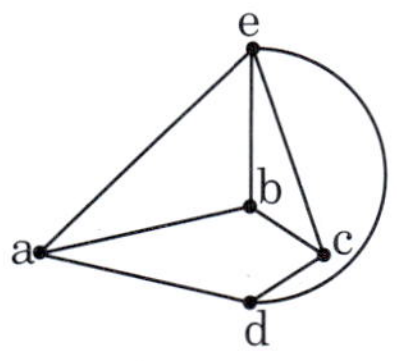

Planarity is an important concept with fine, real-world applications. For example, it figures prominently in designing circuit boards.

The following example answers a problem posed at the beginning of the chapter.

EXAMPLE 8.36

(The Three Houses–Utilities Puzzle) A developer is building three new houses — A, B, and C — on one side of a street. She would like to connect three utilities — electricity (E), telephone (T), and water (W) — to each house. This situation can be modeled by the complete bipartite graph $K_{3,3}$ in Figure 8.96.

Can the developer lay the utility lines without any crossings? To answer this, try to make a planar representation of $K_{3,3}$. Draw eight of the nine edges in $K_{3,3}$ (see the solid edges in Figure 8.97). How can we draw the edge from W to A? Notice that W lies on the circuits W-C-T-B-W and W-C-E-B-W. Consider the regions bounded by them. Since A lies outside both, any edge

Figure 8.96

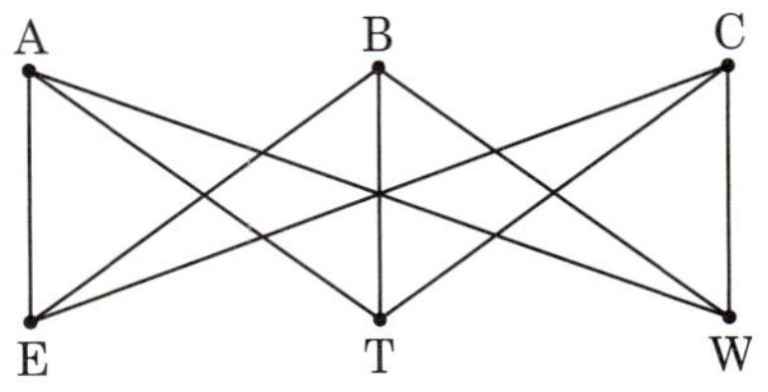

from W to A must cross a boundary (see the broken edges in the figure). Thus $K_{3,3}$ is nonplanar and the puzzle has no solution. The utility lines will cross. (Example 8.41 will prove this algebraically.)

Figure 8.97

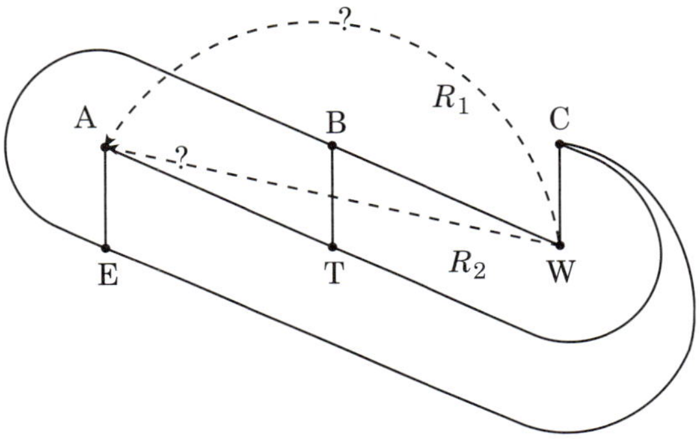

■

An argument like this geometric proof can show that the complete graph K_5 is also nonplanar.

Notice that any planar representation of a connected graph divides the plane into regions, including the unbounded region. For instance, the planar graph in Figure 8.98 divides the plane into five regions, marked 1 through 5. The graph has six edges and and three vertices. Let r denote the number of regions formed by e edges and v vertices. Then $r = e - v + 2$. In fact, this relationship is not a coincidence, but is always true.

Figure 8.98

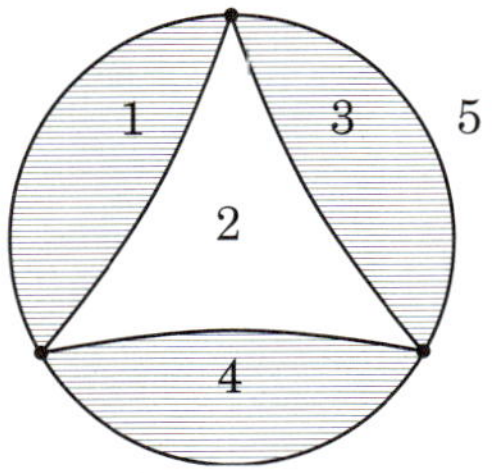

THEOREM 8.11 **(Euler's formula)** Let G be a connected planar graph with e edges and v vertices. Let r be the number of regions formed by a planar representation of G. Then $r = e - v + 2$.

PROOF (by induction on e):
Let P(n): Euler's formula holds for every connected planar graph with n edges.

Basis step Clearly, P(0) is true. Suppose the graph contains one edge. Then there are two possibilities: The edge may or may not be a loop (see Figure 8.99). In the first case, $e = 1$, $v = 1$, and $r = 2$. In the second case, $e = 1$, $v = 2$, and $r = 1$. In both cases, $r = e - v + 2$. Consequently, P(1) is also true.

Figure 8.99

Induction step Assume P(k) is true for some $k \geq 0$; that is, assume the formula holds for every connected planar graph with k edges. Consider a connected planar graph G with $k + 1$ edges, v vertices, and r regions. G either has a cycle or does not.

Case 1 Suppose G does not contain a cycle. Then G must contain a vertex u of degree 1. To find such a vertex, build a path from any vertex until you cannot go any further. The last vertex visited is such a vertex u. (See Figure 8.100.) Let a be the edge incident with u. Let H be the graph obtained by deleting vertex u and edge a from G. H contains $e' = k$ edges and $v' = v - 1$ vertices. Since G is connected, so is H. Therefore, by the inductive hypothesis, the formula holds for H; the number of regions r' formed by H satisfies the formula $r' = e' - v' + 2$. But $r' = r$. So

$$r' = (e - 1) - (v - 1) + 2$$

That is,

$$r = e - v + 2$$

Figure 8.100

Case 2 Suppose G contains a cycle. Let a be an edge in the cycle (see Figure 8.101). Let L be the subgraph obtained by deleting edge a from G. (*Note*: Deleting edge a merges regions R_1 and R_2.) This subgraph contains $e' = k$ edges and $v' = v$ vertices; therefore, by the induction hypothesis, the number of regions r' formed by L is given by $r' = e' - v' + 2$. But $r' = r - 1$, $e' = e - 1$, and $v' = v$. So

$$r - 1 = (e - 1) - v + 2$$

That is,

$$r = e - v + 2$$

Figure 8.101

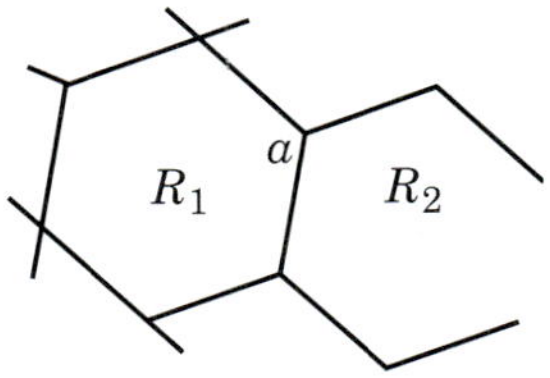

By cases 1 and 2, P($k + 1$) is true. Therefore, by PMI, the formula holds for every connected planar graph. ■

The next two examples illustrate Euler's formula.

EXAMPLE 8.37 Verify Euler's formula for the connected planar graph in Figure 8.102.

Figure 8.102

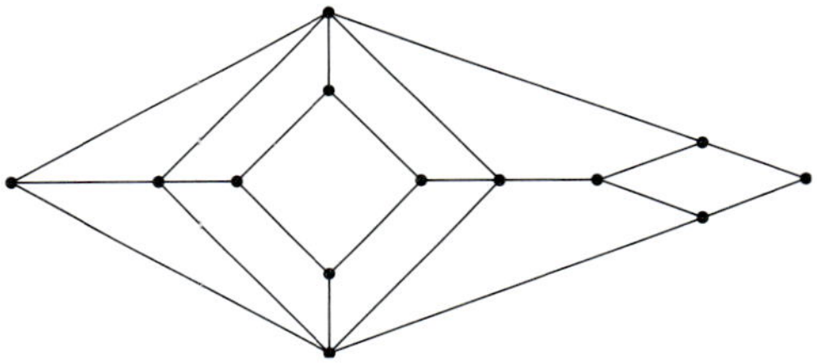

SOLUTION:
The graph contains 22 edges and 13 vertices, which form 11 regions. So $e - v + 2 = 22 - 13 + 2 = 11 = r$. ■

EXAMPLE 8.38 A connected planar graph has 17 edges, dividing the plane into nine regions. How many vertices does the graph have?

SOLUTION:
Here $e = 17$ and $r = 9$. By Euler's formula,

$$v = e - r + 2 = 17 - 9 + 2 = 10$$ ■

Euler's formula can generate simple relationships in special cases. One such formula is given in the following theorem.

THEOREM 8.12 If G is a simple, connected, planar graph that contains e (≥ 2) edges and v vertices, then $e \leq 3v - 6$.

PROOF:
Suppose the plane is divided into r regions by a planar representation of the graph. First, we shall show that $2e \geq 3r$.

If $e = 2$, then $r = 1$. So $2e \geq 3r$.

Suppose $e > 2$. Let s be the total number of edges along the boundaries of the various regions. Every edge appears on the boundary of two regions, so every edge is counted twice in this sum and $s = 2e$. Now every region, including the unbounded region, is bounded by at least three edges. Therefore, $s \geq 3r$ and $2e \geq 3r$.

By Euler's formula,

$$r + v - 2 = e$$

Then

$$3r + 3v - 6 = 3e$$
$$3r + 3v - 6 - e = 2e$$

So

$$3r + 3v - 6 - e \geq 3r$$

That is,

$$e \leq 3v - 6$$ ■

This theorem can prove that K_5 is nonplanar.

EXAMPLE 8.39 Prove that the complete graph K_5 is nonplanar.

PROOF (by contradiction):
Recall that K_5 contains 5 vertices and 10 edges. It is a simple connected graph. If it is planar, $e \leq 3v - 6$, by Theorem 8.12. But $3v - 6 = 3(5) - 6 = 9 < e$, which is a contradiction; so K_5 is not planar. ■

Is the converse of Theorem 8.12 true? That is, if $e \leq 3v - 6$ in a simple connected graph, is it planar? To answer this, recall that in $K_{3,3}$, $v = 6$ and $e = 9$, so $e \leq 3v - 6$. Nonetheless, Example 8.36 showed $K_{3,3}$ is nonplanar; so the converse is false.

In the analysis of planar graphs, the degree of a region becomes useful.

Degree of a Region

For a region R formed by a planar representation of a connected graph, the **degree** of a region R, denoted by deg (R), is the number of edges in a closed path bordering R.

EXAMPLE 8.40 Find the degree of each region formed by the graphs in Figures 8.103 and 8.104.

Figure 8.103

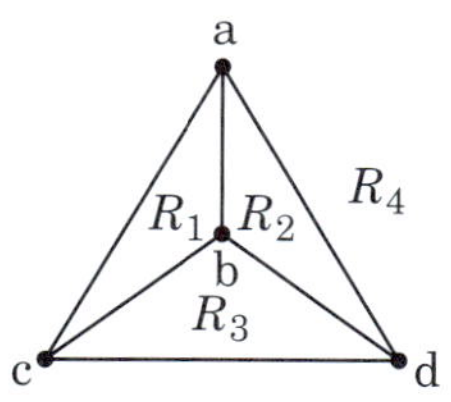

Figure 8.104

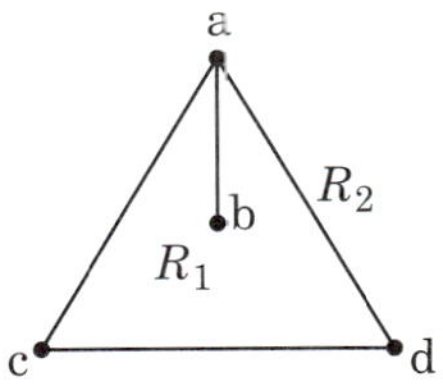

SOLUTION:

- In Figure 8.103, each region is bounded by three edges. So deg $(R_1) =$ deg $(R_2) =$ deg $(R_3) =$ deg $(R_4) = 3$.
- In Figure 8.104, R_2 is bounded by three edges, so, deg$(R_2) = 3$. R_1 is bounded by the closed path a-c-d-a-b-a which contains edge {a, b} twice. So deg $(R_1) = 5$. ■

The next theorem establishes a straightforward relationship between the sum of the degrees of the regions and e.

THEOREM 8.13 The sum of the degrees formed by a planar representation of a connected graph G with e edges equals $2e$; that is, $\sum_i \deg(R_i) = 2e$.

PROOF:
Every edge in a planar representation of G either bounds two regions (see Figure 8.103) or lies within a region (see Figure 8.104), so it is counted twice in computing the degree of the region. So $\sum_i \deg(R_i) = 2e$. ■

Through Euler's formula and Theorem 8.13, the proof that $K_{3,3}$ is nonplanar emerges algebraically.

EXAMPLE 8.41 Prove that $K_{3,3}$ is nonplanar.

PROOF (by contradiction):
Assume $K_{3,3}$ is planar. Since $K_{3,3}$ has six vertices and nine edges, a planar representation must yield $r = e - v + 2 = 9 - 6 + 2 = 5$ regions, by Euler's formula.

If $K_{3,3}$ were planar, each region R_i would be bounded by at least four edges. Then the sum of the degrees of the regions in a planar representation would be at least $5 \cdot 4 = 20$; that is, $\sum_i \deg(R_i) \geq 20$.

By Theorem 8.13, $2e = \sum_i \deg(R_i)$. This yields $18 \geq 20$, which is a contradiction, so $K_{3,3}$ is not planar. ■

Thus both K_5 and $K_{3,3}$ are nonplanar, and any graph containing one of them as a subgraph is also nonplanar. In fact, every nonplanar graph must contain a subgraph derived from K_5 or $K_{3,3}$ through elementary operations defined below.

Homeomorphic Graphs

Let $\{u, w\}$ be an edge in a graph. Make a vertex v of degree two by deleting the edge $\{u, w\}$ and adding two edges $\{u, v\}$ and $\{v, w\}$ incident with v. Now, delete v and replace the two edges with an edge $\{u, w\}$. Such inserting and deleting of vertices of degree two are **elementary operations** on a graph. Two graphs are **homeomorphic** if one can be obtained from the other by a sequence of elementary operations.

For example, the graphs in Figure 8.105 are homeomorphic. Why?

Figure 8.105

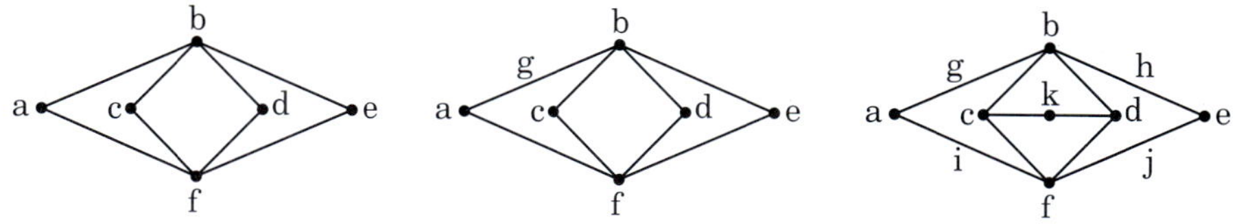

Fortunately, the elementary operations do not affect the planarity of a graph. A criterion for planarity was proved by the Polish mathematician Kazimierz Kuratowski in 1930.

THEOREM 8.14 **(Kuratowski's theorem)** A graph is planar if and only if it does not contain a subgraph homeomorphic to K_5 or $K_{3,3}$. (Equivalently, a graph is nonplanar if and only if it contains a subgraph homeomorphic to K_5 or $K_{3,3}$.) ■

We close this section with an example that illustrates this theorem.

EXAMPLE 8.42 Is the graph in Figure 8.106 planar?

SOLUTION:

The graph in Figure 8.106 contains K_5 as a subgraph (see Figure 8.107); so it is not planar. ■

Although Kuratowski's theorem gives a necessary and sufficient condition for a graph to be planar, it does not provide an algorithm to determine if a graph with n vertices is planar. Such algorithms do exist, however, which can be executed using $O(n)$ time.

Figure 8.106

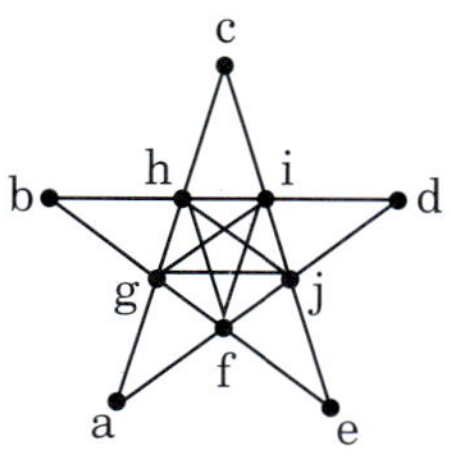

Kazimierz Kuratowski *(1896–1980), a Polish mathematician and son of a well-known lawyer, was born in Warsaw. After completing his secondary education, he studied engineering at the University of Glasgow in Scotland in 1913. When he returned home during the summer of 1914, World War I broke out, so he could not return to Glasgow.*

In 1915, Kuratowski studied mathematics at the University of Warsaw where he was a student of the Polish logician Jan Lukasiewicz (see Section 9.5). He graduated from the University in 1919 and received his Ph.D. in 1921.

In 1927, Kuratowski became professor of mathematics at Lvov Technical University. Six years later, he returned to the University of Warsaw where he had both academic and administrative responsibilities; he worked there until 1966.

In 1928, he joined the editorial board of Fundamenta Mathematicae *and became its editor-in-chief in 1952, a position he held until his death. From 1957 to 1968 he was vice president of the Polish Academy of Sciences, as well as founder and later director of the Institute of Mathematics.*

Kuratowski wrote numerous articles for professional journals, making significant contributions to topology, graph theory, and analysis.

Figure 8.107

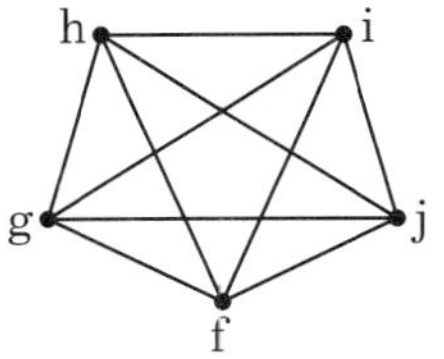

Exercises 8.6

Draw a planar representation of each graph.

1.

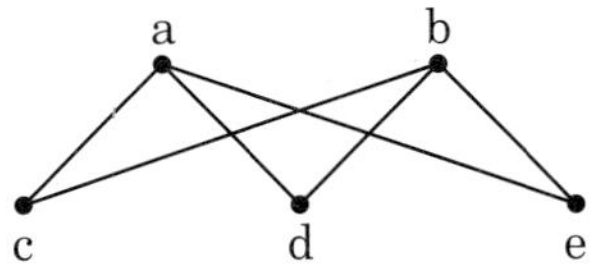

2.

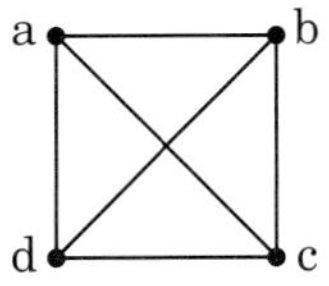

3.

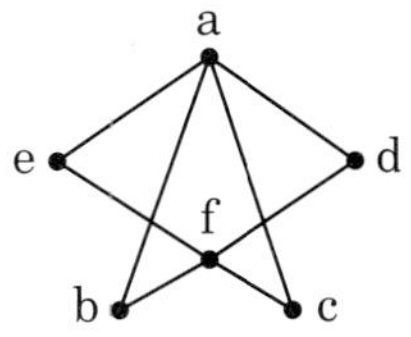

4.

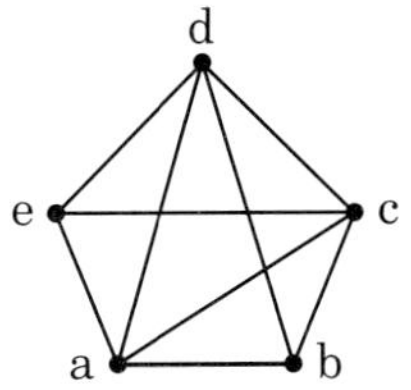

5. Draw a planar representation of the 3-cube.

Are the following graphs planar? If one is, draw a planar representation.

6.

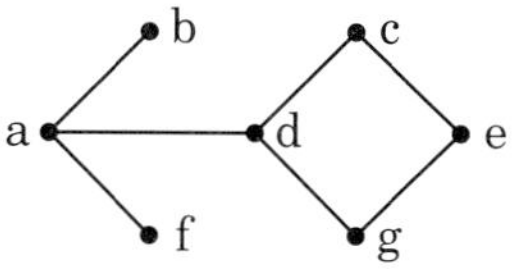

7.

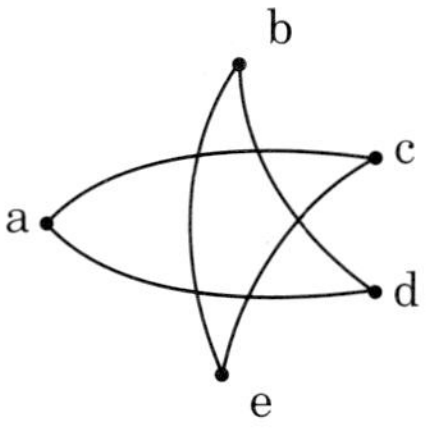

8.

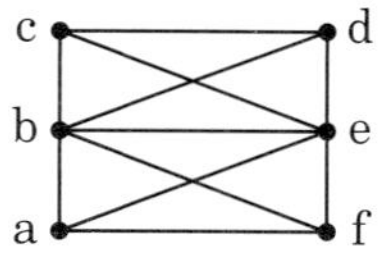

9.

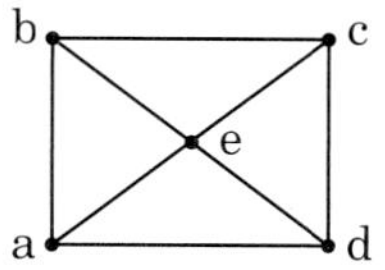

Verify Euler's formula for each connected planar graph.

10.

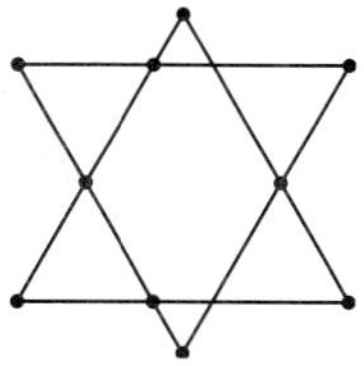

11.

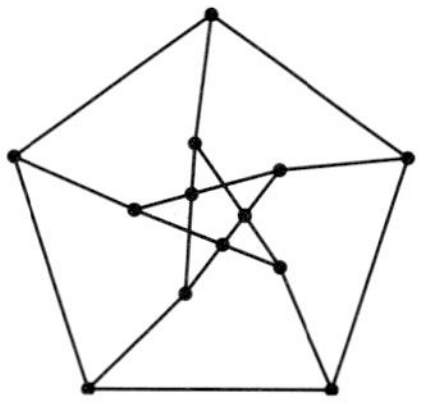

12–13. Verify Theorem 8.12 for the simple, connected, planar graphs in Exercises 10 and 11.

14. A connected, planar graph contains 10 vertices and divides the plane into seven regions. Compute the number of edges in the graph.

15. A connected, planar graph contains 24 edges. It divides the plane into 13 regions. How many vertices does the graph have?

16. Find the maximum number of edges in a simple, connected, planar graph with six vertices. With seven vertices.

17. Find the minimum number of vertices in a simple, connected, planar graph with 12 edges. With 19 edges.

Find the degree of each region formed by these planar graphs.

18.

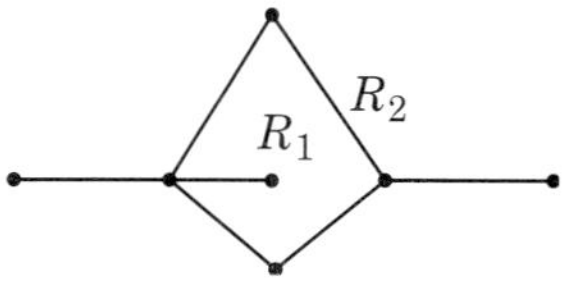

19.

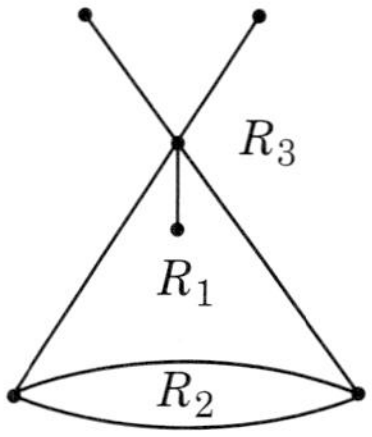

Verify that each graph is nonplanar by finding K_5 or $K_{3,3}$ as a subgraph.

20.

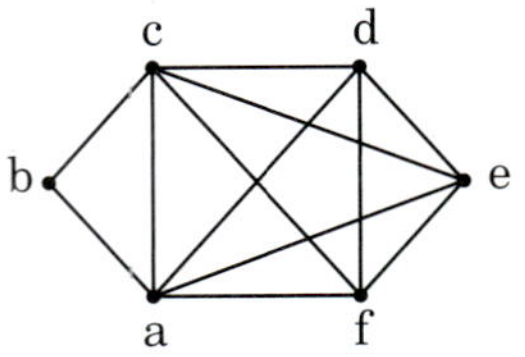

21.

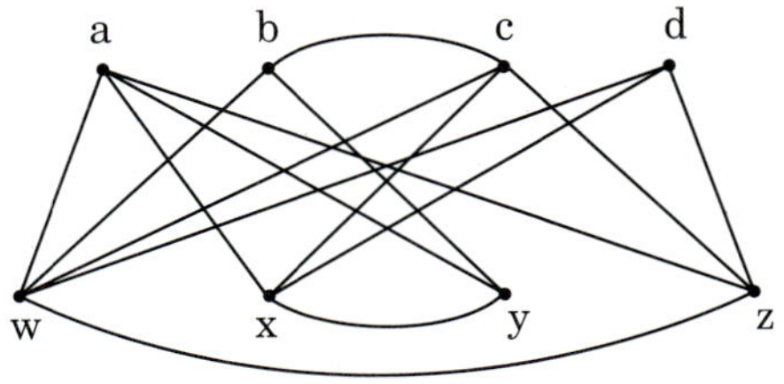

22. For what values of n will the complete graph K_n be planar?

23. For what values of m and n will the complete bipartite graph $K_{m,n}$ be planar?

Find the minimum number of edges that must be removed from each complete graph, so the resulting graph is planar.

24. K_5 **25.** K_6

***26.** Find the minimum number of vertices in a simple, connected, planar graph with e edges.

***27.** Show that the Petersen graph (Figure 8.28) is nonplanar. (*Hint*: Find a subgraph homeomorphic to $K_{3,3}$.)

8.7 Graph Coloring

Have you ever observed the traffic light pattern at intersections and calculated the number of different light patterns? Or have you ever wondered if the final examinations at a college can be scheduled so that no student would have a time conflict? Surprisingly, both problems can be modeled by graphs and solved by a technique called graph coloring.

To begin with, take a careful look at any map—for example, the map of the continental United States. Have you ever wondered how many different colors are needed to color it, so no two adjacent states are assigned the same color? (Two states that meet at a single point are *not* considered adjacent.) For instance, consider the map of a portion of the United States in Figure 8.108. Since it contains 11 states, the map can be colored with 11 colors. But this is inefficient. Can we do better? That is, can we color the map with fewer colors? First, you can verify that it can be done with four colors; no fewer than four will work because the five states around Nevada — Idaho, Oregon, California, Arizona, and Utah — must have three different colors. So with Nevada, at least four colors are needed. Figure 8.109 shows a possible coloring of the map using exactly four colors.

Figure 8.108

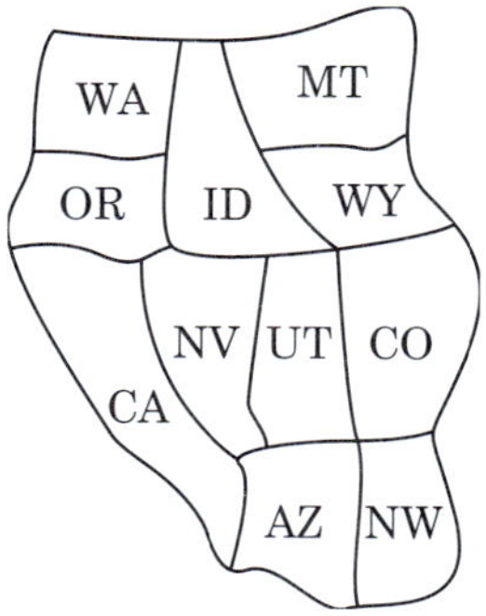

Figure 8.109

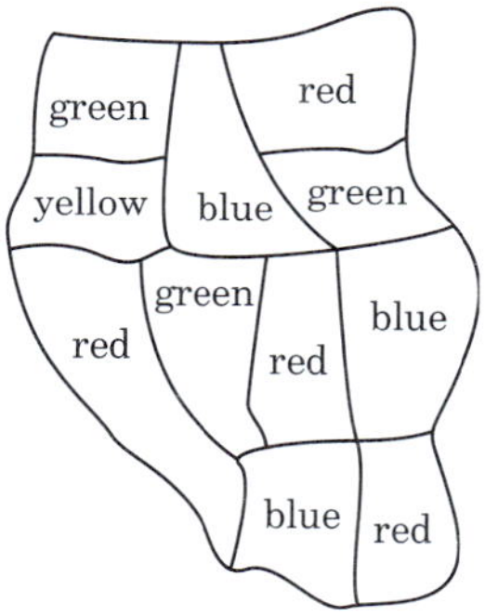

To see how map coloring is a graph-theoretic problem, represent each state by a vertex. Two vertices are adjacent if the states have a common border. Figure 8.110 shows the resulting graph. Thus the map coloring problem is equivalent to assigning a color to each vertex of the graph so that no two adjacent vertices have the same color. Figure 111 shows such a coloring, where "R" represents red; "B", blue; "G", green; and "Y", yellow. (Verify this.)

This discussion brings us to the next definition.

Graph Coloring

A coloring of a simple graph is an assignment of colors to its vertices, so no two adjacent vertices are assigned the same color.

Since a graph with n vertices can be colored with n colors, often we are interested in a coloring that takes the least number of colors. For example, notice that in Figure 8.108 vertices ID, OR, and NV are mutually adjacent; that is, they form a triangular subgraph, so three colors are needed to color them, meaning any coloring of the graph will take at least three colors.

Will three colors suffice? No, and let us see why. Notice that the vertices ID, OR, CA, UT, and AZ of the cycle subgraph must be assigned three colors. Since NV is adjacent to each one of them, it must be assigned a different color, meaning at least four colors are needed to color the graph.

Figure 8.110

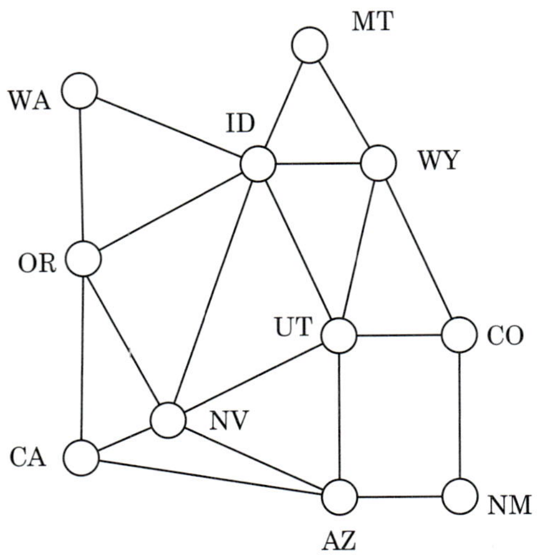

Figure 8.111

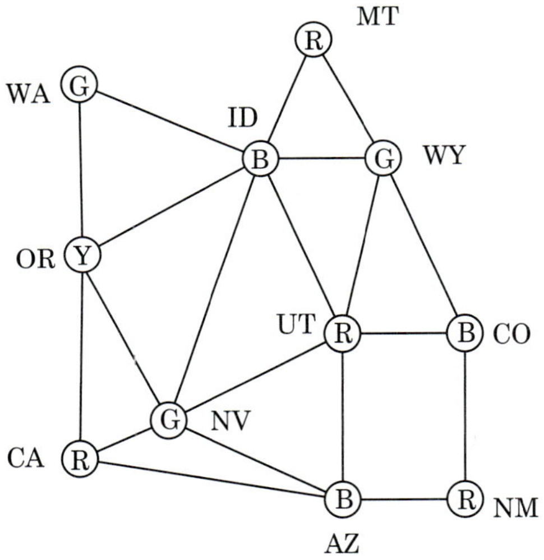

Figure 8.111 shows such a coloring, so the least number of colors needed is four.

Accordingly, we make the next definition.

Chromatic Number

The **chromatic number** of a graph is the minimum number of colors needed for a coloring of the graph.

Thus the chromatic number of the graph in Figure 8.110 is four.

Francis Guthrie *(1831–1899) was born in London, England. After graduating from University College, London, he pursued law and became a barrister.*

Around 1850, after coloring the counties of England on a map with four colors, Guthrie conjectured the four-color problem to his younger brother Frederick. At that time Frederick Guthrie (1833–1866) was a student of Augustus De Morgan, and showed the problem to De Morgan, who in turn communicated it to William Hamilton in 1852. Hamilton did not take any interest in the problem. In 1878, however, the problem was popularized by Arthur Cayley with an announcement at a meeting of The London Mathematical Society.

In 1861, Guthrie was appointed professor of mathematics at Graaff-Reinet College, Cape Colony, South Africa, and in 1876 he moved to South African University, Cape Town.

Besides mathematics, Guthrie was also interested in botany. He is known for his work on the genus Erica.

Guthrie died in Cape Town.

The Four-Color Problem

The map coloring problem pursued above is a special case of the celebrated **four-color problem**, conjectured by Francis Guthrie: *Every planar graph has a coloring using at most four distinct colors.* Its solution eluded the brilliance of mathematicians around the world for over a century. In 1890 the English mathematician Percy John Heawood (1861–1955) of Durham University proved that five colors are sufficient to color any map. In 1968 the problem was solved by O. Ore and J. Stemple, where the number of countries was at most 40. In 1976, two American mathematicians, Kenneth Appel and Wolfang Haken, using an exhaustive computer analysis, answered the problem affirmatively. Thus every planar graph has chromatic number ≤ 4.

The next three examples determine the chromatic numbers of $K_n, K_{m,n}$, and C_n.

EXAMPLE 8.43 Find the chromatic number of the complete graph K_n.

SOLUTION:

Every vertex in K_n is adjacent to each of the remaining $n - 1$ vertices, so exactly n distinct colors are needed for a coloring of K_n. Thus the chromatic number of K_n is n. Figure 8.112 shows a coloring of K_4 and K_5. ■

According to this example, the chromatic number of K_5 is five. Does this contradict the four-color theorem? Certainly not. Since K_5 is nonplanar, the theorem does not apply to K_5.

Figure 8.112

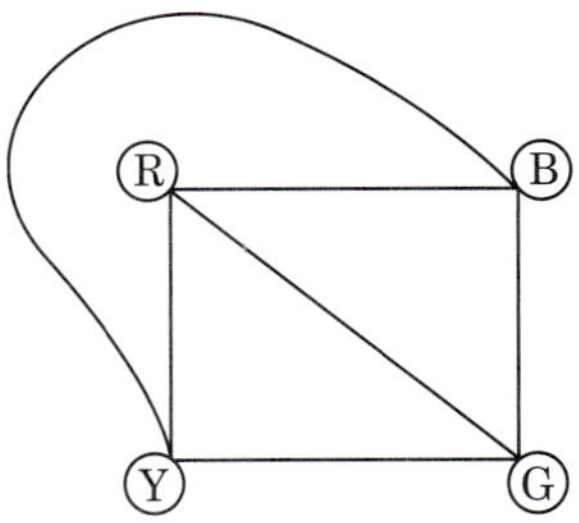

A coloring of K_4

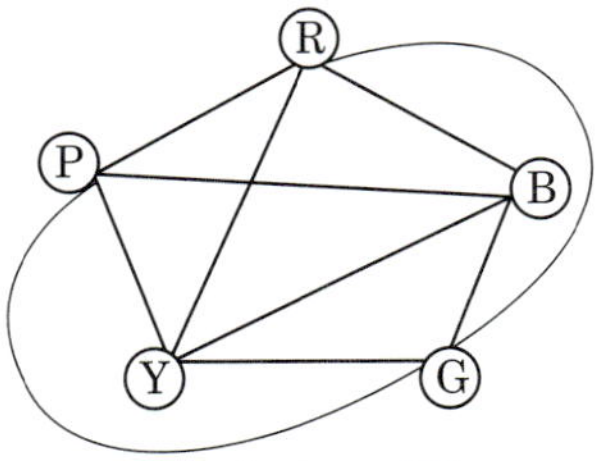

A coloring of K_5

EXAMPLE 8.44 Find the chromatic number of $K_{m,n}$.

SOLUTION:
You may recall that the vertex set in $K_{m,n}$ can be partitioned into two nonempty disjoint subsets A and B such that every vertex in A is adjacent to every vertex in B and vice versa; also, no vertices in A and in B are adjacent. So every vertex in A can be assigned one color (red), and every vertex in B a second color (blue). Thus the chromatic number of $K_{m,n}$ is two. Figure 8.113 shows a coloring of $K_{2,3}$.

Figure 8.113

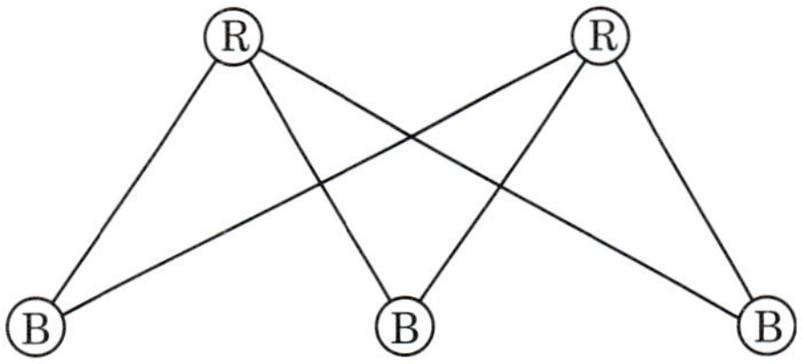

A coloring of $K_{2,3}$ ■

EXAMPLE 8.45 Find the chromatic number of the cycle graph C_n.

SOLUTION:
Suppose n is even and the vertices are $v_1, v_2, \ldots, v_{2n}$. Then the odd-numbered vertices $v_1, v_3, \ldots, v_{2n-1}$ can be assigned red and the even-numbered vertices $v_2, v_4, \ldots, v_{2n}$ blue (see Figure 8.114). So, if n is even, exactly two colors are needed.

Figure 8.114

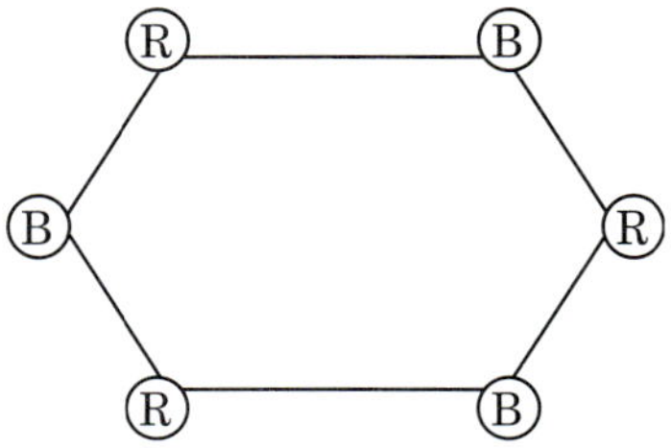

A coloring of C_6

Figure 8.115

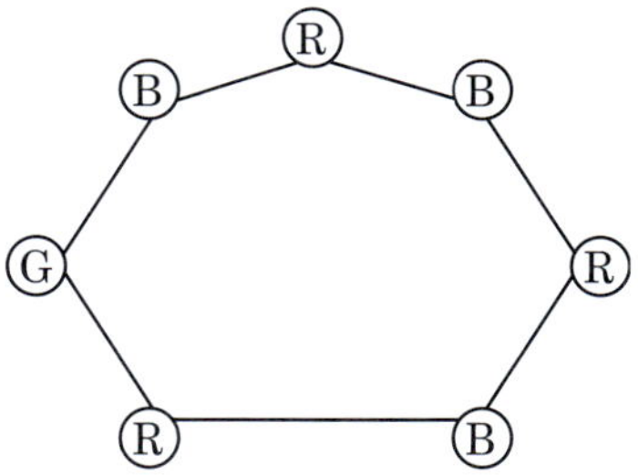

A coloring of C_7

Suppose n is odd and the vertices are $v_1, v_2, \ldots, v_{2n+1}$. If we assign red to the odd-numbered vertices and blue to the even-numbered ones, then v_1 and v_{2n+1}, which are adjacent vertices, would receive the same color. Since this is unacceptable, the chromatic number is ≥ 3. However, we can assign red to $v_1, v_3, \ldots, v_{2n-1}$ and blue to $v_2, v_4, \ldots, v_{2n}$. Since v_{2n+1} cannot be red or blue, it must be assigned a third color, green (see Figure 8.115). So exactly three colors are needed.

Thus the chromatic number of C_n is two if n is even, and three otherwise. ■

A useful observation: Suppose the chromatic number of a subgraph H of graph G is h. Then the chromatic number of G is $\geq h$. For example, the graph in Figure 8.110 contains C_3 as a subgraph with vertices WA, OR, and ID; so, by Example 8.45, the chromatic number of the graph is ≥ 3.

The next example shows how graph coloring is useful in scheduling conflict-free final examinations.

EXAMPLE 8.46 Table 8.7 lists the students taking the various courses at Königsberg College. The registrar would like to develop a conflict-free final exam schedule using as few time slots as possible. How can we help her?

Table 8.7

Course A	Course B	Course C	Course D	Course E	Course F	Course G
Boole	Cantor	Clinton	Boole	Boole	Abel	Abel
Bourbaki	Euler	Euler	Ford	Cantor	Ford	Boole
Cantor	Newton	Gauss	Hamilton	Cauchy	Gauss	Cauchy
Ford	Pascal	Newton	Hardy	Fibonacci	Nobel	Cayley
Hamilton	Russell	Nobel	Pascal	Newton	Russell	Hardy
Williams						

SOLUTION:
First we construct a graph model for the problem. To this end, represent each course by a vertex. Two vertices are adjacent if the corresponding courses are incompatible, that is, they have a common student. Figure 8.116 shows the ensuing graph.

Figure 8.116

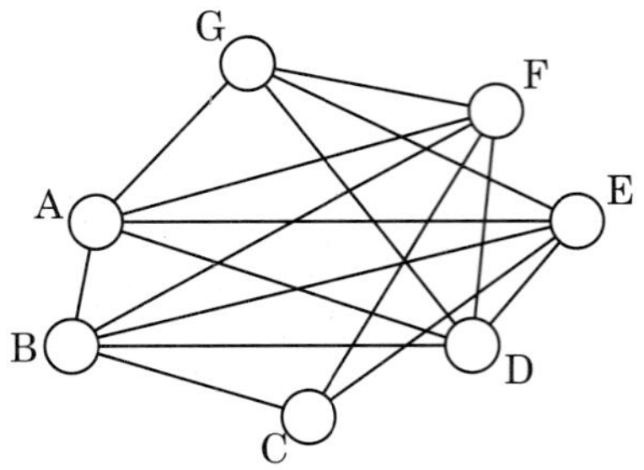

Figure 8.117

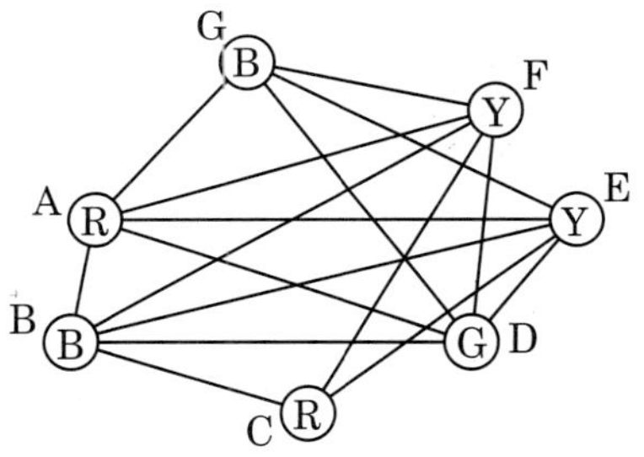

Apply coloring to it. Since the graph contains K_4 as a subgraph (with vertices A, B, D, and E), its chromatic number is ≥ 4. Figure 8.117 shows a coloring with exactly four colors.

How do we interpret this result? Since the chromatic number is four, the final exams can be scheduled conflict-free using four time slots, as Table 8.8 shows.

Table 8.8

Block	1	2	3	4
Course (s)	A, C	B, G	D	E, F

■

We close this section with the next example, which illustrates how graph coloring can be employed to design traffic light patterns at intersections.

EXAMPLE 8.47 Figure 8.118 shows the intersection of two divided avenues, California and New York, where all left and right turns are permitted. The arrows indicate the traffic flows along each avenue. Assuming they are equally heavy in each direction, design a traffic signal pattern for the intersection.*

Figure 8.118

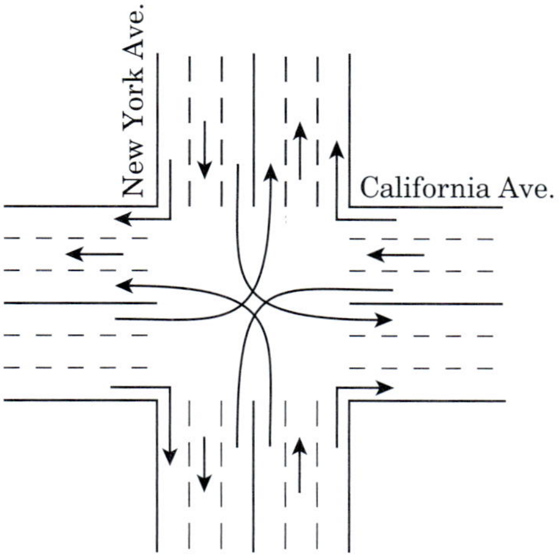

SOLUTION:

Since the four right turns do not interfere with the other traffic flows, they can safely be dropped from our discussion. The remaining traffic directions are pictured in Figure 8.119 and are labeled A through H. We need to develop a traffic pattern so that vehicles can pass through the intersection without interfering with other traffic flows.

Figure 8.119

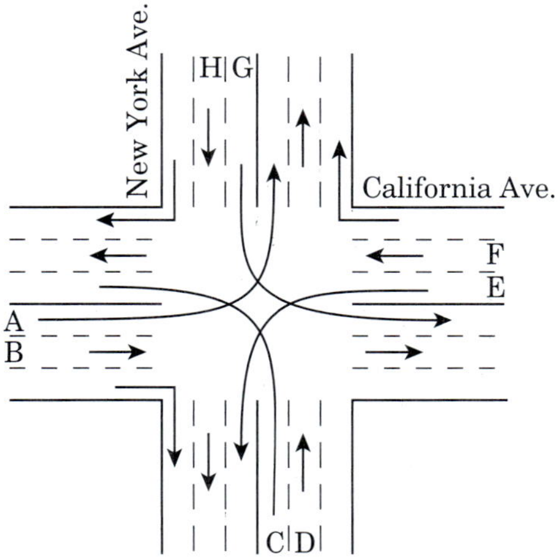

*Based on J. Burling *et al.*, "Using Graphs to Solve the Traffic Light Problem," *FAIM Module*, COMAP, Inc., Lexington, MA, 1989.

Figure 8.120

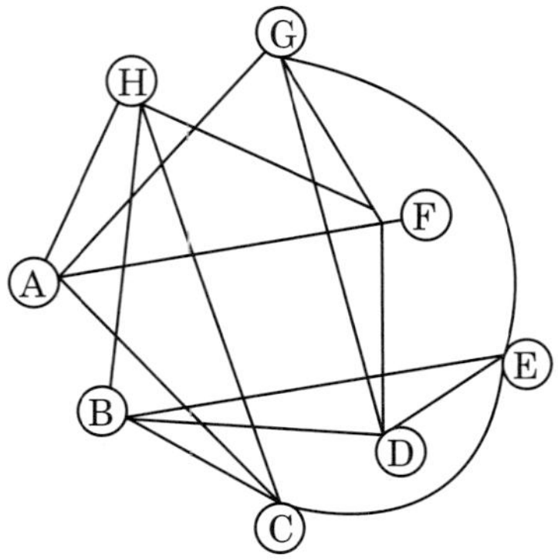

To this end, represent each traffic flow with a vertex. Two vertices are adjacent if the corresponding traffic flows cross each other. For instance, directions C and E intersect, so vertices C and E are adjacent. Figure 8.120 shows the resulting graph.

Now color its vertices. Since the graph contains C_3 as subgraph (verify), its chromatic number is ≥ 3. Is it in fact three? Let us see.

Figure 8.121

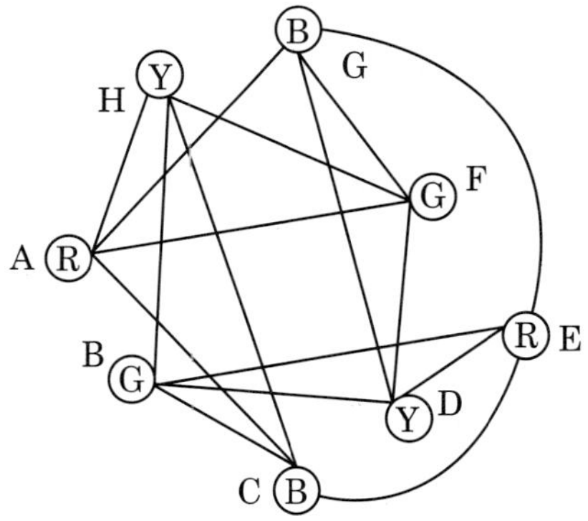

Suppose we color A red, C blue, and H yellow (see Figure 8.121). Then B cannot be blue or yellow, but can be red; so color it red. Then E must be yellow; this forces F to be blue. D, being adjacent to B, F, and H can't be red, blue, or yellow, so it must be assigned a new color, say, green. Thus the graph takes at least four colors.

Figure 8.121 shows a coloring of the graph with exactly four colors, which depicts an efficient way of designing the traffic signal pattern. It consists of four phases:

- Vertices B and F are green, so traffic flows B and F proceed, while others are waiting.
- Vertices D and H are yellow; that is, only traffic directions D and H proceed simultaneously.

- Vertices A and E are red; that is, traffic flows A and E continue at the same time, while others are stopped.
- Vertices C and G are blue, so only traffic directions C and G proceed simultaneously.

See Table 8.9 also.

Table 8.9

Traffic light pattern			
Phase 1	**Phase 2**	**Phase 3**	**Phase 4**
Only B and F proceed.	Only D and H proceed.	Only A and E proceed.	Only C and G proceed.

■

Exercises 8.7

Find the chromatic number of each map or graph.

1.

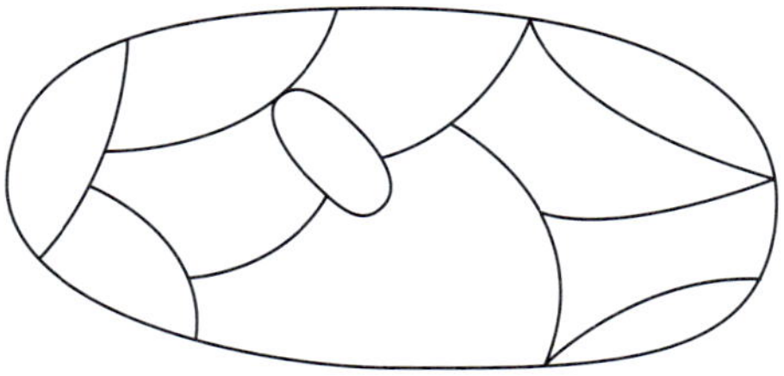

2.

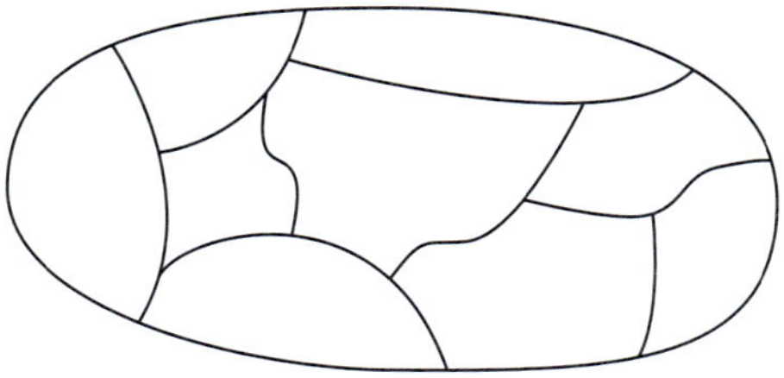

3.

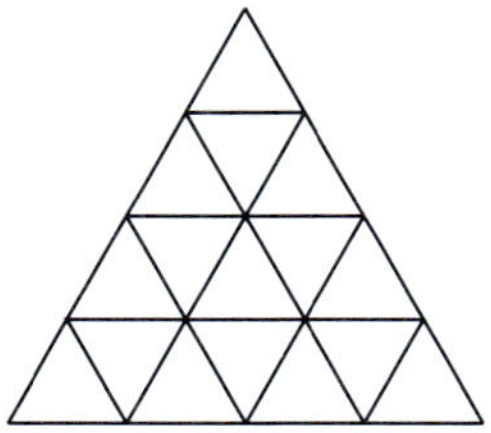

4.

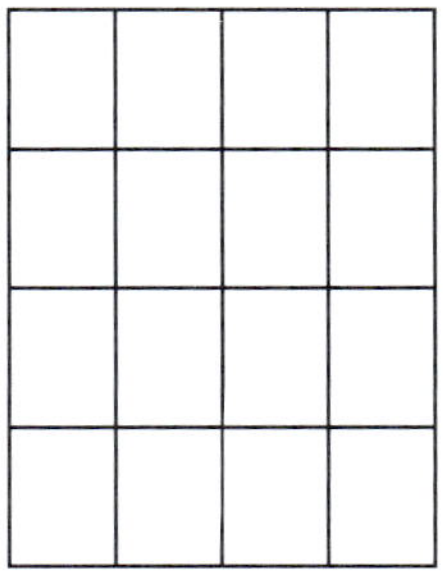

5.

Figure 8.122

Continental United States.

6. Figure 8.5 **7.** Figure 8.18 **8.** Figure 8.56

9. Graph G_1 in Figure 8.73 **10.** Figure 8.102 **11.** Figure 8.106

12. Petersen graph **13.** The Star of David **14.** Wheel graph W_n

15. 3-cube Q_3

16. Characterize graphs with chromatic number 1.

17. Let G be the union of two simple disconnected subgraphs H_1 and H_2 with chromatic numbers m and n, respectively. What can you say about the chromatic number c of G?

In Exercises 18 and 19, schedule conflict-free committee meetings using the smallest number of blocks. Identify such a schedule.

18.

Committee 1	Committee 2	Committee 3	Committee 4	Committee 5
B	A	C	A	B
D	E	D	C	G
E	G	F	E	H
F	H	H	F	I

19.

Committee 1	Committee 2	Committee 3	Committee 4	Committee 5	Committee 6
B	D	A	B	A	C
D	E	C	C	E	G
E	G	D	E	G	H
F	H	F	H	H	I
		H	I		

A zoo curator would like to add new open "islands" for seven species of animals to roam about freely. Unfortunately, some species prey on some others, as given by the incompatibility array in Table 8.10.

Table 8.10

Species	1	2	3	4	5	6	7
1		•		•	•	•	•
2	•		•	•	•	•	•
3		•			•	•	
4	•	•			•	•	
5	•	•	•	•		•	
6	•	•	•	•	•		
7	•	•					

20. Determine the minimum number of islands needed to keep them.

21. Find a possible way of accommodating the animals on those islands.

A pet shop owner just received a shipment of 10 species of tropical fish. Since some species are incompatible, that is, they fight with some other species, they cannot be kept in the same tank. Table 8.11 summarizes the incompatibility of the various species, where a dot in row i and column j indicates species i and j are incompatible.

Table 8.11

Species	1	2	3	4	5	6	7	8	9	10
1			•		•		•	•		•
2			•		•		•		•	
3	•	•		•		•		•		•
4			•			•	•			
5	•	•					•			
6			•	•			•	•		
7	•	•		•	•	•				•
8	•		•			•				
9		•								
10	•		•				•			

22. Determine the minimum number of tanks needed to store the fish.

23. Find a possible way of storing them among those tanks.

Figure 8.123 shows the traffic flows at an exit from a shopping center into a two-way street. (J. Williams, 1992)

24. Represent this information in a graph.

25. Develop a traffic light pattern so that traffic will flow smoothly at the exit.

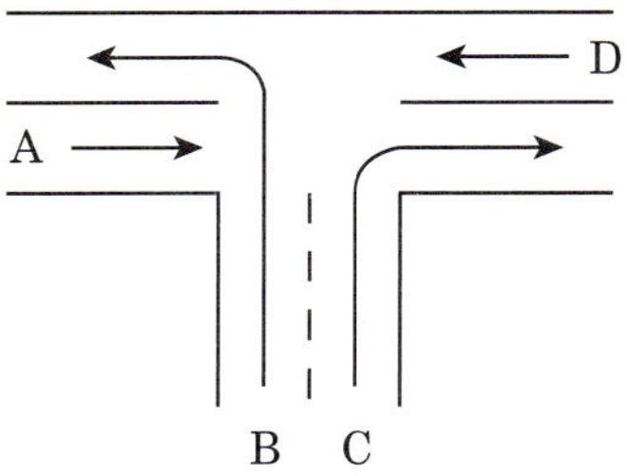

Figure 8.123

Shopping Center

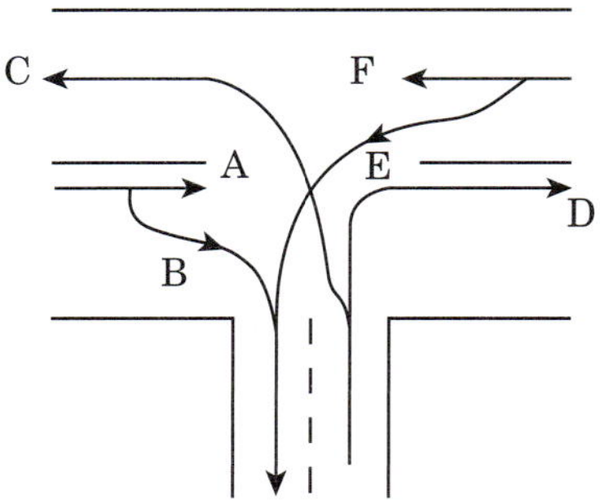

Figure 8.124

Shopping Center

Figure 8.124 shows the traffic flows from a two-way street into a shopping center and from the shopping center into the street.

26–27. Redo Exercises 8.24 and 8.25.

***28.** Let G be a simple graph with n vertices. Let k denote the maximum degree of any vertex in G. Prove that the chromatic number of G is $\leq k+1$.
(*Hint:* Apply induction on n.)

Chapter Summary

This chapter presented the rudiments of graph theory, a fast-growing branch of mathematics. The cornerstone of graph theory was laid by Euler when he solved the Königsberg bridge puzzle in 1736.

Graph

- A **graph** G consists of a nonempty, finite set V of vertices and a set E of edges connecting them: $G = (V, E)$. An edge connecting vertices u and v is denoted by $\{u, v\}$ (page 517).
- A graph with no loops or parallel edges is a **simple graph** (page 518).

Adjacency

- Two vertices v and w in a graph G are **adjacent** if $\{v, w\}$ is an edge in G (page 520).
- The **degree** of a vertex v is the number of edges meeting at v (page 520).
- The **adjacency matrix** of G is an $n \times n$ matrix $A = (a_{ij})$, where $a_{ij} =$ the number of edges from vertex v_i to vertex v_j (page 520).
- Let e denote the number of edges of a graph with n vertices $v_1, v_2, \ldots, v_n$. Then $\sum_{i=1}^{n} \deg(v_i) = 2e$ (page 521).

Special Graphs

- A **subgraph** of a graph $G = (V, E)$ is a graph $H = (V_1, E_1)$, where $V_1 \subseteq V$ and $E_1 \subseteq E$ (page 522).
- A simple graph with n vertices is a **complete graph** K_n if every pair of distinct vertices is connected by an edge (page 524).
- Let $G = (V, E)$ be a simple graph such that $V = V_1 \cup V_2, V_1 \cap V_2 = \emptyset$, and every edge is incident with a vertex in V_1 and a vertex V_2. G is a **bipartite graph**. If every vertex in V_1 is adjacent to every vertex in V_2, G is a **complete bipartite graph**. If $|V_1| = m$ and $|V_2| = n$, then G is denoted by $K_{m,n}$ (page 527).
- A **weighted graph** is a simple graph in which every edge is assigned a positive number, called the **weight** of the edge (page 527).
- An ***r*-regular graph** is a simple graph in which every vertex has the same degree r (page 536).
- The **complement** $G' = (V, E')$ of a simple graph $G = (V, E)$ contains all vertices in G. An edge $\{u, v\} \in E'$ if and only if $\{u, v\} \notin E$ (page 536).

Isomorphic Graphs

- Two simple graphs, $G_1 = (V_1, E_1)$ and $G_2 = (V_2, E_2)$, are **isomorphic** if a bijection $f : V_1 \to V_2$ exists such that $\{u, v\} \in E_1$ if and only if $\{f(u), f(v)\} \in E_2$. The function f is an **isomorphism** between G_1 and G_2. If G_1 and G_2 are isomorphic, they have exactly the same properties (page 541).

- An **isomorphism invariant** is a property shared by isomorphic graphs (page 542).

Paths, Cycles, and Circuits

- A **path of length** n from a vertex v_0 to a vertex v_n is a sequence of vertices v_i and edges e_i of the form v_0-e_1-v_1-$\cdots$-e_n-v_n, where each edge e_i is incident with the vertices v_{i-1} and v_i $(1 \leq i \leq n)$. A **simple** path contains no repeated vertices, except perhaps the endpoints(page 546).
- A path from v_0 to v_n is **closed** if $v_0 = v_n$. Otherwise, it is **open** (page 548).
- A **cycle** is a simple closed path (page 548).
- A **circuit** is a simple closed path with no repeated edges (page 548).

Connected Graph

- A **connected graph** contains a path between any two distinct vertices (page 549).
- The length of a simple path between any two vertices of a connected graph is at most $n - 1$ (page 549).
- If A is the adjacency matrix of a connected graph, the number of paths of length k from a vertex v_i to v_j is given by the ijth entry of A^k, where $1 \leq k \leq n - 1$ (page 549).

Eulerian Graphs

- A path in a connected graph is **Eulerian** if it contains every edge exactly once (page 556).
- A circuit in a connected graph is **Eulerian** if it contains every edge exactly once (page 556).
- A connected graph is **Eulerian** if it contains an Eulerian circuit (page 556).
- A connected graph is Eulerian if and only if every vertex of the graph has even degree (page 557).
- A connected graph contains an Eulerian path if and only if it has exactly two vertices of odd degree (page 561).

Hamiltonian Graphs

- A simple path in a connected graph is **Hamiltonian** if it contains every vertex (page 565).

- A cycle in a connected graph is **Hamiltonian** if it contains every vertex (page 565).
- A connected graph that contains a Hamiltonian cycle is **Hamiltonian** (page 565).

Planar Graph

- A graph is **planar** if it can be drawn in the plane, so no two edges cross (page 576).
- $K_{3,3}$ and K_5 are nonplanar graphs (page 578).
- **Euler's formula** If a connected planar graph with e edges and v vertices partitions the plane into r regions, then $r = e - v + 2$ (page 578).
- **Kuratowski's theorem** A graph is planar if and only if it does not contain a subgraph homeomorphic to K_5 or $K_{3,3}$ (page 583).

Review Exercises

Find the adjacency matrix of each graph.

1.

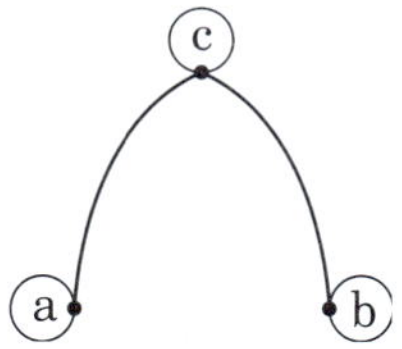

2.

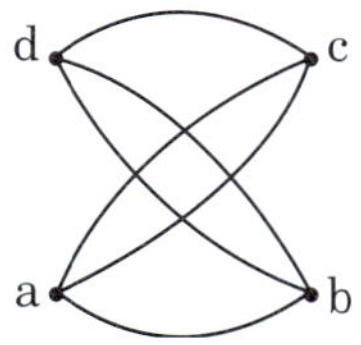

3.

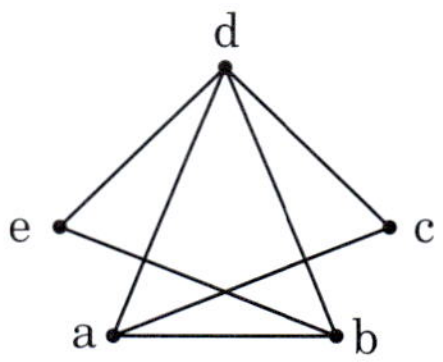

4. Verify Theorem 8.1 for the Petersen graph in Figure 8.28.

Could there be a graph with six vertices of the following degrees?

5. 4, 4, 5, 2, 3, 1

6. 4, 4, 5, 2, 3, 2

Could a hydrocarbon molecule of the following compositions exist? Assume that a carbon atom has degree four.

7. Six carbon atoms and 19 hydrogen atoms.

8. Six carbon atoms and six hydrogen atoms.

9. Find the number of subgraphs of the complete graph K_3.

10. Can there be an n-regular graph with $n \geq 1$ vertices?

11. Let n be a positive integer ≥ 2. What can you say about an $(n-1)$-regular graph with n vertices?

12. Is the complete graph K_4 bipartite? Complete bipartite?

13. Under what conditions will the graph $K_{m,n}$ be regular?

14. Is $K_{m,n}$ a complete graph?

15–17. Find the adjacency list representation of each graph in Exercises 1–3.

In Exercises 18 and 19, are the given graphs isomorphic? Find an isomorphism *f* for any that are.

18.

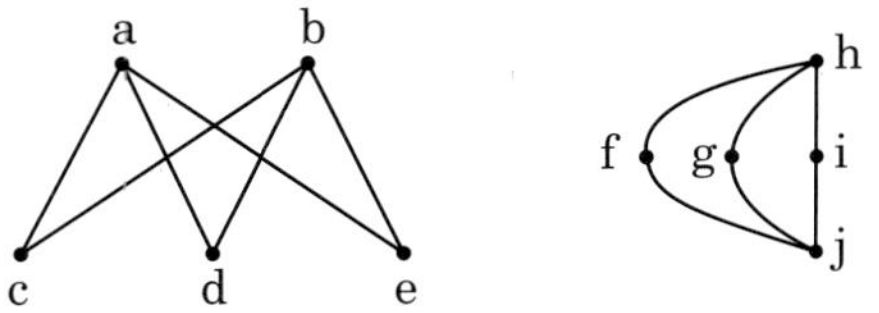

G_1 G_2

19.

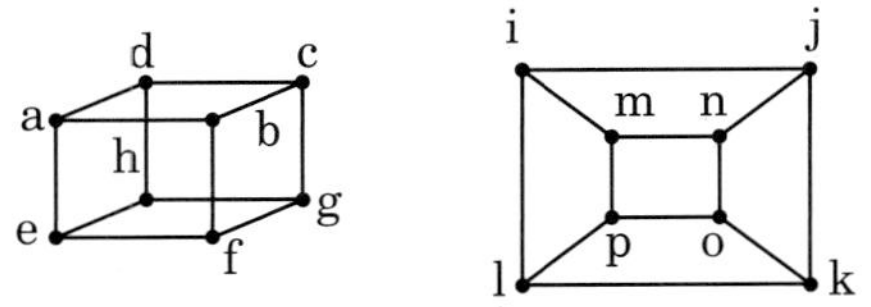

G_1 G_2

Are the following connected graphs?

20.

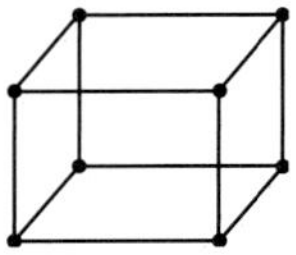

21. $K_{3,4}$

Determine if each graph is Eulerian. If yes, find an Eulerian circuit.

22.

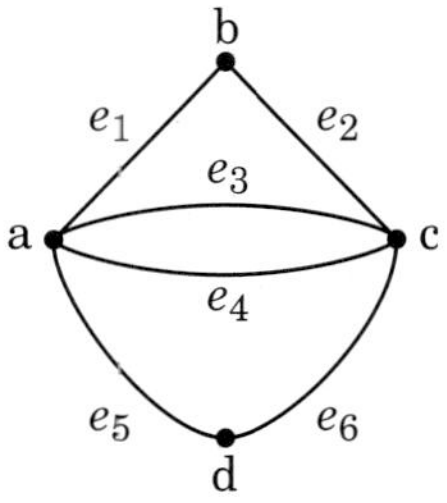

23.

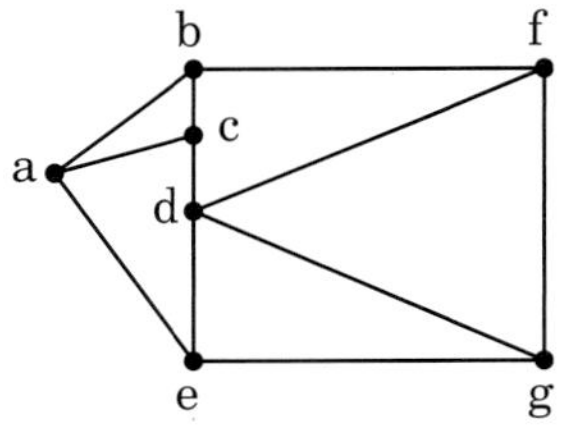

Does each graph contain an Eulerian path, but not an Eulerian circuit? If a graph contains an Eulerian path, find such a path.

24.

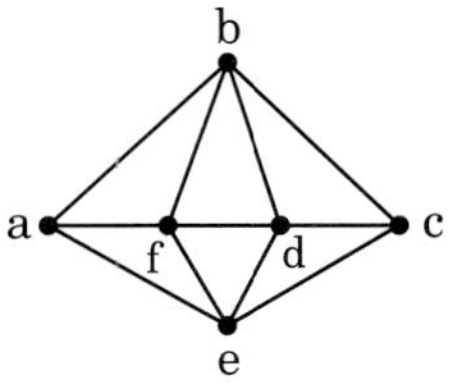

25.

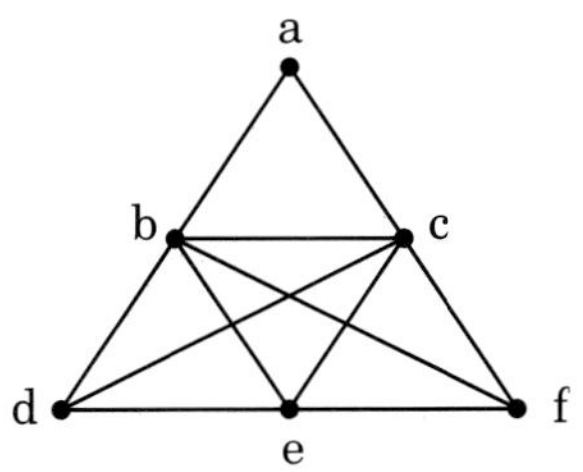

Is the graph with each adjacency matrix Eulerian?

26.
$$\begin{array}{c} \\ a \\ b \\ c \\ d \\ e \end{array}\begin{array}{c} \begin{array}{ccccc} a & b & c & d & e \end{array} \\ \begin{bmatrix} 0 & 0 & 2 & 2 & 0 \\ 0 & 0 & 1 & 1 & 0 \\ 2 & 1 & 0 & 1 & 2 \\ 2 & 1 & 1 & 0 & 2 \\ 0 & 0 & 2 & 2 & 0 \end{bmatrix} \end{array}$$

27.
$$\begin{array}{c} \\ a \\ b \\ c \\ d \\ e \end{array}\begin{array}{c} \begin{array}{ccccc} a & b & c & d & e \end{array} \\ \begin{bmatrix} 0 & 0 & 1 & 1 & 1 \\ 0 & 0 & 1 & 1 & 1 \\ 1 & 1 & 0 & 0 & 0 \\ 1 & 1 & 0 & 0 & 0 \\ 1 & 1 & 0 & 0 & 0 \end{bmatrix} \end{array}$$

28–31. Determine if the graphs in Exercises 24–27 are Hamiltonian.

32. The weights in the graph in Figure 8.125 represent the distances between cities a through e. A saleswoman based at city a would like to visit every other city exactly once and return to the home city, keeping her total travel to a minimum. What route should she take and how far will she travel?

Figure 8.125

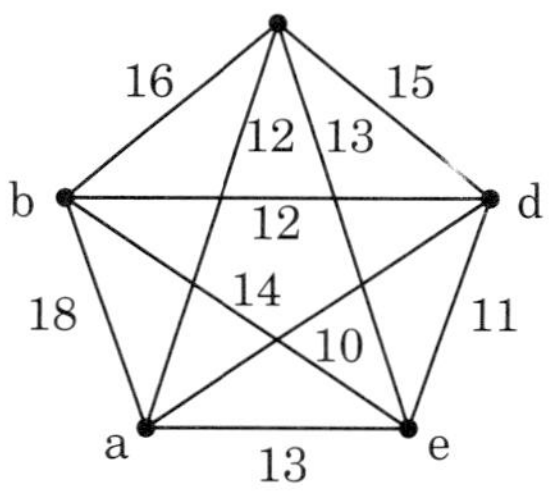

Draw a planar representation of each planar graph.

33.

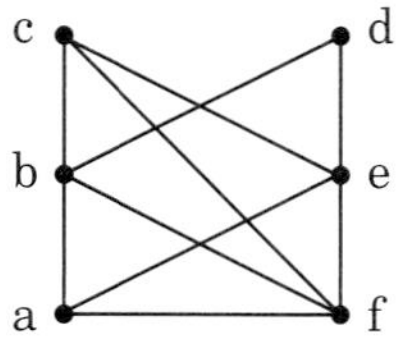

34.

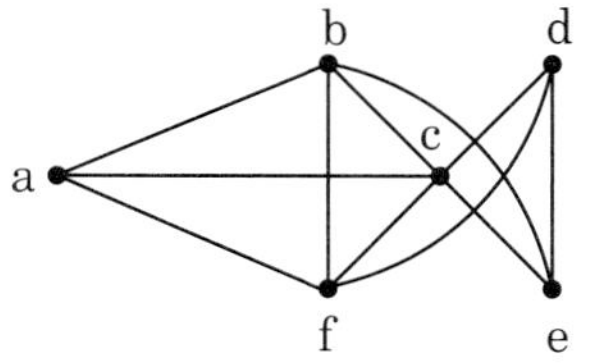

Are the following graphs planar? Draw a planar representation of any that are.

35.

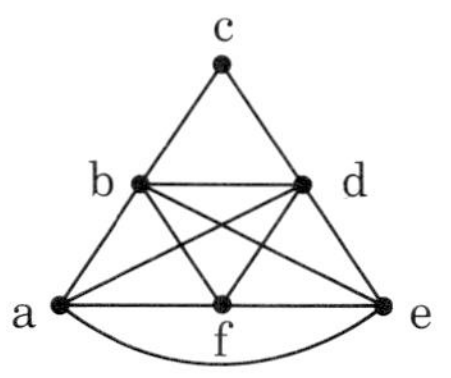

36.

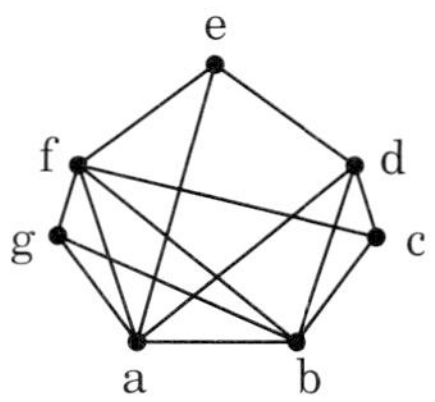

Verify Euler's formula for each connected, planar graph.

37. 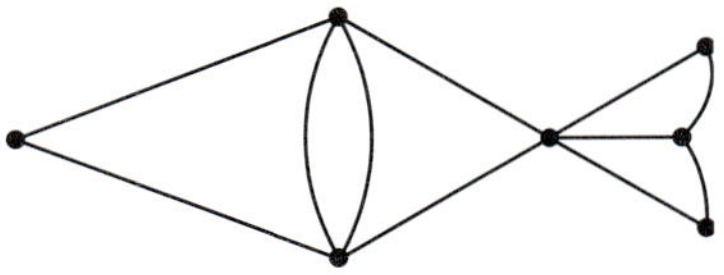

38. 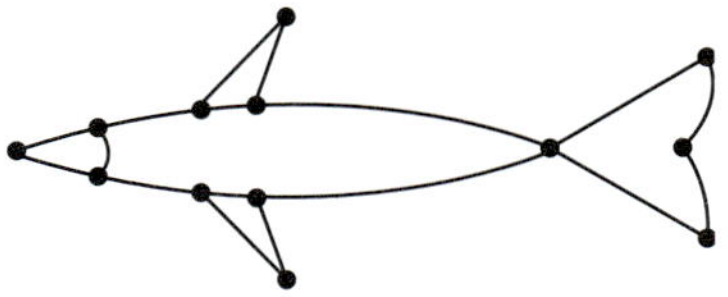

39. A connected, planar graph has seven vertices, and partitions the plane into eight regions. How many edges does it have?

40. A connected, planar graph has 10 edges. It does not divide the plane into smaller regions. Compute the number of its vertices.

Determine if each bipartite graph is planar.

41. $K_{2,2}$ **42.** $K_{3,4}$

Let G be a simple bipartite planar graph with $e \geq 2$ edges and v vertices. Then $e \leq 2v - 4$. Using this fact, show that each bipartite graph is nonplanar.

43. $K_{3,3}$ **44.** $K_{3,4}$ **45.** $K_{3,5}$ **46.** $K_{4,5}$

Supplementary Exercises

1. Find the number of vertices in C_n, W_n, and Q_n. (See Figures 8.16, 8.17, and 8.75.)

2. Find the number of edges in C_n, W_n, and Q_n.

3. For what value(s) of n is C_n a regular graph?

***4.** Prove that if W_n is regular, then $n = 3$.

A graph G with n vertices and e edges is **graceful** if the vertices can be labeled $1, 2, \ldots, n$ in such a way that the values $|i - j|$ are different for every pair of adjacent vertices i and j. For example, the graph in Figure 8.126 is graceful.

Figure 8.126

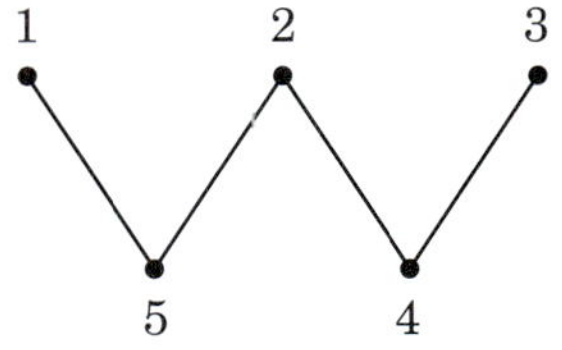

Are the following graphs graceful?

5.

6. 

7. 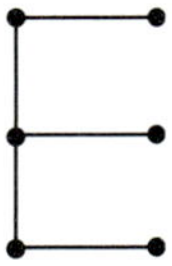

8. Prove that every open path with n vertices is graceful.

9. Prove that $K_{1,n}$ is graceful.

A graph $G = (V, E)$ is **complete n-partite** if the following conditions are satisfied:

- $V = V_1 \cup V_2 \cup \cdots \cup V_n$, where $V_i \neq \emptyset$ for every i and $V_i \cap V_j = \emptyset$ for $i \neq j$; and
- If $u \in V_i$ and $v \in V_j$, then $\{u, v\} \in E$, where $i \neq j$.

10. What can you say about G if $|V_i| = 1$ for every i?

11. Let $|V_i| = v_i$, $1 \leq i \leq n$. Find $|V|$ and $|E|$.

12. Prove that if the degree of every vertex of a regular graph $G = (V, E)$ is odd, $|V|$ is even.

***13.** Prove that two simple graphs are isomorphic if and only if their complements are isomorphic.

A simple graph G is **self-complementary** if G is isomorphic to G'. Figure 8.127 shows a graph G and its complement G'. Verify G and G' are isomorphic; so G is self-complementary.

Figure 8.127

14. Prove that there cannot be a self-complementary graph with three vertices.

15. Draw a self-complementary graph with five vertices.

16. Find the number of edges in a self-complementary graph with n vertices.

***17.** Let n (≥ 2) denote the number of vertices in a self-complementary graph. Show that $n \equiv 0 \pmod 4$ or $n \equiv 1 \pmod 4$.

18. Delete any vertex and edges incident with it in the Petersen graph. Show that the resulting subgraph is Hamiltonian.

19. Find the number of distinct Hamiltonian cycles in K_n, where $n \geq 3$.

***20.** Find the number of distinct Hamiltonian cycles in $K_{n,n}$, where $n \geq 2$.

***21.** Find the number of distinct Hamiltonian paths in $K_{n+1,n}$, where $n \geq 1$.

22. Delete any edge from K_5. Show that the resulting subgraph is planar.

23. Delete any edge from $K_{3,3}$. Show that the resulting subgraph is planar.

Computer Exercises

Let G be a graph with n vertices, labeled 1 through n, where $1 \leq n \leq 10$. Write a program to do each task.

1. Read in n and the various edges $\{i, j\}$ of G, where $1 \leq i, j \leq n$.
   - Print the adjacency matrix.
   - Print the degree of each vertex.
   - Print the linked list representation of G.
   - Determine if G is a simple graph.
   - Determine if G is a complete graph.
2. Let G be a weighted graph. Read in n, the various edges $\{i, j\}$ of the graph, and their weights w.
   - Print the weighted adjacency matrix.
   - Print the adjacency list representation.
3. Read in the adjacency matrix of G. Print its adjacency list representation. Use the adjacency list representation to print the adjacency matrix of G.
4. Read in the adjacency matrix of a simple graph and determine if it is r-regular.
5. Read in two positive integers m and n, where $m, n \leq 10$; the vertex sets V_1 and V_2 of a bipartite graph G, where $|V_1| = m$ and $|V_2| = n$; and the edges $\{i, j\}$ in G, where $i, j \leq 10$. Using the adjacency matrix and adjacency list representation of G, determine if it is the complete bipartite graph $K_{m,n}$.
6. Read in the adjacency matrix of G. Determine if G is connected. Find how many simple paths run from vertex i to vertex j, where $1 \leq i, j \leq 10$ and $i \neq j$.
7. Read in the adjacency matrix of a graph and determine if the graph is Eulerian. If it is not, see if it contains an Eulerian path.
8. Read in the edges $\{i, j\}$ of a graph, where $1 \leq i, j \leq n$. With the linked list representation of the graph, determine if it is Eulerian. If it is not Eulerian, determine if it contains an Eulerian path.
9. Let G be a complete weighted graph whose vertices and weights represent cities and distances between them, respectively. Read in the

weighted adjacency matrix of G. Find a Hamiltonian cycle so that the sum of the weights along the cycle is a minimum.

10. Solve the knights puzzle in Example 8.26. List all moves.

11. Read in the various class lists in Example 8.46. Find a conflict-free final exam schedule for the courses.

12. Read in the adjacency matrix for the 48 states of the continental United States. Assign a coloring to them in such a way that adjacent states receive different colors.

Exploratory Writing Projects

Using library and Internet resources, write a team report on each of the following in your own words. Provide a well-documented bibliography.

1. Explain the applications of graph theory to various disciplines.

2. Describe the history of the traveling salesperson problem. Name a few practical applications.

3. Give a historical background of the four-color problem. Include a biography of K. Appel and W. Haken.

4. Explain the Chinese Postman Problem.

5. Explain how *de Bruijn sequences*, named after Nicolaas G. de Bruijn, can be generated from Eulerian circuits.

6. Explain the relationship between Hamiltonian cycles and the Tower of Brahma.

7. Example 8.11 is an application of *Ramsey's theorem*, developed by the English logician Frank P. Ramsey (1903–1930). The theorem laid the foundation for a branch of combinatorics called *Ramsey theory*. Give a brief introduction to Ramsey theory.

8. Discuss the *Instant Insanity Puzzle*.

9. Discuss the relationship between n-cubes and parallel computers. How can two $(n-1)$-cubes be used to construct an n-cube? Also, discuss the relationship between an n-cube and Gray code.

10. Write an essay on graceful graphs, a term coined by S. W. Golomb of the University of Southern California.

11. Write an essay on the game of *SIM*.

12. Write an essay on the game *DIM*, a three-dimensional variation of SIM invented in 1972 by D. Engel.

13. Describe the game of *Hackenbush*.

14. Write an essay on chromatic polynomials.

15. Write an essay on the Petersen graph.

16. Investigate rook polynomials.

17. Write an essay on matching theory.

18. Write an essay on parallel algorithms.

Enrichment Readings

1. K. Appel and W. Haken, "Every Planar Map is 4-Colorable," *Bulletin of the American Mathematical Society*, Vol. 82 (1976), pp. 711–712.

2. J. A. Bondy and U. S. R. Murty, *Graph Theory with Applications*, Elsevier, New York, 1976.

3. R. A. Brualdi, *Introductory Combinatorics*, 3rd ed., Prentice-Hall, Upper Saddle River, NJ, 1999.

4. N. Cristofides, *Graph Theory: An Algorithmic Approach*, Academic Press, New York, 1975.

5. B. W. Jackson and D. Thoro, *Applied Combinatorics with Problem Solving*, Addison-Wesley, Reading, MA, 1990, pp. 134–200.

6. J. A. McHugh, *Algorithmic Graph Theory*, Prentice-Hall, Englewood Cliffs, NJ, 1990.

7. A. Ralston, "Debruijn Sequences—A Model Example of the Interaction of Discrete Mathematics and Computer Science," *Mathematics Magazine*, Vol. 55 (May 1982), pp. 131–143.

8. S. S. Skieno, *Implementing Discrete Mathematics*, Addison-Wesley, Reading, MA, 1990.

9. K. Thulasiraman and M. N. S. Swamy, *Graphs: Theory and Algorithms*, Wiley, New York, 1992.

10. A. Tucker, *Applied Combinatorics*, 2nd ed., Wiley, New York, 1984, pp. 3–79, 389–410.

11. D. West, *Introduction to Graph Theory*, 2nd ed., Prentice-Hall, Upper Saddle River, NJ, 2001.

12. R. J. Wilson and J. J. Watkins, *Graphs: An Introductory Approach*, Wiley, New York, 1990.

Chapter 9

Trees

An expert problem solver must be endowed with two incompatible qualities — a restless imagination and a pertinent pertinacity.

— HOWARD W. EVES

Trees are the most important class of graphs and they make fine modeling tools. In 1847, the German physicist Gustav Robert Kirchoff used them to solve systems of linear equations for electrical networks. Ten years later, Arthur Cayley studied the isomers of saturated hydrocarbons C_nH_{2n+2} with them. Today trees are widely used in mathematics and computer science, as well as in linguistics and the social sciences.

This chapter presents the concept of a tree, and two necessary and sufficient conditions for a graph to be a tree. In addition, it presents the concept of a spanning tree for a connected graph, algorithms for finding it, and the important class of rooted trees.

Trees facilitate sorting and searching, representing and computing numeric expressions, and constructing efficient coding schemes and algorithms, as will be seen in Sections 9.5 and 9.6.

Some of the interesting problems trees handle well are:

- Can four queens be placed on a 4×4 chessboard, so they are not attacking each other?
- A utility company would like to lay pipelines for natural gas between five towns, a through e, as in Figure 9.1. The weights of the edges represent the costs of building the various pipelines. How cheaply can the company build them, so natural gas can be sent from any town to any other town?
- There are 64 entrants in a singles tennis tournament. The winner of each round is advanced to the next round and plays another. Find the number of matches and rounds played to determine the champion.
- There are eight coins in a collection plate. Although they look identical, one of them is counterfeit and heavier. Identify it, using an equal arm balance and a minimum number of weighings.

***Gustav Robert Kirchoff** (1824–1887), an outstanding German physicist and son of a lawyer, was born in Königsberg, Prussia. After graduating from the local gymnasium, Kirchoff entered the University of Königsberg at the age of 18 and received his doctorate 5 years later.*

In 1848, his teaching career began in Berlin. Two years later he joined the faculty of the University of Breslau, where he met his future research collaborator, the well-known chemist Robert Bunsen. In 1854, they both moved to Heidelberg. While there, with Bunsen's collaboration, Kirchoff made his greatest contributions to science.

In 1875, he accepted the chair of theoretical physics at the University of Berlin, a position he held with great distinction until his death.

Although Kirchoff made significant contributions to every branch of physics, he is best known for his pioneering work in spectroscopy.

Figure 9.1

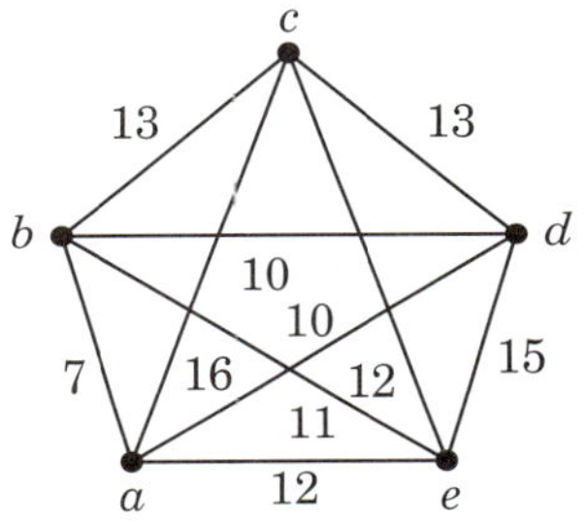

As in graph theory, tree terminology is *not* standardized, so be aware of this when you refer to different texts on the topic.

9.1 Trees

Notice that the graphs in Figures 9.2 and 9.3 are connected; each is **acyclic**, meaning it does not contain a cycle. Such a graph is a tree.

Figure 9.2

Propane C_3H_8.

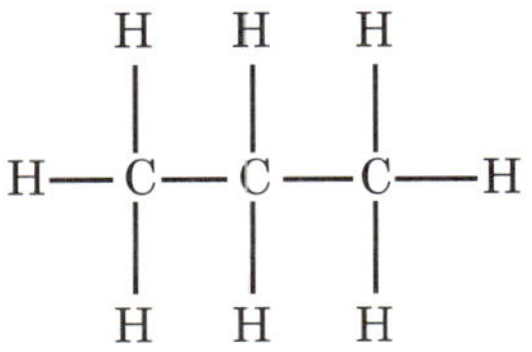

Figure 9.3

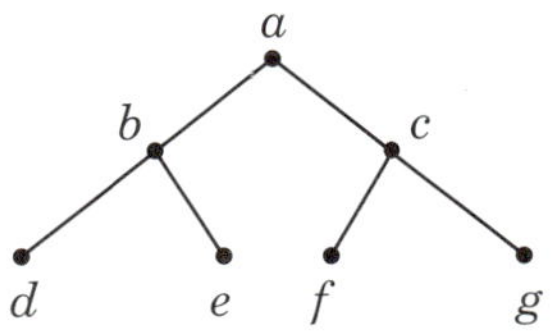

Tree

A connected, acyclic graph is a **tree**.

Trees are everywhere. You have seen several examples of trees in earlier chapters. Figure 1.1 , Figures 5.16–5.19, and Figures 6.7, 6.8, and 6.14 offer fine cases.

The graph in Figure 9.4 is not a tree, since it contains a cycle. The graphs in Figures 9.2 and 9.3 are connected; each is **acyclic**, so each is a tree. The graph in Figure 9.5 is not a tree, either, because it is not connected. Nonetheless, it is a set of disjoint trees, called a **forest**.

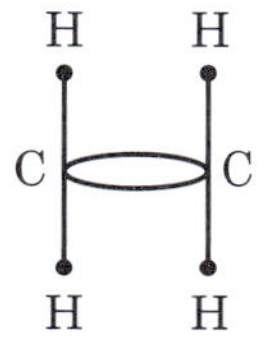

Figure 9.4

Ethylene C_2H_4.

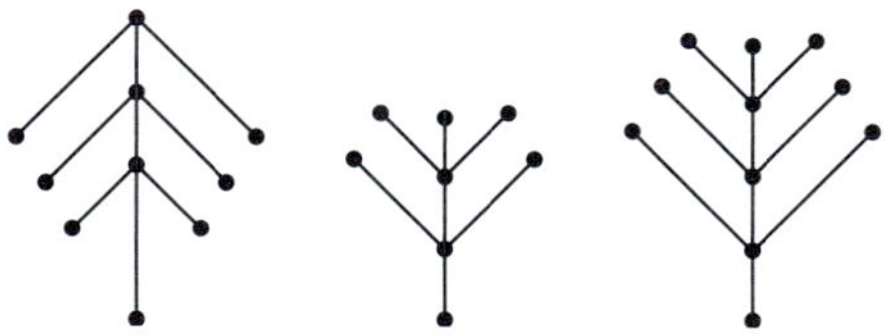

Figure 9.5

A forest.

Figure 9.6 shows the **family tree** of the Bernoullis of Switzerland, the most distinguished family of mathematicians.

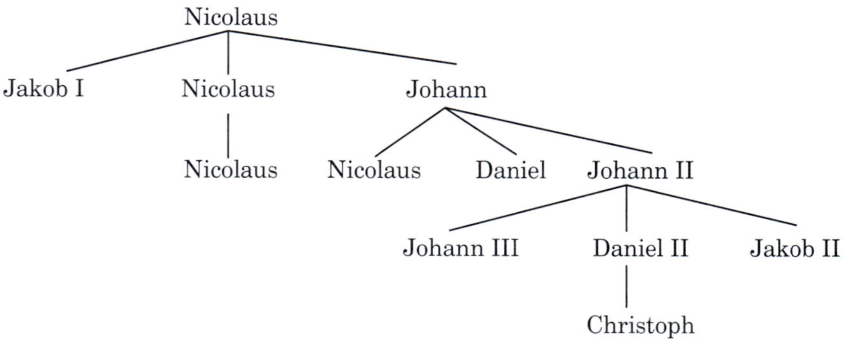

Figure 9.6

The Bernoulli family.

Figure 9.7 depicts a tree of a corporation's top management.

The tree in Figure 9.8 shows a partial hierarchical chart for teams in the National Hockey League (NHL) during the 2002–2003 season.

To qualify as a tree, a connected graph must fulfill path and edge requirements. First we present the path requirement.

Figure 9.7

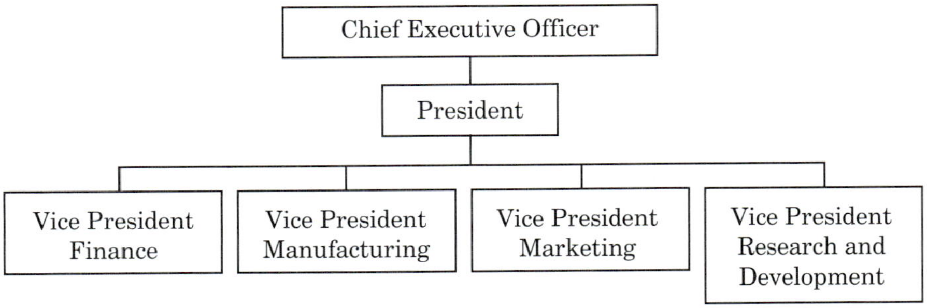

Figure 9.8

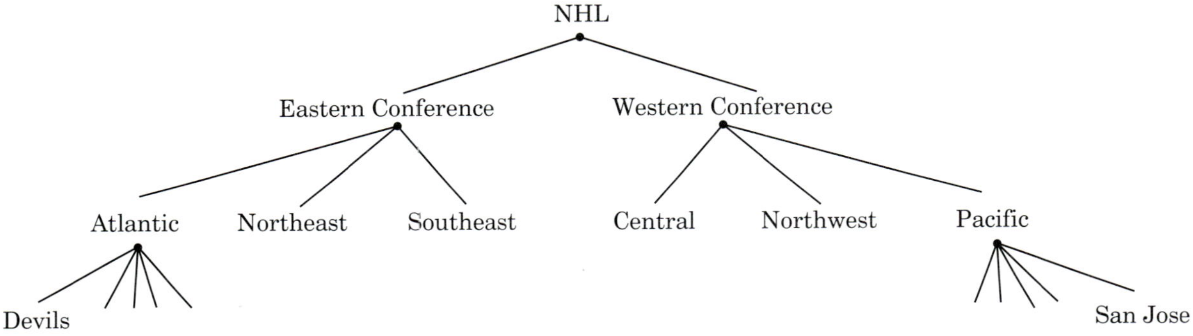

THEOREM 9.1 A connected graph is a tree if and only if there is a unique, simple path between any two vertices.

PROOF:

Let G be a tree, and v and w any two of its vertices. Since G is connected, by Theorem 8.3, a simple path must run between them. If there are two distinct simple paths between them, then one path followed by the other in reverse order would form a cycle. This, however, is impossible since G is a tree; so G contains a unique, simple path between v and w.

Conversely, let G be a graph with a unique, simple path between any two vertices. Clearly, G is connected. Suppose G contains a cycle, and v and w are two vertices in it. Then the cycle can be split into two distinct simple paths between v and w, a contradiction, so G is acyclic. Being connected and acyclic, G is a tree. ■

You may verify that the graph in Figure 9.2 is connected and contains a unique, simple path between any two vertices. Therefore, by Theorem 9.1, it is a tree. The same holds for Figure 9.3.

Next we establish the edge requirement for a connected graph to be a tree.

THEOREM 9.2 A connected graph with n vertices is a tree if and only if it has exactly $n - 1$ edges.

PROOF:
First we prove by strong induction that a tree with n vertices has $n - 1$ edges.

Let P(n): A tree T with n vertices has $n - 1$ edges.

Basis step Suppose T contains one vertex. Since T is acyclic, it is loop-free; consequently, T contains no edges and P(1) is true.

Induction step Suppose the result is true for every tree with k or fewer vertices. Let T be a tree with $k + 1$ vertices and e an edge in T. Deleting e from T yields two connected disjoint graphs, T_1 and T_2. Each is a tree. Suppose T_1 has p vertices; then T_2 has $q = k + 1 - p$ vertices. By the inductive hypothesis, T_1 contains $p - 1$ edges and T_2 contains $q - 1$ edges; so T contains $(p - 1) + (q - 1) + 1 = (p - 1) + (k - p) + 1 = k$ edges.

Thus by strong induction, P(n) is true for every $n \geq 1$.

Conversely, let G be a connected graph with n vertices and $n - 1$ edges. If it were not a tree, it would have a cycle. Remove an edge from this cycle. The resulting graph is still connected. If it is not acyclic, remove an edge from a cycle. Continue this procedure to get an acyclic graph H. Thus H is connected and acyclic; it must be a tree with n vertices. So, by the first part, H has $n - 1$ edges and hence G has more than $n - 1$ edges, which is a contradiction. Thus G is connected and acyclic, and hence a tree. ■

For example, the graph in Figure 9.2 is connected and has 11 vertices, so it must contain 10 edges to be a tree, which is true.

Exercises 9.1

Determine if each graph is a tree.

1. 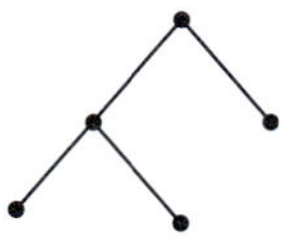

2. 

3. 

4. 

5. 

6. 

7. Is the graph in Figure 8.4b a tree? If not, why?

8. Determine if the Petersen graph in Figure 8.28 is a tree. If it isn't, explain why.

Let n denote the number of vertices of a tree and e the number of edges. Verify that $e = n - 1$ for each tree.

9.  **10.**  **11.** 

12. How many bonds does the hydrocarbon molecule C_nH_{2n+2} have? Assume a carbon molecule has degree four.

13. Let T be a tree with vertices $v_1, \ldots, v_n$. Show that $\sum_{i=1}^{n} \deg(v_i) = 2n - 2$.

14. For what values of n is K_n a tree?

Determine if each complete bipartite graph is a tree.

15. $K_{1,2}$ **16.** $K_{1,3}$ **17.** $K_{2,2}$ **18.** $K_{2,3}$

19. For what values of m and n is $K_{m,n}$ a tree?

20. Let G be an r-regular tree with n vertices. Prove that $n = 1$ or $n = 2$.

Draw all nonisomorphic trees with the given number of vertices n.

21. 2 **22.** 3 **23.** 4 **24.** 5

25. The **eccentricity** of a vertex v in a tree is the length of the longest simple path from v. Find the eccentricity of each vertex of the tree in Figure 9.9.

Figure 9.9

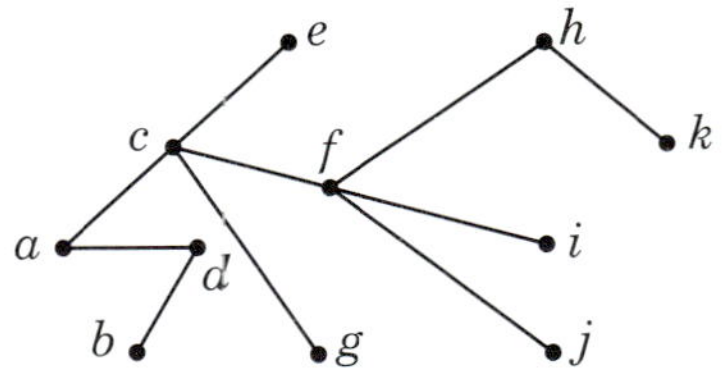

26. The **center** of a tree is a vertex with the least eccentricity. Find the center(s) of the tree in Figure 9.9.

27. Using the adjacency matrix of a connected graph with n vertices, write an algorithm to determine if it is a tree.

9.2 Spanning trees

All connected graphs have trees that span them. We will discuss three ways to find them, as well as solve the 4-queens problem stated at the beginning of the chapter, but first a more pressing practical job needs to be done.

A county's five towns, A through E, are connected by roads (see Figure 9.10), but 2 feet of snow cover them. The county would like to plow as few roads as possible, so one can travel between the towns. Figure 9.11 displays one possible solution.

Figure 9.10

Graph G.

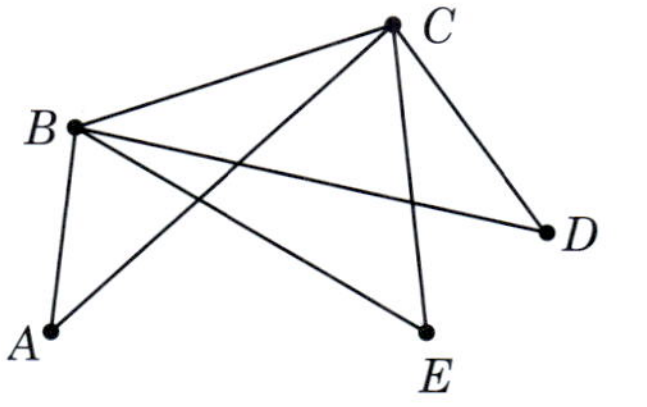

Figure 9.11

Graph H.

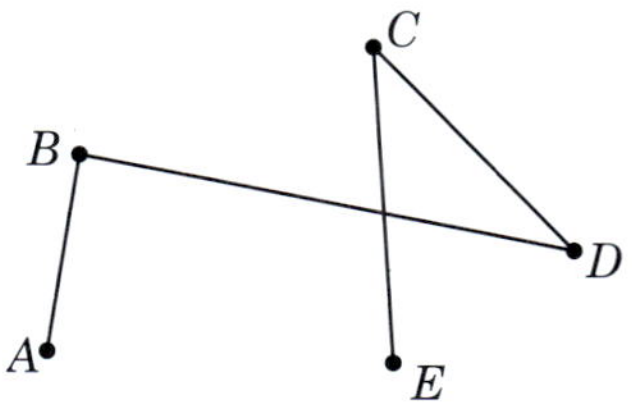

You may notice that the graph H in Figure 9.11 is a subgraph of G. In fact, it is a tree containing every vertex of G. It is called a spanning tree of G.

Spanning Tree

A subgraph H of a connected graph G is a **spanning tree** of G if H is a tree containing every vertex of G.

Using this definition, we can solve the snow-plowing problem by finding a spanning tree for the graph G.

Two additional spanning trees are given in Figure 9.12, indicating that the spanning tree of a graph need not be unique.

Figure 9.12

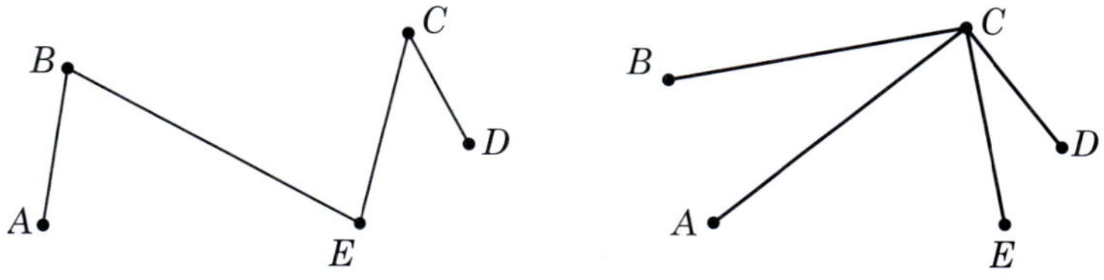

Does every connected graph have a spanning tree? Theorem 9.3 affirms that it does.

Joseph Bernard Kruskal *(1928–) was born in New York City. He graduated from the University of Chicago in 1948 and received his Ph.D. from Princeton in 1954. After being an instructor at Princeton and at the University of Wisconsin, he became an assistant professor at the University of Michigan in 1958. The following year he joined the technical staff at Bell Telephone Labs, a position he still holds. Kruskal has served as visiting professor at Yale, Columbia, and Rutgers.*

THEOREM 9.3 Every connected graph has a spanning tree.

PROOF:
Let G be a connected graph. If G is a tree, we are done.

If G is not a tree, it must contain a cycle. Remove an edge from the cycle. The new graph is still connected. If it is acyclic, then it is a tree and hence a spanning tree. Otherwise, it must have another cycle. Remove an edge from this cycle. Continue this procedure until a subgraph H is acyclic. Since H is both connected and acyclic, it is a tree. H also contains every vertex of G, so it is a spanning tree of G. ■

Three algorithms to find a spanning tree in a connected graph are given below: Kruskal's, the Depth-First Search, and the Breadth-First Search. Kruskal's algorithm represents a special case of an algorithm for a minimal spanning tree that we will examine in the next section. It was developed by the American mathematician Joseph Bernard Kruskal in 1956.

Kruskal's Algorithm for a Spanning Tree

Although the proof of Theorem 9.3 can find a spanning tree, Kruskal's algorithm takes advantage of Theorem 9.2. It is given in Algorithm 9.1.

```
Algorithm Spanning Tree (G,T)
(* G = (V, E) is a connected graph with n vertices and T denotes a
   spanning tree. Numedges denotes the number of edges selected.
      Begin (* algorithm *)
       T ← Ø (* initialize tree *)
       numedges ← 0 (* initialize numedges *)
       while numedges < n − 1 do
       begin (* while *)
         select an edge e in G
         E ← E − {e}
```

```
        if e does not create a cycle in T then
          begin
            T ← T ∪ {e}
            numedges ← numedges + 1
          endif
        else
          discard e
    endwhile
End (* algorithm *)
```

Algorithm 9.1

The next example illuminates this algorithm.

EXAMPLE 9.1 Using Kruskal's algorithm, find a spanning tree for the graph G in Figure 9.13.

Figure 9.13

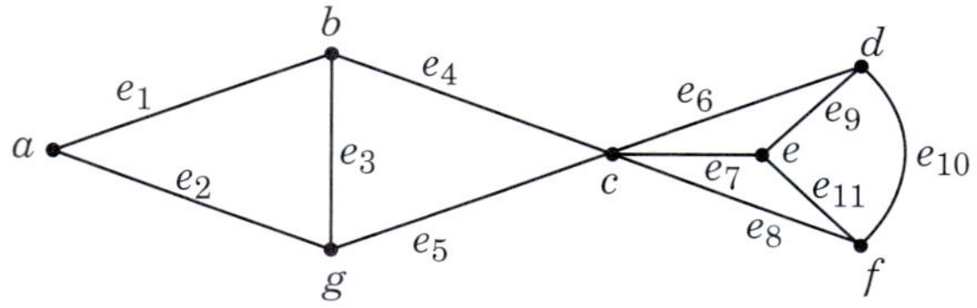

SOLUTION:

We construct a spanning tree for G step-by-step, as in Table 9.1, which shows each step involved. Figure 9.14 depicts the resulting spanning tree.

Table 9.1

Edge in G	Edges selected form a cycle?	Edges in tree	Number of edges in tree	Done?
e_1	No	e_1	1	No
e_2	No	e_1, e_2	2	No
e_3	Yes	e_1, e_2	2	No
e_4	No	e_1, e_2, e_4	3	No
e_5	Yes	e_1, e_2, e_4	3	No
e_6	No	e_1, e_2, e_4, e_6	4	No
e_7	No	e_1, e_2, e_4, e_6, e_7	6	No
e_8	No	$e_1, e_2, e_4, e_6, e_7, e_8$	5	Yes

Figure 9.14

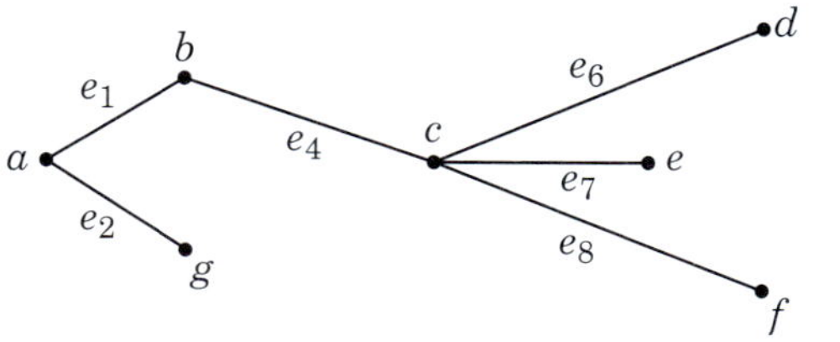

■

A Big-Oh Estimate

To find a big-oh estimate of the total number of comparisons required by Kruskal's algorithm, notice that the **while** loop takes n iterations and line 7 takes e comparisons in the worst case, where e denotes the number of edges in G. So the algorithm takes $O(ne)$ comparisons in the worst case, where $0 \le e \le n(n-1)/2$.

This is unacceptable. Two simple and elegant methods, called Depth-First Search (DFS) and Breadth-First Search (BFS), provide remedies.

The Depth-First Search (DFS) Algorithm

For the DFS algorithm, beginning at a vertex u, traverse the graph as far as possible, marking each vertex visited. Once you reach a vertex from which no more paths are available, backtrack to the last vertex visited; continue traversing in this manner until all vertices are visited.

An outline of this recursive DFS algorithm follows.

- Start at a vertex u.
- Mark it as visited.
- Pick a vertex v adjacent to u and not yet visited.
- Call DFS from v.
- If a vertex w is reached such that all its adjacent vertices have been visited, back up to the last vertex x visited and call DFS from x.
- Stop when you have visited all vertices.

The DFS method, also called **backtracking**, is given in Algorithm 9.2.

```
Algorithm DFS(G,v)
(* G is a connected graph with n vertices. A is an n × 1 boolean matrix;
   Au indicates whether or not vertex u has been visited. This algorithm
   visits all vertices in G, beginning with v. *)
 Begin (* algorithm *)
   Av ← true (* mark vertex v as visited *)
      for each vertex w adjacent to v do
         if Aw = false then (* vertex w not yet visited *)
            DFS(G,w)
 End (* algorithm *)
```

Algorithm 9.2

The following example clarifies this algorithm.

EXAMPLE 9.2 Using the DFS method, find a spanning tree for the graph in Figure 9.13, beginning at vertex a.

SOLUTION:

Step 1 There are two vertices, b and g, adjacent to vertex a, neither visited yet.

Step 2 Choose b. (We shall follow the alphabetic order when a choice exists.) Look at the vertices adjacent to b: a, c, and g. Only a has been visited.

Step 3 Select c. Of the vertices adjacent to c, b has been visited, but not $d, f,$ or g.

Step 4 Choose d. Continuing like this, visit vertices e and f.

Now all vertices adjacent to f have been visited. So backtrack to e. All vertices adjacent to it have been visited. Continue backing up to d and to c. A vertex adjacent to c has not yet been visited, namely g. Visit g. All vertices adjacent to g have been visited. So backtrack to c, to b, and finally to a.

Now all vertices have been visited, and the algorithm terminates. Figure 9.15 shows the resulting spanning tree.

Figure 9.15

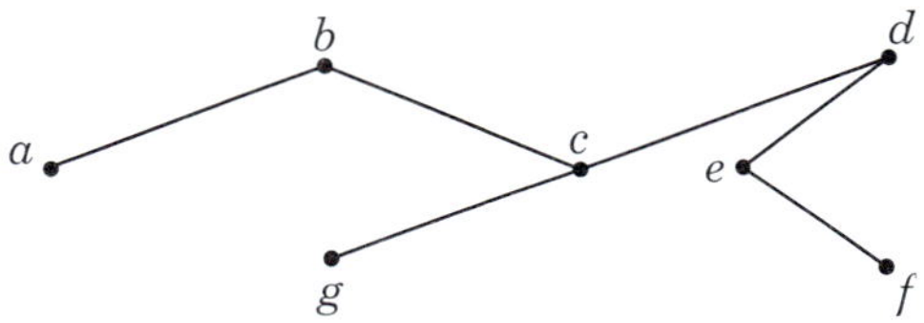

(You can verify this using the adjacency list representation of the graph.) Notice that this spanning tree differs from the one in Figure 9.14. ■

Next we demonstrate how the n-queens problem can be solved using DFS, if a solution exists.

EXAMPLE 9.3 **(The four-queens puzzle)** Place four queens on a 4×4 chessboard so that no two queens attack each other. (A queen attacks another queen if they are in the same row, column, or diagonal.)

SOLUTION:

First, place a queen (Q) in row 1 and column 1 (Figure 9.16). The next queen can be placed in row 3 and column 2 (Figure 9.17). Unfortunately, the third queen cannot be placed in column 3, so backtrack to the previous location and move the queen to the next available location, namely,

Figure 9.16

Q			

Figure 9.17

Q			
	Q		

row 4 and column 2 (Figure 9.18). Now the third queen can be placed in row 2 and column 3 (Figure 9.19). However, this move does not properly

Figure 9.18

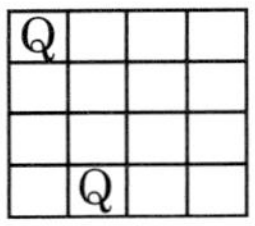

Figure 9.19

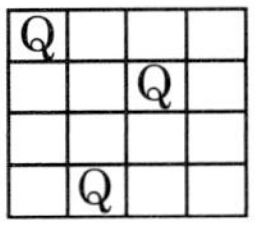

place the fourth queen in the column 4, so backtrack and try to move the third queen. It has no more legal moves. So backtrack to the second queen, which also turns out to have no further legal moves. Therefore, backtrack to the first queen and place it in position (2,1). Continuing as before, we get a solution, as shown in Figure 9.20.

Figure 9.20

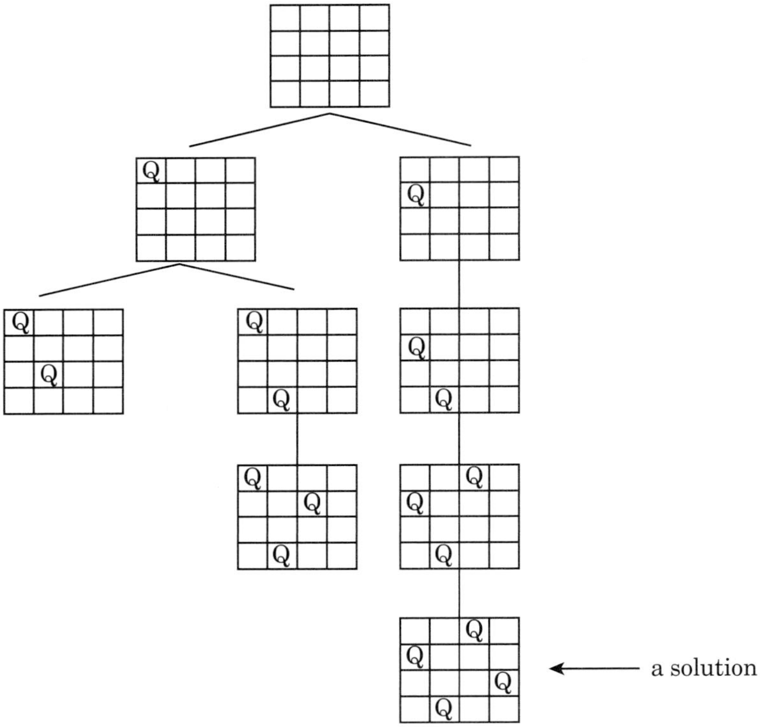

■

Clearly, the backtracking technique can solve the n-queens puzzle, if a solution exists. For instance, try to solve the **eight-queens puzzle**: Place eight queens on an 8×8 chessboard, so no two queens attack each other. There are 92 solutions.

Next we turn our attention to analyzing the DFS algorithm.

An Analysis of the DFS Algorithm

Suppose we apply Algorithm 9.2 to a graph G with n vertices and e edges. If G is represented by its adjacency matrix, it takes $O(n)$ time to identify vertices adjacent to a given vertex. Since at most n vertices are visited, the algorithm takes $O(n^2)$ time for this search.

On the other hand, if G is implemented by adjacency lists (Section 8.2), the DFS examines each vertex at most once. Since there are $2e$ vertices in the adjacency list representation, the algorithm takes $O(e)$ time for the search.

The Breadth-First Search (BFS) Algorithm

The other elegant remedy besides the DFS method to find a spanning tree is the **Breadth-First Search** method. In the BFS method, visit the vertices level by level until all are visited. An outline of the BFS algorithm follows.

- Start at a vertex u.
- Mark it as visited.
- Visit all unvisited vertices adjacent to u.
- Visit all unvisited vertices adjacent to each of them.
- Repeat step 4 until all vertices are visited.

In step 4, we visit all unvisited vertices adjacent to the unvisited vertices found in step 3. The algorithm is given in Algorithm 9.3. (It uses a data structure called a **queue**; omit it if you are not familiar with queues.)

```
Algorithm BFS (G,v)
(* G is a connected graph with n vertices. A is an n × 1 boolean matrix;
   it is initialized to false for every w in G; A_w indicates whether or
   not vertex w has been visited. Q denotes a queue used to store
   vertices adjacent to a given vertex. Beginning with v, the algorithm
   visits all vertices of G using the BFS method. *)
   Begin (* algorithm *)
     A_w ← true
     insert v in Q
     while Q not empty do
     begin (* while *)
       delete a vertex w from Q
       for each vertex x adjacent to w do
         if A_x = false then (* x not yet visited *)
         begin (* if *)
           insert x in Q
           A_x ← true
         endif
     endwhile
   End (* algorithm *)
```

Algorithm 9.3

The following example illustrates this algorithm.

EXAMPLE 9.4 Using the BFS method, construct a spanning tree for the graph in Figure 9.13, beginning at vertex a.

SOLUTION:

Step 1 Beginning at a, there are two vertices adjacent to it, b and g. Visit them.

Step 2 Visit the unvisited vertices adjacent to b. There is only one such vertex, c. There are no unvisited vertices adjacent to g.

Step 3 Visit the unvisited vertices adjacent to c, namely, d, e, and f. There are no more unvisited vertices; so we are done.

The resulting spanning tree appears in Figure 9.21.

Figure 9.21

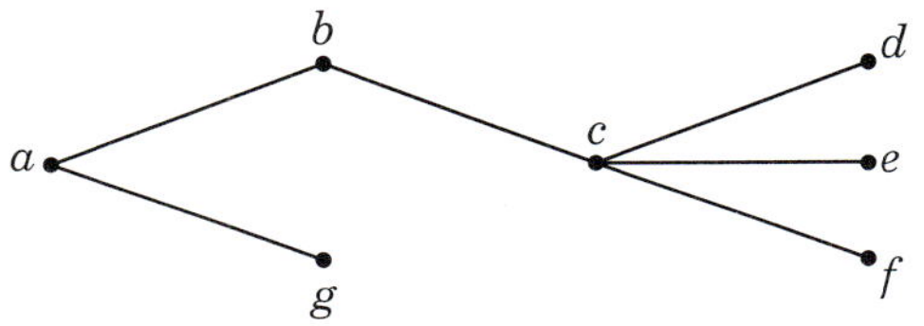

(Notice that this spanning tree differs from the ones obtained earlier.) ■

Next we present an analysis of the BFS method.

An Analysis of the BFS Algorithm

Let G be a graph with n vertices and e edges. If it is represented by its adjacency matrix, the ***while*** loop is executed at most n times. Corresponding to each element in the queue, the ***for*** loop is executed at most n times, so the total time taken is $O(n^2)$.

On the other hand, if G is represented by adjacency lists, the algorithm examines each vertex once. Since the list contains a total of $2e$ vertices, it takes $O(e)$ time for the search.

Notice that the BFS method takes as long as the DFS method in the worst case.

We close this section with yet another refreshing occurrence of Lucas numbers, but first a definition.

The Complexity of a Graph

The **complexity*** of a graph G, denoted by $k(G)$, is the number of distinct spanning trees of G.

*See the author's *Fibonacci and Lucas Numbers with Applications* for a detailed discussion.

For instance, the graph G in Figure 9.22 has three spanning trees, as Figure 9.23 shows, so its complexity is 3; that is, $k(G) = 3$.

Figure 9.22

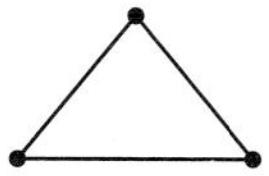

Figure 9.23

Spanning trees of G.

An investigation of the complexity of the wheel graph W_n (see Figure 8.17) yields a surprising dividend.

Lucas and the Wheel Graph

Recall that W_n has $n+1$ vertices; n of them lie on a rim and the remaining vertex (the hub) is connected to every vertex on the rim. In 1969 J. Sedlacek of the University of Calgary, Canada, proved that $k(W_n) = L_{2n} - 2$, where L_m denotes the mth Lucas number; two years later, B. R. Myers of the University of Notre Dame rediscovered the formula.

For example, $k(W_3) = L_6 - 2 = 18 - 2 = 16$; that is, W_3 has 16 different spanning trees. See Figure 9.24.

Figure 9.24

Spanning trees of W_3.

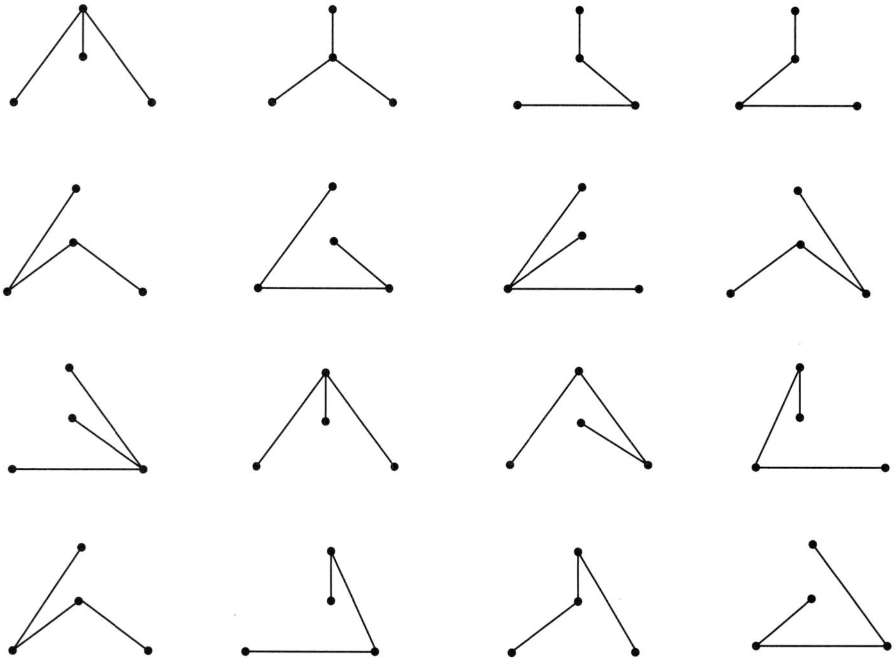

Exercises 9.2

How many edges does a spanning tree of each graph have?

1. K_n **2.** $K_{m,n}$ **3.** Figure 9.1 **4.** The Petersen graph

Using Kruskal's algorithm, construct a spanning tree for each graph, starting at a.

5.

6.

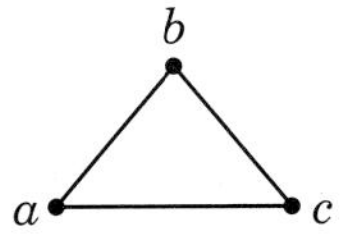

7.

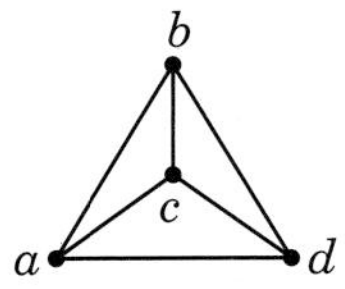

8.

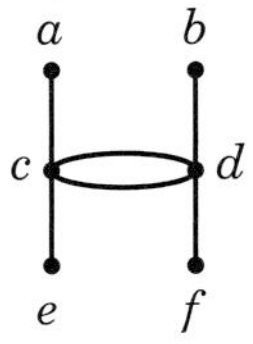

9.

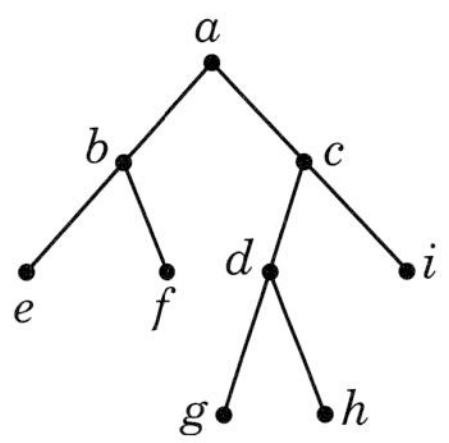

10.

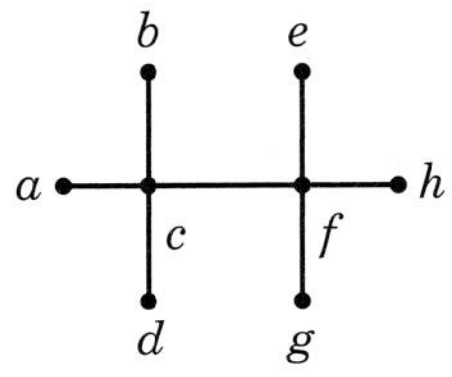

11.

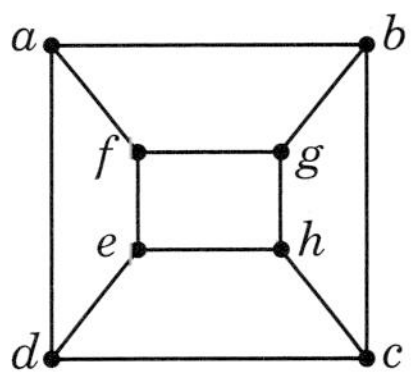

12.

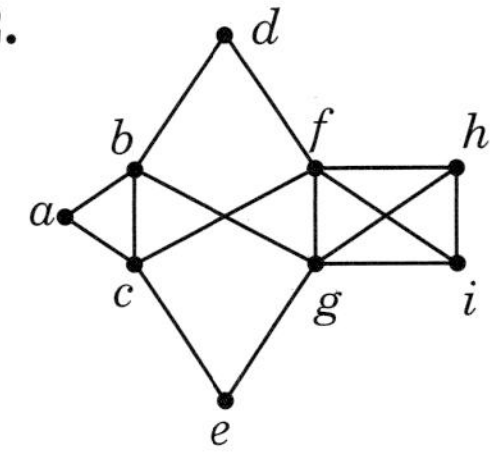

13.

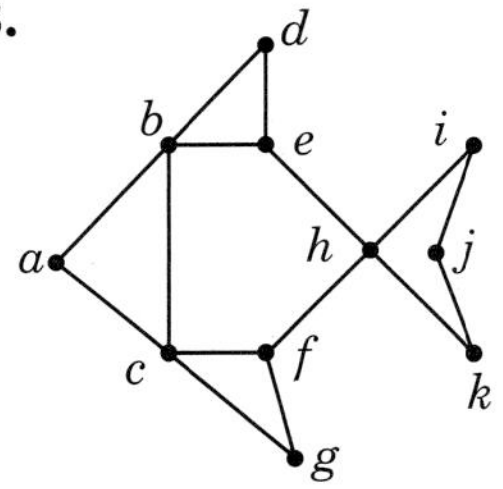

14–22. Using the DFS method, construct a spanning tree for each graph in Exercises 5–13.

23–31. Using the BFS method, construct a spanning tree for each graph in Exercises 5–13.

○ Using the DFS method, construct a spanning tree for each graph with the given adjacency list.

32.

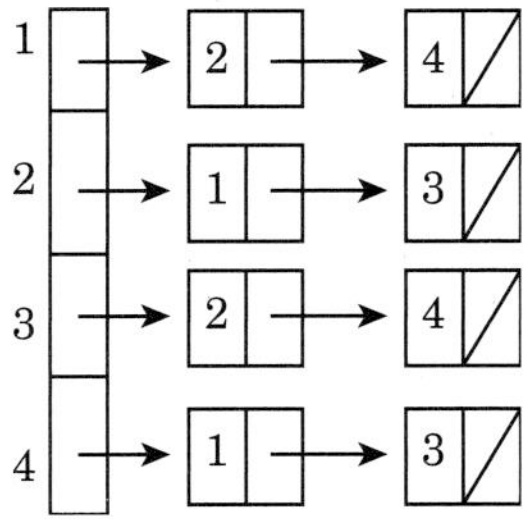

33.

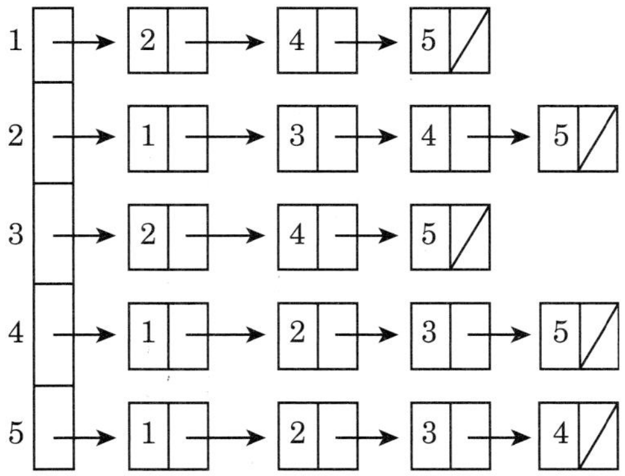

34–35. Using the BFS algorithm, construct a spanning tree for each graph in Exercises 32 and 33.

Using backtracking, solve the n-queens puzzle, if a solution exists, for each value of n.

36. 2 **37.** 3 **38.** 5 **39.** 6

Find a spanning tree for each complete graph.

40. K_2 **41.** K_3 **42.** K_4 **43.** K_5

How many nonisomorphic spanning trees does each complete graph have?

44. K_1 **45.** K_2 **46.** K_3 **47.** K_4

***48.** Using Exercises 40–47, predict the number of nonisomorphic spanning trees for the complete graph K_n.

Find a spanning tree for each complete bipartite graph.

49. $K_{1,1}$ **50.** $K_{2,2}$ **51.** $K_{2,3}$ **52.** $K_{3,3}$

53–56. How many nonisomorphic spanning trees do the complete bipartite graphs in Exercises 49–52 have?

○ Let G be a simple graph with n vertices $v_1, v_2, \ldots, v_n$ and adjacency matrix $A = (a_{ij})_{n\times n}$. Let $B = (b_{ij})_{n\times n}$, where

$$b_{ij} = \begin{cases} -a_{ij} & \text{if } i \neq j \\ \deg(v_i) & \text{if } i = j \end{cases}$$

Let C be the $(n-1)\times(n-1)$ matrix obtained by deleting row 1 and column 1 of B. Then the number of nonisomorphic spanning trees of G is the determinant $|C|$. Using this fact, find the number of nonisomorphic spanning trees for each graph.

57.

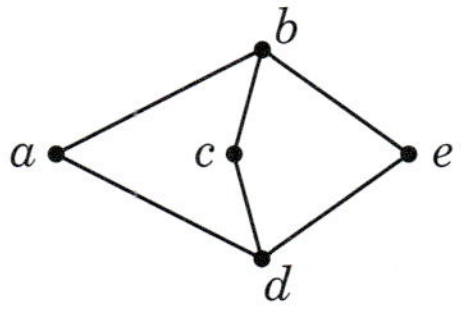

58.

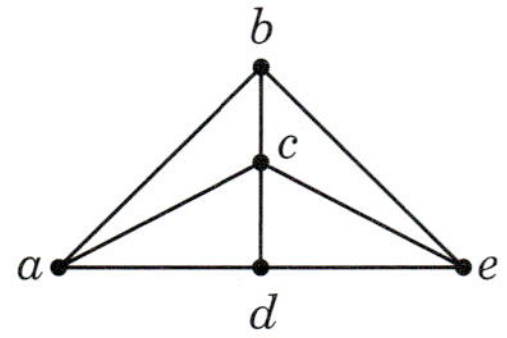

59. K_5

***60.** (**Cayley's formula**) Using the statement preceding Exercises 57–59, prove that the number of nonisomorphic spanning trees for K_n is n^{n-2}.

9.3 Minimal Spanning Trees

Finding a minimal spanning tree in a connected weighted graph can frequently be useful. This section presents two such algorithms, but first an example.

A utility company would like to lay pipelines for natural gas between five towns, a through e, as in Figure 9.25. The weights (in millions) of the edges represent the cost of building the various pipelines. They must be laid in such a way that natural gas can be sent from any town to any other town, holding construction costs to a minimum. In other words, the company needs to find a spanning tree for the weighted graph with the sum of the weights of the tree's edges at a minimum. Such a tree is a minimal spanning tree.

Figure 9.25

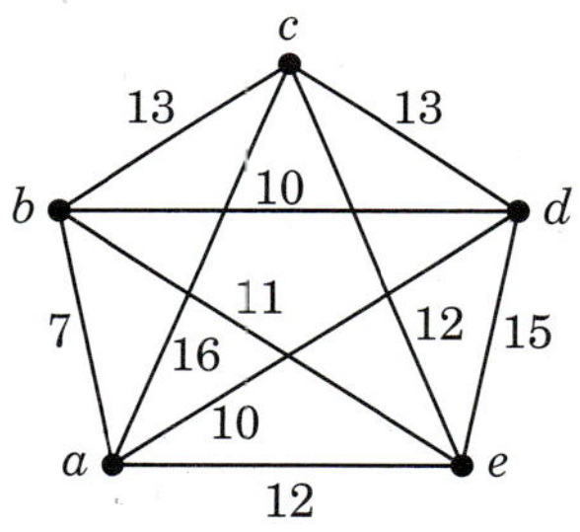

Minimal Spanning Tree

Let G be a connected weighted graph. The **weight** of a spanning tree of G is the sum of the weights of its edges. A **minimal spanning tree** of G weighs the least.

Several algorithms can find a minimal spanning tree T of a connected weighted graph G. Two of them are Kruskal's and Prim's.

Kruskal's Algorithm

Kruskal's algorithm is an extension of his algorithm presented in the preceding section:

- Arrange the edges in G in nondecreasing order of their weights.

- Choose an edge in G with the minimum weight.
- Add an edge of least weight to T if it does not form a cycle with the edges already selected.
- Repeat step 3 until the number of edges selected is $n - 1$, where n denotes the number of vertices in G.

The following example illustrates these steps.

EXAMPLE 9.5 Using Kruskal's algorithm, find a minimal spanning tree for the graph in Figure 9.25.

SOLUTION:

Step 1 Arrange the edges in nondecreasing order of their weights, as in Table 9.2.

Table 9.2

Weight	7	10	10	11	12	12	13	13	15	16
Edge	$\{a,b\}$	$\{a,d\}$	{b, d}	{b, e}	$\{a,e\}$	{c, e}	{b, c}	{c, d}	{d, e}	$\{a,c\}$

Step 2 Choose edge $\{a,b\}$ since it has the smallest weight; include it in T.

Step 3 An edge with the next smallest weight is $\{a,d\}$. It does not form a cycle with the edge in T, so include it in T.

Step 4 Edge $\{b,d\}$ forms a cycle with the edges in T; discard it.

Step 5 Edge $\{b,e\}$ does not produce a cycle with the edges in T; include it in T.

Step 6 Edge $\{a,e\}$ creates a cycle; reject it.

Step 7 Edge $\{c,e\}$ is selected.

Having selected four edges, we are done, by Theorem 9.2. Figure 9.26 shows the resulting spanning tree.

Figure 9.26

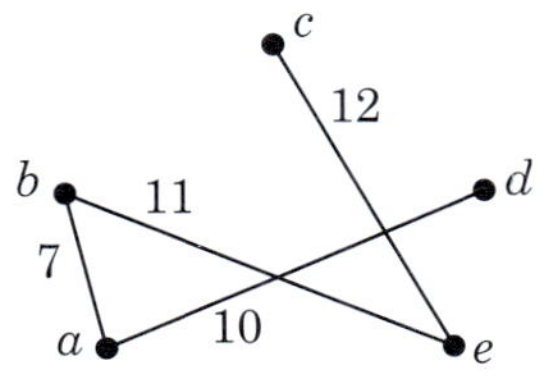

Table 9.3 summarizes these steps, indicating 40 as the weight of a minimal spanning tree, so it would cost $40 million to build the pipelines.

Table 9.3

Edge	Weight	Decision	Spanning tree T	Weight of T	Number of edges in T	Done?
$\{a,b\}$	7	Select		7	1	No
$\{a,d\}$	10	Select		17	2	No
{b, d}	10	Reject	—	17	2	No
{b, e}	11	Select		28	3	No
$\{a,e\}$	12	Reject	—	28	3	No
{c, e}	12	Select	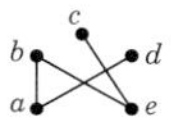	40	4	Yes

■

Algorithm 9.4 presents Kruskal's algorithm.

```
Algorithm Kruskal (G,T)
(* Using Kruskal's algorithm, this algorithm constructs a minimal
   spanning tree T for a connected weighted graph G = (V,E) with n
   vertices and computes the minimum weight. Numedges denotes the
   number of edges in T. *)
   Begin (* algorithm *)
     T ← Ø          (* initialize tree *)
     numedges ← 0  (* initialize numedges in T *)
     weight ← 0    (* initialize weight of T *)
     while (numedges < n− 1) and (E ≠ Ø) do
     begin (* while *)
       select an edge e in E with the smallest weight.
       E ← E − {e} (* delete e from E *)
       if e does not form a cycle in T then
         begin (* if *)
          T ← T ∪{e} (* add e to T *)
          weight ← weight + weight of edge e
          numedges ← numedges + 1
         endif
       else
          discard it
     endwhile
   End (* algorithm *)
```

Algorithm 9.4

Robert Clay Prim *(1921–), an engineer and mathematician, was born in Sweetwater, Texas. He received his B.S. in electrical engineering in 1941 from the University of Texas and his Ph.D. in mathematics from Princeton in 1949.*

After working for a year at the University of Texas, he has held a variety of positions: engineer at the General Electric Co., engineer and mathematician at the U. S. Naval Ordinance Lab, research associate at Princeton, research mathematician and director of mathematics and mechanics at Bell Telephone Labs, vice president for research at Sandia Corporation, and executive director of research at Bell Labs.

Prim's Algorithm

Another method for finding a minimal spanning tree T, **Prim's algorithm**, was discovered by the American engineer Robert Clay Prim in 1957. The following steps outline it.

- Choose an edge with the least weight.
- Include it in T.
- Select an edge of least weight that is incident with a vertex of an edge in T.
- If it does not create a cycle with the edges in T, then include it in T; otherwise, discard it.
- Repeat steps 3 and 4 until T contains $n - 1$ edges.

The next example illustrates these steps.

EXAMPLE 9.6 Using Prim's algorithm, construct a minimal spanning tree for the graph in Figure 9.25.

SOLUTION:
Arrange the edges in nondecreasing order of weights, as in Table 9.2.

Step 1 Choose an edge with the least weight: {a, b}.

Step 2 Include it in T. See Figure 9.27.

Figure 9.27

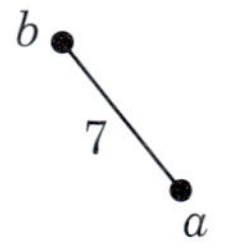

Step 3 Select a lightest edge incident with a or b. Edges $\{a, d\}$ and $\{b, d\}$ are two candidates. Neither of them forms a cycle $\{a, b\}$, so select one of them, say, $\{b, d\}$.

Step 4 Adjoin it to T (see Figure 9.28).

Figure 9.28

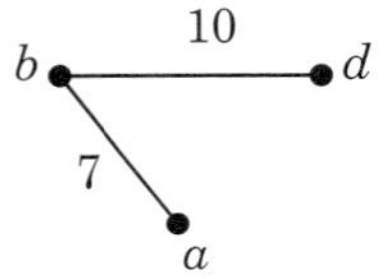

Step 5 Look for a lightest edge incident with a vertex in T: $\{a, d\}$.

Step 6 Edge $\{a, d\}$ creates a cycle in T; reject it.

Step 7 Edge $\{b, e\}$ is the next candidate.

Step 8 Add it to T (see Figure 9.29).

Figure 9.29

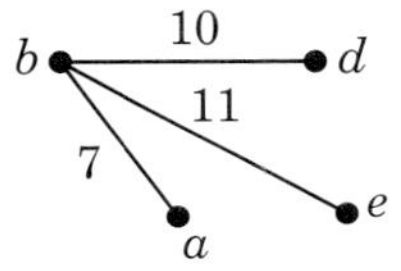

Step 9 Look for a lightest edge incident with a vertex in T; $\{a, e\}$ and $\{c, e\}$ are two possibilities.

Step 10 Edge $\{a, e\}$ forms a cycle in T, so discard it.

Step 11 $\{c, e\}$ does not produce a cycle, so include it in T (see Figure 9.30). T now contains 4 edges; so by Theorem 9.2, T is a minimal spanning tree weighing $7 + 10 + 11 + 12 = 40$.

Figure 9.30

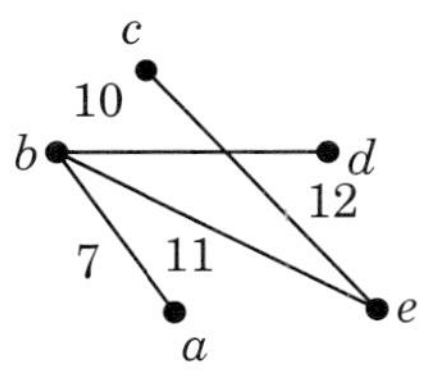

Table 9.4 summarizes these steps.

> *An important observation*: Examples 9.5 and 9.6 indicate that a connected weighted graph may have more than one minimal spanning tree. Nonetheless, they weigh the same.

Table 9.4

Edge	Weight	Decision	Spanning tree T	Weight of T	Number of edges in T	Done?
$\{a,b\}$	7	Select		7	1	No
$\{b,d\}$	10	Select		17	2	No
$\{a,d\}$	10	Reject	—	17	2	No
$\{b,e\}$	11	Select	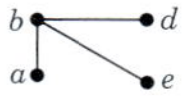	28	3	No
$\{a,e\}$	12	Reject	—	28	3	No
$\{c,e\}$	12	select	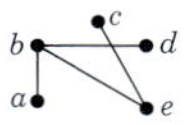	40	4	Yes

■

Algorithm 9.5 gives Prim's algorithm.

```
Algorithm Prim (G,T)
(* Using Prim's algorithm, this algorithm constructs a minimal spanning
   tree T for a connected weighted graph G = (V, E) with n vertices and
   computes the minimum weight. Numedges denotes the number of
   edges in T. *)
   Begin (* algorithm *)
     T ← Ø (* initialize tree *)
     numedges ← 0 (* initialize numedges *)
     weight ← 0 (* initialize weight of T *)
     while (numedges < n − 1) and (E ≠ Ø) do
     begin (* while *)
       select an edge e that is incident with a vertex
       in T that has the least weight.
       E ← E − {e} (* delete edge e from E *)
       if e does not form a cycle in T then
         begin (* if *)
           T ← T ∪ {e} (* adjoin e to T *)
           weight ← weight + weight of edge e
           numedges ← numedges + 1
         endif
        else
         discard it
     endwhile
   end (* algorithm *)
```

Algorithm 9.5

The time complexity of this version of Prim's algorithm is $O(n^3)$, with n the number of vertices in G. However, it can be implemented in $O(n^2)$ time.

To prove Prim's algorithm, induction is used twice as well as a loop invariant (Section 4.5). Unfortunately, the proof, given next, is a bit long.

THEOREM 9.4 Prim's algorithm produces a minimal spanning tree for a connected weighted graph.

PROOF:
Let G be a connected weighted graph with n vertices and T_k its subgraph after k iterations of the **while** loop in Algorithm 9.5. The proof consists of validating three statements: (1) T_k is a tree for every $k \geq 1$; (2) T_{n-1} is a spanning tree; and (3) T_k is contained in a minimal spanning tree.

If the tree T_k is contained in a minimal spanning tree at the end of every iteration, it must be true even when the loop is terminated. Consequently, when we leave the loop, a minimal spanning tree must result.

So we begin with the proof of part 1.

- *To prove that $T_k$is a tree for every $k \geq 1$* (use induction):

Basis step When $k = 1, T_1$ consists of a single vertex. So T_1 is a tree.

Induction step Assume T_k is a tree for any $k \geq 1$. In the $(k+1)$st iteration, when a new edge is added to T_k the resulting subgraph T_{k+1} is still connected and acyclic, so T_{k+1} is also a tree. Thus, by induction, T_k is a tree for every $k \geq 1$.

- *To prove that $T_{n-1}$is a spanning tree*:
  The **while** loop is terminated when $k = n - 1$. The resulting subgraph T_{n-1}, by part 1, is a tree with $n - 1$ edges; so it contains all the vertices of G. Hence T_{n-1} is a spanning tree.

- *To prove that T_k is contained in a minimal spanning tree* (use induction):

Basis step Since T_1 consists of a single vertex, T_1 is contained in every minimal spanning tree.

Induction step Assume T_k is contained in a minimal spanning tree T after k iterations of the loop. We would like to show that T_{k+1} is contained in some minimal spanning tree of G.

Let V be the set of vertices in T_k. In the $(k+1)$st iteration, the algorithm selects an edge $e = \{u, v\}$ of least weight, where $u \in V$ and $v \notin V$, and adds it to T_k. This yields the tree T_{k+1}: $T_{k+1} = T_k \cup \{e\}$.

Case 1 If $e \in T$, clearly T_{k+1} is contained in T.

Case 2 If $e \notin T, T \cup \{e\}$ contains a cycle C. Choose an edge $e' = \{x, y\}$ in C such that $x \in V, y \notin V$, and $e \neq e'$. Then $w(e) \leq w(e')$ by the choice of e, where $w(e)$ denotes the weight of edge e and $w(e')$ the weight of e'.

Let $T' = [T \cup \{e\}] - \{e'\}$. T' is a tree with $n - 1$ edges; so it is a tree, by Theorem 9.2, and hence a spanning one. Then $w(T') \leq w(T)$, where

$w(T')$ denotes the weight of T' and $w(T)$ the weight of T; so T' is also a minimal spanning tree. Besides, it contains T_{k+1}.

Thus, in both cases, T_{k+1} is contained in a minimal spanning tree; therefore, by induction, T_k is contained in a minimal spanning tree for every $k \geq 1$. ■

Kruskal's and Prim's algorithms present fine paradigms of a greedy algorithm. A **greedy algorithm** makes an optimal choice at every step. For instance, each step in the two algorithms sought an edge with minimum weight. Although greedy algorithms may not always yield optimal solutions, they do in these two cases.

Exercises 9.3

Using Kruskal's algorithm, construct a minimal spanning tree for each connected weighted graph.

1.

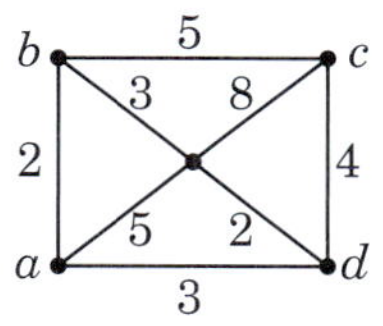

2.

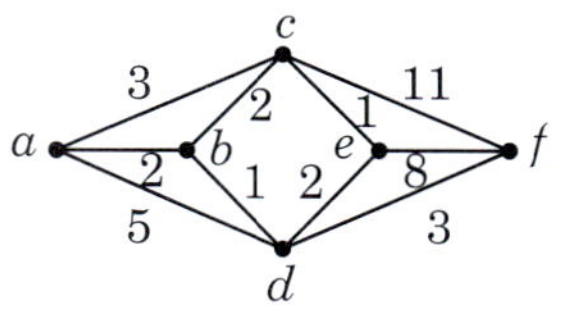

3.

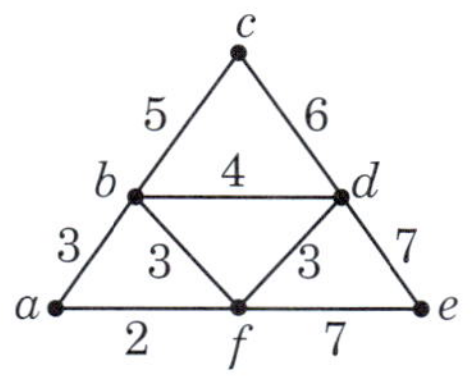

4.

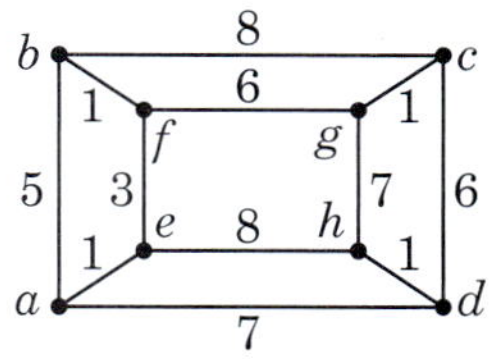

5.

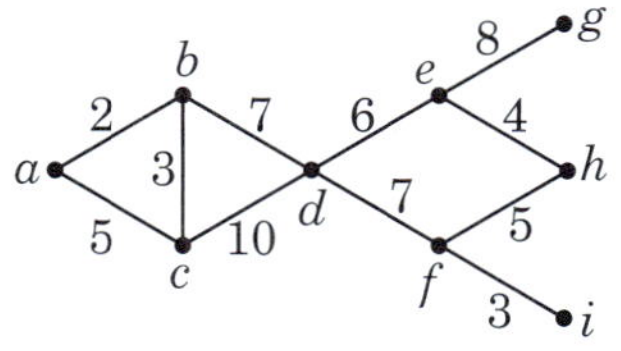

6.

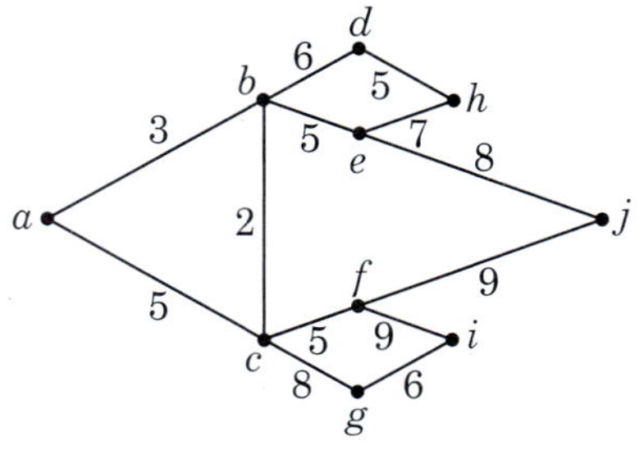

7–12. Using Prim's algorithm, construct a minimal spanning tree for each graph in Exercises 1–6. Compute the weight of each.

○ Using Kruskal's algorithm, create a minimal spanning tree for the connected weighted graph with each adjacency list.

13.

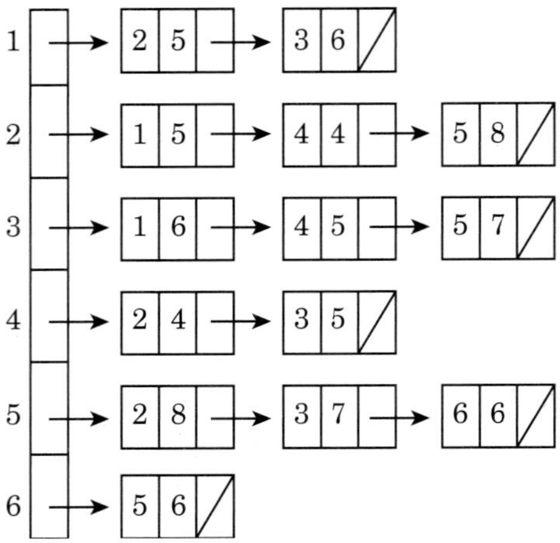

14.

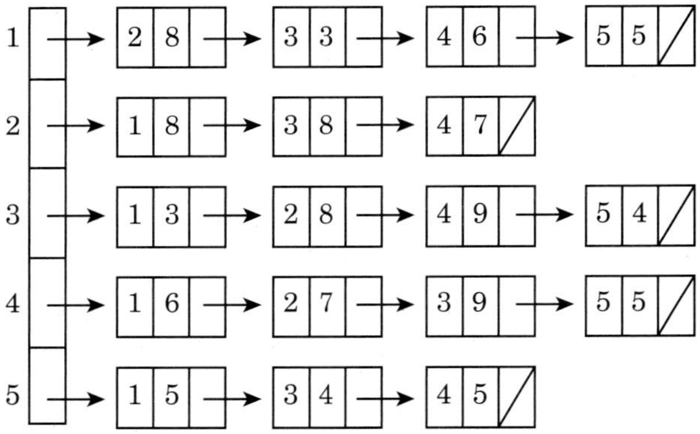

15–16. Redo Exercises 13 and 14 using Prim's algorithm.

Using Kruskal's algorithm, construct a minimal spanning tree for the connected, weighted graph with each modified adjacency matrix. The symbol ∞ in row i and column j indicates the absence of edge $\{i, j\}$; think of it as a number larger than any of the weights in the graph.

17.

$$\begin{array}{c} \\ 1 \\ 2 \\ 3 \\ 4 \end{array} \begin{array}{c} \begin{array}{cccc} 1 & 2 & 3 & 4 \end{array} \\ \begin{bmatrix} \infty & 6 & 3 & 7 \\ 6 & \infty & 5 & 7 \\ 3 & 5 & \infty & 6 \\ 7 & 7 & 6 & \infty \end{bmatrix} \end{array}$$

18.

$$\begin{array}{c} \\ 1 \\ 2 \\ 3 \\ 4 \\ 5 \end{array} \begin{array}{c} \begin{array}{ccccc} 1 & 2 & 3 & 4 & 5 \end{array} \\ \begin{bmatrix} \infty & 8 & 9 & 7 & 7 \\ 8 & \infty & 11 & \infty & 3 \\ 9 & 11 & \infty & 10 & \infty \\ 7 & \infty & 10 & \infty & 5 \\ 7 & 3 & \infty & 5 & \infty \end{bmatrix} \end{array}$$

19–20. Redo Exercises 17 and 18 using Prim's algorithm.

9.4 Rooted Trees

Take a good look at the trees in Figures 9.6, 9.7, and 9.8. Each contains a specially designated vertex called the **root**: *Nicolaus*, *Chief Executive Officer*, and *NHL*, respectively. A tree with a root is a **rooted tree**.

Consider the simple algorithm for making an overseas telephone call:

- Dial the international access code (011).
- Dial the country code.
- Dial the city code.
- Dial the local number.

This algorithm can be used to construct a rooted tree, a portion of which is shown in Figure 9.31. The root of the tree is the vertex 011.

Figure 9.31

A rooted tree.

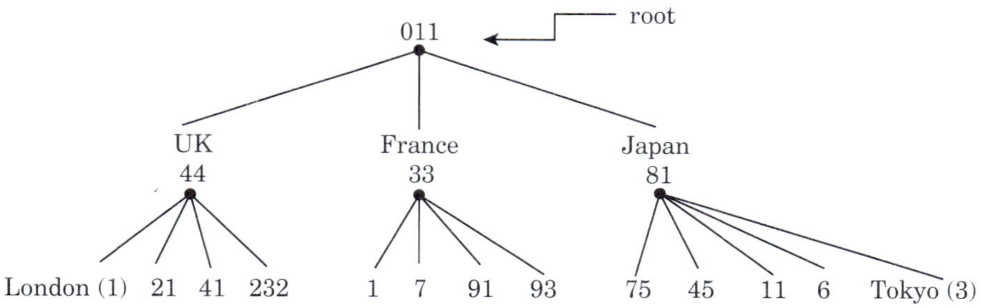

Rooted trees are drawn with the root at the top, especially in computer science; they grow downward. Every "unrooted" tree can be redrawn as rooted and each choice of the root produces a differently shaped tree. For example, the tree in Figure 9.2 when redrawn as a rooted tree appears in Figure 9.32. The spanning tree in Figure 9.26 can be redrawn as differently shaped trees (see Figure 9.33).

Figure 9.32

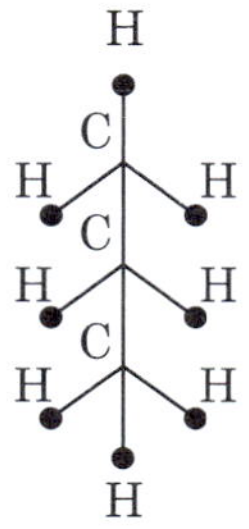

Since a unique, simple path runs between any two vertices in a tree, such a path exists from the root of a rooted tree to any other vertex.

The basic terminology of rooted trees resembles that of a family tree.

Figure 9.33

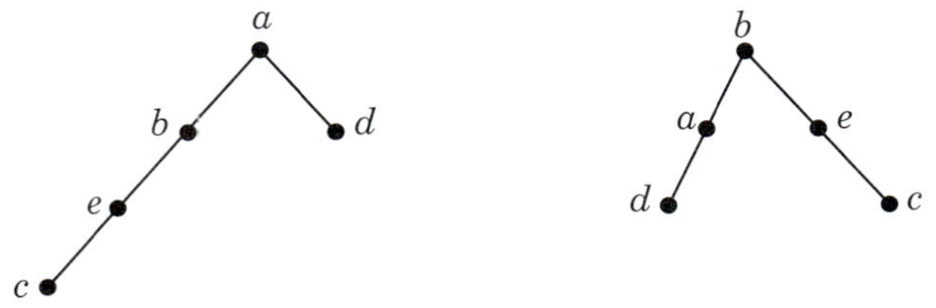

Tree rooted at a Tree rooted at b

Parent, Child, Sibling, Ancestor, Descendant, and Subtree

Let T be a rooted tree with root v_0. Let v_0-v_1- $\cdots$-v_{n-1}-v_n be the path from v_0 to v_n. Then:

- v_{n-1} is the **parent** of v_n.
- v_n is a **child** of v_{n-1}.
- Vertices with the same parent are **siblings**.
- The vertices $v_0, v_1, \ldots, v_{n-1}$ are **ancestors** of v_n.
- The **descendants** of a vertex v are those vertices for which v is an ancestor.
- A vertex with no children is a **leaf** or a **terminal vertex**.
- A vertex that is not a leaf is an **internal vertex**.
- The **subtree** of T rooted at v consists of v, its descendants, and all edges incident with them.

EXAMPLE 9.7 The tree T in Figure 9.34 is rooted at a. Vertex b is the parent of both e and f; so e and f are the children of b.

Figure 9.34

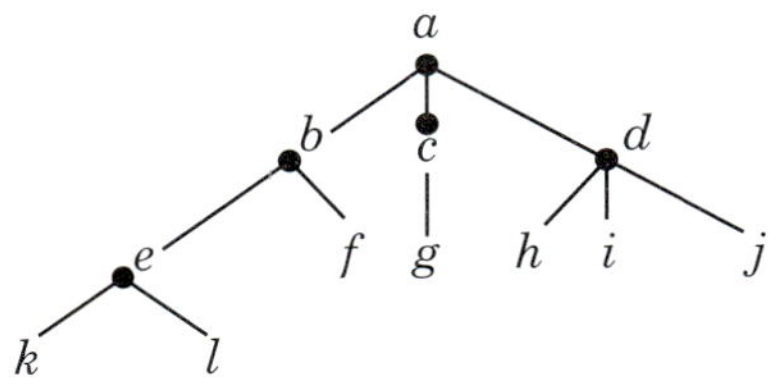

Since b, c, and d have the same parent, they are siblings. Vertices a, b, and e are ancestors of l. Vertices e, f, k, and l are descendants of b. Vertex f has no children, so it is a leaf. Vertices c and e have at least one child, so both are internal vertices. Figure 9.35 displays the subtrees rooted at b and d. ■

Tree structures occur not only in genealogy and hierarchical studies, but in the study of games such as chess, checkers, and tic-tac-toe. In such

Figure 9.35

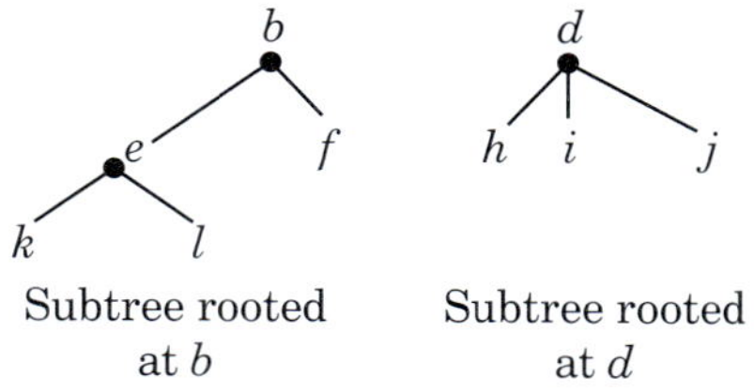

a game, two players take turns for their moves. The root of a **game tree** represents the initial board position and every other vertex represents a subsequent board position.

Figure 9.36, for example, shows a portion of a game tree for tic-tac-toe.

Figure 9.36

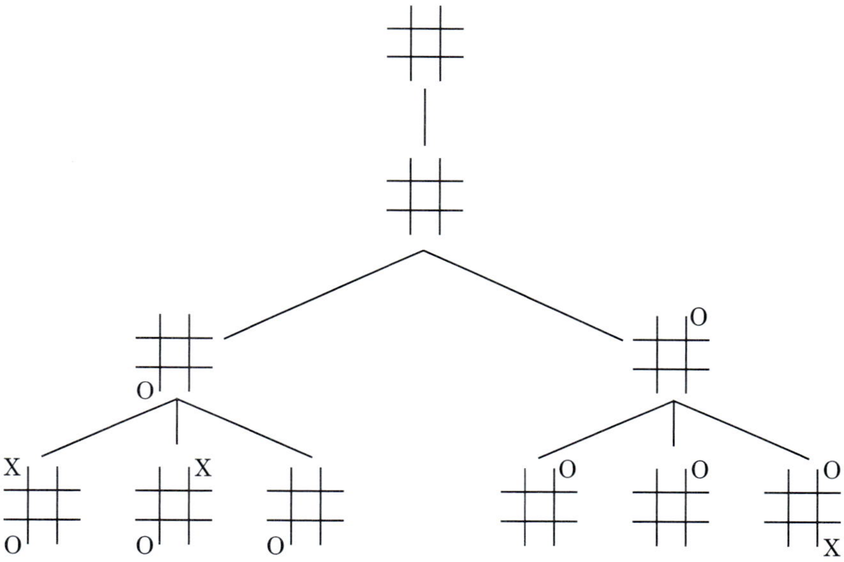

The unique, simple path from the root of a tree to any vertex can define its level and hence its height.

Level and Height

The **level** of a vertex v in a rooted tree, denoted by level (v), is the length of the path from the root to v. The **height** of a rooted tree is the maximum level of any leaf in the tree.

EXAMPLE 9.8 For the rooted tree in Figure 9.34, level $(a) = 0$, level $(b) =$ level $(c) =$ level $(d) = 1$, level $(e) =$ level (f) $=$ level $(j) = 2$, and level $(k)=$ level $(l) = 3$. Since the maximum level of any leaf is 3, the height of the tree is 3. ■

The level of a vertex can also be defined recursively.

A Recursive Definition of the Level of a Vertex v

$$\text{level}\ (v) = \begin{cases} 0 & \text{if } v \text{ is the root} \\ 1 + \text{level (parent of } v) & \text{otherwise} \end{cases}$$

EXAMPLE 9.9 Using the recursive definition, compute the level of vertex k in Figure 9.34.

SOLUTION:

$$\begin{aligned} \text{level}\ (k) &= 1 + \text{level}\ (e) \\ &= 1 + [1 + \text{level}\ (b)] = 2 + \text{level}\ (b) \\ &= 2 + [1 + \text{level}\ (a)] = 3 + \text{level}\ (a) \\ &= 3 + 0 \qquad\qquad\quad = 3 \end{aligned}$$ ■

An important class of rooted trees is ordered rooted trees.

Ordered Rooted Tree

A rooted tree in which the vertices at each level are ordered as the first, second, third, and so on is an **ordered rooted tree**.

EXAMPLE 9.10 As ordered trees, the rooted trees in Figure 9.37 are not the same (Why?). As unordered trees, however, there is no difference between them.

Figure 9.37

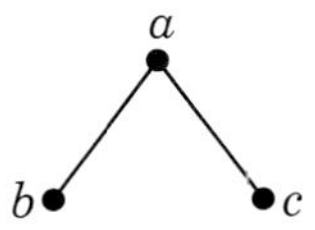

■

When an ordered rooted tree is drawn, vertices at the same level are placed along the same horizontal line. Children from oldest to youngest are arranged from left to right. For example, in Figure 9.37, $b \prec c$, whereas in Figure 9.38, $c \prec b$.

Figure 9.38

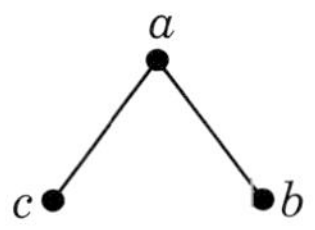

Since only ordered rooted trees are discussed from here on, the term *rooted tree* will imply *ordered rooted tree*.

Rooted trees need not have the same number of children at every internal vertex. Depending on the maximum number of children of any vertex, these trees have special names.

m-ary Tree

A rooted tree is an ***m*-ary tree** if every vertex has at most m children. If $m = 2$, it is a **binary tree**; if $m = 3$, it is a **ternary tree**. An m-ary tree is **full** if every internal vertex has exactly m children.

EXAMPLE 9.11 With the rooted trees in Figure 9.39, every internal vertex in T_1 has at most two children, so T_1 is a binary tree. In fact, it is a full binary tree. Tree T_2

Figure 9.39

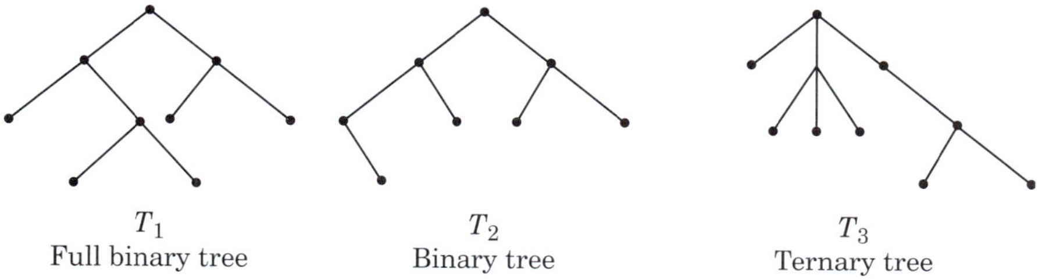

is binary, but not full (Why?). Every vertex in T_3 has at most three children, so T_3 is ternary. It is not, however, a full ternary tree (Why?). ■

m-ary trees help analyze the structures of computer programs.

○ **EXAMPLE 9.12** (optional) A computer program P consists of two subprograms A_1 and A_2. A_1 has three modules B_1, B_2, and B_3; A_2 has two modules B_4 and B_5. In addition, B_1 contains a module C_1, B_2 contains two modules C_2 and C_3, B_4 contains C_4, and B_5 contains C_5, C_6, and C_7. The structure of the program forms the ternary tree in Figure 9.40.

Figure 9.40

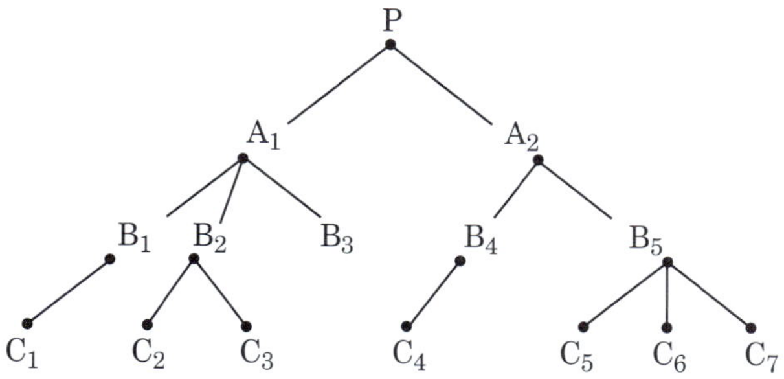

■

m-ary Tree and Partitions

Rooted trees can also find the partitions of a finite set, as the next example illustrates.

EXAMPLE 9.13 Using an m-ary tree, find the partitions of the set $S = \{a, b, c\}$.

SOLUTION:

Build a tree with root $\{a\}$. Use the remaining elements in S to make sets of subsets until each set becomes a partition of S, as Figure 9.41 shows. For

Figure 9.41

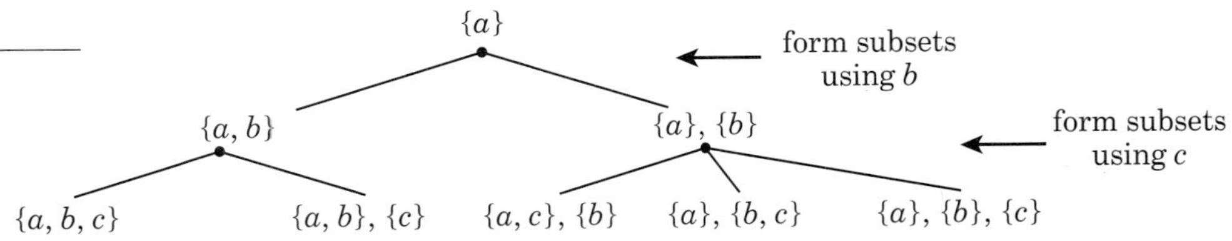

instance, at level 1, element b can be added to the set $\{a\}$ to form a new subset or can form a subset by itself. The leaves of the tree manifest the five partitions of the set. ■

Next we present balanced trees.

Balanced Tree

An m-ary tree of height h is **balanced** if its leaves lie at level $h-1$ or h, that is, if its leaves lie on adjacent levels.

EXAMPLE 9.14 In Figure 9.39, T_1 is balanced since all its leaves lie at level 2 or 3, and $h = 3$. For T_2, $h = 3$ and all its leaves lie at level 2 or 3, so T_2 is also balanced. Tree T_3 is not balanced (Why?). ■

Balanced trees prove extremely useful in sorting and searching, as Section 9.6 demonstrates. Shortening a binary tree, that is, balancing it, reduces the time needed for sorting and searching. In other words, making the tree as *bushy* as possible maximizes the efficiency of the search process.

The next five theorems explore some aspects of m-ary trees.

If you know the number of internal vertices in a full m-ary tree, you can use Theorem 9.5 to compute the total number of vertices.

THEOREM 9.5 A full m-ary tree with i internal vertices has $n = mi + 1$ vertices.

PROOF:
By Theorem 9.2, the tree has $n-1$ edges. Since the tree is full m-ary, each internal vertex has m children, each contributing m edges. So the tree has mi edges. Thus $mi = n - 1$ and $n = mi + 1$. ■

The next theorem presents additional relationships among the number of vertices, leaves, and internal vertices in a full m-ary tree. The proof is fairly straightforward, so we leave it as a routine exercise. (See Exercises 48–50.)

THEOREM 9.6 Let T be a full m-ary tree. Then:

- If T has n vertices, it has $i = \dfrac{n-1}{m}$ internal vertices and $l = \dfrac{(m-1)n+1}{m}$ leaves.
- If T has i internal vertices, it has $l = (m-1)i + 1$ leaves.

- If T has l leaves, it has $i = \dfrac{l-1}{m-1}$ internal vertices and $n = \dfrac{ml-1}{m-1}$ vertices. ■

THEOREM 9.7 An m-ary tree of height h has at most m^h leaves.

PROOF (by induction on h):

Basis step When $h = 0$, the tree contains one vertex, namely, the root. So the number of leaves in the tree is 1. Thus the theorem holds when $h = 0$.

Induction step Assume every m-ary tree of height h has at most m^h leaves. Let T be an m-ary tree of height $h+1$, and T_1 the tree obtained by deleting all leaves of T and the edges adjacent with them. T_1 is an m-ary tree of height h and contains at most m^h leaves by the inductive hypothesis. Since T is m-ary, each of these m^h vertices has a maximum of m children, so T has at most $m \cdot m^h = m^{h+1}$ leaves.

Thus, by induction, the theorem holds for every $h \geq 0$. ■

Theorem 9.8 gives a lower bound for the height of an m-ary tree with l leaves.

THEOREM 9.8 Let h be the height of an m-ary tree with l leaves. Then $h \geq \lceil \log_m l \rceil$.

PROOF:

By Theorem 9.7, $l \leq m^h$, so $h \geq \log_m l$. Since h is an integer, $h \geq \lceil \log_m l \rceil$. ■

In particular, for a binary tree with height h and l leaves, $h \geq \lceil \lg l \rceil$. For example, if the truth table of a compound proposition contains 1024 rows, the proposition is composed of at least $\lceil \lg 1024 \rceil = 10$ simple propositions.

The next theorem gives the exact height of a full, balanced m-ary tree with l leaves.

THEOREM 9.9 If h is the height of a full, balanced m-ary tree T with l leaves, then $h = \lceil \log_m l \rceil$.

PROOF:

By Theorem 9.7, $l \leq m^h$. Since the height of the tree is h, T must contain at least one leaf at level h. Let T_1 be the tree obtained by deleting the leaves at level h and the edges incident with them. Since T is balanced, T_1 has all its leaves at level $h-1$; so, by Theorem 9.7, T_1 has exactly m^{h-1} leaves (see Exercise 54). So $m^{h-1} < l$.

Thus

$$m^{h-1} < l \leq m^h$$

$$h - 1 < \log_m l \leq h$$

Thus

$$h = \lceil \log_m l \rceil$$ ■

The next two examples apply Theorems 9.6 and 9.9.

EXAMPLE 9.15 Compute the height of a full balanced binary tree with 4095 vertices.

SOLUTION:
By Theorem 9.6,

$$l = \frac{(m-1)n+1}{m} = \frac{(2-1)4095+1}{2}$$
$$= 2048$$

By Theorem 9.9, $h = \lceil \lg 2048 \rceil = 11$. ■

We close this section with the next example, which solves Problem 3 from the beginning of the chapter.

EXAMPLE 9.16 There are 64 entrants in a singles tennis tournament. The winner of each round is advanced to the next round. Find the number of matches and the number of rounds played to determine the champion.

SOLUTION:
The tennis tournament can be represented by the full balanced binary tree in Figure 9.42 with its leaves as the initial players, and the root being the champion of the tournament. The internal vertices represent the winners of the various matches. So each level corresponds to a round of the tournament.

Figure 9.42

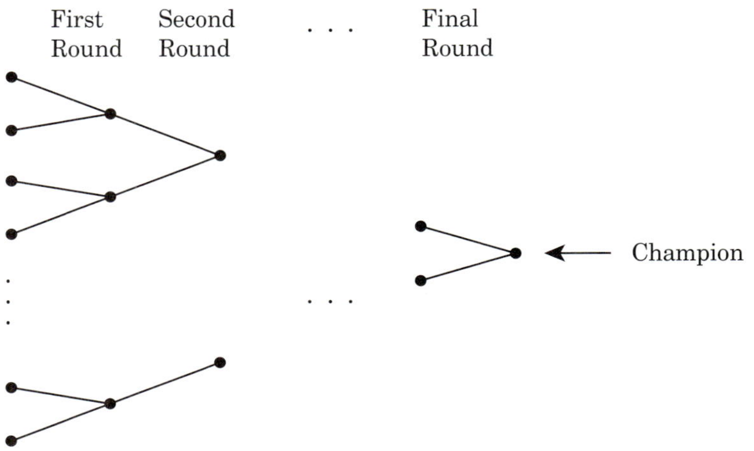

$$\begin{aligned}\text{Number of matches played} &= \text{Number of internal vertices}\\ &= \frac{l-1}{m-1}, \text{ by Theorem 9.6}\\ &= \frac{64-1}{2-1} = 63\\ \text{Number of rounds played} &= \text{height of the tree}\\ &= \lceil \lg 64 \rceil, \text{ by Theorem 9.9}\\ &= 6\end{aligned}$$

■

Exercises 9.4

Using recursion, compute the level of the given vertex in Figure 9.43.

Figure 9.43

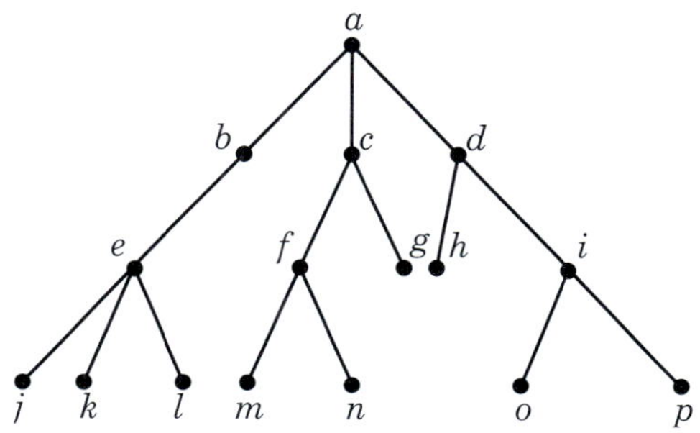

1. Vertex c **2.** Vertex e **3.** Vertex h **4.** Vertex j

Find m for each m-ary tree.

5.

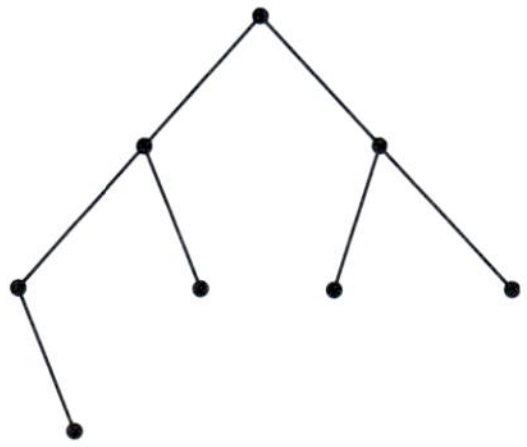

6.

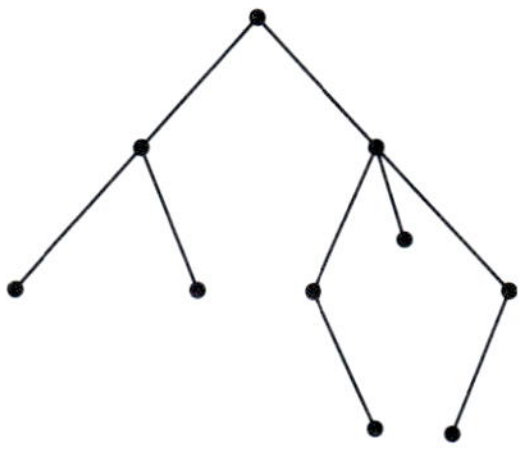

7.

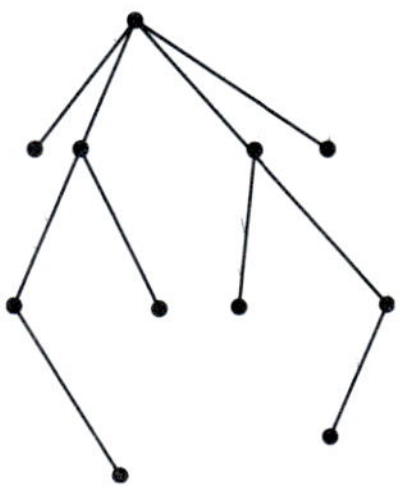

8–10. Is each of the m-ary trees in Exercises 5–7 full?

11–13. Is each of the m-ary trees in Exercises 5–7 balanced?

14. Let $\Sigma = \{0, 1\}$, where $0 \prec 1$. The language Σ^n can be defined as $\Sigma^n = \{wx \mid w \in \Sigma^{n-1}, x \in \Sigma\}$.

Using this definition, display the elements of $\bigcup_{n=0}^{3} \Sigma^n$ in a rooted tree, where the vertices at level k represent the elements of Σ^k.

15. Is the tree in Figure 9.43 binary? Full? Balanced?

16. Draw a full binary tree that is not balanced.

17. Draw a balanced tree that is not full.

18. Determine all simple paths in the maze in Figure 9.44 that a person at gate A can take to exit through gate Z.
(*Hint:* Draw a tree rooted at A.)

Figure 9.44

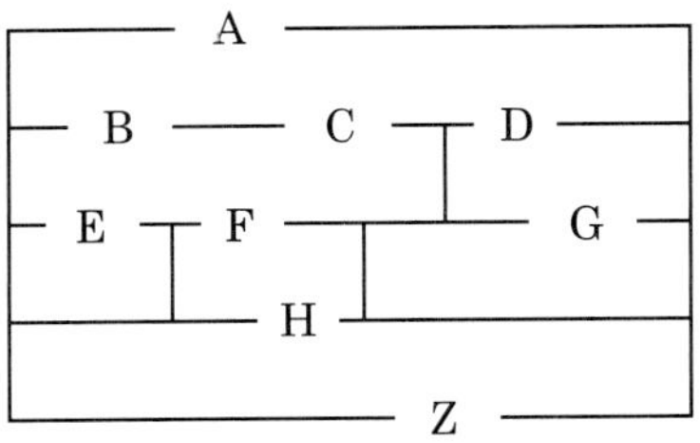

Using a rooted tree, find all partitions of each set.

19. {a, b}

20. $\{a, b, c, d\}$

21. Find the number of vertices of a full ternary tree with four internal vertices.

22. Find the number of leaves of a full 5-ary tree with 156 vertices.

23. How many internal vertices and leaves does a full ternary tree with 121 vertices have?

24. Compute the number of internal vertices and the height of a full and balanced 4-ary tree with 1024 leaves.

25. A full ternary tree has 121 internal vertices. How many leaves does it have?

26. Compute the maximum number of leaves in a full ternary tree of height 5.

Compute the height of each tree.

27. A full balanced binary tree with 10 leaves.

28. A full balanced ternary tree with 10 leaves.

29. A full balanced binary tree with 511 vertices.

30. A full balanced ternary tree with 3280 vertices.

Thirty-two people enter a singles Ping-Pong tournament. Find each.

31. The number of matches played. **32.** The number of rounds played.

An m-ary tree is **complete** if all leaves are at the same level. Are the following trees complete?

33.

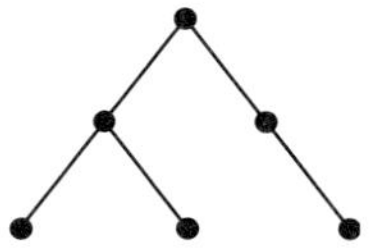

34.

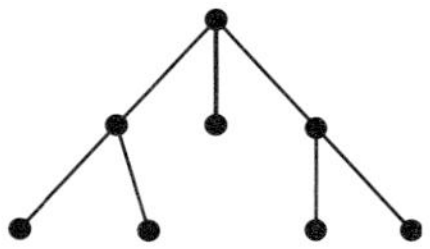

35.

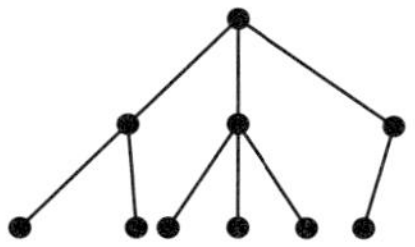

36. Is a complete m-ary tree full?

37. Is a complete m-ary tree balanced?

38. Draw a full binary tree that is *not* complete.

39. Draw a complete binary tree that is *not* full.

40. Draw a binary tree that is both full and complete.

For a full complete m-ary tree of height h, find each.

41. The number of leaves.

42. The number of internal vertices.

43. The number of vertices.

A full complete 6-ary tree has 1296 leaves. Compute each.

44. The number of its internal vertices.

45. Its height.

For a full complete ternary tree with 1093 vertices, find each:

46. Its height. **47.** The number of its leaves.

Prove the following parts of Theorem 9.6.

48. Part 1 **49.** Part 2 **50.** Part 3

Two rooted trees T and T' with vertex sets V and V', respectively, are **isomorphic** if there is a bijection $f : V \to V'$ such that:

- If v_0 is the root of T, then $f(v_0)$ is the root of T'; and
- If $v_1, v_2, \ldots, v_n$ are the children of a vertex $v \in V$ such that $v_1 \prec v_2 \prec \cdots \prec v_n$, then $f(v_1), f(v_2), \ldots, f(v_n)$ are the children of $f(v)$, where $f(v_1) \prec f(v_2) \prec \cdots \prec f(v_n)$.

Are the trees in Exercises 51 and 52 isomorphic?

51.

52.

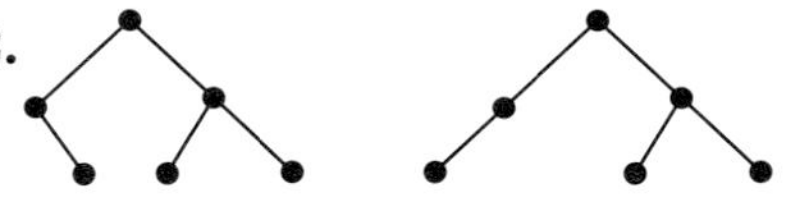

*53. Let T be a full complete m-ary tree with i internal vertices and l leaves, where $m \geq 2$. Prove that $l > i$.

*54. Prove that a full complete m-ary tree with height h has exactly m^h leaves.

9.5 Binary Trees

Binary trees are the most important class of m-ary trees and have a wide range of applications. For instance, they can model tournaments, as in Example 9.16. They also support very well the representation and evaluation of algebraic expressions.

Since a binary tree is a 2-ary tree, every internal vertex has at most two children — the elder is the **left child**; the other is the **right child**. They stand to the left and right of their parent. The subtree at a left child v is the **left subtree** rooted at v, and the subtree rooted at a right child w is the **right subtree** rooted at w.

For example, the left child of b in Figure 9.45 is d; b's right child is e. Vertex d has a left child, but no right. Vertex c has no left child, but does have a right child; its right subtree appears in Figure 9.46.

Figure 9.45

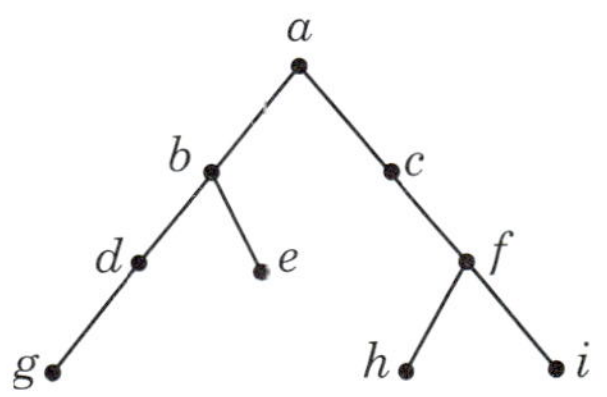

Figure 9.46

Right subtree of c.

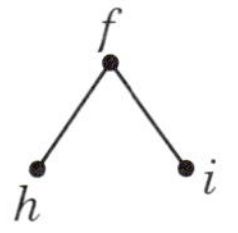

Fibonacci Trees

We are now ready to explore the world of Fibonacci trees T_n, a special class of binary trees, and their close relationship with Fibonacci numbers. **Fibonacci trees** are defined recursively as follows:

- Both T_1 and T_2 are binary trees with exactly one vertex each; and
- T_n is a binary tree with left subtree T_{n-1} and right subtree T_{n-2}, where $n \geq 3$.

Figure 9.47 shows the first six Fibonacci trees, T_1 through T_6.

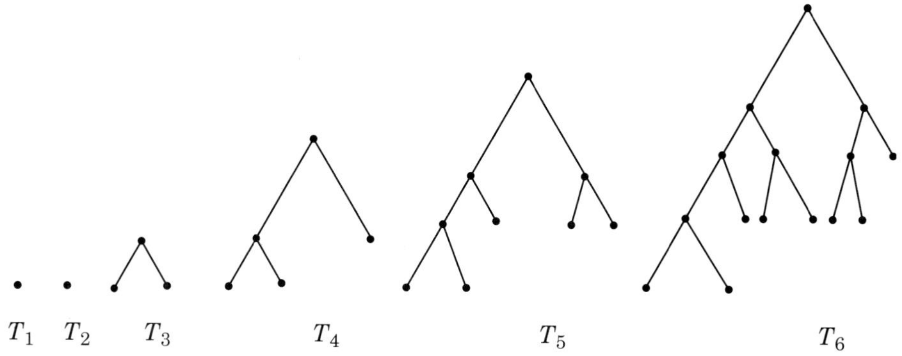

Figure 9.47

Fibonacci trees.

Out of curiosity, we now explore the number of leaves l_n, the number of internal vertices i_n, the number of vertices v_n, and the number of edges e_n of the nth Fibonacci tree T_n. To facilitate our investigation, study the trees in Figure 9.47, collect the needed data, and then summarize them in a table, as in Table 9.5.

Table 9.5

n	**1**	**2**	**3**	**4**	**5**	**6**	...	n
V_n	1	1	3	5	9	15	...	?
l_n	1	1	2	3	5	8	...	?
i_n	0	0	1	2	4	7	...	?
e_n	0	0	2	4	8	14	...	?

Using the table, we conjecture that $l_n = F_n$, $i_n = l_n - 1 = F_n - 1$, $v_n = i_n + l_n = 2F_n - 1$, and $e_n = 2i_n = 2F_n - 2$. They are in fact true. We can confirm them; see Exercises 74–77.

Binary Tree Traversals

An important tree operation is tree traversal, visiting every vertex of a tree in a systematic way. Three elegant methods for traversing a nonempty binary tree exist: **preorder**, **inorder**, **postorder traversals**. Each of them can be defined recursively:

Preorder traversal

- Visit the root.
- Traverse the left subtree in preorder.
- Traverse the right subtree in preorder.

Inorder traversal

- Traverse the left subtree in inorder.

- Visit the root.
- Traverse the right subtree in inorder.

Postorder traversal

- Traverse the left subtree in postorder.
- Traverse the right subtree in postorder.
- Visit the root.

The binary tree in Figure 9.48 has T_1 and T_2 as left and right subtrees. To traverse the tree in preorder, first visit the root v; traverse T_1 in preorder and then traverse T_2 in preorder. The other two traversals can be interpreted similarly.

Figure 9.48

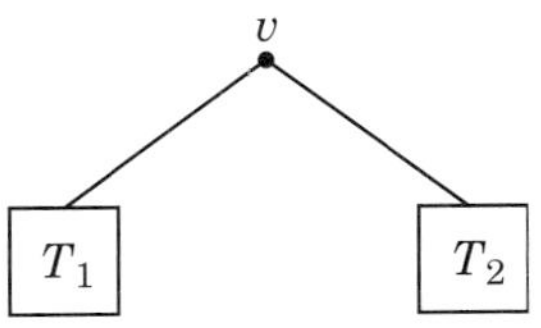

The next three examples illustrate these tree traversals step-by-step.

EXAMPLE 9.17 Give the output from traversing the binary tree in Figure 9.49 in preorder.

Figure 9.49

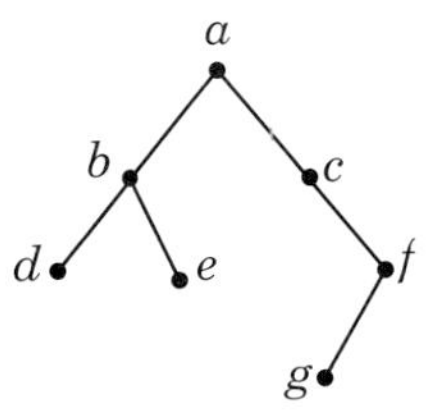

SOLUTION:

1. First visit the root a and output a.
2. Traverse the left subtree rooted at b in preorder (see Figure 9.50). Output b. Now traverse the left subtree rooted at d in preorder (see Figure 9.51). Output d. Traverse the left subtree of d; it is empty. So traverse its right subtree; it is also empty. Therefore, back up to b and traverse its right subtree (see Figure 9.52). Visit e. Output e. It has no subtrees.
3. Backtrack to a. Traverse its right subtree in preorder (see Figure 9.53). Output c. It has no left subtree, so traverse its right subtree (see Figure 9.54). Output f.

Figure 9.50

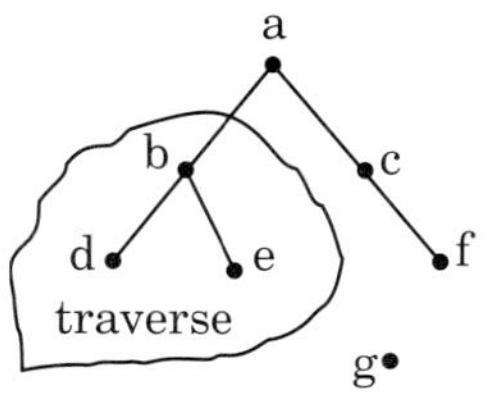

Figure 9.51

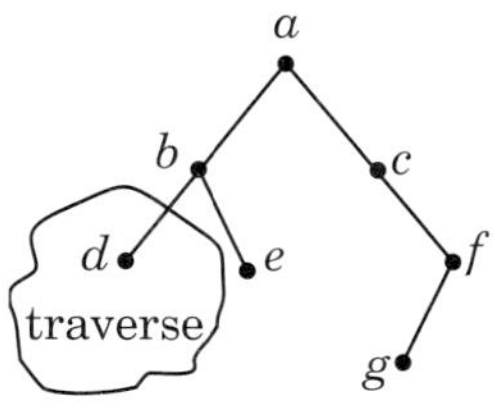

Figure 9.52

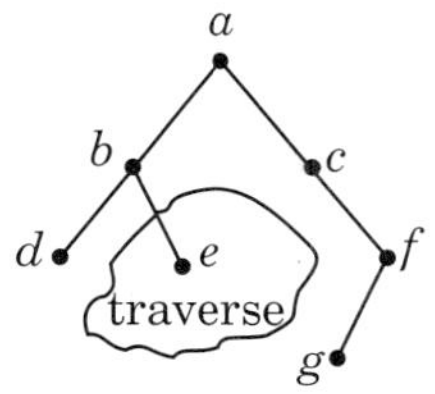

Figure 9.53

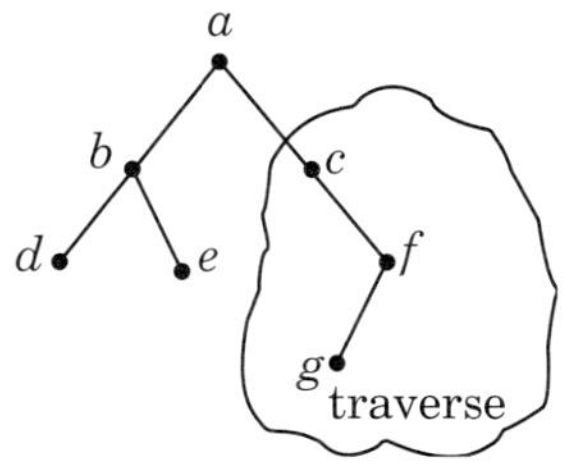

Figure 9.54

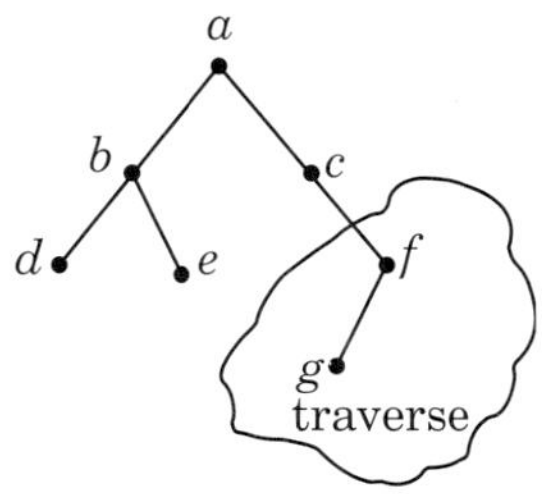

Traverse its left subtree (see Figure 9.55). Output g. It has no subtrees; so backtrack to f and traverse its right subtree in preorder, which is empty.

All vertices have been visited, so the traversal is done. The output is a, b, d, e, c, f, g.

Figure 9.55

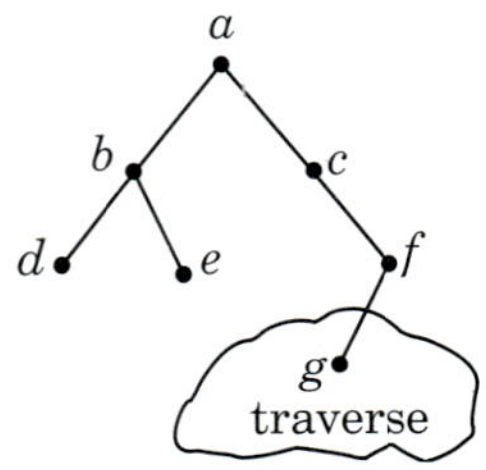

■

EXAMPLE 9.18 Give the output from traversing the binary tree in Figure 9.49 in inorder.

SOLUTION:

1. First traverse the left subtree in inorder (see Figure 9.56). Traverse its left subtree in inorder (see Figure 9.57). Again traverse its left subtree in inorder. It is empty, so visit the root d and output it. Now traverse its right subtree. Since it is empty, back up to b and output it. Traverse its right subtree (see Figure 9.58). Vertex e has no left subtrees, so output e. It has no right subtrees.

Figure 9.56

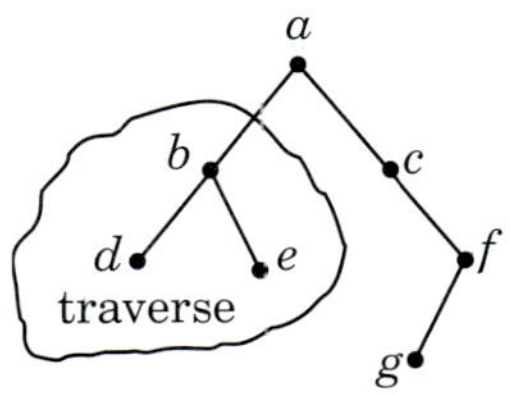

Figure 9.57

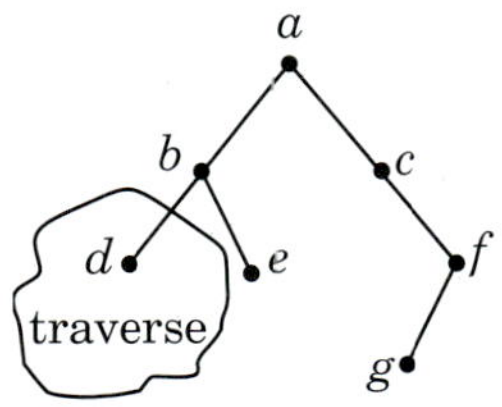

Figure 9.58

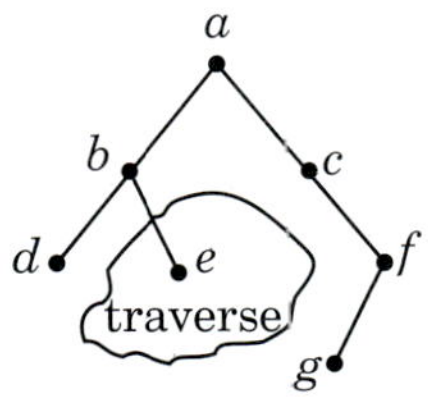

2. Backtrack to a and output a.

3. Now traverse the right subtree in inorder (see Figure 9.59). Since the left subtree of c is empty, output c. Traverse its right subtree (see Figure 9.60). Move to the left and traverse its left subtree (see Figure 9.61). Since g has no left subtree, output g. It also has no right subtree, so back up to f and output it. Since f has no right subtree, the traversal is over. The output is d, b, e, a, c, g, f.

Figure 9.59

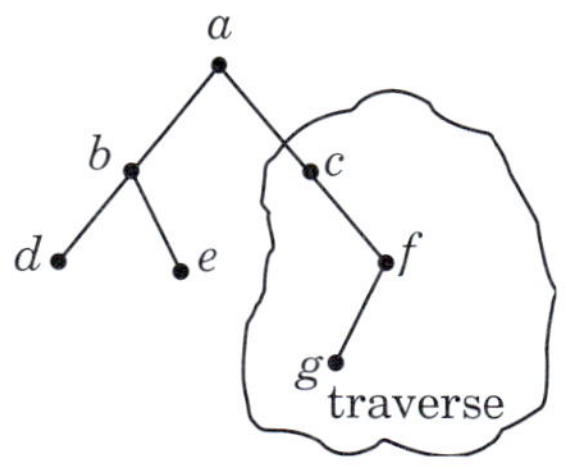

Figure 9.60

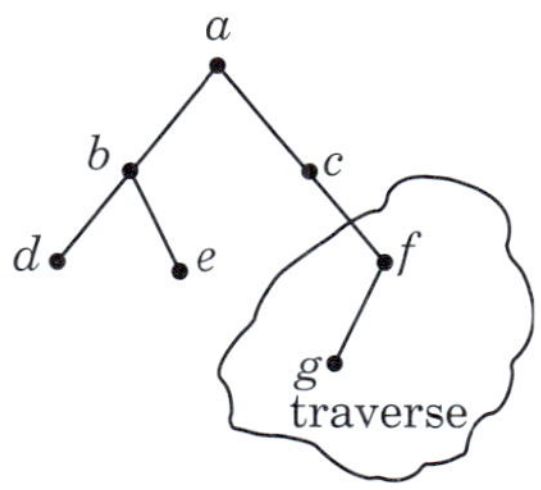

Figure 9.61

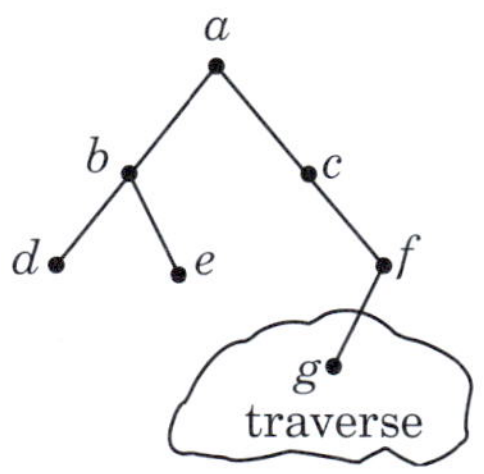

■

EXAMPLE 9.19 Find the output from traversing the binary tree in Figure 9.49 in postorder.

SOLUTION:

The various steps are summarized in Figures 9.62–9.67.

1. Traverse the left subtree in postorder (see Figures 9.62–9.64). The output from this traversal is d, e, b.
2. Traverse the right subtree in postorder (see Figures 9.65–9.67). This traversal yields the output g, f, c.
3. Backtrack to the root a. Output a.

The output from the postorder traversal is d, e, b, g, f, c, a.

Figure 9.62

Traverse the left subtree of a in postorder.

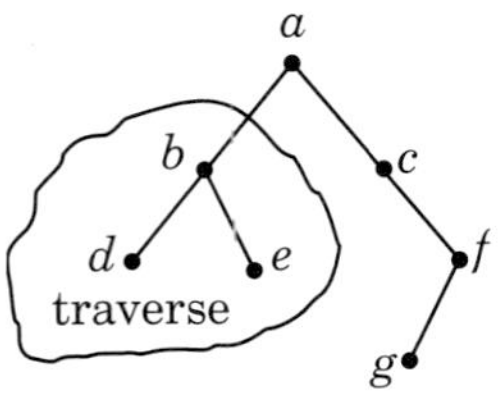

Figure 9.63

d has no subtrees; output d.

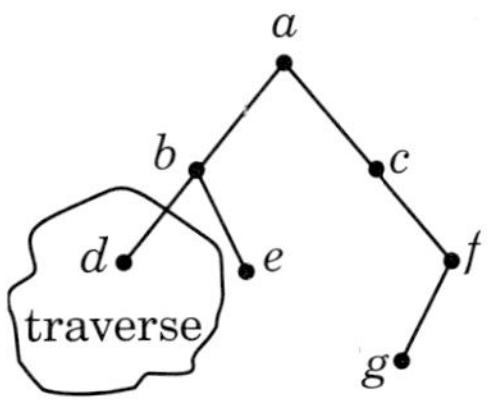

Figure 9.64

Back up to b and traverse its right subtree; e has no subtree, so ouput e. Backtrack to b; output b.

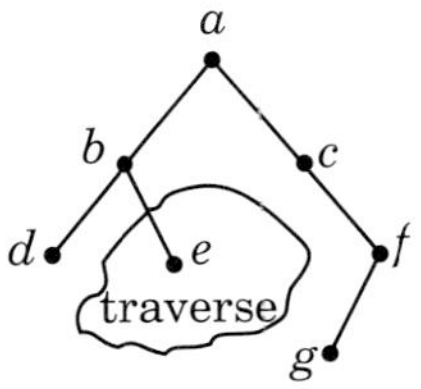

Figure 9.65

c has no left subtree; traverse its right subtree.

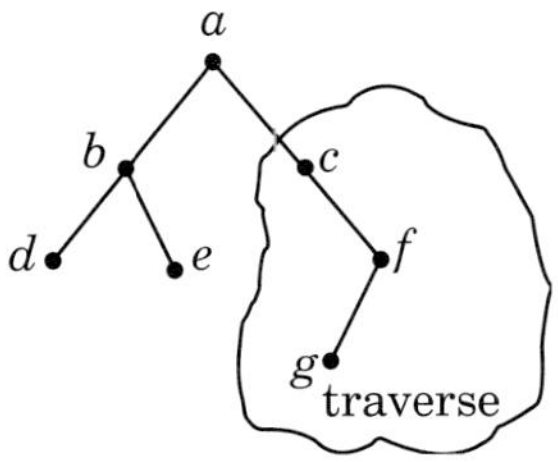

Figure 9.66

Traverse f's left subtree.

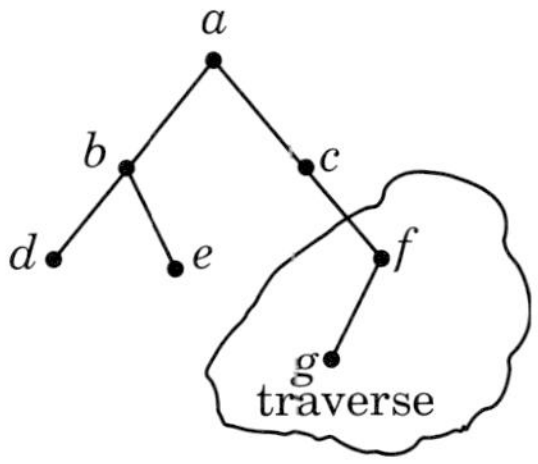

Figure 9.67

g has no subtrees; output *g*. Backtrack to *f*, it has no subtree. Output *f*. Backtrack to *c*: output it.

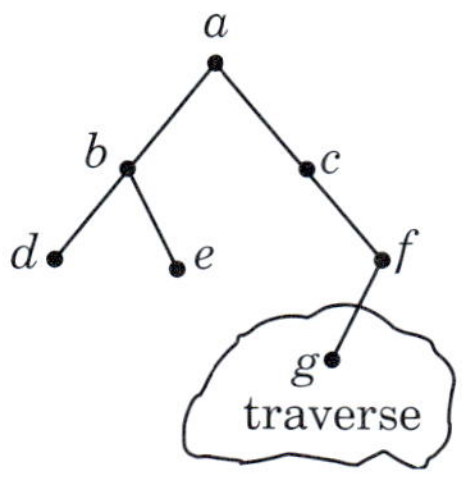

Binary trees are fine tools for representing and evaluating algebraic expressions involving binary operators.

Infix, Prefix, and Postfix Notations

Before discussing how an algebraic expression with only binary operators can be represented by a binary tree, we should examine three standard forms for such expressions. In the familiar form, the operator appears between its two operands. This is the **infix form**. For instance, $a + b$ is an infix expression.

Two alternatives are the **prefix** and **postfix forms**. In the prefix notation, also called the **Polish notation**, the binary operator precedes its two operands; in the postfix notation, also known as the **Reversed Polish Notation (RPN)**, the binary operator follows the two operands. These two notations are attributed to the Polish logician Jan Lukasiewicz.

The three general forms are summarized in Table 9.6. The prefixes *pre*, *in*, and *post* indicate the location of the operator with respect to the operands.

Table 9.6

Notation	General form	Example
Infix	<operand> <operator> <operand>	$a + b$
Prefix	<operator> <operand> <operand>	$+ab$
Postfix	<operand> <operand> <operator>	$ab+$

Some calculators, such as Hewlett-Packard's, use the postfix notation to evaluate numeric expressions. The main advantage of the two Polish notations is that algebraic expressions can be rewritten without parentheses. Invalid expressions can be detected easily, too.

Algebraic expressions involving the binary operators $+$ (addition), $-$ (subtraction), $*$ (multiplication), / (division), and $\uparrow$ (exponentiation) have the following order of precedence (from highest to lowest):

Exponentiation
Multiplication, division
Addition, subtraction

Certainly, parentheses can override this precedence to specify another order in which to perform operations, so parenthesized subexpressions have the highest priority. [Note that $a \uparrow b \uparrow c = a \uparrow (b \uparrow c)$.]

Jan Lukasiewicz *(1878–1956), a Polish logician and philosopher, was born in Lvov. His father was a captain in the Austrian army. After studying mathematics and philosophy, Lukasiewicz earned his Ph.D. in philosophy from the University of Lvov in 1902, where he taught for the next 5 years. In 1915, he accepted an invitation to teach at the University of Warsaw. In 1919, he served as the minister of education in independent Poland, and the following year returned to the University where he was a professor from 1920 to 1939, serving twice as its rector.*

In 1946, while exiled in Belgium, Lukasiewicz accepted a professorship at the Royal Irish Academy, Dublin, where he remained until his death.

A "resourceful and imaginative scholar" and "a gifted and inspiring teacher," he was one of the founding fathers of the Warsaw School of Logic, a member of several scientific societies, and the recipient of numerous honors.

To convert an infix expression into prefix or postfix, look for the subexpression that must be evaluated first, and translate it into the desired form. Now translate the next subexpression that must be evaluated. Continue like this until the whole expression is converted into the desired form. The next example illustrates this technique.

EXAMPLE 9.20 Rewrite the expression $a * (b + c)/d - e$ in prefix and postfix forms.

SOLUTION:

1. *To convert the expression into prefix form:*

- The subexpression $b + c$ must be evaluated first, so translate it into prefix: $+bc$.
- Scanning the expression from left to right indicates the next operation that must be performed is multiplication. The two operands of $*$ are a and $+bc$. So translate this subexpression into prefix: $*a + bc$.
- Now perform division. The two operands of / are $*a + bc$ and d; rewrite this subexpression in prefix: $/ * a + bcd$.
- Finally, perform subtraction. The operands of $-$ are $/ * a + bcd$ and e; convert this subexpression into prefix: $-/ * a + bcde$.

Since the expression is finished, we are done; so the desired prefix expression is $-/ * a + bcde$.

These steps are summarized in Figure 9.68.

2. *To convert the expression into postfix form:*
Use essentially the same steps as above, but convert subexpressions into postfix form.

Figure 9.68

$a * (b + c) / d - e$

$+bc$

$*a+bc$

$/*a+bcd$

$-/*a+bcde$

- Convert $b + c$ into postfix: $bc+$
- Convert $a * (bc+)$ into postfix (parentheses are used for readability): $abc+*$
- Convert $(abc+*)/d$ into postfix: $abc+*d/$
- Convert $(abc+*d/) - e$ into postfix: $abc+*d/e-$. This is the desired postfix expression.

These steps are summarized in Figure 9.69.

Figure 9.69

$a * (b + c) / d - e$

$bc+$

$abc+*$

$abc+*d/$

$abc+*d/e-$

■

Since every binary operator has two operands, every algebraic expression containing only binary operators can be represented by a **binary expression tree**. In such a tree the leaves contain operands and the internal vertices contain operators. Since not all data stored in the vertices are of the same kind, it is a **heterogeneous tree**.

To construct a binary expression tree, store the operator in the root of the tree and represent the first operand by the left subtree and the second operand by the right subtree. Example 9.21 employs this technique.

EXAMPLE 9.21 Represent the expression $a * (b + c)/d - e$ in a binary expression tree.

SOLUTION:

Build the tree from the bottom up, using the order in which the various operations are performed. Construct a binary tree for each expression $b+c$, $a * (b + c)$, $a * (b + c)/d$, and $a * (b + c)/d - e$ successively, as in Figures 9.70–9.73.

Figure 9.70

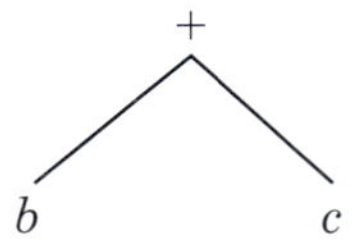

Figure 9.71

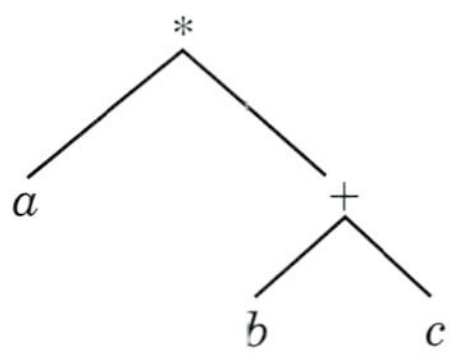

Figure 9.72

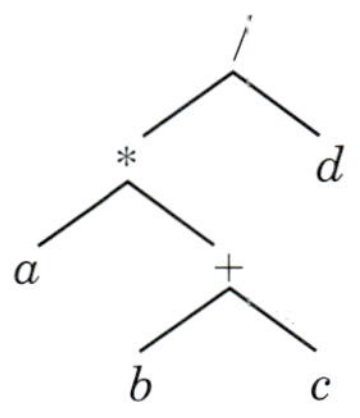

Figure 9.73

Binary expression tree.

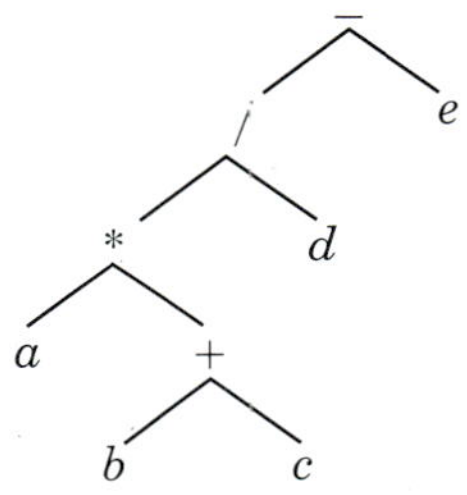

For a computer to evaluate a legal infix expression, it translates it into a Polish expression and then evaluates it. The following example illustrates the second half of this process.

EXAMPLE 9.22 Evaluate the prefix expression $- \uparrow / * 2 + 75429$, where each operand is a single-digit number.

SOLUTION:

In a prefix expression, the binary operator precedes its operands; so scan the expression from left to right until you encounter two successive operands. Then back up to the operand immediately preceding the operator and perform the operation. Repeat this procedure until the expression consists of a single operand. Figure 9.74 summarizes the various steps involved; so the value of the expression is 27.

Figure 9.74

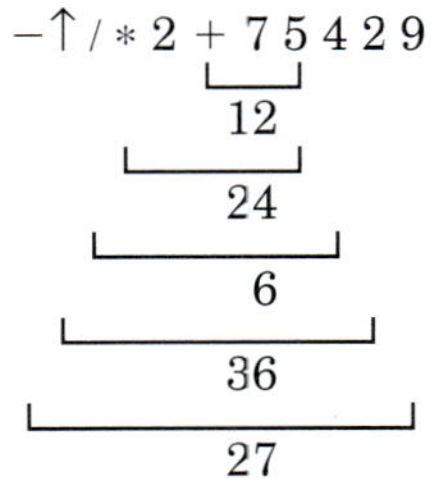

The steps illustrated in Example 9.22 can easily be adapted to evaluate postfix expressions.

> Binary expression trees can evaluate numeric expressions recursively: replace each operator vertex with the value of the expression tree rooted at the vertex. The next example illustrates this method.

EXAMPLE 9.23 Evaluate the expression represented by the binary expression tree in Figure 9.75.

Figure 9.75

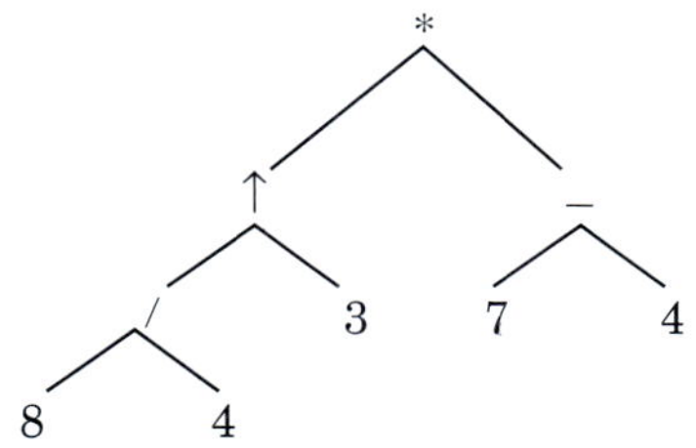

SOLUTION:
First evaluate the subtree rooted at /, giving 2; replace this subtree with 2 (see Figure 9.76). Now evaluate the subtree rooted at ↑, giving 8; replace the subtree rooted at ↑ with 8 (see Figure 9.77). Continue like this until the root of the tree contains an operand. The remaining steps transpire Figures 9.78 and 9.79. The value of the expression is 24.

Figure 9.76

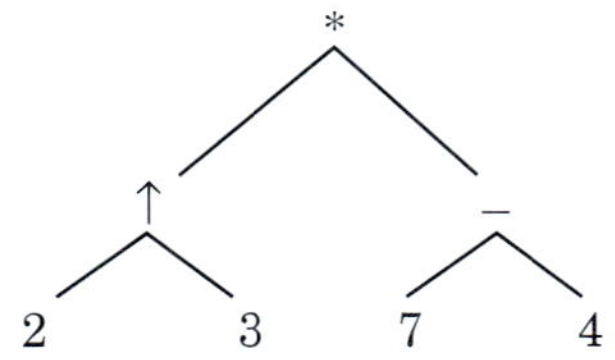

Figure 9.77

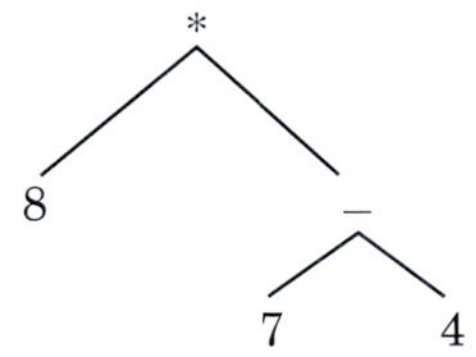

Figure 9.78

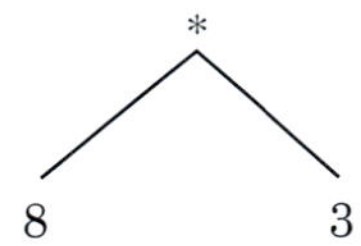

Figure 9.79

■

Suppose you traverse a binary expression tree in preorder. What can you say about the output? In a preorder traversal, the root precedes the subtrees. Since the root of a binary expression tree represents a binary operator and the subtrees represent its operands, the preorder traversal yields the prefix form of the expression. Similarly, the postorder traversal produces the postfix expression.

For example, the binary tree in Figure 9.80 represents the infix expression $[a + (b \uparrow c)] * [d/(e - f)]$. You may verify that preorder traversal yields the prefix expression $* + a \uparrow bc/d - ef$ and postorder traversal produces the postfix expression $abc \uparrow +$def $- /*$.

Figure 9.80

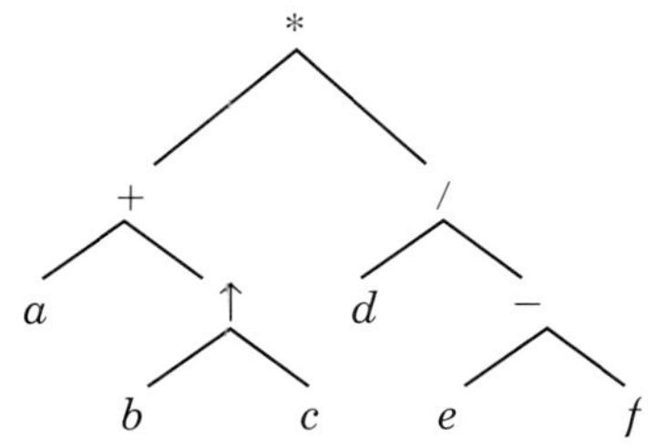

Traversing this tree in inorder yields $a + b \uparrow c * d/e - f$, the same as the infix expression except for parentheses. In general, the inorder traversal of a binary expression tree yields the infix expression without parentheses.

Finally, binary expression trees can represent assignment statements and boolean expressions; the assignment operator $\leftarrow$ commands least priority among the operators. For instance, the statement $x \leftarrow w + y \uparrow z$ is represented by the tree in Figure 9.81. Such binary trees can find the value of the variable on the LHS of the assignment operator $\leftarrow$. With the tree in Figure 9.82, you may verify that the value assigned to x by the statement $x \leftarrow (5 + 6) * (2 \uparrow 3)$ is 88.

Figure 9.81

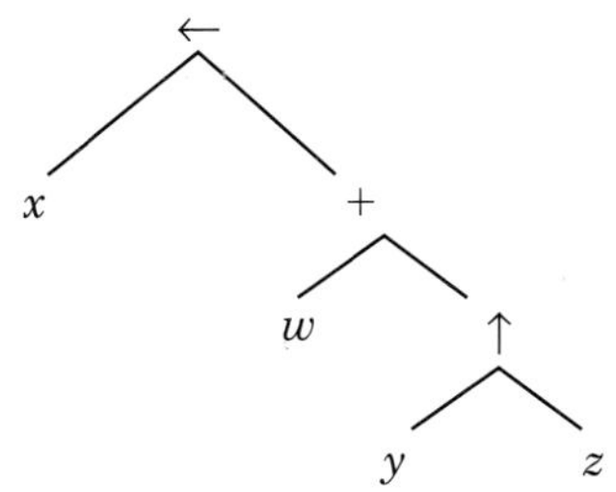

The trees in Figures 9.83 and 9.84 display the boolean expressions $(a < b) \vee (c < d)$ and $a * (b + c) < d \uparrow e - f$, respectively. The **relational operators** $<$, $\leq$, $>$, $\geq$, $=$, and $\neq$ have next-to-last precedence among all operators.

Figure 9.82

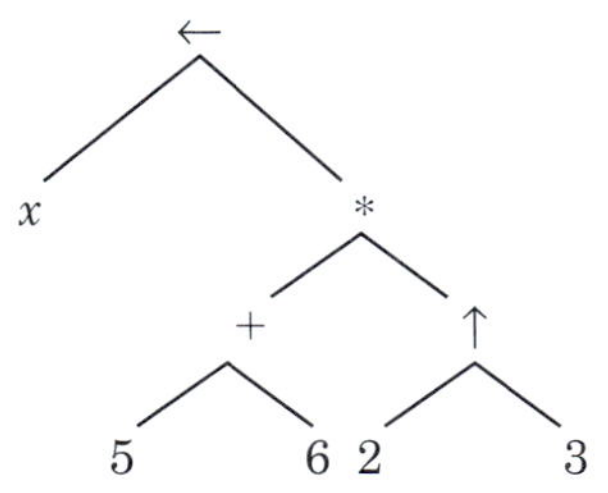

Figure 9.83

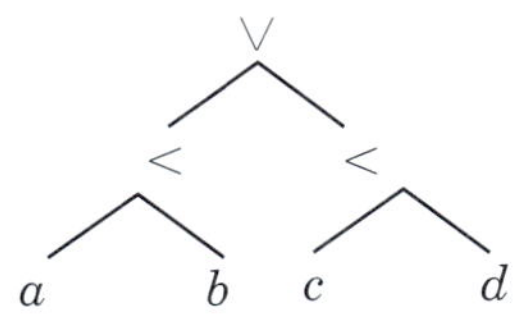

Figure 9.84

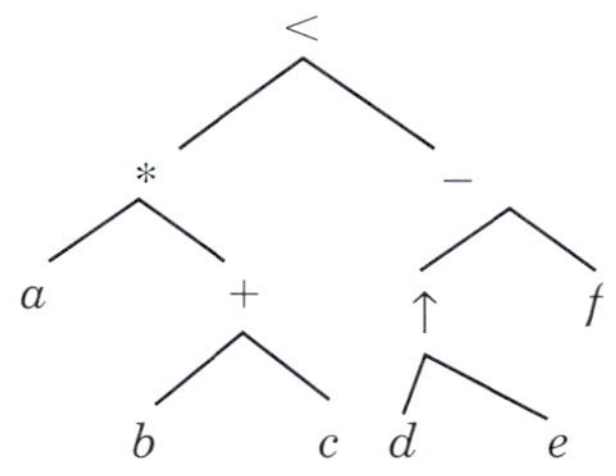

Next we show how parenthesized triangulations are closely related to binary trees.

Binary Trees and Parenthesized Triangulations

Recall from Section 6.6 that the triangulation of a convex $(n + 2)$-gon corresponds to correctly parenthisizing a sequence of $n + 1$ symbols and vice versa. Interestingly, each procedure can be represented in a binary tree. For example, let us return to the parenthesized triangulation in Figure 6.28. Figure 9.85 shows the order of operation employed there; this leads to the binary tree in Figure 9.86 in an obvious way.

Figure 9.85

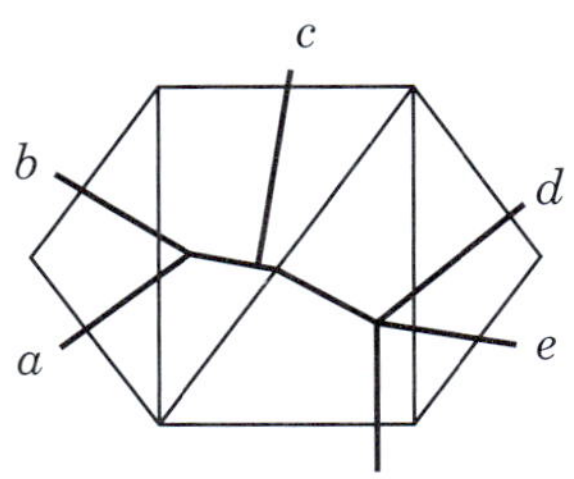

Lukasiewicz discovered an intriguing way to extract the binary number corresponding to a polygonal dissection from its binary tree representation.

Figure 9.86

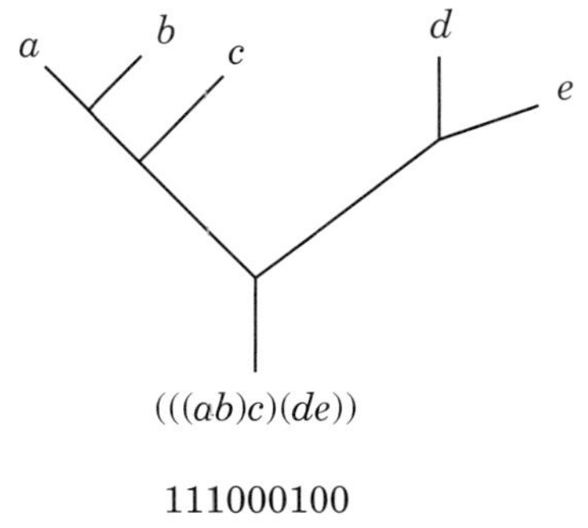

First we label each leaf in Figure 9.86 with a 0 and each internal vertex with a 1, as in Figure 9.87. Next, beginning at the root, we traverse the entire tree. Reading each unvisited vertex, we get the same binary number 111000100 as before.

Figure 9.87

Traversing the binary tree for its binary number.

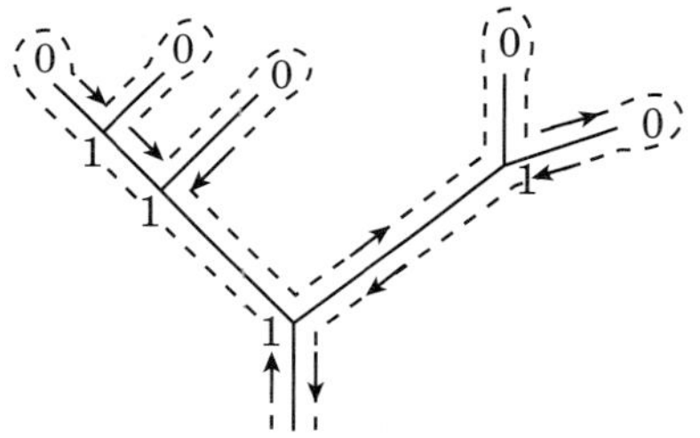

Finally, we observe a close relationship between binary trees and Catalan numbers.

Binary Trees Meet Catalan Numbers

An interesting combinatorial problem is to find the number of nonisomorphic binary trees we can draw with n vertices. For example, there is exactly one binary tree with $n = 1$; there are exactly two distinct binary trees with $n = 2$, exactly five with $n = 3$ (see Figure 9.88), and so on. In general, with n vertices we can draw exactly C_n nonisomorphic binary trees. This should be obvious, because we found earlier that there is a bijection between triangulated convex polygons and binary trees.

Figure 9.88

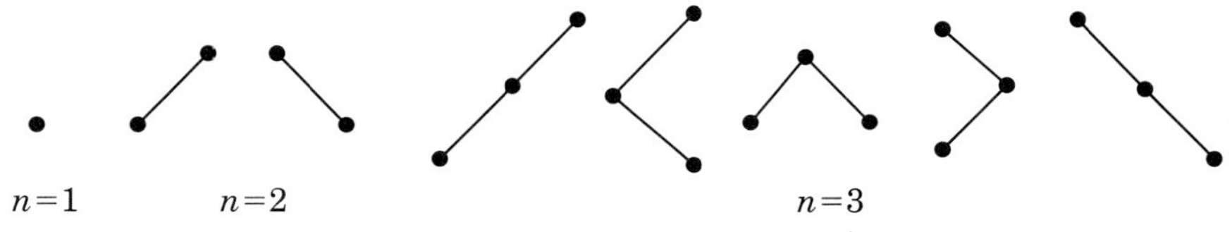

Exercises 9.5

Give the output from traversing each binary tree in preorder.

1.

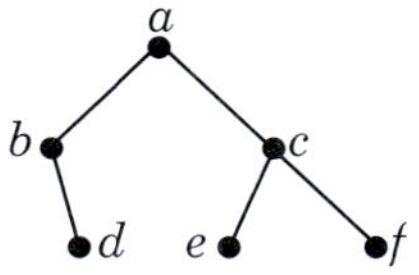

2.

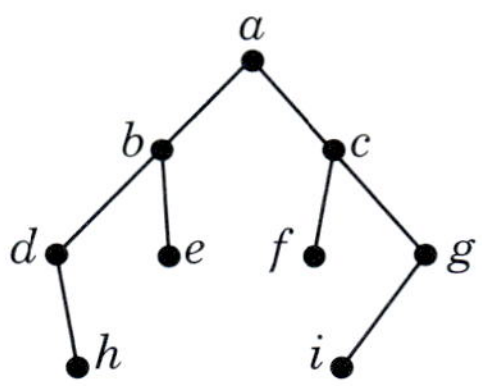

3.

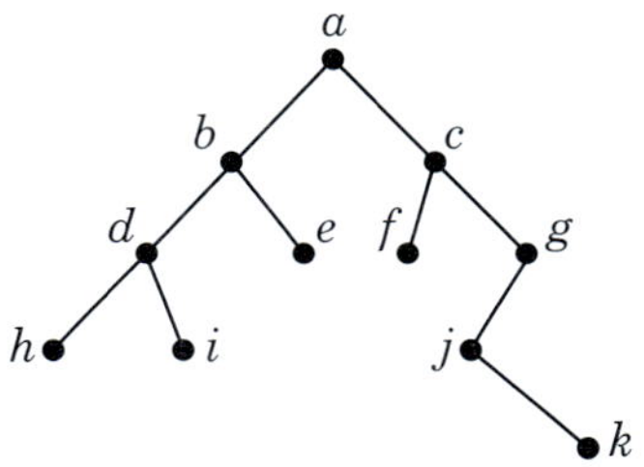

4–6. Give the output from traversing the binary trees in Exercises 1–3 in inorder.

7–9. Give the output from postorder traversing the binary trees in Exercises 1–3.

10–12. Redo Exercises 4–6 using the DFS method.

13–15. Redo Exercises 4–6 using the BFS method.

Rewrite each infix expression in prefix form.

16. $a + b * c/(d - e) \uparrow f$

17. $a \uparrow (b \uparrow c) + d/e - f$

18. $(a + b * c)/(d - e/f) \uparrow g$

19. $a - (b * c + d)/e * f - g \uparrow h$

20–23. Translate the infix expressions in Exercises 16–19 into postfix expressions.

24–27. Construct a binary expression tree for each infix expression in Exercises 16–19.

Rewrite each prefix expression in infix form, supplying parentheses when necessary.

28. $+* \uparrow abc * de$

29. $+a \uparrow /b - cde$

30. $- \uparrow +a * bcd * ef$

31. $* * -a + bcd \uparrow e - fg$

Convert each postfix expression into infix form, supplying parentheses when necessary.

32. $ab - cd - /ef \uparrow *$

33. $abc + d * /e * f-$

34. $ab - cd - /e \uparrow$

35. $ab/cd/efg - + * +$

Evaluate each binary expression, where each operand is a single-digit number.

36. $- \uparrow 2 \uparrow 23 * + 857$

37. $37 * 4 + 5/2 \uparrow$

38. $63/921 + / + 73 - *$

39. $86 + 34 + /5 \uparrow$

Represent each binary expression in a binary expression tree.

40. $a * b + c \uparrow d$

41. $a * b + [c - (d - e)]$

42. $(a + b * c) \uparrow (d/e)$

43. $[(a + b) * c] * [d \uparrow (e \uparrow f)]$

Evaluate each binary expression tree.

44.

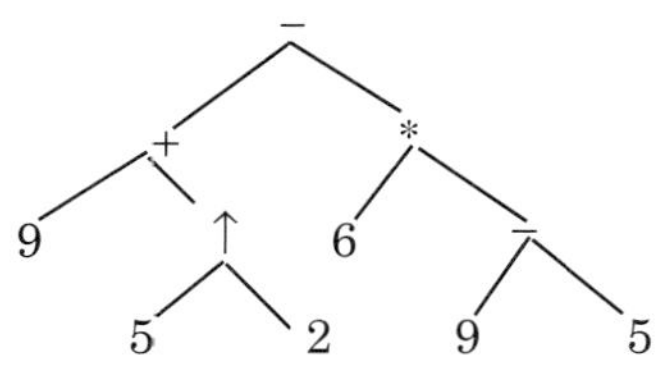

45.

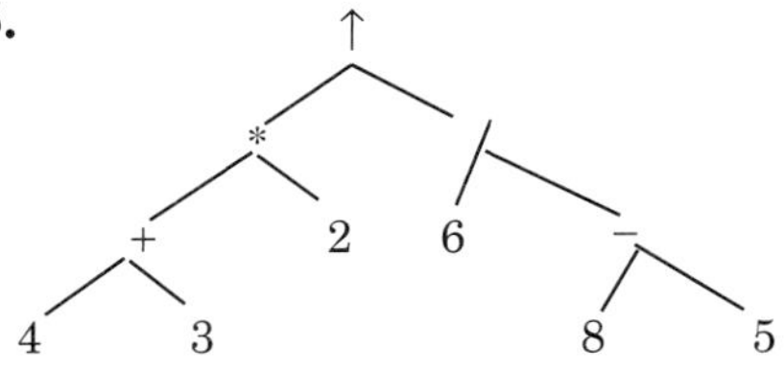

46–47. Traverse the binary trees in Exercises 44 and 45 in preorder.

48–49. Traverse the binary trees in Exercises 44 and 45 in postorder.

50–51. Traverse the binary trees in Exercises 44 and 45 in inorder. (Supply parentheses when needed.)

Represent each assignment statement and boolean expression in a binary expression tree.

52. $x \leftarrow (x + y * z) \uparrow a - b$

53. $x \leftarrow x \uparrow (y \uparrow z)/x + (y - z)$

54. $x + y \,\hat{}\, z < w + y$

55. $(a < b) \vee [(b \leq c) \wedge (d \leq e)]$

56. $x/y + y * z \leq w \uparrow x + z$

57. $(x/y) * z + w \geq (w - y) \uparrow z$

58. How many leaves does a full binary tree with n vertices have?

59. Prove that the number of vertices in a full binary tree is odd.

60. How many leaves does a full binary tree with i internal vertices have?

61. How many vertices does a full binary tree with l leaves have?

Two binary trees, T_1 and T_2, with vertex sets V_1 and V_2 and roots r_1 and r_2 are **isomorphic** if there exists a bijection $f : V_1 \to V_2$ such that:

- $f(r_1) = f(r_2)$;
- Vertices v and w are adjacent in T_1 if and only if $f(v)$ and $f(w)$ are adjacent in T_2; and

- If w is a left (or right) child of a vertex v in T_1, then $f(w)$ is a left (or right) child of $f(v)$.

For example, the binary trees in Figure 9.89 are isomorphic; those in Figure 9.90 are not (Why?).

Figure 9.89

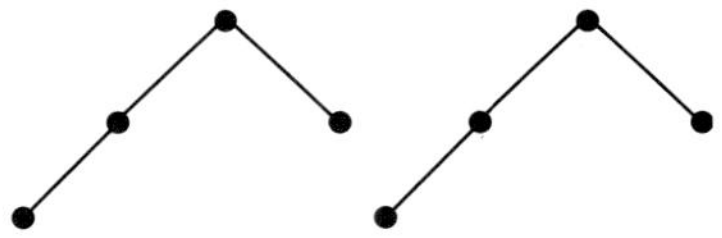

Figure 9.90

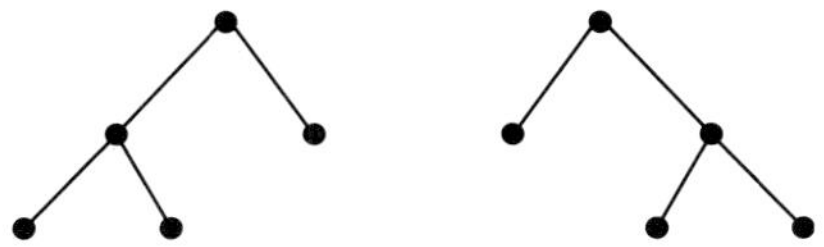

62. Draw all nonisomorphic binary trees with three vertices. With four vertices.

63. Generating functions and the binomial theorem can show* that the number of nonisomorphic binary trees with n vertices is the Catalan number C_n. With this fact, compute the number of binary trees with four vertices. With five vertices.

In Exercises 64–77, T_n denotes the nth Fibonacci tree.

64. Draw the Fibonacci tree T_7.

65. Is T_n a full binary tree?

66. Is T_6 a balanced binary tree?

67. Is T_5 a complete binary tree?

68. For what values of n is T_n a complete binary tree?

Using T_n, define each recursively.

69. The number of leaves l_n.

70. The number of internal vertices i_n.

71. The number of vertices v_n.

72. The height h_n.

73. The number of edges e_n.

Prove each.

74. $l_n = F_n$

75. $i_n = F_n - 1$

76. $v_n = 2F_n - 1$

77. $e_n = 2F_n - 2$

Write an algorithm to traverse a binary tree in:

78. Preorder.

79. Inorder.

80. Postorder.

81. Write an algorithm to evaluate a binary expression tree.

---

*See, S. Sahni, *Concepts in Discrete Mathematics*, 2nd ed., Camelot, Fridley, MN, 1985, pp. 427–431.

9.6 Binary Search Trees

Binary trees that contain items of the same kind are extremely useful. Such trees are **homogeneous trees**. For example, the binary tree in Figure 9.45 is homogeneous, but the one in Figure 9.73 is not.

Suppose a city needs a telephone directory. Since all items in the list are of the same type, a binary tree provides the perfect data structure for such a list. When a family moves into or out of the city, the list can easily be updated with the tree. Such trees efficiently sort a list, search it for a *key*, and eliminate duplicates from the list, as will be seen shortly; but first, we make the following definition.

Binary Search Tree

A **binary search tree** is a homogeneous binary tree such that every item on the left subtree of every vertex v is less than v and every item on its right subtree is greater than v. (Here *less than* refers to any linear or total order.)

EXAMPLE 9.24 In the binary tree in Figure 9.91, using the alphabetic order, every element in the left subtree of each vertex v is less than v, and every element in the right subtree of v is greater than v; so it is a binary search tree. You may verify, however, that Figure 9.92 does not display a binary search tree.

Figure 9.91

A binary search tree.

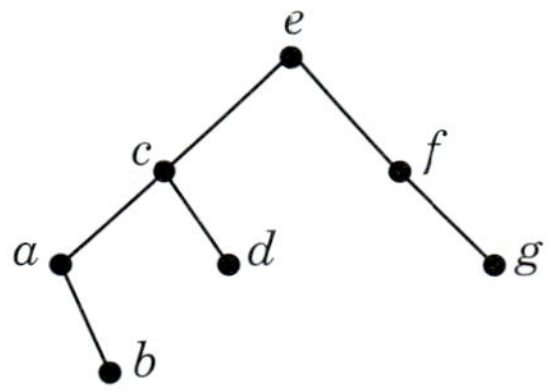

Figure 9.92

Not a binary search tree.

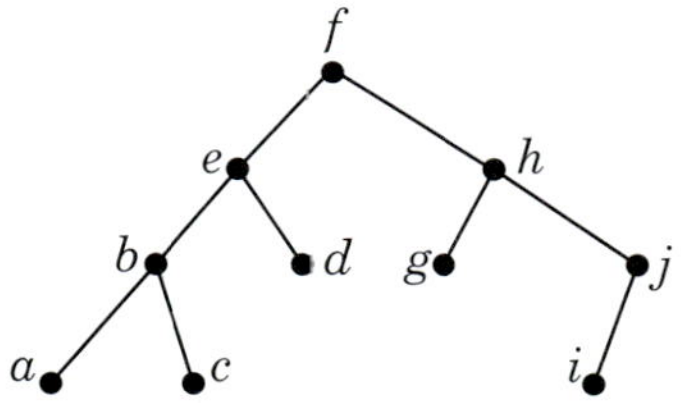

■

A recursive procedure for constructing a binary search tree appears in Algorithm 9.6.

```
Algorithm Binary Search Tree (root)
(* This algorithm constructs a binary search tree using a list of
   distinct items.Info(v) denotes the item stored at vertex v. *)
Begin (* algorithm *)
   store the first item as the root of the tree.
   while there are more data do
   begin (* while *)
     read(item)
     if item < info(root) then
       search the left subtree for the insertion point
     else
       search the right subtree for the insertion point
     insert (item)
   endwhile
End (* algorithm *)
```

Algorithm 9.6

The next example uses this algorithm.

EXAMPLE 9.25 Construct a binary search tree with the three-letter words *hit*, *hat*, *cat*, *rat*, *sat*, *fat*, *mat*, *pat*, *kat*.

SOLUTION:
First store the word *hit* in the root of the binary tree (see Figure 9.93). Since *hat* < *hit*, store *hat* as the left child of *hit* (Figure 9.94). Now *cat* < *hit*; so *cat* should go on the left subtree of *hit*. Since *cat* < *hat*, insert *cat* as the left child of *hat* (Figure 9.95). Notice that *rat* > *hit*; so insert *rat* as the right child of *hit* (Figure 9.96).

Figure 9.93

Insert *hit*.

Figure 9.94

Insert *hat*.

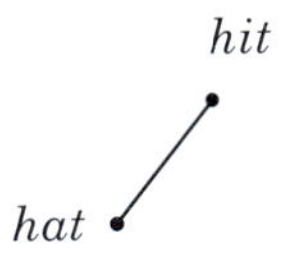

Figure 9.95

Insert *cat*.

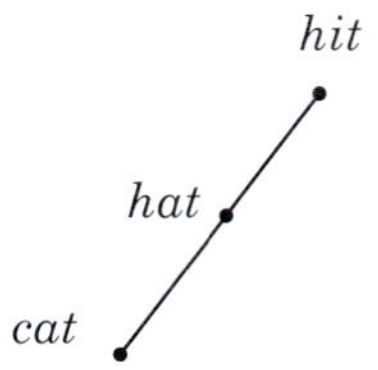

Figure 9.96

Insert *rat*.

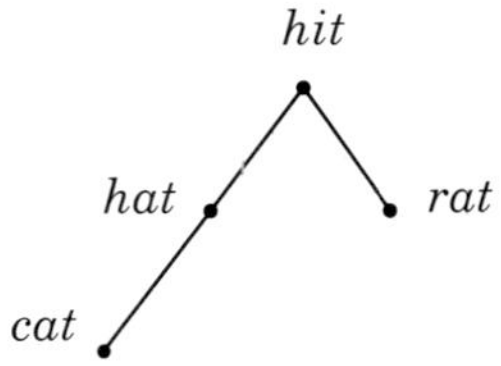

The next word *sat* is greater than both *hit* and *rat*; store it as the right child of *rat* (Figure 9.97). Now *fat* < *hit* and *fat* < *hat*, but greater than *cat*; append it as the right child of *cat* (Figure 9.98). The remaining words similarly fall in place, as in Figures 9.99–9.101. The completed binary search tree stands in Figure 9.101.

Figure 9.97

Insert *sat*.

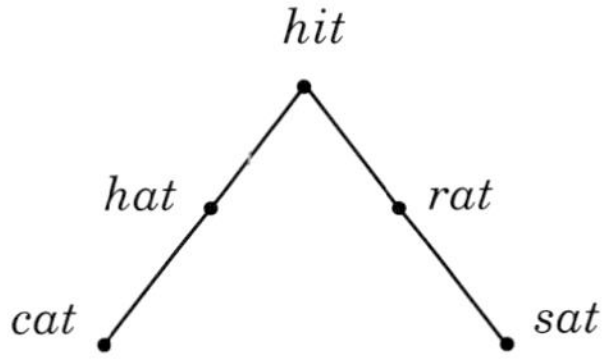

Figure 9.98

Insert *fat*.

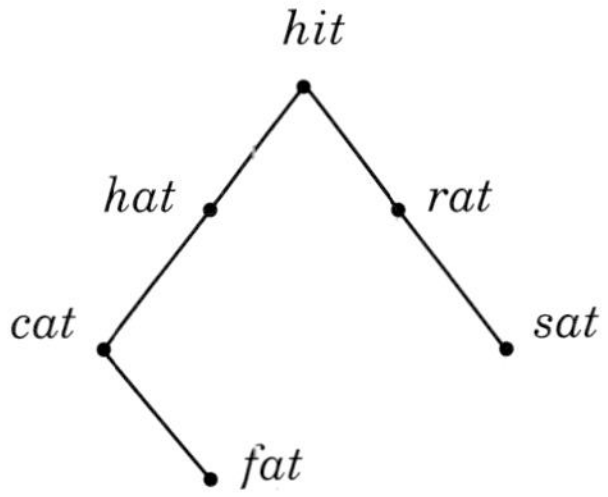

Figure 9.99

Insert *mat*.

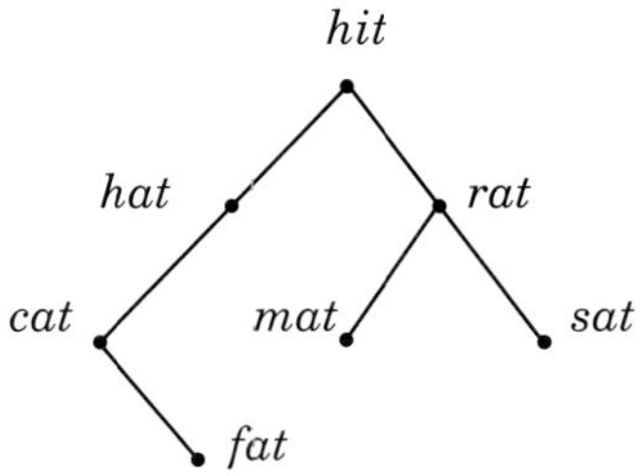

Figure 9.100

Insert *pat*.

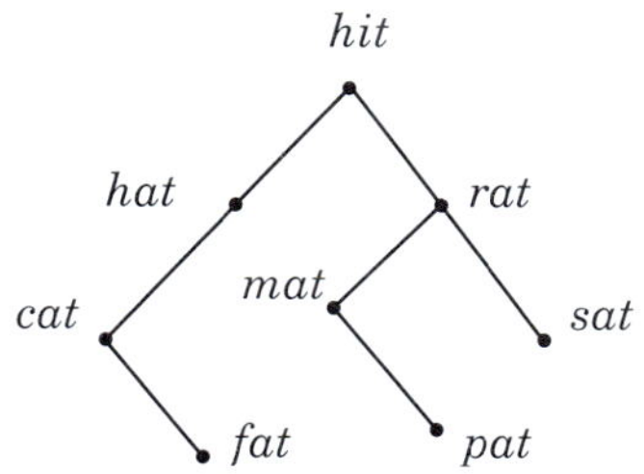

Figure 9.101

Insert *kat*.

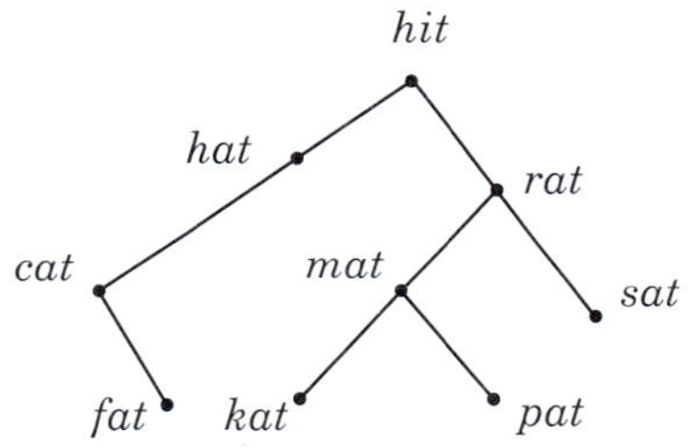

■

The order in which data are inserted in the tree determines its shape. In other words, the same data entered differently will produce differently shaped trees. For instance, the data *kat, rat, hat, cat, hit, pat, fat, sat, mat* produces the tree in Figure 9.102.

Figure 9.102

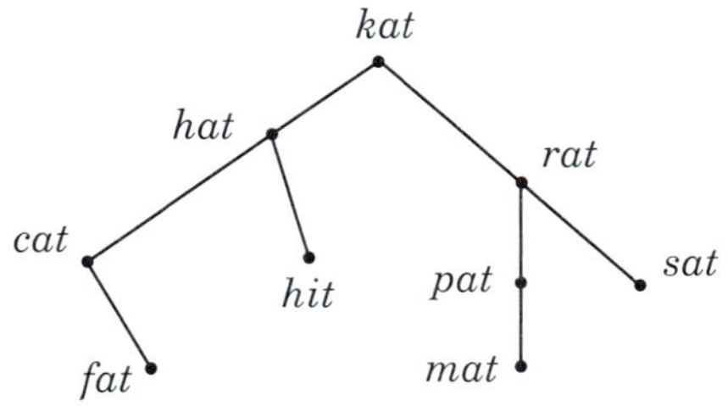

> *An important observation*: If the data in Example 9.25 are stored as an unordered linear list, linear search takes a maximum of nine comparisons to find a given key. On the other hand, searching the binary tree in Figure 9.101 (or 9.102) takes at most four comparisons (Why?), much less than nine.

> More generally, a maximum of n comparisons are needed to locate an item in an unordered list of n items, but if they are stored in a binary search tree of height h, the worst case takes only $h + 1$ comparisons. So making the tree as *bushy* as possible minimizes the search time.

Binary search trees can also be used to eliminate duplicates from a homogeneous list. To accomplish this, build a binary search tree consisting of distinct elements, as the next example illustrates.

EXAMPLE 9.26 Using a binary search tree, eliminate all duplicates in the list 11, 15, 8, 11, 8, 12, 15, 10, 12, 17, 6.

SOLUTION:
With each item, search the tree to determine if it already exists in the tree; if it does, then the item is a duplicate. If it is a new element, insert it in the tree. This procedure finds four duplicates: 11, 15, 8, and 12. The final binary search tree is displayed in Figure 9.103.

Figure 9.103

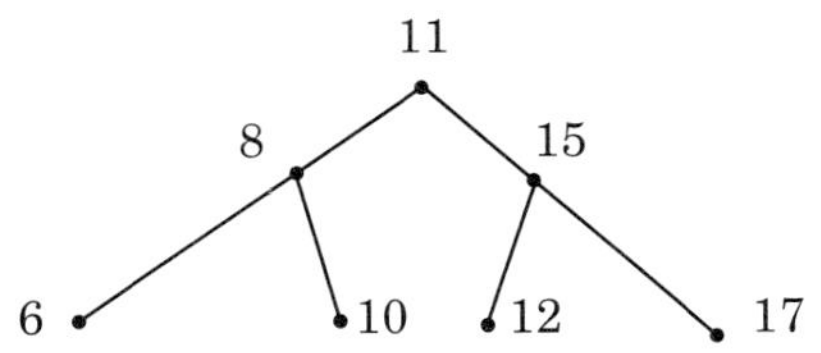

■

These two examples indicate how efficiently binary search trees can handle homogeneous lists.

Exercises 9.6

Construct a binary search tree for each set.

1. i, a, u, o, e **2.** a, e, i, o, u **3.** u, o, i, e, a

4. i, a, e, o, u **5.** 8, 5, 2, 3, 13, 21 **6.** 5, 2, 13, 17, 3, 11

7. *do, re, me, fa, sol, la*

8. *inning, input, output, insect, inroad, inset, insole*

9. *order, ouch, outfit, outing, outcome, outlet, outcry*

10. *canna, coleus, balsam, celosia, dahlia, azalea, tulip*

11–20. Find the maximum number of comparisons needed to locate an item in the binary search trees of Exercises 1–10.

Construct a binary search tree using the words in each phrase or sentence.

21. Fourscore and seven years ago.

22. Ask not what your country can do for you.

23. All that glitters is not gold.

24. Necessity is the mother of invention.

25. Write an algorithm to print the contents of a binary search tree in lexicographic order.

26. **Tournament sort** is a sorting technique that reflects the structure of a tournament. For a list of n items where n is a power of 2, a full, complete binary tree springs from the leaves to the root. For example, consider the list 13, 8, 5, 1, 21, 3, 34, 2. Store the numbers as leaves in a binary tree (Figure 9.104). At each level move up the larger of the siblings to its parent (Figure 9.105). Now the root contains the largest element m. Output it and store 0 in the leaf that contained m. Repeat this procedure until all elements are output. Give the final output.

Figure 9.104

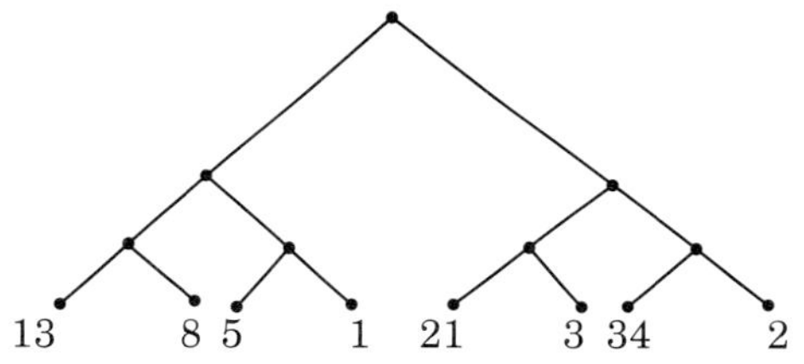

Figure 9.105

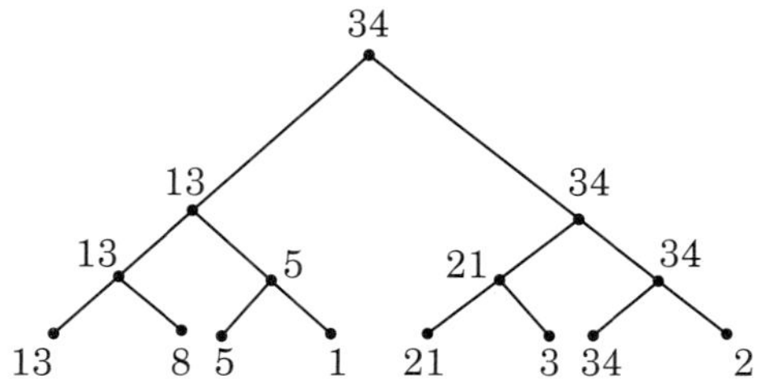

Heapsort is also a sorting method based on binary trees. A **heap** is a balanced homogeneous binary tree such that: (1) all leaves in the lowest level are as far left as possible; (2) info(root) is greater than both info(left child) and info(right child) if they exist; and (3) every subtree is also a heap. To sort a list by this version of heapsort, first store the items level by level from left to right. Beginning with the leftmost internal vertex, build a heap to move the largest element to the root. Output it. Store 0 in the root. Repeat the procedure until all vertices contain 0's. [Heapsort takes $O(n \log n)$ comparisons) in the worst case to sort a list of n elements.] Use the list in Figure 9.106 for Exercises 27 and 28.

***27.** Build the heap from the tree.

***28.** Show how the tree looks after building the second heap.

Figure 9.106

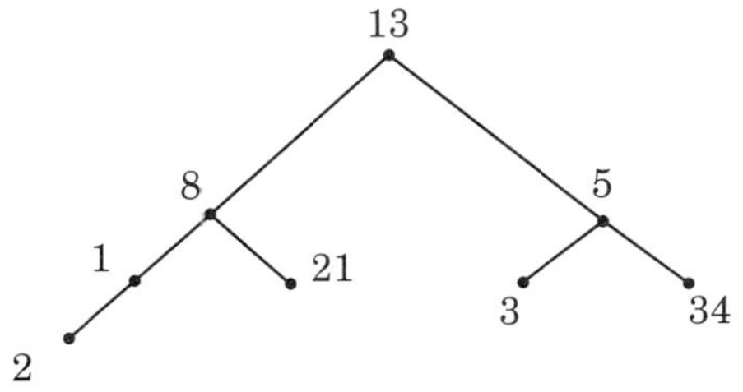

*9.7 Huffman Trees (optional)

One way binary trees are used in the field of communications is through the Huffman coding scheme presented below. The "Huffman Code is one of the fundamental ideas that people in computer science and data communications are using all the time," according to Donald E. Knuth of Stanford University. It is used in computer networks, high-definition televisions, modems, and VCR Plus, a device that automatically programs a VCR.

Suppose we would like to transmit a message over a certain alphabet. We would like to encode it in terms of bits, using an unambiguous coding scheme. The ASCII scheme (Appendix A.1) could translate each letter into a binary word. In ASCII every character is represented by a 7-bit word. Such a system is a **fixed-length code**. Using ASCII the word GRAPH is encoded as 10001111010010100000110100001001000 and the message 1010100100100010001011001111101001010110011 is decoded as THEORY. (You can verify both.)

A fixed-length code has the advantage of being relatively easy to encode and decode, but characters are assigned the same length codes whether or not they appear frequently, which usually wastes both storage space and time.

To rectify this, codes of variable length can be assigned to the symbols in the alphabet. This kind of technique is a **variable-length code**. Characters that occur frequently receive shorter codes than those that occur infrequently. One such system, the **Huffman coding scheme** developed by D. A. Huffman, shortens the encoded messages considerably.

For instance, ASCII encodes the word TREE as 1010100101001010001011000101, 28 bits long, while the Huffman codes in Table 9.7 renders the same word as 111000, a saving of 22 bits.

Table 9.7

Symbol	Code
E	0
R	10
T	11

David Albert Huffman *(1925–1999) was born in Alliance, Ohio. After graduating from Ohio State University in electrical engineering at the age of 18, he joined the Navy. He received his M.S. in electrical engineering from Ohio State in 1949 and his D.Sc. from MIT 4 years later.*

In 1951, while at MIT, in a course on information theory, he and his classmates were given a choice of taking a final examination or writing a term paper on a coding problem. Huffman worked on it for months without much success, but just as he decided to start preparing for the final, a solution came to his mind; the result was the Huffman Code.

Huffman left MIT in 1967 to head the newly created computer science department at the University of California, Santa Cruz.

Although best known for the Huffman Code, he made significant contributions to switching theory, information theory, and picture analysis.

To verify this, scan the encoded message from left to right, using Table 9.7. If the bit you encounter is a 0, it represents the letter E. However, if it is a 1, look at the next bit; if 0, the character is R; otherwise, it is T. See Figure 9.107. Now try it yourself.

Figure 9.107

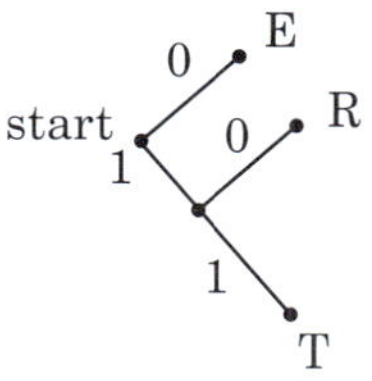

Huffman Algorithm

To illustrate the **Huffman algorithm**, consider the message TERNARY TREE over the alphabet $\{\square, A, E, N, R, T, Y\}$, where the character $\square$ indicates a blank space.

Step 1 Construct a frequency table for the various symbols in the message (Table 9.8).

Table 9.8

Chapter	□	A	E	N	R	T	Y
Frequency	1	1	3	1	3	2	1

Step 2 Find two characters that have the least frequencies, say, □, and A. (Any two of the characters □, A, N, and Y will do.) Concatenate them to form a new symbol □A. The frequency of a newly created symbol is the sum of the frequencies of its components. So the frequency of □A is 2 (Table 9.9). Figure 9.108 shows the resulting symbols and their frequencies. (If the frequencies of the symbols differ, the symbol with the smallest frequency is made the left child; otherwise, preserve the alphabetic order.)

Figure 9.108

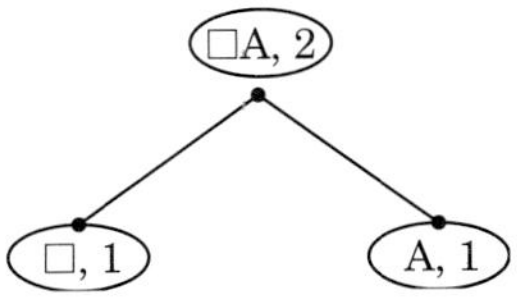

Table 9.9

Symbol	□A	E	N	R	T	Y
Frequency	2	3	1	3	2	1

Step 3 Again look for two symbols of lowest frequencies, namely, N and Y. Concatenate them; the frequency of NY is 2. See Figure 9.109. Five symbols remain in Table 9.10.

Figure 9.109

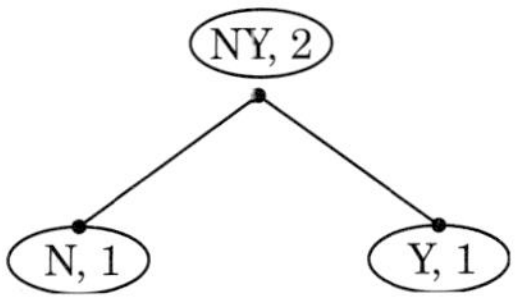

Table 9.10

Symbol	□A	NY	E	R	T
Frequency	2	2	3	3	2

Step 4 Continue like this until one symbol remains: TER□ANY. Tables 9.11–9.14 summarize the steps.

Table 9.11

Symbol	□ANY	E	R	T
Frequency	4	3	3	2

Table 9.12

Symbol	□ANY	TE	R
Frequency	4	5	3

Table 9.13

Symbol	R□ANY	TE
Frequency	7	5

Table 9.14

Symbol	TER□ANY
Frequency	12

The steps of combining symbols to form new symbols can produce a binary tree, as in Figure 9.110. When we drop all symbols and their frequencies except the original characters, a **Huffman tree** remains (Figure 9.111). In this full binary tree, the leaves represent the original characters.

Figure 9.110

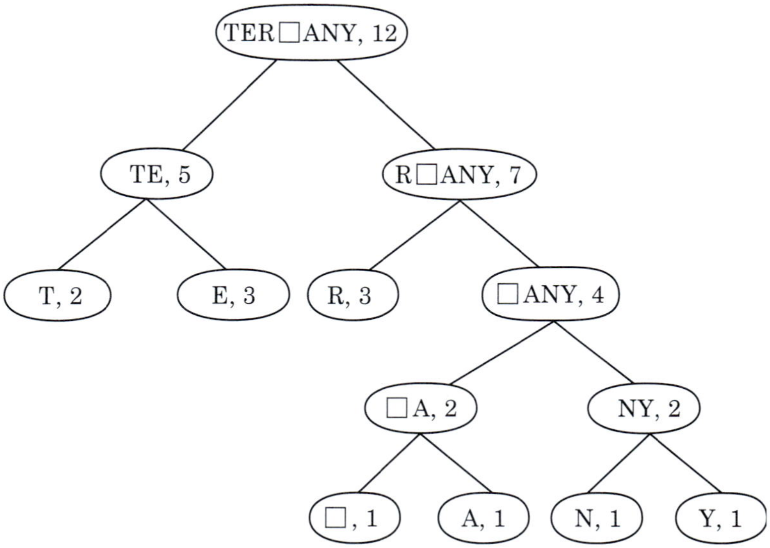

Figure 9.111

Huffman tree.

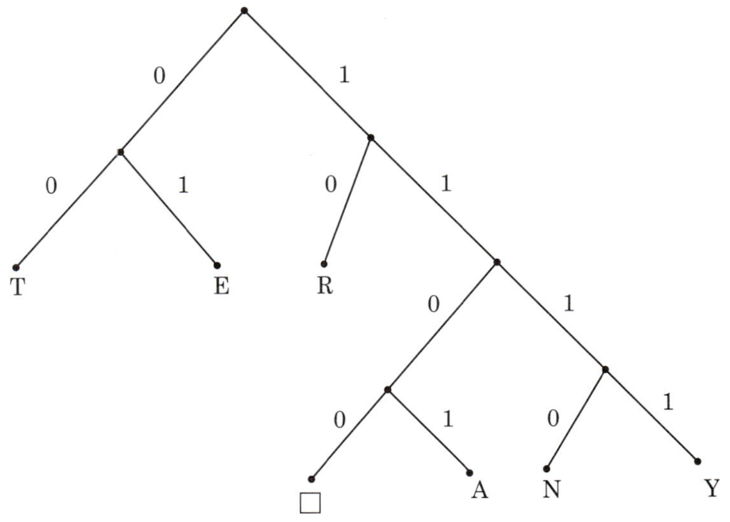

Step 6 Assign a 0 to each edge that leads to a left child and a 1 to each edge that leads to a right child. See Figure 9.111.

Step 7 To find the Huffman code for any character, traverse the path from the root to the corresponding leaf. The sequence of bits along the path is its Huffman code. The codes of the various characters are given in Table 9.15.

Table 9.15

Character	□	A	E	N	R	T	Y
Code	1100	1101	01	1110	10	00	1111

According to Table 9.15, the Huffman code of the message TERNARY TREE is 00011011101101101111110000100101, only 32 bits long, with a saving of 52 bits over ASCII. Interestingly enough, the code for any character does not appear as a prefix in the code for any other character. Such a nonrepetitive code is a **prefix code**. This prefix property guarantees that every message has a unique Huffman code and vice versa.

We close this section with the next example, which illustrates decoding a message using a Huffman code.

EXAMPLE 9.27 Using Table 9.15, decode the message 11011110111111110000100101.

SOLUTION:

Scan the given message from left to right. Since the first bit is 1, the first character can be □, A, N, R, or Y. The second bit is also a 1, so the first character must be □, A, N, or Y. With the third bit 0, the character must be □ or A. The next bit is 1, making the first character A.

Since the fifth bit is 1, the second character must be □, A, N, R, or Y. The sixth bit is also 1: the character must be □, A, N, or Y. With the seventh bit 1 and eighth bit 0, the second character is N.

Continuing like this, you can verify that the original message was ANY TREE. ■

Huffman codes offer a unique and expeditious transmission service, as the previous two examples suggest.

Exercises 9.7

With Table 9.15, encode each word.

1. EARN **2.** EATEN **3.** AERATE **4.** TREATY

Using Table 9.15, decode each message.

5. 111101110110 **6.** 00110100000010

7. 110111110110011011111000 **8.** 110111001001001001110100

Figure 9.112

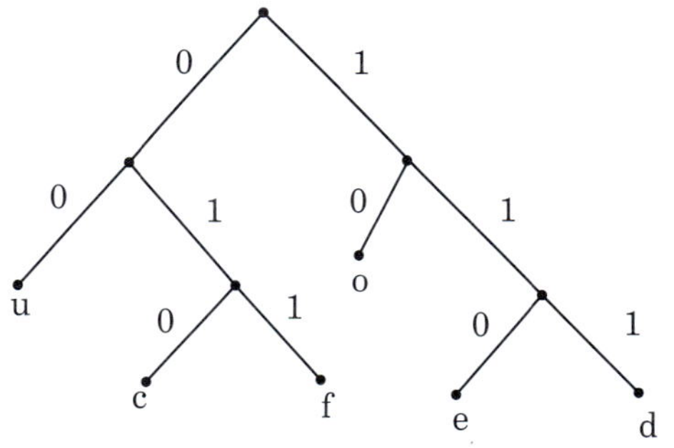

Using the Huffman tree in Figure 9.112, find the Huffman code for the characters in Exercises 9–12.

9. e **10.** u **11.** f **12.** d

With the Huffman tree in Figure 9.112, encode each word.

13. cud **14.** decode **15.** educed **16.** deduced

Using the Huffman tree in Figure 9.112, decode each message.

17. 11110110 **18.** 10111111

19. 01010011011110110 **20.** 011110110111

21. Using the following frequency table, construct a Huffman tree for each character in the alphabet {a, b, c, d, e, f}.

Character	a	b	c	d	e	f
Frequency	4	1	2	3	5	4

22. Using the following frequency table, construct a Huffman tree for the alphabet $\{a, b, c, e, g, l, o, s, u\}$.

Character	a	b	c	e	g	l	o	s	u
Frequency	4	3	2	3	1	2	4	1	5

23. Using Exercise 21, find the Huffman code for the characters a through f.

Using Exercise 21, encode each word.

24. babe **25.** bead **26.** abba **27.** ceded

Using Exercise 21, decode each message.

28. 1101010110 **29.** 00111010110

30. 010011010110 **31.** 10010100111110

With Exercise 22, find a Huffman code for each character.

32. a **33.** c **34.** o **35.** s

Using Exercise 22, encode each word.

36. cabbage **37.** babbage **38.** calculus **39.** caboose

Using Exercise 22, decode each message.

40. 0011001100101010011OO **41.** 1001100110111000001100

42. 1010011011100010001 **43.** 0011010000000001010000100

The **weight** w of a Huffman code, which measures its efficiency, is defined as follows. Let $c_1, c_2, \ldots, c_n$ denote the characters in an alphabet; $f_1, f_2, \ldots, f_n$ their frequencies; and $l_1, l_2, \ldots, l_n$ the lengths of their codes. Then $w = \sum_{i=1}^{n} f_i l_i$. Compute the weight of the Huffman code in:

44. Table 9.15. **45.** Exercise 21. **46.** Exercise 22.

*9.8 Decision Trees (optional)

The wide application of m-ary trees embraces the general decision-making process. Consider this guessing game: Anna thinks of a number

$n \leq 1000$; each time Beena gives an incorrect response, Anna says whether Beena's guess is less than or more than n; by making at most 10 guesses, Beena can find the number n.

Suppose Anna chooses $n = 687$. We shall employ the binary search algorithm to find n. Beena's first guess is $\lfloor (1 + 1000)/2 \rfloor = 500$. Since $500 < n$, Anna says *more*. Since the number must lie between 500 and 1000, Beena computes $\lfloor (500 + 1000)/2 \rfloor = 750$ as her second guess. Anna's response? *Less*. So Beena guesses $\lfloor (500+750) \rfloor = 625$. Anna's response? *More*. Beena's fourth guess is $\lfloor (625 + 750)/2 \rfloor = 687$; she has made the correct guess.

Figure 9.113

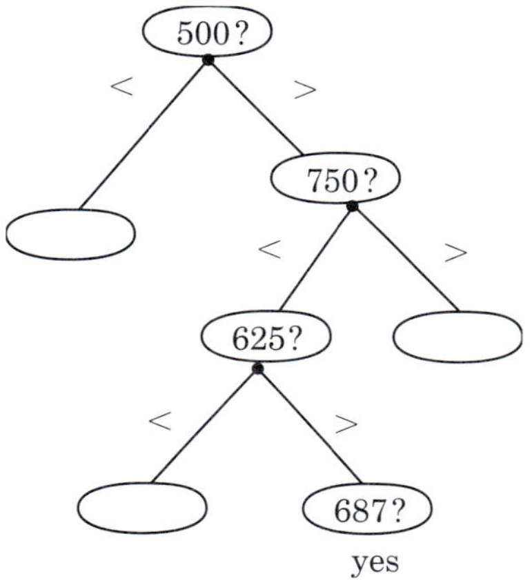

These steps form the binary tree in Figure 9.113, a **decision tree**. At each internal vertex, we select one of two alternatives, thereby selecting a subtree.

The next two examples provide additional decision trees.

EXAMPLE 9.28 Sort three distinct elements a, b, and c.

Case 1 Let $a < b$. Then compare a and c. If $c < a, c < a < b$. On the other hand, if $c > a$, then compare b and c. If $b < c$, then $a < b < c$; otherwise, $a < c < b$.

Case 2 The case $a > b$ can be discussed similarly.

The complete analysis unfolds elegantly into the binary decision tree of Figure 9.114. ■

Notice that this decision tree has six leaves, each representing a possible order. A maximum of three comparisons, which equals the height of the tree, sorts the list.

More generally, for a list of n distinct items, there are $n!$ permutations of n elements, so its binary decision tree contains $n!$ leaves. Therefore, by Theorem 9.8, its height is $\geq \lceil \lg n! \rceil$ and the list will take at least $\lceil \lg n! \rceil$ comparisons to sort.

A more down-to-earth application follows.

Figure 9.114

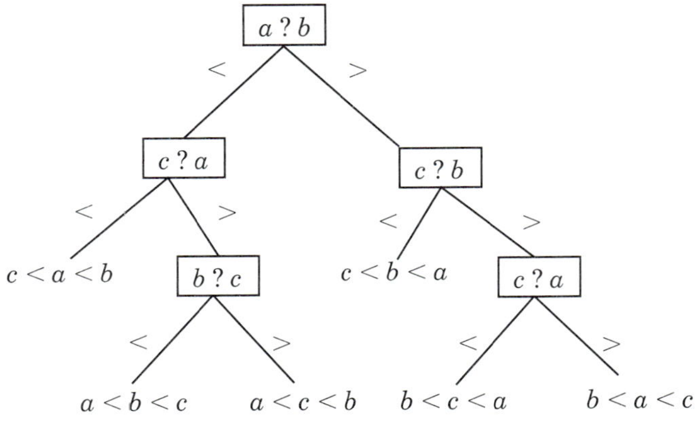

EXAMPLE 9.29 **(The eight-coins puzzle)** Eight coins in a collection plate look identical, but one is counterfeit and heavier. Using an equal-arm balance and a minimum number of weighings, identify the counterfeit coin.

SOLUTION:

First label the coins a through h. Place four coins, say, a through d, on the left side of the balance and the other four on the other side. There are two possible outcomes: either the left side is heavier or the right side is heavier. If the left is heavier, one of the coins a, b, c, or d must be false.

Figure 9.115

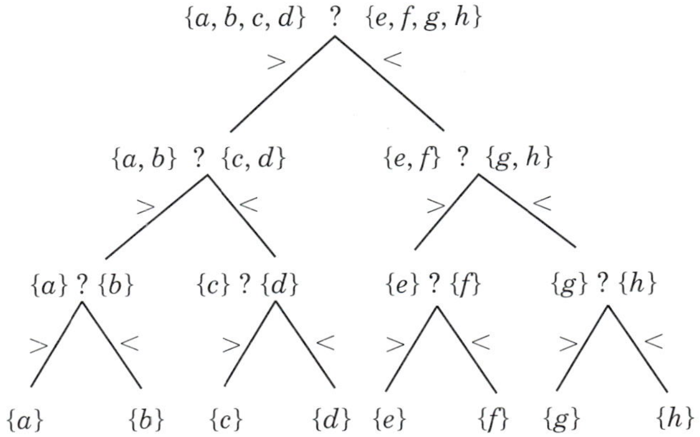

Put two coins, a and b, on the left side of the balance, and c and d on the other side. If the left side is heavier, either a or b is counterfeit. Place a on the left and b on the right. If the left side falls, the counterfeit coin is a; otherwise, it is b.

The possibilities unfold clearly in a decision tree (Figure 9.115). The tree shows that three weighings will identify the false coin.

Can we do better? Can we find the false coin in less than three weighings? Yes. Start with six coins, say, a through f. Place a, b, and c on the left side; d, e, and f on the right. Now there are three possibilities: the left side is heavier, the two sides balance, or the right side is heavier. The complete discussion is summarized in the decision tree in Figure 9.116. It follows from the tree that only two weighings are required.

Figure 9.116

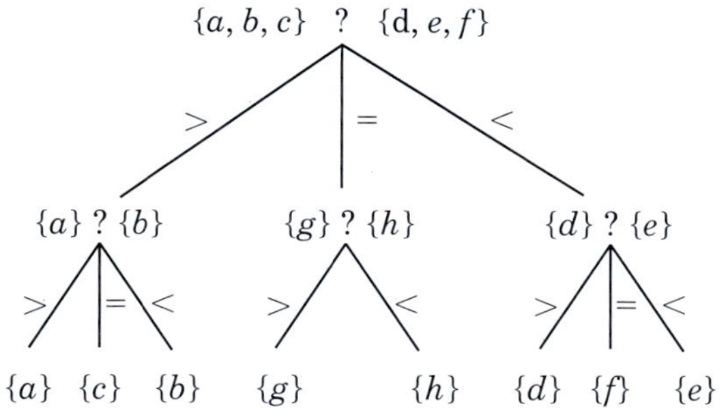

■

More generally, suppose one of n coins is counterfeit and heavier. Let h denote the height of the corresponding ternary tree T. By Theorem 9.8, $h \geq \lceil \log_3 n \rceil$; at least, $\lceil \log_3 n \rceil$ weighings will find the false coin. For example, 8 coins take at least $\lceil \log_3 8 \rceil = 2$ weighings, which agrees with our previous discussion.

Decision trees, as this section has demonstrated, can clarify the reasoning process, producing solutions in a relatively short time.

Exercises 9.8

Find the maximum number of guesses needed to find the positive integer $n \leq N$ for each value of N. (Use the binary search algorithm.)

1. 97 **2.** 243 **3.** 1976 **4.** 3076

Among the N coins in a collection plate, one is counterfeit and heavier. Using an equal-arm balance, find the minimum number of weighings needed to ascertain the counterfeit, for each value of N.

5. 12 **6.** 13 **7.** 28 **8.** 75

9. Four coins, a through d, in a plate look identical, but one is counterfeit and heavier. Using an equal-arm balance and minimum weighings, identify the counterfeit coin and determine if it is lighter or heavier. Display your analysis in a decision tree.

10. Redo Exercise 9 using six coins, a through f.

11. Using Example 9.27, write an algorithm to arrange three distinct elements in lexicographic order.

12. Let n be a positive integer and *key* an arbitrary positive integer $\leq n$. Using binary search, write an algorithm to find key and the number of guesses made.

13. Among seven identical coins lies a heavier counterfeit coin. Write an algorithm to identify the false coin using an equal-arm balance and minimum weighings.

Chapter Summary

This chapter briefly introduced trees, the most important family of graphs. They model isomers of saturated hydrocarbons, hierarchical charts, genealogy, tournaments, and decision-making. They also serve well in evaluating algebraic expressions, as well as in sorting and searching.

Tree

- A **tree** is a connected acyclic graph (page 611).
- A connected graph is a tree if and only if a unique, simple path runs between any two vertices (page 612).
- A connected graph with n vertices is a tree if and only if it has exactly $n - 1$ edges (page 613).

Spanning Tree

- A **spanning tree** of a connected graph contains every vertex of the graph (page 615).
- Every connected graph has a spanning tree (page 616).
- **Kruskal's algorithm** (page 616), the **DFS method** (page 618), and **the BFS method** (page 621) can find spanning trees.
- A **minimal spanning tree** of a connected weighted graph weighs the least (page 626).
- **Kruskal's algorithm** (page 626) and **Prim's algorithm** (page 629) can find minimal spanning trees.

Rooted Tree

- A specially designated vertex in a tree is the **root** of the tree. A tree with a root is a **rooted tree** (page 635).
- The **subtree** rooted at v consists of v, its descendants, and the edges incident with them (page 636).

Level and Height

- The **level** of a vertex is the length of the path from the root to the vertex (page 637).
- The **height** of a tree is the maximum level of any leaf in the tree (page 637).

m-ary Tree

- In an **ordered rooted tree** the children of every vertex are ordered (page 638).
- An ***m*-ary *tree*** is a rooted tree in which every vertex has at most m children. It is **binary** if $m = 2$, and **ternary** if $m = 3$ (page 639).
- An m-ary tree is **full** if every internal vertex has m children(page 639).
- An m-ary tree is **balanced** if all leaves fall on the same level or two adjacent levels (page 640).
- An m-ary tree is **complete** if all leaves lie at the same level (page 645).
- An m-ary tree of height h has at most m^h leaves (page 641).
- For an m-ary tree of height h with l leaves, $h \geq \lceil \log_m l \rceil$ (page 641).
- If it is full and balanced, $h = \lceil \log_m l \rceil$ (page 641).

Binary Tree

- **Preorder**, **inorder**, and **postorder traversal** are three ways to visit every vertex of a binary tree (page 647).
- An algebraic expression with only binary operators can be written in **prefix**, **infix**, or **postfix form**. In prefix form, each operator precedes its operands. The other two forms behave similarly (page 653).
- An algebraic expression containing only binary operators can be represented by a **binary expression tree** (page 655).
- A **binary search tree** is homogeneous with every element in the left subtree of every vertex v less than v and every right subtree element is greater than v (page 664).
- A **Huffman code**, a variable-length code, minimizes the length of encoded messages (page 670).
- A **decision tree** is an m-ary tree in which a decision is made at each internal vertex (page 677).

Review Exercises

Using Kruskal's algorithm, construct a spanning tree, beginning at a, for each graph.

1.

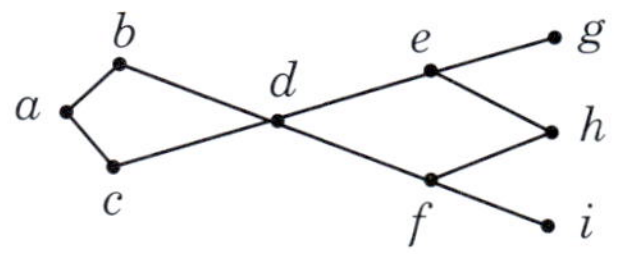

2.

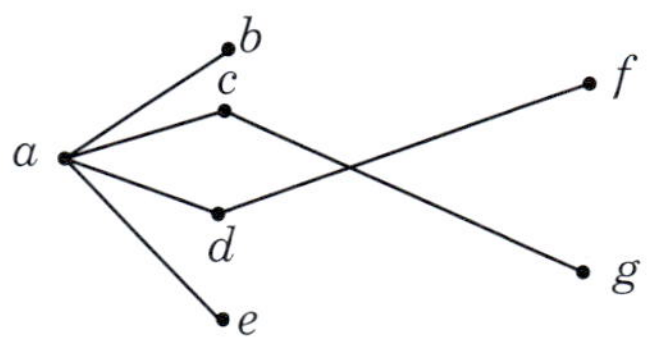

3.

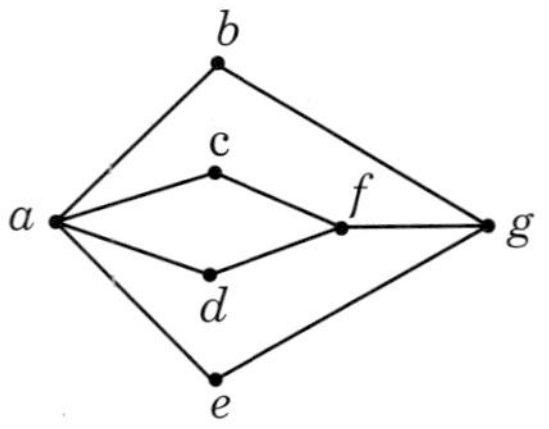

4–6. Using the DFS method, construct a spanning tree for each graph in Exercises 1–3.

7. Using the DFS method, draw a spanning tree for the graph with the following adjacency list representation.

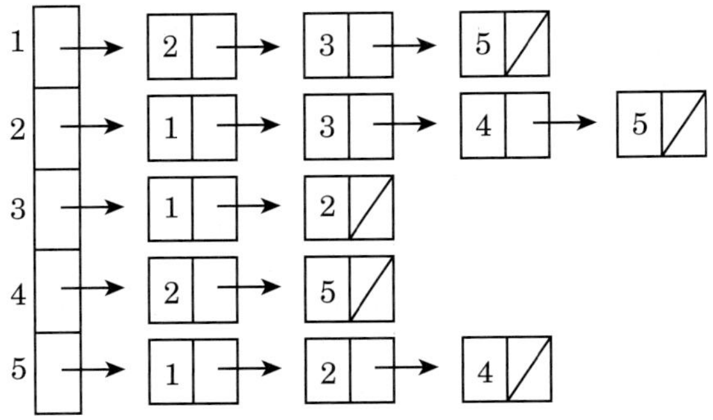

8–11. With the BFS method, construct a spanning tree for each graph in Exercises 1–3 and 7.

12. Using backtracking, solve the six-queens puzzle, if a solution exists.

Using Kruskal's algorithm, construct a minimal spanning tree for each connected weighted graph.

13.

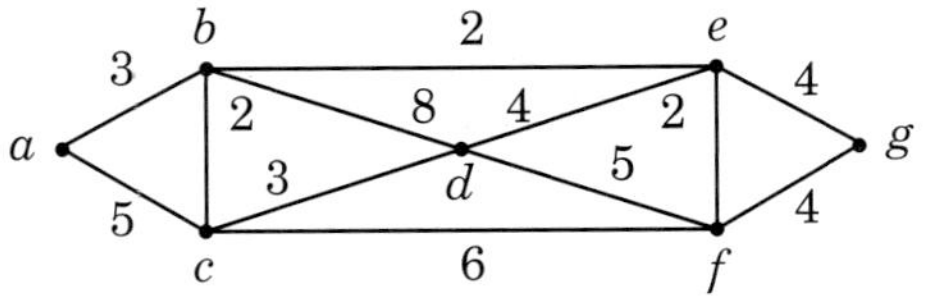

14.

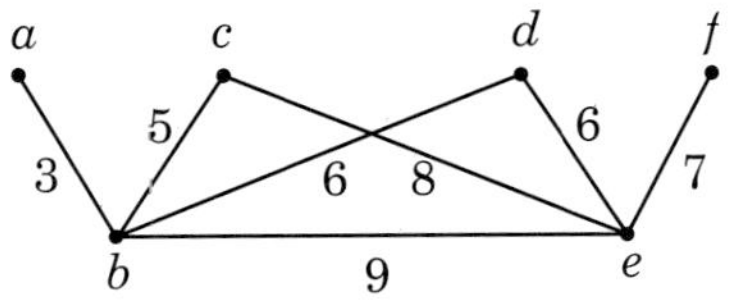

15.

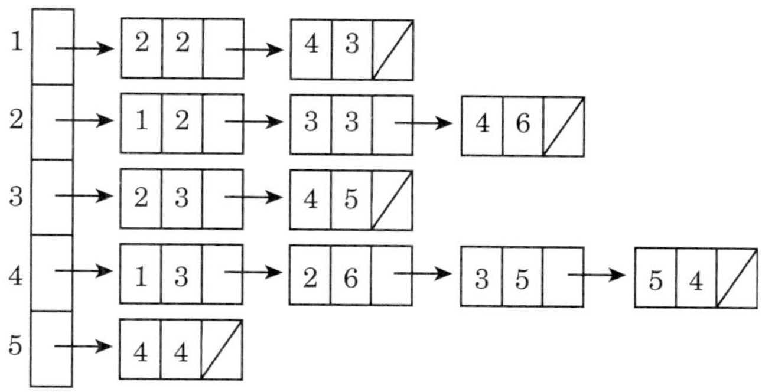

16.

$$\begin{array}{c} \\ 1 \\ 2 \\ 3 \\ 4 \\ 5 \end{array} \begin{array}{c} \begin{array}{ccccc} 1 & 2 & 3 & 4 & 5 \end{array} \\ \begin{bmatrix} \infty & 2 & 3 & 1 & \infty \\ 2 & \infty & 2 & \infty & 6 \\ 3 & 2 & \infty & 3 & 5 \\ 1 & \infty & 3 & \infty & 4 \\ \infty & 6 & 5 & 4 & \infty \end{bmatrix} \end{array}$$

17–20. Using Prim's algorithm, draw a minimal spanning tree for each graph in Exercises 13–16.

A book contains four chapters. Chapters 1 and 4 contain two sections, and Chapters 2 and 3 contain three sections. Section 1 in Chapter 2 contains one subsection while Section 1 in Chapter 3 contains two subsections.

21. Display this information in a rooted tree.

22. Compute the level of *Section 3* in Chapter 2.

23. Compute the level of *Subsection 2* in Chapter 3.

24. How high is the tree?

25. Compute the maximum number of leaves in a ternary tree with height 7.

26. Compute the minimum height of a ternary tree with 1000 leaves.

27. How many vertices does a full 4-ary tree with 15 internal vertices have?

28. How high is a full balanced ternary tree with 9844 vertices?

Seventy-six women enter a singles tennis tournament.

29. How many matches will be played?

30. How many rounds?

If n people enter a tennis tournament and r_n rounds are played to determine the championship, compute each.

31. r_6 **32.** r_7 **33.** r_8 **34.** r_9

35. Define r_n recursively.

36. Predict an explicit formula for r_n.

37. Prove the formula in Exercise 36 inductively.

Give the output from traversing each binary tree in preorder, inorder, and postorder.

38.

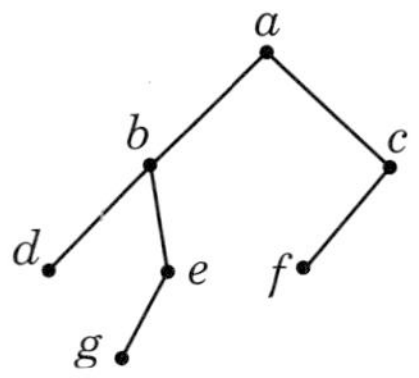

39.

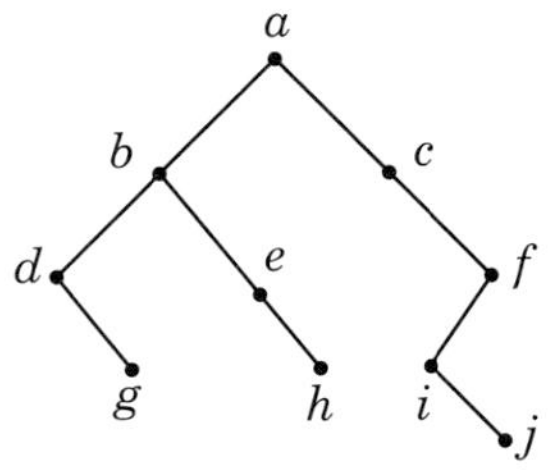

Rewrite each infix expression in prefix form.

40. $a - \left(\dfrac{b}{c+d}\right)^e * f$

41. $a + \dfrac{b}{c}\left(\dfrac{d+e}{f-g}\right) - h^i$

42–43. Translate the expressions in Exercises 40 and 41 into postfix form.

44–45. Represent the expressions in Exercises 40 and 41 in binary expression trees.

Represent each in a binary expression tree.

46. $x \leftarrow x/(y+z), w$

47. $[x \leq (y+z)] \wedge [y \leq (w+z)]$

Find the prefix expression represented by each binary expression tree.

48.

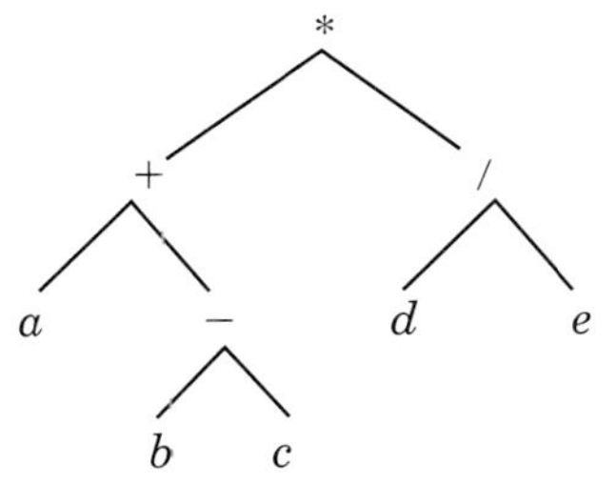

49.

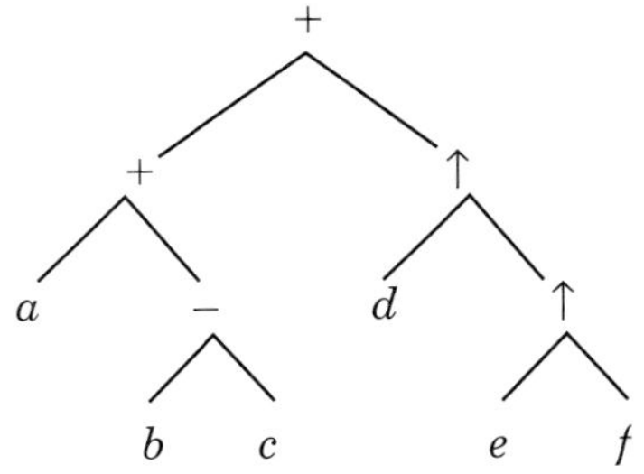

50-51. Find the postfix expressions represented in Exercises 48 and 49.

Evaluate each prefix expression, where each operand is a single-digit number.

52. $- \uparrow - * 35/932 * 88$

53. $34 \uparrow 89 * -3/3 \uparrow$

Evaluate each binary expression tree.

54.

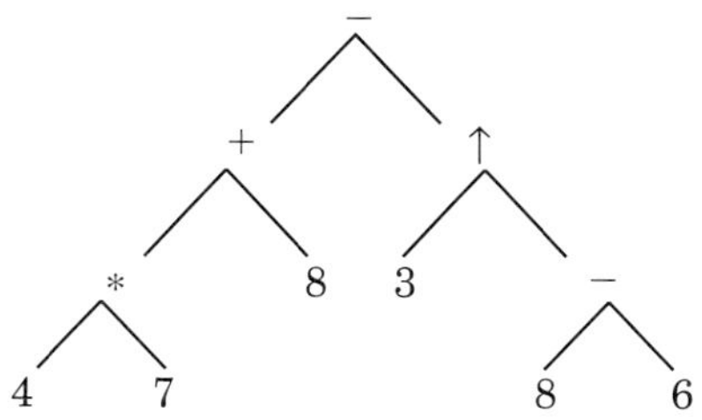

55.

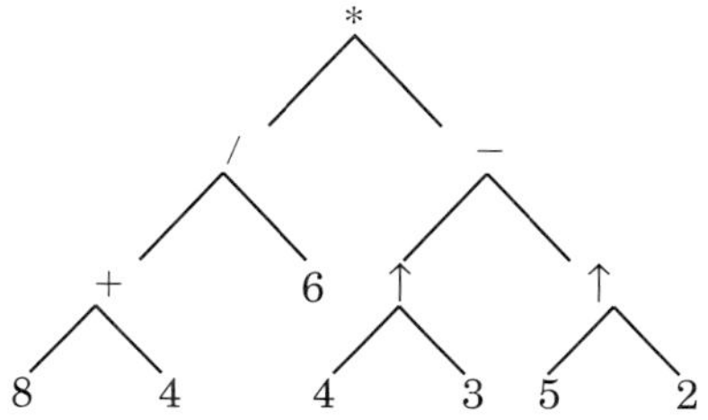

56. Find the number of nonisomorphic binary trees with six vertices.

Using the words in each sentence, construct a binary search tree.

57. Math is not a spectator sport.

58. Too many cooks spoil the broth.

○ **59.** Make a Huffman tree for the alphabet {a, c, d, e, l, n, t, (,), +} with the following frequency table.

Character	a	c	d	e	l	n	t	(	)	+
Frequency	6	4	1	5	1	3	2	1	1	2

○ Using Exercise 59, encode each word.

60. candle

61. canteen

62. $c + d$

63. $(d + e)$

○ With Exercise 59, decode each message.

64. 11010111001001001101

65. 1111010111100111111

○ The n coins in a collection plate look identical, but one is counterfeit and weighs more. Find the minimum number of weighings using an equal-arm balance needed to identify it for each value of n.

66. 43

67. 100

68. 476

69. 2000

Supplementary Exercises

1. Give a counterexample to show that a graph with n vertices and $n-1$ edges need not be a tree.

2. Draw a **graceful tree** with five vertices. With six vertices.

An edge e in a connected graph is a **bridge** if the graph becomes disconnected when e is deleted. The graph in Figure 9.117 has two bridges, namely, $\{a, b\}$ and $\{b, c\}$.

Figure 9.117

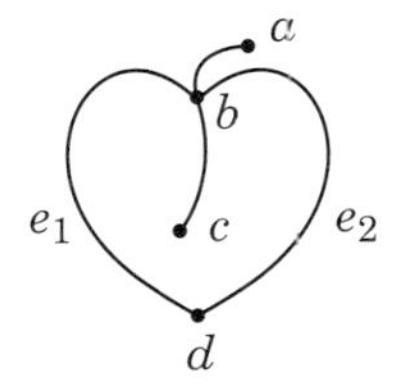

Edges e_1 and e_2 are not bridges (Why?).

3. How many bridges do the trees in Figures 9.2 and 9.3 have?

4. How many bridges does a tree with n vertices have?

5. How many edges does a forest F with n vertices and k trees have?

6. Prove that every tree is a planar graph.

7. Using Exercise 6, prove a tree with n vertices has $n-1$ edges.

○ 8. A certain hydrocarbon molecule contains n carbon atoms. Each carbon atom is bonded to four hydrogen atoms and each hydrogen atom to one carbon atom. Prove that the molecule contains $2n+2$ hydrogen atoms.

9. Prove that any two spanning trees of a connected graph must have the same number of edges.

*10. Let $T = (V, E)$ be a tree with $|V| \geq 2$. Prove that T has at least two vertices with degree 1.

11. Are there trees having two vertices with degree 1? If so, draw such a tree.

*12. Prove that a connected graph G is a tree if and only if every edge is a bridge.

*13. Let $T = (V, E)$ be a tree and $e \in E$. Prove that $(V, E - \{e\})$ is a forest of two trees.

*14. Let $T_1 = (V_1, E_1)$ and $T_2 = (V_2, E_2)$ be two disconnected trees. Let e be an edge connecting a vertex in V_1 with a vertex in V_2. Prove that $T = (V_1 \cup V_2, E_1 \cup E_2 \cup \{e\})$ is a tree.

15. There are two piles of coins and each contains a counterfeit coin. A good coin weighs 2. Pile A contains three coins, one being underweight; pile B contains four coins, one overweight. Determine the minimum number of weighings on an equal-arm balance required to identify the counterfeits. (H. L. Nelson, 1982)

16. Redo Exercise 15 if pile A contains four coins and the combined weight of the two counterfeits is not four. (H. L. Nelson, 1982)

Computer Exercises

Write a program to perform each task.

1. Read in the adjacency matrix A of a connected graph G with n vertices, 1 through n. Determine if G is a tree. If yes, construct a spanning tree for G using Kruskal's algorithm, the DFS method, and the BFS method.

2. Redo program 1 with its adjacency list representation.

3. Read in the modified adjacency matrix of a connected weighted graph G with n vertices, 1 through n, where$1 \leq n \leq 20$. Print the vertices of a minimal spanning tree of G and its weight, using Kruskal's and Prim's algorithms.

4. Redo Program 3, using its adjacency list representation.

5. Read in a positive integer $n \leq 20$ and solve the n-queens problem, if possible.

***6.** Read in a legal numeric prefix expression and evaluate it. Assume that each operand is a single-digit integer.

***7.** Read in a legal numeric postfix expression and evaluate it. Assume that each operand is a single-digit integer.

***8.** Read in a numeric infix expression containing at most 50 single-digit numbers. Convert it into a prefix expression and evaluate it, if legal.

***9.** Read in a numeric infix expression containing at most 50 single-digit numbers. Convert it into a postfix expression and evaluate it, if legal.

***10.** Read in a valid algebraic infix expression. Construct a binary expression tree and traverse it in both preorder and postfix.

***11.** Read in all reserved identifiers in C++ or Java. Print them in lexicographic order.

12. Read in a positive integer $n \leq 15$ and print the number of nonisomorphic binary trees with n vertices.

13. Read in a positive integer $n \leq 5000$ and a *key* $\leq n$. With the binary search method, determine the *key*. Print the number of comparisons needed to locate the key and the value of $\lceil \lg n \rceil$.

14. Among seven coins lies a heavier counterfeit coin. Read in their weights and identify the counterfeit coin with a minimum number of weighings on an equal-arm balance.

15. Eight coins look identical, but one is a counterfeit of a different weight. Read in their weights. With the least number of weighings on an equal-arm balance, identify the false coin and determine if it is lighter or heavier.

Exploratory Writing Projects

Using library and Internet resources, write a team report on each of the following in your own words. Provide a well-documented bibliography.

1. Explain how to represent the National Hockey League (NHL) and the National Basketball Association (NBA) by trees.
2. Write an essay on the applications of trees to the various disciplines.
3. Describe how Kruskal developed his algorithm.
4. Explain how J. Sedlacek and B. R. Myers established that $k(W_n) = L_{2n} - 2$.
5. Explain how Prim developed his algorithm.
6. Explain how decision trees are used in computer science and management science. Give examples.
7. Discuss the *knapsack problem*, which is related to scheduling.
8. Explain labeled trees.
9. Write an essay on *game trees*, trees used in the analysis of games such as backgammon, chess, checkers, and tic-tac-toe.
10. Discuss the card game of *solitaire* and its relationship to trees.
11. Write an essay on *Steiner trees.*

Enrichment Readings

1. A. V. Aho, *et al., Data Structures and Algorithms,* Addison-Wesley, Reading, MA, 1983, pp. 75–196, 230–292.
2. T. H. Carmen *et al., Introduction to Algorithms,* McGraw-Hill, New York, 1990.
3. D. I. A. Cohen, *Basic Techniques of Combinatorial Theory*, Wiley, New York, 1978, pp. 228–282.

4. R. P. Grimaldi, *Discrete and Combinatorial Mathematics: An Applied Introduction,* 4th ed., Addison-Wesley, Reading, MA, 1999, pp. 547–590.

5. E. Horowitz and S. Sahni, *Fundamentals of Computer Algorithms*, Computer Science Press, Rockville, MD, 1978, pp. 152–370.

6. R. L. Kruse, *Data Structures & Program Design*, 2nd ed., Prentice Hall, Englewood Cliffs, NJ, 1987, pp. 318–498, 512–517.

7. J. A. McHugh, *Algorithmic Graph Theory*, Prentice-Hall, Englewood Cliffs, NJ, 1990.

8. S. Sahni, *Concepts in Discrete Mathematics*, 2nd ed., Camelot, Fridley, MN, 1985, pp. 391–448.

9. S. S. Skieno, *Implementing Discrete Mathematics*, Addison-Wesley, Reading, MA, 1990.

10. A. Tucker, *Applied Combinatorics*, 2nd ed., Wiley, New York, NY, 1984, pp. 80–122.

11. R. J. Wilson and J. J. Watkins, *Graphs: An Introductory Approach,* Wiley, New York, 1990, pp. 185–214.

Chapter 10

Digraphs

It is the man not the method that solves the problem.

—H. MASCHKE

Digraphs, a special class of graphs mentioned in Chapter 7, are geometrical representations of finite relations. Besides the basic terminology of digraphs, this chapter explores weighted digraphs and dags. Both finite relations and digraphs can be implemented in a computer using adjacency matrices and linked lists.

Digraphs have applications to computer science, linguistics, genetics, the social sciences, sports, and management science.

Some of the interesting problems digraphs can handle are:

- How can the adjacency matrix of a digraph determine if there exists a path from every vertex to every other vertex?
- How can the adjacency matrix of the digraph of a round-robin tournament identify the champion? The second-best player? The third-best player, and so on?
- How can digraphs represent algebraic and boolean expressions, as well as assignment statements?
- A computer company has a manufacturing plant at location a and a warehouse at location z. The possible streets a truck driver can take from a to z are given by the digraph in Figure 10.1. The arrows signify one-way streets and the weights, distances in miles. Find the shortest route he could take.

10.1 Digraphs

Vertices with indegrees and outdegrees and the reachability factor stand out among the basic features of digraphs. Matrices also play a significant role.

Figure 10.1

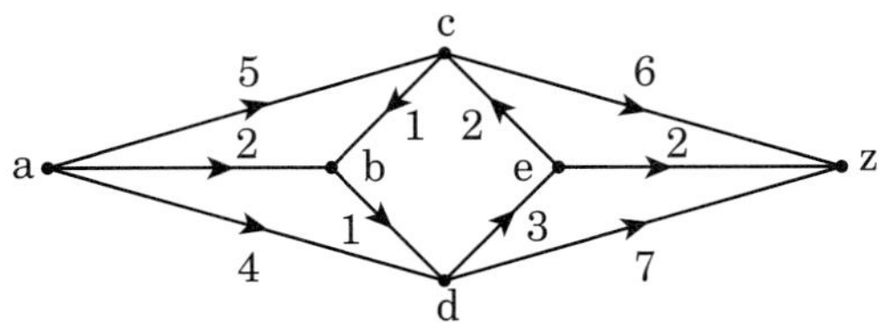

As Section 7.2 showed, finite relations can be modeled by digraphs. For instance, the relation $R = \{(x,y)|\, x \text{ is a factor of } y \text{ and } xy \text{ is an even integer}\}$ on the set $V = \{1,2,3,6,8\}$ appears as the digraph in Figure 10.2. It consists of a set V of vertices and a set E of directed edges.

Figure 10.2

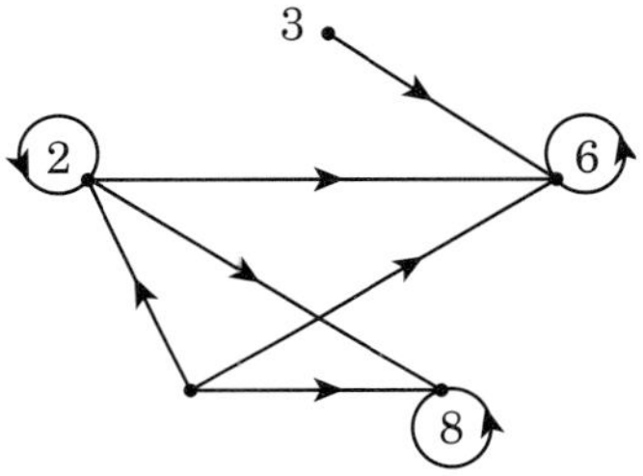

From such observations, a more general definition emerges.

Digraph

A **digraph** (or **di**rected **graph**) $D = (V,E)$ consists of a nonempty finite set V of points (called **vertices** or **nodes**) and a set E of **directed edges** joining them. A directed edge from vertex x to vertex y is denoted by the ordered pair (x,y). The vertex x is the **initial vertex** and y the **terminal vertex** of the edge (x,y). Vertex x is an **immediate predecessor** of vertex y and y an **immediate successor** of x.

The definition indicates $E \subseteq V \times V$. A digraph may have loops (see Figure 10.2), but no parallel edges. At most one edge exists between any two vertices.

Every edge in a digraph is unidirectional, like a one-way street. Consequently, digraphs can represent street maps.

Fibonacci and Digraphs

We now present a fascinating application of Fibonacci numbers to digraphs. Recall from Chapter 3 that a *lattice point* on the cartesian plane is a point (x,y), where both x and y are integers.

The next example, proposed as a problem in 1970 by R. C. Drake of North Carolina A & T University at Greensboro, deals with digraphs whose vertices are lattice points. The solution, based on the one given in the same year by L. Carlitz of Duke University, uses generating functions.

EXAMPLE 10.1 (optional) Let $f(n)$ denote the number of paths from $(0,0)$ to $(n,0)$ on the cartesian plane. Each path consists of directed edges of one or more of the four types in Table 10.1. For example, the vertex on the path following $(k,0)$ can be $(k,1)$ or $(k+1,0)$, and that following $(k,1)$ can be $(k+1,1)$ or $(k+1,0)$. Find a formula for $f(n)$.

Table 10.1

Type	**1**	**2**	**3**	**4**
Initial point	$(k,0)$	$(k,0)$	$(k,1)$	$(k,1)$
End point	$(k,1)$	$(k+1,0)$	$(k+1,1)$	$(k+1,0)$

SOLUTION:

Notice that there are $f(1) = 2 = F_3$ paths from $(0,0)$ to $(1,0)$; $f(2) = 5 = F_5$ paths from $(0,0)$ to $(2,0)$; and $f(3) = 13 = F_7$ paths from $(0,0)$ to $(3,0)$. They are displayed in Figure 10.3. (The arrows are omitted for convenience.)

Figure 10.3

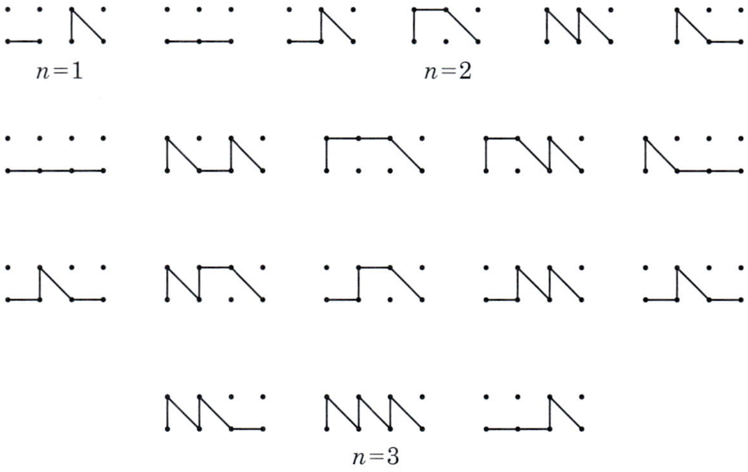

Let $f_2(n)$ denote the number of paths ending with a line segment of type 2 and $f_4(n)$ the number of paths ending with a line segment of type 4. Then

$$f_2(n+1) = f_2(n) + f_4(n) = f(n) \text{ and}$$

$$f_4(n+1) = f(0) + f(1) + \cdots + f(n) = \sum_{k=0}^{n} f(k)$$

Therefore,

$$f(n+1) = f_2(n+1) + f_4(n+1) = f(n) + \sum_{k=0}^{n} f(k)$$

Then $f(1) = f(0) + f(0) = 2f(0)$. But $f(1) = 2$, so $f(0) = 1$.

Let $F(x)$ denote the generating function of the numbers $f(n)$. Then

$$\begin{aligned} F(x) &= \sum_{n=0}^{\infty} f(n)x^n = f(0) + \sum_{n=1}^{\infty} f(n)x^n = 1 + \sum_{n=0}^{\infty} f(n+1)x^{n+1} \\ &= 1 + \sum_{n=0}^{\infty} \left[f(n) + \sum_{k=0}^{n} f(k) \right] x^{n+1} \\ &= 1 + xF(x) + \frac{x}{1-x}F(x) \end{aligned}$$

So

$$F(x) = \frac{1-x}{1-3x+x^2}$$

In 1971, V. E. Hoggatt, Jr., of then San Jose State College, showed that $F(x)$ is the generating function of the Fibonacci numbers F_{2n+1}. That is,

$$\frac{1-x}{1-3x+x^2} = \sum_{n=0}^{\infty} F_{2n+1}x^n$$

Thus, as expected, $f(n) = F_{2n+1}$, where $n \geq 0$. ■

The concept of the degree of a vertex for graphs extends to digraphs also.

Degree of a Vertex

Let v be a vertex of a digraph. The **indegree** of v, denoted by indeg(v), is the number of directed edges terminating at v. The **outdegree** of v, denoted by outdeg(v), is the number of directed edges leaving v. The **degree** of v, or deg(v), is the sum of its indegree and outdegree. A vertex with indegree 0 is a **source** and with outdegree 0 a **sink**.

The next example clarifies these basic terms.

EXAMPLE 10.2 Find the indegree, outdegree, and degree of each vertex of the digraph in Figure 10.2. Identify any source(s) and sink(s).

SOLUTION:

The indegrees, outdegrees, and degrees of the various vertices are given in Table 10.2. Since indeg(1) = 0 = indeg(3), vertices 1 and 3 are sources. Since the outdegree of no vertex is zero, the digraph contains no sinks.

Table 10.2

	Vertex				
	1	**2**	**3**	**6**	**8**
Indegree	0	2	0	4	3
Outdegree	3	3	1	1	1
Degree	3	5	1	5	4

■

The next example applies the basic features of a digraph in computer science.

EXAMPLE 10.3 The flowchart in Figure 10.4 used to find the minimum (min) of three integers x, y, and z is actually a digraph with 11 vertices. The vertices represent operations; the edges, flows of execution.

Figure 10.4

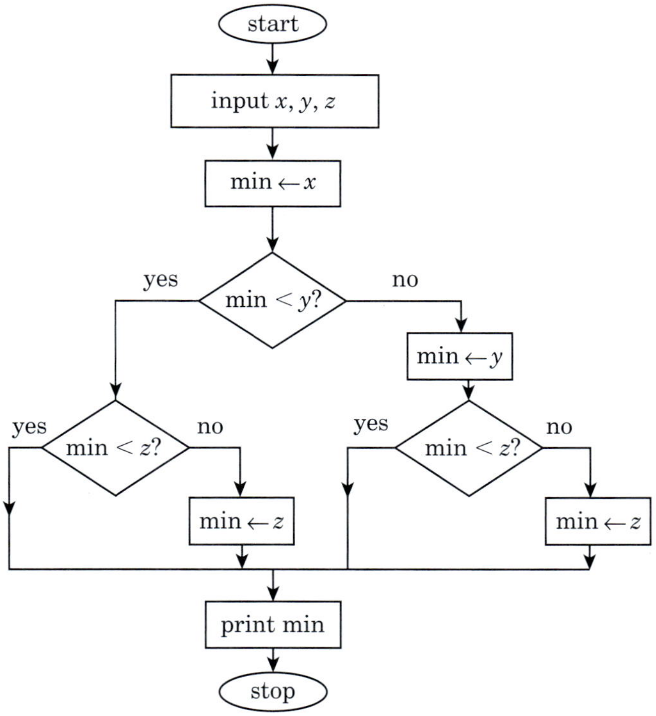

■

The adjacency matrix of a digraph D parallels that of an ordinary graph. Suppose D contains n vertices $v_1, v_2, \ldots, v_n$. The adjacency matrix of D is the matrix $A = (a_{ij})_{n \times n}$, where a_{ij} equals 1 if a directed edge runs from v_i to v_j and 0 otherwise. (Notice that A is a boolean matrix.) The next example illustrates this.

EXAMPLE 10.4 Find the adjacency matrix of the digraph in Figure 10.2.

SOLUTION:
The adjacency matrix of the digraph is

$$
\begin{array}{c} \\ \\ A = \end{array}
\begin{array}{c} \\ 1 \\ 2 \\ 3 \\ 6 \\ 8 \end{array}
\begin{array}{c} \begin{array}{ccccc} 1 & 2 & 3 & 6 & 8 \end{array} \\ \begin{bmatrix} 0 & 1 & 0 & 1 & 1 \\ 0 & 1 & 0 & 1 & 1 \\ 0 & 0 & 0 & 1 & 0 \\ 0 & 0 & 0 & 1 & 0 \\ 0 & 0 & 0 & 0 & 1 \end{bmatrix} \end{array}
\begin{array}{c} \\ \\ \\ \leftarrow \\ \\ \\ \end{array}
\begin{array}{c} \text{row sums} \\ 3 \\ 3 \\ 1 \\ 1 \\ 1 \end{array}
$$

$$
\begin{array}{cccccc} & & & \uparrow & & \\ \text{column sums} & 0 & 2 & 0 & 4 & 3 \end{array}
$$
■

> In this example, the row sums are the outdegrees and the column sums are the indegrees of the vertices. This always happens.

Recall that the degrees of the vertices of a graph total twice the number of edges. Digraphs manifest a similar characteristic.

THEOREM 10.1 Let D be a digraph with vertices $v_1, v_2, \ldots, v_n$ and e edges. Then $\sum_{i=1}^{n} \text{indeg}(v_i) = e = \sum_{i=1}^{n} \text{outdeg}(v_i)$.

PROOF:
Since indeg(v_i) denotes the number of edges terminating at v_i, each is counted exactly once in $\sum_{i=1}^{n} \text{indeg}(v_i)$. Similarly, $\sum_{i=1}^{n} \text{outdeg}(v_i) = e$. ■

For example, the digraph in Figure 10.2 contains $e = 9$ edges. Using the adjacency matrix in Example 10.3, the sum of the indegrees = the sum of the row sums = 9 = the sum of the column sums = the sum of the outdegrees, thus verifying Theorem 10.1.

In Example 10.5, a digraph depicts a more mundane aspect of the world, meriting the designation *tournament*.

EXAMPLE 10.5 Every team in a round-robin tournament plays every other team and no ties are allowed. The results of such a tournament by five teams, a through e, are:

a beat b, c, e; b beat c; c beat d;
d beat a, b, e; e beat both b and c.

A digraph clearly arranges this information in Figure 10.5. A directed edge from vertex x to vertex y indicates team x beat team y. (You may verify Theorem 10.1 for this digraph also.)

Figure 10.5

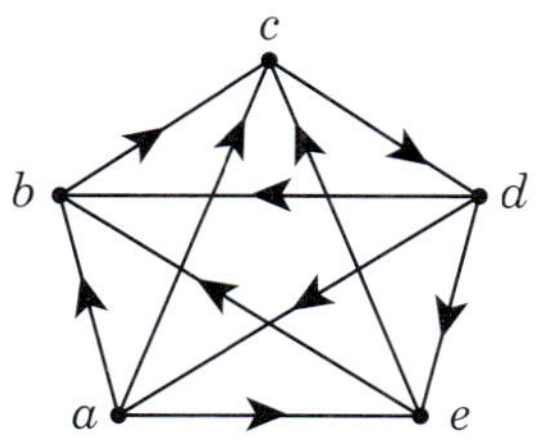

■

This digraph is loop-free and contains exactly one edge between any two distinct vertices. It is a dominance digraph or a tournament.

Dominance Digraph

A loop-free digraph with exactly one directed edge between any two distinct vertices is a **dominance digraph** or a **tournament**.

A dominance digraph can serve in less mundane areas, too, as we can see by an example from the natural sciences.

EXAMPLE 10.6 Suppose a single inherited characteristic of individuals is determined by the gene types A and a, where A denotes the dominant gene and a the recessive gene. So an individual may be type AA, Aa, or aa. (Types Aa and aA are the same.) Of these types, AA dominates both Aa and aa, and Aa dominates aa. This information falls nicely into a dominance digraph, as Figure 10.6 shows.

Figure 10.6

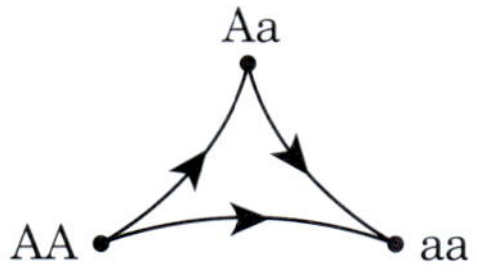

■

The next example shows how we can interpret the outdegrees and indegrees of vertices in a tournament.

EXAMPLE 10.7 The adjacency matrix of the dominance digraph in Figure 10.5 is

$$\begin{array}{c} \\ a \\ b \\ c \\ d \\ e \\ \\ \text{column sums} \end{array} \begin{array}{c} \begin{array}{ccccc} a & b & c & d & e \end{array} \\ \begin{bmatrix} 0 & 1 & 1 & 0 & 1 \\ 0 & 0 & 1 & 0 & 0 \\ 0 & 0 & 0 & 1 & 0 \\ 1 & 1 & 0 & 0 & 1 \\ 0 & 1 & 1 & 0 & 0 \end{bmatrix} \\ \begin{array}{ccccc} & & \uparrow & & \end{array} \\ \begin{array}{ccccc} 1 & 3 & 3 & 1 & 2 \end{array} \end{array} \begin{array}{c} \\ \\ \\ \leftarrow \\ \\ \\ \\ \\ \end{array} \begin{array}{c} \text{row sums} \\ 3 \\ 1 \\ 1 \\ 3 \\ 2 \\ \\ \\ \end{array}$$

The outdegree of each vertex—each row sum—gives the number of wins; its indegree—each column sum—gives the losses. For instance, outdeg(d) = 3 and d has three wins; indeg(e) = 2 and e has two losses. ■

The other characteristics of graphs, paths, and cycles presented in Chapter 8, apply to digraphs also.

Paths, Reachability, and Cycles

Let v_0 and v_n be two vertices in a digraph. A (**directed**) **path of length** $\boldsymbol{n}$ from v_0 to v_n is a sequence of vertices v_i in the form v_0-v_1-$\cdots$-v_n. (Since a digraph contains no parallel edges, labels of edges need not be listed.) The path is **closed** if $v_0 = v_n$ and **open** otherwise. A path containing no repeated vertices, except perhaps the endpoints, is a **simple** path. A simple, closed path is a **cycle**. If a directed path runs from v_0 to v_n, v_n is **reachable** from v_0.

For example, Figure 10.1 has a directed path of length 4 from a to z, namely, a-c-b-d-z. The path b-d-e-c-b is a closed one, whereas a-b-d-e is an open one. Vertex e is reachable from vertex a, but not from z. The path c-b-d-e-c is a cycle.

Suppose there is a directed path v_0-v_1-$\cdots$-v_n from v_0 to v_n in a tournament. This implies that v_0 beat v_1, v_1 beat $v_2, \ldots,$ and v_{n-1} beat v_n. Then v_0 is said to have an $\boldsymbol{n}$**-stage win** over v_n. The following example elucidates this.

EXAMPLE 10.8 The tournament in Figure 10.5 contains three simple directed paths from a to e: a-e, a-b-c-d-e, and a-c-d-e. Vertex e is reachable from vertex a. The length of the path a-c-d-e is three, so team a had a three-stage win over e, a beat c, c beat d, and d beat e.

The concept of reachability bears significance in computer science. To illustrate this in the flow chart in Figure 10.4, if a vertex is not reachable from the start vertex, the corresponding instruction will not be executed for any data. So either the program will not work or the instruction is superfluous.

The same problem can arise in programs consisting of several modules. If a certain module is not reachable, control will not pass to it and hence it will never be invoked.

Is there some way to determine if a vertex v_j in a digraph is reachable from a vertex v_i? A simple definition will help answer this.

Strongly Connected Graph

A digraph D is **strongly connected** if every vertex in D is reachable from every other vertex.

Theorem 10.3 answers the question posed above, which is analogous to Theorem 8.6.

THEOREM 10.3 Let A be the adjacency matrix of a digraph D with vertices $v_1, v_2, \ldots, v_n$. Let $R = A \vee A^{[2]} \vee \cdots \vee A^{[n-1]} = (r_{ij})_{n \times n}$. Then vertex v_j is reachable from vertex v_i if and only if $r_{ij} > 0$ for every $i \neq j$. ($A^{[k]}$ denotes the kth boolean power of A, as defined in Section 7.1.) ■

According to Theorem 10.3, the digraph D is strongly connected if and only if $r_{ij} > 0$ for every $i \neq j$. The matrix R is the **reachability matrix** of the digraph.

The next example illustrates Theorem 10.3.

EXAMPLE 10.9 Is the digraph D in Figure 10.5 strongly connected?

SOLUTION:

By Example 10.7,

$$A = \begin{bmatrix} 0 & 1 & 1 & 0 & 1 \\ 0 & 0 & 1 & 0 & 0 \\ 0 & 0 & 0 & 1 & 0 \\ 1 & 1 & 0 & 0 & 1 \\ 0 & 1 & 1 & 0 & 0 \end{bmatrix}$$

You may verify that

$$A^{[2]} = \begin{bmatrix} 0 & 0 & 1 & 1 & 0 \\ 0 & 0 & 0 & 1 & 0 \\ 1 & 1 & 0 & 0 & 1 \\ 0 & 1 & 1 & 0 & 1 \\ 0 & 0 & 1 & 1 & 0 \end{bmatrix}, \quad A^{[3]} = \begin{bmatrix} 1 & 1 & 1 & 1 & 1 \\ 1 & 1 & 0 & 0 & 1 \\ 0 & 1 & 1 & 0 & 1 \\ 0 & 0 & 1 & 1 & 0 \\ 1 & 1 & 0 & 1 & 1 \end{bmatrix}$$

and

$$A^{[4]} = \begin{bmatrix} 1 & 1 & 1 & 1 & 1 \\ 0 & 1 & 1 & 0 & 1 \\ 0 & 1 & 1 & 1 & 0 \\ 1 & 1 & 1 & 1 & 1 \\ 1 & 1 & 1 & 0 & 1 \end{bmatrix}$$

Then

$$R = A \vee A^{[2]} \vee A^{[3]} \vee A^{[4]} = \begin{bmatrix} 1 & 1 & 1 & 1 & 1 \\ 1 & 1 & 1 & 1 & 1 \\ 1 & 1 & 1 & 1 & 1 \\ 1 & 1 & 1 & 1 & 1 \\ 1 & 1 & 1 & 1 & 1 \end{bmatrix}$$

(You can verify this also.) Since every element off the main diagonal of R is nonzero, the digraph is strongly connected, as can easily be seen. ■

Digraphs exhibit fundamental features of any graph — vertices, paths, adjacency matrices — and can solve a range of problems, as this section demonstrates.

We close this section with a fascinating application of digraphs to ancient Sanskrit poetry and the theory of communications.

The Teleprinter's Problem*

The ten syllabic Sanskrit word *yamátárájabhánasalagám* was used nearly a thousand years ago as an aid for remembering certain rhythms. This nonsensical magic word contains all possible arrays of short and long beats. For example, the first three syllables, *ya*, *má*, and *tá* represent short, long, and long beats; the second through fourth syllables, *má*, *tá*, and *rá* have long, long, and long beats, and so on. Suppose we represent a short beat by 0 and a long beat by 1 as done by Sherman K. Stein of the University of California at Davis, in 1961. Then the magic word can be translated into the 10-bit word 0111010001. This magic word contains another hidden treasure, as we shall see shortly.

Interestingly, this binary word contains all eight possible binary triplets: 011, 111, 110, 101, 010, 100, 000, and 001. They manifest an obvious and interesting pattern: Every two consecutive triplets overlap in bits of two each, as Figure 10.7 shows. In fact, the two ends, 001 and 011, also overlap in two bits:

```
0 0 1
  0 1 1
```

Figure 10.7

```
0 0 1
  1 1 1
    1 1 0
      1 0 1
        0 1 0
          1 0 0
            0 0 0
              0 0 1
0 1 1 1 0 1 0 0 0 1
```

*Based on S. K. Stein, "The Mathematician as an Explorer," *Scientific American*, Vol. 205 (May 1961), pp. 149–158.

Accordingly, using the technique of *wrap around*, we can recover all triplets by storing the bold bits (in Figure 10.7) around a circle, as Figure 10.8 shows. The circular array is called a *memory wheel*. Starting at any bit and counting three bits at a time in the counterclockwise direction, we can recover the various triplets. (In fact, either direction will yield all triplets.)

Figure 10.8

A memory wheel.

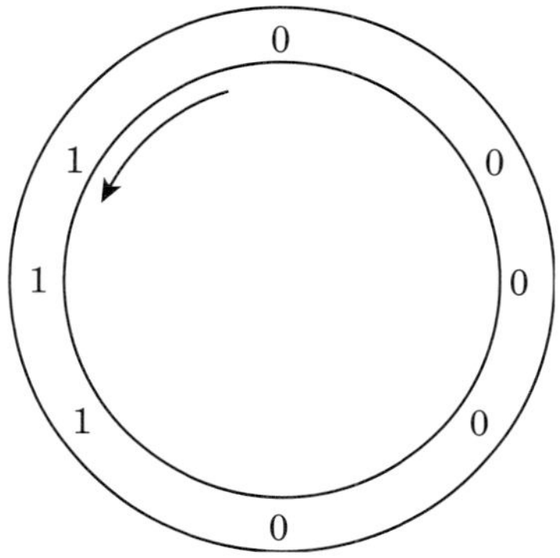

Figure 10.9

A rotating drum.

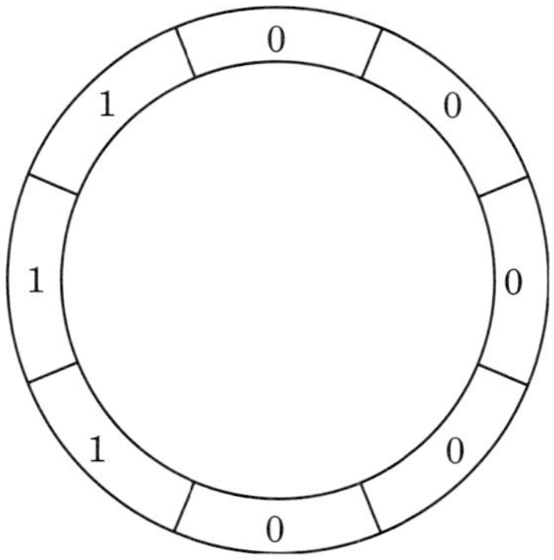

Computers use memory wheels to search for storage locations in their memory drums. The drum in Figure 10.9, for instance, is divided into eight segments of two types, denoted by 0's and 1's. The position of the drum is determined by three consecutive segments, that is, by a binary triplet. Clearly, this is exactly the same as the memory wheel problem, so it is also known as the **rotating drum problem**.

In 1882, a French telegraph engineer, Émile Baudot, used these wheels to transmit information in terms of 5-bit words. In 1961, D. A. Kerr, also a telegraph engineer, observed in *Scientific American* that a memory wheel appears in a Western Electric tape teletypewriter used very early in this century . . . , and it is known here as the combination-wheel and seeker mechanism. The combination wheel is, of course, a mechanically coded memory wheel, and the seekers form a device for locating the desired quintuplet on its periphery, thus indexing the desired character for printing.

Using the memory wheel in Figure 10.8, we can compress the 10-bit magic word into an 8-bit word: 01110100; we can also obtain it by linearly arranging the bold bits in Figure 10.7 from left to right. We can recover all eight binary triplets from this using wrap around. Such a binary word (sequence) is called a **de Bruijn sequence**, after the Dutch mathematician Nicolaas G. de Bruijn (1918–). More generally, we make the following definition.

de Bruijn Sequence

A 2^n-bit word (sequence) w is a **de Bruijn sequence** if every n-bit word can be obtained as a subword (subsequence) of w by cyclically moving one bit at a time from left to right, beginning with the leftmost n-bit word.

Symbolically, let $w = b_0b_1 \dots b_{2^n-1}$. Then every n-bit word s can be obtained as $s = b_0b_1 \dots b_{i+n-1}$ for some integer i, where $b_{2^n+j} = b_j$ and $0 \le i, j < 2^n$.

For example, let $w = b_0b_1 \dots b_7 = 01110100$ and $s = 001$. Then $s = b_7b_8b_9 = b_7b_0b_1 = 001$; see Figure 10.8.

The fact that the last two bits in each word are the same as the first two bits in another word enables us to draw the digraph in Figure 10.10, where the vertices represent the triplets and there is a directed edge from x to y if the last two bits in x are the same as the first two bits in y.

Figure 10.10

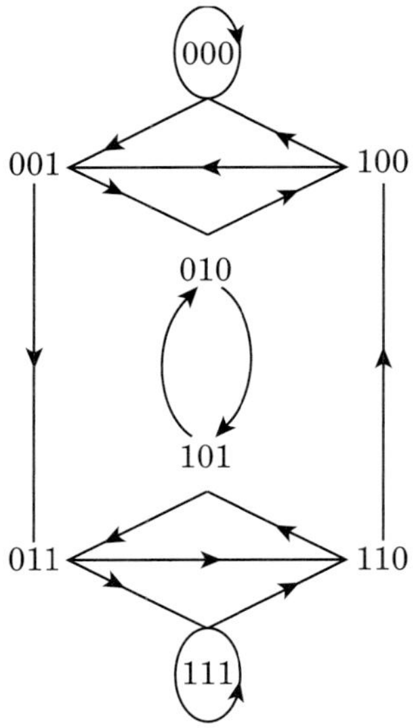

This digraph has the interesting property that indeg(x) $= 2 =$ outdeg(x) for each vertex x. So the **underlying graph**, obtained by dropping the arrowheads, is Eulerian. In addition, the digraph is Hamiltonian. (We could have guessed this because of the cyclic nature of the de Bruijn sequence.) 000 → 100 → 010 → 101 → 110 → 111 → 011 → 001 → 000 is a Hamiltonian cycle.

Interestingly, in 1946, I. J. Good of Great Britain showed yet another way of representing the de Bruijn sequence 01110100 in a digraph. To see this, first observe that the eight binary triplets yield four overlapping couplets: 11, 10, 01, and 00 (see Figure 10.7). Use them as vertices; there is a directed edge from vertex x to vertex y if the second bit in x is the same as the first bit in y. For example, there is a directed edge from **1**1 to **1**0; it is labeled 1**1**0 to indicate that it runs from 11 to 10. The resulting digraph in Figure 10.11 is called a **de Bruijn graph**. The edges yield the various binary triplets. This digraph is also Hamiltonian.

Figure 10.11

de Bruijn graph.

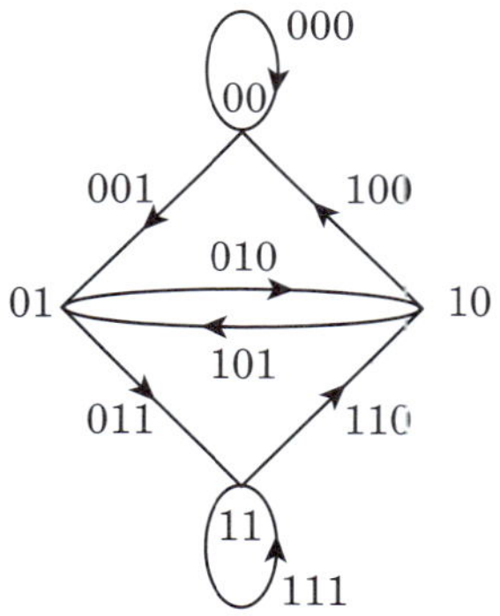

In his study of memory wheels, the Dutch engineer K. Posthumus found that there is exactly one wheel for binary couplets, two for binary triplets, 16 for binary quadruplets (4 bits), and 2048 for binary quintuplets (5 bits). He then conjectured that there are $2^{2^{n-1}-n}$ different memory wheels for binary n-tuples. In 1946, de Bruijn established his conjecture.

An Intriguing By-Product*

The de Bruijn sequence has an intriguing by-product. To see this, consider the de Bruijn Hotel with eight guest rooms. Each guest receives a three-bit code to enter his room. Each door has a keypad with two pushbuttons, one for 0 and the other for 1. When the correct sequence of the three bits is entered, regardless of what bit was entered earlier, the door will open. Suppose a burglar wishes to enter a room. Find the minimum number of bits he needs to enter to be certain that the door will open.

*Based on S. B. Maurer and A. Ralston, *Discrete Algorithmic Mathematics*, Addison-Wesley, Reading, MA, 1991.

The burglar is indeed lucky; he does not have to enter all eight three-bit words, a total of 24 bits. The de Bruijn sequence 01110100 comes to his rescue. Since he does not have the luxury of wrap around, the sequence needs to be stretched linearly to 10 bits: 0111010001. (This is the binary code for the Sanskrit word.) Thus, by entering 0's and 1's in that order, a total of 10 bits, he is guaranteed that the door will open.

Before concluding this section, we should note that the de Bruijn sequence and memory wheels can be extended to m-ary alphabets, where $m \geq 2$. Figure 10.12, for example, shows five such memory wheels; beginning with the innermost, they represent binary couplets, binary triplets, ternary couplets, binary quadruplets, and 4-ary couplets, respectively.

Figure 10.12

Five memory wheels.

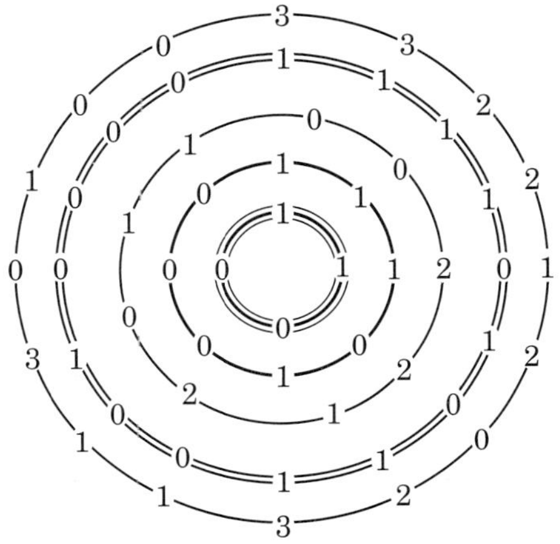

Exercises 10.1

Find the indegree, outdegree, and degree of every vertex in each graph.

1.

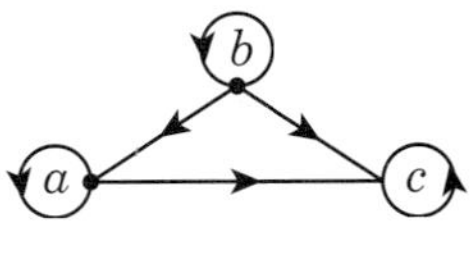

2.

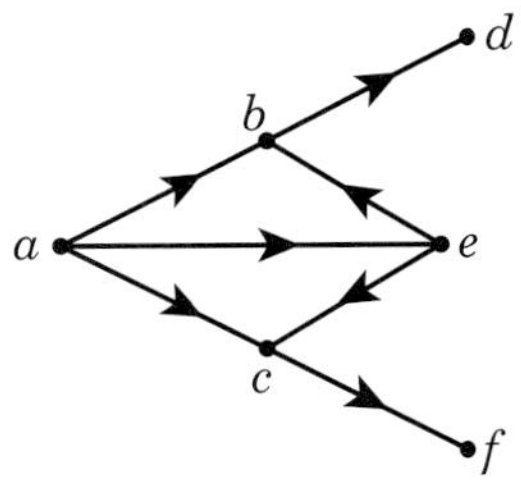

3–4. Identify any source(s) and sink(s) in Exercises 1 and 2.

5–6. Find the adjacency matrices of the graphs in Exercises 1 and 2.

7. Verify Theorem 10.1 using Exercise 1.

8. Verify Theorem 10.1 using Exercise 2.

9. How many edges does a dominance digraph with n vertices have?

The adjacency matrix of the digraph of a round-robin tournament played by five teams, a through e, is

$$\begin{array}{c} \\ a \\ b \\ c \\ d \\ e \end{array} \begin{array}{c} \begin{array}{ccccc} a & b & c & d & e \end{array} \\ \begin{bmatrix} 0 & 0 & 1 & 0 & 1 \\ 1 & 0 & 0 & 1 & 1 \\ 0 & 1 & 0 & 0 & 0 \\ 1 & 0 & 1 & 0 & 0 \\ 0 & 0 & 1 & 1 & 0 \end{bmatrix} \end{array}$$

Find each.

10. The number of wins by a.

11. The number of wins by b.

12. The number of losses by c.

13. The number of losses by e.

14. The champion of the tournament.

15. The second-best team of the tournament.
(*Hint:* Find the team with the maximum number of one- or two-stage wins.)

Use the digraph in Exercise 2 to answer Exercises 16–18.

16. Is vertex a reachable from vertex e?

17. Is vertex f reachable from vertex e?

18. Is the digraph strongly connected?

Find the reachability matrix of the digraph with each adjacency matrix.

19. $$\begin{array}{c} \\ a \\ b \\ c \end{array} \begin{array}{c} \begin{array}{ccc} a & b & c \end{array} \\ \begin{bmatrix} 1 & 0 & 1 \\ 1 & 1 & 1 \\ 0 & 0 & 1 \end{bmatrix} \end{array}$$

20. $$\begin{array}{c} \\ a \\ b \\ c \\ d \end{array} \begin{array}{c} \begin{array}{cccc} a & b & c & d \end{array} \\ \begin{bmatrix} 1 & 1 & 0 & 0 \\ 1 & 0 & 1 & 0 \\ 1 & 1 & 1 & 0 \\ 1 & 0 & 1 & 0 \end{bmatrix} \end{array}$$

21–22. Are the digraphs in Exercises 19 and 20 strongly connected?

23. A row of the adjacency matrix of a digraph is zero. Prove that the digraph is not strongly connected.

24. A column of the adjacency matrix of a digraph is zero. Prove that the digraph is not strongly connected.

25–26. The **underlying graph** G of a digraph D is obtained by deleting all arrows from its edges. Find the underlying graph in Exercises 1 and 2.

27–30. A digraph is **weakly connected** if its underlying graph is connected. Determine if the digraphs in Exercises 1, 2, 19, and 20 are weakly connected.

31. Is a strongly connected digraph also weakly connected?

32. Draw a weakly connected graph that is not strongly connected.

Using the adjacency matrix of a weakly connected digraph with vertices 1 through n, what can you say about each vertex, where $1 \leq i, j \leq n$?

33. Vertex i if row i is zero.

34. Vertex j if column j is zero.

35. Construct the de Bruijn sequence for binary couplets.

36. Construct the memory wheel for binary couplets.

One of the two distinct de Bruijn sequences for binary triplets is 01110100.

37. Find the other de Bruijn sequence.

38. List the binary triplets resulting from it.

39. Represent the de Bruijn sequence in Exercise 37 in a memory wheel.

40. Construct a de Bruijn sequence for binary quadruplets. (There are 16 different sequences.)

41. Represent the de Bruijn sequence in Exercise 40 in a memory wheel.

42. Represent the de Bruijn sequence in Exercise 41 in a de Bruijn digraph.

Suppose de Bruijn Hotel has 16 guest rooms. Each guest receives a three-bit code to enter his room. Each door has a keypad with two pushbuttons, one for 0 and the other for 1. When the correct sequence of the four bits is entered, regardless of what bit was entered earlier, the door will open. Suppose a burglar wishes to enter a room.

43. Find the minimum number of bits he needs to enter to be certain that the door will open.

44. Find the corresponding binary word.

45. Count the number of 0's and 1's in the de Bruijn sequence for binary couplets, triplets, and quadruplets.

46. Using Exercise 45, make a conjecture about the number of 0's and 1's in a de Bruijn sequence for binary words.

In 1934, M. H. Martin developed an algorithm for constructing a de Bruijn sequence for binary n-tuples. Begin with the n-bit word consisting of all 0's. Successively append the larger of the bits 0 and 1 that does not lead to a duplicate n-tuple. Using this method, construct a de Bruijn sequence for each.

47. Binary couplets

48. Binary triplets

49. Binary quadruplets

10.2 Dags

A dag is another special digraph. It represents and evaluates assignment statements and algebraic expressions containing repeating subexpressions, as well as their prefix and postfix forms, with great efficiency.

Dag

A digraph that contains no cycles is a **dag** (**d**irected **a**cyclic **g**raph).

The next example illuminates this definition.

EXAMPLE 10.10 Being acyclic, the digraphs D_1 and D_2 in Figure 10.13 are dags. Since D_3 contains a cycle, *a-b-c-d-a*, it is *not* a dag.

You may verify that the digraphs in Figures 10.1, 10.2, and 10.4 are not dags (Why?).

Figure 10.13

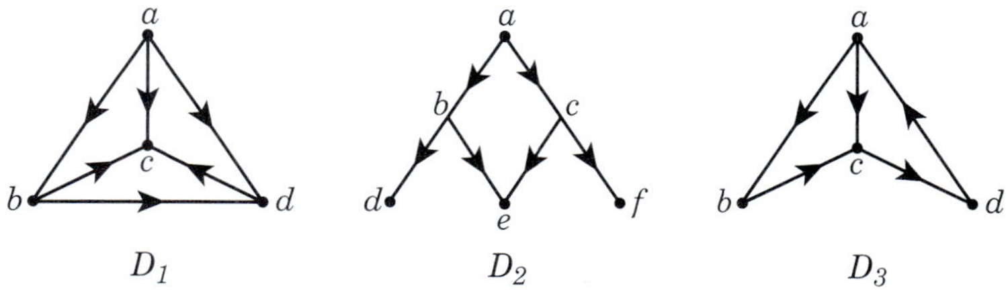

■

Dags are used in both computer science and management science. For instance, the Hasse diagram for various tasks t_1–t_{13} in building a house, given by Table 7.6, can translate into the dag in Figure 10.14. In fact, every

Figure 10.14

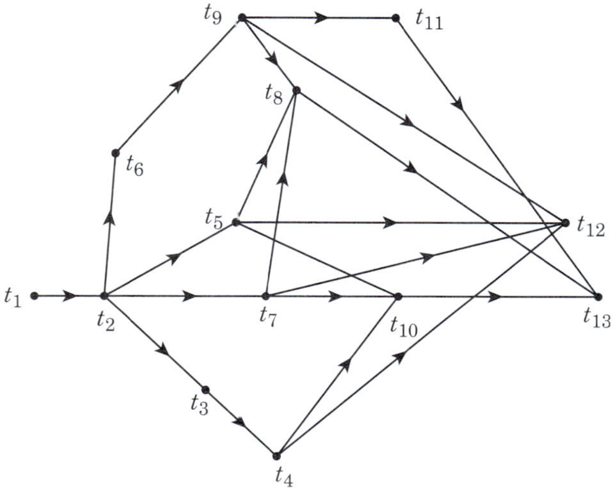

Hasse diagram can be considered a dag, provided you put the arrows back on the edges.

> Dags also can represent assignment statements and algebraic expressions with repeating subexpressions.

For instance, consider the assignment statement, $x \leftarrow x + y - y * x$. It contains repeating subexpressions: x three times and y twice. Store each once (Figure 10.15). The arrows pointing to them indicate the operations to be performed. The digraph with vertices $+$, x, and y represents the subexpression $x + y$, while the digraph with vertices $*$, x, and y represents $x * y$.

Figure 10.15

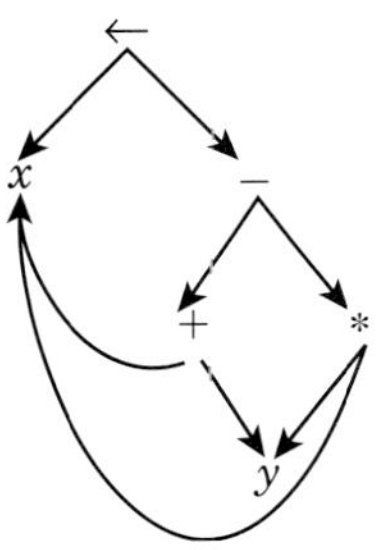

In the dag in Figure 10.15, each interior vertex represents an operator, the source the assignment operator, and each sink an operand. Unlike a tree, a vertex for a repeating subexpression has more than one parent. Vertex x, for instance, has three parents: $\leftarrow$, $+$, and $*$.

Compare the binary tree representation of the assignment statement in Figure 10.16.

Figure 10.16

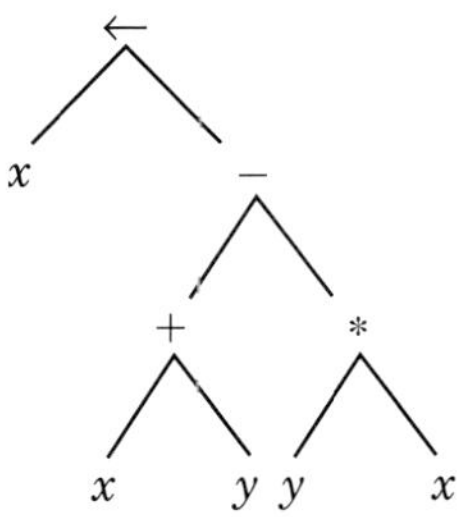

The next example illustrates this technique using a more complex algebraic expression.

EXAMPLE 10.11 Represent the expression $[(a - b * c)/(d - b * c)] + (d - b * c) \uparrow e$ in a dag.

SOLUTION:
The expression contains two common subexpressions: $b * c$ and $d - b * c$. Since the operator $+$ has the least priority in the expression, the source of the dag is $+$. Build the dag step-by-step like a binary expression tree, but with one exception: do not duplicate subtrees. Since the second operand for the minus sign in $d - b * c$ is $b * c$, draw a directed edge from vertex $-$ to the root of the tree that represents $b * c$, vertex $*$ (Figure 10.17). The first operand for $\uparrow$, $d - b * c$, has already appeared in Figure 10.17, so draw an arrow from $\uparrow$ to the source of the dag that represents $d - b * c$. Figure 10.18 shows the completed dag. (Again, compare the corresponding binary expression tree in Figure 10.19.) ■

Figure 10.17

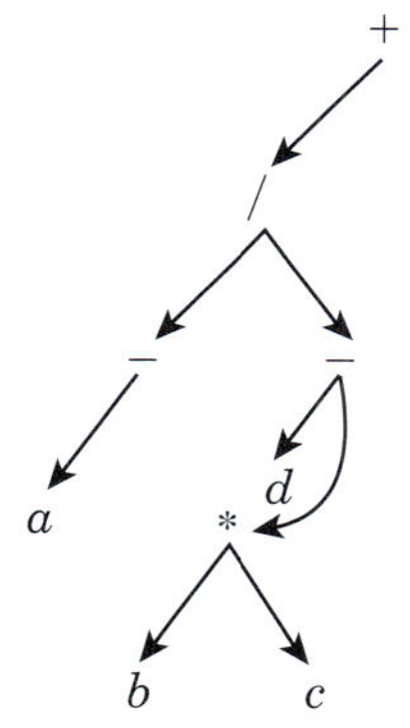

Figure 10.18

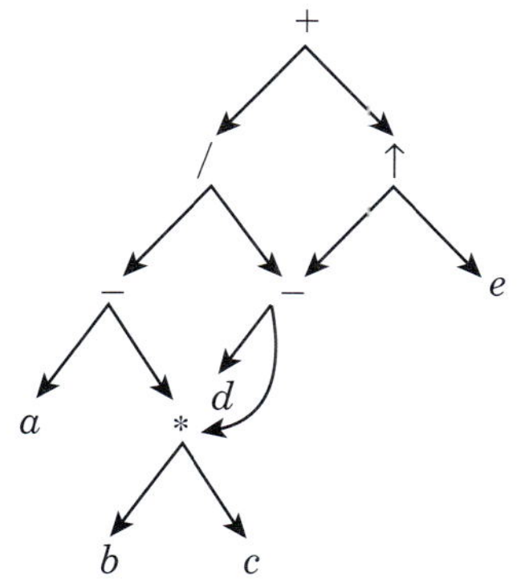

The technique for evaluating binary expression trees can be applied for dags representing arithmetic expressions: replace each operator vertex with the value of the subexpression it represents. Repeat this procedure until the source contains an operand, as the next example demonstrates.

Figure 10.19

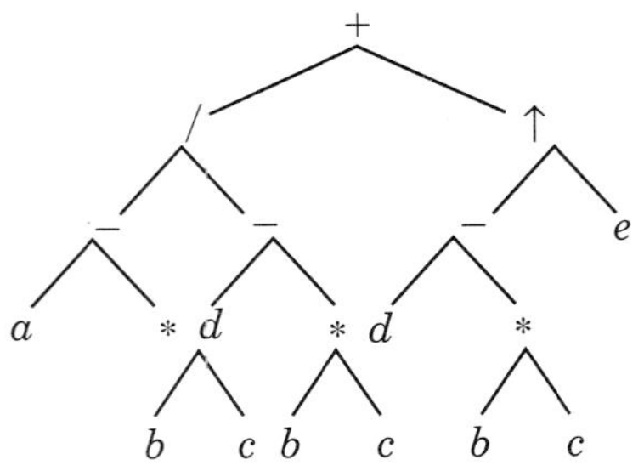

■

EXAMPLE 10.12 Evaluate the arithmetic expression represented by the dag in Figure 10.20.

Figure 10.20

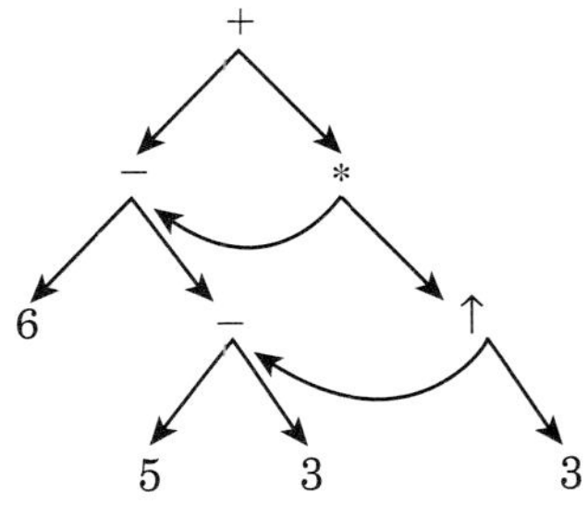

SOLUTION:
The procedure unfolds in Figures 10.21–10.25, reaching the value of the expression: 36 (see Figure 10.25).

Figure 10.21

Evaluate 5 − 3.

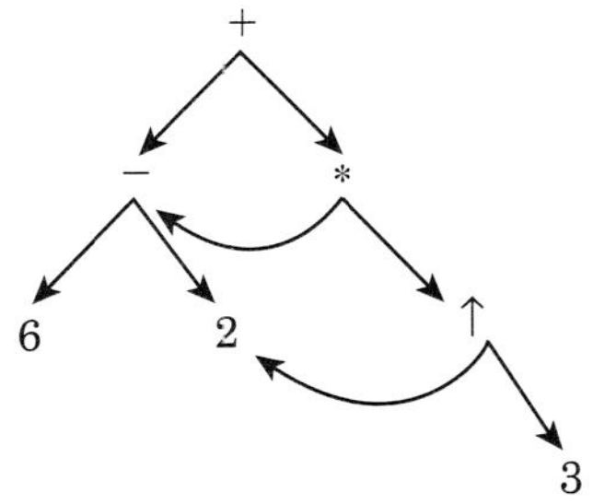

Figure 10.22

Evaluate 2 ↑ 3.

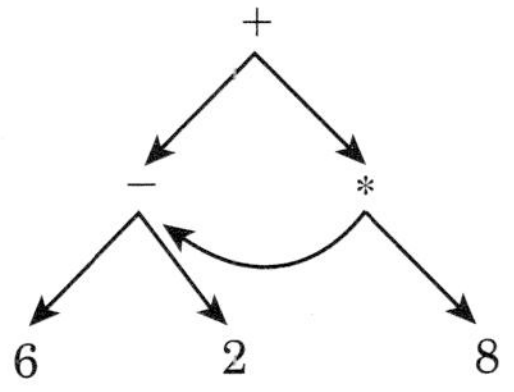

Figure 10.23

Evaluate $6 - 2$.

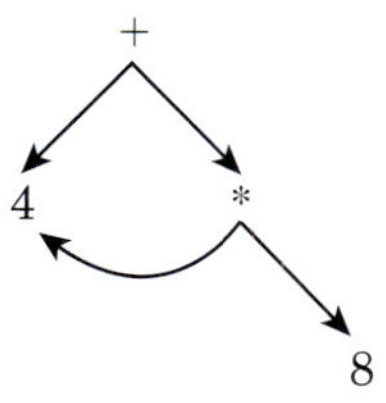

Figure 10.24

Evaluate $4 * 8$.

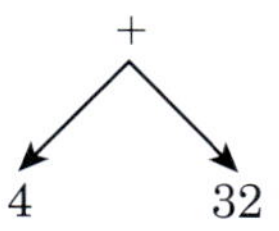

Figure 10.25

Evaluate $4 + 32$.

■

Binary tree traversals can also work with dags representing binary expressions. This makes sense because the outdegree of each operator vertex is two.

Example 10.13, for instance, finds the postfix form of an algebraic expression using a dag.

EXAMPLE 10.13 Find the postfix form of the expression represented by the dag in Figure 10.18.

SOLUTION:

In a postfix expression each operator immediately follows its operands. Again the steps unfold in Figures 10.26–10.31, ending with the postfix expression $abc* - dbc* - /dbc* - e\uparrow +$. The binary expression tree in Figure 10.19 will verify the result.

Figure 10.26

Convert $b * c$ to postfix.

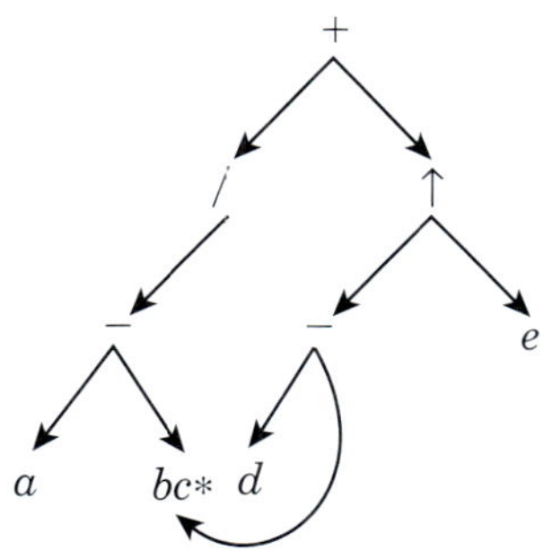

Figure 10.27

Convert $a - (bc*)$ to postfix.

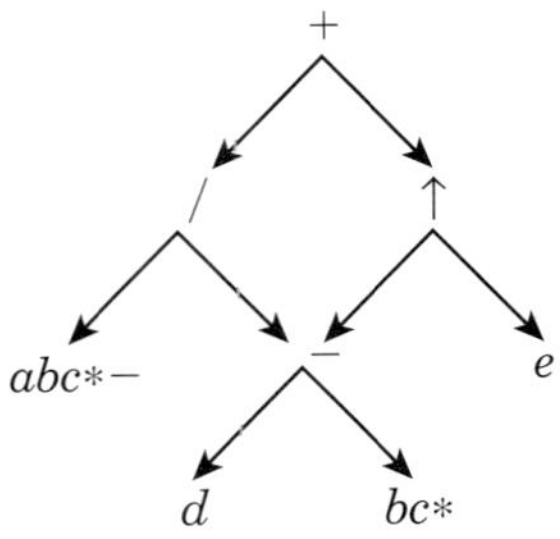

Figure 10.28

Convert $d - (bc*)$ to postfix.

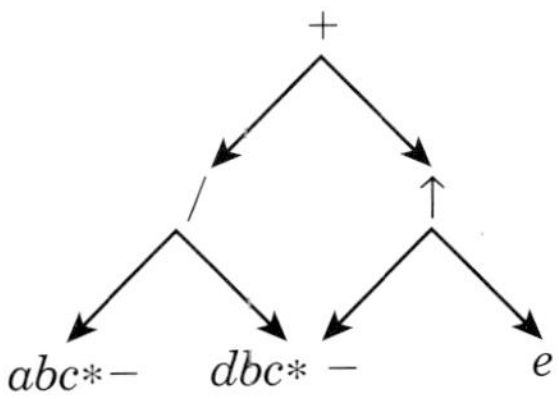

Figure 10.29

Convert $(abc*-)/(dbc*-)$ to postfix.

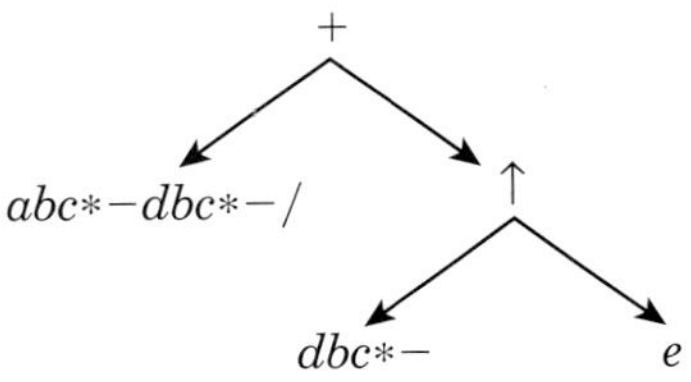

Figure 10.30

Convert $(dbc*-) \uparrow e$ to postfix.

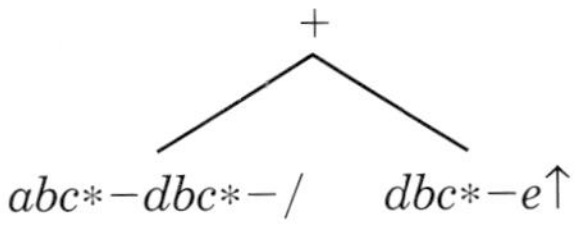

Figure 10.31

Convert $(abc* - dbc* - /) + (dbc* - e\uparrow)$ to postfix.

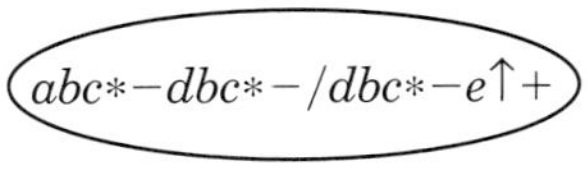

■

Dags are used extensively in a variety of contexts from management science to algebra.

Exercises 10.2

Determine if each digraph is a dag.

1.

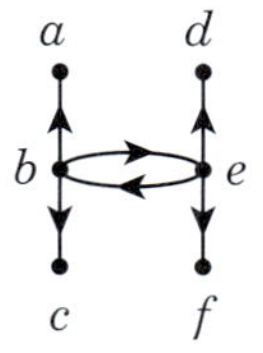

2.

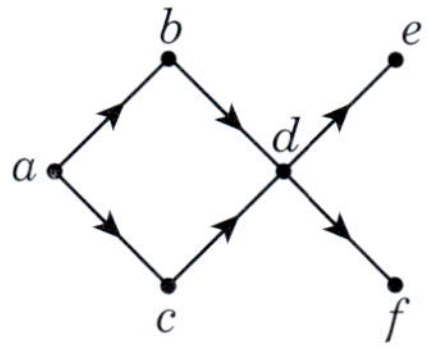

3.

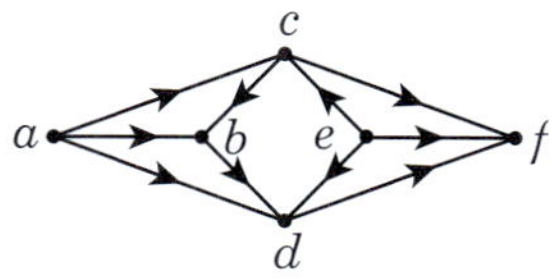

4.

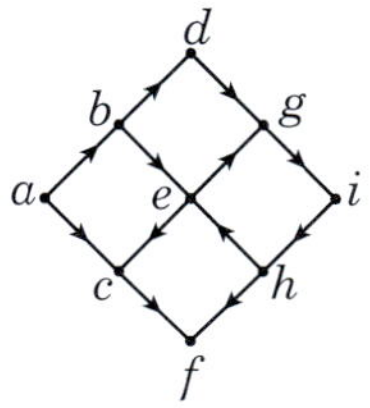

5–8. Identify the source(s) and sink(s) in Exercises 1–4.

9–12. Find the source(s) and sink(s) of the digraphs in Exercises 1–4 using their adjacency matrices.

Let $A = (a_{ij})$ be the adjacency matrix of a digraph with n vertices. Then D is a dag if and only if the main diagonal of the boolean matrix $R = A \vee A^{[2]} \vee \cdots \vee A^{[n]}$ is zero. Using this fact, determine if the digraphs in Exercises 13–17 are dags.

13.

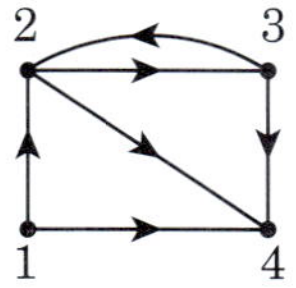

14.

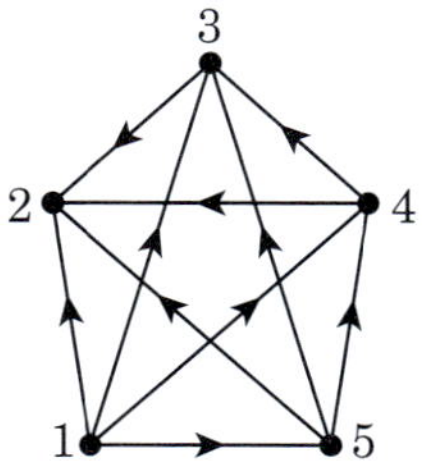

15.

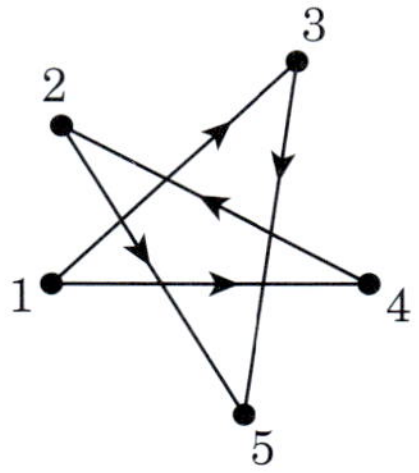

16.
$$\begin{array}{c} \\ a \\ b \\ c \\ d \end{array} \begin{array}{c} \begin{array}{cccc} a & b & c & d \end{array} \\ \begin{bmatrix} 0 & 1 & 1 & 1 \\ 0 & 0 & 1 & 1 \\ 0 & 0 & 0 & 1 \\ 0 & 1 & 0 & 0 \end{bmatrix} \end{array}$$

17.
$$\begin{array}{c} \\ a \\ b \\ c \\ d \\ e \end{array} \begin{array}{c} \begin{array}{ccccc} a & b & c & d & e \end{array} \\ \begin{bmatrix} 0 & 1 & 1 & 0 & 0 \\ 0 & 0 & 1 & 0 & 0 \\ 0 & 0 & 0 & 0 & 0 \\ 1 & 0 & 1 & 0 & 1 \\ 0 & 1 & 0 & 0 & 0 \end{bmatrix} \end{array}$$

18. Represent the subprojects A through F in Table 7.7 (Exercises 7.9) as a dag.

19. Represent the tasks A through G in Table 7.8 (Exercises 7.9) as a dag.

Draw a dag for each assignment statement and algebraic expression.

20. $x \leftarrow x + y + 1$

21. $x \leftarrow x + y + y \uparrow z$

22. $x \leftarrow (x + y) * (x + y + z)$

23. $x \leftarrow (x - y) \uparrow (x - y + z) * w$

24. $(a + b) \uparrow c + [(a + b)/d]$

25. $a + [(a + b) * c] \uparrow (c - b)$

26. $(a/b + b * c) - (b * c) \uparrow (c * d)$

27. $\{a \uparrow (b + c) - [a \uparrow (b + c) + d]\} * [(d + b + c) + c]$

Evaluate the arithmetic expression represented by each dag.

28.

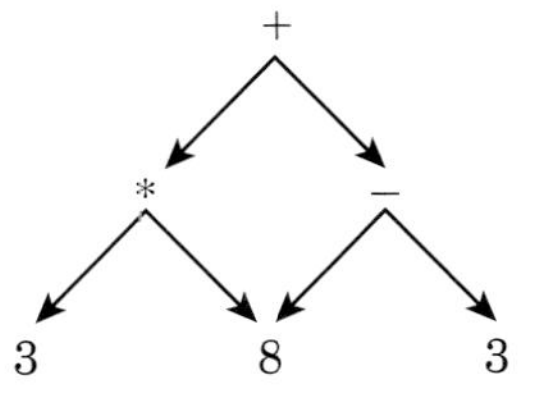

29.

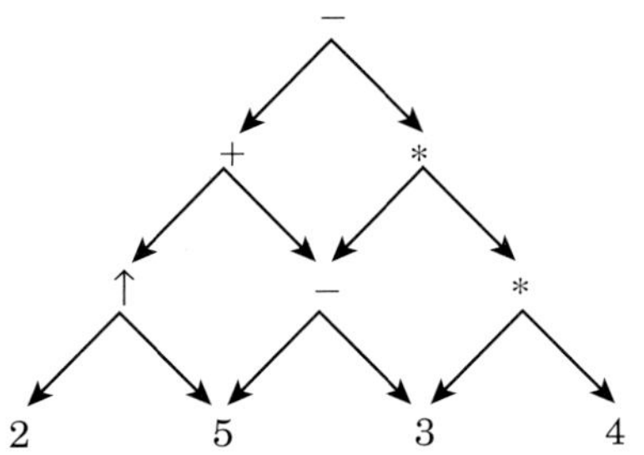

30.

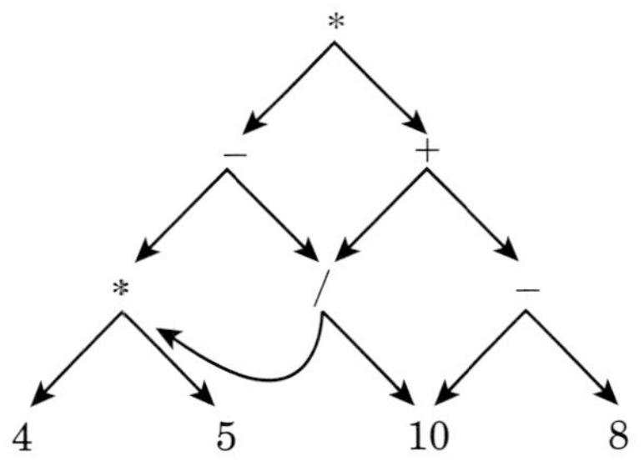

31.

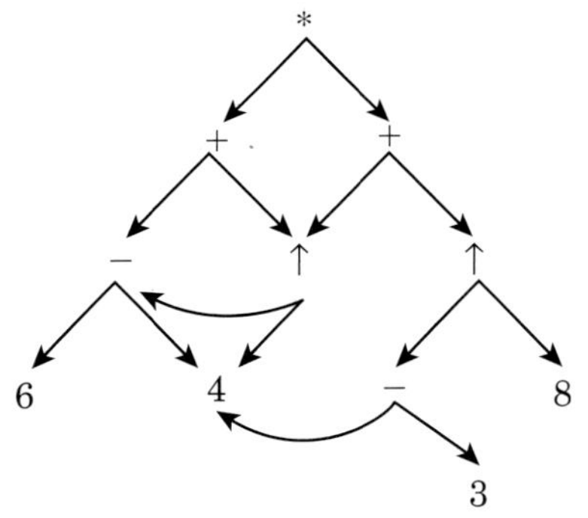

Find the postfix expression represented by each dag.

32.

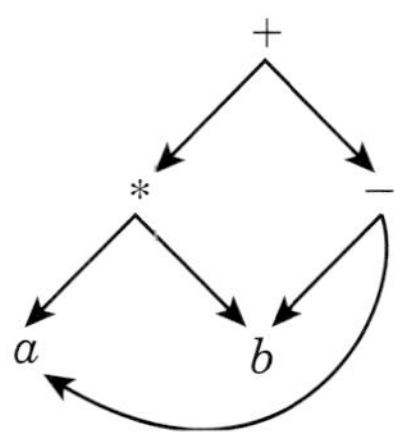

33.

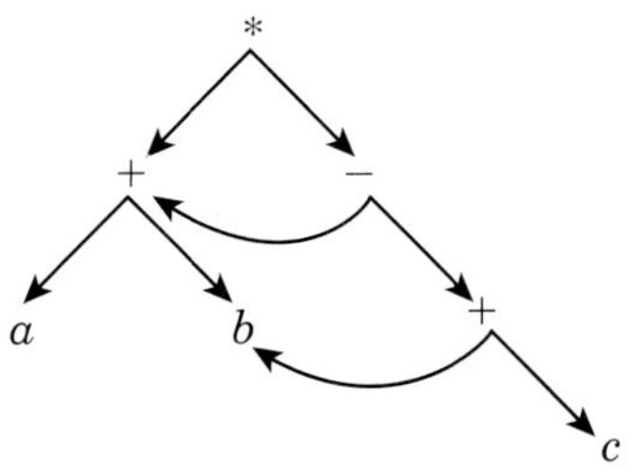

34.

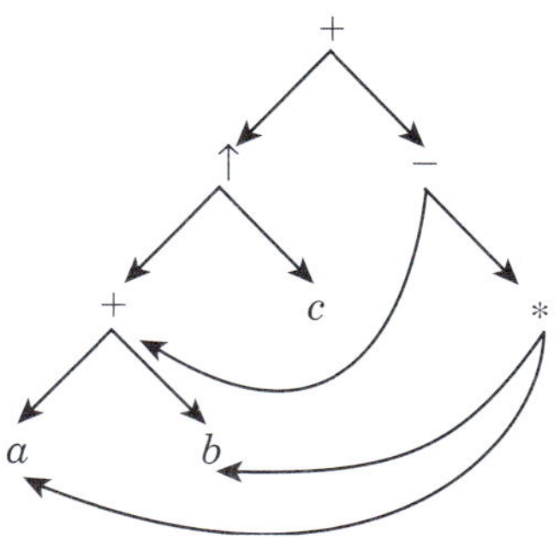

35.

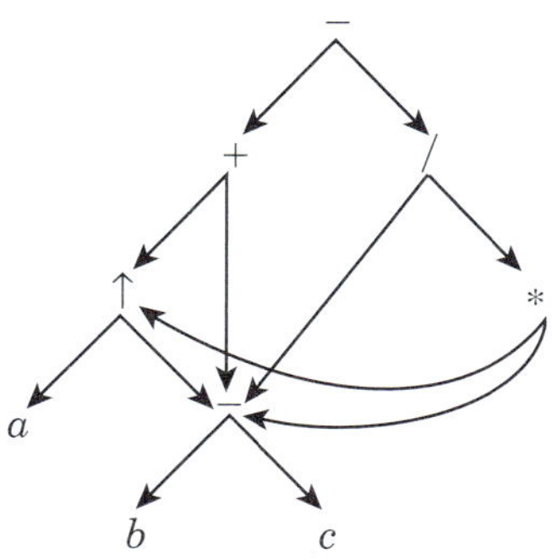

36–39. Find the prefix expression represented by the dags in Exercises 32–35.

40–43. Find the infix expression represented by the dags in Exercises 32–35. Insert parentheses as needed.

Prove each.

***44.** Every dag has a source.

***45.** Every dag has a sink.

***46.** The vertices of a dag can be topologically sorted. (*Hint:* Use induction.)

47–49. Using Exercise 46, topologically sort the vertices of the dags in Figure 10.13 as well as in Exercises 18 and 19.

10.3 Weighted Digraphs

Just as there are weighted graphs, there are weighted digraphs. This section presents them and an algorithm for finding a shortest path from a unique source to any other vertex.

Weighted Digraph

A **weighted digraph** is a digraph with a weight assigned to each directed edge. The weight of the edge (x,y) is denoted by $\mathrm{w}(x,y)$.

For example, consider the weighted digraph in Figure 10.32. The weight of the edge (a,c) is 5; that is, $w(a,c) = 5$. Similarly, $w(d,e) = 3$.

The **weight** of a path in a weighted digraph is the sum of the weights of the edges along the path. For instance, the weight of the path *a-c-b-d-z* in Figure 10.32 is $5 + 1 + 1 + 7 = 14$, and that of *c-b-d-e-z* is 7.

The weights in a weighted digraph may represent measurements such as distances between cities, transportation costs, or completion times for various jobs on a project.

Figure 10.32

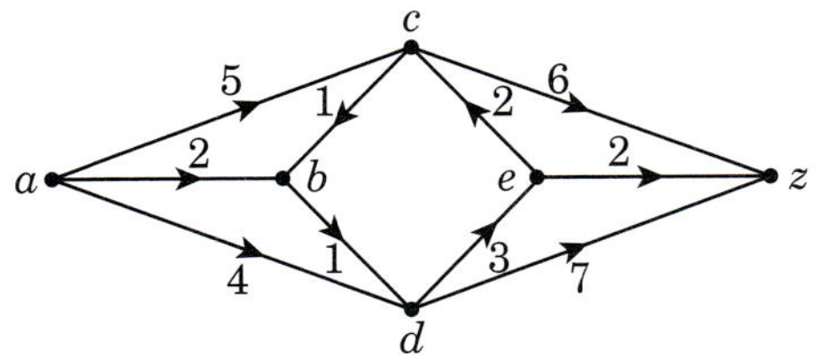

Weighted Adjacency Matrix

The **weighted adjacency matrix** W of a digraph with n vertices v_i is defined as $(w_{ij})_{n\times n}$ where

$$w_{ij} = \begin{cases} \infty & \text{if there is no edge from } v_i \text{ to } v_j \\ w(i,j) & \text{otherwise} \end{cases}$$

Here the infinity symbol ∞ denotes a number greater than the largest of all weights in the digraph.

For instance, the weighted adjacency matrix of the digraph in Figure 10.32 is

$$W = \begin{array}{c} \\ a \\ b \\ c \\ d \\ e \\ z \end{array} \begin{array}{c} \begin{array}{cccccc} a & b & c & d & e & z \end{array} \\ \begin{bmatrix} \infty & 2 & 5 & 4 & \infty & \infty \\ \infty & \infty & \infty & 1 & \infty & \infty \\ \infty & 1 & \infty & \infty & \infty & 6 \\ \infty & \infty & \infty & \infty & 3 & 7 \\ \infty & \infty & 2 & \infty & \infty & 2 \\ \infty & \infty & \infty & \infty & \infty & \infty \end{bmatrix} \end{array}$$

The next example finds the shortest route in the fourth problem posed at the beginning of the chapter.

EXAMPLE 10.14 A computer company has a manufacturing plant at location a and a warehouse at location z. The possible streets a truck driver can take from a to z are given by the digraph in Figure 10.32. The arrows indicate one-way streets and the weights, the distances in miles. Find the shortest route the driver could take from a to z.

SOLUTION:

Vertex a is the source of the digraph. To find a shortest route, list all simple paths from a to z, compute their weights, and select a path with the least weight:

- a-c-z; weight $= 11$
- a-c-b-d-e-z; weight $= 12$
- a-c-b-d-e-c-z; weight $= 18$
- a-c-b-d-z; weight $= 14$

Edsger Wybe Dijkstra *(1930–2002), a computer science educator and mathematician, was born in Rotterdam, The Netherlands. After graduating from the University of Leyden in 1951, he received his doctorate from Leyden in 1956, a Ph.D. from Amsterdam in 1959, and an honorary D.Sc. from Queen's University, Belfast, in 1976.*

After 6 years at the Mathematics Center at Amsterdam, Dijkstra taught mathematics at the Technical University, Nuenen, until 1973. For the next 11 years, he was a research fellow at Burroughs Corporation, Nuenen. In 1984, he became the Schlumberger Centennial chair in computer science at the University of Texas, Austin.

A distinguished fellow of the British Computer Society, Dijkstra was honored with the prestigious Alan M. Turing award in 1972. He has made outstanding contributions to operating systems and programming languages.

- a-b-d-e-c-z; weight $= 14$
- a-b-d-e-z; weight $= 8$
- a-b-d-z; weight $= 10$
- a-d-e-c-z; weight $= 15$
- a-d-e-z; weight $= 9$
- a-d-z; weight $= 11$

Clearly, the path with the minimum weight is a-b-d-e-z and the minimum distance to be traveled is 8 miles. ■

This example leads us to the next definition.

Shortest Path

A path of least weight from vertex u to vertex v in a weighted digraph is a **shortest path** from u to v.

The shortest path from a to z in Example 10.14 is a-b-d-e-z, but the technique used to identify it and to compute the minimum weight is certainly not the best, especially with a digraph of many vertices and edges. An efficient method was developed in 1959 by the Dutch mathematician Edsger W. Dijkstra, a pioneer in the art of computer programming. **Dijkstra's algorithm** can find a shortest path from any vertex to any vertex in the digraph, if it exists, as the next example demonstrates. Nonetheless, assume a single source a.

EXAMPLE 10.15 Using Dijkstra's algorithm, find a shortest path from source a to z in the weighted digraph in Figure 10.32.

SOLUTION:

Step 0 Let V be the set of all vertices in the digraph except a. Let S be the set of all vertices in the digraph whose shortest distances from a are already known. Make $S = \{a\}$, since vertex a is the closest vertex to itself. Let D be a one-dimensional array such that D_u denotes the length of a shortest path from a to u passing through the vertices in S. Initially define D as follows:

$$D_u = \begin{cases} 0 & \textit{if } u = a \\ w(a,u) & \text{if edge } (a,u) \text{ exists} \\ \infty & \text{if edge } (a,u) \text{ does not exist} \end{cases}$$

Thus, initially, $V = \{b,c,d,e,z\}$, $S = \{a\}$, $V - S = \{b,c,d,e,z\}$, and

$$\begin{array}{cccccccc} & & a & b & c & d & e & z \\ D & = & [0 & 2 & 5 & 4 & \infty & \infty] \end{array}$$

Step 1 Select a vertex v in $V - S$ with D_v at a minimum: $v = b$.

Step 2 Add v to S and update $V - S$: $S = \{a, b\}, V - S = \{c, d, e, z\}$.

Step 3 Update the array D of distances of vertices x in $V - S$, using v as a vertex along the path from a to x. That is, check if a shorter path runs from a to each vertex x in $V - S$ passing through v; if so, update D_x:

$$D_x \leftarrow \min\{D_x, D_v + w(v,x)\}^{\dagger}$$

See Figure 10.33. To update D_x, make the following definitions:

$$n + \infty = \infty = \infty + n$$
$$\min\{n, \infty\} = n \text{ and } \min\{\infty, \infty\} = \infty$$

Figure 10.33

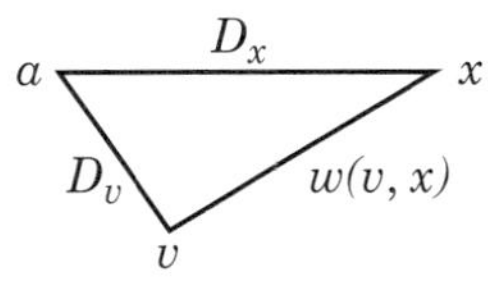

Figure 10.34

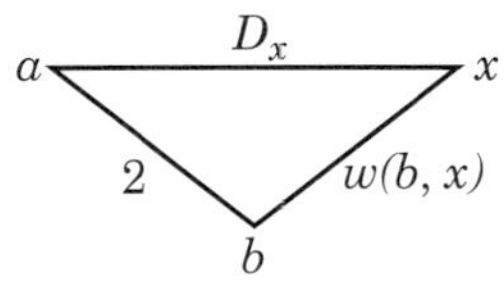

†$\min\{x,y\}$ denotes the minimum of x and y.

Thus (see Figure 10.34):

$$
\begin{aligned}
D_c &= \min\{D_c, D_b + w(b,c)\} = \min\{5, 2+\infty\} = 5\\
D_d &= \min\{D_d, D_b + w(b,d)\} = \min\{4, 2+1\} = 3\\
D_e &= \min\{D_e, D_b + w(b,e)\} = \min\{\infty, 2+\infty\} = \infty\\
D_z &= \min\{D_z, D_b + w(b,z)\} = \min\{\infty, 2+\infty\} = \infty
\end{aligned}
$$

Consequently, $D = \begin{array}{cccccc} a & b & c & d & e & z \\ [0 & 2 & 5 & 3 & \infty & \infty] \end{array}$.

Repeat steps 1–3 until $|V - S| = 1$.

Step 4 Find a vertex v in $V - S$ with minimum D_v. Choose $v = d$.

Step 5 Add d to S and update $V - S$: $S = \{a, b, d\}$, $V = \{c, e, z\}$.

Step 6 Update the array D of distances from a to each vertex x in $V - S$, using $v = d$ as a vertex along the path from a to x (see Figure 10.35): $D_c = \min\{5, 3+\infty\} = 5$; $D_e = \min\{\infty, 3+3\} = 6$; and $D_z = \min\{\infty, 3+7\} = 10$.
Thus,

$$
D = \begin{array}{cccccc} a & b & c & d & e & z \\ [0 & 2 & 5 & 3 & 6 & 10] \end{array}
$$

Figure 10.35

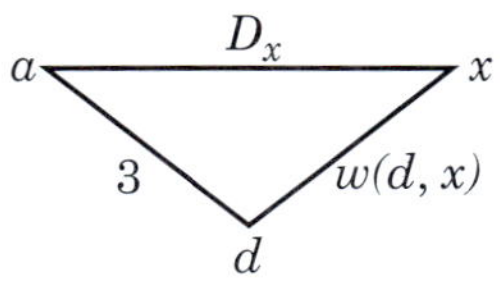

Since $|V - S| \neq 1$, continue with step 1.

Step 7 Find a vertex v in $V - S$ with minimum D_v; choose $v = c$.

Step 8 Update S and $V - S$: $S = \{a, b, d, c\}$, $V - S = \{e, z\}$.

Step 9 Update D (Figure 10.36):

$$
D_e = \min\{6, 5+\infty\} = 6 \text{ and } D_z = \min\{10, 5+6\} = 10
$$

Figure 10.36

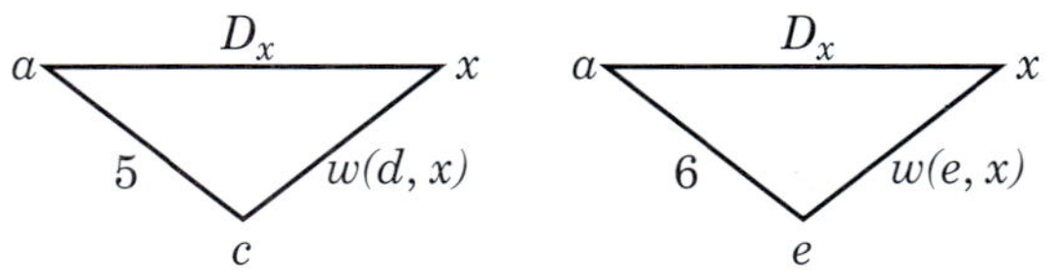

Thus,

$$\begin{array}{cccccc} a & b & c & d & e & z \end{array}$$
$$D = \begin{bmatrix} 0 & 2 & 5 & 3 & 6 & 10 \end{bmatrix}$$

Step 10 Again $|V - S| \neq 1$, so continue with step 1. Find a vertex v in $V - S$ with minimum D_v; choose $v = e$.

Step 11 Update S and $V - S$: $S = \{a, b, d, c, e\}, V - S = \{z\}$.

Step 12 Update D (see Figure 10.36): $D_z = \min\{10, 6 + 2\} = 8$.

Thus,

$$\begin{array}{cccccc} a & b & c & d & e & z \end{array}$$
$$D = \begin{bmatrix} 0 & 2 & 5 & 3 & 6 & 8 \end{bmatrix}$$

Step 13 Since $V - S = \{z\}$, the process is over. The length of the shortest path from a to any vertex v is D_v. For example, $D_z = 8$, which agrees with the conclusion in Example 10.14. Since $D_e = 6$, the shortest path from a to e is 6; the corresponding path, *a-b-d-e*.

Table 10.3 lists the above steps.

Table 10.3

Steps	v	S	$V - S$	$\|V - S\| = 1$?	D_b	D_c	D_d	D_e	D_z
0	—	$\{a\}$	$\{b, c, d, e, z\}$	No	2	5	4	∞	∞
1–3	b	$\{a, b\}$	$\{c, d, e, z\}$	No	2	5	3	∞	∞
4–6	d	$\{a, b, d\}$	$\{c, e, z\}$	No	2	5	3	6	10
7–9	c	$\{a, b, d, c\}$	$\{e, z\}$	No	2	5	3	6	10
10–12	e	$\{a, b, d, c, e\}$	$\{z\}$	Yes	2	5	3	6	8

■

Although the method illustrated in this example provided the weights of shortest paths from a to every other vertex in the digraph, it did not find the shortest paths. To do this, we use a one-dimensional **predecessor array** P of vertices, where P_v denotes the immediate predecessor of v in the shortest path from a to v. Initially, make $P_v = a$ for every vertex $v \neq a$:

	a	b	c	d	e	z
P	.	a	a	a	a	a

Every time a shorter path to a vertex x passing through v appears, update P_x to v. In other words, when updating array D, update P, too:

```
If Dv - w(v,x) < Dx then
   Px ← v.
```

Table 10.4 shows array P after steps 1–3, 4–6, 7–9, and 10–12. (*Note*: The cell P_a could be omitted.)

A shortest path from a to any vertex v is obtained by finding the immediate predecessors of v using the last row in Table 10.4. For instance, to find a shortest path from a to z, notice that e is the immediate predecessor of z, d immediately precedes e, b immediately precedes d, and a immediately precedes b. Thus a shortest path is a-b-d-e-z, which matches the path obtained in Example 10.14.

Table 10.4

Step	P_a	P_b	P_c	P_d	P_e	P_z
0	—	a	a	a	a	a
1–3	—	a	a	b	a	a
4–6	—	a	a	b	d	d
7–9	—	a	a	b	d	d
10–12	—	a	a	b	d	e

A pseudocode for Dijkstra's algorithm unfolds in Algorithm 10.1.

```
Algorithm Dijkstra (W,D,P)
(* This algorithm computes the length of a shortest path from
   a unique source 1 to every other vertex of a connected
   weighted digraph with V = {1,2,...,n}. W = (wij) denotes
   the weighted adjacency matrix of the digraph. S denotes the
   set of vertices whose shortest distances from the source
   are known. D and P are one-dimensional arrays containing
   (n - 1) cells, labeled 2 through n; Di denotes the shortest
   distance from the source to vertex i and Pi denotes the
   immediate predecessor of vertex i along the shortest path. *)
   Begin (* algorithm *)
     (* Initialize S, V, D, and P. *)
     S ← {1}
     V ← {2,3,...,n}
     for i = 2 to n do
     begin (* for *)
       Di ← w1i
       Pi ← 1
     endfor
     for i = 1 to n - 1 do
     begin (* for *)
       find a vertex j in V such that Dj is a minimum.
       S ← S ∪ {j} (* add vertex j to S *)
       V ← V - {j} (* update the set of remaining vertices *)
       for each vertex k in V do
       begin (* for *)
         if Dj + wjk < Dk then
           (* A shorter path runs from the source to
              vertex k through j *)
         begin (* if *)
           (* update the minimum distance Dk *)
           Dk ← Dj + wjk
```

```
            (* update the immediate predecessor of vertex k *)
            P_k ← j
         endif
      endfor
   endfor
End (* algorithm *)
```

Algorithm 10.1

EXAMPLE 10.16 Apply Dijkstra's algorithm to find a shortest path from vertex a to vertex e in the weighted digraph of Figure 10.37 and its length.

Figure 10.37

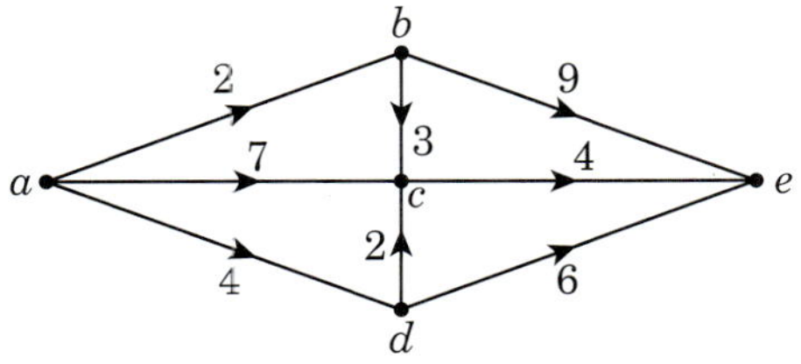

SOLUTION:

Table 10.5 summarizes the various steps of the algorithm. In line 1, initialize the sets S and V and the arrays D and P. In line 2, find a vertex j in V such that D_j is a minimum, namely, $j = b$; update S, V, D, and P. Is $i = 4$? No. So in line 3, again find a vertex j in V such that D_j is a minimum, namely, $j = d$; update S, V, D, and P. Is $i = 4$? No. Therefore, in line 4, look for a vertex j for which D_j is a minimum, namely, $j = c$; again, update S, V, D, and P. By now, $i = 4$. Exit the ***for*** loop and the algorithm terminates.

Table 10.5

					D				**P**			
Steps	i	j	S	V	D_b	D_c	D_d	D_e	P_b	P_c	P_d	P_e
1–7	1	—	$\{a\}$	$\{b,c,d,e\}$	2	7	4	∞	a	a	a	a
8 – 21	2	b	$\{a,b\}$	$\{c,d,e\}$	2	5	4	11	a	b	a	b
8 – 21	3	d	$\{a,b,d\}$	$\{c,e\}$	2	5	4	10	a	b	a	d
8 – 21	4	c	$\{a,b,d,c\}$	$\{e\}$	2	5	4	9	a	b	a	c

According to the table, the shortest distance from the source to vertex e is $D_e = 9$. The immediate predecessor of e along a shortest path is c; that of c is b, and that of b is a. So a shortest path from vertex a to vertex e is a-b-c-e. ■

Next we establish the correctness of Dijkstra's algorithm.

THEOREM 10.3 Dijkstra's algorithm yields the weight of a shortest path from the source to any vertex in a weighted digraph.

PROOF (by PMI):
Let D be a weighted digraph, as in Algorithm 10.1. Let $D_i(\text{v})$ denote the weight of a shortest path from the source to any vertex v after i iterations of the ***for*** loop (lines 8–21).

Basis step When $i = 0$, the loop is not entered and no iterations take place. $S = \{1\}$, and $D_0(v) = D_v$ if v is adjacent to the source and ∞ otherwise. Thus the basis step works by default.

Induction step Assume $D_i(v)$ is the weight of a shortest path from the source to any vertex $v \in S$ after i iterations. In the $(i + 1)$st iteration, let j be the vertex added to S (and hence deleted from V), where D_j is a minimum. If $v \in S$, then $D_i(v)$ is the weight of a shortest path from the source to v, by the inductive hypothesis. Therefore, $D_{i+1}(v) = D_i(v)$. If $v \notin S$, then $v \in V$. In line 15, check for a shorter path to v through j. If $D_j + w_{jv} < D_v$, then $D_{i+1}(v) = D_j + w_{jv}$. If $D_j + w_{jv} \geq D_v$, then $D_{i+1} = D_i(v) = D_v$. In either case, $D_{i+1}(v)$ is the weight of a shorter path to v after $i + 1$ iterations.

Therefore, by induction, $D_i(v)$ is the weight of a shortest path to any vertex v at the end of i iterations for every $i \geq 0$.

Consequently, when the *for* loop ends, $D_{n-1}(v)$ gives the weight of a shortest path from the source to any vertex v in D. This establishes the validity of the algorithm.

(A similar argument proves the algorithm also yields a shortest path to any vertex v.) ■

We close this section with an analysis of the algorithm.

An Analysis of Dijkstra's Algorithm

The number of additions and comparisons measures the computational complexity of Dijkstra's algorithm. The *for* loop (lines 8–21) is executed $n - 1 = \Theta(n)$ times. Each iteration takes at most $n - 1$ additions and $n - 1$ comparisons: a total of $2(n - 1) = \Theta(n)$ operations. Thus, by Theorem 4.15, Dijkstra's algorithm in the worst case takes $\Theta(n^2)$ operations to find a shortest path from the source to any vertex in a weighted digraph. (Dijkstra's algorithm is also a greedy algorithm.)

Exercises 10.3

Using the weighted digraph in Figure 10.38, compute the length of each path in Exercises 1–4.

1. *a-c-b-e* **2.** *a-b-e* **3.** *e-c-b-e* **4.** *a-c-b-e-g*

Figure 10.38

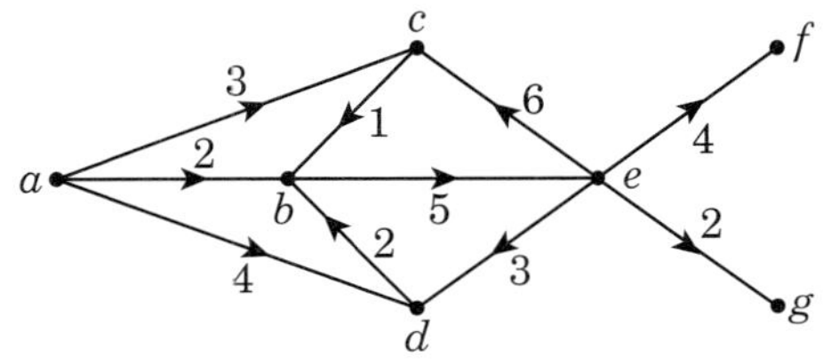

Find the weighted adjacency matrix of each weighted digraph.

5.

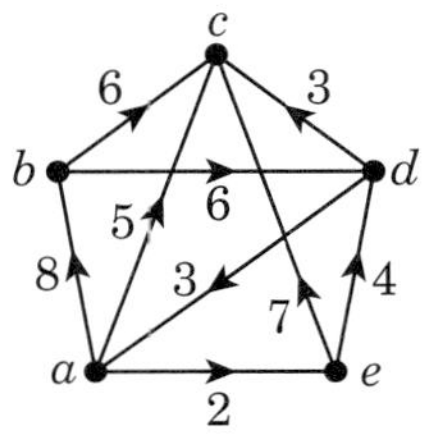

6.

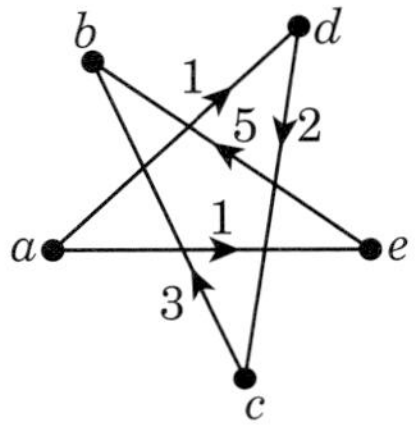

Use Figure 10.38 for Exercises 7–10.

7. List all possible paths from vertex a to vertex f and their lengths.

8. Find the shortest path from vertex a to vertex f and its length.

9. Redo Exercise 7 with vertices a and g.

10. Redo Exercise 8 with vertices a and g.

Using Figure 10.32, compute the length of a shortest path from vertex a to each vertex.

11. Vertex b **12.** Vertex c **13.** Vertex d **14.** Vertex e

15–18. Redo Exercises 11–14 with Table 10.3.

19–22. Using the predecessor array P in Table 10.4, redo Exercises 11–14.

With Table 10.5, find the shortest distance in the weighted digraph of Figure 10.37 from the source to each vertex.

23. Vertex b **24.** Vertex c **25.** Vertex d

26–28. Use Table 10.5 to find the shortest path from the source to the vertices in Exercises 23–25.

29–31. Using Figure 10.37, compute the length of the paths in Exercises 26–28.

Using Dijkstra's algorithm, find a shortest path and its length from vertex a to the other vertices in Exercises 32–35.

32.

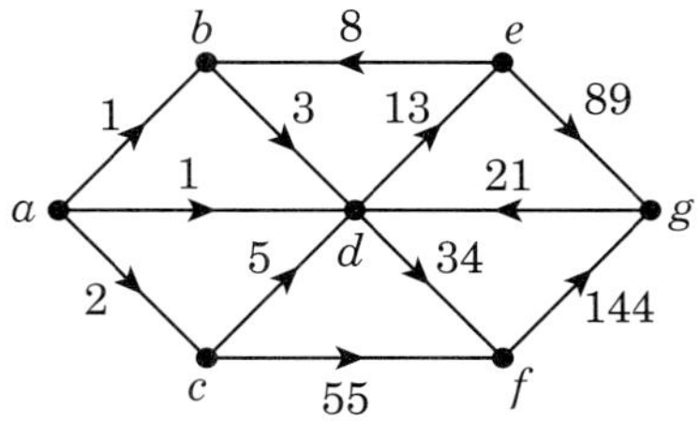

33.

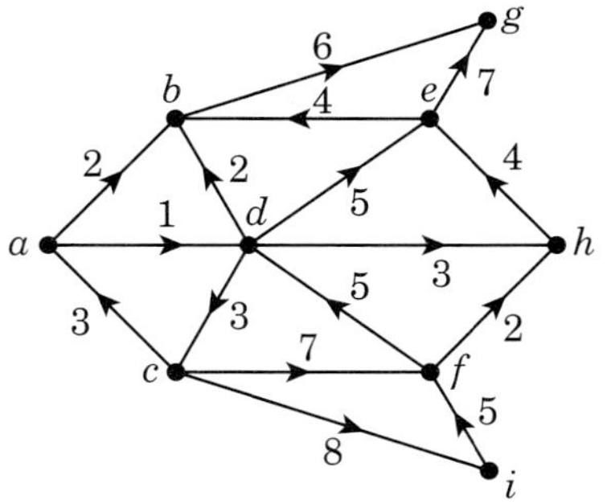

34.

	a	b	c	d	e	f
a	∞	3	5	∞	∞	∞
b	∞	∞	∞	2	∞	∞
c	∞	∞	∞	3	∞	∞
d	4	∞	∞	∞	3	6
e	∞	5	∞	∞	∞	∞
f	∞	∞	4	∞	∞	∞

35.

	a	b	c	d	e	f	g
a	∞	∞	2	5	∞	∞	∞
b	3	∞	∞	∞	∞	5	∞
c	∞	4	∞	∞	7	∞	∞
d	∞	6	∞	∞	∞	8	8
e	∞	∞	∞	∞	∞	∞	∞
f	∞	∞	6	∞	∞	∞	∞
g	∞	6	∞	∞	∞	∞	∞

Algorithm 10.2, developed by R. W. Floyd in 1962, can also find the weight of a shortest path from the source to any vertex in a weighted digraph.

```
Algorithm Floyd (D,W)
(* D is a connected weighted graph with weighted adjacency matrix
   W = (wij)n×n. The algorithm computes the weight of a shortest path
   from the unique source 1 to any other vertex.  *)
0. Begin (* Floyd *)
1.   for i = 1 to n do
2.     for j = 1 to n do
3.       for k = 1 to n do
4.         if wji + wik < wjk then (* a shorter path runs from
                  vertex j to vertex k passing through vertex i *)
5.           wjk ← wji + wik
6. End (* Floyd *)
```

Algorithm 10.2

36–37. Using Floyd's algorithm, find the weight of a shortest path from the source a to the other vertices in Exercises 32 and 33.

38. Give a big-oh estimate of the number of operations (additions and comparisons in line 4) required by Floyd's algorithm in the worst case.

39. Modify Floyd's algorithm to determine a shortest path from the source to the other vertices in the digraph D.

40–41. Use the algorithm in Exercise 39 to find a shortest path from the source to the other vertices in Exercises 32 and 33.

***42.** Prove Floyd's algorithm.

Robert W. Floyd *(1936–), a computer scientist and an educator, was born in New York City. He graduated from the University of Chicago at the age of 17 and received a B.S. in physics 5 years later. After working at Westinghouse in Elmira, New York, for a year, Armour Research Foundation in Chicago for 6 years, and Computer Associates in Wakefield, Massachusetts, for three years, Floyd joined the computer science faculty at Carnegie-Mellon University in 1965. Three years later, he moved to Stanford, where he chaired the computer science department during 1974–1977.*

A fellow of the American Academy of Arts and Sciences, Floyd received the Alan M. Turing award from the American Association of Computing Machinery in 1978.

Chapter Summary

The edges in a digraph are never parallel, but always directed. It pictorially represents a finite relation.

Digraph

- A **digraph** $D = (V, E)$ consists of a finite set of vertices V joined by a set of directed edges E. A directed edge from vertex x to vertex y is denoted by (x, y) (page 692).
- The **indegree** of a vertex v, denoted by indeg(v), is the number of edges terminating at v; its **outdegree**, denoted by outdeg(v), is the number of edges leaving v. A vertex with indegree 0 is a **source**; a vertex with outdegree 0 is a **sink** (page 694).
- The **adjacency matrix** of a digraph is $A = (a_{ij})$, where a_{ij} = number of directed edges from vertex v_i to vertex v_j (page 695).
- Let D be a digraph with vertices $v_1, v_2, \ldots, v_n$ and e edges. Then $\sum_{i=1}^{n} \text{indeg}(v_i) = e = \sum_{i=1}^{n} \text{outdeg}(v_i)$ (page 696).
- A loop-free digraph with exactly one edge between any two distinct vertices is a **dominance digraph** or a **tournament** (page 697).

Reachability

- A vertex v_j is **reachable** from a vertex v_i if there is a directed path from v_i to v_j (page 698).

- If every vertex in a digraph is reachable from every other vertex, the digraph is **strongly connected** (page 698).
- Let $A = (a_{ij})$ be the adjacency matrix of a digraph D with vertices $v_1, v_2, \ldots, v_n$. Let $R = A \vee A^{[2]} \vee \cdots \vee A^{[n-1]} = (r_{ij})$. Then vertex v_j is reachable from vertex v_i if and only if $r_{ij} > 0$, where $i \neq j$. R is the **reachability matrix** of the digraph (page 699).

de Bruijn Sequence

- A **de Bruijn sequence** for a binary alphabet is a 2^n-bit sequence that contains every n-bit word as a subsequence in a cyclic fashion (page 702).

Dag

- A **dag** is a digraph with no directed cycles (page 707).
- Dags pictorially represent assignment statements and algebraic expressions with repeating subexpressions; they help find the prefix and postfix forms of such expressions (page 707).

Weighted Digraph

- A **weighted digraph** has a weight w for every edge (page 715).
- The **weight** of a directed path is the sum of the weights of the edges along the path (page 715).
- The **weighted adjacency matrix** of a digraph (V, E) is $W = (w_{ij})$, where

$$w_{ij} = \begin{cases} \infty & \text{if } (v_i, v_j) \notin E \\ \text{weight of edge } (i, j) & \text{otherwise} \end{cases}$$ (page 716).

- A **shortest path** from u to v in a weighted digraph weighs the least (page 717).
- Dijkstra's algorithm finds a shortest path and its length from the source to any vertex in a weighted digraph (page 717).

Review Exercises

Use the digraph in Figure 10.39 to answer Exercises 1–7.

1. Find the indegree and outdegree of each vertex.

2. Find the adjacency matrix.

Figure 10.39

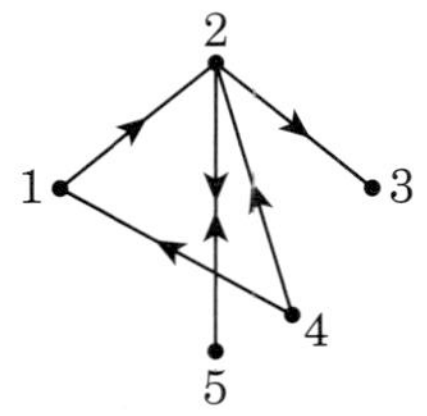

3. Is vertex 3 reachable from vertex 5?

4. Is the digraph strongly connected?

5. Find the reachability matrix R of the digraph.

6. Is every entry of R positive, except those on the main diagonal?

7. Is the digraph weakly connected?

In 1953, E. N. Gilbert of Bell Telephone Laboratories developed a method for constructing a de Bruijn sequence for nonzero binary n-tuples: Start with any nonzero n-bit word x. If the first two digits in x are the same, then append a 0 to it; otherwise, append a 1. Drop the first digit in the new word to produce the next n-bit word. Using this technique successively, produce all nonzero n-bit words from each word.

8. 01 **9.** 010 **10.** 1011

Using Gilbert's algorithm, check if all nonzero binary quintuples can be generated beginning with the given word.

11. 01011 **12.** 11111

13. Using Exercises 11 and 12, determine if Gilbert's algorithm works for $n = 5$.

14. List all possible ternary couplets.

15. Construct a de Bruijn sequence for ternary couplets. (There are 24 such choices.)

16. List all possible 4-ary couplets.

17. Construct a de Bruijn sequence for 4-ary couplets. (There are 16 such couplets.)

Figure 10.40

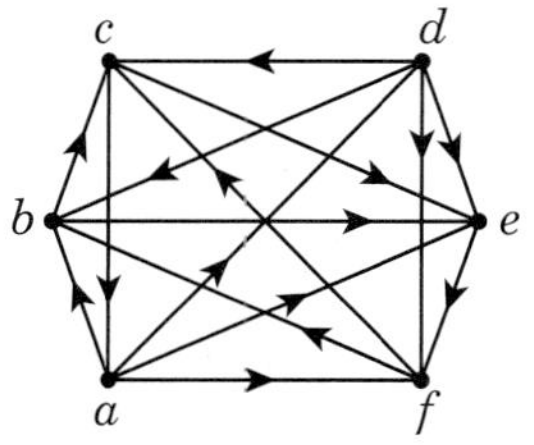

The digraph in Figure 10.40 shows the results from a round-robin tournament of six players, a through f. Find each.

18. The number of wins by player e.

19. The number of losses by player f.

20. The champion of the tournament.

21. The second-best player.

22–23. Are the digraphs in Figures 10.39 and 10.40 dags?

24. Represent the courses given by Table 7.1 in a dag.

Represent each assignment statement and algebraic expression in a dag.

25. $x \leftarrow (x+y)+(y+z)$

26. $x \leftarrow [x+(x+y)]*z$

27. $[(a*b)\uparrow c]+[(b-a)+d]$

28. $\{[(a*b)+c]+d\}\uparrow[c/(b+a)]$

Evaluate the arithmetic expression represented by each dag.

29.

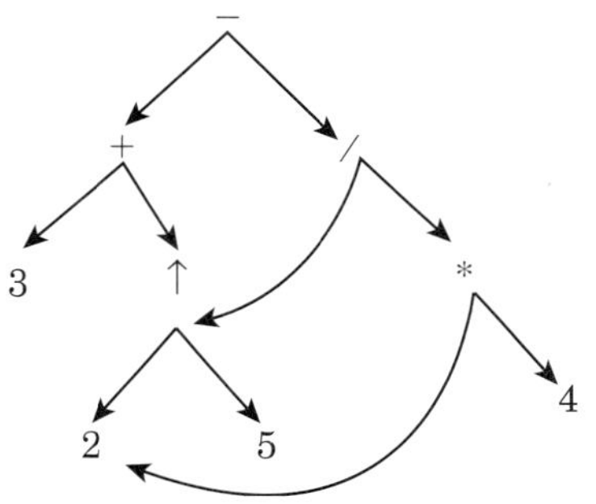

30.

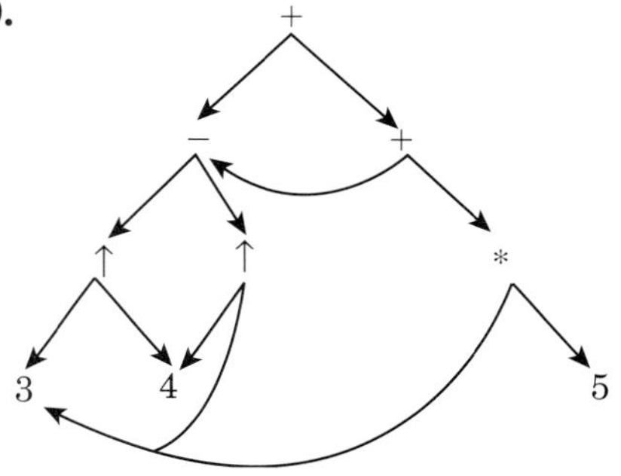

31–34. Find the postfix expressions represented by the dags in Exercises 27–30.

35–38. Find the prefix expressions represented by the dags in Exercises 27–30.

39–40. Find the infix expressions represented by the dags in Exercises 29 and 30. Provide parentheses as needed.

Use the weighted digraph in Figure 10.41 to answer Exercises 41–46. Compute the length of each path.

41. a-b-e-d

42. a-d-g-h

43. a-c-f-e

44. a-c-e-g-h

45. Find the shortest path from vertex a to vertex g.

46. Find the shortest path from vertex e to vertex h.

47–48. With Dijkstra's algorithm, find a shortest path from vertex a to vertex g and its length in the weighted digraphs in Figures 10.42–10.43.

Figure 10.41

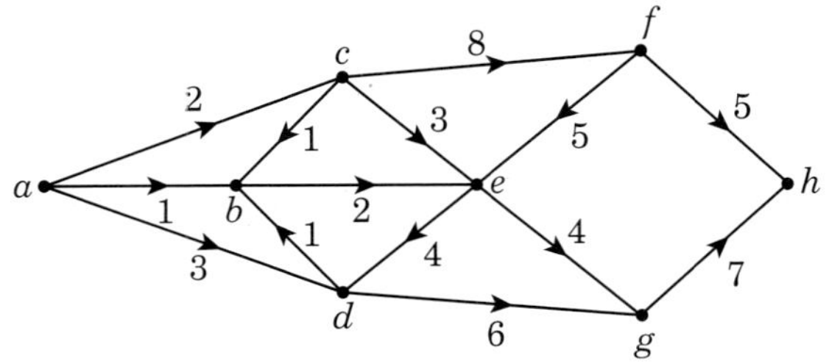

Figure 10.42

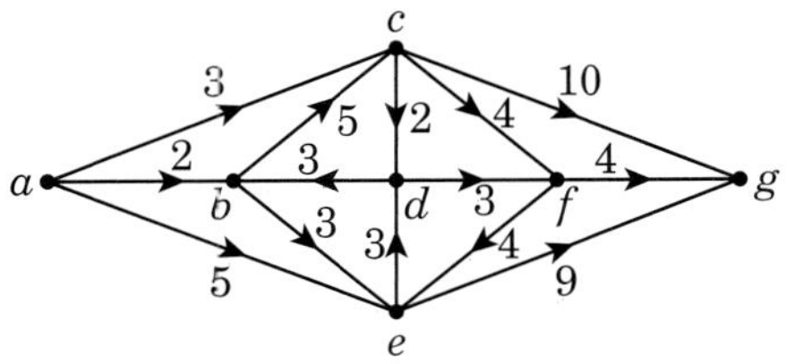

Figure 10.43

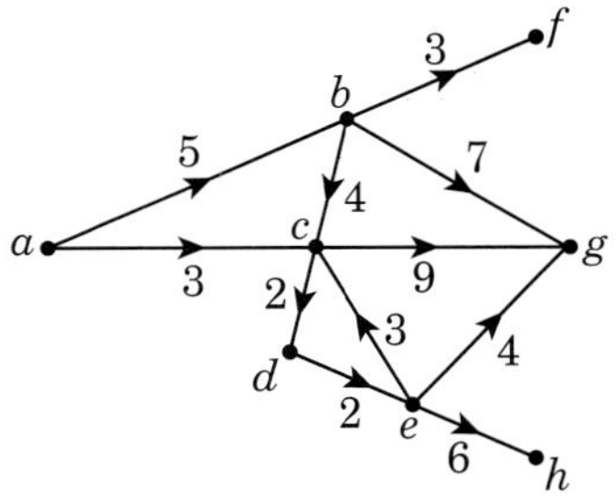

Supplementary Exercises

1. Find the sum of the indegrees and outdegrees of a vertex in a tournament with n vertices.
2. Find the maximum number of edges in a digraph with n vertices and no parallel edges.
3. Find the maximum number of edges in a loop-free digraph with n vertices and no parallel edges.
4. Identify the underlying graph of a tournament with n vertices.
5. How many edges are in a tournament?
6. How many different tournaments (simple digraphs) with n vertices are there?

***7.** Prove that every tournament has a Hamiltonian path, that is, a simple path containing every vertex.

8. Let V be a set of n elements. Let R be an irreflexive binary relation on V such that either aRb or bRa, but *not* both, for every $a, b \in V$. Characterize the digraph of R.

9. A **root** in a dag D is a vertex v such that every vertex ($\neq v$) is reachable from v. Which dags in Figure 10.13 are rooted?

Give a Hamiltonian path in each digraph, if it exists.

10. Figure 10.4

11. Figure 10.5

12. Figure 10.6

13. D_1 in Figure 10.13

A graph G is **strongly orientable** if a direction can be assigned to each edge in such a way that the resulting digraph is strongly connected. Such an assignment of directions is a **strong orientation** of G. The graphs in Figure 10.44 show the two-way streets in three towns. Determine if each is strongly orientable; when yes, give a strong orientation.

14.

15.

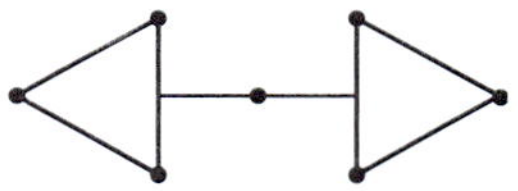

16.

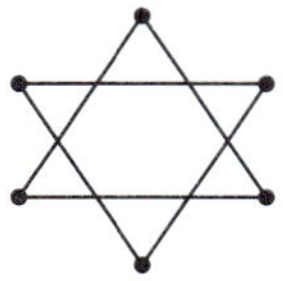

***17.** Using Exercises 14–16, predict a condition for a graph to be strongly orientable.

Computer Exercises

Let n be a positive integer ≤ 20. Write a program to perform each task.

1. Read in the adjacency matrix for a round-robin tournament of n players, 1 through n. Determine the champion and the second-best team of the tournament.

2. Read in the adjacency matrix of a digraph with n vertices, 1 through n.

- Is vertex j reachable from vertex i, where $1 \leq i, j \leq n$, $i \neq j$?
- Find the reachability matrix of the digraph.
- Is the digraph strongly connected?

3. Read in the weighted adjacency matrix of a weighted digraph with n vertices, 1 through n. Use Dijkstra's algorithm to compute the length of a shortest path from vertex 1 to vertex n and print the corresponding path.

Exploratory Writing Projects

Using library and Internet resources, write a team report on each of the following in your own words. Provide a well-documented bibliography.

1. Investigate the various applications of digraphs to the various disciplines.
2. Explain the *Program Evaluation and Review Technique* (PERT) for scheduling the various tasks in a large project.
3. Examine the teleprinter's problem.

Enrichment Readings

1. A. V. Aho *et al.*, *Data Structures and Algorithms*, Addison-Wesley, Reading, MA, 1983, pp. 198–229.
2. R. K. Ahuja *et al.*, *Network Flows*, Prentice-Hall, Englewood Cliffs, NJ, 1993.
3. J. A. Bondy and U. S. R. Murty, *Graph Theory with Applications*, Elsevier, New York, 1976.
4. T. H. Cormen, *Introduction to Algorithms*, McGraw-Hill, New York, 1990.
5. L. Ford and D. Fulkerson, *Flows and Networks*, Princeton University Press, Princeton, NJ, 1962.
6. R. P. Grimaldi, *Discrete and Combinatorial Mathematics: An Applied Introduction*, 4th ed., Addison-Wesley, Reading, MA, 1999, pp. 324–480.
7. E. Horowitz and S. Sahni, *Fundamentals of Data Structures*, Computer Science Press, Potomac, MD, 1976, pp. 301–334.
8. S. Sahni, *Concepts in Discrete Mathematics*, 2nd ed., Camelot, Fridley, MN, 1985, pp. 379–397.
9. R. J. Wilson and J. J. Watkins, *Graphs: An Introductory Approach*, Wiley, New York, 1990, pp. 80–111.

Chapter 11

Formal Languages and Finite-State Machines

Time as he grows old teaches many lessons.

—AESCHYLUS

The study of finite-state machines began with the neural networks investigations of Warren S. McCulloch and Walter Pitts in 1943. Today paradigms of finite-state constructs can be seen everywhere: turnstiles, traffic signal controllers, automated teller machines, automated telephone service, garage door openers, household appliances, and coin-operated machines such as vending machines and slot machines.

Finite-state machines significantly assist the study of formal languages; for example, a machine can be designed (or a program developed) that determines if an input string over the alphabet {a,b} contains *abba* as a substring. (The string *babaabbaba* does, while *abaaba* does not.) This type of machine produces no output, but instead tells whether or not the input string has a certain property. (Example 11.26 explores this.)

Some machines, however, produce output values. For instance, adding two binary numbers requires the input of two numbers, and yields their sum as the output (see Example 11.42). Such a machine, a finite-state automaton, is described in Section 11.4.

Since all finite-state automata must recognize particular languages, formal languages and types of grammars become important. This chapter explores formal languages, and how automata and formal languages are related, as well as other interesting questions such as:

- How do we determine if a string of characters contains a certain substring?
- How do we simulate an automatic teller machine?

- Can we develop a program that accepts two binary numbers, adds them bit by bit, and outputs their sum?
- What sort of languages are accepted by finite-state automata?

11.1 Formal Languages

We now continue our study of formal languages, begun in Section 2.1. The language of sets plays an important role in the study, as we saw in Chapter 2.

You may recall that an alphabet Σ is a finite set of symbols; and a word (or string) over Σ is a finite sequence of symbols from Σ.

How do we determine whether or not two words over Σ are equal? To this end, we make the following definition.

Equality of Words

Two words $x = x_1x_2\dots x_m$ and $y = y_1y_2\dots y_n$ over Σ are **equal**, denoted by $x = y$, if $m = n$ and $x_i = y_i$ for every i. Thus two words are equal if they contain the same number of symbols and the corresponding symbols are the same.

For example, if $01z = xy0$, then $x = 0$, $y = 1$, and $z = 0$. Also, $011 \neq 001$.

The length of a word w is the number of symbols in it. A word of length zero is the empty word, denoted by the lowercase Greek letter λ; it contains no symbols.

Again recall that Σ^* denotes the set of words over Σ. (Σ^* can be defined recursively. See Exercise 23.) A language over Σ is a subset of Σ^*; it may be finite or infinite.

EXAMPLE 11.1 Let $\Sigma = \{x, y, z, +, -, *, /, \uparrow, (,)\}$, where $*$ denotes multiplication and $\uparrow$ denotes exponentiation. Define a language L over Σ recursively as follows:

- $x, y, z \in L$.
- If u and v are in L, then so are $(+u)$, $(-u)$, $(u + v)$, $(u - v)$, $(u * v)$, (u/v), and $(u \uparrow v)$.

Then L consists of all fully and legally parenthesized algebraic expressions in x, y, and z. For instance, $((((x * (y \uparrow z)) - (y * z)) + x) \uparrow z)$ is a fully parenthesized and well-formed algebraic expression. Note that Σ^* contains nonsensical expressions such as $) + x\,(\,/\,y^* \uparrow$ also. ■

EXAMPLE 11.2 (optional) Let $\Sigma = \{\_, \$, a, \dots, z, 0, \dots, 9, \}$. Define the language L of legal identifiers in Java recursively.

SOLUTION:

An identifier in Java begins with a letter, underscore (_), or the dollar sign ($), followed by any number of alphanumeric characters (letters or digits). (See the syntax diagram in Figure 11.1.) A letter, an underscore, or $ by itself is a valid identifier. It can also be followed by a letter or a digit; that is, if $x \in L$ and $y \in \Sigma$, then $xy \in L$. Thus the language L can be defined recursively as follows:

- _ (underscore), $, and every letter of the English alphabet are in L.
- If $x \in L$ and $y \in \Sigma$, then $xy \in L$.

Figure 11.1

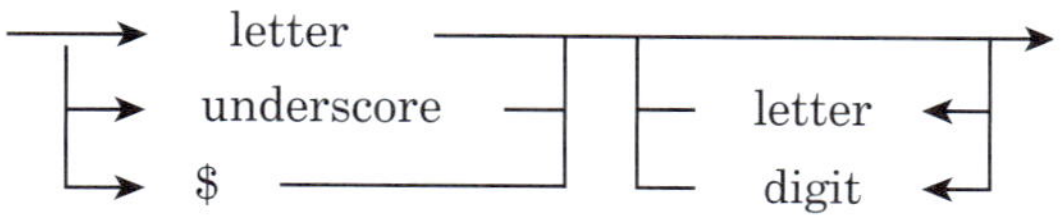

■

EXAMPLE 11.3 (optional) The alphabet Σ for Java on a computer system that uses the ASCII character set consists of the blank character, the upper- and lower-case letters, digits, arithmetic and relational operators, special characters, and control characters. So Java is a subset of Σ^*, consisting of all words over Σ that are recognizable by a Java compiler. ■

Since both $\emptyset$ and $\{\lambda\}$ are subsets of Σ^*, both are languages by definition. The language $\emptyset$ is the **empty language**. The language $\{\lambda\}$ is denoted by the upper case Greek letter Λ. We emphasize that $\emptyset \neq \Lambda$, since $|\emptyset| = 0$, whereas $|\Lambda| = 1$. However, if $\Sigma = \emptyset$, $\Sigma^* = \Lambda$. Why?

> Suppose an alphabet Σ contains at least one element a. Then $L = \{\text{a, aa, aaa}, \dots\}$ is an infinite language over Σ. Since $L \subseteq \Sigma^*$, Σ^* is also infinite. Thus, if $\Sigma \neq \emptyset$, Σ^* contains infinitely many words, each of finite length (see Exercise 29).

Let z be the concatenation of the words x and y; that is, $z = xy$. Then x is a **prefix** of z and y a **suffix** of z. For instance, consider the word $z =$ *readability* over the English alphabet; $x =$ read is a prefix of z and $y =$ ability is a suffix of z. Since $x = \lambda x = x\lambda$, every word is a prefix and a suffix of itself; further, λ is both a prefix and a suffix of every word.

The operations of union and intersection can be applied to languages also; after all, languages are also sets. To this end, we extend the definition of concatenation of strings to languages.

Concatenation of Languages

Let A and B be any two languages over Σ. The **concatenation** of A and B, denoted by AB, is the set of all words ab with $a \in A$ and $b \in B$. That is, $AB = \{ab \mid a \in A \wedge b \in B\}$.

The next two examples illustrate this definition.

EXAMPLE 11.4 Let $\Sigma = \{0, 1\}, A = \{0, 01\}$, and $B = \{\lambda, 1, 110\}$. Find the concatenations AB and BA.

SOLUTION:

- AB consists of strings of the form ab with $a \in$ A and $b \in$ B. So

$$\begin{aligned} AB &= \{0\lambda, 01, 0110, 01\lambda, 011, 01110\} \\ &= \{0, 01, 0110, 01, 011, 01110\} \\ &= \{0, 01, 011, 0110, 01110\} \end{aligned}$$

- $$\begin{aligned} BA &= \{ba \mid b \in B \wedge a \in A\} \\ &= \{\lambda 0, \lambda 01, 10, 101, 1100, 11001\} \\ &= \{0, 01, 10, 101, 1100, 11001\} \end{aligned}$$ ■

Tree diagrams are useful in finding the various strings in the concatenation of two finite languages. Figure 11.2, for example, shows the different ways of obtaining the elements in AB for the languages in Example 11.4.

Figure 11.2

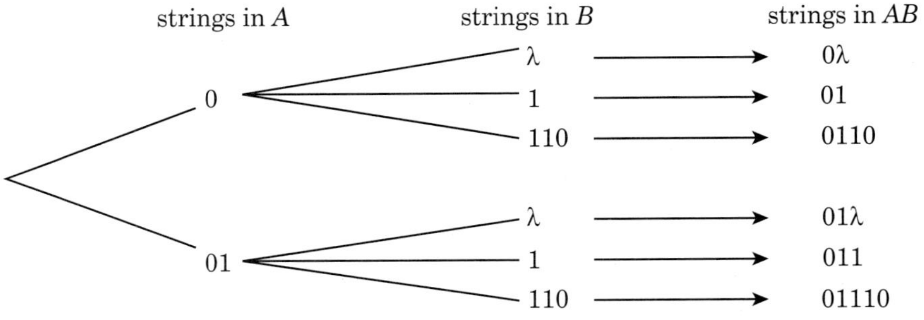

> Two interesting points arise from this example:
>
> (1) $AB \neq BA$.
>
> (2) Further, $|AB| = 5 \leq 6 = 2 \cdot 3 = |A| \cdot |B|$; whereas $|BA| = 6 = 3 \cdot 2 = |B| \cdot |A|$. This is so since the word 01 in AB can be obtained in two ways. Therefore, all we can say in general is, if A and B are finite languages, then $|AB| \leq |A| \cdot |B|$.

EXAMPLE 11.5 (optional) In the programming language QUICKBASIC, a numeric variable name must begin with a letter followed by either a period or an alphanumeric character. (QUICKBASIC does not distinguish between upper and lowercases in variable names.) Let $A = \{a, b, \ldots, z\}$ and $B = \{a, \ldots, z, 0, \ldots, 9, .\}$. The concatenation AB gives all numeric variable names containing exactly two characters, namely:

$$aa,\ ab,\ \ldots,\ a0,\ \ldots,\ a9,\ a.$$
$$ba,\ bb,\ \ldots,\ b0,\ \ldots,\ b9,\ b.$$
$$\vdots$$
$$za,\ zb,\ \ldots,\ z0,\ \ldots,\ z9,\ z.$$

■

EXAMPLE 11.6 Let A be a language over Σ. Identify the languages $A\emptyset$ and $A\Lambda$.

SOLUTION:

- $A\emptyset = \{ab \mid a \in A \wedge b \in \emptyset\}$. Since $\emptyset$ contains no elements, no concatenations ab can be performed; therefore, $A\emptyset = \emptyset$. (Similarly, $\emptyset A = \emptyset$.)
- $A\Lambda = A\{\lambda\} = \{a\lambda \mid a \in A\} = \{a \mid a \in A\} = A$. (Similarly, $\Lambda A = A$.) ■

We are now ready to study some properties of the concatenation operation on languages.

THEOREM 11.1 Let A, B, C, and D be any languages over an alphabet Σ. Then:

(1) $A\emptyset = \emptyset = \emptyset A$ (2) $A\Lambda = A = \Lambda A$ (3) $A(BC) = (AB)C$

(4) $A(B \cup C) = AB \cup AC$ (5) $(B \cup C)A = BA \cup CA$

(6) $A(B \cap C) \subseteq AB \cap AC$ (7) $(B \cap C)A \subseteq BA \cap CA$

(8) If $A \subseteq B$ and $C \subseteq D$, then $AC \subseteq BD$.

PROOF:

We already proved parts 1 and 2 in Example 11.6. We now shall prove parts 4 and 6, and leave the other parts as exercises.

(4) *To prove that* $A(B \cup C) = AB \cup AC$:

We need to show that (a) $A(B \cup C) \subseteq AB \cup AC$ and (b) $AB \cup AC \subseteq A(B \cup C)$.

- *To prove that* $A(B \cup C) \subseteq AB \cup AC$:
  Let $x \in A(B \cup C)$. Then x is of the form yz, where $y \in A$ and $z \in B \cup C$. If $z \in B$, then $yz \in AB$ and hence $yz \in AB \cup AC$. If $z \in C$, then $yz \in AC$ and therefore $yz \in AB \cup AC$. Thus, in both cases $x = yz \in AB \cup AC$. Consequently, $A(B \cup C) \subseteq AB \cup AC$.

- *To prove that* $AB \cup AC \subseteq A(B \cup C)$:
  Let $x \in AB \cup AC$. Suppose $x \in AB$. Then $x = ab$ for some $a \in A$ and $b \in B$. Since $b \in B$, b also belongs to $B \cup C$. So $x = ab \in A(B \cup C)$. Similarly, if $x \in AC$, then also $x \in A(B \cup C)$. Thus in both cases $x \in A(B \cup C)$. Consequently, $AB \cup AC \subseteq A(B \cup C)$.

Therefore, by parts (a) and (b), $A(B \cup C) = AB \cup AC$.

(6) *To prove that* $A(B \cap C) \subseteq AB \cap AC$:
Let $x \in A(B \cap C)$. Then $x = yz$ for some element $y \in A$ and $z \in B \cap C$. Since $z \in B \cap C$, $z \in B$ and $z \in C$. So yz belongs to both AB and AC, and hence $yz \in AB \cap AC$; in other words, $x \in AB \cap AC$. Thus $A(B \cap C) \subseteq AB \cap AC$. ■

The next example verifies parts (4) and (6) of this theorem.

EXAMPLE 11.7 Let $\Sigma = \{a, b, c\}$, $A = \{a, ab\}$, $B = \{b, ab\}$, and $C = \{\lambda, bc\}$. Verify that (1) $A(B \cup C) = AB \cup AC$ and (2) $A(B \cap C) \subseteq AB \cap AC$.

SOLUTION:

$$AB = \{ab, aab, abb, abab\}$$

$$AC = \{a\lambda, abc, ab\lambda, abbc\} = \{a, ab, abc, abbc\}$$

$$AB \cup AC = \{a, ab, aab, abb, abc, abab, abbc\}$$

$$AB \cap AC = \{ab, abb\}$$

(1) $$B \cup C = \{\lambda, b, ab, bc\}$$

Then $$A(B \cup C) = \{a\lambda, ab, aab, abc, ab\lambda, abb, abab, abbc\}$$
$$= \{a, ab, aab, abb, abc, abab, abbc\}$$
$$= AB \cup AC$$

(2) Since $B \cap C = \emptyset$, $A(B \cap C) = \emptyset$ and hence $A(B \cap C) \subseteq AB \cap AC$. ■

If the languages A and B are the same, then AB is often denoted by A^2. Thus A^2 consists of words obtained by concatenating each word in A with every word in A: $A^2 = \{xy \mid x, y \in A\}$. More generally, let $n \in \mathbb{N}$. Then A^n consists of all words obtained by $n - 1$ concatenations of words in A. In particular, Σ^n denotes the set of words obtained by $n - 1$ concatenations of symbols in Σ, that is, words of length n.

EXAMPLE 11.8 Let $\Sigma = \{a, b, c\}$, $A = \{a, ab, bc\}$, and $B = \{a, bc\}$. Find Σ^2, A^2, and B^3.

SOLUTION:

- $\Sigma^2 = \{xy \mid x, y \in \Sigma\} = \{aa, ab, ac, ba, bb, bc, ca, cb, cc\}$
- $A^2 = \{xy \mid x, y \in A\} = \{aa, aab, abc, aba, abab, abbc, bca, bcab, bcbc\}$
- $B^2 = \{aa, abc, bca, bcbc\}$

So $B^3 = \{aaa, aabc, abca, abcbc, bcaa, bcabc, bcbca, bcbcbc\}$ ■

Stephen Cole Kleene *(1909–1994) was born in Hartford, Connecticut. His father was an economics professor and his mother, a poet. In 1930, he graduated from Amherst College and 4 years later received his Ph.D. in mathematics from Princeton.*

After teaching for 6 years at the University of Wisconsin, Madison, he joined the faculty at Amherst College for a year. For the next 4 years he served in the U.S. Naval Reserve. In 1946, he returned to the Madison campus and in 1964 became the Cyrus C. MacDuffee Professor of Mathematics and Computer Science. He served as Chairman of the Department of Mathematics, Acting Director of the Mathematics Research Center, and Dean of the College of Letters and Science.

Kleene was awarded an honorary Doctor of Science by Amherst College in 1970, the Steele Prize by the American Mathematical Society in 1983, and the National Medal of Science in 1990.

Kleene contributed significantly to the theory of recursive functions and the theory of automata.

Note: It follows by part 8 of Theorem 11.1 that if $A \subseteq B$, then $A^2 \subseteq B^2$ (Why?). More generally, it can be shown by induction that if $A \subseteq B$, then $A^n \subseteq B^n$ for every $n \in \mathbb{N}$.

Finally, from any language A over Σ, we can construct a new language A^* using the various powers of A. First we define $A^0 = \Lambda$.

Kleene Closure

Let A be a language over an alphabet Σ. Then $A^* = \bigcup_{n=0}^{\infty} A^n$ is the **Kleene closure** of A, in honor of the American logician Stephen Kleene. A^* consists of strings obtained by an arbitrary number of concatenations of words from A. $*$ is the **Kleene operator**.

The following example illustrates this definition.

EXAMPLE 11.9 Let $A = \{0\}$, $B = \{11\}$, $C = \{000\}$, and $\Sigma = \{0, 1\}$. Find their Kleene closures.

SOLUTION:

- Since $A = \{0\}$, $A^n = \{0^n\}$. So $A^* = \bigcup_{n=0}^{\infty} A^n = \{0^n \mid n \geq 0\}$. In other words, A^* consists of strings of zero or more 0's.
- Since $B = \{11\} = \{1^2\}$, $B^2 = BB = \{1111\} = \{1^4\}$. So $B^3 = BB^2 = \{1^2\}\{1^4\} = \{1^6\}$. Thus, in general, $B^n = \{1^{2n}\}$. Thus $B^* = \bigcup_{n=0}^{\infty} B^n = \{1^{2n} \mid n \geq 0\}$. It consists of words of 1's of even length.
- Since $C = \{0^3\}$, $C^n = \{0^{3n}\}$. So $C^* = \{0^{3n} \mid n \geq 0\}$, the set of strings of 0's whose lengths are divisible by 3.

- The Kleene closure is the set of all possible words over Σ, namely, Σ^*. (This explains why we denoted it by Σ^* from the beginning of the section.) ■

We now turn to a few properties satisfied by the Kleene operator. We shall prove one of them and leave the others as exercises. Property 6 is a bit hard to prove, so we omit it; properties 4 and 5 require induction.

THEOREM 11.2 Let A and B be any languages over an alphabet Σ. Then:

(1) $\Lambda \subseteq A^*$ (2) $A \subseteq A^*$

(3) $A^*A^* = A^*$ (4) If A $\subseteq$ B, then $A^* \subseteq B^*$.

(5) $(A^*)^* = A^*$ (6) $(A \cup B)^* = (A^* \cup B^*)^* = (A^*B^*)^*$

PROOF:

(3) **To prove that $\mathbf{A^*A^* = A^*}$:**

- *To prove that* $A^* \subseteq A^*A^*$: Since $\Lambda \subseteq A^*$, $A^*\Lambda \subseteq A^*A^*$ by Theorem 11.1. But $A^*\Lambda = A^*$ by Theorem 11.1. So

$$A^* \subseteq A^*A^* \tag{11.1}$$

- *To prove that* $A^*A^* \subseteq A^*$: Let $x \in A^*A^*$. Then $x = yz$ with $y, z \in A^*$. Since $y, z \in A^*$, $y \in A^m$ and $z \in A^n$ where $m, n \in W$. So $yz \in A^mA^n = A^{m+n}$. But $A^{m+n} \subseteq A^*$, so $x = yz \in A^*$. Thus

$$A^*A^* \subseteq A^* \tag{11.2}$$

Thus, by set inclusions (11.1) and (11.2), $A^*A^* = A^*$. ■

An interesting observation: For any language A, $A \subseteq A^*$. That is, when we apply the Kleene operator * on A, the resulting language A^* contains A. However, if we apply * to A^*, we find that $(A^*)^* = A^*$; so we do not get a new language. This explains why A^* is called the Kleene closure of A.

We conclude this section with an example involving both concatenation and the Kleene operators.

EXAMPLE 11.10 Identify each language over $\Sigma = \{a, b\}$.

(1) $\{a, b\}^*\{b\}$ (2) $\{a\}\{a, b\}^*$ (3) $\{a\}\{a, b\}^*\{b\}$ (4) $\{a, b\}^*\{b\}^*$

SOLUTION:

(1) $\{a, b\}^*$ consists of all possible words over Σ including λ, whereas $\{b\}$ contains just one word, namely, b. Therefore, the language $\{a,b\}^*\{b\}$ consists of words over Σ that have b as a suffix.

(2) Similarly, $\{a\}\{a, b\}^*$ consists of words that have a as a prefix.

(3) {a}{a, b}*{b} consists of words that begin with a and end in b.
(4) Every element in {b}* consists of a finite number of b's. Therefore, {a, b}*{b}* consists of strings followed by a finite number of b's. Notice that this is different from {a, b}*{b} (Why?). ■

Exercises 11.1

In Exercises 1–4, a language L over $\Sigma = \{a, b\}$ is given. Find five words in each language.

1. $L = \{x \in \Sigma^* \mid x \text{ begins with and ends in } b.\}$

2. $L = \{x \in \Sigma^* \mid x \text{ contains exactly one } b.\}$

3. L is defined recursively as follows: (i) $\lambda \in L$ (ii) $x \in L \to xbb \in L$

4. L is defined recursively as follows: (i) $\lambda \in L$ (ii) $x \in L \to axb \in L$

Define each language L over the given alphabet recursively.

5. The language L of all palindromes over $\Sigma = \{a, b\}$. (A **palindrome** over Σ is a word that reads the same both forwards and backwards. For instance, *abba* is a palindrome.)

6. $L = \{a^n b^n \mid n \in N\}$, $\Sigma = \{a, b\}$

7. $L = \{0, 00, 10, 100, 110, 0000, 1010, \ldots\}$, $\Sigma = \{0, 1\}$

8. L = set of binary representations of positive integers, $\Sigma = \{0,1\}$

9. $L = \{1, 11, 111, 1111, 11111, \ldots\}$, $\Sigma = \{0, 1\}$

10. $L = \{x \in \Sigma^* \mid x = b^n a b^n,\ n \geq 0\}$, $\Sigma = \{a, b\}$

11. L = set of words over $\Sigma = \{0, 1\}$ with prefix 00

12. L = set of words over $\Sigma = \{0, 1\}$ with suffix 11

Mark each as true or false.

13. Every language over an alphabet is infinite.

14. If $\Sigma = \emptyset$, then $\Sigma^* = \emptyset$.

∘ **15.** C++ is a finite language.

16. Every language is a set.

Using Example 11.1, determine if each is a well-formed and fully parenthesized arithmetic expression.

17. $(((x + y)/(((x - y) * z) \uparrow z))$

18. $(x \uparrow ((y - x) \uparrow (-z)))$

19. $(y + (z \uparrow (+x))/(-x))$

20. $((x - (y \uparrow z)) * (x + (y \uparrow (+z))))$

21. Define the set of words S over an alphabet Σ recursively. (*Hint*: Use concatenation.)

22. Define the language L of all binary representations of nonnegative integers recursively.

23. Let Σ be an alphabet. Define Σ^* recursively. (*Hint:* Use concatenation.)

○ **24.** Define recursively the set S of integers acceptable in Java.

Arrange the binary words of each length in increasing order.

25. Length two.

26. Length three.

A **ternary word** is a word over the alphabet $\{0, 1, 2\}$. Arrange the ternary words of each length in increasing order.

27. Length one

28. Length two

***29.** Let Σ be a nonempty alphabet. Prove that Σ^* is infinite. (*Hint*: Assume Σ^* is finite. Since $\Sigma \neq \emptyset$, it contains an element a. Let $x \in \Sigma^*$ with largest length. Now consider xa.)

Let $A = \{a, bc\}$ and $B = \{\lambda, ab, bc\}$. Find each concatenation.

30. AB

31. BA

32. A^2

33. A^3

Let $A = \{a, ab\}$, $B = \{a, b, ab\}$, $C = \{c\}$, *and* $D = \{c, bc\}$. Verify each.

34. $A\Lambda = A$

35. $\Lambda A = A$

36. $A(B \cup C) = AB \cup AC$

37. $(B \cup C)A = BA \cup CA$

38. $A(B \cap C) = AB \cap AC$

39. $(B \cap C)A = BA \cap CA$

40. If $A \subseteq B$ and $C \subseteq D$, then $AC \subseteq BD$.

Mark each as true or false, where A and B are arbitrary finite languages.

41. $\Lambda = \emptyset$

42. $A\emptyset = \emptyset$

43. $A\emptyset = \emptyset A$

44. $A\Lambda = \Lambda$

45. $A\Lambda = \Lambda A$

46. $| A \times B | = | B \times A |$

47. $|AB| = | BA |$

Find three words belonging to each language over $\sigma = \{0, 1\}$.

48. $\{0\}^*$

49. $\{0\}\{1\}^*$

50. $\{0\}^*\{1\}$

51. $\{0\}\{11\}^*\{1\}$

52. $\{0\}^*\{1\}^*$

53. $\{01\}^*$

54. $\{0\}\{0, 1\}^*\{1\}$

55. $\{0\}^*\{1\}^*\{0\}^*$

Prove each, where A, B, and C are arbitrary languages over Σ and $x \in \Sigma$.

56. $\|x^n\| = n\|x\|$ for every $n \geq 0$.

57. If $A \subseteq B$, then $A^n \subseteq B^n$ for every $n \geq 0$.

58. If $A \subseteq B$, then $A^* \subseteq B^*$.

59. $(A^*)^* = A^*$

60. $\emptyset A = \emptyset$

61. $\Lambda A = A$

62. $\Lambda \subseteq A^*$

63. $A \subseteq A^*$

64. $A(B \cap C) \subseteq AB \cap AC$

65. $(B \cup C)A = BA \cup CA$

66. $(B \cap C)A \subseteq BA \cap CA$

67. $(A^*B^*)^* = (B^*A^*)^*$

68. $(A^* \cup B^*)^* = (A \cup B)^*$

11.2 Grammars

Words in a natural language such as English or French can be combined in several ways. Some combinations form valid sentences, while others do not. The **grammar** of a language is a set of rules that determines whether or not a sentence is considered valid. For instance, *The milk drinks child quickly*, although meaningless, is a perfectly legal sentence.

The sentences in a language may be nonsensical, but must obey the grammar. Our discussion deals with only the **syntax** of sentences (the way words are combined), and not with the **semantics** of sentences (meaning). Although listing the rules that govern a natural language such as English is extremely complex, specifying the rules for subsets of English is certainly possible.

The next example introduces such a language.

EXAMPLE 11.11 The sentence *The child drinks milk quickly*, has two parts: a subject, *The child*, and a predicate *drinks milk quickly*. The subject consists of the definite article *The* and the noun *child*. The predicate, on the other hand, consists of the verb *drinks* and the object phrase *milk quickly*; the object phrase in turn has the object *milk* and the adverb *quickly*. This structure of the sentence can appear as a sequence of trees (Figures 11.3–11.7), with the **derivation tree** of the sentence in Figure 11.7.

Figure 11.3

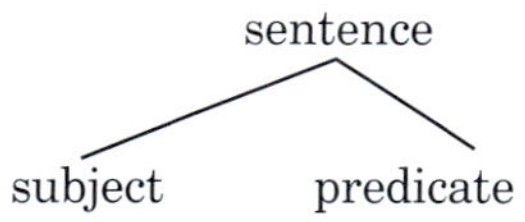

Figure 11.4

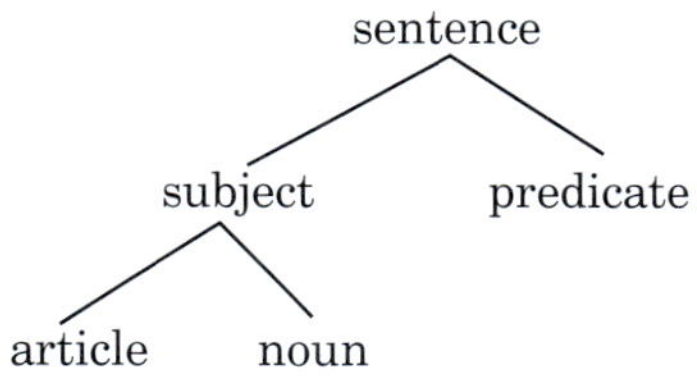

The derivation tree exhibits certain characteristics:

- Each leaf represents a word, a **terminal symbol**. The set of terminal symbols is $T = \{\text{the, child, drinks, milk, quickly}\}$.

Figure 11.5

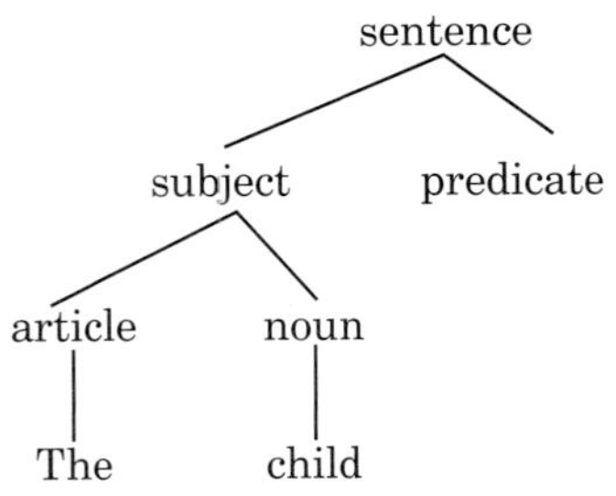

Figure 11.6

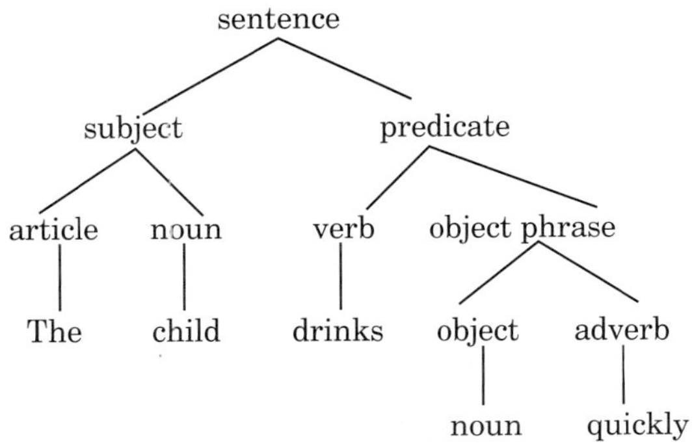

Figure 11.7

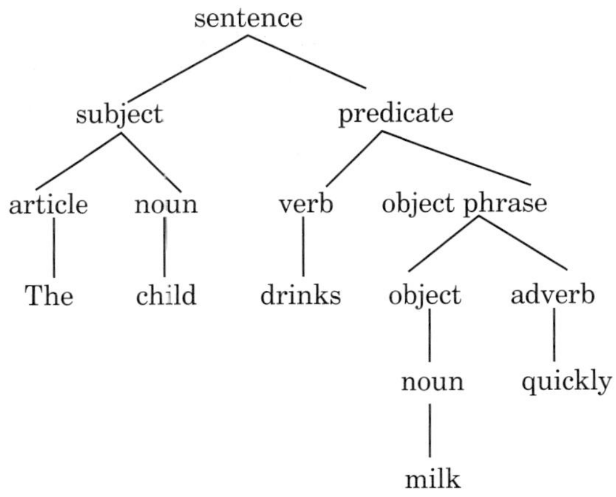

- Each internal vertex represents a grammatical class, a **nonterminal**. The set of nonterminals is $N = \{\text{sentence, subject, predicate, article, noun, object phrase, object, verb, adverb}\}$. A nonterminal symbol is enclosed within **angle brackets**, $\langle$ and $\rangle$. For instance, the nonterminal "subject" is denoted by $\langle\text{subject}\rangle$.
- The root of the tree represents the nonterminal symbol $\langle\text{sentence}\rangle$ called the **start symbol**, denoted by σ.

Certain rules can generate the above sentence. Every rule, called a **production rule** or a **substitution rule**, is of the form $w \to w'$ where $w \in N$

and w' may be a terminal symbol, a nonterminal symbol, or a combination of both.

The production rules of the above sentence are:

$$\begin{aligned}
\langle\text{sentence}\rangle &\to \langle\text{subject}\rangle\langle\text{predicate}\rangle\\
\langle\text{subject}\rangle &\to \langle\text{article}\rangle\langle\text{noun}\rangle\\
\langle\text{article}\rangle &\to \textit{the}\\
\langle\text{noun}\rangle &\to \textit{child}\\
\langle\text{noun}\rangle &\to \textit{milk}\\
\langle\text{predicate}\rangle &\to \langle\text{verb}\rangle\langle\text{object phrase}\rangle\\
\langle\text{verb}\rangle &\to \textit{drinks}\\
\langle\text{object phrase}\rangle &\to \langle\text{object}\rangle\langle\text{adverb}\rangle\\
\langle\text{object}\rangle &\to \langle\text{noun}\rangle\\
\langle\text{adverb}\rangle &\to \textit{quickly}
\end{aligned}$$

■

The production rules specify the arrangement of words in a sentence: the **syntax** of the language. They produce syntactically correct sentences (which can be meaningless). For instance, the sentence, *The milk drinks child quickly* makes no sense but is syntactically valid. Figure 11.8 shows the derivation tree of this sentence.

Figure 11.8

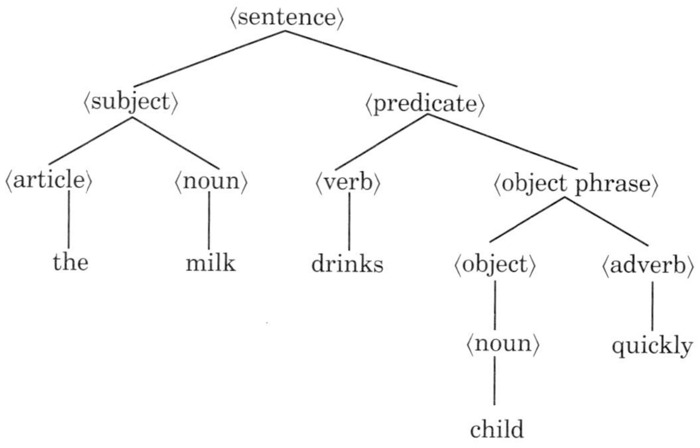

Determining whether a program is syntactically correct is of the utmost importance in computer science. Before executing a program, the compiler checks the syntax of each sentence (or expression) by constructing derivation trees. (This process is **parsing**, and the corresponding derivation tree is a **parse tree**.)

We now turn to present the definition of a phrase-structure grammar.

Phrase-Structure Grammar

A **phrase-structure grammar** (or simply a **grammar**) G bears four features:

- A finite set N of **nonterminal symbols**;
- A finite set T of **terminal symbols**, where $N \cap T = \emptyset$;
- A finite subset P of $[(N \cup T)^* - T^*] \times (N \cup T)^*$; each element of P is called a **production**;
- A **start symbol** σ belonging to N;

The grammar G is denoted by $G = (N, T, P, \sigma)$.

These features meet certain requirements:

- The start symbol σ is nonterminal.
- No symbol can be both terminal and nonterminal.
- Every production has at least one nonterminal symbol on its LHS, because $P \subseteq [(N \cup T)^* - T^*] \times (N \cup T)^*$. Also, P is a binary relation from $(N \cup T)^* - T^*$ to $(N \cup T)^*$.
- If $(w, w') \in P$, we then write $w \to w'$; since $w \in (N \cup T)^* - T^*$, w contains at least one nonterminal symbol; but $w' \in (N \cup T)^*$; so it may contain terminal symbols, nonterminals, or both.

Grammars not only produce natural languages, but also formal ones, as the next two examples demonstrate.

EXAMPLE 11.12 Let $N = \{A, B, \sigma\}$, $T = \{a, b\}$, and $P = \{\sigma \to aA,\ A \to bA,\ A \to a\}$. Then $G = (N, T, P, \sigma)$ is a grammar. Notice that the production $A \to bA$ is recursive. ■

EXAMPLE 11.13 Let $N = \{A, \sigma\}$, $T = \{a, b\}$, and $P = \{\sigma \to a\sigma,\ \sigma \to Aa,\ A \to b\}$. Then $G = (N, T, P, \sigma)$ is a grammar. Again notice that the production $\sigma \to a\sigma$ is recursive. ■

Next we define the language generated by a grammar.

Derivation and Language

Let $G = (N, T, P, \sigma)$ be a grammar. If $w = x\alpha y$ and $w' = x\beta y$ are any two words in $(N \cup T)^*$, and if there exists a production $\alpha \to \beta$, then the word w' is said to be **directly derivable** from w; we then write $w \Longrightarrow w'$. If there is a finite sequence of words $w_0, w_1, \ldots, w_n$ in $(N \cup T)^*$ such that

$w_0 \Longrightarrow w_1, w_1 \Longrightarrow w_2, \ldots, w_{n-1} \Longrightarrow w_n$, then w_n is **derivable** from w_0. The finite sequence of steps, $w_0 \Longrightarrow w_1 \Longrightarrow \cdots \Longrightarrow w_n$, is a **derivation** of w_n from w_0.

The set of words in T^* derivable from σ by G is the **language generated** by G, denoted by $L(G)$.

The next two examples illustrate these definitions.

EXAMPLE 11.14 Identify the language $L(G)$ generated by the grammar in Example 11.12.

SOLUTION:
Since the grammar contains exactly one production involving σ, namely, $\sigma \to aA$, start with it to find every word in the language. Now select the next production: $A \to bA$ or $A \to a$. The production $A \to a$ produces exactly one word, a. $A \to bA$ chosen n times, produces $\sigma \Longrightarrow aA \Longrightarrow abA \Longrightarrow ab^2A \Longrightarrow \cdots \Longrightarrow ab^nA$. Now $A \to a$ yields the word ab^na and, when $n = 0$, this yields $a\lambda a = aa$. (*Note*: $b^0 = \lambda$, the null word.) Every word derivable from σ fits the form ab^na, where $n \geq 0$. In other words, $L(G) = \{ab^na \mid n \geq 0\}$. ■

Example 11.14 illustrates that a grammar G can determine if it generates a string in the language $L(G)$. With some difficulty, the language could be described. Again with some difficulty, and a lot of patience and practice, a grammar G that generates a given language can be found, as Example 11.15 demonstrates.

EXAMPLE 11.15 Define a grammar $G = (N, T, P, \sigma)$ that generates the language $L = \{a^nb^n \mid n \geq 1\}$.

SOLUTION:
Since every word in L must contain the same number of a's and b's, G must contain a production of the form $\sigma \to$ aAb. Consequently, to produce a new word from aAb containing the same number of a's and b's requires another production A $\to$ aAb. From these two productions, we can derive all strings of the form $\mathrm{a}^n\mathrm{Ab}^n$ (Verify this.). All that remains to be done to define the grammar is the production A $\to \lambda$ to terminate the recursive procedure. Thus, $N = \{\sigma, \mathrm{A}\}$, $T = \{\mathrm{a}, \mathrm{b}, \lambda\}$, and $P = \{\sigma \to \mathrm{aAb}, \mathrm{A} \to \mathrm{aAb}, \mathrm{A} \to \lambda\}$. ■

The first two production rules in this example look quite similar, except for the start symbol, and can be combined into a single production, $\sigma \to \mathrm{a}\sigma\mathrm{b}$. The production rules $\sigma \to$ aAb and A $\to \lambda$ can yield the word ab, so the third production is $\sigma \to$ ab. Thus $P' = \{\sigma \to \mathrm{a}\sigma\mathrm{b}, \sigma \to \mathrm{ab}\}$ is an additional production set that yields the same language. In other words, the grammars $G = (N,\ T,\ P, \sigma)$ and $G' = \{N', T', P', \sigma)$ generate the same language L, where $N' = \{\sigma\}$ and $T' = \{\mathrm{a}, \mathrm{b}\}$. Thus $L(G) = L(G')$, so the grammars G and G' are **equivalent**. Our conclusion: *The grammar that generates a language need not be unique.*

John W. Backus *(1924–) was born in Philadelphia. He received his B.S. and M.S. in mathematics from Columbia University. After joining IBM in 1950, he became instrumental in the development of FORTRAN and ALGOL (**ALGO**rithmic **L**anguage). He received the W. W. McDowell Award from The* Institute of Electrical and Electronics Engineers *(IEEE) in 1967, the National Medal of Science in 1975, the A. M. Turing Award from the* Association for Computing Machinery *in 1977, the Harold Pender Award from the University of Pennsylvania in 1983, and an honorary doctorate from York University, England, in 1985.*

Peter Naur *(1928–), a computer scientist and prolific writer, was born in Frederiksberg, Denmark. After receiving his M.A. in astronomy from Copenhagen University in 1949, he spent the next two years at Cambridge University, England, where he used the EDSAC, one of the earliest computers, to pursue astronomy. He received his Ph.D. in astronomy from Copenhagen in 1957.*

From 1953 to 1959, he consulted for the design of the first Danish computer, the DASK. Beginning around 1964, he became increasingly involved in datalogy *(a word he coined), the study of data and data processes. In 1963, Naur was given the Hagemanns Gold Medal and three years later the Rosenhjaer Prize.*

Backus-Normal Form

The most widely used notation for describing the syntax of programming languages is the **Backus-Normal Form** (**BNF**), developed by John Backus, who described ALGOL 60 with it. Peter Naur edited the ALGOL 60 report, which appeared in 1963, so the BNF notation is also called the **Backus–Naur Form**.

In BNF, the production symbol $\rightarrow$ is denoted by ::=; thus the production $w \rightarrow w'$ is written as $w ::= w'$. Production rules with the same LHS are combined by separating their RHS with vertical bars. For instance, the productions $w \rightarrow w_1, w \rightarrow w_2, \ldots, w \rightarrow w_n$ become $w ::= w_1 \mid w_2 \mid \ldots \mid w_n$. (You may read the vertical bar as *or*.) Nonterminal symbols have angle brackets around them.

EXAMPLE 11.16 Study the following production rules:

$$\begin{aligned}
\langle\text{sentence}\rangle &\to \langle\text{subject}\rangle\langle\text{predicate}\rangle\\
\langle\text{subject}\rangle &\to \langle\text{article}\rangle\langle\text{noun}\rangle\\
\langle\text{predicate}\rangle &\to \langle\text{verb}\rangle\langle\text{object}\rangle\\
\langle\text{object}\rangle &\to \langle\text{article}\rangle\langle\text{noun}\rangle\\
\langle\text{article}\rangle &\to a\\
\langle\text{article}\rangle &\to \textit{the}\\
\langle\text{noun}\rangle &\to \textit{hare}\\
\langle\text{noun}\rangle &\to \textit{tortoise}\\
\langle\text{noun}\rangle &\to \textit{race}\\
\langle\text{verb}\rangle &\to \textit{beats}\\
\langle\text{verb}\rangle &\to \textit{wins}
\end{aligned}$$

BNF shortens these rules:

$$\begin{aligned}
\langle\text{sentence}\rangle &::= \langle\text{subject}\rangle\langle\text{predicate}\rangle\\
\langle\text{subject}\rangle &::= \langle\text{article}\rangle\langle\text{noun}\rangle\\
\langle\text{predicate}\rangle &::= \langle\text{verb}\rangle\langle\text{object}\rangle\\
\langle\text{object}\rangle &::= \langle\text{article}\rangle\langle\text{noun}\rangle\\
\langle\text{article}\rangle &::= a \mid \textit{the}\\
\langle\text{noun}\rangle &::= \textit{hare} \mid \textit{tortoise} \mid \textit{race}\\
\langle\text{verb}\rangle &::= \textit{beats} \mid \textit{wins}
\end{aligned}$$

■

EXAMPLE 11.17 The grammar for the language of correctly nested parentheses contains one production:

$$\langle\text{nested parentheses}\rangle ::= \lambda \mid (\langle\text{nested parentheses}\rangle)$$

where λ denotes the null string. [Using this definition, you may verify that (()) and ((())) are valid nested parentheses, whereas (() and (())) are not.] ■

EXAMPLE 11.18 (optional) An integer is a string of digits preceded by an optional sign, + or −. Using BNF, it can be defined as follows:

$$\begin{aligned}
\langle\text{integer}\rangle &:: = \langle\text{signed integer}\rangle \mid \langle\text{unsigned integer}\rangle\\
\langle\text{signed integer}\rangle &:: = \langle\text{sign}\rangle \mid \langle\text{unsigned integer}\rangle\\
\langle\text{sign}\rangle &:: = + \mid -
\end{aligned}$$

$$\langle\text{unsigned integer}\rangle ::= \langle\text{digit}\rangle \mid \langle\text{digit}\rangle\langle\text{unsigned integer}\rangle$$
$$\langle\text{digit}\rangle ::= 0 \mid 1 \mid 2 \mid 3 \mid 4 \mid 5 \mid 6 \mid 7 \mid 8 \mid 9$$

For instance, 234, +234, and −234 are valid integers. Figure 11.9 shows the derivation tree for the integer +234.

Figure 11.9

Derivation tree for the integer +234.

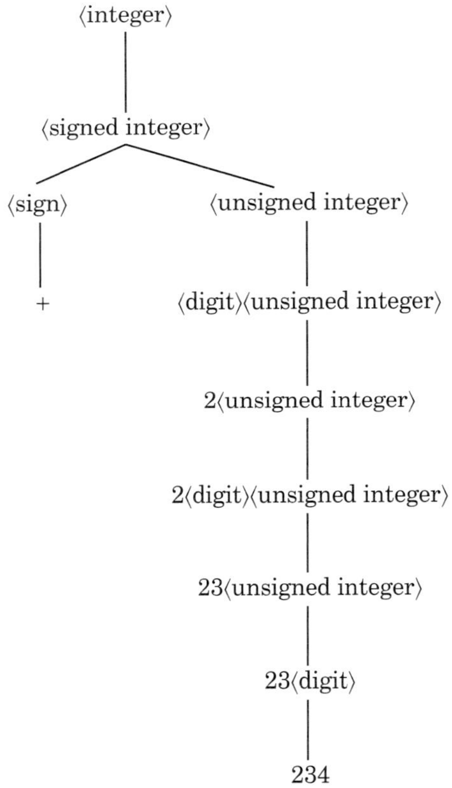

The grammar defined in this example is $G = (N, T, P, \sigma)$, where:

- $N = \{\langle\text{integer}\rangle, \langle\text{signed integer}\rangle, \langle\text{unsigned integer}\rangle, \langle\text{sign}\rangle, \langle\text{digit}\rangle\}$,
- $T = \{+, -, 0, 1, 2, 3, 4, 5, 6, 7, 8, 9\}$,
- The production rules are:

$$\langle\text{integer}\rangle \to \langle\text{signed integer}\rangle \mid \langle\text{unsigned integer}\rangle$$
$$\langle\text{signed integer}\rangle \to \langle\text{sign}\rangle \mid \langle\text{unsigned integer}\rangle$$
$$\langle\text{sign}\rangle \to + \mid -$$
$$\langle\text{unsigned integer}\rangle \to \langle\text{digit}\rangle \mid \langle\text{digit}\rangle\langle\text{unsigned integer}\rangle$$
$$\langle\text{digit}\rangle \to 0 \mid 1 \mid 2 \mid 3 \mid 4 \mid 5 \mid 6 \mid 7 \mid 8 \mid 9$$

- The start symbol σ is $\langle$integer$\rangle$.

Grammars are categorized by the productions that define them.

Context-Sensitive, Context-Free, and Regular Grammars

Let $G = (N, T, P, \sigma)$ be a grammar. Let $A, B \in N$ and α, α', $\beta \in (N \cup T)^*$. Notice that α, α', and β could be the null word.

- Any phrase-structure grammar is **type 0**.
- G is **context-sensitive** (or **type 1**) if every production is of the form $\alpha \mathrm{A} \alpha' \to \alpha \beta \alpha'$.
- G is **context-free** (or **type 2**) if every production is of the form $\mathrm{A} \to \alpha$.
- G is **regular** (or **type 3**) if every production is of the form $\mathrm{A} \to t$ or $\mathrm{A} \to t\mathrm{B}$, where $t \in T$.

In a context-sensitive grammar, β can replace A in the word $\alpha \mathrm{A} \alpha'$ only when A lies between α and α'. In a context-free grammar, the LHS of every production is a single nonterminal symbol A, which α can replace. In a regular grammar, the LHS of every production consists of a single nonterminal symbol A and the RHS consists of a terminal symbol t or a terminal symbol t followed by a nonterminal symbol B; t or tB can always replace A. (In tB, the nonterminal must be on the RHS of the terminal symbol t.)

A regular grammar is also context-free and a context-free grammar is also context-sensitive. The Venn diagram in Figure 11.10 shows the **Chomsky hierarchy** of the various grammars, named in honor of Noam Chomsky, who developed the theory of formal languages.

Figure 11.10

Chomsky hierarchy of grammars.

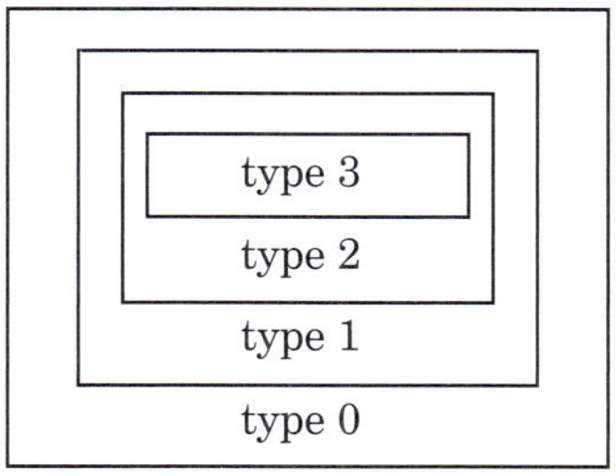

Context-Sensitive, Context-Free, and Regular Languages

A language $L(G)$ is **context-sensitive, context-free,** or **regular** if the grammar G is context-sensitive, context-free, and regular, respectively.

The next five examples clarify these definitions.

(Avram) Noam Chomsky *(1928–), a linguist, writer, and political activist, was born in Philadelphia, as the son of a Hebrew scholar. At 10 he proofread the manuscript of his father's edition of a 13th century Hebrew grammar. "This backdoor introduction to 'historical linguistics' had considerable impact on his future" (The New York Times Magazine). The young Chomsky, however, was more passionate about politics than about grammar.*

On graduating from Central High School in Philadelphia in 1945, Chomsky entered the University of Pennsylvania and received his B.A. in 1949 and M.A. 2 years later.

Chomsky received his Ph.D. in linguistics from the University of Pennsylvania in 1955 and joined the faculty at the Massachusetts Institute of Technology.

His first book, Syntactic Structures *(1957), developed from his notes for an introductory course in linguistics, triggered the Chomskyan revolution in linguistics "by disputing traditional ideas about language development." Chomsky is considered the father of the theory of formal languages.*

In 1966, Chomsky became the Ferrari P. Ward Professor of Modern Languages and Linguistics. He had been a visiting professor at Columbia, Princeton, and the University of California at Los Angeles and at Berkeley.

A recipient of numerous awards and honorary degrees, including the Kyoto prize in Basic Sciences in 1988, Chomsky was named one of the thousand "makers of the twentieth century" by the London Times.

EXAMPLE 11.19 Every production of the grammar G in Example 11.12 is A $\to$ t or A $\to$ tB, so G is a regular grammar. Consequently, $L(G) = \{\text{ab}^n\text{a} \mid n \geq 0\}$ is a regular language. (See also Example 11.14.) ■

EXAMPLE 11.20 In Example 11.13, the RHS of the production $\sigma \to \text{Aa}$ contains the terminal symbol a on the right of the nonterminal symbol A, so G is not regular. However, since every production appears as $w \to \alpha$ where $w \in N$ and $\alpha \in (N \cup T)^*$, G is context-free; thus $L(G)$ is a context-free language. ■

EXAMPLE 11.21 (optional) Not every production of the grammar G in Example 11.18 is of the form A $\to$ t or A $\to$ tB. For instance, the production $\langle$unsigned integer$\rangle$ ::= $\langle$digit$\rangle\langle$unsigned integer$\rangle$ is not of either form.

The production rules, however, can be rewritten as follows:

$$\langle\text{integer}\rangle ::= +\langle\text{unsigned integer}\rangle$$

$$\langle\text{integer}\rangle ::= -\langle\text{unsigned integer}\rangle$$

$$\langle\text{unsigned integer}\rangle ::= 0\langle\text{unsigned integer}\rangle \mid \ldots \mid 9\langle\text{unsigned integer}\rangle$$

$$\langle\text{unsigned integer}\rangle ::= 0 \mid 1 \mid 2 \mid 3 \mid 4 \mid 5 \mid 6 \mid 7 \mid 8 \mid 9$$

Clearly, the form A → t or A → tB always results. So this grammar G for the set L(G) of integers is regular. Thus the set of integers is a regular language and hence context-free. ■

EXAMPLE 11.22 Consider the grammar $G = (N, T, P, \sigma)$, where $N = \{A, B, \sigma\}$, $T = \{\mathrm{a}, \mathrm{b}\}$, and $P = \{\sigma \to \mathrm{a}\sigma\mathrm{b}, \sigma \to \mathrm{aAb}, \mathrm{aAb} \to \mathrm{aBb}, \mathrm{A} \to \mathrm{a}, \mathrm{B} \to \mathrm{b}, \mathrm{A} \to \lambda, \mathrm{B} \to \lambda\}$. In the production aAb → aBb, A can be replaced with B only if A is surrounded by a and b. Notice that $L(G) = \{\mathrm{a}^m\mathrm{b}^m, \mathrm{a}^m\mathrm{b}^{m+1}, \mathrm{a}^{m+1}\mathrm{b}^m \mid m \geq 1\}$. ■

EXAMPLE 11.23 The grammar $G = (N, T, P, \sigma)$ in Example 11.15 is context-free, so $L(G) = \{\mathrm{a}^n\mathrm{b}^n \mid \mathrm{n} \geq 1\}$ is a context-free language. Example 11.53 will demonstrate that G is not regular. ■

A language $L(G)$ may contain words derivable from σ in more than one way. Accordingly, we make the following definition.

Ambiguous Grammar

A grammar G is **ambiguous** if a string in $L(G)$ has more than one derivation tree.

The next two examples present ambiguous grammars.

EXAMPLE 11.24 The following grammar G defines the syntax of simple algebraic expressions:

$$\langle\text{expression}\rangle ::= \langle\text{expression}\rangle\langle\text{sign}\rangle\langle\text{expression}\rangle \mid \langle\text{letter}\rangle$$

$$\langle\text{sign}\rangle ::= + \mid -$$

$$\langle\text{letter}\rangle ::= \mathrm{a} \mid \mathrm{b} \mid \mathrm{c} \mid \ldots \mid \mathrm{z}$$

This grammar can produce the expression a − b + c two ways, as the derivation trees in Figure 11.11 show. As a result, G is an ambiguous grammar. ■

○ **EXAMPLE 11.25** (optional) The following are simplified production rules for an *if–then statement* S:

$$S ::= \text{if } \langle\text{expression}\rangle \text{ then } \langle\text{statement}\rangle \mid$$
$$\text{if } \langle\text{expression}\rangle \text{ then } \langle\text{statement}\rangle \text{ else } \langle\text{statement}\rangle$$

$$\langle\text{expression}\rangle ::= E_1 \mid E_2$$

$$\langle\text{statement}\rangle ::= S_1 \mid S_2 \mid \text{ if } \langle\text{expression}\rangle \text{ then } \langle\text{statement}\rangle$$

To see that these rules produce an ambiguous grammar, notice that the if–then statement

$$\textit{If } E_1 \textit{ then if } E_2 \textit{ then } S_1 \textit{ else } S_2 \qquad (11.3)$$

Figure 11.11

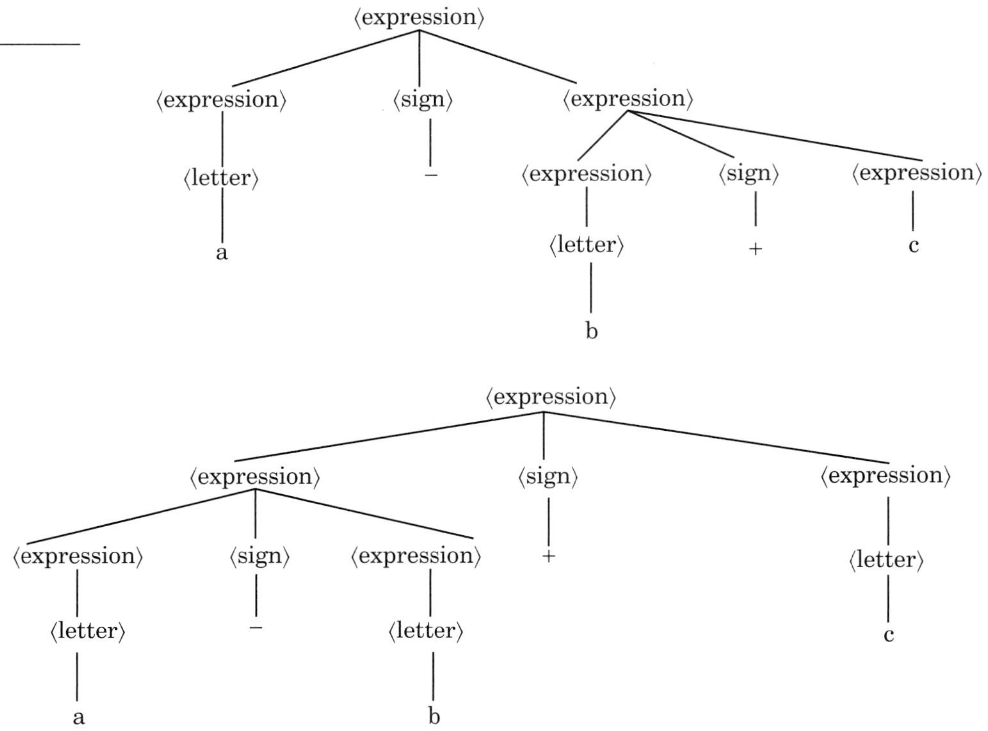

can be interpreted in two ways:

(i) If E_1 then (if E_2 then S_1 else S_2), or
(ii) If E_1 then (if E_2 then S_1) else S_2.

Using indentation, these possibilities can be displayed as follows:

```
(i)  if E1 then                     (ii)  if E1 then
        if E2 then                            if E2 then
           S1                                    S1
        else                              else
           S2                                S2
```

Accordingly, statement (11.3) can be generated by two distinct derivation trees (see Figure 11.12).

To avoid this confusion, each **else** is paired with the nearest **if**. Consequently, statement (i) is the correct interpretation of statement (11.3). If you would like statement (11.3) to mean statement (ii), you have

Figure 11.12

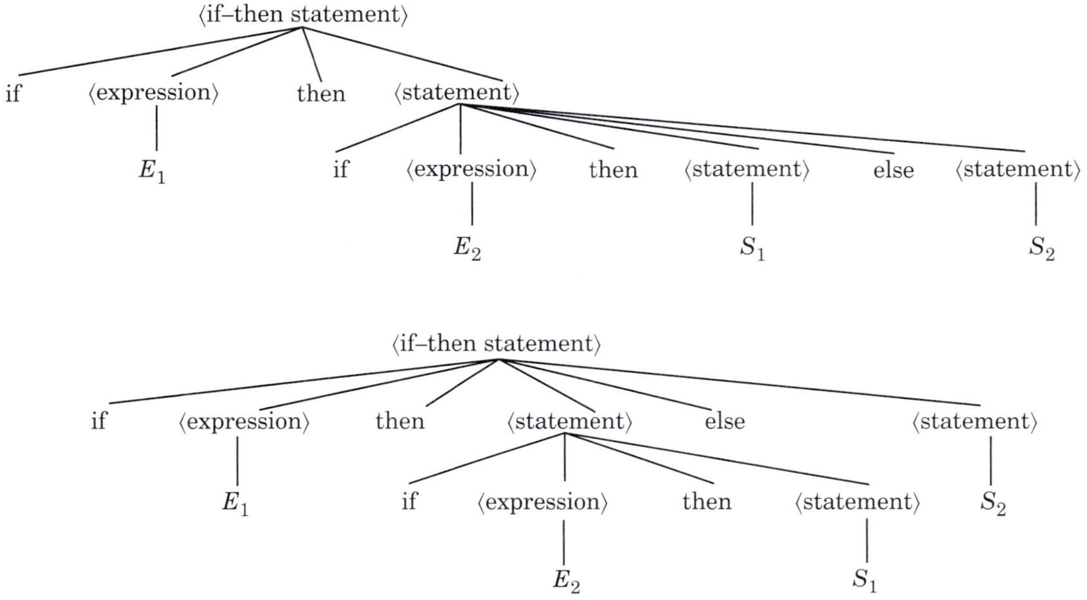

two options:

```
if E1 then
   if E2 then
      S1
   else
else
   S2
```

```
if E1 then
          begin
            if E2 then
              S1
          end
else
    S2
```

■

The way a grammar produces its language of terminal and nonterminal symbols determines whether it is regular, context-free, or context-sensitive. The BNF notation facilitates such a differentiation.

Exercises 11.2

In the grammar $G = (N, T, P, \sigma)$, $N = \{\langle\text{sentence}\rangle, \langle\text{noun phrase}\rangle, \langle\text{verb}\rangle, \langle\text{object phrase}\rangle, \langle\text{article}\rangle, \langle\text{noun}\rangle\}$, $T = \{$a, the, cat, dog, chicken, milk, drinks, eats$\}$, $\sigma = \langle\text{sentence}\rangle$ and the production rules are:

$$\langle\text{sentence}\rangle \rightarrow \langle\text{noun phrase}\rangle\langle\text{verb}\rangle\langle\text{object phrase}\rangle$$

$$\langle\text{noun phrase}\rangle \rightarrow \langle\text{article}\rangle\langle\text{noun}\rangle$$

$$\langle\text{article}\rangle \rightarrow a \mid the$$

$$\langle\text{noun}\rangle \rightarrow \textit{cat} \mid \textit{dog} \mid \textit{chicken} \mid \textit{milk}$$

$$\langle\text{verb}\rangle \rightarrow \textit{drinks} \mid \textit{eats}$$

$$\langle\text{object phrase}\rangle \rightarrow \langle\text{article}\rangle\langle\text{noun}\rangle$$

Determine if each is a valid sentence in L(G).

1. The cat drinks the milk.
2. A chicken eats the dog.
3. The dog swallows the cat.
4. The chicken drinks a rabbit.

Construct a derivation tree for each sentence in $L(G)$.

5. The cat eats the chicken.
6. A dog drinks the milk.

With the grammar in Example 11.12, construct a derivation tree for each word in $L(G)$.

7. aa
8. aba
9. $\mathrm{ab^2a}$
10. $\mathrm{ab^3a}$

Determine if each word belongs to the language generated by the grammar in Example 11.13.

11. aba
12. abba
13. $\mathrm{a^3ba}$
14. $\mathrm{a^2b^3a^4}$

Use the grammar $G = (N, T, P, \sigma)$, where $N = \{A, \sigma\}$, $T = \{\mathrm{a}, \mathrm{b}\}$, and $P = \{\sigma \rightarrow \mathrm{a}\sigma, \sigma \rightarrow \mathrm{aA}, \mathrm{A} \rightarrow \mathrm{b}\}$, to answer Exercises 15–23.
Draw a derivation tree for each word in $L(G)$.

15. ab
16. $\mathrm{a^2b}$
17. $\mathrm{a^3b}$
18. $\mathrm{a^4b}$

Do the following words belong to $L(G)$?

19. aba
20. abba
21. $\mathrm{a^3b}$
22. $\mathrm{a^5b}$

23. Identify the language $L(G)$.

Consider the grammar $G = (N, T, P, \sigma)$, where $N = \{\sigma\}$, $T = \{\mathrm{a}, \mathrm{b}\}$, and $P = \{\sigma \rightarrow \mathrm{a}\sigma\mathrm{b}, \sigma \rightarrow \mathrm{ab}\}$. Determine if each word belongs to $L(G)$.

24. abba
25. abab
26. $\mathrm{a^2b^2}$
27. $\mathrm{a^3b^3}$

28. Identify the language $L(G)$.

Find the language generated by each grammar $G = (N, T, P, \sigma)$ where:

29. $N = \{\sigma, \mathrm{A}, \mathrm{B}\}$, $T = \{\mathrm{a}, \mathrm{b}\}$, $P = \{\sigma \rightarrow \mathrm{aA}, \mathrm{A} \rightarrow \mathrm{Bb}, \mathrm{A} \rightarrow \mathrm{a}, \mathrm{B} \rightarrow \mathrm{b}\}$

30. $N = \{\sigma, \mathrm{A}, \mathrm{B}\}$, $T = \{\mathrm{a}, \mathrm{b}\}$, $P = \{\sigma \rightarrow \mathrm{aAa}, \mathrm{A} \rightarrow \mathrm{bBb}, \sigma \rightarrow \lambda, \mathrm{A} \rightarrow \mathrm{a}, \mathrm{B} \rightarrow \mathrm{a}, \mathrm{B} \rightarrow \mathrm{b}\}$

Develop a grammar that generates each language over $\{0, 1\}$.

31. $\{1, 11, 1111, 11111111, \ldots\}$

32. $\{0, 00, 10, 100, 110, 0000, 1010, \ldots\}$

33. The set of words with prefix 00.

34. The set of words with suffix 11.

35. The set of binary representations of positive integers.

Create a grammar to produce each language over {a,b}.

36. $\{b^n ab^n \mid n \geq 0\}$ **37.** $\{a^n b \mid n \geq 1\}$ **38.** $\{a^n ba \mid n \geq 1\}$

39. $\{a^m b^n \mid m, n \geq 1\}$ **40.** The set of palindromes.

○ Using Example 11.18, draw the derivation tree for each integer.

41. 234 **42.** −234

○ **43.** An identifier in Java is a letter, underscore, or \$, followed by any number of alphanumeric characters. With BNF, define the grammar for a Java identifier.

Use the grammar in Exercise 43 to see if each string is a valid Java identifier.

○ **44.** catch 22 **45.** 20/20 **46.** algorist **47.** three roots

Construct a derivation tree for each identifier.

○ **48.** result2 **49.** value **50.** R2D2 **51.** math

The production rules of a grammar for simple arithmetic expressions are:

$$\begin{aligned}\langle\text{expression}\rangle &::= \langle\text{digit}\rangle \mid (\langle\text{expression}\rangle) \mid +(\langle\text{expression}\rangle) \mid \\ &\quad -(\langle\text{expression}\rangle) \mid \langle\text{expression}\rangle\langle\text{operator}\rangle\langle\text{expression}\rangle \\ \langle\text{digit}\rangle &::= 0 \mid 1 \mid 2 \mid 3 \mid 4 \mid 5 \mid 6 \mid 7 \mid 8 \mid 9 \\ \langle\text{operator}\rangle &::= + \mid - \mid * \mid / \mid\uparrow\end{aligned}$$

Use this grammar to answer Exercises 52–59.
Determine if each is a valid arithmetic expression.

52. $2 * 3 + 4$ **53.** $-(3 * 4 \uparrow 5)$ **54.** $3+\uparrow 7$ **55.** $6 + 5/8*$

Construct a derivation tree for each expression.

56. $3 + 5 * 6$ **57.** $5 + (4 \uparrow 3)$ **58.** $(5 + 3) - 7/4$ **59.** $-(3 \uparrow (5 + 2))$

A number in ALGOL (excluding the exponential form) is defined as follows:

$$\begin{aligned}\langle\text{number}\rangle &::= \langle\text{decimal number}\rangle \mid \langle\text{sign}\rangle\langle\text{decimal number}\rangle \\ \langle\text{decimal number}\rangle &::= \langle\text{unsigned integer}\rangle \mid .\langle\text{unsigned integer}\rangle \mid \\ &\quad \langle\text{unsigned integer}\rangle.\langle\text{unsigned integer}\rangle\end{aligned}$$

$$\langle\text{unsigned integer}\rangle ::= \langle\text{digit}\rangle \mid \langle\text{unsigned integer}\rangle\langle\text{digit}\rangle$$
$$\langle\text{digit}\rangle ::= 0 \mid 1 \mid 2 \mid 3 \mid 4 \mid 5 \mid 6 \mid 7 \mid 8 \mid 9$$
$$\langle\text{sign}\rangle ::= + \mid -$$

Use this grammar to answer Exercises 60–67.

○ Determine if each is a valid ALGOL number.

60. 234 **61.** 2.34 **62.** 234. **63.** .234

Draw a derivation tree for each ALGOL number.

○ **64.** −3.76 **65.** +376 **66.** .376 **67.** 0.23

For Exercises 68–73, use the following definition of a simple algebraic expression:

$$\begin{aligned}
\langle\text{expression}\rangle &::= \langle\text{term}\rangle \mid \langle\text{sign}\rangle\langle\text{term}\rangle \mid \\
&\quad\ \langle\text{expression}\rangle\langle\text{adding operator}\rangle\langle\text{term}\rangle \\
\langle\text{sign}\rangle &::= + \mid - \\
\langle\text{adding operator}\rangle &::= + \mid - \\
\langle\text{term}\rangle &::= \langle\text{factor}\rangle \mid \\
&\quad\ \langle\text{term}\rangle\langle\text{multiplying operator}\rangle\langle\text{factor}\rangle \\
\langle\text{multiplying operator}\rangle &::= * \mid / \\
\langle\text{factor}\rangle &::= \langle\text{letter}\rangle \mid (\langle\text{expression}\rangle) \mid \langle\text{expression}\rangle \\
\langle\text{letter}\rangle &::= \text{a} \mid \text{b} \mid \text{c} \mid \ldots \mid \text{z}
\end{aligned}$$

○ Determine if each is a legal expression.

68. $a + b * (c/d)$ **69.** $a + b + c$ **70.** $-a * b/c + d$ **71.** $((a - b) + c)$

○ Construct a derivation tree for each expression.

72. $(a * b) + c/d$ **73.** $a * (b + c/d)$

74. Use BNF to define a grammar for the language of well-formed parentheses (wfp).

Use the grammar in Exercise 74 to see if each is a valid sequence of parentheses.

75. (()) **76.** ()(()) **77.** (()()) **78.** ()()()

79. Figures 11.13 and 11.14 diagram the syntax for an unsigned integer and an unsigned number, respectively. Define the grammar for an unsigned number in BNF.

Figure 11.13

Figure 11.14

○ Using the grammar in Exercise 79, check if each is a valid unsigned number.

80. 177.76 **81.** .1776 **82.** 1776. **83.** 17.76E-2

11.3 Finite-State Automata

This section presents an abstract model of a machine that accepts input values, but produces no output values.

Often the question arises whether or not a word over an alphabet is acceptable. For example, is 2R2D an acceptable identifier or is 17.06 a valid real number in C++? Finite-state automata can model the steps in determining if a given word exists in a language. Accordingly, finite-state automata, also known as **language recognizers**, play a central role in the development of compilers.

Before we study the definition, we present a simple example of a language recognizer.

EXAMPLE 11.26 Determining if an input string over the alphabet {a, b} contains *abba* as a substring involves the following five steps:

Step 0 If the first symbol in the string is *a*, move to step 1 and look for the character *b*. Otherwise, no progress has been made.

Step 1 If the next character is *b*, the substring *ab* has occurred, so go to step 2 and look for another *b*. Otherwise, the symbol *b* is still missing, so stay in step 1.

Step 2 If the next symbol is *b*, the substring *abb* exists; go to step 3; if *a*, return to step 1.

Step 3 If the next symbol is *a*, the given input string contains the substring *abba*; otherwise, return to step 0 and start all over again.

Step 4 Once the substring *abba* has occurred in the input string, any sequence of a's and b's may follow.

These steps can be represented by a digraph (see Figure 11.15), each vertex representing a step. Exactly two edges, labeled a or b, leave each vertex.

Figure 11.15

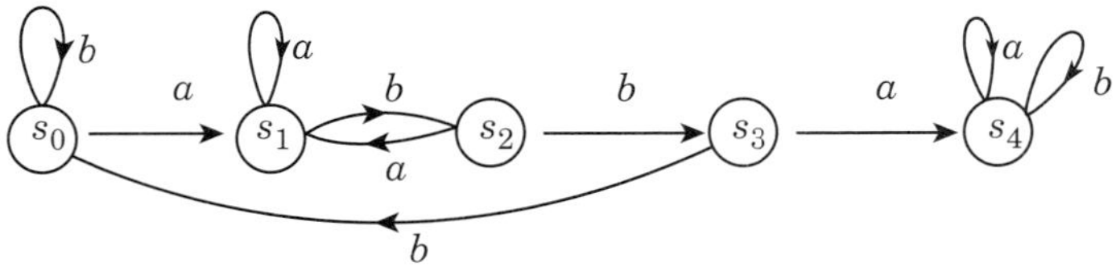

To determine the action required from a given step, simply follow the directed edges from the corresponding vertex. For example, at vertex s_3 (step 3) if the next input symbol is a, move to vertex s_4 (step 4); otherwise, return to vertex s_0 (step 0). The other (labeled) edges are interpreted similarly.

The digraph indicates a string contains *abba* as a substring if and only if the directed path the string determines terminates at vertex s_4. The string *abab* determines the path s_0-s_1-s_2-s_1-s_2, which does not end at s_4; consequently, *abab* is not acceptable. On the other hand, the string *ababbab* determines the path s_0-s_1-s_2-s_1-s_2-s_3-s_4-s_4, which terminates at s_4; so the string does have the desired property. ■

The digraph in Figure 11.15 displays a **finite-state automaton**. (*Automaton* is the singular form of *automata*.) Its five vertices, s_0 through s_4, are the **states** of the automaton. Since the whole process begins at s_0 (step 0), s_0 is the **initial state**. A string is acceptable, that is, contains *abba* as a substring, if and only if its path ends at s_4; accordingly, s_4 is an **accepting state**.

The digraph shows the transition of the machine between states. For example, if the automaton is at state s_2 and the input symbol is a, the automaton switches its state to s_1. The digraph is the **transition diagram** of the finite-state automaton.

The initial state is customarily identified by an arrow pointing to it and an accepting state by two concentric circles, as Figure 11.16 shows. The transition diagram appears in Figure 11.17.

Figure 11.16

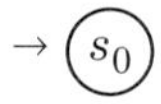

The initial state

An accepting state

Figure 11.17

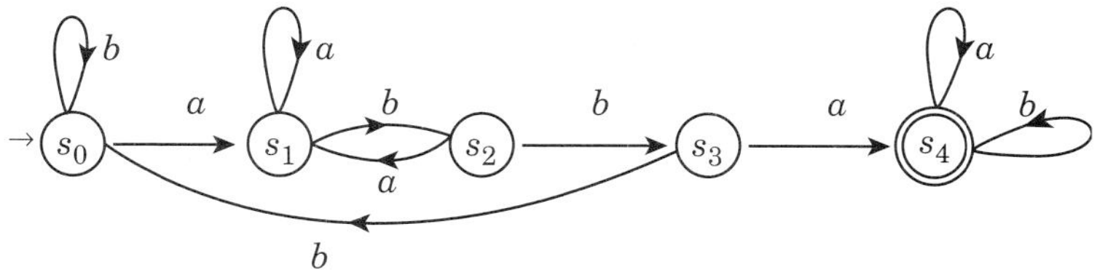

Each state s_i and an input symbol determine a unique state s_j. So we can define a function $f : S \times I \to S$ as follows, where $S = \{s_0, s_1, s_2, s_3, s_4\}$, the set of states, and $I = \{a, b\}$, the input alphabet:

$f(s_0, \text{a}) = s_1$	$f(s_0, \text{b}) = s_0$	$f(s_1, \text{a}) = s_1$	$f(s_1, \text{b}) = s_2$
$f(s_2, \text{a}) = s_1$	$f(s_2, \text{b}) = s_3$	$f(s_3, \text{a}) = s_4$	$f(s_3, \text{b}) = s_0$
$f(s_4, \text{a}) = s_4$	$f(s_4, \text{b}) = s_4$		

The function f is the **transition function** of the finite-state automaton. It can also be defined by the **transition table** in Table 11.1.

Table 11.1

State	Input symbol	
	a	**b**
s_0	s_1	s_0
s_1	s_1	s_2
s_2	s_1	s_3
s_3	s_4	s_0
s_4	s_4	s_4

We are now ready to define a finite-state automaton.

Finite-State Automaton

A **finite-state automaton** (FSA), M, manifests five characteristics:

- A finite set, S, of **states** of the automaton.
- A specially designated state, s_0, called the **initial state**.
- A subset A of S, consisting of the **accepting states** (or **final states**) of the automaton.
- A finite set, I, of **input symbols**.
- A function $f : S \times I \to S$, called the **transition function** or the **next-state function**.

In symbols, $M = (S, A, I, f, s_0)$.

For instance, for the FSA in Example 11.26, $S = \{s_0, s_1, s_2, s_3, s_4\}$, $A = \{s_4\}$, $I = \{\text{a}, \text{b}\}$, and the transition function f is defined by Table 11.1.

New York City subway commuters use an FSA everyday, as the next example shows.

EXAMPLE 11.27* A turnstile in the subway entrance contains four arms at waist level (Figure 11.18). Initially, it is locked so that the arms cannot be moved. Depositing a token into the slot, however, unlocks it and allows the arms to rotate through one quarter of a complete turn, so the commuter passes through the turnstile.

Figure 11.18

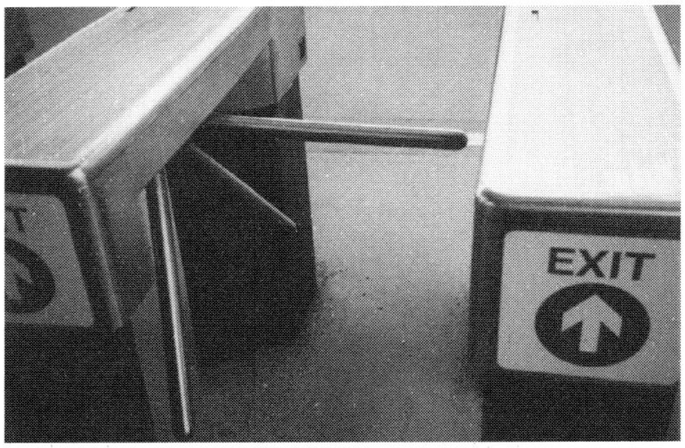

The turnstile has two states: locked (l) and unlocked (u). Depositing a token (t) shifts the turnstile from the locked state to the unlocked state and no matter how many times the commuter inputs t, the turnstile remains in the same state. Pushing (p), the arms, takes the turnstile back to the locked state. Once it is in the locked state, it remains there regardless of how many times the commuter pushes the arms; that is, regardless of the number of times he inputs p into the device.

The turnstile exemplifies an FSA. Figure 11.19 shows its transition diagram.

Figure 11.19

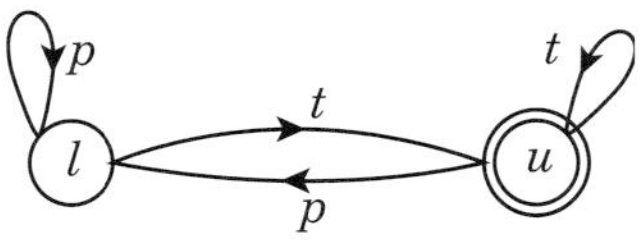

■

The next two examples draw transition diagrams of FSAs from their algebraic definitions.

EXAMPLE 11.28 Draw the transition diagram of the FSA $M = (S, A, I, f, s_0)$, where $S = \{s_0, s_1, s_2\}$, $A = \{s_2\}$, $I = \{\text{a}, \text{b}\}$, and the transition function f is defined by

$$f(s_0, \text{a}) = s_1, \quad f(s_0, \text{b}) = s_0, \quad f(s_1, \text{a}) = s_2,$$
$$f(s_1, \text{b}) = s_0, \quad f(s_2, \text{a}) = s_2, \quad f(s_2, \text{b}) = s_0.$$

*Based on B. Hayes, "On the Finite-State Machine, A Minimal Model of Mousetraps, Ribosomes, and the Human Soul," *Scientific American*, Vol. 249 (Dec. 1983), pp. 20–28, 178.

SOLUTION:

The FSA contains three states — s_0, s_1, and s_2 — with s_2 the only accepting state. Since there are two input symbols, exactly two edges leave each vertex. Draw a directed edge from state s_i to state s_j if there is an input symbol x such that $f(s_i, x) = s_j$; then label the edge x. For example, since $f(s_1, b) = s_0$, a directed edge runs from s_1 to s_0 labeled b. Figure 11.20 shows the resulting transition diagram.

Figure 11.20

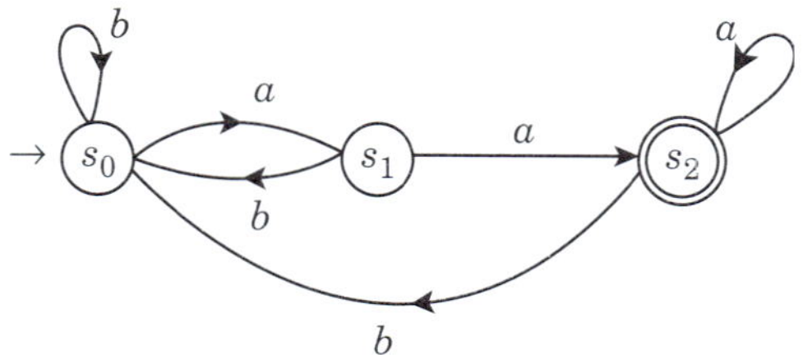

■

EXAMPLE 11.29 Draw the transition diagram of the FSA $M = (S,\ A,\ I,\ f,\ s_0)$, where $S = \{s_0, s_1, s_2, s_3, s_4\}$, $A = \{s_2\}$, $I = \{\text{a, b, c}\}$, and f is defined by Table 11.2.

Table 11.2

S \ I	a	b	c
s_0	s_1	s_2	s_3
s_1	s_4	s_2	s_3
s_2	s_1	s_4	s_3
s_3	s_1	s_2	s_4
s_4	s_4	s_4	s_4

SOLUTION:

The automaton contains five states, with s_2 the only accepting one. Since there are three input symbols, three edges originate from every state. Draw a directed edge from state s_i to state s_j if there exists an input symbol x such that $f(s_i, x) = s_j$. For instance, $f(s_1, c) = s_3$, so a directed edge labeled c runs from state s_1 to state s_3. Figure 11.21 displays the resulting transition diagram, where, for convenience, the three loops at s_4 appear as a single loop with labels a, b, and c.

Figure 11.21

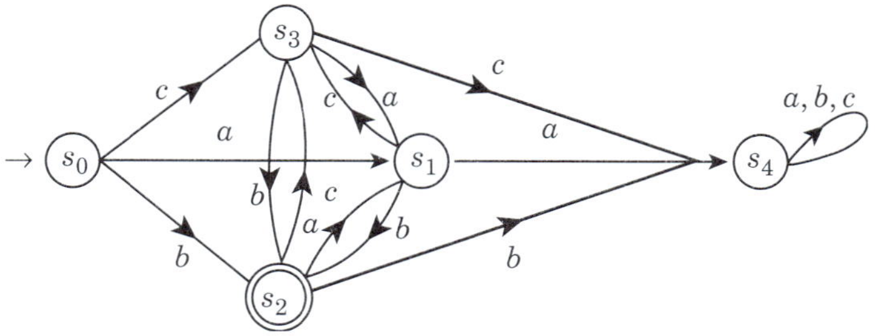

■

Suppose a string is input into an FSA. If the path it determines ends at an accepting state, the string is **accepted** (or **recognized**) by the automaton; otherwise, it is **rejected** by the automaton.

EXAMPLE 11.30 Determine if the strings a^3b^2ab and ab^3a are accepted by the FSA in Figure 11.17.

SOLUTION:

First, find the path determined by the string and check if it terminates at s_4, the accepting state. (Recall that a^3b^2ab = aaabbab.) Begin at the initial state, s_0. When a is input, move to state s_1. Every time a is input, remain there, so the path defined by aaa is s_0-s_1-s_1-s_1. When b is input, transfer to state s_2. The path obtained thus far is s_0-s_1-s_1-s_1-s_2. Now b moves to s_3 and a to s_4, yielding the path s_0-s_1-s_1-s_1-s_2-s_3-s_4. Once in s_4, remain there no matter what the input is. Thus the path determined by the given string is s_0-s_1-s_1-s_1-s_2-s_3-s_4-s_4. Since it terminates at s_4, the FSA accepts the given word.

Notice that the path determined by the string ab^3a is s_0-s_1-s_2-s_3-s_0-s_1, and it does not end at the accepting state s_4; consequently, the automaton rejects the string. ■

Two different FSAs may accept the same language over an alphabet. This occurrence requires that we make a new definition.

Equivalent Finite-State Automata

The set of words accepted by an FSA, M, is the **language accepted** (or **recognized**) by M and is denoted by $L(M)$. Two finite-state automata, M and M', are **equivalent** if they recognize the same language: $L(M) = L(M')$.

EXAMPLE 11.31 Identify the language $L(M)$ accepted by the automaton M in Figure 11.20.

SOLUTION:

Look for paths beginning at s_0 and terminating at s_2. $L(M)$ consists of all words over {a, b} that end in aa. ■

EXAMPLE 11.32 By Example 11.31, the automaton in Figure 11.20 accepts the language of words over {a, b} ending in aa. You may verify that the FSA in Figure 11.22 accepts the same language. Consequently, the automata in Figures 11.20 and 11.22 are equivalent.

Figure 11.22

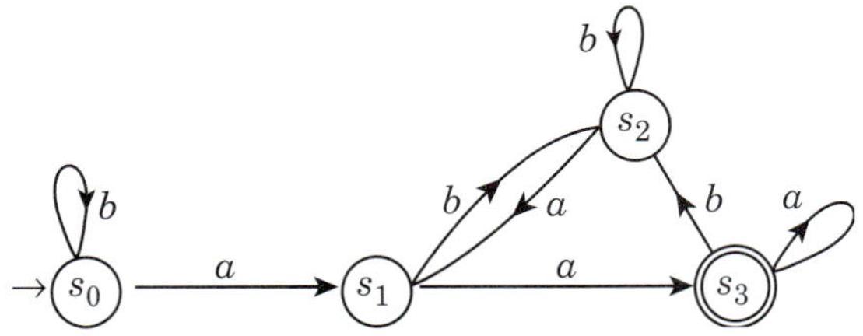

■

The next four examples build FSAs with desired properties, as Example 11.26 did.

EXAMPLE 11.33 Design an FSA that accepts words over $I = \{a, b\}$ containing an even number of a's.

SOLUTION:
Every word over I contains either an even number of a's (E) or an odd number of a's (O), so the automaton has two states, E and O, E being the accepting state. Initially, the number of a's in the word is zero, an even integer; E is the initial state of the automaton. If the automaton is at E and an a is input, it moves to state O. If it is at O and an a is input, it moves to state E. Figure 11.23 shows the transition diagram of the FSA.

Figure 11.23

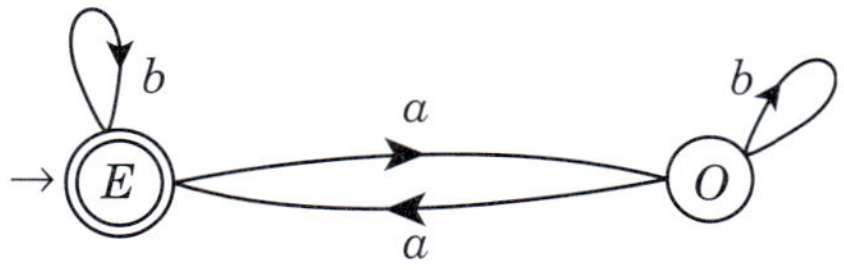

A word over I has **even parity** if it contains an even number of a's and **odd parity** if an odd number. Since the automaton in Example 11.33 determines whether a word has even or odd parity, it is called a **parity-check machine**. ■

EXAMPLE 11.34 Design an FSA accepting words over $\{a, b\}$ that begin with aa and end in bb.

SOLUTION:
We build the automaton step by step:

Step 0 Initially, the automaton is at the initial state s_0.

Step 1 If the first symbol is a, move to state s_1 from s_0 and wait for the next symbol. But if the first symbol is b, the word is not acceptable (state s_2). See Figure 11.24.

Figure 11.24

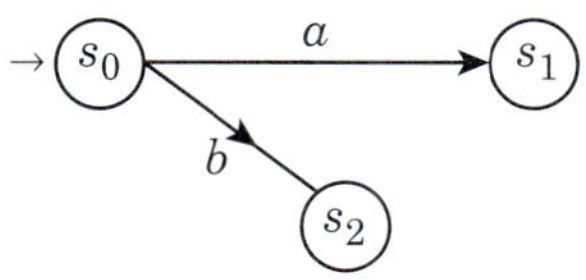

Step 2 If the input symbol at s_1 is a, move to state s_3 and determine whether the string ends with bb. On the other hand, if the input symbol at s_1 is b, move to s_2 to trap such unacceptable words. Once at s_2, remain there no matter what the input symbol is. See Figure 11.25.

Figure 11.25

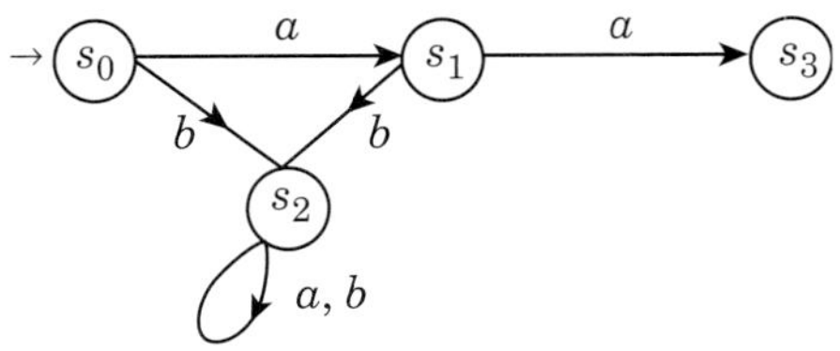

Step 3 Every word that triggers a move from s_0 to s_3 begins with aa. Any number of a's can follow it (see the loop at s_3 in Figure 11.26). However, if b follows the word, move to a new state s_4, as in Figure 11.26.

Figure 11.26

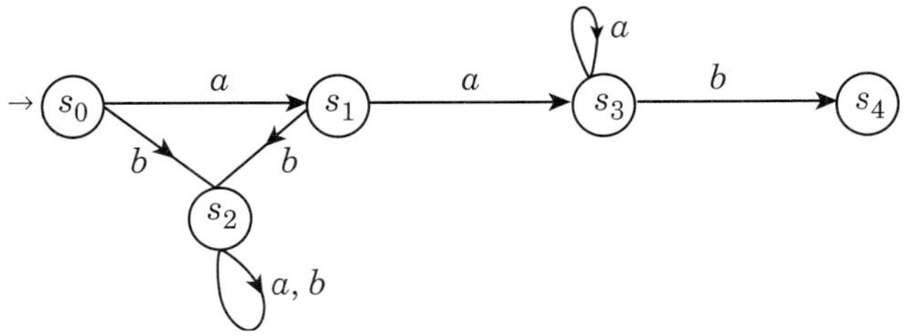

Step 4 If the input symbol at s_4 is a, return to s_3 and look for the pair bb. But if it is b, move to a new state s_5. See Figure 11.27.

Figure 11.27

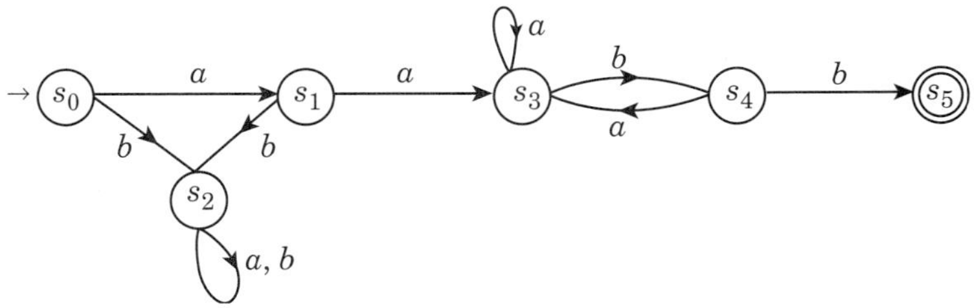

Step 5 Once at s_5, any number of b's may occur. However, if the input symbol at s_5 is a, return to s_3 to look for bb. Since words ending in bb are acceptable, s_5 is the accepting state. These six steps create the FSA in Figure 11.28.

Figure 11.28

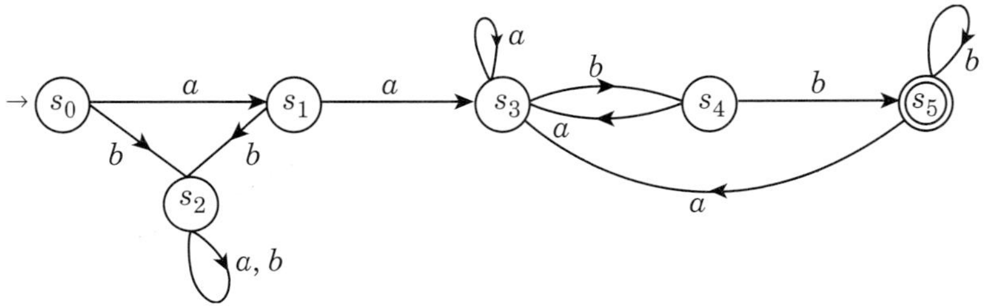

■

EXAMPLE 11.35 (optional) An identifier in a programming language consists of a letter followed by any number of alphanumeric characters (Section 11.1). Design an FSA that recognizes such legal identifiers.

SOLUTION:
Let I denote the set of all characters in the alphabet recognizable by a compiler. Let l denote a letter, d a digit, and n any nonalphanumeric character. The automaton will have three states: s_0, s_1, and s_2. State s_2 traps all invalid strings. (Accordingly, it is called a **trap state** or a **dump state**.) The resulting automaton appears in Figure 11.29.

Figure 11.29

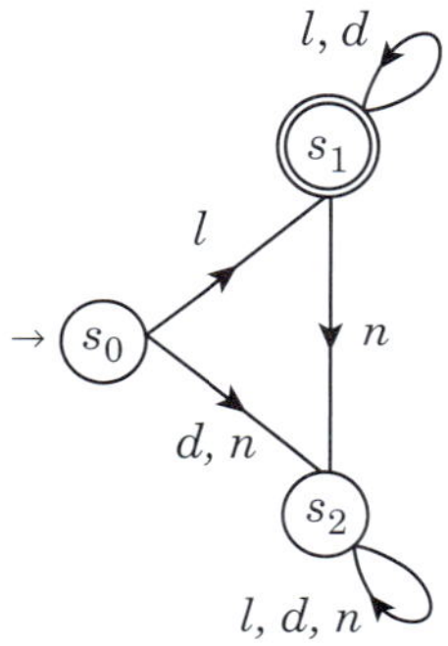

■

The FSA in Figure 11.29 can be translated into an algorithm which determines if a sequence of characters is a legal identifier. See Algorithm 11.1.

```
Algorithm identifier
(* This algorithm determines whether a sequence of characters is a
   valid identifier, using the FSA in Figure 11.29. All characters
   are read from the same input line. Symbol denotes an arbitrary
   character; state denotes an arbitrary state; state0, state1,
   and state2 denote the various states of the FSA. state2 is a
   dump state. *)
 Begin (* algorithm *)
  state ← state0      (* initialize state *)
  read(symbol)
  while not at the end of the current line
  begin
  case state of
    state0:  if symbol is a letter then
               state ← state1
             else     (* invalid sequence; dump it. *)
               state ← state2
    state1:  if symbol is a letter or a digit then
               state ← state1
             else
               state ← state 2
    state2:  (* do nothing; stay there. *)
 read(symbol)
endwhile
```

```
if state = state1 then
   the sequence is a valid identifier
else
   the sequence is an invalid identifier
End (* algorithm *)
```

Algorithm 11.1

With a trap state, an FSA can simulate an automatic teller machine, or ATM, which is widely used because it allows bank customers to make transactions without human intervention, as the following example demonstrates.

EXAMPLE 11.36 After a bank customer inserts his bank card into the ATM, it requests him to input his secret identification number (ID). Suppose the ID is 234. Design an FSA that models the ATM.

SOLUTION:
The input to the automaton contains three digits d. It has five states: s_0 (the initial state, waiting for the first digit in the ID), s_1 (the first digit is correct; now waiting for the second digit), s_2 (the second digit is correct; waiting for the third digit), s_3 (the third digit is correct), and s_4 (the trap state that captures all invalid ID's). The ensuing FSA is shown in Figure 11.30.

Figure 11.30

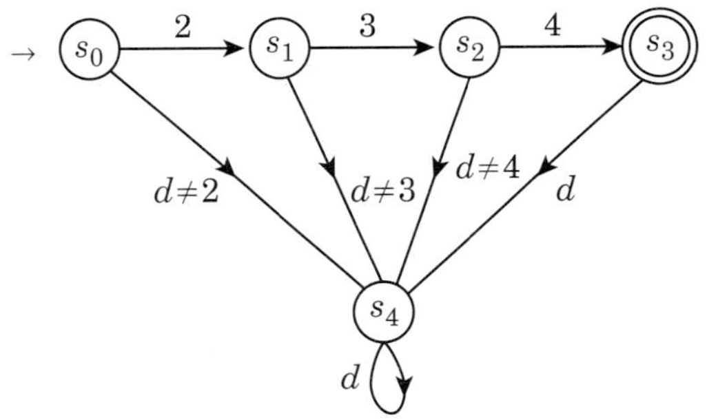

■

The salient characteristics of an FSA have emerged through its many applications to ATMs, programming languages, parity checks, and subway turnstiles. Every FSA manifests an input set, a transition function, and a finite number of states.

Exercises 11.3

Using the FSA in Figure 11.17, identify the directed paths determined by each input string.

1. a^3b **2.** abab **3.** ab^3 **4.** a^2b^3a

With the FSA in Figure 11.21, identify the directed path determined by each word:

5. abcab **6.** caba^2 **7.** a^{2}bc^3 **8.** ab^{2}c^3

Determine if each word is acceptable by the FSA in Figure 11.17.

9. ab^3 **10.** a^{2}b^{2}a^2 **11.** a^{3}b^{2}a^3 **12.** ab^{4}ab^{2}ab

Determine if the FSA in Figure 11.21 recognizes each word.

13. abcabc **14.** abacbc **15.** ab^{4}c^3 **16.** ab^{5}c^6

Draw the transition diagram of the FSA, $M = (S, A, I, f, s_0)$, where $I = \{\text{a,b}\}$, and:

17. $S = \{s_0, s_1, s_2\}$, $A = \{s_2\}$

$$f(s_0, \text{a}) = s_0 \quad f(s_0, \text{b}) = s_1 \quad f(s_1, \text{a}) = s_0 \quad f(s_1, \text{b}) = s_2$$
$$f(s_2, \text{a}) = s_0 \quad f(s_2, \text{b}) = s_2$$

18. $S = \{s_0, s_1, s_2, s_3\}$, $A = \{s_3\}$

$$f(s_0, \text{a}) = s_1 \quad f(s_0, \text{b}) = s_0 \quad f(s_1, \text{a}) = s_1 \quad f(s_1, \text{b}) = s_2$$
$$f(s_2, \text{a}) = s_1 \quad f(s_2, \text{b}) = s_3 \quad f(s_3, \text{a}) = s_1 \quad f(s_3, \text{b}) = s_0$$

19. $S = \{s_0, s_1, s_2, s_3\}$, $A = \{s_2\}$

S \ I	f: a	f: b
s_0	s_0	s_1
s_1	s_1	s_2
s_2	s_2	s_3
s_3	s_3	s_3

20. $S = \{s_0, s_1, s_2, s_3, s_4\}$, $A = \{s_3\}$

S \ I	f: a	f: b
s_0	s_1	s_4
s_1	s_4	s_2
s_2	s_3	s_4
s_3	s_3	s_3
s_4	s_4	s_4

Construct a transition table for each FSA.

21.

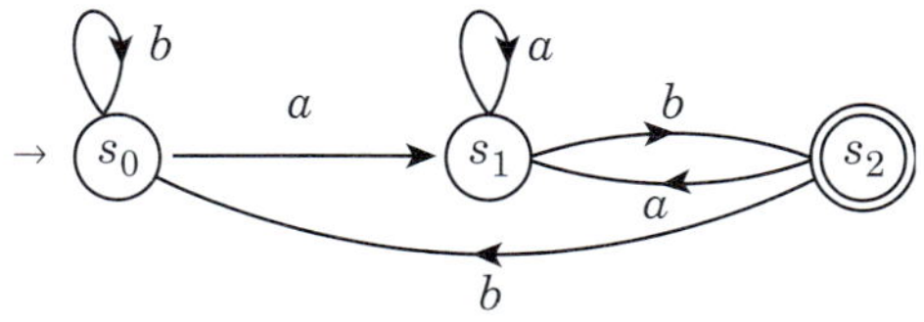

22.

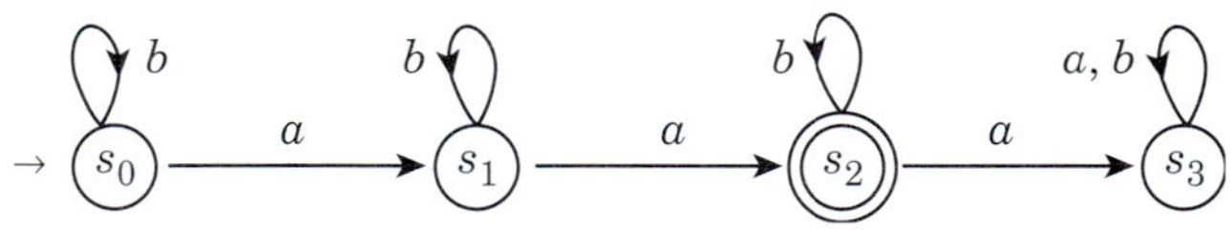

23.

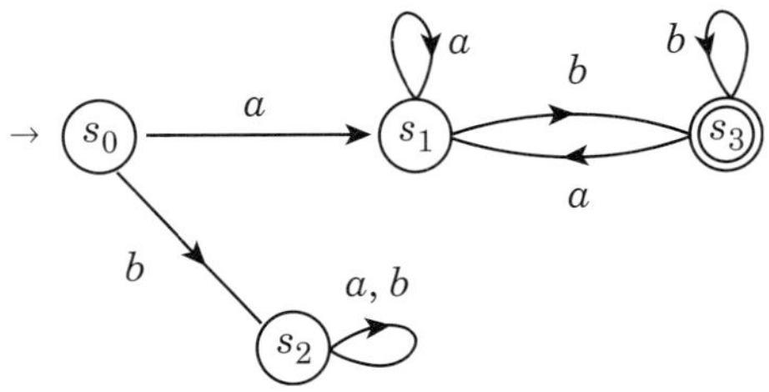

24.

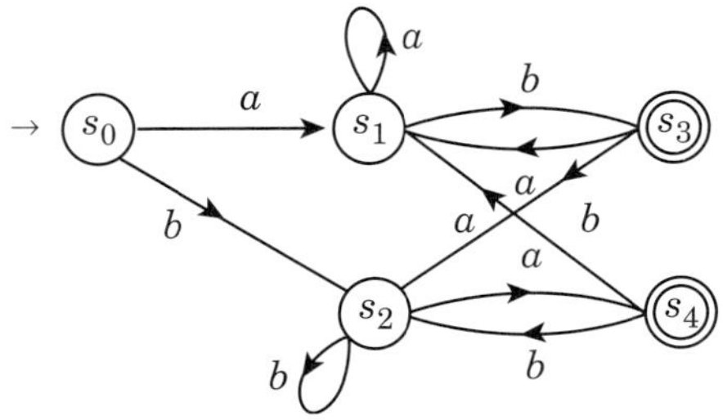

Characterize the language recognized by the FSAs in Exercises 25–35.

25.

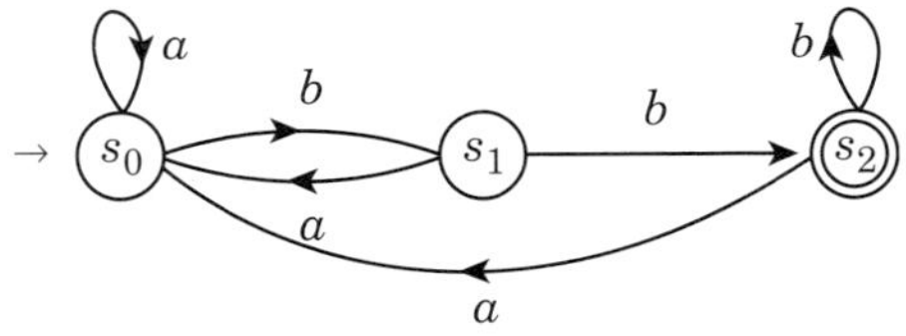

26.

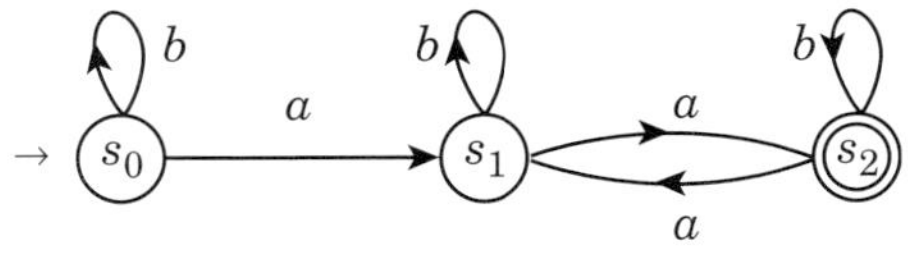

27–34. The finite-state automata in Exercises 17–24.

***35.**

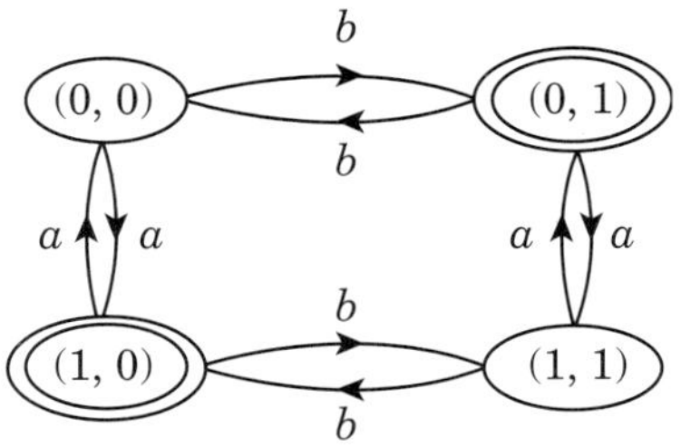

Let m denote the number of a's in a string. Design an FSA that accepts strings over $\{\mathrm{a}, \mathrm{b}\}$ which:

36. Contain exactly one a.

37. Begin with aa.

38. Contain aba as a substring.

39. Contain aaa as a substring.

40. Begin with aa or bb.

41. Contain $baab$ as a substring.

42. Have $m \equiv 0 \pmod 3$).

43. Have $m \equiv 2 \pmod 3$).

44. Simulate an automatic teller machine by means of an FSA that accepts 1776 as a valid identification number.

45. Design an FSA to model an automatic teller machine that accepts 23 or 45 as a valid identification number.

○ **46.** An integer is a nonempty string of digits, preceded by an optional sign (+ or −). See the syntax diagram in Figure 11.31. Design an FSA that recognizes integers.

Figure 11.31

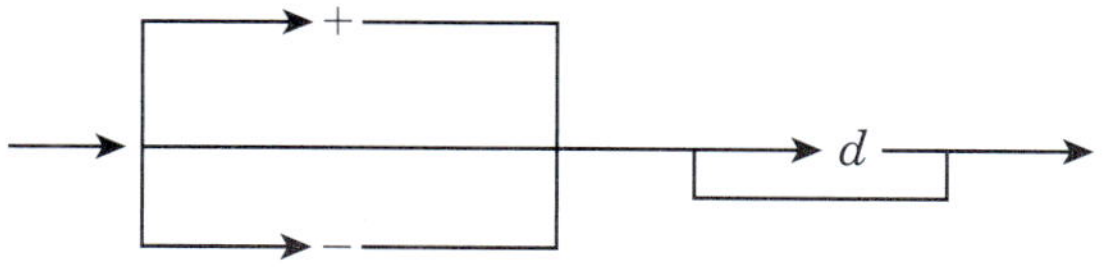

○ **47.** A real number, excluding the exponential form, consists of an optional sign (+ or −) followed by one or more digits, a decimal point, and one or more digits. (See the syntax diagram in Figure 11.32.) Design an FSA that recognizes such real numbers.

Figure 11.32

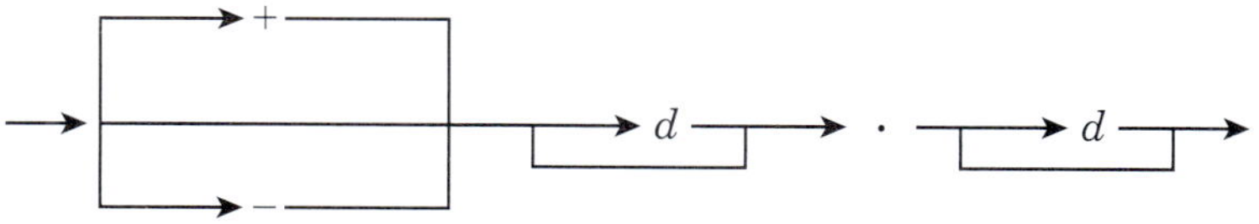

48. Write an algorithm to implement an automatic teller machine as an FSA that accepts 234 as a valid identification number.

○ **49.** Write an algorithm to determine if a sequence of characters represents a valid integer.

○ **50.** Write an algorithm to determine if a sequence of characters represents a valid real number. Exclude the exponential form.

11.4 Finite-State Machines

As a generalization of FSAs, finite-state machines abstractly model computing machines. In an FSA, movements from state s_i to state s_j depend on the input at s_i, and no output emerges. But as a finite-state machine moves from state s_i to state s_j, an output does emerge. Consequently, a finite-state machine possesses two features not required of an FSA: a finite set O of output symbols and an output function $g : S \times I \to O$, where I is the input alphabet. (An accepting state cannot exist here because a word is not being checked for certain characteristics.) The output depends on two things: the current state and the input symbol.

Finite-State Machine

A **finite-state machine** (FSM), M, bears six characteristics:

- A finite set, S, of **states**;
- A finite **input alphabet**, I;
- A finite set, O, of **output symbols**;
- A **transition function**, $f : S \times I \to S$;
- An **output function**, $g : S \times I \to O$;
- An **initial state**, s_0.

In symbols, $M = (S, I, O, f, g, s_0)$.

In this definition, the output function g depends on both the state of the machine and the current input. Such FSMs are called **Mealy machines**, after George H. Mealy, who introduced them in 1955. (Another type of FSM appears in the Supplementary Exercises.)

EXAMPLE 11.37 Let $S = \{s_0, s_1, s_2\}$, $I = \{a, b\}$, and $O = \{0, 1\}$. Define functions $f : S \times I \to S$ and $g : S \times I \to O$ by means of Table 11.3. For example, $f(s_0, b) = s_1$, $f(s_2, b) = s_1$, $g(s_0, b) = 1$, and $g(s_2, b) = 1$.

Table 11.3

	f		g	
S \ I	a	b	a	b
s_0	s_0	s_1	0	1
s_1	s_1	s_2	1	0
s_2	s_2	s_1	1	1

Then $M = (S, I, O, f, g, s_0)$ is an FSM with transition function f and output function g. Table 11.3 is the **transition table** of the machine. ■

Like an FSA, an FSM can be represented by a **transition diagram**, with one main difference: every directed edge (s_j, s_k) has two labels. One indicates the input symbol i; the other the output o from entering i into state s_j. For instance, if $f(s_j, i) = s_k$ and $g(s_j, i) = 0$, the directed edge (s_j, s_k) is labeled $i/0$.

The next example illustrates how to draw transition diagrams of FSMs.

EXAMPLE 11.38 Draw the transition diagram for the FSM in Example 11.37.

SOLUTION:
The FSM has three states — s_0, s_1, and s_2; and two input symbols — a and b; two output symbols–0 and 1. Two input symbols produce exactly two outgoing edges for each state. Each directed edge $(s_j,\ s_k)$ in the diagram is labeled i/o, where $f(s_j, i) = s_k$ and $g(s_j, i) = o$. For instance, since $f(s_0, a) = s_0$ and $g(s_0, a) = 0$, a loop exists at s_0 labeled $a/0$. And because $f(s_0,\text{b}) = s_1$ and $g(s_0,\text{b}) = 1$, the edge (s_0, s_1) is labeled $b/1$. The other directed edges carry similar labels. Figure 11.33 shows the transition diagram produced by this process.

Figure 11.33

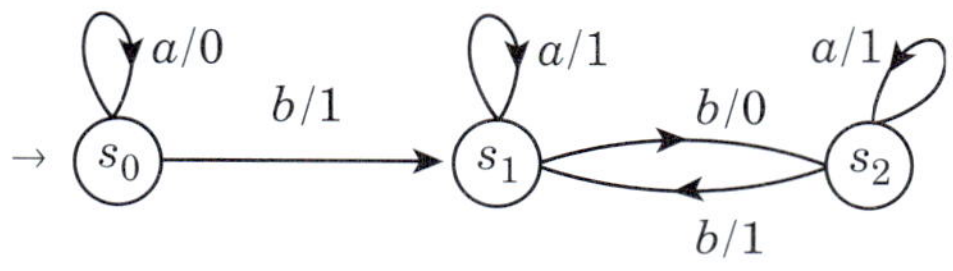

■

The transition diagram of an FSM can generate the transition table, as the following example demonstrates.

EXAMPLE 11.39 Construct the transition table of the FSM in Figure 11.34.

Figure 11.34

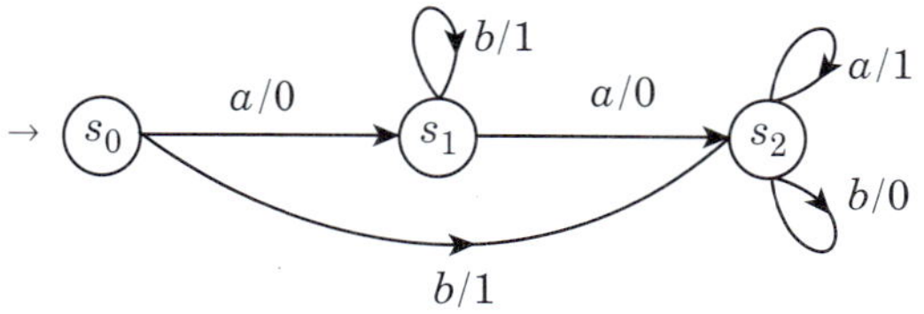

SOLUTION:
From the transition diagram, $f(s_0,\text{a}) = s_1$, $f(s_0,\text{b}) = s_2$, $f(s_1,\text{a}) = s_2$, $f(s_1,\text{b}) = s_1$, $f(s_2,\text{a}) = s_2$, and $f(s_2,\text{b}) = s_2$; also $g(s_0,\text{a}) = 0$, $g(s_0,\text{b}) = 1$, $g(s_1,\text{a}) = 0$, $g(s_1,\text{b}) = 1$, $g(s_2,\text{a}) = 1$, and $g(s_2,\text{b}) = 0$. These values generate the transition table in Table 11.4.

Table 11.4

	f		g	
S \ I	**a**	**b**	**a**	**b**
s_0	s_1	s_2	0	1
s_1	s_2	s_1	0	1
s_2	s_2	s_2	1	0

■

Suppose we input the string $x = x_1x_2 \ldots x_n$ into an FSM. Suppose further that there exist states s_{i-1} and s_i, and an output y_i such that $f(s_{i-1}, x_i) = s_i$ and $g(s_{i-1}, x_i) = y_i$ for every i. Then $y_1y_2 \ldots y_n$ is the output **produced** by the machine for the input x.

EXAMPLE 11.40 Find the output of the FSM in Figure 11.33 for the input string *abbaba*.

SOLUTION:
Start at state s_0. When a is input, stay at s_0 with output 0. When the next symbol b is input, move to s_1 and produce the output 1. When the third symbol b is input at s_1, move to s_2 and output 0. Continuing like this yields the output 010111. ■

The next two examples present FSMs useful in electronics. These machines have limited memory: at each state they must remember the previous input.

EXAMPLE 11.41 Let $I = O = \{0, 1\}$. A **unit delay machine**, an FSM $M = (S, I, O, f, g, s_0)$, delays an input string by unit time. When the string $x_1x_2 \ldots x_n$ is input, it produces $0x_1x_2 \ldots x_n$ as the output. Construct such a machine.

SOLUTION:
Since each state has two possible outputs, each has two outgoing edges. The machine must certainly have an initial state s_0. With the first output always 0, both edges leaving s_0 must yield 0. The machine must remember whether the previous input was 0 or 1; this requires two additional states, s_1 and s_2. If the previous input was 0, the machine moves to state s_1 and outputs 0; if it was 1, it moves to state s_2 and outputs 1. Figure 11.35 shows the transition diagram of this FSM.

Figure 11.35

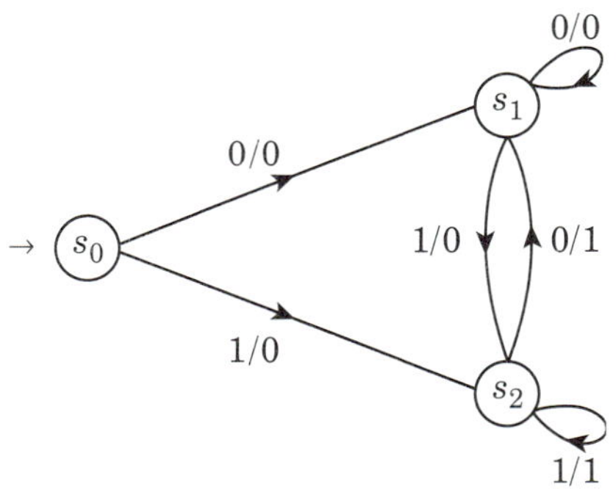

For instance, the input 101110 yields the output 010111 (Verify this.), which has lost the trailing zero of the input. By appending a 0, however, to the input, that is, by inputing 1011100, the desired output results: 0101110. Deleting the leading 0 yields an exact copy of the input. ■

EXAMPLE 11.42 Design an FSM that adds two binary integers, x and y.

SOLUTION:
Assume, for convenience, x and y contain the same number of bits, and the leftmost bits are zeros. Thus, let $x = (x_nx_{n-1} \ldots x_1x_0)_{\text{two}}$ and $y = (y_ny_{n-1} \ldots y_1y_0)_{\text{two}}$, where $x_n = y_n = 0$. Add the corresponding bits x_i and

y_i from right to left, as usual. Adding x_i and y_i yields a sum bit z_i and a carry bit c_i:

$$z_i = (x_i + y_i) \bmod 2 \text{ and } c_i = (x_i + y_i) \text{ div } 2$$

For instance, adding the bits 1 and 1 gives the sum bit 0 and the carry bit 1. Tables 11.5 and 11.6 display the sum and carry bits for paired values of x_i and y_i.

Table 11.5

Sum bits.

		y_i = **0**	y_i = **1**
x_i	0	0	1
	1	1	0

Table 11.6

Carry bits.

		y_i = **0**	y_i = **1**
x_i	0	0	0
	1	0	1

Any two binary numbers can be added if the pairs 00, 01, 10, and 11 can be. When two bits x_i and y_i are added, the carry is 0 or 1. Consequently, a machine can be manufactured with two states: $c0$ (carry is 0) and $c1$ (carry is 1). Since at first the carry is 0, $c0$ is the initial state of the machine (Figure 11.36).

Figure 11.36

Since four bit-pairs exist, exactly four edges leave each state. Tables 11.5 and 11.6 can find the state following a given state and the output from a given input. For instance, if at state $c0$ and input 11, output 0 and move to state $c1$ (Figure 11.37). If at state $c1$ and input 10, output 0 and remain at state $c1$ (Figure 11.38). Continuing like this produces the transition diagram in Figure 11.39.

Figure 11.37

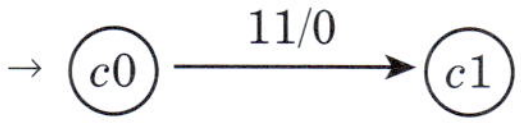

Figure 11.38

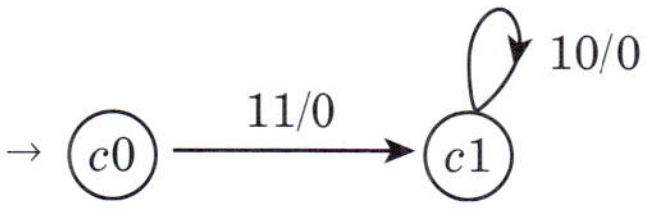

Figure 11.39

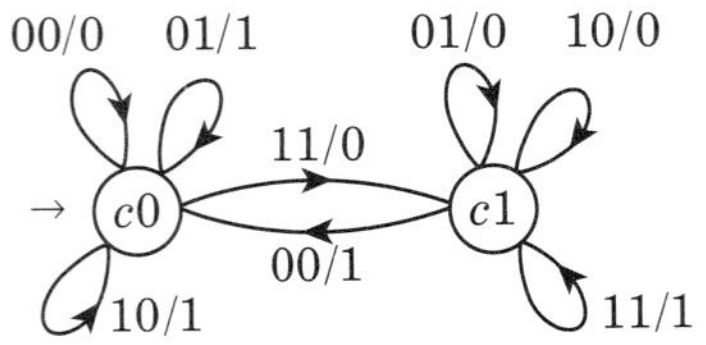

■

Finally, every FSA is a special case of an FSM. To see this, label all incoming edges to each accepting state with output 1 and all incoming edges to each nonaccepting state with output 0. Consequently, an input string is **accepted** by the FSM if and only if the last output of the machine is 1, as the following example illustrates.

EXAMPLE 11.43 Example 11.26 showed that the FSA in Figure 11.40 accepts a string over {a, b} if and only if the string contains *abba* as a substring. To convert the automaton into an FSM, add an output to every edge. Each incoming edge to the accepting state s_4 is labeled with output 1, and every incoming edge to other edges 0. The resulting FSM appears in Figure 11.41.

Figure 11.40

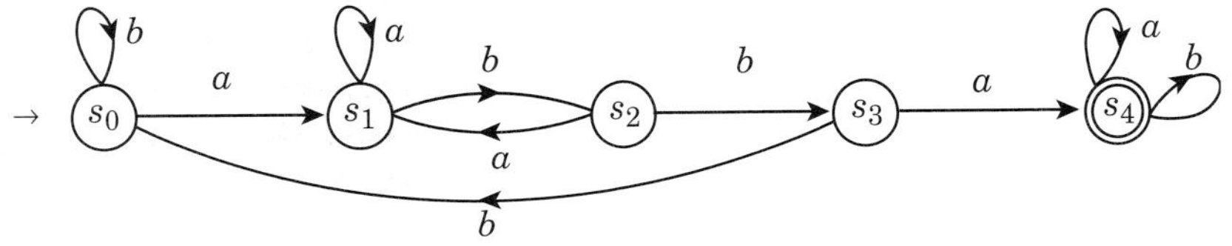

Figure 11.41

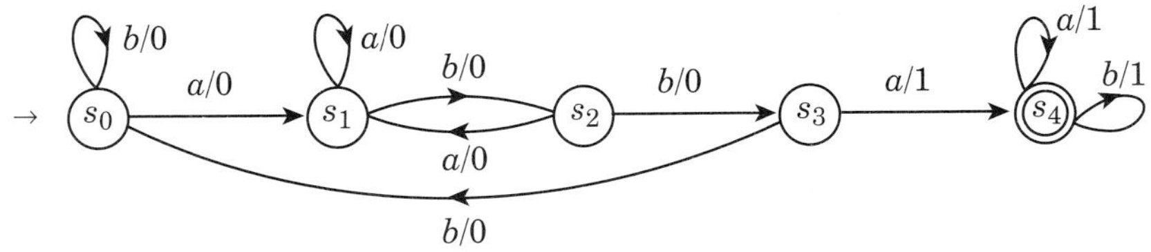

According to Example 11.30, the word a^3b^2ab is accepted by the automaton in Figure 11.40. The machine in Figure 11.41 verifies this: the substring a^3 takes the machine from s_0 to s_1 and it outputs 0 three times, b^2 takes it to s_3 and it outputs 0 twice, a takes it from s_3 to s_4 and it outputs 1; b takes the machine from s_4 to itself and it outputs 1. With the last output 1, the string is accepted by the FSM, as expected. ■

As this example indicates, FSMs like Mealy machines add output to the FSA configuration. This means that we can use them in such fields as electronics, in addition to using their transition tables and diagrams as definitional models.

Exercises 11.4

Using the FSM in Figure 11.33, evaluate each.

1. $f(s_1, \text{a})$ **2.** $f(s_2, \text{b})$ **3.** $f(s_0, \text{b})$ **4.** $f(s_2, \text{a})$

5. $g(s_1, \text{b})$ **6.** $g(s_2, \text{b})$ **7.** $g(s_0, \text{b})$ **8.** $g(s_2, \text{a})$

Draw the transition diagram of the FSM with each transition table.

9.

S \ I	f: a	f: b	g: a	g: b
s_0	s_0	s_1	1	0
s_1	s_1	s_2	0	0
s_2	s_0	s_1	1	1

10.

S \ I	f: a	f: b	g: a	g: b
s_0	s_1	s_1	0	1
s_1	s_1	s_2	1	0
s_2	s_1	s_2	0	1

11.

S \ I	f: a	f: b	g: a	g: b
s_0	s_1	s_1	0	0
s_1	s_1	s_2	0	1
s_2	s_3	s_2	0	1
s_3	s_3	s_1	1	0

12.

S \ I	f: a	f: b	g: a	g: b
s_0	s_1	s_2	1	0
s_1	s_2	s_2	0	1
s_2	s_2	s_3	0	0
s_3	s_2	s_3	1	1

Construct a transition table for each FSM.

13.

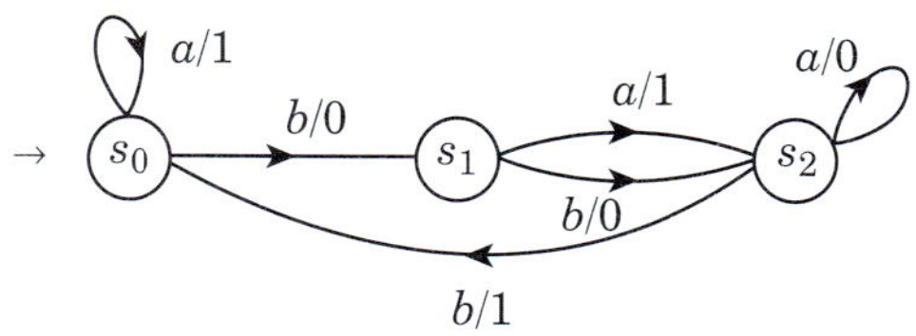

14.

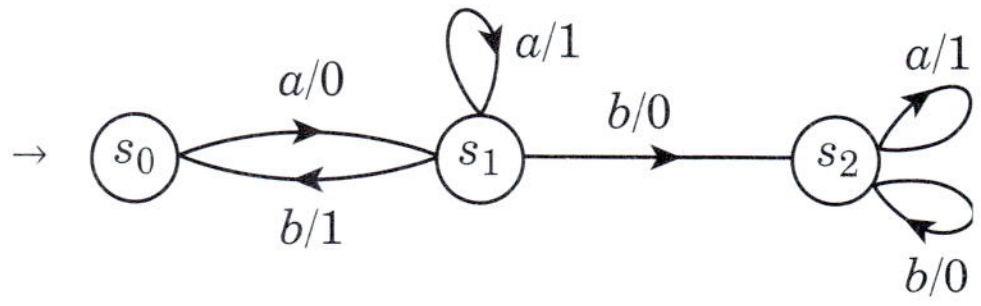

15.

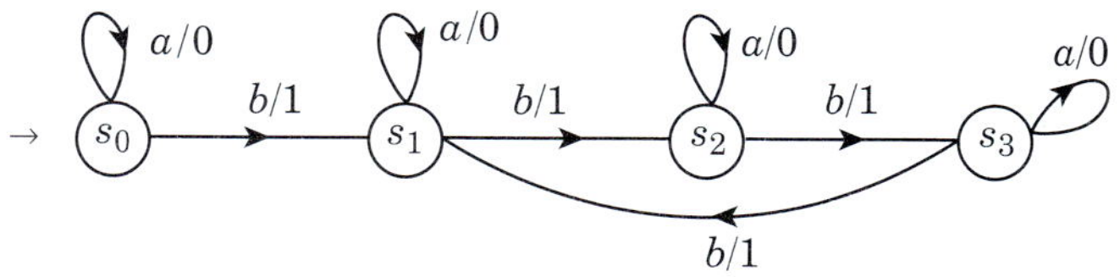

16.

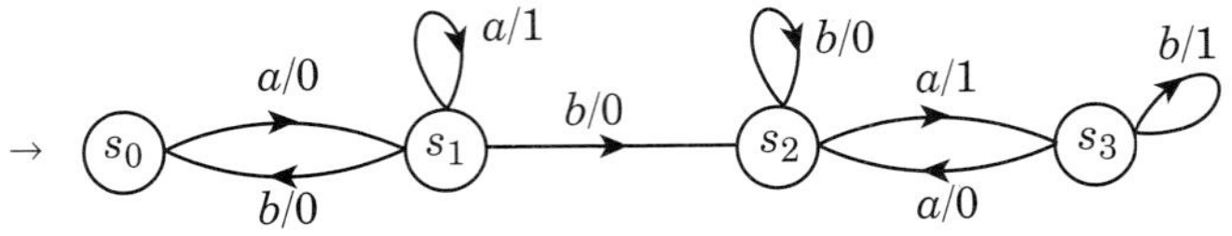

Using the FSM in Figure 11.33, find the output from each input string.

17. *abba* **18.** *baab* **19.** a^2b^3a **20.** $a^3b^2ab^3$

Using the unit delay machine in Figure 11.35, find the output of each input string.

21. 1101 **22.** 1111 **23.** 0000 **24.** 101110

25. With a transition table, define the transition function f and the output function g of the FSM for binary addition in Figure 11.39.

Using the FSM in Figure 11.39, compute the sum of each pair of binary numbers.

26. 1001 0110	**27.** 00111 10010	**28.** 1011 0110	**29.** 11011 10101

30. Redraw Figure 11.20 as the transition diagram of an FSM.

31–34. Redraw the transition diagram of each automaton in Exercises 17–20 of Section 11.3 as that of an FSM.

Determine if the input string in Exercises 35–38 is accepted by the FSM in Figure 11.42.

Figure 11.42

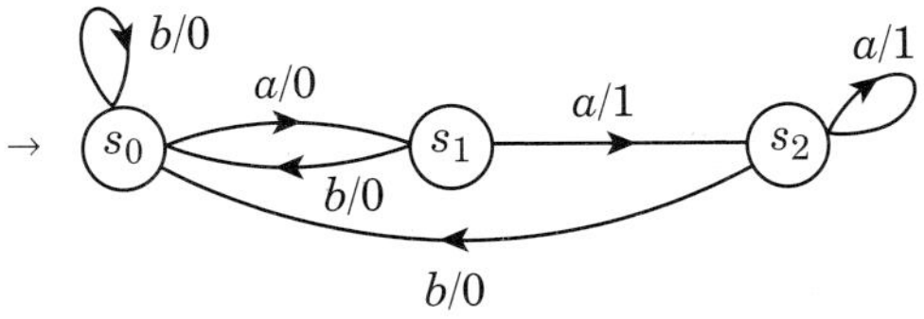

35. *abba* **36.** *aabb* **37.** a^3 **38.** b^3a^4

39. Identify the language accepted by the FSM in Figure 11.42.

Design an FSM accepting strings over $\{a, b\}$ that:

40. Contain *aa* as a substring. **41.** Contain exactly one *a*.

With x an input symbol and s an arbitrary state of an FSM $M = (S, I, O, f, g, s_0)$, define $g(s, x)$ in each case.

42. $f(s, x)$ is an accepting state. **43.** $f(s, x)$ is a nonaccepting state.

11.5 Deterministic Finite-State Automata and Regular Languages

Is the language accepted by an FSA context-sensitive? Or is it context-free, regular, or something else? This section provides a definitive answer to these questions.

In an FSA $M = (S, A, I, f, s_0)$, where $|I| = m$, exactly m outgoing edges leave every state s_i, each labeled with a unique element of I. Besides, since $f : S \times I \to S$, every state–input pair yields a unique state; in other words, every state–input pair uniquely determines the next state.

For the automaton in Figure 11.15, the pair (s_2, a) determines the state s_1, whereas the pair (s_2, b) determines the state s_3. Accordingly, the automata in Section 11.3 are called **deterministic finite-state automata** (DFSA).

This determinism suggests that the language accepted by a DFSA is indeed regular, as the next example demonstrates.

EXAMPLE 11.44 By Example 11.31 the language $L(M)$ accepted by the DFSA in Figure 11.43 consists of words over $\{\text{a}, \text{b}\}$ ending in aa. Employing it, a regular grammar $G = (N, T, P, \sigma)$ can be constructed. Choose $\{\text{a}, \text{b}\}$ as the set of terminal symbols: $T = \{\text{a}, \text{b}\}$. Choose the states as the nonterminal symbols: $N = \{s_0, s_1, s_2\}$. Select the initial state s_0 as the start symbol: $\sigma = s_0$.

Figure 11.43

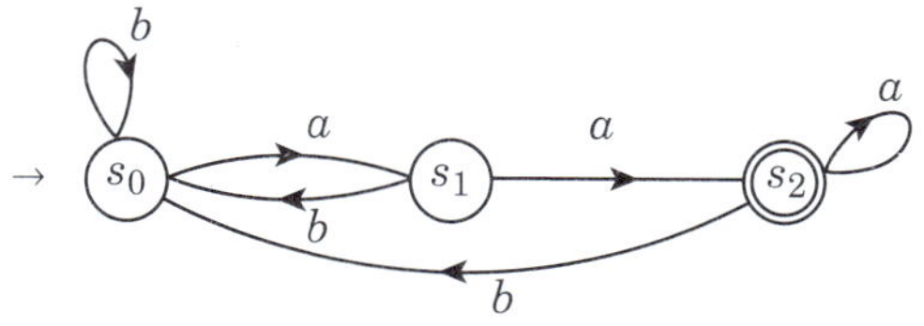

Define the two productions rules:

- If there is an edge labeled x from state s_i to state s_j, define the production $s_i \to \text{x}s_j$. The various productions obtained this way are:

$$s_0 \to as_1, \quad s_0 \to bs_0, \quad s_1 \to as_2,$$
$$s_1 \to bs_0, \quad s_2 \to as_2, \quad \text{and} \quad s_2 \to bs_0.$$

- If there is an edge labeled x from state s_i to an accepting state, induce the production $s_i \to x$. Two additional productions can be obtained by this method:

$$s_1 \to a \quad \text{and} \quad s_2 \to a$$

The grammar $G = (N, T, P, \sigma)$ where N, T, P, and σ are defined as above is clearly regular, therefore $L(G)$ is a regular language. You may verify that $L(G)$ consists of strings over T ending in aa. Thus $L(M) = L(G)$. ■

This example leads us to a fundamental result whose proof resembles that in Example 11.44.

THEOREM 11.3 The language accepted by a DFSA is regular.

PROOF:
Let $M = (S, A, I, f, s_0)$ be a DFSA and $L(M)$ denote the language accepted by the automaton. We shall construct a regular grammar G using the machine M and show that $L(G) = L(M)$.

To construct the grammar $G = (N, T, P, \sigma)$, choose $N = S$ as the set of states, $T = I$ as the input alphabet, and $\sigma = s_0$ as the initial state. Define the productions P this way:

Let s_i and s_j be any two states, and x any input symbol. If $f(s_i, x) = s_j$, define the production $s_i \to xs_j$; if $f(s_i, x) = s_j$, an accepting state, include the production $s_i \to x$. Clearly, G is a regular grammar.

To prove that $L(M) \subseteq L(G)$:

Let $x = x_1x_2 \dots x_n$ be a string accepted by the automaton M; that is, let $x \in L(M)$. Then the transition diagram of the automaton contains a directed path s_0-s_1-s_2-$\cdots$-s_n, where s_n is an accepting state. Correspondingly these production rules follow:

$$\begin{aligned} s_0 &\to x_1 s_1 \\ s_1 &\to x_2 s_2 \\ &\vdots \\ s_{i-1} &\to x_i s_0 i \\ &\vdots \\ s_{n-1} &\to x_n \qquad \text{(Note: } s_n \text{ is an accepting state.)} \end{aligned} \tag{11.4}$$

and the derivation of the string x:

$$\begin{aligned} &\Longrightarrow x_1 s_1 \\ &\Longrightarrow x_1 x_2 s_2 \\ &\vdots \\ &\Longrightarrow x_1 x_2 \dots x_{n-1} s_{n-1} \\ &\Longrightarrow x_1 x_2 \dots x_{n-1} x_n \end{aligned} \tag{11.5}$$

since $s_{n-1} \to x_n$. Thus $x \in L(G)$, so $L(M) \subseteq L(G)$.

Conversely, let $x = x_1x_2 \dots x_n \in L(G)$. Then it must have a derivation of the form (11.4). Correspondingly, the transition diagram of the automaton M must contain a directed path, s_0-s_1-s_2-$\cdots$-s_n. The string determined by this path is $x = x_1x_2 \dots x_n$. Since the last production in the derivation

(11.4) is $s_{n-1} \to x_n$, s_n must be an accepting state, thus $x \in L(M)$ and hence $L(G) \subseteq L(M)$.

Thus $L(M) = L(G)$. In other words, the language accepted by the DFSA is regular. ■

This proof provides an elegant method for finding the regular language accepted by a DFSA. We demonstrate it again in the next example.

EXAMPLE 11.45 Find the grammar of the regular language accepted by the parity check machine in Example 11.33.

SOLUTION:

Using the transition diagram in Figure 11.23, $N = \{E, O\}$, $T = \{a, b\}$, $S = \{E\}$, and the production rules are:

$$E \to \text{a}O, \quad E \to \text{b}E, \quad O \to \text{a}E, \quad O \to \text{b}O, \quad E \to \text{b}, \quad \text{and} \quad O \to \text{a}$$

The regular grammar defined by the parity check machine M is $G = (N, T, P, S)$. [So $L(G) = L(M) =$ the set of strings over T containing an even number of a's.] ■

Finally, is the converse of Theorem 11.3 true? With G a regular grammar, does a DFSA exist such that $L(M) = L(G)$? The next two sections will give us an answer.

Exercises 11.5

Determine if each is a DFSA.

1.

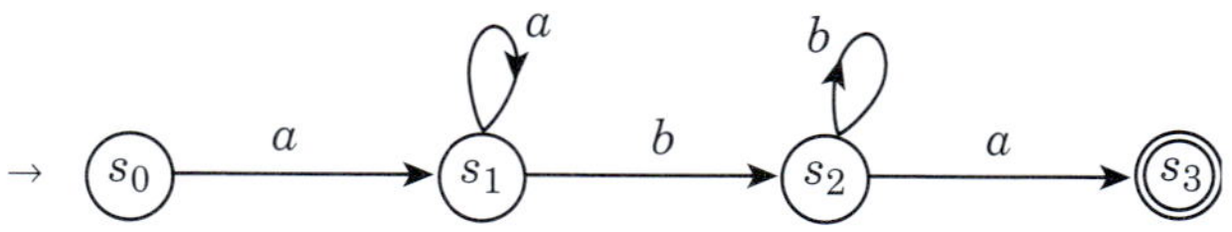

2.

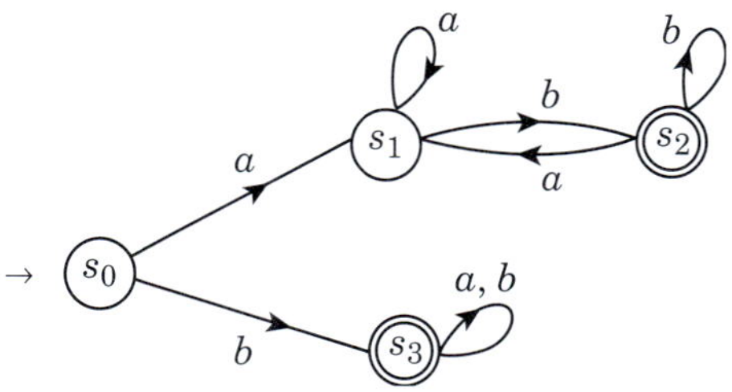

3.

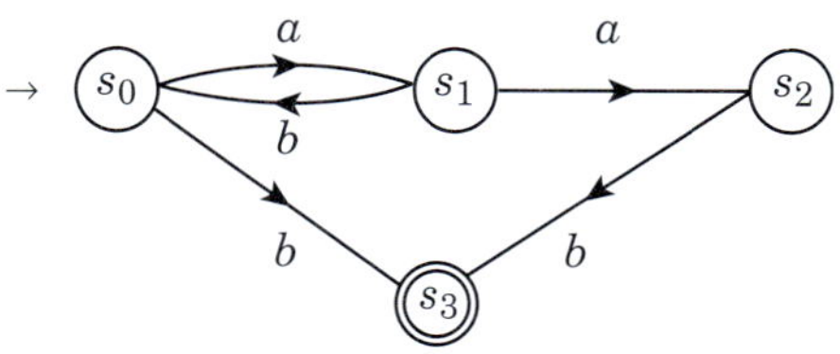

4.

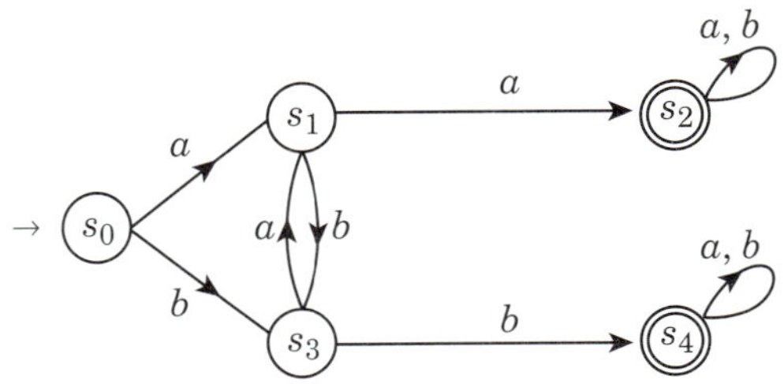

Write the regular grammar defined by the DFSA in each figure.

5. Figure 11.17

6. Figure 11.28

7–14. Construct the regular grammar defined by each DFSA in Exercises 17–24 of Section 11.3.

By making a DFSA, define a regular grammar $G = (N, T, P, \sigma)$ that generates the language consisting of strings over {a, b} that:

15. Contain exactly one a.

16. Contain at least one a.

17. Begin with aa.

18. End with bb.

19. Contain aba as a substring.

20. Contain aaa as a substring.

21. Begin with aa or bb.

22. Contain $baab$ as a substring.

11.6 Nondeterministic Finite-State Automata

We ended the preceding section with a question: For a regular grammar G, is there a DFSA M such that $L(G) = L(M)$? The obvious temptation is to simply reverse the steps in Example 11.44 (or Theorem 11.3) to look for it. Let's see what happens if we do so.

EXAMPLE 11.46

With the regular grammar $G = (N, T, P, \sigma)$, where $N = \{A, \sigma\}$, $T = \{a, b\}$, and $P = \{\sigma \to a\sigma,\ \sigma \to aA,\ A \to b\}$, let us see what happens if we reverse the steps in Theorem 11.3 in order to construct a DFSA $M = (S, A, I, f, s_0)$. Then $I = T = \{a, b\}$ and $s_0 = \sigma$. Corresponding to the productions $\sigma \to a\sigma$ and $\sigma \to aA$, there must be two states, namely, σ and A; besides, by virtue of the production $A \to b$, an accepting state F must exist. Thus S must be $\{\sigma, A, F\}$.

Use the productions to draw the edges in the transition diagram of the automaton: If $s_i \to xs_j$, draw an edge from state s_i to s_j and label it x; if $s_i \to x$, draw an edge from s_i to the accepting state F and label it x. The diagram in Figure 11.44 results.

Unfortunately, it is not a DFSA for two reasons: (1) A state, σ, has two outgoing edges with the same label a; (2) not every state, namely A and F, has two edges with different labels. Thus reversing the steps illustrated in Example 11.44 does *not* yield a DFSA.

Figure 11.44

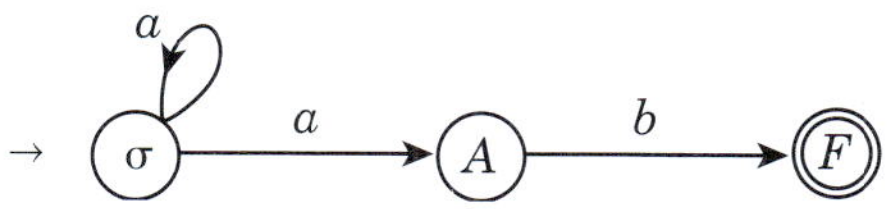

But, fortunately, we have another option. The automaton in Figure 11.44 is a nondeterministic finite-state automaton. "Nondeterministic" means that each state–input pair may determine more than one state. For instance, the pair (σ, a) determines two states, σ and A. If a is input at state σ, two choices exist for the next state: remain at σ or move to A.

We can now move to the following definition.

Nondeterministic Finite-State Automata

A **nondeterministic finite-state automaton** (NDFSA) M exhibits five characteristics:

- A finite set S of **states**;
- A specially designated state σ, called the **initial state**;
- A subset A of S consisting of the **accepting states** (or **final states**) of the automaton;
- A finite set I of **input symbols**;
- A function $f : S \times I \to P(S)$, called the **transition function** (or the **next-state function**). [*Note*: $P(S)$ denotes the power set of S.]

In symbols, $M = (S, A, I, f, \sigma)$.

In an NDFSA, each state–input pair is linked with a set of states, not necessarily a unique state; it can be the null set. A NDFSA can be represented by a transition diagram and a transition table can define a transition function, as the next two examples illustrate.

EXAMPLE 11.47 For the NDFSA in Figure 11.44, $S = \{\sigma, A, F\}$ and $A = \{F\}$. The transition table in Table 11.7 defines the transition function.

Table 11.7

S \ I	a	b
σ	$\{\sigma, A\}$	$\emptyset$
A	$\emptyset$	$\{F\}$
F	$\emptyset$	$\emptyset$

■

EXAMPLE 11.48 The NDFSA $M = (S, A, I, f, \sigma)$, where $S = \{\sigma, \mathrm{A}, \mathrm{B}, \mathrm{C}\}$, $A = \{\mathrm{F}\}$, $I = \{\mathrm{a}, \mathrm{b}\}$, and f is defined by Table 11.8. Its transition diagram is given in Figure 11.45.

Table 11.8

$S \backslash I$	a	b
σ	$\{\sigma, A\}$	$\emptyset$
A	$\{A\}$	$\{B, F\}$
B	$\{F\}$	$\{B\}$
F	$\emptyset$	$\emptyset$

Figure 11.45

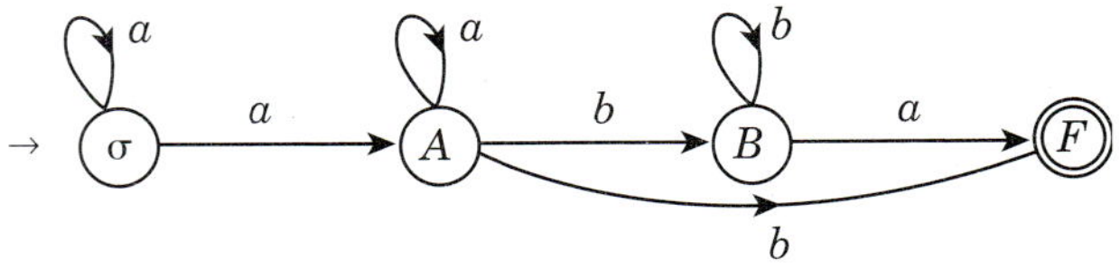

The definition of a string accepted by an FSA can be extended to NDFSA as well.

Equivalent Nondeterministic Finite-State Automata

A string is **accepted** or **recognized** by a NDFSA $M = (S, A, I, f, s_0)$ if a directed path runs from the initial vertex s_0 to an accepting state that generates the string. The language of all strings accepted by M is $L(M)$. Two NDFSAs are **equivalent** if they accept the same language.

The next two examples illustrate the definition of $(L(M))$.

EXAMPLE 11.49 The word $\mathrm{a}^3\mathrm{b}$ is accepted by the NDFSA in Figure 11.44 since the corresponding path, σ-σ-σ-A-F, ends at an accepting state F. Notice that $L(M) = \{\mathrm{a}^n\mathrm{b} \mid n \geq 1\}$. ■

EXAMPLE 11.50 The string $\mathrm{a}^2\mathrm{b}^3\mathrm{a}$ is accepted by the NDFSA in Figure 11.45. Two paths generate it, σ-σ-A-B-B-B-F and σ-A-A-B-B-B-F. The automaton accepts strings $\mathrm{a}^m\mathrm{b}$ and $\mathrm{a}^m\mathrm{b}^n\mathrm{a}$, where $m, n \geq 1$. Thus $L(M) = \{\mathrm{a}^m\mathrm{b}, \mathrm{a}^m\mathrm{b}^n\mathrm{a} \mid m, n \geq 1\}$. ■

The question we posed at the beginning of this section can be partially answered now.

THEOREM 11.4 Every regular language is accepted by an NDFSA.

PROOF:

Let $G = (N, T, P, \sigma)$ be a regular grammar. Through essentially the same steps as in Example 11.46, make a suitable NDFSA $M = (S, A, I, f, s_0)$ such that $L(G) = L(M)$. Select $I = T$, $s_0 = \{\sigma\}$, and N as the set of nonaccepting states of M. Since the grammar contains productions of the form $s_i \to x$, introduce an accepting state F; choose $S = N \cup \{F\}$ and $A = \{F\}$. Finally, since every production of G is $s_i \to xs_j$ or $s_i \to x$, the transition function $f : S \times I \to P(S)$ follows: $f(s_i, x) = \{s_j \mid s_i \to xs_j\} \cup \{F \mid s_i \to x\}$.

As in Theorem 11.3, it can be shown that $L(G) = L(M)$. (Complete the proof.) ■

Although nondeterministic finite-state automata have been defined, an explicit answer to the question posed earlier has yet to surface: Given a regular grammar G, does there exist a DFSA such that $L(G) = L(M)$? We will answer this in the next section.

Exercises 11.6

Draw the transition diagram of the NDFSA $M = (S, A, I, f, s_0)$, where:

1. $S = \{s_0, s_1, s_2)$, $A = \{s_2\}$

S \ I	a	b
s_0	$\{s_1\}$	$\{s_0\}$
s_1	$\{s_1\}$	$\{s_1, s_2\}$
s_2	$\emptyset$	$\emptyset$

2. $S = \{s_0, s_1, s_2\}$, $A = \{s_1\}$

S \ I	a	b
s_0	$\{s_1\}$	$\{s_0\}$
s_1	$\{s_2\}$	$\{s_1, s_2\}$
s_2	$\emptyset$	$\emptyset$

3. $S = \{s_0, s_1, s_2, s_3), A = \{s_2\}$

S \ I	a	b
s_0	$\{s_0, s_1\}$	$\{s_3\}$
s_1	$\{s_1, s_2\}$	$\{s_1\}$
s_2	$\{s_2\}$	$\{s_3\}$
s_3	$\{s_3\}$	$\{s_3\}$

4. $S = \{s_0, s_1, s_2, s_3\}, A = \{s_2\}$

S \ I	a	b
s_0	$\{s_0, s_1\}$	$\{s_3\}$
s_1	$\{s_1, s_2\}$	$\{s_0\}$
s_2	$\emptyset$	$\emptyset$
s_3	$\{s_1\}$	$\{s_3\}$

5. $S = \{s_0, s_1, s_2, s_3, s_4)$, $A = \{s_2, s_3\}$

S \ I	a	b
s_0	$\{s_0, s_1\}$	$\{s_4\}$
s_1	$\{s_1, s_2\}$	$\{s_1, s_3\}$
s_2	$\emptyset$	$\emptyset$
s_3	$\emptyset$	$\emptyset$
s_4	$\{s_4\}$	$\{s_4\}$

6. $S = \{s_0, s_1, s_2, s_3, s_4, s_5)$, $A = \{s_2, s_5\}$

S \ I	a	b
s_0	$\{s_0, s_1\}$	$\{s_4\}$
s_1	$\{s_1, s_2\}$	$\{s_3\}$
s_2	$\{s_2\}$	$\{s_2\}$
s_3	$\{s_3\}$	$\{s_3\}$
s_4	$\{s_3\}$	$\{s_4, s_5\}$
s_5	$\{s_5\}$	$\{s_5\}$

Construct a transition table for each NDFSA.

7.

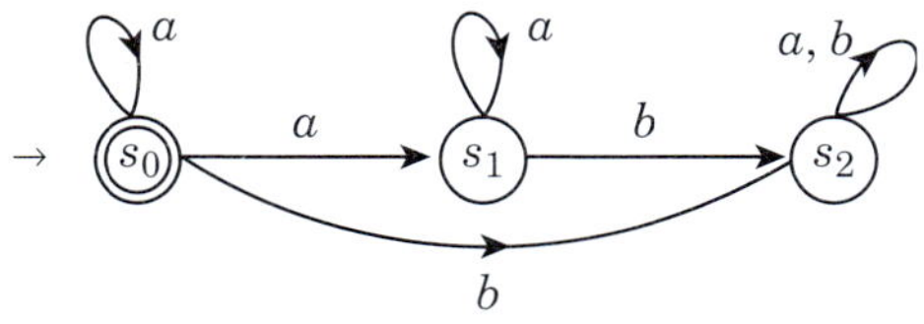

8.

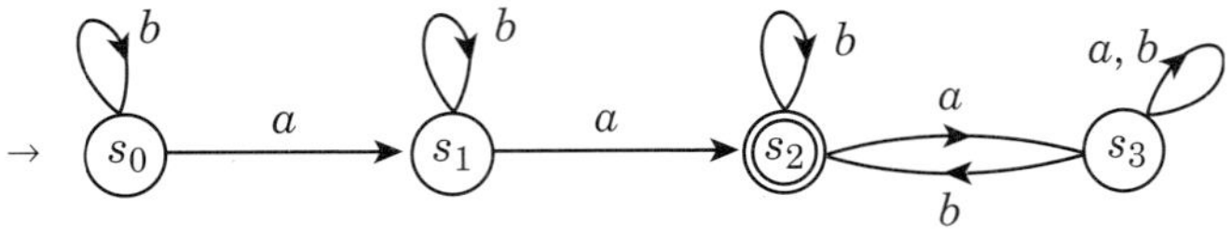

9.

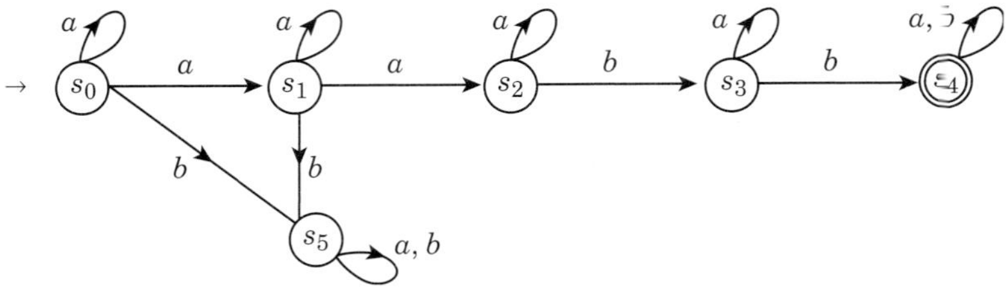

10.

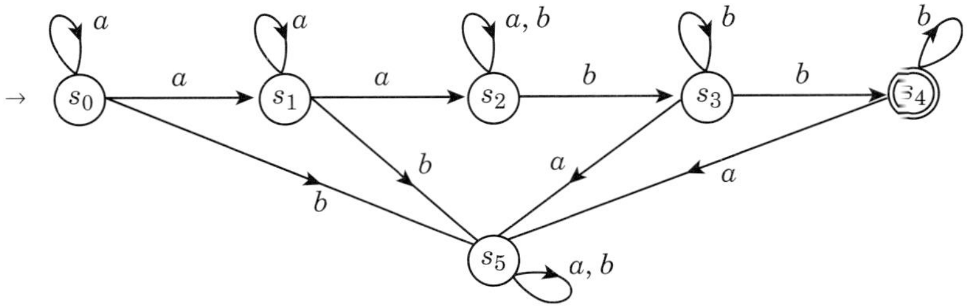

Does the NDFSA in Figure 11.45 accept each string? Identify a path defined by any accepted string.

11. ab^2a **12.** abab

13. a^3b **14.** ab^2ab

Is each string accepted by the NDFSA in Exercise 1? Give a path for accepted strings.

15. a^2b **16.** ab^2a

17. a^3b^3 **18.** $(ab)^3$

Does the NDFSA in Exercise 10 accept each string? Show a path that defines any accepted string.

19. abba **20.** $(ab)^3$

21. a^2b^2 **22.** $a^4b^2ab^3$

Construct a NDFSA that accepts the language generated by the regular grammar $G = (N, T, P, \sigma)$, where:

23. $N = \{\sigma, A, B\}$, $T = \{a, b\}$, and $P = \{\sigma \to aA, A \to aA, A \to bB, B \to bB, A \to a\}$

24. $N = \{\sigma, A, B\}$, $T = \{a, b\}$, and $P = \{\sigma \to aA, \sigma \to bA, A \to aB, \sigma \to b, B \to b\}$

25. $N = \{\sigma, \text{A}, \text{B}, \text{C}, \text{D}\}$, $T = \{\text{a}, \text{b}\}$, and $P = \{\sigma \to \text{b}\sigma,\ \sigma \to \text{aA},\ \text{A} \to \text{aA},\ \text{A} \to \text{bB},\ \text{B} \to \text{aA},\ \text{B} \to \text{bC},\ \text{C} \to \text{aD},\ \text{C} \to \text{b}\sigma,\ \text{D} \to \text{aD},\ \text{D} \to \text{bD},\ \text{C} \to \text{a}\}$

26. $N = \{\sigma, \text{A}, \text{B}, \text{C}\}$, $T = \{\text{a}, \text{b}\}$, and $P = \{\sigma \to \text{b}\sigma,\ \sigma \to \text{aA},\ \text{A} \to \text{aA},\ \text{A} \to \text{bB},\ \text{B} \to \text{aA},\ \text{B} \to \text{bC},\ \text{C} \to \text{aA},\ \text{C} \to \text{b}\sigma,\ \text{B} \to \text{b}\}$

Create a NDFSA that accepts the regular language over $\{\text{a}, \text{b}\}$ of strings that:

27. Contain exactly one a.

28. Contain at least one a.

29. Begin with aa.

30. End with bb.

31. Contain aba as a substring.

32. Contain a^3 as a substring.

33. Begin with aa or bb.

34. Contain ba^2b as a substring.

***35.** Begin with aa, but not end in bb.

***36.** Begin with aa and end in bb.

11.7 Automata and Regular Languages

The preceding two sections demonstrated that the language accepted by a DFSA is regular and that every regular language is accepted by an NDFSA. This section shows that every NDFSA is equivalent to a DFSA, which answers affirmatively our question about the existence of a possible DFSA M such that $L(G) = L(M)$. Every regular language is, in fact, accepted by a suitable DFSA.

The next two examples illustrate step by step how to construct a DFSA equivalent to a given NDFSA.

EXAMPLE 11.51 Consider the regular grammar $G = (N, T, P, \sigma)$, where $N = \{A, \sigma\}$, $T = \{\text{a}, \text{b}\}$, and $P = \{\sigma \to \text{a}\sigma, \sigma \to \text{aA}, \text{A} \to \text{b}\}$. The NDFSA $M = (S, A, I, f, s_0)$ that accepts $L(G)$ is shown in Figure 11.46 (same as Figure 11.44). By Example 11.49, $L(M) = \{\text{a}^n b \,|\, n \geq 1\}$. Using M, we shall construct the DFSA $M' = (S', A', I', f', s'_0)$ which accepts $L(G)$:

Figure 11.46

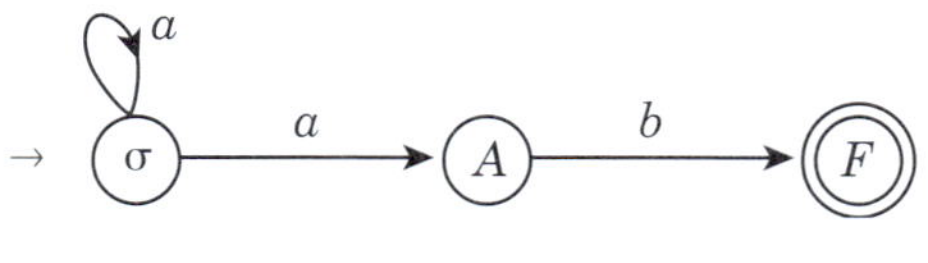

Step 1 Choose $I' = I = \{\text{a}, \text{b}\}$, $s'_0 = \{s_0\} = \{\sigma\}$, and $S' = P(S)$. The various states in M' are subsets of S. If there are n states in M, there can be 2^n states in M', so the states of M' are:

$$\emptyset,\ \{\sigma\},\ \{A\},\ \{F\},\ \{\sigma, A\},\ \{\sigma, F\},\ \{A, F\},\ \text{and}\ \{\sigma, A, F\}$$

Step 2 The accepting states of M' are those states of M' that contain an accepting state of M. They are $\{F\}$, $\{\sigma,F\}$, $\{A,F\}$, and $\{\sigma,A,F\}$.

Step 3 Let $X = \{s_1, s_2, \ldots, s_m\}$ be a state in M'. An input symbol x leads from state X to state Y, where $Y = \bigcup_{i=1}^{m} f(s_i,x)$. In other words, an edge labeled x runs from state X to state Y if $Y = \bigcup_{i=1}^{m} f(s_i,x)$.

Figure 11.46 produces all possible transitions:

$$f(\emptyset,\text{a}) = \emptyset \quad f(\emptyset,\text{b}) = \emptyset \quad f(\sigma,\text{a}) = \{\sigma,A\} \quad f(\sigma,\text{b}) = \emptyset$$
$$f(A,\text{a}) = \emptyset \quad f(A,\text{b}) = \{F\} \quad f(F,\text{a}) = \emptyset \quad f(F,\text{b}) = \emptyset$$

Since $f(\emptyset,\text{a}) = \emptyset = f(\emptyset,\text{b})$, edges run from $\emptyset$ to itself labeled a and b. Since $f(\sigma,\text{a}) = \{\sigma, A\}$ and $f(\sigma,\text{b}) = \emptyset$, an edge labeled a goes from $\{\sigma\}$ to $\{\sigma, A\}$ and an edge b from $\{\sigma\}$ to $\emptyset$. Similarly, there is an edge labeled a from $\{A\}$ to $\emptyset$, an edge b from $\{A\}$ to $\{F\}$, and two edges a and b from $\{F\}$ to $\emptyset$.

Since $f(\sigma,\text{a}) \cup f(A,\text{a}) = \{\sigma, A\} \cup \emptyset = \{\sigma, A\}$, an edge labeled a runs from $\{\sigma, A\}$ to $\{\sigma, A\}$. Also, $f(\sigma,\text{b}) \cup f(\text{A},\text{b}) = \emptyset \cup \{F\} = \{F\}$, so an edge b goes from $\{\sigma,A\}$ to $\{F\}$. Similarly, there are edges labeled a and b from $\{\sigma, F\}$ to $\{\sigma, A\}$ and $\emptyset$, respectively; edges a and b from $\{A, F\}$ to $\emptyset$ and $\{F\}$, respectively; and edges a and b from $\{\sigma, A, F\}$ to $\{\sigma, A\}$ and $\{F\}$, respectively.

These results appear in the transition table in Table 11.9.

Table 11.9

S' \ I'	a	b
$\emptyset$	$\emptyset$	$\emptyset$
$\{\sigma\}$	$\{\sigma,A\}$	$\emptyset$
$\{A\}$	$\emptyset$	$\{F\}$
$\{F\}$	$\emptyset$	$\emptyset$
$\{\sigma,A\}$	$\{\sigma,A\}$	$\{F\}$
$\{\sigma,F\}$	$\{\sigma,A\}$	$\emptyset$
$\{A,F\}$	$\emptyset$	$\{F\}$
$\{\sigma,A,F\}$	$\{\sigma,A\}$	$\{F\}$

Figure 11.47 shows the resulting DFSA.

Since the states $\{A\}$, $\{\sigma, F\}$, $\{A, F\}$, and $\{\sigma, A, F\}$ cannot be reached from the initial state $\{\sigma\}$, they can be dropped out to yield the simplified DFSA M' in Figure 11.48.

From this transition diagram, $L(M') = \{\text{aa}^n\text{b} \mid n \geq 0\} = \{\text{a}^n\text{b} \mid n \geq 1\} = L(G)$. Thus the automata M and M' are equivalent, so the NDFSA is the same as the DFSA. ■

EXAMPLE 11.52 Construct a DFSA $M' = (S', A', I', f', s'_0)$ equivalent to the NDFSA $M = (S, A, I, f, s_0)$ in Example 11.50. Recall that $L(M) = \{\text{a}^m\text{b}, \text{a}^m\text{b}^n\text{a} \mid m, n \geq 1\}$. The key steps lie below. (Fill in the details.)

Figure 11.47

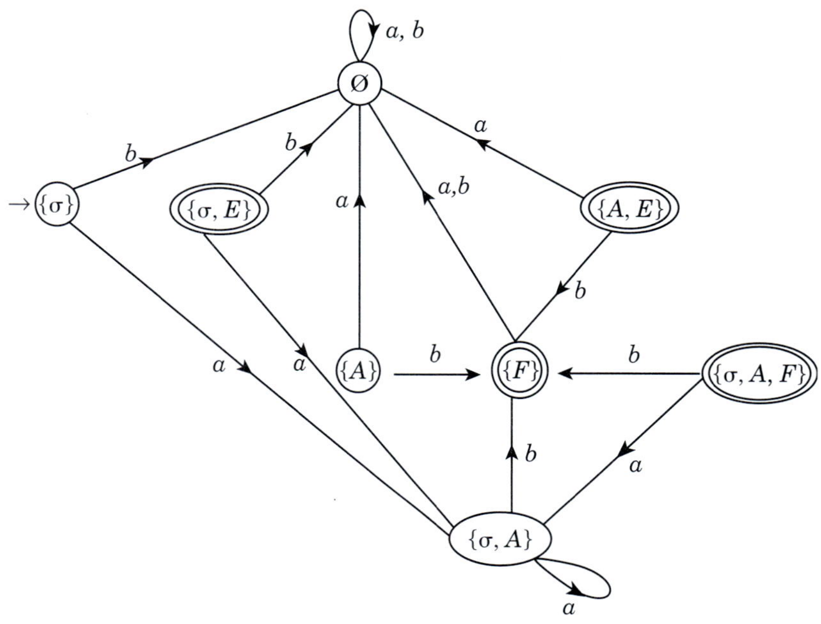

Figure 11.48

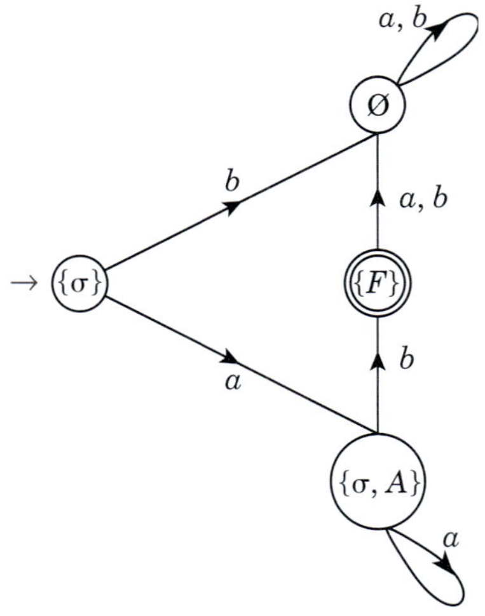

SOLUTION:

Step 1 Select $I' = I = \{a,b\}$, $s_0' = \{s_0\} = \{\sigma\}$, and $S' = P(S)$. The states of M' are Ø, $\{\sigma\}$, $\{A\}$, $\{B\}$, $\{F\}$, $\{\sigma,A\}$, $\{\sigma,B\}$, $\{\sigma,F\}$, $\{A,B\}$, $\{A,F\}$, $\{B,F\}$, $\{\sigma,A,B\}$, $\{\sigma,A,F\}$, $\{\sigma,B,F\}$, $\{A,B,F\}$, and $\{\sigma,A,B,F\}$.

Step 2 The accepting states of M′ are $\{F\}$, $\{\sigma,F\}$, $\{A,F\}$, $\{B,F\}$, $\{\sigma,A,F\}$, $\{\sigma,B,F\}$, $\{A,B,F\}$, and $\{\sigma,A,B,F\}$.

Step 3 The transition table of the DFSA is Table 11.10.

Table 11.10

S \ I	a	b
$\emptyset$	$\emptyset$	$\emptyset$
$\{\sigma\}$	$\{\sigma, A\}$	$\emptyset$
$\{A\}$	$\{A\}$	$\{B, F\}$
$\{B\}$	$\{F\}$	$\{B\}$
$\{F\}$	$\emptyset$	$\emptyset$
$\{\sigma, A\}$	$\{\sigma, A\}$	$\{B, F\}$
$\{\sigma, B\}$	$\{\sigma, A, F\}$	$\{B\}$
$\{\sigma, F\}$	$\{\sigma, A\}$	$\emptyset$
$\{A, B\}$	$\{A, F\}$	$\{B, F\}$
$\{A, F\}$	$\{A\}$	$\{B, F\}$
$\{B, F\}$	$\{F\}$	$\{B\}$
$\{\sigma, A, B\}$	$\{\sigma, A, F\}$	$\{B, F\}$
$\{\sigma, A, F\}$	$\{\sigma, A\}$	$\{B, F\}$
$\{\sigma, B, F\}$	$\{\sigma, A, F\}$	$\{B\}$
$\{A, B, F\}$	$\{A, F\}$	$\{B, F\}$
$\{\sigma, A, B, F\}$	$\{\sigma, A, F\}$	$\{B, F\}$

Step 4 The table indicates the states $\{\sigma, B\}$, $\{\sigma, F\}$, $\{A, B\}$, $\{\sigma, A, B\}$, $\{\sigma, B, F\}$, $\{A, B, F\}$, and $\{\sigma, A, B, F\}$ are not reachable from any state, so they are *not* the initial state $\{\sigma\}$. Delete the corresponding rows from the table. It is now obvious from the table that the states $\{A\}$, $\{A, F\}$, $\{\sigma, A, F\}$ also cannot be reached from $\{\sigma\}$; delete those rows also from the table.

The resulting transition diagram of the DFSA M' appears in Figure 11.49.

Figure 11.49

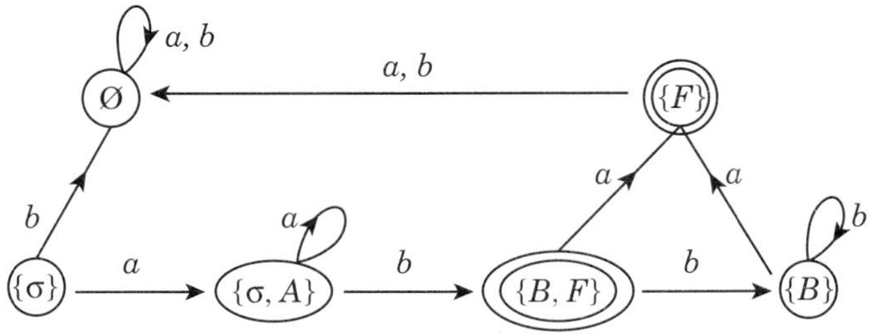

From the diagram, it follows that $L(M') = \{a^m b, a^m b^n a \mid m, n \geq 1\} = L(M)$. Thus M and M' are equivalent automata. As in the previous example, we have shown that the equivalency between an NDFSA and a DFSA. ■

The techniques illustrated in the two previous examples can be generalized to arrive at the following result. (The proof is a bit complicated, so we omit it.)

THEOREM 11.5 Every NDFSA is equivalent to a DFSA. ■

The next theorem follows from Theorems 11.3, 11.4, and 11.5.

THEOREM 11.6 A language is regular if and only if it is accepted by a DFSA. ■

As Theorem 11.6 indicates, a DFSA can define a regular grammar and vice versa. Each is a characterization of the other.

> We now look for an example of a simple-looking language that is *not* regular.

EXAMPLE 11.53 Show that the language $L = \{a^n b^n \mid n \geq 1\}$ is not regular.

PROOF (by contradiction):
Suppose L is regular. Then, by Theorem 11.6, a DFSA M exists such that $L(M) = L$. Suppose M has m states. Since the string $x = a^{m+1}b^{m+1} \in L$, x is accepted by the DFSA. Let P be the path corresponding to x; it ends at an accepting state F.

Figure 11.50

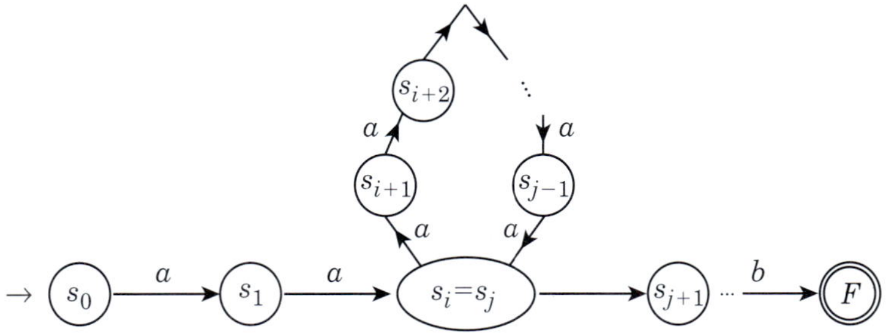

The path corresponding to the substring a^{m+1} contains $m+1$ states. But, since only m states exist, by the pigeonhole principle at least two of the $m+1$ states, say, s_i and s_j, where $i < j$, must be the same; consequently, there must be a directed cycle at s_i, each edge labeled a (see Figure 11.50). Let l be the length of the cycle. The path s_0-s_1-$\cdots$-s_i-s_{j+1}-s_{j+2}-$\cdots$-F generates the string $x' = a^{m+1-l}b^{m+1}$. Since this path ends at F (an accepting state), x' is accepted by the automaton; so $x' \in L$. This is a contradiction, since x' does not contain the same number of a's and b's. Thus L is not a regular language. ■

It follows by this example that the set of well-formed nested parentheses is also *not* a regular language. (Why?)

These discussions lead us to a powerful conclusion: Regular languages are accepted by DFSAs.

Exercises 11.7

1–6. Construct a DFSA equivalent to each NDFSA in Exercises 1–4, 7, 8 of Section 11.6. Eliminate all unreachable states.

7–8. Design DFSAs equivalent to the NDFSAs in Exercises 23 and 24 of Section 11.6. Eliminate all unreachable states.

Let L be the language recognized by an FSA and $L^R = \{x_n \dots x_1 \mid x_1 \dots x_n \in L\}$. Construct an NDFSA that accepts L^R from each FSA in Exercises 9–16. (*Hint:* Reverse the directions of the edges; switch the roles of the initial state and the accepting states.)

9. Figure 11.20

10. Exercise 17 in Section 11.3

11. Exercise 18 in Section 11.3

12. Exercise 36 in Section 11.3

13. Exercise 37 in Section 11.3

14. Exercise 38 in Section 11.3

15. Exercise 40 in Section 11.3

16. Exercise 41 in Section 11.3

17–24. Identify the language $L(M)$ accepted by the FSA in Exercises 9–16.

25–32. Construct a DFSA equivalent to the NDFSA in Exercises 9–16.

Chapter Summary

The abstract models of computing machines with limited capabilities are DFSA, FSM, and NDFSA. An automaton checks if a given input string has a desired property and produces no output values. An FSM, on the other hand, yields an output value corresponding to each input.

Formal Language

- A **formal language** over an alphabet Σ is a subset of Σ^*, the set of all possible words over Σ (page 734).
- The **concatenation** of two languages A and B over Σ consists of words ab with $a \in A$ and $b \in$ B (page 736).
- $A^n = \{a_1 a_2 \dots a_n \mid a_i \in A\}$, where $A^0 = \Lambda$ (page 739).
- $A^* = \bigcup_{n=0}^{\infty} A^n$ is the **Kleene closure** of a language A (page 739).

Grammar

- A **grammar** $G = (N, T, P, \sigma)$ consists of a finite set N of **nonterminal symbols**, a finite set T of **terminal symbols**, a finite set P of **production rules**, and a **start symbol** σ (page 746).

- A word $w' = x\beta y$ is **directly derivable** from a word $w = x\alpha y$ if a production $\alpha \to \beta$ exists; we then write $w \Rightarrow w'$. A word w_n is **derivable** from w_1 if there exists a finite sequence of **derivations**, $w_1 \Rightarrow w_2, w_2 \Rightarrow w_3, \ldots, w_{n-1} \Rightarrow w_n$. The language derivable from σ is the **language generated** by G, denoted by $L(G)$ (page 746).
- In BNF, each production $w \to w'$ is written as $w ::= w'$; each nonterminal symbol w is enclosed by angle brackets, as in $\langle w \rangle$; and production rules with the same left-hand sides are combined with vertical bars into a single rule (page 748).
- A **type 0** grammar has phrase-structure (page 751).
- In a **context-sensitive** (**type 1**) grammar, every production is of the form $\alpha A \alpha' \to \alpha\beta\alpha'$. (page 751)
- In a **context-free** (**type 2**) grammar, every production is of the form $A \to \alpha$. (page 751).
- In a **regular** (**type 3**) grammar, every production is of the form A$\to t$ or $A \to t$B (page 751).
- A language $L(G)$ is **context-sensitive**, **context-free**, or **regular** according as whether G is context-sensitive, context-free, or regular (page 751).
- An **ambiguous** language contains a word that has more than one derivation tree (page 753).
- The language accepted by a DFSA is regular (page 780).

Finite-State Automaton (FSA)

- A **FSA** $M = (S, A, I, f, s_0)$ consists of a finite set S of **states**, a finite set A of **accepting states**, a finite set I of **input symbols**, a **transition function** $f : S \times I \to S$, and an **initial state** s_0. Every state–input pair yields a unique next-state of the automaton (page 761).
- A **transition table** defines the transition function. (page 761).
- A **transition diagram** can represent a DFSA. The initial state s_0 is identified by drawing an arrow toward it; an accepting state by two concentric circles around it (page 762).
- An input string is **accepted** by an automaton M if and only if the string traces a path that ends at an accepting state. The **language** $L(M)$ **accepted** by M consists of all words recognized by it (page 764).
- Two automata M and M' are **equivalent** if $L(M) = L(M')$ (page 764).

Finite-State Machine (FSM)

- An **FSM** $M = (S, I, O, f, g, s_0)$ consists of a finite set S of states, a finite set I of **input symbols**, a finite set O of **output symbols**, a **transition function** $f : S \times I \to S$, an **output function** $g : S \times I \to O$, and an **initial state** s_0. Every state–input pair produces a next-state and an output value (page 772).
- A **transition table** can define the transition and output functions of an FSM (page 772).
- A **transition diagram** also can define an FSM (page 772).

Nondeterministic Finite-State Automaton (NDFSA)

- An **NDFSA** $M = (S, A, I, f, \sigma)$ consists of a finite set S of states, a subset A of S of accepting states, a finite set I of input symbols, a transition function $f : S \times I \to P(S)$, and an initial state σ. A state–input pair may be paired with zero, one, or more states (page 783).
- Every regular language is accepted by a NDFSA (page 784).
- Every NDFSA is equivalent to a DFSA (page 787).
- Every regular language is accepted by a DFSA (page 787).

Review Exercises

Let $A = \{\lambda, \text{a}, \text{bc}\}$ and $B = \{\text{a}, \text{ab}\}$. Find each.

1. AB **2.** BA **3.** A^3 **4.** B^3

Find three words belonging to each language over {a,b,c}.

5. $\{\text{a,b}\}\{\text{c}\}^*$ **6.** $\{\text{a}\}\text{b}^*\{\text{c}\}^*$

7. $\{\text{ab}\}\{\text{ab}\}^*$ **8.** $\{\text{b}\}\{\text{a,b,c}\}^*\{\text{b}\}$

A grammar $G = (N, T, P, \sigma)$ has N = {⟨noun phrase⟩,⟨verb⟩,⟨adjective⟩, ⟨noun⟩, ⟨article⟩}, T = {a,the,chicken,wolf,cabbage,eats,walks,reliable, discreet, gracious}, σ = ⟨sentence⟩, and the production rules are:

⟨sentence⟩ → ⟨noun phrase⟩⟨verb⟩⟨noun phrase⟩

⟨noun phrase⟩ → ⟨article⟩⟨noun⟩ | ⟨article⟩⟨adjective⟩⟨noun⟩

⟨article⟩ → *the* | *a*

⟨noun⟩ → *chicken* | *wolf* | *cabbage*

⟨adjective⟩ → *reliable* | *discreet* | *gracious*

⟨verb⟩ → *eats* | *walks*

Determine if each is a valid sentence in $L(G)$.

9. The gracious chicken walks the wolf.

10. The reliable wolf eats a chicken.

Make a derivation tree for each sentence.

11. The discreet wolf eats the cabbage.

12. The reliable cabbage walks the gracious chicken.

Using the grammar $G = (N, T, P, \sigma)$ where $N = \{\sigma, A, B\}$, $T = \{\mathrm{a,b}\}$, and $P = \{\sigma \to \mathrm{b}\sigma,\ \sigma \to \mathrm{a}A,\ A \to \mathrm{a}B,\ A \to \mathrm{b}\sigma,\ B \to \mathrm{a}B,\ B \to \mathrm{b}B,\ A \to \mathrm{a},\ B \to \mathrm{a},\ B \to \mathrm{b}\}$, determine if each string belongs to $L(G)$.

13. $\mathrm{ab^3a}$ **14.** $\mathrm{(ab)^3}$ **15.** $\mathrm{aba^2b}$ **16.** $\mathrm{ab^2a^4}$

Construct a parse tree for each string.

17. $\mathrm{ba^2b}$ **18.** $\mathrm{b^2a^3b}$ **19.** $\mathrm{aba^2}$ **20.** $\mathrm{a^2b^2a^2}$

Develop a grammar that generates each language over {a,b}.

21. $\{\mathrm{b}^n \mid n \geq 1\}$ **22.** $\{\mathrm{a}^n\mathrm{ba}^n \mid n \geq 0\}$ **23.** $\{\mathrm{b}^{2n+1} \mid n \geq 0\}$ **24.** $\{\mathrm{ab}^n\mathrm{a} \mid n \geq 0\}$

With the grammar below, construct parse trees for the simple **while statements** in Exercises 25 and 26.

⟨while statement⟩ ::= while ⟨expression⟩ do ⟨statement⟩

⟨statement⟩ ::= ⟨assignment statement⟩ | ⟨while statement⟩ | λ

⟨assignment statement⟩ ::= ⟨variable⟩ := ⟨expression⟩

⟨variable⟩ ::= $a \mid b \mid c \mid \ldots \mid z$

⟨expression⟩ ::= ⟨variable⟩⟨sign⟩⟨variable⟩ |
⟨variable⟩⟨operator⟩⟨variable⟩

⟨operator⟩ ::= $= \mid \neq \mid < \mid \leq \mid > \mid \geq$

⟨sign⟩ ::= $+ \mid -$

○ **25.** While $x \geq y$ do $x := y + z$.

○ **26.** While $x \geq y$ do while $y < z$ do $a := b + c$.

27. Draw the transition diagram of the *DFSAM* $= (S, \mathrm{A}, I, f, s_0)$, where $S = \{s_0, s_1, s_2, s_3, s_4\}$, $A = \{s_3\}$, $I = \{\mathrm{a,b}\}$, and f is defined by Table 11.11.

28. Redo Exercise 27 with $A = \{s_1, s_3\}$ and f defined by Table 11.12.

Construct the transition table for each DFSA.

Table 11.11

S \ I	a	b
s_0	s_1	s_4
s_1	s_4	s_2
s_2	s_4	s_3
s_3	s_3	s_3
s_4	s_4	s_4

Table 11.12

S \ I	a	b
s_0	s_1	s_2
s_1	s_1	s_1
s_2	s_4	s_3
s_3	s_3	s_3
s_4	s_4	s_4

29.

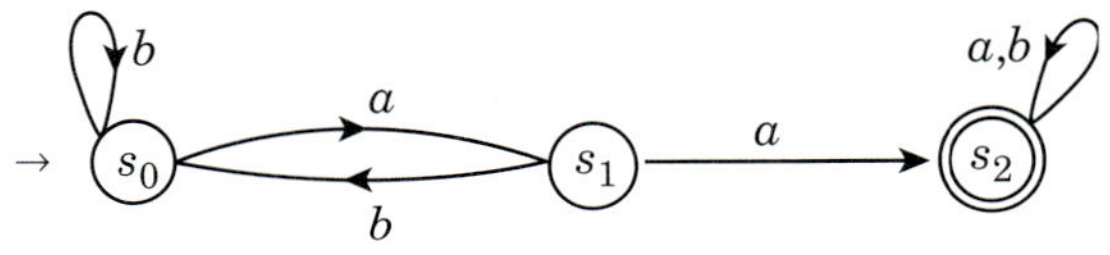

30.

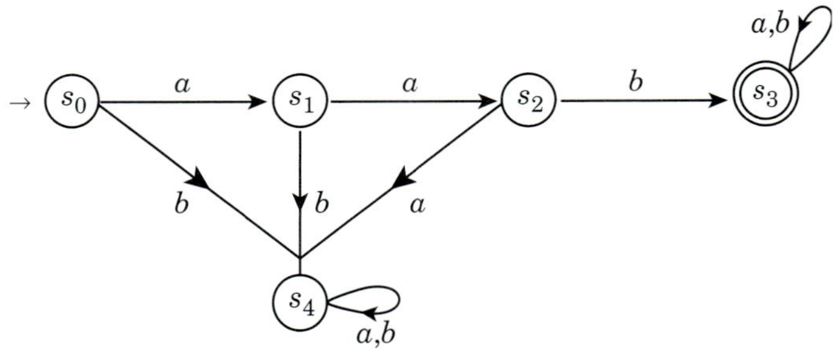

31–34. Identify the language $L(M)$ accepted by the automata in Exercises 27–30.

Design a DFSA that accepts strings over {a, b} that:

35. Begin with *aaa*.

36. Contain *abb* as a substring.

37.

	f		g	
S \ I	a	b	a	b
s_0	s_0	s_1	1	0
s_1	s_2	s_2	0	0
s_2	s_0	s_3	1	0
s_3	s_3	s_2	1	0

38.

	f		g	
S \ I	a	b	a	b
s_0	s_0	s_1	0	0
s_1	s_2	s_3	0	1
s_2	s_3	s_2	1	1
s_3	s_3	s_2	1	0

Construct a transition table for each FSM.

39.

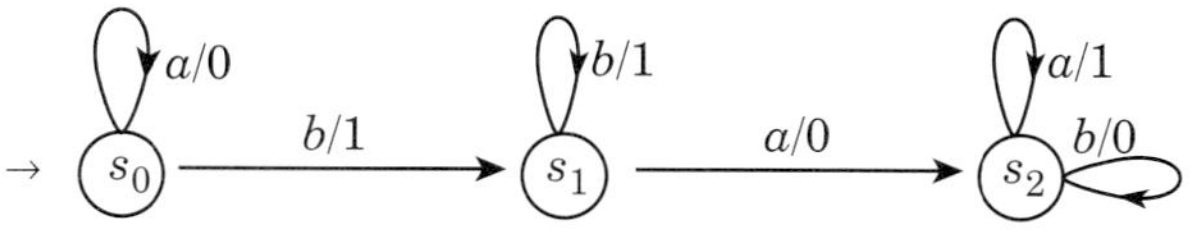

40.

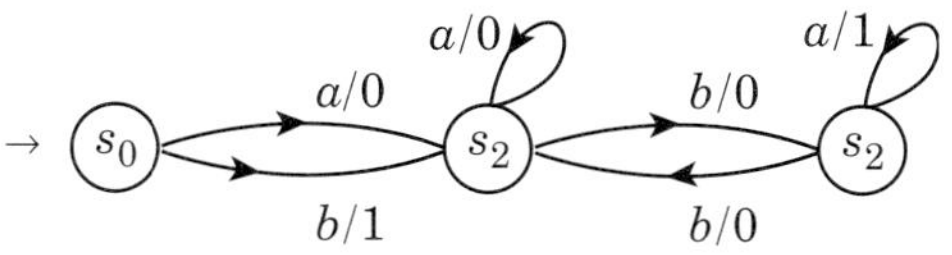

Using the FSM in Figure 11.33 (Example 11.38), find the output from each input string.

41. a^2b^2aba **42.** aba^2ba **43.** ab^3ab **44.** $a^2b^3a^2$

45–46. Redraw the DFSAs in Exercises 27 and 29 as FSMs.

Design an FSM to accept string over {a, b} that:

47. Contain ab^2 as a substring. **48.** Begin with a or b^2.

49–50. Compose the regular grammar defined by the DFSA in Exercises 27–28.

Draw the transition diagram of the NDFSA $M = (S, A, I, f, s_0)$, where $I = \{a,b\}$ and:

51. $S = \{s_0, s_1, s_2), A = \{s_1\}$

S \ I	a	b
s_0	$\{s_1\}$	$\{s_0\}$
s_1	$\{s_2\}$	$\{s_1, s_2\}$
s_2	$\{s_2\}$	$\{s_2\}$

52. $S = \{s_0, s_1, s_2\}, A = \{s_1\}$

S \ I	a	b
s_0	$\{s_1\}$	$\{s_0\}$
s_1	$\{s_0, s_2\}$	$\{s_1\}$
s_2	$\{s_1\}$	$\{s_0, s_2\}$

Construct a transition table for each NDFSA.
Determine if the NDFSA in Exercise 52 accepts each input string.

53.

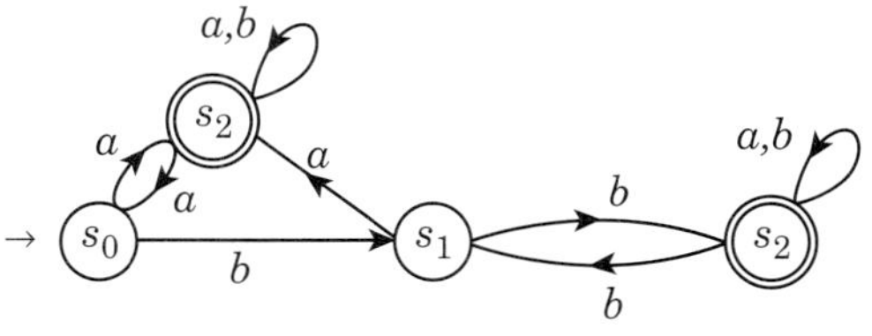

54.

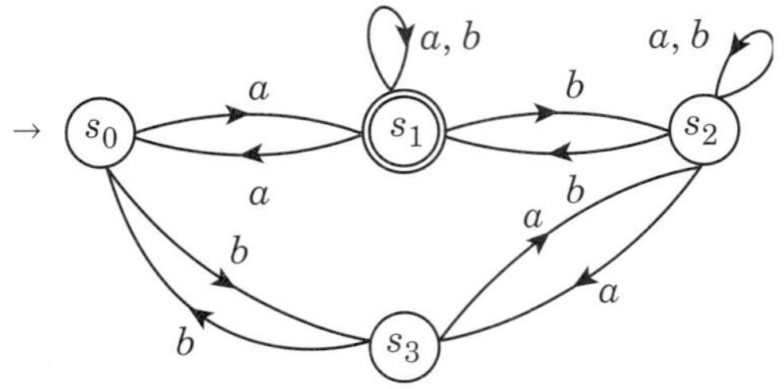

55. a^3 **56.** ab^2ab^4 **57.** a^2b^3 **58.** a^3b^4

59–62. Determine if each input string in Exercises 55–58 is accepted by the NDFSA in Exercise 54.

Create a NDFSA that accepts the language $L(G)$ generated by the regular grammar $G = (N, T, P, \sigma)$, where:

63. $N = \{\sigma, A, B\}$, $T = \{A, B\}$, and $P = \{\sigma \to \text{b}\sigma,\ \sigma \to \text{a}A,\ A \to \text{a}B,\ A \to \text{b}A,\ A \to \text{b}B,\ B \to \text{a}B,\ B \to \text{b}B,\ \sigma \to \text{a},\ A \to \text{b}\}$

64. $N = \{\sigma, A, B\}$, $T = \{\text{a,b}\}$, and $P = \{\sigma \to \text{a}A,\ \sigma \to \text{b}\sigma,\ A \to \text{a}\sigma,\ A \to \text{a}B,\ A \to \text{b}A,\ B \to \text{a}B,\ B \to \text{b}\sigma,\ B \to \text{b}A,\ \sigma \to \text{a},\ A \to \text{b}\}$

65–66. Construct a DFSA equivalent to each NDFSA in Exercises 51 and 52.

67–68. What languages do the DFSAs in Exercises 65 and 66 accept?

Let A and B be any languages over a finite alphabet Σ. Prove each.

***69.** $(A \cup B^*)^* = (A^* \cup B)^*$ ****70.** $(A \cup B)^* = (A^* \cup B^*)^* = (A^*B^*)^*$

Supplementary Exercises

Let m denote the number of a's and n the number of b's in a string over {a,b}. Design an FSA that accepts strings with the given properties.

1. $m \equiv 1 \pmod 2$ and $n \equiv 1 \pmod 2$.

2. $m \equiv 0 \pmod 3$ and $n \equiv 1 \pmod 3$.

3. $m \equiv 0 \pmod 2$ and $n \equiv 1 \pmod 2$, or $m \equiv 1 \pmod 2$ and $n \equiv 0 \pmod 2$.

4. Design an FSA that accepts positive integers n divisible by 3.

○ ***5.** Using the syntax diagram in Figure 11.51 for a real number, design an FSA to recognize valid real numbers.

***6.** The Roman numerals M, D, C, L, X, V, and I have values 1000, 500, 100, 50, 10, 5, and 1, respectively. In the strict additive notation no numeral with a smaller value precedes a numeral with a larger value. For instance, 19 is written as XVIIII in lieu of the shorter

Figure 11.51

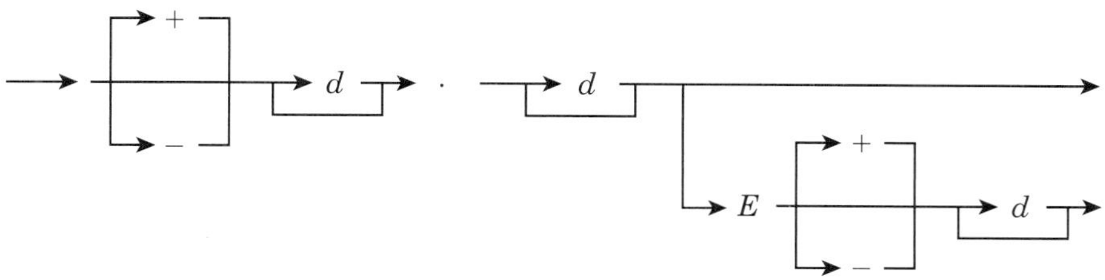

representation XIX and MMLXXVI, unlike MCMXCVI, is a well-formed sequence. Excepting M's, C, X, or I should not appear more than four times in the same sequence, and D, L, or V no more than once. This makes sense since CCCCC = D, XXXXX = L, and so on. Design an FSA to recognize the language of such well-formed sequences of additive Roman numerals.

Develop a grammar that generates each language over {a,b}.

***7.** The set of words that begin and end with a.

***8.** The set of words that begin with aa and end with bb.

9. Using BNF, define a grammar for the language of well-formed nested parentheses.

10. Use productions instead of BNF to define the grammar in Exercise 9.

A **Moore machine** $M = (S, I, O, f, g, s_0)$, named after Edward Moore who introduced it in 1956, is an FSM consisting of a finite set S of states, a finite set I of input symbols, a finite set O of output symbols, a transition function $f : S \times I \to S$, an output function $g : S \to O$, and an initial state s_0. Draw a transition diagram for the Moore machine defined by each transition table.

11.

	f input		**g**
s	**0**	**1**	
s_0	s_0	s_1	1
s_1	s_3	s_2	0
s_2	s_2	s_3	1
s_3	s_0	s_1	1

12.

	f input		**g**
s	**0**	**1**	
s_0	s_1	s_3	0
s_1	s_1	s_2	1
s_2	s_3	s_2	1
s_3	s_1	s_2	0

Construct a transition table for each Moore machine.

13.

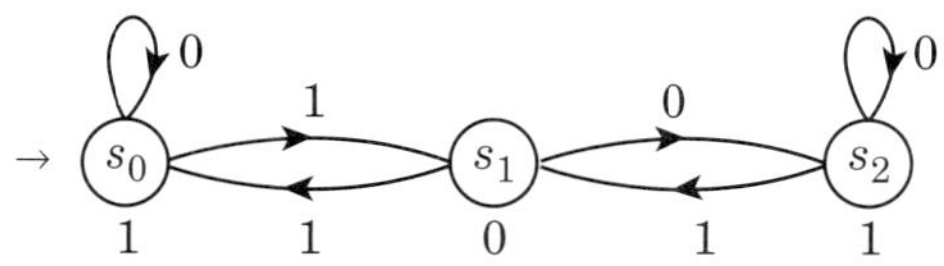

Edward Forrest Moore *(1925–) was born in Baltimore, Maryland. He graduated from Virginia Polytechnic Institute in 1947 and received his Ph.D. in mathematics from Brown 3 years later. After teaching at the University of Illinois for a year, he joined the technical staff at Bell Telephone Labs. In 1966, he joined the faculty of the University of Wisconsin, Madison, and taught there until his retirement in 1985.*

Moore has made outstanding contributions to the logical design of switching circuits, automata theory, graph theory, and database management.

14.

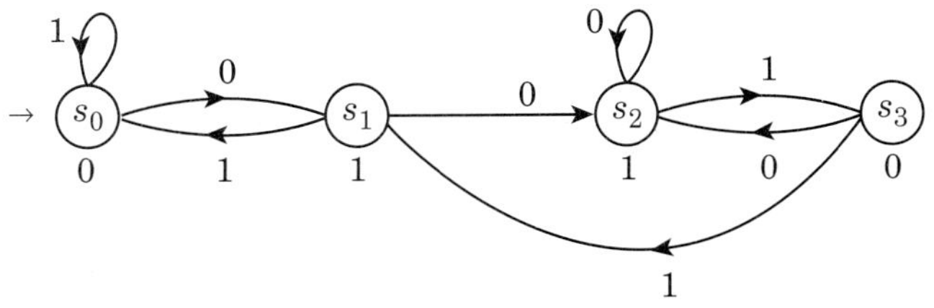

The output generated by the Moore machine $M = (S, I, O, f, g, s_0)$ for the input string $a_1a_2 \ldots a_m$ is $g(s_0)g(s_1)\ldots g(s_m)$, where $s_i = f(s_{i-1}, a_i)$ and $1 \le i \le m$. Find the output produced by the machine in Exercise 13 for each input.

15. 011 **16.** 1010 **17.** 10001 **18.** 1101101

19. Let L be a regular language. Prove that $L^R = \{x_n \ldots x_1 \mid x_1 \ldots x_n \in L\}$ is also regular.

****20.** An FSM $M = (S, I, O, f, g, s_0)$ is **simply minimal** if no output rows in its transition table are identical. If $|S| = n$, $|I| = m$, and $|O| = p$, how many simply minimal FSMs are possible?

Computer Exercises

Write a program to do each task, where $\Sigma = \{a,b\}$.

1. Determine if a string over Σ:

- Begins with aa.
- Contains exactly one a.
- Ends with bb.
- Contains at least one a.
- Contains aba as a substring.
- Has its number of a's congruent to 1 mod 3.

- Has an even number of a's and b's.
- Has both its number of a's and b's congruent to 1 mod 3.

2. Let m denote the number of a's and n the number of b's in a string over Σ. Read in a word over Σ and see if it has:

- $m \equiv 0 (\text{mod } 5)$
- $m \equiv 3 (\text{mod } 5)$
- $m \equiv 0 (\text{mod } 3)$ or $m \equiv 1 (\text{mod } 3)$
- $m \equiv 0 (\text{mod } 3)$ or $m \equiv 2 (\text{mod } 3)$
- $m \equiv 0 (\text{mod } 3)$ and $n \equiv 1 (\text{mod } 3)$
- $m \equiv 0 (\text{mod } 3)$ and $n \equiv 2 (\text{mod } 3)$

3. For a DFSA with n (≤ 10) states, labeled 1 through n, read in its number of states n and transition table. Read in a sequence of input strings over Σ and determine if each is accepted by the DFSA.

4. Implement the unit delay machine in Example 11.41.

5. Read in two binary numbers and use the FSM in Example 11.42 to compute their sum.

○ **6.** By means of the syntax diagram in Figure 11.31, determine if a string of characters represents a valid integer.

○ **7.** Ascertain with the syntax diagram in Figure 11.32 whether a string of characters represents a valid real number. (Excluding the exponential form.)

8. Read in the number of states n, the transition table, and a set of input strings for an FSM with n (≤ 10) states, labeled 1 through n. Print the output produced by each input string.

9. Using a DFSA with n (≤ 10) states, labeled 1 through n, read its number of states n and transition table. Determine the corresponding regular grammar.

Exploratory Writing Projects

Using library and Internet resources, write a team report on each of the following in your own words. Provide a well-documented bibliography.

1. Discuss how BNF rules are used to define programming languages such as C++ and Java.

2. Discuss Turing machines and Church's thesis.

3. Explain how vending machines, slot machines, and garage door openers can be modeled by FSAs.

4. Write an essay on Kleene closure.

5. Write an essay on different types of FSMs and their applications.

Enrichment Readings

1. W. J. Barnier, "Finite-State Machines as Recognizers," *The UMAP Module 671* (1986), pp. 209–232.
2. B. Hayes, "On the Finite-State Machine, a Minimal Model of Mousetraps, Ribosomes, and the Human Soul," *Scientific American*, Vol. 249 (Dec. 1983), pp. 19–28, 178.
3. J. E. Hopcroft and J. D. Ullman, *Introduction to Automata Theory, Languages, and Computation*, Addison-Wesley, Reading, MA, 1979.
4. Z. Kohavi, *Switching and Finite Automata Theory*, 2nd ed., McGraw-Hill, New York, 1978.
5. P. Linz, *An Introduction to Formal Languages and Automata*, D. C. Heath, Lexington, MA 1990.
6. J. C. Martin, *Introduction to Languages and the Theory of Computation*, 2nd ed., McGraw-Hill, New York, 1997.
7. M. Sipser, *Introduction to the Theory of Computation*, PWS, Boston, 1997.
8. W. A. Wulf *et al.*, *Fundamental Structures of Computer Science*, Addison-Wesley, Reading, MA, 1981, pp. 1–64.

Chapter 12

Boolean Algebra and Combinatorial Circuits

Mathematics is music for the mind; music is mathematics for the soul.

—ANONYMOUS

George Boole's classic *An Investigation of the Laws of Thought*, published in 1854, led to the development of two closely related areas of mathematics: symbolic logic and the mathematical system called boolean algebra. Chapter 1 of Boole's book demonstrated how valuable the symbols and the laws of logic are in investigating how we reason in order to reach conclusions. That material will help us in studying this chapter.

Until the late 1930s boolean algebra did not seem to have many useful applications. In 1938, Claude E. Shannon, while working at the Massachusetts Institute of Technology, used boolean algebra to analyze electrical circuits, thus opening the door for a world of applications of boolean algebra. Since then, boolean algebra has played a central role in the design, analysis, and simplification of electronic devices, including digital computers.

In this chapter we will address some of the interesting problems that boolean algebra handles well:

- Three switches for a light fixture are in a hallway. If the light is on, it can be turned off by flipping one of the switches off. On the other hand, if it is off, flipping one of the tongues turns it on. What does the circuit look like?
- Design a circuit to compute the sum of two 3-bit numbers.
- How can boolean algebra simplify a circuit while maintaining the circuit's capabilities?
- Electronic devices display digits by lighting up a maximum of seven line segments in the adjacent configuration. What kind of circuit will accept the binary-coded decimal expansion of a decimal digit and light up a segment?

Claude Elwood Shannon *(1916–2001), a pioneer in artificial intelligence and information theory, was born in Gaylord, Michigan. After graduating from the University of Michigan in 1936, he earned his M.S. in 1937 and Ph.D. in applied mathematics in 1940 from MIT.*

Shannon was a research mathematician at the Bell Telephone Labs from 1941 to 1957 and became professor of electrical engineering at MIT in 1958. In 1948, he coined the word ***bit*** *for* ***b****inary* ***d****igit.*

He received honorary doctorates from many universities: Michigan, Pittsburgh, Princeton, Northwestern, Edinburgh, Oxford, Carnegie-Mellon, Tufts, and Pennsylvania. He also received numerous honors and awards, including the Nobel Peace Prize, Morris Liebmann Memorial Award, Ballantine Medal, Medal of Honor from the Institute of Electrical and Electronics Engineers (IEEE), and National Medal of Science. Shannon made outstanding contributions to the theory of switching circuits, information theory, cryptography, and artificial intelligence.

12.1 Boolean Algebra

A boolean algebra is a **mathematical system**; it consists of a nonempty set S with one or more operations defined on S, and a set of axioms that the elements of S satisfy.

A mathematical system can be thought of as a skeleton, like a human skeleton. Whether people are black or white, Caucasian or Chinese, their skeletons have common characteristics. Likewise, concrete examples of a mathematical system share common properties. When we study mathematical systems, we need to study properties common to all examples of such systems. Real numbers form an important number system.

Before formally defining a boolean algebra, we examine two concrete examples. Look for the properties common to both because they will help us define a boolean algebra.

EXAMPLE 12.1 Let U be an arbitrary set and $P(U)$ its power set. Let A, B, and C be any three elements of $P(U)$. You may recall from Chapter 2 that the union, intersection, and complementation operations on $P(U)$ satisfy the following:

Commutative properties

- $A \cup B = B \cup A$
- $A \cap B = B \cap A$

Associative properties

- $A \cup (B \cup C) = (A \cup B) \cup C$
- $A \cap (B \cap C) = (A \cap B) \cap C$

Distributive properties

- $A \cap (B \cup C) = (A \cap B) \cup (A \cap C)$
- $A \cup (B \cap C) = (A \cup B) \cap (A \cup C)$

Identity properties

- $A \cup \emptyset = A$
- $A \cap U = A$

Complement properties

- $A \cup A' = U$
- $A \cap A' = \emptyset$ ■

The next example is a bit sophisticated.

EXAMPLE 12.2 Let D_{30} be the set of positive factors of 30: $D_{30} = \{1, 2, 3, 5, 6, 10, 15, 30\}$. Define three operations $\oplus$, $\odot$, and $'$ on D_{30} as follows:

$$a \oplus b = \text{lcm}\{a, b\}$$

$$a \odot b = \text{gcd}\{a, b\}^{\dagger}$$

$$a' = \frac{30}{a}$$

For instance,

$$2 \oplus 3 = \text{lcm}\{2, 3\} = 6 = 3 \oplus 2; \qquad 6 \oplus 10 = \text{lcm}\{6, 10\} = 30 = 10 \oplus 6;$$

$$2 \odot 3 = \text{gcd}\{2, 3\} = 1 = 3 \odot 2; \qquad 6 \odot 10 = \text{gcd}\{6, 10\} = 2 = 10 \odot 6;$$

$$6' = \frac{30}{6} = 5; \qquad 5' = \frac{30}{5} = 6;$$

$$6 \oplus 6' = 6 \oplus 5 = \text{lcm}\{6, 5\} = 30; \qquad 5 \odot 5' = 5 \odot 6 = \text{gcd}\{5, 6\} = 1.$$

These operations satisfy the following properties:

Commutative properties

- $a \oplus b = b \oplus a$
- $a \odot b = b \odot a$

Associative properties

- $a \oplus (b \oplus c) = (a \oplus b) \oplus c$
- $a \odot (b \odot c) = (a \odot b) \odot c$

Distributive properties

- $a \odot (b \oplus c) = (a \odot b) \oplus (a \odot c)$
- $a \oplus (b \odot c) = (a \oplus b) \odot (a \oplus c)$

Identity properties

- $a \oplus 1 = a$
- $a \odot 30 = a$

---

†Recall that lcm $\{a, b\}$ denotes the least common multiple of a and b, and gcd $\{a, b\}$ the greatest common divisor of a and b.

Complement properties

- $a \oplus a' = 30$
- $a \odot a' = 1$ ■

Take a close look at both examples. What do they have in common? In each case, a nonempty set B contains two special elements — Ø and U in Example 12.1; 1 and 30 in Example 12.2. Furthermore, both examples contain two binary operators and a unary operator, satisfying the same 10 properties. These commonalities suggest a mathematical system called a boolean algebra, defined below.

Boolean Algebra

A **boolean algebra** consists of a nonempty set B containing two distinct elements 0 and 1, two binary operators $+$ and $\cdot$, and a unary operator $'$ satisfying the following conditions for all x, y, and z in B:

Commutative laws

- $x + y = y + x$
- $x \cdot y = y \cdot x$

Associative laws

- $x + (y + z) = (x + y) + z$
- $x \cdot (y \cdot z) = (x \cdot y) \cdot z$

Distributive laws

- $x \cdot (y + z) = (x \cdot y) + (x \cdot z)$
- $x + (y \cdot z) = (x + y) \cdot (x + z)$

Identity laws

- $x + 0 = x$
- $x \cdot 1 = x$

Complement laws

- $x + x' = 1$
- $x \cdot x' = 0$

In symbols, the boolean algebra is denoted by $\langle B, +, \cdot, ', 0, 1\rangle$.

We clarify a few points before we go any further:

- The operations $+$, $\cdot$, and $'$ are called **sum**, **product**, and **complementation**, respectively. (For instance, in Example 12.1, the binary operators are ∩ and ∪; the unary operator is complementation.) These operators are generic symbols: the operator $+$ does not stand for addition nor the operator $\cdot$ for multiplication.
- Since $+$ and $\cdot$ are binary operators, both $x + y$ and $x \cdot y$ belong to B for all x and y in B; since $'$ is a unary operator, x' belongs to B for every $x \in B$.
- The elements 0 and 1 are the **zero element** and the **unit element**, respectively. Again, they are generic symbols for the zero element and the unit element, respectively; they need not be the familiar numbers zero and one. In Example 12.1, the zero element is Ø and the unit element is U; in Example 12.2, they are 1 and 30, respectively.

- When it is clear from the context, we designate the boolean algebra $\langle B, +, \cdot, ', 0, 1\rangle$ as the boolean algebra B for convenience.
- The operator $\cdot$ in $x \cdot y$ is usually omitted for convenience; thus $xy = x \cdot y$. No parentheses appear when there is no danger of confusion. For instance, $(xy) + (xz) = xy + xz$ and $x + y + z = x + (y + z) = (x + y) + z$.
- Precedence rules govern evaluating expressions in boolean algebra:

  First, parenthesized subexpressions are evaluated.

  Complementation has the highest priority among the operators, followed by $\cdot$ and then $+$.

  For example, $xy + zx' = (xy) + [z(x')]$.
- The 10 axioms are paired off in two columns. In each pair, an axiom can be obtained from the other by swapping $+$ with $\cdot$, and 0 with 1. These are **dual axioms**. For instance, the dual of $x + x' = 1$ (axiom 9) is $x \cdot x' = 0$ (axiom 10). (In a boolean algebra, the dual of every true statement is also true. This property is the **principle of duality**.)

According to the definition, a boolean algebra contains at least two elements, the zero and the unit. Consequently we can ask: Does any boolean algebra contain exactly two elements? The next example shows that answer is yes.

EXAMPLE 12.3 Let $B = \{0, 1\}$. Define the operations $+$, $\cdot$, and $'$ on B as follows:

$$\begin{array}{cccc} 0+0=0 & 0+1=1 & 1+0=1 & 1+1=1 \\ 0\cdot 0=0 & 0\cdot 1=0 & 1\cdot 0=0 & 1\cdot 1=1 \\ 0'=1 & 1'=0 & & \end{array}$$

[Think of the operators $+$ and $\cdot$ as the *or* ($\vee$) and the *and* ($\wedge$) operators, respectively.] By Theorem 7.1, the commutative, associative, and distributive laws are satisfied.

Clearly, $x + 0 = x$ and $x \cdot 1 = x$, for every $x \in B$; therefore, 0 is the zero element and 1 is the unit element. Besides,

$$0 + 0' = 0 + 1 = 1 \quad \text{and} \quad 1 + 1' = 1 + 0 = 1$$
$$0 \cdot 0' = 0 \cdot 1 = 0 \quad \text{and} \quad 1 \cdot 1' = 1 \cdot 0 = 0$$

Consequently, the complement laws are also satisfied. Thus $\langle B, +, \cdot, ', 0, 1\rangle$ is a boolean algebra.

Some fundamental facts about boolean algebras come from the above axioms. We will find them useful in later discussions.

THEOREM 12.1 **(Unique Identities)** The zero element and the unit element of a boolean algebra B are unique.

PROOF:
The zero element in B will be proved unique; the other half will be left as a routine exercise.

Suppose there are two zero elements, 0_1 and 0_2, in B. Since 0_2 is a zero element, $0_1 + 0_2 = 0_1$. Likewise, since 0_1 is a zero element, $0_2 + 0_1 = 0_2$. But $0_2 + 0_1 = 0_1 + 0_2$, by the commutative law. Therefore, $0_1 = 0_1 + 0_2 = 0_2 + 0_1 = 0_2$, so the zero element is unique. ■

THEOREM 12.2 **(Unique Complement)** The complement of every element in a boolean algebra is unique.

PROOF:
Let x be an arbitrary element in a boolean algebra. Then by the complement laws, $x + x' = 1$ and $xx' = 0$. If x has a second complement y, $x + y = 1$ and $xy = 0$.

To show that $y = x'$ (the reason for each step is given on the RHS):

$y = y1$	identity law
$= y(x + x')$	complement law
$= yx + yx'$	distributive law
$= xy + x'y$	commutative law
$= 0 + x'y$	complement law
$= xx' + x'y$	complement law
$= x'x + x'y$	commutative law
$= x'(x + y)$	distributive law
$= x'1$	complement law
$= x'$	identity law

Thus the complement of every element in a boolean algebra is unique. ■

The next example uses Theorem 12.2.

EXAMPLE 12.4 With the boolean algebra D_{30} in Example 12.2, verify each.

(1) $6 \oplus 6 = 6$ (2) $5 \oplus 30 = 30$ (3) $(5')' = 5$
(4) $3 \oplus (3 \odot 5) = 3$ (5) $(3 \oplus 5)' = 3' \odot 5'$ (6) $(5 \odot 6)' = 5' \oplus 6'$

SOLUTION:

(1) $6 \oplus 6 = \text{lcm}\{6, 6\} = 6$

(2) $5 \oplus 30 = \text{lcm}\{5, 30\} = 30$

(3) $5' = \dfrac{30}{5} = 6$

So $(5')' = 6' = \dfrac{30}{6} = 5$

(4) $3 \odot 5 = \text{gcd}\{3, 5\} = 1$

So $3 \oplus (3 \odot 5) = 3 \oplus 1$
$= \text{lcm}\{3, 1\} = 1$

(5) $3 \oplus 5 = \text{lcm}\{3, 5\} = 15$

$$\begin{aligned}\text{So } (3 \oplus 5)' = 15' &= \frac{30}{15} \\ &= 2 \\ 3' \odot 5' &= 10 \odot 6 \\ &= \gcd\{10, 6\} = 2 \\ &= (3 \oplus 5)'\end{aligned}$$

(6) $5 \odot 6 = \gcd\{5, 6\} = 1$

$$\begin{aligned}\text{So } (5 \odot 6)' = 1' &= \frac{30}{1} \\ &= 30 \\ &= 5 \oplus 6 \\ &= \text{lcm}\{5, 6\} \\ &= 5' \oplus 6'\end{aligned}$$

■

We now establish a few more properties of boolean algebras.

THEOREM 12.3 Let x and y be arbitrary elements in a boolean algebra $\langle B, +, \cdot, \ ', 0, 1\rangle$. Then:

Idempotent laws

- $x + x = x$
- $xx = x$

Boundedness laws

- $x + 1 = 1$
- $x0 = 0$

Involution laws

- $(x')' = x$
- $0' = 1$
- $1' = 0$

Absorption laws

- $x + xy = x$
- $x(x + y) = x$

De Morgan's laws

- $(x + y)' = x'y'$
- $(xy)' = x' + y'$

PROOF:

- *To prove that* $x + x = x$:

$$\begin{aligned} x + x &= (x + x)1 && \text{identity law} \\ &= (x + x)(x + x') && \text{complement law} \\ &= xx + xx' + xx + xx' && \text{distributive law} \\ &= x + xx' && \text{idempotent law} \\ &= x + 0 && \text{identity law} \\ &= x && \text{identity law} \end{aligned}$$

- *To prove that* $x + 1 = 1$:

$$x + 1 = x + (x + x') \qquad \text{complement law}$$

$$
\begin{aligned}
&= (x + x) + x' && \text{associative law} \\
&= x + x' && \text{idempotent law} \\
&= 1 && \text{complement law}
\end{aligned}
$$

- *To prove that* $(x')' = x$:

Since x' is the complement of x, $x + x' = 1$ and $xx' = 0$. Using the commutative laws, these equations can be rewritten as

$$x' + x = 1 \quad \text{and} \quad x'x = 0 \tag{12.1}$$

Since x' is also an element of B, it has a complement $(x')'$. Therefore,

$$x' + (x')' = 1 \quad \text{and} \quad x'(x')' = 0 \tag{12.2}$$

Equations (12.1) imply x is a complement of x'; by Equations (12.2), $(x')'$ is also a complement of x'. But, by Theorem 12.2, the complement of x' is unique; therefore, $(x')' = x$.

- *To prove that* $x + xy = x$:

$$
\begin{aligned}
x + xy &= x1 + xy && \text{identity law} \\
&= x(1 + y) && \text{distributive law} \\
&= x(y + 1) && \text{commutative law} \\
&= x1 && \text{boundedness law} \\
&= x
\end{aligned}
$$

- *To prove that* $(x + y)' = x'y'$:
  By the complement laws, we must show that $(x + y) + x'y' = 1$ and $(x + y)(x'y') = 0$.

$$
\begin{aligned}
(1)\ (x + y) + x'y' &= x + (y + x'y') && \text{associative law} \\
&= x + (y1 + x'y') && \text{identity law} \\
&= x + [y(x + x') + x'y'] && \text{complement law} \\
&= x + [(yx + yx') + x'y'] && \text{distributive law} \\
&= x + [xy + (x'y + x'y')] && \text{commutative law} \\
&= (x + xy) + (x'y + x'y') && \text{associative law} \\
&= x(1 + y) + x'(y + y') && \text{distributive law} \\
&= x1 + x'(y + y') && \text{boundedness law} \\
&= x1 + x'1 && \text{complement law} \\
&= x + x' && \text{identity law} \\
&= 1 && \text{complement law}
\end{aligned}
$$

$$
\begin{aligned}
(2)\ (x + y)(x'y') &= (x'y')(x + y) && \text{commutative law} \\
&= (x'y')x + (x'y')y && \text{distributive law} \\
&= x(x'y') + (x'y')y && \text{commutative law} \\
&= (xx')y' + x'(y'y) && \text{associative law}
\end{aligned}
$$

$= (xx')y' + x'(yy')$	commutative law
$= 0y' + x'0$	complement law
$= y'0 + x'0$	commutative law
$= 0 + 0$	boundedness law
$= 0$	identity law

Thus, by parts 1 and 2, $x'y'$ is the complement of $x + y$; that is, $(x + y)' = x'y'$. ■

A close relationship exists between the boolean algebras D_{30} and $P(U)$, where $U = \{a, b, c\}$. They are displayed by the Hasse diagrams in Figures 12.1 and 12.2, respectively, which have the same structure. This is *not* surprising because the algebras are **isomorphic**. [Find an isomorphism $f: D_{30} \to P(U)$ that preserves the operations; that is, $f(x \oplus y) = f(x) \cup f(y), f(x \odot y) = f(x) \cap f(y)$, and $f(x') = (f(x))'$.]

Figure 12.1

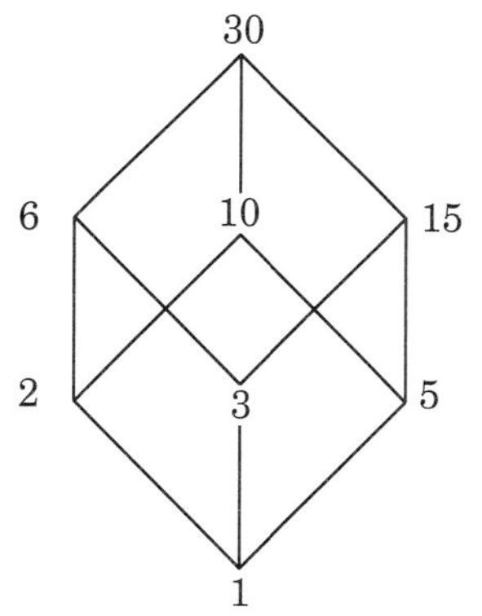

Figure 12.2

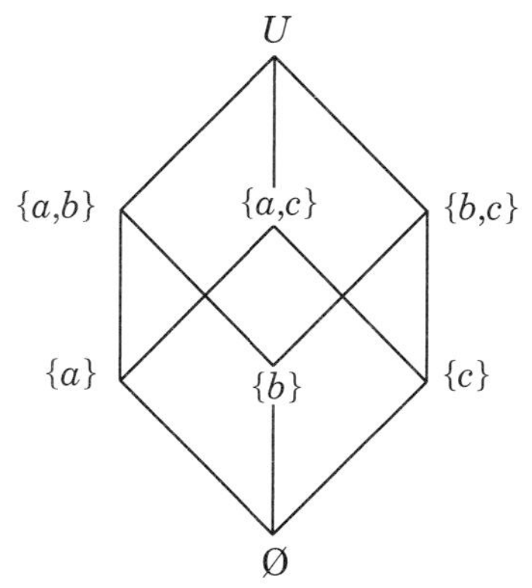

Any boolean algebra, then, has at least two elements, as well as two binary operators and a unary one that adhere to various laws.

Exercises 12.1

Using Example 12.2, evaluate each.

1. $6 \oplus 10$ **2.** $6 \odot 10$ **3.** $2 \oplus (3 \oplus 5)$ **4.** $(2 \oplus 3) \oplus 5$

5. $3 \odot (5 \odot 6)$ **6.** $(3 \odot 5) \odot 6$ **7.** $(3 \oplus 6)'$ **8.** $3' \odot 6'$

9. $(5 \odot 10)'$ **10.** $5' \oplus 10'$ **11.** $2 \odot 3 \oplus 5$ **12.** $2 \oplus 3 \odot 5$

The set $D_{70} = \{1, 2, 5, 7, 10, 14, 35, 70\}$ of positive factors of 70 is a boolean algebra under the operations $\oplus, \odot$, and $'$ defined by $x \oplus y = \text{lcm}\{x, y\}$, $x \odot y = \gcd\{x, y\}$, and $x' = 70/x$. Compute each.

13. $7 \oplus 5$ **14.** $2 \odot 7$ **15.** $(5 \oplus 7)'$ **16.** $5' \odot 7'$

17. $(7 \odot 2)'$ **18.** $7' \oplus 2'$ **19.** $10 \oplus 10$ **20.** $7 \odot 7$

Using the boolean algebra D_{70}, verify each.

21. $(5')' = 5$ **22.** $7 \oplus (7 \odot 5) = 7$

23. $5 \odot (5 \oplus 7) = 5$ **24.** $(5 \oplus 7)' = 5' \odot 7'$

25. With the boolean algebras in Examples 12.1, 12.2, and 12.3, and D_{70}, predict the number of elements in a finite boolean algebra.

26. Define the operations $+$, $\cdot$, and $'$ on $B = \{0, 1\}$ as follows: $x + y = \max\{x, y\}$, $x \cdot y = \min\{x, y\}$, $0' = 1$, and $1' = 0$. Is $\langle B, +, \cdot, ', 0, 1\rangle$ a boolean algebra?

Determine if $\langle S, +, \cdot, ', 0, 1\rangle$ is a boolean algebra for each subset S of the boolean algebra D_{30}.

27. $\{1, 2, 15, 30\}$ **28.** $\{1, 6, 10, 30\}$ **29.** $\{1, 3, 5, 30\}$ **30.** $\{1, 5, 6, 30\}$

Is $\langle S, \cup, \cap, ', \emptyset, U\rangle$ a boolean algebra for each subset S of $P(U)$, where $U = \{a, b, c\}$?

31. $\{\emptyset, \{a\}, \{b, c\}, U\}$ **32.** $\{\emptyset, \{a\}, \{b\}, \{a, b\}, U\}$

33. $\{\emptyset, \{b\}, \{a, c\}, U\}$ **34.** $\{\emptyset, \{c\}, \{a, b\}, U\}$

35. Define the operations $+$, $\cdot$, and $'$ on $B = \{a, b, 0, 1\}$ so that $\langle B, +, \cdot, ', 0, 1\rangle$ would be a boolean algebra.

Find the dual of each boolean property.

36. $0' = 1$ **37.** $x(x + y) = x$

38. $(x + y)z = xz + yz$ **39.** $(xy)' = x' + y'$

Prove algebraically.

40. The unit element in a boolean algebra is unique.

41. $xx = x$ **42.** $x0 = 0$ **43.** $0' = 1$

44. $1' = 0$ **45.** $x(x + y) = x$ **46.** $(xy)' = x' + y'$

47. $x + y = (x'y')'$ **48.** $xy = (x' + y')'$ **49.** $(x + y)z = xz + yz$

50. $xy' + x'y = (x + y)(xy)'$ **51.** $(xy) + z = (x + z)(y + z)$

***52.** Let $U = \{a, b, c\}$. Define a suitable function $f : D_{30} \to P(U)$ that preserves the operations; that is, $f(x \oplus y) = f(x) \cup f(y)$, $f(x \odot y) = f(x) \cap f(y)$, and $f(x') = (f(x))'$.

12.2 Boolean functions

This section introduces the concept of a boolean function and illustrates the possible equality of boolean expressions. We also see that through the laws of boolean algebra, boolean expressions can be rewritten in a standard form. Since the design and analysis of electronic devices rely on the two-element boolean algebra $\langle B, +, \cdot, ', 0, 1\rangle$, discussed in Example 12.3, in the rest of this chapter we will employ only this boolean algebra. The topics in this section closely resemble the topics in Section 1.1, so you will find it beneficial to review that section before we continue.

We begin with a couple of definitions.

Boolean Function

A **boolean variable** assumes the value 0 or 1. A **boolean function** is a function $f : B^n \to B$, where $B^n = B \times B \times \cdots \times B$, the cartesian product of B with itself to n factors. Let $(x_1, x_2, \ldots, x_n) \in B^n$. Then the value of the boolean function f at $(x_1, x_2, \ldots, x_n)$ is denoted by $f(x_1, x_2, \ldots, x_n)$, although technically it should be $f((x_1, x_2, \ldots, x_n))$. Since $\text{codom}(f) = B$, the value of the boolean function for any input value $(x_1, x_2, \ldots, x_n)$ is either 0 or 1. [Recall that codom(f) denotes the codomain of the function.]

Boolean functions can be defined by **logic tables**. Tables 12.1 and 12.2 define two boolean functions from B^2 to B. The value $f(x, y)$ is given for every combination of values of x and y. (Notice the similarity between these tables and the truth tables for the disjunction and conjunction of two propositions.)

Table 12.1

x	y	$f(x,y)$
0	0	0
0	1	1
1	0	1
1	1	1

Table 12.2

x	y	$f(x,y)$
0	0	0
0	1	0
1	0	0
1	1	1

Boolean functions can also be defined by **boolean expressions** made up of boolean variables and boolean operators. For instance, if x, y, and z are boolean variables, then xy, $(xy)'$, $x+yz$, and $x+(xy)'$ are boolean expressions.

Next we define a boolean expression recursively.

A Recursive Definition of a Boolean Expression

Let $x_1, x_2, \ldots, x_n$ be n boolean variables. Then:

- **Basis clause** 0, 1, $x_1, x_2, \ldots, x_n$ are boolean expressions.
- **Recursive clause** If E_1 and E_2 are boolean expressions, then so are $(E_1), E_1', E_1E_2$, and $E_1 + E_2$.

(The symbols 0 and 1 denote bits. Parentheses are dropped when no confusion results.)

The next example demonstrates this definition.

EXAMPLE 12.5 Verify that $x + (yz)'$ is a boolean expression, where the variables x, y, and z are boolean variables.

SOLUTION:
Since y and z are boolean variables, they are boolean expressions by the basis clause. Consequently, yz and hence $(yz)'$ are boolean expressions by the recursive clause. Since x is a boolean variable, x is also a boolean expression; therefore, by the recursive clause, $x + (yz)'$ is a boolean expression. ■

The following example shows how to evaluate boolean expressions for given values of boolean variables.

EXAMPLE 12.6 Evaluate the boolean expression $x+(yz)'$ for the triplets (0, 1, 0) and (0, 1, 1).

SOLUTION:
When $x = 0$, $y = 1$, and $z = 0$:

$$x + (yz)' = 0 + (1 \cdot 0)' = 0 + 0' = 0 + 1 = 1$$

When $x = 0$, $y = 1 = z$:

$$x + (yz)' = 0 + (1 \cdot 1)' = 0 + 1' = 0 + 0 = 0$$

■

When are boolean expressions equal? The next definition gives us an answer.

Equality of Boolean Expressions

Two boolean expressions $E_1(x_1,x_2,\ldots,x_n)$ and $E_2(x_1,x_2,\ldots,x_n)$ are **equal**, denoted by $E_1(x_1,x_2,\ldots,x_n) = E_2(x_1,x_2,\ldots,x_n)$, if they yield the same value for all $(x_1,x_2,\ldots,x_n)$ in B^n.

When the boolean expressions E_1 and E_2 are fairly simple, logic tables can determine if they are equal. This procedure is quite similar to verifying the logical equivalence of two compound statements, as the next example illustrates.

EXAMPLE 12.7 Using a logic table, verify that $(x+y)' = x'y'$.

SOLUTION:

As in propositional logic, constructing Table 12.3 shows that the columns headed by $(x+y)'$ and $x'y'$ are identical; therefore, $(x+y)' = x'y'$.

Table 12.3

x	y	$x+y$	$(x+y)'$	x'	y'	$x'y'$
0	0	0	1	1	1	1
0	1	1	0	1	0	0
1	0	1	0	0	1	0
1	1	1	0	0	0	0

↑ identical columns ↑ ■

The boolean expression $x+(yz)'$ defines the boolean function $f : B^3 \to B$, where $f(x,y,z) = x + (yz)'$. Then:

$$f(0,0,0) = 0 + (0 \cdot 0)' = 0 + 0' = 0 + 1 = 1$$

and

$$f(1,1,1) = 1 + (1 \cdot 1)' = 1 + 1' = 1 + 0 = 1$$

The output values of the function for various input values (x,y,z) are shown in the logic table of Table 12.4.

Table 12.4

x	y	z	yz	$(yz)'$	$f(x,y,z) = x + (yz)'$
0	0	0	0	1	1
0	0	1	0	1	1
0	1	0	0	1	1
0	1	1	1	0	0
1	0	0	0	1	1
1	0	1	0	1	1
1	1	0	0	1	1
1	1	1	1	0	1

EXAMPLE 12.8 Construct a logic table for the boolean function

$$f(x,y,z) = (x+y+z)(xyz)'$$

SOLUTION:

Just as we build up truth tables for complex propositions from simple ones, we build up the boolean expression $(x+y+z)(xyz)'$ from boolean subexpressions in steps: $x, y, z, x+y, x+y+z, xy, xyz, (xyz)'$, and finally $(x+y+z)(xyz)'$. The table contains a column for each subexpression; fill in the various columns using the definitions of sum, product, and complement. Table 12.5 is the resulting table.

Table 12.5

x	y	z	$x+y$	$(x+y)+z$	xy	xyz	$(xyz)'$	$f(x,y,z) = (x+y+z)(xyz)'$
0	0	0	0	0	0	0	1	0
0	0	1	0	1	0	0	1	1
0	1	0	1	1	0	0	1	1
0	1	1	1	1	0	0	1	1
1	0	0	1	1	0	0	1	1
1	0	1	1	1	0	0	1	1
1	1	0	1	1	1	0	1	1
1	1	1	1	1	1	1	0	0

■

Every boolean expression in n boolean variables defines a boolean function from B^n to B. On the other hand, given a boolean function $f : B^n \to B$, can we find a boolean expression $E(x_1, x_2, \ldots, x_n)$ such that $f(x_1, x_2, \ldots, x_n) = E(x_1, x_2, \ldots, x_n)$? Yes! Such a boolean expression E **defines** the boolean function f.

A system for finding such an expression lies below.

EXAMPLE 12.9 Find a boolean expression that defines the boolean function f in Table 12.6.

Table 12.6

x	y	$f(x,y)$
0	0	0
0	1	1
1	0	1
1	1	0

SOLUTION:

To find a boolean expression E that yields the same values as the function f, look at each row that corresponds to the functional value 1. From each such row construct a boolean subexpression that has the value 1, and has the value 0 for any other combination of x and y. The desired boolean expression is obtained by taking the sum of all such subexpressions.

- When $x = 0$ and $y = 1$, the value of the function is 1 (see row 2); so the value of the expression must be 1 when $x = 0$ and $y = 1$, that is, when $x' = 1 = y$. One such expression is $E_1 = x'y$. When $x = 0 = y$, or $x = 1$ and $y = 0$, or $x = 1 = y$, its value is 0.
- When $x = 1$ and $y = 0$, the value of the function is again 1 (see row 3). Therefore, the value of the expression must be 1 when $x = 1$ and $y = 0$, that is, when $x = 1 = y'$. Again, one such function is $E_2 = xy'$. When $x = 0 = y$, or $x = 0$ and $y = 1$, or $x = 1 = y$, its value is 0.

The desired boolean expression E is the sum of subexpressions E_1 and E_2: $E = E_1 + E_2 = x'y + xy'$. Thus, $f(x, y) = x'y + xy'$.

Let us now verify this. The subexpression $E_1 = x'y$ has the value 1 when $x = 0$ and $y = 1$; otherwise, it is 0. The subexpression $E_2 = xy'$ has the value 1 when $x = 1$ and $y = 0$; otherwise, it is 0. Therefore, the expression $E = E_1 + E_2$ has the value 1 either when $x = 0$ and $y = 1$, or when $x = 1$ and $y = 0$. This is precisely when the function has the value 1. $E = x'y + xy'$ does indeed work.

Notice that the boolean expression $x'y + xy'$ is the sum of two boolean subexpressions $x'y$ and xy', and each subexpression is of the form st', with s and t boolean variables. Such an expression is called a **minterm** in s and t, defined next.

Minterm

A **minterm** in n variables $x_1, x_2, \ldots, x_n$ is a boolean expression $y_1 y_2 \ldots y_n$ where each y_i is either x_i or x'_i. A **literal** is a boolean variable or its complement. Thus each y_i is a literal, and a minterm is the product of n literals.

A minterm contains a literal y_i corresponding to each variable x_i; y_i is either x_i or x'_i. In two variables, four minterms are possible: xy, xy', $x'y$, and $x'y'$. The boolean expressions xyz, xyz', and $xy'z$ are a few minterms in three variables.

In Example 12.9, the boolean expression E defining the function f is a sum of minterms. In fact, the technique illustrated there always yields a sum of minterms, called the **disjunctive normal form (DNF)** of the boolean function. Consequently, the DNF of the boolean function in Example 12.9 is $x'y + xy'$. It is unique except for the order in which the minterms appear in the sum and the order of literals in each minterm.

The next four examples find the DNFs of boolean functions.

EXAMPLE 12.10 Find the DNF of the boolean function defined by Table 12.7.

SOLUTION:

Step 1 Locate the rows in the table that yield the functional value 1. They are rows 4, 6, 7, and 8.

Table 12.7

x	y	z	$f(x,y,z)$
0	0	0	0
0	0	1	0
0	1	0	0
0	1	1	1
1	0	0	0
1	0	1	1
1	1	0	1
1	1	1	1

Step 2 Build a minterm $y_1 y_2 y_3$ corresponding to each row, where

$$y_i = \begin{cases} x_i & \text{if } x_i = 1 \\ x_i{}' & \text{if } x_i = 0 \end{cases}$$

and $1 \le i \le 3$. The minterm corresponding to row 4 is $x'yz$. The minterms for rows 6, 7, and 8 are $xy'z$, xyz', and xyz, respectively.

Step 3 Add the minterms in step 2 to get the DNF of the given function.

$$\begin{aligned} f(x,y,z) &= x'yz + xy'z + xyz' + xyz \\ &= xyz + x'yz + xy'z + xyz' \end{aligned}$$

■

EXAMPLE 12.11 Construct the DNF of the boolean function $f(x,y,z) = x(y+z)$.

SOLUTION:

First, set up the logic table for the boolean function f. (See Table 12.8.) As before, find a minterm corresponding to each value 1 of the function and add the minterms to produce the DNF:

$$f(x,y,z) = xyz + xy'z + xyz'$$

Table 12.8

x	y	z	$y+z$	$f(x,y,z) = x(y+z)$
0	0	0	0	0
0	0	1	1	0
0	1	0	1	0
0	1	1	1	0
1	0	0	0	0
1	0	1	1	1
1	1	0	1	1
1	1	1	1	1

■

From Examples 12.10 and 12.11 unfolds Algorithm 12.1 for constructing the DNF of a boolean function.

```
Algorithm DNF(f)
(* This algorithm finds the DNF of a boolean function. *)
  Begin (* algorithm *)
    if the function f is given by a boolean expression then
      construct a logic table for the function f
    DNF ← 0 (* initialize DNF *)
    for each row in the table do
      if the corresponding functional value is 1 then
        begin (* if *)
          construct a minterm
          DNF ← DNF + minterm
        endif
  End (* algorithm *)
```

Although we will usually find it more difficult, we can use the laws of boolean algebra to uncover the DNF of a boolean function elegantly, as the next two examples illustrate.

EXAMPLE 12.12 Using the laws of boolean algebra, construct the DNF of the boolean function $f(x,y,z) = x(y+z)$.

SOLUTION:

$x(y+z)$	$= xy + xz$	distributive law
	$= xy \cdot 1 + xz \cdot 1$	identity law
	$= xy(z+z') + xz(y+y')$	complement law
	$= xyz + xyz' + xzy + xzy'$	distributive law
	$= xyz + xyz' + xyz + xy'z$	commutative and associative laws
	$= xyz + xyz' + xy'z$	idempotent law

Thus

$$f(x,y,z) = xyz + xy'z + xyz'$$

This is the DNF obtained in Example 12.11. (A logic table can verify the equality of the two boolean expressions. Verify this.) ■

EXAMPLE 12.13 Find the DNF of the boolean function $f(x,y,z) = x + yz$ using the laws of boolean algebra.

SOLUTION:

$f(x,y,z)$	$= x + yz$	
	$= x \cdot 1 + 1 \cdot yz$	identity law
	$= x[yz + (yz)'] + (x + x')yz$	complement law
	$= xyz + x(yz)' + xyz + x'yz$	distributive law
	$= xyz + x(yz)' + x'yz$	idempotent law

$= xyz + x(y' + z') + x'yz$	DeMorgan's law
$= xyz + xy' + xz' + x'yz$	distributive law
$= xyz + xy' \cdot 1 + xz' \cdot 1 + x'yz$	identity law
$= xyz + xy'(z + z') + xz'(y + y') + x'yz$	complement law
$= xyz + xy'z + xy'z' + xz'y + xz'y' + x'yz$	distributive law
$= xyz + xy'z + xy'z' + xyz' + xy'z' + x'yz$	commutative and associative laws
$= xyz + xy'z + xy'z' + xyz' + x'yz$	idempotent law
$= xyz + x'yz + xy'z + xyz' + xy'z'$	commutative and associative laws

[Alternatively, you could use the fact that $x + yz = x \cdot 1 \cdot 1 + 1 \cdot yz = x(y + y')(z + z') + (x + x')yz$.] ■

Every boolean function can be defined by means of a boolean expression, and every boolean expression by means of the boolean operators $+$, $\cdot$, and $'$. Consequently, every boolean function can be defined using these three boolean operators. Such a set of binary and unary operators is said to be functionally complete.

Functional Completeness

A set of boolean operators is **functionally complete** if every boolean function can be defined using them.

The triad of boolean operators $\{+, \cdot, '\}$ forms a functionally complete set, but can fewer operators do so? By De Morgan's law, $(x + y)' = x'y'$, so $x + y = [(x + y)']' = (x'y')'$. Consequently the boolean operator $+$ can be defined using the operators $\cdot$ and $'$. Thus $\{\cdot, '\}$ is a functionally complete set of two boolean operators. In fact, such a set can have just one binary operator. We define two such operators next.

NAND and NOR

The binary operators **NAND** (not and) and **NOR** (not or), represented by $\uparrow$ and $\downarrow$ respectively, are defined as follows:

$$x \uparrow y = \begin{cases} 0 & \text{if } x = 1 = y \\ 1 & \text{otherwise} \end{cases}$$

$$x \downarrow y = \begin{cases} 1 & \text{if } x = 0 = y \\ 0 & \text{otherwise} \end{cases}$$

(These definitions and the ones presented in Section 1.1 bear similarities.)

EXAMPLE 12.14 Verify that $\{\uparrow\}$ is functionally complete.

SOLUTION:

Since $\{\cdot, {}'\}$ is functionally complete, it suffices to show that both operators $\cdot$ and $'$ can be expressed in terms of $\uparrow$. First, notice that

$$(xy)' = x \uparrow y \tag{12.3}$$

So

$$(xx)' = x \uparrow x$$

But $xx = x$, by the idempotent law; so

$$x' = x \uparrow x \tag{12.4}$$

By the involution law,

$$\begin{aligned} xy &= [(xy)']' \\ &= (xy)' \uparrow (xy)' \qquad \text{by Equation (12.4)} \\ &= (x \uparrow y) \uparrow (x \uparrow y) \end{aligned} \tag{12.5}$$

by Equation (12.3).

Equations (12.4) and (12.5) prove that $\{\uparrow\}$ is functionally complete. ■

With functionally complete operators, boolean functions can arise through recursively defined expressions, as well as through logic tables. We can also use such tables to form the DNF of a boolean function.

Exercises 12.2

1. Evaluate the boolean expression $x(yz' + y'z)$ at the ordered triplets $(1, 0, 1)$ and $(1, 1, 1)$.

2. Evaluate the boolean expression $(x + y + z)(x + y' + z)$ at the triplets in Exercise 1.

3. How many constant boolean functions can be defined from B^n to B, with B a two-element boolean algebra?

4. Find the number of boolean functions that can be defined from B^n to B, where B is a two-element boolean algebra.

Determine if each is a boolean expression, where each variable is boolean.

5. $((xy')')'$ **6.** $x' + yz$ **7.** $(xy + y'z')'$ **8.** $x(yz')'$

Construct a logic table for each boolean function defined by each boolean expression.

9. $(x+y')(x'+y)$

10. $x(y'z+yz')$

11. $xy+y'z+yz'$

12. $(x+y'+z)(x'+y+z')$

13. $(x+y'+z)(xy'z)$

14. $xyz+(xyz)'$

15. $xyz+x(yz)'$

16. $x'yz'+x'(yz)'$

Using a logic table, verify each.

17. $x+xy=x$

18. $x(x+y)=x$

19. $(x+y)'=x'y'$

20. $(xy)'=x'+y'$

21. $(x+y)'\neq x'+y'$

22. $(xy)'\neq x'y'$

23. Is the equality relation on the set of boolean expressions in n variables an equivalence relation?

24. List all minterms in two boolean variables x and y.

25. Give all minterms three boolean variables x, y, and z can generate.

26. How many minterms can n boolean variables produce?

Find the DNFs of the boolean functions in Exercises 27–34.

27.

x	y	$f(x,y)$
0	0	0
0	1	0
1	0	0
1	1	1

28.

x	y	$f(x,y)$
0	0	0
0	1	1
1	0	1
1	1	1

29.

x	y	$f(x,y)$
0	0	1
0	1	0
1	0	0
1	1	1

30.

x	y	$f(x,y)$
0	0	1
0	1	1
1	0	1
1	1	0

31.

x	y	z	$f(x,y,z)$
0	0	0	0
0	0	1	0
0	1	0	0
0	1	1	0
1	0	0	1
1	0	1	0
1	1	0	0
1	1	1	1

32.

x	y	z	$f(x,y,z)$
0	0	0	0
0	0	1	0
0	1	0	0
0	1	1	1
1	0	0	1
1	0	1	0
1	1	0	1
1	1	1	0

33.

x	y	z	$f(x,y,z)$
0	0	0	0
0	0	1	1
0	1	0	1
0	1	1	0
1	0	0	1
1	0	1	0
1	1	0	0
1	1	1	0

34.

x	y	z	$f(x,y,z)$
0	0	0	0
0	0	1	0
0	1	0	0
0	1	1	1
1	0	0	0
1	0	1	1
1	1	0	1
1	1	1	0

Find the DNF of each boolean function.

35. $f(x,y) = x + y$ **36.** $f(x,y) = x + xy'$ **37.** $f(x,y) = (x + y)xy'$

38. $f(x,y,z) = x + yz'$ **39.** $f(x,y,z) = y(x + z)$ **40.** $f(x,y,z) = (x+y)xyz$

41–46. Using the laws of boolean algebra, find the DNF of each boolean function in Exercises 35–40.

Evaluate each boolean expression.

47. $1 \uparrow (0 \uparrow 1)$ **48.** $1 \downarrow (1 \downarrow 0)$ **49.** $0 \uparrow (1 \downarrow 1)$

50. $1 \downarrow (0 \uparrow 1)$ **51.** $(0 \uparrow 1) \downarrow (0 \uparrow 1)$ **52.** $(1 \downarrow 0) \uparrow (1 \downarrow 1)$

Construct a logic table for each boolean expression.

53. $(x \uparrow y) \uparrow (x \uparrow y)$ **54.** $(x \downarrow y) \downarrow (x \downarrow y)$ **55.** $(x \uparrow y) \downarrow (x \uparrow y)$

56. $(x \downarrow y) \uparrow (x \downarrow y)$ **57.** $(x \uparrow x) \downarrow (y \uparrow y)$ **58.** $(x \downarrow x) \downarrow (y \downarrow y)$

Using the laws of boolean algebra, find the DNF of each boolean function.

59. $f(x,y) = x \uparrow y$ **60.** $f(x,y) = x \downarrow y$

Verify each.

61. $x + y = (x \uparrow x) \uparrow (y \uparrow y)$ **62.** $xy = (x \uparrow y) \uparrow (x \uparrow y)$

63. $x + y = (x \downarrow y) \downarrow (x \downarrow y)$ **64.** $xy = (x \downarrow x) \downarrow (y \downarrow y)$

Give a counterexample to disprove each statement.

65. $x \uparrow (y \uparrow z) = (x \uparrow y) \uparrow z$ **66.** $x \downarrow (y \downarrow z) = (x \downarrow y) \downarrow z$

Verify that each set of boolean operators is functionally complete.

67. $\{+, '\}$ **68.** $\{\downarrow\}$ (*Hint:* See Example 12.14.)

Use the following definition of the binary operator **XOR**, denoted by $\oplus$, for Exercises 69–81.

$$x \oplus y = \begin{cases} 1 & \text{if exactly one of the bits } x \text{ and } y \text{ is } 1 \\ 0 & \text{otherwise} \end{cases}$$

Evaluate each.

69. $1 \oplus (1 \oplus 1)$

70. $1 \oplus (0 \oplus 1)$

71. $(1 \oplus 0) \oplus 1$

72. $1 \uparrow (0 \oplus 1)$

73. $1 \downarrow (0 \oplus 1)$

74. $(1 \uparrow 0) \oplus (1 \downarrow 1)$

75. Is $\{\oplus\}$ functionally complete?

76. Find the DNF of the boolean function $f(x,y) = x \oplus y$.

Prove each.

77. $x \oplus x = 0$

78. $x \oplus x' = 1$

79. $x \oplus y = y \oplus x$

80. $x \oplus y = (x + y)(xy)'$

81. $x \oplus (y \oplus z) = (x \oplus y) \oplus z$

The dual of a DNF is called the **conjunctive normal form** (**CNF**). It is the product of maxterms; a **maxterm** in n variables $x_1, x_2, \ldots, x_n$ is $y_1 + y_2 + \cdots + y_n$, where each y_i is a literal x_i or x_i'.

82–87. With a procedure like the one in Examples 12.9 and 12.10, find the CNF of each boolean function in Exercises 35–40.

88. Write an algorithm to find the CNF of a boolean function f.

12.3 Logic Gates

The laws of boolean algebra help us construct DNFs and also give us an advantage in studying more practical electrical circuits, the mainstay of every electronic device. Elemental components of a circuit perform the fundamental boolean operations, and circuits can be designed to perform specific tasks.

In electronic circuits, the bits 0 and 1 indicate a low and a high voltage, respectively. Within the circuits, tiny, simple devices called **gates** accept one or more input signals (that is, bits) and produce a unique output signal (again, a bit). Each gate performs a unique logical operation on bits and hence can be considered a function, as defined below.

Gate

A **gate** is a function from B^n to B, where B denotes the boolean algebra $\{0, 1\}$.

The three fundamental gates — AND, OR, and NOT gates — carry out the three basic operations of product, sum, and complement, respectively.

AND Gate

An **AND gate** receives two arbitrary bits x and y as input signals and produces a unique output xy, where

$$xy = \begin{cases} 1 & \text{if } x = 1 = y \\ 0 & \text{otherwise} \end{cases}$$

The output xy is also denoted by $x \wedge y$.

Logic gates are drawn with standard symbols developed by the Institute of Electrical and Electronics Engineers. For instance, Figure 12.3 shows an AND gate.

Figure 12.3

An AND gate.

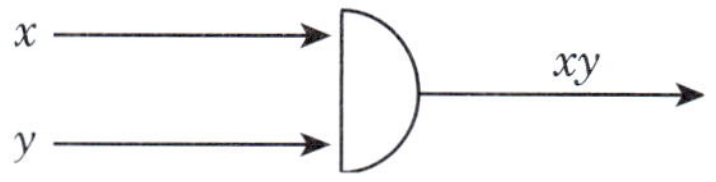

The outputs of an AND gate corresponding to the possible combinations of inputs appear in the **logic table** in Table 12.9, which approximates the truth table for the conjunction of two propositions. It follows from Table 12.9 that the output from an AND gate is 1 if and only if both inputs are 1.

Table 12.9

Logic table for an AND gate.

Inputs		Output
x	y	xy
0	0	0
0	1	0
1	0	0
1	1	1

More generally, n inputs into an AND gate yield an output of 1 if and only if every input is 1.

Many real-world situations fit AND gate models, as the next two examples demonstrate.

EXAMPLE 12.15 A tractor-mower will mow the lawn only if the tractor engine is on and the mower engaged. Figure 12.4 delineates these restrictions.

Figure 12.4

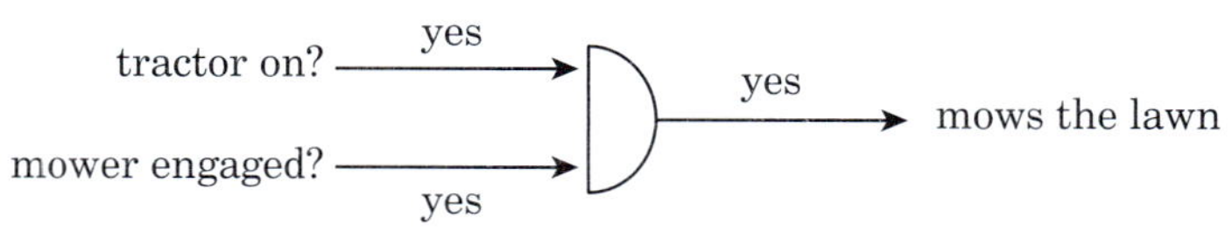

■

EXAMPLE 12.16 Another AND gate from everyday life appears in Figure 12.5. A laser printer will print the output from a personal computer only if the ON LINE and READY indicators are on.

Figure 12.5

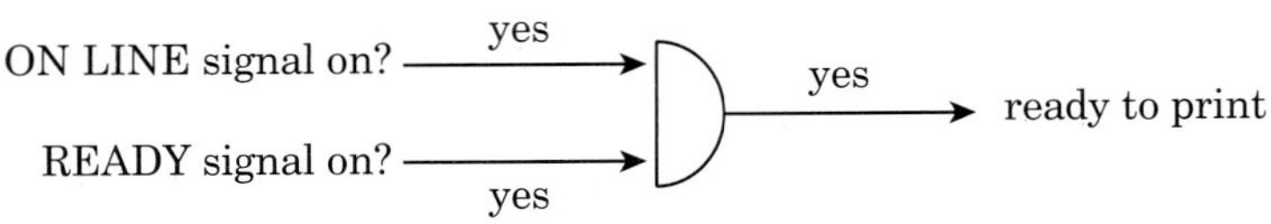

■

Next we define an OR gate.

OR Gate

An **OR gate** receives two bits x and y as input signals and produces a unique output $x+y$, where

$$x+y=\begin{cases}1 & \text{if } x=1 \text{ or } y=1\\ 0 & \text{otherwise}\end{cases}$$

The output $x+y$ is also denoted by $x \vee y$.

Figure 12.6 shows an OR gate, and Table 12.10 gives its logic table. According to the table, the output from an OR gate is 1 if at least one of the input symbols is 1. (Again, similarities exist between this table and the one for the disjunction of two propositions.)

Figure 12.6

An OR gate.

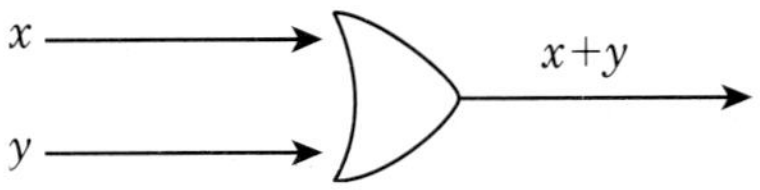

Table 12.10

Logic table for an OR gate.

Inputs		Output
x	y	$x+y$
0	0	0
0	1	1
1	0	1
1	1	1

More generally, n inputs into an OR gate will output 1 if and only if at least one input symbol is 1.

NOT Gate

A **NOT gate** receives a bit x and produces an output x':

$$x' = \begin{cases} 0 & \text{if } x = 1 \\ 1 & \text{otherwise} \end{cases}$$

A NOT gate, also known as an **inverter**, produces an output opposite from the input, as Figure 12.7 and Table 12.11 indicate. (A NOT gate resembles the logic operator NOT.)

Figure 12.7

A NOT gate.

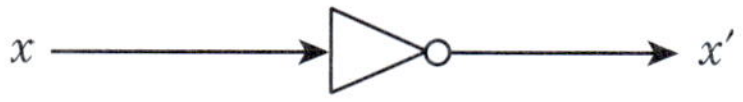

Table 12.11

Logic table for a NOT gate.

Input x	**Output** x'
0	1
1	0

The next example presents two NOT gates from everyday life.

EXAMPLE 12.17

One type of light bulb, often placed along walkways and driveways, contains a sensor and lights up automatically when there is no sunlight. This situation can be modeled by a NOT gate (see Figure 12.8).

Figure 12.8

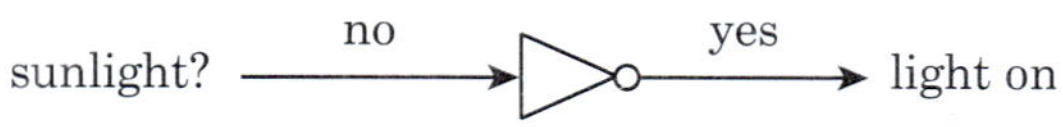

Another NOT gate condition is shown in Figure 12.9. Power plants are often equipped with two generators, one main and one backup. The backup generator works only if the main generator fails.

Figure 12.9

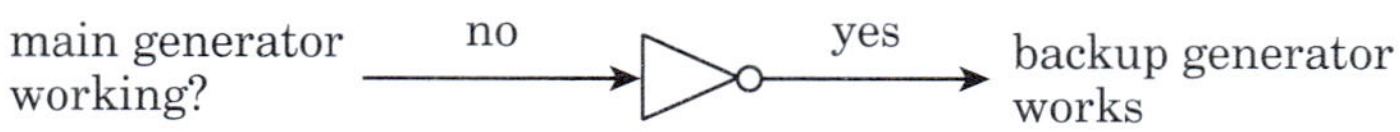

■

Two additional gates, **NAND** and **NOR**, prove useful.

NAND Gate

A **NAND gate** receives two bits x and y as input signals and produces a unique output denoted by $x \uparrow y$, where

$$x \uparrow y = \begin{cases} 0 & \text{if } x = 1 = y \\ 1 & \text{otherwise} \end{cases}$$

Figure 12.10 displays a NAND gate.

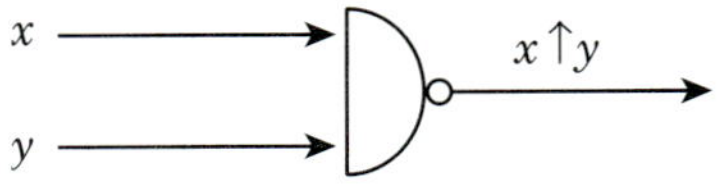

Figure 12.10
A NAND gate.

The next example provides an interesting instance of a NAND gate from everyday life.

EXAMPLE 12.18 Modern cars are equipped with alarms to warn forgetful drivers. Assume the car key is in the ignition. If the door on the driver's side is not closed or if her shoulder belt is not buckled up, a sound alarm will go off automatically. (See Figure 12.11.)

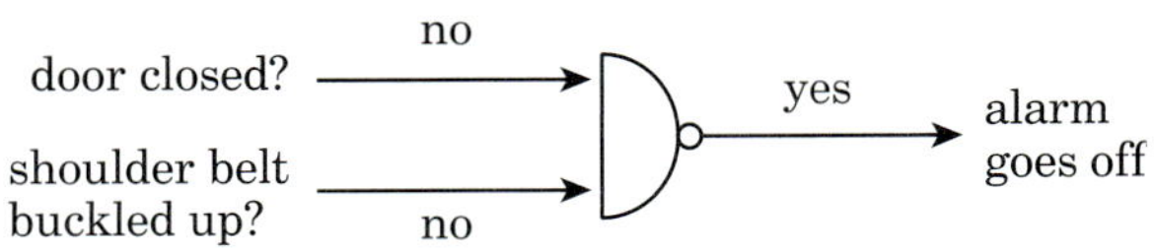

Figure 12.11

■

NOR Gate

A **NOR gate** receives two input signals x and y and produces a unique output denoted by $x \downarrow y$, where

$$x \downarrow y = \begin{cases} 1 & \text{if } x = 0 = y \\ 0 & \text{otherwise} \end{cases}$$

A NOR gate is pictured in Figure 12.12.

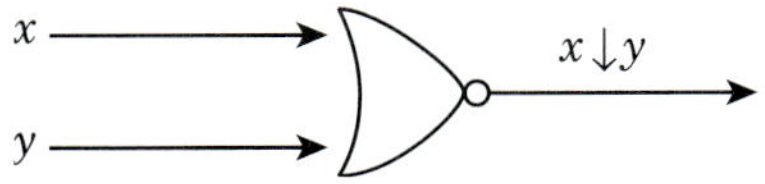

Figure 12.12
A NOR gate.

The next example presents a real-world application of a NOR gate.

EXAMPLE 12.19 Modern bathroom circuits are required for safety reasons to have "ground fault interrupt" circuit breakers to terminate the flow of electricity. These circuit breakers turn power off automatically if they detect a short-circuit. Circuit breakers carry test buttons, so the user can confirm that the breaker is working. Power is on unless a short-circuit is detected or a test indicates a dead breaker. This situation can be represented by a NOR gate, as Figure 12.13 shows.

Figure 12.13

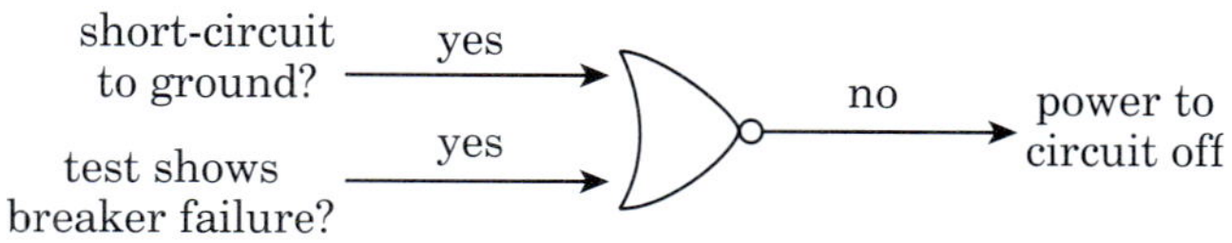

■

The five logic gates introduced carry out basic operations of easy practical value, yielding predictable output from known input. AND gates yield products, OR sums, and NOT complements. NAND and NOR gates produce exactly the opposite outputs of AND and OR gates.

Exercises 12.3

Find the output, if the bits 1, 1, and 0 are input into each gate.

1. AND gate **2.** OR gate

Compute the NAND gate output from inputing each pair of bits.

3. 0,0 **4.** 0,1 **5.** 1,0 **6.** 1,1

7–10. Redo Exercises 3–6 with a NOR gate.

Construct a logic table for each gate.

11. $x \uparrow y$ **12.** $x \downarrow y$

Find the DNF of each boolean function.

13. $f(x,y) = x \uparrow y$ **14.** $f(x,y) = x \downarrow y$

15. $f(x,y,z) = x \uparrow (y \uparrow z)$ **16.** $f(x,y,z) = (x \uparrow y) \uparrow z$

17. Suppose the bits x, y, and z are input into a NAND gate. List the possible output in a logic table.

18. Redo Exercise 17 for a NOR gate.

Mark each statement as true or false.

19. $x \uparrow y = y \uparrow x$ **20.** $x \downarrow y = y \downarrow x$

21. $x \uparrow (y \uparrow z) = (x \uparrow y) \uparrow z$ **22.** $x \downarrow (y \downarrow z) = (x \downarrow y) \downarrow z$

12.4 Combinatorial Circuits

Logic gates can be used to construct **combinatorial circuits** for a variety of applications in electrical engineering and electronics. A combinatorial circuit has no memory, but produces a unique output for every combination of input signals. The output does not depend on previous inputs or the state of the system. (On the other hand, outputs from **sequential circuits** depend on previous inputs and the state of the system. The transition diagrams of the FSMs discussed in Section 11.4 illustrate sequential circuits.)

Figure 12.14

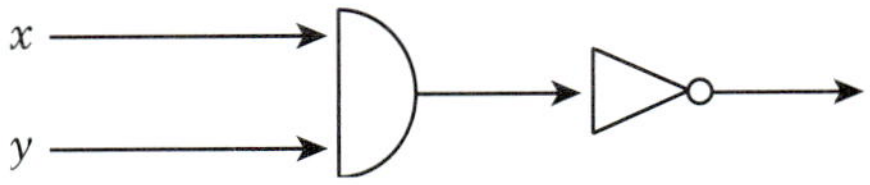

Figure 12.15

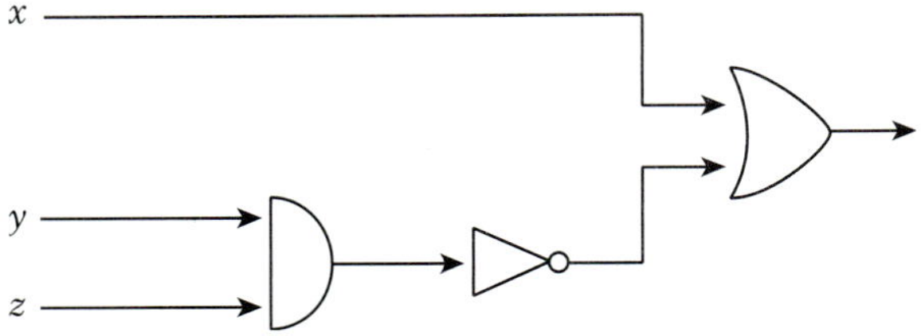

Figures 12.14 and 12.15 display combinatorial circuits. The one in Figure 12.14 contains two gates (an AND gate and a NOT gate), whereas Figure 12.15 contains three gates.

In Examples 12.20 and 12.21, we find, step by step, the output from a combinatorial circuit.

EXAMPLE 12.20 Find the output of the combinatorial circuit in Figure 12.14.

SOLUTION:

The bits x and y are input to an AND gate, yielding the output xy (see Figure 12.16). This output is sent to an inverter, yielding $(xy)'$ (see Figure 12.17). The output of the circuit equals the boolean expression $(xy)'$.

Figure 12.16

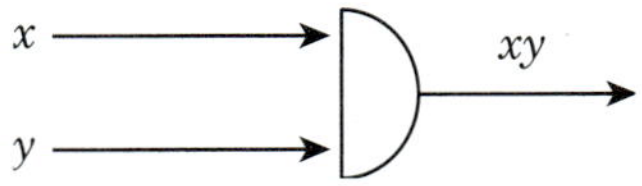

Figure 12.17

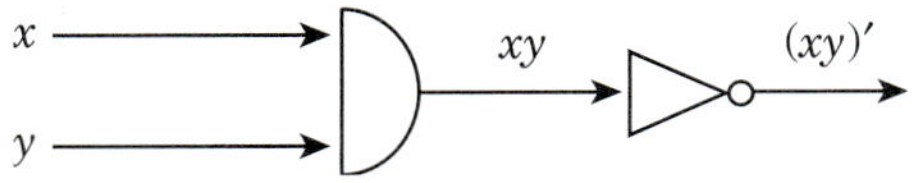

■

EXAMPLE 12.21 What output does the combinatorial circuit in Figure 12.15 produce?

SOLUTION:
As in the preceding example, the circuit could be built in steps, computing the output at each stage. But, for convenience, we can combine the various steps into one as shown in Figure 12.18.

Figure 12.18

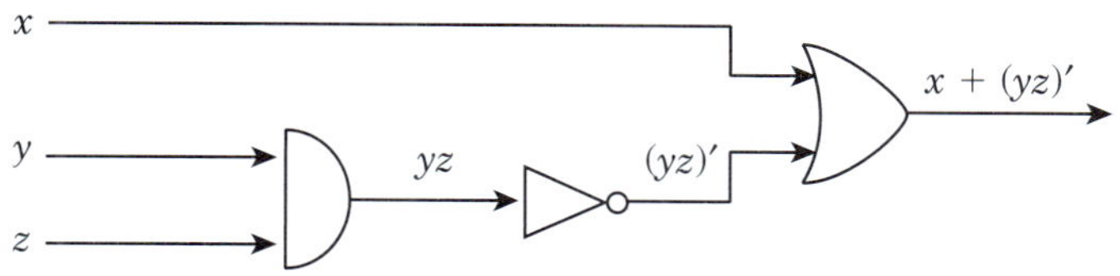

The output from the AND gate is yz. It enters an inverter, which outputs $(yz)'$. Then x and $(yz)'$ are input to an OR gate, for the final output of $x + (yz)'$. ■

A logic table helps to determine the output corresponding to every combination of input signals to a combinatorial circuit, as the next example shows.

EXAMPLE 12.22 Construct a logic table for the combinatorial circuit in Figure 12.15.

SOLUTION:
Since an input can be 0 or 1, eight combinations are possible, so the table contains eight rows, as in Chapter 1. Columns headed by x, y, z, yz, $(yz)'$,

Table 12.12

Input					Final output
x	y	z	yz	yz'	$x + (yz)'$
0	0	0	0	1	1
0	0	1	0	1	1
0	1	0	0	1	1
0	1	1	1	0	0
1	0	0	0	1	1
1	0	1	0	1	1
1	1	0	0	1	1
1	1	1	1	0	1

and $x + (yz)'$ coincide with Figure 12.18. Now, with the definitions of the AND, NOT, and OR gates, fill in the columns. Table 12.12, the same as Table 12.4, results. (Once again, the close relationship between building truth tables and logic tables is striking.) ■

In a combinatorial circuit, the same signal can be input for more than one gate. In such a case, branching serves well. For instance, the combinatorial circuit in Figure 12.19 has y as an input to both the AND and the OR gates. You can verify that the output of this circuit is $xy + y$.

Figure 12.19

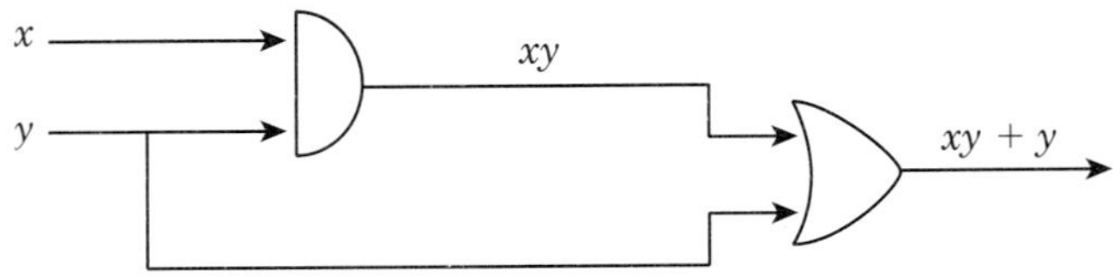

The output produced by a combinatorial circuit is a boolean expression. On the other hand, with a boolean expression E, we can form a combinatorial circuit that yields E as the output, as the next example illustrates.

EXAMPLE 12.23 Make a combinatorial circuit that yields the boolean expression $(x+y)z+z'$ as its output.

SOLUTION:
To construct the circuit successively in steps, begin with the *innermost* subexpression $x + y$. The corresponding combinatorial circuit is shown in Figure 12.20. Its output $x + y$ is ANDed with z to yield $(x + y)z$, as the corresponding circuit in Figure 12.21 indicates. Now devise an inverter for z (see Figure 12.22); it yields z'.

Figure 12.20

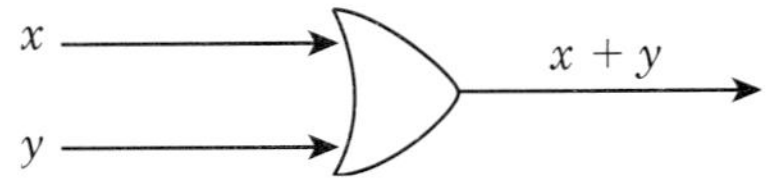

Figure 12.21

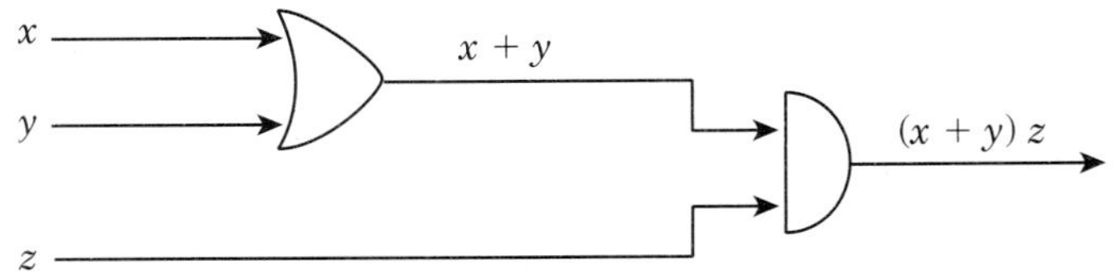

Figure 12.22

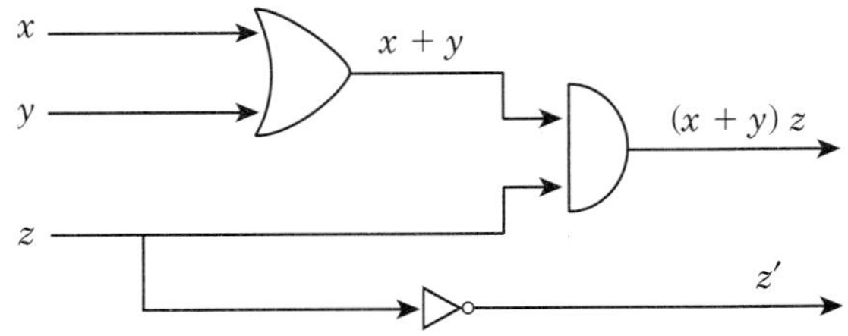

Figure 12.23

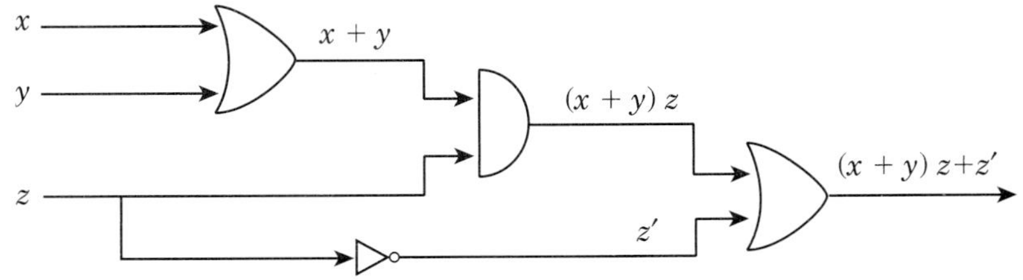

Finally, the subexpressions $(x+y)z$ and z' can combine to yield the given boolean expression. Consequently, the required combinatorial circuit is completed by ORing the outputs $(x+y)z$ and z', as in Figure 12.23. ■

Since a combinatorial circuit comprises logic gates, we can build it up using AND, OR, and NOT gates. In other words, {AND, OR, NOT} is a functionally complete set of gates. So are {AND, NOT}, {OR, NOT}, {NAND}, and {NOR}, according to the previous section.

Example 12.24 devises a combinatorial circuit for a desired output.

EXAMPLE 12.24 Construct a combinatorial circuit that produces $(x+y+z)(xyz)'$ as its output.

SOLUTION:

The given boolean expression can evolve in gradual steps: $x+y+z$, xyz, $(xyz)'$, and $(x+y+z)(xyz)'$. In similar succession, the corresponding combinatorial circuits unfold in Figures 12.24–12.27. Figure 12.27 shows the required circuit.

Figure 12.24

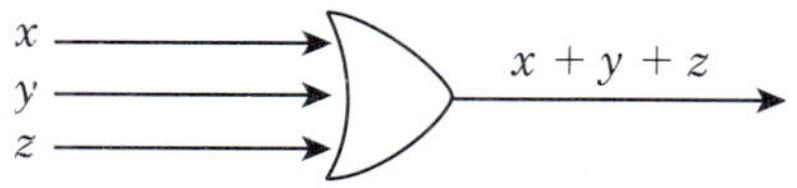

Figure 12.25

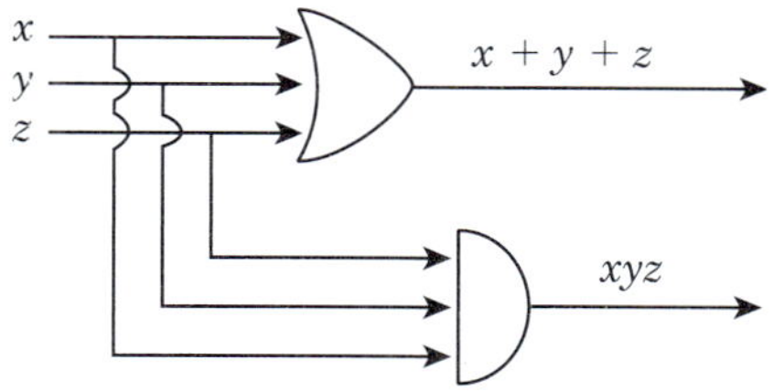

Figure 12.26

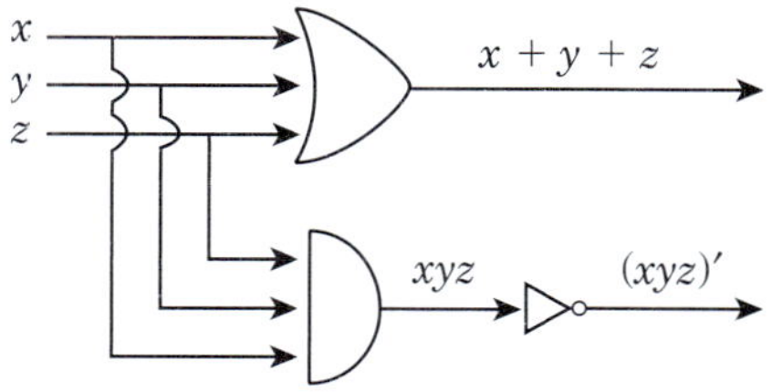

Figure 12.27

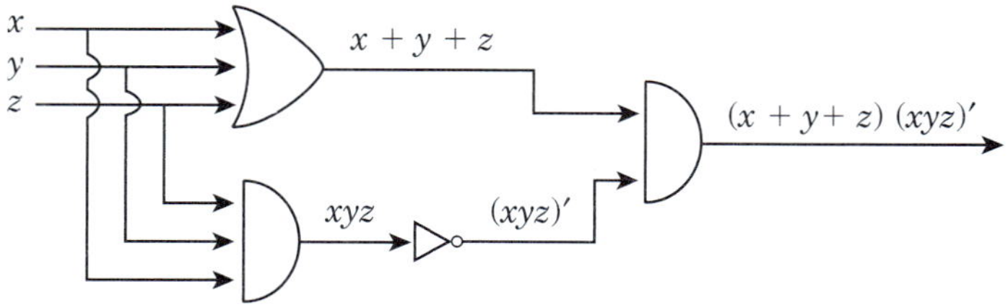

■

In the next example we design a combinatorial circuit used in everyday life.

EXAMPLE 12.25

Light fixtures in hallways are usually controlled by two or more switches. Suppose there are three such switches for a light fixture and the light is on. When the toggle of one of the switches is flipped, the light is turned off. On the other hand, if the light is off, then flipping one of the toggles turns the light on. Design a combinatorial circuit for this.

SOLUTION:

Let x, y, and z denote the three switches and $f(x,y,z)$ the output of the required combinatorial circuit. Let $f(x,y,z) = 1$ denote the state that the light is on. Arbitrarily assume the light is off when all switches are off, that is, when $x = y = z = 0$; in other words, $f(0,0,0) = 0$.

If one switch is turned on, the light goes on: $f(0,0,1) = f(0,1,0) = f(1,0,0) = 1$.

Now suppose one more switch is turned on. The light is turned off. Consequently, $f(0,1,1) = f(1,0,1) = f(1,1,0) = 0$.

Finally, suppose the third switch is also turned on. Then the light comes on again: $f(1,1,1) = 1$.

These functional values can be arranged in a logic table (see Table 12.13), and the DNF of the function is $f(x,y,z) = xyz + xy'z' + x'yz' + x'y'z$. The combinatorial circuit is shown in Figure 12.28.

Table 12.13

x	y	z	$f(x,y,z)$
0	0	0	0
0	0	1	1
0	1	0	1
1	0	0	1
0	1	1	0
1	0	1	0
1	1	0	0
1	1	1	1

Figure 12.28

A combinatorial circuit for a light fixture controlled by three switches.

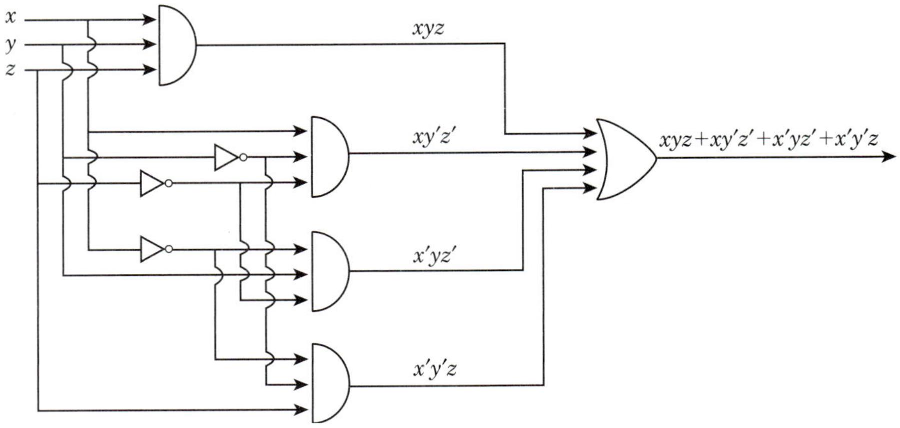

■

Some circuits can perform binary addition. We can find the sum of any two binary numbers using Table 12.14, where s denotes the sum bit and c the carry bit. To add 1011_{two} and 1110_{two}, use the familiar addition algorithm. Add the corresponding bits from right to left, propagating each carry to the left. (Initially, the carry bit is 0.) The final sum is 11001_{two}, as Figure 12.29 shows. (Base two is omitted for convenience.) Similarly, $111_{\text{two}} + 1001_{\text{two}} = 10000_{\text{two}}$ (see Figure 12.30).

Table 12.14

Logic table for a half-adder.

Input x	Input y	Output s	Output c
0	0	0	0
0	1	1	0
1	0	1	0
1	1	0	1

Figure 12.29

```
   (1)(1)(1)(0)  ← carry bits

     1  0  1  1
     1  1  1  0
  ---------------
  1  1  0  0  1
```

Figure 12.30

```
   (1)(1)(1)(1)  ← carry bits

        1  1  1
     1  0  0  1
  ---------------
  1  0  0  0  0
```

Half-Adder

Suppose you would like to design a circuit that will add any two bits x and y. Notice from Table 12.14 that $c = xy$ and $s = xy' + x'y$ (the DNF of s). But $xy' + x'y = (x + y)(xy)'$ by Exercise 50 in Section 12.1, so $s = (x + y)(xy)'$. As a result, the circuit that yields the sum bit s and the carry bit c lies in Figure 12.31. Being a **half-adder**, it adds only two bits without considering the carry from a possible previous addition. It contains two AND gates, an OR gate, and a NOT gate.

Figure 12.31

A half-adder.

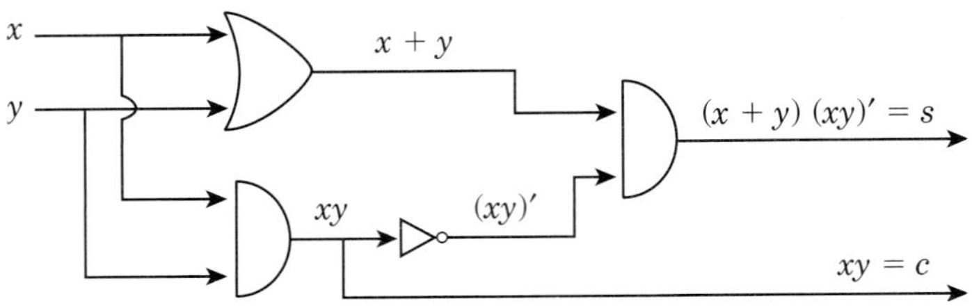

Full-Adder

Full-adders find the sum of any two binary numbers. A **full-adder** accepts three bits: the two bits x_i and y_i in the addends and the carry bit c_i, where $i \geq 1$. After accepting these three bits, the full-adder outputs the sum bit s_i and the new carry bit c_{i+1}. Table 12.15 lays out the values of s_i and c_{i+1} for various values of x_i, y_i, and c_i. According to the table,

$$s_i = x_i y_i c_i + x_i y_i' c_i' + x_i' y_i c_i' + x_i' y_i' c_i$$

and

$$c_{i+1} = x_i y_i c_i + x_i y_i c_i' + x_i y_i' c_i + x_i' y_i c_i$$

Consequently, a full-adder can be made with AND, OR, and NOT gates.

Table 12.15

Logic table for a full-adder.

Input			Output	
x_i	y_i	c_i	s_i	c_{i+1}
0	0	0	0	0
0	0	1	1	0
0	1	0	1	0
0	1	1	0	1
1	0	0	1	0
1	0	1	0	1
1	1	0	0	1
1	1	1	1	1

But two half-adders will work well together as a full-adder. To find the sum bit s_i and the carry bit c_{i+1}, input x_i and y_i into a half-adder H_1; it produces a sum bit s and a carry bit c. The bits s and c_i are then input to another half-adder H_2. Their sum is the desired sum bit s_i. The carry bits from H_1 and H_2 are sent to an OR gate, which outputs the new carry bit c_{i+1}. This full-adder appears in Figure 12.32.

Figure 12.32

A full-adder.

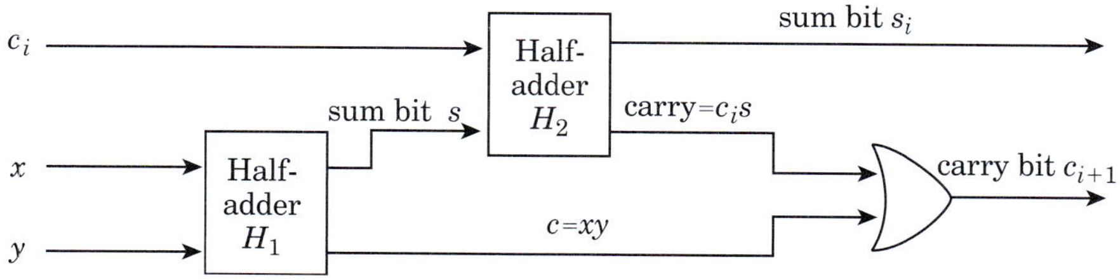

We close this section by showing how we can use half- and full-adders to compute the sum of two n-bit numbers.

EXAMPLE 12.26 Using a half-adder and full-adder, design a circuit that computes the sum of two 3-bit numbers $x = x_2x_1x_0$ and $y = y_2y_1y_0$.

SOLUTION:

A half-adder accepts two bits outputing the sum and carry bits. On the other hand, a full-adder accepts three bits to produce the sum and carry bits. The circuit in Figure 12.33 produces the sum $s = s_3s_2s_1s_0$.

Figure 12.33

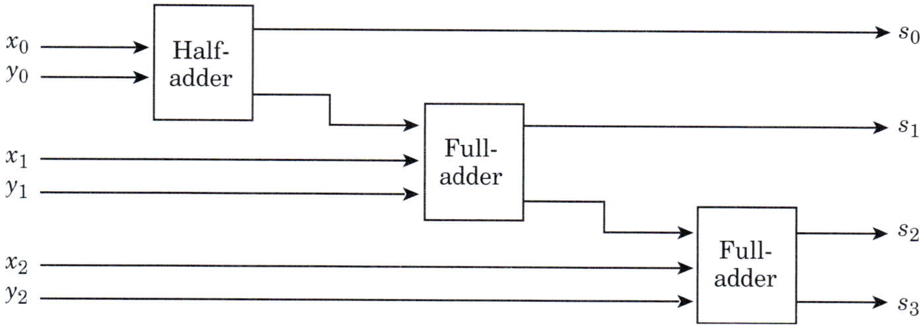

■

The range of possible combinatorial circuits expands as logic gates from half- and full-adders describe many electrical systems through the production of bit sums.

Exercises 12.4

When will the combinatorial circuit for each boolean expression produce 1 as the output?

1. x' **2.** $x + y$ **3.** xy

Find the output produced by the combinatorial circuits in Exercises 4–13.

4.

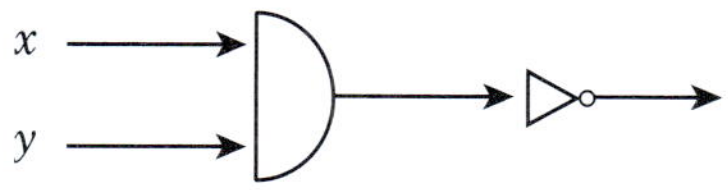

5.

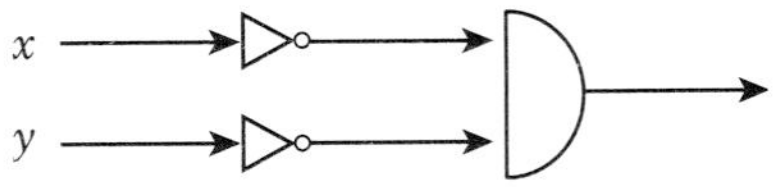

6.

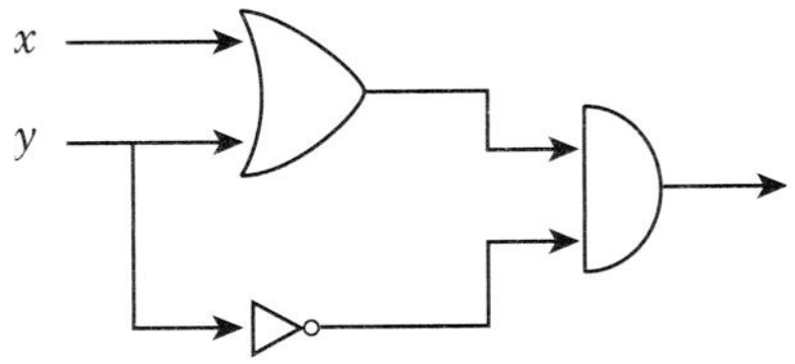

7.

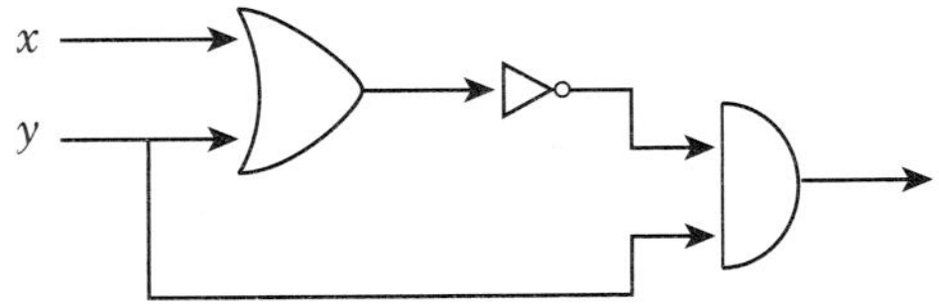

8.

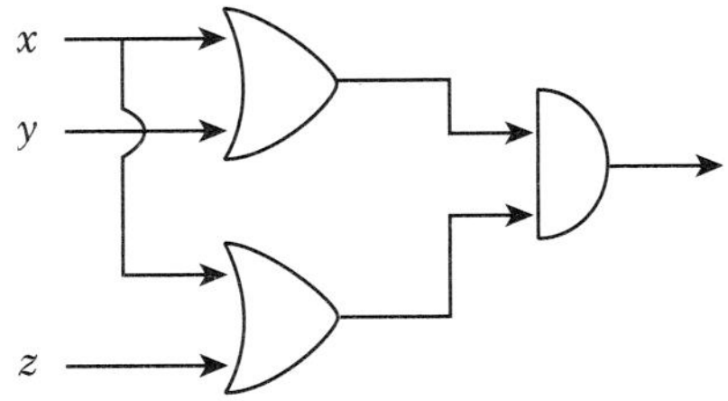

9.

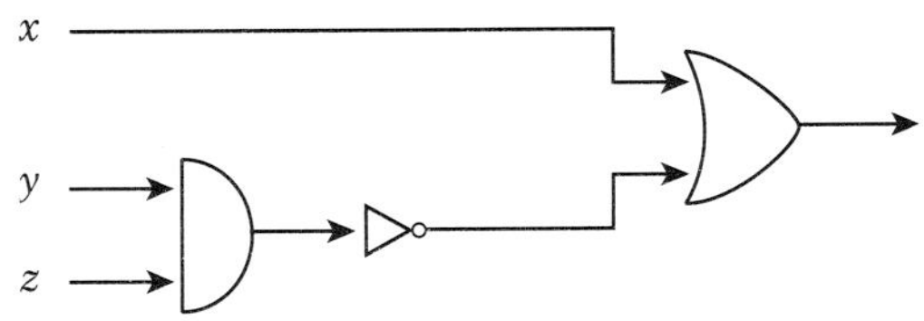

10.

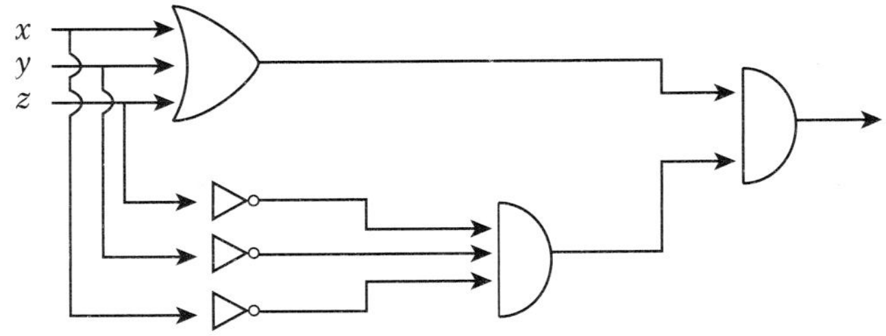

11.

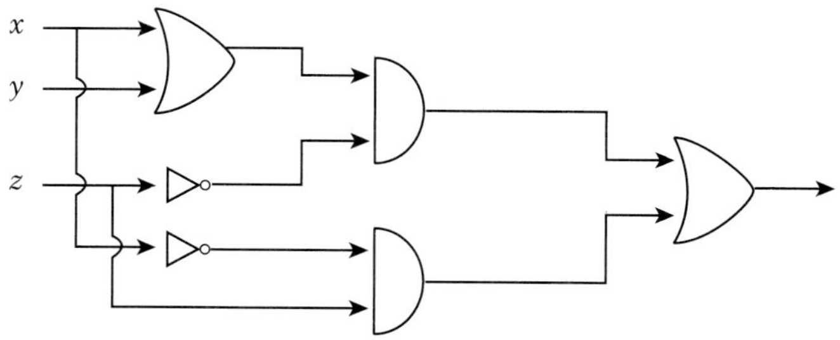

12.

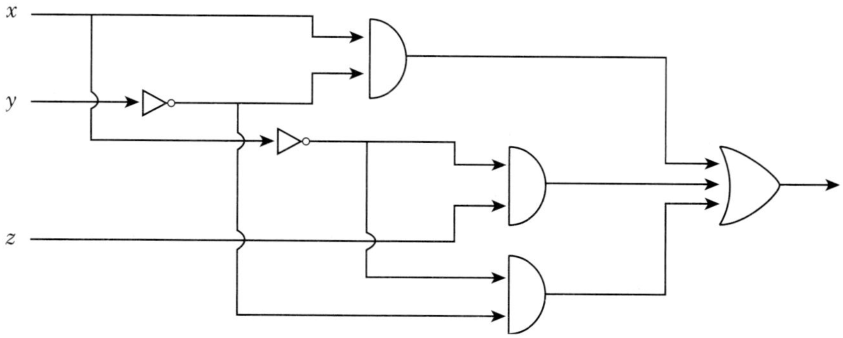

13.

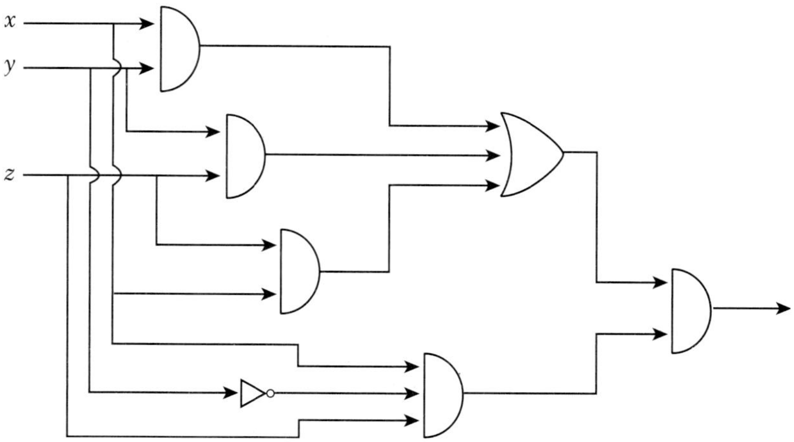

14–17. Devise a logic table for each circuit in Exercises 6–9.

18–21. Construct a combinatorial circuit for each boolean expression in Exercises 9, 10, 14, and 15 in Section 12.2.

Using only NAND gates, design a combinatorial circuit that receives x and y as input signals and outputs:

22. x' **23.** $x + y$ **24.** xy

25–27. Redo Exercises 22–24, using only NOR gates.

28. Make a combinatorial circuit for a hallway light fixture controlled by two switches x and y. Assume the light is off when both switches are.

By means of the circuit in Figure 12.33, add each pair of binary numbers.

29. 101, 100 **30.** 110, 011 **31.** 110, 101 **32.** 101, 111

Using the full-adder in Figure 12.32, verify each algebraically.

33. $s_i = x_iy_ic_i + x_iy_i'c_i' + x_i'y_ic_i' + x_i'y_i'c_i$

34. $c_{i+1} = x_iy_ic_i + x_iy_ic_i' + x_iy_i'c_i + x_i'y_ic_i$.

Design a half-adder with:

35. NAND gates. **36.** NOR gates.

12.5 Minimization of Combinatorial Circuits

With the set of gates {AND, OR, NOT} functionally complete, every combinatorial circuit can be represented by a boolean expression. So simplifying such expressions amounts to simplifying, or minimizing, circuits. Boolean algebraic laws can reduce boolean expressions, as two examples with three variables will demonstrate. However, Karnaugh maps give us an easier method. The DNFs of expressions of up to four variables will shrink through this graphic procedure.

Section 12.2 explored boolean expressions that yield the same value no matter how we combine the values of variables. Consequently, the corresponding combinatorial circuits yield the same output for the same set of input values. Accordingly, we make the following definition.

Equivalent Combinatorial Circuits

Two combinatorial circuits are **equivalent** if the corresponding boolean expressions are equal.

EXAMPLE 12.27 Are the combinatorial circuits C_1 and C_2 in Figure 12.34 equivalent?

Figure 12.34

Combinatorial circuit C_1 Combinatorial circuit C_2

SOLUTION:

First, find the boolean expressions representing the two circuits: $E_1 = x'y'$ and $E_2 = (x+y)'$, respectively. Since $E_1 = E_2$ by De Morgan's law, the two combinatorial circuits are equivalent. ■

Circuit C_1 contains three gates; C_2, only two. Accordingly, C_2 is **simpler** than C_1.

Through two methods, algebraic and graphical, we can find a circuit C_2 equivalent to but simpler than a given circuit C_1. Since a gate in a combinatorial circuit corresponds to a boolean operator, the key to both procedures lies in locating a boolean expression with fewer boolean operators and literals.

We can simplify a boolean expression through the laws of boolean algebra, as the next two examples demonstrate. However, we can use this method successfully only if we know the laws of boolean algebra well, so review them as often as needed.

EXAMPLE 12.28 Find a simpler combinatorial circuit equivalent to the one in Figure 12.35.

Figure 12.35

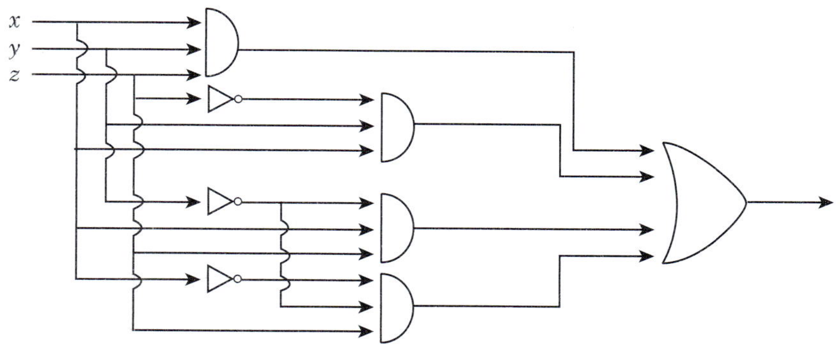

SOLUTION:

The boolean expression represented by the circuit is $f(x,y,z) = xyz + xyz' + xy'z + x'y'z$. Simplify it as much as possible, justifying every step:

$$\begin{aligned} f(x,y,z) &= xyz + xyz' + xy'z + x'y'z \\ &= xy(z+z') + y'z(x+x') \\ &= xy1 + y'z1 \\ &= xy + y'z \end{aligned}$$

Consequently, we can replace the given circuit by the much simpler version in Figure 12.36.

Figure 12.36

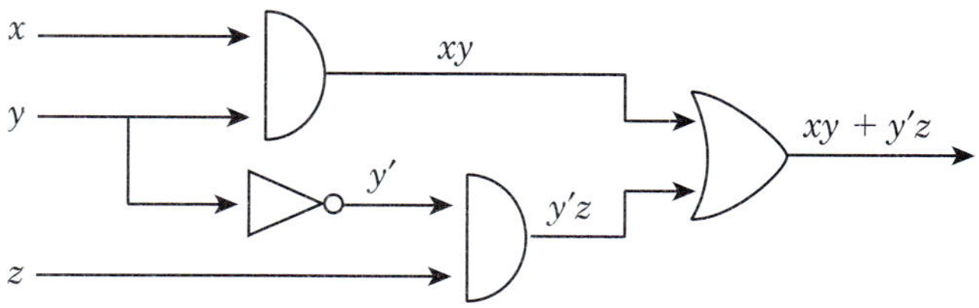

■

EXAMPLE 12.29 Find a simpler combinatorial circuit equivalent to the one represented by the boolean function $f(x,y,z) = xyz + xyz' + xy'z + x'yz + xy'z'$ (see Example 12.13 also).

SOLUTION:

$$\begin{aligned}
f(x,y,z) &= xyz + xyz' + xy'z + x'yz + xy'z' \\
&= xyz + xyz' + xy'z + x'yz + (xy'z' + xy'z') \\
&= xyz + (xy'z + xy'z') + (xyz' + xy'z') + x'yz \\
&= xyz + xy'(z + z') + xz'(y + y') + x'yz \\
&= xyz + xy'1 + xz'1 + x'yz \\
&= xyz + xy' + xz' + x'yz \\
&= xyz + x(y' + z') + x'yz + xyz \\
&= xyz + x(yz)' + yz(x + x') \\
&= x[yz + (yz)'] + yz1 \\
&= x1 + yz \\
&= x + yz
\end{aligned}$$

Therefore, the simpler two-gate circuit in Figure 12.37 can replace the original nine-gate one, saving seven gates.

Figure 12.37

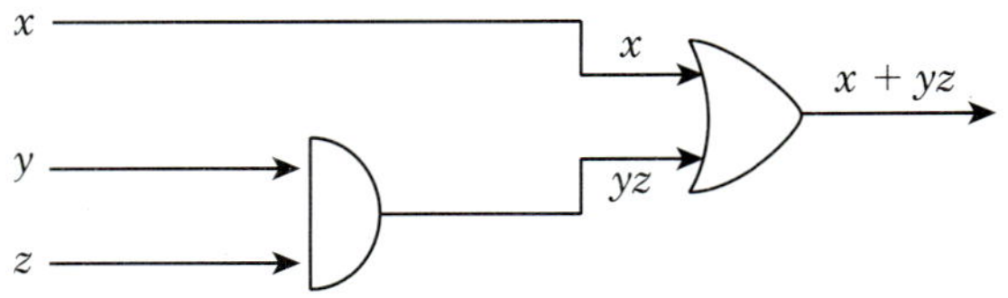

■

Simplifying a boolean expression can be arduous and frustrating. To a large extent, success depends on grouping the various terms properly and then ingeniously applying the laws of boolean algebra. But there is another way.

Maurice Karnaugh developed the graphical method of the **Karnaugh map** at Bell Laboratories in 1953 to simplify the DNFs of boolean expressions. The essence of the Karnaugh map lies in grouping minterms that differ by exactly one literal.

For instance, the minterms $xy'z$ and $xy'z'$ differ in exactly one literal and their sum can shrink: $xy'z + xy'z' = xy'(z + z') = xy'1 = xy'$ (see also Example 12.29).

The Karnaugh map has a rectangular grid of squares. Each square stands for a possible minterm in the DNF of the boolean expression that represents

Maurice Karnaugh *(1924–), a physicist, was born in New York City. After graduating from the City College of New York in 1948, he received his Ph.D. in physics from Yale in 1952.*

Karnaugh was a member of the research staff at Bell Telephone Labs from 1952 to 1966, and a research and development manager at the Federal Systems Division of AT&T during the next 4 years. In 1970, he joined the research staff at IBM. His research interests include knowledge-based computer systems.

the circuit. Each contains a 1 if the corresponding minterm exists in the expression.

A 2×2 Karnaugh map can help simplify the DNF of a boolean expression in two variables x and y. Since four possible minterms are feasible—xy, xy', $x'y'$, and $x'y$—the map consists of four squares. Label the squares so the minterms in any two adjacent squares in each row and column differ by exactly one literal; two such squares are **adjacent**. For instance, the squares xy and xy' are adjacent, whereas the squares xy and $x'y'$ are not. The resulting arrangement appears in Table 12.16.

Table 12.16

	y	y'
x	xy	xy'
x'	$x'y$	$x'y'$

EXAMPLE 12.30 Find the Karnaugh map for each boolean expression.

(1) $xy + x'y + x'y'$ (2) $xy' + x'y$

SOLUTION:

Place a 1 in the square corresponding to each minterm and leave the other squares blank, as shown in Tables 12.17 and 12.18, respectively.

Table 12.17

Karnaugh map for $xy + x'y + x'y'$.

	y	y'
x	1	
x'	1	1

Table 12.18

Karnaugh map for $xy' + x'y$.

	y	y'
x	1	
x'	1	1

■

The sum of minterms in adjacent squares can be simplified. By drawing a loop around such blocks containing 1's, always beginning with the largest block, we can always reduce the minterms corresponding to the blocks, as Example 12.31 suggests.

EXAMPLE 12.31 Using a Karnaugh map, simplify the boolean expressions in Example 12.30, if possible.

SOLUTION:

(1) First, loop off adjacent squares containing 1's. There are two such blocks in Table 12.19. Therefore,

$$
\begin{aligned}
xy + x'y + x'y' &= (xy + x'y) + (x'y + x'y') \\
&= (x + x')y + x'(y + y') \\
&= 1y + x'1 \\
&= y + x' \\
&= x' + y
\end{aligned}
$$

(With a little practice, we can read this answer directly from the map.)

Table 12.19

	y	y'
x	1	
x'	1	1

(2) The squares corresponding to the minterms xy' and $x'y$ are not adjacent, so each forms a block by itself in Table 12.20. Consequently, the expression $xy' + x'y$ cannot be simplified.

Table 12.20

	y	y'
x		1
x'	1	

■

A Karnaugh map for three variables x, y, and z expands to a 2×4 rectangular grid. Each of the eight squares in Table 12.21 corresponds to a possible minterm in x, y, and z. Again, two squares are **adjacent** if the corresponding minterms differ by exactly one literal. For instance, the squares $xy'z$ and xyz are adjacent. To see this geometrically, cut out the grid, bend, and glue the two shorter edges to form the cylinder in Figure 12.38. Two adjacent squares share a boundary.

Table 12.21

	yz	yz'	$y'z'$	$y'z$
x	xyz	xyz'	$xy'z'$	$xy'z$
x'	$x'yz$	$x'yz'$	$x'y'z'$	$x'y'z$

Figure 12.38

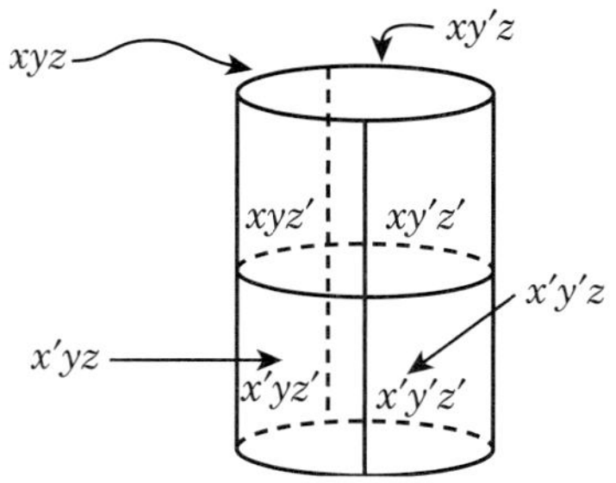

To simplify a sum of minterms in x, y, and z, identify blocks of minterms that can be combined by drawing loops around them. Always begin with the largest block and use the minimum number of blocks. This time a block may be a 1×1, 1×2, 1×4, 2×2, or a 2×4 rectangle. An example of each possibility appears in Figure 12.39.

Figure 12.39

A few possible blocks in a 2×4 Karnaugh map.

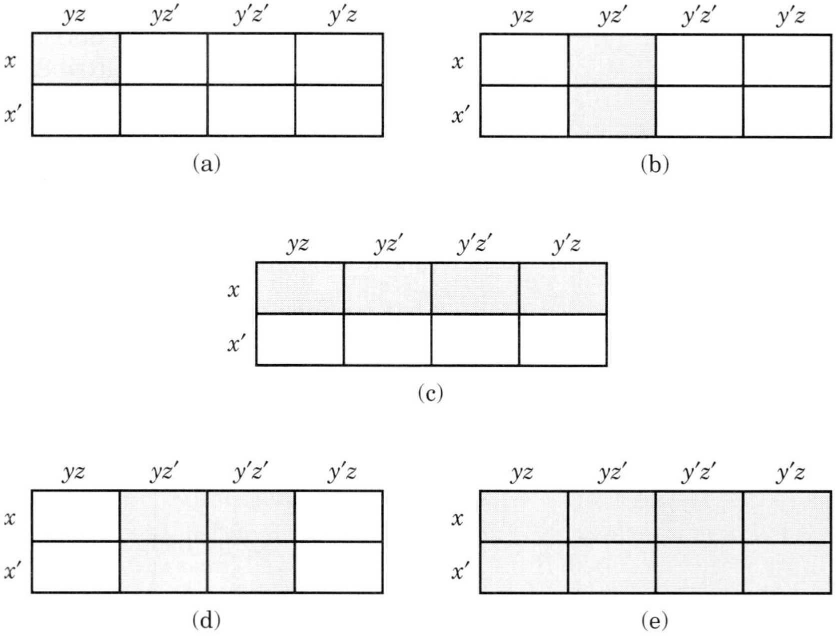

Notice that the sum of the minterms in the block in Figure 12.39d can be simplified as follows:

$$\begin{aligned} xyz' + xy'z' + x'yz' + x'y'z' &= yz'(x + x') + y'z'(x + x') \\ &= yz'1 + y'z'1 \\ &= yz' + y'z' \\ &= (y + y')z' \\ &= 1z' \\ &= z' \end{aligned}$$

The next example provides two additional exercises in simplifying boolean expressions.

EXAMPLE 12.32 Simplify each boolean expression with a Karnaugh map.

$$(1)\ E_1 = xyz + xyz' + xy'z + x'y'z$$
$$(2)\ E_2 = xyz + xyz' + xy'z + x'yz + xy'z'$$

SOLUTION:

Figures 12.40 and 12.41 show the Karnaugh maps for the two expressions.

Figure 12.40

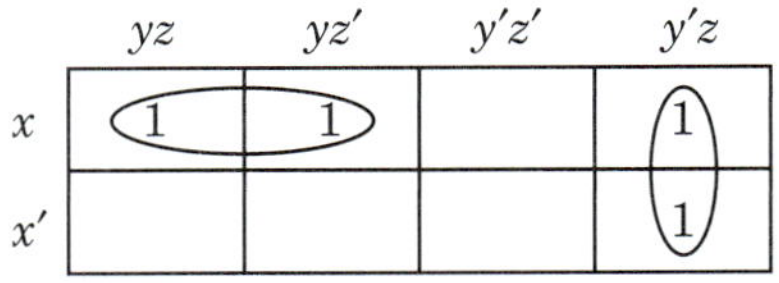

Figure 12.41

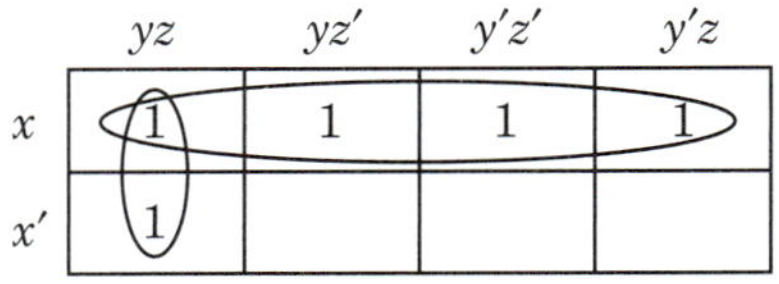

(1) From Figure 12.40,

$$\begin{aligned} E_1 &= (xyz + xyz') + (xy'z + x'y'z) \\ &= xy(z + z') + y'z(x + x') \\ &= xy1 + y'z1 \\ &= xy + y'z \qquad \text{(also see Example 12.28)} \end{aligned}$$

(2) Since a 1 occurs in both loops, count the corresponding minterm xyz twice using the idempotent law:

$$\begin{aligned} E_2 &= (xyz + xyz' + xy'z' + xy'z) + (xyz + x'yz) \\ &= x[y(z+z') + y'(z'+z)] + (x+x')yz \\ &= x(y1 + y'1) + 1yz \\ &= x(y+y') + yz \\ &= x + yz \qquad \text{(also see Example 12.29)} \end{aligned}$$

Again, with a little experience and patience, we can read these expressions directly from the maps. ■

Finally, a Karnaugh map for four variables w, x, y, and z encompasses the 4×4 grid of Table 12.22. Each of the 16 squares represents a possible minterm in w, x, y, and z. Geometrically, the grid can be cut, bent, and glued to form the doughnut-shaped surface called a **torus**. Two squares are adjacent if they share a border on the torus. The cells $wxy'z$ and $wxyz$ are adjacent; $wxyz$ and $wx'yz'$ are *not*.

Table 12.22

Karnaugh map for $xy + z'y + x'y'$.

	yz	yz'	$y'z'$	$y'z$
wx	$wxyz$	$wxyz'$	$wxy'z'$	$wxy'z$
wx'	$wx'yz$	$wx'yz'$	$wx'y'z'$	$wx'y'z$
$w'x'$	$w'x'yz$	$w'x'yz'$	$w'x'y'z'$	$w'x'y'z$
$w'x$	$w'xyz$	$w'xyz'$	$w'xy'z'$	$w'xy'z$

As was done with two- and three-variable expressions, place a 1 in the square corresponding to each minterm in the boolean expression and loop off the 1's into blocks of minterms that can be combined, always looking for the largest block first and using as few blocks as possible. The blocks useful for minimization are 1×1, 1×2, 1×4, 2×2, 2×4, and 4×4. Four such blocks are highlighted in Figure 12.42.

For example, the sum of the minterms in Figure 12.42a can be simplified:

$$\begin{aligned} wxyz + wxy'z + w'xyz + w'xy'z &= wxz(y+y') + w'xz(y+y') \\ &= wxz + w'xz \\ &= (w+w')xz \\ &= xz \end{aligned}$$

We conclude this section with an example that illustrates how to simplify four-variable boolean expressions using a Karnaugh map.

Figure 12.42

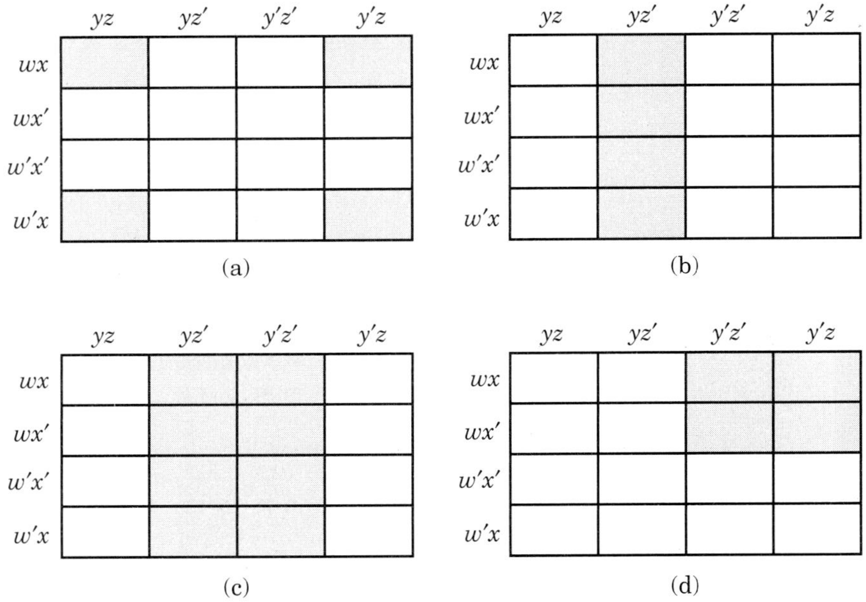

EXAMPLE 12.33 Using a Karnaugh map, simplify each boolean expression.

(1) $E_1 = wxyz + wxyz' + wxy'z' + wxy'z + w'xyz + w'xy'z$

(2) $E_2 = wxy'z' + wx'y'z' + wx'y'z + w'x'yz + w'x'yz' + w'xyz + w'xyz'$

SOLUTION:

(1) Place a 1 in the square for each minterm in the expression. Loop off each block, beginning with the largest. The Karnaugh map is shown in Figure 12.43. Accordingly, the expression E_1 can be pruned:

Figure 12.43

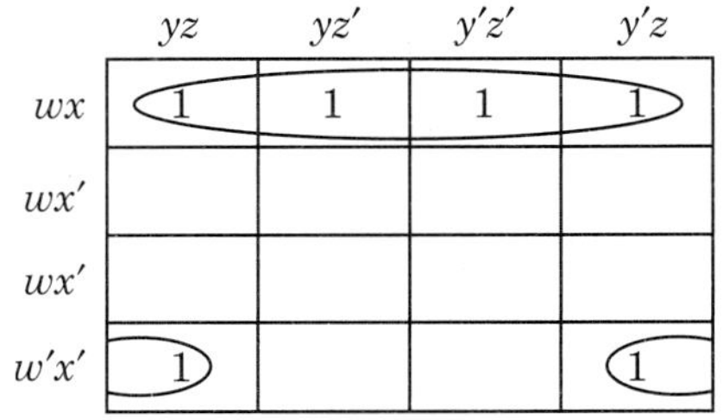

$$\begin{aligned} E_1 &= (wxyz + wxyz' + wxy'z' + wxy'z) + (w'xyz + w'xy'z) \\ &= wx[y(z + z') + y'(z' + z)] + w'xz(y + y') \\ &= wx(y + y') + w'xz \\ &= wx + w'xz \end{aligned}$$

Figure 12.44

	yz	yz'	$y'z'$	$y'z$
wx			1	
wx'			1	1
$w'x'$	1	1		
$w'x$	1	1		

(2) Using the Karnaugh map in Figure 12.44, we have:

$$\begin{aligned}E_2 &= (wxy'z' + wx'y'z') + (wx'y'z' + wx'y'z)\\ &\quad + (w'x'yz + w'x'yz' + w'xyz + w'xyz')\\ &= wy'z'(x + x') + wx'y'(z + z') + [w'x'y(z + z') + w'xy(z + z')]\\ &= wy'z' + wx'y' + (w'x'y + w'xy)\\ &= wy'z' + wx'y' + w'y(x + x')\\ &= wy'z' + wx'y' + w'y\end{aligned}$$

■

These cases suggest that Karnaugh maps simplify boolean expressions more easily than algebraic laws, especially when variables are few.

Exercises 12.5

Simplify each boolean expression using the laws of boolean algebra.

1. $xy + xy'$ **2.** $x(x + y) + xy'$ **3.** $(x + y)xy'$ **4.** $xy + xy' + x'y'$

5. $x'yz + x'y'z' + x'yz' + x'y'z$ **6.** $xy'z' + x'y'z' + xy'z + x'y'z$

7. $(x + y)(x + y + z)xy$ **8.** $(x + y + z)xyz$

9. $(x + y)(y + z)(z + x)$ **10.** $(xy + yz + zx)xyz$

11. $(x + y)(x' + y)(x + y')$ **12.** $(x + y' + z)(x + y + z')xy'z'$

13. $(x + y)(y + z)(z + x)xyz$ **14.** $(x + yz)(y + zx)(z + xy)$

15. $wxyz + w'xy'z' + wxyz' + w'xy'z$ **16.** $wx'yz + wx'yz' + w'x'yz' + w'xyz'$

Find the boolean expression represented by each Karnaugh map.

17.

	y	y'
x	1	1
x'		

18.

	y	y'
x	1	
x'	1	

19.

	y	y'
x		1
x'	1	

20.

	y	y'
x	1	
x'		1

Display each sum of minterms in a Karnaugh map.

21. $xy + x'y'$ **22.** $x'y + xy'$

Using a Karnaugh map, simplify each sum of minterms.

23. $xy + xy'$ **24.** $xy + xy' + x'y'$

Find the boolean expression represented by each Karnaugh map.

25.

	yz	yz'	$y'z'$	$y'z$
x	1	1	1	
x'			1	

26.

	yz	yz'	$y'z'$	$y'z$
x	1			
x'	1	1	1	

27.

	yz	yz'	$y'z'$	$y'z$
x		1	1	
x'		1	1	

28.

	yz	yz'	$y'z'$	$y'z$
x	1			
x'	1	1		

Using a Karnaugh map, simplify each boolean expression.

29. $xy'z + xy'z'$ **30.** $xyz + xy'z + x'yz + x'y'z$

31. $xy'z' + xy'z + x'y'z' + x'y'z$ **32.** $xyz + xyz' + x'y'z' + x'y'z$

33–36. Using a Karnaugh map, simplify the boolean expressions in Exercises 25–28.

Find the boolean expression represented by each Karnaugh map.

37.

	yz	yz'	$y'z'$	$y'z$
wx		1	1	
wx'		1	1	
$w'x'$		1	1	
$w'x$		1	1	

38.

	yz	yz'	$y'z'$	$y'z$
wx	1			1
wx'				
$w'x'$				
$w'x$	1			1

39.

	yz	yz'	$y'z'$	$y'z$
wx	1			
wx'	1			
$w'x'$			1	1
$w'x$			1	1

40.

	yz	yz'	$y'z'$	$y'z$
wx	1	1	1	
wx'			1	1
$w'x'$				
$w'x$	1			

Represent each sum of minterms in a Karnaugh map.

41. $wxy'z + w'xyz$ **42.** $wxyz + wxy'z + w'xyz + w'xy'z$

43. $wxy'z + wx'y'z + w'xy'z + w'x'y'z$ **44.** $wx'yz' + wx'y'z' + w'x'yz' + w'x'y'z'$

45–48. Using a Karnaugh map, simplify the boolean expressions in Exercises 37–40.

Using a Karnaugh map, simplify each boolean expression.

49. $wxyz + wx'yz + w'x'yz + w'xyz$

50. $wx'yz' + wx'y'z' + w'x'yz' + w'x'y'z'$

51. $wx'yz + wx'yz' + wx'y'z' + w'x'y'z' + w'xy'z' + w'xy'z$

52. $wxyz + wxyz' + wxy'z' + wxy'z + wx'y'z + w'x'y'z + w'xy'z$

12.6 Don't Care Conditions

Occasionally, combinatorial circuits may occur that do not accept certain combinations of input signals. In other words, a boolean function f can be generated for which some value combinations never occur as input. Consequently, we *don't care* about corresponding output values of f in such cases; they remain **unspecified**. Such functions are **incompletely specified**, and such input values are **don't care conditions**.

Consequently, to simplify the sum of minterms of such a combinatorial circuit with a Karnaugh map, place a d in each cell corresponding to a don't care condition. This simply means that the corresponding value of the function can be arbitrarily assigned: the minterm may or may not enter the simplification process. Whenever the term can help minimization, its d counts as a 1 in the otherwise standard procedure.

Three examples will illustrate this technique.

EXAMPLE 12.34 Simplify the boolean expression E represented by the Karnaugh map in Figure 12.45.

Figure 12.45

	yz	yz'	$y'z'$	$y'z$
wx	d	d	d	d
wx'	d	d		
$w'x'$	1	1		
$w'x$	1			1

SOLUTION:

First, draw a loop around each block as usual. Since the top row contains no 1's, exclude it in the minimization process; think of its cells as blanks. The resulting blocks are shown in Figure 12.46. You may verify that:

Figure 12.46

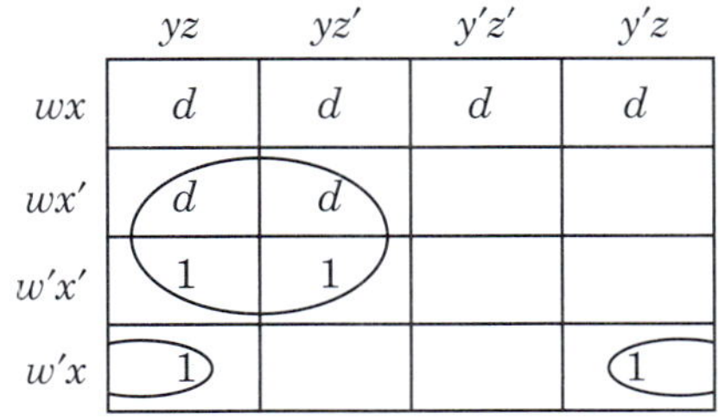

$$E = (wx'yz + wx'yz' + w'x'yz + w'x'yz') + (w'xyz + w'xy'z)$$
$$= x'y + w'xz$$

Note: If the top row counts as a block, $E = wx + x'y + w'xz$. Thus, $wx + x'y + w'xz = x'y + w'xz$ for all combined values of the variables for which the expression E is specified. As a result, the solution depends on a judicious choice of d's. ■

The next two examples provide interesting applications of don't care conditions.

EXAMPLE 12.35 A system to represent decimal numbers is to replace each decimal digit with its binary version. Four bits are needed to encode the 10 digits, as Table 12.23 shows. With Table 12.23, 345 can be encoded as 001101000101. This binary representation is the **binary coded decimal (BCD)expansion**. Although there are 16 possible arrangements of four bits, only 10 of them appear in the table. The others (see Table 12.24) are never used, meaning those BCD expansions manifest don't care conditions.

Table 12.23

Decimal digit	Binary representation
0	0000
1	0001
2	0010
3	0011
4	0100
5	0101
6	0110
7	0111
8	1000
9	1001

Table 12.24

Decimal number	Binary representation
10	1010
11	1011
12	1100
13	1101
14	1110
15	1111

We would like to design a combinatorial circuit that will accept four input signals w, x, y, and z, and output 1 if they represent the BCD expansion of a decimal prime number < 10 (see Figure 12.47). That is, we would like to develop a boolean function f such that $f(w,x,y,z)$ equals 1 if $(wxyz)_{\text{two}}$ is the BCD expansion of a prime number and equals 0 otherwise.

Combining Tables 12.23 and 12.24, Table 12.25 reveals the logic table of the function f. Since the numbers 10 through 15 do not apply, they generate

Figure 12.47

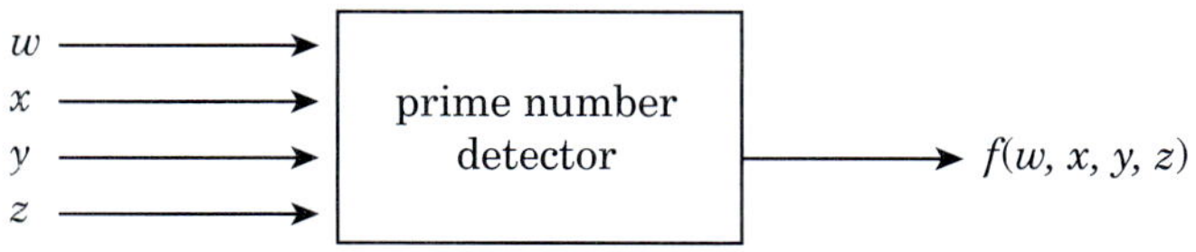

Table 12.25

Decimal value	BCD input w	x	y	z	Output
0	0	0	0	0	0
1	0	0	0	1	0
2	0	0	1	0	1
3	0	0	1	1	1
4	0	1	0	0	0
5	0	1	0	1	1
6	0	1	1	0	0
7	0	1	1	1	1
8	1	0	0	0	0
9	1	0	0	1	0
10	1	0	1	0	d
11	1	0	1	1	d
12	1	1	0	0	d
13	1	1	0	1	d
14	1	1	1	0	d
15	1	1	1	1	d

(Rows 10–15: don't care conditions)

don't care conditions, signaled by the d's in the last column. (Recall that d-cells crop up in the map only if helpful in minimization.) The blocks are shown in Figure 12.48. It follows that:

$$
\begin{aligned}
f(w,x,y,z) &= (wxyz + wxy'z) + (wx'yz + wx'yz' + w'x'yz + w'x'yz') \\
&\quad + (w'xyz + w'xy'z) \\
&= wxz(y+y') + wx'y(z+z') + w'x'y(z+z') + w'xz(y+y') \\
&= wxz + wx'y + w'x'y + w'xz
\end{aligned}
$$

Figure 12.48

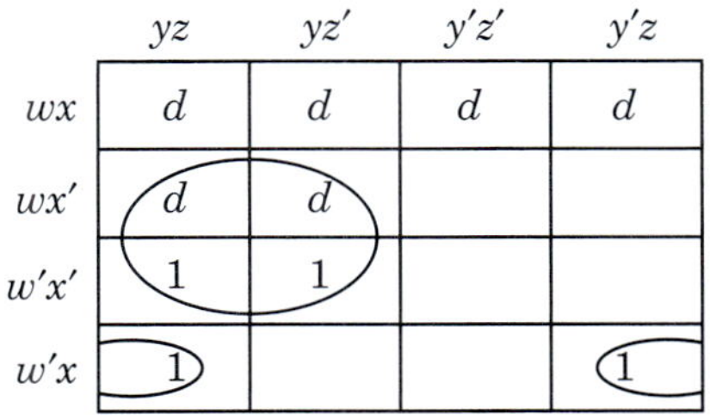

$$= xz(w + w') + x'y(w + w')$$
$$= xz + x'y$$

Figure 12.49 gives the circuit.

Figure 12.49

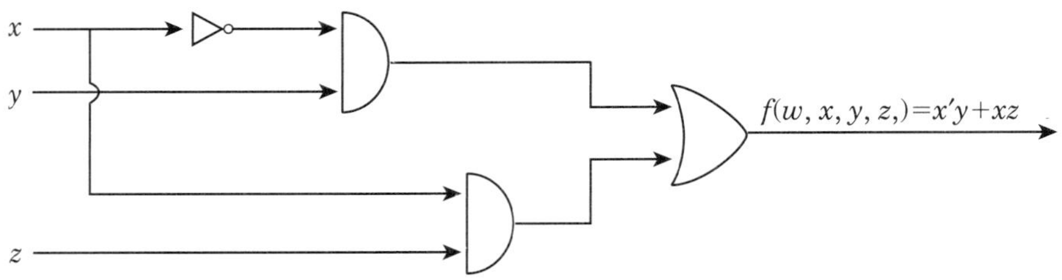

■

The next example presents a familiar application of don't care conditions to everyday life.

EXAMPLE 12.36

(Digital Displays) Electronic devices such as modern calculators, microwave ovens, and video cassette recorders display digits by lighting up a maximum of seven line segments, labeled a through g in Figure 12.50. Figure 12.51 provides the display strategy: the digit 0 lights up the segments a, b, c, d, e, and f; 6 lights up the segments c, d, e, f, and g; and so on.

Figure 12.50

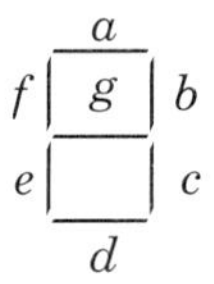

Figure 12.51

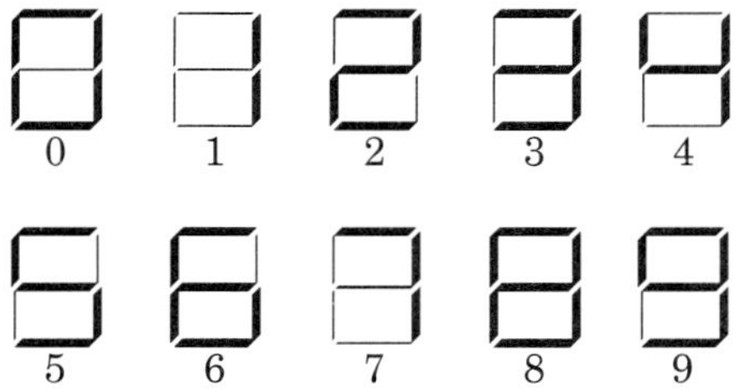

Design a combinatorial circuit to accept the BCD expansion $(wxyz)_{\text{two}}$ of a decimal digit and light up a segment s. In other words, develop a boolean expression to define the boolean function $f(w,x,y,z) = s$.

As in Example 12.35, 16 arrangements are available. However, since there are only 10 digits, six arrangements are don't care conditions; they are listed in Table 12.26. (Verify them.)

For instance, suppose we wish to find a minimal boolean expression for the boolean function $f(w,x,y,z) = a$. Table 12.26 gives the corresponding

Table 12.26

Decimal digit	BCD input *w*	*x*	*y*	*z*	Inputs *a*	*b*	*c*	*d*	*e*	*f*	*g*
0	0	0	0	0	1	1	1	1	1	1	0
1	0	0	0	1	0	1	1	0	0	0	0
2	0	0	1	0	1	1	0	1	1	0	1
3	0	0	1	1	1	1	1	1	0	0	1
4	0	1	0	0	0	1	1	0	0	1	1
5	0	1	0	1	1	0	1	1	0	1	1
6	0	1	1	0	0	0	1	1	1	1	1
7	0	1	1	1	1	1	1	0	0	0	0
8	1	0	0	0	1	1	1	1	1	1	1
9	1	0	0	1	1	1	1	0	0	1	1

DNF, $a = w'x'y'z' + w'x'yz' + w'x'yz + w'xy'z + w'xyz + wx'y'z' + wx'y'z$, which leads to the Karnaugh map in Figure 12.52. The cartographical conclusion, $a = w + x'y + x'y'z' + w'xz$, yields the desired circuit in Figure 12.53.

Figure 12.52

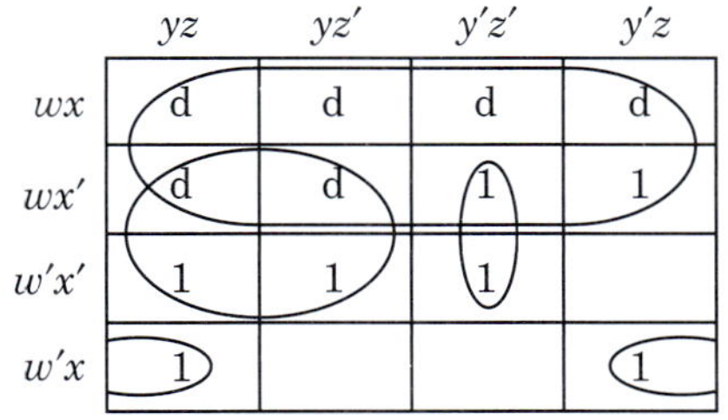

Figure 12.53

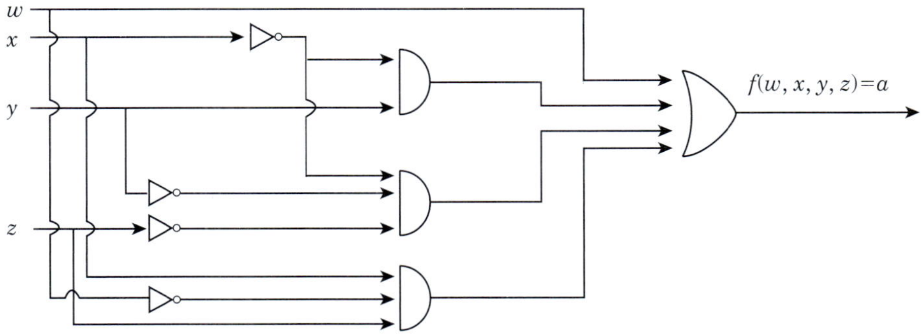

In accommodating through d-cells, the don't care conditions of incompletely specified functions in Karnaugh maps emerge as strong workhorses in the search for combinatorial circuits that mimic boolean expressions.

Exercises 12.6

Simplify the boolean expression represented by each Karnaugh map.

1.

	yz	yz'	$y'z'$	$y'z$
x	1	d	d	d
x'			1	

2.

	yz	yz'	$y'z'$	$y'z$
x	d		1	d
x'		d	1	1

3.

	yz	yz'	$y'z'$	$y'z$
wx			1	1
wx'			d	d
$w'x'$	d	d	d	d
$w'x$	1	1	d	d

4.

	yz	yz'	$y'z'$	$y'z$
wx	1	d		
wx'	d	1		
$w'x'$			d	d
$w'x$			1	1

5.

	yz	yz'	$y'z'$	$y'z$
wx	d	1	1	1
wx'				
$w'x'$				d
$w'x$	1		1	d

6.

	yz	yz'	$y'z'$	$y'z$
wx		d		
wx'	1	d	d	1
$w'x'$			1	d
$w'x$	d	1	1	d

Simplify the boolean expression defined by each table.

7.

x	y	z	f
0	0	0	1
0	0	1	1
0	1	0	0
0	1	1	0
1	0	0	d
1	0	1	d
1	1	0	1
1	1	1	1

8.

x	y	z	f
0	0	0	1
0	0	1	d
0	1	0	0
0	1	1	0
1	0	0	d
1	0	1	1
1	1	0	1
1	1	1	1

Using the BCD expansions of the decimal numbers 0 through 9, develop a boolean function $f(w, x, y, z)$ such that the decimal value of $(wxyz)_{\text{two}}$ is divisible by:

9. Two

10. Three

11. Two and three

12. Two or three

13. Seven or more

14. Not more than five, but not one

15–22. Find the DNFs of the boolean functions in Exercises 7–14.

23–28. Generate a minimal boolean expression for each function b through g in Example 12.36.

Chapter Summary

Boolean operators and the laws of boolean algebra serve well in designing and simplifying electronic circuits.

Boolean Algebra

- A **boolean algebra** $\langle B, +, \cdot, {}', 0, 1\rangle$ comprises a set B of at least two distinct elements 0 and 1, two binary operators + and ·, and a unary operator $'$ that satisfies the commutative, associative, distributive, identity, and complement laws (page 806).
- The zero element 0 and the unit element 1 of a boolean algebra are unique (page 806).
- Every element x of a boolean algebra has a unique complement x' (page 808).
- Two important laws of boolean algebra are (see Theorem 12.3):

  Absorption laws: $x + xy = x$, $x(x + y) = x$

  DeMorgan's laws: $(x + y)' = x'y'$, $(xy)' = x' + y'$ (page 809).

Boolean Function

- A **boolean variable** takes the value 0 or 1 (page 813).
- A **boolean function** f is a function $f: B^n \to B$ (page 813).
- Boolean functions can be defined by logic tables or boolean expressions (page 813).

Boolean Expression

- A **boolean expression** over the boolean variables $x_1, x_2, \ldots, x_n$ is defined recursively:

  $0, 1, x_1, x_2, \ldots, x_n$ are boolean expressions.

  If E_1 and E_2 are boolean expressions, so are (E_1), E_1', E_1E_2, and $E_1 + E_2$.

  Only expressions thus obtained are boolean. (page 814)
- Two boolean expressions are **equal** if they yield the same value for every combination of values of the variables (page 815).

Disjunctive Normal Form (DNF)

- A **literal** is a boolean variable or its complement (page 817).

- A **minterm** in n boolean variables $x_1, x_2, \ldots, x_n$ is a boolean expression $y_1 y_2 \ldots y_n$, where $y_i = x_i$ or x_i', for every i (page 817).
- The DNF of a boolean function is a sum of minterms that do not repeat (page 817).
- The DNF of a boolean function can be constructed using its logic table and the laws of boolean algebra (page 819).

Functionally Complete Boolean Operators

- A set of boolean operators is **functionally complete** if every boolean function can be defined in terms of them (page 820).
- The sets $\{+, \cdot, '\}, \{+, '\}, \{\cdot, '\}, \{\uparrow\}$ and $\{\downarrow\}$ are functionally complete (page 821).

Logic Gates

- A **gate** is a boolean function from B^n to B (page 824).
- An **AND gate** accepts two or more boolean values and outputs their boolean product (page 825).
- An **OR gate** accepts two or more boolean values and outputs their boolean sum (page 826).
- A **NOT gate** (or an **inverter**) accepts a boolean value and outputs its complement (page 827).
- A **NAND gate** accepts two boolean values x and y and outputs 0 if and only if $x = y = 1$ (page 828).
- A **NOR gate** accepts two boolean values x and y and outputs 1 if and only if $x = y = 0$ (page 828).

Combinatorial Circuit

- A **combinatorial circuit** produces a unique boolean value for every combination of boolean input values. It has no memory; it can be built using AND, OR, and NOT gates (page 830).
- The output of a combinatorial circuit can be described by a boolean function (page 832).

Adders

- A **half-adder** circuit adds two bits, yielding their sum and carry bits (page 835).

- A **full-adder** circuit accepts two bits and a carry bit from previous addition, yielding their sum and carry bits. It consists of two half-adders and an OR gate (page 836).

Minimization of a Combinatorial Circuit

- Two combinatorial circuits are **equivalent** if the corresponding boolean expressions are equal (page 840).
- The laws of boolean algebra and Karnaugh maps can simplify combinatorial circuits (page 840).

Don't Care Conditions

- A combination of boolean values that is not a valid input into a combinatorial circuit is a **don't care condition** (page 851).
- A don't care condition triggers a d in the corresponding cell of a Karnaugh map and counts as 1 if it helps simplify the boolean expression (page 851).

Review Exercises

Let D_n denote the set of positive factors of a positive integer n. Define the operations $\oplus$, $\odot$, and $'$ on D_n as

$$x \oplus y = \text{lcm}\,\{x, y\}, \qquad x \odot y = \gcd\,\{x, y\}, \quad \text{and} \quad x' = n/x$$

Determine if $\langle D_n, \oplus, \odot, ', 1, n\rangle$ is a boolean algebra for each value of n.

1. 2 **2.** 6 **3.** 12 **4.** 42

***5.** Using Exercises 1–4, predict when $\langle D_n, \oplus, \odot, ', 1, n\rangle$ will be a boolean algebra.

The set D_{210} of positive factors of 210 is a boolean algebra under the operations $\oplus$, $\odot$, and $'$ defined as follows:

$$x \oplus y = \text{lcm}\,\{x, y\}, \quad x \odot y = \gcd\,\{x, y\}, \quad \text{and} \quad x' = 210/x$$

Compute each.

6. $21' \odot 35'$ **7.** $7' \odot 15'$ **8.** $5 \odot (10 \oplus 30)$ **9.** $6 \oplus (10 \odot 15)$

In the boolean algebra $\langle B, +, \cdot, ', 0, 1\rangle$, find the dual of each statement.

10. $(1 + x)(x + y) = x + y$ **11.** $xy + x'y = y$

Construct a logic table for each boolean function.

12. $f(x,y,z) = x + y'z'$

13. $f(x,y,z) = (x + y + z)(x'y'z')$

Using a logic table, verify each statement.

14. $(x + z)(y + z) = z + xy$

15. $x + x'yz = x + yz$

Find the DNF of each boolean function *f* in Exercises 16–19.

16.

x	y	z	$f(x,y,z)$
0	0	0	1
0	0	1	0
0	1	0	0
0	1	1	0
1	0	0	0
1	0	1	0
1	1	0	0
1	1	1	1

17.

x	y	z	$f(x,y,z)$
0	0	0	0
0	0	1	0
0	1	0	0
0	1	0	1
1	0	0	0
1	0	1	1
1	1	0	1
1	1	1	0

18. $f(x,y,z) = x + x'yz$

19. $f(x,y,z) = xy + yz + xz$

20–21. Using the laws of boolean algebra, find the DNF of each boolean function in Exercises 18 and 19.

Evaluate each.

22. $1 \uparrow (0 \downarrow 1)$

23. $0 \downarrow (1 \uparrow 0)$

Set up a logic table for each boolean expression.

24. $(x \uparrow x) \uparrow (y \uparrow y)$

25. $(x \downarrow x) \downarrow (y \downarrow y)$

Find the DNF of each boolean function.

26. $f(x,y) = (x \uparrow x) \uparrow y$

27. $f(x,y) = (x \downarrow y) \uparrow x$

Find the output produced by each combinatorial circuit.

28.

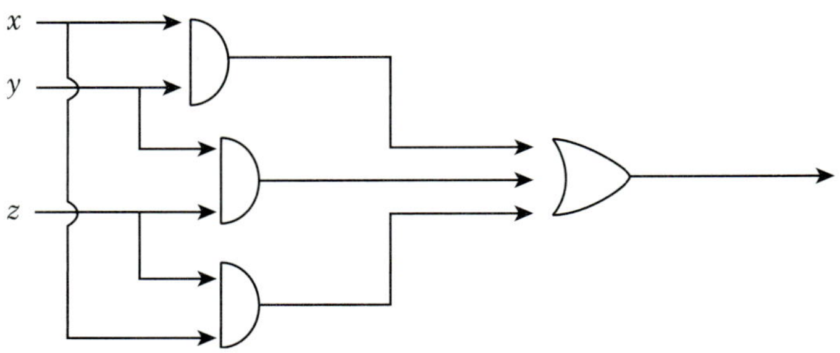

29.

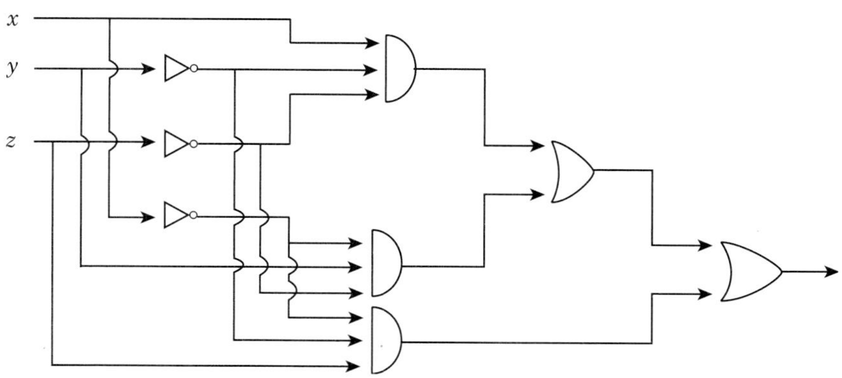

Construct a combinatorial circuit represented by each boolean expression.

30. $xy' + yz' + zx'$

31. $xyz + x'y'z'$

Using the circuit in Figure 12.33, compute the sum of each pair of binary numbers.

32. 110, 111

33. 111, 111

Represent each boolean expression in a Karnaugh map.

34. $xyz + xyz' + x'yx + x'y'z'$

35. $wxyz + wxy'z + w'xyz + wx'yz' + w'xy'z + wx'y'z' + w'x'yz' + w'x'y'z'$

Simplify each sum of minterms, if possible.

36. $xyz + xyz' + xy'z + x'yz + x'yz' + x'y'z$

37. $xy'z + x'yz + xy'z' + x'yz' + x'y'z + x'y'z'$

38. $wxyz' + wx'yz + wxy'z' + wx'yz' + wx'y'z + wx'y'z'$

39. $wxyz + wxyz' + wx'yz + w'xy'z + wx'yz' + w'xy'z' + w'x'y'z + w'x'y'z'$

Simplify the boolean expression represented by each Karnaugh map.

40.

	yz	yz'	$y'z'$	$y'z$
x	1	1	d	d
x'			d	d

41.

	yz	yz'	$y'z'$	$y'z$
x			1	1
x'	d	d	d	d

42.

	yz	yz'	$y'z'$	$y'z$
wx	1	1	1	1
wx'			1	d
$w'x'$			d	d
$w'x$			d	1

43.

	yz	yz'	$y'z'$	$y'z$
wx	1	d	d	d
wx'		1	d	d
$w'x'$			1	d
$w'x$				1

Simplify each boolean expression defined by the tables in Exercises 44 and 45.

44.

x	y	z	f
0	0	0	0
0	0	1	0
0	1	0	1
0	1	1	1
1	0	0	0
1	0	1	1
1	1	0	d
1	1	1	1

45.

x	y	z	f
0	0	0	1
0	0	1	d
0	1	0	1
0	1	1	1
1	0	0	1
1	0	1	d
1	1	0	1
1	1	1	0

Using the BCD expansions of decimal numbers 0 through 9, develop a boolean function $f(w,x,y,z)$ such that the decimal value of $(wxyz)_{\text{two}}$ is:

46. A composite number.

47. Divisible by five.

48. At least seven.

49. Divisible by three or four.

50–53. Find the DNF of each boolean function in Exercises 46–49.

Supplementary Exercises

1. Let p and q be distinct prime numbers, $n = pq$, and $S = \{1, p, q, n\}$. Define the operations $\oplus$, $\odot$, and $'$ on S as follows:

$$x \oplus y = \text{lcm}\,\{x,y\}, \quad x \odot y = \text{gcd}\,\{x,y\}, \quad x' = n/x$$

Is $\langle S, \oplus, \odot, ', 1, n\rangle$ a boolean algebra?

2. Let $\langle B, +, \cdot, ', 0, 1\rangle$ be a boolean algebra and $A \subseteq B$. Prove that $\langle A, +, \cdot, ', 0, 1\rangle$ is a boolean algebra if and only if these two conditions are satisfied:

- $1 \in A$.
- If $x, y \in$ A, then $xy' \in A$.

3. Let x be any element of a boolean algebra B. Let $x^{(n)}$ denote the expression $(\ldots(x')'\ldots)'$, where there are n (≥ 1) primes. Prove that

$$x^{(n)} = \begin{cases} x & \text{if } n \text{ is even} \\ x' & \text{if } n \text{ is odd} \end{cases}$$

Let $x_1, x_2, \ldots, x_n$ be any elements of a boolean algebra B. By induction, prove the following **DeMorgan's laws.**

4. $(x_1 + x_2 + \cdots + x_n)' = x_1' x_2' \ldots x_n'$

5. $(x_1 x_2 \cdots x_n)' = x_1' + x_2' + \cdots + x_n'$

Find the dual of each boolean expression.

6. $(x+y)+x'y'$
7. $(x+y)(xy)'(x+y')$
8. $[(x'+y')+(x')'(y')']'$
9. $[(x'+y')(x'y')'(x'+(y')')]'$
10. $(x+y)(y'+0)(x1')$
11. $\{(x'+y')[(y')'+0'][x'(1')']\}'$
12. Using Exercises 6–11, predict the dual of the boolean expression $E(x_1,\ldots,x_n)$.

A boolean expression $E(x_1,\ldots,x_n)$ is **self-dual** if $E(x_1,\ldots,x_n) = [E(x'_1,\ldots,x'_n)]'$. Determine if each expression is self-dual.

13. x
14. $x+yz$
15. xyz
16. $xy+yz'$

Computer Exercises

Write a program to perform each task.

1. Let n be the product of k distinct prime numbers, where $1 \le k \le 5$. Then $\langle D_n, \oplus, \odot, ', 1, n\rangle$ is a boolean algebra, where the operations $\oplus$, $\odot$ and $'$ are defined below:

$$x \oplus y = \text{lcm}\,\{x,y\}, \quad x \odot y = \text{gcd}\,\{x,y\}, \quad x' = n/x$$

   (a) Read in n and find the elements of D_n.

   Read in three elements x, y, and z of D_n and compute each.

   (b) $x \oplus y$ **(c)** $x \odot y$ **(d)** x'

   (e) $x \oplus y \oplus z$ **(f)** $x \odot y \odot z$ **(g)** $x \oplus (y \odot z)$

2. Construct a logic table for the boolean expressions $x+y, xy, x \uparrow y$, and $x \downarrow y$.
3. Set up a logic table for each boolean expression.

   (a) $(x+y+z)(xyz)'$ **(b)** $(x+y+z)(xy'+yz'+zx')$

   (c) $x \uparrow (y \uparrow z)$ **(d)** $x \downarrow (y \downarrow z)$

4. Determine if the given boolean expressions are equal.

   (a) $x+xy, x$ **(b)** $x \uparrow (y \uparrow z), (x \uparrow y) \uparrow z$

   (c) $(x+y)', x'y'$ **(d)** $(x \uparrow y) \uparrow (x \uparrow y), xy$

5. Read in the logic table for a boolean function $f(x,y,z)$. Find the DNF of the function.
6. Read in each boolean function $f(x,y,z)$. For each one, create the logic table for f and the DNF of the function.

(a) $f(x,y,z) = x + y' + z'$ **(b)** $f(x,y,z) = x \uparrow (y \downarrow z)$

(c) $f(x,y,z) = x(y + z)$ **(d)** $f(x,y) = (x \uparrow y) \downarrow (x \uparrow z)$

7. Read in two n-bit binary numbers. Compute their sum bit by bit.

8. Read in the logic table for a boolean expression $E = E(x,y,z)$.

(a) Construct its DNF.

(b) Construct the corresponding Karnaugh map.

9. Redo program 8 if $E = E(w,x,y,z)$.

10. Draw a table containing the binary representations of the integers 0 through 15.

11. Let n be a positive integer ≤ 10. Find a boolean function f of the variables w, x, y, and z such that $f(w,x,y,z) = 1$ if n is a prime number and 0 otherwise, where $(wxyz)_{\text{two}}$ denotes the BCD expansion of n.

(a) Print the logic table for f.

(b) Build the DNF of the function.

(c) Draw the Karnaugh map for the expression $f(w,x,y,z)$.

12. Read in a decimal digit d. Use the configuration in Figure 12.50 to print the line segments a through g that must light up to display the digit d. Print the corresponding DNFs, too.

Exploratory Writing Projects

Using library and Internet resources, write a team report on each of the following in your own words. Provide a well-documented bibliography.

1. Develop number applications of the logic operations NOT, AND, OR, NOR, and NAND to everyday life.

2. Describe C. Shannon's contributions to the theory of switching circuits.

3. Describe the *Quine–McCluskey method*, originally developed by W. V. Quine (1908–2000) and modified by E. J. McCluskey, Jr. (1929–), used for the simplification of boolean functions.

Enrichment Readings

1. J. C. Abbot, *Sets, Lattices, and Boolean Algebras,* Allyn and Bacon, Boston, 1969.

2. B. H. Arnold, *Logic and Boolean Algebra,* Prentice-Hall, Englewood Cliffs, NJ, 1962.

3. H. G. Flegg, *Boolean Algebra and Its Applications,* Wiley, New York, 1964.

4. K. G. Gopolan, *Introduction to Digital Microelectronic Circuits*, Irwin, Chicago, 1996.

5. F. E. Hohn, *Applied Boolean Algebra,* 2nd ed., Macmillan, New York, 1966.

6. J. E. Whitesitt, *Boolean Algebra and Its Applications,* Addison-Wesley, Reading, MA, 1961.

7. J. F. Wakerly, *Digital Design: Principles and Practices*, 2nd ed., Prentice-Hall, Englewood Cliffs, NJ, 1994.

8. G. E. Williams, *Boolean Algebra with Computer Applications,* McGraw-Hill, New York, 1970.

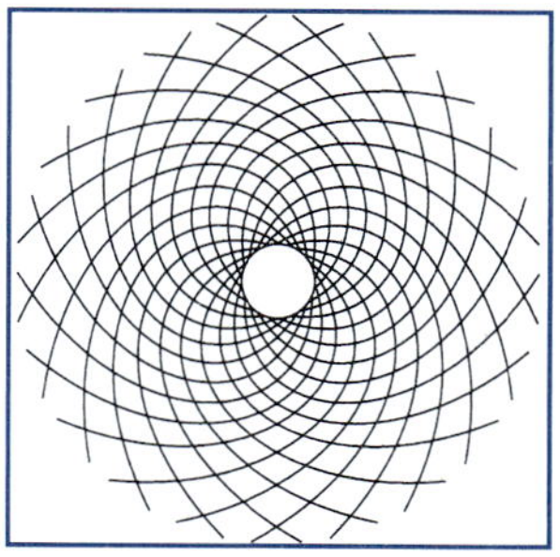

Appendix A

A.1 ASCII Character Set

Left Digit(s) \ *Right Digit*	ASCII *0*	*1*	*2*	*3*	*4*	*5*	*6*	*7*	*8*	*9*
3			□	!	"	#	$	%	&	'
4	(	)	*	+	,	−	.	/	0	1
5	2	3	4	5	6	7	8	9	:	;
6	<	=	>	?	@	A	B	C	D	E
7	F	G	H	I	J	K	L	M	N	O
8	P	Q	R	S	T	U	V	W	X	Y
9	Z	[	1/8	]	∧	—	´	a	b	c
10	d	e	f	g	h	i	j	k	l	m
11	n	o	p	q	r	s	t	u	v	w
12	x	y	z	{	|	}	˜			

Codes 00-31 and 127 are nonprintable control characters. The symbol □ denotes the *blank* character.

A.2 Determinants*

Determinants were invented by the Japanese mathematician Seki Kowa (1642–1708). However, German mathematician Gottfried Wilhelm Leibniz (1646–1716) is credited with their creation, although his contribution (in 1693) came 10 years later than Kowa's. Leibniz developed determinants while solving systems of linear equations.

*Based on T. Koshy, *College Algebra and Trigonometry with Applications*, McGraw-Hill, New York, 1986, pp. 423–432.

Determinant

A **determinant** is a function from the set of square matrices to $\mathbb{R}$, the set of real numbers. The determinant of a square matrix A is denoted by $\det A$ or $|A|$. If A is an $n \times n$ matrix, $\det A$ is called a determinant of **order** n. (Determinants of nonsquare matrices are *not* defined.)

For example, the determinant of the matrix $A = (a_{ij})_{2\times 2}$ is written as

$$\det A = \begin{vmatrix} a_{11} & a_{12} \\ a_{21} & a_{22} \end{vmatrix}$$

using vertical bars and is defined by

$$\det A = a_{11}a_{22} - a_{21}a_{12}$$

For instance,

$$\begin{vmatrix} 2 & 3 \\ 4 & -5 \end{vmatrix} = 2 \cdot (-5) - 4 \cdot 3 = -22$$

If we know how to evaluate 2×2 determinants, we can then evaluate higher-order determinants.

Minor and Cofactor

Let (a_{ij}) be a square matrix of order n. The determinant of the matrix obtained by deleting row i and column j is the **minor** of the element a_{ij}, denoted by M_{ij}. The **cofactor** A_{ij} of a_{ij} is defined by $A_{ij} = (-1)^{i+j} M_{ij}$.

$A_{ij} = +M_{ij}$ if $i + j$ is even, and $-M_{ij}$ if $i + j$ is odd. The signs associated with each cofactor A_{ij} can be displayed in a checkerboard pattern like this 3×3 matrix:

$$\begin{bmatrix} + & - & + \\ - & + & - \\ + & - & + \end{bmatrix}$$

The signs alternate for each row and column, beginning with + for the first row and first column.

EXAMPLE A.1 For the matrix

$$A = \begin{bmatrix} 1 & -2 & 3 \\ 4 & -5 & 6 \\ -7 & 8 & -9 \end{bmatrix}$$

compute M_{13}, A_{13}, M_{32}, and A_{32}.

SOLUTION:

$$M_{13} = \begin{vmatrix} 4 & -5 \\ -7 & 8 \end{vmatrix} = 4 \cdot 8 - (-7) \cdot (-5) = -3$$

$$A_{13} = (-1)^{1+3} M_{13} = -3$$

$$M_{32} = \begin{vmatrix} 1 & 3 \\ 4 & 6 \end{vmatrix} = 1 \cdot 6 - 4 \cdot 3 = -6$$

$$A_{32} = (-1)^{3+2} M_{32} = +6$$

■

Expansion of *det A*

Let A_{ij} denote the cofactor of the element a_{ij} of the matrix $A = (a_{ij})_{n \times n}$. Then the determinant of A is defined by

$$\begin{aligned} \det A &= a_{i1}A_{i1} + a_{i2}A_{i2} + \cdots + a_{in}A_{in} \qquad \text{(expansion by row } i\text{)} \\ &= a_{1j}A_{1j} + a_{2j}A_{2j} + \cdots + a_{nj}A_{nj} \qquad \text{(expansion by column } j\text{)} \end{aligned}$$

($\det A$ can be expanded by any row or any column.)

EXAMPLE A.2 Evaluate $\det A$ for the matrix A in Example A.1.

SOLUTION:

Expanding $\det A$ by the first row,

$$\begin{aligned} \det A &= +1 \begin{vmatrix} -5 & 6 \\ 8 & -9 \end{vmatrix} - (-2) \begin{vmatrix} 4 & 6 \\ -7 & -9 \end{vmatrix} + 3 \begin{vmatrix} 4 & -5 \\ -7 & 8 \end{vmatrix} \\ &= +1(45 - 48) + 2(-36 + 42) + 3(32 - 35) \\ &= 0 \end{aligned}$$

We can expand $\det A$ by any other row or any column to verify this. ■

Suppose that every element in a row or column of a square matrix A is zero. Expanding $\det A$ by that row or column shows that $\det A = 0$ (Verify this.). Theorem A.1 confirms this.

THEOREM A.1 If every element in a row or column of a square matrix A is zero, then $\det A = 0$. ■

Evaluating a determinant of a large order is not easy unless it contains many zeros. However, the following theorems can make evaluation easier.

THEOREM A.2 Let B be the matrix obtained by interchanging any two rows or columns of a square matrix A. Then $\det B = -\det A$.

PROOF:
To prove this theorem for the matrix $A = (a_{ij})_{2\times 2}$, interchange the rows of A:

$$B = \begin{bmatrix} a_{21} & a_{22} \\ a_{11} & a_{12} \end{bmatrix}$$

and

$$\begin{aligned} \det A &= a_{21}a_{12} - a_{11}a_{22} \\ &= -(a_{11}a_{22} - a_{21}a_{12}) \\ &= -\det A \end{aligned}$$

■

THEOREM A.3 If any two rows or columns of a square matrix A are identical, $\det A = 0$.

PROOF:
Let B be the matrix obtained by interchanging two identical rows or columns of A. By Theorem A.2, $\det B = -\det A$. But since $A = B$, $\det B = \det A$. Therefore, $\det A = -\det A$ and hence $\det A = 0$. ■

The following example illustrates this.

EXAMPLE A.3 Evaluate $\det A$ if

$$A = \begin{bmatrix} 2 & 3 & 5 \\ -8 & 4 & 7 \\ 2 & 3 & 5 \end{bmatrix}$$

SOLUTION:
Since rows 1 and 3 are the same, $\det A = 0$, by Theorem A.3. Expanding $\det A$ by column 1 confirms this:

$$\begin{aligned} \det A &= 2\begin{vmatrix} 4 & 7 \\ 3 & 5 \end{vmatrix} - (-8)\begin{vmatrix} 3 & 5 \\ 3 & 5 \end{vmatrix} + 2\begin{vmatrix} 3 & 5 \\ 4 & 7 \end{vmatrix} \\ &= 2(20-21) + 8(15-15) + 2(21-20) \\ &= -2 + 0 + 2 \\ &= 0 \end{aligned}$$

as expected. ■

The next theorem shows that we can factor out common factors from a row or column.

THEOREM A.4 Let B be the matrix produced by multiplying one row or column of a square matrix A by a real number k. Then $\det B = k \cdot \det A$.

PROOF:
Let $A = (a_{ij})_{n\times n}$ and B be the matrix derived from multiplying its ith row by k. Expanding $\det B$ by row i,

$$\begin{aligned}\det B &= (ka_{i1})A_{i1} + (ka_{i2})A_{i2} + \cdots + (ka_{in})A_{in}\\ &= k(a_{i1}A_{i1} + a_{i2}A_{i2} + \cdots + a_{in}A_{in})\\ &= k \cdot \det A\end{aligned}$$

■

EXAMPLE A.4 Using Theorem A.4, evaluate $\det A$, where

$$A = \begin{bmatrix} 3 & -6 & 9\\ 4 & -5 & 6\\ -7 & 8 & -9\end{bmatrix}$$

SOLUTION:
Factor out 3 from row 1:

$$\det A = 3\begin{vmatrix} 1 & -2 & 3\\ 4 & -5 & 6\\ -7 & 8 & -9\end{vmatrix}$$

Now factor out 3 from column 3:

$$\det A = 9\begin{vmatrix} 1 & -2 & 1\\ 4 & -5 & 2\\ -7 & 8 & -3\end{vmatrix}$$

Expand by row 1:

$$\begin{aligned}\det A &= 9\left[1\begin{vmatrix} -5 & 2\\ 8 & -3\end{vmatrix} + 2\begin{vmatrix} 4 & 2\\ -7 & -3\end{vmatrix} + 1\begin{vmatrix} 4 & -5\\ -7 & 8\end{vmatrix}\right]\\ &= 9[1(15-16) + 2(-12+14) + 1(32-35)]\\ &= 0\end{aligned}$$

■

THEOREM A.5 Let B be the matrix obtained by adding k times a row or column to another row or column of a square matrix A. Then $\det B = \det A$.

PROOF:
We shall prove the theorem for an arbitrary matrix $A = (a_{ij})_{2\times 2}$. Let B be the matrix obtained by adding k times row 2 to row 1:

$$B = \begin{vmatrix} a_{11} + ka_{21} & a_{12} + ka_{22}\\ a_{21} & a_{22}\end{vmatrix}$$

$$\begin{aligned}\det A &= (a_{11} + ka_{21})a_{22} - a_{21}(a_{12} + ka_{22}) \\ &= (a_{11}a_{22} - a_{21}a_{12}) + k(a_{21}a_{22} - a_{21}a_{22}) \\ &= \det A + k \cdot 0 \\ &= \det A\end{aligned}$$

Similarly, if B is the matrix obtained by adding k times row 1 to row 2, then $\det B = \det A$. ■

EXAMPLE A.5 Evaluate $\det A$ with Theorem A.5, where

$$A = \begin{bmatrix} 1 & 2 & -3 \\ 4 & -5 & 6 \\ -7 & 8 & -9 \end{bmatrix}$$

SOLUTION:
Keep the 1 in column 1 to make the other two entries in column 1 (4 and -7) 0's. To this end, add -4 times row 1 to row 2 and 7 times row 1 to row 3:

$$\det A = \begin{vmatrix} 1 & 2 & -3 \\ 0 & -13 & 18 \\ 0 & 22 & -30 \end{vmatrix}$$

Expanding $\det A$ by column 1, we see that

$$\det A = 1[(-13) \cdot (-30) - 22 \cdot 18] = -6$$ ■

The above properties used simultaneously can speed up evaluating determinants, as the next example shows.

EXAMPLE A.6 Evaluate $\det A$ if

$$A = \begin{bmatrix} -3 & 1 & 2 & -1 \\ 3 & 0 & -5 & 6 \\ 2 & 4 & 0 & 8 \\ -1 & 2 & 0 & 3 \end{bmatrix}$$

SOLUTION:
Use Theorems A.4 and A.5 to evaluate $\det A$. Factor out 2 from row 3:

$$\det A = 2\begin{vmatrix} -3 & 1 & 2 & -1 \\ 3 & 0 & -5 & 6 \\ 1 & 2 & 0 & 4 \\ -1 & 2 & 0 & 3 \end{vmatrix}$$

Now use the 1 in column 1 to make the remaining entries in the column zeros. To this end, add 3 times row 3 to row 1 and −3 times row 3 to row 2; add row 3 to row 4:

$$\det A = 2\begin{vmatrix} 0 & 7 & 2 & -11 \\ 0 & -6 & -5 & -6 \\ 1 & 2 & 0 & 4 \\ 0 & 4 & 0 & 7 \end{vmatrix}$$

Expand this determinant by column 1:

$$\det A = 2\begin{vmatrix} 7 & 2 & -11 \\ -6 & -5 & -6 \\ 4 & 0 & 7 \end{vmatrix}$$

Expand the new determinant by row 3:

$$\begin{aligned} \det A &= 2\left[4\begin{vmatrix} 2 & -11 \\ -5 & -6 \end{vmatrix} - 0 + 7\begin{vmatrix} 7 & 2 \\ -6 & -5 \end{vmatrix}\right] \\ &= 2[4(-12-55) - 0 + 7(-35+12)] \\ &= -858 \end{aligned}$$

■

Exercises A.2

Evaluate each determinant.

1. $\begin{vmatrix} 2 & 0 \\ 0 & 3 \end{vmatrix}$ **2.** $\begin{vmatrix} -3 & 2 \\ 5 & -7 \end{vmatrix}$ **3.** $\begin{vmatrix} a & -b \\ b & a \end{vmatrix}$

4. $\begin{vmatrix} 1 & 0 & 0 \\ 0 & 2 & 0 \\ 0 & 0 & 3 \end{vmatrix}$ **5.** $\begin{vmatrix} 3 & 5 & 0 \\ 6 & 7 & 0 \\ 2 & 4 & 0 \end{vmatrix}$ **6.** $\begin{vmatrix} 2 & 3 & 5 \\ 1 & 4 & 7 \\ 6 & 9 & 15 \end{vmatrix}$

7. $\begin{vmatrix} a & b & 1 \\ b & c & 1 \\ c & a & 1 \end{vmatrix}$ **8.** $\begin{vmatrix} 1+x & 1 & 1 \\ 1 & 1+y & 1 \\ 1 & 1 & 1+z \end{vmatrix}$ **9.** $\begin{vmatrix} a & b & c \\ b & c & a \\ c & a & b \end{vmatrix}$

10. $\begin{vmatrix} 1 & 1 & 1 \\ a & b & c \\ a^2 & b^2 & c^2 \end{vmatrix}$ **11.** $\begin{vmatrix} a & b & 0 & 0 \\ c & d & 0 & 0 \\ 0 & 0 & e & f \\ 0 & 0 & g & h \end{vmatrix}$ **12.** $\begin{vmatrix} -1 & 0 & 1 & 1 \\ 0 & 1 & -1 & -1 \\ 1 & -2 & 3 & -4 \\ -2 & 4 & -6 & 8 \end{vmatrix}$

Prove each.

13. Let A and B be two 2×2 matrices. Then $|AB| = |A| \cdot |B|$.

14. Let A be any square matrix of order n. Then $|kA| = k^n|A|$.

15. Let A be an $n \times n$ matrix such that $AA^{\mathrm{T}} = I_n$. Then $|A| = \pm 1$.

16. Let $A = (a_{ij})_{n\times n}$ where $a_{ij} = 0$ for $j > i$. (Such a matrix is called an **upper triangular matrix**.) Then $|A| = a_{11}a_{22}\ldots a_{nn}$.

Solve each equation for $f(x)$.

17. $\begin{vmatrix} f(x) & 1 & x \\ f(a) & 1 & a \\ f(b) & 1 & b \end{vmatrix} = 0$

18. $\begin{vmatrix} f(x) & 1 & x & x^2 \\ f(a) & 1 & a & a^2 \\ f(b) & 1 & b & b^2 \\ f(c) & 1 & c & c^2 \end{vmatrix} = 0$

A.3 Exponential and Logarithmic Functions*

Exponential functions and their closest relatives, the logarithmic functions, form one of the most important classes of functions. They have a wide range of fine applications to computer science, life sciences, management science, physical sciences, and social sciences.

For example, an exponential function occurs in the natural sciences: *E. coli* are microscopic unicellular bacteria that multiply by halving into two bacteria about every 20 minutes under ideal conditions. Suppose initially there is one bacterium and 20 minutes is unit time. Then there will be two bacteria in one unit time, four in two units, eight in three units, and so on. In general, there will be $f(x) = 2^x$ bacteria in x units of time, assuming none dies during the period. This function belongs to the large class of exponential functions, defined below.

Exponential Function

Let a be a positive real number different from 1 and x an arbitrary real number. Then the function $f(x) = a^x$ is called an **exponential function with base** a.

If $a = 1$ in this definition, the function becomes the constant function $f(x) = 1$, which holds little practical interest. As a result, it is excluded from the definition.

Figure A.1 shows the graphs of several exponential functions with different bases a, where $a > 1$. They grow faster and more steeply as a increases. We can use such exponential functions to describe growths of populations; consequently, their graphs are called **growth curves**.

Figure A.2 shows some exponential graphs $y = a^x$, where $0 < a < 1$. Each can be obtained by flipping the corresponding graph $y = a^x$, where $a > 1$, about the y-axis.

---

*Based on T. Koshy, *College Algebra and Trigonometry with Applications*, McGraw-Hill, New York, pp. 204–241.

Figure A.1

Growth curves $y = a^x$, $a > 1$.

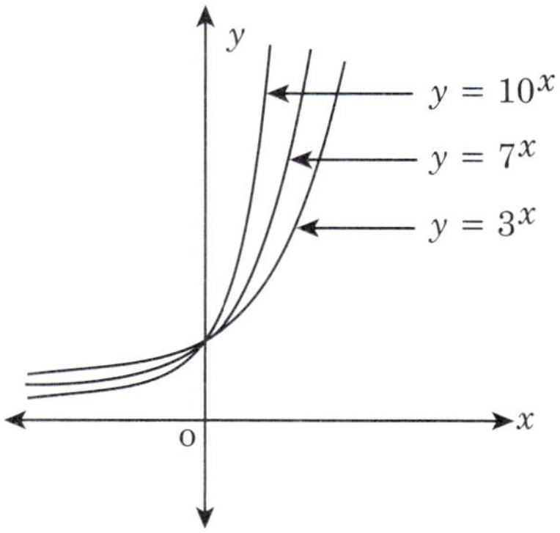

Figure A.2

Decay curves $y = a^x$, $0 < a < 1$.

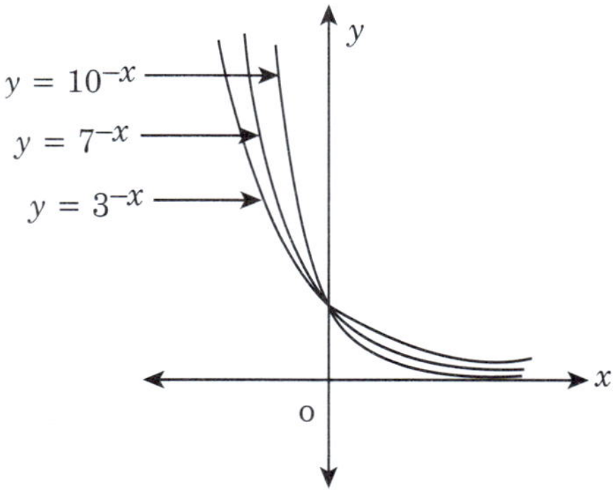

Here the y-values decrease as x increases. Such graphs are used to study the decay of radioactive substances and are called **decay curves**.

The important properties of the exponential function $f(x) = a^x$ are listed below:

- $\text{Dom}(f) = (-\infty, \infty)$; $\text{range}(f) = (0, \infty)$.
- The y-intercept of the graph is 1.
- The exponential function is bijective.
- If $a > 1$, then f is an **increasing function;** that is, if $x_1 > x_2$ then $f(x_1) > f(x_2)$. If $0 < a < 1$, then f is a **decreasing function**; that is, if $x_1 > x_2$ then $f(x_1) < f(x_2)$.
- The x-axis is a horizontal asymptote for the graph.
- The graphs $y = a^x$ and $y = a^{-x}$ are reflections of each other about the y-axis.

The next two examples explore some applications of the exponential function.

EXAMPLE A.7

Consider the world population (in billions) given by Table A.1. We can use these to draw a smooth curve as in Figure A.3. The graph shows that the world population is growing exponentially.

Table A.1

Year t	1	1000	1650	1750	1800	1850	1900	1950	1970	1990	2000
Population $P(t)$	0.25?	0.3	0.5	0.7	0.9	1.1	1.6	2.5	3.6	5.2	6.1

Figure A.3

The exponential growth of the world population.

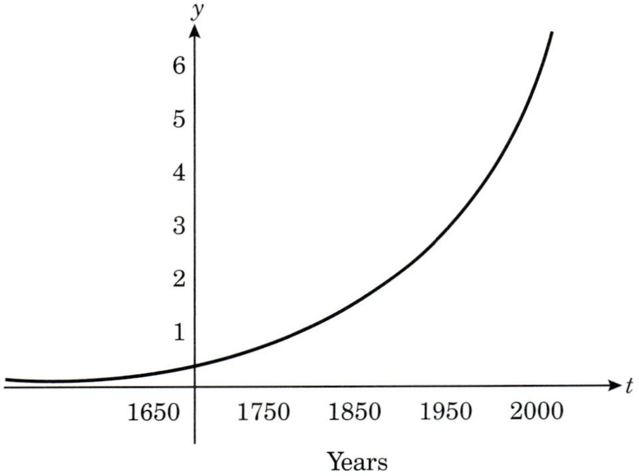

■

EXAMPLE A.8*

Suppose a bowl of nutrient broth contains 1,000,000 viable bacteria per millimeter. Table A.2 shows the number of viable bacteria per millimeter in the broth after an agent has been added to kill them. The population at time t is only one-tenth of that at time $t - 10$. Let $N(t)$ denote the number of viable bacteria at time t. Then $N(t) = 10^{6-0.1t}$. If t can take on all possible values (at least theoretically), then the graph of $N(t)$ looks like the decay curve in Figure A.4.

Table A.2

Time t (in minutes)	0	10	20	30	40	50	60
Number of bacteria $N(t)$	10^6	10^5	10^4	10^3	10^2	10	1

■

An extremely useful exponential function is $f(x) = e^x$, where e is the irrational number 2.718281828.... (The letter e for the base was chosen in

*Based on R. M. Thrall *et al.* (ed.), "Extermination of Bacteria," *Some Mathematical Models in Biology*, The University of Michigan Press, Ann Arbor, MI, 1967, pp. PE3.1–PE3.2.

Figure A.4

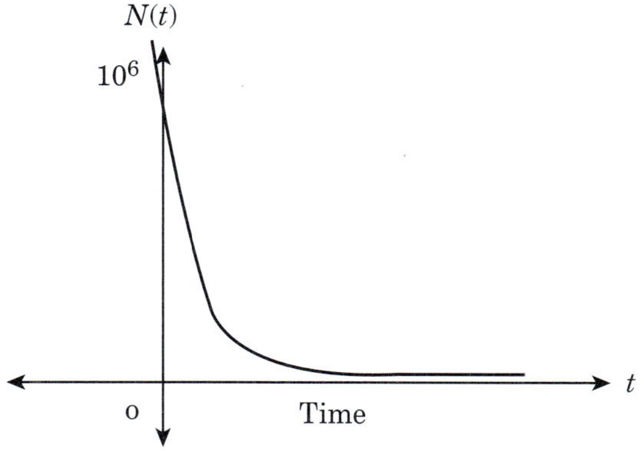

honor of Euler.) It is established in calculus that

$$e = \lim_{n\to\infty}\left(1 + \frac{1}{n}\right)^n$$

The graphs of the functions $f(x) = e^x$ and $g(x) = e^{-x}$ appear in Figure A.5.

Figure A.5

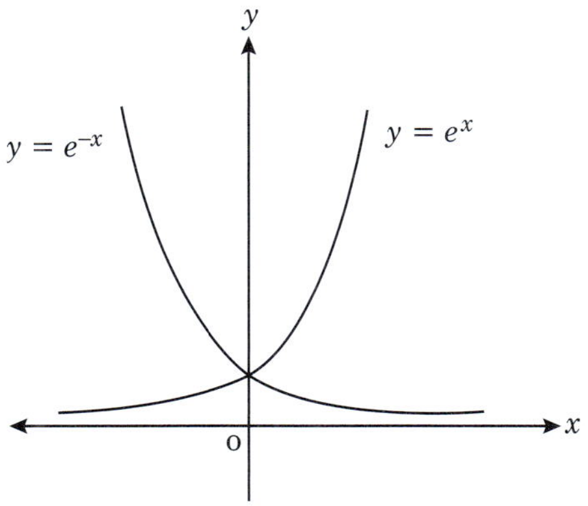

Logarithms

Logarithms were invented by the Scottish mathematician John Napier (1550–1617), to facilitate numeric computations. He published a summary of his results in 1614.

Logarithmic Function

In the exponential function $y = a^x$, where $a > 0$ and $a \neq 1$, the exponent x is the **logarithm of y to the base a**; it is denoted by $\log_a(y)$ or simply $\log_a y$. Thus,

$$\log_a y = x \text{ if and only if } y = a^x$$

The function $g(x) = \log_a x$ is the **logarithmic function** with base a.

With the base a always positive, $y = a^x$ is positive for all values of x. As a result, $x = \log_a y$ is defined only if y is positive. In other words, only logarithms of positive numbers are defined. Thus the domain of the logarithmic function is $(0,\infty)$, and the range is $(-\infty, \infty)$.

Since the exponential function is bijective with domain $(-\infty, \infty)$ and range $(0, \infty)$, it has an inverse function. The next theorem shows the inverse is indeed the logarithmic function.

THEOREM A.6 The logarithmic function $g(x) = \log_a x$ is the inverse of the exponential function $f(x) = a^x$.

PROOF:
By definition $a^{\log_a x} = x$ for every $x > 0$, and that $\log_a(a^x) = x$ for every x. Then

$$(f \circ g)(x) = f(g(x)) = f(\log_a x) = a^{\log_a x} = x$$

and

$$(g \circ f)(x) = g(f(x)) = g(a^x) = \log_a(a^x) = x$$

So the logarithmic function g is the inverse of the exponential function f. ■

According to this theorem, we can obtain the graph of $y = \log_a x$ by reflecting that of $y = a^x$ about the line $y = x$, as shown in Figure A.6.

Figure A.6

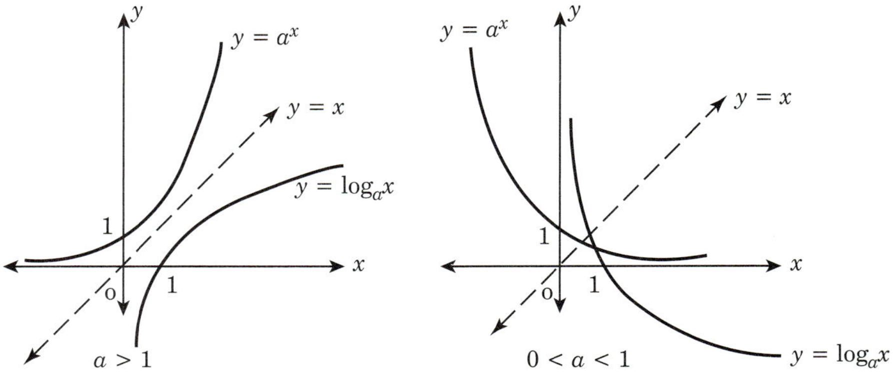

Below are listed the important properties of the logarithmic function $g(x) = \log_a x$.

- $\text{Dom}(g) = (0,\infty)$; $\text{range}(g) = (-\infty,\infty)$.
- The x-intercept of the logarithmic graph is 1.
- The logarithmic function is bijective.
- If $a > 1$, g is an increasing function; if $0 < a < 1$, g is a decreasing function.
- The y-axis is a vertical asymptote.
- The functions $f(x) = a^x$ and $g(x) = \log_a x$ are inverses of each other; their graphs are mirror images of each other about the line $y = x$.
- The graphs of $y = \log_a x$ and $y = \log_{1/a} x$ are reflections of each other about the x-axis.

Frequently Used Bases

Three bases of logarithms, 10, e, and 2, are frequently used. Logarithms to the base 10 are called **common logarithms** or **Briggsian logarithms**, after the English mathematician Henry Briggs (1561–1631). Their base is usually omitted:

$$\log x = \log_{10} x$$

Logarithms to the base e are **natural logarithms** and abbreviated as ln:

$$\ln x = \log_e x$$

Logarithms to the base 2, which we often use in computer science, appear as lg:

$$\lg x = \log_2 x$$

The fundamental properties of logarithms are given in the next theorem. We can prove each using the definition of a logarithm.

THEOREM A.7 Let a, b, x, and y be any positive real numbers such that $a \neq 1$, $b \neq 1$, and n any real number. Then:

(1) $a^{\log_a x} = x$

(2) $\log_a(xy) = \log_a x + \log_a y$

(3) $\log_a \dfrac{x}{y} = \log_a x - \log_a y$

(4) $\log_a(x^n) = n \log_a x$

(5) $\log_a 1 = 0$

(6) $\log_a a = 1$

(7) $\log_a (a^x) = x$

(8) $\log_a x = \log_a y$ if and only if $x = y$

(9) $\log_b x = \dfrac{\log_a x}{\log_a b}$ ■

The next three examples illustrate some useful exponential and logarithmic functions.

EXAMPLE A.9 The pH of a solution, which measures its acidity or alkalinity, is defined as the negative logarithm of the hydrogen concentration x in moles per liter of the solution; that is, pH $= -\log x$. At 25°C, the pH of pure water is 7; if the pH of a solution falls below 7, the solution is acidic, and above 7 it is alkaline. Compute the pH of a solution where the concentration of hydrogen ions is 6.7×10^{-4}.

SOLUTION:

$$\begin{aligned} \text{pH of the solution} &= -\log(6.7 \times 10^{-4}) \\ &= -\log 6.7 - \log(10^{-4}) \\ &= -\log 6.7 + 4 \\ &\approx -0.8261 + 4 \\ &= 3.1739 \end{aligned}$$ ■

EXAMPLE A.10 The world population (in billions) at any time t (in years) is given by $p(t) = 6.1e^{0.0167t}$, where $p(0)$ denotes the population in 2000. At the given rate, when will the population double the 2000 level?

SOLUTION:
Population in 2000 $= p(0) = 6.1$ billion. We need the value of t when $p(t) = 12.2$ (Figure A.7). Then

$$6.1e^{0.0167t} = 12.2$$

Divide by 6.1:

$$e^{0.0167t} = 2$$

Then

$$\begin{aligned} \ln e^{0.0167t} &= \ln 2 \\ 0.0167t &= \ln 2 \\ t &= \frac{\ln 2}{0.0167} \\ &\approx 42 \text{ years} \end{aligned}$$ ■

Figure A.7

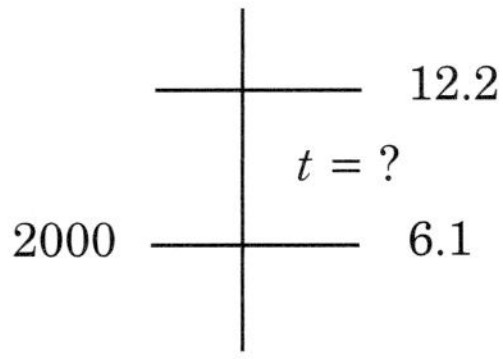

EXAMPLE A.11 The **half-life** of a radioactive substance is the time required for one-half of that substance to decay. The amount of ^{11}C, an isotope of carbon, present at a future time t (in months) is given by $A(t) = 100e^{-0.0338t}$. Find the half-life of the material.

SOLUTION:
Original amount $= A(0) = 100$ grams. The half-life is the value of t when $A(t) = \frac{1}{2}A(0) = 50$ (Figure A.8). Then

$$100e^{-0.0338t} = 50$$

Divide by 100:

$$e^{-0.0338t} = \frac{1}{2}$$

Then

$$\begin{aligned} \ln e^{-0.0338t} &= \ln \frac{1}{2} \\ -0.0338t &= -\ln 2 \\ t &= \frac{\ln 2}{0.0338} \\ &\approx 20.5 \text{ months} \end{aligned}$$

Figure A.8

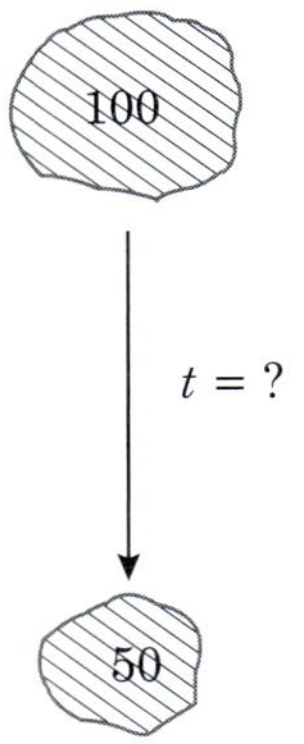

■

Exercises A.3

Prove each, where f denotes the exponential function.

1. $f(x+y) = f(x) \cdot f(y)$

2. $f(x-y) = \dfrac{f(x)}{f(y)}$

3. f is injective.

4. The logarithmic function $g(x) = \log_a x$ is injective.

Prove, by contradiction, that each is an irrational number.

5. log 2

6. log 3

7. The concentration of hydrogen ions in a solution is 3.76×10^{-8}. Compute the pH of the solution.

8. The pH of a solution is 5.3575. Compute its hydrogen ion concentration.

The intensity of sound the human ear can hear is measured in *decibels*, named after Alexander Graham Bell (1847–1922), the American scientist who invented the telephone. The number of decibels in a sound of intensity I is given by

$$B = 10 \log \frac{I}{I_0}$$

where I_0 is the standard intensity of 10^{-12} watts per square meter ($\mathrm{W/m^2}$). Find the number of decibels in:

9. Threshold of hearing (10^{-12} $\mathrm{W/m^2}$)

10. Normal conversation (10^{-6} $\mathrm{W/m^2}$)

11. Street traffic (10^{-5} $\mathrm{W/m^2}$)

12. Two sounds of intensity I_1 and I_2 have decibels B_1 and B_2, respectively. Show that $B_1 - B_2 = 10 \log \dfrac{I_1}{I_2}$.

13. The magnitude of a star is a measure of its brightness: the brighter the star, the smaller the magnitude. If b_1 and b_2 denote the magnitudes of two stars, B_1 and B_2 their respective brightnesses, $b_1 - b_2 = -2.512 \log \dfrac{B_1}{B^2}$. The brightest stars, Aldebaran and Sirius, have magnitudes 1 and -1.6, respectively. How much brighter is Sirius than Aldebaran?

The absolute magnitude M of a star, its apparent magnitude m, and the stellar distance d (in parsecs) from the earth are related by the formula $m = M + 5 \log 0.1d$.

14. Compute the stellar distance of Aldebaran if its absolute magnitude is -0.2 and its apparent magnitude 0.86.

15. The difference $m - M$ is the **distance modulus** of the star. The distance modulus of Sirius is -2.86. Find its stellar distance.

16. With **Rubik's Cube**, a popular game, the goal is to turn its faces to one particular configuration. The minimum number of turns required from some position to the desired state is $1 + \log_{15}\left(\dfrac{4.3 \times 10^{19}}{18}\right)$. Compute it. (C. Kluepfel, 1982)

A.4 Generating Permutations and Combinations

Sections 6.2 through 6.5 presented various types of permutations and combinations, as well as formulas for computing the number of r-permutations and r-combinations of a finite set S. In this section, we will study algorithms which enumerate them.

Generating Permutations

For convenience, we choose $S = \{1, 2, \dots, n\}$. We shall present an algorithm for generating the various permutations of elements of the set. The algorithm is based on **lexicographic ordering**, the same type used to arrange words in a dictionary.

Lexicographic Order

In lexicographic ordering, a permutation $A = a_1 a_2 \dots a_n$ is **less than** (or **precedes**) a permutation $B = b_1 b_2 \dots b_n$ if:

- $a_1 < b_1$ or
- $a_i = b_i$ for $1 \le i \le k - 1$, and $a_k < b_k$.

This is precisely the order we use to alphabetize words (of the same length) in the dictionary. For instance, the word *compute* precedes the word *permute* and the word *estate* precedes the word *esteem*.

EXAMPLE A.12 Consider the permutation $a_1a_2a_3a_4 = 2134$ and $b_1b_2b_3b_4 = 2143$ of the set $\{1, 2, 3, 4\}$. They agree in the first two positions: $a_1 = b_1$ and $a_2 = b_2$. But they differ in the third positions: $a_3 \ne b_3$. Since $a_3 < b_3$, the permutation 2134 precedes the permutation 2143 in lexicographic ordering. ■

Recall that there are $4! = 24$ permutations of the set $\{1, 2, 3, 4\}$. They are listed in Figure A.9 in lexicographic order columnwise, beginning with 1234. (Can you find a pattern for enumerating them?)

Figure A.9

1234	1423	2314	3124	3412	4213
1243	1432	2341	3142	3421	4231
1324	2134	2413	3214	4123	4312
1342	2143	2431	3241	4132	4321

Suppose we have an algorithm to generate the next larger permutation $b_1 b_2 \dots b_n$ from a given permutation $a_1 a_2 \dots a_n$. Then, beginning with the permutation $123 \dots n$, we can invoke it to generate the remaining $n - 1$ permutations. The procedure to generate the next larger permutation from a given permutation in lexicographic order is described below.

If $a_{n-1} < a_n$, switch them to obtain the new permutation $a_1a_2 \ldots a_{n-2}$ a_na_{n-1}. Then the permutation $a_1a_2 \ldots a_{n-2}a_{n-1}a_n$ precedes the permutation $a_1a_2 \ldots a_{n-2}a_na_{n-1}$ in lexicographic ordering.

For example, consider the permutation 2143; it succeeds the permutation 2134 in the listing in Figure A.9.

On the other hand, suppose $a_{n-1} > a_n$. (Then by switching them we cannot obtain a larger permutation.) So we look at a_{n-2}. If $a_{n-2} < a_{n-1}$, find the smaller of the elements a_{n-1} and a_n that is larger than a_{n-2}; place it in position $n - 2$; and now arrange a_{n-2} and the remaining elements in increasing order.

For instance, consider the permutation $a_1a_2a_3a_4 = 2341$ in Figure A.9. Here $a_3 > a_4$, but $a_2 < a_3$; the smaller of a_3 and a_4 that is larger than $a_2 = 3$ is $a_3 = 4$; so hold 4 in position 2 to get 24__; now arrange the remaining elements $a_2 = 3$ and $a_4 = 1$ in ascending order to yield the next larger permutation 2413 (see Figure A.9). If $a_{n-2} > a_{n-1}$, we need to look at the elements a_{n-3} through a_n.

More generally, all we need to do is the following. From right to left, find the first pair of elements a_i, a_{i+1} such that $a_i < a_{i+1}$, where $a_{i+1} > a_{i+2} > \ldots > a_n$. Then find the smallest of the elements a_{i+1}, $a_{i+2}, \ldots, a_n$, say, a_j, that is larger than a_i. Place a_j in position i. Now arrange the elements a_i, $a_{i+1}, \ldots, a_{j-1}$, a_j, $a_{j+1}, \ldots, a_n$ in increasing order in positions $i + 1$ through n.

The following example illustrates this procedure.

EXAMPLE A.13 Find the next permutation larger than $a_1a_2a_3a_4 = 3421$.

SOLUTION:

The first pair a_i, a_{i+1} from right to left for which $a_i < a_{i+1}$ is the pair a_1, a_2. So find the smallest of the elements $a_2 = 4$, $a_3 = 2$, and $a_4 = 1$ that is larger than $a_1 = 3$. Clearly, it is $a_2 = 4$. So we place 4 in position 1. Now we arrange the elements 3, 2, and 1 in increasing order: 123. Consequently, the next larger permutation is 4123. (See Figure A.9.) ■

The above discussion can be translated into an algorithm for finding the next larger permutation that follows a given permutation $a_1a_2 \ldots a_n$. It is presented in Algorithm A.1.

```
Algorithm next-permutation (a1a2...an)
  (* This algorithm finds the permutation larger than the
     given permutation a1a2...an. Assume it is not the
     largest permutation n(n - 1)...321. *)
0. Begin (* algorithm *)
     (* From right to left, find the first pair ai, ai+1 for
        which ai < ai+1. *)
1. i ← n - 1
```

```
2. While a_i > a_i+1 do (* continue searching to the left *)
3.    i ← i - 1
      (* When we exit the loop, we will have found an i for
      which a_i < a_i+1. Find the smallest of the elements a_i+1
      through a_n, say, a_k, that is larger than a_i. Since
      a_i+1 > a_i+2 > ··· > a_n, scan from right to left to find
      the element a_k that is larger than a_i. *)
4. k ← n
5. While a_k < a_i do (* continue scanning to the left. *)
6.    k ← k - 1
     (* a_k is the smallest of the elements a_i+1 through a_n
        that is larger than a_i. Swap a_i and a_k. *)
7. swap a_i and a_k
8. sort the elements in positions i + 1 through n into
   ascending order.
9. End (* algorithm *)
```

Algorithm A.1

EXAMPLE A.14 Using Algorithm A.1, enumerate in lexicographic order all permutations of elements of the set $\{1, 2, 3\}$.

SOLUTION:
We begin with the basic permutation $a_1a_2a_3 = 123$. Since $a_2 < a_3$ and the least of the elements(s) a_3 that is greater than a_2 is a_3, switch 2 and 3 to get the next larger permutation 132. Since $1 < 3$ and the smaller of 3 and 2 that is larger than 1 is 2, swap 1 and 2, and sort the elements 3 and 1 to yield the permutation 213. Continuing like this, we get the remaining permutations in lexicographic order: 123, 132, 213, 231, 312, 321. (Also see the tree diagram in Figure A.10.)

Figure A.10

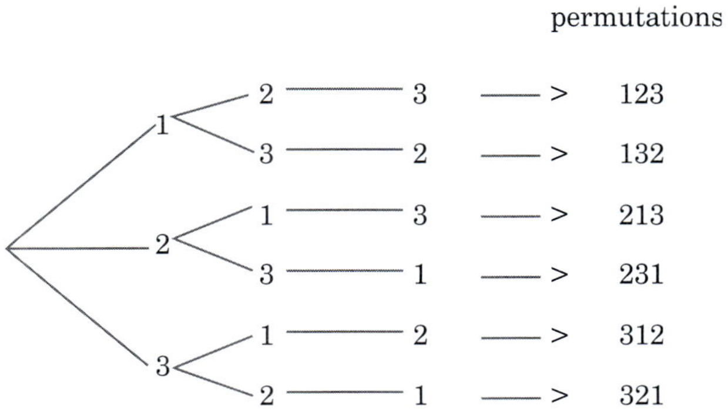

■

Generating Combinations

How can we generate the r-combinations of set $\{1, 2, \ldots, n\}$, where $0 < r < n$? First we list the various combinations in increasing numerical order.

For instance, there are ten 3-combinations of the set $S = \{1, 2, 3, 4, 5\}$; in lexicographic order, they are:

$$123,\ 124,\ 125,\ 134,\ 135,\ 145,\ 234,\ 235,\ 245,\ 345$$

How can we find the next larger 3-combination from a given 3-combination? For instance, consider the 3-combination 134 of the set S. The last element (from right) is less than 5 $(= n)$. So we increment it by 1 to get the next larger combination 135.

EXAMPLE A.15 Using the set $S = \{1, 2, 3, 4, 5\}$, find the combination following $a_1a_2a_3 = 135$ in lexicographic order.

SOLUTION:
Notice that $a_3 = 5$, the largest in S. So look at a_2; $a_2 < 5$; so increment a_2 by 1 to get 4. Replace a_3 by 1 more than the current a_2: $a_3 \leftarrow a_2 + 1 = 4 + 1 = 5$. Thus the next larger combination is 145. ■

EXAMPLE A.16 Find the combination that follows in lexicographic order the 3-combination 145 of the set $\{1, 2, 3, 4, 5\}$.

SOLUTION:
Let $a_1a_2a_3 = 145$. Clearly, $a_3 = 5$ and $a_2 = 4$. Since $a_2 < 5$, if we increment it by 1 to get $4 + 1 = 5$, the new a_3 should be $5 + 1$; unfortunately, it does not exist in the set. So go to $a_1 = 1$, which is less than 5. (Notice that $a_3 = 5 - 3 + 3$; $a_2 = 4 = 5 - 3 + 2$; but $a_1 = 1 \neq 5 - 3 + 1$. Therefore, a_1 is the first element from the right such that $a_i \neq n - r + i$.) Increase a_1 by $1 : a_1 \leftarrow a_1 + 1$; so $a_1 \leftarrow 2$. Now assign $a_2 = a_1 + 1 = 3$ and $a_3 = a_1 + 2 = 4$. The resulting combination is 234. ■

> Thus, to find the combination that follows the r-combination $a_1a_2 \ldots a_r$, we proceed as follows. From right to left, we find the first element a_i such that $a_i \neq n - r + i$. Increment a_i by $1 : a_i \leftarrow a_i + 1$. (Using the new a_i) assign the values $a_i + 1, a_i + 2, \ldots$ to $a_{i+1}, a_{i+2}, \ldots, a_r$, respectively; that is, $a_j \leftarrow a_i + j - i$, where $i + 1 \leq j \leq r$.

EXAMPLE A.17 Find the combination that follows in lexicographic ordering the 3-combination 245 of the set $\{1, 2, 3, 4, 5\}$.

SOLUTION:
Here $n = 5$, $r = 3$, and $n - r = 2$. Let $a_1a_2a_3 = 245$. From right to left, find the first a_i such that $a_i \neq n - r + i = 2 + i$. Clearly, $a_3 = 5 = 2 + 3$, $a_2 = 4 = 2 + 2$; but $a_1 = 2 \neq 2 + 1$. Therefore, the first such a_i is a_1. Update a_1 as $a_1 + 1 : a_1 \leftarrow 2 + 1 = 3$. Now assign the value $a_1 + j - 1$ to a_j for the remaining positions j, namely, 2 and 3:

$$\text{When } j = 2,\ a_2 = a_1 + 2 - 1 = 3 + 1 = 4$$

$$\text{When } j = 3,\ a_3 = a_1 + 3 - 1 = 3 + 2 = 5$$

The resulting combination is 345. ■

These discussions lead us to Algorithm A.2.

```
Algorithm next-combination (a1a2...ar)
(* This algorithm finds the combination that follows the
   r-combination a1a2...ar of the set {1,2,...,n}. Assume
   the given combination is not the largest combination
   (n - r + 1)...(n - 1)n. *)
0. Begin (* next-combination *)
      (* Find the first ai from right to left for which
         ai ≠ n - r + i. *)
1. i ← r
2. while ai = n - r + i do (* continue scanning *)
3. i ← i - 1
4. ai ← ai + 1 (* update ai *)
   (* update the values of ai+1, ai+2,...,ar *)
5. for j = i + 1 to r do
6. aj ← ai + j - i
7. End (* next-combination *)
```

Algorithm A.2

EXAMPLE A.18 Using Algorithm A.2, find the combination that follows in lexicographic order the 4-combination 1345 of the set $\{1, 2, 3, 4, 5\}$.

SOLUTION:

Here $n = 5$, $r = 4$, and $n - r = 1$. Let $a_1a_2a_3a_4 = 1345$. Beginning with a_4, compute a_i and $n - r + i$ until $a_i \neq n - r + i$:

$$a_4 = 5 = 1 + 4 \qquad a_3 = 4 = 1 + 3$$
$$a_2 = 3 = 1 + 2 \qquad a_1 = 1 \neq 1 + 1$$

Therefore, $i = 1$. Consequently, increment the value of a_1 by 1:

$$a_1 \leftarrow a_1 + 1 = 2.$$

Update the values of a_2, a_3, and a_4: $a_2 \leftarrow a_1 + 1 = 3$, $a_3 \leftarrow a_1 + 2 = 4$, $a_4 \leftarrow a_1 + 3 = 5$. Thus the resulting combination is 2345. ■

Exercises A.4

Find the next permutation larger than each permutation in lexicographic ordering, if it exists.

1. 1432 **2.** 2341 **3.** 4132 **4.** 4321

5. 21345 **6.** 21354 **7.** 21543 **8.** 35421

Find the next three permutations that follow each permutation in lexicographic order.

9. 213 **10.** 1324 **11.** 2314 **12.** 2413

13. 3412 **14.** 13245 **15.** 23514 **16.** 45213

Using the next-permutation algorithm, enumerate all permutations of elements of each set.

17. $\{1, 2\}$ **18.** $\{1, 2, 3\}$

Find the next r-combination larger than each combination of the corresponding set, if possible.

19. 124, $\{1, 2, 3, 4\}$ **20.** 134, $\{1, 2, 3, 4\}$

21. 1245, $\{1, 2, 3, 4, 5\}$ **22.** 234, $\{1, 2, 3, 4\}$

23. 2456, $\{1, 2, 3, 4, 5, 6\}$ **24.** 3456, $\{1, 2, 3, 4, 5, 6\}$

Using the next-combination algorithm, generate all r-combinations of the set $\{1, 2, 3, \ldots, n\}$ for each pair of values of r and n.

25. $r = 2, n = 3$ **26.** $r = 1, n = 3$ **27.** $r = 2, n = 4$ **28.** $r = 3, n = 4$

29. Recall that a byte is an 8-bit word. There are 256 bytes and they can be arranged in lexicographic order: 00000000, 00000001, 00000010, ..., 11111110, 11111111. Write an algorithm to find the byte that follows a given byte $a_1a_2\ldots a_8$ in lexicographic order.

30. A ternary word is a word over the alphabet $\{0, 1, 2\}$. Write an algorithm to find the ternary word that follows the ternary word $a_1a_2a_3a_4$ in lexicographic order. For example, the ternary word that follows 0222 is 1000.

Using the next-combination algorithm, find all subsets of each set.

31. $\{1\}$ **32.** $\{1, 2\}$ **33.** $\{1, 2, 3\}$ **34.** $\{1, 2, 3, 4\}$

A.5 The Multinomial Theorem

The binomial theorem enables us to expand powers of the binomial $x + y$. It can be restated in a seemingly different way as follows:

$$(x+y)^n = \sum_{r=0}^{n} \binom{n}{r} x^{n-r}y^r = \sum_{r=0}^{n} \frac{n!}{r!(n-r)!}x^{n-r}y^r$$

$$= \sum_{\substack{i,j\ge 0\\ i+j=n}} \frac{n!}{i!j!}x^i y^j$$

The binomial coefficient $\frac{n!}{i!j!}$ can be interpreted as the number of ways of dividing a set of size n into two disjoint subsets of sizes i and j. This section generalizes this idea: We would like to find the expansion of $(x_1 + x_2 + \cdots + x_k)^n$.

Trinomial Theorem

To this end, let us consider the special case $(x+y+z)^n$. Clearly,

$$(x+y+z)^n = (x+y+z)(x+y+z)\cdots(x+y+z) \tag{A.1}$$

to n factors. Therefore, every term in the expansion of $(x+y+z)^n$ is of the form $Cx^iy^jz^k$, where i, j, and k are nonnegative integers, $i+j+k=n$, and C is the coefficient to be determined. Thus, the various terms are obtained by assigning values to i, j, and k, such that they are nonnegative integers and their sum is n.

The coefficient C is the number of ways i x's, j y's, and k z's can be selected from the n factors on the RHS of Equation (A.1). The i x's can be selected in $\binom{n}{i}$ ways. Now $n-i$ factors are left. So the j y's can be selected in $\binom{n-i}{j}$ ways. Now $n-i-j$ factors are left. Therefore, the k z's can be selected in $\binom{n-i-j}{k}$ ways. Thus, by the multiplication principle,

$$\begin{aligned}
C &= \binom{n}{i}\binom{n-i}{j}\binom{n-i-j}{k} \\
&= \frac{n!}{i!(n-i)!}\cdot\frac{(n-i)!}{j!(n-i-j)!}\cdot\frac{(n-i-j)!}{k!(n-i-j-k)!} \\
&= \frac{n!}{i!j!k!0!}, \quad \text{since } n = i+j+k. \\
&= \frac{n!}{i!j!k!}
\end{aligned}$$

Thus we have the following theorem.

THEOREM A.8 **(Trinomial Theorem)** Let x, y, and z be any real numbers, and n any whole number. Then

$$(x+y+z)^n = \sum_{\substack{i,j,k\geq 0 \\ i+j+k=n}} \frac{n!}{i!j!k!}x^iy^jz^k \tag{A.2}$$

■

EXAMPLE A.19 Using the trinomial theorem, expand $(x+y+z)^2$.

SOLUTION:
By the trinomial theorem,

$$(x+y+z)^2 = \sum_{\substack{i,j,k \geq 0 \\ i+j+k=2}} \frac{2!}{i!j!k!} x^i y^j z^k$$

To find the various terms in the expansion, let us list the possible combinations of values of i, j, and k, and the corresponding coefficients and terms in a table, as in Table A.3.

Table A.3

i	j	k	$\frac{2!}{i!j!k!}$	$x^i y^j z^k$	Term
2	0	0	$\frac{2!}{2!0!0!} = 1$	x^2	x^2
0	2	0	$\frac{2!}{0!2!0!} = 1$	y^2	y^2
0	0	2	$\frac{2!}{0!0!2!} = 1$	z^2	z^2
1	1	0	$\frac{2!}{1!1!0!} = 2$	xy	$2xy$
0	1	1	$\frac{2!}{0!1!1!} = 2$	yz	$2yz$
1	0	1	$\frac{2!}{1!0!1!} = 2$	zx	$2zx$

Adding up the terms in the last column,

$$(x+y+z)^2 = x^2 + y^2 + z^2 + 2xy + 2yz + 2zx$$ ■

EXAMPLE A.20 Find the coefficient of $x^2y^3z^4$ in the expansion of $(x+y+z)^9$.

SOLUTION:
Here $i = 2, j = 3, k = 4$, and $n = 9$. Therefore, the required coefficient is given by

$$\frac{n!}{i!j!k!} = \frac{9!}{2!3!4!} = 1260$$ ■

The proof used to develop the trinomial theorem can be generalized to k variables $x_1, x_2, \ldots, x_k$ in an obvious way. The resulting theorem, called the **multinomial theorem**, is given below.

THEOREM A.9 **(The Multinomial Theorem)** Let $x_1, x_2, \ldots, x_k$ be any k real variables and n any nonnegative integer. Then

$$(x_1 + x_2 + \cdots + x_k)^n = \sum \frac{n!}{i_1!i_2!\ldots i_k!} x_1^{i_1} x_2^{i_2} \ldots x_k^{i_k}$$

where $i_1, i_2, \ldots, i_k \geq 0$ and $i_1 + i_2 + \cdots + i_k = n$. ■

EXAMPLE A.21 Using the multinomial theorem, find the coefficient of $x_1x_2{}^2x_3{}^3x_4{}^4x_5{}^5$ in the expansion of $(x_1+x_2+x_3+x_4+x_5)^{15}$.

SOLUTION:

$$\begin{aligned}\text{Required coefficient} &= \frac{15!}{1!2!3!4!5!}\\ &= 37{,}837{,}800\end{aligned}$$

■

The coefficient $\dfrac{n!}{i_1!i_2!\ldots i_k!}$ in the multinomial theorem, called the **multinomial coefficient**, is denoted by $\binom{n}{i_1, i_2, \ldots, i_k}$. For instance,

$$\binom{5}{3,0,1,1} = \frac{5!}{3!0!1!1!} = 20$$

Just as the binomial coefficient $\binom{n}{i,j}$ denotes the number of ways a set of size n can be divided into two disjoint subsets of sizes i and $j = n - i$, the multinomial coefficient can also be interpreted in a similar way, as stated in the following theorem.

THEOREM A.10 The number of ways of dividing a set S of size n into k mutually disjoint ordered subsets $S_1, S_2, \ldots, S_k$ of sizes $i_1, i_2, \ldots, i_k$, respectively, is given by the multinomial coefficient

$$\binom{n}{i_1, i_2, \ldots, i_k}$$

where $i_1, i_2, \ldots i_k \geq 0$ and $i_1 + i_2 + \cdots + i_k = n$.

PROOF:

The i_1 elements of the subset S_1 can be selected in $\binom{n}{i_1}$ ways. This leaves $n - i_1$ elements in $S - S_1$. Therefore, the i_2 elements of S_2 can be selected in $\binom{n-i_1}{i_2}$ ways. Similarly, the i_3 elements of S_3 can be selected in $\binom{n-i_1-i_2}{i_3}$ ways, and so on. The i_k elements of S_k can be chosen in $\binom{n-i_1-i_2-\ldots-i_{k-1}}{i_k}$ ways. Consequently, by the multiplication principle, the number of ways of choosing the mutually disjoint subsets S_1, $S_2, \ldots, S_k$ is

$$\binom{n}{i_1}\binom{n-i_1}{i_2}\binom{n-i_1-i_2}{i_3}\cdots\binom{n-i_1-i_2-\cdots-i_{k-1}}{i_k}$$

$$= \frac{n!}{i_1!(n-i_1)!} \cdot \frac{(n-i_1)!}{i_2!(n-i_1-i_2)!} \cdots \frac{(n-i_1-\cdots i_{k-1})}{i_k!(n-i_1-\cdots-i_k)!}$$

$$= \frac{n!}{i_1!i_2!\ldots i_k!} \qquad (Note: n = i_1 + i_2 + \cdots + i_k)$$

$$= \binom{n}{i_1, i_2, \ldots, i_k}$$

This concludes the proof. ■

EXAMPLE A.22 Find the number of ways of dividing a set of size five into three mutually disjoint ordered subsets of sizes 2, 1, and 2.

SOLUTION:

By Theorem A.10, the number of ways of dividing the set in the desired way is given by the multinomial coefficient

$$\binom{5}{2, 1, 2} = \frac{5!}{2!1!2!}$$
$$= 30$$

■

EXAMPLE A.23 Find the number of ways of dividing the set $S = \{a, b, c\}$ into three mutually disjoint ordered subsets S_1, S_2, and S_3 of sizes 1, 1, and 1, respectively, and list them.

SOLUTION:

$$\text{Number of possibilities} = \binom{3}{1, 1, 1} = \frac{3!}{1!1!1!} = 6$$

They are listed in Table A.4.

Table A.4

S_1	S_2	S_3
{a}	{b}	{c}
{a}	{c}	{b}
{b}	{a}	{c}
{b}	{c}	{a}
{c}	{a}	{b}
{c}	{b}	{a}

An important observation: The division {{a}, {b}, {c}} is different from the division {{a}, {c}, {b}}. However, they form the same partition. ■

Exercises A.5

Evaluate each multinomial coefficient.

1. $\binom{5}{2, 2, 1}$ **2.** $\binom{6}{1, 2, 3}$ **3.** $\binom{7}{2, 3, 2}$ **4.** $\binom{8}{0, 3, 5}$

5. $\binom{8}{2,1,3,2}$ **6.** $\binom{10}{1,2,3,4}$ **7.** $\binom{10}{3,1,4,2}$ **8.** $\binom{10}{1,4,4,1}$

Find the coefficient of each.

9. x^2yz^2 in the expansion of $(x+y+z)^5$

10. xy^2z^3 in the expansion of $(x+y+z)^6$

11. yz in the expansion of $(x+y-z)^2$

12. xy^2 in the expansion of $(x-y-z)^3$

13. xy^2z^5 in the expansion of $(x+y+z)^8$

14. xy^3z^3 in the expansion of $(x-y+2z)^8$

Expand each.

15. $(x+y-\mathrm{z})^2$ **16.** $(x-y-z)^2$ **17.** $(x-y+z)^3$

18. $(x-y-z)^3$ **19.** $(x+2y+z)^3$ **20.** $(x-2y+2z)^3$

Consider the various ways of dividing the set $S=\{\mathrm{a,b,c,d}\}$ into three mutually disjoint ordered subsets S_1, S_2, and S_3 with sizes 1, 1, and 2, respectively.

21. Find the number of possibilities.

22. List the possibilities.

Find the number of ways of dividing a set of size n into k mutually disjoint ordered subsets of sizes i_1, $\mathrm{i}_2, \ldots, i_k$ in each case.

23. $n=10$, three subsets with sizes 2, 3, and 5.

24. $n=10$, four subsets with sizes 2, 2, 3, and 3.

25. $n=12$, six subsets with sizes 1, 1, 2, 2, 3, and 3.

26. $n=15$, four subsets with sizes 1, 1, 5, and 8.

27. Find the sum of the multinomial coefficients $\binom{n}{i_1, i_2, \ldots, i_k}$.

It follows by Theorem A.10 that the number of permutations of n items of which i_1 are alike, i_2 are alike, ..., and i_k are alike is given by the multinomial coefficient $\binom{n}{i_1, i_2, \ldots, i_k}$. Using this fact, compute each.

28. The number of ways of scrambling the letters of the word ABRA-CADABRA.

29. The number of ways of scrambling the letters of the word TINTINNABULATION.

30. The number of binary words of length 10 and containing exactly three 1's and seven 0's.

31. The number of bytes containing exactly five 0's.

32. The number of ternary words of length 12 and containing three 0's, four 1's, and five 2's.

Find the number of terms in the expansion of each.

33. $(x+y+z)^3$

34. $(x+y+z)^{10}$

35. $(w+x+y+z)^{12}$

36. $(2w-3x+4y-4z)^{15}$

Find the coefficient of each.

37. $x^2y^2z^3$ in the expansion of $(x+y+z)^7$

38. xy^3z in the expansion of $(x+2y-z)^5$

Using the multinomial theorem, expand each.

39. $(x-2y+z)^3$

40. $(x+y-z)^4$

Find the number of terms in the expansion of each.

41. $(x+y-z)^6$

42. $(x-y-z)^8$

43. $(2x+3y-4z)^9$

44. $(w+x-y-z)^{10}$

Find the number of ways of dividing a set with size 15 into each.

45. Four mutually disjoint subsets with sizes 7, 2, 3, and 3.

46. Five mutually disjoint subsets with sizes 3, 6, 2, 1, and 3.

A.6 The Greek Alphabet

A	α	alpha	N	ν	nu
B	β	beta	Ξ	ξ	xi
Γ	γ	gamma	O	o	omicron
Δ	δ	delta	Π	π	pi
E	ϵ	epsilon	P	ρ	rho
Z	ζ	zeta	$\sum$	σ	sigma
H	η	eta	T	τ	tau
Θ	θ	theta	Υ	υ	upsilon
I	τ	iota	Φ	ϕ	phi
K	κ	kappa	X	χ	chi
Λ	λ	lambda	Ψ	ψ	psi
M	μ	mu	Ω	ω	omega

A.7 Web Sites

The following Web sites provide valuable information for further exploration and enrichment. The status of a Web site could change with time, so it may not exist when you look for it; if it does not, use a search engine to locate a similar Web site.

1. Wilhelm Ackermann
   www-gap.dcs.st-and.ac.uk/~history/Mathematicians/Ackermann.html
2. John Backus
   www.digitalcentury.com/encyclo/update/backus.html
   www-gap.dcs.st-and.ac.uk/~history/Mathematicians/Backus.html
   www.acm.org/awards/turing_citations/backus.html
   www.cs.nyu.edu/cs/faculty/shashaoutofmind/backus.html
3. George Boole
   www-gap.dcs.st-and.ac.uk/~history/Mathematicians/Boole.html
   www.digitalcentury.com/encyclo/update/boole.html
   homepages.enterprise.net/rogerp/george/boole.html
4. Georg Cantor
   www-groups.dcs.st-and.ac.uk/~history/Mathematicians/Cantor.html
   www.treasure-troves.com/bios/CantorGeorg.html
   www.aug.edu/dvskel/JohnsonSU97.html
5. Arthur Cayley
   www-gap.dcs.st-and.ac.uk/~history/Mathematicians/Cayley.html
   scienceworld.wolfram.com/biography/Cayley.html
   www.stetson.edu/~efriedma/periodictable/html/C.html
   www.optisyn.com/presentations/Cayley.html
   www.math.ukans.edu/~engheta/bio/cayley.html
   www.geometry.net/Biographer/Cayley.html
6. Edsger Dijkstra
   www.digidome.nl/edgser_wybe_dijkstra.html
   www.acm.org/classics/oct95/
   www.cs.utexas/users/EWD/obituary.html
   news.com.com/2100-1001-949023.html
7. Leonhard Euler
   www.maths.tcd.ie/pub/HistMath/People/Euler/RouseBall/RB_Euler.html
   www.shu.edu/html/teaching/math/reals/history/euler.html
8. Pierre de Fermat
   www.maths.tcd.ie/pub/HistMath/People/Fermat/RouseBall/RB_Fermat.html
9. Karl F. Gauss
   www-gap.dcs.st-and.ac.uk/~history/Mathematicians/Gauss.html
   www.english.upenn.edu/~jlynch/FrankenDemo/People/gauss.html
   scienceworld.wolfram.com/biography/Gauss.html
   www.indiana.edu/~intell/gauss.html
   www.brown.edu/Students/OHJC/hm4/gauss.html

10. Christian Goldbach
www-gap.dcs.st-and.ac.uk/~history/Mathematicians/Goldbach.html
www.geocities.com/Heartland/Hills/7972/math-goldbach.html
www.wikipedia.org/wiki/Christian_Goldbach
plus.maths.org/issue11/news/Goldbach

11. Sir William R. Hamilton
www.chembio.uoguelph.ca/educmat/chm386/rudiment/tourclas/hamilton.html
scienceworld.wolfram.com/biography/HamiltonWilliamRowan.html
www.iaste.com/hall_of_fame/hamilton.html

12. Maurice Karnaugh
www.informatik.uni-trier.de/~ley/db/indices/a-tree/k/Karnaugh:Maurice.html
www.maxmon.com/library.html
www-cse.stanford.edu/classes/cs103a/h9BooleanAlgebra.pdf

13. Alfred B. Kempe
www-gap.dcs.st-and.ac.uk/~history/Mathematicians/Kempe.html
mappa.mundi.net/locus_014
www.uwinnipeg.ca/~ooellerm/guthrie/FourColor.html
www.mathsyear2000.org/explorer/morphing/13usedownload.shtml

14. Stephen C. Kleene
www.library.wisc.edu/libraries/Math/kleen.html
www-groups.dcs.st-and.ac.uk/~history/Mathematicians/Kleene.html
www.dcs.ed.ac.uk/homes/als/lics/newsletters/19.html
www.student.math.uwaterloo.ca/~cs462/Hall/kleene.html

15. Donald E. Knuth
sunburn.stanford.edu/~knuth
www-cs-staff.stanford.edu/~knuth/index.html
www.digitalcentury.com/encyclo/update/knuth.html
laurel.actlab.utexas.edu/~cynbe/muq/muf3_20.html

16. Kazimierz Kuratowski
www-groups.dcs.st-and.ac.uk/~history/Mathematiticains/Kuratowski.html
www.stetson.edu/~efriedma/periodictable/html/Kr.html

17. Gabriel Lame
www-groups.dcs.st-and.ac.uk/~history/Mathematicians/Lame.html
www.bath.ac.uk/~ma0dmp/Lamelife.html

18. Edmund Landau
www.ma.huji.ac.il/~landau/landuniv.html
www-groups.dcs.st-and.ac.uk/~history/Mathematicians/Landau.html
www.ard.huji.ac.il/publications/25years/chap13.html

19. Pierre-Simon Laplace
www-groups.dcs.st-and.ac.uk/~history/Mathematicians/Laplace.html
www.maths.tcd.ie/pub/HistMath/People/Laplace/RouseBall/RB_Laplace.html
www. stetson.edu/~efriedma/periodictable/html/La.html

20. Gottfried W. Leibniz
www.maths.tcd.ie/pub/HistMath/People/Leibniz/RouseBall/RB_Leibniz.html

21. Jan Lukasiewicz
www-groups.dcs.st-and.ac.uk/~history/Mathematiticains/Lukawiewicz.html
www.hpmuseum.org/rpn.html
www.wikipedia.com/wiki/Jan_Lukawiewicz

22. George H. Mealy
www.informatik.uni-trier.de/~ley/db/indices/a-tree/m/Mealy:George_H=.html

23. Marin Mersenne
www-groups.dcs.st-and.ac.uk/~history/Mathematiticains/Mersenne.html
www2.andrews.edu/~calkins/math/biograph/biomerse.html

24. Blaise Pascal
members.aol.com/KatherenaE/private/Philo/Pascal/pascal.html
www.maths.tcd.ie/pub/HistMath/People/Pascal/RouseBall/RB_Pascal.html
www.cs.washington.edu/homes/jbaer/classes/blaise/blaise.html
www.math.sfu.ca/histmath/Europe/17thCenturyAD/Balise.html
www-groups.dcs.st-and.ac.uk/~history/Mathematiticains/Pascal.html

25. Bertrand Russell
www.mcmaster.ca/russdocs/russell.html
plato.stanford.edu/entries/russell
desktop12.cis.macmaster.cal/~bertrand

26. Claude E. Shannon
www-groups.dcs.st-and.ac.uk/~history/Mathematiticains/Shannon.html
www.bell-labs.com/news/2001/february/26/1.html
www.digitalcentury.com/encyclo/update/shannon.html

27. Alan M. Turing
www.turing.org.uk/turing
ei.cs.vt.edu/~history/Turing.html
www-groups.dcs.st-and.ac.uk/~history/Mathematiticains/Turing.html

28. Alexandre-Theophile Vandermonde
www-groups.dcs.st-and.ac.uk/~history/Mathematiticains/Vandermonde.html

29. John Venn
www-groups.dcs.st-and. ac.uk/~history/Mathematicians/Venn.html
userwww.sfsu.edu/~rsauzier/Venn.html

30. Boolean Algebra
educ.queensu.ca/~compsci/units/BoolLogic/titlepage.html
www.bit.umkc.edu/vu/course/cs281/lectures/boolean-algebra/boolean-algebra.html

ever.phys.ualberta.ca/~gingrich/phys395/notes/node121.html
www.maxmon.com/1847ad.html
www.yale.edu/ynhti/curriculum/units/1989/7/89.07.07.x.html

32. Finite-State Machines
www.cs.brown.edu/people/jes/book/BOOK/node10.html
classwww.gsfc.nasa.gov/CAGESite/pages/cage_gpf_fsm.html
www.cs.arizona.edu/classes/cs352/summer02/fsa.html
www.c3.lanl.gov/mega-math/workbk/machine/mabkgd.html
www.beigarath.demon.co.uk/java/fsme.html
members.aol.com/asakharov/fsm.html

33. Fuzzy Sets
www.answermath.com/fuzzymath.html
sun16.cecs.missouri.edu/index.html
news:comp.ai.fuzzy

34. Generating Functions
www.cs.wpi.edu/~cs504/s00m/notes/ln/1999/class06/class06.html
www.cs.wpi.edu/~cs504/s00m/classes/class05/Class05.html
msl.cs.uiuc.edu/~lavalle/cs576/projects/wmchan
www.eco.rug.nl/gauss/GAUSS00/mhonarc.db
www.maths.surrey.ac.uk/personal/st/d.fisher/MS103/MS105-5.mws

35. Graph Theory
www.utm.edu/departments/math/graph
www.c3.lanl.gov/mega-math/workbk/graph/graph.html
www-groups.dcs.st-and.ac.uk/~history/Mathematicians/Erdos.html
www.nada.kth.se/~viggo/problemlist/compendium.html
www.shodor.org/interactive/lessions/frac1.html
www.scism.sbu.ac.uk/law/Section3/chapter3/s3c3int.html
www.cs.cmu.edu/~cburch/survey/recurse
www.nd.edu/~cholak/computability/computability.html

36. Hilbert's Paradoxes
www.wordsmith.demon.co.uk/paradoxes
eluzions.com/Puzzles/Logic/Paradoxes.shtml
www.c3.lanl.gov/mega-math/workbk/infinity/infinity.html
www.cs.tpu.ee/~jaagup/uk/fmm/math/1.2.5.3.html

37. The Four-Color Problem
www.math.gatech.edu/~thomas/FC/fourcolor.html
www.cs.uidaho.edu/~casey931/mega-math/gloss/math/4ct.html
www-groups.dcs.st-and.ac.uk/~history/HistTopics/The_four_colour
_theorem.html

References

1. A. V. Aho *et al.*, *Compilers: Principles, Techniques, and Tools*, Addison-Wesley, Reading, MA 1986.
2. V. S. Alagar, *Fundamentals of Computing: Theory and Practice,* Prentice-Hall, Englewood Cliffs, NJ, 1989.
3. K. Appel and W. Haken, "Every Planar Graph is 4-colorable," *Bulletin of the American Mathematical Society*, Vol. 82 (Sept. 1976), pp. 711–712.
4. R. G. Archibald, *An Introduction to the Theory of Numbers*, 3rd edition, Wiley, New York, 1972.
5. V. Bain, "An Algorithm for Drawing the n-cube," *The College Mathematics Journal*, Vol. 29 (Sept. 1998), pp. 320–322.
6. C. Baltus, "A Truth Table on the Island of Truthtellers and Liars," *Mathematics Teacher*, Vol. 94 (Dec. 2001), pp. 730–732.
7. W. J. Barneir, "Finite-State Machines as Recognizers," *The UMAP Journal*, 7:3 (1986), pp. 209–232.
8. W. Barnier and J. B. Chan, *Discrete Mathematics with Applications*, West, St. Paul, MN, 1989.
9. B. Barwell, Solution to Problem 702, *J. Recreational Mathematics*, Vol. 12:1 (1979–1980), p. 67–68.
10. B. Barwell, Solution to Problem 1046, *J. Recreational Mathematics*, Vol. 15:1 (1981–1982), pp. 70–72.
11. R. Bellman *et al., Algorithms, Graphs, and Computers*, Academic Press, New York, 1970.
12. M. Bellmore and G. L. Nemhauser, "The Traveling Salesman Problem," *Operations Research*, Vol. 16 (1968), pp. 538–558.
13. S. J. Bezsuska, Solution to Problem 791, *J. Recreational Mathematics*, 12:4 (1979–1980), p. 311.
14. B. Bissinger, "Ask Marilyn," *Parade Magazine* (April 25, 1993), p. 15.
15. M. L. Bittinger, *Logic and Proof*, Addison-Wesley, Reading, MA 1972.
16. W. G. Brown, "Historical Note on a Recurrent Combinatorial Problem," *The American Mathematical Monthly*, Vol. 72 (Nov. 1965), pp. 973–977.

17. A. V. Boyd and M. J. Glencorss, "Dissecting a Circle by Chords through n Points," *Mathematics Teacher*, Vol. 84 (April 1991), pp. 318–319.
18. J. Burling *et al.,* "Using Graphs to Solve the Traffic Light Problem," *FAIM Module*, COMAP, Inc., Lexington, MA, 1989.
19. D. Burns, Problem 596, *J. Recreational Mathematics*, Vol. 10:1 (1977–1978), p. 52.
20. W. H. Bussey, "Origin of Mathematical Induction," *The American Mathematical Monthly*, Vol. 24 (May 1917), pp. 199–207.
21. Calendar Problems, *Mathematics Teacher*, Vol. 83 (Oct. 1990), p. 550.
22. Calendar Problems, *Mathematics Teacher*, Vol. 85 (Dec. 1992), p. 736.
23. Calendar Problems, *Mathematics Teacher*, Vol. 79 (April 1986), p. 274.
24. Calendar Problems, *Mathematics Teacher*, Vol. 79 (Nov. 1986), p. 627.
25. D. M. Campbell, "The Computation of Catalan Numbers," *Mathematics Magazine*, Vol. 57 (Sept. 1984), pp. 195–208.
26. L. Carlitz, Solution to Problem B-180, *The Fibonacci Quarterly*, Vol. 8:5 (Dec. 1970), pp. 547–548.
27. M. Caudill, "Using Neural Nets: Fuzzy Decisions," *AI Expert*, Vol. 5 (April 1990), pp. 59–64.
28. M. Charosh, Problem 1160, *J. Recreational Mathematics*, Vol. 15:1 (1983–1984), p. 58.
29. E. F. Codd, "A Relational Model of Data for Large Shared Databanks," *Communications of the ACM*, Vol. 13 (1970), pp. 377–387.
30. F. Cohen and J. L. Selfridge, "Not Every Integer Is the Sum or Difference of Two Prime Powers," *Mathematics of Computation*, Vol. 29 (1975), p. 79.
31. E. Comfort, Solution to Problem 596, *J. Recreational Mathematics*, Vol. 11:1 (1978–1979), p. 66.
32. J. W. Cortada, *Historical Dictionary of Data Processing: Biographies*, Greenwood Press, New York, 1987.
33. M. Coughlin and C. Kerwin, "Mathematical Induction and Pascal's Problem of the Points," *Mathematics Teacher*, Vol. 78 (May 1985), pp. 376–380.
34. T. Crilly, "A Victorian Mathematician," *The Mathematical Gazette*, Vol. 79 (July 1995), pp. 259–262.
35. P. Cull and E. F. Ecklund, Jr., "Towers of Hanoi and Analysis of Algorithms," *The American Mathematical Monthly*, Vol. 92 (June–July 1985), pp. 407–420.
36. N. Deo, *Graph Theory with Applications to Engineering and Computer Science*, Prentice Hall, Englewood Cliffs, NJ, 1974.
37. R. C. Drake, Problem B-180, *The Fibonacci Quarterly*, Vol. 8:1 (Feb. 1970), p. 106.
38. L. R. Duffy, "The Duffinian Numbers," *J. Recreational Mathematics*, Vol. 12:2 (1979–1980), pp. 112–115.
39. R. Euler, Problem 1551, *J. Recreational Mathematics*, Vol. 19:2, 1987, p. 151.
40. H. Eves, Problem E579, *The American Mathematical Monthly*, Vol. 50 (June–July 1943), p. 386.
41. H. Eves, *An Introduction to the History of Mathematics*, 3rd edition, Holt, Rinehart and Winston, New York, 1969.

42. A. Filz, Problem 1046, *J. Recreational Mathematics*, Vol. 14:1 (1981–1982), p. 64.
43. T. Fletcher, Problem 602, *J. Recreational Mathematics*, Vol. 10:1 (1978–1979), p. 52.
44. H. G. Forder, "Some Problems in Combinatorics," *The Mathematical Gazette*, Vol. 45 (1961), pp. 199–201.
45. A. J. Friedland, *Puzzles in Math & Logic*, Dover, New York, 1970.
46. J. A. Gallian and S. Winters, "Modular Arithmetic in the Market Place," *The American Mathematical Monthly*, Vol. 95 (June–July 1988), pp. 548–551.
47. M. Gardner, "Mathematical Games," *Scientific American*, Vol. 219 (Sept. 1968), pp. 218–230.
48. M. Gardner, "Catalan Numbers: An Integer Sequence That Materializes in Unexpected Places," *Scientific American*, Vol. 234 (June 1976), pp. 120–125.
49. M. Gardner, *Mathematical Circus*, Knopf, New York, NY, 1979.
50. M. Gardner, *Mathematical Puzzles and Diversions*, The University of Chicago Press, Chicago, IL, 1987.
51. M. Gardner, "Ask Marilyn," *Parade Magazine* (April 18, 1993), p. 12.
52. S. W. Golomb, "Pairings and Groupings," *Johns Hopkins Magazine*, Vol. 45:2 (April 1993), p. 7.
53. R. L. Graham *et al.*, *Concrete Mathematics*, Addison-Wesley, Reading, MA, 1990.
54. J. J. Gray, "Arthur Cayley (1821–1895)," *The Mathematical Intelligencer*, Vol. 17:4 (1995), pp. 62–63.
55. T. M. Green, "Pascal's Pizza," *Mathematics Teacher*, Vol. 81 (Sept. 1988), p. 445, 454.
56. R. P. Grimaldi, *Discrete and Computational Mathematics*, 4th ed., Addison-Wesley, Reading, MA, 1999.
57. A. Guckin *et al.*, *The Euler Circuit Project*, COMAP, Inc., Lexington, MA, 1989.
58. S. Gudder, *A Mathematical Journey*, McGraw-Hill, New York, 1976.
59. B. Hamilton, *Brainteasers and Mindbenders*, Freeside, New York, 1992.
60. D. K. Hanson *et al.*, "Matching, Derangements, and Rencontres," *Mathematics Magazine*, Vol. 56 (Sept. 1983), pp. 224–229.
61. F. Harray and J. S. Maybee (eds.), *Graphs and Applications*, Wiley, New York, 1985.
62. B. Hayes, "On the Finite-State Machine, A Minimal Model of Mousetraps, Ribosomes, and the Human Soul," *Scientific American*, Vol. 249 (Dec. 1983), pp. 20–28, 178.
63. L. Henkin, "On Mathematical Induction," *The American Mathematical Monthly*, Vol. 67 (April 1960), pp. 323–338.
64. V. E. Hoggatt, Jr., *Fibonacci and Lucas Numbers*, Houghton Mifflin, Boston, 1963.
65. V. E. Hoggatt, Jr., and S. L. Basin, "A Primer on the Fibonacci Sequence, Part II," *The Fibonacci Quarterly*, Vol. 1:2 (April 1963), pp. 61–68.

66. V. E. Hoggatt, Jr., "Some Special Fibonacci and Lucas Generating Functions," *The Fibonacci Quarterly*, Vol. 9:2 (April 1971), pp. 121–133.

67. T. C. Hu, *Combinatorial Algorithms*, Addison-Wesley, Reading, MA, 1982.

68. K. K. Huang, Solution to Problem 1160, *J. Recreational Mathematics*, Vol. 16:1 (1983–1984), p. 69.

69. R. V. Jean, "The Fibonacci Sequence," *The UMAP Journal*, Vol. 5:1 (1984), pp. 23–47.

70. J. T. Johnson, "Fuzzy Logic," *Popular Science*, Vol. 237 (July 1990), pp. 87–89.

71. R. Johnsonbaugh, *Discrete Mathematics*, 5th ed., Prentice-Hall, Upper Saddle River, NJ, 2001.

72. E. Just, "A Note on the nth Term of the Fibonacci Sequence," *Mathematics Magazine*, Vol. 44 (Sept.–Oct. 1971), p. 199.

73. M. Karnaugh, "The Map Method for Synthesis of Combinational Logic Circuits," *Transactions of the AIEE*, Part I, Vol. 72:9 (Nov. 1953), pp. 593–599.

74. F. H. Kierstead, Jr., Problem 791, *J. Recreational Mathematics*, Vol. 11:4 (1978–1979), p. 302.

75. F. H. Kierstead, Jr., Problem 1014, *J. Recreational Mathematics*, Vol. 14:4 (1981–1982), p. 309.

76. M. Keith and T. Carver, "The Ultimate Perpetual Calendar," *J. Recreational Mathematics*, Vol. 22:4 (1990), pp. 280–282.

77. D. E. Knuth, "Algorithms," *Scientific American*, Vol. 243 (April 1977), pp. 63–80.

78. D. E. Knuth, "Algorithmic Thinking and Mathematical Thinking," *The American Mathematical Monthly*, Vol. 92 (March 1985), pp. 170–181.

79. B. Kolman *et al.*, *Discrete Mathematical Structures*, 4th ed., Prentice-Hall, Upper Saddle River, NJ, 2000.

80. T. Koshy, *Finite Mathematics and Calculus with Applications*, Goodyear, Pacific Palisades, CA, 1979.

81. T. Koshy, *Fibonacci and Lucas Numbers with Applications*, Wiley, New York, 2001.

82. T. Koshy, *Elementary Number Theory with Applications*, Harcourt/Academic Press, Boston, 2002.

83. B. Kosko and S. Isaka, "Fuzzy Logic," *Scientific American*, Vol. 269 (July 1993), pp. 76–81.

84. J. B. Kruskal, "On the Shortest Spanning Subtree of a Graph and the Traveling Salesman Problem," *Proceedings of the American Mathematical Society*, Vol. 1 (1956), pp. 48–50.

85. E. A. Kuehls, "The Truth-Value of $\{\forall, \exists, P(x,y)\}$: A Graphical Approach," *Mathematics Teacher*, Vol. 43 (Nov. 1970), pp. 260–261.

86. N. J. Kuenzi and B. Prielipp, Problem 4026, *School and Science Mathematics*, Vol. 85 (Dec. 1985), pp. 714–716.

87. L. J. Lander *et al.*, "A Survey of Equal Sums of Like Powers," *Mathematics of Computation*, Vol. 21 (1967), p. 446.

88. E. L. Lawler *et al.* (eds.), *The Traveling Salesman Problem*, Wiley, New York, 1986.

89. C. T. Long, "On Pigeons and Problems," *Mathematics Teacher*, Vol. 81 (Jan. 1988), pp. 28–30, 64.

90. E. Maier, "Counting Pizza Pieces and Other Combinatorial Problems," *Mathematics Teacher*, Vol. 81 (Jan. 1988), pp. 22–26.

91. C. L. Mallows, "Conway's Challenge Sequence," *The American Mathematical Monthly*, Vol. 98 (Jan. 1991), pp. 5–20.

92. L. E. Mauland, "An Exercise with Polygonal Numbers," *Mathematics Teacher*, Vol. 78 (May 1985), pp. 340–344.

93. S. B. Maurer and A. Ralston, *Discrete Algorithmic Mathematics*, Addison-Wesley, Reading, MA, 1991.

94. M. Martelli, "The Farmer and the Goose — a Generalization," *Mathematics Teacher*, Vol. 86 (March 1993), pp. 202–203.

95. W. S. McCulloch and W. Pitts, "A Logical Calculus of the Ideas Imminent in Nervous Activity," *Bulletin of Mathematical Biophysics*, Vol. 5 (1943), pp. 115–133.

96. W. A. Miller, "Polynomial Numbers and Recursion," *Mathematics Teacher*, Vol. 83 (Oct. 1990), pp. 555–558.

97. B. R. Myers, "Number of Spanning Trees in a Wheel," *IEEE Transactions on Circuit Theory*, CT-18 (March 1971), pp. 280–281.

98. H. L. Nelson, Problem 702, *J. Recreational Mathematics*, Vol. 11:1 (1978–1979), p. 36.

99. H. L. Nelson, "Two Counterfeits," *J. Recreational Mathematics*, Vol. 15:1 (1982–1983), p. 65.

100. J. C. Nichols, Solution to Problem 602, *J. Recreational Mathematics*, Vol. 11:1 (1978–1979), p. 75.

101. G. Polya, *Mathematical Discovery*, combined ed., Wiley, New York, 1981.

102. R. C. Prim, "Shortest Connection Networks and Some Generalizations," *Bell System Technical Journal*, Vol. 36 (1957), pp. 1389–1401.

103. I. Vun and P. Belcher, "Catalan Numbers," *Mathematical Spectrum*, Vol. 29:3 (1996–1997), pp. 3–5.

104. A. Wayne, Solution to Problem E579, *The American Mathematical Monthly*, Vol. 51 (March 1944), p. 165.

105. A. Ralston, "De Bruijn Sequences — A Model Example of the Interaction of Discrete Mathematics and Computer Science," *Mathematics Magazine*, Vol. 55 (May 1982), pp. 131–143.

106. B. Recaman, "The Games of Ham," *J. Recreational Mathematics*, Vol. 10 (1977–1978), pp. 251–253.

107. J. V. Roberti, "The Indirect Method," *Mathematics Teacher*, Vol. 80 (Jan. 1987), pp. 41–43.

108. K. H. Rosen, *Discrete Mathematics and Its Applications*, 4th ed., McGraw-Hill, New York, 1999.

109. K. A. Ross and C. R. B. Wright, *Discrete Mathematics*, 3rd ed., Prentice-Hall, Englewood Cliffs, NJ, 1992.

110. G. L. Ritter *et al.*, "An Aid to the Superstitious," *Mathematics Teacher*, Vol. 70 (May 1977), pp. 456–457.

111. S. Sahni, *Concepts in Discrete Mathematics*, 2nd ed., Camelot, Fridley, MN, 1985.

112. B. J. Schwartz, Solution to Problem 1014, *J. Recreational Mathematics*, Vol. 14:4 (1981–1982), p. 309.

113. J. Sedlacek, "On the Skeletons of a Graph or Digraph," *Proceedings of the Calgary International Conference of Combinatorial Structures and their Applications*, Gordon & Breach, New York, pp. 387–391.

114. D. E. Shasha, *Out of their Minds: The Lives and Discoveries of 15 Great Computer Scientists*, Copernicus, New York, 1995, pp. 89–101.

115. D. R. Sherbert, "Difference Equations with Applications," *UMAP Module 322*, Arlington, MA, 1980.

116. R. M. Smullyan, *What is the name of this book?*, Prentice-Hall, Englewood Cliffs, NJ, 1978.

117. R. M. Smullyan, *Alice in Puzzle-Land: A Carrollian Tale for Children Under Eighty*, Penguin Books, New York, 1982.

118. R. M. Smullyan, "Leaps of Logic," *Discover* (March 1993), p. 96.

119. S. K. Stein, "The Mathematician as an Explorer,"*Scientific American*, Vol. 204 (May 1961), pp. 149–158.

120. S. K. Stein, *Mathematics: The Man-Made Universe*, W. H. Freeman, San Francisco, CA, 1969.

121. A. Sterrett, "Gambling Doesn't Pay," *Mathematics Teacher*, Vol. 60 (March 1967), pp. 210–214.

122. P. Stevens, *Patterns in Nature*, Little Brown, Boston, 1974.

123. D. R. Stone, "A Different Prime Proof," *Mathematics Teacher*, Vol. 83 (Jan. 1990), p. 63.

124. A. S. Tanenbaum, *Structured Computer Organization*, Prentice-Hall, Englewood Cliffs, NJ, 1976, pp. 420–423.

125. *The Official LSAT PrepBook*, Law Services, Newtown, PA, 1991.

126. R. M. Thrall, "Insulin Requirements as a Linear Process in Time," *Some Mathematical Models in Biology* (R. F. Baum, ed.), The University of Michigan Press, Ann Arbor, MI, 1967, pp. 0L2.1–0L2.4.

127. P. M. Tuchinsky, "International Standard Book Numbers," *The UMAP Journal*, Vol. 6:1 (1985), pp. 41–53.

128. A. Tucker, *Applied Combinatorics*, Wiley, New York, NY, 1980.

129. T. Tymoczko, "Computers, Proofs, and Mathematics: A Philosophical Investigation of the Four-Color Proof," *Mathematics Magazine*, Vol. 53 (May 1980), pp. 131–138.

130. J. H. van Lint and R. M. Wilson, *A Course in Combinatorics*, Cambridge University Press, New York, 1992.

131. M. vos Savant, *Ask Marilyn*, St. Martin Press, New York, p. 228.

132. S. Warshall, "A Theorem on Boolean Matrices," *J. of the Association of Computing Machinery*, Vol. 9 (1962), pp. 11–12.

133. J. Williams, "Graph Coloring Used to Model Traffic Lights," *Mathematics Teacher*, Vol. 85 (March 1992), pp. 212–214.

134. R. J. Wilson and J. J. Watkins, *Graphs: An Introductory Approach*, Wiley, New York, 1990.

135. D. Wood, "Towers of Brahma and Hanoi Revisited," *J. Recreational Mathematics*, Vol. 14:1 (1981–1982), pp. 17–24.

136. R. V. Young (ed.), *Notable Mathematicians*, Gale Research, Detroit, MI, 1997.

Solutions to Odd-Numbered Exercises

Chapter 1 The Language of Logic

Exercises 1.1 (p. 17)

1. yes **3.** no **5.** F **7.** T

9. $1 + 1 \neq 0$ **11.** $\sim p \vee q$ **13.** $\sim p \wedge (q \vee r)$ **15.** T

17. F **19.** T **21.** F

23.

p	q	$\sim p$	$\sim q$	$\sim p \vee \sim q$
T	T	F	F	F
T	F	F	T	T
F	T	T	F	T
F	F	T	T	T

25.

p	q	$p \vee q$	$\sim q$	$(p \vee q) \vee \sim q$
T	T	T	F	T
T	F	T	T	T
F	T	T	F	T
F	F	F	T	T

27. F **29.** T

31.

p	q	p XOR q
T	T	F
T	F	T
F	T	T
F	F	F

33. If two lines are perpendicular to the same line, then they are parallel.

35. If $x = 1$, then $x^2 = 1$.

37. *converse*: If Paris is in England, then London is in France.
inverse: If London is not in France, then Paris is not in England.
contrapositive: If Paris is not in England, then London is not in France.

39. $\sim q \vee p \to r$ **41.** $\sim p \wedge \sim q \leftrightarrow \sim r$ **43.** no

45. yes **47.** T **49.** F

51. T **53.** T

55.

p	q	$p \wedge q$	$\sim p$	$p \wedge q \to \sim p$
T	T	T	F	F
T	F	F	F	T
F	T	F	T	T
F	F	F	T	T

57.

p	q	$p \vee q$	$p \wedge q$	$p \vee q \leftrightarrow p \wedge q$
T	T	T	T	T
T	F	T	F	F
F	T	T	F	F
F	F	F	F	T

59. yes **61.** yes **63.** yes **65.** no

67. $(p \wedge q) \to ((\sim p) \vee (\sim q))$ **69.** $(p \to q) \leftrightarrow ((\sim p) \vee q)$

71. $(A' \wedge C) \vee (B' \vee C') \vee (A \wedge B)$

73.

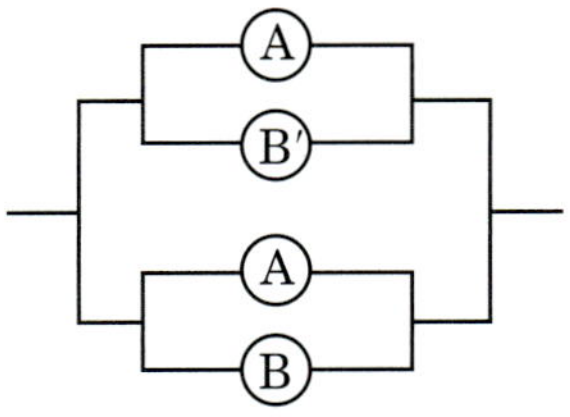

75.

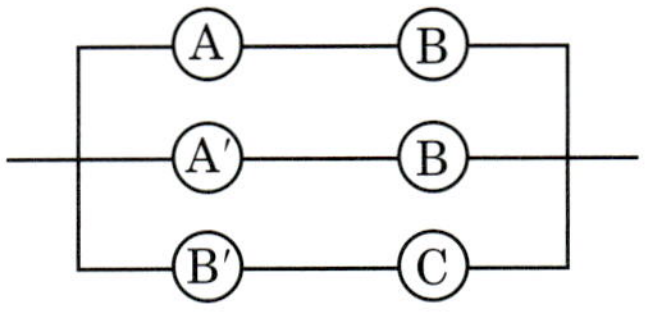

77. Ellen must use computer 1.

Exercises 1.2 (p. 29)

1. F

3.

p	$\sim p$	$\sim(\sim p)$
T	F	T
F	T	F

identical

5.

p	$p \vee p$
T	T
F	F

7.

p	q	$p \vee q$	$q \vee p$
T	T	T	T
T	F	T	T
F	T	T	T
F	F	F	F

9.

p	q	$p \to q$	$\sim(p \to q)$	$\sim q$	$p \wedge \sim q$
T	T	T	F	F	F
T	F	F	T	T	T
F	T	T	F	F	F
F	F	T	F	T	F

↑——identical——↑ ($\sim(p \to q)$ and $p \wedge \sim q$)

11.

p	q	r	$q \wedge r$	$p \wedge (q \wedge r)$	$p \wedge q$	$(p \wedge q) \wedge r$
T	T	T	T	T	T	T
T	T	F	F	F	T	F
T	F	T	F	F	F	F
T	F	F	F	F	F	F
F	T	T	T	F	F	F
F	T	F	F	F	F	F
F	F	T	F	F	F	F
F	F	F	F	F	F	F

↑——identical——↑ ($p \wedge (q \wedge r)$ and $(p \wedge q) \wedge r$)

13.

p	q	r	$p \vee q$	$(p \vee q) \to r$	$p \to r$	$q \to r$	$(p \to r) \wedge (q \to r)$
T	T	T	T	T	T	T	T
T	T	F	T	F	F	F	F
T	F	T	T	T	T	T	T
T	F	F	T	F	F	T	F
F	T	T	T	T	T	T	T
F	T	F	T	F	T	F	F
F	F	T	F	T	T	T	T
F	F	F	F	T	T	T	T

↑——identical——↑ ($(p \vee q) \to r$ and $(p \to r) \wedge (q \to r)$)

15. T **17.** F **19.** yes **21.** no **23.** yes **25.** yes

27.

p	q	p NOR q
T	T	F
T	F	F
F	T	F
F	F	T

29. T **31.** T **33.** T **35.** F

37. T **39.** T **41.** F

43. $\sim(\sim p \vee q) \equiv \sim(\sim p) \wedge \sim q$
$\equiv p \wedge \sim q$

45. $\sim(p \wedge \sim q) \equiv \sim p \vee \sim(\sim q)$
$\equiv \sim p \vee q$

47. $p \rightarrow \sim q \equiv \sim p \vee \sim q$
$\equiv \sim(p \wedge q)$

49. $p \wedge q$ **51.** $p \vee q$ **53.** $p \wedge \sim q$

55. $(A \wedge B') \vee [(A \wedge B) \vee C] \equiv A \vee C$

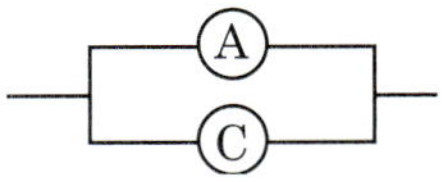

57.

p	$\sim p$	$p\|p$
T	F	F
F	T	T

59.

p	q	$p \wedge q$	$p\|q$	$(p\|q)\|(p\|q)$
T	T	T	F	T
T	F	F	T	F
F	T	F	T	F
F	F	F	T	F

↑—identical—↑

61.

p	q	$p \rightarrow q$	$p\|p$	$(p\|p)\|(p\|p)$	$q\|q$	$((p\|p)\|(p\|p))\|(q\|q)$
T	T	T	F	T	F	T
T	F	F	F	T	T	F
F	T	T	T	F	F	T
F	F	T	T	F	T	T

↑—identical—↑

63. $p \text{ XOR } q \equiv (((p|p)|(q|q))|(((p|q)|(p|q))|((p|q)|(p|q))))|$
$(((p|p)|(q|q))|(((p|q)|(p|q))|((p|q)|(p|q))))$

65. 1 **67.** 1 **69.** 0.3 **71.** 0.7 **73.** 0 **75.** 0.5

77. $\begin{cases} 1-x & \text{if } 0 \le x \le 1/2 \\ x & \text{otherwise} \end{cases}$

Exercises 1.3 (p. 36)

1. F **3.** T **5.** T **7.** F **9.** F **11.** T

13. $(\exists x)(P(x) \wedge Q(x))$ **15.** $(\exists x)(P(x) \vee \sim Q(x))$ **17.** $(\exists x)\,(x^2 \leq 0)$

19. Some super-computers are not manufactured in Japan.

21. $(\forall x)(\forall y)(xy > 0)$ **23.** $(\forall x)(\exists y)(xy = x)$

25. F **27.** T

29. The square of every integer is nonnegative.

31. There are integers x and y such that $x + y = 7$.

33. There is an integer x such that $y - x = y$ for every integer y.

35. T **37.** T **39.** T **41.** T **43.** F **45.** T **47.** T

49. T **51.** T **53.** T **55.** F **57.** F **59.** T **61.** T

63. T **65.** F **67.** T

Exercises 1.4 (p. 45)

1. $p \rightarrow q$
$\sim q$
______
$\therefore \sim p$

3.

p	q	$p \vee q$	$p \rightarrow (p \vee q)$
T	T	T	T
T	F	T	T
F	T	T	T
F	F	F	T

5. invalid

7. valid

9. valid

11. valid

13. q is false.

15. r is true.

17. The program is running.

19. Carol is a baby.

21. Benjamin to Cindy and Aaron to Daphne; they are 34, 27, 28, and 29 years old, respectively.

23. blue

25. B

27. A is a knight and B a knave.

29. yes

31. knight

33. "I am red."

35. Kitty is guilty.

37. At least one cowgirl will escape injury.

Exercises 1.5 (p. 54)

1. yes

3. yes

5. Let x and y be any two even integers. Then $x = 2m$ and $y = 2n$ for some integers m and n. Then $x + y = 2m + 2n$
$= 2(m + n)$
which is also an even integer.

11. Let x be any even integer and y any odd integer. Then $x = 2m$ and $y = 2n + 1$ for some integers m and n. Then $xy = (2m)(2n + 1) = 2(2mn + m)$, an even integer.

15. Let x be any integer. Assume it is not even; it is odd and is of the form $2m + 1$. Then
$x^2 = (2m + 1)^2 = 4m^2 + 4m + 1$
$= 2(2m^2 + 2m) + 1$
which is an odd integer. So the given hypothesis is false and the result follows.

19. Suppose $\sqrt{2}$ is not an irrational number; that is, $\sqrt{2}$ is a rational number. Let $\sqrt{2} = a/b$, where a and b have no positive common factors except 1. Then $(a/b)^2 = 2$ or $a^2 = 2b^2$. $\therefore$ 2 is a factor of a^2 and hence of a. Then $a = 2m$ for some integer m. $\therefore (2m)^2 = 2b^2$ or $b^2 = 2m^2$. $\therefore$ 2 is a factor of b^2 and hence of b. Consequently, 2 is a factor of both a and b, which contradicts the assumption.

25. proof: Every integer n is of the form $3k$, $3k + 1$, or $3k + 2$.
case 1 Let $n = 3k$. Then
$n^3 - n = (3k)^3 - (3k)$
$= 27k^3 - 3k = 3(9k^3 - k)$
case 2 Let $n = 3k + 1$. Then
$n^3 - n = (3k + 1)^3 - (3k + 1)$
$= (27k^3 + 27k^2 + 9k + 1)$
$-(3k + 1)$
$= 27k^3 + 27k^2 + 6k$
$= 3(9k^3 + 9k^2 + 2k)$

7. Let $x = 2m$ be any even integer. Then $x^2 = (2m)^2 = 2(2m^2)$ is also an even integer.

9. Let x be any odd integer. Then $x = 2m + 1$ for some integer m.
$\therefore x^2 = (2m + 1)^2$
$= 2(2m^2 + 2m) + 1$
which is an odd integer.

13. Let $x = 4k + 1$. Then
$x^2 = (4k + 1)^2 = 16k^2 + 8k + 1$
$= 4(4k^2 + 2k) + 1 = 4m + 1$,
where $m = 4k^2 + 2k$ is an even integer.
$\therefore x^2$ is also an integer of the same form.

17. Let x and y be any two integers. Assume that the given conclusion is false; that is, assume that both x and y are odd integers. Then $x = 2m + 1$ and $y = 2n + 1$ for some integers m and n. $\therefore xy = (2m + 1)(2n + 1)$
$= 2(2mn + m + n) + 1$, an odd integer. This negates the given hypothesis and so the result follows.

21. Assume $\sqrt{p}$ is a rational number a/b, where a and b have no positive common factors except 1. Then $\sqrt{p} = a/b$, $(a/b)^2 = p$ or $a^2 = pb^2$. Consequently, p is a factor of a^2 and hence of a. So let $a = mp$. Then $(mp)^2 = pb^2$ or $b^2 = pm^2$. As before, this shows p is a factor of b. Thus p is a common factor of a and b, a contradiction.

23. proof:
case 1 n is an even integer, say, $2m$. Then $n^2 + n = (2m)^2 + 2m = 2(2m^2 + m)$, an even integer.
case 2 n is an odd integer, say, $2m + 1$.

case 3 Let $n = 3k + 2$. Then
$$n^3 - n = (3k+2)^3 - (3k+2) = (27k^3 + 54k^2 + 36k + 8) - (3k+2) = 27k^3 + 54k^2 + 33k + 6 = 3(9k^3 + 18k^2 + 11k + 2)$$
Thus, in every case, $n^3 - n$ is divisible by 3.

29. **proof:** Choose $x = 1$. Clearly, $x^2 = x$.

31. **proof:** Let a be any nonzero integer.
Then $1729a^3 = a^3 + (12a)^3 = (9a)^3 + (10a)^3$
Since a is arbitrary, $1729a^3$ has infinitely many choices.

33. 0 **35.** 2

37. $x = 2$, $P(x)$: $x^2 = 4$.

39. Canceling $a - b$ is invalid since $a = b$.

43. **proof:** Let $a \cdot b = 0$ and $a \neq 0$. Since $a \cdot 0 = 0$, $a \cdot b = a \cdot 0$. Canceling a from both sides, we get $b = 0$.

Then
$$n^2 + n = (2m+1)^2 + (2m+1) = (4m^2 + 4m + 1) + (2m + 1) = 2(2m^2 + 3m + 1),$$
again, an even integer.

27. **proof:**
case 1 Let $x, y \geq 0$. Then
$|x \cdot y| = xy = |x| \cdot |y|$
case 2 Let $x < 0$, $y < 0$. Then
$|x| = -x$ and $|y| = -y$.
$|x \cdot y| = xy = (-x) \cdot (-y) = |x| \cdot |y|$
case 3 Let $x \geq 0$, $y < 0$. Then
$|x \cdot y| = -(xy) = x \cdot (-y) = |x| \cdot |y|$
case 4 Let $x < 0$, $y \geq 0$.
This case is similar to case 3.

41. **proof:** Since $a < b$, there is a positive real number x such that $a + x = b$.
$\therefore (a + x) + c = b + c$
That is, $(a + c) + x = b + c$
$\therefore a + c < b + c$

45. 11, since $2047 = 23 \cdot 89$.

Review Exercises (p. 58)

1.

p	q	$p \vee q$	$\sim q$	$(p \vee q) \wedge (\sim q)$
T	T	T	F	F
T	F	T	T	T
F	T	T	F	F
F	F	F	T	F

3.

p	q	r	$q \rightarrow r$	$p \rightarrow (q \rightarrow r)$
T	T	T	T	T
T	T	F	F	F
T	F	T	T	T

5. T **7.** F

9. $(w < x) \wedge (y < z) \rightarrow (w + y < x + z)$

p	q	r	$q \to r$	$p \to (q \to r)$
T	F	F	T	T
F	T	T	T	T
F	T	F	F	T
F	F	T	T	T
F	F	F	T	T

11. yes **13.** no

15. $[A \wedge (A' \vee B)] \vee (A \wedge B')$

17. T **19.** T **21.** T

23. F **25.** T

27. F **29.** F

31. *converse* : If $x+z < y+z$, then $x < y$.
inverse : If $x \geq y$, then $x+z \geq y+z$.
contrapositive: If $x+z \geq y+z$, then $x \geq y$.

33. If $x \leq 3$ and $x \geq -3$, then $|x| \leq 3$.

35. yes **37.** no **39.** yes **41.** yes **43.** T **45.** T **47.** F

49. $(A' \wedge B) \vee (A' \wedge B') \vee (A \wedge B') \equiv A' \vee B'$

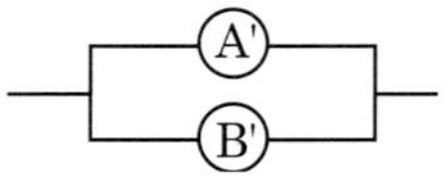

51. invalid **53.** F **55.** T **57.** F **59.** F **61.** F

63. **proof:** Let x and $x+1$ be two consecutive integers.
case 1 Let x be even. Then $x = 2m$ for some integer m. Then $x(x+1) = 2m(2m+1)$
$= 2[m(m+1)]$,
an even integer.
case 2 Let x be odd. Then $x = 2m+1$ for some integer m.
$\therefore\ x(x+1) = (2m+1)(2m+2)$
$= 2[(m+1)(2m+1)]$,
again an even integer.

65. **proof:** $n^4 - n^2 = [(n-1)n(n+1)]n$ contains a product of three consecutive integers. Therefore, by cases, it is divisible by 3.

67. **indirect proof:** Assume that the given conclusion is false, that is, $a \leq 6$ and $b \leq 6$. Then $a+b \leq 6+6$, that is, $a+b \leq 12$, which contradicts the hypothesis.

69. **proof:** Assume that $a^2 = b^2$ and $a \neq b$. Since $a^2 - b^2 = (a-b)(a+b) = 0$ and $a - b \neq 0, a+b = 0$. $\therefore\ a = -b$.

71. 0.5 **73.** 0.5 **75.** 0 **77.** 0.3

79. Since $t(p) = x, t(p') = 1 - x$. Then $t(p \vee p') = \max\{x, 1-x\}$.

Suppose $t(p \vee p') = 1$. Then $\max\{x, 1-x\} = 1$. Suppose $0 \leq x \leq 1/2$. Then $\max\{x, 1-x\} = 1 - x = 1$, so $x = 0$; that is, $t(p) = 0$. On the other hand, let $1/2 < x \leq 1$. Then $\max\{x, 1-x\} = x = 1$; so $t(p) = 1$. In either case, $t(p) = 0$ or 1.

Conversely, let $t(p) = 0$ or 1. Then $t(p \vee p') = \max\{t(p), t(p')\} = \max\{t(p), 1 - t(p)\}$. If $t(p) = 0$, then $t(p \vee p') = \max\{0, 1\} = 1$; on the other hand, if $t(p) = 1$, then also $t(p \vee p') = \max\{1, 0\} = 1$. In both cases, $t(p \vee p') = 1$.

Thus $t(p \vee p') = 1$ if and only if $t(p) = 0$ or $t(p) = 1$.

81. **proof:**
$$\begin{aligned} t((p \vee q)') &= 1 - t(p \vee q) \\ &= 1 - \max\{t(p), t(q)\} \\ &= 1 - \max\{x, y\} \\ &= 1 - \begin{cases} y & \text{if } x \le y \\ x & \text{otherwise} \end{cases} \\ &= \begin{cases} 1-y & \text{if } x \le y \\ 1-x & \text{otherwise} \end{cases} \\ t(p' \wedge q') &= \min\{t(p), t(q')\} \\ &= \min\{1-x, 1-y\} \\ &= \begin{cases} 1-y & \text{if } x \le y \\ 1-x & \text{otherwise} \end{cases} \end{aligned}$$
Thus $t((p \vee q)') = t(p') \wedge t(q')$.

Supplementary Exercises (p. 62)

1. *converse* : If $-a < x < a$, then $|x| < a$.
inverse : If $|x| \ge a$, then $x \ge a$ or $x \le -a$.
contrapositive: If $(x \le -a)$ or $(x \ge a)$, then $|x| \ge a$.

3. $\sim q$ **5.** $\sim p \vee \sim q$ **7.** yes

9. $\sim [(\forall x)(\forall y)(xy = yx) \equiv (\exists x)(\exists y)(xy \ne yx)$

11. $\sim [(\forall x)(\exists y)(\exists z)(x + y = z) \equiv (\exists x)(\forall y)(\forall z)(x + y \ne z)$

13. **proof:** Let $K = (n^2 + n)(n^2 + n + 1)(n^2 + n + 2)$
and $L = (n^2 - n)(n^2 - n + 1)(n^2 - n + 2)$.
$$\begin{aligned} 2n(3n^4 + 7n^2 + 2) &= 6n^5 + 14n^3 + 4n \\ &= (4n^3 + 6n)n^2 + (2n^4 + 8n^2 + 4)n \\ &= (n^2 + n)(n^4 + 2n^3 + 4n^2 + 3n + 2) \\ &\quad - (n^2 - n)(n^4 - 2n^3 + 4n^2 - 3n + 2) \\ &= K - L \end{aligned}$$
Since K is the product of three consecutive integers, it is divisible by 3. But $n^2 + n = n(n+1)$ is divisible by 2. So $n^2 + n + 2$ is divisible by 4. Thus K is divisible by 24. Likewise, L is also divisible by 24. Therefore, $K - L$ is divisible by 24. Thus $n(3n^4 + 7n^2 + 2)$ is divisible by 12.

15. 40

17.
$$\begin{aligned} t(p \vee p') &= \max\{t(p), t(p')\} \\ &= \max\{t(p), 1 - t(p)\} = 1 \end{aligned}$$
only if $t(p) = 0$ or $t(p) = 1$.

19. Not a three-valued tautology.

21. Is a three-valued tautology.

23.

p	q	$p \vee q$	$(p \vee q)'$	p'	q'	$p' \wedge q'$
0	0	0	1	1	1	1
0	u	u	u	1	u	u
0	1	1	0	1	0	0
u	0	u	u	u	1	u
u	u	u	u	u	u	u
u	1	1	0	u	0	0
1	0	1	0	0	1	0
1	u	1	0	0	u	0
1	1	1	0	0	0	0

↑——identical——↑

Chapter 2 The Language of Sets

Exercises 2.1 (p. 76)

1. {April, August} **3.** {Jan, March, May, July, Aug, Oct, Dec}

5. $\{x \in \mathbf{Z} | 0 < x < 5\}$ **7.** $\{x | x$ is a member of the United Nations$\}$

9. yes **11.** yes **13.** F

15. T **17.** F **19.** F **21.** T **23.** T **25.** T **27.** T

29. F **31.** $\{\emptyset, \mathrm{A}\}$ **33.** 2^n

35. $\{3, 6, 9\}$ **37.** $\{2, 3, 4, 6, 8, 9\}$

39. $\{\emptyset, \{1\}, \{2\}, \{3\}, \{1,3\}\}$ **41.** b, b^2, bab, b^3, ba^2b **43.** $\lambda, b, a^2, a^2b, aba$

45. 3 **47.** 5 **49.** 00, 01, 10, 11 **51.** 0, 1, 2

53. Consider the implication $x \in \emptyset \to x \in A$. Since the hypothesis is false, this is a true implication. $\therefore\ \emptyset \subseteq A$.

55. Let x be an arbitrary element in A. Since $A \subseteq B$, $x \in B$. Since $B \subseteq C$, $x \in C$. Thus, every element in A is also in C. $\therefore A \subseteq C$.

Exercises 2.2 (p. 93)

1. {a, b, c, g, j, k} **3.** {d, h} **5.** {a, b, c, f, g, i, j, k}

7. {a, c, d, e, f, h, i, j, k} **9.** {a} **11.** {a, b, g}

13. {w, y, z} **15.** {a, b} **17.** {x, y}

19. $\{(b,x),(c,x)\}$

21. $\emptyset$

23. $\{(b,x),(b,z),(c,x),(c,z)\}$

25. $\{(b,x,x),(b,x,z),(c,x,x),(c,x,z)\}$

27. T

29. F

31. F

33. T

35. F

37. F

39. $A = \{a\}, B = \emptyset, C = \{a\}$

41. $A = \{a\}, B = \{b\}, C = \{a,b\}$

43. no

45. no

47. Let $x \in (A')'$. Then $x \notin A'$, so $x \in A$. Thus $(A')' \subseteq A$. Conversely, let $x \in A$. Then $x \notin A'$, so $x \in (A')'$. $\therefore$ $A \subseteq (A')'$. Thus $(A')' = A$.

49.
$$\begin{aligned} A \cap (A \cup B) &= (A \cap A) \cup (A \cap B) \\ &= A \cup (A \cap B) \\ &= A, \text{ since } A \cap B \subseteq A. \end{aligned}$$

51.
$$\begin{aligned} A \oplus A &= (A - A) \cup (A - A) \\ &= \emptyset \cup \emptyset = \emptyset \end{aligned}$$

53.
$$\begin{aligned} A \oplus B &= (A - B) \cup (B - A) \\ &= (B - A) \cup (A - B) \\ &= B \oplus A \end{aligned}$$

55.
$$\begin{aligned} (A \cup B \cup C)' &= [(A \cup B) \cup C]' \\ &= (A \cup B)' \cap C' \\ &= (A' \cap B') \cap C' \\ &= A' \cap B' \cap C' \end{aligned}$$

57. $A \cap (A - B) = A - B$

59. $(A - B') - (B - A') = \emptyset$

61. $A \cup B - (A \cap B)' = A \cap B$

63. $(A \cap B)' \cup (A \cup B') = U$

65. $(A' \cup B')' \cup (A' \cap B) = B$

67.
$$A \cup \left(\bigcap_{i \in I} B_i\right) = \bigcap_{i \in I} (A \cup B_i)$$
$$A \cap \left(\bigcup_{i \in I} B_i\right) = \bigcup_{i \in I} (A \cap B_i)$$

69. $\emptyset$

71. {Angelo 0.4, Bart 0.7, Cathy 0.6, Dan 0.7, Elsie 0.2, Frank 0.6}

73. $\emptyset$

75. {Angelo 0.4, Bart 0.7, Cathy 0.6}

77. {(Angelo, Dan) 0.3, (Angelo, Elsie) 0.4, (Angelo, Frank) 0.4, (Bart, Dan) 0.3, (Bart, Elsie) 0.7, (Bart, Frank) 0.4, (Cathy, Dan) 0.3, (Cathy, Elsie) 0.6, (Cathy, Frank) 0.4}

79. {(Angelo, Angelo) 0.4, (Angelo, Bart) 0.4, (Angelo, Cathy) 0.4, (Bart, Angelo) 0.4, (Bart, Bart) 0.7, (Bart, Cathy) 0.6, (Cathy, Angelo) 0.4, (Cathy, Bart) 0.6, (Cathy, Cathy) 0.6}

81. **proof:** By De Morgan's law in ordinary sets, it suffices to show that $d_{(A \cap B)'}(x) = d_{A' \cup B'}(x)$ for every element x.

Let $x \in (A \cap B)'$. Then

$$d_{(A\cap B)'}(x) = 1 - d_{(A\cap B)}(x) = 1 - \min\{d_A(x), d_B(x)\} = \begin{cases} 1 - d_A(x) & \text{if } d_A(x) \le d_B(x) \\ 1 - d_B(x) & \text{otherwise} \end{cases}$$

On the other hand, let $x \in A' \cup B'$. Then

$$d_{A'\cup B'}(x) = \max\{d_{A'}(x), d_{B'}(x)\} = \begin{cases} 1 - d_A(x) & \text{if } d_A(x) \le d_B(x) \\ 1 - d_B(x) & \text{otherwise} \end{cases}$$

Thus $d_{(A\cap B)'}(x) = d_{A'\cup B'}(x)$ for every element x, so $(A\cap B)' = A'\cup B'$ for fuzzy sets.

Exercises 2.3 (p. 98)

1. 01010101 **3.** 11110001 **5.** $\{s_0, s_3\}$ **7.** $\{s_0, s_2, s_3\}$

9. $\{\emptyset, \{s_0\}, \{s_1\}, \{s_0, s_1\}$ **11.** 10010010 **13.** 01001001

15. 01001000 **17.** 11011010 **19.** 10010011

21. 00000000 **23.** 00000001 **25.** 01001001

Exercises 2.4 (p. 102)

1. 26 **3.** 7 **5.** $3a$ **7.** $a+2b$ **9.** $a+2b$ **11.** $a-b$

13. 0 **15.** b **17.** a **19.** b **21.** $b-a$ **23.** 50

25. 366 **27.** 134 **29.** 1007 **31.** 1244 **33.** 55 **35.** 10

37. 13 **39.** 230 **41.** 60

43. {Ø, {a}, {b}, {c}, {a, b}, {b, c}, {c, a}, {a, b, c}}

45. $4b$ **47.** b^2 **49.** $|U| - |A \cup B \cup C|$

51.

$$\begin{aligned} |A_1 \cup A_2 \cup A_3 \cup A_4| &= |A_1 \cup (A_2 \cup A_3 \cup A_3)| \\ &= |A_1| + |A_2 \cup A_3 \cup A_4| - |A_1 \cap (A_2 \cup A_3 \cup A_3)| \\ &= |A_1| + [|A_2| + |A_3| + |A_4| - |A_2 \cap A_3| \\ &\quad -|A_2 \cap A_4| - |A_3 \cap A_4| - |A_2 \cap A_3 \cap A_4|] \\ &\quad -|A_1 \cap (A_2 \cup A_3 \cup A_3)| \qquad (1) \end{aligned}$$

$$\begin{aligned} |A_1 \cap (A_2 \cup A_3 \cup A_3)| &= |(A_1 \cap A_2) \cup (A_1 \cap A_3) \cup (A_1 \cap A_4)| \\ &= |A_1 \cap A_2| + |A_1 \cap A_3| + |A_1 \cap A_4| \\ &\quad -|(A_1 \cap A_2) \cap (A_1 \cap A_3)| \\ &\quad -|(A_1 \cap A_2) \cap (A_1 \cap A_4)| - |(A_1 \cap A_3) \cup (A_1 \cap A_4)| \\ &\quad +|(A_1 \cap A_2) \cap (A_1 \cap A_3) \cap (A_1 \cap A_4)| \end{aligned}$$

$$= |A_1 \cap A_2| + |A_1 \cap A_3| + |A_1 \cap A_4| - |A_1 \cap A_2 \cap A_3| - |A_1 \cap A_2 \cap A_4| - |A_1 \cap A_3 \cap A_4| + |A_1 \cap A_2 \cap A_3 \cap A_4| \quad (2)$$

Substituting (2) in (1), we get the desired result:

$$|\bigcup_{i=1}^{4} A_i| = \sum_{i=1}^{4} |A_i| - \sum_{1 \le i < j \le 4} |A_i \cap A_j| + \sum_{1 \le i < j < k \le 4} |A_i \cap A_j \cap A_k| - |\bigcap_{i=1}^{4} A_i|$$

Exercises 2.5 (p. 107)

1. $1, 2, 4, 8$ **3.** $e, e^e, e^{e^e}, e^{e^{e^e}}$ **5.** λ, b^2, b^4, b^6 **7.** $\mathbf{Z}^+$

9. set of even integers

11. 1) $0 \in L$
2) $x \in L \to 1x, xx \in L$

13. 1) $a \in L$
2) $x \in L \to bxb \in L$

15. 1) $b \in L$
2) $x \in L \to bx \in L$

17. 1) $a \in L$
2) $x \in L \to axa \in L$

19. yes

21. no **23.** 5 **25.** 42

27. ( )( )( ),(( )( )),(( ))( ),( )(( )),((( )))

29. yes **31.** no

33. no, since (( ))(( )) cannot be generated.

35. 1) $\lambda \in \Sigma^*$
2) $x \in \Sigma^*, y \in \Sigma \to xy, yx \in \Sigma^*$

Review Exercises (p. 111)

1. {b, y, z} **3.** {b, f, x, y} **5.** {b, c, y} **7.** {b, f, x, y}

9. {a, c, d, f}, {b, d, e, f, g} **11.** {b, e, k}

13. {Ø,{Ø},{{Ø}},{Ø,{Ø}}}

15. no **17.** 01000100 **19.** 01100110 **21.** 54%

23. 21 **25.** 7 **27.** 2871

29. $3, \lg 3, \lg\lg 3, \lg\lg\lg 3$

31. $1, \sqrt{3}, \sqrt{1 + 2\sqrt{3}}, \sqrt{1 + 2\sqrt{1 + 2\sqrt{3}}}$

33. 1) $1 \in S$
2) $x \in S \to 1 + 2x \in S$

35. 1) $\lambda \in S$
2) $x \in S \to bxa \in S$

37. a, ab, ba, abb, bab

39. 1) $a, b \in L$
2) $x \in L \to ax, bx \in L$

41. 1) $1 \in L$
2) $x \in L \to 0x0 \in L$

43. yes

45. Let $x \in A \cup (B \cap C)$. Then $x \in A$ or $x \in B \cap C$. $\therefore x \in A$ or ($x \in B$ and $x \in C$). $\therefore x \in A \cup B$ and $x \in A \cup C$. Thus, $x \in (A \cup B) \cap (A \cup C)$. So $A \cup (B \cap C) \subseteq (A \cup B) \cap (A \cup C)$. Similarly, $(A \cup B) \cap (A \cup C) \subseteq A \cup (B \cap C)$ Hence the result.

47. $A \cap (B - C) = A \cap (B \cap C') = (A \cap B) \cap C' = (A \cap B) - C = (A \cap B) - (A \cap C)$

49. $\emptyset$

51. $\emptyset$

53. {(Mike, Jean) 0.6, (Mike, June) 0.5, (Andy, Jean) 0.3, (Andy, June) 0.3, (Jeff, Jean) 0.7, (Jeff, June) 0.5}

55. As ordinary sets, $(A - B)' = (A \cap B')' = A' \cup B$. So is remains to show that $d_{(A-B)'}(x) = d_{A' \cup B}(x)$ for every element x.

$$
\begin{aligned}
d_{(A-B)'}(x) &= 1 - d_{A-B}(x) \\
&= 1 - d_{A \cap B'}(x) \\
&= 1 - \min\{d_A(x), d_{B'}(x)\} \\
&= 1 - \min\{d_A(x), 1 - d_B(x)\} \\
&= \begin{cases} 1 - d_A(x) & \text{if } d_A(x) \le 1 - d_B(x) \\ d_B(x) & \text{otherwise} \end{cases}
\end{aligned}
$$

Similarly,

$$
d_{A' \cup B}(x) = \begin{cases} 1 - d_A(x) & \text{if } d_A(x) \le 1 - d_B(x) \\ d_B(x) & \text{otherwise} \end{cases}
$$

Thus $d_{(A-B)'}(x) = d_{A' \cup B}(x)$ for every element x; so $(A - B)' = A' \cup B$ as fuzzy sets.

Supplementary Exercises (p. 113)

1. $(A - B) \cap (A - C)$

$$
\begin{aligned}
&= (A \cap B') \cap (A \cap C') \\
&= A \cap B' \cap C' \\
&= A \cap (B' \cap C') \\
&= A \cap (B \cup C)' \\
&= A - (B \cup C)
\end{aligned}
$$

3. $(A \cap B) \oplus (A \cap C)$

$$
\begin{aligned}
&= [(A \cap B) - (A \cap C)] \cup [(A \cap C) - (A \cap B)] \\
&= [(A \cap B) \cap (A \cap C)'] \cup [(A \cap C) \cap (A \cap B)'] \\
&= [(A \cap B) \cap (A' \cup C)'] \cup [(A \cap C) \cap (A' \cup B')] \\
&= [(A \cap B \cap A') \cup (A \cap B \cap C')] \cup [(A \cap C \cap A') \cup (A \cap C \cap B')] \\
&= \emptyset \cup (A \cap B \cap C') \cup \emptyset \cup (A \cap C \cap B') \\
&= (A \cap B \cap C') \cup (A \cap C \cap B') \\
&= A \cap [(B \cap C') \cup (C \cap B')] \\
&= A \cap [(B - C) \cup (C - B)] \\
&= A \cap (B \oplus C)
\end{aligned}
$$

5. $A \cap B \cap C$ **7.** $A' \cap B'$ **9.** 266

11. 1) $\lambda \in L$
2) If $x \in L$, then $0x1 \in L$.

13. Let $x = x_1 \ldots x_n$ and $y = y_1 \ldots y_n$.

$$
\begin{aligned}
xy &= x_1 \ldots x_n y_1 \ldots y_n \\
(xy)^{\mathrm{R}} &= y_n \ldots y_1 x_n \ldots x_1 \\
&= y^{\mathrm{R}} x^{\mathrm{R}}
\end{aligned}
$$

15. By Exercise 14, it suffices to show that $(xx^{\mathrm{R}})^{\mathrm{R}} = xx^{\mathrm{R}}$.
$(xx^{\mathrm{R}})^{\mathrm{R}} = (x^{\mathrm{R}})^{\mathrm{R}} x^{\mathrm{R}}$, by Exercise 13
$= xx^{\mathrm{R}}$, since $(x^{\mathrm{R}})^{\mathrm{R}} = x$.

Chapter 3 Functions and Matrices

Exercises 3.1 (p. 123)

1. $-40°$F **3.** 9.8 **5.** 5 **7.** \$75
9. \$84.60 **11.** 6 **13.** 7 **15.** combinatorics
17. & **19.** Z **21.** 43 **23.** 122
25. -23 **27.** 41 **29.** azalea **31.** aroma
33. 5 **35.** undefined **37.** 6 **39.** $1 + p + q$
41. 5 **43.** $x^2 + 2x + 2$

45. **proof:** Let $y \in f(A \cup B)$. Then there is an element $x \in A \cup B$ such that $y = f(x)$. If $x \in A$, then $y = f(x) \in f(A)$; if $x \in B$, then $y = f(x) \in f(B)$. In either case, $y \in f(A) \cup f(B)$. $\therefore f(A \cup B) \subseteq f(A) \cup f(B)$.

Conversely, let $y \in f(A) \cup f(B)$. Then either $y \in f(A)$ or $y \in f(B)$. If $y \in f(A)$, then there is an element $a \in A$ such that $y = f(a)$. Since $a \in A \cup B$, $y = f(a) \in f(A \cup B)$. If $y \in f(B)$, then similarly, $y \in f(A \cup B)$. Thus, in both cases, $y \in f(A \cup B)$. $\therefore f(A) \cup f(B) \subseteq f(A \cup B)$.

47. proof: Let $y \in f(A) - f(B)$. Then $y \in f(A)$ and $y \notin f(B)$. Then there is an element $a \in A$ such that $y = f(A)$; since $y \neq f(B), a \notin B$. $\therefore a \in A - B$. Consequently, $y = f(a) \in f(A - B)$. $\therefore f(A) - f(B) \subseteq f(A - B)$.

Exercises 3.2 (p. 134)

1. n **3.** $n+1$ **5.** 6 **7.** 9 **9.** -4

11. 7 **13.** $\{-1, 0\}$ **15.** 1538 **17.** 1435 **19.** 97

21. 358 **23.** $h(\mathrm{a}) = h(\mathrm{c}) = h(\mathrm{d}) = 1, h(\mathrm{b}) = h(\mathrm{e}) = 0$

25. $h(\mathrm{b}) = h(\mathrm{c}) = h(\mathrm{g}) = 1, h(\mathrm{a}) = h(\mathrm{d}) = h(\mathrm{e}) = h(\mathrm{f}) = 0$

27. {c,e,g,h} **29.** {b,d,f,h} **31.** Thursday **33.** Monday

35. T **37.** T **39.** Wednesday **41.** Saturday

43. Wednesday **45.** 1 **47.** 2 **49.** April 7 **51.** April 26

53. Let $n = 2k + 1$. Then $\lfloor n/2 \rfloor = \lfloor k + 1/2 \rfloor = k$

55. Let $n = 2k + 1$. Then $\lfloor n^2/4 \rfloor = \lfloor k^2 + k + 1/4 \rfloor = k^2 + k = (4k^2 + 4k)/4 = (n^2 - 1)/4$

57. case 1 Let $n = 2k + 1$. Then $\lfloor n/2 \rfloor = k$ and $\lceil n/2 \rceil = k + 1$. $\therefore \lfloor n/2 \rfloor + \lceil n/2 \rceil = k + (k + 1) = 2k + 1 = n$

case 2 Let $n = 2k$. Then $\lfloor n/2 \rfloor = k = \lceil n/2 \rceil$. So $\lfloor n/2 \rfloor + \lceil n/2 \rceil = k + k = 2k = n$.

59. case 1 Let $x \in \mathbf{Z}$. Then $\lceil x \rceil = x = -(-x) = -\lfloor -x \rfloor$.

case 2 Let $x \notin \mathbf{Z}$. Then $x = k + x'$, where $k \in \mathbf{Z}$ and $0 < x' < 1$. $-\lfloor -x \rfloor = -(-k - 1) = k + 1 = \lceil x \rceil$

61. case 1 Let $x \in A \cap B$. Then $x \in A$, $x \in B$, and $x \in A \cup B$. So, $f_{A \cap B}(x) = f_A(x) = f_B(x) = f_{A \cup B}(x) = 1$ $\therefore$ RHS $= 1 + 1 - 1 = 1 =$ LHS.

case 2 Let $x \notin A \cap B$. Then $x \notin A$ or $x \notin B$. If $x \notin A$ and $x \in B$, then RHS $= 0 + 1 - 0 = 1 = 1 =$ LHS. If $x \notin A$ and $x \notin B$, then RHS $= 0 + 0 - 0 = 0 =$ LHS. If $x \in A$ and $x \notin B$, then RHS $= 1 + 0 - 0 = 1 =$ LHS.

63. case 1 Let $x \in A \cap B$. Then $x \in A$, $x \in B$, and $x \notin A \oplus B$. $\therefore$ RHS $= 1 + 1 - 2 = 0 =$ LHS.

case 2 Let $x \notin A \cap B$. Then $x \notin A$ or $x \notin B$. If $x \notin A$ and $x \in B$, then RHS $= 0 + 1 - 0 = 1 =$ LHS. If $x \notin A$ and $x \notin B$, then RHS $= 0 + 0 - 0 = 0 =$ LHS. If $x \in A$ and $x \notin B$, then RHS $= 1 + 0 - 0 = 1 =$ LHS.

65. **case 1** Let $x \geq y$. Then $\max\{x, y\} = x$ and $\min\{x, y\} = y$. $\max\{x, y\} - \min\{x, y\} = x - y = |x - y|$

case 2 Let $x < y$. Then $\max\{x, y\} = y$ and $\min\{x, y\} = x$. $\max\{x, y\} - \min\{x, y\} = y - x = |y - x| = |x - y|$

Exercises 3.3 (p. 141)

1. yes **3.** no **5.** yes **7.** no **9.** no

11. no **13.** yes **15.** no **17.** no **19.** no

21. no **23.** no, not injective. **25.** no, not surjective. **27.** yes

29. 153 **31.** 190 **33.** 12

35.

A	B	C	D	E	F	G	H	I	J	K	L	M
AL	AZ	CA			FL							MA

N	O	P	Q	R	S	T	U	V	W	X	Y	Z
NY	OH	MI										

37. **proof:** Let $f : A \to A$ such that $f(a) = a\ \forall a \in A$. Since f is the identity function on A, it is bijective. $\therefore A \sim A$.

41. **proof:** Let $f : [a,b] \to [c,d]$ defined by
$f(x) = c + \dfrac{d-c}{b-a}(x-a)$.
Notice that $f(a) = c$ and $f(b) = d$.
To prove that f is injective: Let $\alpha, \beta \in [a,b]$ such that $f(\alpha) = f(\beta)$. Then
$c + \dfrac{d-c}{b-a}(\alpha - a) = c + \dfrac{d-c}{b-a}(\beta - a)$
This yields $\alpha = \beta$; so f is injective.
To prove that f is surjective: Let $\beta \in [c,d]$. Choose $\alpha = a + \dfrac{b-a}{d-c}(\beta - c)$. You may verify that $a \leq \alpha \leq b$ and $f(\alpha) = \beta$. $\therefore f$ is surjective.
Thus f is bijective and hence $[a,b] \sim [c,d]$.

39. **proof:** Since $A \sim B$, there is a bijection $f : A \to B$. Let $g : A \times \{1\} \to B \times \{2\}$ defined by $g(a,1) = (f(a),2)$. Clearly, g is well-defined.
To show that g is injective: Let a' and a'' be any two elements in A such that $g(a',1) = g(a'',1)$. Then $(f(a'),2) = (f(a''),2)$. $\therefore a' = a''$, since f is injective. So, $(a',1) = (a'',1)$ and hence g is injective.
To show that g is surjective: Let $(b,2)$ be any element in $B \times \{2\}$. Since f is surjective, there is an element $a \in A$ such that $f(a) = b$. Then $x = (a,1) \in A \times \{2\}$ and
$g(x) = g(a,1) = (f(a),2) = (b,2)$.
$\therefore g$ is surjective.
Thus g is a bijection and hence the result.

43. **proof:** Let $f : Z \to \mathbb{N}$ defined by

$$f(n) = \begin{cases} 2n+1 & \text{if } n \geq 0 \\ -2n & \text{if } n < 0 \end{cases}$$

You may verify that f is injective.
To prove that f is surjective: Let $m \in \mathbb{N}$.
case 1 Let m be even. Then choose $x = -m/2$. Clearly, $x \in Z$ and $f(x) = -2(-m/2) = m$.
case 2 Let m be odd. Then choose $x = \dfrac{m-1}{2}$. Again, $x \in Z$. Also $f(x) = 2\left(\dfrac{m-1}{2}\right) + 1 = m$. $\therefore f$ is surjective. Thus f is bijective and hence $Z \sim \mathbb{N}$.

49. **proof:** Let $A_1, A_2, \ldots$ be a countable number of countable sets.
To show that $A = \cup A_i$ *is countable*: Since each A_i is countable, its elements can be listed as $a_{i1}, a_{i2}, \ldots$. Now list all elements a_{ij} in A with $i + j = 2$, then list all elements a_{ij} with $i + j = 4$, and so on. This procedure establishes a systematic procedure for listing all elements in A.
$\therefore A$ is countable.

45. **proof:** Let $A = \{a_1, a_2, \ldots, a_n, \ldots\}$ be an infinite set. Then $B = \{a_1, a_2, a_4, \ldots\}$ is a proper subset of A. The function $f : A \to B$ defined by $f(a_i) = a_{2i}$ is a bijection (verify). $\therefore A \sim B$.
Conversely, assume that A is finite, $B \subset A$, and $A \sim B$. Since A is finite and $B \subset A$, $|B| < |A|$. But since $A \sim B$, $|A| = |B|$. This is a contradiction. $\therefore A$ is infinite.

47. **proof:** Let $f : Q^+ \to \mathbb{N} \times \mathbb{N}$ defined by $f(m/n) = (m, n)$, where m and n are relatively prime. Clearly, f is bijective.
$\therefore Q^+ \sim \mathbb{N} \times \mathbb{N}$ and hence Q^+ is countable.

51. **proof:** Let $|\Sigma| = n$. Let Σ_i denote the set of all words containing i symbols over Σ. Clearly, Σ_i is countable (in fact finite) and $|\Sigma_i| = n^i$. Since $\Sigma^* = \bigcup_{i \in W} \Sigma_i$ is a countable union of countable sets, Σ^* is countable.

Exercises 3.4 (p. 149)

1. No. of digits in the integer $m = 11$
No. of distinct digits available $n = 10$
$\therefore$ Result follows by the pigeonhole principle.

3. Apply the pigeonhole principle with $m = 6$ and $n = 5$.

5. Choose $m = 8$ and $n = 7$.

7. Choose $m = 7$ and $n = 6$.

9. Apply the generalized pigeonhole principle with $m = 12{,}305$ and $n = 13$.

11. Let $x = 0 \cdot a_1a_2 \cdots a_i\overline{b_1b_2 \cdots b_j}$.

$\therefore\ 10^i x = a_1a_2 \cdots a_i \cdot \overline{b_1b_2 \cdots b_j}$

$10^{i+j}x = a_1a_2 \cdots a_ib_1b_2 \cdots b_j.\overline{b_1b_2 \cdots b_j}$

$\therefore\ 10^{i+j}x - 10^i x = a_1a_2 \cdots a_ib_1b_2 \cdots b_j - a_1a_2 \cdots a_i$

$$\therefore \quad x = \frac{a_1a_2 \cdots a_ib_1b_2 \cdots b_j - a_1a_2 \cdots a_i}{10^{i+j} - 10^i}$$

which is a rational number.

13. Join the midpoints of the sides to form four small congruent

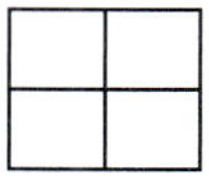

squares, each of side 1/2.
By the PHP, at least two of the five points must lie inside a small square. Since its diagonal is $\sqrt{2}/2$ units long, the distance between the points must be $\sqrt{2}/2$.

17. Let $|X| = m$, $|Y| = n$, and $S_y = \{f^{-1}(y) | y \in Y\}$. Then $\{S_y\}$ partitions X into b blocks, where $b \leq n$. Suppose $|S_y| < m/b\ \forall y$. Since X is a disjoint union of the blocks S_y,

$$m = |x| = \sum_{i=1}^{b} |S_y| < b\left(\frac{m}{b}\right) = m,$$

a contradiction. Therefore, one of the blocks S_y has at least m/b elements. In other words, there is an element $t \in Y$ such that $f^{-1}(t)$ has at least m/b elements.

Since $b \leq n$ and $m > kn$, $\dfrac{m}{b} > \dfrac{kn}{n} = k$. So $|f^{-1}(t)| > k$. Hence the result. (Note that $k = 1$ yields the PHP.)

15. Divide the triangle into nine small congruent triangles:

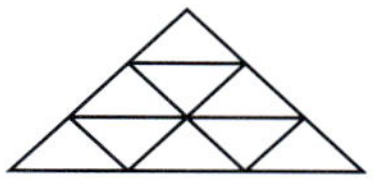

By the PHP, at least two of the ten points must lie inside the same small triangle. Since each side is 1/3 units long, the result follows.

19. Let S be a finite set. Assume $|S|$ is not unique, say, $|S| = m$ and $|S| = n$.
To show that $m = n$: Suppose $m > n$. Then, by the PHP, no injective function exists on S. This is a contradiction, since the identity function 1_S is injective. $\therefore m \leq n$. Similarly, $n \leq m$. Thus $m = n$.

Exercises 3.5 (p. 155)

1. 10 **3.** $4x^2 - 4x + 2$ **5.** -3 **7.** -5

9. 0

11. 0

13. yes

15.

x	1	2	3	4
$g(x)$	b	d	c	a

17. yes

19. not injective; not invertible

21. yes

23. not injective; not invertible

25. T

27. F

29. T

31. T

33. T

35. 4

37. 16

39. $(1_Y \circ f)(x) = 1_Y(f(x)) = f(x)$
$\therefore 1_Y \circ f = f$

41. Let z be any element in Z. Since g is surjective, there exists an element y in Y such that $g(y) = z$. Since f is surjective, there exists an element x in X such that $f(x) = y$. Then $g(f(x)) = g(y) = z$. That is, $(g \circ f)(x) = z$
$\therefore g \circ f$ is surjective.

43. 1) To show that 1_X is injective:
Let $1_X(x) = 1_X(x')$. Then
$$x = x'$$
$\therefore 1_X$ is injective.
2) *To show that 1_X is surjective:*
Let x be any element in X. Since $1_X(x) = x$, 1_X is clearly surjective.
Thus, 1_X is bijective.

45. **proof:** Let $z \in Z$. Since $g \circ f$ is surjective, there is an element $x \in X$ such that $(g \circ f)(x) = z$, that is, $g(f(x)) = z$. Let $y = f(x)$. Then $y \in Y$ and $g(y) = z$. Thus, given any element $z \in Z$, there is an element $y \in Y$ such that $g(y) = z$. $\therefore g$ is surjective.

47. Let x be any element in X and let $f(x) = y$. Then
$$(f^{-1} \circ f)(x) = f^{-1}(f(x)) = f^{-1}(y) = x$$
$\therefore f^{-1} \circ f = 1_X$

49. **proof:** *To prove that f^{-1} is injective:*
Let $f^{-1}(y_1) = f^{-1}(y_2)$. Then $f(f^{-1}(y_1)) = f(f^{-1}(y_2))$. That is,
$$(f \circ f^{-1})(y_1) = (f \circ f^{-1})(y_2)$$
$$1_Y(y_1) = 1_Y(y_2)$$
$$y_1 = y_2$$
$\therefore f^{-1}$ is injective.
To show that f^{-1} is surjective:
Let $x \in X$. Then $f(x) \in Y$. Let $y = f(x)$.
$f^{-1}(y) = f^{-1}(f(x)) = (f^{-1} \circ f)(x) = 1_X(x) = x$. $\therefore f^{-1}$ is surjective.
Thus f^{-1} is bijective.

51. **proof:**
$$\begin{aligned}(g \circ f) \circ (f^{-1} \circ g^{-1}) &= g \circ (f \circ (f^{-1} \circ g^{-1}))\\ &= g \circ (f \circ f^{-1}) \circ g^{-1}\\ &= g \circ (1_Y \circ g^{-1})\\ &= g \circ g^{-1}\\ &= 1_Z\end{aligned}$$

Similarly, $(f^{-1} \circ g^{-1}) \circ (g \circ f) = 1_X$.
$\therefore (g \circ f)^{-1} = f^{-1} \circ g^{-1}$

53. $$\begin{aligned}(h \circ (g \circ f))(x) &= h((g \circ f)(x)) \\ &= h(g(f(x))) \\ &= (h \circ g)(f(x)) \\ &= ((h \circ g) \circ f)(x)\end{aligned}$$
$\therefore h \circ (g \circ f) = (h \circ g) \circ f$

55. Let $A \sim B$. Then there exists a bijection $f : A \to B$. Since f is a bijection, $f^{-1} : B \to A$ exists and is also a bijection. $\therefore B \sim A$.

57. **proof:**
To show that $f^{-1}(S \cup T) \subseteq f^{-1}(S) \cup f^{-1}(T)$:
Let $x \in f^{-1}(S \cup T)$. Then $f(x) \in S \cup T$.
$\therefore f(x) \in S$ or $f(x) \in T$. In other words, $x \in f^{-1}(S)$ or $x \in f^{-1}(T)$.
$\therefore x \in f^{-1}(S) \cup f^{-1}(T)$. Thus $f^{-1}(S \cup T) \subseteq f^{-1}(S) \cup f^{-1}(T)$.
By retracing the steps, it can be shown that
$f^{-1}(S) \cup f^{-1}(T) \subseteq f^{-1}(S \cup T)$.
Hence the result.

Exercises 3.6 (p. 162)

1. 21 **3.** 5 **5.** 20 **7.** 21 **9.** 135 **11.** 19

13. $\sum_{i=1}^{12} (2k - 1)$ **15.** $\sum_{k=1}^{11} k(k+1)$ **17.** T

19. $S = (a_{m+1} - a_m) + (a_{m+2} - a_{m+1}) + \cdots + (a_n - a_{n-1}) = a_n - a_m$

21.
$$\sum_{i=1}^{n}(i+1)^2 - \sum_{i=1}^{n} i^2 = 2\sum_{i=1}^{n} i + \sum_{i=1}^{n} 1$$
$$(n+1)^2 - 1 + \left(\sum_{i=2}^{n} i^2 - \sum_{i=2}^{n} i^2\right) = 2\sum_{i=1}^{n} i + n$$
$$n^2 + n = 2\sum_{i=1}^{n} i$$
$$\therefore \sum_{i=1}^{n} i = \frac{n(n+1)}{2}$$

23. 28 **25.** 350 **27.** 255 **29.** 15

31. 24 **33.** 1 **35.** 10 **37.** 210

39. 28 **41.** 6 **43.** 480, 480 **45.** 613

47. $a_{1j} + a_{2j} + a_{3j}$ **49.** $a_{11} + a_{12} + a_{21} + a_{22} + a_{31} + a_{32}$

51. $2a_1 + 2a_2 + 2a_3$ **53.** $|A_1 \cap A_2| + |A_1 \cap A_3| + |A_2 \cap A_3|$

Exercises 3.7 (p. 172)

1. $x = -1,\ y = -4,\ z = -4$

3. $\begin{bmatrix} -2 & 3 \\ 0 & -4 \end{bmatrix}$

5. $\begin{bmatrix} 0 & 3 & 2 \\ -1 & 2 & -4 \\ -2 & 5 & -6 \end{bmatrix}$

7. $\begin{bmatrix} -3 & -2 & 5 \\ 0 & 1 & 3 \end{bmatrix}$

9. $\begin{bmatrix} 0 & 4 & -10 \\ 0 & 0 & -2 \end{bmatrix}$

11. $\begin{bmatrix} 2 & -6 & 13 \\ 0 & 4 & 9 \end{bmatrix}$

13. $\begin{bmatrix} 3 & 4 & -13 \\ 0 & 6 & 7 \end{bmatrix}$

15. $p = r, q = s$

17. $m = n$

19. $n = p = r, q = s$

21. $n = p, q = r$

23. 1470, 875, 1890

25. 2025, 1250, 2710

27. proof:
$$\begin{aligned} A + B &= (a_{ij}) + (b_{ij}) \\ &= (a_{ij} + b_{ij}) \\ &= (b_{ij} + a_{ij}) \\ &= (b_{ij}) + (a_{ij}) \\ &= B + A \end{aligned}$$

29. proof:
$$\begin{aligned} A + O &= (a_{ij}) + (0) \\ &= (a_{ij} + 0) \\ &= (a_{ij}) \\ &= A \end{aligned}$$
Similarly, $O + A = A$.

31. proof:
$$\begin{aligned} c(A + B) &= c(a_{ij} + b_{ij}) \\ &= (ca_{ij} + cb_{ij}) \\ &= c(a_{ij}) + c(b_{ij}) \\ &= cA + cB \end{aligned}$$

33. Let $A = (a_{ij})$, $B = (b_{ij})$, and $C = (c_{ij})$.

$$BC = \begin{bmatrix} \sum b_{1k}c_{k1} & \sum b_{1k}c_{k2} \\ \sum b_{2k}c_{k1} & \sum b_{2k}c_{k2} \end{bmatrix}$$

$$\begin{aligned} A(BC) &= \begin{bmatrix} \sum a_{1t}(\Sigma b_{1k}c_{k1}) & \sum a_{1t}(\Sigma b_{1k}c_{k2}) \\ \sum a_{2t}(\sum b_{2k}c_{k1}) & \sum a_{2t}(\sum b_{2k}c_{k2}) \end{bmatrix} \\ &= \begin{bmatrix} \sum(\sum a_{1t}b_{t1})c_{k1} & \sum(\sum a_{1t}bt_{t1})c_{k2} \\ \sum(\sum a_{2t}b_{t2})c_{k1} & \sum(\sum a_{2t}b_{t2})c_{k2} \end{bmatrix} \\ &= (AB)C \end{aligned}$$

35. proof: $(A + B)C$
$$= \begin{bmatrix} (a_{11} + b_{11})c_{11} + (a_{12} + b_{12})c_{21} & (a_{11} + b_{11})c_{12} + (a_{12} + b_{12})c_{22} \\ (a_{21} + b_{21})c_{11} + (a_{22} + b_{22})c_{21} & (a_{21} + b_{21})c_{12} + (a_{22} + b_{22})c_{22} \end{bmatrix}$$

$$= \begin{bmatrix} a_{11}c_{11}+a_{12}c_{21} & a_{11}c_{12}+a_{12}c_{22} \\ a_{21}c_{11}+a_{22}c_{21} & a_{21}c_{12}+a_{22}c_{22} \end{bmatrix} + \begin{bmatrix} b_{11}c_{11}+b_{12}c_{21} & b_{11}c_{12}+b_{12}c_{22} \\ b_{21}c_{12}+b_{22}c_{12} & b_{21}c_{12}+b_{22}c_{22} \end{bmatrix}$$

$$= AB + AC$$

37. $\begin{bmatrix} a & d & f \\ b & e & g \\ c & f & h \end{bmatrix}$

39. Let $A = (a_{ij})$.

$$A^{\mathrm{T}} = (a_{ji})$$
$$\therefore (A^{\mathrm{T}})^{\mathrm{T}} = (a_{ij}) = A$$

41. Let $A = (a_{ij})$ and $B = (b_{ij})$.

$$AB = \begin{bmatrix} a_{11}b_{11} + a_{12}b_{21} & a_{11}b_{12} + a_{12}b_{22} \\ a_{21}b_{11} + a_{22}b_{21} & a_{21}b_{12} + a_{22}b_{22} \end{bmatrix}$$

$$(AB)^{\mathrm{T}} = \begin{bmatrix} a_{11}b_{11} + a_{12}b_{21} & a_{21}b_{11} + a_{22}b_{21} \\ a_{11}b_{12} + a_{12}b_{22} & a_{21}b_{12} + a_{22}b_{22} \end{bmatrix}$$

$$B^{\mathrm{T}}A^{\mathrm{T}} = \begin{bmatrix} b_{11} & b_{21} \\ b_{12} & b_{22} \end{bmatrix} \begin{bmatrix} a_{11} & a_{21} \\ a_{12} & a_{22} \end{bmatrix} = \begin{bmatrix} b_{11}a_{11} + b_{21}a_{12} & b_{11}a_{21} + b_{21}a_{22} \\ b_{12}a_{11} + b_{22}a_{12} & b_{12}a_{21} + b_{22}a_{22} \end{bmatrix} = (AB)^{\mathrm{T}}$$

43. $(ABC)^{\mathrm{T}} = [A(BC)]^{\mathrm{T}}$
$= (BC)^{\mathrm{T}}A^{\mathrm{T}}$, by Exercise 41
$= (C^{\mathrm{T}}B^{\mathrm{T}})A^{\mathrm{T}}$, by Exercise 41
$= C^{\mathrm{T}}B^{\mathrm{T}}A^{\mathrm{T}}$

45. $$AB = \frac{1}{17}\begin{bmatrix} 1 & -2 & 0 \\ 3 & 1 & -1 \\ 1 & 2 & -3 \end{bmatrix} \begin{bmatrix} 1 & 6 & -2 \\ -8 & 3 & -1 \\ -5 & 4 & -7 \end{bmatrix} = \frac{1}{17}\begin{bmatrix} 17 & 0 & 0 \\ 0 & 17 & 0 \\ 0 & 0 & 17 \end{bmatrix} = I_3$$

47. $\begin{bmatrix} x \\ y \end{bmatrix}$

49. $\begin{bmatrix} 2 & 3 \\ 4 & 5 \end{bmatrix} \begin{bmatrix} x \\ y \end{bmatrix} = \begin{bmatrix} 4 \\ 6 \end{bmatrix}$

51. $$\begin{bmatrix} x \\ y \end{bmatrix} = \begin{bmatrix} 2 & 3 \\ 4 & 5 \end{bmatrix}^{-1} \begin{bmatrix} 4 \\ 6 \end{bmatrix} = \frac{1}{2}\begin{bmatrix} 5 & -3 \\ -4 & 2 \end{bmatrix} \begin{bmatrix} 4 \\ 6 \end{bmatrix} = \frac{1}{2}\begin{bmatrix} 2 \\ -4 \end{bmatrix} = \begin{bmatrix} -1 \\ 2 \end{bmatrix}$$

so $x = -1, y = 2$.

53. $f(2) = \begin{bmatrix} 2 & 1 \\ 1 & 1 \end{bmatrix}$

55. $f(4) = \begin{bmatrix} 5 & 3 \\ 3 & 2 \end{bmatrix}$

57. **proof:** Since B is an inverse of A, $AB = I = BA$. Likewise, $AC = I = CA$. Then $B = BI = B(AC) = (BA)C = IC = C$

59. **proof:** $(AB)(B^{-1}A^{-1}) = A[B(B^{-1}A^{-1})]$
$= A[(BB^{-1})A^{-1}]$
$= A(IA^{-1})$
$= AA^{-1} = I$
Similarly, $(B^{-1}A^{-1})(AB) = I$.
$\therefore (AB)^{-1} = B^{-1}A^{-1}$

Review Exercises (p. 177)

1. 558 **3.** 761 **5.** Tuesday **7.** July

9. Wednesday **11.** Monday **13.** April 1 **15.** March 28

17. no **19.** yes **21.** no **23.** 8

25. 0 **27.** 0 **29.** 5540 **31.** 5

33.

A	B	C	D	E	...	K	L
CBS1AA	CBA3BA		AQX5CD		...	NBC4GK	VPS3SL

M	N	O	P	...	X	Y	Z
NCR4SK	CNN1TK						ABC5ZZ

35.

0	1	2	3	4	5
AQZ5CD	CBS1AA	CNN1TK	VPS3SL	NBC4GK	CBA3BA

6	7	8	9
NCR4SK	ABC5ZZ		

37. Use $m = 19, n = 6$. **39.** -27

41. -39 **43.** 1 **45.** 36

47. F **49.** F **51.** 3.4, 4.6

53. 3.4, 4.5 **55.** 2, 3.5 **57.** 1, 1, 2, 3

59. 1, 1, 2, 3 **61.** 25 **63.** 72

65. **proof:** Let $b_1, b_2 \in B$ such that $g(b_1) = g(b_2)$. Then $f(g(b_1)) = f(g(b_2))$; that is, $(f \circ g)(b_1) = (f \circ g)(b_2)$. Thus $1_B(b_1) = 1_B(b_2)$; that is, $b_1 = b_2$. $\therefore$ g is injective.

Supplementary Exercises (p. 179)

1. $(A+B)^{\mathrm{T}} = (a_{ij}+b_{ij})^{\mathrm{T}}$
$= (a_{ji}+b_{ji})$
$= (a_{ji}) + (b_{ji})$
$= A^{\mathrm{T}} + B^{\mathrm{T}}$

3. $(AA^{\mathrm{T}})^{\mathrm{T}} = (A^{\mathrm{T}})^{\mathrm{T}}A^{\mathrm{T}}$
$= AA^{\mathrm{T}}$, since $(A^{\mathrm{T}})^{\mathrm{T}} = A$.
So AA^{T} is symmetric.

5. **proof** (by contradiction): Suppose each of the numbers is less than their average A. Then

$$\sum_{i=1}^{n} a_i < nA = n\left(\frac{1}{n}\sum_{i=1}^{n} a_i\right) = \sum_{i=1}^{n} a_i,$$

a contradiction.

7. 63.

9. 134,217,728

11. 65,536

13. **proof:** $a_k = a + (k-1)d$, $k \geq 1$

$$\begin{aligned} S_n &= \sum_{k=1}^{n} a_k \\ &= \sum_{k=1}^{n} [a + (k-1)d \\ &= \sum_{k=1}^{n} a + d\sum_{k=1}^{n}(k-1) \\ &= na + d\cdot\frac{(n-1)n}{2} \\ &= \frac{n}{2}[2a + (n-1)d] \end{aligned}$$

15. Let $S_n = a + ar + \cdots + ar^{n-1}$

$$rS_n = ar + ar^2 + \cdots + ar^{n-1} + ar^n$$

$$\therefore rS_n - S_n = ar^n - a$$

$$S_n = \frac{a(r^n-1)}{r-1} (r \neq 1)$$

17. Let $X = \{\mathrm{a,b,c,d,e,g}\}$, $A = \{\mathrm{a,b,c}\}$, $\mathrm{B} = \{\mathrm{b,c,d,e}\}$, and $Y = \{0,3,5\}$. Let $f : X \to Y$ defined by $f(\mathrm{a}) = 0$, $f(\mathrm{b}) = f(\mathrm{c}) = 3$, $f(\mathrm{d}) = f(\mathrm{e}) = f(\mathrm{g}) = 5$. Then $A \cap B = \{\mathrm{b,c}\}$, $f(A) = \{0,3\}$, $f(B) = \{3,5\}$, $f(A\cap B) = \{3\} = f(A)\cap f(B)$. Nonetheless, f is not injective.

19. **proof:** Let $a_1 = \cdots = a_n$.

$$\text{Then } \sum_{i=1}^{n} a_i - n + 1 = 2n - n + 1 = n + 1$$

So, by Exercise 18, if $n+1$ pigeons occupy n pigeonholes, one pigeonhole must contain at least two pigeonholes.

21.

$$\begin{aligned} a_n &= \sum_{i=n+1}^{\infty} a_i \\ &= a_{n+1} + \sum_{i=n+2}^{\infty} a_i \\ &= a_{n+1} + a_{n+1} \\ &= 2a_{n+1} \\ \therefore\quad a_{n+1} &= \frac{1}{2}a_n \end{aligned}$$

Chapter 4 Induction and Algorithms

Exercises 4.1 (p. 188)

1. yes

3. 12; 5

5. -3; 5

7. $\{0, 1\}$

9. $\{0, 1, 2, 3, 4, 5, 6\}$

11. **proof** (by contradiction): Assume there is an integer n such that $0 < n < 1$. Let $S = \{a|0 < a < 1\}$. Since $n \in S$, $S \neq \emptyset$. Therefore, by the WOP, S has a least element ℓ, where $0 < \ell < 1$. $\therefore 0 < \ell^2 < \ell$ and hence $0 < \ell^2 < 1$. So $\ell^2 \in S$, where $\ell^2 < \ell$, a contradiction.

13. **proof** (by contradiction): Assume there is an integer $n' \in S$ such that $n' < n_0 + \ell^* - 1$. Then $n_0 + \ell^* - 1 < \ell^*$. Since $n' - n_0 + 1 \in S^*$ and $n' - n_0 + 1 < \ell^*$, ℓ is not the least element of S^*, a contradiction.

15. **proof** (by contradiction): Let $D = S - T$. Then $D \cap T = \emptyset$ and $D \cup T = S$. To prove that $S = T$, suffices to show that $D = \emptyset$. Assume $D \neq \emptyset$. Since S is well-ordered, so is D. So D contains a least element d. Then $d \neq a$, since $d \in D$, $a \in T$, and $D \cap T = \emptyset$. Since $d \neq a$ and $S = \{a, a+1, \ldots\}$, $d - 1$ must lie in S. Clearly, $d - 1 \notin D$. So $d - 1 \in T$. $\therefore (d-1) + 1 = d \in T$, a contradiction. Thus $D = \emptyset$ and hence $S = T$.

Exercises 4.2 (p. 195)

1. yes

3. no

5. no, 3 is a counterexample.

7. 4

9. 4

11. $2 = 2 \cdot 28 + (-3) \cdot 18$

13. $1 = (-8) \cdot 28 + 15 \cdot 15$

15. 12

17. 39/18

19. choose $a = 3, b = 5, c = 7$.

21. April 23

23. April 13

25. 4

27. 16

29. 5; 6; 10; 12

31. **proof:** Since $a|b$ and $a|c$, there are positive integers d and d' such that $b = da$ and $c = d'a$. Then $b - c = da - d'a = (d - d')a$ $\therefore a|(b - c)$.

33. **proof:** By the division algorithm, there exists a quotient q such that $a = bq + r$. Since $d' = \gcd\{b, r\}, d'|b$ and $d'|r$. $\therefore d'|a$. Thus $d'|a$ and $d'|b$. $\therefore d'| \gcd\{a, b\}$; that is, $d'|d$.

35. **proof:** Let $d = \gcd\{a, b\}$ and $d' = \gcd\{b, a + b\}$. Then $d|a$ and $d|b$. $\therefore d|(a + b)$ and hence $d|d'$. Since $d' = \gcd\{b, a + b\}$, $d'|b$ and $d'|(a + b)$. $\therefore d'|[(a + b) - b]$, that is, $d'|a$. Thus $d'|a$ and $d'|b$, and hence $d'|d$. $\therefore d = d'$.

37. Assume $p|ab$ and $p \nmid a$. Since $p \nmid a$, $\gcd\{p, a\} = 1$. Then there are integers s and t such that $1 = sp + ta$. $\therefore b = spb + tab$. Since $p|spb$ and $p|tab$, $p|b$.

39. Since $\gcd\{a, b\} = d$, by Theorem 4.6, there are integers s and t such that $d = sa + tb$. $\therefore 1 = s(a/d) + t(b/d)$. Thus, $\gcd\{a/d, b/d\} = 1$, and hence a/d and b/d are relatively prime.

43. Let n and $n+1$ be any two consecutive integers. Clearly, one of them must be even. Suppose it is n. Then $2|n$. But n is a prime, so $n = 2$ and hence $n + 1 = 3$. The case $n + 1$ is even yields $n = 1$, which is a contradiction.

47. We have $p = q + 3$. Suppose q is even. Then $q = 2$ and $p = 5$. On the other hand, let q be odd. Then $q + 3$ is an even prime > 6, which is impossible.

41. Let $d = \gcd\{\gcd\{a, b\}, c\}$ and $d' = \gcd\{a, \gcd\{b, c\}\}$. Then $d|$ $\gcd\{a, b\}$ and $d|c$. $\therefore d|a$, $d|b$, and $d|c$. So $d|a$, and $(d|b$ and $d|c)$. $\therefore d|a$ and $d|\gcd\{b, c\}$.

45. Because $p^2 + 8$ is prime, p must be odd. By the division algorithm, p is of the form $3n, 3n - 1$, or $3n + 1$. If $p = 3n \pm 1$, then $p^2 + 8 = 9n^2 \pm 6n + 9$ is divisible by 3. So $p = 3$ and hence $n = 1$ and hence $p = 3$. Then $p^3 + 4 = 31$ is a prime.

49. False. Choose $a = 3, b = 4, c = 9$.

51. $E_7 = 510511 = 19 \cdot 26869$ is composite.

Exercises 4.3 (p. 205)

1. 13

3. 1022

5. 10000110100_{two}

7. 3360_{eight}

9. 15_{eight}

11. 72_{eight}

13. $1\text{D}_{\text{sixteen}}$

15. 75_{sixteen}

17. 110110_{two}

19. 1000110111_{two}

21. 11010_{two}

23. $87\text{EC}_{\text{sixteen}}$

25. 10100_{two}

27. 3071730_{two}

29. 110, 1011, 10110, 11011, 101010

31. 0, 1

33. 2

35. 5

37. 1) $0 \in S$.
2) If $x \in S$, then $0x, 1x \in S$.

39. 1) $1 \in S$.
2) If $x \in S$, then $x0, 1x, x1 \in S$.

41. 165

43. 101

45. $10\text{EF}_{\text{sixteen}}$

Exercises 4.4 (p. 221)

1. 666

3. $\sum_{i=1}^{12} i = \frac{12 \cdot 13}{2} = 78$

(*Note*: In Exercises 5-13, $P(n)$ denotes the given statement.)

5. Basis step: When $n = 1$, LHS $= 1 = 1^2 =$ RHS. $\therefore P(1)$ is true.

Induction step: Assume $P(k)$ is true: $\sum_{i=1}^{k}(2i-1) = k^2$

Then $\sum_{i=1}^{k}(2i-1) + (2k+1) = k^2 + (2k+1) = (k+1)^2$

$\therefore P(k+1)$ is true. Thus, the given result follows by PMI.

7. Basis step: When $n = 1$, LHS $= 1^3 = 1 =$ RHS. $\therefore P(1)$ is true.

Induction step: Assume $P(k)$ is true: $\sum_{i=1}^{k} i^3 = \left[\frac{k(k+1)}{2}\right]^2$

Then

$$\begin{aligned}\sum_{i=1}^{k} i^3 + (k+1)^3 &= \left[\frac{k(k+1)}{2}\right]^3 + (k+1)^2 \\ &= (k+1)^2\left[\frac{k^2+4k+4}{4}\right] \\ &= \frac{(k+1)^2(k+2)^2}{4} \\ &= \left[\frac{(k+1)(k+2)}{2}\right]^2\end{aligned}$$

$\therefore P(k+1)$ is true. Thus the result is true by induction.

9. Basis step: When $n = 1$, $n^2 + n = 1 + 1 = 2$ is divisible by 2. $\therefore P(1)$ is true.

Induction step: Assume $P(k)$ is true: $k^2 + k$ is divisible by 2. Let $k^2 + k = 2m$ for some positive integer m. Then

$$\begin{aligned}(k+1)^2 + (k+1) &= k^2 + 3k + 2 = (k^2 + k) + 2(k+1) \\ &= 2m + 2(k+1)\end{aligned}$$

which is divisible by 2. $\therefore P(k+1)$ is true. Thus the result follows by PMI.

11. Basis step: When $n = 2, P(2) = 1 = \frac{2(2-1)}{2}$. $\therefore P(2)$ is true.

Induction step: Assume $P(k)$ is true. Consider a new point Q. By joining Q to each of the original k points, we get k new lines.

$$\begin{aligned}\therefore \text{Total no. of points with } k+1 \text{ points} &= \frac{k(k-1)}{2} + k = \frac{k^2+k}{2} \\ &= \frac{(k+1)[(k+1)-1]}{2}\end{aligned}$$

$\therefore P(k+1)$ is true. Thus the result follows by PMI.

13. Basis step: Clearly, $P(1)$ is true.

Induction step: Assume $P(k)$ is true: If $p|a^k$, then $p|a$. Suppose $p|a^{k+1}$. Since $a^{k+1} = a^k \cdot a, p|(a^k \cdot a)$. Then $p|a^k$ or $p|a$, by Exercise 33 in Section 4.2. Since $p|a^k$ by the IH, $P(k+1)$ is true. Thus the result follows by induction.

15. 28,335

17. $\begin{cases} n^2/4 & \text{if } n \text{ is even} \\ (n^2-1)/4 & \text{otherwise} \end{cases}$

19. $x = \sum_{i=1}^{n} (2i-1) = n^2$

21. $n(n+1)/2$

23. $n(n+1)(n+2)/6$

25. $n(n+1)(2n+1)/6$

27. $(n!)^2$

29. $2^{n^2(n+1)}$

31. $\text{sum} = \sum_{s=0}^{i} p^s = \dfrac{p^{i+1}-1}{p-1}$

33. $\text{sum} = \left(\sum_{s=0}^{i} p^s\right)\left(\sum_{s=0}^{j} q^s\right)\left(\sum_{s=0}^{k} r^s\right)$

$$= \frac{p^{i+1}-1}{p-1} \cdot \frac{q^{j+1}-1}{q-1} \cdot \frac{r^{k+1}-1}{r-1}$$

35. $\sum_{i=1}^{n} i = \dfrac{n(n+1)}{2}$

37. $\sum_{i=1}^{n} \sum_{j=1}^{i} j = \dfrac{n(n+1)(n+2)}{6}$

39. Recall that the k-th term of the geometric sequence $a, ar, ar^2, \ldots$ is $a_k = ar^{k-1}$. $\therefore$ no. of grains on the last square $= n^2$-th term in the sequence 1, 2, 2^2, 2^3, …. Here $a = 1, r = 2$, and $k = n^2$. $\therefore$ no. of grains on the last square $= 1\cdot(2^{n^2-1}) = 2^{n^2-1}$. (We could prove this PMI also.)

41. **case 1** Let n be even. Then

$$\begin{aligned} a_n &= 0+1+1+\cdots \\ &\quad + (n/2-1) + (n/2-1) + n/2 \\ &= 2[1+2+\cdots+(n/2-1)] + n/2 \\ &= \frac{2(n/2-1)(n/2+1)}{2} + \frac{n}{2} = \frac{n^2}{4} \end{aligned}$$

case 2 Let n be odd. Then

$$\begin{aligned} a_n &= 0+1+1+\cdots+(n-1)/2 \\ &\quad +(n-1)/2 \\ &= 2[1+2+\cdots+(n-1)/2] \\ &= \frac{2[(n-1)/2][(n+1)/2]}{2} \\ &= \frac{n^2-1}{4} \end{aligned}$$

43. 93

45. 249

47. $8t_n + 1 = 8n(n+1)/2 + 1$
$= 4n(n+1) + 1$
$= (2n+1)^2$

49. LHS
$= [(n-1)n/2]^2 + [n(n+1)/2]^2$
$= n^2[(n-1)^2 + (n+1)^2]/4$
$= n^2(n^2+1)/2 = t_{n^2} = \text{RHS}$

51. **basis step:** When $n = 0$, LHS $= A \cup B_1 =$ RHS. So $P(1)$ is true.

induction step: Assume $P(k)$ is true: $A \cup (\bigcap_{i=1}^{k} B_i) = \bigcap_{i=1}^{k} (A \cup B_i)$. Then

$$A \cup \left(\bigcap_{i=1}^{k+1} B_i\right) = A \cup \left(\bigcap_{i=1}^{k} B_i \cap B_{k+1}\right) = [A \cup \left(\bigcap_{i=1}^{k} B_i\right)] \cap (A \cup B_{k+1})$$

$$= \bigcap_{i=1}^{k} (A \cup B_i) \cap (A \cup B_{k+1}) = \bigcap_{i=1}^{k+1} (A \cup B_i)$$

$\therefore$ $P(k) \to P(k+1)$, so the result follows by PMI.

53. **basis step:** When $n = 1$, LHS $= p_1 =$ RHS. $\therefore$ $P(1)$ is true.
induction step: Assume $P(k)$ is true. Then

$$\begin{aligned}\sim(p \wedge \cdots \wedge p_k \wedge p_{k+1}) &\equiv \sim[(p_1 \wedge \cdots \wedge p_k) \wedge p_{k+1} \\ &\equiv \sim(p_1 \wedge \cdots \wedge p_k) \vee \sim p_{k+1} \\ &\equiv (\sim p_1 \vee \cdots \sim p_k) \vee \sim p_{k+1}\end{aligned}$$

$\therefore$ $P(k+1)$ is true. Thus the result follows by PMI.

55. **basis step:** Clearly, $P(2)$ is true.
induction step: Assume $P(k)$ is true. By the division algorithm, $k = 3q + r$, where $0 \le r < 3$. Then $k+1 = 3q + r + 1$.
case 1 If $r = 0$, then $k+1 = 3q+1 = 3(q-1)+4$. Therefore, a change of $3q+1$ cents can be paid with $q-1$ 3-cent coins and two 2-cent coins.
case 2 If $r = 1$, then $k+1 = 3q+2$. Clearly, a change of $3q+2$ coins can be paid with q 3-cent coins and one 2-cent coin.
case 3 If $r = 2$, then $k+1 = 3(q+1)$. Such a change can be paid with $(q+1)$-cent coins. Thus $P(k+1)$ is true.
Hence result follows by PMI.

57. **proof:** Suppose $q(n)$ satisfies the conditions of the second principle. Let $p(n) = q(n_0) \wedge q(n_0+1) \wedge \cdots \wedge q(n)$, where $n \ge n_0$. Since $q(n_0)$ is true, $p(n_0)$ is true.

Assume $p(k)$ is true, where $k \ge n_0$. By condition (2), $q(k+1)$ is true. Therefore, $p(k+1) = q(n_0) \wedge \cdots \wedge q(k) \wedge q(k+1)$ is true. So $p(k) \to p(k+1)$. So by the first principle, $p(n)$ is true $\forall n \ge n_0$. Thus $q(n)$ is true $\forall n \ge n_0$.

59. $n(3n^4 + 7n^2 + 2)/12$

Exercises 4.5 (p. 234)

1. When $n = 0$, $a_0 = 1 = x^0$. $\therefore$ $P(0)$ is true. Assume $P(k)$ is true: $a_k = x^k$. Then

$$\begin{aligned} a_{k+1} &= a_k * x, && \text{by step 4} \\ &= x^k \cdot x, && \text{by the inductive hypothesis} \\ &= x^{k+1} \end{aligned}$$

$\therefore$ $P(k+1)$ is true. Thus $P(n)$ is a loop invariant by PMI.

3. Let $\gcd\{a,b\} = d$. When $n = 0$, $\gcd\{x_n, y_n\} = \gcd\{x_0, y_0\} = \gcd\{x, y\}$. $\therefore P(0)$ is true. Assume $P(k)$ is true: $\gcd\{x_k, y_k\} = \gcd\{x, y\} = d$. Then $x_k = y_{k-1}, y_k = r_{k-1}, r_k = x_k \bmod y_k$, and $x_k = q_k y_k + r_k$. (1)
Since $d|x_k$ and $d|y_k$, $d|r_k$.
To show that $P(k+1)$ is true: $\gcd\{x_{k+1}, y_{k+1}\} = d$.
Let $\gcd\{x_{k+1}, y_{k+1}\} = d'$. (2)
$x_{k+1} = y_k$, $y_{k+1} = r_k$. $\therefore d|x_{k+1}$ and $d|y_{k+1}$. So $d|d'$. (3)
From (1), $d'|x_{k+1}$ and $d'|y_{k+1}$. $\therefore d'|y_k$ and $d'|r_k$. $\therefore d'|x_k$ by (1). Thus $d'|x_k$ and $d'|y_k$. $\therefore d'|d$. (4)
Thus by (3) and (4), $d = d'$. Thus $P(n)$ is a loop invariant.

5. When $n = 0$, $s_0 = x = x + 0$. $\therefore P(0)$ is true. Assume $P(k)$ is true: $s_k = x + k$. Then $s_{k+1} = s_k + 1 = (x+k) + 1 = x + (k+1)$. $\therefore P(k+1)$ is true. Thus $P(n)$ is a loop invariant.

7. 3

9. 4

11. 7, 18, 19, 23, 53

13. 7, 18, 19, 23, 53

15.
```
Algorithm factorial(n)
Begin (* algorithm *)
  fact ← 1
  i ← 1
  while i ≤ n do
    fact ← fact * i
End (* algorithm *)
```

17.
```
Algorithm product(A,B,C)
Begin (* algorithm *)
  for i = 1 to n do
  for j = 1 to n do
            n
    cij  =  Σ  aitbtj
           t=1
End (* algorithm *)
```

19.
```
Algorithm sum(X)
Begin (* algorithm *)
  sum ← 0
  for i = 1 to n do
    sum ← sum + xi
End (* algorithm *)
```

21.
```
Algorithm max(X)
Begin (* algorithm *)
  max ← x1
  for i = 2 to n do
  if xi > max then
    max ← xi
End (* algorithm *)
```

23.
```
Algorithm print(X)
Begin (* algorithm *)
  i ← 1
  flag ← true
  while (i ≤ n) and (flag) do
  if si = sn+1−i then
    i ← i + 1
  else
    flag ← false
End (* algorithm *)
```

25.
```
Algorithm palindrome (S)
Begin (* algorithm *)
  for i = 1 to n do
    write (xi)
End (* algorithm *)
```

(In Exercises 26–36, $P(n)$ denotes the statement that the algorithm works correctly.)

27. Initially, flag = true, and $i = j = 1$.
basis step: When $n = 1$, if $a_{11} = b_{11}$, flag is still true, so $A = B$; otherwise, flag = false (line 10), so $A \neq B$. In both cases, $P(1)$ is true.

induction step: Assume $P(k)$ is true. To show that $P(k+1)$ is true, it suffices to show that the algorithm works correctly for the $(k+1)$st rows and columns of A and B.
When $j = k+1$, then inner loop (lines 6–10) checks whether or not $a_{i(k+1)} = b_{i(k+1)}$. So as i varies from 1 to $k+1$, the outer loop (lines 4–12) checks if the $(k+1)$st columns of A and B are equal. When $i = k+1$, the inner loop checks if the $(k+1)$st rows of A and B are equal. Thus, $P(k) \to P(k+1)$. So, by induction, the algorithm works correctly $\forall n \geq 1$.

29. Initially, flag = true, and $i = j = 1$.
basis step: When $n = 1$, if $a_{11} \leq b_{11}$, flag is still true, so $A = B$; otherwise, flag = false (line 10), so $A \leq B$. In both cases, $P(1)$ is true.
induction step: Assume $P(k)$ is true. To show that $P(k+1)$ is true, it suffices to show that the algorithm works correctly for the $(k+1)$st rows and columns of A and B.
When $j = k+1$, then inner loop (lines 6–10) checks whether or not $a_{i(k+1)} \leq b_{i(k+1)}$. So as i varies from 1 to $k+1$, the outer loop (lines 4–12) checks if the $(k+1)$st column of A is $\leq$ that of B. When $i = k+1$, the inner loop checks if the $(k+1)$st row of A is $\leq$ that of B.
Therefore, $P(k) \to P(k+1)$. So, by induction, the algorithm works correctly $\forall n \geq 1$.

31. basis step: When $n = 1$, $A = [a_{11}]$ and $B = [b_{11}]$. So $c_{11} = a_{11}b_{11}$ by line 3. $\therefore P(1)$ is true.
induction step: Assume $P(k)$ is true, that is, the algorithm works correctly for any two $k \times k$ matrices. In other words, line 3 gives the correct value of c_{ij} for $1 \leq i,j \leq k$. Now suppose $n = k+1$. By line 4, $\sum_{t=1}^{k+1} a_{it}b_{tj} = \sum_{t=1}^{k+1} a_{it}b_{tj} + a_{i(k+1)}b_{(k+1)j}$ gives the correct value of c_{ij}, by the inductive hypothesis. Therefore, the nested **for** loops give the correct value of c_{ij} $\forall i, j$. $\therefore P(k) \to P(k+1)$. So, by induction, the algorithm works correctly $\forall n \geq 1$.

33. basis step: When $n = 1$, lines 2–4, are skipped. So the algorithm returns the correct value of *min* form line 1. $\therefore P(1)$ is true.
induction step: Assume $P(k)$ is true. Suppose the algorithm is invoked with $n = k+1$ elements. As i varies from 2 through k, the **for** loop finds the minimum of the first k numbers. When $n = k+1$, this value is compared to x_{k+1} and the minimum of x_1 through x_{k+1} is returned in line 4.
$\therefore P(k) \to P(k+1)$. So, by induction, the algorithm works correctly $\forall n \geq 1$.

35. basis step: When $n = 1$, $x_{n-i+1} = x_{1-1+1} = x_1$ is printed. $\therefore P(1)$ is true.
induction step: Assume $P(k)$ is true. Suppose the algorithm is invoked with $n = k+1$ elements. When $i = 1$, x_{k+1} is printed. Then as

i varies from 2 through k, $x_k, x_{k-1}, \ldots, x_1$ are printed, by the inductive hypothesis. $\therefore P(k) \to P(k+1)$. So, by induction, the algorithm works correctly $\forall n \geq 1$.

37. 2, 3, 5, 6, 8, 13

39. **basis step:** When $n = 1$, the **for** loop is skipped. So $P(1)$ is true by default.

induction step: Assume $P(k)$ is true. Suppose the algorithm is invoked with $n = k + 1$ elements. When $i = 2$, the algorithm places the largest of the $k + 1$ elements in the correct position. This leaves a sublist with k elements. By the inductive hypothesis, the algorithm correctly sorts it. $\therefore P(k) \to P(k+1)$. So, by induction, the algorithm works correctly $\forall n \geq 1$.

Exercises 4.6 (p. 246)

1. $O(n)$ **3.** $O(n^3)$ **5.** $O(\lg n)$

7. $O(n \lg n)$ **9.** $O(n^3)$ **11.** $O(n^2)$

13. $2^n = 2 \cdot 2 \cdot \cdots \cdot 2$ (n times) $\leq 2 \cdot 3 \cdot \cdots \cdot n = O(n!)$

15. $\sum_{i=1}^{n} i^k \leq n^k + \cdots + n^k = n(n^k) = O(n^{k+1})$

17. $\sum_{i=1}^{n} i(i+1) = \sum_{i=1}^{n} i^2 + \sum_{i=1}^{n} i = \frac{n(n+1)(2n+1)}{6} + \frac{n(n+1)}{2} = O(n^3)$

19. $O(n^2)$ **21.** $O(n^3)$

23. $\Omega(n)$ **25.** $\Omega(n^3)$

27. $\Omega(\lg n)$ **29.** $\Omega(n^3)$

31. $\Omega(n^2)$

33. $(3n)! = 3 \cdot 6 \cdot 9 \cdots (3n) = 3^n n! \geq 3^n 2^n$, where $n \geq 4$ $= \Omega(6^n)$

35. sum $= n^2 = \Omega(n^2)$

37. $2n + 3 \geq 2n = \Omega(n)$

39. $2n^3 - 3n^2 + 4n \geq 2n^3$, where $n \geq 2$ $= \Omega(n^3)$

41. $3\lg n + 2 \geq 3\lg n = \Omega(\lg n)$

43. True, since sum $= n^2/4$ if n is even and $(n^2 - 1)/4$.

45. Since $f_1(n) = O(n(g(n))$, $|f_1(n)| \leq A|g(n)|$ for some constant $A > 0$. $f_2(n) = kf_1(n)$. $\therefore |f_2(n)| = k|f_1(n)| \leq kA|g(n)| = C|g(n)|$, where $C = kA$. $= O(g(n))$

47. Since $f(n) = O(h(n))$, $|f(n)| \leq A|h(n)|$ for some constant A. Similarly, $|g(n)| \leq B|h(n)|$ for some constant B. Then $|(f+g)(n)| = |f(n) + g(n)| \leq |f(n)| + |g(n)| \leq A|h(n)| + B|h(n)|$

49. $|f(n)| \leq A|g(n)|$ for some constant $A > 0$. $|g(n)| \leq B|h(n)|$ for some constant $B > 0$.

$\therefore\ |f(n)| \leq AB|h(n)|$
$= C|h(n)|$, where $C = AB$.
$= O(h(n))$

$\leq C|h(n)| + C|h(n)|$
$\leq 2C|h(n)|$
$= O(h(n))$
where $C = \max\{A, B\}$

51. Suppose $f(n) = \Theta(g(n))$. Then $f(n) = O(g(n))$, $f(n) = \Omega(g(n))$. So $|f(n)| \leq B|g(n)|$ and $|f(n)| \geq A|g(n)|$ for some constants A and B. Thus $A|g(n)| \leq |f(n)| \leq B|g(n)|$.
Conversely, let $A|g(n)| \leq |f(n)| \leq B|g(n)|$. Since $A|g(n)| \leq |f(n)|$, $f(n) = \Omega(g(n))$. Likewise $f(n) = O(g(n))$. So $f(n) = \Theta(g(n))$.

Exercises 4.7 (p. 251)

1. No. of entries in each matrix $= n^2$
$\therefore$ no. of additions required $= n^2$
$= O(n^2)$

3. $s_n = n(n-1)$

5. No. of additions $= n = O(n)$

7. $a_n = n = O(n)$

9. $c_n = n - 1 = O(n)$

11. 37, 2

13. $O(n)$

Review Exercises (p. 254)

1. 2 **3.** 3 **5.** 1024 **7.** 2989

9. 11110101_{two} **11.** 2305_{two} **13.** 100001_{two} **15.** $\text{D033}_{\text{sixteen}}$

17. 265_{eight} **19.** 463_{eight} **21.** $\text{B5}_{\text{sixteen}}$ **23.** 133_{sixteen}

25. n^2

27. $a_n = \begin{cases} n(n+2)/4 & \text{if } n \text{ is even} \\ (n+1)^2/4 & \text{otherwise} \end{cases}$

29. **proof:** Let $d = \gcd\{a-b, a+b\}$. Then $d|(a-b)$ and $d|(a+b)$; so $d|[(a-b)+(a+b)]$. That is, $d|2a$. Similarly, $d|2b$. But a and b are relatively prime. So $d|2$. $\therefore\ d = 1$ or $d = 2$.

31. Clearly, $P(1)$ is true. Assume $P(k)$ is true; that is, $3|(k^3 - k)$.
$(k+1)^3 - (k+1) = (k^3 + 3k^2 + 3k + 1) - (k+1)$
Since $3|(k^3 - k)$, the RHS is divisible by 3. $\therefore\ P(k+1)$ is true. Thus $P(n)$ is true $\forall n \geq 1$.

33. When $n = 1$, LHS $= \frac{1}{(2-1)(2+1)} = \frac{1}{3} =$ RHS. $\therefore\ P(1)$ is true.

Assume $P(k)$ is true: $\sum_{i=1}^{k} \frac{1}{(2i-1)(2i+1)} + \frac{k}{2k+1}$. Then

$$\begin{aligned}\sum_{i=1}^{k+1} \frac{1}{(2i-1)(2i+1)} &= \sum_{i=1}^{k} \frac{1}{(2i-1)(2i+1)} + \frac{1}{(2k+1)(2k+3)}\\ &= \frac{k}{2k+1} + \frac{1}{(2k+1)(2k+3)} = \frac{k(2k+3)+1}{(2k+1)(2k+3)}\\ &= \frac{(k+1)(2k+1)}{(2k+1)(2k+3)} = \frac{k+1}{(k+1)+1}\end{aligned}$$

$\therefore$ $P(k) \to P(k+1)$. So the result holds by induction.

35. Let $P(n)$: We must select at least $2n+1$ socks to ensure n matching pairs. By the PHP, $P(1)$ is true. Assume $P(k)$ is true, that is, we must select at least $2k+1$ socks to ensure k matching pairs. Add one matching pair. This ensures $k+1$ matching pairs by selecting $(2k+1)+2 = 2(k+1)+1$ socks. $\therefore$ $P(k) \to P(k+1)$. Thus $P(n)$ is true $\forall n \geq 1$.

37. $\frac{5(n+1)n^2}{2}$ **39.** $2^{n+2} - n - 4$ **41.** $3^{n(n+1)(n+2)/6}$ **43.** $(n+1)! - 1$

45. $\sum_{i=1}^{m} \lfloor 5^i/n \rfloor$, where m is the largest integer such that $5^m \leq n$.

47.
$$t_{19} + t_{20} + t_{21} + t_{22} + t_{23} + t_{24} = t_{25} + t_{26} + t_{27} + t_{28}$$
$$t_{29} + t_{30} + t_{31} + t_{32} + t_{33} + t_{34} + t_{35} = t_{36} + t_{37} + t_{38} + t_{39} + t_{40}$$

49.
$$\begin{aligned}t_n + t_{n-1}t_{n+1} &= n(n+1)/2 + (n-1)n/2 \cdot (n+1)(n+2)/2\\ &= [n(n+1)/2][1 + (n^2+n-2)/2]\\ &= [n(n+1)/2][n(n+1)/2]\\ &= t_n^2\end{aligned}$$

Supplementary Exercises (p. 256)

1.
$$\begin{aligned}(m^2-n^2)^2 + (2mn)^2 &= (m^4 - 2m^2n^2 + n^4) + 4m^2n^2\\ &= m^4 + 2m^2n^2 + n^4 = (m^2+n^2)^2\end{aligned}$$

3. $t_k = k(k+1)/2$. Let $n = m(m+1)/2$. Then

$$\begin{aligned}(2k+1)^2 n + t_k &= \frac{(2k+1)^2 \cdot m(m+1)}{2} + \frac{k(k+1)}{2}\\ &= \frac{[(2k+1)^2(m^2+m) + k(k+1)]}{2}\\ &= \frac{[(2k+1)^2m^2 + (2k+1)^2m) + k(k+1)]}{2}\\ &= \frac{[(2k+1)m + k][(2k+1)m + (k+1)]}{2} = \frac{N(N+1)}{2}\end{aligned}$$

where $N = (2k+1)m + k$ $\therefore$ $(2k+1)^2 n + t_k$ is a triangular number.

5. **proof:** Suppose 111 is a perfect square a^2 in some base b, so $a^2 = b^2 + b + 1 < (b+1)^2$. Then $(b+1/2)^2 = b^2 + b + 1/4 < b^2 + b + 1$. That is, $(b+1/2)^2 < a^2 < (b+1)^2$. This yields $b + 1/2 < a < b + 1$; that is, a lies between $b + 1/2$ and $b + 1$, which is impossible. Thus 111 cannot be a square in any base.

7. 25 and 36 are Duffinian, but 18 and 43 are not.

9. $\sigma(n) = \sum_{\substack{d|n \\ d \neq n}} d + n = s + n$. $\therefore$ None of the n's factors, except 1, divides s if and only if none divides $\sigma(n)$. Thus n is Duffinian if and only if none of n's factors, except 1, divides $\sigma(n)$.

11. In Exercise 59 in Section 4.4, we established that $S_n - S_{n-1} = n(3n^4 + 7n^2 + 2)/12$. Since both S_n and S_{n-1} are positive integers, it follows that $n(3n^4 + 7n^2 + 2)/12$ is also an integer. Consequently, $12 | n(3n^4 + 7n^2 + 2)$.

13. n^3 **15.** Georgia

Chapter 5 Recursion

Exercises 5.1 (p. 273)

1. 1, 4, 7, 10 **3.** 1, 2, 3, 4 **5.** 1, 1, 2, 4

7. 1, 3, 4, 7, 11, 18 **9.** 3

11. $A(0) = 1000$
$A(n) = 1.015A(n-1), n \geq 1$

13. \$1024

15. $a_1 = 1$
$a_n = a_{n-1} + 3, n \geq 2$

17. $a_0 = 0$
$a_n = 2a_{n-1} + 3, n \geq 1$

19. 00000, 00001, 00010,00011,00100,00101,00110,01000,01001,01010, 01011,01100,01101

21. $S_1 = A_1$
$S_n = S_{n-1} \cup A_n, n \geq 2$

23. $S_0 = 1$
$S_n = 2S_{n-1}, n \geq 1$

25. $a_1 = a$
$a_n = ra_{n-1}, n \geq 2$

27. 91

29. 91

31. $f(99) = f(f(110)) = f(100)$
$= f(f(111)) = f(101)$
$= 91$

33. Let k be the smallest integer such $90 \le x + 11k \le 100$. Then

$$\begin{aligned} f(x) &= f(f(x+11)) \\ &= f(f(f(x + 11 \cdot 2))) \\ &\vdots \\ &= f^{k+1}(x + 11k) \end{aligned}$$

Since $90 \le x + 11k \le 100$ and $k \ge 1$, $f(x + 11k) = 91$, by Exercise 43. $\therefore f^{k-1}(x + 11k) = f^k(91)$
Since $f(91) = 91$ by Exercise 32, $f^k(91) = 91$. $\therefore f(x) = 91$ for $0 \le x < 90$.

35. 429

37. 1, 2, 1.66666667, 1.5, 1.66666667, 1.6, 1.625, 1. 61538462, 1.61904762, 1.61764706, 1.61818182, 1.61797753

39.
$$\begin{aligned} 2F_{n-2} + F_{n-3} &= F_{n-2} + (F_{n-2} + F_{n-3}) \\ &= F_{n-2} + F_{n-1} \\ &= F_n \end{aligned}$$

41. **proof** (by PMI)**:** When $n = 1$, $F_{5n} = F_5 = 5$ is divisible by 5. So $P(1)$ is true. Assume $P(k)$ is divisible by 5: F_{5k} is divisible by 5.

$$\begin{aligned} F_{5(k+1)} &= F_{5k+5} = F_{5k+4} + F_{5k+1} \\ &= F_{5k+2} + 2F_{5k+3} \\ &= F_{5k+2} + 2(F_{5k+1} + F_{5k+2}) \\ &= 3F_{5k+2} + 2F_{5k+1} \\ &= 3(F_{5k+1} + F_{5k}) + 2F_{5k+1} \\ &= 3F_{5k} + 5F_{5k+1} \end{aligned}$$

which is clearly divisible by 5. $\therefore$ $P(k+1)$ is true. Thus the result follows by induction.

43. **proof** (by PMI): When $n = 1$, $F_1 = F_2 = 1 \le 2$. $\therefore P(1)$ is true. Assume $P(i)$ is true for every $i \le k$. Then

$$\begin{aligned} F_{k+1} &= F_k + F_{k-1} \\ &\le 2^k + 2^{k-1} \\ &\le 2^k + 2^k \\ &= 2^{k+1} \end{aligned}$$

$\therefore$ $P(k+1)$ is true. Thus the result follows by induction.

45. $|A| = -1$. $|A^n| = F_{n+1}F_{n-1} - F_n^2$. But $|A^n| = |A|^n = (-1)^n$

47. **proof** (by PMI): Let $P(n)$ denote the given statement. When $n = 1$, RHS $= 3 + \sum_{k=1}^{0} L_k = 3 + 0 = 3 = L_2 =$ LHS. So $P(2)$ is true. Assume $P(i)$ is true: $L_{2i} = 3 + \sum_{k=1}^{2i-1} L_k$. To show that $P(i+1)$ is true:

$L_{2i+2} = L_{2i} + L_{2i+1}$, by definition.

$$= \left(3 + \sum_{k=1}^{2i-2} L_k\right) + L_{2i+1}$$

$$= \left(3 + \sum_{k=1}^{2i-2} L_k\right) + (L_{2i-1} + L_{2i})$$

$$= \left(3 + \sum_{k=1}^{2i} L_k\right)$$

$\therefore$ $P(i+1)$ is true. Thus $P(n)$ is true $\forall n \in N$.

49. Since α and β are solutions of the equation $x^2 = x + 1$, $\alpha + \beta = 1$.

$$b_2 = \frac{\alpha^2 - \beta^2}{\alpha - \beta} = \alpha + \beta = 1$$

51. $u_1 = \alpha + \beta = 1$, by Exercise 49.

53. Since α and β are solutions of the equation $x^2 = x + 1, \alpha^2 = \alpha + 1$ and $\beta^2 = \beta + 1$.

$$\begin{aligned} u_{n-1} + u_{n-2} &= (\alpha^{n-1} + \beta^{n-1}) + (\alpha^{n-2} + \beta^{n-2}) \\ &= \alpha^{n-2}(\alpha + 1) + \beta^{n-2}(\beta + 1) \\ &= \alpha^{n-2} \cdot \alpha^2 + \beta^{n-2} \cdot \beta^2 \\ &= \alpha^n + \beta^n = u_n \end{aligned}$$

55. 2

57. $a_1 = 1$
$a_n = a_{n-1} + n, n \geq 2$

59. $a_1 = 1, a_2 = 3$
$a_n = a_{n-1} + a_{n-2}, n \geq 3$

61. $a_1 = 2, a_2 = 3$
$a_n = a_{n-1} + a_{n-2}, n \geq 3$
Let $b_n = a_{n-2}$. Then $b_3 = a_1 = 2$ and $b_4 = a_2 = 3$. Also, $b_{n-1} + b_{n-2} = b_n$. $\therefore$ $b_n = F_n$, $n \geq 3$. Thus, $a_n = F_{n+2}, n \geq 1$

63. 15

65. 3

67. 7

69. **proof** (by PMI): When $n = 0$,

$$\begin{aligned} \text{LHS} &= A(2,0) = A(1,1) \\ &= A(0, A(1,0)) = 1 + A(1,0) \\ &= 1 + A(0,1) \\ &= 1 + 2 = 3 = \text{RHS} \end{aligned}$$

$\therefore$ $P(0)$ is true. Assume $P(k)$: $A(2,k) = 3 + 2k$. Then

$$\begin{aligned}A(2,k+1) &= A(1, A(2,k))\\ &= A(1, 3+2k)\\ &= A(0, A(1, 2+2k))\\ &= 1 + A(1, 2+2k)\\ &= 1 + A(0, A(1, 1+2k))\\ &= 2 + A(1, 1+2k)\\ &= 2 + A(0, A(1, 2k))\\ &= 3 + A(1, 2k)\\ &= 3 + (2k+2)\\ &= 3 + 2(k+1)\end{aligned}$$

$\therefore$ $P(k+1)$ is true. Thus the result follows by PMI.

71. **proof** (by PMI): When $n = 0$,

$$\begin{aligned}\text{LHS} &= A(3,0) = A(2,1)\\ &= A(1, A(2,0)) = A(1,3)\\ &= 2 + 3 = 2^3 - 3 = \text{RHS}\end{aligned}$$

$\therefore$ $P(0)$ is true. Assume $P(k)$:

$$A(3,k) = 2^{k+3} - 3$$

$$\begin{aligned}A(3,k+1) &= A(2, A(3,k))\\ &= A(2, 2^{k+3} - 3)\\ &= 3 + 2(2^{k+3} - 3)\\ &= 2^{k+4} - 3\end{aligned}$$

$\therefore$ $P(k+1)$ is true. Thus the result follows by PMI.

Exercises 5.2 (p. 282)

1. $s_n = 2^n, n \geq 1$

3. $a_n = n(n+1)/2 + 1, n \geq 1$

5. $a_n = 2n(n+1), n \geq 0$

7. $s_n = n(n+1)(2n+1)/6, n \geq 1$

9. When $n = 0, s_0 = 2^0 = 1$. $\therefore$ $P(0)$ is true. Assume $P(k)$ is true: $s_k = 2^k$. Then $s_{k+1} = 2s_k = 2(2^k) = 2^{k+1}$. $\therefore$ $P(k+1)$ is true. Consequently the result follows by PMI.

11. When $n = 0$, $a_0 = 0 \cdot 1/2 + 1 = 1$. $\therefore$ $P(0)$ is true. Assume $P(k)$ is true: $a_k = k(k+1)/2+1$. Then $a_{k+1} = a_k + (k+1) = [k(k+1)/2+1] +$ (k + 1) $=$ $(k+1)(k+2)/2 + 1$. $\therefore$ $P(k+1)$ is true. Thus the result follows by PMI.

13. When $n = 0, a_0 = 2(0)(0+1) = 0$. $\therefore P(0)$ is true. Assume $P(k)$ is true: $a_k = 2k(k+1)$. Then $a_{k+1} = a_k + 4(k+1) = 2k(k+1) + 4(k+1) = 2(k+1)(k+2)$ $\therefore P(k+1)$ is true. Thus the result is true by PMI.

15. When $n = 1$, $s_1 = 1 \cdot 2 \cdot 3/6 = 1$. $\therefore P(1)$ is true. Assume $P(k)$ is true: $s_k = k(k+1)(2k+1)/6$. Then $s_{k+1} = s_k + (k+1)^2 = k(k+1)(2k+1)/6 + (k+1)^2 = (k+1)(k+2)(2k+3)/6$. $\therefore P(k+1)$ is true. Thus the result follows by PMI.

17. When $n = 0$, $A(0) = 1000(1.08)^0 = 1000$. So $P(0)$ is true. Now assume $P(k)$ is true: $A(k) = 1000(1.08)^k$. Then $A(k+1) = (1.08)A(k) = 1000(1.08)^{k+1}$.
$\therefore P(k+1)$ is true. Thus the given result follows by PMI.

19. When $n = 1, f(1) = 1 \cdot 2/2 + 1 = 2$. So, $P(1)$ is true. Now assume $P(k)$ is true: $f(k) = k(k+1)/2 + 1$. Then $f(k+1) = f(k) + (k+1) = k(k+1)/2 + 1 + (k+1) = (k+1)(k+2)/2 + 1$. $\therefore P(k+1)$ is true. Thus the given result follows by PMI.

21. **proof** (by PMI): Let $P(n)$ denote the statement that a_n is a solution of the recurrence relation. Clearly, $P(1)$ is true. Assume $P(k)$ is true for an arbitrary $k \geq 1$: $a_k = a_0 + \sum_{i=1}^{k} f(i)$. Then

$$a_{k+1} = a_k + f(k+1) = [a_0 + \sum_{i=1}^{k} f(i)] + f(k+1) = a_0 + \sum_{i=1}^{k+1} f(i)$$

23. **proof** (by PMI): Let $P(n)$ denote the statement that a_n is a solution of the recurrence relation. Clearly, $P(1)$ is true. Assume $P(k)$ is true:

$a_k = c^k a_0 + \sum_{i=1}^{k} c^{k-i} f(i)$. Then

$$\begin{aligned} a_{k+1} = ca_k + f(k+1) &= c[c^k a_0 + \sum_{i=1}^{k} c^{k-i} f(i)] + f(k+1) \\ &= c^{k+1} a_0 + \sum_{i=1}^{k+1} c^{k+1-i} f(i) \end{aligned}$$

$\therefore P(k+1)$ is true. Thus the result follows by PMI.

25. **case 1** Let $n = 1$. Then the statement $x \leftarrow x + 1$ is executed zero times. So $a_1 = 0$.
case 2 Let $n > 1$ and even. When n is even, $\lfloor n/2 \rfloor = n/2$.
$\therefore a_n = a_{n-1} + n/2$
case 3 Let $n > 1$ and odd. When n is even, $\lfloor n/2 \rfloor = (n-1)/2$.
$\therefore a_n = a_{n-1} + (n-1)/2$

27. $a_1 = 1$
$a_n = a_{n-1} + \lceil n/2 \rceil, n \geq 2$

29. $a_n = \begin{cases} n(n+2) & \text{if } n \text{ is even} \\ (n+1)^2/4 & \text{otherwise} \end{cases}$

31. $a_n = n(n+1)/2, n \geq 1$

33. $a_n = n(n+1)(2n+1)/6, n \geq 1$

35. $a_n = n(n+1)(n+2)/6, n \geq 1$

37. $a_n = [n(n+1)/2]^2, n \geq 1$

39. $t_n = n(n+1)/2$

41.

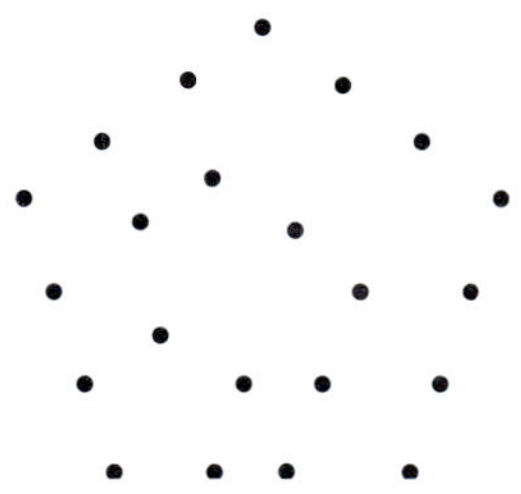

43. $p_n = n(3n-1)/2$

45. $h_n = \begin{cases} 1 & \text{if } n = 1 \\ h_{n-1} + 4(n-1) + 1 & \text{if } n \geq 2 \end{cases}$

47. $p_n + t_n - n = n(3n-1)/2 + n(n+1)/2 - n$
$= 2n^2 - n = (2n-1)n = h_n$

49. $T_1 = 1$
$T_n = T_{n-1} + n(n+1)/2, n \geq 2$

51. We shall prove by PMI that $T_n = n(n+1)(n+2)/6 \; \forall n \geq 1$. When $n = 1$, $T_1 = 1 \cdot 2 \cdot 3/6 = 1$, which is true. So the formula works when $n = 1$.
Assume it works for an arbitrary positive integer k. Using the recurrence relation,

$$T_{k+1} = T_k + (k+1)(k+2)/2 = k(k+1)(k+2)/6 + (k+1)(k+2)/2 = (k+1)(k+2)(k+3)/6$$

Thus, by PMI, the formula holds for every $n \geq 1$.

53.
$$\begin{aligned} S_n &= S_{n-1} + n^2 \\ &= S_{n-2} + (n-1)^2 + n^2 \\ &= S_{n-2} + (n-2)^2 + (n-1)^2 + n^2 \\ &\;\vdots \\ &= S_1 + 2^2 + 3^2 + \cdots + n^2 \\ &= 1^2 + 2^2 + 3^2 + \cdots + n^2 \\ &= n(n+1)(2n+1)/6 \end{aligned}$$

So we conjecture that $S_n = n(n+1)(2n+1)/6$, where $n \geq 1$.

55. 1

57. 2

59. $a_0 = 1, a_1 = 2$
$a_n = a_{n-1} + a_{n-2}, n \geq 2$

61. $a_n = \begin{cases} 1 & \text{if } n = 0 \\ (1 + a_{n-1})/k & \text{if } n \geq 1 \end{cases}$

63. **proof** (by PMI): Let $P(n)$ denote the given statement. Clearly $P(0)$ is true. So assume $P(m)$ is true for any $m \geq 0$. Then

$$\begin{aligned} a_{m+1} &= (1 + a_m)/k \\ &= 1 + \frac{k^m + k - 2}{k^m(k-1)k} = \frac{k^{m+1} + k - 2}{k^{m+1}(k-1)} \end{aligned}$$

Thus $P(k)$ implies $P(k+1)$, so the result follows by PMI.

Exercises 5.3 (p. 296)

1. yes

3. yes

5. no

7. yes

9. $a_n = 2(-1)^n + 2^n, n \geq 0$

11. $a_n = 3(-2)^n + 2 \cdot 3^n, n \geq 0$

13. $a_n = F_{n+2}, n \geq 0$

15. $L_n = \alpha^n + \beta^n$

17. $a_n = 2 \cdot 3^n - n \cdot 3^n, n \geq 0$

19. $a_n = 2^n - 3^n + n \cdot 3^n, n \geq 0$

21. $a_n = 2^n + (-2)^{n+1} + 2 \cdot 3^n - (-4)^n, n \geq 0$

23. $a_n = 3 \cdot 2^n - n2^n + n^2 2^n + 2 \cdot 3^n, n \geq 0$

25. $a_n^{(p)} = an + b$

27. $a_n^{(p)} = an^2 + bn + c$

29. $a_n^{(p)} = (an + b)2^n$ if 2 is a characteristic root; otherwise, $a_n^{(p)} = n^m(an + b)2^n$.

31. $a_n^{(p)} = an^2 2^n$

33. $a_n^{(p)} = n^2(an^2 + bn + c)2^n$

35. $a_n = 2^n - 1, n \geq 0$

37. $a_n = 11 \cdot 4^n - 11 \cdot 3^n - n \cdot 3^{n+1}, n \geq 0$

39. $a_n = 1 + n(n+1)/2, n \geq 0$

41. Since r_n and s_n are solutions of equation (9), $r_n = ar_{n-1} + br_{n-2}$ and $s_n = as_{n-1} + bs_{n-2}$. $\therefore r_n + s_n = a(r_{n-1} + s_{n-1}) + b(r_{n-2} + s_{n-2})$. Thus a_n is a solution of (9).

43. Since α is a root of $x^3 - ax^2 - bx - c = 0$ with multiplicity 3,

$$x^3 - ax^2 - bx - c = (x - \alpha)^3 = x^3 - 3\alpha x^2 + 3\alpha^2 x - \alpha^3$$

$$\therefore 3\alpha = a,\ 3\alpha^2 = -b, \text{ and } \alpha^3 = c \qquad (1)$$

$$\text{Also, } \alpha^3 = a\alpha^2 + b\alpha + c \qquad (2)$$

a) Multiply equation (2) by α^{n-3}: $\alpha^n = a\alpha^{n-1} + b\alpha^{n-2} + c\alpha^{n-3}$ (3)
$\therefore \alpha^n$ is a solution.

b) When $a_n = n\alpha^n$,

$$\begin{aligned}
&aa_{n-1} + ba_{n-2} + ca_{n-3} \\
&\quad = a(n-1)\alpha^{n-1} + b(n-2)\alpha^{n-1} + c(n-3)\alpha^{n-3} \\
&\quad = n(a\alpha^{n-1} + b\alpha^{n-2} + c\alpha^{n-3}) - (a\alpha^{n-1} + 2b\alpha^{n-2} + 3c\alpha^{n-3}) \\
&\quad = n\alpha^n - (3\alpha^n - 6\alpha^n + 3\alpha^n), \text{ by equations (1)} \\
&\quad = a_n = n\alpha^n
\end{aligned}$$

$\therefore n\alpha^n$ is a solution.

c) When $a_n = n^2\alpha^n$,

$$\begin{aligned}
&aa_{n-1} + ba_{n-2} + ca_{n-3} \\
&\quad = a(n-1)^2\alpha^{n-1} + b(n-2)^2\alpha^{n-1} + c(n-3)^2\alpha^{n-3}
\end{aligned}$$

$$= n^2(a\alpha^{n-1} + b\alpha^{n-2} + c\alpha^{n-3}) - 2n(a\alpha^{n-1} + 2b\alpha^{n-2} + 3c\alpha^{n-3})$$
$$+ (a\alpha^{n-1} + 4b\alpha^{n-2} + 9c\alpha^{n-3})$$
$$= n^2\alpha^n - 2n(3\alpha^n - 6\alpha^n + 3\alpha^n) + (3\alpha^n - 12\alpha^n + 9\alpha^n), \text{ by (1)}$$
$$= a_n = n^2\alpha^n$$

$\therefore n^2\alpha^n$ is a solution.

Exercises 5.4 (p. 306)

1. $\dfrac{2}{x-1} - \dfrac{1}{x+3}$ **3.** $\dfrac{2}{1+2x} + \dfrac{3}{1-3x}$ **5.** $-\dfrac{8}{13(2+3x)} + \dfrac{7x+4}{13(x^2+1)}$

7. $\dfrac{1-x}{x^2+2} + \dfrac{2x}{x^2+3}$ **9.** $\dfrac{x-1}{x^2+1} + \dfrac{2x+1}{x^2-x+1}$ **11.** $a_n = 2^n, n \ge 0$

13. $a_n = 2n - 1, n \ge 1$ **15.** $a_n = 3(-2)^n - 2^n, n \ge 0$

17. $a_n = 5 \cdot 2^n - 3^n, n \ge 0$ **19.** $a_n = F_{n+3}, n \ge 0$

21. $a_n = (2n+3)2^n, n \ge 0$ **23.** $a_n = 3(-2)^n + 2^n - 3^n, n \ge 0$

25. $a_n = 2^n + 3n \cdot 2^n - 3^n, n \ge 0$ **27.** $a_n = n \cdot 2^{n+1} - n^2 2^n, n \ge 0$

29. $a_n = (3n^2 - n + 2)(-1)^n - 2^{n+1}, n \ge 0$

Exercises 5.5 (p. 314)

1. 2 **3.** 13 **5.** 2

7. 7 **9.** $a_n = F_n - 1$

11. **Algorithm Fibonacci(n)**

```
Begin (* algorithm *)
  if (n = 1) or (n = 2) then
    answer ← 1
  else
    begin
      previous ← 1
      current ← 1
      for i = 3 to n do
      begin
        answer ← previous +
                 current
        previous ← current
        current ← answer
      endfor
    endelse
End (* algorithm *)
```

13. yes

15. 2, 3, 5, 7, 9, 11,17, 19

17. **Algorithm Lucas (n)**

```
(* This algorithm computes
the nth Lucas number using
recursion *)
Begin (* Lucas *)
  if n = 1 then
    Lucas ← 1
  else if n = 2 then
      Lucas ← 3
    else
      Lucas ← Lucas(n-1) +
              Lucas (n-2)
End (* Lucas *)
```

19. **Algorithm sum (X,i,n)**

```
(* This algorithm computes the
sum of the numbers xi through
xn in a list X from left to
right. *)
Begin (* sum *)
  if i = n then
    sum ← xi
  else
    sum ← xi + sum (X,i+1,n)
End (* sum *)
```

21. **Algorithm product (X,i,n)**

```
(* This algorithm computes
the product of the numbers xi
through xn in a list X from
left to right *)
Begin (* product *)
  if i = n then
    product ← xi
  else
    product ← xi * product
                (X,i+1,n)
End (* product *)
```

23. **Algorithm maximum (X,i,n,max)**

```
(* This algorithm determines
the maximum of the elements
xi through xn of a list X. It
is returned in the variable
max. *)
Begin (* maximum *)
  if i ≤ n then
  begin (* if *)
    if i = 1 then
      max ← x1
    else if xi > max then
            max ← xi
    maximum (X,i+1,n,max)
  endif
End (* maximum *)
```

25. **Algorithm print in order (X,i,n)**

```
(* This algorithm prints the
elements xi through xn of a
list X using recursion *)
Begin (* print in order *)
  if i = n then
     write(xn)
  else
  begin
    write(xi)
    print in order(X,i+1,n)
  endelse
End (* print in order *)
```

27. See Algorithm 9 in Section 6.

29. **Algorithm palindrome (w,i,j,flag)**

```
(* This algorithm, using
recursion, determines if a
word w of length j - i + 1 is a
palindrome or not. *)
Begin (* palindrome *)
  if i = j then
    flag ← true
  else if wi = wj then
         begin
           flag ← true
           palindrome (w,i+1,
             j-1, flag)
         endif
       else
         flag ← false
End (* palindrome *)
```

31. **Algorithm bubble sort (X,n)**

```
(* This algorithm sorts a list
X of n elements. Sorted is
a boolean variable. Swap is
an algorithm which swaps two
elements. *)
Begin (* bubble sort *)
  sorted ← false
  for j = 1 to n do
    if xj > xj+1 then
      begin
        swap(xj,xj+1)
        sorted ← false
      endif
  if not sorted then
    bubble sort (X,n-1)
End (* bubble sort *)
```

33. 4, 5, 6, 7, 8, 11, 13

Exercises 5.6 (p. 318)

In Exercises 1–29, $P(n)$ denotes the following statement: The algorithm works correctly.

1. proof: When $n = 0$, the algorithm returns the value 1 from line 2. So it works correctly when $n = 0$.

Assume $P(k)$ is true, that is, assume that the algorithm returns the value $k!$ for an arbitrary integer $k \geq 0$. Suppose we invoke the algorithm with $n = k + 1$. Then factorial $= (k + 1)$ * factorial $= (k + 1) * k! = (k + 1)!$.
$\therefore P(k + 1)$ is true. Thus, by induction, the algorithm works correctly $\forall n \geq 0$.

3. proof: Let b_n denote the number of moves needed. When n = 1, disk 1 is moved from X to Z and $b_1 = 1$ (line 4). So $P(1)$ is true.

Assume $P(k)$ is true. Suppose the algorithm is invoked with $n = k+1$. By line 8, it correctly transfers the top $n - 1 = k$ disks from X to Y; count $= b_k$. Disk $k + 1$ is moved from X to Z (line 9); count $= b_k + 1$. By the IH, the algorithm moves the k disks at Y to Z using b_k moves (line 11). Now count $= (b_k + 1) + b_k = 2b_k + 1$. Thus $P(k) \rightarrow P(k + 1)$. Thus the result follows by PMI.

5. proof (by strong induction): Initially, low = 1 and high = n. When $n = 1$, mid = 1 (line 4). If key $= x_1$, then the list contains *key*. If key $\neq x_1$, the algorithm is invoked in line 9 or 11. In either case, the condition in line 1 is not satisfied; so the algorithm returns the value false for **found**.

Assume $P(i)$ is true $\forall i \leq k, k \geq 1$. Suppose the algorithm is invoked with $n = k + 1$; then low = 1, high $= k + 1$, and mid $= \lfloor (k + 2)/2 \rfloor$. If key $= X_{\text{mid}}$, $P(k + 1)$ is true. Otherwise, line 9 or line 11 is executed. In either case, the algorithm is invoked for a sublist containing $< k$ elements. By the IH, the algorithm works correctly in each case. So $P(k) \rightarrow P(k + 1)$. Thus the result follows by PMI.

7. proof: When $n = 1$, sum $= x_1$ by line 2. So $P(1)$ is true. Assume $P(k)$ is true for some $k \geq 1$. Suppose the algorithm is invoked for a list with $k + 1$ elements. If $i < k + 1$, then the algorithm computes the correct sum of the first k items by the hypothesis. When $i = k + 1$, line 2 is excuted. Then sum $= x_{k+1}$ is added to the previous sum in line 4. $\therefore P(k + 1)$ is true. Thus, by PMI, the result holds $\forall n \geq 1$.

9. proof: When $n = 1$, product $= x_1$ by line 2. So $P(1)$ is true. Assume $P(k)$ is true for an arbitrary $k \geq 1$. Suppose the algorithm is invoked for a list with $k+1$ elements. If $i < k+1$, then the algorithm correctly computes the product of the first k elements by the inductive hypothesis. When $i = k + 1$, line 2 is executed. Then product $= x_{k+1}$ * previous product in line 4. $\therefore P(k + 1)$ is true. Thus the result follows by PMI.

11. **proof:** When $n = 1$, the list contains one element. So $P(1)$ is true. Now assume $P(k)$ is true. Consider a list X with $k+1$ elements. When $i = 1$, max $= x_1$ from line 3. In line 7, the algorithm is invoked for a list with k elements. By the inductive hypothesis, the correct maximum value of max and the k elements is returned in line 7. So $P(k + 1)$ is true. Thus the result follows by PMI.

13. **proof:** When $n = 1$, x_1 is printed by line 2. So $P(1)$ is true. Assume $P(k)$ is true. Consider a list with $k + 1$ elements. Then lines 4 and 5 are executed. In line 5, x_1 is printed. The remaining k elements are printed in order by line 6, by the inductive hypothesis. $\therefore P(k + 1)$ is true. Thus, by PMI, $P(n)$ is true $\forall n \geq 1$. Hence the result.

15. Same as Example 4.36.

17. **proof** (by strong induction): When $n = 1, i = 1 = j$; flag = true from line 2. $\therefore P(1)$ is true. Assume $P(i)$ is true $\forall i \leq k$. Consider a word with $k + 1$ elements.
case 1 If $w_1 = w_{k+1}$, then flag = true by line 5. Further, the algorithm is invoked with $i = 2$ and $j = k$ for a word with $k - 1$ elements. By the inductive hypothesis, the algorithm works correctly for such a word.
case 2 If $w_1 \neq w_{k+1}$, then flag = false by line 9. Then also the algorithm works correctly.
Thus in both cases, $P(k+1)$ is true. $\therefore$ By PMI, the result holds $\forall n \geq 1$.

19. 0 **21.** 3 **23.** $\lg n$

25. Clearly, the algorithm works when $n = 1$ or $n = 2$. So $P(1)$ and $P(2)$ are true. Assume $P(k)$ is true for any $k \geq 2$. Suppose the algorithm is invoked with $n = k+1$ (≥ 3). In line 9, the previous number (F_{k-1}) and the current number (F_k) are added to yield F_{k+1}. So $P(k) \to P(k + 1)$. Thus the result follows by PMI.

Exercises 5.7 (p. 329)

1. $O(n^2)$ **3.** $O(n^2)$ **5.** $O(n^2)$ **7.** $O(n2^n)$

9. 3 **11.** 3 **13.** 5 **15.** 5

17. $b_n = n, n \geq 0$ **19.** $O(n^2)$ **21.** $O(n^3)$

23. $O(n^2)$ **25.** $O(n^4)$

27. $b_n = \begin{cases} 0 & \text{if } n = 1 \\ b_{n-1} + (n-1) & \text{if } n \geq 2 \end{cases}$ **29.** $O(n^2)$

31. $c_n = nb + 1, n \geq 0$ **33.** $c_n = [n(n + 1)/2 - 1]b, n \geq 1$

35. $f(n) = (n - 1)^2 + 1, n \geq 1$ **37.** $f(n) = \lfloor \lg n \rfloor + 1$

39. $f(n) =$ (no. of bits in the binary representation of n) + (no. of multiplications)

$$\begin{aligned} &\le \text{(no. of bits)} + \text{(no. of bits in } n-1) \\ &= (\lfloor \lg n \rfloor + 1) + \lfloor \lg n \rfloor \\ &= 2\lfloor \lg n \rfloor + 1 \\ &= O(\lg n) \end{aligned}$$

41. $O(n^3)$

43. $c_n = (a+1)n - 1 = O(n)$

45.
$$\begin{aligned} f(n) &= af(n/b) + g(n) \\ &= a^2 f(n/b^2) + ag(n/b) + g(n) \\ &= a^3 f(n/b^3) + a^2 g(n/b^2) + ag(n/b) + g(n) \\ &\;\;\vdots \\ &= a^k f(n/b^k) + \sum_{i=0}^{k-1} a^i g(n/b^i) \\ &= a^k f(1) + \sum_{i=0}^{k-1} a^i g(n/b^i) \end{aligned}$$

47. Let $n = 2^k$. By Exercise 45,

$$\begin{aligned} a_n &= 2 \cdot 0 + \sum_0^{k-1} 2^i (n/2^i) \\ &= 0 + nk = n\lg n \\ &= O(n\lg n) \end{aligned}$$

49. By Exercise 48,

$$\begin{aligned} f(n) &= da^k + cn\left[\sum_0^{k-1} (a/b)^i\right] \\ &= 0 + cn\left(\sum_0^{k-1} 1\right) = ckn \\ &= O(n\lg n) \end{aligned}$$

51. $c_n = 2n - 2$

53. $5a + 12b, 7a + 19b$

55. Since n is a power of 2,

$$t(n) = \begin{cases} a & \text{if } n = 1 \\ 2t(n/2) + bn & \text{otherwise} \end{cases}$$

$\therefore$ By Exercise 45,

$$\begin{aligned} t(n) &= 2^k t(1) + \sum_{i=0}^{k-1} 2^i \frac{bn}{2^i} \\ &= a2^k + kbn \\ &= an + kbn \\ &= Cn + Ckn, \text{ where } C = \max\{a, b\} \\ &\le Ckn \\ &= O(n\lg n), \text{ since } k = \lg n \end{aligned}$$

57. By Exercise 56,

$$\begin{aligned} f(n) &= 2cn^2 + (d-2c)n \\ &\le Cn^2 + Cn \\ &\le 2Cn^2 \\ &= O(n^2) \end{aligned}$$

where $C = \max\{2c, d-2c\}$.

59. $h_n = \begin{cases} 1 & \text{if } n = 1 \\ h_{n-1} + 1/n & \text{if } n \ge 2 \end{cases}$

61. **proof** (by induction): Let $P(m)$ be the given statement. Since $h_{2^0} = h_1 = 1 \ge 1 + 0/2$, $P(0)$ is true.

Assume $P(k)$ is true: $h_{2^k} \ge 1 + k/2$. Then

$$\begin{aligned} h_{2^{k+1}} &= \sum_{i=1}^{2^{k+1}} \frac{1}{i} \\ &= \sum_{i=1}^{2^k} \frac{1}{i} + \frac{1}{2^k+1} + \cdots + \frac{1}{2^{k+1}} \\ &\ge \left(1 + \frac{k}{2}\right) + \frac{1}{2^k+1} + \cdots + \frac{1}{2^{k+1}} \\ &\ge \left(1 + \frac{k}{2}\right) + 2^k \cdot \frac{1}{2^{k+1}} \\ &= \left(1 + \frac{1}{k}\right) + \frac{1}{2} = 1 + \frac{k+1}{2} \end{aligned}$$

$\therefore P(k) \to P(k+1)$. So the result follows by PMI.

63.
$$\begin{aligned} h_n &= \sum_{i=1}^{n} \frac{1}{i} \le \int_1^n \frac{dx}{x} \\ &= \ln x \Big|_1^n \\ &= \ln n - \ln 1 = \ln n - 0 \\ &= \ln n = (\ln 2)\lg n \\ &= O(\lg n) \end{aligned}$$

65. **proof:**

case 1 Let n be a power of b, say, $n = b^k$. Then

$$\begin{aligned} f(n) &= af(n/b) + c \\ &= a^2 f(n/b^2) + ac + c \\ &= a^3 f(n/b^3) + a^2c + ac + c \\ &\vdots \\ &= a^k f(n/b^k) + a^{k-1}c + \cdots + a^2c + ac + c \\ &= a^k d + c \sum_{i=0}^{k-1} a^i, \qquad \text{(1) since } f(n/b^k) = f(1) = d. \end{aligned}$$

Since $a = 1, f(n) = d + ck = d + c\log_b n$

$$= O(\lg n) \qquad (\text{Note: } \log_b n = \log_2 n \cdot \log_b 2)$$

case 2 Let n not be a power of b.
Suppose that $b^k < n < b^{k+1}$ for some positive integer k. Since f is nondecreasing, this implies that $f(n) \le f(b^{k+1})$.

If $a = 1$, then by subcase (1) above,

$$f(b^{k+1}) = d + c(k+1) = (c+d) + ck$$
$$\le (c+d) + c\log_b n, \quad \text{since } k < \log_b n.$$
$$\therefore \quad f(n) \le (c+d) + c\log_b n$$
$$= O(\lg n)$$

67. Using Exercise 45 with $g(n) = cn^2$ and $n = b^k$,

$$f(n) = a^k f(1) + \sum_{i=0}^{k-1} a^i (cn^2/b^{2i})$$
$$= a^k d + cn^2 \sum_{i=0}^{k-1} (a/b^2)^i$$
$$= a^k d + cn^2 \sum_{i=0}^{k-1} 1, \quad \text{if } a = b^2$$
$$= n^2 d + cn^2 k, \quad \text{since } a^k = b^{2k} = n^2$$
$$= n^2 d + cn^2 \log_b n$$

69. If $a = b^2$, by Exercise 67,

$$f(n) = n^2 d + cn^2(\log_2 n)(\log_b 2) = n^2 d + Cn^2 \lg n, \text{ where } C = c\log_b 2.$$
$$= O(n^2 \lg n)$$

If $a \ne b^2$, then by Exercise 68, $f(n) = An^2 + Ba^k$. If $a < b^2$, $a^k < b^{2k}$; that is, $a^k < n^2$.

$$\therefore \quad f(n) \le An^2 + Bn^2$$
$$= (A+B)n^2 = O(n^2)$$

If $a > b^2$, then $a^k > b^{2k}$; that is, $a^k > n^2$.

$$\therefore \quad f(n) \le Aa^k + Ba^k$$
$$= (A+B)a^k = (A+B)a^{\log_b n}$$
$$= (A+B)n^{\log_b a} = O(n^{\log_b a})$$

Review exercises (p. 334)

1. 7

3. $a_1 = 0 = a_2$
$a_n = a_{n-1} + a_{n-2} + 1, n \geq 3$

5. $h(n) = 2n(n+1), n \geq 1$

7. $a_n = n(n+1)(n+2)/3, n \geq 1$

9. $a_n = 2^n - 1, n \geq 1$

11. Let $P(n)$: $a_n = n(n+1)(n+2)/3$. When $n = 1$, $a_1 = 1 \cdot 2 \cdot 3/3 = 2$. So $P(1)$ is true. Assume $P(k)$ is true. Then $a_{k+1} = a_k + (k+1)(k+2) = k(k+1)(k+2)/3 + (k+1)(k+2) = (k+1)(k+2)[k/3+1] = (k+1)(k+2)(k+3)/3$. Thus $P(k) \rightarrow P(k+1)$. So the result follows by PMI.

13. Let $P(n)$: $a_n = 2^n - 1$. Since $a_1 = 2^1 - 1 = 1$, $P(1)$ is true. Assume $P(k)$ is true: $a_k = 2^k - 1$. Then $a_{k+1} = a_k + 2^k = (2^k - 1) + 2^k = 2^{k+1} - 1$. So $P(k) \rightarrow P(k+1)$. Thus the result follows by PMI.

15. $a_n = F_{n+2}$

17. $a_n = \frac{6}{5}(-1)^n + \frac{14}{5}2^n - \frac{28}{25}(-2)^n + \frac{28}{25}3^n, n \geq 0$

19. $a_n = -\frac{333}{144} \cdot 3^n + \frac{173}{144} \cdot 7^n + \frac{5}{12}n + \frac{10}{9}, n \geq 0$

21. $a_n = \lfloor (n+1)/2 \rfloor$
$\leq (n+1)/2$
$= O(n)$

23. $c_n = (n-1)n/2$

25. 8

27. $c_1 = 0$
$c_n = c_{n-1} + 3n, n \geq 2$

29. $c_n = 3n(n+1)/2$
$= O(n^2)$

31. 9

33. $b_n = 2n$

35. $a_1 = 2, a_2 = 4, a_3 = 7$
$a_n = a_{n-1} + a_{n-2} + a_{n-3}, n \geq 4$

37. $a_n = F_n$

39. **Algorithm sum (n)**

```
Begin (* sum *)
  if n = 1 then
    sum ← 1
  else
    sum = sum(n-1) + (2n-1)
End (* sum *)
```

41. Let $P(n)$: The algorithm works correctly $\forall\, n \geq 1$. When $n = 1$, sum $= 1$, by line 2. So $P(1)$ is true.

Assume $P(k)$ is true. Suppose $n = k+1$. Then, by line 4, sum $=$ sum $(k) + (2k+1)$, which is the sum of the first k odd positive integers, by the IH. $\therefore P(k) \rightarrow P(k+1)$. Thus the result follows by PMI.

43. $c_1 = 0$
$c_n = c_{n-1} + 2, n \geq 2$

45. $c_n = O(n)$

47. **proof** (by PMI): When $n = 1$, LHS $= 1 =$ RHS; so the result is true when $n = 1$. Now assume it is true for an for an arbitrary positive integer k. Then

$$\sum_{i=1}^{k+1} F_{2i-1} = \sum_{i=1}^{k} F_{2i-1} + F_{2k+1}$$
$$= F_{2k} + F_{2k+1} = F_{2(k+1)}$$

So, by PMI, the result is true $\forall n \geq 1$.

49. **proof** (by PMI): When $n = 1$, LHS $= L_1 = 1 = L_3$; so the result is true when $n = 1$. Now assume it is true for an arbitrary positive integer k. Then

$$\sum_{i=1}^{k+1} L_i = \sum_{i=1}^{k} L_i + L_{k+1}$$
$$= (L_{k+2} - 3) + L_{k+1}$$
$$= L_{k+3} - 3$$

Thus, the result is true $\forall n \geq 1$.

51. **proof** (by PMI): When $n = 1$, LHS $= 3 =$ RHS; so the result is true when $n = 1$. Now assume it is true for an arbitrary positive integer k. Then

$$\sum_{i=1}^{k+1} L_{2i} = \sum_{i=1}^{k+1} L_{2i} + L_{2k+2} = (L_{2k+1} - 1) + L_{2k+2}$$
$$= L_{2k+3} - 1$$

Thus, by PMI, the result is true $\forall n \geq 1$.

53. $$F_{n+1}^2 - F_{n-1}^2 = (F_{n+1} + F_{n-1})(F_{n+1} - F_{n-1})$$
$$= L_n F_n = F_{2n}$$

55. Let $P(n)$ be the given statement. When $n = 2$, RHS $= F_2 x + F_1 = x + 1$ $= x^2 =$ LHS. $\therefore P(2)$ is true.
Assume $P(k)$ is true:

$$x^k = xF_k + F_{k-1}. \text{ Then}$$
$$x^{k+1} = x^2 F_k + xF_{k-1}$$
$$= (x+1)F_k + xF_{k-1}$$
$$= x(F_k + F_{k-1}) + F_k$$
$$= xF_{k+1} + F_k$$

Thus $P(k) \to P(k+1)$. So the result follows by PMI.

59. $$A(n) = (n+1) + \frac{2}{n}\sum_{i=2}^{n-1} A(i)$$

$$nA(n) = n(n+1) + 2\sum_{i=2}^{n-1} A(i)$$

$$\therefore (n-1)A(n-1) = (n-1)n + 2\sum_{i=2}^{n-2} A(i)$$

$$nA(n) - (n-1)A(n-1) = 2n + 2A(n-1)$$
$$nA(n) = 2n + (n+1)A(n-1)$$

$$\begin{aligned}\frac{A(n)}{n+1} &= \frac{A(n-1)}{n} + \frac{2}{n+1}\\ &= \frac{A(n-2)}{n-1} + \frac{2}{n} + \frac{2}{n+1}\\ &\vdots\\ &= \frac{A(1)}{2} + \frac{2}{3} + \frac{2}{4} + \cdots + \frac{2}{n+1}\\ &= \frac{0}{2} + \frac{2}{3} + \cdots + \frac{2}{n+1}\\ &= 2\sum_{i=3}^{n+1}\frac{1}{i}\end{aligned}$$

Supplementary Exercises (p. 338)

1. $t_1 = 1$
$t_n = t_{n-1} + n(n+1)/2,\ n \geq 2$

3. $f_n^2 - 2f_n + 2 = (2^{2^n}+1)^2 - 2(2^{2^n}+1) + 2$
$= 2^{2^{n+1}} + 2^{2+1^n} + 1 - 2^{2+1^n} - 2 + 2$
$= 2^{2^{n+1}} + 1 = f_{n+1}$

5. (solution by V. G. Feser)
a) Consider the sequence $\{a_n\}$, where $a_{n+2} = a_{n+1} + a_n$. Then $a_{n+3} = a_{n+2} + a_{n+1} = a_{n+2} + (a_{n+2} - a_n)$. Thus $a_n + a_{n+3} = 2a_{n+2}$ (1)
$\therefore$ The sequence satisfies the given recurrence relation.
b) Using the given conditions, $a_4 = 2a_3 - a_1 = 2a_3 - 1$
$a_5 = 5$
$a_6 = 2a_5 - a_3 = 10 - a_3$
$\vdots$
$a_{12} = 290 - 70a_3 = 144$
$\therefore a_3 = 2$ and $a_4 = 3$. Since $a_5 = 2a_4 - a_2 = 5, a_2 = 1$. Thus the sequence $\{a_n\}$ satisfies the given conditions and the recurrence relation, so it must be the Fibonacci sequence; that is, $a_n = F_n$.

7. $S_n = (3n^2+5)n^3/8$ **9.** $a_n = a_{n+5},\ n \in \mathbf{Z}$ **11.** 8, 8

13. $f(n) = \phi(n) + \cdots + \phi(2) + \phi(1)$. Since $\phi(n)$ is even for $n > 2$ and $\phi(2) + \phi(1) = 1 + 1 = 2$ is even, $f(n)$ is even.

15. $f_1 = 2, f_2 = 3$
$f_n = f_{n-1} + f_{n-2},\ n \geq 3$

17. $f(n,1) = 1$
$$f(n,n) = \begin{cases} 1 & \text{if } n = 0 \text{ or } 1\\ 0 & \text{otherwise}\end{cases}$$
$f(n,k) = f(n-2,k-1) + f(n-1,k)$

Chapter 6 Combinatorics

Exercises 6.1 (p. 349)

1. 1,317 **3.** 1,449 **5.** 312 **7.** 18

9. 200 **11.** $36^{\lceil n/2 \rceil}$ **13.** 3^n **15.** 1000

17. 0 **19.** 260 **21.** 100,000 **23.** 0

25. 1,757,600 **27.** 0 **29.** 1,572,120,576 **31.** 186,624

33. 36,504 **35.** 32 **37.** 128 **39.** 192

41. 16 **43.** 24 **45.** 2^{n^2} **47.** 364

49. 9 **51.** 1,053 **53.** $1 + m + m^2$

55. $\dfrac{m^{n+1} - 1}{m - 1}$ **57.** $m!$ **59.** $\dfrac{n!}{(n-m)!}$

61. Let $m = p_1^{a_1} \dots p_m^{a_m}$ and $n = q_1^{b_1} \dots q_m^{b_m}$. Since $\gcd\{m, n\} = 1$, $mn = p_1^{a_1} \dots p_m^{a_m} q_1^{b_1} \dots q_m^{b_m}$ is the prime factorization of mn. $\therefore$ $\tau(mn) = (a_1 + 1) \dots (a_m + 1)(b_1 + 1) \dots (b_m + 1) = \tau(m)\tau(n)$

Exercises 6.2 (p. 358)

1. 5 **3.** 60 **5.** F

7. T **9.** F **11.** F

13. F **15.** F **17.** 20

19. 362,880 **21.** 40,320 **23.** 358,800

25. 1 **27.** 562,432 **29.** 208,827,062,548

31. 103,680 **33.** 1,693,440 **35.** 60,480

37. 120 **39.** 6 **41.** 6 **43.** 7

45.
$$\begin{aligned} P(n-1, r) + rP(n-1, r-1) &= \frac{(n-1)!}{(n-r-1)!} + \frac{r(n-1)!}{(n-r)!} \\ &= \frac{(n-1)!(n-r+r)}{(n-r)!} \\ &= \frac{n!}{(n-r)!} = P(n, r) \end{aligned}$$

47. $(n+1)! - n! = (n+1)n! - n!$

$$= (n+1-1)n! = n(n!)$$

49.

```
Algorithm permutations (n,r,answer)
(* This algorithm computes P(n,r) where 0 ≤ r ≤ n and the result
is returned in the variable answer. The recursive factorial
subalgorithm is invoked by the algorithm.*)
Begin (* permutations *)
answer ← factorial(n)/factorial (n - r)
End (* permutations *)
```

51. $n! = n(n-1)\ldots 2.1$

$= 2n[(n-1)\ldots 3]$

$> 2n$ if $n \geq 4$

$\therefore\ (n!)! > (2n)!,\quad$ if $n \geq 4$

Exercises 6.3 (p. 364)

1. 44 **3.** 1, 854 **5.** 1 **7.** 265 **9.** $d_n = = (-1)^n$

11. **proof:** $D_n = (n-1)(D_{n-1} + D_{n-2})$ Suppose n is odd. Then $n-1$ is even. $\therefore (n-1)(D_{n-1} + D_{n-2}) = D_n$ is even.

13. 2

15. 8

17. $b_0 = 0 = b_1$
$b_n = b_{n-1} + b_{n-2} + 2,\ n \geq 2$

19. 14

21. 66

23. 4

25. 8

27. $c_0 = 0$
$c_n = c_{n-1} + 2,\ n \geq 1$

29. 10

31. 16

33. $c_n = 2n = O(n)$

35. $a_n = (n-3)(n-2)$
$= O(n^2)$

37. 0.368879

39. 0.367879

41. $$p_n - p_{n-1} = \frac{D_n}{n!} - \frac{D_{n-1}}{(n-1)!} = \frac{D_n - nD_{n-1}}{n!} = \frac{(-1)^n}{n!},\ \text{by equation (4)}$$

43. $D_n = (n-1)(D_{n-1} + D_{n-2})$

$$\frac{D_n}{n!} = \frac{(n-1)}{n!}(D_{n-1} + D_{n-2})$$

$$p_n = \frac{n-1}{n}p_{n-1} + \frac{1}{n}p_{n-2}$$

45. Notice that $p_0 = 1$ and $p_1 = 0$. Let $g_n = p_n - p_{n-1}$. Then $g_1 = -1$. By Exercise 43,

$$\begin{aligned} g_n &= \left(-\frac{1}{n}\right) g_{n-1} = \left(-\frac{1}{n}\right)\left(-\frac{1}{n-1}\right) g_{n-2} \\ &\vdots \\ &= \left(-\frac{1}{n}\right)\left(-\frac{1}{n-1}\right)\cdots\left(-\frac{1}{2}\right) g_1 \\ &= \frac{(-1)^n}{n!} \end{aligned}$$

47. **Algorithm derangements (n)**

```
(* This algorithm computes the number of derangements of n items using
the alternate recursive definition. *)
Begin (* derangements *)
    if n = 0 then
     derangements ← 1
    else if n is odd then
           derangements ← n * derangements(n-1) - 1
        else
           derangements ← n * derangements(n-1) + 1
End (* derangements *)
```

Exercises 6.4 (p. 372)

1. 36 **3.** 84 **5.** 16,170 **7.** 45

9. $$\begin{aligned} f_n &= f_{n-1} + n \\ &= f_{n-2} + (n-1) + n \\ &\vdots \\ &= f_0 + (1 + 2 + \ldots + n) \\ &= 1 + n(n+1)/2, \; n \geq 0 \end{aligned}$$

11. yes

13. 78

15. No. of gifts sent on the nth day $= \sum_{i=1}^{n} i = n(n+1)/2 = C(n+1, 2)$

17. any nonnegative integer.

19. 8 **21.** $C(n,r)$ **23.** 243 **25.** 260 **27.** 957

29. 42 **31.** 40 **33.** 105 **35.** 5

37. $$\begin{aligned} \text{RHS} &= \frac{n}{r} \cdot \frac{(n-1)!}{(r-1)!(n-r)!} \\ &= \frac{n(n-1)!}{[r(r-1)!](n-r)!} \\ &= \frac{n!}{r!(n-r)!} = \text{LHS} \end{aligned}$$

39. $$\begin{aligned} \text{RHS} &= C(n-1, r) + C(n-1, r-1) \\ &= \frac{(n-1)!}{r!(n-r-1)!} + \frac{(n-1)!}{(r-1)!(n-r)!} \\ &= \frac{(n-1)!(n-r+r)}{r!(n-r)!} \\ &= \frac{n!}{r!(n-r)!} = \text{LHS} \end{aligned}$$

41. Let $P(n): g_n = C(n,0) + C(n,2) + C(n,4)$. Then $C(0,0) + C(0,2) + C(0,4) = 1 + 0 + 0 = 1 = g_0$.

$\therefore$ $P(0)$ is true.

Assume $P(k)$ is true:

$$g_k = C(k,0) + C(k,2) + C(k,4)$$

$$\begin{aligned} g_{k+1} &= g_k + C(k,1) + C(k,3) \\ &= C(k,0) + C(k,2) + C(k,4) + C(k,1) + C(k,3) \\ &= C(k,0) + [C(k,1) + C(k,2)] + [C(k,3) + C(k,4)] \\ &= C(k+1,0) + C(k+1,2) + C(k+1,4) \end{aligned}$$

$\therefore$ $P(k) \to P(k+1)$. Thus the result follows by PMI.

43. 2

45. $$A(n,r) = \begin{cases} 0 & \text{if } r = 0 \\ 0 & \text{if } r = n \\ A(n-1,r) + A(n-1,r-1) + 1 & \text{if } 0 < r < n \end{cases}$$

47. $$\begin{aligned} \text{RHS} &= C(2n,n) - C(2n,n-1) \\ &= \frac{(2n)!}{n!n!} - \frac{(2n)!}{(n-1)!(n+1)!} \\ &= \frac{(2n)!(n+1-n)}{n!(n+1)!} \\ &= \frac{(2n)!}{n!(n+1)!} = C_n = \text{LHS} \end{aligned}$$

49. $C_{n-1} = \dfrac{(2n-2)!}{(n-1)!n!}$, by definition

$$\begin{aligned} \text{RHS} &= \frac{2(2n-1)}{(n+1)} C_{n-1} \\ &= \frac{2(2n-1)}{(n+1)} \cdot \frac{(2n-2)!}{(n-1)!n!} \\ &= \frac{(2n)(2n-1)(2n-2)!}{n!(n+1)!} \\ &= \frac{(2n)!}{n!(n+1)!} = C = \text{LHS} \end{aligned}$$

51. $$\begin{aligned} \text{RHS} &= 4C_{n-1} - \frac{6}{n+1} C_{n-1} \\ &= \frac{2(2n-1)}{n+1} C_{n-1} \\ &= C_n, \text{ by Exercise 49} \\ &= \text{LHS} \end{aligned}$$

53. 42

55. Algorithm combinations (n,r, answer)

```
(* This algorithm computes C(n,r) using recursion *)
  Begin (* combinations *)
    if r = 0 then
      combinations ← 1
    else if r = n then
          combinations ← 1
        else
          combinations ← combinations(n-1,r) + combinations (n-1,r-1)
  End (* combinations *)
```

57. 3

59. 6

61. $2^n - 2$

Exercises 6.5 (p. 384)

1. 5040 **3.** 28 **5.** 93 **7.** 1 **9.** 8

11. 60 **13.** 10 **15.** 84 **17.** {a, a, a, a} **19.** 495

21. 8008 **23.** 120 **25.** 1365 **27.** 495 **29.** 3

31. 4 **33.** (3,1,6), (3,2,5), (4,1,5) **35.** 18

Exercises 6.6 (p. 396)

1. 56 **3.** 112

5. $x^4 + 4x^3y + 6x^2y^2 + 4xy^3 + y^4$

7. $32x^5 - 80x^4 + 80x^3 - 40x^2 + 10x - 1$

9. 6 **11.** 17,920 **13.** 10 **15.** 35

17. $\binom{n}{\lfloor n/2 \rfloor}$ **19.** F_n **21.** 1 **23.** 8

25. 5 **27.** 52

29.
$$\begin{aligned} 4^{2n} + 10n - 1 &= (5-1)^{2n} + 10n - 1 \\ &= \sum_{r=0}^{2n} \binom{2n}{r} 5^{2n-r}(-1)^r + 10n - 1 \\ &= \sum_{r=0}^{2n-2} \binom{2n}{r} 5^{2n-r}(-1)^r - 10n + 1 + 10n - 1 \\ &= \sum_{r=0}^{2n-2} \binom{2n}{r} 5^{2n-r}(-1)^r \\ &= 5^{2n} - \binom{2n}{1} 5^{2n-1} + \cdots + \binom{2n}{2n-2} 5^2 \end{aligned}$$

which is clearly divisible by 25.

31. **proof:** $(1+x)^n = \sum_{r=0}^{n} \binom{n}{r} x^r$
Let $x = 2$. Then
$$3^n = \sum_{r=0}^{n} 2^r \binom{n}{r}$$

33. **proof:** $(1+x)^{2n} = (x+1)(1+x)^n$
The coefficient of x^{n+1} on the LHS is $\binom{2n}{n+1}$ and that on the RHS is $\sum_{r=1}^{n} 2^r \binom{n}{i-1}\binom{n}{i}$. Hence the result.

35. Let $S = \sum_{k=0}^{n} (a + kd) \binom{n}{k}$ (1)

Reverse the sum on the RHS:

$$S = \sum_{k=0}^{n} [a + (n-k)d] \binom{n}{n-k} = \sum_{k=0}^{n} [a + (n-k)d] \binom{n}{k} \qquad (2)$$

Adding equations (1) and (2),

$$2S = \sum_{k=0}^{n} (2a + nd) \binom{n}{k} = (2a + nd) \sum_{k=0}^{n} \binom{n}{k}$$
$$= (2a + nd) \cdot 2^n$$
$$\therefore\ S = (2a + nd)2^{n-1}$$

37. proof: By Exercise 36,
$\binom{n}{0} < \binom{n}{1} < \cdots < \binom{n}{r-1} < \binom{n}{r}$, where $r < \frac{n+1}{2}$. $\therefore \binom{n}{r}$ is maximum when r is the largest integer $< (n+1)/2$.
case 1 Let n be odd, say, $n = 2k+1$. Then $(n+1)/2 = k+1$. So r must be $k = (n-1)/2 = \lfloor n/2 \rfloor$.
case 2 Let n be even, say, $n = 2k$. Then $(n+1)/2 = (2k+1)/2 = k+1/2$. So r must be $k = n/2 = \lfloor n/2 \rfloor$.
Thus, by cases (1) and (2), $\binom{n}{r}$ is maximum when $r = \lfloor n/2 \rfloor$.

39. proof (by PMI): Let $P(n)$ denote the given statement. When $n = 1$, LHS $= n =$ RHS. $\therefore$ $P(1)$ is true. Assume $P(k)$ is true, where $k \geq 1$: $\sum_{i=1}^{k} i \binom{k}{i} = k \cdot 2^{k-1}$

$$\sum_{i=1}^{k+1} i \binom{k}{i} = \sum_{i=1}^{k+1} i \left[\binom{k}{i} + \binom{k}{i-1} \right]$$
$$= \sum_{r=1}^{k+1} i \binom{k}{i} + \sum_{r=1}^{k+1} i \binom{k}{i-1}$$
$$= \sum_{i=1}^{k} i \binom{k}{i} + (k+1) \binom{k}{k+1} + \sum_{i=1}^{k+1} (i-1) \binom{k}{i-1} + \sum_{i=1}^{k+1} \binom{k}{i-1}$$
$$= \sum_{i=1}^{k} i \binom{k}{i} + \sum_{i=1}^{k} i \binom{k}{i} + \sum_{i=0}^{k} \binom{k}{i}$$
$$= k \cdot 2^{k-1} + k \cdot 2^{k-1} + 2^k$$
$$= (k+1)2^k.$$

$\therefore$ $P(k+1)$ is true. Thus, by induction, $P(n)$ is true $\forall n \geq 1$.

41. proof: $n(1+x)^{n-1} = \sum_{r=1}^{n} \binom{n}{r} r x^{r-1}$. Let $x = 1$. Then $n \cdot 2^{n-1} = \sum_{r=1}^{n} r \binom{n}{r}$.

43. proof (by induction): Let $P(n)$: $(x+y)^n = \sum_{r=0}^{n} \binom{n}{r} x^{n-r}y^r$. When $n=0$, LHS $= 1 =$ RHS. So $P(0)$ is true. Assume $P(k)$ is true for some $k \geq 0$:

$$(x+y)^k = \sum_{r=0}^{k} \binom{k}{r} x^{k-r}y^r \qquad (1)$$

Then
$$\begin{aligned}
(x+y)^{k+1} &= (x+y)^k(x+y) \\
&= \left[\sum_{r=0}^{k} \binom{k}{r} x^{k-r}y^r\right](x+y) \\
&= \sum_{r=0}^{k} \binom{k}{r} x^{k+1-r}y^r + \sum_{r=0}^{k} \binom{k}{r} x^{k-r}y^{r+1} \\
&= \left[\binom{k}{0} x^{k+1} + \sum_{r=1}^{k} \binom{k}{r} x^{k+1-r}\right] + \left[\sum_{r=0}^{k-1} \binom{k}{r} x^{k-r}y^{r+1} \right. \\
&\qquad \left. + \binom{k}{k} y^{k+1}\right] \\
&= \binom{k+1}{0} x^{k+1} + \sum_{r=1}^{k} \binom{k}{r} x^{k+1-r}y^r + \sum_{r=1}^{k} \binom{k}{r-1} x^{k+1-r}y^r \\
&\qquad + \binom{k+1}{k+1} y^{k+1} \\
&= \binom{k+1}{0} x^{k+1} + \sum_{1}^{k} \left[\binom{k}{r} + \binom{k}{r-1}\right] x^{k+1-r}y^r \\
&\qquad + \binom{k+1}{k+1} y^{k+1} \\
&= \binom{k+1}{0} x^{k+1} + \sum_{r=1}^{k} \binom{k+1}{r} x^{k+1-r}y^r + \binom{k+1}{k+1} y^{k+1} \\
&= \sum_{r=0}^{k} \binom{k+1}{r} x^{k+1-r}y^r
\end{aligned}$$

$\therefore$ $P(k+1)$ is true. Thus, by PMI, $P(n)$ is true $\forall n \geq 0$.

45. proof: RHS $= \binom{n}{r}\binom{n-r}{m-r} = \dfrac{n!}{r!(n-r)!} \cdot \dfrac{(n-r)!}{(m-r)!(n-m)!}$
$= \dfrac{n!}{m!(n-m)!} \cdot \dfrac{m!}{r!(m-r)!} = \binom{n}{m}\binom{m}{r} =$ LHS

47. proof: $(1+x)^{m+n} = (1+x)^m(x+1)^n = \left[\sum_{r=0}^{m} \binom{m}{r} x^r\right]\left[\sum_{r=0}^{n} \binom{n}{r} x^{n-r}\right]$

Coefficient of x^r on the LHS $= \binom{m+n}{r}$

Coefficient of x^r on the RHS $= \sum_{i=0}^{r} \binom{m}{i}\binom{n}{r-i}$

Hence the result.

49. proof: Let $P(n)$ denote the given statement, where $n \geq 2$. When $n = 2$, LHS $= \binom{2}{2} = 1 = \binom{3}{3} =$ RHS. $\therefore P(2)$ is true. Assume $P(k) : \sum_{i=2}^{k} \binom{i}{2} = \binom{k+1}{3}$, where $k \geq 2$.

$$\text{Then } \sum_{i=2}^{k+1} \binom{i}{2} = \sum_{i=2}^{k} \binom{i}{2} + \binom{k+1}{2} = \binom{k+1}{3} + \binom{k+1}{2}$$
$$= \binom{k+2}{3}$$

$\therefore$ $P(k+1)$ is true. Thus, by PMI, the result holds $\forall n \geq 2$.

51. proof: Let $P(n)$ denote the given statement, where $n \geq 3$. When $n = 3$, LHS $= \binom{3}{3} = 1 = \binom{4}{4} =$ RHS. $\therefore$ $P(3)$ is true. Assume $P(k)$: $\sum_{i=3}^{k} \binom{i}{3} = \binom{k+1}{4}$, where $k \geq 3$.

$$\text{Then} \quad \sum_{i=3}^{k+1} \binom{i}{3} = \sum_{i=3}^{k} \binom{i}{3} + \binom{k+1}{3}$$
$$= \binom{k+2}{4}$$

$\therefore$ $P(k+1)$ is true. Thus, by PMI, the formula holds $\forall n \geq 3$.

Exercises 6.7 (p. 408)

1. 74 **3.** 533 **5.** 266 **7.** 21 **9.** 30

11. Let S be the set of solutions with $x \leq 3$, $y \leq 4$, and $z \leq 4$. Then $|S| = C(3+11-1, 11) = 78$. Let $p_1 : x > 3$, $p_2 : y > 4$, and $p_3 : z > 5$. Then:

$N(p_1) =$ no. of solutions with $x \geq 4 = C(3+7-1, 7) = 36$
$N(p_2) =$ no. of solutions with $y \geq 5 = C(3+6-1, 6) = 28$
$N(p_3) =$ no. of solutions with $z \geq 6 = C(3+5-1, 5) = 21$
$N(p_1p_2) =$ no. of solutions with $x \geq 4$ and $y \geq 5 = C(3+2-1, 2) = 6$
$N(p_1p_3) =$ no. of solutions with $x \geq 4$ and $z \geq 6 = C(3+1-1, 1) = 3$
$N(p_2p_3) =$ no. of solutions with $y \geq 5$ and $z \geq 6 = C(3+0-1, 0) = 1$
$N(p_1p_2p_3) =$ no. of solutions with $x \geq 4$, $y \geq 5$, and $z \geq 6$
$= C(3-4-1, -4) = 0$

$\therefore$ answer $= N(p'_1p'_2p'_3) = 78 - (36+28+21) + (6+1+3) - 0 = 3$

13. When $n = 5$, LHS $= 600 = 5(5!) =$ RHS; when $n = 6$, LHS $= 3600 = 5(6!) =$ RHS.

Exercises 6.8 (p. 416)

1. $\frac{1}{13}$ **3.** $\frac{2}{13}$ **5.** $\frac{1}{36}$ **7.** $\frac{1}{12}$ **9.** $\frac{9}{1000000}$

11. $\frac{1890}{1000000}$ **13.** $\frac{98901}{1000000}$ **15.** $\frac{1}{17}$ **17.** $\frac{13}{102}$ **19.** $\frac{2}{429}$

21. 0 **23.** $\frac{2}{5}$ **25.** 0 **27.** $\frac{39}{95}$ **29.** $\frac{9}{19}$

Exercises 6.9 (p. 425)

1. $\frac{2}{11}$ **3.** $\frac{5}{13}$ **5.** $\frac{1}{4}$ **7.** $\frac{1}{4}$ **9.** 0 **11.** $\frac{7}{15}$ **13.** $\frac{1}{2}$

15. $\frac{1}{2}$ **17.** $\frac{1}{16}$ **19.** 0 **21.** $\frac{1}{17}$ **23.** 0 **25.** $\frac{15}{91}$ **27.** $\frac{20}{273}$

29. $-\$1.75$ **31.** $-92.11¢$ **33.** $\frac{125}{324}$ **35.** $\frac{5}{324}$ **37.** $\frac{425}{432}$ **39.** $\frac{256}{625}$ **41.** $\frac{624}{625}$

43. $E = \sum_{k=0}^{n} kC(n,k)p^k q^{n-k} = \sum_{k=1}^{n} kC(n,k)p^k q^{n-k}$

By the binomial theorem, $n(1+x)^{n-1} = \sum_{k=1}^{n} kC(n,k)x^{k-1}$. Let $x = p/q$.

Then

$$\frac{n(p+q)^{n-1}}{q^{n-1}} = \sum_{k=1}^{n} kC(n,k)\,\frac{p^{k-1}}{q^{k-1}} = \sum_{k=1}^{n} kC(n,k)\,\frac{p^{k-1}}{q^{k-1}}$$

Since $p+q=1$ in a Bernoulli trial, this yields

$$np = \sum_{k=1}^{n} kC(n,k)p^k q^{n-k} = E$$

Review Exercises (p. 429)

1. 1,038 **3.** 1,454 **5.** 1,572,120,576 **7.** 436,700,160

9. 1,542,636,576 **11.** 0 **13.** 80 **15.** 1,296

17. 2,401 **19.** 103,680 **21.** 39,916,800 **23.** 59,049

25. 176 **27.** 2,598,960 **29.** 3,744 **31.** 34,650

33. 20

35. $h(n) = h(n-1) +$ number of handshakes made by the nth couple
$= h(n-1) + 4n$, where $h(0) = 0$.

37. 84 **39.** 26 **41.** 165 **43.** -330

45. $x^6 + 6x^5y + 15x^4y^2 + 20x^3y^3 + 15x^2y^4 + 6xy^5 + y^6$

47. $243x^5 - 405x^4y + 270x^3y^2 - 90x^2y^3 + 15xy^4 - y^5$

49. **proof** (by strong induction): Let $P(n)$ denote the given statement. When $n = 0$, $r = 0$. Then $C(n,r) = C(0,0) = 1$ is an integer. So $P(0)$ is true.
Assume $P(i)$ is true $\forall i \leq k$, where $k \geq 0$. By Theorem 6.12, $C(k+1,r) = C(k,r-1) + C(k,r)$. Both numbers on the RHS are integers, by the inductive hypothesis. So their sum is also an integer. $\therefore$ $P(k+1)$ is true. Thus, by PMI, the result holds $\forall n \geq 0$.

51. No. of gifts sent on the nth day $= \sum_{i=1}^{n} i = n(n+1)/2 = C(n+1,2)$

53. **proof:** Consider a permutation of a set of n elements. Suppose exactly k of them occupy their natural positions. None of the remaining $n-k$ elements occupy their natural positions; the number of such arrangements is D_{n-k}. But the original k elements can be chosen in $C(n,k)$ different ways. Therefore, the number of permutations with exactly k elements in their natural positions is $C(n,k)D_{n-k}$.

55. 2^n **57.** $A(n,1) = n-1$

59. **proof** (by PMI): Let $A(n,r) = C(n,r)-1$. When $n = 1$, RHS $= C(n,1) - 1 = n-1 = A(n,1) =$ LHS. So $P(1)$ is true. Assume $P(k)$ is true: $A(k,r) = C(k,r) - 1$. Then

$$A(k+1,r) = A(k,r-1) + A(k,r) + 1 = [C(k,r-1) - 1] + [C(k,r) - 1] + 1$$
$$= [C(k,r-1) + C(k,r)] - 1 = C(k+1,r) - 1 = A(k+1,r)$$

$\therefore$ $P(k) \rightarrow P(k+1)$. Thus the result follows by PMI.

61. \$375,000 **63.** 0.40951

Supplementary Exercises (p. 432)

1. Let k be the least of the r consecutive positive integers. Their product is $k(k+1)\dots(k+r-1)$. Since $\dfrac{k(k+1)\dots(k+r-1)}{r!} = C(k+r-1,r)$ is an integer, $k(k+1)\dots(k+r-1)$ is divisible by $r!$.

3. $C_n = C(2n,n)/(n+1)$ is an integer; so $n+1 \mid C(2n,n)$.

5. $\text{RHS} = \dfrac{C(2n+1,n)}{2n+1} = \dfrac{(2n+1)!}{n!(n+1)!(2n+1)} = \dfrac{(2n)!}{n!(n+1)!} = C_n = \text{LHS}$

7. 13; 80

9. We shall use the following identities:

1. $n\binom{n-1}{k-1} = k\binom{n}{k}$ 2. $\sum_{k=0}^{n}\binom{n}{k} = 2^n$ 3. $\sum_{k=1}^{n} k\binom{n}{k} = n2^{n-1}$

$$\begin{aligned}\sum_{k=0}^{n}\binom{n}{k}k^2 &= \sum_{k=0}^{n}\binom{n}{k}k{\cdot}k = \sum_{k=0}^{n} n\binom{n-1}{k-1}k = n\sum_{k=0}^{n}\binom{n-1}{k-1}(k-1+1)\\ &= n\sum_{k=0}^{n}\binom{n-1}{k-1}(k-1) + n\sum_{k=0}^{n}\binom{n-1}{k-1}\\ &= n\sum_{k=0}^{n}(n-1)\binom{n-2}{k-2} + \sum_{k=1}^{n-1}\binom{n-1}{k-1}\\ &= n(n-1)\sum_{k=2}^{n-2}\binom{n-2}{k-2} + n\sum_{k=1}^{n-1}\binom{n-1}{k-1}\\ &= n(n-1)\sum_{k=0}^{n-2}\binom{n-2}{k} + n\sum_{k=0}^{n-1}\binom{n-1}{k}\\ &= n(n-1)\cdot 2^{n-2} + n\cdot 2^{n-1} = n(n+1)2^{n-2}\end{aligned}$$

11.
$$\begin{aligned}\sum_{d=1}^{n-1}\sum_{j=0}^{d}\sum_{k=0}^{n-d-1}\binom{j+k}{j} &= \sum_{d=1}^{n-1}\sum_{j=0}^{d}\binom{j+n-d}{j+1}\\ &= \sum_{d=1}^{n-1}\left[\binom{n+1}{d+1} - 1\right]\\ &= \sum_{r=0}^{n-1}\binom{n+1}{r} - \binom{n+1}{0} - \binom{n+1}{1}\\ &\quad - \binom{n+1}{n+1} - (n-1)\\ &= 2^{n+1} - 1 - (n+1) - 1 - (n-1)\\ &= 2^{n+1} - 2n - 2\end{aligned}$$

13. no, no

15. $f(n,k) = \begin{cases} n & \text{if } k = 1\\ 0 & \text{if } k = n\\ f(n-2,k-1) + f(n-1,k) & \text{if } 1 < k < n\end{cases}$

17. By Exercise 16, $f(n,k) = C(n-k+1,k)$.
∴ Total no. of subsets that do not contain consecutive integers is given by

$$\sum_{k=0}^{n} f(n,k) = \sum_{k=0}^{n} C(n-k+1,k) = \sum_{k=0}^{n} C((n+2)-k-1,k) = F_{n+2}$$

19. F_{n+2}

Chapter 7 Relations

Exercises 7.1 (p. 441)

1. $\begin{bmatrix} 1 & 1 \\ 0 & 0 \end{bmatrix}$

3. $\begin{bmatrix} 1 & 1 \\ 0 & 0 \end{bmatrix}$

5. $\begin{bmatrix} 1 & 1 \\ 1 & 0 \end{bmatrix}$

7. $\begin{bmatrix} 1 & 0 \\ 0 & 0 \end{bmatrix}$

9. $\begin{bmatrix} 0 & 0 & 1 \\ 0 & 0 & 0 \\ 1 & 1 & 0 \end{bmatrix}$

11. $\begin{bmatrix} 0 & 0 & 1 \\ 0 & 0 & 0 \\ 1 & 1 & 0 \end{bmatrix}$

13. $\begin{bmatrix} 1 & 0 & 1 \\ 0 & 0 & 1 \\ 1 & 1 & 0 \end{bmatrix}$

15. $\begin{bmatrix} 0 & 1 & 0 \\ 0 & 0 & 0 \\ 0 & 1 & 1 \end{bmatrix}$

17. $\begin{bmatrix} 1 & 1 & 1 \\ 0 & 0 & 0 \\ 1 & 1 & 1 \end{bmatrix}$

19. n^2

21. $(2n-1)n^2$

23. $(2p-1)mn$

25. $\begin{bmatrix} 0 & 0 \\ 1 & 1 \end{bmatrix}$

27. $\begin{bmatrix} 0 & 0 \\ 0 & 1 \end{bmatrix}$

29. $\begin{bmatrix} 1 & 0 \\ 1 & 1 \end{bmatrix}$

31. $\begin{bmatrix} 1 & 1 \\ 0 & 0 \end{bmatrix}$

33. A

35. O

Note: In Exercises 37–48, let $A = (a_{ij})$, $B = (b_{ij})$, and $C = (c_{ij})$.

37. $$\begin{aligned} A \wedge A &= (a_{ij}) \wedge (a_{ij}) \\ &= (a_{ij} \wedge a_{ij}) \\ &= (a_{ij}) = A \end{aligned}$$

39. $$\begin{aligned} A \wedge B &= (a_{ij}) \wedge (b_{ij}) \\ &= (a_{ij} \wedge b_{ij}) \\ &= (b_{ij} \wedge a_{ij}) \\ &= (b_{ij}) \wedge (a_{ij}) = B \wedge A \end{aligned}$$

41. $$\begin{aligned} \text{LHS} &= (a_{ij}) \wedge [(b_{ij}) \wedge (c_{ij})] \\ &= (a_{ij}) \wedge (b_{ij} \wedge c_{ij}) \\ &= (a_{ij} \wedge (b_{ij} \wedge c_{ij})) \\ &= ((a_{ij} \wedge b_{ij}) \wedge c_{ij}) \\ &= [(a_{ij}) \wedge (b_{ij})] \wedge (c_{ij}) \\ &= \text{RHS} \end{aligned}$$

43. $$\begin{aligned} \text{LHS} &= (a_{ij}) \wedge [(b_{ij}) \vee (c_{ij})] \\ &= (a_{ij}) \wedge (b_{ij} \vee c_{ij}) \\ &= (a_{ij} \wedge (b_{ij} \vee c_{ij})) \\ &= ((a_{ij} \wedge b_{ij}) \vee (a_{ij} \wedge c_{ij})) \\ &= ((a_{ij}) \wedge (b_{ij})) \vee (a_{ij} \wedge c_{ij}) \\ &= \text{RHS} \end{aligned}$$

45.

```
Algorithm meet (A,B,C)
(* Let A = (aij)n×n and
   B = (aij)n×n. C is the meet of A and B.)
   Begin (* meet *)
     for i = 1 to n do
       for j = 1 to n do
         cij ← aij ∧ bij
   End (* meet *)
```

47.

```
Algorithm boolean product (A,B,C)
(* C is the boolean product of A = (a_ij)_m×p and B = (b_ij)_p×n *)
 Begin (* boolean product *)
   for i = 1 to m do
     for j = 1 to p do
       c_ij ← (a_i1 ∧ b_1j) ∨ (a_i2 ∧ b_2j) ··· ∨ (a_ip ∧ b_pj)
 End (* boolean product *)
```

Exercises 7.2 (p. 448)

1. $\{(1,2),(1,4),(1,8),(3,4),(3,8),(5,8)\}$ **3.** $\{(1,4),(3,2)\}$

5. $\{(1,2)\}$ **7.** $A; B$ **9.** $\{1,3\}; \{2,4\}$ **11.** $\{1\}; \{2\}$

13. $\begin{bmatrix} 1 & 1 & 1 \\ 0 & 1 & 1 \\ 0 & 0 & 1 \end{bmatrix}$ **15.** $\begin{bmatrix} 0 & 1 & 0 \\ 1 & 0 & 0 \\ 0 & 0 & 0 \end{bmatrix}$ **17.** $\begin{bmatrix} 1 & 0 & 0 \\ 0 & 0 & 0 \\ 0 & 0 & 0 \end{bmatrix}$

19.
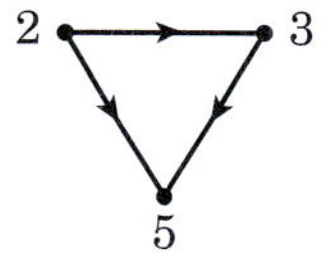

21.
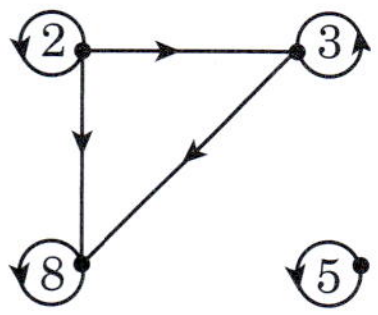

23. yes **25.** no

27. yes **29.** yes

31. yes **33.** yes

35. $\{(a,b),(b,c),(c,b)\}$

37.
$$\begin{array}{c} \\ a \\ b \\ c \end{array} \begin{array}{c} \begin{array}{ccc} a & b & c \end{array} \\ \begin{bmatrix} 0 & 1 & 0 \\ 0 & 0 & 1 \\ 0 & 1 & 0 \end{bmatrix} \end{array}$$

39. 0000, 0001, 0011, 0010, 0110, 0111, 0101, 0100, 1100, 1101, 1111, 1110, 1010, 1011, 1001, 1000

41. b-b-b, b-c-a, b-c-d **43.** 12 **45.** no **49.** no **51.** 2^{m^2}

53.

```
Algorithm print relation (R,A)
Begin (* print *)
  for i = 1 to n do
    for j = 1 to n do
      if a_ij = 1 then
         write ((i,j))
End (* print *)
```

Exercises 7.3 (p. 453)

1.

	Tail	head	next
1	2	1	5
2	3	1	1
start 3	1	3	0
↳ 4	1	2	6
5	3	2	3
6	2	3	2

3.
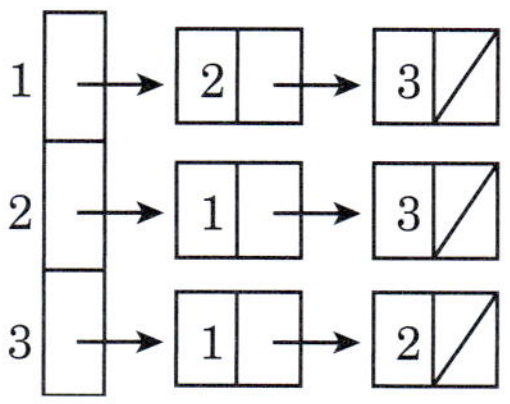

5.

$$\begin{array}{c} \\ 1 \\ 2 \\ 3 \end{array}\begin{array}{c} \begin{array}{ccc} 1 & 2 & 3 \end{array} \\ \begin{bmatrix} 0 & 1 & 0 \\ 0 & 1 & 0 \\ 1 & 1 & 1 \end{bmatrix} \end{array}$$

7.

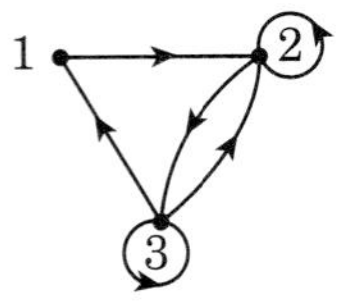

9.

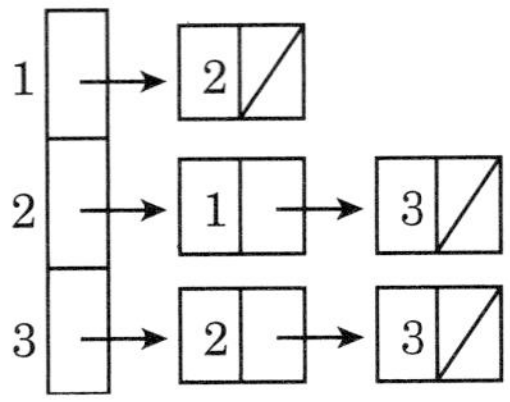

11.

```
Algorithm adjacency list (R,list)
(* List is the array of pointers pointing to each linked list
   in the adjacency list representation. *)
Begin (* algorithm *)
   initialize list
   while there are more elements in R do
   begin (* while *)
      read a pair (i,j)
      create a node
      info(node) ← j
      link(node) ← nil
      insert the node
   endwhile
End (* algorithm *)
```

13.

```
Algorithm adjacency matrix (A,list)
(* List is the array of pointers. *)
Begin (* algorithm *)
   (* initialize A with zeros *)
   for i = 1 to n do
      for j = 1 to n do
         a_ij ← 0
   (* traverse the adjacency list *)
   for i = 1 to n do
   begin (* for *)
      (* traverse each linked list *)
      k ← link(i)
      while k ≠ nil do
      begin (* while *)
         j ← info(node(k))
         a_ij ← 1
         k ← link(node(j))
      endwhile
   endfor
End (* algorithm *)
```

Exercises 7.4 (p. 459)

1. symmetric, antisymmetric, transitive

3. symmetric, antisymmetric, transitive

5. yes **7.** no **9.** yes

11. no **13.** no **15.** no

17. reflexive, symmetric, antisymmetric, transitive

19. reflexive, antisymmetric

21. R is not symmetric if there is an $a \in A$ such that aRb, but $b \not R a$.

23. $\{(a,a),(b,b),(c,c)\}$ **25.** $\{(a,a),(a,b),(b,b),(c,c)\}$

27. $\{(a,a),(a,b),(b,b),(b,c),(c,c)\}$ **29.** $\{(a,b)\}$

31. $\{(a,b),(b,a)\}$ **33.** $\{(a,a)\}$

35. $\begin{bmatrix} 1 & 1 & 0 \\ 0 & 1 & 1 \\ 1 & 0 & 1 \end{bmatrix}$ **37.** $\begin{bmatrix} 0 & 0 & 1 \\ 1 & 1 & 0 \\ 0 & 1 & 0 \end{bmatrix}$

39. If R contains only elements of the form (a,a).

41. no **43.** no **45.** yes **47.** $m_{ii} = 0\ \forall i$.

49. yes **51.** yes **53.** no **55.** $m_{ij} = 1 \rightarrow m_{ji} = 0\ \forall i,j$.

57. 4 **59.** 2 **61.** 4

63. Let $M_R = (x_{ij})$ and $M_R^{[2]} = (y_{ij})$. Suppose R is not transitive. Then there must be elements $a_i, a_j, a_k \in A$ such that $a_i R a_j$ and $a_j R a_k$, but $a_i \not R a_k$; that is, $x_{ij} = 1 = x_{jk}$, but $x_{ik} = 0$. Since $x_{ij} = 1 = x_{jk}$ and $y_{ik} = (x_{i1} \wedge x_{1k}) \vee \ldots \vee (x_{ij} \wedge x_{jk}) \vee \ldots \vee (x_{in} \wedge x_{nk}), y_{ik} = 1$. So $y_{ik} > x_{ik}$, which contradicts the assumption that $M_R^{[2]} \leq M_R$. $\therefore R$ is transitive.

Exercises 7.5 (p. 469)

1. $\{(a,a),(a,b),(a,c),(b,b),(b,c),(c,a)\}; \{(a,b),(b,b)\}$

3. $\leq; \emptyset$

5. $\{(a,a),(a,b),(b,b),(b,c),(b,d),(c,a),(d,a)\}; \{(a,b),(b,c)\}$

7. $\{(a,a),(a,b),(b,c),(c,a),(c,b)\}$

9. $\{(a,a),(a,b),(a,c),(b,c),(c,c)\}$

11. $\emptyset$

13. $\begin{bmatrix} 1 & 1 & 1 \\ 0 & 0 & 0 \\ 1 & 0 & 1 \end{bmatrix}$ **15.** $\begin{bmatrix} 1 & 1 & 1 \\ 0 & 1 & 0 \\ 0 & 1 & 1 \end{bmatrix}$

17. R^2 = communicate through an intermediary; R^n = communicate through $n-1$ intermediaries.

19. $\{(a,a),(b,a),(c,b),(c,c)\}$

21. $\begin{bmatrix} 0 & 0 & 1 \\ 1 & 1 & 0 \\ 1 & 1 & 0 \end{bmatrix}$

23. $\begin{bmatrix} 1 & 0 & 0 \\ 1 & 0 & 0 \\ 0 & 1 & 1 \end{bmatrix}$

25. $\begin{bmatrix} 1 & 0 & 1 \\ 0 & 1 & 0 \\ 0 & 1 & 0 \end{bmatrix}$

27. $\{(1,a),(2,b),(3,b)\}$

29. R

31. $\{(a,1),(a,2),(a,3),(b,1),(b,3)\}$

33. $\{(a,3)\}$

35. $\{(1,a),(1,b),(2,a),(2,b),(3,b)\}$

37. $\{(2,b)\}$

39. Let $M_{R'} = (a_{ij})_{n\times n}$ and $(M_R)' = (b_{ij})_{n\times n}$. Then $a_{ij} = 1$ iff $(i,j) \in R'$; that is, iff $b_{ij} = 1$. So $a_{ij} = 1$ iff $b_{ij} = 1$. Thus $M_{R'} = (M_R)'$.

41. Let $(x,y) \in (R^{-1})^{-1}$. Then $(y,x) \in R^{-1}$ and hence $(x,y) \in R$. $\therefore (R^{-1})^{-1} \subseteq R$. Conversely, let $(x,y) \in R$. Then $(y,x) \in R^{-1}$ and hence $(x,y) \in (R^{-1})^{-1}$. $\therefore R \subseteq (R^{-1})^{-1}$. Thus $(R^{-1})^{-1} = R$.

43. Suppose $R \subseteq S$. Let $(x,y) \in R^{-1}$. Then $(y,x) \in R$. $\therefore (y,x) \in S$ and hence $(x,y) \in S^{-1}$. Thus $R^{-1} \subseteq S^{-1}$.

45. Let $(x,y) \in (R \cap S)'$. Then $(x,y) \notin (R \cap S)$. So $(x,y) \notin R$ or $(x,y) \notin S$, that is, $(x,y) \in R'$ or $(x,y) \notin S'$. Thus $(x,y) \in (R' \cup S')$ and hence $(R \cap S)' \subseteq (R' \cup S')$. Similarly, $(R' \cup S') \subseteq (R \cap S)'$. Hence the result.

47. Let $(x,y) \in (R \cap S)^{-1}$. Then $(y,x) \in R \cap S$. So $(y,x) \in R$ and $(y,x) \in S$, and hence $(x,y) \in R^{-1}$ and $(x,y) \in S^{-1}$. Thus $(x,y) \in (R^{-1} \cap S^{-1})$. $\therefore (R \cap S)^{-1} \subseteq (R^{-1} \cap S^{-1})$. Similarly, $(R^{-1} \cap S^{-1}) \subseteq (R \cap S)^{-1}$. Hence the result.

49. Assume R is symmetric. Let $(x,y) \in R'$. Then $(x,y) \notin R$, so $(y,x) \notin R$ since R is symmetric. $\therefore (y,x) \in R'$. Thus R' is symmetric. Conversely, Let R' be symmetric. Then $(R')'$ is symmetric. But $(R')' = R$. $\therefore R$ is symmetric.

51. Let R be symmetric and $(x,y) \in R^{-1}$. Then $(y,x) \in R$. Since R is symmetric, $(x,y) \in R$. $\therefore R^{-1} \subseteq R$. So, by Exercise 43, $(R^{-1})^{-1} \subseteq R^{-1}$; that is, $R \subseteq R^{-1}$. $\therefore R^{-1} = R$. Conversely, let $R^{-1} = R$ and $(x,y) \in R$. Then $(x,y) \in R^{-1}$. $\therefore (y,x) \in R^{-1}$. Thus R is symmetric.

53. Let R and S be symmetric. Let $(x,y) \in R \cap S$. Then $(x,y) \in R$ and $(x,y) \in S$. $\therefore (y,x) \in R$ and $(y,x) \in S$. Thus $(y,x) \in R \cap S$ and hence $R \cap S$ is symmetric.

55. The relations $R = \{(a,b)\}$ and $S = \{(b,c)\}$ on $\{a,b,c\}$ are transitive, but $R \cup S = \{(a,b),(b,c)\}$ is not.

57.

```
Algorithm union (MR∪S, MR,MS)
(* Let MR = (aij)m×n, MS = (bij)m×n, and MR∪S = (cij)m×n. *)
Begin (* algorithm *)
  for i = 1 to m do
    for j = 1 to n do
      cij ← aij ∨ bij
End (* algorithm *)
```

59.

```
Algorithm inverse (MR,MR-1)
(* Let MR = (aij)m×n and MR-1 = (cij)m×n.*)
Begin (* algorithm *)
  for i = 1 to m do
    for j = 1 to n do
      if aij=1 then
        cij ← 0
          else
            cij ← 1
End (* algorithm *)
```

61.

```
Algorithm composition (X,Y,Z)
(* X = MR, Y = MS, and Z = MR⊙S *)
Begin (* algorithm *)
  for i = 1 to m do
    for j = 1 to p do
      zij ← (xij ∧ yij) ∨ ··· ∨ (xin ∧ ynj)
End (* algorithm *)
```

Exercises 7.6 (p. 475)

1. $\{(a,a)\}$ **3.** $\{(a,a),(b,b),(c,c)\}$ **5.** $\emptyset$

7. $\{(a,a),(b,b),(c,c)\}$ **9.** $A \times A$

11. $\{(a,b),(a,c),(a,d),(b,c),(d,b),(d,c)\}$

Exercises 7.7 (p. 481)

1. $\{(a,a),(a,b),(b,a),(b,b)\}$ **3.** $\{(b,a),(b,b),(b,c),(c,a),(c,b),(c,c)\}$

5. $\{(a,a),(a,b)\}$ **7.** $\{(a,b),(a,c),(b,c)\}$

9. $A \times A$ **11.** $\{(a,a),(a,b),(a,c),(b,b),(c,b),(c,c)\}$

13. $\begin{bmatrix} 0 & 1 & 1 \\ 0 & 0 & 1 \\ 0 & 0 & 0 \end{bmatrix}$ **15.** $\begin{bmatrix} 0 & 1 & 0 \\ 0 & 1 & 1 \\ 1 & 1 & 0 \end{bmatrix}$ **17.** $\begin{bmatrix} 1 & 0 & 0 & 0 \\ 0 & 1 & 0 & 1 \\ 0 & 1 & 0 & 1 \\ 0 & 1 & 0 & 1 \end{bmatrix}, \begin{bmatrix} 1 & 0 & 0 & 0 \\ 0 & 1 & 0 & 1 \\ 0 & 1 & 0 & 1 \\ 0 & 1 & 0 & 1 \end{bmatrix}$

19. $\{(a,b),(a,c),(b,c)\}$

21. $\{(a,b),(a,c),(b,a),(b,b),(b,c),(c,b),(c,c)\}$

23. $\{(a,a),(b,b),(b,d),(c,b),(c,d),(d,b),(d,d)\}$

25. $\{(a,a),(b,a),(b,b),(c,c)\}$ **27.** $\{(a,a),(b,a),(b,b),(b,c),(c,b),(c,c)\}$

29. $\{(a,a),(a,b),(b,b),(c,c)\}$ **31.** $\{(a,a),(a,b),(a,c),(b,b),(b,c),(c,c)\}$

33. $\{(a,a),(a,b),(b,b),(b,c),(c,a),(c,c)\}$ **35.** $\geq$

37. $\{(a,b),(a,c),(b,a),(b,c),(c,a),(c,b)\}$

39. $\{(a,a),(a,c),(b,b),(c,a),(c,b)\}$

41. $\{(a,b),(a,c),(b,a),(b,c),(c,a),(c,b)\}$

43. $\{(a,a),(a,c),(b,b),(b,c),(c,b),(c,c)\}$

45. Assume R is reflexive. Let $(a,a) \in \Delta$. Since R is reflexive, $(a,a) \in R$. $\therefore \Delta \subseteq R$. Conversely, let $\Delta \subseteq R$. Let $a \in A$. Since $(a,a) \in \Delta$, $(a,a) \in R$. Thus R is reflexive.

47. $R \cup \Delta$ is reflexive, by Exercise 46. Let S be a reflexive relation such that $R \subseteq S \subseteq R \cup \Delta$. Since $R \subseteq S$, $R \cup \Delta \subseteq S \cup \Delta$. Since S is reflexive, $\Delta \subseteq S$; so $S \cup \Delta \subseteq S$. But $S \subseteq R \cup \Delta$. $\therefore S = R \cup \Delta$.

49. $R \cup R^{-1}$ is symmetric, by Exercise 48. Let S be a symmetric relation such that $R \subseteq S \subseteq R \cup R^{-1}$. Since $R \subseteq S$, $R^{-1} \subseteq S^{-1}$ by Exercise 61 in Section 7.5. Further, since S is symmetric, $S^{-1} = S$. $\therefore R^{-1} \subseteq S$. Thus $R \subseteq S$ and $R^{-1} \subseteq S$. $\therefore R \cup S^{-1} \subseteq S \cup S$; that is, $R \cup R^{-1} \subseteq S$. $\therefore S = R \cup R^{-1}$.

Exercises 7.8 (p. 490)

1. no **3.** yes **5.** no

7. yes **9.** no **11.** $\{a,b\}$

13. $\{d\}$ **15.** {A, B, C, F, H} **17.** {J}

19. {a, b} **21.** {aaa, aab, aba, abb, baa, bab, bba, bbb}

23. {a, b, c} **25.** {aa, ab, ac, ba, bb, bc, ca, cb, cc}

27. $\{\{a\},\{b\},\{c\}\}$ **29.** {{A, B, C, F, H}, {D, E}, {G}, {I}, {J}}

31. $\{(a,a),(b,b),(b,c),(c,b),(c,c),(d,d)\}$

33. 2 **35.** 15

37. 2 **39.** 15

41. $R = \{(a,a),(a,b),(b,a),(b,b),(c,c)\}$ and $S = \{(a,a),(a,c),(b,b),(c,a),(c,c)\}$ are equivalence relations on $\{a,b,c\}$, but $R \cup S = \{(a,a),(a,b),(a,c),(b,a),(b,b),(c,a),(c,c)\}$ is not.

43. Monday

45. Tuesday

47. Let $a \equiv b \pmod m$ and $c \equiv d \pmod m$. Then $a - b = lm$ and $c - d = km$ for some $l,k \in \mathbf{Z}$. Then $(a-b)+(c-d) = lm + km$, that is, $(a+c)-(b+d) = (l+k)m$. $\therefore a + c \equiv (b+d) \pmod m$.

49. By the division algorithm, $a = bm + r$ for some $b \in \mathbf{Z}$. $\therefore a \equiv r \pmod m$.

51. By the division algorithm, $a = d_1m + r_1$ and $b = d_2m + r_2$, where $0 \le r_1, r_2 < m$. Then $a - b = (d_1 - d_2)m + (r_1 - r_2)$. $\therefore a \equiv b \pmod m$ if and only if $r_1 \equiv r_2 \pmod m$.

53. Let $N = \sum_{i=0}^{n} a_i 10^{n-i}$ be the decimal expansion of N. Since $10 \equiv 1 \pmod 9$, $10^k \equiv 1 \pmod 9$ by Exercise 48. Thus $a_i 10^{n-i} \equiv a_i \forall i$ by Exercise 48.
$\therefore N \equiv \sum_{i=0}^{n} a_i \pmod 9$. Consequently, N is divisible by 9 if and only if $\sum_{i=0}^{n} a_i$ is divisible by 9.

55. 0

57. 8

59. 4

61. 15, 621

Exercises 7.9 (p. 503)

1. no **3.** yes **5.** yes **7.** yes

9. no **11.** no **13.** yes **15.** yes

17. yes

19. $(2,6) \preceq (3,5)$

21. $(2,3,2),(2,3,3),(2,2,4)$

23. luxury, maximum, moment, neglect, neighbor, neophyte

25. discount, discourse, discreet, discrete, discretion, diskette

27.

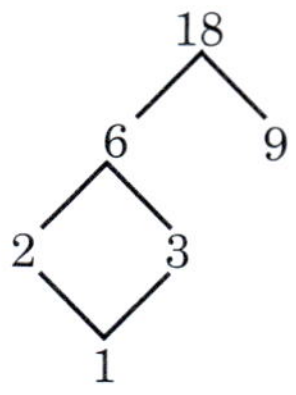

29.

31.

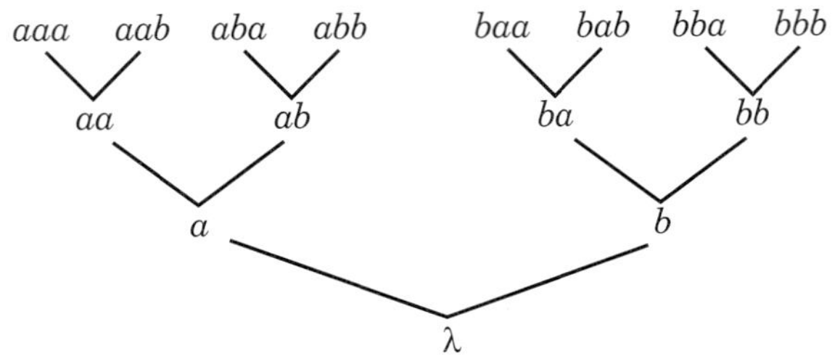

33. $c, d; a$ **35.** none; none **37.** 18; 1 **39.** none; none

41. 18, 1 **43.** F **45.** F **47.** F

49. $(\mathbf{Z}, \leq)$ **51.** $(\mathbf{Z}, \leq)$ **53.** a, b, c, d, e **55.** a, c, b, d, e, f

57. $1, 2, 3, 6, 9, 18$ **59.** $t_1, t_2, t_3, t_4, t_5, t_6, t_7, t_9, t_8, t_{10}, t_{11}, t_{12}, t_{13}$

61. 26; C, G, B, E, A, F, D

63. Suppose there are two greatest elements, a and a' in the poset. Since a is a greatest element, $a' \preceq a$. Likewise, $a \preceq a'$. Since $\preceq$ is antisymmetric, $a = a'$.

65. Let a_1 be any element in A. If a_1 is not maximal, there must be an element a_2 such that $a_1 \prec a_2$. If a_2 is maximal, then we are done. If it is not, there must be an element a_3 in A such that $a_2 \prec a_3$. If a_3 is not maximal, continue this procedure. Since A is finite, this procedure must terminate with some element a_n. Thus $a_1 \prec a_2 \prec \ldots \prec a_n$. So a_n is a maximal element. Hence the result.

Review Exercises (p. 508)

1. symmetric **3.** $\{(1,3), (2,1), (3,1), (3,3)\}$

5. $\{(1,2), (1,3), (2,1), (2,3), (3,1), (3,3)\}$

7. $\{(1,1), (2,1), (2,2), (2,3), (3,2)\}$ **9.** $\{(1,1), (2,2), (2,3)\}$

11. $\{(1,1), (1,2), (1,3), (2,2), (2,3), (3,2), (3,3)\}$

13. $\{(1,1), (1,2), (2,2), (3,2)\}$

15. $\begin{bmatrix} 0 & 0 & 1 \\ 1 & 0 & 1 \\ 1 & 0 & 1 \end{bmatrix}$ **17.** $\begin{bmatrix} 1 & 1 & 1 \\ 0 & 1 & 1 \\ 0 & 1 & 1 \end{bmatrix}$

19. $\{(1,1), (1,2), (1,3), (2,2), (2,3), (3,2), (3,3)\}$

21. $\begin{bmatrix} 1 & 1 & 0 \\ 0 & 1 & 0 \\ 0 & 1 & 0 \end{bmatrix}$ **23.** $A \times A$ **25.** 2

27. no **29.** $\{a, c\}$

31. $\{a, c\}$

33. $\{\{a, c\}, \{b, d\}\}$

35. $\{(2,2), (2,4), (3,3), (4,2), (4,4), (7,7)\}$

37. 877

39. T

41. T

43. T

45. T

47. T

49. $\lambda, 0, 00, 01, 1, 10, 11, 000, 001, 010, 011, 100, 101, 110, 111$

51. CS 100, CS 150, CS 200, CS 250, CS 300, CS 350, CS 400, CS 450

53. none; a

55. Let $(a,b) \in (R \cap S)^2$. Then there exists an element c in A such that $(a,c) \in R \cap S$ and $(c,b) \in R \cap S$. $\therefore (a,c) \in R$ and $(c,b) \in R$, so $(a,b) \in R^2$. Similarly, $(a,b) \in S^2$. Thus $(a,b) \in R^2 \cap S^2$. So $(R \cap S)^2 \subseteq R^2 \cap S^2$.

57. Suppose R is antisymmetric. Let $(a,b) \in R \cap R^{-1}$. Then aRb and $aR^{-1}b$. Since $aR^{-1}b$, bRa. Thus aRb and bRa. So $a = b$. Thus $(a,a) \in R \cap R^{-1} \subseteq \Delta$. Conversely, let $R \cap R^{-1} \subseteq \Delta$. Suppose aRb and bRa. Since bRa, $aR^{-1}b$. Thus aRb and $aR^{-1}b$. $\therefore a(R \cap R^{-1})b$. Since $R \cap R^{-1} \subseteq \Delta$, $(a,b) \in \Delta$. So $a = b$ and R is antisymmetric.

Supplementary Exercises (p. 511)

1. Sunday

3. Recall that $\binom{p}{k} = \dfrac{p!}{k!(p-k)!}$ is an integer. When $0 < k < p, 0 < p-k < p$. Since p is a prime, none of the factors of $k!$ or $(p-k)!$ can divide p. So $p \,\Big|\, \binom{p}{k}$.

5. By the binomial theorem, $(a+b)^p = \sum_{r=0}^{p} \binom{p}{r} a^{p-r} b^r = a^p + \sum_{r=1}^{p-1} \binom{p}{r} a^{p-r} b^r + b^p \equiv a^p + 0 + b^p \equiv a^p + b^p \pmod{p}$

7. 2

9. Let a be any integer. Then, by the division algorithm, $a \equiv 0, 1,$ or 2 modulo 3. If $a \equiv 0 \pmod 3$, then $a(a+1)(a+2) \equiv 0 \pmod 3$. Likewise, $a(a+1)(a+2) \equiv 0 \pmod 3$ when $a \equiv 1$ or 2 modulo 3. Thus, in every case, the product is congruent to 0 modulo 3.

11. 5, 13

13. By Euler's theorem, $a^{\phi(b)} \equiv 1 \pmod b$. Since $b \mid \phi(a)$, $b^{\phi(a)} \equiv 1 \pmod b$. Therefore, $a^{\phi(b)} + b^{\phi(a)} \equiv 1 \pmod b$. Likewise, $a^{\phi(b)} + b^{\phi(a)} \equiv 1 \pmod a$.

Since a and b are relatively prime, it follows that $a^{\phi(b)} + b^{\phi(a)} \equiv 1 \pmod{ab}$.

15. The set has 2^{n^2} relations (pigeons) on it. Using adjacency matrices, $2^{(n-1)n/2+n} = 2^{n(n+1)/2}$ of them are symmetric relations (pigeonholes). Suppose each relation is assigned to its symmetric closure. Then, by the GPHP, at least one of the holes must contain $\lfloor (2^{n^2}-1)/2^{n(n+1)/2} \rfloor + 1 = \lfloor 2^{n(n-1)/2} - 1/2^{n(n+1)/2} \rfloor + 1 = 2^{n(n-1)/2} - 1 + 1 = 2^{n(n-1)/2}$ pigeons; that is, at least $2^{n(n-1)/2}$ relations must have the same symmetric closure.

Chapter 8 Graphs

Exercises 8.1 (p. 533)

1. no **3.** no **5.** 3; 4 **7.** 4; 4; 4

9.

$$\begin{array}{c} \\ a \\ b \\ c \end{array}\begin{array}{c} \begin{array}{ccc} a & b & c \end{array} \\ \begin{bmatrix} 0 & 2 & 1 \\ 2 & 0 & 1 \\ 1 & 1 & 0 \end{bmatrix} \end{array}$$

11. 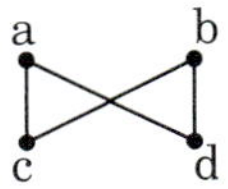

13. The graph is made up of two disjoint components.

15. Sum of the degrees $= 6 + 4 + 4 = 14 = 2 \cdot 7 = 2$ (no. of edges)

17. 4 **19.** yes **21.** 13 **23.** 12 **25.** 15 **27.** 45

29. $(a_{ij})_{n \times n}$ where $a_{ij} = \begin{cases} 1 & \text{if } i \neq j \\ 0 & \text{otherwise} \end{cases}$

31. yes; $V_1 = \{a, c\}, V_2 = \{b, d, e\}$

33. mn

35. By Theorem 8.1, $\sum_{i=1}^{n} v_i = 2e$

$$nm \le \sum_{i=1}^{n} v_i \le nM$$

That is, $nm \le 2e \le nM$

$$m \le \frac{2e}{n} \le M$$

37. 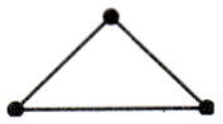

39. 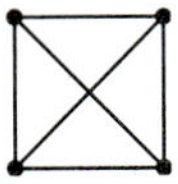

41. 

43. yes, $n-1$

45. By Exercise 44, $e = \dfrac{nr}{2}$.

$\therefore nr$ is even.

47. no

49. 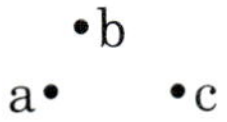

51. Let A be the adjacency matrix of G. Then A' is the adjacency matrix of G'.

53. $K'_{m,n}$ consists of a loop at each vertex and a path connecting the vertices in each vertex set.

55.

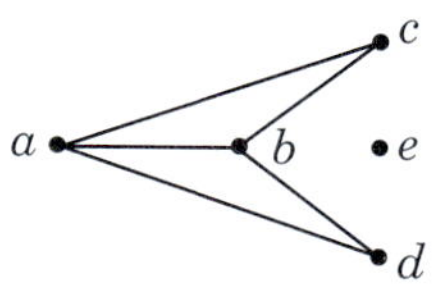

57. K_5

59. K_6

61. K_n

63.

	team					
round	**a**	**b**	**c**	**d**	**e**	**x**
1	b	a	d	c	x	e
2	c	e	a	x	b	d
3	d	x	e	a	c	b
4	e	d	x	b	a	c
5	x	c	b	e	d	a

65. Let $P(e)$ denote the given proposition. When $e = 1$, a graph consisting of a loop at a vertex satisfies the condition. $\therefore P(1)$ is true.

Assume $P(k)$ is true: $\sum_{i=1}^{n} \deg(v_i) = 2k$. Add an edge between vertices v_s and v_t, where $1 \le s < t \le n$. Then both $\deg(v_s)$ and $\deg(v_t)$ are increased by 1.

$\therefore \sum_{i=1}^{n} \deg(v_i) = 2k + 2 = 2(k+1)$. So $P(k+1)$ is true. Thus the result follows by PMI.

Exercises 8.2 (p. 539)

1.

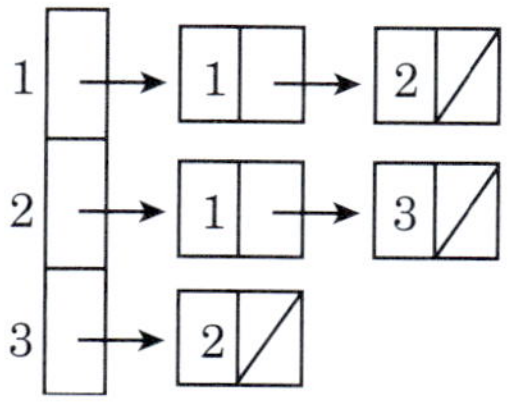

3.

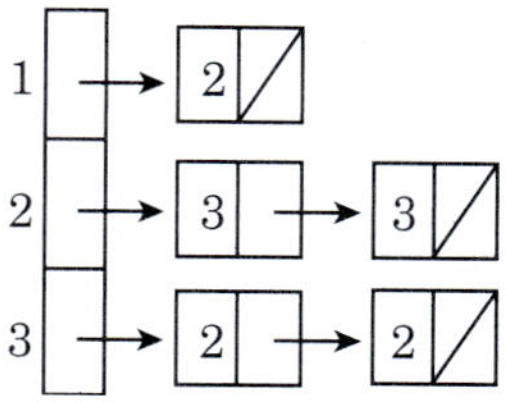

5.

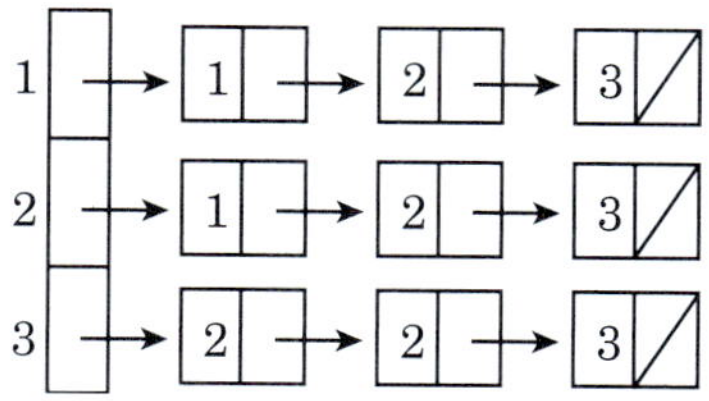

7.

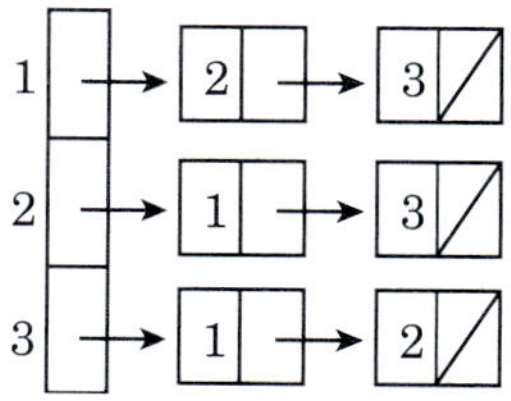

9.

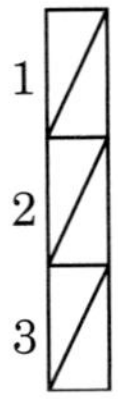

11.

$$\begin{array}{c} \\ 1 \\ 2 \\ 3 \end{array}\begin{array}{c} \begin{array}{ccc} 1 & 2 & 3 \end{array} \\ \begin{bmatrix} 1 & 1 & 0 \\ 1 & 0 & 1 \\ 0 & 1 & 0 \end{bmatrix} \end{array}$$

13.

$$\begin{array}{c} \\ 1 \\ 2 \\ 3 \end{array}\begin{array}{c} \begin{array}{ccc} 1 & 2 & 3 \end{array} \\ \begin{bmatrix} 0 & 1 & 0 \\ 1 & 0 & 2 \\ 0 & 2 & 0 \end{bmatrix} \end{array}$$

15.

$$\begin{array}{c} \\ 1 \\ 2 \\ 3 \end{array}\begin{array}{c} \begin{array}{ccc} 1 & 2 & 3 \end{array} \\ \begin{bmatrix} 1 & 1 & 1 \\ 1 & 1 & 1 \\ 1 & 1 & 1 \end{bmatrix} \end{array}$$

17.

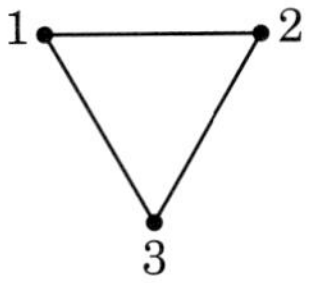

19.

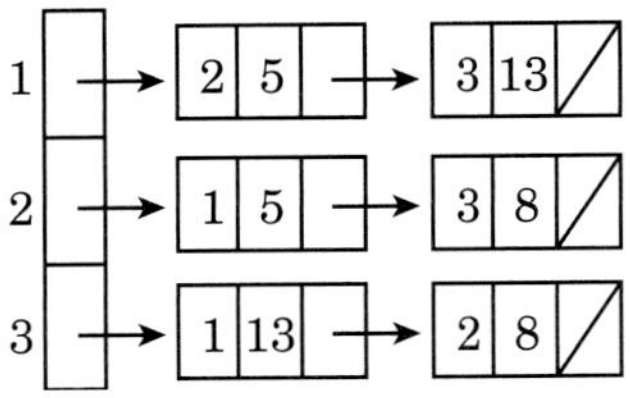

21.

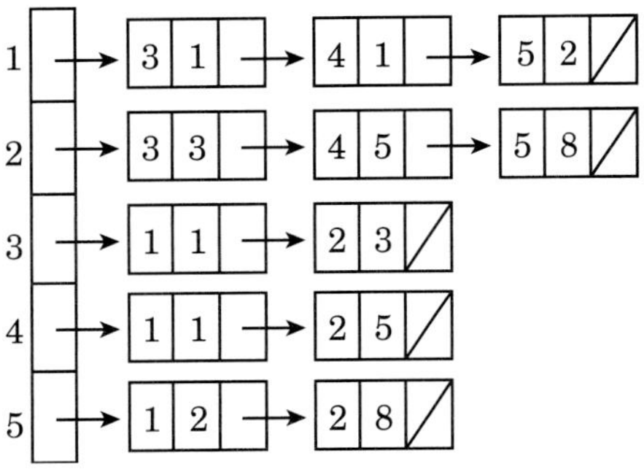

Exercises 8.3 (p. 544)

1. no

3. yes; $f(\mathrm{a}) = \mathrm{b}, f(\mathrm{b}) = \mathrm{f}, f(\mathrm{c}) = \mathrm{e}, f(\mathrm{d}) = \mathrm{g}$

5. yes; $f(\mathrm{a}) = \mathrm{f}, f(\mathrm{b}) = \mathrm{e}, f(\mathrm{c}) = \mathrm{g}, f(\mathrm{d}) = \mathrm{h}$

7. no

9. no

11. yes; $f(\mathrm{a}) = \mathrm{i}, f(\mathrm{b}) = \mathrm{h}, f(\mathrm{c}) = \mathrm{g},$
$f(\mathrm{d}) = \mathrm{f}, f(\mathrm{e}) = \mathrm{j}$

13. no

15. 1) Every graph is isomorphic to itself. ∴ The relation is reflexive.

2) Let $f : G_1 \to G_2$ be an isomorphism. Then $f^{-1} : G_2 \to G_1$ is also an isomorphism.

3) Let $f : G_1 \to G_2$ and $g : G_2 \to G_3$ be isomorphisms. Then $g \circ f : G_1 \to G_3$ is also an isomorphism. ∴ The relation is transitive. Thus the relation is an equivalence relation.

Exercises 8.4 (p. 554)

1. 2

3. 4

5. 7

7. 1

9. b-a-d-b, b-a-c-b, b-d-c-b **11.** no **13.** no **15.** yes

17. 5 **19.** 10 **21.** 45 **23.** 0

25. 10 **27.** 0 **29.** 10 **31.** 2

33. 0

35. start — ab — a — aA — A — AB — aA — AC — A — aA — C — cC
(second branch from AC:) C — cC — A — aA

37. Initially, the farmer takes $a_2, a_4, \ldots, a_{2n}$ across the river and then returns. She takes $a_3, a_5, \ldots, a_{2n-1}$ across and brings $a_2, a_4, \ldots, a_{2n}$ back to the original side. She takes a_1 and a_{2n+1} to the other side, and then returns. Finally, she takes $a_2, a_4, \ldots, a_{2n}$ to the other side.

39. **Algorithm graph (G,A,connected)**

```
(* This algorithm determines if the graph G with n vertices is
connected using Theorems 8.4 and 8.5, and its adjacency matrix A. The
boolean variable connected denotes whether or not G is connected. *)
Begin (* algorithm *)
  (* find S = Σ A^i *)
  B ← A (* initialize *)
  S ← A (* initialize *)
  for i = 2 to n - 1 do
  begin (* for *)
    B ← AB
    S ← S + B
  endfor
  (* check if any entry of S is nonzero. *)
  i ← 1
  j ← 1
  done ← false
  connected ← true
  while (not done) and (i ≤ n) and (j ≤ n) do
    if s_ij = 0 then
      begin (* graph is not connected *)
        connected ← false
        done ← true
      endif
    else
      begin (* else *)
        i ← i + 1
        j ← j + 1
      endelse
End (* algorithm *)
```

Exercises 8.5 (p. 573)

1. no **3.** no **5.** yes

7. a-e_1-b-e_2-c-e_3-a-e_4-c-e_5-a

9. a-e_1-b-e_2-a-e_5-c-e_6-a-e_7-c-e_3-b-e_4-c-e_9-d-e_8-a

11. a-e_1-b-e_4-d-e_7-e-e_3-b-e_5-c-e_6-d-e_2-a

13. no

15. contains an Eulerian circuit.

17. contains an Eulerian circuit.

19. no

21. none

23. n odd

25. $n = 2$

27. r even

29. contains a Hamiltonian cycle.

31. contains a Hamiltonian cycle.

33. not a connected graph.

35. contains a Hamiltonian cycle.

37. contains a Hamiltonian cycle.

39. contains neither

41. no Hamiltonian cycle; contains a Hamiltonian path, a-b-c-e-d.

43. $n \geq 3$

45. contains a Hamiltonian path.

47. neither

49. $m = n$

51. T_n is Eulerian $\forall n \geq 1$.

53. S_n is not Eulerian if $n \geq 2$.

55.

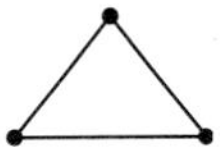

57.

59. a-c-b-d-e-a, 53

61. none

63. none

65. 1-15-12-4-5-11-16-9-7-2-14-13-3-6-10-17-8

67. **Algorithm Eulerian path (G,A)**

```
(* G is a connected graph with n vertices and with adjacency matrix
A = (aij)n×n. Using Theorem 8.8, this algorithm determines if G
contains an Eulerian path. Odd is a counter that keeps track of the
number of odd vertices in G. *)
Begin (* algorithm *)
  odd ← 0 (* initialize the counter *)
  flag ← false (* boolean flag to exist when odd > 2 *)
  i ← 1 (* initialize row index *)
  while (i ≤ n) and (not flag ) do (* compute each row sum *)
  begin (* while *)
    sum ← 0
    for j = 1 to n do
      sum ← sum + aij
    if sum is odd then (* update the counter *)
      odd ← odd + 1
    if odd > 2 then (* exit the loop *)
      flag ← true
  endwhile
```

```
  if odd = 2 then
    graph has an Eulerian path
  else
    graph does not have an Eulerian path
End (* algorithm *)
```

Exercises 8.6 (p. 584)

1.

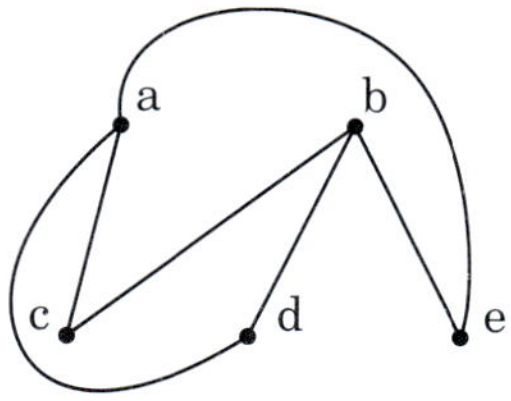

3.

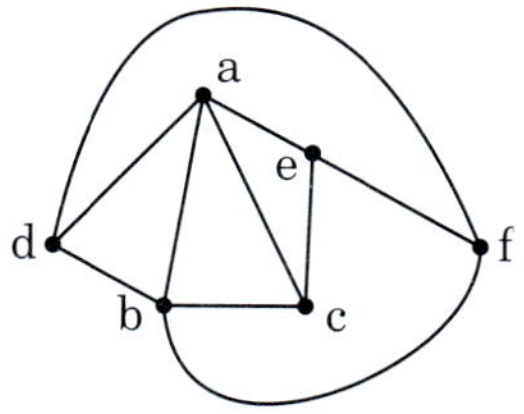

5.

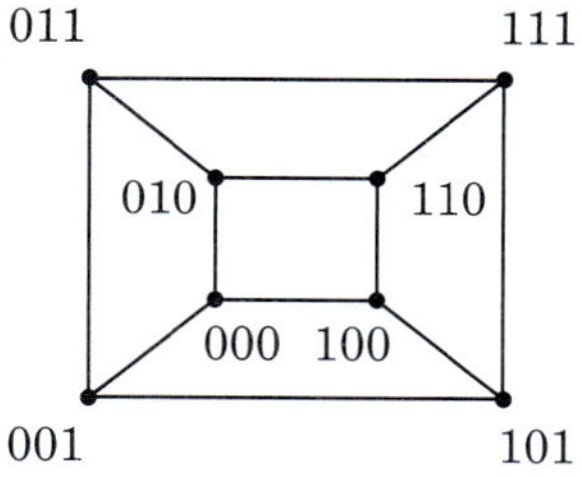

7.

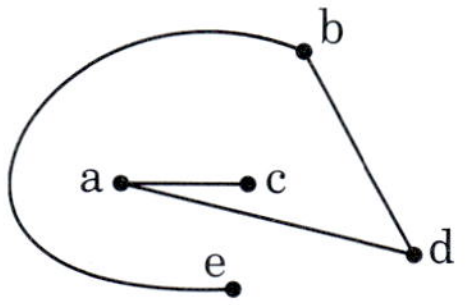

9.

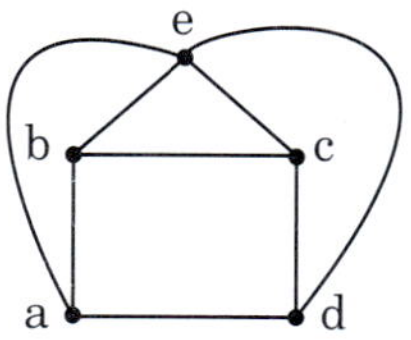

11. $e = 25, v = 15, r = 12$
$r = e - v + 2$

13. $e = 25, v = 15, 3v - 6 = 39 \geq e$

15. 13 **17.** 6, 9 **19.** 5; 2; 7

21.

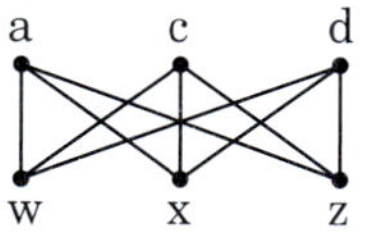

23. $m, n \leq 2$

25. 3

27.

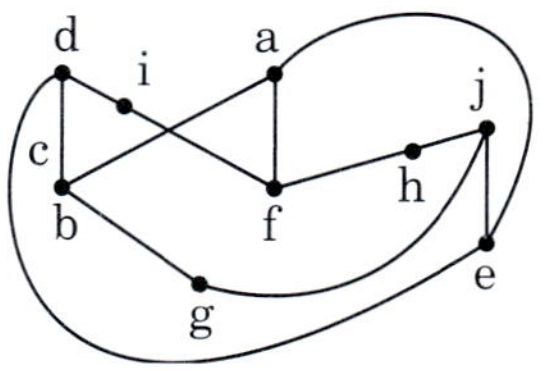

Exercises 8.7 (p. 595)

1. 4 **3.** 2 **5.** 4 **7.** 2 **9.** 2 **11.** 5 **13.** 3 **15.** 2

17. $\min\{m, n\} \leq \max\{m, n\}$

19. 1, 6; 2, 3; 4; 5

21. 1, 3; 2; 4, 7; 5; 6

23. 1, 2, 4; 3, 5; 6; 7, 8, 9

25. A, D; B, C

27. A, B, F; C, D; E

Review Exercises (p. 601)

1.
$$\begin{array}{c} \\ a \\ b \\ c \end{array}\begin{array}{c} \begin{array}{ccc} a & b & c \end{array} \\ \begin{bmatrix} 1 & 0 & 1 \\ 0 & 1 & 1 \\ 1 & 1 & 1 \end{bmatrix} \end{array}$$

3.
$$\begin{array}{c} \\ a \\ b \\ c \\ d \\ e \end{array}\begin{array}{c} \begin{array}{ccccc} a & b & c & d & e \end{array} \\ \begin{bmatrix} 0 & 1 & 1 & 1 & 0 \\ 1 & 0 & 0 & 1 & 1 \\ 1 & 0 & 0 & 1 & 0 \\ 1 & 1 & 1 & 1 & 0 \\ 0 & 1 & 0 & 1 & 0 \end{bmatrix} \end{array}$$

5. no **7.** no **9.** 17 **11.** K_n **13.** $m = n$

15.

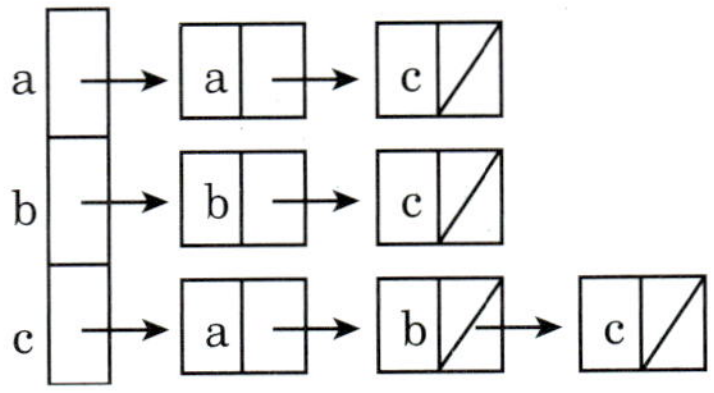

17.

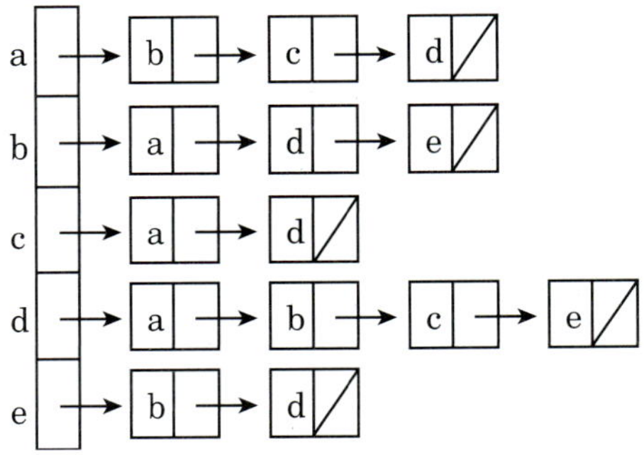

19. yes; $f(\text{a}) = \text{i}, f(\text{b}) = \text{j}, f(\text{c}) = \text{k}, f(\text{d}) = \text{l}, f(\text{e}) = \text{m}, f(\text{f}) = \text{n}, f(\text{g}) = \text{o}, f(\text{n}) = \text{p}$

21. yes **23.** no **25.** no **27.** no **29.** yes **31.** no

33.

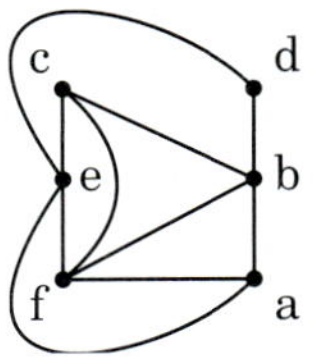

35. no

37. $e = 11, v = 7, r = 6; r = e - v + 2$

39. 13

41. yes

43. Assume $K_{3,3}$ is planar. For $K_{3,3}$, $v = 6$ and $e = 9$. Then $2v - 4 = 12 - 4 = 8 < e$, is a contradiction. $\therefore K_{3,3}$ is nonplanar.

45. Assume $K_{3,5}$ is planar. Then $2v - 4 = 16 - 4 = 12 < e$, which is a which contradiction. $\therefore K_{3,5}$ is nonplanar.

Supplementary Exercises (p. 604)

1. $n, n+1, 2^n$ **3.** $n \geq 3$ **5.** yes **7.** yes

9. Label the vertex of degree n with 1 and the others $2, 3, \ldots, (n+1)$:

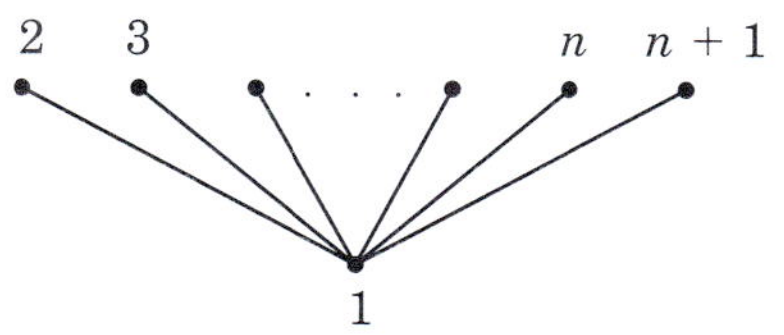

11. $\sum v_i, \sum_{i \neq j} v_i v_j$

13. Let $G = (V_1, E_1)$ and $H = (V_2, E_2)$ be two simple graphs. Assume $f: G \to H$ is an isomorphism. Let $G' = (V_1, E'_1)$ and $H' = (V_2, E'_2)$ be their complements. Let $g: G' \to H'$ defined by $g(u) = f(u)$. Since $|E_1| = |E_2|$, $|E'_1| = |E'_2|$. To show that g preserves the adjacency relationship, assume $\{u, v\} \in E'_1$. Then $\{u, v\} \in E_1$. So $\{f(u), f(v)\} \notin E_2$; that is, $\{g(u), g(v)\} \in E'_2$. Thus if $\{u, v\} \in E'_1$, then $\{g(u), g(v)\} \in E'_2$. $\therefore g$ is an isomorphism.

Conversely, let G' and H' be isomorphic. Then $(G')' = G$ and $(H')' = H$ are isomorphic, by the first part.

15. C_5

17. Suppose the graph contains e edges. Then $e = C(n, 2)/2 = n(n-1)/4$. $\therefore 4|n$ or $4|(n-1)$, that is, $n \equiv 0 \pmod 4$ or $n \equiv 1 \pmod 4$.

19. $(n-1)!/2$

21. $(n!)^2/2$

23.

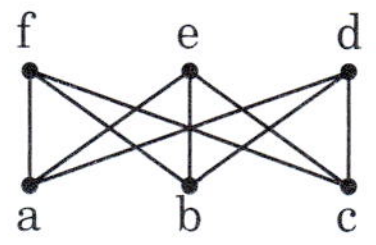

Delete a-d. The ensuing subgraph is planar:

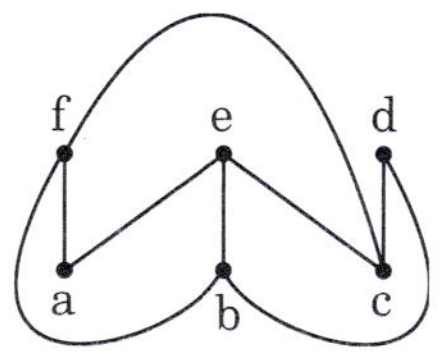

Chapter 9 Trees

Exercises 9.1 (p. 613)

1. yes **3.** no **5.** no **7.** no; contains cycles.

9. $n = 8, e = 7 = n - 1$ **11.** $n = 12, e = 11 = n - 1$

13. By Theorem 8.1, $\sum_{i=1}^{n} \deg(v_i) = 2$(no. of edges). But, by Theorem 9.2, number of edges $= n - 1$. $\therefore \sum_{i=1}^{n} \deg(v_i) = 2(n-1) = 2n - 2$.

15. yes **17.** no **19.** $m = 1$ or $n = 1$

21. 

23. 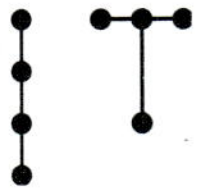

25. 4, 6, 3, 5, 4, 4, 4, 5, 5, 5, 6

27. Algorithm tree (A,n)

```
(* Using Theorem 9.2, this algorithm determines if a graph
with adjacency matrix A = (aij)n×n is a tree. Flag is a boolean
variable and stores the value true if the graph is a tree and false
otherwise. *)
Begin (* tree *)
   flag ← false (* initially assume that the graph is not a tree *)
   (* compute the sum of the degrees of the vertices *)
   sum ← 0
   For i = 1 to n do
     For j = 1 to n do
       sum ← sum + aij
   If sum = 2n - 2 then
     flag ← true
End (* tree *)
```

Exercises 9.2 (p. 624)

1. $n - 1$ **3.** 4

5. 

7. 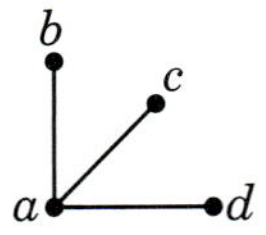

9. 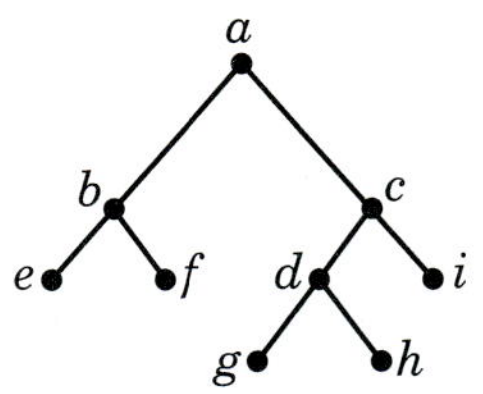

11. 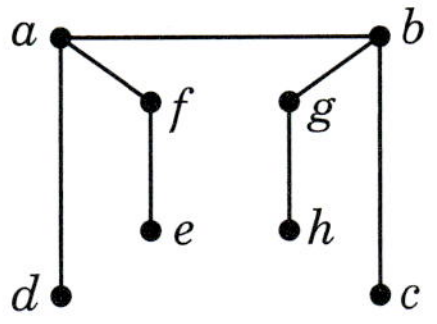

13. 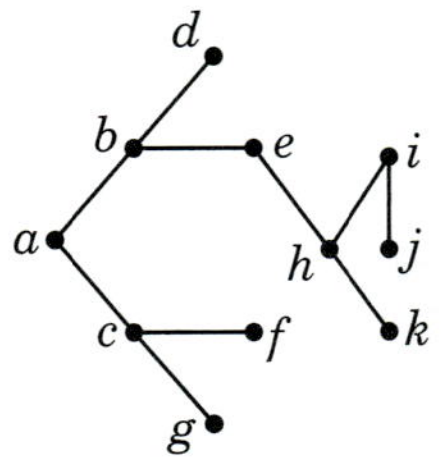

15. 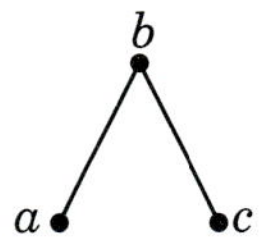

17. 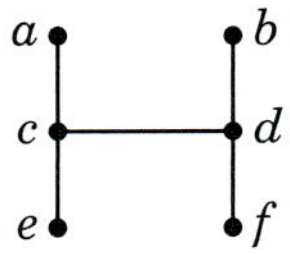

19. 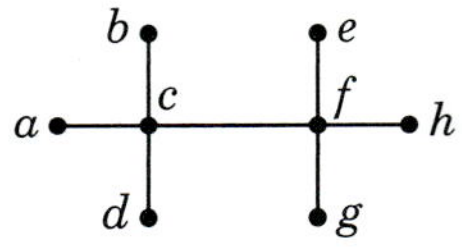

21.

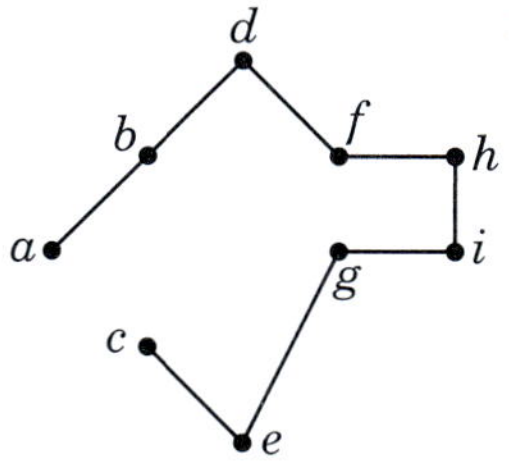

23.

25.

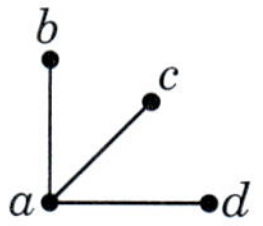

27.

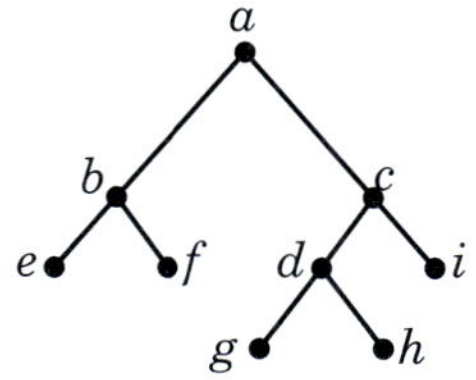

29.

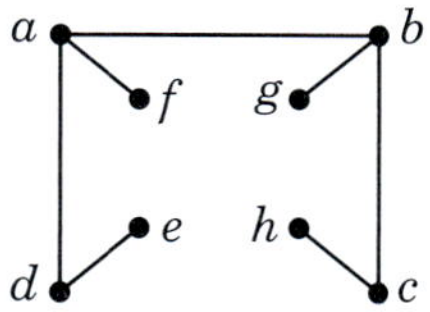

31.

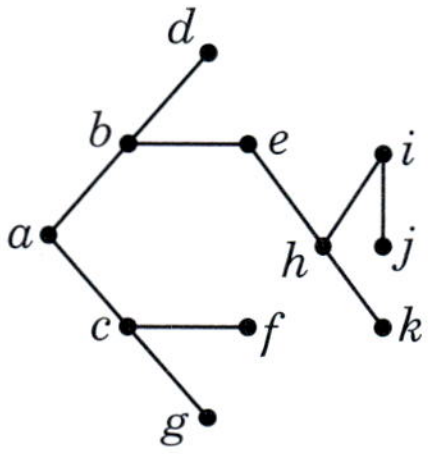

33.

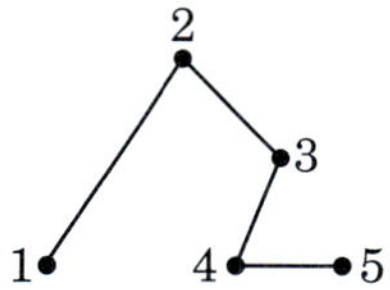

35.

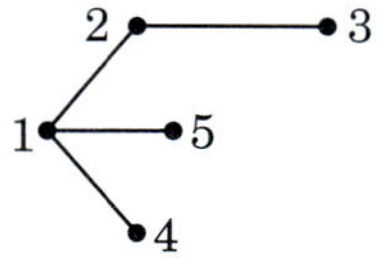

37. no solution

39.

			Q		
Q					
				Q	
	Q				
					Q
		Q			

41.

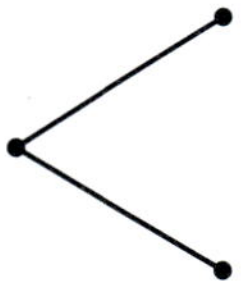

43.

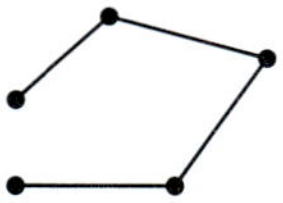

45. 1

47. 16

49.

51.

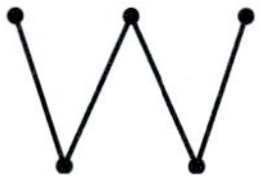

53. 1

55. 9

57. 12

59. 125

Exercises 9.3 (p. 633)

1.

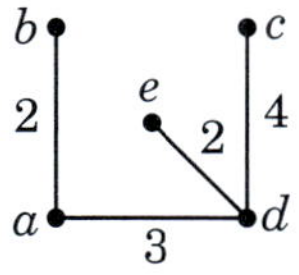

3.

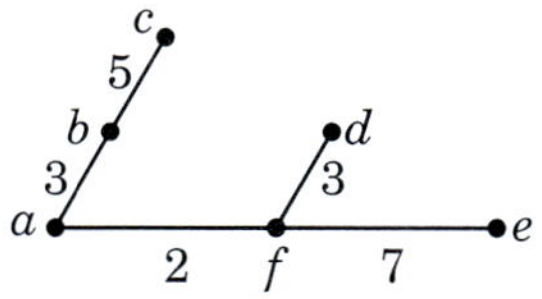

5.

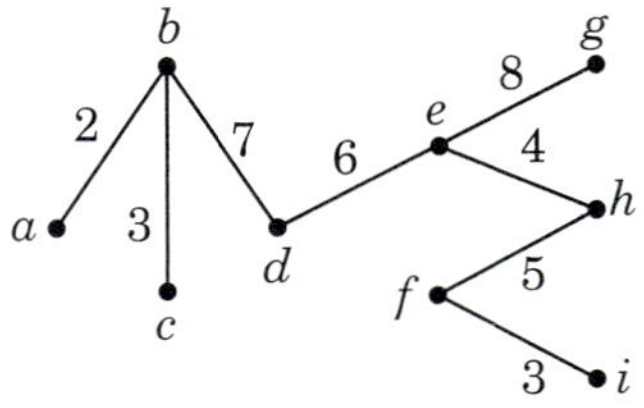

7.

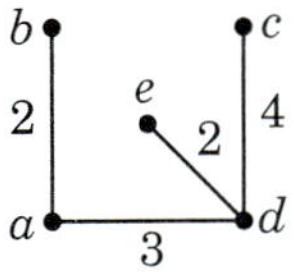

9.

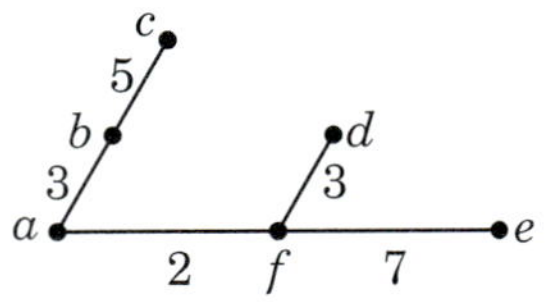

11.

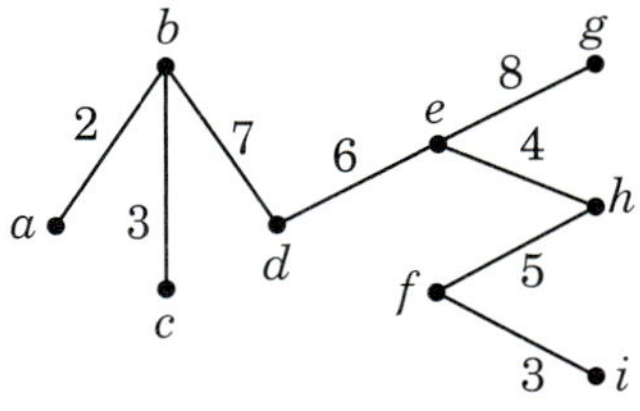

13.

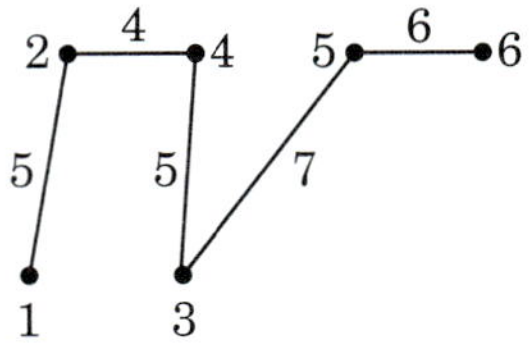

15.

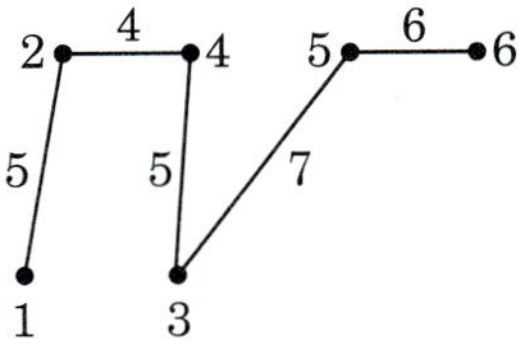

17.

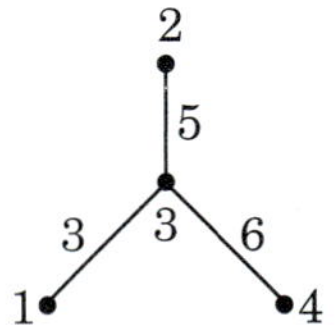

19.

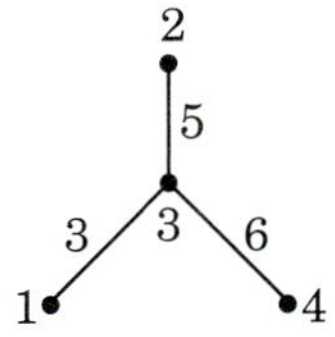

Exercises 9.4 (p. 643)

1. 1

3. 2

5. $m = 2$

7. $m = 4$

9. no

11. yes

13. no

15. no, no, yes

17.

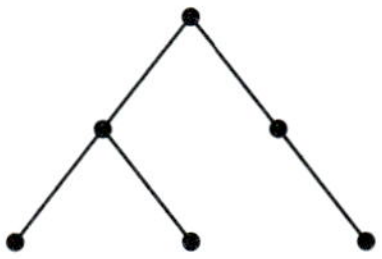

19.

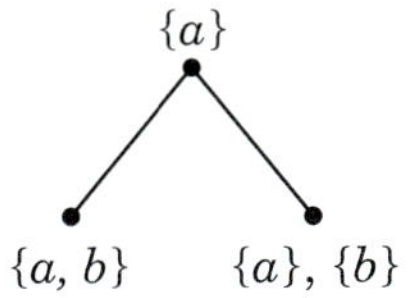

21. 13

23. 40, 81

25. 243

27. 4

29. 8

31. 31

33. yes

35. yes

37. yes

39.

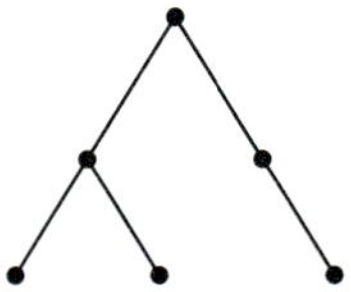

41. m^h

43. $\dfrac{m^{h+1} - 1}{m - 1}$

45. 4

47. 729

49. $l = n - i = (mi + 1) - i = (m - 1)i + 1$

51. yes

53. Since $m \geq 2$, $m \cdot m^h \geq 2 \cdot m^h$. That is, $m^{h+1} \geq 2m^h$, so $m^h(m - 1) > m^h - 1$.
$\therefore m^h > (m^h - 1)/(m - 1)$. That is, $l > i$, by Exercise 41.

Exercises 9.5 (p. 661)

1. a, b, d, c, e, f

3. $a, b, d, h, i, e, c, f, g, j, k$

5. $d, h, b, e, a, f, c, i, g$

7. d, b, e, f, c, a

9. $h, i, d, e, b, f, k, j, g, c, a$

11. $a, b, d, h, e, c, f, g, i$

13. a, b, c, d, e, f

15. $a, b, c, d, e, f, g, h, i, j, k$

17. $- + \uparrow a \uparrow bc/def$

19. $- - a + *bcd * ef \uparrow gh$

21. $abc \uparrow\uparrow de/ + f-$

23. $abc * d + e/f * gh \uparrow --$

25.

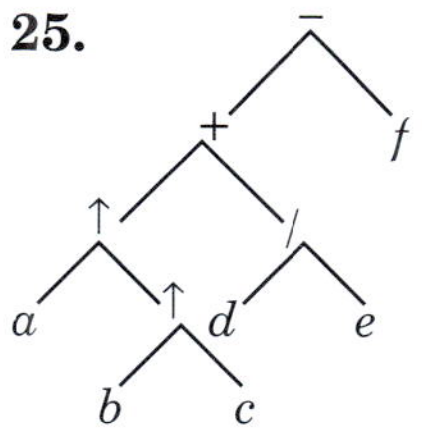

27.

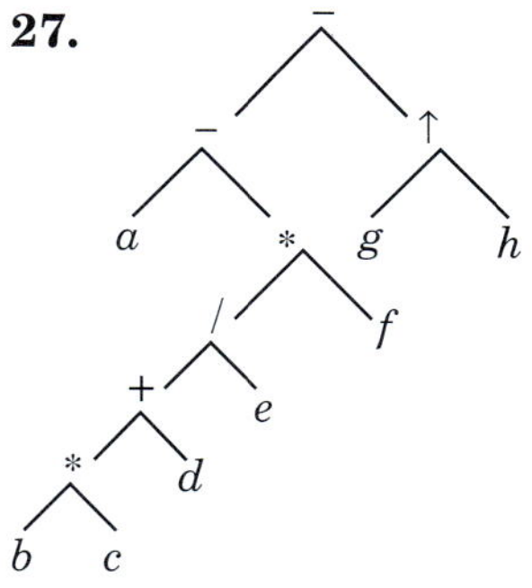

29. $a + [b/(c - d)] \uparrow e$

31. $[a - (b + c)] * d * e \uparrow (f - g)$

33. $a/[(b + c) * d * e - f$

35. $\dfrac{a}{b} + \left(\dfrac{c}{d}\right) * [e + (f - g)]$

37. 22

39. 32

41.

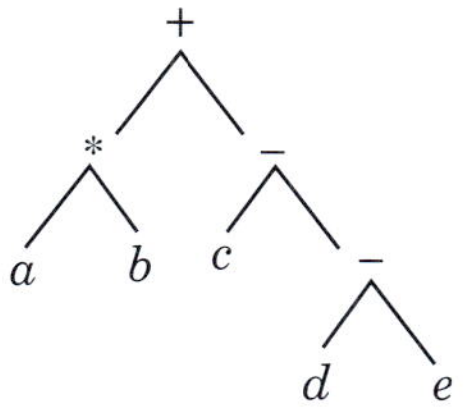

43.

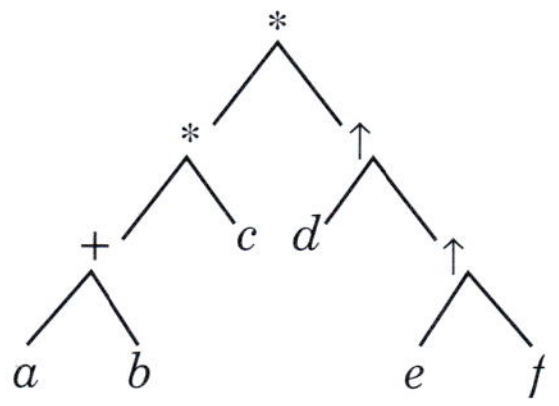

45. 196

47. $\uparrow * + 432/6 - 85$

49. $43 + 2 * 685 - / \uparrow$

51. $[(4 + 3) * 2] \uparrow [6/(8 - 5)]$

53.

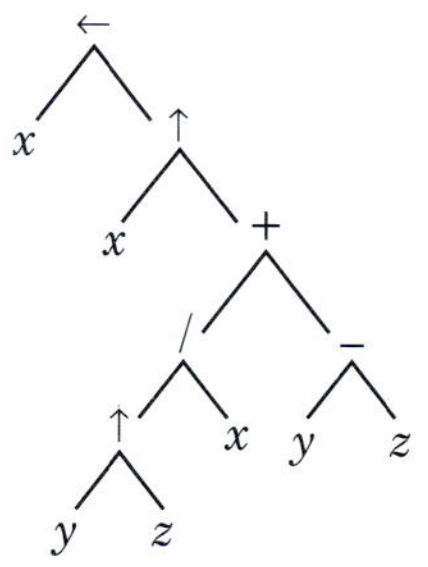

55.

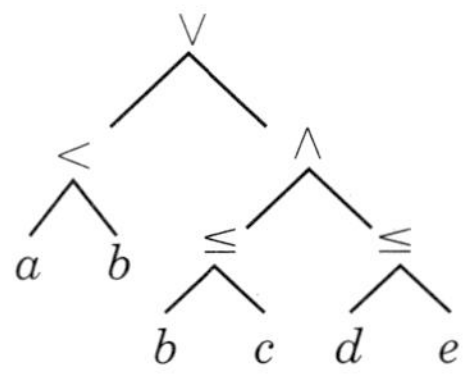

57.

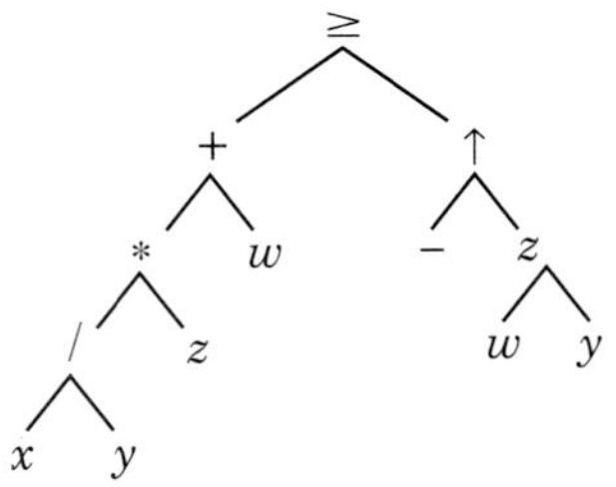

59. Let l denote the number of leaves and n the number of vertices in a full binary tree. Then, by Exercise 58, $l = \frac{n+1}{2}$. Since l is an integer, n must be odd.

61. $n = 2l - 1$ **63.** 14; 42 **65.** yes if $n \geq 3$ **67.** no

69. $l_1 = 1 = l_2$
$l_n = l_{n-1} + l_{n-2}, n \geq 3$

71. $v_n = \begin{cases} 1 & \text{if } n = 1 \text{ or } 2 \\ v_{n-1} + v_{n-2} + 1 & \text{if } n \geq 3 \end{cases}$

73. $e_1 = 0 = e_2$
$e_n = e_{n-1} + e_{n-2} + 2, n \geq 3$

75. Let $a_n = i_n + 1$. Then, by Exercise 70, $a_1 = 1 = a_2$ and $a_n = a_{n-1} + a_{n-2}, n \geq 3$. So $i_n = a_n - 1 = F_n - 1$.

77. Let $a_n = e_n + 1$. Then, by Exercise 73, $a_1 = 1 = a_2$ and $a_n = a_{n-1} + a_{n-2} + 1$, where $n \geq 3$. By Exercise 76, $e_n = a_n - 1 = (2F_n - 1) - 1 = 2F_n - 2$.

79. **Algorithm inorder (vertex)**

```
(* Prints a binary tree using
recursion *)
Begin (*inorder*)
  if tree ≠ Ø then
    begin (* if *)
      inorder (left subtree)
      print (vertex)
      inorder (right subtree)
    endif
End (* inorder *)
```

81. **Algorithm eval (vertex)**

```
(* This algorithm evaluates a
binary expression tree using
recursion. *)
Begin (* eval *)
  if tree ≠ Ø then
    begin (* if *)
      eval(left subtree)
      eval(right subtree)
      perform the operation at
      the vertex
    endif
End (* eval *)
```

Exercises 9.6 (p. 668)

1.

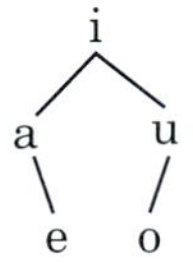

3.

5.

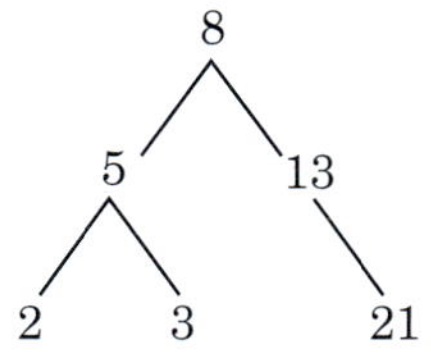

7.

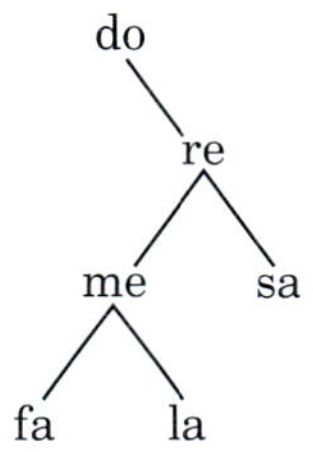

9.

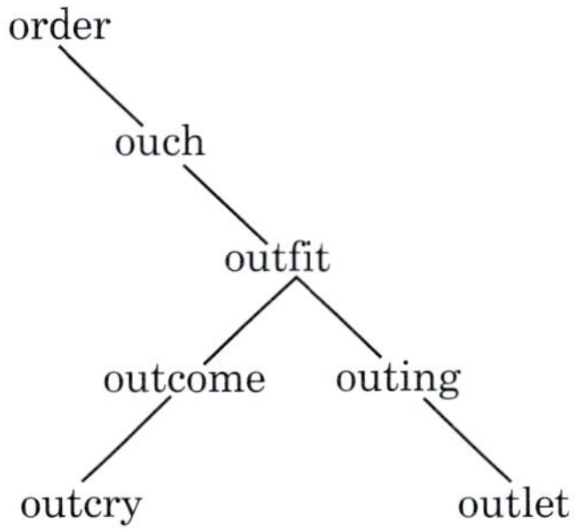

11. 3 **13.** 5 **15.** 3 **17.** 4 **19.** 5

21.

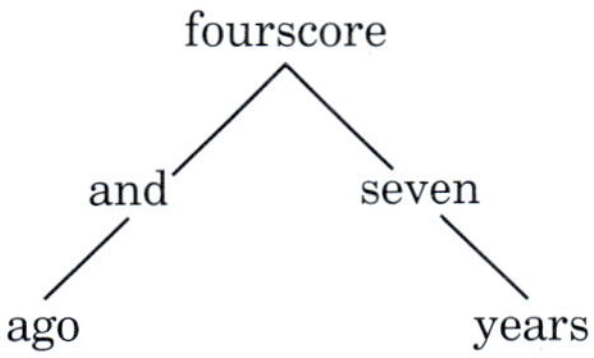

23.

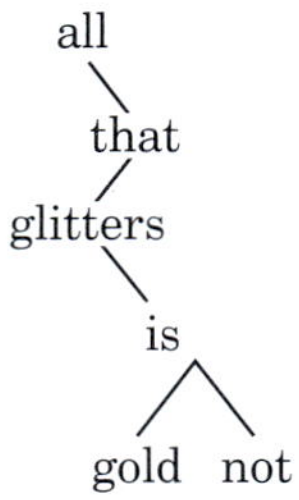

25.

```
Algorithm traverse (vertex)
    (* This algorithm prints the
    contents of a binary search tree
    rooted at vertex in lexicographic
    order using preorder traversal. *)
      Begin (* traverse *)
        If tree ≠ Ø then
        begin (* if *)
          traverse (left subtree)
          print (vertex)
          traverse (right subtree)
        endif
      End (* traverse *)
```

27.

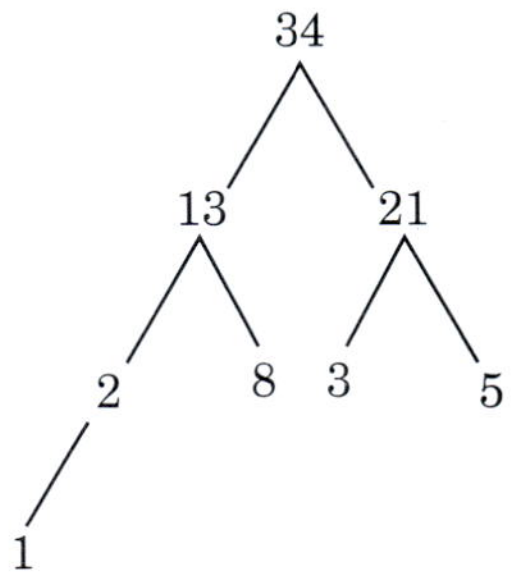

Exercises 9.7 (p. 675)

1. 011101101110
3. 1101011011010001
5. YEAR
7. AN ANT
9. 110
11. 011
13. 01000111
15. 11011100010110111
17. doe
19. coffee
21.

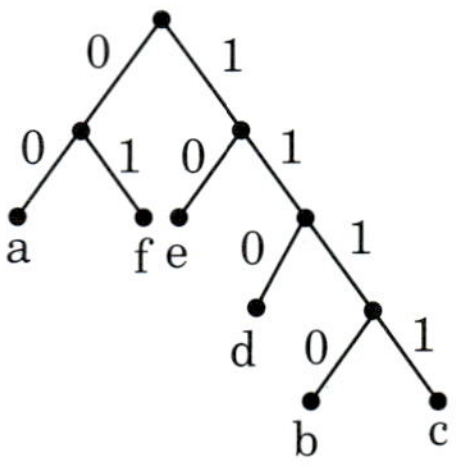

23. 00, 1110, 1111, 110, 10, 01
25. 111000111010
27. 11111011010110
29. abed
31. efface
33. 1100
35. 0001
37. 0011010010011010000100
39. 11001010011111110001100
41. eclogue
43. baggage
45. 47

Exercises 9.8 (p. 679)

1. 7
3. 11
5. 3
7. 4
9.

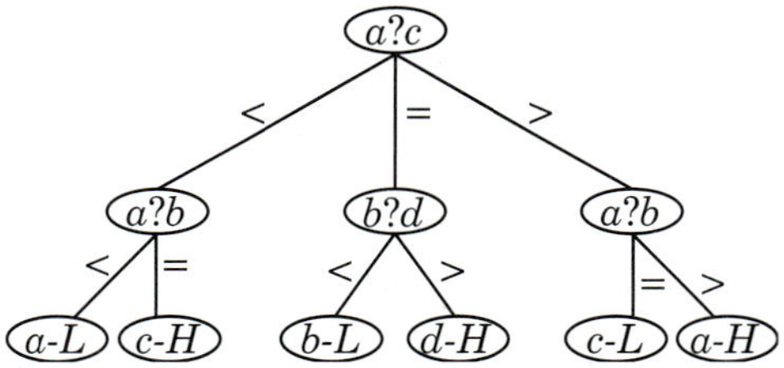

11. **Algorithm sort (a,b,c)**

```
(* This algorithm sorts the
elements a, b, and c in
lexico-graphic order. *)
Begin (* sort *)
  if a < b then
    if c < a then
      print (c,a,b)
    else if b < c then
          print(a,b,c)
        else
          print(a,c,b)
```

13. **Algorithm counterfeit coin(a,b,c,d,e,f,g)**

```
(* There are seven coins
a through g; one of them
is heavier. This algorithm
identifies the false coin using
an equal arm balance and a
minimum number of weighings. *)
Begin (* counterfeit coin *)
  if a + b + c < d + e + f then
    if d < e then
      e is heavier
```

```
        else if c < b then
              print(c,b,a)
           else if c < a then
              print(b,c,a)
            else
              print(b,a,c)
End (* sort *)
```

```
   else if d = e then
         f is heavier
       else
         d is heavier
 else if a + b + c = d + e + f then
       g is heavier
     else if a < b then
           b is heavier
         else if a = b then
               c is heavier
             else
               a is heavier
End (* counterfeit coin *)
```

Review Exercises (p. 681)

1.

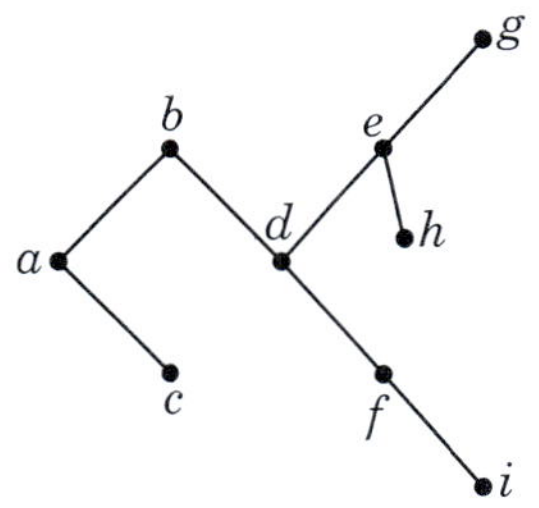

3.

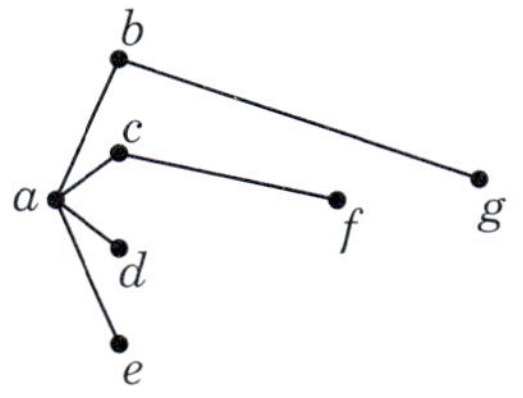

5.

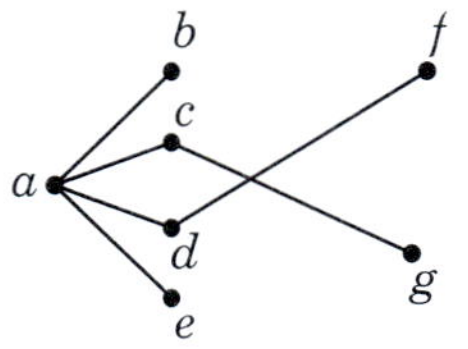

7.

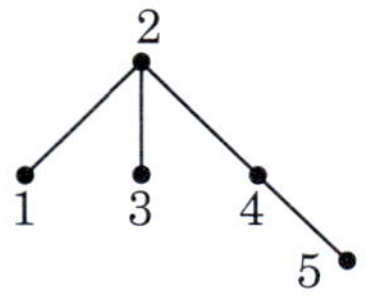

9.

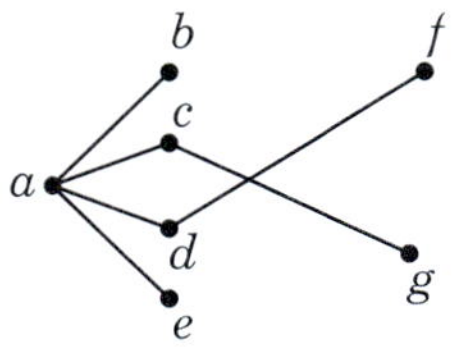

11.

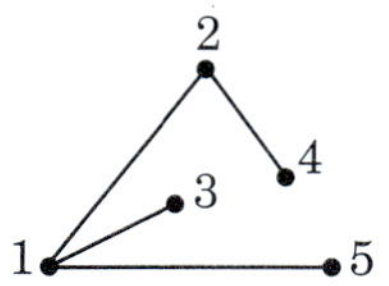

13.

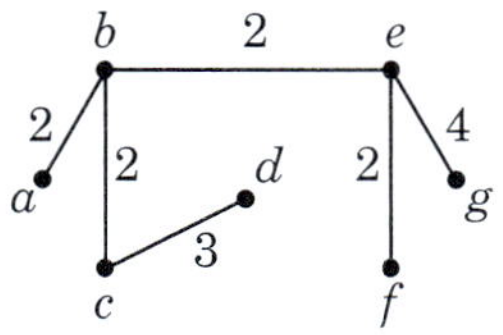

15.

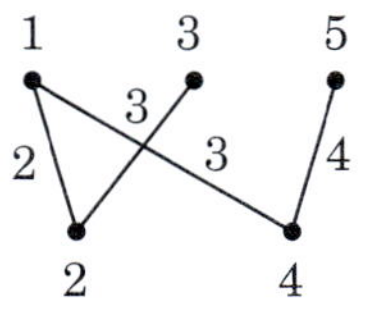

17.

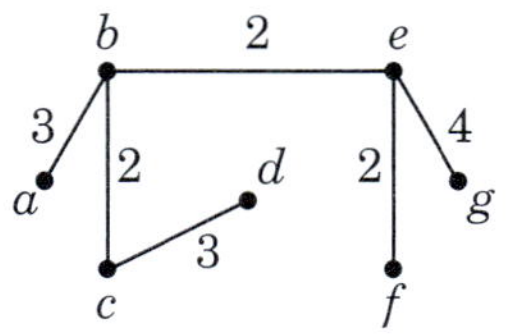

19.

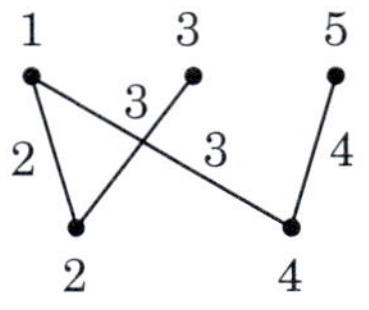

21.

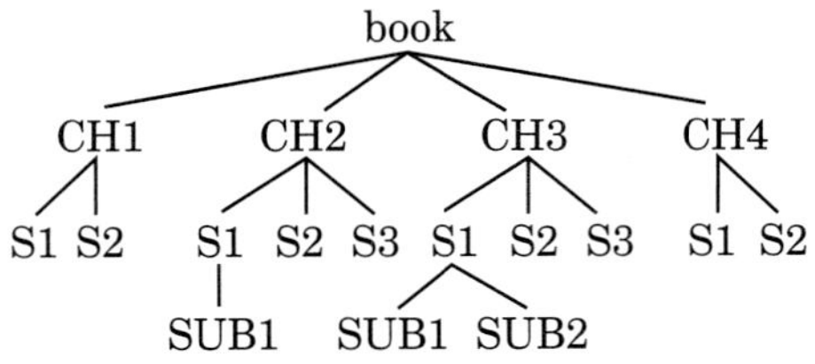

23. 3 **25.** 2187 **27.** 61 **29.** 75 **31.** 3 **33.** 3

35. $r_n = \begin{cases} 0 & \text{if } n = 1 \\ r_{\lceil n/2 \rceil} + 1 & \text{if } n \geq 2 \end{cases}$

37. **proof** (by PMI): Let $P(n) : r_n = \lceil \lg n \rceil$. When $n = 1$, $r_1 = 0 = \lceil \lg 1 \rceil$. $\therefore P(1)$ is true. Assume $P(k)$ is true; that is, assume that $r_k = \lceil \lg k \rceil$. Then $r_{k+1} = r_{\lceil (k+1)/2 \rceil} + 1 = \lceil \lg \lceil (k+1)/2 \rceil \rceil + 1 = \lceil \lg (k+1) \rceil$. $\therefore P(k+1)$ is true. Thus the result follows by induction.

39. *abdgehcfij*, *dgbehacijf*, *gdhebjifca*

41. $- + a * / bc / + de - fg \uparrow hi$

43. $abc/de + fg - / * + hi \uparrow -$

45.

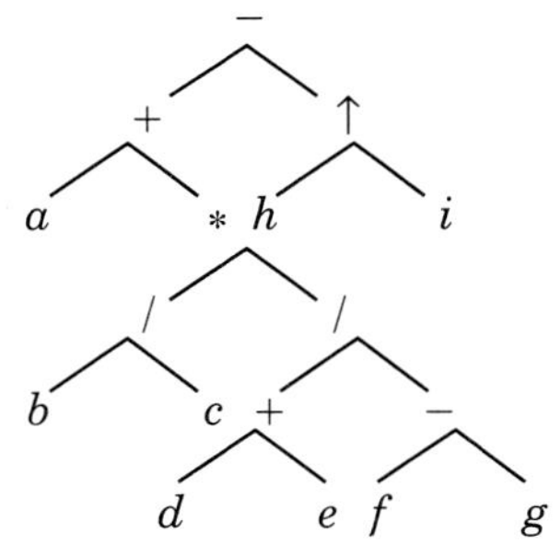

47.

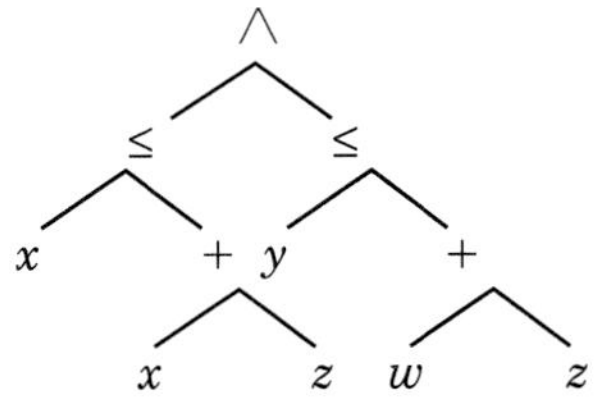

49. $+ + a - bc \uparrow d \uparrow ef$

51. $abc - + def \uparrow\uparrow +$

53. 27

55. 78

57.

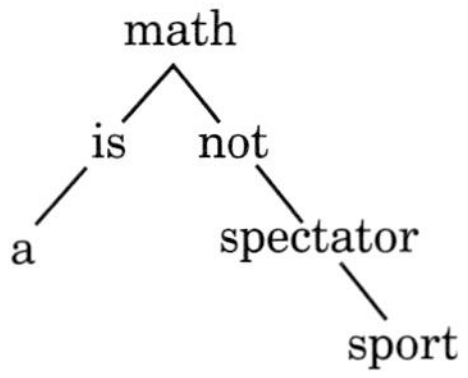

59.

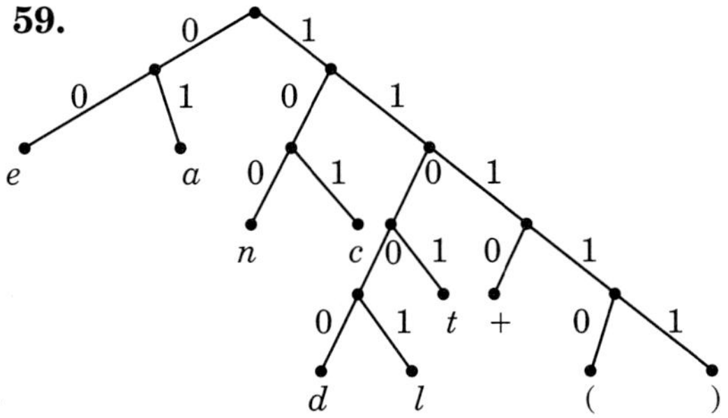

61. 10101100011010000100

63. 11110110001110001111 1

65. $(c + a)$ **67.** 5 **69.** 7

Supplementary Exercises (p. 686)

1. 

3. 10, 6

5. Let e be the number of edges in F. Let $T_i = (V_i, E_i)$ be a tree in F, $1 \le i \le k$. Let $|V_i| = v_i$ and $|E_i| = e_i$. Then $\sum v_i = n$ and $e_i = v_i - 1$. $\therefore e = \sum e_i = \sum(v_i - 1) = \sum v_i - k = n - k$.

7. Since a tree is planar, the number r of regions formed is 1. By Euler's formula, the is number of edges e of the tree is given by $e = r + v - 2 = 1 + n - 2 = n - 1$.

9. Let T and T' be two spanning trees of a connected graph G with n vertices. Assume T and T' contain e and e' edges respectively. Then, by Theorem 9.2, $e = n - 1 = e'$.

11.

13. Since e is a bridge, $(V, E\text{-}\{e\})$ is disconnected with two subgraphs, each being a tree. So it is a forest of two trees.

15. 3 weighings

Chapter 10 Digraphs

Exercises 10.1 (p. 704)

1. 2, 1, 3; 2, 3, 1; 4, 4, 4

3. no sources or sinks.

5. $\begin{array}{c} \\ a \\ b \\ c \end{array}\begin{array}{c} \begin{matrix} a & b & c \end{matrix} \\ \begin{bmatrix} 1 & 0 & 1 \\ 1 & 1 & 1 \\ 0 & 0 & 1 \end{bmatrix} \end{array}$

7. $\sum \text{indeg}(v_i) = 2 + 1 + 3 = 6 = e = \sum \text{outdeg}(v_i)$

9. $C(n, 2)$

11. 3

13. 2

15. b

17. yes

19. $\begin{bmatrix} 1 & 1 & 1 \\ 1 & 1 & 1 \\ 1 & 1 & 1 \end{bmatrix}$

21. yes

23. Let $A = (a_{ij})_{n \times n}$ be the adjacency matrix of a digraph. Suppose row k is zero; that is, $a_{kj} = 0\ \forall k$. Then row k of every power of $A^{[m]}$ is zero, so row k of the reachability matrix is zero. $\therefore$ By Theorem 10.3, the digraph is not strongly connected.

25. 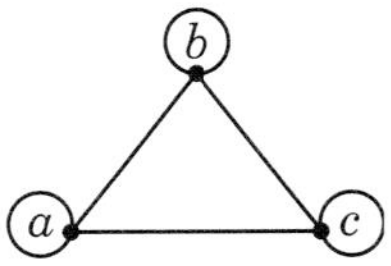

27. yes

29. yes

31. yes

33. sink

35. 0011

37. 00101110

39. 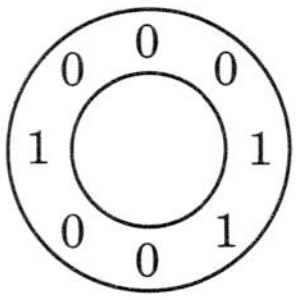

41. 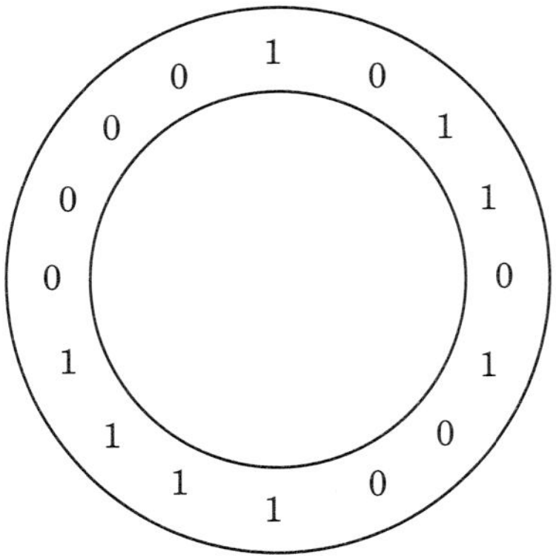

43. 16

47. 0011

49. 0000111101100101

Exercises 10.2 (p. 713)

1. no

3. yes

5. no sources; a, c, d, f

7. a; no sinks.

9. no sources; a, c, d, f

11. a and e; f

13. no

15. yes

17. yes

19. 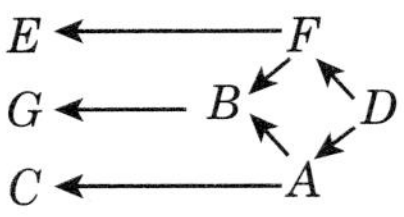

21. 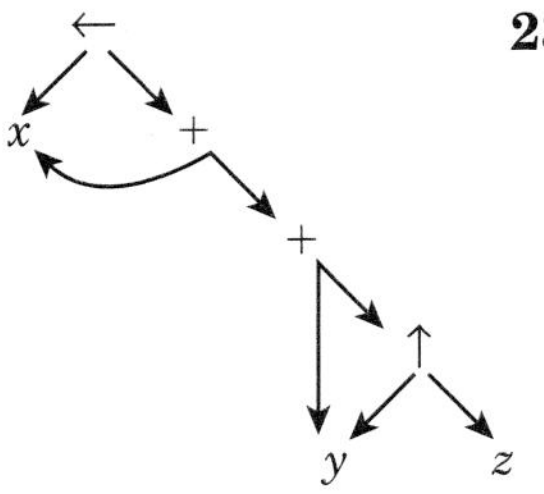

23. 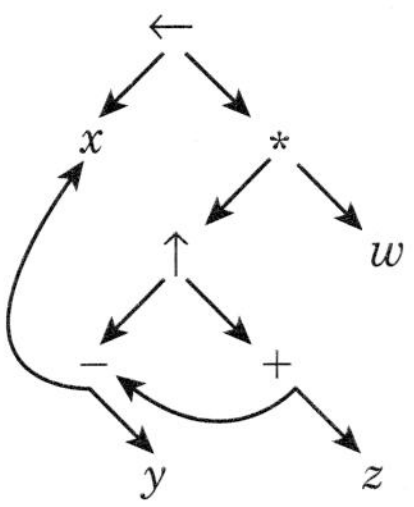

25. 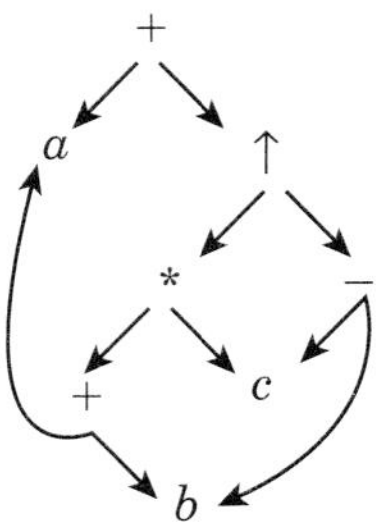

27. 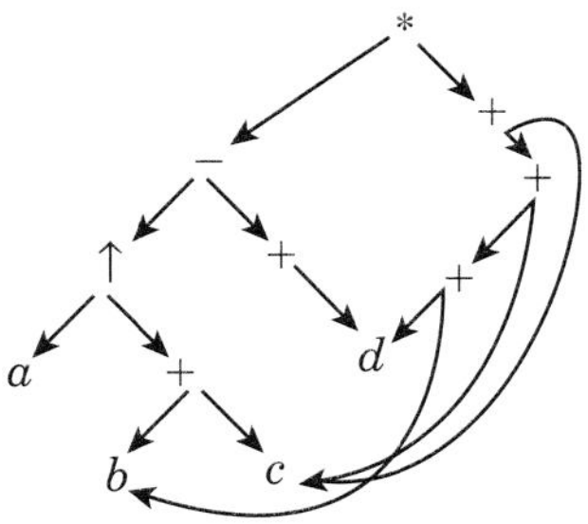

29. 10

31. 306

33. $ab + ab + bc + -*$

35. $abc- \uparrow bc - +bc - abc- \uparrow bc - */-$

37. $* + ab - +ab + bc$

39. $-+ \uparrow a - bc - bc/ - bc * \uparrow a - bc - bc$

41. $(a+b)*((a+b)-(b+c))$

43. $((a \uparrow (b-c))+(b-c))-(b-c)/((a \uparrow (b-c)*(b-c))$

45. Suppose dag D contains n vertices and no sinks. Then outdeg$(v) \geq 1\ \forall v$. Since outdeg$(v) \geq 1$, v has an immediate successor v_1. Since outdeg$(v_1) \geq 1$, v_1 has an immediate successor v_2. Continuing like this, a directed path results: v-v_1-v_2-$\cdots$-v_i. If two vertices along this path are the same, then D would be cyclic, a contradiction. If no two vertices are identical, the procedure can be continued until we get a path continuing all vertices in D : v-v_1-v_2-$\cdots$-v_i-$\cdots$-v_j. Since outdeg$(v_j) \geq 1$, v_j has an immediate successor and it must be one of the vertices in the path. So the path and hence D contains cycle, a contradiction. $\therefore D$ contains a source.

47. $t_1, t_2, t_3, t_4, t_5, t_6, t_7, t_9, t_8, t_{10}, t_{11}, t_{12}, t_{13}$

49. E, G, B, F, C, A, D

Exercises 10.3 (p. 723)

1. 9

3. 12

5.

$$\begin{array}{c|ccccc} & a & b & c & d & e \\ \hline a & 0 & 8 & 5 & 0 & 2 \\ b & 0 & 0 & 6 & 6 & 0 \\ c & 0 & 0 & 0 & 0 & 0 \\ d & 3 & 0 & 3 & 0 & 0 \\ e & 3 & 0 & 7 & 4 & 0 \end{array}$$

7. a-c-b-e-f, 13; a-b-e-f, 11; a-b-e-d-b-e-f, 21; a-d-b-e-f, 15

9. a-c-e-g, 11; a-c-b-e-g, 11; a-b-e-g, 9; a-d-b-e-g, 13

11. 2 **13.** 3 **15.** 2 **17.** 3 **19.** 2 **21.** 3

23. 2 **25.** 4 **27.** a-b-c **29.** 2 **31.** 4

33. a-b, 2; a-d-c, 4; a-d, 1; a-d-e, 6; a-d-c-f, 11; a-b-g, 8; a-d-h, 4; a-d-c-i, 12

35. a-c-b, 6; a-c, 2; a-d, 5; a-c-e, 9; a-d-f, 13; a-d-g, 13

37. a-b, 2; a-c, 4; a-d, 1; a-d-e, 6; a-d-c-f, 11; a-b-g, 8; a-d-h, 4; a-d-c-i, 12

39.
```
Algorithm Floyd (D,W,P)
(* P is a one-dimensional array such that Pi is the immediate
predecessor of vertex i along a shortest path from source 1. *)
Begin (* Floyd *)
  (* initialize P. *)
  for i = 1 to n do
    Pi ← 1
  for i = 1 to n do
    for j = 1 to n do
      for k = 1 to n do
        if wji + wik < wjk then
```

```
begin (* if *)
  wjk ← wji + wik
  Pk ← i
endif
End (* Floyd *)
```

41. a-b, a-c, a-d, a-d-e, a-d-c-f, a-b-g, a-d-h, a-d-c-i

Review Exercises (p. 727)

1. 1, 3, 1, 0, 1; 1, 2, 0, 2, 1

3. yes

5. $\begin{bmatrix} 0 & 1 & 1 & 0 & 1 \\ 0 & 1 & 1 & 0 & 1 \\ 0 & 0 & 0 & 0 & 0 \\ 1 & 1 & 1 & 0 & 1 \\ 0 & 1 & 1 & 0 & 1 \end{bmatrix}$

7. yes

9. 010, 101, 011, 111, 110, 100, 001

11. no

13. no

15. 221201100

17. 33, 32, 22, 21, 12, 20, 02, 23, 31, 11, 13, 11, 13, 30, 01, 10, 00, 03

19. 3

21. d

23. no

25.

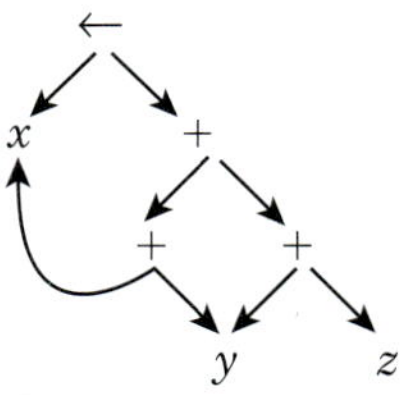

27.

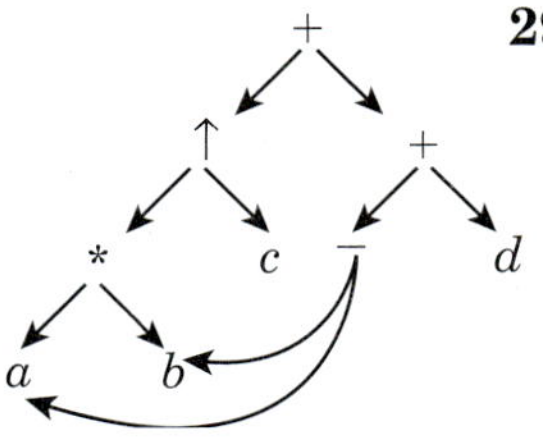

29. 31

31. $ab * c \uparrow ba - d + +$

33. $325 \uparrow + 25 \uparrow 24 * / -$

35. $+ \uparrow * abc + -bad$

37. $- + 3 \uparrow 25 / \uparrow 25 * 24$

39. $(3 + 2 \uparrow 5) - 2 \uparrow 5/(2 * 5)$

41. 7

43. 15

45. a-b-e-g

47. a-c-f-g; 11

Supplementary Exercises (p. 730)

1. $n - 1$

3. $n^2 - 2n$

5. $C(n, 2)$

7. **proof** (by PMI): Suppose the tournament has n vertices. When $n = 2$, the tournament has a single edge and the edge is a Hamiltonian path.

Now assume that every tournament with n vertices has a Hamiltonian path. Let T be a tournament with $n + 1$ vertices, $v_1, v_2, \ldots, v_{n+1}$. Delete v_{n+1} from T. This yields a tournament with n vertices; by the IH, it has a Hamiltonian path, say, v_1-v_2-$\cdots$-v_n.

Returning to the tournament T, if v_n-v_{n+1} is an edge in T, then v_1-v_2-$\cdots$-v_n-v_{n+1} is a Hamiltonian path in T; on the other hand, if v_{n+1}-v_1 is an edge in T, then v_{n+1}-v_1-v_2-$\cdots$-v_n is a Hamiltonian path in T. Otherwise, there must be a positive integer i such that both v_i-v_{n+1} and v_{n+1}-v_{i+1} are edges in T. Then v_1-v_2-$\cdots$ v_i-v_{n+1}-v_{i+1}-$\cdots$-v_n is a Hamiltonian path in T.

Thus, by PMI, the result is true $\forall n \geq 2$; that is, every tournament with $n \geq 2$ vertices has Hamiltonian path.

9. In both D_1 and D_2, a is a root. In D_3, every vertex is a root.

11. a-e-c-d-b

13. a-b-d-c

15. no

17. A graph is strongly orientable iff it is connected and has no bridges.

Chapter 11 Formal Languages and Finite-State Machines

Exercises 11.1 (p. 741)

1. $b, bb, bab, bbabb, bbabb$

3. $\lambda, b^2, b^4, b^6, b^6$

5. $\lambda, a, b \in L$;
$x \in L \Rightarrow axa, bxb \in L$.

7. $0 \in L; x \in L \Rightarrow 1x, xx \in L$.

9. $1 \in L; x \in L \Rightarrow xx \in L$.

11. $00 \in L; x \in L \Rightarrow x0, x1 \in L$.

13. F

15. F

17. no

19. invalid

21. $\lambda \in S; x \in S, y \in S \Rightarrow xy \in S$.

23. $\lambda \in \Sigma^*; x \in \Sigma^*$,
$y \in \Sigma \Rightarrow xy \in \Sigma^*$.

25. 00, 01, 10, 11

27. 0, 1, 2

29. **proof** (by contradiction): Assume Σ^* is finite. Since $\Sigma \neq \emptyset$, it contains an element a. Let x be a word of largest length in Σ^*. Then $l(xa) = l(x) + l(a) > l(x)$. $\therefore xa \in \Sigma^*$ and is longer than x, which is a contradiction.

31. $\{a, bc, aba, abbc, bca, bcbc\}$

33. $\{a^3, a^2bc, abca, a(bc)^2, bca^2, bcabc, bcbca, (bc)^3\}$

35. $\Lambda A = \{\lambda\}\{a, bc\} = \{a, bc\} = A$

37. $(B \cup C)A = \{aa, aab, ba, bab, ca, cab, aba, abab\}$
$= BA \cup CA$

39. $(B \cap C)A = \emptyset = BA \cap CA$

41. F **43.** T **45.** T **47.** T

49. 0, 01, 011 **51.** $01, 01^3, 01^5$ **53.** λ, 01, 0101 **55.** λ, 0, 1, 01

57. **proof** (by PMI): Since $\Lambda \subseteq \Lambda$, the result is true when $n = 0$. Assume it is true for an arbitrary integer $k \geq 0$. Then $A^{k+1} \subseteq AB^{k+1} \subseteq BB^{k+1} = B^{k+1}$. Thus the result follows by PMI.

59. Since $A \subseteq A^*$, $A^* \subseteq (A^*)^*$. Conversely, let $x \in (A^*)^*$. Then $x \in (A^*)^m$ for some integer $m \geq 0$. So $x = y^m$, where $y \in A^*$. Since $y \in A^*$, $x = y^y \in A^*$. So $(A^*)^* \subseteq A^*$. Thus $(A^*)^* = A^*$.

61. $\Lambda A = \{\lambda\}A = \{\lambda a | a \in A\} = \{a | a \in A\} = A$

63. Since $A^1 = A$, the result follows.

65. Let $x \in (B \cup C)A$. Then $x = ya$, where $y \in B \cup C$ and $a \in A$. If $y \in B$, then $ya \in BA$, so $x \in BC \cup CA$. If $y \in C$, then $ya \in CA$, so $x \in BC \cup CA$. Thus, in both cases, $(B \cup C)A \subseteq BC \cup CA$.

Conversely, let $x \in BA \cup CA$. If $x \in BA$, then $x = ba$, where $b \in B$ and $a \in A$. Since $b \in B \cup C$, $x \in (B \cup C)A$. Similarly, if $x \in CA$, then also $x \in (B \cup C)A$. Again in both cases, $BC \cup CA \subseteq (B \cup C)A$. Thus $(B \cup C)A = BC \cup CA$.

67. $(A^*B^*)^* = (A \cup B)^* = (B \cup A)^* = (B^*A^*)^*$

Exercises 11.2 (p. 755)

1. yes **3.** no **5.**

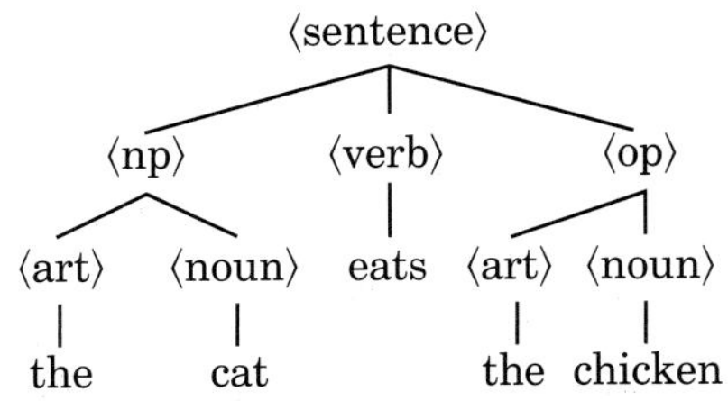

7.

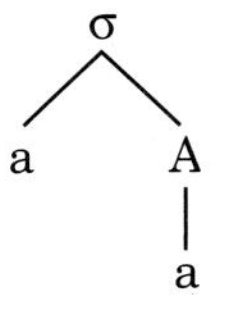

9.

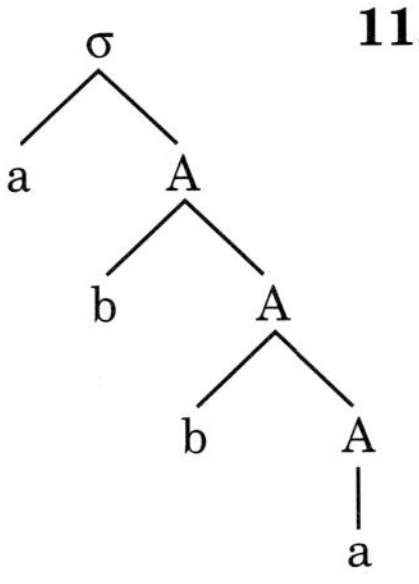

11. yes **13.** yes

15.

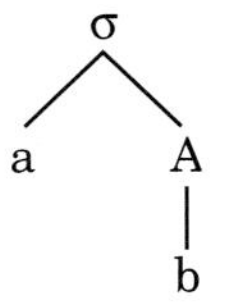

17.

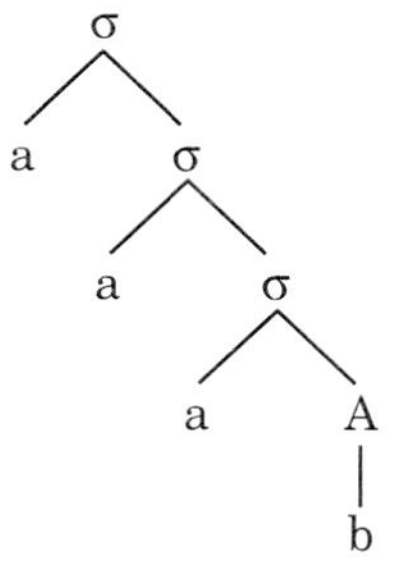

19. no

21. yes

23. $\{a^n b \mid n \geq 1\}$

25. no

27. yes

29. {aa, abb}

31. $\sigma \to \sigma\sigma$, $\sigma \to 1$

33. $\sigma \to \sigma 0$, $\sigma \to \sigma 1$, $\sigma \to 00$

35. $\sigma \to \sigma 0$, $\sigma \to 1\sigma$, $\sigma \to 1$

37. $\sigma \to \mathrm{a}\sigma$, $\sigma \to \mathrm{aA}$, $\mathrm{A} \to \mathrm{b}$

39. $\sigma \to \mathrm{aAB}$, $\mathrm{A} \to \mathrm{Aa}$, $\mathrm{B} \to \mathrm{Bb}$, $\mathrm{A} \to \mathrm{a}$, $\mathrm{B} \to \mathrm{b}$

41.

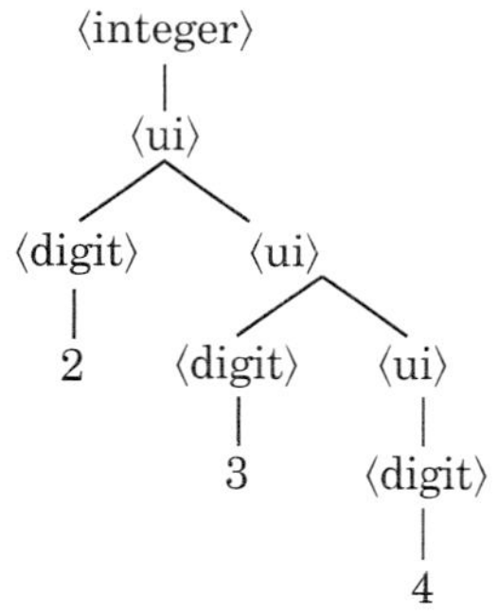

43.

⟨id⟩ ::= ⟨letter⟩ | ⟨alfa⟩
⟨alfa⟩ ::= ⟨letter⟩|⟨letter⟩⟨alfa⟩|⟨ui⟩|⟨ui⟩⟨alfa⟩
⟨ui⟩ ::= ⟨digit⟩|⟨digit⟩⟨ui⟩
⟨letter⟩ ::= a|b|c| . . . |z
⟨digit⟩ ::= 0|1|2| . . . |9

45. no

47. no

49.

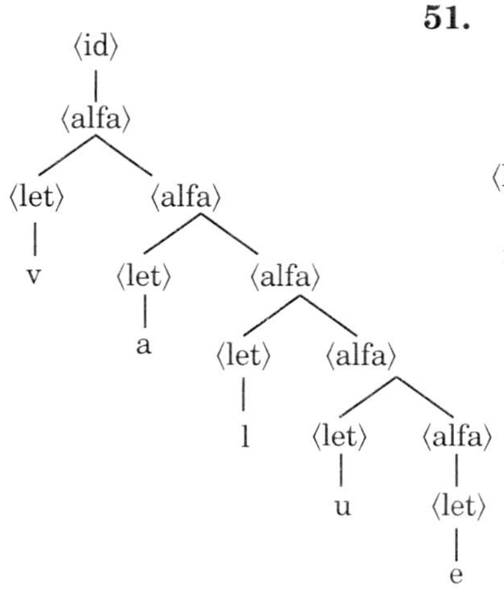

51.

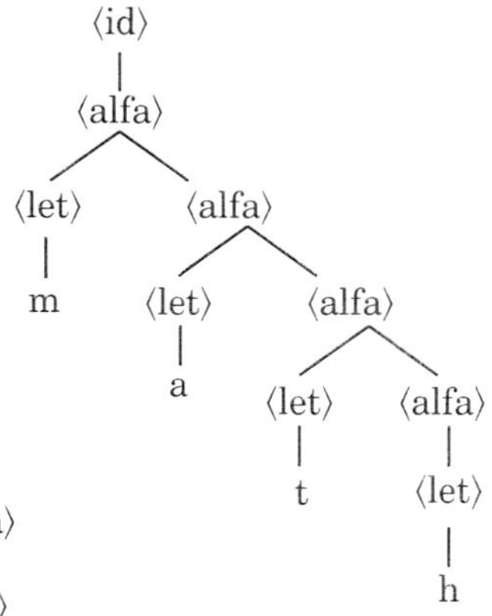

53. yes

55. no

57.

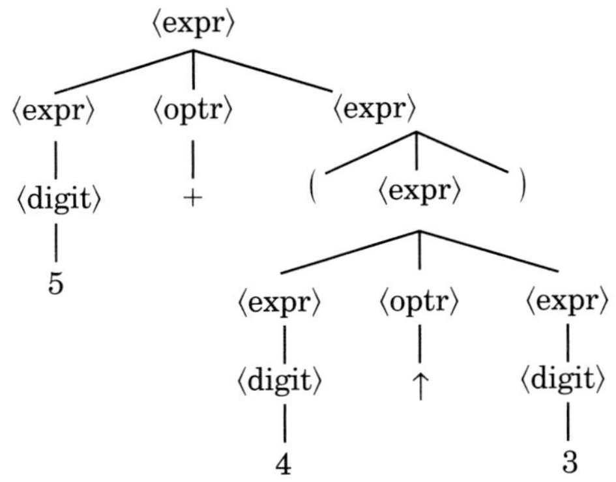

59.

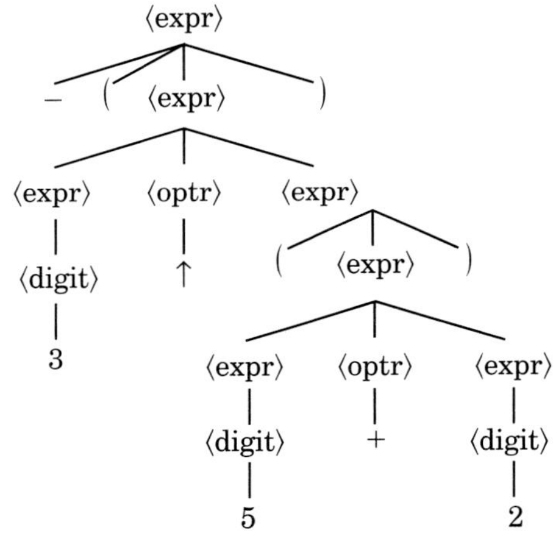

61. yes

63. yes

65.

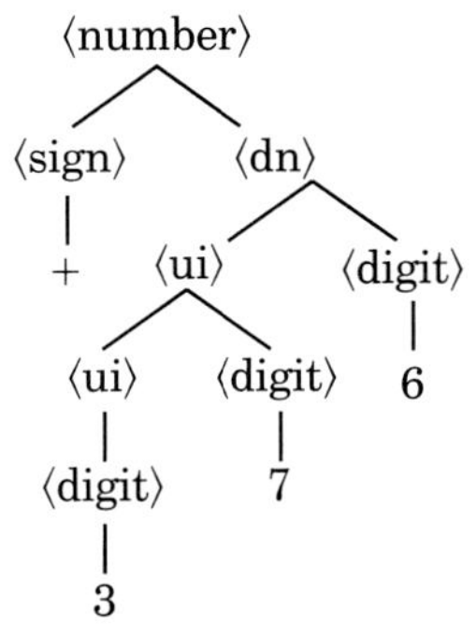

67.

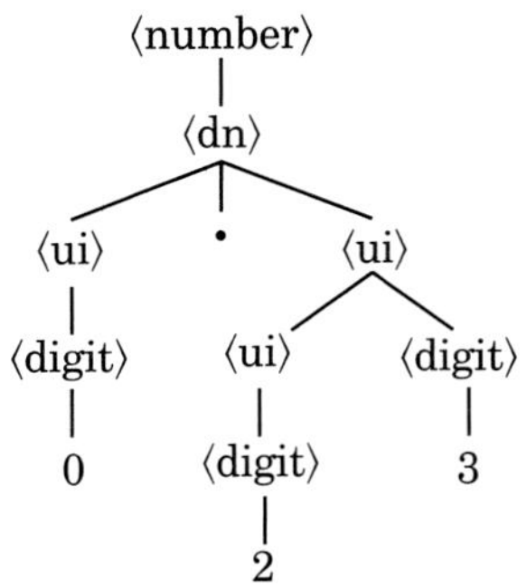

69. yes

71. yes

73.

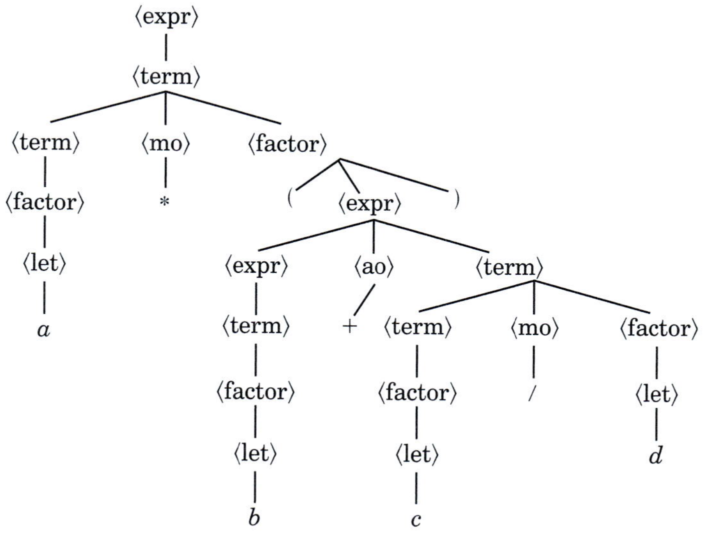

75. yes

77. yes

79. $\langle \text{un} \rangle ::= \langle \text{ui} \rangle \cdot \langle \text{ui} \rangle \text{E} \langle \text{ui} \rangle \,|\, \langle \text{ui} \rangle \cdot \langle \text{ui} \rangle \text{E} \langle \text{sign} \rangle \langle \text{ui} \rangle$
$\langle \text{ui} \rangle ::= \langle \text{digit} \rangle \,|\, \langle \text{digit} \rangle \langle \text{ui} \rangle$
$\langle \text{sign} \rangle ::= + \,|\, -$

81. no

83. yes

Exercises 11.3 (p. 768)

1. s_0-s_1-s_1-s_1-s_2

3. s_0-s_1-s_2-s_3-s_0

5. s_0-s_1-s_2-s_3-s_1-s_2

7. s_0-s_1-s_4-s_4-s_4-s_4-s_4

9. no

11. yes

13. no

15. no

17.

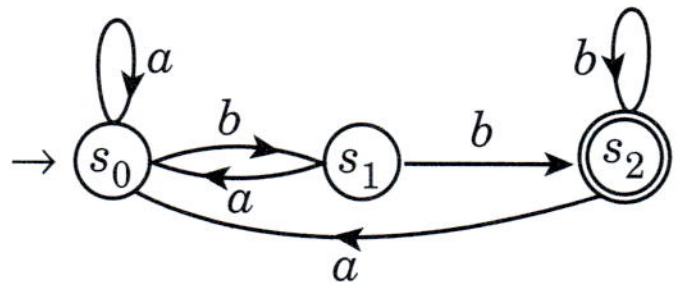

19.

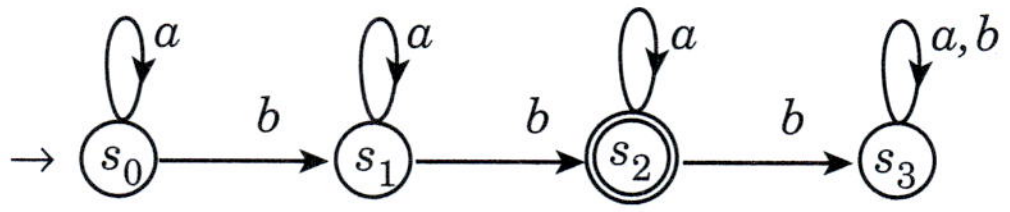

21.

S \ I	f: **a**	**b**
s_0	s_1	s_0
s_1	s_1	s_2
s_2	s_1	s_0

23.

S \ I	f: **a**	**b**
s_0	s_1	s_2
s_1	s_1	s_3
s_2	s_2	s_2
s_3	s_1	s_3

25. Words ending in bb.

27. Words ending in bb.

29. Words containing exactly two b's.

31. Words ending in ab.

33. Words beginning with a and ending in b.

35. Words containing an odd number of a's and an odd number of b's.

37.

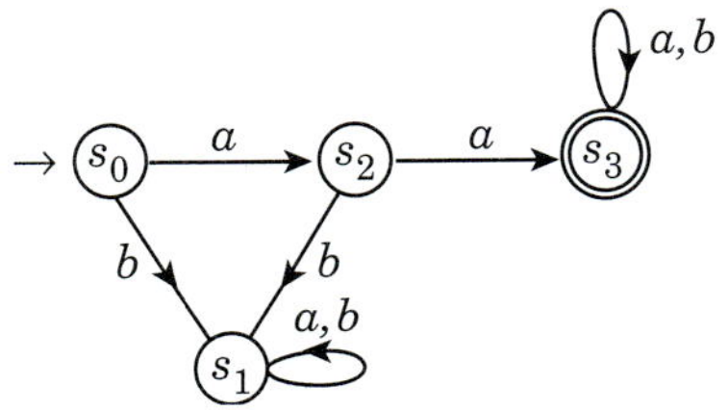

39.

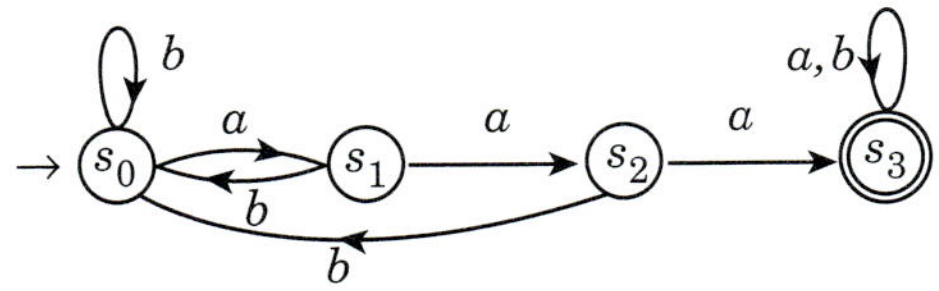

41.

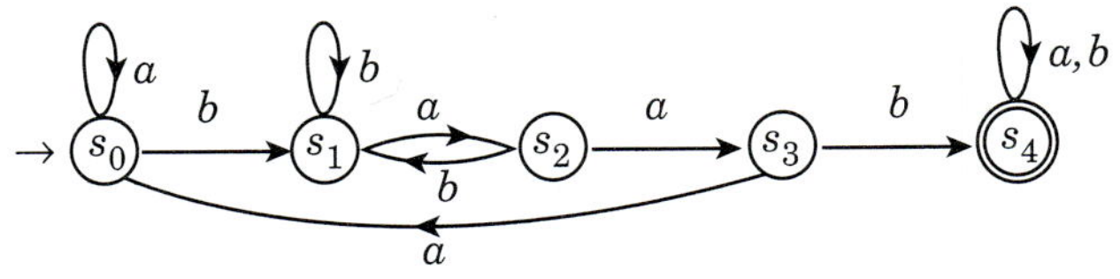

43.

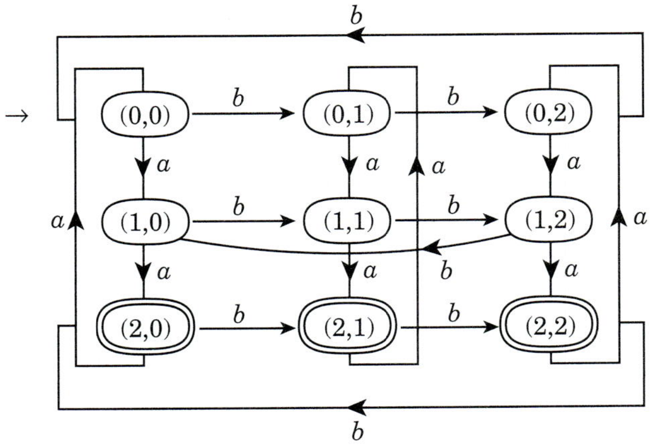

45.

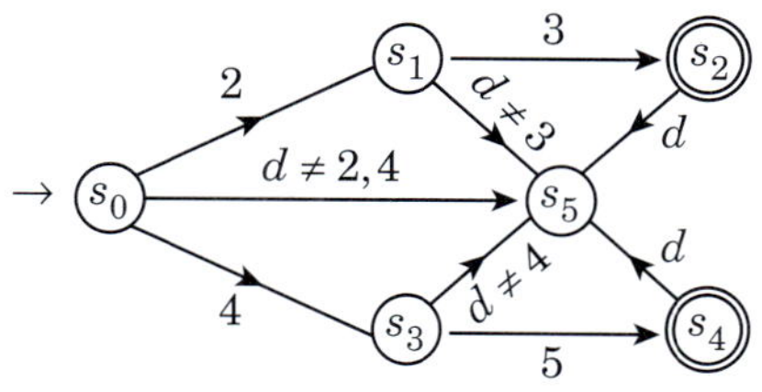

47.

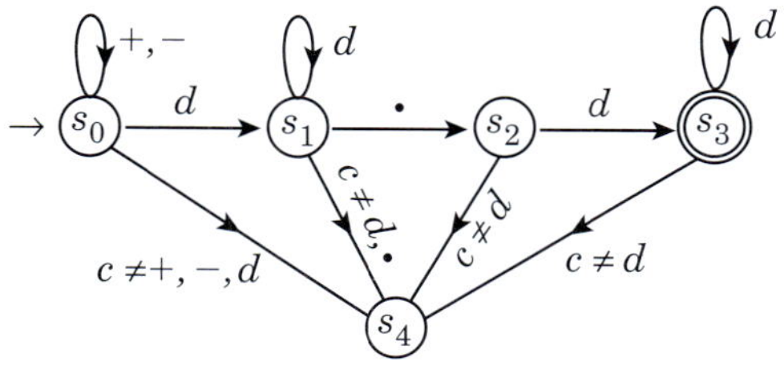

Exercises 11.4 (p. 777)

1. s_1

3. s_1

5. 0

7. 1

9.

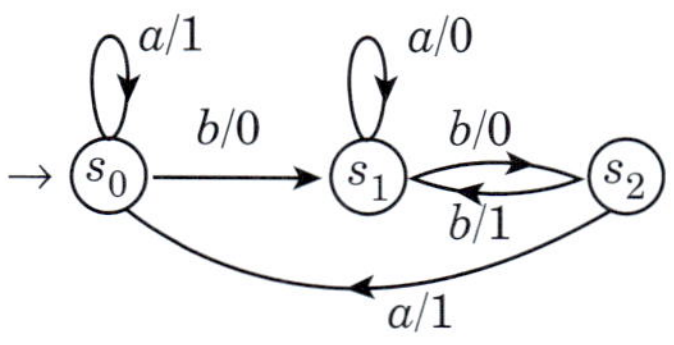

11.

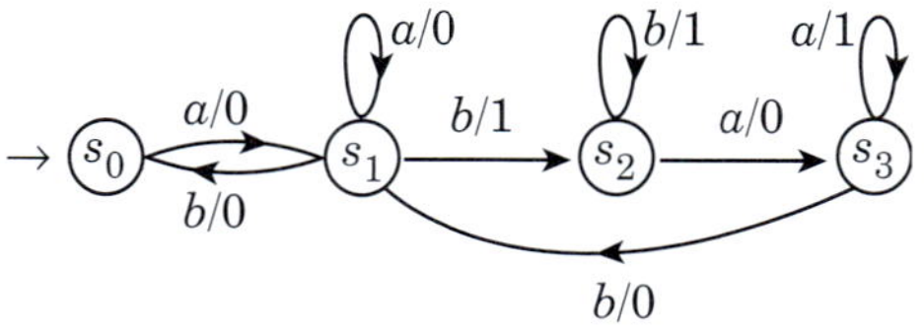

13.

	f		*g*	
S \ I	**a**	**b**	**a**	**b**
s_0	s_0	s_1	1	0
s_1	s_2	s_2	1	0
s_2	s_2	s_0	0	1

15.

	f		g	
S \ I	a	b	a	b
s_0	s_0	s_1	0	1
s_1	s_1	s_2	0	1
s_2	s_2	s_3	0	1
s_3	s_3	s_1	0	1

17. 0101 **19.** 001011 **21.** 0110 **23.** 0000

25.

	f				g			
S \ I	00	01	10	11	00	01	10	11
C0	C0	C0	C0	C1	0	1	1	0
C1	C0	C1	C1	C1	1	0	0	1

27. 11001 **29.** 110000

31.

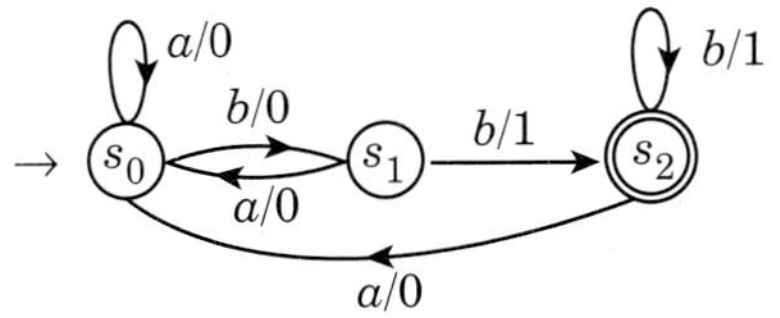

33.

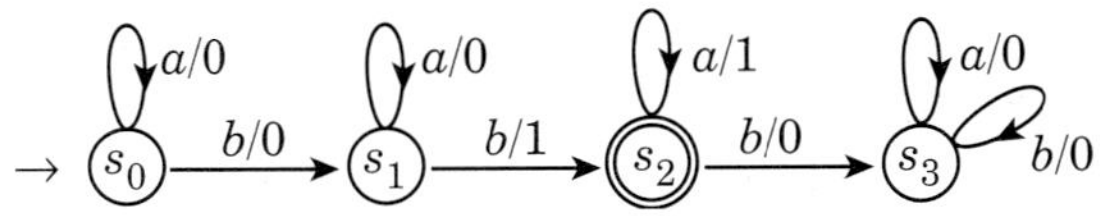

35. no

37. yes

39. Words ending in aa.

41.

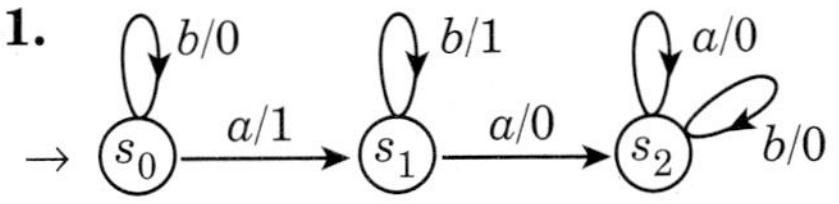

43. $g(s, x) = 0$

Exercises 11.5 (p. 781)

1. no

3. no

5. $s_0 \to as_1$, $s_0 \to bs_0$, $s_1 \to as_1$, $s_1 \to bs_2$, $s_2 \to as_1$, $s_2 \to bs_3$, $s_3 \to as_4$, $s_3 \to bs_0$, $s_3 \to a$, $s_4 \to a$, $s_4 \to b$

7. $s_0 \to as_0$, $s_0 \to bs_1$, $s_1 \to as_0$, $s_1 \to bs_2$, $s_2 \to as_0$, $s_2 \to bs_2$, $s_1 \to b$, $s_2 \to b$

9. $s_0 \to as_0$, $s_0 \to bs_1$, $s_1 \to as_1$, $s_1 \to bs_2$, $s_2 \to as_2$, $s_2 \to bs_3$, $s_3 \to as_3$, $s_3 \to bs_3$, $s_2 \to b$, $s_3 \to a$, $s_3 \to b$

11. $s_0 \to as_1, s_0 \to bs_0, s_1 \to as_1$, $s_1 \to bs_2$, $s_2 \to as_1$, $s_2 \to bs_0$, $s_1 \to b$

13. $s_0 \to as_1$, $s_0 \to bs_2$, $s_1 \to as_1$, $s_1 \to bs_3$, $s_2 \to as_2$, $s_2 \to bs_2$, $s_3 \to as_1$, $s_3 \to bs_3$, $s_1 \to b$, $s_3 \to b$

15. $s_0 \to as_1$, $s_0 \to bs_0$, $s_1 \to as_2$, $s_1 \to bs_1$, $s_2 \to as_2$, $s_2 \to bs_2$, $s_1 \to a$, $s_2 \to b$

17. $s_0 \to as_2$, $s_0 \to bs_1$, $s_1 \to as_1$, $s_1 \to bs_1$, $s_2 \to as_3$, $s_2 \to bs_1$, $s_3 \to as_3$, $s_3 \to bs_3$, $s_2 \to a$, $s_3 \to a$, $s_3 \to b$

19. $s_0 \to as_1$, $s_0 \to bs_0$, $s_1 \to as_1$, $s_1 \to bs_2$, $s_2 \to as_3$, $s_2 \to bs_0$, $s_3 \to as_3$, $s_3 \to bs_3$, $s_2 \to a$, $s_3 \to a$, $s_3 \to b$

21. $s_0 \to as_1$, $s_0 \to bs_2$, $s_1 \to as_4$, $s_1 \to bs_3$, $s_2 \to as_3$, $s_2 \to bs_4$, $s_3 \to as_3$, $s_3 \to bs_3$, $s_4 \to as_4$, $s_4 \to bs_4$, $s_4 \to a$, $s_4 \to b$

Exercises 11.6 (p. 785)

1.

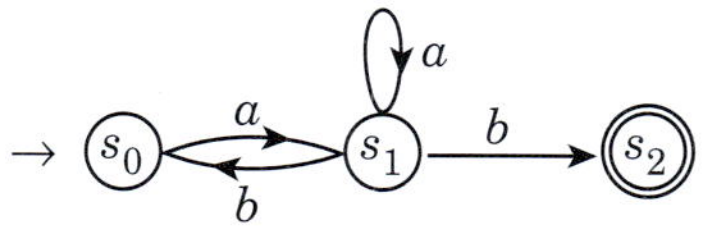

3.

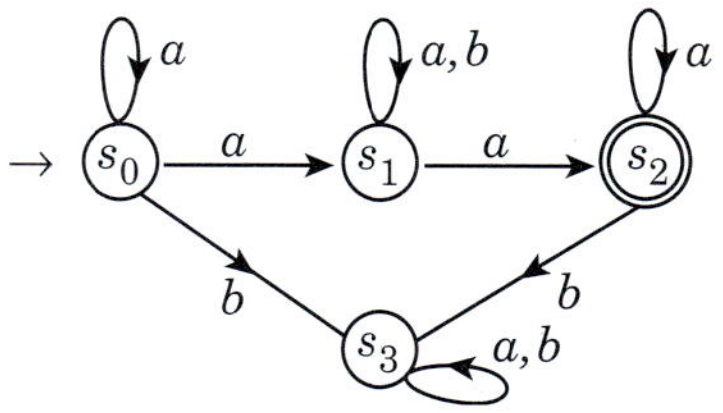

5.

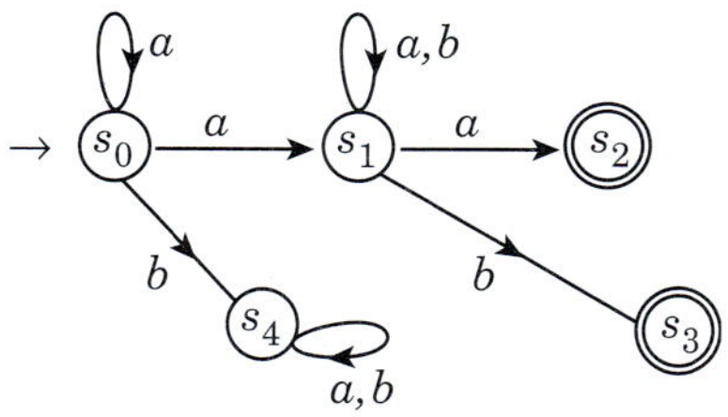

7.

S \ I	a	b
s_0	$\{s_0, s_1\}$	$\{s_2\}$
s_1	$\{s_1\}$	$\{s_2\}$
s_2	$\{s_2\}$	$\{s_2\}$

9.

S \ I	**a**	**b**
s_0	$\{s_0, s_1\}$	$\{s_5\}$
s_1	$\{s_1, s_2\}$	$\{s_5\}$
s_2	$\{s_2\}$	$\{s_3\}$
s_3	$\{s_3\}$	$\{s_4\}$
s_4	$\{s_4\}$	$\{s_4\}$
s_5	$\{s_5\}$	$\{s_5\}$

11. yes; σ-A-B-B-F

13. yes; σ-A-A-A-F

15. yes; s_0-s_1-s_1-s_2

17. no

19. no

21. yes; s_0-s_1-s_2-s_3-s_4

23.

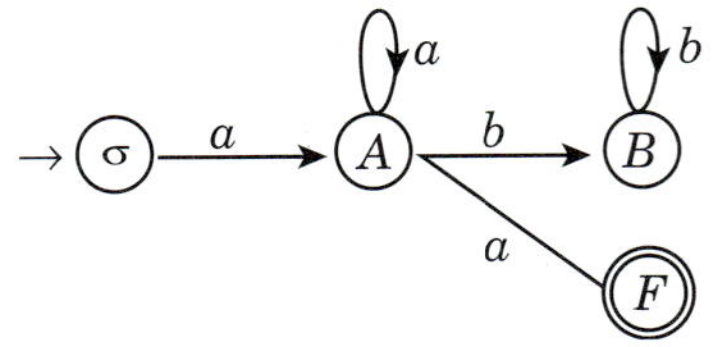

25.

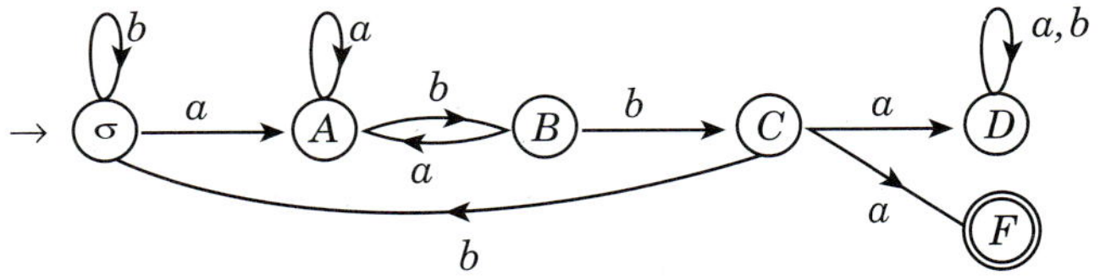

27.

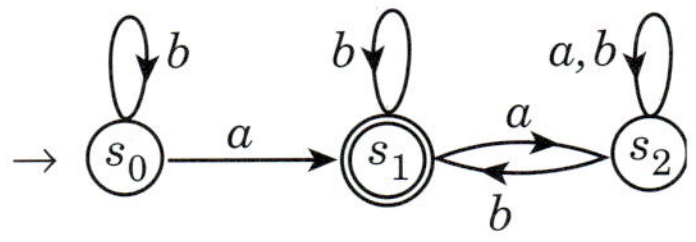

29.

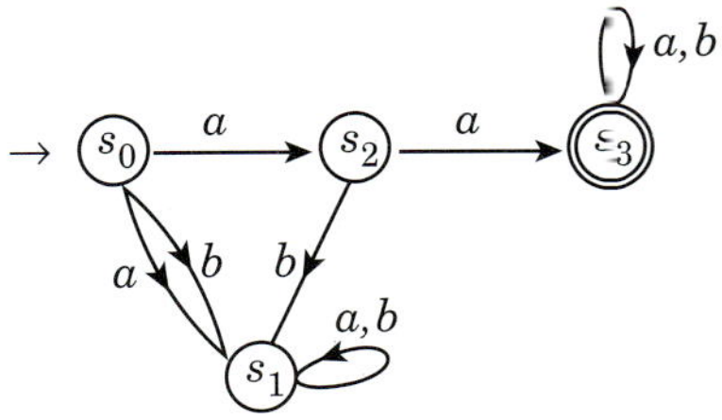

31.

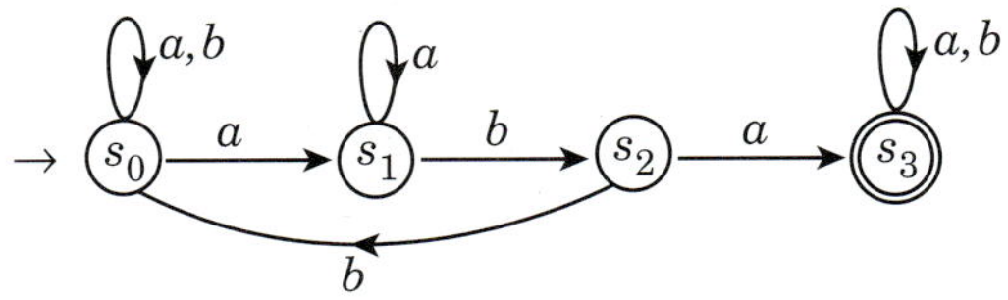

33.

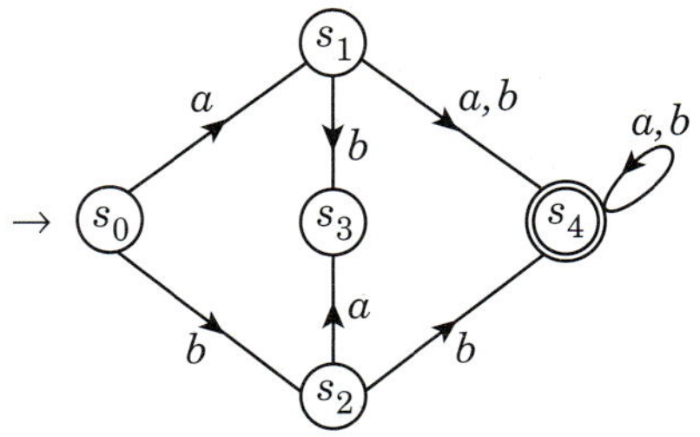

35.

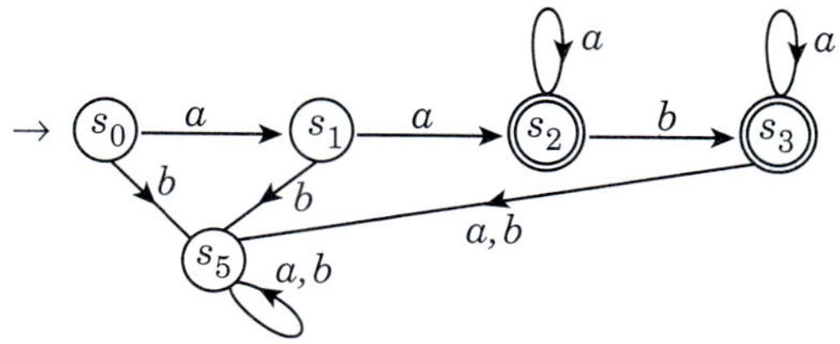

Exercises 11.7 (p. 792)

1.

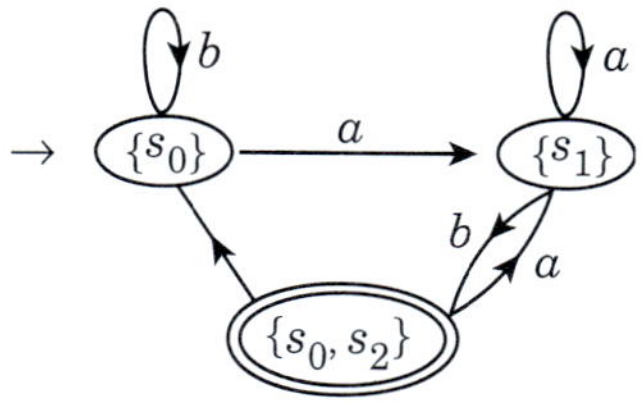

3.

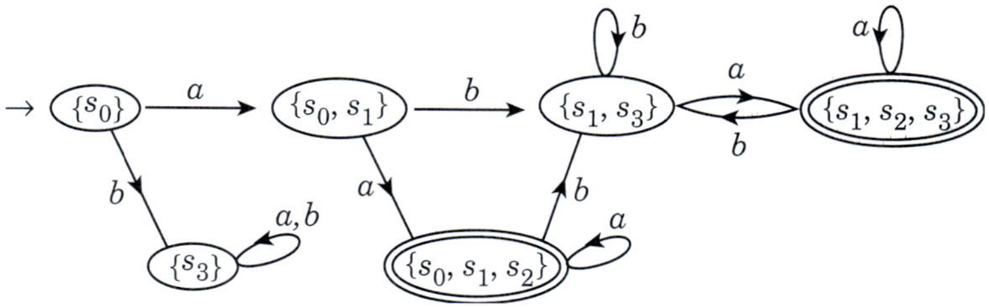

5.

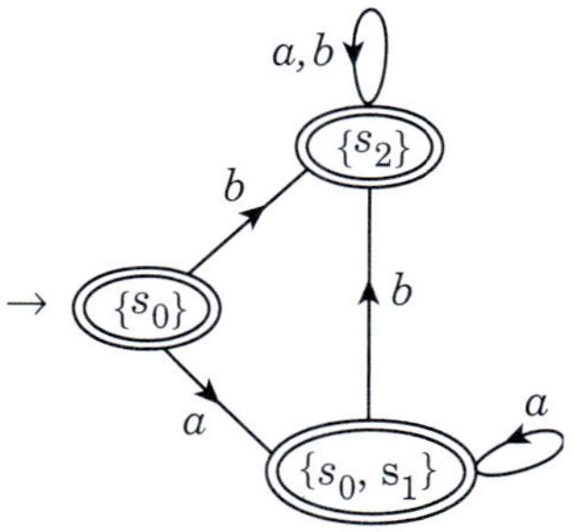

7.

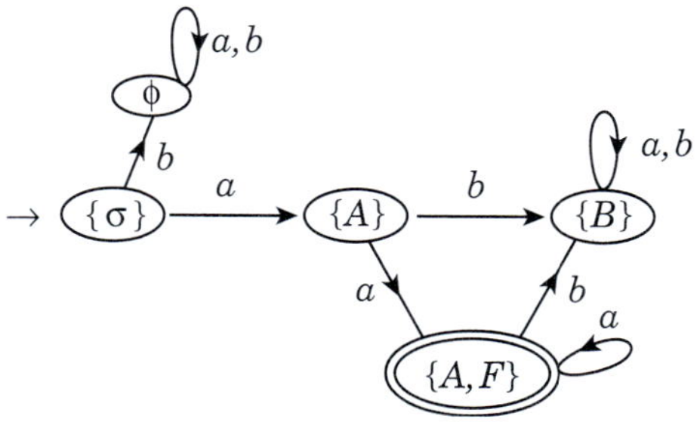

9.

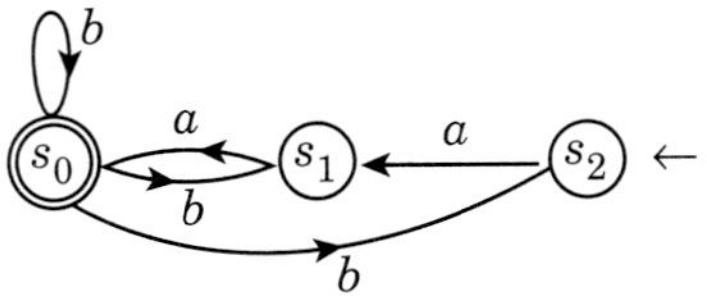

11.

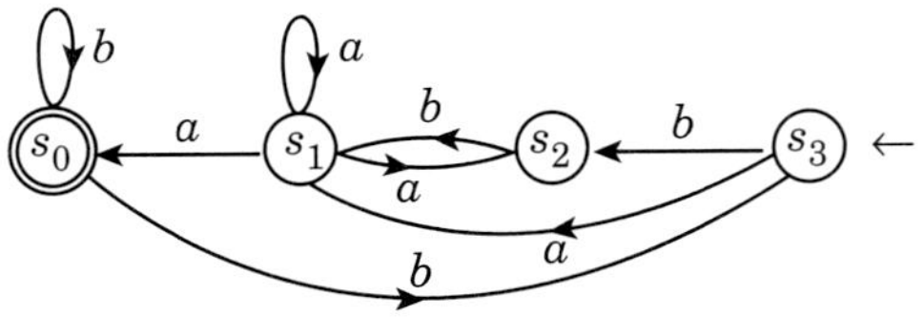

13.

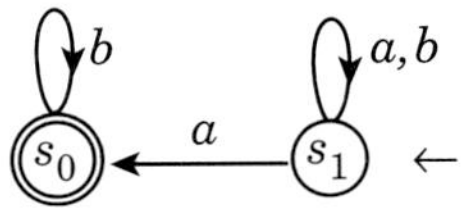

15.

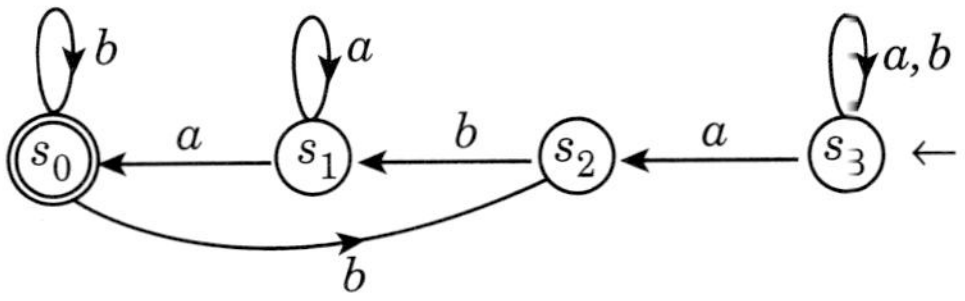

17. Words beginning with aa. **19.** Words beginning with bba.

21. Words with at least one a. **23.** Words containing aba as a substring.

25.

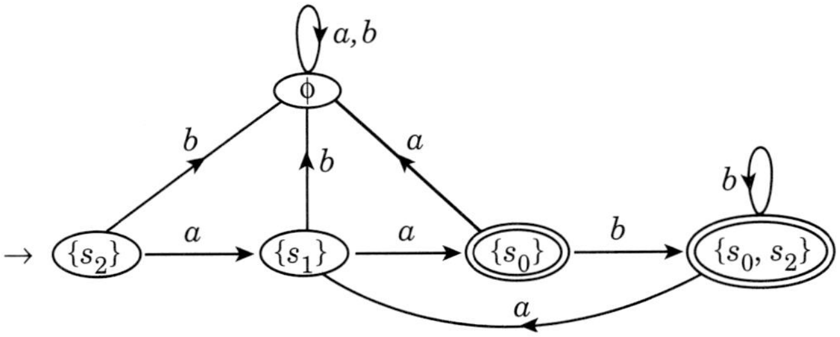

27.

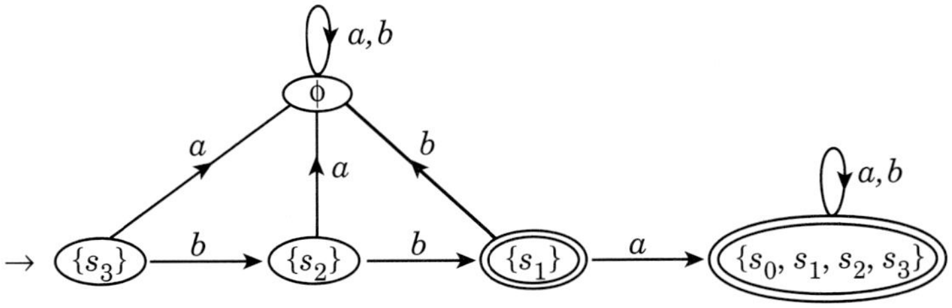

29.

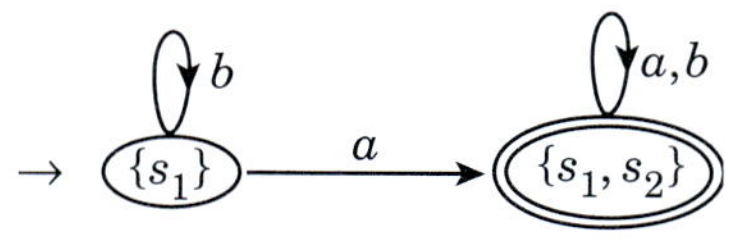

31.

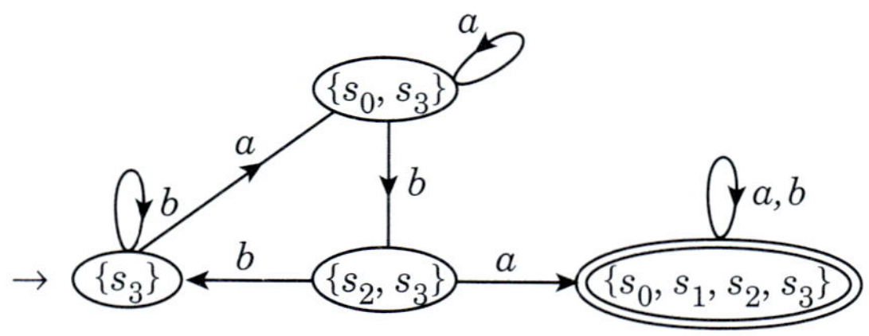

Review Exercises (p. 794)

1. {a, aa, ab, aab, bca, bcab}

3. $\{\lambda$, a, aa, bc, abc, bca, bcbc$\}$, $\{\lambda$, a, aa, bc, abc, bca, bcbc, a^3,abca, bca^2, bcbca, a^2bc, abcbc, bcabc, bcbcbc$\}$

5. a, b, ac

7. ab, abab, ababab

9. yes

11.

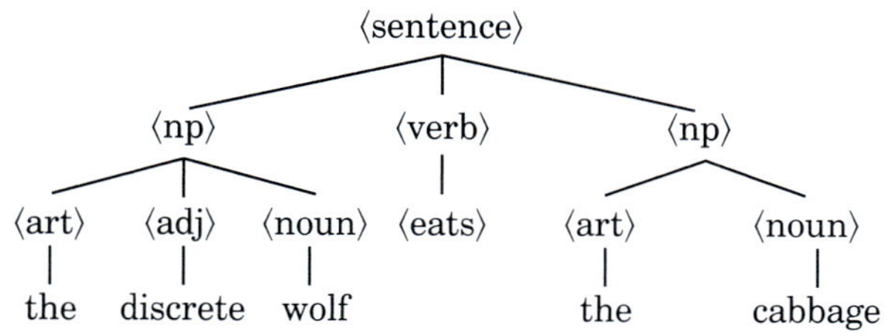

13. no

15. yes

17.

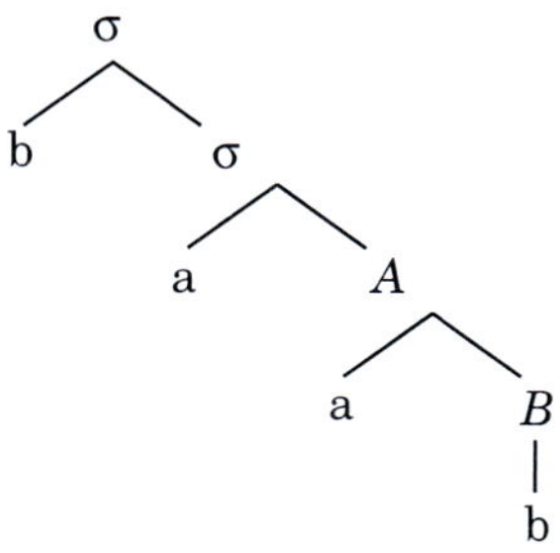

19.

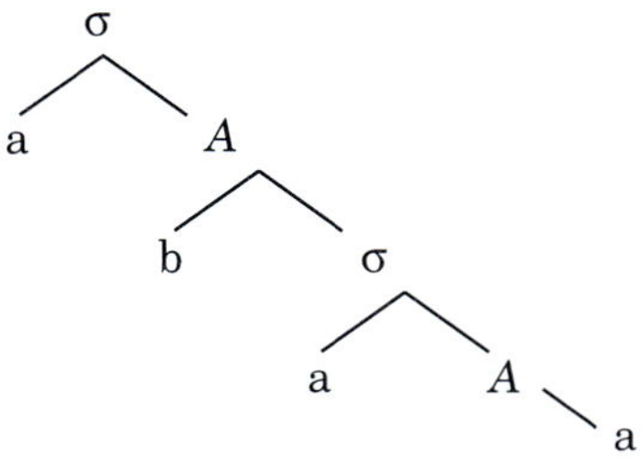

21. $\sigma \to b\sigma, \sigma \to b$

23. $\sigma \to b\sigma b, \sigma \to b$

25.

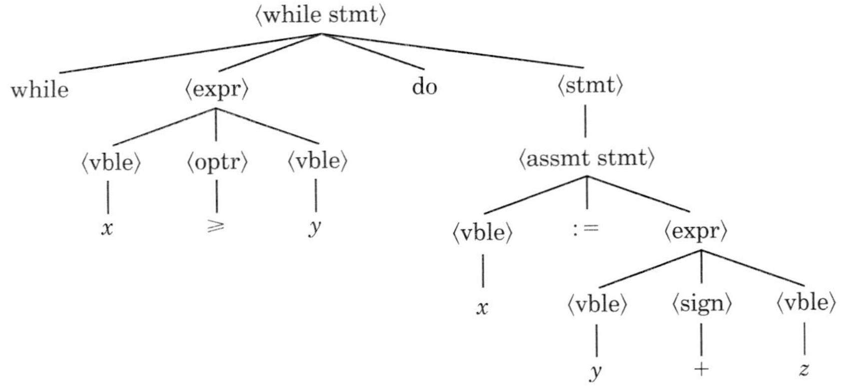

27.

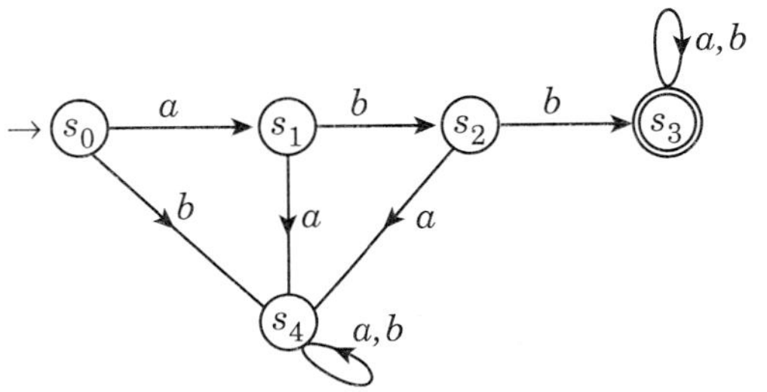

29.

S \ I	a	b
s_0	s_1	s_0
s_1	s_2	s_0
s_2	s_2	s_2

31. Words beginning with *abb*. **33.** Words containing *aa*.

35.

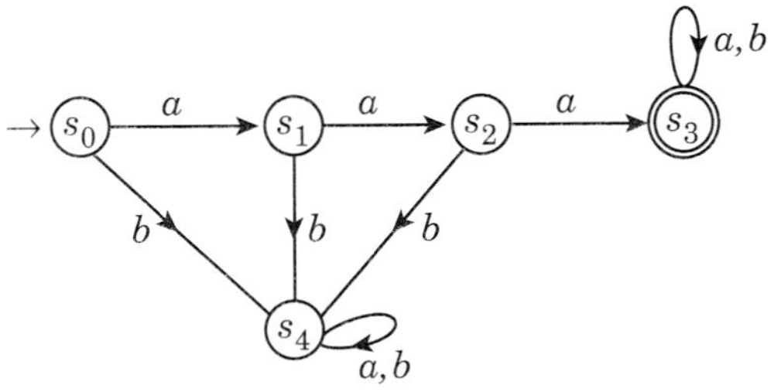

37.

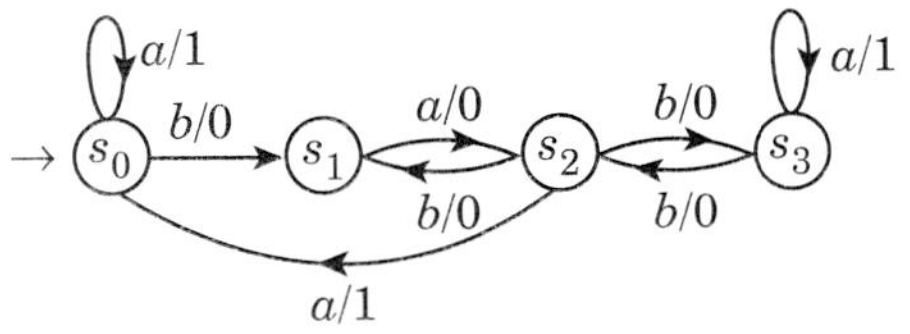

39.

	f		*g*	
S \ *I*	**a**	**b**	**a**	**b**
s_0	s_0	s_1	0	1
s_1	s_2	s_1	0	1
s_2	s_2	s_2	1	0

41. 0010111

43. 010110

45.

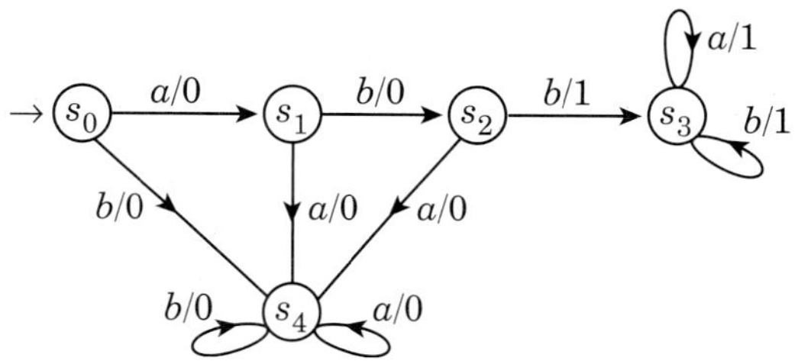

47.

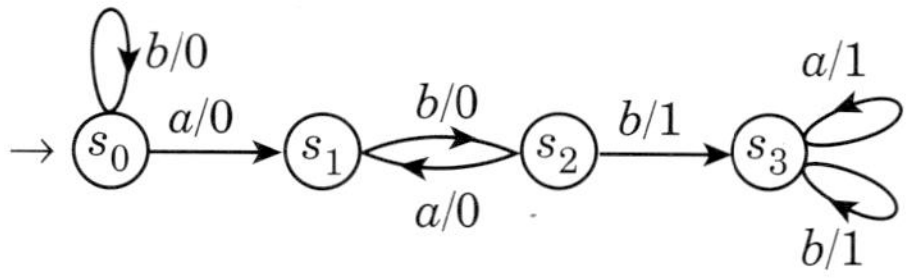

49. $\sigma \to \mathrm{a}A, \sigma \to \mathrm{b}C, A \to \mathrm{b}B, A \to \mathrm{a}C,$
$B \to \mathrm{a}C, C \to \mathrm{a}C, C \to \mathrm{b}C, B \to \mathrm{b}D,$
$D \to \mathrm{a}D, D \to \mathrm{b}D, D \to \mathrm{a}, D \to \mathrm{b}$

51.

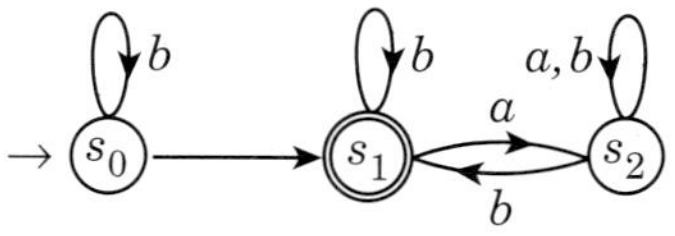

53.

S \ *I*	**a**	**b**
s_0	$\{s_1\}$	$\{s_2\}$
s_1	$\{s_0, s_1\}$	$\{s_1\}$
s_2	$\{s_1\}$	$\{s_3\}$
s_3	$\{s_3\}$	$\{s_2, s_2\}$

55. yes

57. no

59. yes

61. yes

63.

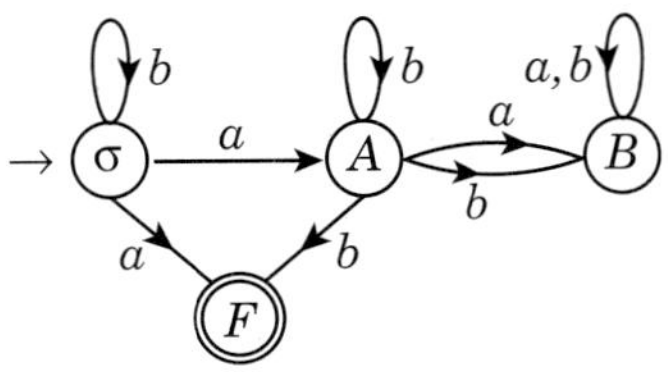

65.

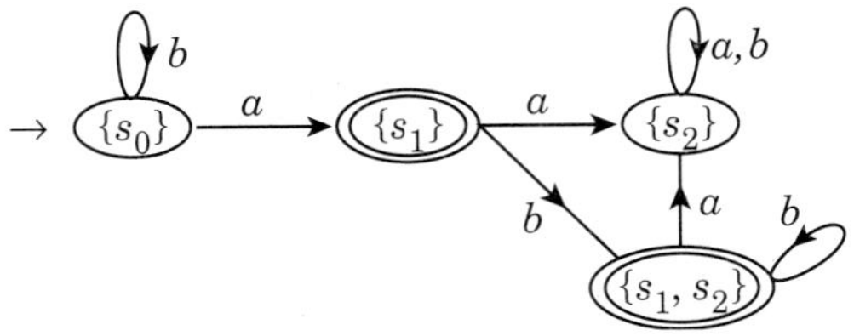

67. Words containing exactly one a.

69. By Exercise 68 in Section 11.1, $(A \cup B^*)^* = (A \cup B)^* = (A^* \cup B)^*$.

Supplementary Exercises (p. 798)

1.

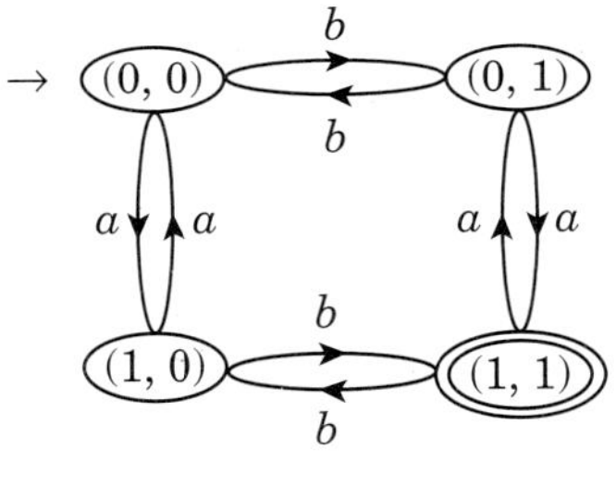

3.

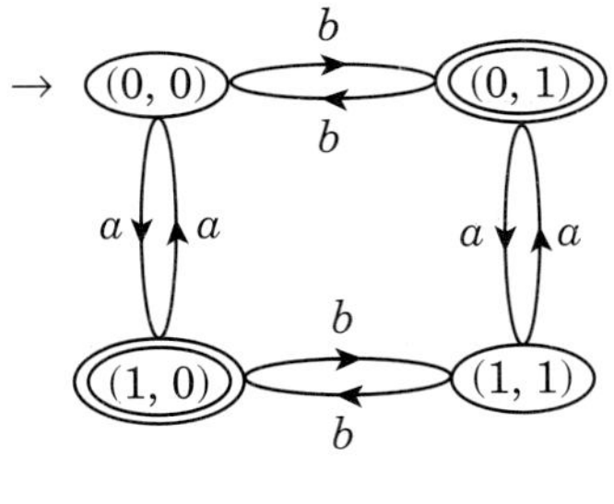

5.

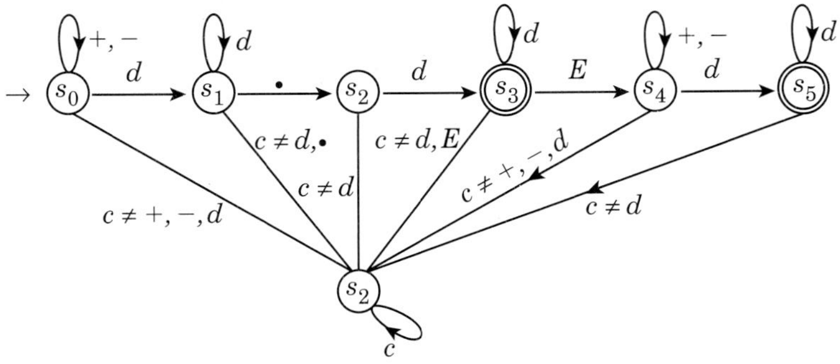

7. $\sigma \to aA$, $\sigma \to bC$, $A \to aA$, $A \to bB$, $B \to aA$, $B \to bB$, $C \to aC$, $C \to bC$, $B \to b$

9. $\langle \text{wfnp} \rangle ::= \lambda | \langle \text{wfnp} \rangle$

11.

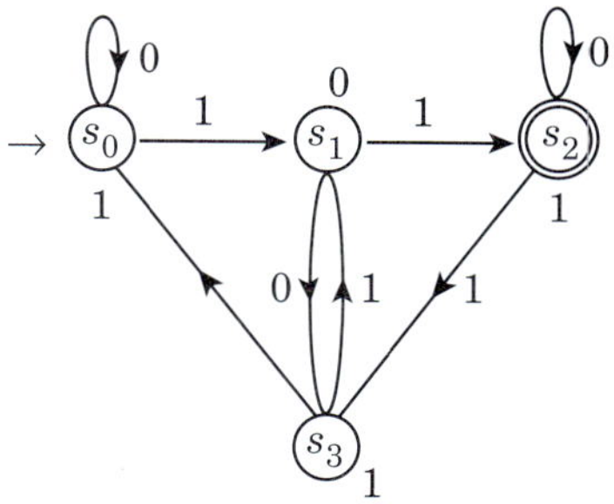

13.

	f input		*g*
s	0	1	
s_0	s_0	s_1	1
s_1	s_2	s_0	0
s_2	s_2	s_1	1

15. 1101

17. 101110

19. Suppose $x = x_1x_2 \ldots x_n \in L$. Then there is a DFSA A that excepts x, so there is a path that begins at s_0 and ends at an accepting state s_n. Reversing the path yields the string $x_nx_{n-1} \ldots x_1$. So making s_n the initial state and s_0 the accepting state yields a NDFSA A' that accepts L^R. (If A' contains more than one accepting state, introduce a new state s; then corresponding to every incoming edge to an accepting state, add an edge terminating at s.) Now the result follows by Theorems 11.3 and 11.4.

Chapter 12 Boolean Algebra and Combinatorial Circuits

Exercises 12.1 (p. 812)

1. 30 **3.** 30 **5.** 1 **7.** 5 **9.** 6

11. 5 **13.** 35 **15.** 2 **17.** 70 **19.** 10

21. $5' = 70/5 = 14$
$\therefore (5')' = 14' = 70/14 = 5$

23. $5 \odot (5 \oplus 7) = 5 \odot 35 = 5$

25. 2^n **27.** yes **29.** no **31.** yes **33.** yes

35.

+	0	1	*a*	*b*
0	0	1	a	b
1	1	1	1	1
a	a	1	a	1
b	b	1	1	b

·	0	1	*a*	*b*
0	0	0	0	0
1	0	1	a	b
a	0	a	a	0
b	0	b	0	b

$0' = 1, 1' = 0, a' = b, b' = a.$

37. $x + xy = x$

39. $(x + y)' = x'y'$

41. $xx = xx + 0 = xx + xx'$
$= x(x + x') = x1 = x$

45. $x(x+y) = xx+xy = x+xy = x$

49. $(x + y)z = z(x + y)$
$= zx + zy = xz + yz$

51. $(x+z)(y+z) = x(y+z)+z(y+z)$
$= xy + xz + zy + zz$
$= xy + xz + zy + z$
$= xy + (z + zx) + zy$
$= xy + (z + zy)$
$= xy + z$

43. By complement law, $0 + 0' = 1$. But by identity law, $0 + 0' = 0'$. $\therefore 0' = 1$.

47. $(x+y)' = x'y'$. $\therefore [(x+y)']' = (x'y')'$ i.e. $x + y = (x'y')'$

Exercises 12.2 (p. 821)

1. 1, 0 **3.** 2 **5.** yes **7.** yes

9.

x	y	x'	y'	$x + y'$	$x' + y$	$(x + y')(x' + y)$
0	0	1	1	1	1	1
0	1	1	0	0	1	0
1	0	0	1	1	0	0
1	1	0	0	1	1	1

11.

x	y	z	xy	y'	$y'z$	z'	yz'	$xy + y'z + yz'$
0	0	0	0	1	0	1	0	0
0	0	1	0	1	1	0	0	1
0	1	0	0	0	0	1	1	1
0	1	1	0	0	0	0	0	0
1	0	0	0	1	0	1	0	0
1	0	1	0	1	1	0	0	1
1	1	0	0	0	0	1	1	1
1	1	1	1	0	0	0	0	1

13.

x	y	z	y'	$x + y' + z$	$xy'z$	$(x + y' + z)(xy'z)$
0	0	0	1	1	0	0
0	0	1	1	1	0	0
0	1	0	0	0	0	0
0	1	1	0	1	0	0
1	0	0	1	1	0	0
1	0	1	1	1	1	1
1	1	0	0	1	0	0
1	1	1	0	1	0	0

15.

x	y	z	yz	xyz	$(yz)'$	$x(yz)'$	$xyz + x(yz)'$
0	0	0	0	0	1	0	0
0	0	1	0	0	1	0	0
0	1	0	0	0	1	0	0
0	1	1	1	0	0	0	0
1	0	0	0	0	1	1	1
1	0	1	0	0	1	1	1
1	1	0	0	0	1	1	1
1	1	1	1	1	0	0	1

17.

x	y	xy	$x + xy$
0	0	0	0
0	1	0	0
1	0	0	1
1	1	1	1

↑—— identical ——↑ (x and $x + xy$)

19.

x	y	$x + y$	$(x + y)'$	x'	y'	$x'y'$
0	0	0	1	1	1	1
0	1	1	0	1	0	0
1	0	1	0	0	1	0
1	1	1	0	0	0	0

↑—— identical ——↑ ($(x + y)'$ and $x'y'$)

21.

x	y	$x + y$	$(x + y)'$	x'	y'	$x' + y'$
0	0	0	1	1	1	1
0	1	1	0	1	0	1
1	0	1	0	0	1	1
1	1	1	0	0	0	0

↑—— not identical ——↑ ($(x + y)'$ and $x' + y'$)

23. yes

25. $xyz, xyz', xy'z, xy'z', x'yz, x'yz', x'y'z, x'y'z'$

27. xy

29. $xy + x'y'$

31. $xyz + xy'z'$

33. $xy'z' + x'yz' + x'y'z$

35. $xy + x'y + xy'$

37. xy'

39. $xyz + xyz' + x'yz$

41. $x+y = x1 + 1y = x(y+y') + (x+x')y$
$= xy + xy' + xy + x'y$
$= (xy + xy) + xy' + x'y$
$= xy + xy' + x'y$

43. $(x+y)xy' = xxy' + yxy'$
$= xy' + x(yy')$
$= xy' + x0$
$= xy'$

45. $y(x+z) = yx + yz = xy + yz$
$= xy(z+z') + (x+x')yz$
$= xyz + xyz' + xyz + x'yz$
$= xyz + xyz' + x'yz$

47. 0

49. 1

51. 0

53.

x	y	$x\uparrow y$	$(x\uparrow y)\uparrow(x\uparrow y)$
0	0	1	0
0	1	1	0
1	0	1	0
1	1	0	1

55.

x	y	$x\uparrow y$	$(x\uparrow y)\downarrow(x\uparrow y)$
0	0	1	0
1	0	1	0
1	0	1	0
1	1	0	1

57.

x	y	$x \uparrow x$	$y \uparrow y$	$(x \uparrow x) \downarrow (y \uparrow y)$
0	0	1	1	0
0	1	1	0	0
1	0	0	1	0
1	1	0	0	1

59. $xy' + x'y + x'y'$

61.

x	y	$x + y$	$x \uparrow x$	$y \uparrow y$	$(x \uparrow x) \uparrow (y \uparrow y)$
0	0	0	1	1	0
0	1	1	1	0	1
1	0	1	0	1	1
1	1	1	0	0	1

↑ identical ↑ ($x + y$ and $(x \uparrow x) \uparrow (y \uparrow y)$)

63.

x	y	$x + y$	$x \downarrow y$	$(x \downarrow y) \downarrow (x \downarrow y)$
0	0	0	1	0
0	1	1	0	1
1	0	1	0	1
1	1	1	0	1

↑ identical ↑ ($x + y$ and $(x \downarrow y) \downarrow (x \downarrow y)$)

65. (0,1,1)

67. Since $\{\uparrow\}$ is functionally complete, $x' = x \uparrow x$ and $x + y = (x \uparrow x) \uparrow (y \uparrow y)$, $\{+, '\}$ is also functionally complete.

69. 1

71. 0

73. 0

75. no

77.

x	$x \oplus x$
0	0
1	0

79.

x	y	$x \oplus y$	$y \oplus x$
0	0	0	0
0	1	1	1
1	0	1	1
1	1	0	0

81.

x	y	z	$y \oplus z$	$x \oplus (y \oplus z)$	$x \oplus y$	$(x \oplus y) \oplus z$
0	0	0	0	0	0	0
0	0	1	1	1	0	1
0	1	0	1	1	1	1
0	1	1	0	0	1	0
1	0	0	0	1	1	1
1	0	1	1	0	1	0
1	1	0	1	0	0	0
1	1	1	0	1	0	1

identical

83. $(x + y)(x + y')$

85. $(x + y + z)(x + y + z')(x + y' + z')$

87. $(x+y+z)(x+y+z')(x+y'+z)(x+y'+z')(x'+y+z)(x'+y+z')(x'+y'+z)$

Exercises 12.3 (p. 829)

1. 0 **3.** 1 **5.** 1 **7.** 1 **9.** 0

11.

x	y	$x \uparrow y$
0	0	1
0	1	1
1	0	1
1	1	0

13. $xy' + x'y + x'y'$

15. $xyz + x'yz + x'y'z + x'yz' + x'y'z'$

17.

x	y	z	NAND
0	0	0	1
0	0	1	1
0	1	0	1
0	1	1	1
1	0	0	1
1	0	1	1
1	1	0	1
1	1	1	0

19. true

21. false

Exercises 12.4 (p. 837)

1. $x = 0$ **3.** $x = y = 1$ **5.** $x'y'$ **7.** $(x + y)'y$

9. $x + (yz)'$ **11.** $(x + y)z' + x'z$ **13.** $(xy + yz + zx)xy'z$

15.

x	y	$x + y$	$(x + y)'$	$(x + y)'y$
0	0	0	1	0
0	1	1	0	0
1	0	1	0	0
1	1	1	0	0

17.

x	y	z	yz	$(yz)'$	$x + (yz)'$
0	0	0	0	1	1
0	0	1	0	1	1
0	1	0	0	1	1
0	1	1	1	0	0
1	0	0	0	1	1
1	0	1	0	1	1
1	1	0	0	1	1
1	1	1	1	0	1

19.

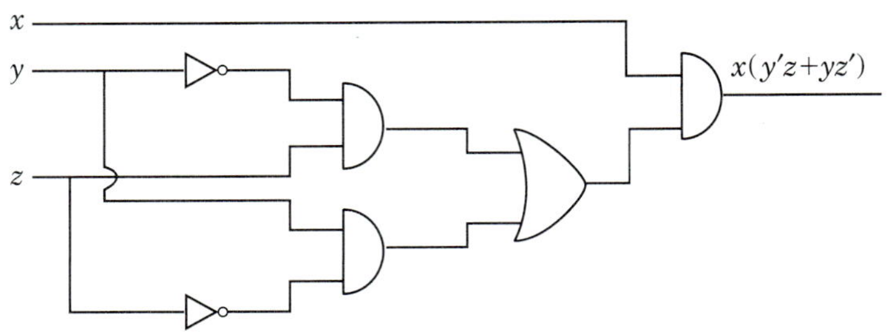

21.

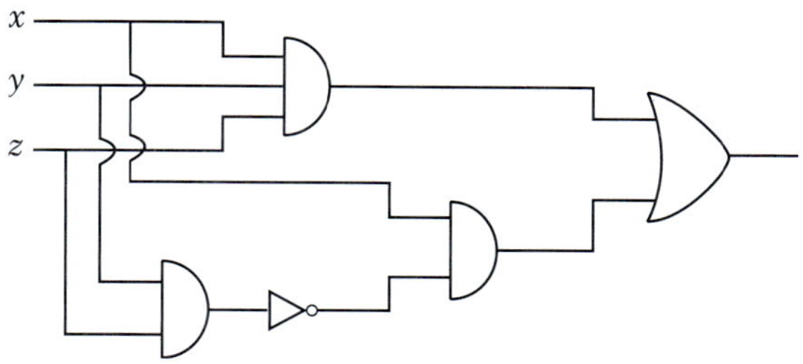

23.

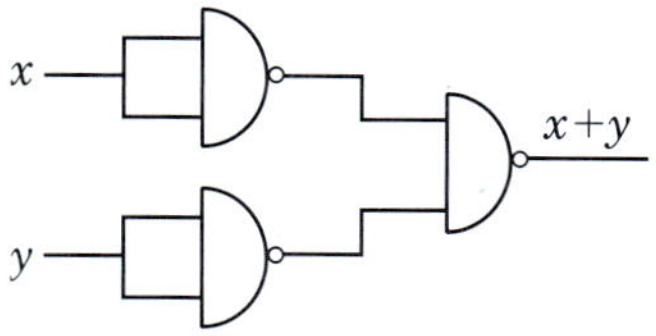

25.

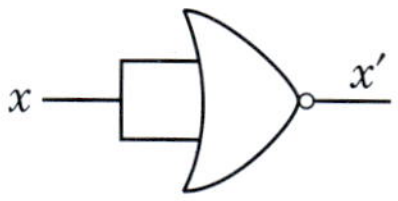

27.

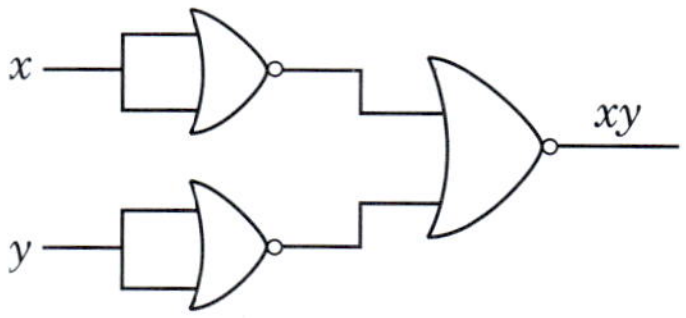

29. 1001 **31.** 1011

33. $s = (x + y)(xy)' = xy' + x'y$

$$
\begin{aligned}
\therefore s_i = sc_i + s'c_i &= (xy' + x'y)c_i' + (xy' + x'y)'c_i\\
&= xy'c_i' + x'yc_i' + (x' + y)(x + y')c_i\\
&= xy'c_i' + x'yc_i' + (xx' + x'y' + xy + yy')c_i\\
&= xy'c_i' + x'yc_i' + (0 + x'y' + xy + 0)c_i\\
&= xyc_i + xy'c_i' + x'yc_i' + x'y'c_i
\end{aligned}
$$

35.

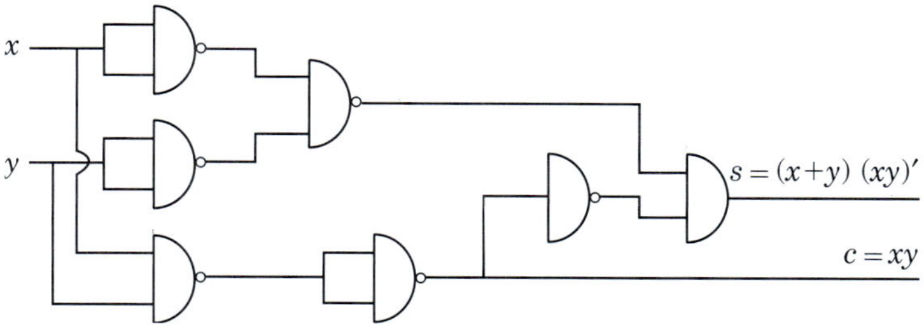

Exercises 12.5 (p. 849)

1. x **3.** xy' **5.** x'

7. xy **9.** $xy + yz + zx + xyz$ **11.** xy

13. xyz **15.** $x(wy + w'y')$ **17.** x

19. $xy' + x'y$

21.

	y	y'
x	1	
x'		1

23. x

25. $xyz + xyz' + xy'z' + x'y'z'$

27. $xyz' + xy'z' + x'yz' + x'y'z'$

29. xy'

31. y'

33. $xy + y'z'$

35. z'

37. $wxyz' + wxy'z' + wx'yz' + wx'y'z' + w'x'yz' + w'x'y'z' + w'xyz' + w'xy'z'$

39. $wxyz + wx'yz + w'x'y'z' + w'x'y'z + w'xy'z' + w'xy'z$

41.

	yz	yz'	$y'z'$	$y'z$
wx				1
wx'				
$w'x'$				
$w'x$	1			

43.

	yz	yz'	$y'z'$	$y'z$
wx				1
wx'				1
$w'x'$				1
$w'x$				1

45. z'

47. $wyz + w'y'$

49. yz

51. $wx + y'z$

Exercises 12.6 (p. 856)

1. $x + y'z'$

3. $y' + w'y$

5. $wx + xyz + w'xy'$

7. $x + y'$

9. z'

11. $xyz' + w'x'y'z'$

13. $wx' + xyz$

15. $x'y'z' + x'y'z + xyz' + xyz$

17. $w'x'y'z' + w'x'yz' + w'xy'z' + w'xyz' + wx'y'z' + wx'yz'$

19. $w'x'y'z' + w'xyz'$

21. $w'xyz + wx'y'z' + wx'y'z + wx'yz'$

23. $yz + y'z' + wx' + w'x'$

25. $yz' + w'x'y + x'y'z' + w'xy'z$

27. $xy'z + xyz' + w'x + y'z'$

Review Exercises (p. 859)

1. yes

3. no

5. When n is a product of distinct primes.

7. 2

9. 30

11. $(x + y)(x' + y) = y$

13.

x	y	z	$x+y$	$x+y+z$	x'	y'	$x'y'$	z'	$x'y'z'$	f
0	0	0	0	0	1	1	1	1	1	0
0	0	1	0	1	1	1	1	0	0	0
0	1	0	1	1	1	0	0	1	0	0
0	1	1	1	1	1	0	0	0	0	0
1	0	0	1	1	0	1	0	1	0	0
1	0	1	1	1	0	1	0	0	0	0
1	1	0	1	1	0	0	0	1	0	0
1	1	1	1	1	0	0	0	0	0	0

15.

x	y	z	yz	x'	$x'y'z$	$x+x'yz$	$x+yz$
0	0	0	0	1	0	0	0
0	0	1	0	1	0	0	0
0	1	0	0	1	0	0	0
0	1	1	1	1	1	1	1
1	0	0	0	0	0	1	1
1	0	1	0	0	0	1	1
1	1	0	0	0	0	1	1
1	1	1	1	0	0	1	1

↑ identical ↑

17. $x'yz' + xy'z + xyz'$

19. $xyz + xyz' + xy'z + x'yz$

21. $xyz + xyz' + xy'z + x'yz$

23. 0

25.

x	y	$x \downarrow x$	$y \downarrow y$	$(x\downarrow x)\uparrow(y\downarrow y)$
0	0	1	1	0
0	1	1	0	1
1	0	0	1	1
1	1	0	0	1

27. $xy + xy' + x'y + x'y'$

29. $xy'z' + x'yz' + x'y'z$

31.

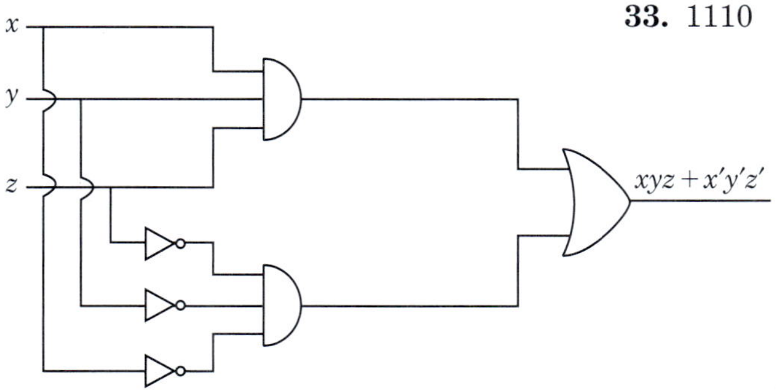

33. 1110

35.

	yz	yz'	$y'z'$	$y'z$
wx	1			1
wx'		1	1	
$w'x'$		1	1	
$w'x$	1			1

37. $x' + y'$ **39.** $wy + w'y'$ **41.** y'

43. $wx + wyz' + x'y'z' + y'z$

45. $x' + z'$ **47.** $wy + xy'z + w'x'y'z'$

49. $wy' + yz + xyz' + x'y'z'$

51. $w'x'y'z' + w'xy'z + wx'yz'$

53. $wx'y'z' + w'xyz + w'x'y'z + w'x'y'z' + w'xyz' + w'x'yz$

Supplementary Exercises (p. 862)

1. yes

3. **proof** (by PMI): Clearly, $x^{(1)} = x'$. So assume $x^{(k)}$ equals x if k is even and x' if k is odd, where $k \geq 1$. Then:

$$x^{(k+1)} = \left(x^{(k)}\right)' = \begin{cases} x' & \text{if } k \text{ is even} \\ (x')' & \text{otherwise} \end{cases}$$

$$= \begin{cases} x' & \text{if } k \text{ is even} \\ x & \text{otherwise} \end{cases}$$

Thus the result follows by induction.

5. **proof** (by strong induction): The result is clearly true when $n = 1$. So assume it is true for any boolean expression with $2, 3, 4, \ldots,$ or k variables. Then $(x_1x_2 \ldots x_kx_{k+1})' = [(x_1x_2 \ldots x_k)x_{k+1}]' = [(x_1x_2 \ldots x_k)' + x_{k+1}] = (x_1' + x_2' + \cdots + x_k') + x_{k+1}' = x_1' + x_2' + \cdots + x_k' + x_{k+1}'$. Thus the result follows by PMI.

7. $xy + (x + y)' + xy'$

9. $[x'y' + (x' + y')' + x'(y')']'$

11. $\{x'y' + [(y')'1'] + [x' + (0')']\}$

13. yes

15. no

Appendix A

Exercises A.2 (p. 873)

1. 6

3. $a^2 + b^2$

5. 0

7. $ab+bc+ca-a^2-b^2-c^2$ **9.** $(a+b+c)(ab+bc+ca-a^2-b^2-c^2)$

11. $(ad-bc)(eh-fg)$ **13.** $|AB|=(ad-bc)(eh-fg)=|A|\cdot|B|$

15. $|A|\cdot|A|=|A|\cdot|A^{\mathrm{T}}|=|A\cdot A^{\mathrm{T}}|=|I_n|=1$, so $|A|=\pm 1$.

17. $f(x)=\dfrac{x-b}{a-b}f(a)+\dfrac{x-a}{b-a}f(b)$

Exercises A.3 (p. 881)

1. $f(x+y)=a^{x+y}=a^x\cdot a^y=f(x)\cdot f(y)$

3. Let $f(x)=f(y)$. Then $a^x=a^y$, so $x=y$. Thus $f(x)=f(y)$ implies $x=y$; so f is injective.

5. Suppose $\log 2=a/b$, where a and b are relatively prime integers and $a<b$. Then $2=10^{a/b}$; that is, $2^b=10^a$. So $5^a=2^{b-a}$. This implies $2|5$, which is a contradiction. Thus log 2 is an irrational number.

7. $\mathrm{pH}=-\log(3.76\times 10^{-8})\approx 7.4248$

9. 0 decibals **11.** 70 decibals

13. 10.84006 **15.** 2.67917 parsecs

Exercises A.4 (p. 887)

1. 2134 **3.** 4213 **5.** 21354 **7.** 23145

9. 231, 312, 321 **11.** 2341, 2413, 2431

13. 3421, 4123, 4132 **15.** 23541, 24135, 24153

17. 12, 21 **19.** 134 **21.** 1345 **23.** 3456

25. 12, 13, 23 **27.** 12, 13, 14, 23, 24, 34

29. **Algorithm next byte ($a_1a_2\cdots a_8$)**

```
(* This algorithm finds byte that follows byte a1a2 ... a8 in
lexicographic order. *)
Begin (* algorithm *)
  i ← 8
  while bi = 1 do (* continue until bi = 0 *)
  begin (* while *)
    bi ← 0
    i ← i - 1
  endwhile
  bi ← 0 (* update bi *)
End (* algorithm *)
```

31. $\emptyset$, {1} **33.** $\emptyset$, {1}, {2}, {3}, {1, 2}, {1, 3}, {2, 3}, {1, 2, 3}

Exercises A.5 (p. 892)

1. 30 **3.** 210 **5.** 1680 **7.** 12,600

9. 30 **11.** -2 **13.** 168

15. $x^2 + y^2 + z^2 + 2xy - 2yz - 2zx$

17. $x^3 - y^3 + z^3 - 3x^2y + 3x^2z + 3xy^2 + 3y^2z + 3xz^2 - 3yz^2 - 3xyz$

19. $x^3 + 8y^3 + z^3 + 6x^2y + 3x^2z + 12xy^2 + 12y^2z + 3xz^2 + 6yz^2 + 6xyz$

21. 12 **23.** 2520 **25.** 3,326,400 **27.** k^n

29. 12,080,096,000 **31.** 56 **33.** 10

35. 455 **37.** 210

39. $x^3 + 8y^3 + z^3 - 6x^2y + 3x^2z + 12xy^2 - 12y^2z + 3xz^2 - 6yz^2 - 6xyz$

41. 10 **43.** 165 **45.** 1,801,800

Credits

Cover image (sunflower head) courtesy of Runk, Schoenberger/Grant Heilman Photography, Inc.

Text from The Divine Proportion by H. E. Huntley (Mineola, New York: Dover Publications, 1970) reprinted courtesy of the publisher.

Coffee pot illustration (Figure 1.7) by Michael Crawford, New York.

All biographical portrait illustrations by Nancy Crumpton.

Index

A

ABRACADABRA, 378–379, 387–388
Absolute value, 33, 126
Absorption laws, 83
Accepted string, 764
Accepting state, automaton, 760, 761, 783
Ackermann, Wilhelm, 278
  function, 278
Acyclic graphs, 610–611
  See also Dags
Adder
  full, 836–837
  half, 835–836
Addition
  in base *b*, 199–201
  of functions, 123
  of generating functions, 300–301
  of matrices, 166
  rule (principle), 43, 99, 344–345, 415
Adjacency
  edges and, 520
  graphs and, 520
  list representation, 452–453, 538
  matrix, 444–445, 520–522, 528, 716
Adjacent squares, in Karnaugh maps, 843
Algebra
  See also Boolean algebra
  origin of word, 96
ALGOL (ALGOrithmic Language), 748
Algorithms
  addition, 199–200
  binary multiplication, 203
  binary search, 228–230, 250–251, 311–312, 327–328
  binary search tree, 665
  breath-first search (BFS), 621–622
  bubble sort, 230–233, 248–249, 317–318
  complexities of, 247–252, 319–333, 420–422
  connectivity, 471–475, 481
  correctness of, 224–237
  correctness of recursive, 316–319
  defined, 96
  depth-first search (DFS), 618–621
  Dijkstra, 717–723
  disjunctive normal form (DNF), 819
  divide and conquer, 228, 327–329
  division, 185–191
  equivalence relation, 490
  Euclidean, 191–197, 322–323
  Eulerian circuit, 561
  Eulerian graph, 559–560
  exponentiation, 318–319
  factorial, 226
  Fibonacci, 310
  findmax, 251
  Floyd, 725
  Gilbert's, 728
  greatest common divisor (gcd), 311
  greedy, 633
  handshake, 308–309
  Horner's, 336
  Huffman, 671–675
  identifier, 767–768
  insertion sort, 236–237
  Kruskal's, 616–617, 626–628
  legally paired sequence, 106–107
  linear search, 227–228, 249–250, 316–317, 319–320, 421–422
  merge, 312–313
  merge sort, 313–314, 328–329
  minmax, 252
  multiplication, 225
  next-combination, 887
  next-permutation, 884–885
  next-subset, 96–97
  nondecimal bases (base-*b* expansion), 197–207
  origin of word, 96
  Prim's, 628–633
  product, 171
  recursive, 307–333
  selection sort, 233–234
  space complexity, 247
  spanning tree, 616–617
  subsets, 102
  time complexity, 247–248
  topological sort, 501
  tower, 309
  Warshall's, 477–481
Alphabet, 75
  Greek, 894
  input, 772
Alphanumeric character, 348
Ambiguous grammar, 753–755
Ancestors, 636
And
  Boolean, 438
  conjunctions and use of, 5
AND gate, 825–826
Angle brackets, 744
Antisymmetric relations, 456–458
Appel, Kenneth, 589
Arguments, 38
  valid and invalid, 39–49
Aristotle, 1, 2
Arithmetic, fundamental theorem of, 218–221
Arithmetic mean, 54
Arithmetic sequence, 180
Around the World puzzle, 564–565
Array, predecessor, 720

ASCII, 670
  character set, 867
Assignment operator, 7
Assignment statement, 7
Associated linear homogeneous recurrence relations with constant coefficients (ALHRRWCCs), 287, 293–296
Associative law
  Boolean algebra, 806
  of logic, 22
  of sets, 83
Associative properties, 470, 804, 805
Automata. *See* Finite-state automata (FSA)
Average-case time complexity, 248, 420–422
Axioms
  defined, 185
  dual, 807

B
Bachmann, Paul Gustav Heinrich, 237, 238
Backtracking, 618
Backus, John W., 748
Backus-Naur Form, 748
Backus-Normal Form (BNF), 748–751
Balanced trees, 640–643
Barber paradox, 42–43
Base-*b* expansion, 197–207
Basis clause, 105, 262, 814
Basis step, 209
Begin, 18
Bell, Eric T., 397
Bell numbers, 397, 492
Bernoulli, Jacob I., 212, 422
  inequality, 212
  trials, 422–425
Best-case time complexity, 247
Biconditional statements, 14–15
Big-oh estimate, 618
Big-oh notation, 237–239
Big-omega notation, 243–245
Big-theta notation, 245
Bijection function, 137–138
Binary alphabet, 75
Binary coded decimal (BCD) expansion, 852–854
Binary digits, 95
Binary expression trees, 655
Binary multiplication, 201–203
Binary numbers, 394–395
Binary operators, 12, 31, 806
Binary predicate, 33–34
Binary relation, 443–444
Binary search algorithm, 228–230, 250–251, 311–312, 327–328
Binary search tree algorithm, 665
Binary search trees, 664–668
Binary subtraction, 203–205
Binary trees, 639, 646–663
  Catalan numbers and, 660
  parenthesized triangulations and, 659–660
Binet, Jacques Phillipe Marie, 277
Binet form of Fibonacci number, 276–277, 290, 304–305
Binomial coefficient, 366, 386, 889
Binomial expansion, 388
Binomial probability, 423
Binomial theorem, 386–399, 888–889
Bipartite graphs, 526
Birthday paradox, 182, 412
Bits
  operations, 438
  string, 95
Boole, George, 1, 4, 803
Boolean algebra
  applications, 803
  complement, 442, 806
  defined, 806–811
  examples, 804–806
  expressions, 7, 814–817
  expression, self-dual, 863
  matrices, 438–443
  operators, 12, 806
  power, 441
  precedence rules, 807
  product, 439–440, 806
  sum, 806
  unit element, 806
  variables, 2, 813
  zero element, 806
Boolean functions, 134
  conjunctive normal form (CNF), 824
  defined, 813–814
  disjunctive normal form (DNF), 817–820
  equality, 815–817
  functional completeness, 820
  minterm, 817–820
  NAND and NOR, 820–821
  recursive definition, 814
Bound variables, 33
Breath-first search (BFS) algorithm, 621–622
Bridge, 686
Briggs, Henry, 879
Briggsian logarithms, 879
Bubble sort algorithm, 230–233, 248–249, 317–318
Bus topology, 529
Bytes, 348–349

C
Cabbage-goat-wolf puzzle, 552–554
Cantor, Georg, 67, 68
  diagonalization process, 141
Card dealing, 132–133
Cardinality of sets, 98–102, 140
Carroll, Lewis, 44, 45
Carry, 199
Cartesian plane, 88
Cartesian products, 87–89
Catalan, Eugene Charles, 108
  numbers, 108, 388–390, 393, 395–396, 660
  parenthesization problem, 394
Cayley, Arthur, 164, 519, 609
  formula, 626
Ceiling function, 126–130
Center, tree, 614
Characteristic equation, 287–288
Characteristic function, 131–132
Characteristic roots, 288
Chessboard puzzle
  eight queen puzzle, 620
  four queen puzzle, 619–620
  knight's tour problem, 558–560, 567–571
  two queen puzzle, 133–134
Child, 636
  left and right, 646
Chomsky, (Avram) Noam, 751, 752
  hierarchy of grammar, 751
Chromatic number, 588
Chuck-a-luck game, 423–424
Circle problem, 370–372
Circuit, combinatorial, 830–840
  minimization of, 840–850
Circuits, graph, 548
  Eulerian, 556
Circuits, sequential, 830
Class, equivalence, 486–490
Closed interval, 73
Closed-open interval, 73

Closed path, 548
Closure
  Kleene, 739–741
  reflexive, 482
  symmetric, 482
  transitive, 475–482
Coconuts and monkey problem, 437, 493
Code-a-bar system, 493
Codes, gray, 446–447
Coding scheme, Huffman, 670
Codomain of functions, 118
Coefficients
  binomial, 366, 386, 889
  multinomial, 891
Cofactor, 868
Collision, 138–139
Coloring, graph, 586–598
Column vector, matrix, 165
Combinations, 365–375
  generating, 383–384, 885–887
  with repetitions, 379–381
Combinatorial circuits, 830–840
  minimization of, 840–850
Combinatorics
  binomial theorem, 386–399
  combinations, 365–375
  defined, 343
  derangements, 360–365
  fundamental counting principles, 344–351
  generalized inclusion/exclusion principle, 399–409
  permutations, 351–360
  repetitions, 375–386
Common difference, 180
Common factor, 191
Common ratio, 180
Communication link, 2-stage, 550
Commutative law
  Boolean algebra, 806
  of logic, 22
  of sets, 83
Commutative properties, 804, 805
Comparable elements, 495
Complements
  Boolean, 442, 806
  of graphs, 536
  properties, 805, 806
  relation, 469
  of sets, 81
Complete bipartite graphs, 527
Complete graphs, 524–526
Completeness,
  functional, 820
Complete n-partite graph, 605
Complexities of algorithms, 247–252
Complexities of graphs, 622–623
Complexities of recursive algorithms, 319–333
Components, 5
Composite numbers, 189–190
Composition
  function, 150–153
  of relations, 463–464
Compound propositions, 5
Computer operations, with sets, 94–98
Computer representations
  of graphs, 538–539
  of relations, 449–454
Concatenation, 76
  of languages, 736–739
Conclusion, 9
Concurrent lines, 266
Conditional probability, 417–419
Conditional statements,
  defined, 9
Congruence
  to b modulo m, 484
  relation, 484–486
Conjunction rule, 43
Conjunctions, 5–7
  symbol for, 5
Conjunctive normal form (CNF), 824
Connected digraphs
  strongly, 698–700
  weakly, 705
Connected graphs, 549–554
Connectives, 5
Connectivity relation, 471–475, 481
Constructive existence proof, 52–53
Context-free grammar, 751
Context-free language, 751–753
Context-sensitive grammar, 751
Context-sensitive language, 751–753
Contingencies, 16
Contradictions, 16
  proof by, 51
Contrapositive, 11–12
  proof of, 51
Contrapositive law, 22, 43
Converse, 11–12
Correctness of algorithms, 224–237
Correctness of recursive algorithms, 316–319
Correct program, 224–225
Correspondence, one-to-one, 137–138
Countable sets, 140–141
Counterexample, 53
Counting principles, 344–351
Cycle
  digraphs, 698
  graphs, 526
  Hamiltonian, 565–566
  of a path, 446, 548
Cyclic permutations, 355–356

D

Dags (directed acyclic graphs), 707–715
Databases, 464–466
Data field, 451
  de Bruijn, Nicolaas G., 702
  graph, 703
  sequence, 702–704
Decay curves, 875
Decimal expansion, 197
Decision trees, 676–680
Degree of a region, 581–582
Degree of vertex, 520, 694–698
De Moivre, Abraham, 298, 299
De Morgan, Augustus, 23, 207
De Morgan's law of logic
  description of, 22–24
  for negating quantifiers, 35–36
De Morgan's law
  of sets, 83
Dependent events, 419–420
Depth-first search (DFS) algorithm, 618–621
Derangements, 360–365
  counting, 404–408
Derivation, language and, 746–747
Derivation trees, 743–744
Descartes, René, 87
Detachment, law of, 43
Determinants, 867–874
Deterministic finite-state automata (DFSA), 779–782
Diagonal, matrix, 166
Diagrams
  Hasse, 496–499
  transition, 760, 772
  tree, 6
  Venn, 72–73

Difference
between sets, 80
symmetric, 82
Differential and Integral Calculus, The (De Morgan), 23
Digraphs (directed graphs)
closed, 698
dags, 707–715
de Bruijn sequence, 702–704
degree of vertex, 694–698
features of, 691–692
open, 698
relations and, 443–447
Fibonacci numbers and, 692–694
n-stage win, 698
paths, reachability, and cycles, 698
strongly connected, 698–700
teleprinter's problem, 700–702
weakly connected, 705
weighted, 715–725
Dijkstra, Edsger Wybe, 717
algorithm, 717–723
Dirac, Gabriel Andrew, 571
theorem, 570
Directed acyclic graphs. *See* Dags
Directed edges, 692
Directed graphs. *See* Digraphs
Directly derivable, 746
Direct proofs, 50
Dirichlet, Peter Gustav Lejeune, 144, 145
Dirichlet box principle. *See* Pigeonhole principle
Disconnected graphs, 549
Discrete probability, 409–427
Disjoint sets, 71
Disjunctions
defined, 7–8
exclusive, 8, 18
inclusive, 8
symbol for, 7
Disjunctive normal form (DNF), 817–820
Distributive law
Boolean algebra, 806
of logic, 22
of sets, 83
Distributive properties, 804, 805
Div function, 132–134
Divide and conquer algorithm, 228, 327–329
Dividend, 186
Divisibility properties, 189–197
Division algorithm, 185–191
Divisor, 186
greatest common divisor (gcd), 191
Doctrine of Chances, The (De Moivre), 299
Domain
of functions, 118
of relations, 443
Domination law
of logic, 21
of sets, 83
Don't care conditions, 851–856
Dot product, 493
Double complementation law, 83
Double negation law, 22
Dual axioms, 807
Duality, principle of, 807
Dudeney, Henry Ernest, 40–41
Duffinian, 257
Dump state, 767

E

Easter Sunday, 135, 195
Eccentricity, tree, 614
Edges, graph, 445, 517
adjacent, 520
directed, 692
parallel, 518
Eight coins puzzle, 678
Eight Queens puzzle, 620
Elementary operations, graph, 583
Elements
comparable, 495
greatest and least, 500
matrix, 165
maximal, 499
minimal, 499
noncomparable, 495
of sets, 68
Elements (Euclid), 192
Ellipsis, 73
Empty language, 735
Empty relation, 455
Empty sets, 70
Empty word, 75
Endpoints, path, 546
Endwhile, 18
Equality
of Boolean expressions, 815–817
of functions, 123
of generating functions, 299–300
of matrices, 166
of relations, 455
of sets, 70
of words, 734–735
Equivalence class, 486–490
Equivalence relations, 482–484
Equivalence statement, 483
Equivalent combinatorial circuits, 840–850
Equivalent finite-state automata, 764–768
Equivalent nondeterministic finite-state automata, 784–785
Equivalent switching networks 25–26
Erdös, Paul, 147
theorem, 146–147
Euclid, 191
Euclidean algorithm, 191–195, 322–323
Euler, Leonhard, 53, 72, 393, 516–517
circuit, 556
formula, 578–581
graphs, 556–564
path, 556
phi-function, 196
theorem, 512
Even parity, 765
Event(s)
defined, 410
dependent and independent, 419–420
mutually exclusive, 414–415
probability of, 410–414
Eves, Howard, 67, 391
Excluded middle, law of, 28
Exclusive disjunction, 8, 18
Existence proof, 52–53
Existential quantifiers, 32
Expansion
base-b, 197–207
binary coded decimal (BCD), 852–854
binomial, 388
decimal, 197
Expected value, 419–420
Experiment, 409
Exponential function, 125, 874–877
Expressions,
Boolean, 7, 814–817, 863

F
Factorial, 108, 226, 308
Fallacy, 39
Family tree, 611
Fermat, Pierre-Simon de, 4, 5
  Last Theorem, 4, 145, 511
  number, 338
  primes, 53
Fibonacci, Leonardo, 268, 269
Fibonacci numbers, 269–271
  algorithm, 310
  Binet form of, 276–277, 290, 305–305
  complexity of iterative version, 320–321
  digraphs and, 692–694
  paraffins and, 523–524
  permutations and, 356–358
  recursive, 310
Fibonacci trees, 646–647
Final state, automaton, 761, 783
Finite sequences, 158
Finite sets, 73–74
  number of partitions of, 490
Finite-state automata (FSA)
  applications, 759–761
  defined, 761–764
  deterministic, 779–782
  equivalent, 764–768, 784–785
  languages accepted, 787–792
  nondeterministic, 782–787
Finite-state machines (FSMs), 771
  applications, 733
  characteristics, 772–778
  Mealy, 772
  Moore, 799–800
  simply minimal, 800
  unit delay, 774
Fixed-length code, 670
Floor function, 126–130
Floyd, Robert W., 725, 726
  algorithm, 725
Football pools, 426–427
Forest, 611
For loop, 233, 381–382
Formal languages. *See* Languages
Formal power series, 299
Formulas
  Cayley's, 626
  Euler's, 578–581
  midpoint, 145
  molecular, 519
  recursion, 262
  structural, 519–520
  well-formed, 107
  Zeller's, 492
FORTRAN, 748
Four-color problem, 589–595
Four Queens puzzle, 619–620
Free variables, 33
Friday-the-13th, 485–486
Friedberg, S. H., 325
Full-adder, 836–837
Full tree, 639
Functional completeness, 820
Functions
  absolute value, 126
  Ackermann, 278
  applications, 117–118
  bijection, 137–138
  Boolean, 134, 813–824
  cardinality, 140
  ceiling, 126–130
  characteristic, 131–132
  codomain of, 118
  composition, 150–153
  defined, 118–120, 448
  div, 132–134
  domain of, 118
  equality of, 123
  Euler's phi, 196
  exponential, 125, 874–877
  floor, 126–130
  generating, 298–307, 383–384
  graphing, 122–123
  greatest integer, 127
  growth of, 237–247
  hashing, 138–139
  identity, 136
  injection, 136–137
  inverse, 153–155
  least integer, 127
  linear, 125
  logarithmic, 125–126, 877–879
  McCarthy's 91, 275
  mod, 132–134
  next-state, 761, 783
  one-to-one, 136–137
  one-to-one correspondence, 137–138
  onto, 137
  order, most common, 239–241
  output, 772
  piecewise definition, 121–123
  pigeonhole principle and, 144–150
  polynomial, 125
  product of, 123
  properties of, 136–143
  quadratic, 125
  range of, 120–121
  recursive definition of, 262–278
  sum of, 123
  surjection, 137
  transition, 761, 772, 783
Fundamental theorem of arithmetic, 218–221
Fuzzy decisions, 28–29
Fuzzy logic, 26–28
Fuzzy sets, 91–92
Fuzzy subsets, 91–92

G
Game tree, 637
Gates
  AND, 825–826
  defined, 824
  logic, 824–829
  NAND, 828
  NOR, 828–829
  NOT, 827
  OR, 826
  XOR, 18
Gauss, Karl Friedrich, 145, 222, 484
Generalized inclusion/exclusion principle (GIEP), 399–409
Generalized pigeonhole principle, 147–149
Generating combinations, 383–384, 885–887
Generating functions, 298–307, 383–384
Generating permutations, 352–358, 883
Geometric mean, 54
Geometric sequence, 180
Gilbert, E. N., 728
  algorithm, 728
Goldbach, Christian, 4, 5
  conjecture, 4
Golden ratio, 179, 276
Graceful graph, 604
Graceful tree, 686
Grammars, 743–759
  ambiguous, 753–755
  Chomsky hierarchy, 751
  context-sensitive, context-free, and regular, 751
  Type 0, 751
  Type 1, 751
  Type 2, 751
  Type 3, 751

Graphing functions, 122–123
Graphs
  See also Digraphs (directed graphs)
  acyclic, 610–611
  adjacency and incidence, 520
  bipartite, 526
  coloring, 586–598
  complement of, 536
  complete, 524–526
  complete bipartite, 527
  complete n-partite, 605
  complexity, 622–623
  computer representations of, 538–539
  connected, 549–554
  cycle, 526
  de Bruijn, 703
  degree of vertex, 520
  disconnected, 549
  edges, 445, 517
  Eulerian, 556–564
  graceful, 604
  Hamiltonian, 565–567
  homeomorphic, 583
  isomorphic, 541–544
  nodes, 445, 517
  nonplanar, 582
  paths, 446–447, 546–547
  Petersen, 530–531
  planar, 576–586
  planar representation, 576
  regular, 536
  self-complementary, 605
  simple, 518–520
  square, 575
  sub, 522–523
  underlying, 702, 705
  undirected, 517
  union of, 532–533
  vertex, 445, 517
  weighted, 527–528
  wheel, 526, 623
Gray, Frank, 447
Gray codes, 446–447, 566
Greatest common divisor (gcd), 191
  recursive, 311
Greatest element, 500
Greatest integer function, 127
Greedy algorithm, 633
Growth curves, 874
Guthrie, Francis, 589

H

Haken, Wolfang, 589
Half-adder, 835–836
Half-life, 881
HAM (Hampton Court) game, 572–573
Hamilton, William Rowan, 564, 565
  Around the World puzzle, 564–565
  graph, 565–567
Hamming, Richard W., 180
  distance, 179, 446
Handshake problem, 264–265, 280, 308–309, 431, 524–526, 533
Harmonic mean, 351
Harmonic number, 332
Hash function, 138–139
Hasse, Helmut, 496, 497
  diagrams, 496–499
Heap, 669
Heapsort, 669
Heawood, Percy John, 589
Height, tree, 637
Heterogeneous trees, 655
Hexagonal number, 284
Hilbert, David, 74
  hotel paradoxes, 74–75
Hoare, C. Anthony R., 316
Homeomorphic graphs, 583
Homogeneous trees, 664
Horner, William G., 207
  algorithm, 336
  method, 207
Huffman, David Albert, 671
  algorithm, 671–675
  coding scheme, 670
  trees, 670–675
Huygens, Christian, 419
Hypothesis, 9
  inductive, 209

I

Ibn Ezra, Rabbi, 366
Icosian calculus, 565
Idempotent law
  of logic, 21
  of sets, 83
Identity function, 136
Identity law
  Boolean algebra, 806
  of logic, 21
  of sets, 83
Identity matrices, 166
Identity properties, 805
If-and-only-if statement, 14–15
If statements, Boolean expressions in, 7
If-then-else statements
  composition, 152
  piecewise definition, 122
If-then statements, 753–754
Image, function, 119
Immediate predecessor, 692
Immediate successor, 692
Implication conversion law, 22
Implications, 9–11
  converse, inverse, and contrapositive of, 11–12
Incidence, graph, 520
Inclusion/exclusion principle, 99–101, 345, 414–415
  generalized, 399–409
Inclusive disjunction, 8
Indegree, vertex, 694
Independent events, 419–420
Index, summation, 158, 160–161
Indexed sets, 86
Indirect proofs, 51
Induction. *See* Mathematical induction
Induction step, 209
Inductive definitions, 263
Inductive hypothesis, 209
Inequality,
  mathematical induction and, 212
Inference rules, 43
Inferential form, 38
Infinite sequences, 158
Infinite sets, 73–74
Infinity symbol, 74
Infix notation, 653
Initial state
  finite-state automaton, 760, 761, 783
  finite-state machine, 772
Initial vertex, 692
Injection function, 136–137
Inorder traversal, tree, 647–648
Input alphabet, 772
Input symbols, 761, 783
Insertion sort algorithm, 236–237
Interior points, 477
Internal vertex, 636
International Standard Book Number (ISBN), 509

Intersections
  relations, 462–463
  of sets, 78–80
Intervals, 73
Invalid arguments, 39–49
Invariant(s)
  isomorphism, 542–544
  loops, 225–227
Inverse/invertible, 11–12
  functions, 153–155
  of a matrix, 174
  relation, 469
Inverse law
  of logic, 21
  of sets, 83
Invertible, 153
Investigation of the Laws of Thought, An (Boole), 1, 4, 803
Irreflexive relation, 461
Isaka, Satoru, 26, 27
Island of Knights and Knaves example, 12–14, 39–40
Isolated vertex, 520
Isomorphic graphs, 541–544
Isomorphic trees, 645
Isomorphism
  defined, 541
  invariants, 542–544
Iteration method, 226, 279–282, 320–321
Iverson, Kenneth E., 127

J
Java, 735
Join, Boolean, 439

K
Kaplansky, Irving, 285, 290
Karnaguth, Maurice, 842, 843
  maps, 842–850
Keno, 420
al-Khowarizmi, Abu-Abdullah Muhammed ibn-Musa, 96
Kirchoff, Gustav Robert, 609, 610
Kleene, Stephen Cole, 739
  closure, 739–741
  operator, 739
Knight's tour problem, 558–560, 567–570
Knuth, Donald Ervin, 244
Königsberg bridge puzzle, 516–517
Kosko, Bart, 26, 27
Kowa, Seki, 867
Kronecker, Leopold, 68, 162
Kronecker's delta, 162
Kruskal, Joseph Bernard, 616
  algorithm for minimal spanning tree, 626–628
  algorithm for spanning tree, 616–617
Kuratowski, Kazimierz, 584
  theorem, 583

L
Lagrange, Joseph Louis, 159
Lambda, use of, 75
Lamé, Gabriel, 325
  theorem, 323–324
Landau, Edmund, 237, 238
  symbol, 237
Language recognizers. *See* Finite-state automata
Languages
  accepted-recognized, 764, 787–792
  Backus-Normal Form (BNF), 748–751
  concatenation of, 736–739
  context-sensitive, context-free, and regular, 751–753
  derivation and, 746–746
  empty, 735
  equality of words, 734–735
  grammars, 743–759
  grammars, ambiguous, 753–755
  Kleene closure, 739–741
  recognized by deterministic finite-state automata, 779–782
  of sets, 75
Laplace, Pierre-Simon, 410
Lattice-walking, 377–378
Laws, of logic
  associative, 22
  commutative, 22
  contrapositive, 22, 43
  De Morgan's, 22–24, 35–36
  detachment, 43
  distributive, 22
  domination, 21
  double negation, 22
  examples using, 24–25
  excluded middle, 28
  idempotent, 21
  identity, 21
  implication conversion, 22
  inverse, 21
  reductio ad absurdum, 22
Laws, of sets
  absorption, 83
  associative, 83
  commutative, 83
  De Morgan's, 83
  distributive, 83
  domination, 83
  double complementation, 83
  idempotent, 83
  identity, 83
  inverse, 83
Leaf, 636
Leap year, 130
Least element, 500
Least integer function, 127
Left child, 646
Left subtree, 646
Leibniz, Gottfried Wilhelm, 1, 3, 72, 867
Length
  of a path, 446, 546
  of a word, 75
Level, tree, 637, 638
Lexicographic order, 495–496, 883–885
Linear combination, 194
Linear function, 125
Linear homogeneous recurrence relations with constant coefficients (LHRRWCCs), 287–293
Linearly ordered sets, 495
Linear nonhomogeneous recurrence relations with constant coefficients (LNHRRWCCs), 293–298
Linear probing, 139
Linear search algorithm, 227–228, 249–250, 316–317, 319–320, 421
Linked field, 451
Linked lists, 449–450
Listing method, 68–69
Literal, 817
Local area networks (LANs), 529
Logarithmic function, 125–126, 877–879
Logarithms
  Briggsian, 879
  common, 879
  natural, 879
Logic
  See also Laws, of logic
  arguments, 38–49

Logic *(continued)*
compared with set operations, 82–86
defined, 1
gates, 824–829
predicate, 36
proof methods, 49–56
propositional, 36
propositions, 2–20
quantifiers, 32–38
tables, 813–814, 825–827
variables, 2
Logical equivalence, 20–32
defined, 20
equivalent switching networks, 25–26
fuzzy decisions, 28–29
fuzzy logic, 26–28
Logic of Chance, The (Venn), 72
Loops
digraphs/graphs, 446
invariant, 225–227
Lower limit, summation, 158
Lucas, François-Edouard-Anatole, 273, 274
number, 273, 524, 623
theorem, 512
Lukasiewicz, Jan, 62, 653, 654, 659

M

Machines. *See* Finite-state machines
Magic constant, 222
Magic square, 222
Maps, Karnaguth, 842–850
Master Mind game, 430
Mathematical Analysis of Logic (Boole), 4
Mathematical induction
defined, 207
description of, 207–224
strong version of, 218
weak version of, 208–209
Mathematical system, 804
Matrix (matrices)
addition, 166
adjacency, 444–445, 520–522
Boolean, 438–443
defined, 165
discovery of, 164
equality of, 166
identity, 166
inverse, 174
invertible, 174
multiplication, 167–172
negative, 167
reachability, 699
submatrices of, 391–393
subtraction, 167
sum, 166
symmetric, 174
telecommunications/networks and, 528–532
transpose, 175
upper triangular, 874
weighted adjacency, 528, 716
zero, 166
Maurocylus, Francesco, 207
Maximal element, 499
Maxterm, 824
McCarthy, John, 275
91 function, 275
McCulloch, Warren S., 733
Mealy, George H., 772
machines, 772
Mean
arithmetic, 54
geometric, 54
harmonic, 351
Meet, Boolean, 439
Members of sets, 68
Memory wheels, 701–702
Merge algorithm, 312–313
Merge sort algorithm, 313–314, 328–329
Mersenne, Marin, 56
primes, 56
Midpoint formula, 145
Minimal element, 499
Minimal spanning tree, 626–634
Minor, 868
Minterm, 817–820
m-nary tree, 639–640
complete, 645
Mod function, 132–134
Mod operator, 485
Modulus, 484
Molecular formula, 519
Moore, Edward Forrest, 799, 800
machine, 799–800
Multinomial coefficient, 891
Multinomial theorem, 890–892
Multiplication
in base b, 201
of generating functions, 300–301
of matrices, 167–172
nested, 207
principle, 345–347, 418–419
scalar, 167–170
shifting and binary, 201–203
Mutually exclusive
events, 414–415
tasks, 344

N

NAND (not and), 30, 820–821
NAND gate, 828
Napier, John, 877
n-ary predicate, 33
Naur, Peter, 748
n-cube, 566
Negation, 8–9
Negative of matrices, 167
Nested multiplication, 207
Networks, 528–530
Newton's identity, 398
Next-state function, 761, 783
Next-subset algorithm, 96–97
Nil pointer, 451
Nodes, 445, 450–451, 517, 692
Noncomparable elements, 495
Nonconstructive existence proof, 52, 53
Nondecimal bases, 197–207
Pascal's triangle and, 390–391
Nondeterministic finite-state automata (NDFSA), 782–787
Nonterminal symbol, 744, 746
NOR (not or), 30, 820–821
NOR gate, 828–829
Notations
big-oh, 237–239
big-omega, 243–245
big-theta, 245
infix, 653
Polish, 653
postfix, 653
prefix, 653
reversed Polish, 653
set-builder, 69
summation, 158–161
NOT gate, 827
n-stage win, 698
n-tuple, 86
Null sets, 70
Null word, 75
Numbers
Bell, 397, 492
binary, 394–395
Catalan, 108, 388–390, 393, 395–396, 660

chromatic, 588
composite, 189–190
Fermat, 338
Fibonacci, 269–271, 276–277, 290, 304–305, 310, 356–358, 523–524, 692–694
game, 420
harmonic, 332
hexagonal, 284
Lucas, 273, 524, 623
perfect, 124
pentagonal, 284
polygonal, 216–218
prime, 189–190
square pyramidal, 285
Stirling, 278, 375, 490
tetrahedral, 285, 382
triangular, 216, 285, 382

O

Odd parity, 765
One's complement, 203
One-to-one correspondence function, 137–138
One-to-one function, 136–137
Onto function, 137
Open-closed interval, 73
Open interval, 73
Open path, 548
Operators
Boolean, 12, 806
Kleene, 739
relational, 658–659
Or
Boolean, 438
exclusive, 8
inclusive, 8
symbol for, 7
ORD, 136
Ordered pair, 86
Ordered rooted trees, 638
Ordered sets, 86
linearly, 495
partial and total, 493–506
Order functions, most common, 239–241
Orderings
lexicographic, 495–496, 883–885
linear, 495
partial and total, 493–506
Order of precedence, 15
Ordinal number, 121
Ore, Oystein, 571, 589
theorem, 570
OR gate, 826

Outcome, 409, 410
Outdegree, vertex, 694
Output function, 772
Output symbols, 772

P

Palindrome, 108, 741
Paradox, 4
birthday, 182, 412
Russell's, 42–43, 69
Parallel edges, 518
Parent, 636
Parentheses, legally paired, 106–107
Parenthesization problem, Catalan's, 394
Parenthesized triangulations, 659–660
Parity, 433
-check machine, 765
even, 765
odd, 765
Parse trees, 745
Parsing, 745
Partial fraction decomposition rule, 301–306
Partially ordered sets (poset), 494–503
Partial orderings, 493–506
Partitions of finite sets, 490
Partitions of sets, 89–91
Pascal, Blaise, 370, 371
identity, 370
triangle, 386–391
Pascal (computer language), 71
Paths
closed, 548
in digraphs, 698
in digraphs and relations, 446–447
endpoints, 546
Eulerian, 556
Hamiltonian, 565–566
interior points, 477
length, 446, 546
open, 548
shortest, 717–723
simple, 546, 548
Pentagonal number, 284
Pentagrams, 519
Perfect number, 124
Permutations, 351–360
cyclic, 355–356
generating, 883
with repetitions, 376–379

Petersen, Julius, 530, 532
graphs, 530–531
Phrase-structure grammar, 746
Piecewise definition, 121–123
Pigeonhole principle
description of, 144–150
generalized, 147–149
Pitts, Walter, 733
Pizza problem, 368
Planar graphs, 576–586
Planar representation, 576
Pointer field, 451
Polish notation, 653
reversed, 653
Polya, G., 378
Polygonal number, 216–218
Polygons, triangulation of convex, 275
Polynomial function, 125
Poset. *See* Partially ordered sets
Postfix notation, 653
Postorder traversal, tree, 648
Power
Boolean, 441
chain of order, 569
cycle of order, 576
sets, 72–73
Precedence relation, 459
Precedence rules, 807
Predecessor array, 720
Predecessor function (PRED), 142
Predicate logic, 36
Predicates, 33–34
Prefix, 735
Prefix code, 674
Prefix notation, 653
Premise, 9
Preorder traversal, tree, 647
Prim, Robert Clay, 628, 629
algorithm, 628–633
Prime circle of order, 551–552
Prime numbers
divisibility and, 189–190
Fermat, 53
Mersenne, 56
twin, 195
Wilson, 512
Principia Mathematica (Russell and Whitehead), 42
Principle of duality, 807
Principle of inclusion/exclusion, 99–101, 345
Principle of mathematical induction (PMI). *See* Mathematical induction

Principles of counting, 344–351
Principles of Empirical Logic, The (Venn), 72
Product
  Boolean, 439–440, 806
  Cartesian, 87–89
  dot, 493
  of functions, 123
  of matrices, 170
  symbol, 163
  of two functions, 242–243
Production rule, 744–745, 746
Program, correct, 224–225
Proof by cases, 52
Proofs
  by cases, 52
  by contradiction, 51
  by contrapositive, 51
  counterexample, 53
  direct, 50
  existence, 52–53
  indirect, 51
  trivial, 50
  vacuous, 49–50
Proper factor, 124, 189
Proper subset, 70
Propositional logic, 36
Propositions, 2–32
  compound, 5
  simple, 5
Propositional forms. *See* Statement forms

Q
Quadratic function, 125
Quantifiers
  defined, 32
  existential, 32
  how to symbolically right, 32–33
  negation of, 35–36
  truth values of, 33–35
  universal, 32
Quarternions, system of, 565
Queen chessboard puzzle
  eight, 620
  four, 619–620
  two, 133–134
Queue, 621
QUICKBASIC, 737
Quicksort, 316
Quotient, 186

R
Rabbits, calculating reproductive rates of, 268–269, 286
Range
  of functions, 120–121
  of relation, 443
r-combination, 366–367
Reachability, 698
  matrix, 699
Recognized string, 764
Recurrence relation, 262
  solving, 278–298
Recursion (recursive)
  algorithms, 307–333
  clause, 105, 262, 814
  defined, 261
  definition of Boolean expressions, 814
  definition of functions, 262–278
  definition of sets, 104–109
  derangements, 362–365
  formula, 262
  generating functions, 298–307
  linear homogeneous recurrence relations with constant coefficients, 287–293
  linear nonhomogeneous recurrence relations with constant coefficients, 293–298
  relations, 466–468
  solving by iteration, 278–286
Reductio ad absurdum law, 22
Reflexive closure, 482
Reflexive property, 143
Reflexive relation, 455–456
Region, degree of a, 581–582
Regular
  grammar, 751
  graph, 536
  language, 751–753
Rejected string, 764
Relational operators, 658–659
Relations
  antisymmetric, 456–458
  applications, 437
  binary, 443–444
  Boolean matrices, 438–443
  closure, 475–482
  complement, 469
  composition, 463–464
  computer representations of, 449–454
  connectivity, 471–475
  congruence, 484–486
  digraphs and, 443–449
  empty, 455
  equality, 455
  equivalence, 482–493
  intersection, 462–463
  inverse, 469
  irreflexive, 461
  operations on, 461–471
  partial and total orderings, 493–506
  paths, 446–447
  precedence, 459
  properties of, 454–461
  recursive, 466–468
  reflexive, 455–456
  symmetric, 456–459
  transitive, 459
  transitive closure, 475–482
  union, 462–463
Relative complement of sets, 80
Relatively prime, 194–195
Remainder, 186
Reversed Polish Notation (RPN), 653
Right child, 646
Right subtree, 646
Ring topology, 529
Root, dag, 731
Rooted trees, 635–646
  ordered, 638
Rotating drum problem, 701
Round-robin tournaments, 526
Row vector, matrix, 165
r-permutation, 352
Rubik's cube, 882
Russell, Bertrand, 42, 68
  paradox, 42–43, 69

S
Sample space, 409, 410
Scalar multiplication, 167–170
Search algorithms, 227–230
Selections, 379
Selection sort, 233–234
Self-complementary graph, 605
Self-dual Boolean expressions, 863
Semantics, 743
Sequences
  defined, 157
  finite and infinite, 158
  summation notation and, 158–161
  terms of, 157
Sequential circuits, 830
Set-builder notation, 69
Sets
  See also Laws, of sets; Subsets
  cardinality, 98–104, 140

Cartesian products, 87–89
compared with logic operations, 82–86
concepts, 67–78
countable, 140–141
countably infinite, 140–141
defined, 68–69
disjoint, 71
empty, 70
equal, 70
finite and infinite, 73–74
fuzzy, 91–92
indexed, 86
linearly ordered, 495
null, 70
operations, 78–94
operations (computer), 94–98
ordered, 86
partially ordered, 494
partitions, 89–91
power, 72–73
recursively defined, 104–109
uncountable, 141
universal, 70–71
Set theory, 67, 68
Shannon, Claude Elwood, 803, 804
Sheffer, Henry M., 31
Sheffer stroke, 31–32
Shifting and binary multiplication, 201–203
Shortest path, 717–723
Siblings, 636
SIM, 531–532
Simmons, Gustavus J., 531
Simple graphs, 518–520
union of, 532–533
Simple path, 546, 548
Simple propositions, 5
Simplification rule, 43
Simply minimal, 800
Sink, 694
Smullyan, Raymond, 39–40
Sorting
algorithms, 230–234
heap, 669
topological, 500–503
tournament, 669
Source, 694
Space complexity, 247
Spanning trees, 614–634
Square graph, 575
Square pyramidal number, 285
Star of David, 549
Star-ring topology, 529–530
Star topology, 528
Start symbol, 744, 746
Statements,
defined, 2–3
States
automaton, 760, 761, 767, 783
for finite-state machine, 772
Stein, Sherma K., 700
Stemple, J., 589
Stifel, Michel, 366
Stirling, James, 278
numbers of the second kind, 278, 375, 490
Strings
accepted/recognized, 764
rejected, 764
Strongly connected digraphs, 698–700
Strongly orientable, 731
Strong orientation, 731
Structural formula, 519–520
Subgraphs, 522–523
Submatrices of amatrix, 391–393
Subsets
defined, 69
fuzzy, 91–92
proper, 70
Substitution rule, 744–745
Subtraction
binary, 203–205
of matrices, 167
Subtrees, 636
left and right, 646
Successive squaring, 331
Successor function (SUCC), 142
Suffix, 735
Sum, 806
of functions, 123
of matrices, 166
of two functions, 241–242
Summation notation, 158–161
Surjection function, 137
Surjections, counting, 402–404
Switching networks, 16–17
equivalent, 25–26
Sylvester, James Joseph, 164
Symbolic Logic (Carroll), 44–45
Symbolic Logic (Venn), 72
Symmetric closure, 482
Symmetric difference, 82
Symmetric of a matrix, 174
Symmetric relations, 456–459
Syntactic Structures (Chomsky), 752
Syntax, 743, 745

T

Tables
See also Truth tables
logic, 813–814, 825–827
transition, 761, 772
Tau function, 350
Tautology, 16
Teleprinter problem, 700–702
Term, sequence, 157
Terminal clause, 105, 262
Terminal symbol, 743, 746
Terminal vertex, 636, 692
Ternary tree, 639
Ternary word, 78, 349
Tetrahedral numbers, 285, 382
Theorem, 49
See also under name of
Théorie Analytique des Probabilités (Laplace), 410
Three Houses-Utilities puzzle, 577–578
Three-valued logic, 62
Time complexity, 247–248
Topological index, 523
Topological sorting, 500–503
Topology
bus, 529
ring, 529
star, 528
star-ring, 529–530
Torus, 847
Total orderings, 493–506
Tournament sort, 669
Tournaments, round-robin, 526
Tower of Brahma, 265, 281–282, 309–310, 321
Tower of Hanoi, 265
Trails, 422
Traité de Méchanique Céleste (Laplace), 410
Transition diagram, 760, 772
Transition function, 761, 772, 783
Transition table, 761, 772
Transitive closure relation, 475–482
Transitive property, 55, 78
Transitive relation, 459
Transpose of a matrix, 174
Trap state, 767
Traveling salesperson problem, 571–573
Traversals, tree, 647–653
inorder, 647–648
postorder, 648
preorder, 647

Treatise on Differential Equations (Boole), 4
Treatise on the Calculus of Finite Differences (Boole), 4
Trees
acyclic graphs, 610–611
ancestors, 636
applications, 609
balanced, 640–643
binary, 639, 646–663, 659–661
binary expression, 655
binary search, 664–668
center, 614
child, 636, 646
decision, 676–680
derivation, 743–744
descendants, 636
diagram, 6
eccentricity, 614
edge requirements, 612–613
family, 611
Fibonacci, 646–647
full, 639
game, 637
graceful, 686
height and level, 637–638
heterogeneous, 655
homogeneous, 664
Huffman, 670–675
internal vertex, 636
isomorphic, 645
leaf, 636
minimal spanning, 626–634
m-nary, 639–640
m-nary, complete, 645
ordered rooted, 638
parent, 636
parse, 745
path requirements, 611–612
rooted, 635–646
siblings, 636
spanning, 614–634
sub, 636, 646
terminal vertex, 636
ternary, 639
traversals, 647–653
Triangular numbers, 216, 285, 382
Triangulation of convex polygons, 275
Trinomial theorem, 889–890
Trivial proofs, 50
Truth tables
for biconditional statements, 14–15
for conjunctions, 6–7
for disjunctions (inclusive), 8
for implications, 10–11
for Island of Knights and Knaves example, 12–14
for negation, 9
Truth values, 3–5
Twelve Days of Christmas, 221, 373, 431
Twin primes, 195
Two Queens puzzle, 133–134
Two's complement, 203
Type 0 grammar, 751
Type 1 grammar, 751
Type 2 grammar, 751
Type 3 grammar, 751

U
Unary operator, 12
Unary predicate, 33
Uncountable set, 141
Underlying graph, 702, 705
Undirected graph, 517
Union
of graphs, 532–533
relations, 462–463
of sets, 78
Unit delay machines, 774
Unit element, 806
Universal quantifiers, 32
Universal sets, 70–71
Universe of discourse (UD), 33
Upper limit, summation, 158
Upper triangular matrix, 874

V
Vacuously true, 11
Vacuous proofs, 49–50
Valid arguments, 39–49
Value
absolute, 33, 126
expected, 419–420
truth, 3–5
Vandermonde, Alexandre-Théophile, 568, 569
identity, 398–399
Variables
Boolean, 2, 813
logic, 2
Venn, John, 72
diagrams, 72–73
Vertex (vertices)
degree of, 520, 694–698
digraph, 445, 692
graph, 445, 517
initial, 692
interior, 478
internal, 636
isolated, 520
terminal, 636, 692
Von Ettinghausen, Andreas, 366
Von Segner, Johann Andreas, 393

W
Warshall, Stephen, 477
algorithm, 477–481
Weakly connected digraphs, 705
Web sites, 895–898
Weierstrass, Karl, 68
Weight, of spanning trees, 626
Weighted adjacency matrix, 528, 716
Weighted digraphs, 715–725
Weighted graphs, 527–527
Well-formed formula, 107
Well-ordering principle, 186
Wheel graphs, 526, 623
While loops, 618, 622
boolean expressions in, 7
correctness, 225
While statements, 795
Whitehead, Alfred North, 42
Wiles, Andrew J., 4
Williams, Ben Ames, 493
Wilson, John, 512
prime, 512
theorem, 512
Word(s)
empty, 75
equality of, 734–735
length, 75
null, 75
ternary, 78, 349
Worst-case time complexity, 248

X
XOR, 18
xy-plane, 88

Y
Yager, Ronald R., 28
method, 28–29
Yashima, 530–531

Z
Zadeh, Lotfi A., 91, 92
Zeller, Christian Julius Johannes, 492
formula, 492
Zero element, 806
Zero matrices, 166

Applications Index

Item	*Page*
Automatic teller machine	768
Automobile license plate numbers	347
Bank check ID number	493
A binary puzzle	198
The birthday paradox	182, 412
The cabbage-goat-wolf puzzle	552
Card dealing	132
Casino games	132
Coconuts and monkey	493
Code-a-bar system	493
The circle problem	370
Circuit breakers	829
Conflict-free scheduling	591
Counting derangements	404
Counting surjections	402
Cycloparaffins	524
Databases	464
Day of the week	135, 492
De Bruijn sequences	702
Derangements	360
Digital displays	854
Easter Sunday	135, 195
The eight-coins puzzle	678
EQUIVALENCE in FORTRAN	483
Fibonacci numbers	269
Fibonacci trees	646
Friday-the-thirteenth	485
The four-queens puzzle	619
Fuzzy logic in decision-making	28
The game of Yashima	530
Gene types	697
The golden ratio	276
Gray codes	446
The handshake problem	264, 533
Hashing	138
Huffman code	674
Hydrocarbon molecules	519
Insulin requirements	173
International Standard Book No.	509
Knights on a 3 H 3 chessboard	558
Knight's tour	567
Lattice points	145
Lattice walking	377
Legally paired parentheses	106
Light fixtures in hallways	834
Local area networks	529
Logic puzzles	39–42
Lucas numbers	273
Lucas numbers and the wheel graph	623
Magic squares	222
Mass. state lottery	411
National Football League	91
National Hockey League	164, 612
n-cube	566
A nondecimal puzzle	205
Number of leap years	130
Number of partitions	490, 492
Paraffins	523
Parity-check machine	765
The pizza problem	368
Post-office function	128
Roman numerals	798
Round-robin tournament	526
SIM	531
Sunflower seeds	271
Survey analysis	100
Switching circuits	16
Symbolic logic in decision-making	39
Tasks in building a house	494
Telecommunications	528
The teleprinter's problem	700
The three houses-utilities puzzle	577
Tic-tac-toe	637
Tractor-mower	825
Traffic light patterns	592
Tower of Brahma	265
Triangulations of a an n-gon	275, 393
Turnstiles	762
Twelve days of Christmas	221, 373
Two-queens puzzle	133
Unit-delay machine	774
United Parcel Service	492
Water bill	121
Well-formed formulas	107

Algorithms Index

Algorithm	*Page*	*Algorithm*	*Page*
Next-subset algorithm	96	Polynomial evaluation	336
Subsets of a set	102	Connectivity relation	474
Legally paired sequence	106	Warshall's algorithm	481
Matrix product	171	Equivalence relation	490
Prime number algorithm	190	Topological sort	501
Euclid's algorithm	193	Horner's algorithm	336
Base-b representation	198	Eulerian graph	559
Addition in base b	199	Eulerain circuit	561
Binary multiplication	203	Minimal spanning tree	616
Integer multiplication	225	Depth-first search	618
Factorial algorithm	226	Breadth-first search	621
Linear search	227, 316	Kruskal's algorithm	628
Binary search	229, 301	Prim's algorithm	631
Bubble sort	232, 317	Inorder traversal	647
Selcetion sort	233, 321	Preorder traversal	647
Insertion sort	236	Postorder traversal	648
Maximum value in a list	251	Binary search tree	665
Minimum and maximum in a list	252	Huffman algorithm	671
Handshakes	308	Dijkstra's algorithm	721
Tower of Brahma	309	Valid identifier	767
Fibonacci	310, 320	Disjunctive normal form	819
Merge sort	314	Next-permutation	884
Exponentiation	318	Next-combination	887

List of Biographical Sketches

Name and dates	*Page*
1. Aristotle (384–322 B.C.)	(2)
2. Leibniz, Baron Gottfried Wilhelm (1646–1716)	(3)
3. Boole, George (1815–1864)	(4)
4. Fermat, Pierre-Simon de (1601–1665)	(5)
5. Goldbach, Christian (1690–1764)	(5)
6. De Morgan, Augustus (1806–1871)	(23)
7. Kosko, Bart	(27)
8. Isaka, Satoru	(27)
9. Dudeney, Henry Ernest (1857–1930)	(40)
10. Russell, Bertrand Arthur William (1872–1970)	(42)
11. Carroll, Lewis (1832–1898)	(45)
12. Cantor, Georg (1845–1918)	(68)
13. Venn, John (1834–1923)	(72)
14. Hilbert, David (1862–1943)	(74)
15. Descartes, René (1596–1650)	(87)
16. Zadeh, Lotfi, A. (1921–)	(92)
17. Iverson, Kenneth E. (1920–)	(127)
18. Dirichlet, Gustav Peter Lejeune (1805–1859)	(145)
19. Erdös, Paul (1913–1996)	(147)
20. Lagrange, Joseph Louis (1736–1813)	(159)
21. Cayley, Arthur (1821–1895)	(164)
22. Sylvester, James Joseph (1814–1897)	(164)
23. Hamming, Richard Wesley (1915–1998)	(180)
24. Euclid	(192)
25. Bernoulli, Jacob I. (1654–1705)	(212)
26. Bachmann, Paul Gustav Heinrich (1837–1920)	(238)
27. Landau, Edmund (1877–1938)	(238)
28. Knuth, Donald Ervin (1938–)	(244)
29. Fibonacci, Leonardo (1170?-1250?)	(269)
30. Lucas, François-Edouard-Anatole (1842–1891)	(274)